CAMBRIDGE STUDIES IN BIOLOGICAL &
EVOLUTIONARY ANTHROPOLOGY 33

The Primate Fossil Record

The Primate Fossil Record is the first comprehensive treatment of primate paleontology in more than 20 years. Profusely illustrated and up to date, it captures the complete history of the discovery and interpretation of primate fossils. The chapters range from primate origins to the advent of anatomically modern humans. Each emphasizes three key components of the record of primate evolution: history of discovery, taxonomy of the fossils, and evolution of the adaptive radiations they represent. *The Primate Fossil Record* summarizes objectively the many intellectual debates surrounding the fossil record and provides a foundation of reference information on the last two decades of astounding discoveries and worldwide field research for physical anthropologists, paleontologists and evolutionary biologists.

WALTER HARTWIG is Associate Professor of Anatomy at Touro University College of Osteopathic Medicine in northern California. He has conducted paleontological field research in South America and Africa, and has authored over 40 scientific articles and book chapters on comparative anatomy, primate evolution and the history of sciences. Professor Hartwig is also founder and director of ibob.org, a non-profit organization dedicated to improving medical care, education and scientific research in underdeveloped countries.

The Primate Fossil Record

Edited by

WALTER CARL HARTWIG
TOURO UNIVERSITY

CAMBRIDGE UNIVERSITY PRESS
Cambridge, New York, Melbourne, Madrid, Cape Town, Singapore, São Paulo

Cambridge University Press
The Edinburgh Building, Cambridge CB2 8RU, UK

Published in the United States of America by Cambridge University Press, New York

www.cambridge.org
Information on this title: www.cambridge.org/9780521663151

First published 2002
Reprinted 2004
This digitally printed version 2008

A catalogue record for this publication is available from the British Library

Library of Congress Cataloguing in Publication data

The primate fossil record / edited by Walter Carl Hartwig.
p. cm.
Includes bibliographical references and index.
ISBN 0 521 66315 6
1. Primates, Fossil. I. Hartwig, Walter Carl, 1964–
QE882.P7 P75 2002
569'.8 – dc21 2001037847

ISBN 978-0-521-66315-1 hardback
ISBN 978-0-521-08141-2 paperback

To **F. CLARK HOWELL**, mentor and visionary

and **ELWYN LAVERNE SIMONS**, who began modern
study of the primate fossil record

Contents

Contributors

K. CHRISTOPHER BEARD
Vertebrate Paleontology Section
Carnegie Museum of Natural History
Pittsburgh, PA 15213, USA

DAVID R. BEGUN
Department of Anthropology
University of Toronto
Toronto, ONT, M5S 3G3
Canada

BRENDA R. BENEFIT
Department of Sociology and Anthropology
New Mexico State University
Las Cruces, NM 88003, USA

HERBERT H. COVERT
Department of Anthropology
University of Colorado–Boulder
Boulder, CO 80309, USA

MARIAN DAGOSTO
Department of CMS Biology
Northwestern University
Chicago, IL 60611, USA

HOLLY DUNSWORTH
Department of Anthropology
Pennsylvania State University
University Park, PA 16802, USA

DANA L. DUREN
School of Biomedical Sciences
Kent State University
Kent, OH 44242, USA

JOHN G. FLEAGLE
Anatomical Sciences
State University of New York
Stony Brook, NY 11794, USA

DANIEL L. GEBO
Department of Anthropology
Northern Illinois University
DeKalb, IL 60115, USA

LAURIE R. GODFREY
Department of Anthropology
University of Massachusetts
Amherst, MA 01003, USA

GREGG F. GUNNELL
Museum of Paleontology
University of Michigan
Ann Arbor, MI 48109, USA

TERRY HARRISON
Department of Anthropology
New York University
New York, NY 10003, USA

WALTER C. HARTWIG
Department of Basic Sciences
Touro University College of Osteopathic Medicine
Mare Island, Vallejo, CA 94592, USA

INES HOROVITZ
Section of Mammalogy
Natural History Museum of Los Angeles County
Los Angeles, CA 90007, USA

NINA G. JABLONSKI
Department of Anthropology
California Academy of Sciences
San Francisco, CA 94115, USA

WILLIAM L. JUNGERS
Anatomical Sciences
State University of New York
Stony Brook, NY 11794, USA

JAY KELLEY
Department of Oral Biology
College of Dentistry
University of Illinois–Chicago
Chicago, IL 60612, USA

ROSS D. E. MACPHEE
Division of Vertebrate Zoology
American Museum of Natural History
New York, NY 10024, USA

MONTE L. MCCROSSIN
Department of Sociology and Anthropology
New Mexico State University
Las Cruces, NM 88003, USA

HENRY M. MCHENRY
Department of Anthropology
University of California–Davis,
Davis, CA 95116, USA

D. JEFFREY MELDRUM
Department of Biological Sciences
Idaho State University
Pocatello, ID 83202, USA

ERICA M. PHILLIPS
Department of Anthropology
Pennsylvania State University
University Park, PA 16802, USA

DAVID R. PILBEAM
Peabody Museum
Harvard University
Cambridge, MA 02138, USA

DAVID TAB RASMUSSEN
Department of Anthropology
Washington University
St. Louis, MO 63130, USA

KENNETH D. ROSE
Department of Cell Biology and Anatomy
Johns Hopkins University School of Medicine
Baltimore, MD 21205, USA

ALFRED L. ROSENBERGER
Department of Public Programs
National Museum of Natural History
Smithsonian Institution
Washington, DC 20560, USA

FRED H. SMITH
Department of Anthropology
Northern Illinois University
DeKalb, IL 60115, USA

MARCELO F. TEJEDOR
Facultad de Ciencias Naturales
Sede Esquel, Universidad Nacional de la Pat
Esquel, Prov. Chubut, 9200, Argentina

ALAN WALKER
Department of Anthropology
Pennsylvania State University
University Park, PA 16802, USA

STEVEN C. WARD
Department of Anatomy
NE Ohio University College of Medicine
Rootstown, OH 44272, USA

TIM D. WHITE
Laboratory of Human Evolutionary Studies
Museum of Vertebrate Zoology
and Department of Integrative Biology
University of California–Berkeley
Berkeley, CA 94720, USA

Preface

The idea for this book arose from discussions with Alfred L. Rosenberger, F. Clark Howell, Eric Delson and John Fleagle in 1998. I was anxious for the revision of Fred Szalay and Eric Delson's seminal reference work *Evolutionary History of the Primates* (Academic Press, 1979). Twenty years after publication this work was still the standard reference work for the primate fossil record, but many new and important fossils had been discovered in the meantime. I realized that only Eric and Fred could update their own reference work effectively, so I polled several colleagues to see if we could mount a group effort independent of them. To my delight the 31 contributors agreed to the effort immediately and enthusiastically. Our goal was to document the record usefully and objectively, and in a manner that would serve as a referential starting point for study of the history, classification and interpretation of the primate fossil record.

Once assembly of the book began in earnest it was clear that a comprehensive volume of all factual information and historical chronicle of the primate fossil record would be much larger than the practical word limit to which Cambridge and I had agreed. In some ways this may signal the end of large-scale primate evolution reference works in a printed medium. But I believe it also compelled the contributors to distill the essential information of history, taxonomy and interpretation from the vast accumulation reflected in the bibliography. My role in all of this was merely to build the whole out of the parts. As this required me to navigate through every sentence of every chapter, I alone am responsible for any errors or omissions in the final product.

Walter Carl Hartwig
Mare Island
Vallejo, California

Acknowledgements

Several scholars from around the world provided illustrations directly to me for this project. Masanaru Takai of the Primate Research Institute in Inuyama generously provided illustrations of La Venta fossils and of *Branisella* from Bolivia. Masato Nakatsukasa and Hidemi Ishida of Kyoto University generously provided illustrations of *Samburupithecus*, *Nacholapithecus* and *Neosaimiri* fossils. David Lordkipanidze of the Georgian State Museum in Tbilisi generously provided illustrations of the Dmanisi hominid fossils. Emmanuel Gheerbrant of the Muséum National d'Histoire Naturelle in Paris generously provided illustrations of *Oligopithecus rogeri*. Salvador Moyà-Solà of Institut Paleontologia Miguel Crusafont generously provided illustrations of *Dryopithecus*. Glenn Conroy of Washington University generously provided illustrations of *Otavipithecus*. Fred Anapol of University of Wisconsin–Milwaukee generously provided illustrations of *Carlocebus*.

I relied upon the advice and expertise of many people over the time it has taken to complete this work. For responding promptly and in some cases repeatedly to my unsolicited inquiries I thank F. Clark Howell, Alfred L. Rosenberger, Glenn Conroy, B. Holly Smith, Ellen Miller, Mark Teaford, Jeffrey T. Schwartz, Jack Fooden, Brigitte Senut, Meave Leakey, Leslie Aiello and, of course, the contributors themselves.

For advice during the proposal phase of this book I thank Blake Edgar, Bill Woodcock, William Curtis, Nina Jablonski, Kirk Jensen and especially Christie Henry of University of Chicago Press. For help producing the manuscript the editor thanks Ms. Vicki Woodman, whose work saved weeks of production time, Mr. Alex Perez and Ms. Sahskkia Saballos of TUCOM. I thank my faculty colleagues at TUCOM, particularly Barbara Kriz, for their encouragement and support.

Dr. Tracey Sanderson of Cambridge University Press provided expert guidance and infinite patience during the development of this book. I learned very quickly that no matter how much you believe in a project, without a visionary acquisitions editor it will always be just a project. During her absence Sarah Jeffery kept everything moving forward.

I owe perhaps the largest debt of thanks to the copy-editor, Anna Hodson, whose deep knowledge of primate evolution and expert reading of the manuscript make her an editor's dream. I cannot express how much her interest in this volume meant to me, and to the quality of the final product.

Finally, I thank my wife YeunShin Lee for her unwavering support of my ideas, large and small, then and now.

Abbreviations

ACM	Amherst College Museum
AD	Anatomy Department (University of Witwatersrand, South Africa)
AL	Afar Locality
AMNH	American Museum of Natural History
ANS	Academy of Natural Sciences, Philadelphia
BMNH	British Museum of Natural History
BNM	Basel Natural History Museum
BPI	Bernard Price Institute
BSM	Bavaria Science Museum
CAN	Cangalongue Cave (Angola)
CGM	Cairo Geological Museum
CENDIA	Centro Dominicano de Investigaciones Antropológicas
CM	Carnegie Museum of Natural History
CORD	Cordoba
CT	South African Museum, Cape Town, South Africa
DPC	Duke University Primate Center
DU	Duke University
FLM	Fundación Miguel Lillo
FMNH	Field Museum of Natural History
FSL	Faculty of Science, University of Lyon
GEN	Geological Institute, Russian Academy of Sciences
GIN	Geological Institute, Russian Academy of Sciences
GMH	Geiseltal Museum Halle, Germany
GSI	Geological Survey of India
GSP	Geological Society of Pakistan
GZC	Glib Zegdou Collection, Oran University
HCRP	Hominid Corridor Research Project
IGC-UFMG	Instituto de Geociencias, Universidade Federal de Minas Gerais
IGF	Institute of Geosciences, Firenze (University of Florence)
IGM	Geological Museum of INGEOMINAS, Bogotá, Colombia
IPMC	Instituto Paleontologico Miguel Crusafont
IPS	Instituto Paleontologico, Sabadell
IRSNB	Institut Royal des Sciences Naturelles Belge (Belgium)
IVPP	Institute of Vertebrate Paleontology and Paleoanthropology, Beijing

KA Kromdraai
KNM Kenya National Museum
KOAN Koanaka, Botswana
LACM Los Angeles County Museum
LAET Laetoli, Tanzania
LGPUT Laboratory of Geology and Paleontology, University of Thessalonika
LPS, LPX Luoding Paleontological Survey, Guangdong, China
LUVP Lahore University Department of Vertebrate Paleontology (now Panjab University)
Ma megannum, equivalent to 1 million geochronologic years
MACN Museo de Ciencias Naturales, Buenos Aires, Argentina
MCZ Museum of Comparative Zoology, Harvard University
MLP Museo La Plata, La Plata, Argentina
MNHM Museum National d'Histoire Naturelle, Paris
MNHNH Museo Nacional de Historia Natural, La Habana
MTA Maden Tetkik ve Arama (Ankara, Turkey)
Munich AS Deutsches Museum, Munich, Germany
NAP Napak (Kenya National Museum)
NBV Nihewan Basin Vertebrate
NMMP National Museum of Myanmar – Primate
OLD Olduvai
PA Province Anhui (China)
PEN Paleontological Institute, Russian Academy of Sciences
PIN Paleontological Institute, Russian Academy of Sciences
PMU Peking Medical Union
PQ Phosphorites of Quercy (France)
PSS Mongolian Academy of Sciences
PU Princeton University
QD Quercy District (France)
rcyrbp radiocarbon years before "present" (i.e., 1950)
SAM South African Museum, Cape Town, South Africa
SB Skurweberg (South Africa)
SDSM San Diego State Museum
SDSNH San Diego Society of Natural History
SGOPV Museo Nacional de Historia Natural, Santiago, Chile
SK Swartkrans (South Africa)
SNJ Sangiran, Java
SNM Staatliches Museum für Naturkunde, Stuttgart, Germany
STS Sterkfontein
TCH Tchuia Cave (Angola)
TF See Geological Survey Division, Department of Mineral Resources, Bangkok, Thailand
THR Natural History Museum, Rabat (Morocco)
TM Transvaal Museum
TMM Texas Memorial Museum
TMP Transvaal Museum, Pretoria
TQ Taqah
Tvl Transvaal, South Africa
UCM University of Colorado Museums
UCMP University of California Museum of Paleontology
UF Florida Museum of Natural History, University of Florida
UL University of Lyon
UM University of Michigan
UMP Uganda Museum, Paleontology
USGS United States Geological Survey
USNM United States National Museum of Natural History
USNMM United States National Museum of Natural History, Division of Mammals
USTL Université Montpellier Sciences et Techniques du Languedoc
UTBEG University of Texas Bureau of Economic Geology
UW University of Wyoming
UZM Universitets Zoologist Museum, University of Copenhagen
YGSP Yale Geological Survey of Pakistan (Peabody Museum, Yale University)
YPM Yale Peabody Museum
YV Yunnan Vertebrate

1 | Introduction to *The Primate Fossil Record*

WALTER CARL HARTWIG

Fossil primates have been known to science for nearly 200 years. Discoveries are scrutinized and inspire further exploration. They attract a flow of information from many different sources including geology, biology, anthropology and popular media. Periodically, a census of the record itself makes a useful reference. What follows is our attempt at such a census.

This book honors and departs from the seminal 1979 reference, *Evolutionary History of the Primates*, by Fred Szalay and Eric Delson. The last two decades have provided a tremendous new inventory of fossils and the interpretive debates that inevitably attend them. The time has come for a reference work to complement textbooks such as *Primate Adaptation and Evolution* (Fleagle, 1988, 1999), *Primate Evolution* (Conroy, 1990) and *Primate Origins and Evolution* (Martin, 1990). To keep *The Primate Fossil Record* to a reasonable size the chapters emphasize the history of fossil discovery rather than comprehensive comparative anatomy. The goal of this book is to reference the primate fossil record according to the history of its study, the taxonomic groups to which the fossils have been referred, and its implications for understanding primate and human evolution.

Organization of *The Primate Fossil Record*

The book is organized into chapter units according to a combination of time (geological), place and taxonomy. These composite "units" of the record contain all currently recognized fossil genera, and enable the history of their discovery to be presented with a minimum amount of redundancy. Perhaps a better approach would be to construct the ultimate systematic paleontology of the Order Primates, in the profound manner of *Evolutionary History of the Primates*, so that readers could locate fossils according to a master evolutionary classification. Our volume departs from this approach because we explicitly wish to avoid promotion of one interpretive phylogeny, or means of deriving one, over another. The chapter authors are charged with objectively accounting for an assigned geological/geographical/taxonomic unit of the primate fossil record. Although we each have strong opinions about the best method of interpreting those units, we recognize that interpretive differences are the essential fabric of the field. We are obliged, willingly, to represent all of the rich variety of interpretation that has been cogently argued in the last two decades.

The chapters reviewing the fossil record use the following framework:

Introduction
History of discovery and debate
Taxonomic framework
Evolution

The anatomical "nuts-and-bolts" are found in the illustrations of the fossils themselves and under the **Taxonomic framework** subheading. Anatomical description is limited in favor of illustration. The narrative emphasis in each chapter, rather, is **History of discovery and debate**. Without doubt, interpretations of the fossil record are functions of when the fossils were found and of who described them. We emphasize the **History of discovery and debate** section because it provides readers with the full context necessary to make their own judgements about the raw data (Fleagle & Hartwig, 1997). Although classifications are presented in the **Taxonomic framework** section of each chapter, we stress that they are means to the end of putting the genera into context, rather than ends in themselves. We encourage students to arrive at their own taxonomic frameworks by using this book to assemble the primary literature and understand what has been debated until now. For this same reason very few phylogenetic trees are presented, because the intention is to chronicle the fossil record rather than to promote a particular interpretation. Finally, the **Evolution** section is developed to account for the realms of interpretation that the fossils have impacted since their discovery. Because each unit of the fossil record presents a different spectrum of information and implication, the domains of phylogeny, functional morphology and paleobiology are not weighted equally across all chapters.

The book as a whole also includes one topical chapter on primate origins and introductory chapters for each major adaptive radiation. Because the depth of the fossil record differs for each of these major adaptive radiations, these introductory essays vary in emphasis. For example, Bert Covert synthesizes the biology of the earliest primates in a style predicated by the quantity and tremendous diversity of Eocene fossils. By contrast, David Pilbeam pivots his introduction to the fossil record of hominoid primates on a current burning issue of how fossil anatomy should be interpreted relative to genetic data. The styles of these authors reflect where they expect our knowledge of primate evolution to grow.

Literature citations form a key database in any reference work. In this book we separate the references that describe new fossil material (**Primary references**) from analytical, interpretive or related references. A master bibliography of all cited works is presented alphabetically by author in the back of the book. The **Primary references** of the fossil record are listed at the end of each chapter, in chronological order under each genus. Ordering the references chronologically reinforces the historical attributes of the record and its interpretation. Some **Primary references** lists are more comprehensive than others, depending upon the extent of the fossil record for that radiation. Eocene primates are so numerous that the list is limited to the earliest citations; by contrast, fossil New World monkeys are so scarce that their **Primary references** lists can include every published description.

This book accounts the human fossil record but does not analyze it exhaustively. No other fossil group is so thoroughly studied and, as a result, treated so extensively in the research and popular literature. A special effort has been made by the chapter authors for this unit to render the hominid fossil record within practical limits and the implications of the fossils themselves.

This framework is applied to each of the review chapters, but not at the expense of customizing the coverage to the attributes of each unit of the fossil record. Readers will notice, for example, that each contributor brings a different style of description and itemization. Species of Miocene hominoids are scrutinized more closely than species of adapiforms. Likewise, the criteria for recognizing a valid taxon vary from one part of the record to the next. Although the units used to define chapter boundaries result in almost no overlap of coverage, one species, *Afrotarsius chatrathi*, is reviewed in two chapters (as a potential tarsiiform and as a potential early anthropoid); conversely, one species, *Plesiopithecus teras*, is not accepted as part of any chapter unit (see Chapter 6). Sticklers for uniformity may be disappointed, but we believe the differences between chapters better reflect the contours of the record itself.

A brief introduction to the record itself

Twenty years before publication of *The Origin of Species* researchers on three continents risked their scientific reputations by arguing that primates had indeed lived in antediluvian times and that some went "extinct" along with other mammals in a catastrophic biblical flood (Hartwig, 1995*b*). Since that time the explicit mission to discover fossil primates (especially fossil humans) has fueled fieldwork and inspired the distinction of primate evolution as its own field of study. In this book we try to separate the bounty of knowledge gained by virtue of this venture from the diversions of belief that the very same zeal instills.

Certain patterns of understanding pertain to the fossil record throughout its history. Two in particular are worth mention in this introduction. One, fossil primates have always been interpreted in light of how they might relate to living ones. Whether this tendency stems from scientific insight or myopia, a framework of closely related forms and conservative phylogenies has persevered. Two, we discoverers of fossil primates irresistibly ascribe superlatives, novelties and profound implications to virtually every scrappy fossil we have the providence to find. In forming their own judgements students are advised to regard the former pattern a challenge and the latter pattern not at all.

The history of discovery of the primate fossil record can be viewed through a chronology of its key unit, the genus. Table 1.1 presents a "history-gram" of fossil non-human primate genera, arbitrarily divided into six geographic/taxonomic units. From this perspective it is easy to see the increase in inventory of genera in the last two decades compared to previous decades. It should also be easy to see why the subject of early anthropoid evolution has received more attention in the last decade (e.g., Fleagle & Kay, 1994) than in all previous decades combined. Just as a fossil must have known provenience to be informative, the information it generates has its own, equally vital, historical context.

The primate fossil record is expanded by discovery, and this book is unabashedly "discovery-centric". Competing interpretations, cladograms and phylogenies endure, fade away, or transform with new discoveries. No single book-length accounting of the primate fossil record will satisfy everyone's sense of proper emphasis or database management. We hope that this treatment appeals, at the very least, to a common love of discovery.

Table 1.1. Chronology of discovery of fossil non-human primate genera, updated from Fleagle & Hartwig (1997)

Date	North American adapids + omomyids	Old World prosimians + tarsiiforms	Early anthropoids + catarrhines
1990–	*Chipetaia* Rasmussen, 1996	*Guangxilemur* Qi & Beard, 1998	*Bahinia* Jaeger et al., 1999
	Hesperolemur Gunnell, 1995	*Xanthorhysis* Beard, 1998	*Siamopithecus* Chaimanee et al., 1997
	Sphacorhysis Gunnell, 1995	*Wadilemur* Simons, 1997	*Tabelia* Godinot & Mahboubi, 1994
	Wyomomys Gunnell, 1995	*Wailekia* Ducroq et al., 1995	*Eosimias* Beard et al., 1994
	Ageitodendron Gunnell, 1995	*Aframonius* Simons et al., 1995	*Arsinoea* Simons, 1992
	Tatmanius Bown & Rose, 1991	*Barnesia* Thalmann, 1994	*Algeripithecus* Godinot & Mahboubi, 1992
	Yaquius Mason, 1990	*Rencunius* Gingerich et al., 1994	*Serapia* Simons, 1992
		Adapoides Beard et al., 1994	*Plesiopithecus* Simons, 1992
		Shizarodon Gheerbrant et al., 1993	
		Omanodon Gheerbrant et al., 1993	
		Djebelemur Hartenberger & Marandat, 1992	
		Asiomomys Wang & Li, 1990	
		Babakotia Godfrey et al., 1990	
1980–1990	*Jemezius* Beard, 1987	*Buxella* Godinot, 1988	*Proteopithecus* Simons, 1989
	Steinius Bown & Rose, 1984	*Panobius* Russell & Gingerich, 1987	*Catopithecus* Simons, 1989
		Sinoadapis Wu & Pan, 1985	*Biretia* Bonis et al., 1988
		?*Afrotarsius* Simons & Bown, 1985	?*Afrotarsius* Simons & Bown, 1985
		Cryptadapis Godinot, 1984	*Qatrania* Simons & Kay, 1983
		Nycticeboides Jacobs, 1981	
		Kohatius Russell & Gingerich, 1980	
1970–1980	*Aycrossia* Bown, 1979	*Sivaladapis* Gingerich & Sahni, 1979	
	Strigorhysis Bown, 1979	*Altanius* Dashzeveg & McKenna, 1977	
	Gazinius Bown, 1979	*Donrussellia* Szalay, 1976	
	Arapahovius Savage & Waters, 1978	*Cercamonius* Gingerich, 1975	
	Copelemur Gingerich & Simons, 1977	*Microadapis* Szalay, 1974	
	Mahgarita Wilson & Szalay, 1976	*Agerinia* Crusafont-Pairo and Golpe-Posse, 1973	
	Pseudotetonius Bown, 1974		
1960–1970	*Rooneyia* Wilson, 1966	*Komba* Simpson, 1967	*Aegyptopithecus* Simons, 1965
	Ekgmowechashala MacDonald, 1963	*Mioeuoticus* Leakey, 1962	*Oligopithecus* Simons, 1962
	Cantius Simons, 1962	*Lushius* Chow, 1961	
1950–1960	*Chlororhysis* Gazin, 1958		
	Anemorhysis Gazin, 1958		
	Ourayia Gazin, 1958		
	Utahia Gazin, 1958		
	Stockia Gazin, 1958		
1940–1950	*Macrotarsius* Clark, 1941	*Pachylemur* Lamberton, 1948	
	Loveina Simpson, 1940	*Progalago* MacInnes, 1943	
		Teilhardina Simpson, 1940	
1930–1940	*Dyseolemur* Stock, 1934	*Europolemur* Weigelt, 1933	*Amphipithecus* Colbert, 1937
	Chumashius Stock, 1933	*Indraloris* Lewis, 1933	
		Hoanghonius Zdansky, 1930	
1920–1930			*Pondaungia* Pilgrim, 1927
1910–1920	*Tetonius* Matthew, 1915	*Anchomomys* Stehlin, 1916	*Parapithecus* Schlosser, 1910
	Absarokius Matthew, 1915	*Periconodon* Stehlin, 1916	*Propliopithecus* Schlosser, 1910
	Uintanius Matthew, 1915	*Nannopithex* Stehlin, 1916	*Moeripithecus* Schlosser, 1910
		Pseudoloris Stehlin, 1916	
1900–1910	*Shoshonius* Granger, 1910	*Archaeoindris* Standing, 1909	*Apidium* Osborn, 1908
	Trogolemur Matthew, 1909	*Mesopropithecus* Standing, 1905	
	Smilodectes Wortman, 1903	*Pronycticebus* Grandidier, 1904	
1890–1900		*Hadropithecus* Lorenz, 1899	
		Paleopropithecus Grandidier, 1899	
		Archaeolemur Filhol, 1895	
		Megaladapis Forsyth-Major, 1894	
1880–1890			
1870–1880	*Pelycodus* Cope, 1875	*Protoadapis* Lemoine, 1878	
	Washakius Leidy, 1873	*Leptadapis* Gervais, 1876	
	Anaptomorphus Cope, 1872	*Necrolemur* Filhol, 1873	
	Hemiacodon Marsh, 1872	*Palaeolemur* Delfortrie, 1873	
	Notharctus Leidy, 1870		
1860–1870	*Omomys* Leidy, 1869	*Caenopithecus* Rütimeyer, 1862	
1850–1860		*Adapis* Cuvier, 1821[a]	
1840–1850		*Microchoerus* Wood, 1846	
1836–1840			

Table 1.1. (*cont.*)

Date	New World monkeys	Old World monkeys	Hominoids + pliopithecoids
1990–	*Proteropithecia* Kay *et al.*, 1998	*Parapresbytis* Kalmykov & Maschenko, 1992	*Egarapithecus* Moyà-Solà *et al.*, 2001
	Patasola Kay & Meldrum, 1997		*Orrorin* Senut *et al.*, 2001
	Nuciruptor Meldrum & Kay, 1997		*Equatorius* Ward *et al.*, 1999
	Caipora Cartelle & Hartwig, 1996		*Nacholapithecus* Ishida *et al.*, 1999
	Antillothrix MacPhee *et al.*, 1995		*Samburupithecus* Ishida & Pickford, 1997
	Chilecebus Flynn *et al.*, 1995		*Morotopithecus* Gebo *et al.*, 1997
	Lagonimico Kay, 1994		*Kamoyapithecus* Leakey *et al.*, 1995
	Laventiana Rosenberger *et al.*, 1991		*Otavipithecus* Conroy *et al.*, 1992
	Paralouatta Rivero & Arredondo, 1991		
	Carlocebus Fleagle, 1990		
1980–1990	*Soriacebus* Fleagle *et al.*, 1987	*Microcolobus* Benefit & Pickford, 1986	*Kalepithecus* Harrison, 1988
	Aotus dindensis Setoguchi & Rosenberger, 1987	*Rhinocolobus* M.G. Leakey, 1982	*Lufengpithecus* Wu, 1987
	Mohanamico Luchterhand *et al.*, 1986		*Heliopithecus* Andrews & Martin, 1987
	Micodon Setoguchi & Rosenberger, 1985		*Simiolus* Leakey & Leakey, 1987
			Nyanzapithecus Harrison, 1986
			Afropithecus Leakey & Leakey, 1986
			Turkanapithecus Leakey & Leakey, 1986
			Laccopithecus Wu & Pan, 1984
1970–1980	*Tremacebus* Hershkovitz, 1974		*Dionysopithecus* Li, 1978
	Stirtonia Hershkovitz, 1970		*Platodontopithecus* Li, 1978
			Micropithecus Fleagle & Simons, 1978
			Dendropithecus Andrews & Simons, 1977
			Ouranopithecus de Bonis & Melentis, 1977
			Anapithecus Kretzoi, 1975
			Rangwapithecus Andrews, 1974
			Graecopithecus von Koenigswald, 1972
1960–1970	*Branisella* Hoffstetter, 1969	*Paracolobus* R. Leakey, 1969	*Mabokopithecus* von Koenigswald, 1969
		Victoriapithecus von Koenigswald, 1969	*Ankarapithecus* Ozansoy, 1965
		Paradolichopithecus Necrasov *et al.*, 1961	*Kenyapithecus* L. Leakey, 1962
			Plesiopliopithecus Zapfe, 1961
1950–1960	*Xenothrix* Williams & Koopman, 1952		*Epipliopithecus* Zapfe & Hürzeler, 1957
	Cebupithecia Stirton & Savage, 1951		
	Neosaimiri Stirton, 1951		
	Dolichocebus Kraglievich, 1951		
1940–1950		*Gorgopithecus* Broom & Robinson, 1949	
		Cercopithecoides Mollett, 1947	
1930–1940		*Dinopithecus* Broom, 1937	*Gigantopithecus* von Koenigswald, 1935
		Parapapio Jones, 1937	*Proconsul* Hopwood, 1933
			Limnopithecus Hopwood, 1933
1920–1930		*Procynocephalus* Schlosser, 1924	
1910–1920		*Prohylobates* Fourtau, 1918	*Sivapithecus* Pilgrim, 1910
		Libypithecus Stromer, 1913	
1900–1910			*Griphopithecus* Abel, 1902
1890–1900	*Homunculus* Ameghino, 1891		*Paidopithex* Pohlig, 1895
1880–1890		*Dolichopithecus* Depéret, 1889	
1870–1880			*Oreopithecus* Gervais, 1872
1860–1870			
1850–1860			*Dryopithecus* Lartet, 1856
1840–1850			*Pliopithecus* Gervais, 1849
1836–1840	*Protopithecus* Lund, 1838	*Mesopithecus* Wagner, 1839	

[a] First recognized as primate in 1859.

2 | The origin of primates

DAVID TAB RASMUSSEN

Introduction

Charles Darwin, renowned for convincingly demonstrating that organic evolution occurred and for discovering that natural selection was the primary mechanism of evolutionary change, also discovered the phenomenon of phylogeny. He showed that organisms not only change through time (which several prominent evolutionists had believed before him), but that one species could split and *diverge* into two or more descendant species. Lamarck, for example, believed each species on Earth to have undergone separate evolutionary trajectories from independent origins. In contrast, from Darwin's earliest notebooks on the subject it was clear that he was keenly interested in common ancestral origins and evolutionary divergences among genealogically related organisms. This is why his famous book of 1859 is called *On the Origin of Species* and only subtitled *by Means of Natural Selection*. The origin of species by splitting of lineages leads to a diverging tree of evolutionary relatedness among organisms, now called a phylogeny. It was immediately clear to Darwin that the traditional Linnean system of classification had worked well because it more or less overlaid the natural nested hierarchy of relatedness among organisms.

Darwin pointed out that with the discovery of phylogeny the rationale for fitting organisms into the Linnean hierarchy became obvious: each taxon – whether a class, order, family or genus – should represent a natural genealogical group. The classificatory system "must be, as far as possible, genealogical in arrangement, – that is, the co-descendants of the same form must be kept together in one group, separate from the co-descendants of any other form; but if the parent-forms are related, so will be their descendants, and the two groups together will form a larger group" (Darwin, 1871: 188). Darwin endorsed Linnaeus's order Primates as just such a genealogical group, and argued forcefully that humans belonged in the order. Thus it was that the order Primates was recognized as a natural phylogenetic grouping, not just an arbitrary assemblage of similar organisms useful for classification purposes (Table 2.1). The initial primate must have originated by splitting off and diverging from another mammalian species. From Darwin's insight on, questions about the ancestral primate and the selective pressures that gave rise to the earliest primates entered the realm of scientific scrutiny.

The Order Primates

Primates lack any one shared specialization that is as clear and unambiguous as the wing shared by all bats or the paddling tail of porpoises. The list of apparent specializations that characterize most (but not all) primates and which are found uncommonly among other mammals are the following (Cartmill, 1972, 1992; Wible & Covert, 1987; Martin, 1990):

Primates have grasping hands and feet usually with opposable big toes and thumbs.

The primitive mammalian claw has been modified in primates to form a flat nail, although some species retain grooming claws on one or two digits, and the marmosets and tamarins have re-evolved a claw from the primitive primate nail.

Primates further depart from the primitive mammalian pattern in having their eyes placed forward on the face so that the visual fields overlap, yielding stereoscopic or binocular vision. Primates also share specialized traits of the retina and the visual cortex of the brain.

Primates tend to have larger brains than would be expected for mammals within their size range.

Primates are further characterized by having singleton births (or very small litters), and their offspring grow up relatively slowly when compared to non-primates of comparable size.

Finally, primates share what may be one truly unique specialization of the basicranium: the auditory bulla (the rounded bony case protecting the underside of the inner and middle ears) is formed by ballooning of the petrosal bone rather than being constructed from an entirely different center of ossification.

The first cohesive theory of primate origins was developed by the British anatomists Sir Grafton Elliot Smith (1913) and Frederic Wood Jones (1916). They proposed that arboreal life in and of itself was the key context in which natural selection would favor primate attributes: grasping hands and feet obviously functioned to hold onto branches, running and leaping in an arboreal environment required binocular vision for judging distance, sophisticated brains were

Table 2.1. The living mammalian orders

Monotremata	echidnas, platypus
Marsupialia[a]	opossums, kangaroos, koalas, etc.
Afrosoricida	tenrecs, golden moles
Insectivora	hedgehogs, shrews, moles, etc.
Macroscelidea	elephant shrews
Scandentia	tree shrews
Primates	primates
Dermoptera	colugos
Chiroptera	bats
Rodentia	rodents
Lagomorpha	rabbits, pikas
Edentata	sloths, anteaters, armadillos
Pholidota	pangolins
Tubulidentata	aardvark
Hyracoidea	hyraxes
Proboscidea	elephants
Sirenia	manatees, dugongs
Carnivora	dogs, cats, bears, mongooses, hyenas, etc.
Pinnipedia	seals, sea lions, walruses
Cetacea	whales, porpoises, dolphins
Perissodactyla	tapirs, rhinoceroses, horses
Artiodactyla	pigs, hippopotami, camels, deer, antelope, etc.

[a] May be two or up to seven orders.

required to process the complex three-dimensional space of the canopy, while the sense of smell was of diminished value in the shifting air of the treetops. According to Elliot Smith and Wood Jones, the simple step of departing from a terrestrial habitat in favor of an arboreal one catalyzed the suite of selective pressures that would yield an ancestral primate. This elegant theory of primate origins was widely adopted and remained unchallenged for half a century.

In the early 1970s the arboreal theory was tackled head on by anthropologist Matt Cartmill (1972, 1974*c*) who pointed out what afterwards seems so obvious: many arboreal mammals never evolved primate-like attributes. Among the non-primate orders that contain species largely or wholly arboreal are Marsupialia, Dermoptera, Scandentia, Chiroptera, Rodentia, Edentata, Pholidota, Hyracoidea and Carnivora. A primate-like trait shows up here or there among arboreal mammals, but never the whole set of primate traits, and never in a consistent pattern that would allow arboreality *per se* to be credited as the key selective pressure.

Primate relatives

By the time Cartmill challenged the arboreal hypothesis, new information and ideas were available concerning the closest living relatives of primates, and about the fossil record of primates and their kin. The arboreal theory of primate origins made the best sense if primates sprang right out of a basal, primitive insectivore-like animal foraging through leaf litter on the forest floor. However, it had long been suspected that the living orders of mammals most closely related to Primates might be ones that are all partly or largely arboreal: Scandentia, Chiroptera, Dermoptera, plus an extinct group called Plesiadapiformes (orders that are sometimes collectively referred to as archontans). If ancestral archontans were already experimenting with arboreality, then this suggests that the key innovation leading to primate origins involved some other particular adaptive shift in an arboreal setting.

The history of interpreting the relationship of tree shrews (Scandentia) to primates is long and complex (Gregory, 1910, 1913; Jenkins, 1974; Luckett, 1980; Martin, 1990). At times the tree shrews have been included within the order Primates, a position most prominently espoused by Wilfrid E. Le Gros Clark (e.g., 1959) who listed several specialized characters linking tree shrews to the primates (see also Martin, 1990). Le Gros Clark considered the possibility that the shared attributes arose convergently but dismissed that option because of what he interpreted as a detailed pattern of resemblance. Robert D. Martin (1968*a*, 1968*b*) was largely responsible for changing the consensus view on tree shrews, throwing them out of Primates based in part on reproductive and developmental characters. Since the 1970s, tree shrews have been placed typically in their own order, Scandentia (Butler, 1972). Ironically, more recent analyses suggest that the closest living relative of Primates may indeed be Scandentia (Wible & Covert, 1987; Wible & Martin, 1993; Wible & Zeller, 1994), meaning that Le Gros Clark would be correct in his phylogeny, if not widely followed in his choice of Linnean categorical level at which to express the link.

The tree shrew situation is simple compared to the tangled history of how primates have been viewed in relation to Plesiadapiformes, an extremely diverse, extinct group of mammals usually classified together at approximately ordinal level (Gingerich, 1976*a*; Gunnell, 1989; Martin, 1990; Rose, 1995*a*). Originally described on the basis of fossil teeth and jaws from the Paleocene and Eocene of Europe and North America, plesiadapiforms were classified as primates based on their molar structure. Several of the better known plesiadapiforms really do not conform to our view of the order Primates in terms of cranial construction (long low heads lacking postorbital bars), in the structure of the anterior teeth (enlarged, procumbent incisors superficially resembling those of rodents, sometimes separated by diastemata from the cheek teeth) and in limb morphology (lack of prehensile hand and foot structures, and the presence of claws). However, given the dental similarities and the lack of any more obvious place to classify the diverse plesiadapiforms, they were usually put in Primates (requiring researchers to refer to the non-plesiadapiform primates as "euprimates", or true primates). With plesiadapiforms classified as primates, the earliest known fossil plesidapiform, *Purgatorius* (which preceded the earliest euprimate by several million years) gained some notoriety as the world's earliest primate (van Valen & Sloan, 1965; Clemens, 1974; Buckley, 1997). Research by Cartmill (1972, 1974*c*) and others on the assemblage of characters shared by euprimates led to increasing skepticism that plesiadapiforms could be accommodated in the order. Like the tree shrews, the plesiadapiforms have been placed in their own order, Plesiadapiformes. The outcome of these developments is that anthropological interest in models of primate origins that involve the origination of a combined plesiadapiform–euprimate clade has declined (Szalay, 1981), especially in light of the superordinal phylogenetic uncertainties.

Comparable taxonomic flip-flopping has also characterized assessments of primate relationships to Dermoptera and Chiroptera. Based on similarities in the retina and the brain's visual system, Pettigrew (1989, 1991) concluded that a traditional bat suborder, the Megachiroptera (fruit bats and kin) were actually the closest relatives of the primates (see also Martin, 1986). This view has met with criticism (Wible & Novacek, 1988; Baker *et al.*, 1991). The order Dermoptera has also been promoted as the closest living relatives of the primates (by way of the suspect primate–plesiadapiform link) based on studies finding that colugos share cranial specializations (Kay *et al.*, 1990, 1992) and claimed postcranial specializations (Beard, 1990, 1993) with a plesiadapiform family, Paromomyidae. The arguments have been effectively rebutted (Krause, 1991; Runestad & Ruff, 1995; Hamrick *et al.*, 1999; Lemelin, 2000; Bloch & Silcox, 2001).

The phylogenetic arguments about the superordinal relationships of primates are important because if primate affinities could be resolved with confidence this might allow more precise assessments of the sequence of trait acquisition that characterized the adaptive origin of the true primates (Martin, 1990). However, it is fair to say that at this juncture we really do not know if primates are more closely related to Scandentia, Plesiadapiformes, Chiroptera or Dermoptera (MacPhee, 1993*a*; Adkins & Honeycutt, 1991; Allard *et al.*, 1996). These four orders and Primates are conveniently lumped together as "archontans" in what may be a true clade but which for lack of unambiguous evidence is often used as an informal grouping (Martin, 1990; MacPhee, 1993*b*; Lemelin, 2000). Trying to unravel the superordinal relations of Primates is not uniquely vexing; most of the other living orders have proven equally intractable (e.g., Novacek, 1992*a*; Stanhope *et al.*, 1998).

Simultaneous work on early fossil primates has shown that the primate package of specializations was present unambiguously in the early Eocene (55 million years ago). Eocene primates had a petrosal bulla, orbital convergence and the development of a postorbital bar, relatively large brains by Eocene standards, olfactory structures consistent with those of living primates, grasping hands and feet, and nails (Conroy & Rose, 1983; Dagosto, 1988, 1993; Franzen, 1988; Szalay & Dagosto, 1988; Rose, 1995*a*; Hamrick & Alexander, 1996; Anemone & Covert, 2000). At the same time, categorization of all early primates as sharing certain general similarities should not obscure the observation that there is great variability among early primates for some of these traits, such as the structure of the grasping hand and digits (Godinot, 1992*b*). The earliest fossil euprimates constitute a significant shift from any known archontan morphology and this significant shift must be accounted for in any viable theory of primate origins. The evidence now suggests that euprimates originated at a time and place currently unsampled or poorly sampled by the fossil record, so the intermediate stages between primates and their immediate ancestors remain unknown. When the primate adaptive package does appear in the fossil record, primates occur alongside other arboreal mammals that lack the assemblage of primate traits, more evidence that Elliot Smith's arboreal theory is wrong.

New models of primate origins

Cartmill (1972, 1974*c*, 1992) proposed an alternative model of primate origins called the visual predation hypothesis. Based on detailed comparative studies, Cartmill found that animals relying on vision to hunt and track prey items before capturing them with the extremities – like cats and raptorial birds – have high degrees of optic convergence and adept prehension in their extremities. Cartmill also emphasized that while orbital convergence provides stereoscopy it also reduces parallax, which actually diminishes depth perception at a distance. Extreme convergence seems to be most useful "in animals that needed a wide field of stereoscopic vision at close range" (Cartmill, 1992: 107). He proposed that the ancestral primate was an insect-hunting specialist that hunted in the fine branches of forest canopy or undergrowth. To Cartmill, the key primate innovation was not the move into the trees *per se*, it was that an arboreal mammal came to rely on visually directed predation to capture arthropods or other animal prey.

This hypothesis predicted that the earliest primate did not depend heavily on fruit, flowers or other plant matter, even though cheirogaleids and lorisids (often promoted as models of early primates) do just that (Charles-Dominique & Martin, 1970; Perret, 1995). Primatologist Robert Sussman, who had studied living prosimians in Madagascar, knew that most prosimians depended to some extent on the characteristic rich products of angiosperms (Sussman & Raven, 1978; Sussman, 1991, 1995, 1999). Sussman also noted that the primate-like visual system of megachiropteran bats could not be explained by the visual predation hypothesis because they feed almost exclusively on angiosperm products. In addition, Sussman was struck by the coincidence that the origin and radiation of euprimates followed directly upon a significant evolutionary upheaval among angiosperms in the Paleocene, including the first development of tropical forests with closed angiosperm canopies and the evolution of new, large fruit types (Sussman, 1995). He proposed that the new availability of rich, abundant fruits and flowers in the terminal branches of tropical forests provided a windfall resource that was taken advantage of by the earliest primate ancestors. According to Sussman, utilization of angiosperm products required sophisticated vision to spot colorful flowers and fruits at a distance and yet also demanded significant olfactory abilities to assess fruit and flowers up close. Grasping hands and feet were favored in order to negotiate the fine terminal branches that were the only non-flying access to the angiosperm products.

Both Cartmill's and Sussman's models are elegant and are supported by a fair share of comparative data. They both also have some difficulties (Cartmill, 1992; Crompton, 1995). In an attempt to directly test their predictions, I conducted a short field study on a small, prosimian-like, nocturnal marsupial of the Neotropics, *Caluromys derbianus*, a member of the family Didelphidae, which includes among other forms the familiar Virginia opossum (*Didelphis*) of North America. *Caluromys* differs from other didelphids in ways that resemble primates: it has the relatively largest brain of any didelphid, the largest eyes, the highest ratio of orbit size to snout length (perhaps reflecting a more primate-like reliance on vision rather than olfaction), digital proportions and manipulative behavior similar to those of small prosimians, relatively small litters of only two to four infants, relatively slow development, and unusually agile behavior and skillful climbing (Biggers, 1967; Phillips & Jones, 1968; Hunsaker & Shupe, 1977; Eisenberg & Wilson, 1981; Steiner, 1981; Rasmussen, 1990*a*; Lemelin, 1996). Thus, *Caluromys* potentially offered a test of whether its natural history was based on harvesting angiosperm products or visually tracking and grabbing insect prey.

The results of the field study revealed that *Caluromys* was significantly more arboreal than other didelphids (Rasmussen, 1990*a*). While *Caluromys* often rested on medium or small branches, it spent fully half of its foraging and locomotor time in fine terminal branches. It foraged for fruit at the tips of fine branches, which swayed and shook under the animal's weight, by hanging on with prehensile hands and feet while pulling branch tips with a hand to the front of the face for inspection. Fruit held in the hands was often consumed while hanging upside down supported by the hindlimbs and the semi-prehensile tail. These behavioral observations matched quite closely the predictions generated from Sussman's angiosperm hypothesis. At the same time, *Caluromys* adeptly snatched moths and other insects from the substrate or directly from the air using both hands while foraging in angiosperm terminal branches. Comparable two-handed insect-catching has been reported in the primate *Cheirogaleus* (Wright & Martin, 1995). Rather than refuting the predictions of one or both models, *Caluromys* seemed to fit both.

The behavior of *Caluromys* in the context of angiosperm utilization and insect capture suggested modifications to the models. Angiosperms and insects have a long history of coevolution, and a mammal that finds fresh flowers and fruit also finds insects. An animal the size of *Caluromys* (200–400 g) or a small primate crashing through the terminal branches of an angiosperm would shake up insects as certainly as it would find fruit or flowers. One of the weaknesses of the visual predation hypothesis was that it did not provide an ecological rationale for why an ancestral primate alone among insect eaters would come to specialize on visual predation of its prey. In crashing terminal branches, visually directed predation may be the only viable method of acquiring insects. So the question of "Why did a primate ancestor become a visual predator?" can be answered by drawing on Sussman's ecological model – it did so because it was foraging in the demanding environment of the terminal branches for the combined windfall of angiosperm products and associated insects.

This revised angiosperm–insect hypothesis finds support in the work of other primatologists. Crompton (1995: 26) evaluated the locomotor adaptations of primates that are principal predators of animal prey versus those that rely largely on angiosperm products, and in the end he regarded "insectivory as more probably opportunistic than specialized in the common primate ancestor". In a broad synthesis of primate origins and evolution, Martin (1990: 659) analyzed variations on the theme of the arboreal primate and concluded that: "In the tropical forests inhabited by the earliest primates, small fruits would doubtless have been concentrated in the fine branch zone of shrubs and trees and arthropod prey . . . could have been collected in an opportunistic fashion."

New comparative studies

Subsequent laboratory research on hand use and locomotion in primates and other mammals has shown that *Caluromys* is primate-like in morphology and behavior (Lemelin, 1996; Larson *et al*., 2000). In both the hands and feet, *Caluromys* and a smaller arboreal didelphid, *Marmosa*, exhibit longer digits in relation to metacarpal and metatarsal length than do any other didelphids (Lemelin, 1999). These long-digited proportions reflect greater prehension in *Caluromys*, which resembles the proportions and prehension of cheirogaleid prosimians. Lemelin (1999) interpreted the hand proportions of *Caluromys* and cheirogaleids as functioning to locomote in fine branches and vines. He suggested that locomoting on small-diameter branches and catching moving prey with the hands "were interdependent factors in the evolution of prehensile extremities of early primates" (Lemelin, 1999: 173).

The functional morphology of the hands in utilizing small-diameter supports has been further clarified by Hamrick (1998; see also Lemelin, 2000), who showed that small platyrrhines utilizing small-diameter supports have larger apical pads and more complex epidermal ridge development than do species utilizing larger-diameter supports, which have acquired secondarily derived claws. More recently, Hamrick (2001) showed that the distinctive pattern of long digits and relatively short metacarpals which is essential for fine prehension was already present in Eocene primates. Hamrick and Lemelin both concluded that prehension in a small-branch milieu was an important part of the adaptive profile of early primates.

The fossil record

Alongside the work on living mammals, we need new fossils of very early primates. More than just a few teeth or jaws are required – the primate assemblage of traits cannot be fully assessed in fossils without having crania and partial hand and foot skeletons. Both of those wish-list items are fairly rare in the early Eocene record. Currently, the best candidate for the title of world's earliest known euprimate is *Altiatlasius*, a small mammal from the Paleocene of Morocco (Sigé *et al.*, 1990), known only from several isolated teeth. While these do indeed look quite a bit like the teeth of true primates, so did the molars of plesiadapiforms and several other extinct groups that with more fossil material eventually proved to be non-primate. In any case, simply identifying a fossil as the earliest record of a primate does not mean that it is the actual point of evolutionary origins of the order.

The sudden, widespread appearance of full-blown euprimates occurs at the base of the Eocene in the northern hemisphere, and this is the beginning of the story of the rest of this book. The abrupt appearance suggests a migration event from an unknown source, probably coinciding with climatic shifts towards greater warmth and tropicality near the Paleocene–Eocene transition. The most likely candidates as source areas for the ancestral primate are Africa or the Indian subcontinent (Krause & Maas, 1990). In North America, the earliest euprimates are *Cantius* and *Teilhardina*, followed shortly by *Steinius* and others (Gingerich, 1986, 1993*a*; Rose & Bown, 1991). In Europe, *Teilhardina* also appears, this time alongside *Donrussellia* and *Cantius* (Godinot, 1978; Godinot *et al.*, 1987). Interestingly, the very earliest undoubted euprimate species are also known only from jaws and teeth, but they have earned their "undoubted" status by being clearly related in a stratigraphic context to younger species of the same genera that are represented by relatively complete fossil remains, including crania and limb bones. The best known of these earliest genera is *Cantius*, represented by crania and partial skeletal remains in stratigraphic levels significantly higher than the first dental appearance of the taxon in the fossil record (Rose & Walker, 1985; Covert, 1988; Gebo, 1988).

The adaptations of early primates are becoming clearer. They were small animals with grasping hands, which were particularly adapted for maneuvering in small branches. They were stereoscopic. Judging from the shape of cusps and crests, the earliest euprimates were animals that could have fed on both insects and fruits. All have trenchant slicing crests and high trigonids useful for puncturing and tearing insect exoskeletons They also have broader, flatter talonids than what are found in most true insect specialists, suggesting a fruit, nut or seed component to the diet. On a continuum of crestiness to bulbousness, *Teilhardina* lies in the direction of emphasizing insects over plants, while *Cantius* is probably more of a fruit-eater that also took insects. The fossil record at this coarse level of resolution indicates the presence of an early Eocene radiation of small-bodied, arboreal fruit- and insect-eaters moving among the fine branches, perhaps the terminal ones, where they would find an abundance of invertebrates attracted to rich new radiations of angiosperm products.

The earliest fossil primates and the evolution of prosimians

3 | The earliest fossil primates and the evolution of prosimians: Introduction

HERBERT H. COVERT

Introduction

Since the exclusion of the Plesiadapiformes from the order Primates following critical taxonomic discussions during the 1960s, 1970s and 1980s (Martin, 1968*a*; Cartmill, 1972, 1974*c*; Wible & Covert, 1987), it has been widely recognized that the Omomyoidea and Adapoidea represent the earliest well-known primate adaptive radiations. Their adaptive radiations have been shown during the past decade to be even broader than previously recognized. For example, in recent years approximately a dozen new adapoid species have been identified from the Eocene of North Africa and the Middle East (Hartenberger & Marandat, 1992; Gheerbrant *et al.*, 1993; Simons *et al.*, 1995*c*; Simons, 1997*a*; Simons & Miller, 1997), new adapoids and omomyoids have been described from Asia (Beard & Wang, 1991; Beard *et al.*, 1994; Gingerich *et al.*, 1994) and new taxa continue to be described from North American and European Eocene sediments (Bown & Rose, 1991; Beard *et al.*, 1992*a*; Godinot *et al.*, 1992; Williams & Covert, 1994; Gunnell, 1995*a*, 1995*b*; Köhler & Moyà-Solà, 1999). These new discoveries add to the importance of studying these creatures for both phylogenetic and adaptive reasons. Phylogenetic debates of omomyoid–adapoid relationships, and how they are related to the more recent prosimian and anthropoid radiations continue (Covert, 1988; Beard *et al.*, 1991, 1994; Covert & Williams, 1994; Fleagle & Kay, 1994; Rose *et al.*, 1994; Simons & Rasmussen, 1994*b*; Dagosto *et al.*, 1996). Recent studies of the adaptations of the earliest primate radiations continue to challenge the long held notion that adapoids were essentially lemur-like creatures and that omomyoids were essentially tarsier-like creatures (Gebo *et al.*, 1991; Covert & Hamrick, 1993; Dagosto, 1993; Covert, 1995, 1997; Anemone & Covert, 2000). Furthermore, such studies reveal that the range of earliest primate adaptations overlaps that of extant prosimians.

The 1990s saw a number of changes in our understanding of early primate evolution as noted above. The most dramatic series of new discoveries are those documenting that the anthropoid adaptive radiation was well under way by the late Eocene and that tarsiers may have diverged from other primates before the middle Eocene. Eocene anthropoids were initially documented at the Fayum Depression in Egypt at the late Eocene locality L-41 (Simons, 1989, 1990, 1995*b*; Simons & Rasmussen, 1994*a*). Eocene anthropoids have also been recovered in Algeria (Godinot & Mahboubi, 1992), in China (Beard *et al.*, 1994, 1996; Gebo *et al.*, 2000*a*), and in Thailand (Chaimanee *et al.*, 1997). Middle Eocene tarsiers have been recovered from two regions in China (Beard, 1998*a*; Beard et al., 1994). There are a number of fascinating aspects of these recent discoveries. For example, the early presumed anthropoids from Asia nearly bracket the omomyoids and adapoids in terms of estimated body mass. When combined with the late Eocene anthropoids of Africa evidence for a significant early adaptive radiation by this group is clear. In addition, it appears that anthropoids and tarsiers both must have diverged from other primates by the beginning of the Eocene.

By the end of the Eocene the vast majority of adapoids and omomyoids had apparently become extinct. Yet new discoveries have recently added to the evidence that these creatures lived on in reduced numbers into the Oligocene of Europe (Köhler & Moyà-Solà, 1999), into the Oligocene and even the Miocene of Asia (Beard, 1998*a*), and the late Oligocene of North America. Finally, it is possible that one of the omomyoids gave rise to the tarsiids and eosimiids of China by the early Eocene and that the lemuroids and lorisoids are derived from one of the adapoids. There is, however, no consensus on which of the omomyoids and adapoids gave rise to the more recent occurring members of the haplorhine and strepsirhine radiations.

Classification

As noted in the introduction, the Adapoidea and the Omomyoidea are the earliest well-known primates. Members of both superfamilies are characterized by a series of derived traits diagnostic of the order Primates including the presence of a petrosal bulla, a postorbital bar, orbital convergence, an opposable hallux, a flattened nail on the hallux, and some anterior elongation of the calcaneus (Figs. 3.1–3.3). While only a few species are known from specimens illustrating all of these traits, the widespread distribution of these diagnostic features indicates that they were present universally throughout this radiation of creatures.

For much of the past century these primates were allocated to either the Adapidae or the Omomyidae with the former

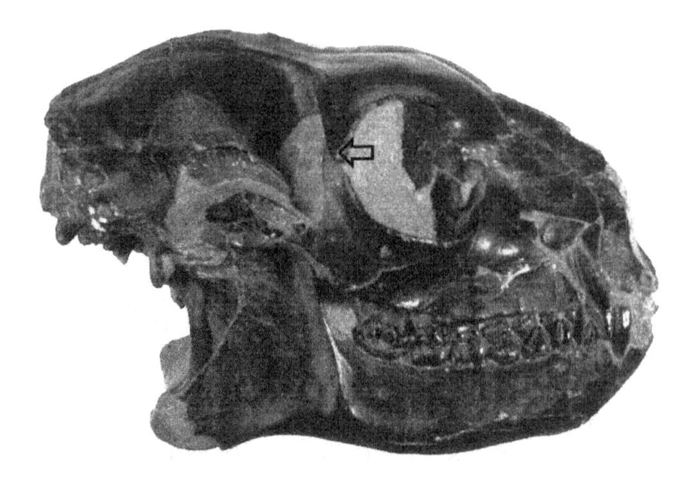

Fig. 3.1 *Smilodectes gracilis*, a middle Eocene adapoid. Lateral view of the skull. The arrow denotes the postorbital bar, a diagnostic trait for the primate order. All omomyoids and adapoids known from cranial material have a postorbital bar.

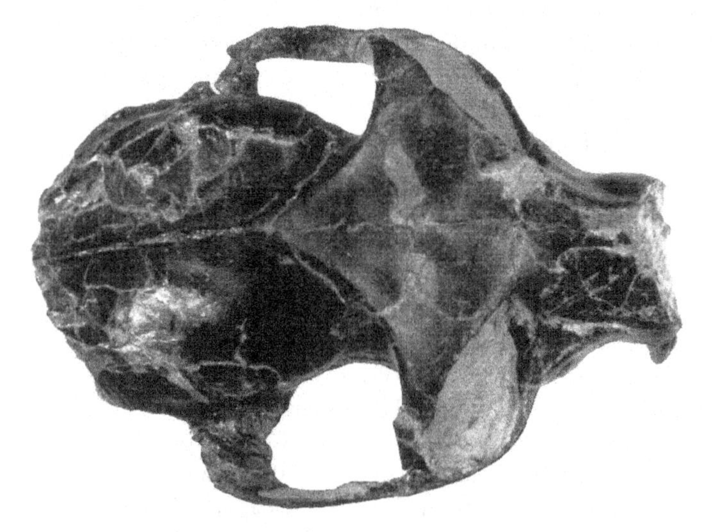

Fig. 3.2 *Notharctus tenebrosus*, a middle Eocene adapoid. Superior view of the skull, which has has convergent orbits, a diagnostic trait for the primate order. All omomyoids and adapoids known from cranial material have convergent orbits.

being grouped in one way or another with the strepsirhine primates and the latter with the tarsiids. It has become apparent during the last 20 years, however, that the diversity within the taxa traditionally placed in the adapids and omomyids is too great to be accommodated by a familiar rank. Thus, these groups are now granted superfamily rank (Table 3.1). Furthermore, the adapoids are placed in the semiorder Strepsirhini because they are characterized by a handful of derived tarsal features that are diagnostic to this group (Beard *et al.*, 1988; Covert, 1988; Dagosto, 1988). Omomyoids are placed in the semiorder Haplorhini because they are characterized by a handful of derived basicranial traits diagnostic to this group (Ross, 1994; Kay *et al.*, 1997*b*).

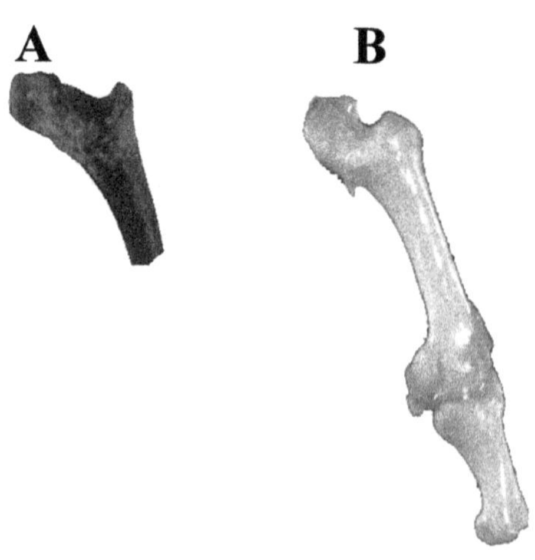

Fig. 3.3 *Omomys carteri*, a middle Eocene omomyoid (A, proximal hallucial metatarsal) and *Mirza*, a modern lemur (B, hallucial metatarsal of and proximal phalanx). Note that the proximal articular surface of the hallucial metatarsal is a saddle-shaped joint, a diagnostic trait for the primate order. This features allows for an opposable hallux.

Taxonomic diversity

Both the adapoids and omomyoids include a substantial number of taxa and the number continues to grow. It is interesting to see how their taxonomic composition has changed as conceptualized by Fleagle (1988, 1999) in the first two editions of his influential text, *Primate Adaptation and Evolution*. As denoted in Table 3.2, Fleagle (1988) recognized 66 species in 26 adapoid genera and 70 species in 34 omomyoid genera. In a single decade this changed to 88 species in 37 adapoid genera and 93 species in 41 omomyoid genera, a 33% increase in specific diversity. While this in part might represent a more liberal approach to classification at the specific level, it is more likely a reflection of the continued intensification of fieldwork. At present then, more than 180 species of adapoids and omomyoids are recognized and the vast majority of these taxa are known solely from Eocene deposits. This clearly documents a very successful adaptive radiation simply in terms of taxonomic diversity.

Adaptive diversity

We now also have a much better appreciation of early primate adaptive diversity. This reflects not only the dramatic improvement of the fossil record but also a much better understanding of the ecology and behavior of the modern prosimians (Martin *et al.*, 1974; Charles-Dominique et al., 1980; Alterman *et al.*, 1995; Fleagle, 1999). Such information provides a rich context in which to analyze the adaptation of the adapoids and omomyoids. What is an adaptive radiation? According to Fleagle (1999: 571) adaptive radiation is defined as "a group of closely related organisms that have evolved morphological and behavioral features

Table 3.1. General classification of the earliest primates

Order Primates
Semiorder Strepsirhini
Superfamily Adapoidea
Family Notharctidae
Subfamily Notharctinae
Genera: *Cantius, Copelemur, Notharctus, Smilodectes, Pelycodus, Hesperolemur*
Subfamily: Cercamoniinae
Genera *Donrussellia, Protoadapis, Europolemur, Periconodon, Caenopithecus,Pronycticebus, Cercamonius, Anchomomys, Huerzeleria, Buxella, Agerinia, Panobius, Mahgarita, Djebelemur, Aframonius, Omanodon, Shizarodon, Wadilemur*
Family Adapidae
Genera *Adapis, Cryptadapis, Microadapis, Leptadapis, Adapoides*
Family Sivaladapidae
Genera *Hoanghonius*
Family *Incertae sedis*
Genera *Azibius, Lushius, Rencunius, Wailekia*
Semiorder Haplorhini
Superfamily Omomyoidea
Family Omomyidae
Subfamily Anaptomorphinae
Genera *Teilhardina, Anaptomorphus, Gazinius, Tetonius, Pseudotetonius, Absarokius, Tatmanius, Strigorhysis, Aycrossia, Trogolemur, Sphacorhysis, Anemorhysis, Tetonoides, Arapahovius, Chlororhysis, Washakius, Shoshonius, Dyseolemur, Loveina*
Subfamily Omomyinae
Genera *Omomys, Chumashius, Steinius, Uintanius, Jemezius, Macrotarsius,Hemiacodon, Yaquius, Ourayia, Wyomomys, Ageitodendron, Utahia, Stockia, Chipetaia, Asiomomys*
Family Microchoeriidae
Genera *Nannopithex, Pseudoloris, Necrolemur, Microchoerus*
Family *Incertae sedis*
Genera *Ekgmowechashala, Rooneyia, Kohatius*
Semiorder *Incertae sedis*
Genera *Altanius, Altiatlasius*

Modified from Covert (1997) and Fleagle (1999). For a more detailed classification see Fleagle (1999).

Table 3.2. The change in taxonomic composition of Adapoidea, Omomyoidea, extinct Lorisoidea and extinct Tarsiidae as reported by Fleagle (1988) and Fleagle (1999). The numbers in the columns report the number of genera and species recognized for each taxon

	1988	1999
Adapoidea		
Notharctinae	5, 20	6, 25
Cercamoniinae[a]	—	18, 42
Adapinae	14, 38	5, 12
Sivaladapidae	3, 4	3, 4
Adapidae *incertae sedis*	4, 4	5, 5
Subtotal	26, 66	37, 88
Omomyoidea		
Anaptomorphinae	13, 34	19, 47
Omomyinae	15, 20	16, 24
Microchoeridae	4, 14	4, 20
Omomyoidea *incertae sedis*	2, 2	2, 2
Subtotal	34, 70	41, 93
Lorisoidea		
Galagidae	3, 6	3, 7
Lorisidae	2, 3	2, 3
Pleisiopithecidae[a]	NR	1, 1
Subtotal	5, 9	6, 11
Tarsiidae	2, 2	3, 4
Total	67, 147	87, 196

[a] Please note that the Cercamoniinae were included with the Adapinae in 1988 and that the Plesiopithecidae were unknown in 1988.

enabling them to exploit different ecological niches". While it is quite difficult, if not impossible, to reconstruct a complete behavioral profile for a group of extinct creatures certain attributes can be documented. These include a consideration of taxonomy, body mass, diet, locomotion and activity cycles. Body mass, diet, locomotion and activity cycle have traditionally been estimated or hypothesized for extinct primates using a comparative method. As outlined by Kay & Cartmill (1977), Kay & Covert (1984) and Anthony & Kay (1993), various comparative methods have been utilized and some approaches appear to be more effective than others. Specifically, one must carefully consider the functional relationship between a specific morphological feature and the adaptive role of interest. If one finds that there is a consistent or nearly consistent association of the adaptive role with the specific morphological feature (but not necessarily the reverse) this adaptive role can be attributed to extinct animals that are characterized by this specific morphological trait. When such information is gathered for a group of closely related creatures, the omomyoids and the adapoids, for example, one can describe the breadth of their adaptive radiations. A sound understanding of the biology of the extant animals that are used as the reference sample is essential. The earliest primates most closely resemble the prosimian primates among extant forms.

A great deal of information has been gathered on the biology of prosimians during the past thirty years (Martin *et al.*, 1974; Doyle & Martin, 1979; Charles-Dominique *et al.*, 1980; Tattersall, 1982; Kappeler & Ganzhorn, 1993; Alterman *et al.*, 1995; Fleagle, 1999). In addition, we have learned that the modern prosimians express a rich diversity of social behavior, dietary specialization and locomotor behavior. Adapoid primates were long known as "lemur-like" and omomyoids "tarsier-like". Because lemurs are so diverse in their adaptations "lemur-like" does not provide a specific statement about the biology of an extinct primate group. In contrast, the description "tarsier-like" does have a specific meaning because all modern-day tarsiers are small, nocturnal, acrobatic leaping faunivores. As reviewed below, adapoids are somewhat lemur-like in the diversity of their adaptive radiation (although this does not allow one to understand anything specific about the adaptation of any given adapoid species) and the vast majority of omomyoids were not tarsier-like.

Extant prosimians

Taxonomic diversity

Today we recognize approximately 24 genera and 54 species of extant prosimians (Fleagle, 1999). The majority of these taxa are restricted to the island of Madagascar. There are 7 genera and 14 species residing in Africa and 3 genera and 8 species in Asia. How does the omomyoid and adapoid diversity compare? During the Eocene North America was home to 41 genera and 82 species of prosimians and Europe was home to 22 genera and 69 species of prosimians. At present 12 genera and species of prosimians are known from the Eocene of Asia whereas approximately 10 genera and species are known from the Eocene and early Oligocene of Africa. As previously noted, our knowledge of early primates is quite limited in regards to African and Asian deposits thus the diversity identified is likely to grow dramatically in the next few years. The diversity of both North American and European Eocene primates is much greater than the modern diversity of prosimians. These numbers are not directly comparable, of course, because the Eocene includes nearly 20 million years of Earth history. Just the same, however, the taxonomic diversity seen during the Eocene for primates reflects that they were experiencing a series of successful adaptive radiations.

Body mass estimation

Extant prosimian primates range in body mass from about 40 g for the smallest mouse lemurs to nearly 7 kg for the indri (Fleagle, 1999). Thus these creatures are small to medium size mammals. It is interesting to note, however, that during the recent past, Madagascar was home to more than 15 additional species of prosimians with estimated body masses ranging from about 10 kg to about 200 kg, a group of medium to quite large mammals! Eocene primates overlap comfortably with the extant prosimians (excluding the subfossil Malagasy forms). Body mass has been estimated for the vast majority of these Eocene creatures based on dental regressions and the range of body mass estimates reported here comes from Gingerich *et al.* (1982) and Fleagle (1999). In addition, a number of researchers have recently generated regressions for estimating body mass for extinct primates based on various skeletal measurements (e.g., Dagosto & Terranova, 1992; Payseur *et al.*, 1999).

While there is substantial overlap the vast majority of omomyoids were smaller than the vast majority of adapoids. Omomyoids appear to have ranged in body mass from about 35 g to 1775 g with over 75% of the species being smaller than 400 g. Omomyoids show a body size range and distribution that is quite similar to that of the extant African and Asian prosimians. Adapoids appear to have ranged in body mass from about 100 g to 7 kg with over 75% of the species being larger than 1 kg. Adapoids show a body size range and distribution that is quite similar to that of the extant Malagasy primates.

Dietary specialization

Extant prosimians show an impressive array of dietary specializations. Tarsiers, as stated above, are faunivores and have never been observed eating plant matter in the wild. A number of other prosimians are primarily insectivorous including some of the small galagos, the slender loris and the angwantibo. Both the needle-clawed bushbaby and the fork-marked lemur specialize on gum whereas the indriids, sportive lemurs and gentle lemurs have diets that include a preponderance of foliage. Many of the other prosimians have largely frugivorous diets while a few species including the ring-tailed lemur and some of the mouse lemurs are omnivores. Using comparative methods focusing primarily on anatomical correlates to specific diets among extant prosimians a number of researchers have offered hypotheses about the diets of omomyoids and adapoids (e.g., Covert, 1986, 1997; Strait, 1991, 1997). The majority of these creatures appear to have been primarily frugivorous including small and moderate sized omomyoids such as *Necrolemur* and *Absarokius* and medium and large sized adapoids such as *Cantius*, *Pelycodus* and *Cercamonius*. A number of medium and large sized adapoids including *Smilodectes*, *Notharctus*, *Adapis* and *Leptadapis* appear to have been primarily folivorous creatures. Finally, a few of the smaller omomyoids including *Teilhardina* and *Pseudoloris* and a few of smallest adapoids including *Anchomomys* and *Donrussellia* appear to have been primarily insectivorous creatures.

Locomotor specialization

Extant prosimians also show an impressive array of locomotor adaptations although all are primarily arboreal creatures with only the ring-tailed lemur spending up to one-third of its time on the ground. Tarsiers and some of the smaller galagos are acrobatic leapers, often moving from vertical substrate to vertical substrate. The indriids, the sportive lemur and the gentle lemur are also acrobatic leapers. The majority of extant prosimians can be described as active arboreal quadrupeds. In contrast, all of the lorises are highly

Table 3.3. Limb indices for modern prosimians and two Eocene adapoids. For the extant forms the mean for each index is followed by the range. All of the modern forms with an intermembral index mean below 65 are acrobatic leapers

Taxa	Intermembral index	Brachial index	Crural index
Modern prosimians			
Eulemur fulvus	70.5 (69.2–71.7)	95.0 (92.8–97.8)	104.3 (101.8–106.7)
Varecia	70.8 (70.2–71.7)	91.7 (91.5–91.8)	97.1 (94.5–98.7)
Hapalemur	62.4 (61.0–64.4)	96.9 (93.1–102.7)	112.7 (105.9–116.3)
Lepilemur	63.0 (58.9–65.7)	91.8 (87.4–96.8)	113.2 (109.2–117.8)
Avahi	56.8 (56.0–57.7)	85.1 (83.3–86.3)	118.5 (117.5–120.0)
Propithecus	60.5 (59.2–62.6)	86.3 (85.2–87.2)	106.3 (103.0–109.3)
Indri indri	64.7 (63.6–65.7)	87.4 (85.2–87.2)	120.0 (119.0–121.1)
Eocene adapoids			
Smilodectes gracilis[a] (USNM 25686)	60.8	83.0	89.0
Notharctus pugnax[a] (CM 11910)	62.9	84.5	97.5

[a] Note that *Notharctus* and *Smilodectes* both have very low intermembral indices. Also note that these Eocene forms have a lower crural index than all of the extant forms.

specialized slow-moving arboreal quadrupeds. A number of researchers have used comparative methods focusing primarily on anatomical correlates to specific modes of locomotion among extant prosimians to develop hypotheses about the locomotor behaviors of omomyoids and adapoids (e.g., Gregory, 1920; Simpson, 1940; Szalay & Dagosto, 1980; Dagosto, 1983, 1985, 1993; Godinot & Dagosto, 1983; Covert, 1986, 1995; Rose & Walker, 1985; Gebo, 1988; Gebo *et al.*, 1991; Anemone & Covert, 2000). These Eocene primates varied significantly in locomotor behavior. *Adapis* shares a number of distinctive anatomical traits with the lorises and apparently was a specialized slow-moving arboreal quadruped. Some omomyoids such as *Absarokius* and *Omomys* appear to have been active arboreal quadrupeds with a locomotor adaptation most similar to the dwarf lemurs although both have anatomical features that hint at a higher frequency of leaping for these Eocene primates. Some of the adapoids such as *Cantius* and *Pronycticebus* were active arboreal quadrupeds with a locomotor adaptation most similar to the brown lemurs. Other adapoids such as *Notharctus* and particularly *Smilodectes* appear to have been specialized leapers more similar to the indriids although they do not show any specializations for vertical clinging (Table 3.3). Finally, at least one of the omomyoids, *Necrolemur*, appears to have had a leaping adaptation similar to the most acrobatic leaping galagos and tarsiers, although, again, it does not show any specializations for vertical clinging.

Activity cycle

All of the extant prosimians that inhabit regions of Africa and Asia are nocturnal. The Malagasy primates show a range of activity cycles with some being predominantly diurnal creatures such as the ring-tailed lemur, indris and sifakas, and others are nocturnal creatures such as the aye-aye, mouse lemurs, wooly lemurs and sportive lemurs. Yet others are cathemeral, active in the day during certain times of the year and active at night during others (red-bellied and mongoose lemurs). Each of the nocturnal primates has eyes (and orbits) that are larger than those of diurnal forms. As with diet and locomotor adaptations researchers have compared Eocene prosimians with extant forms to make inferences about behavior. Based on orbit size it appears that most omomyoids were nocturnal. Species that preserve sufficient portions of the orbits that appear to have been nocturnal include *Necrolemur*, *Shoshonius*, *Omomys* and *Tetonius*. The only omomyoid that appears to have had an orbit size consistent with diurnality is *Rooneyia*. In contrast to the omomyoid condition, the vast majority of adapoids appear to have been diurnal including *Cantius*, *Notharctus*, *Smilodectes*, *Adapis* and *Leptadapis*. The only adapoid that appears to have had an orbit size consistent with nocturnality is *Pronycticebus*.

Behavioral profile

Omomyoids and adapoids show an impressive diversity of adaptations in regards to body size, diet, locomotion and activity cycle and thus it is not particularly productive to describe them as "tarsier-like" (since most omomyoids are not like tarsiers in diet or locomotion) or as "lemur-like" (since this description fails to capture the range of adaptive diversity shown by the adapoids). Just the same, when the adaptive radiations of these creatures are considered together it is apparent that there is extensive overlap with the adaptive radiation exhibited by extant prosimians. Another fruitful way to summarize their adaptive diversity is to consider the adaptations of some of the better represented genera. Table 3.4 provides simple adaptive profiles for two omomyoids, *Omomys carteri* and *Necrolemur antiquus*, and three adapoids, *Notharctus tenebrosus*, *Smilodectes gracilis* and *Adapis parisiensis*. Both

Table 3.4. Adaptive profiles for two omomyoids (top row) and three adapoids (bottom row)

Taxon	*Omomys carteri*		*Necrolemur antiquus*
Body mass	170–290 g		320 g
Diet	fruit and insects		frugivorous
Locomotion	active arboreal runner and leaper		active arboreal leaper
Activity pattern	nocturnal		nocturnal
Taxon	*Notharctus tenebrosus*	*Smilodectes gracilis*	*Adapis parisiensis*
Body mass	3100 g	1950 g	1500 g
Diet	folivorous	folivorous	folivorous
Locomotion	active arboreal leaper and runner	active arboreal leaper and runner	slow, deliberate arboreal quadruped
Activity pattern	diurnal	diurnal	diurnal

of these omomyoids were small animals, although both are substantially larger than extant tarsiers. Both were also presumably nocturnal because both have relatively very large orbits. They appear to have differed, however, in dietary and locomotor behaviors. The molars of *Necrolemur* are low crowned with poorly developed shear features (characteristic of frugivory) whereas the molars of *Omomys* are higher crowned with fairly well developed shear features (characteristic of diet with significant amounts of insects). The skeleton of *Omomys* closely resembles that of *Mirza*, an active arboreal runner and leaper, differing only in features of the hip and knee that indicate a slightly higher frequency of leaping in *Omomys* (Figs. 3.4, 3.5). The skeleton of *Necrolemur* resembles that of *Tarsius* in a number of hindlimb and foot features consistent with a specialized leaping adaptation. All of the adapoids in Table 3.4 appear to have been diurnal folivores because they have relatively small orbits, moderate body masses, and well-developed shearing features on their molars. They do show variation in locomotor adaptations, however. As identified by Dagosto (1983) *Adapis* closely resembles lorises in a range of forelimb and hindlimb features that indicate slow, deliberate arboreal quadrupedalism. In contrast, both *Smilodectes* and *Notharctus* show features consistent with frequent leaping behavior in an arboreal milieu. The adaptive profiles of these three adapoids clearly demonstrate that the expression "lemur-like" fails to describe accurately their adaptive behaviors.

Phylogeny

Significantly, much of the debate about phylogenetic relationships between the earliest primates and more recent forms essentially revolves around our understanding of the relationships among modern primates. A number of scholars have argued that extant tarsiers and anthropoids (haplorhines) are more closely related to one another than either is to the "non-tarsier" prosimians (strepsirhines) and that this must be considered to determine the relationships among adapoids, omomyoids and more recent primates (e.g., Rosenberger & Szalay, 1980; Szalay, 1976; Cartmill & Kay, 1978; Ross, 1994; Kay *et al.*, 1997*a*). Each of these scholars has argued that anthropoids and tarsiers are derived from an omomyoid. Others, however, argue that the haplorhine/strepsirhine dichotomy is far from certain (e.g., Gingerich, 1980*b*; Rasmussen, 1986; Simons & Rasmussen, 1994*a*, 1994*b*) and these authors have concluded that anthropoids are derived from an adapoid. In addition, the past few years have seen the description of a number of new anthropoid primates that has added fuel to this debate (e.g., Simons, 1989, 1990; Beard *et al.*, 1994, 1996; Simons & Rasmussen, 1994*a*, 1994*b*; Chaimanee *et al.*, 1997). Paradoxically, the eosimiids of China have been argued to provide strong support for an omomyoid origin for anthropoids whereas *Siamopithecus* and the late Eocene anthropoids from the Fayum have been argued to be adapoid-like. While it is safe to say that there is no consensus on this issue at present I find that Kay *et al.*'s (1997*a*) argument the most compelling to date. They provide sufficient cladistic evidence supporting the haplorhine/strepsirhine dichotomy among modern primates and argue that the division between these groups dates to before the beginning of the Eocene. Furthermore, they consider adapoids as basal strepsirhines and omomyoids as basal haplorhines. The eosimiids are linked to other early anthropoids on the basis of a handful of dental traits forming a plausible intermediate form between an omomyoid and the later occurring anthropoids from the Fayum.

As reviewed by Rose (1995*a*) the oldest omomyoids and adapoids (from the earliest Eocene) closely resemble one another in dental morphology and differ only in minor skeletal and cranial features indicating that they likely shared a common ancestor from the middle or late Paleocene. In addition, as noted by Fleagle (1999), two of the earliest known primates, *Altanius* and *Altiatlasius*, known solely from dental material, cannot be attributed to either the omomyoids or adapoids with confidence (although they are more likely to be omomyoids given their tiny size). Moreover, Beard (1998*a*) argued that *Altiatlasius* might even be a member of the basal anthropoid radiation. All of this reflects the fact that cranial and skeletal material is often needed to separate out primitive members of various primate groups.

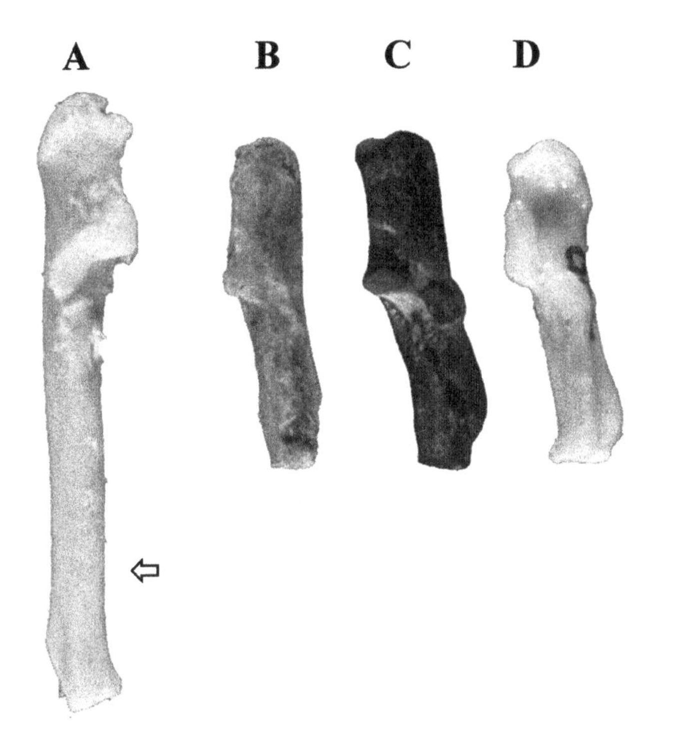

Fig. 3.4 The calcaneus of *Galago senegalensis* (A), *Omomys carteri* (B, C) and *Mirza* (D). Note that the calcaneus of *Omomys* closely resembles that of *Mirza* and lacks the dramatic anterior elongation (arrow) characteristic of galagos and tarsiers. This is one of the many features in which omomyoids differ from tarsiers.

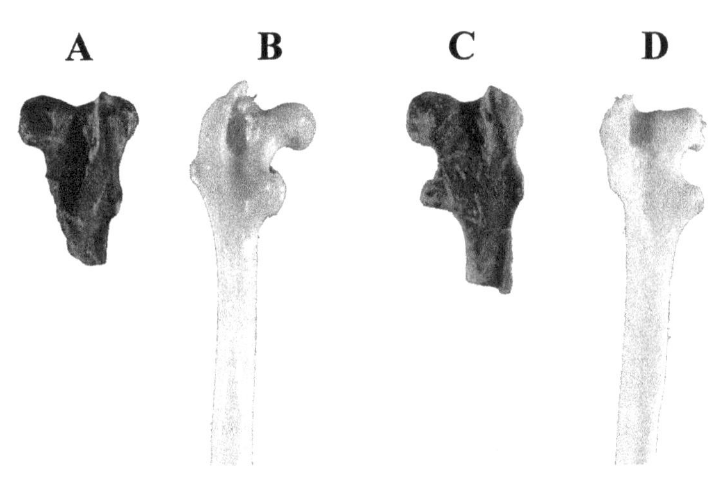

Fig. 3.5 Posterior view of the proximal femur of *Omomys carteri* (A, C), *Mirza* (B) and *Tarsius* (D). Note that the proximal femur of *Omomys* is intermediate in morphology between these extant primates, being somewhat cylindrical in shape. This is a feature in which several omomyoids resemble *Tarsius*.

The divergence of the major subdivisions of Haplorhini, the tarsiers, anthropoids and omomyoids, appears to date back to our earliest records of this group. Thus, at present, the earliest well-known primates, early Eocene adapoids and omomyoids, clearly document that the origin of the order dates back to the Paleocene.

It has been recognized for a number of years that while adapoids are extinct in North America and Europe by the early Oligocene they reappear in Asia during the late Miocene. These Miocene primates, sivaladapids, have been a puzzle because there is a 20 million year or so gap in the adapoid fossil record. Beard (1998*c*) has described new material of late Eocene *Hoanghonius* material from China as resembling the sivaladapids in a number of interesting dental traits and that these similarities are suggestive of a close phylogenetic relationship. Finally, while it is widely assumed that the extant tooth-combed primates are derived from the adapoid radiation, none of the adapoids clearly shows a series of traits linking them exclusively to the more recent strepsirhines.

Extinct tarsiers, galagos and lorises

Omomyoids and adapoids are known from an impressive fossil record documenting broad and successful adaptive radiations for both of these superfamilies. In contrast to this situation, very little is known about the evolutionary history of the prosimian primates postdating the time of the split between the Malagasy forms and the galagos and lorises or between the tarsiers and strepsirhine primates.

Beard *et al.* (1994) and Beard (1998*a*) have recently described new tarsier material from the middle Eocene of China. *Tarsius eocaenus* is known from isolated dental material from southern Jiangsu Province and *Xanthorhysis tabrumi* is known from a single, well preserved mandible from the southern Shanxi Province. This material, while limited, illustrates striking similarity to modern tarsiers and indicates that this clade has been separate from other primates since early in the Eocene. Unfortunately, only a single molar is known for tarsiers (from the Miocene of Thailand) that existed between the occurrence of these Eocene forms and modern tarsiers. The fossil record for the Asian lorises is extremely poor, consisting solely of one late Miocene species, *Nycticeboides simpsoni*, from Pakistan.

The fossil record for African prosimians is only slightly better than that for the Asian forms. Both galagos and lorises are known from the early Miocene of East Africa and galagos are also known from Pliocene to recent deposits. These creatures appear to be fairly similar to the modern African prosimians although Gebo (1989) has noted that the Miocene galagos lack the extreme elongation of the calcaneus and navicular seen among the modern members of this group. While there is a record of a radiation of large to very large Malagasy primates from the recent past, nothing is known about Pliocene, Miocene or earlier radiations.

Where to next?

The past thirty years has seen a dramatic improvement of our understanding of the evolutionary biology of the earliest primates. This reflects a vastly enlarged fossil record, sampling a much wider geographic area for the remains of early primates, a better understanding of the phylogenetic relationships among extant primates, and a significantly improved understanding of the ecology and behavior of extant prosimians. The next thirty years promises to illuminate further our understanding of the earliest chapters of primate evolution as field work in North America and Europe continues and that in Africa and Asia continues to increase in intensity and covers an ever-increasing geographic range.

Research in Paleocene deposits in Africa and Asia may allow us to discover potential common ancestors for the omomyoids and adapoids (*Altanius* from Asia and *Altiatlasius* from Africa are our best candidates to be representatives of such a stem group at present). Additional research in Eocene and Oligocene deposits in sub-Saharan Africa and Madagascar may shed light on the origins of the modern Malagasy primates specifically and about the last common ancestor of the tooth-combed primates in general. It is also likely that the adaptive radiations of early primates in Africa and Asia will be discovered to rival, or even surpass, those that have been well documented to have occurred in North America and Europe during the Eocene epoch. Material from Africa and Asia is also likely to improve dramatically our understanding of the past adaptive radiations of galagos, tarsiers and lorises.

Acknowledgements

First, I thank Dr. Walter Hartwig for the invitation to contribute this chapter. I thank K.C. Beard, M. Dagosto, J. Fleagle, D. Gebo, M. Hamrick, R.F. Kay, C. Ross, E. Simons and B. Williams for informative discussions and comments on the morphology of early primates. I thank the following persons and institutions for access to skeletal material of extant and extinct primates in their collections: the American Museum of Natural History, G. Musser and R.H. Tedford; the National Museum of Natural History (Smithsonian Institution), R. Thorington and R. Emry; Yale Peabody Museum, J.H. Ostrom; the Carnegie Museum of Natural History, K.C. Beard; the Museum of Paleontology at the University of Michigan, P.D. Gingerich and G. Gunnell; the Museum of Comparative Zoology at Harvard University, F.A. Jenkins, Jr. and M. Rutzmoser; Museum of Paleontology at the University of California, Berkeley, P. Holroyd and H. Hutchison; the U.S. Geological Survey (Denver), T.M. Bown and K. McKinney; the University of Wyoming Geological Museum, J.A. Lillegraven; the Duke University Primate Center, E.L. Simons and P. Chatrath; the Museum National d'Histoire Naturelle, Paris, D.E. Russell: Naturhistorisches Museum, Basel, B. Engesser; and the Laboratoire de Paléontologie, University of Montpellier, M. Godinot. I am particularly grateful for the generosity of P. Robinson of the University of Colorado Museum for access to fossil primates.

4 | Adapiformes: Phylogeny and adaptation

DANIEL L. GEBO

Introduction

Adapiformes is a highly diversified and biogeographically dispersed group of extinct primates. These primates date from the beginning of the early Eocene (55 Ma) and thus represent one of the two earliest branches of euprimates. Adapiform primates inhabited four continents (North America, Europe, Asia and Africa) and survived at least 47 Ma from the early Eocene until the late Miocene. As a matter of history, the very first fossil primate to be described was *Adapis parisiensis* (Cuvier, 1821), although Cuvier mistakenly believed "ad-apis" (toward the sacred bull Apis) to have artiodactyl affinities (Simons, 1972). Later in the nineteenth century, Delfortrie (1873), Gervais (1872) and Filhol (1873) would correctly align *Adapis* with primates. The cercamoniine *Caenopithecus lemuroides* (Rütimeyer, 1862) has the distinction of being the very first adapiform to be recognized. Today, adapiform primates are represented by at least 30 genera and about 80 species (see Godinot, 1998). They are well represented in terms of skulls, jaws and body parts, although many genera are still systematically controversial.

There have been several alternative proposals for higher-level adapiform taxonomy. Some workers recognize only one family, Adapidae (Gregory, 1920; Simons, 1972; Gingerich, 1977*a*; Szalay & Delson, 1979; Covert, 1986). Others utilize multiple family approaches, i.e., Notharctidae and Adapidae (Stehlin, 1912; Gazin, 1958; Schwartz & Tattersall, 1985; Franzen, 1987, 1994*b*; Rose *et al.*, 1994); Notharctidae, Adapidae, Sivaladapidae and Petrolemuridae (Groves, 1989); or Notharctidae, Adapidae and Sivaladapidae (Godinot, 1998). This review uses a three-family approach although a four- or five-family approach may best reflect adapiform phylogeny.

History of discovery and debate

The discovery and description of adapiform primates has a long history with the earliest finds occurring more than 100 years ago: *Adapis* (Cuvier, 1821), *Caenopithecus* (Rütimeyer, 1862), *Notharctus* (Leidy, 1870), *Pelycodus* (Cope, 1875), *Protoadapis* (Lemoine, 1878), *Smilodectes* (Wortman, 1903), *Pronycticebus* (Grandidier, 1904) and *Anchomomys* (Stehlin, 1916). Many of these early descriptions aligned these taxa with artiodactyls or intermediate forms between primates and ungulates. Some even believed these fossils to be carnivorous ungulates (Leidy, 1872). In contrast, Ludwig Rütimeyer (1862) was ahead of his time is assessing the primate status of his fossils from Egerkingen, Switzerland (*Caenopithecus*). Later Othniel Charles Marsh (1872), Edward Drinker Cope (1873, 1882*a*), Eugène Delfortrie (1873), Henri Filhol (1873), Major Charles Immanuel Forsyth-Major (1901), Henry Fairfield Osborn (1902), Guillaume Grandidier (1904), Hans Georg Stehlin (1912) and William King Gregory (1920) would clear up these early misconceptions about adapiform primates. Thus, the early problems centered on the primate status of these "new" adapiform fossils. Later, the problem would switch to evolutionary concerns – in short, how do these fossil primates link up with living primates as well as to other extinct primates? Are lemurs the living descendants of adapiform primates? Did adapiforms give rise to anthropoid primates? These questions and their phylogenetic debates continue to this day.

The most significant publication in understanding adapiform primates may well be Gregory's 1920 monograph "On the structure and relations of *Notharctus*, an American Eocene primate". His anatomical interpretation of the skeletons of *Notharctus* begins the modern era of understanding adapiform adaptation and phylogeny. Gregory's (1920) detailed comparison of *Notharctus* with living lemurs provided the first thorough connection of Adapiformes and Lemuriformes. Later views concerning primate evolution (e.g., Le Gros Clark, 1959; Simons, 1972; Gingerich, 1975*a*) would suggest other similarities to lemurs. In fact, very few shared derived characteristics or special resemblances do indeed link these two groups (see Cartmill & Kay, 1978; Rasmussen & Nekaris, 1998). Adapiforms are very primitive euprimates (Table 4.1).

However, most of the anatomical features shared by Lemuriformes (living lemurs and lorises) and Adapiformes are primitive primate features, which are therefore non-diagnostic. However, in some anatomical areas like the distal tibia (Dagosto, 1985) and the upper ankle joint (Gebo, 1986; Dagosto, 1988), adapiforms are very specialized and highly diagnostic from other primates. In fact, these morphological regions link all known adapiform primates with all tooth-comb lemurs (Lemuriformes), including the subfossil lemurs. One could say that lemurs and adapiforms form a clade that could be called primates with sloping ankles ("Angulartalarformes"), although we normally use the term Strepsirhini (Pocock, 1918).

Rasmussen & Nekaris (1998) prefer to not use the taxonomic term "Strepsirhini" with adapiforms since the

Table 4.1. Primitive characteristics

Skull	Dentition	Postcranium
a postorbital bar	early adapiforms possess a 2–1–4–-3 dental formula	a grasping hallux and pollex (a prominent 1st metatarsal with a large peroneal tubercle and a sellar shaped distal entocuneiform facet)
obliquely facing orbits	spatulate incisors that are slightly procumbent	nails (broad distal phalanges)
a petrosal bulla with a ring-like ectotympanic within the bulla	robust canines	good elbow mobility (a round capitulum with a wide zona conoidea separating the capitulum from the trochlea)
a relatively long and broad snout with many nasal turbinals	a two-rooted P_2	a robust humerus with a prominent brachial flange and deltopectoral crest
	an elevated trigonid	a spherical humeral head that rises above the tubercles
		a short radius
	lack of hypocone	a long hindlimb (femur and tibia)
	an unfused mandibular symphysis	a ball-like femoral head
		a large lesser trochanter
		a proximally placed and prominent third trochanter
		a high lateral patellar rim with a deep patellar groove
		good foot inversion capabilities (calcaneocuboid joint pivot and subtalar joint)
		an unfused fibula

characters that define living strepsirhine primates are soft-tissue characters and cannot be easily identified in the fossil record. Thus, a wet nose with lateral slits, a philtrum (median cleft), a frenulum (tethered upper lip), long vibrissae and a connection between the rhinarium and the vomeronasal organ cannot be identified among adapiform primates, who may not be strepsirhines in their soft tissue nasal anatomy. Thus, Rasmussen & Nekaris (1998) would prefer an alternative taxonomic term for the adapiform–lemuriform clade.

On the other hand, Rosenberger *et al.* (1985) argue that the spacing, occlusion and size–shape relationship of the upper incisors link notharctines with lemuriforms, suggesting at least some hard-tissue correlations with wet noses. Rosenberger & Strasser (1988) and Rosenberger *et al.* (1985) argue that the upper-incisor morphology of *Notharctus tenebrosus* and that of lemuriforms is shared derived while the incisor–canine morphology of *Adapis* is autapomorphic.

Beard (1988*b*) also examined the front teeth and central gap of lemuriforms and adapiforms in an attempt to document hard-tissue correlations with wet-nose strepsirhinism. He argued that strepsirhinism is shared among a variety of non-primate mammals and may also be inferred for omomyids (non-strepsirhine primates). Thus, strepsirhinism cannot be an adaptive innovation that separates early Haplorhini from Strepsirhini. The only osteological feature associated with the strepsirhine nasal condition is the relatively wide median gap between the upper central incisor roots. Beard (1988*b*) agrees with Rosenberger *et al.* (1985) in suggesting that adapiforms were anatomically strepsirhine but argues that several omomyids (haplorhine primates; e.g., *Pseudoloris*, *Necrolemur* and *Rooneyia*) must also be considered anatomically strepsirhine. In the end, no matter what soft-tissue anatomy exists in adapiform fossil primates, most researchers like the phylogenetic connection between adapiform and lemuriform primates and this grouping requires a taxonomic term that reflects this association (i.e., a sister taxa relationship).

Another key evolutionary question surrounding the interpretation of adapiforms concerns their possible connection to anthropoid evolution (Gidley, 1923; Gingerich, 1975*b*, 1977*a*, 1980*b*, 1984*b*, 1995; Franzen, 1994*b*; Rasmussen, 1994; Simons, 1995*b*). Gingerich (1980*b*) has produced a long list of anatomical features supporting an adapiform–anthropoid phylogenetic connection. For example, small vertically implanted spatulate lower incisors, I_2 larger than I_1, a lower premolar honing facet for the upper canine, fusion of the mandibular symphysis, sexual dimorphism, non-elongated tarsals, and an unfused tibia and fibula have been used to support an adapiform–anthropoid affinity (but see Covert & Williams, 1994). Other anatomical areas such as the basicrainium (Gingerich, 1980*b*; Rasmussen, 1990*b*) and molar anatomy have also been used to support an adapiform–anthropoid phyletic link (Gingerich, 1977*a*, 1980*b*; Kay, 1980; Rasmussen, 1986; Rasmussen & Simons, 1988; Franzen, 1994*b*; Simons, 1995*b*), as has large body size (Gingerich, 1980*b*). Currently, cercamoniine adapiforms are believed to best bridge the morphological gap between adapiforms and Anthropoidea. Rasmussen (1994), in particular, has suggested that when the adapiform *Periconodon* is better known, it will possess postorbital closure, identifying it as an anthropoid primate. Although there appear to be several interesting similarities between adapiform primates and anthropoids in the upper dentition, the cranial and postcranial evidence is decidedly against this evolutionary interpretation (e.g., Szalay, 1975*a*; Cartmill & Kay, 1978; Rosenberger & Szalay, 1980; Beard *et al.*, 1988; Kay *et al.*, 1997*a*; Ross *et al.*, 1998).

Taxonomy

Systematic framework

Order Primates Linnaeus, 1758
- Suborder Strepsirhini É. Geoffroy Saint-Hilaire, 1812
  - Infraorder Adapiformes Hoffstetter, 1977
    - Family Notharctidae Trouessart, 1879
      - Subfamily Notharctinae Trouessart, 1879
        - Genera *Cantius* Simons, 1962
          - *Copelemur* Gingerich & Simons, 1977
          - *Hesperolemur* Gunnell, 1995
          - *Notharctus* Leidy, 1870
          - *Pelycodus* Cope, 1875
          - *Smilodectes* Wortman, 1903
      - Subfamily Cercamoniinae Gingerich, 1975
        - Genera *Aframonius* Simons *et al.*, 1995
          - *Agerinia* Crusafont-Pairo & Golpe-Posse, 1973
          - *Anchomomys* Stehlin, 1916
          - *Barnesia* Thalmann, 1994
          - *Buxella* Godinot, 1988
          - *Caenopithecus* Rütimeyer, 1862
          - *Cercamonius* Gingerich, 1975
          - *Donrussellia* Szalay, 1976
          - *Europolemur* Weigelt, 1933
          - *Mahgarita* Wilson & Szalay, 1976
          - *Periconodon* Stehlin, 1916
          - *Pronycticebus* Grandidier, 1904
          - *Protoadapis* Lemoine, 1878
          - *Wadilemur* Simons, 1997
    - Family Adapidae Trouessart, 1879
      - Subfamily Adapinae Trouessart, 1879
        - Genera *Adapis* Cuvier, 1821
          - *Adapoides* Beard *et al.*, 1994
          - *Cryptadapis* Godinot, 1984
          - *Leptadapis* Gervais, 1876
          - *Microadapis* Szalay, 1974
          - *Palaeolemur* Delfortrie, 1873
    - Family Sivaladapidae Thomas & Verma, 1979
      - Subfamily Sivaladapinae Thomas & Verma, 1979
        - Genera *Guangxilemur* Qi & Beard, 1998
          - *Hoanghonius* Zdansky, 1930
          - *Indraloris* Lewis, 1933
          - *Rencunius* Gingerich *et al.*, 1994
          - *Sinoadapis* Wu & Pan, 1985
          - *Sivaladapis* Gingerich & Sahni, 1979
          - *Wailekia* Ducrocq *et al.*, 1995
    - Problematic Taxa
          - *Amphipithecus* Colbert, 1937
          - *Djebelemur* Hartenberger & Marandat, 1992
          - *Lushius* Chow, 1961
          - *Omanodon* Gheerbrandt *et al.*, 1993
          - *Panobius* Russell & Gingerich, 1987
          - *Pondaungia* Pilgrim, 1927
          - *Shizarodon* Gheerbrandt *et al.*, 1993

Family Notharctidae

Subfamily Notharctinae

The notharctines are essentially restricted to North America with only *Cantius* (*C. eppsi*; Cooper, 1932, and *C. savagei*; Gingerich, 1977*a*) occuring in Europe and in North America. Notharctines possess relatively long, broad snouts with a small infraorbital foramen, implying a decrease in vibrissae and tactile sense importance. Notharctines possess small orbital diameters suggesting a diurnal lifestyle. Spatulate lower incisors, I_1 smaller than I_2, interlocking canines, a honed premolar, sexually dimorphic canines, reduced or absent paraconids, and a free ectotympanic ring within the bulla are features that help diagnose this group. Notharctines retain four premolars and a lacrimal bone within the orbit. Most species possess unfused mandibles although *Notharctus* exhibits a fused mandible. The internal carotid enters the bulla posterolaterally. The size of the promontory and stapedial arteries in notharctines is variable (see Gingerich, 1973; MacPhee & Cartmill, 1986; Gunnell, 1995*b*) but they generally retain a large stapedial branch. Shearing coefficients for M_2 for *Notharctus* and *Smilodectes* (possessing the highest shearing coefficients) suggest a folivorous diet (Covert, 1986). All species of *Cantius* and *Pelycodus* appear to have been frugivorous in this measure, while *Copelemur praetutus* is more folivorous (Covert, 1986). Notharctines possess some of the most complete and best-known postcranial specimens and their locomotion has been likened to that of living lemurs and indriids.

GENUS *Cantius* Simons, 1962*a* (Fig. 4.1)
INCLUDED SPECIES *C. abditus, C. angulatus, C. antediluvius, C. eppsi, C. frugivorus, C. mckennai, C. nuniensis, C. ralstoni, C. savagei, C. torresi, C. trigonodus*

TYPE SPECIES *Cantius angulatus* Cope, 1875
TYPE SPECIMEN Type specimen is lost; AMNH 55511 is the neotype (Cope, 1875; Simons, 1962*a*; Beard, 1988*a*)

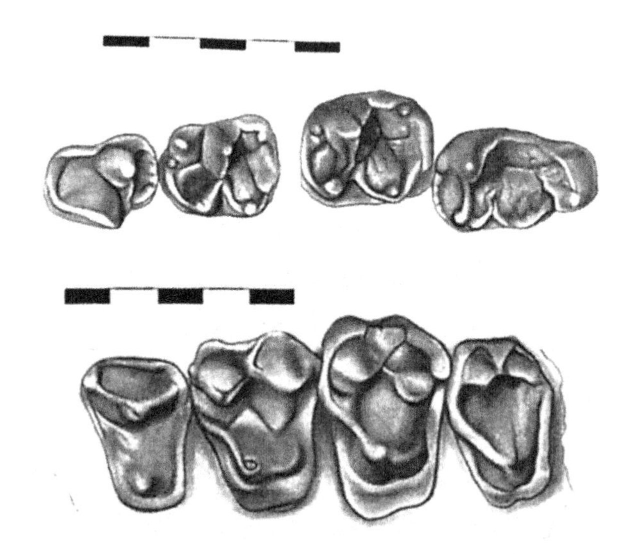

Fig. 4.1 *Cantius trigonodus*. Upper (USGS 9516, bottom) and lower (USGS 9525, above) dentition. Scales in mm.

AGE AND GEOGRAPHIC RANGE Early Eocene, North America and Europe

ANATOMICAL DEFINITION

All of the original specimens of this genus were first placed with the genus *Pelycodus* (Cope, 1875). In 1962, Simons named *Cantius* and later Gingerich & Simons (1977) reanalyzed the Big Horn Basin collections (then *Pelycodus*) and sunk all species except *Pelycodus jarrovii* into *Cantius* (Fig. 4.1). *Cantius torresi* (earliest Wasatchian biochron Wa-0) is the oldest species of this lineage (Gingerich, 1986, 1995). *Cantius torresi* differs from all other species of *Cantius* in that it is smaller (1100–1200 g), possesses relatively square upper molars, and has shorter and relatively broader lower premolars and molars. P_4 is distinctive in having a broad trigonid and talonid (Gingerich, 1986). P_4 is similar in form to the fourth premolars found among the early Omomyidae, suggesting little differentiation from the ancestral euprimate. The upper molars of *Cantius* have distinct paracones, metacones and prominent protocones with small but distinct paraconules and metaconules. There is a "nannopithex fold" or a partially developed postprotocingulum. M_3 is unreduced and it possesses an elongated heel. *Cantius torresi* has marked canine dimorphism (Gingerich, 1995) as do other notharctines (Gregory, 1920; Gingerich, 1979; Krishtalka *et al*., 1990). Godinot (1998) believes that *C. torresi* may be derived relative to the slightly younger *C. ralstoni* in that *C. torresi* possesses shorter and broader P_{3-4}, and more squared upper molars. This suggests two immigrations of *Cantius* into North America by Godinot (1992*a*), although Gingerich (1995) disagrees. In *Cantius*, the mandibular symphysis is unfused and the rostrum is relatively short. The lacrimal canal is anterior to the orbital margin in contrast to *Notharctus*, *Smilodectes* and *Hesperolemur* (Gunnell, 1995*b*). Of all of the adapiforms, *Cantius* in particular has been used to demonstrate ancestor–descendant relationships (Gingerich & Simons, 1977; O'Leary, 1996). The continuous and gradual development over time of pseudohypocones, mesostyles and mandibular fusion, in addition to changes in body size, has played an essential role in understanding notharctine evolution (Gingerich & Schoeninger, 1977; Gingerich & Simons, 1977; Maas & O'Leary, 1996). The application of the stratophenetic approach to notharctines in the Bighorn Basin, Wyoming, has also played a key role in understanding the evolution of North American adapiforms (Gingerich, 1976*b*).

GENUS *Copelemur* Gingerich & Simons, 1977
INCLUDED SPECIES *C. australotutus, C. praetutus, C. tutus*

TYPE SPECIES *Copolemur tutus* Cope, 1877
TYPE SPECIMEN Type specimen is lost; AMNH 16206 is the topotype, a dentary with M_{1-2} (Cope, 1877; Gingerich & Simons, 1977)
AGE AND GEOGRAPHIC RANGE Early Eocene, North America
ANATOMICAL DEFINITION

Gingerich & Simons (1977) reallocated early Eocene (Wasatchian) Bighorn Basin specimens from *Pelycodus* to *Copelemur*. *Copelemur* differs from other notharctines in having an open talonid with a more distinct entoconid notch (Gingerich & Simons, 1977). Most are smaller with less well-developed pseudohypocones and mesostyles compared to contemporary species of *Cantius*. They lack symphyseal fusion. *Copelemur* differs from *Cantius* in having a distinct entoconid notch on P_4–M_3, relatively narrower molars, an anteriorly shifted paraconid on M_2 and M_3, and a lengthened premetacristid on M_2 and M_3 (Covert, 1990). *Copelemur* differs from *Smilodectes* in having less compressed lower premolars and more distinct paraconids (Gingerich & Simons, 1977). Beard (1988*a*) restricts *Copelemur* to the southern western basins in New Mexico and Wyoming (San Juan, Green River and Washakie Basins) and believes that *Copelemur* and *Smilodectes* are sister taxa. He lists "reduced, mesioinferiorly placed paraconids on M_1–M_2; relatively greatly developed paracristids on M_1–M_3; protoconids and metaconids that are relatively close together on M_1–M_3; a relatively deep hypoflexid on M_3; and a simply constructed, relatively long and narrow hypoconulid lobe on M_3" (Beard, 1988*a*: 465). In contrast, Covert (1985) lists several dental synapomorphies that unite *Copelemur* to *Notharctus* and *Smilodectes*.

GENUS *Hesperolemur* Gunnell, 1995
INCLUDED SPECIES *H. actius*

SPECIES *Hesperolemur actius* Gunnell, 1995
TYPE SPECIMEN SDSNH 35233, compressed skull
AGE AND GEOGRAPHIC RANGE Middle Eocene, North America

ANATOMICAL DEFINITION

Hesperolemur actius (Gunnell, 1995*b*) was the latest notharctine to live in North America. *Hesperolemur* is known from southern Californian (Sespe Formation) and is thought to be an immigrant taxon. It is a mid-Eocene (Uintan) primate that differs from other notharctines in that (1) it possess an a fused ectotympanic anulus (the anterior third) along the internal lateral wall of the bulla; and (2) it lacks bony canals that transmit the internal carotid (but see Rose *et al.*, 1999). *Hesperolemur* also lacks a stapedial artery like that of *Mahgarita* (Gunnell, 1995*b*). *Hesperolemur* possesses a premolariform P^4, a pseudohypocone (protocone fold) and a well-developed metaconule (Gunnell, 1995b). It lacks lower molar paraconids and is unlike other notharctines in this feature. It has a smaller sagittal crest and less massive zygomatics compared to adapids. Dental measurements suggest a body size similar to that of *Pelycodus jarrovii* (4500 g, Fleagle, 1999). Rose *et al.* (1999) dispute the generic distinctiveness of *Hesperolemur* stating that "H." *actius* differs little from *Cantius*. They believe this specimen to be a distinct species (*Cantius actius*) but their reassessment of the basicranium provides a decidedly different interpretation from that of Gunnell (1995*b*).

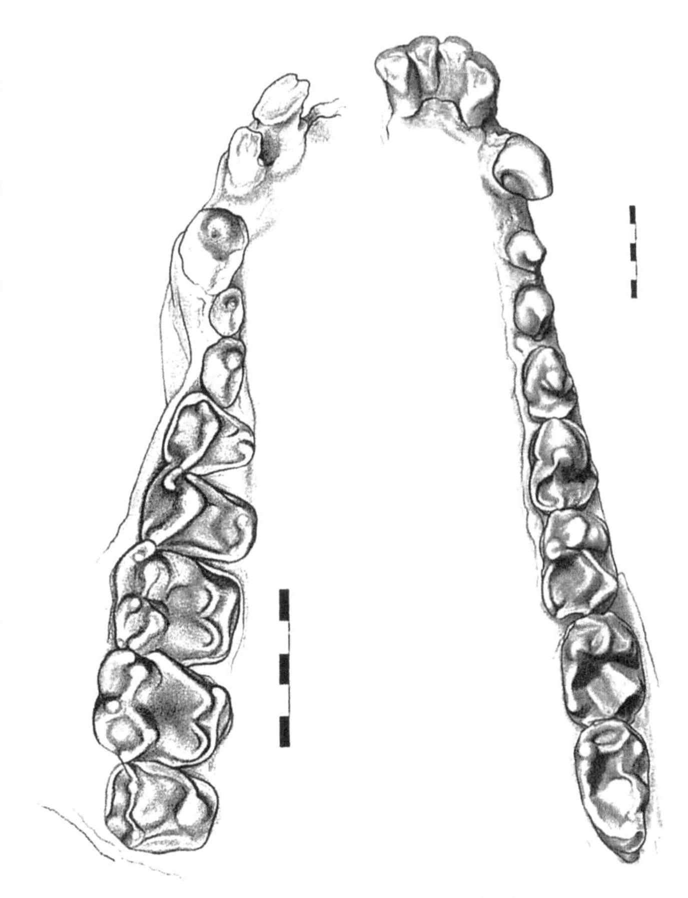

Fig. 4.2 *Notharctus tenebrosus*. Upper (left) and lower (right) dentition (AMNH 127167). Scales in mm.

GENUS *Notharctus* Leidy, 1870 (Figs. 4.2, 4.3)
INCLUDED SPECIES *N. pugnax, N. robustior, N. tenebrosus, N. venticolus*

TYPE SPECIES *Notharctus tenebrosus* Leidy, 1870
TYPE SPECIMEN USNM 3752, mandible with C-M_3
AGE AND GEOGRAPHIC RANGE Late to middle Eocene, North America
ANATOMICAL DEFINITION

Notharctus represents the best known of all adapiform primates (Gregory, 1920; Gazin, 1958; Gingerich, 1979). It is an early late Eocene to middle Eocene (late Wasatchian to Bridgerian) form that is last known from early Uintan sediments (Turnbull, 1972). *Notharctus* was a large adapiform, 4 to 7 kg in size (Fleagle, 1999). *Notharctus* has well-developed pseudohypocones and mesostyles, reduced paraconids, and in general more complicated upper molars than *Cantius* (Fig. 4.2). *Notharctus* differs from *Cantius* in having (1) distinct entoconid notches on P_4–M_3; (2) narrower lower molars; (3) an anteriorly shifted and reduced paraconid on M_2 and M_3; (4) a lengthened premetacristid on M_2 and M_3; (5) narrower upper molars; (6) a distobucally running postparacrista from the paracone; (7) a premetacrista running mesiobuccally from the metacone; (8) a rectangular upper M^3; (9) a reduced height for P_4; and (10) a fused symphysis (Covert, 1990). *Notharctus* possesses a small braincase, with pronounced nuchal and sagittal crests (Fig. 4.3). The face and nose are long and the orbits small, suggesting diurnality (Kay & Cartmill, 1977). The lacrimal bone is at the edge of the orbit rather than being anteriorly located as in extant lemurs (Fleagle, 1999). *Notharctus* is known to be sexually dimorphic in canine size (Krishtalka *et al.*, 1990; Alexander, 1994). *Notharctus* had a highly folivorous diet (Szalay & Delson, 1979; Covert, 1986). There are many relatively complete skeletons of *Notharctus* and many have commented on its locomotor pattern (e.g., Gregory, 1920; Dagosto, 1993).

GENUS *Pelycodus* Cope, 1875
INCLUDED SPECIES *P. danielsae, P. jarrovii*

TYPE SPECIES *Pelycodus jarrovii* Cope, 1875
TYPE SPECIMEN Type specimen is lost; AMNH 16298 is the neotype, a mandible with M_{1-2} (Gingerich & Simons, 1977)
AGE AND GEOGRAPHIC RANGE Early Eocene, North America and Europe
ANATOMICAL DEFINITION

Cope (1875) initially named *Pelycodus jarrovii*, a rare and poorly known primate from the Wasatchian Land-Mammal Age of New Mexico (San Juan Basin) and Wyoming (Bighorn Basin; Rose & Bown, 1984). Matthew (1915*a*) conducted the first comprehensive study of *Pelycodus* but since the holotype was lost, he designated a neotype from the Bighorn Basin (Wyoming). This specimen became the

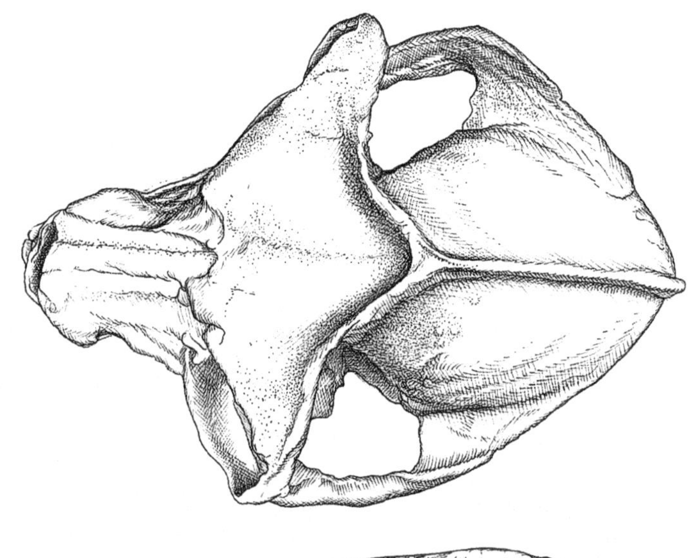

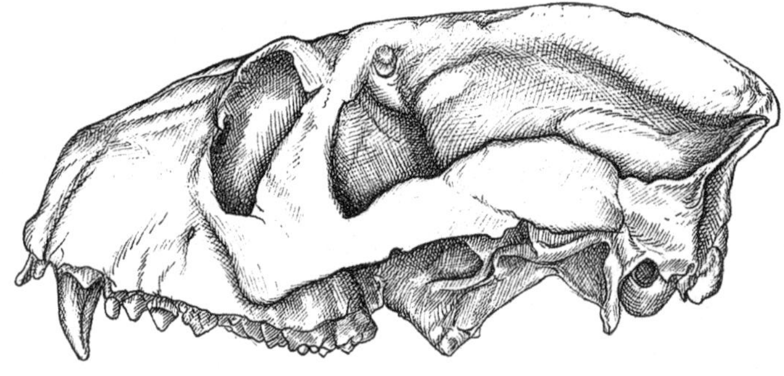

Fig. 4.3 *Notharctus tenebrosus*. Skull in dorsal and lateral views (AMNH 127167).

concept of *P. jarrovii* (Rose & Bown, 1984). Gingerich & Haskin (1981) would later sink *Pelycodus* into *Cantius* (Simons, 1962*a*), due to the fact that Matthew's neotype represented *P. abditus* (now *Cantius abditus*). *Pelycodus jarrovii* differs from *Cantius* in being relatively larger (Rose & Bown, 1984). It possesses large and broad molar basins and squared upper molars with widely separated paracones and metacones. No mesostyle is present. The trigonids are anteroposteriorly compressed. According to Rose & Bown (1984), this last feature separates *Pelycodus* from contemporary *Cantius*. Covert (1990) notes that *Pelycodus* differs from *Cantius* in (1) lacking or having an extremely reduced paraconid on M_2; (2) lacking a hypoconulid on M_1 and M_2; and (3) lacking styles on its upper molars.

GENUS *Smilodectes* Wortman, 1903
INCLUDED SPECIES *S. gingerichi, S. gracilis, S. mcgrewi*

TYPE SPECIES *Smilodectes gracilis* Wortman, 1903
TYPE SPECIMEN YPM 11800, left mandible with P_4, M_1 and part of P_3 (Marsh, 1871; Wortman, 1903)
AGE AND GEOGRAPHIC RANGE Middle Eocene, North America
ANATOMICAL DEFINITION
Smilodectes was named by Wortman in 1903 and it is a middle Eocene (Bridgerian) adapiform from North America. Gazin's (1958) study of skull morphology shows that *Smilodectes* is clearly a distinct genus from that of *Notharctus*. It differs from *Notharctus* in being smaller and in possessing a relatively shorter and broader skull. The frontal bone is more rounded (Gingerich, 1979). *Smilodectes* lacks mandibular fusion and its lower molar paraconids are reduced to a crest or lost (Gingerich, 1979). *Smilodectes* differs from *Cantius* in lacking a paraconid on M_2 and M_3, and having a cristid obliqua that runs mesiolingually (Covert, 1990). *Smilodectes* was a diurnal and folivorous primate.

Subfamily Cercamoniinae

The cercamoniine adapiforms are very diverse in terms of morphology, dietary adaptations, body size and geographical distribution compared to other adapiforms. They represent an Old World radiation (with the exception of *Mahgarita* in North America). Cercamoniines represent 14 genera with 11 taxa coming from Europe. One genus is found in North America and two are from Africa (one shared with Europe). *Donrussellia* is by far the most primitive of this group and *Donrussellia magna* has features suggesting a close relationship to *Cantius* (Godinot, 1992*a*, 1998). The cercamoniine *Anchomomys* represents the smallest known adapiform (~50 g). Cercamoniines are anatomically similar to notharctines in many features of their skull, teeth and body. They differ in that the hypocone develops from the lingual cingulum, they lack a nannopithex fold, the M_3 talonid lacks an expanded lobe of the hypoconulid and the M_3 talonid is wider than its trigonid (Szalay & Delson, 1979). Cercamoniines possess fused mandibles. *Pronycticebus* and *Caenopithecus* had relatively larger orbits compared to other Notharctidae suggesting nocturnal lifestyles for these two taxa. In terms of diet, *Donrussellia, Periconodon, Anchomomys, Buxella, Pronycticebus, Microadapis* and *Agerinia* were likely insectivorous on the basis of shearing quotients and their body size (Covert, 1986). *Protoadapis* was frugivorous (Covert, 1986) as was *Barnesia* (Thalmann, 1994) while *Aframonius* and *Caenopithecus* were more folivorous.

GENUS *Aframonius* Simons et al., 1995
INCLUDED SPECIES *A. diedes*

TYPE SPECIES *Aframonius diedes* Simons et al., 1995
TYPE SPECIMEN CGM 42202, left dentary P_3–M_2
AGE AND GEOGRAPHIC RANGE Late Eocene, Africa
ANATOMICAL DEFINITION
Simons et al. (1995*c*) named a new cercamoniine adapid from the Fayum (Locality 41, Egypt). *Aframonius dieides* is a late Eocene or early Oligocene adapiform from Africa. It is the largest primate (about 1600 g) at L-41 with a relatively long and shallow dentary. Symphyseal fusion is present in

older adults. *Aframonius* was folivorous. Dentally, this form possesses (1) small vertical incisors with $I_2 > I_1$; (2) dimorphic canines; (3) a single-rooted reduced P_2 with honing; (4) P_{3-4} with a complex talonid shelf; (5) lower molars showing cresting with relatively long and narrow crowns; (6) molars lacking a paraconid; (7) the entoconid in the posterolingual corner of the talonid basin; (8) M_1 and M_2 with a very small hypoconulid; and (9) M_3 with a well-developed hypoconulid lobe (Simons *et al.*, 1995*c*). The paracristid, postprotocristid and the cristid obliqua are prominent. Molar length increases posteriorly. *Aframonius* is similar to *Protoadapis* and *Europolemur* in size and molar form. Godinot (1998) believes *Aframonius* may be similar to *Caenopithecus* and a part of the Adapidae (Caenopithecinae) rather than a cercamoniine.

GENUS *Agerinia* Crusafont-Pairo & Golpe-Posse, 1973
INCLUDED SPECIES *A. roselli*

TYPE SPECIES *Agerinia roselli* Crusafont-Pairo & Golpe-Posse, 1973
TYPE SPECIMEN Ager No.1, left mandibular fragment with M_{2-3}
AGE AND GEOGRAPHIC RANGE Early Eocene, Europe
ANATOMICAL DEFINITION
Agerinia roselli (late early Eocene, Rhenanian, Spain) differs from *Protoadapis* in having P_3 and P_4 subequal in height and in lacking a distinct anteroposteriorly oriented cristid obliqua on P_4. The lower molars are low-crowned with crests. The trigonid is sealed off lingually. *Agerinia* differs from *Caenopithecus* in having a larger metaconid on P_4 and in lacking a metastylid as well as an entoconid (Szalay & Delson, 1979). Its robust mandible suggests omnivory or frugivory (Szalay & Delson, 1979). *Agerinia* may be a pronycticebine according to Godinot (1998).

GENUS *Anchomomys* Stehlin, 1916
INCLUDED SPECIES *A. crocheti, A. gaillardi, A. milleri, A. pygmaeus, A. quercyi*

TYPE SPECIES *Anchomomys gaillardi* Stehlin, 1916
TYPE SPECIMEN UL No. L46bis, left M^{1-3}/M_{1-3}
AGE AND GEOGRAPHIC RANGE Middle Eocene, Europe, Africa
ANATOMICAL DEFINITION
Anchomomys (Stehlin, 1916) represents the smallest genus of an adapiform primate (body size ~50 g; Conroy, 1987). *Anchomomys* is from the middle Eocene (middle Bartonian) of Europe, primarily. In France and Switzerland, *Anchomomys* shows a lineage in that the hypocone diminishes in size over time as does the transverse breadth of the upper molars (Godinot, 1998). *Anchomomys* is likely the most insectivorous of all adapiform primates. *Anchomoys crocheti* is believed to be the sister taxon to *Buxella* (Godinot, 1988). The lineage of *Anchomomys* survives into the upper Eocene and it represents the longest-lived genus within the Anchomomyini of Europe (Godinot, 1988). Simons (1997*a*) recently described *Anchomomys milleri* from Africa (Fayum, Locality 41, Egypt), extending the known geographic range of this genus. *Anchomomys milleri* is smaller than *Wadilemur* and *Aframonius*, two recently described adapiforms from Africa.

GENUS *Barnesia* Thalmann, 1994
INCLUDED SPECIES *B. hauboldi*

TYPE SPECIES *Barnesia hauboldi* Thalman, 1994
TYPE SPECIMEN GMH CeV-4338, crushed skull with upper right P^3–M^3 and upper left canine, P^2–M^3
AGE AND GEOGRAPHIC RANGE Middle Eocene, Europe
ANATOMICAL DEFINITION
Barnesia hauboldi (Thalmann, 1994) comes from the Geiseltal Valley (Germany) and is middle Eocene in age. *Barnesia* is similar dentally to the cercamoniines (*Protoadapis, Agerinia, Europolemur, Mahgarita* and *Pronycticebus*) in possessing a premolariform P^4 and the lack of a postprotocone fold. *Europolemur* appears to be the most similar in upper dental morphology to *Barnesia*. M^2 in *Barnesia* differs from *Europolemur* in being wider with additional conules on the continuous cingulum (Thalmann, 1994). *Barnesia* was primarily a frugivorous adapiform (Thalmann, 1994).

GENUS *Buxella* Godinot, 1988
INCLUDED SPECIES *B. magna, B. prisca*

TYPE SPECIES *Buxella prisca* Godinot, 1988
TYPE SPECIMEN BUX 80.69 (USTL Montpellier), isolated M_1
AGE AND GEOGRAPHIC RANGE Middle Eocene, Europe
ANATOMICAL DEFINITION
Buxella prisca and *B. magna* were named in 1988 by Godinot from isolated teeth. Godinot (1988) placed formerly attributed species of *Anchomomys* at Bouxwiller (a middle Eocene Alsacian locality) into the genus *Buxella*. This genus is closely related to *Anchomomys* and *Periconodon* (Godinot, 1988, 1998). All three genera are small adapiforms within the Anchomomyini. The upper molar morphology of *Buxella* is similar to that of *Anchomomys* (Godinot, 1988). There is no metaconule and the paraconule is very small. The hypocone is small and the lingual cingulum is discontinuous. The M_1 morphology of *Buxella* is also very similar to that of *Anchomomys* (Godinot, 1998). The metaconid and protoconid are about the same height and the paraconid is very low and small. The paraconid is connected to the premetacristid by a sinuous crest. Godinot (1988) suggests that *Buxella* and *Periconodon* are short-lived genera relative to *Anchomomys*. The derived characteristics that distinguish *Buxella* from other members of the Anchomomyini are a well-developed premetacristid on the lower molars, the absence of a metaconule and a reduced paraconule on the upper molars (Godinot, 1988). *Anchomomys* represents the sister taxon to *Buxella* and these

two lineages are united by an elevated trigonid that is constrained laterally on their M_1.

GENUS *Caenopithecus* Rütimeyer, 1862
INCLUDED SPECIES *C. lemuroides*

TYPE SPECIES *Caenopithecus lemuroides* Rütimeyer, 1862
TYPE SPECIMEN BNM Ef 383, right maxilla with M^1–M^3
AGE AND GEOGRAPHIC RANGE Middle Eocene, Europe
ANATOMICAL DEFINITION
Caenopithecus lemuroides has had a long history (Rütimeyer, 1862). Godinot (1998) places *Caenopithecus* within the Adapidae under the Caenopithecinae, a subfamily that includes *Caenopithecus*, *Adapoides*, *Mahgarita* and possibly *Aframonius*. In contrast, Szalay & Delson (1979) place *Caenopithecus* within the Adapinae, as does Conroy (1990). Franzen (1994*b*) and Fleagle (1999) place *Caenopithecus* within the Cercamoniinae, a view I follow here. *Caenopithecus lemuroides* is a large adapiform (3.5 kg; Fleagle, 1999) with a fused symphysis and a shortened snout. *Caenopithecus* has a non-molarized P^4 with small hypocones, mesotyles and mesostylids on its molars (Godinot, 1998). *Caenopithecus lemuroides* shares upper molar mesostyles and lower molar hypoconulids with *Hesperolemur* (Gunnell, 1995*b*). The mesostyles and mesostylids are autapomorphic relative to adapines. It shares with other adapids a broad trigon, non-reduced third molars (upper and lower), a deep groove between metaconid and entoconid, and a high cristid obliqua (Godinot, 1998). P_2 is reduced while the other premolars are premolariform in shape. *Caenopithecus* has lost a lower incisor and its dental formula appears to be 2.1.3.3/1.1.3.3. Orbit diameter is relatively large in *Caenopithecus*, like that of *Pronycticebus*, and the creature is thus considered to be nocturnal (Thalmann *et al.*, 1989; Franzen, 1994*b*). Its known body anatomy suggest cercamoniine rather than adapid affinities. *Caenopithecus* was a folivorous primate.

GENUS *Cercamonius* Gingerich, 1975
INCLUDED SPECIES *C. brachyrhynchus*

TYPE SPECIES *Cercamonius brachyrhynchus* Gingerich, 1975
TYPE SPECIMEN BNM Qv 619, left dentary, P_4–M_2 with alveoli for C, P_{2-3} and M_3 (Stehlin, 1912; Gingerich, 1975*b*)
AGE AND GEOGRAPHIC RANGE Late Eocene, Europe
ANATOMICAL DEFINITION
Gingerich (1975*b*) named *Cercamonius* from the Phosphorites du Quercy of southern France. The original specimen was described by Stehlin (1912) as *Protoadapis brachyrhynchus*. *Cercamonius* is from the late Bartonian in the late middle Eocene. *Cercamonius* resembles *Protoadapis* and *Notharctus* in lower molar morphology (Gingerich, 1975*b*). Its mandible is shorter and deeper. *Cercamonius* lacks a P_1 and P_2 is reduced to a single root. *Cercamonius* possesses large interlocking canines and is similar in size to *Notharctus robustior* (about 7 kg; Gingerich, 1975*b*). Gingerich (1975*b*) has suggested that *Cercamonius* may be closedly related to Old World anthropoids. Both Szalay & Delson (1979) and Godinot (1998) have placed this material within *Protoadapis*.

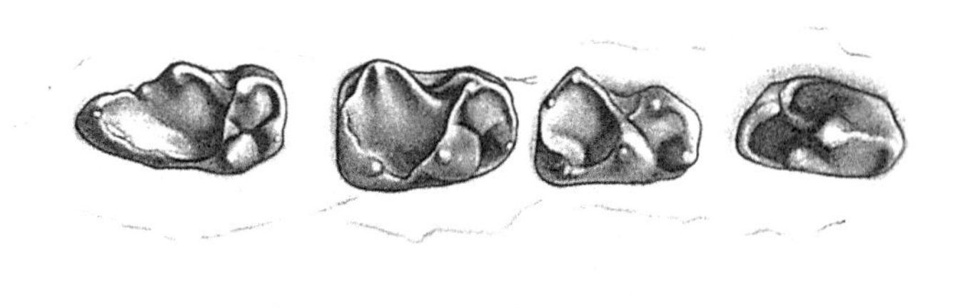

Fig. 4.4 *Donrussellia gallica*. Lower dentition of (composite). Scale in mm.

GENUS *Donrussellia* Szalay, 1976 (Fig. 4.4)
INCLUDED SPECIES *D. gallica, D. louisi, D. magna, D. provincialis, D. russelli*

TYPE SPECIES *Donrussellia gallica* Russell *et al.*, 1967
TYPE SPECIMEN MNHN Av 5755, left M_2 (Russell *et al.*, 1967; Szalay, 1976)
AGE AND GEOGRAPHIC RANGE Early Eocene, Europe
ANATOMICAL DEFINITION
Teeth associated with *Donrussellia* (early Eocene, Neustrian, France) were first referred to *Teilhardina gallica* (Russell *et al.*, 1967; Szalay, 1976). Szalay (1976), Savage *et al.* (1977) and Szalay & Delson (1979) all placed this genus within the Omomyidae. With better material (Fig. 4.4), this genus was placed within the Adapiformes (Godinot, 1978). *Donrussellia provincialis* is the most primitive adapiform known (Godinot, 1992*a*; Rose *et al.*, 1994; but see Gingerich, 1986). *Donrussellia magna* has features suggesting a close relationship to *Cantius* (Godinot, 1992*a*, 1998). *Donrussellia* is similar in size to *Teilhardina*, but has a P_4 that is transversely wide, less inflated metaconids relative to *Teilhardina*, large third molars and a metaconule that is not clearly defined and part of the protocrista (Rose *et al.*, 1994). *Donrussellia* has low-crowned, squared molars with low conules, high cusped premolars, and was probably an omnivore according to Szalay & Delson (1979). *Donrussellia* has a double-rooted P_2, a feature that has never been known to occur in living or fossil haplorhines. *Donrussellia* is advanced over *Teilhardina* in that the cristid obliqua is directed toward the apex of the metaconid, the postmetaconule crista is absent and the postprotocrista and metacone are connected (Godinot, 1992*a*).

GENUS *Europolemur* Weigelt, 1933
INCLUDED SPECIES *E. dunaifi, E. klatti, E. koenigswald*

TYPE SPECIES *Europolemur klatti* Weigelt, 1933
TYPE SPECIMEN GMH 232/3656, crushed cranium and mandible
AGE AND GEOGRAPHIC RANGE Middle Eocene, Europe
ANATOMICAL DEFINITION
Europolemur is a middle Eocene (Rhenanian) adapiform from

Germany. The crushed skull of *Europolemur* (Franzen, 1987) shows the orbits to be small, the postorbital bar and zygomatic arch to be robust, and the mid-skull moderate in height. Thus, *Europolemur* is more like *Notharctus* and *Adapis* than *Pronycticebus*. Its orbital diameter is similar to diurnal prosimians (Franzen, 1994*b*). *Europolemur* lacks a sagittal crest and the massive zygomatics of European adapids. *Europolemur* has a 2.1.4.3 dental formula. It has a well-defined hypoconulid on M_1 and M_2, and sharp crests on its molars. The upper molars have cingula and hypocones, except for M^3 (Szalay & Delson, 1979). The upper incisors are small, low-crowned, and spatulate. The canines are moderately large. P_1 is very small when present. *Europolemur* possesses non-molarized fourth premolars and a sharp crista obliqua (a crest running from the protocone to the metacone) (Franzen, 1994*b*). *Europolemur* was a frugivorous/ insectivorous primate (Szalay & Delson, 1979). Its body proportions are distinctly non-*Adapis*-like as well (Franzen, 1994*b*).

GENUS *Mahgarita* Wilson & Szalay, 1976
INCLUDED SPECIES *M. stevensi*

TYPE SPECIES *Mahgarita stevensi* Wilson & Szalay, 1976
TYPE SPECIMEN TMM 41578-9, crushed skull with C–M^3
AGE AND GEOGRAPHIC RANGE Late Eocene, North America
ANATOMICAL DEFINITION
Mahgarita has been a controversial genus (Szalay & Delson, 1979; Rasmussen, 1990*b*). This taxon is of Duchesnean age (early late Eocene) and is likely an immigrant from Asia (Gingerich, 1980*a*; Beard *et al.* 1994; Godinot, 1998). Thalmann (1994) and Godinot (1998) place *Mahgarita* within the Adapidae, while Franzen (1994*b*) and Fleagle (1999) prefer the Cercamoniinae and the Notharctidae, respectively. P_1 is lost and P_2 is extremely reduced in this taxon. The dental formula is 2.1.3.3. The twinned paraconid and metaconid on M_3 and a postprotocone fold are missing in *Mahgarita* (Szalay & Delson, 1979). *Mahgarita* is dentally similar to *Europolemur klatti* (Szalay & Delson, 1979). A crushed skull is known for this taxon but the middle ear anatomy has been controversial (Wilson & Szalay, 1976; Szalay & Delson, 1979; Rasmussen, 1990*b*).

GENUS *Periconodon* Stehlin, 1916
INCLUDED SPECIES *P. helleri, P. helveticus, P. huerzeleri, P. jaegeri, P. lemoinei*

TYPE SPECIES *Periconodon helveticus* Stehlin, 1916
TYPE SPECIMEN BNM Ef 366, left maxilla with M^{1-2} (Rütimeyer, 1890; Stehlin, 1916)
AGE AND GEOGRAPHIC RANGE Middle Eocene, Europe
ANATOMICAL DEFINITION
Periconodon is from the early middle Eocene of Western Europe (Stehlin, 1916; Szalay & Delson, 1979). *Periconodon* has been perceived as an omomyid in the past (Simons, 1962*a*). It possesses a well-developed hypocone, a prominent pericone and a P^3 with a reduced protocone (Szalay & Delson, 1979). A faint metaconule is present. P_4 is molariform and the lower molars show reduced paracristids. Shared derived characters uniting species of *Periconodon* include a pericone on the upper molars bunodont lower molars and a continuous crest between the protoconid and the metaconid (Godinot, 1988). Conroy (1990) estimates *P. huerzeleri* to be between 286 and 344 g. Its small size suggests an insectivorous diet, but the bunodont molars imply frugivory. *Periconodon* likely consumed both insects and fruit.

GENUS *Protoadapis* Lemoine, 1878
INCLUDED SPECIES *P. angustidens, P. brachyrhynchus, P. curvicuspidens, P. ignoratus, P. muechelnensis, P. recticuspidens, P. weigelti*

TYPE SPECIES *Protoadapis curvicuspidens* Lemoine, 1878
TYPE SPECIMEN MNHN AL-5179, right mandible M_{2-3}
AGE AND GEOGRAPHIC RANGE Middle Eocene, Europe
ANATOMICAL DEFINITION
Protoadapis is an early middle Eocene adapiform (Godinot, 1998). *Protoadapis* has a large canine, P_3 taller than P_4 (with a metaconid) with small hypoconulids (Szalay & Delson, 1979). *Protoadapis* has inflated protoconid and metaconid cusps separated by a V-shaped trigonid notch which distinguishes it from *Pronycticebus* (Szalay & Delson, 1979). *Protoadapis* has an unfused mandible with a 2.1.4.3 dental formula. *Protoadapis louisi* probably represents the common ancestor of *Periconodon* and later species of *Protoadapis* (Gingerich, 1977*b*). Other species of *Protoadapis* may have given rise to *Pronycticebus* and *Europolemur* (Szalay & Delson, 1979). The reviews by Szalay & Delson (1979) and Godinot (1998) attempt to sort out the complicated taxonomic history of *Protoadapis*.

GENUS *Pronycticebus* Grandidier, 1904 (Fig. 4.5)
INCLUDED SPECIES *P. gaudryi, P. neglectus*

TYPE SPECIES *Pronycticebus gaudryi* Grandidier, 1904
TYPE SPECIMEN MNHH 1893-11, cranium and mandible
AGE AND GEOGRAPHIC RANGE Middle Eocene, Europe
ANATOMICAL DEFINITION
Pronycticebus is a late Bartonian (late middle Eocene) adapiform from Europe. Many have commented upon *Pronycticebus* (Grandidier, 1904; Le Gros Clark, 1934; Simons, 1972; Szalay & Delson, 1979; Godinot, 1998). *Pronycticebus gaudryi* has a talonid basin that is broader than in *Protoadapis* and *Europolemur* (Godinot, 1998). Third molars (upper and lower) are elongated in this genus and this feature helps to distinguish *Pronycticebus* from *Protoadapis* and *Europolemur*. P^4 is also very broad (transversely) relative to *Europolemur* and *Protoadapis* while M^1 has a larger hypocone and no lingual cingulum. M^2 has a mid-sized hypocone and an incomplete lingual cingulum. M^3 has a complete lingual

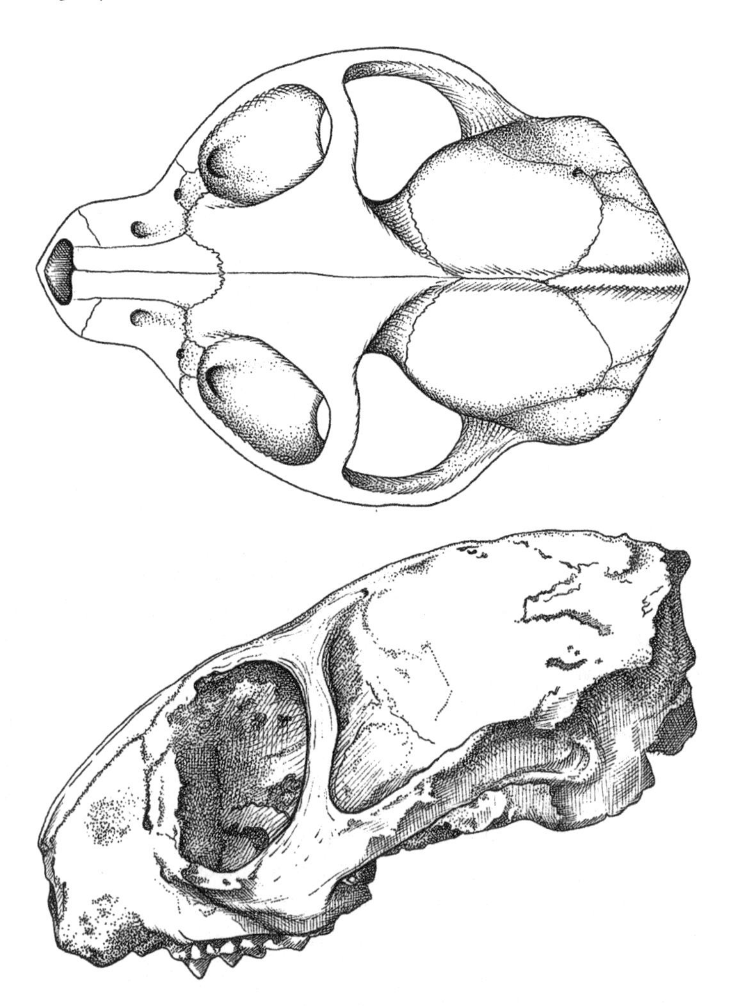

Fig. 4.5 *Pronycticebus gaudryi*. Skull in dorsal and lateral views (MNHN 1893-11).

cingulum and no hypocone (Godinot, 1998). The skull (Fig. 4.5) has a short muzzle and weak crests (Simons, 1962*a*). Sloping of the nasal region makes this taxon unusual compared to other adapiform skulls. *Pronycticebus* has large orbits (Simons, 1962*a*; Martin, 1990; Godinot, 1998) suggesting nocturnality. The maxilla is broad at P^2 compared to the maxillary broadening at M^1 seen among other adapiforms (Godinot, 1998).

Godinot (1998) places *Pronycticebus* in its own subfamily and suggests that it might share several similarities with *Plesiopithecus* from the Fayum (Simons & Rasmussen, 1994*a*). *Agerinia* and *Pronycticebus* also share several dental similarities according to Godinot (1998). These similarities include a large talonid compared to the trigonid, a narrow trigonid with the protoconid pointing anteriorly and molar reduction in size from M^1 to M^3. *Pronycticebus* has a longer M_3 than *Agerinia*. Godinot (1988) wonders if one mandible of this taxon might possess a tooth-comb. The mixed characteristics of *Pronycticebus* do not allow a clear link with ceramoniines according to Godinot (1998).

GENUS *Wadilemur* Simons, 1997
INCLUDED SPECIES *W. elegans*

TYPE SPECIES *Wadilemur elegans* Simons, 1997
TYPE SPECIMEN CGM 42211, right mandible with P_3–M_3
AGE AND GEOGRAPHIC RANGE Late Eocene, Africa
ANATOMICAL DEFINITION
Wadilemur elegans (Simons, 1997*a*) represents one of the three known adapiforms from a southern continent (all from Locality 41, Fayum, Egypt). Simons (1997*a*) describes this taxon as being very similar to *Microcebus* and *Anchomomys* and places *Wadilemur* within the Cercamoniinae (Tribe Anchomomyini). *Wadilemur* has crescentric ridges that connect the lower molar cusps. A continuous crest runs from the protoconid to the metaconid encircling the anterior fovea (Simons, 1997*a*). The lower molars lack a paraconid and increase in size distally. M_3 has a hypoconulid. The metaconids are shifted posteriorly relative to the protoconid. Lastly, the lower molars are notched externally between the trigonid and talonid basins. The jaw of *Wadilemur* is about half the length of *Aframonius* (P_3–M_2 length).

Family Adapidae

Subfamily Adapinae

The Adapidae are chiefly European although their origin may stem from Asia (*Adapoides*). *Adapis parisiensis* is the last surviving member of the Adapidae in Europe just before the Grande Coupure, the late Eocene–Oligocene boundary. All taxa have a 2.1.4.3 dental formula. *Adapis parisiensis* and *Leptadapis magnus* were folivorous adapiforms (Covert, 1986). The anterior dentition of *Adapis* differs remarkably from that of *Leptadapis*. The skulls of *Adapis* and *Leptadapis* show large sagittal crests and wide zygomatic regions associated with powerful chewing muscles (temporalis and masseter). *Microadapis* lacked a fused mandible. The postcranial anatomy of this group is highly divergent from other adapiforms. Climbing and quadrupedalism appear to be the dominant modes of locomotion (i.e., non-leaping adapiforms).

GENUS *Adapis* Cuvier, 1821 (Figs. 4.6–4.7)
INCLUDED SPECIES *A. bruni, A. collinsonae, A. parisiensis, A. sudrei*

TYPE SPECIES *Adapis parisiensis* Cuvier, 1821
TYPE SPECIMEN GY 685, fragmentary skull and lower jaw (crushed)
AGE AND GEOGRAPHIC RANGE Late Eocene, Europe
ANATOMICAL DEFINITION
Adapis parisiensis has a 2.1.4.3 dental formula, broad spatulate upper and lower incisors, and premolars that become progressively larger and more molariform (Fig. 4.6). The molars are sharply crested with small hypocones on M^1 and M^2, and M_3 is elongated with a well-developed

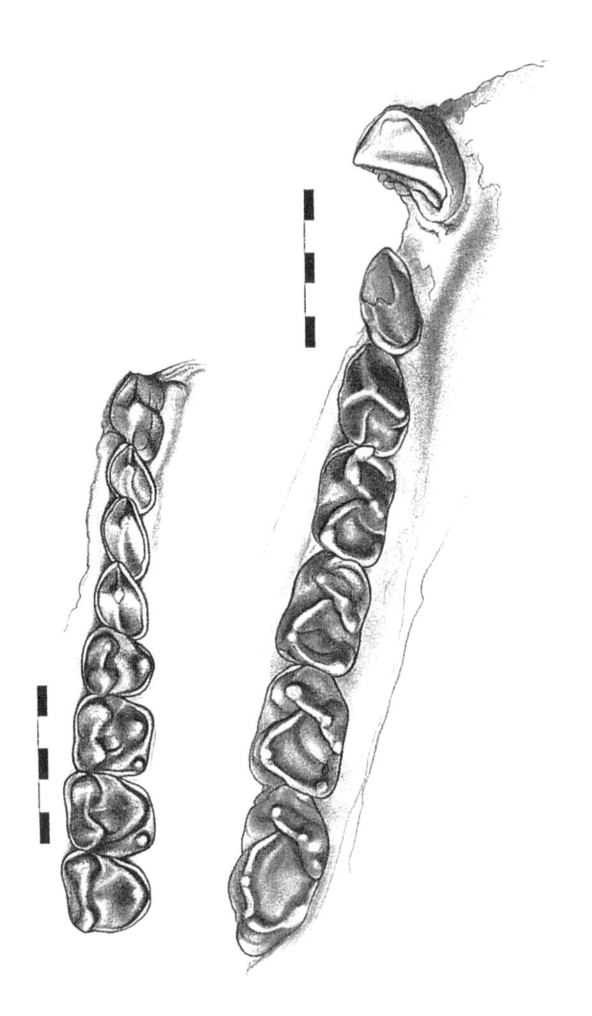

Fig. 4.6 Lower dentition (right) of *Leptadapis magnus* (QD 34); upper dentition (left) of *Adapis parisiensis* (unnumbered palate). Scales in mm.

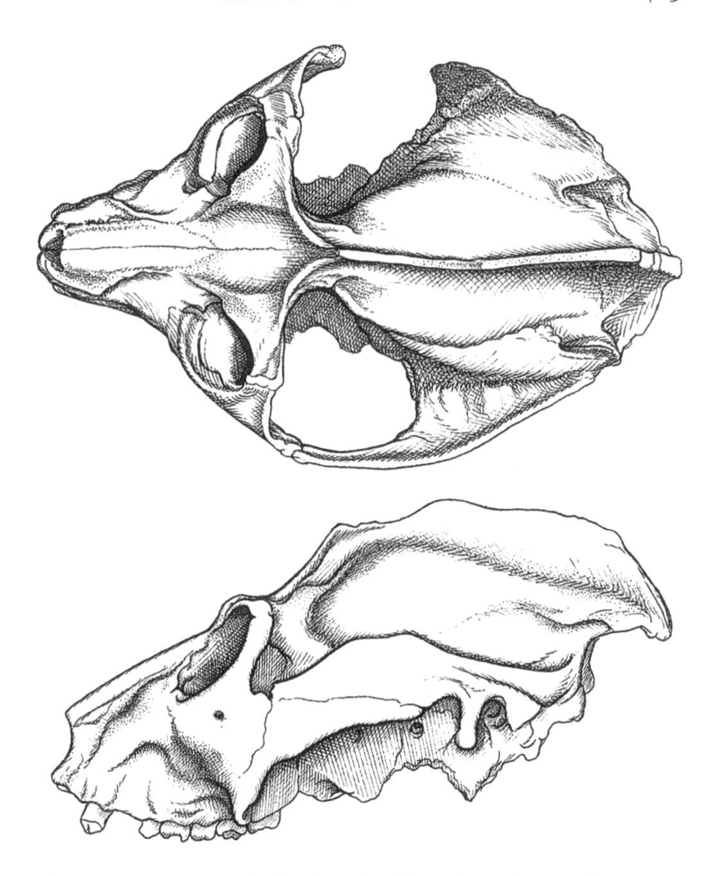

Fig. 4.7 *Adapis parisiensis*. Skull in dorsal and lateral view (PQ 1700).

hypoconulid. Gingerich (1975*a*) notes similarities between the anterior dentition of *Adapis* and that of *Hapalemur*, suggesting a similar incisor–canine functional unit and the possibility of evolving the lemuriform tooth-comb from this functional complex. The canine is reduced in *Adapis* and forms a cutting edge or cropping mechanism with the incisors (Gingerich, 1975*a*). In *A. sudrei*, the mandible is unfused, while *A. parisiensis* possesses a fused symphysis. The skulls of *Adapis* are characterized by large sagittal crests and robust zygomatics (Fig. 4.7). The braincase is small but it possesses large olfactory bulbs. The orbits are small (Kay & Cartmill, 1977; Lanèque, 1993), a derived condition from that of *Leptadapis* (Lanèque, 1993). Lanèque (1992*a*, 1992*b*) has examined the skulls attributed to *Adapis* and concluded that three groups or taxa are present. This interpretation contrasts with the suggestion of sexual dimorphism within *Adapis* as expressed by Gingerich (1981*a*) and Szalay & Delson (1979), and suggests a much more heterogenetic sample from earlier interpretations. *Adapis* has less frontation than do lemurs, with more upward-facing orbits, but greater convergence than lemurs and indrids. Lanèque (1993) suggests that adapines were very visually oriented primates.

GENUS *Adapoides* Beard et al., 1994
INCLUDED SPECIES *A. troglodytes*

TYPE SPECIES *Adapoides troglodytes* Beard et al., 1994
TYPE SPECIMEN IVPP V11023, right mandibular fragment with M_2–M_3
AGE AND GEOGRAPHIC RANGE Middle Eocene, Asia
ANATOMICAL DEFINITION
In 1994 Beard and colleagues named the earliest Asian adapiform from the middle Eocene fissure fillings from Shanghuang (Jiangsu Province, southeastern China). The mammalian faunas at Shanghuang represent Irdinmanhan and early Sharamurunian land mammal ages. The Irdinmanhan can be correlated with the Bridgerian and early Uintan of North America (about 46 Ma; Beard et al., 1994). *Adapoides troglodytes* is small and similar in size to *Microadapis sciureus* (Beard et al., 1994). Its lower molars are longer and narrower than in *Leptadapis* and they lack metastylids. Lower molars exhibit a high, continuous crest between the protoconid and metaconid and a deep talonid notch. Upper molars show buccal crests that form a single mesiodistally oriented shearing surface, a weak or absent metaconule and well-developed hypocones. This form suggests that adapines immigrated into Europe from Asia (Beard et al., 1994) and not from Africa (Franzen, 1987). Although Asian, this species shows no special resemblance to the Sivaladapidae.

GENUS *Cryptadapis* Godinot, 1984
INCLUDED SPECIES *C. laharpei, C. tertius*

TYPE SPECIES *Cryptadapis tertius* Godinot, 1984
TYPE SPECIMEN SNB 2273 (Laboratory of Paleontology, University of Montpellier), maxilla with P^4–M^1).
AGE AND GEOGRAPHIC RANGE Late Eocene, Europe
ANATOMICAL DEFINITION
Cryptadapis tertius is similar in size to *Adapis parisiensis* and is late Eocene in age (Godinot, 1984, 1998). *Cryptadapis* has transversely elongated upper and lower molars with strong metastylids on all lower molars (Godinot, 1984). A crest connects the protocone and hypocone like that of notharctines. M^3 has a small hypocone. P_4 is molarized with a metaconid and metastylid. *Cryptadapis* was folivorous.

GENUS *Leptadapis* Gervais, 1876 (Fig. 4.6)
INCLUDED SPECIES *L. assolicus, L. capellae, L. leenhardti, L. magnus, L. priscus, L. ruetimeyeri*

TYPE SPECIES *Leptadapis magnus* Filhol, 1874
TYPE SPECIMEN MNHN QU 11002, skull (Filhol, 1874; Gervais, 1876)
AGE AND GEOGRAPHIC RANGE Middle Eocene, Europe
ANATOMICAL DEFINITION
Leptadapis is a middle to late Eocene adapid from western Europe. Species range in size from 1300 to 4000 g (Fleagle, 1999). The incisors and canines of *Leptadapis* are very different from those of *Adapis* (Szalay, 1974). The canines are well-developed, robust and caniniform in structure (Fig. 4.6). The incisors are not part of an anterior dentition functional complex like that of *Adapis* (Gingerich, 1975*a*). *Leptadapis* has a large protocone on P^3 and a reduced P_2 (Szalay & Delson, 1979). Its dental formula is 2.1.4.3. "*Leptadapis* has relatively larger orbits, weaker frontation and weaker obital convergence than *Adapis*" (Lanèque, 1993: 287).

GENUS *Microadapis* Szalay, 1974
INCLUDED SPECIES *M. lynnae, M. sciureus*

TYPE SPECIES *Microadapis sciureus* Stehlin, 1916
TYPE SPECIMEN BNM Eh 750, left C–M_3
AGE AND GEOGRAPHIC RANGE Middle Eocene, Europe
ANATOMICAL DEFINITION
Szalay (1974) and Szalay & Delson (1979) referred the original specimen which was attributed to *Adapis sciureus* by Stehlin (1916) to *Microadapis sciureus* (but see Gingerich, 1977*b*). *Microadapis* is a middle Eocene adapid. Schwartz & Tattersall (1982) allied this form with that of *Smilodectes* but Covert (1990) disapproves of this notion. It lacks an entoconid notch on its lower molars and a squared talonid heel on M_3, features that characterize *Smilodectes* (Covert, 1990). *Microadapis* possesses a paraconid shelf and lacks a paraconid and a cristid obliqua on M_3. Molars are low-crowned with a large hypocone, as is the canine (Szalay & Delson, 1979). The upper molars have well-developed paraconules and metaconules. *Microadapis* has non-molarized fourth premolars and a 2.1.4.3 dental formula. The symphysis is unfused.

GENUS *Palaeolemur* Delfortrie, 1873
INCLUDED SPECIES *P. betillei*

TYPE SPECIES *Palaeolemur betillei* Delfortrie, 1873
TYPE SPECIMEN MNHN BOR 613, skull
AGE AND GEOGRAPHIC RANGE Late Eocene, Europe
ANATOMICAL DEFINITION
Palaeolemur betillei was originally described by Delfortrie in 1873, but was subsequently placed within *Adapis* by Simons (1972) and Gingerich (1981*a*). However, Godinot (1986, 1998) has recently resurrected *Palaeolemur* as a valid genus. Lanèque (1992*a*, 1992*b*) has also shown that this skull is much smaller in several dimensions to be subsumed within the genus *Adapis*.

Family Sivaladapidae

Subfamily Sivaladapinae

Although a few sivaladapids are known in the Eocene of Asia at present, this group is best known as a late surviving branch of Miocene adapiforms. With recent discoveries from China and Thailand, this view is changing. Sivaladapids appear to be a diverse group of adapiforms that weathered cooling in the northern hemisphere by shifting southward toward the Asian equator (Qi & Beard, 1998). *Rencunius* may well represent the stem taxon for this group, while *Periconodon* may be the nearest outgroup (Qi & Beard, 1998). The Sivaladapidae are united by three dental features: (1) strong lower molar hypoconulids; (2) lower molar hypoconulids twinned with entoconids; and (3) a strong and continuous lingual cingulum on upper molars (Qi & Beard, 1998). The Miocene sivaladapids are known for their large body sizes, robust and fused mandibles, and highly crested teeth adapted for folivory. Sivaladapids were the last surviving members of Adapiformes.

GENUS *Guangxilemur* Qi & Beard, 1998
INCLUDED SPECIES *G. tongi*

TYPE SPECIES *Guangxilemur tongi* Qi & Beard, 1998
TYPE SPECIMEN IVPP V11652, left M^2
AGE AND GEOGRAPHIC RANGE Late Eocene, Asia
ANATOMICAL DEFINITION
Guangxilemur tongi is a late Eocene sivaladapine from China, probably equivalent in age to *Wailekia* from southern Thailand (Qi & Beard, 1998). This form is known only from a M^2 and a broken canine. The upper molar is larger than *Hoanghonius, Rencunius* and *Wailekia*. It differs from *Hoanghonius* and *Rencunius* in showing greater external shearing crests, a well-developed parastyle and mesostyle,

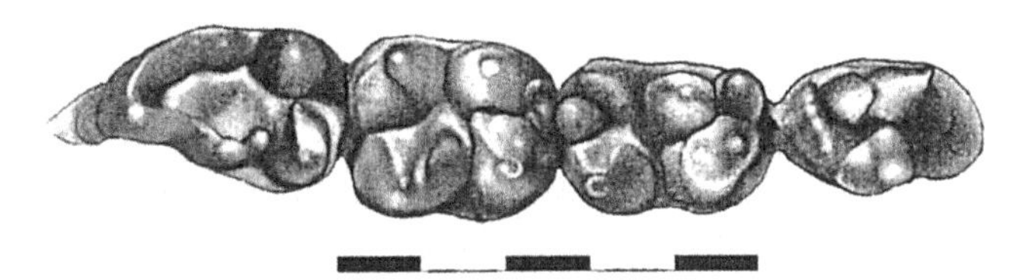

Fig. 4.8 *Hoanghonius stehlinii*. Lower dentition (composite). Scale in mm.

and weaker conules (Qi & Beard, 1998). It differs from *Sivaladapis* and *Sinoadapis* in that it possesses a large pericone and a hypocone on the lingual cingulum. The preprotocrista joins the preparacrista just lingual to the parastyle. The buccal shearing crests are well developed and connect the parastyle with the paracone, mesostyle and metacone. Among adapiforms only *Periconodon*, *Hoanghonius* and *Rencunius* possess both a pericone and a hypocone (Qi & Beard, 1998). *Guangxilemur* differs from *Periconodon*, *Hoanghonius* and *Rencunius* in its greater development of external shearing crests and associated stylar structures. *Guangxilemur* also lacks the molar conules found in these three taxa. The Miocene sivaladapines show even greater development of external shearing crests.

GENUS *Hoanghonius* Zdansky, 1930 (Fig. 4.8)
INCLUDED SPECIES *H. stehlini*

TYPE SPECIES *Hoanghonius stehlini* Zdansky, 1930
TYPE SPECIMEN University of Uppsala, Shanxi Province, China, no number, mandibular fragment with M_{2-3}.
AGE AND GEOGRAPHIC RANGE Middle Eocene, Asia
ANATOMICAL DEFINITION
Hoanghonius stehlini is believed to be a sivaladapid (Gingerich, 1980*a*; Rasmussen & Simons, 1988; Beard, 1998*c*), rather than an omomyid (Szalay & Delson, 1979). Zdansky (1930) first named this taxon from northern China and it dates to the late middle Eocene. *Hoanghonius* (Fig. 4.8) possesses a large lingual cingulum, a pericone, a twinned entoconid–hypoconulid, and a reduced paracristid (Szalay & Delson, 1979). The upper molars possess a pre- and postprotocrista, paraconules and metaconules, and a mesostyle. M_3 has a hypoconulid lobe. Gingerich (1977*a*) and Szalay & Delson (1979) have suggested similarities to the anthropoid *Oligopithecus*. Beard (1998*c*) has recently recovered more specimens of *H. stehlini* and has noted a double-rooted P_2 and upper and lower molar features that link this genus with other sivaladapids. The only known postcranial specimen for a sivaladapid is attributed to *Hoanghonius* and this first metatarsal is very similar in its morphology to that of *Cantius* (Gebo *et al.*, 1999).

GENUS *Indraloris* Lewis, 1933
INCLUDED SPECIES *I. himalayensis*

TYPE SPECIES *Indraloris himalayensis* Pilgrim, 1932
TYPE SPECIMEN Geological Survey of India (Calcutta) No. D. 237, right mandibular ramus with M_1
AGE AND GEOGRAPHIC RANGE Late Miocene, Asia
ANATOMICAL DEFINITION
Indraloris himalayensis is a late Miocene adapiform from northern India (Pilgrim, 1932; Lewis, 1933; Gingerich & Sahni, 1979; Chopra & Vasishat, 1980). Its upper molars lack a hypocone. The lower molars possess a trigonid that is mesiodistally constricted with lophodont crests and twinned hypoconulids and entoconids. P_4 is molarized. The canines are large and interlocking and the symphysis is fused (Gingerich & Sahni, 1979; Szalay & Delson, 1979). The dental adaptations for *Indraloris* indicate a folivorous diet. Wu & Wang (1988) suggest that *Indraloris* is possibly congeneric with *Sinoadapis*.

GENUS *Rencunius* Gingerich *et al.*, 1994
INCLUDED SPECIES *R. zhoui*

TYPE SPECIES *Rencunius zhoui* Gingerich *et al.*, 1994
TYPE SPECIMEN IVPP 5312, left dentary with P_4–M_2
AGE AND GEOGRAPHIC RANGE Late middle Eocene, Asia
ANATOMICAL DEFINITION
Rencunius zhoui is a late middle Eocene adapiform from China (Gingerich *et al.*, 1994). This genus is similar to *Hoanghonius* in that it has twinned entoconids and hypoconids on M_2 but with larger and more separated cusps. *Rencunius* differs from *Hoanghonius* in having higher and more inflated M_2 crowns, and the retention of a small paraconid on M_2. M^1 has a large pericone and hypocone, and larger, more centrally placed paraconules and metaconules. There exists a flexure of enamel running along the posterolingual side of the paracone. A small centrocrista exists within the trigon basin.

GENUS *Sinoadapis* Wu & Pan, 1985
INCLUDED SPECIES *S. carnosus*

TYPE SPECIES *Sinoadapis carnosus* Wu & Pan, 1985
TYPE SPECIMEN PA 855, two mandibular fragments, left P_4–M_2 and right M_1–M_3 (Pan, 1988)
AGE AND GEOGRAPHIC RANGE Late Miocene, Asia
ANATOMICAL DEFINITION
Sinoadapis carnosus comes from the late Miocene site ($\sim$ 8 Ma) of Lufeng, China (Pan, 1988), where more than 380 specimens have been recovered (Wu & Pan, 1985; Pan, 1988). *Sinoadapis* is distinguished by its large size and its lower P_4, which is long and highly molarized. The canines show marked sexual dimorphism. The mandible is unfused. The dental formula is 2.1.3.3. The incisors are spatulate with I2 > I1. The upper and lower dentitions are similar to *Sivaladapis* (Pan, 1988). The hypoconulid is twinned with the entoconid, and these cusps are separated by a deep notch. *Sinoadapis* shows distinct swellings of enamel on the lingual cingula in precisely the areas occcupied by the pericone and hypocone in *Guangxilemur* (Qi & Beard, 1998). Thomas & Verna (1979) suggested that the later

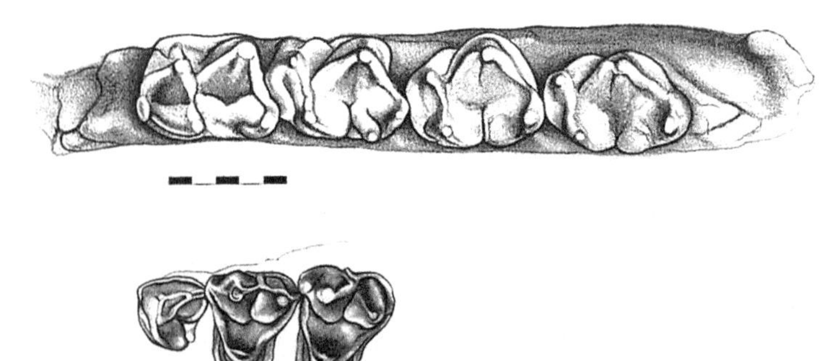

Fig. 4.9 *Sivaladapis nagrii*. Upper (LUVP 14506, bottom) and lower (LUVP 14500, above) dentition. Scales in mm.

sivaladapids secondarily lost pericones and hypocones on their upper molars and *Sinoadapis* and *Guangxilemur* corroborate this view. *Sinoadapis* was frugivorous (Pan, 1988). *Sinoadapis* was the last surviving member of the Sivaladapidae.

GENUS *Sivaladapis* Gingerich & Sahni, 1979 (Fig. 4.9)
INCLUDED SPECIES *S. nagrii, S. palaeindicus*

TYPE SPECIES *Sivaladapis palaeindicus* Gingerich & Sahni, 1979
TYPE SPECIMEN Geological Survey of India (Calcutta) No. D. 224 associated right, P_4 and M_1^{-2} or M_2^{-3} (Pilgrim, 1932; Gingerich & Sahni, 1979)
AGE AND GEOGRAPHIC RANGE Middle Miocene, Asia
ANATOMICAL DEFINITION
Sivaladapis palaeindicus is found in the Siwaliks (India and Pakistan) in the middle Miocene (Chinji faunal zone) at about 13–14 Ma. *Sivaladapis nagrii* is late Miocene (9 Ma) in the Nagri faunal zone. The best-known specimens are associated with *S. nagrii* (Gingerich & Sahni, 1979, 1984). The incisors have spatulate crowns and are slightly procumbent (Gingerich & Sahni, 1979), the canines are large, and the mandible is fused. The dental formula is 2.1.3.3. P_2 has slight honing. The upper premolars are progressively more molarized with P^4 being three-rooted and fully molarized (Gingerich & Sahni, 1984). The paracone, metacone and protocone are prominent with a completely closed trigon with pre- and postcristas. A sharp ectoloph exists. There are distinct para- and mesostyles. A prominent lingual cingulum surrounds the protocone. All of the upper molars have high sharp cusps. No hypocone is present. The lower incisors are small, vertically implanted and spatulate in shape. They are high-crowned. P_2 is single-rooted, high-crowned and caniniform with no accesory cusps (Gingerich & Sahni, 1984). Gingerich & Sahni (1984) suggest canine dimorphism for *S. nagrii* . P_4 is highly molarized but lacks an entoconid. The lower molars are high-crowned with crests (Fig. 4.9). The protoconid and metaconid are large and separated by a protolophid. The paraconid is reduced and crest-like. The entoconid is sharp and pointed. The hypoconulid is twinned with the entoconid and these two cusps are separated by a notch (Gingerich & Sahni, 1979). The molars appear to be highly specialized for puncturing and shearing. *Sivaladapis* was between 4 and 7 kg in body size suggesting that it was a specialized arboreal folivore. *Sivaladapis* differs from *Indraloris* in having longer, relatively narrow lower molars with a much larger hypoconulid (Gingerich & Sahni, 1979). *Sivaladapis* shares with *Sinoadapis* narrower trigonids, large and high hypoconulids that are deeply separated from the entoconid, and no postprotocrista (Godinot, 1998).

GENUS *Wailekia* Ducrocq et al., 1995
INCLUDED SPECIES *W. orientale*

TYPE SPECIES *Wailekia orientale* Ducrocq et al., 1995
TYPE SPECIMEN Dept. of Mineral Resources collections, Bangkok TF 2632, right lower jaw with M_2–M_3 and alveoli for lower C and P_2–M_1
AGE AND GEOGRAPHIC RANGE Late Eocene, Asia
ANATOMICAL DEFINITION
Wailekia orientale (late Eocene, Krabi Basin, southern Thailand) was originally described as an anthropoid, being morphologically similar to *Oligopithecus* and placed within the Propliopithecidae (Ducrocq et al., 1995). More recently, Gingerich et al. (1994), Godinot (1998) and Qi & Beard (1998) all believe that this species is in fact a sivaladapid and that it is morphologically very similar to *Hoanghonius stehlini* (Qi & Beard, 1998). *Wailekia* retains a two-rooted P_2 (a non-haplorhine characteristic). *Wailekia* is a relatively large primate at 1.5 kg (Ducrocq et al., 1995). On its lower molars, *Wailekia* shows a large notch separating the metaconid and entoconid, a twinned entoconid and hypoconulid, and an absence of a paraconid. The preprotocristid and premetacristid unite mesially and close the trigonid basin lingually. The long paralophid and closed trigonid basin are similar to *Hoanghonius* and *Oligopithecus* (Godinot, 1998). The cristid obliqua is low and short. The mandible is unfused at the symphysis with a slender mandibular corpus. The coronoid process slopes low (Godinot, 1998). The diet of *Wailekia* is believed to be frugivorous.

Problem taxa

Asia

Amphipithecus mogaungensis (Colbert, 1937) and *Pondaungia cotteri* (Pilgrim, 1927) have long suffered under enigmatic systematics. These 40-million-year-old fossil primates from Burma (Pondaung Formation) have been attributed to adapiform and to anthropoid primates (Szalay, 1970; Simons, 1972; Ba Maw et al., 1979; Ciochon et al., 1985; Ciochon & Holroyd, 1994; Chaimanee et al., 2000*a*). Ducrocq (1998) and Ciochon et al. (1999) have recently attributed anthropoid status to these two taxa (see Chapter 9).

Lushius qinlinensis (Chow, 1961), a late Eocene fossil from China (maxilla fragment with P^4–M^3), could be either an omomyid or an adapiform primate (Szalay & Delson, 1979). Its large size (~2.9 kg; Fleagle, 1999) and buccal shearing suggest folivory and a connection to adapiforms, while its reduced M^3 is similar to anaptomorphines (Szalay & Delson, 1979).

Panobius afridi (two isolated molars from the Kuldana Formation, Pakistan) has been described as an early or middle Eocene adapiform primate by Russell & Gingerich (1987). *Panobius* is a small primate (~130 g; Fleagle, 1999) and is thought to be similar to *Donrussellia*. Fleagle (1999) lists this form as a cercamoniine.

Africa

The tiny (100 g; Fleagle, 1999) early Eocene *Djebelemur martinezi* (Chambi, Tunisia) described by Hartenberger & Marandat (1992) shows similarities to adapiform primates. Godinot (1994) finds many similarities to cercamoniines and to early anthropoids in the left mandible (P_3–M_3) and the few isolated teeth known for *Djebelemur*. In 1997, Hartenberger *et al.* continued to support the adapiform affinities for *Djebelemur*, while Godinot (in Hartenberger *et al.*, 1997) argued that *Djebelemur* is a "simiiforme" (anthropoid) primate.

Likewise, *Omanodon minor* (two lower and five upper teeth) and *Shizarodon dhofarensis* (one and a half lower molars) from the early Oligocene of Oman (Gheerbrant *et al.*, 1993) have been described as being similar to anchomomyins and cheirogaleids. Both are tiny primates (100–200 g; Fleagle, 1999). Both taxa have been allocated to Adapiformes (Gheerbrant *et al.*, 1993). Simons (1998) also believes these taxa to be adapiforms and Fleagle (1999) lists them among the cercamoniines. In contrast, Godinot (1998) notes that both taxa have anthropoid affinities and thus cannot be adapiform primates.

Evolution of Adapiformes

Paleoenvironment

Adapiformes is a northern hemisphere adaptive radiation with only three species known from the southern continent of Africa. Until the late Eocene, a land bridge connected North America and Europe while the southern continents were isolated by water barriers. The collision of the Afro-Arabian plates in the mid-Miocene produced the first stable land bridge to a southern continent. Thus, during the Eocene and Oligocene, adapiforms must have crossed a water barrier, traveled across an island arc, or crossed short-lived or unstable land bridges to arrive in Africa before the mid-Miocene, given their present biogeographic distribution.

The Eocene epoch represents the warmest time period of the Tertiary. Paleobotanical evidence indicates that the northern continents were covered with tropical rainforests with strong Indo-Malay affinities (Conroy, 1990). Seasonality was not marked until the period of rapid cool-off toward the late Eocene and early Oligocene. In fact, the mid-Eocene (49-47 Ma) represents the "Golden Age" of adapiform diversity. Europe, in particular, shows the greatest number of taxa at this time. In contrast with the mid-Eocene, the Eocene–Oligocene boundary documents a severe reduction in adapiform species, an event correlated with cooling in the northern hemisphere. By the Oligocene, adapiforms are gone from Europe, very rare in North America, and not present in Africa (given its poorer record). Only the late surviving sivaladapids last until the late Miocene in southern Asia. Thus, southern Asia represents an important refugium for rainforest-adapted primates like the adapiforms (Beard, 1998*a*).

Size and evolution

Adapiform primates are generally above 1 kg in body size (see Conroy, 1987; Fleagle, 1999). The largest adapiform is *Notharctus robustior*, weighing about 7 kg. Only five genera (*Anchomomys*, *Periconodon*, *Agerinia*, *Donrussellia* and *Adapoides*) are less than 500 g (Gingerich, 1980*a*; Conroy, 1987), with *Anchomomys gaillardi* representing the smallest adapiform primate at 50 g (Conroy, 1987). Gingerich (1977b) has shown that four adapiform lineages in Europe become larger over time, while three others reduce in size. In North America, *Cantius* increases in body size through time while *Copelemur*, *Smilodectes*, *Cantius frugivorus* and *Notharctus tenebrosus* are reduced from their ancestral precursors (Gingerich & Simons, 1977; Gingerich, 1979). Of all the adapiform lineages, only the anchomomyines are tiny. There are several morphological trends accompanying these size changes. For example, symphyseal fusion occurred independently in three different European lineages (*Caenopithecus*, *Cercamonius* and *Adapis*) and in the North American *Notharctus* (Gingerich, 1977*b*), while shearing crests evolve independently in sivaladapids and adapids (Gingerich, 1977*a*).

Biostratigraphy

Using good stratigraphic information and large collections from the Bighorn Basin, Gingerich & Simons (1977) plotted log length × width of molars. They noted a single lineage of *Cantius* increasing in size over time and noted the progressive enlargement of mesostyles and hypocones in this lineage. Gingerich & Simons (1977) also noted several smaller lineages evolving from *Cantius* at various stratigraphic levels. Later Gingerich (1977*b*, 1979, 1980*a*) would analyze *Notharctus* and *Smilodectes* as well as the European adapiforms in a similar manner, albeit with less stratigraphic precision. This stratophenetic approach (Gingerich & Schoeninger, 1977) has been highly influential in analyzing North American fossil primates (e.g., Bown & Rose, 1987). These analyses provide little doubt that *Cantius* (early Eocene) represents the earliest adapiform in North America. In the early Eocene of Europe, *Donrussellia*, *Cantius* and *Protoadapis* represent the earliest adapiforms (Gingerich, 1980*a*). Thus, the Notharctidae encompasses the earliest known adapiforms. In 1998, Godinot attempted to apply Gingerich's stratophenetic approach in Europe, but found that in several localities different genera overlap in size, making this approach very difficult. He noted that "a purely stratodimensional diagram no longer allows us to separate the lineages and show phylogenetic relationships (which are quite complex in an area as large as Europe)" (Godinot, 1998: 226).

Adaptations

Cranium

Adapiforms generally have long noses and jaws with fairly broad snouts. They have a relatively large ethmoid region with numerous ethnoturbinates, implying a well-developed olfactory sense. They have small orbit sizes relative to skull length (see Kay & Cartmill, 1977; Gingerich, 1980*a*). Small orbits imply a diurnal lifestyle for these primates with one exception. *Pronycticebus* has large orbits and likely was active at night. Adapiform orbits are obliquely oriented to the midline of the skull suggesting the right and left visual fields overlapped. Stereoscopic or three-dimensional vision is inferred from the spatial position and orientations of adapiform orbits. All adapiform skulls have postorbital bars surrounding the lateral orbit and postorbital constriction.

Relative brain size for the known adapiforms such as *Smilodectes*, *Notharctus* and *Adapis* is lower than most living primates (mean encephalization quotients are in the range 0.39–0.53 for adapiforms compared to 0.67–1.89 for living prosimians; Gingerich & Martin, 1981; Conroy, 1990: 107). MacLarnon (1995, 1996) has noted that the spinal cord is also small in notharctines. She suggested that adapiform coordination (neural control) was not as sophisticated as that in living lemurs.

In terms of cranial anatomy, the lacrimal bone is located within the orbit and a free ring-like ectotympanic is within the petrosal auditory bulla. The carotid artery enters the bulla posterolaterally before dividing into the promontory and stapedial arteries. Comparatively well-developed promontory and stapedial arteries are present in adapiforms with the stapedial artery usually larger than the promontory (Conroy, 1990). In *Adapis*, the bony canal for the stapedial artery is four times the cross-sectional area of the canal for the promontory artery (Gingerich, 1980*a*). In *Notharctus*, the promontory artery is more than twice the size of the stapedial artery (Gingerich, 1973). Thus, the size of the stapedial artery, a character that has been used to distinguish adapiforms from tarsiiforms (Gregory, 1920; Szalay, 1975*b*), varies widely among taxa (Gingerich, 1973; MacPhee & Cartmill, 1986; Gunnell, 1995*b*).

The infraorbital foramen, through which the maxillary branches of the trigeminal nerve exit, is small. This would suggest that the nose and whiskers were less important sensory structures (Kay & Cartmill, 1977; Gingerich, 1980*a*). In terms of chewing musculature, some adapiform taxa have very prominent zygomatic regions (e.g., *Adapis*). Sagittal crests vary from being well developed among the adapids or *Hesperolemur* to being weakly developed among *Smilodectes* and *Notharctus*. Well-developed zygomatic regions and sagittal crests imply well-developed jaw musculature for chewing and suggest more frugivorous or folivorous diets among adapiforms.

Diet and dentition

Adapiforms span a large range of diets. The smallest forms, like *Anchomomys*, possess teeth indicative of insectivory, having tall and pointy teeth. According to Covert (1986), *Donrussellia*, *Periconodon*, *Microadapis*, *Agerinia* and *Pronycticebus* may also have been insectivorous. Most of the other adapiforms are larger than 1 kg and possess dentitions best adapted for frugivory and folivory. Kay (1984) has shown that primates above 1 kg prefer a diet of fruits and leaves. Since metabolic rate scales inversely with mass, only the larger adapiforms have metabolic rates low enough to survive a diet of leaves, a dietary item low in carbohydrates (energy) but high in protein. Covert (1986) has measured adapiform molar crest heights (shearing coefficients), a feature indicative of leaf-eating or folivory. He notes that adapids like *Adapis* and *Leptadapis* possess high shearing coefficients as do *Notharctus*, *Smilodectes* and *Caenopithecus*. The sivaladapines are also highly folivorous. Shearing coefficients for *Cantius*, *Copelemur* and *Pelycodus* on the other hand suggest frugivory (Covert, 1986). Gingerich (1980*a*) also noted that the anterior dentition of *Adapis* was like the short-tusked condition in callitrichids, primates that use their front teeth to harvest gums.

Body

Since the initial description of notharctine tarsals by Matthew in 1915 and Gregory's 1920 opus on *Notharctus*, the postcranium of adapiforms has received considerably more attention (Napier & Walker, 1967; Simons, 1969; Decker & Szalay, 1974; Szalay & Delson, 1979; von Koeningswald, 1979; Godinot & Jouffroy, 1982; Conroy & Rose, 1983; Dagosto, 1983, 1986, 1988, 1993; Rose & Walker, 1985; Covert, 1985, 1986, 1988; Gebo, 1985, 1987*a*, 1988; Franzen, 1987, 1988, 1994*b*; Beard & Godinot, 1988; Szalay & Dagosto, 1988; Thalmann *et al.*, 1989; Gebo *et al.*, 1991; Godinot, 1991; Dagosto & Terranova, 1992; Moyà-Solà & Köhler, 1993*a*; Covert & Williams, 1994; Bacon & Godinot, 1998). Although we have complete skeletons of *Notharctus* and a few other adapiformes, most of the adapiform postcranial material is represented by isolated and unassociated remains. This leaves many questions concerning the bodies of a large number of adapiform primates. On the other hand, family-level adaptations in the body among living primates are highly accentuated suggesting that one good representative might be adequate to understand a broad level of adaptation within a primate group or lineage. I would suggest that adapiform primates easily fall within this morphological pattern. Thus, I can explain more about the body of *Notharctus* than I can for *Cantius*, but in reality our inferences about locomotor adaptation between the two differ only in subtle distinctions. As I will show below, notharctines and cercamoniines are very similar in their body adaptations and both are linked within Notharctidae. In contrast, the body adaptations within the Adapidae are very different from that of the Notharctidae. At present, we know too little about the Sivaladapidae to make any significant comments. We currently have postcranial remains from the following adapiforms: Notharctidae (*Cantius, Copelemur, Pelycodus, Notharctus, Smilodectes, Anchomomys, Cercamonius, Donrussellia, Europolemur, Caenopithecus, Pronycticebus*); Adapidae (*Adapis, Leptadapis, Palaeolemur*); and Sivaladapidae (*Hoanghonius*). This represents only 45% (15/33) of the known genera.

Notharctidae

Upper body

The thorax of *Notharctus* is long and deep, like that of a lemur (Gregory, 1920), although the ribs are shorter and more robust. There are 7 cervical, 12 thoracic, 8 lumbar and 3 sacral vertebrae according to Gregory (1920). *Notharctus* had a long tail with at least 19 caudal vertebrae. Gregory noted that the cervical foramina and canals for the vertebral artery, suboccipital nerve and spinal cord were smaller relative to lemurs and stated that the size decrease in arterial and nerve foramina in *Notharctus* is "in harmony with its smaller brain" (Gregory, 1920: 110).

Gregory (1920) reported that the thoracic vertebrae were very lemur-like. He noted that the transverse processes are longer in notharctines than in lemurs, suggesting better developed erector spinae muscles. The lumbar vertebrae are elongated (Simons, 1972) with stout neural spines being directed forward in *Notharctus* and in lemurs (Gregory, 1920). Rose & Walker's (1985) morphological assessments of the lumbar vertebrae of *Cantius* show no significant differences from that of lemurs as well. In sum, *Notharctus* possessed a horizontal or pronograde orientation to its back like that of living lemurs (Gregory, 1920).

Gregory (1920), Thalmann *et al.* (1989), Franzen (1987) and Covert (1985) have all found similarities in the scapula and clavicle of notharctids to those of living lemuriforms, particularly lemurids and indrids. These similarities imply well-developed musculature for shoulder and arm mobility for climbing. I would add that the scapula of notharctines possesses a broader, flatter, and laterally shorter acromion process as well as a broader coracoid process relative to lemurs. At the shoulder, the glenoid facet is pear-shaped and oval like that of a lemur (but without the flattened or caudal notch).

The notharctid humeral head faces posteriorly, suggesting fore-and-aft movement capabilities. The humeral head is round with the greater and lesser tuberosities lying just below the humeral head. This type of a shoulder suggests good mobility for a variety of raised-arm activities (Gebo, 1987*a*). *Notharctus* and *Smilodectes* possess humeral heads that narrow distally (Dagosto, 1993; Schmitt, 1996), a feature correlated with vertical clinging in living lemurs. Godinot's (1992*a*) assessment of humeri attributed to *Protoadapis* and another unnamed cercamoniine suggests a lemur-like locomotor pattern, including vertical clinging and leaping. The lesser tuberosity (attachment for the subscapularis muscle) is very prominent in notharctids. This is important for medial rotation of the humerus or lateral rotation of the trunk, which are both important during climbing. The deltopectoral crest is longer distally than in living prosimians, implying powerful deltoid and pectoralis muscles. The brachioradialis flange (origin for brachialis) is also very broad and long, extending well proximally compared to living prosimians (Gregory, 1920; Dagosto, 1993). A very large brachialis muscle implies powerful pronation and flexion of the forearm at the elbow. All of these features suggest well-muscled shoulders and upper arms among the notharctids.

At the elbow joint, notharctids are characterized by a mediolaterally broad and cylindrically shaped trochlea, a wide zona conoidea (the gap separating the trochlea from the capitulum), a round or ovoid capitulum and a shallow olecranon fossa posteriorly (Gregory, 1920; Gebo, 1987*a*; Thalmann *et al.*, 1989; Dagosto, 1993). *Smilodectes* and *Notharctus* possess shorter and higher trochleas than *Cantius* and *Pronycticebus* (Szalay & Dagosto, 1980; Gebo, 1987*a*; Dagosto, 1993). Szalay & Dagosto (1980) have suggested that shorter and higher trochleas are associated with vertical clinging primates. The capitulum extends below the trochlea in *Cantius*, *Notharctus* and *Smilodectes*, a feature similar to the living indrids and *Lepilemur*, also vertical clinging species. The shallow olecranon fossa and the olecranon process (attachment site of the triceps) imply a forearm that could not be fully extended (Rose & Walker, 1985; Gebo, 1987*a*). In contrast, the round capitulum and wide zona conoidea (gap) suggest a very mobile radius, forearm and wrist (Rose, 1993*a*). Thus, the elbow joint of notharctids shows enhanced rotational mobility but was unable to fully extend (Rose & Walker, 1985). The elbow anatomy of *Pronycticebus* is similar to that of North American notharctines and *Europolemur* (Thalmann *et al.*, 1989).

The radius and ulna of notharctines are generally similar to living lemurs only shorter. Compared to lemurs, the distal radius is wider, while the ulna is bowed and exhibits a shorter styloid process, and a longer olecranon process (Gregory, 1920; Rose & Walker, 1985; Dagosto, 1993). I would add that the radius of notharctines is shorter, the radial head much larger, the radial tubercle more distally located and the distal joint surface broader than in living lemurs. Notharctids also have a low brachial index (a short radius relative to their humerus; Dagosto, 1993).

The wrist of notharctines is quite generalized compared to other living primates (Beard & Godinot, 1988). Beard & Godinot (1988) note that notharctines possess a small lunate and capitate, a capitate that articulates only with the third metacarpal, and a triangular-shaped trapezium. All are likely primitive for primates. *Smilodectes* and *Notharctus* lacked fusion of the centrale and scaphoid (Beard & Godinot, 1988). The carpal anatomy of *Smilodectes* shows several similarities to living lemurs but differs in two important aspects (Beard & Godinot, 1988). First, *Smilodectes* and *Notharctus* do not possess an oblique articulation between the centrale and the trapezoid, the derived midcarpal anatomy of lemuriforms. Second, *Smilodectes* and haplorhine primates share a similar ulnocarpal joint, unlike that of living lemurs. Beard & Godinot (1988) suggest that *Notharctus* may have utilized more palmigrade quadrupedalism than *Smilodectes* on the basis of its known wrist morphology. Hamrick (1996) states that *Notharctus*, *Smilodectes* and *Adapis* all share features of the wrist (a tall pisiform, a mediolaterally flat midcarpal joint and an unexpanded thumb) that are functionally related to arboreal quadrupedalism. He further notes that the adapiform from Messel shares some features with vertical cling and leaping primates.

Notharctus has relatively long hands and fingers (Jouffroy *et al.*, 1991; Godinot, 1992*a*; Dagosto, 1993). The third digit is longest in *Pronycticebus* (Thalmann *et al.*, 1989), as it is in *Notharctus* (Gregory, 1920). The distal phalanges of notharctines are nail-bearing and narrow and long like indrids (Dagosto, 1993). Franzen (1987) notes that the phalanges of *Europolemur* are like living lemurs. In general, notharctine hands are mobile but not hypermobile like lorises.

Lower body

The innominates of *Cantius*, *Notharctus*, *Smilodectes* and *Europolemur* are generally similar in morphology (Gregory, 1920; Rose & Walker, 1985). They all possess short and wide ilia but a longer ischium than living lemurs (but see *Daubentonia*; Rose & Walker, 1985). A long ischium indicates more powerful but less speedy action by the hamstrings. Short ischia are normally associated with leaping lemurs (Rose & Walker, 1985). *Cantius* has a relatively narrower ilium than *Notharctus* (Rose & Walker, 1985). Fleagle & Anapol (1992) note that the degree of dorsal extension of the ilium in notharctines is similar to lemurids that use vertical supports. The gluteal fossa in notharctines is shallow compared to lemurs but the wide ilia indicates gluteal muscle expansion (Gregory, 1920). The acetabulum is deep and hemispherical, being high and thinner dorsally and posteriorly like that of leaping primates (Schultz, 1969). The anterior inferior spine (rectus femoris attachment site) is very pronounced in notharctine pelves (Gregory, 1920). The pelvis of notharctines is narrower than in lemurs with less lateral bending of the proximal ilia. The pelvic inlet is restricted suggesting smaller newborns with relatively smaller brains (Gregory, 1920). Gregory (1920: 86) believed "The whole configuration of the pelvis of *Notharctus* indicates that the animal was an arboreal quadruped which did not sit fully upright but leaped about on all fours among the branches."

Notharctids have long hindlimbs. *Notharctus* and *Smilodectes* have a lower intermembral index (about 60) compared to *Pronycticebus* (less than 73; Dagosto, 1993). A low index implies longer legs and a value of 60 is similar to vertical clinging and leaping prosimians (Napier & Walker, 1967). A long femur with powerful thigh muscles implies a frequent leaping primate. McArdle (1981) and Anemone (1990, 1993) have noted that frequent leaping primates often have a certain combination of hindlimb indices or proportions. In contrast, Runestad (1994) noted that notharctines have short humeri, not long femora, the exact opposite pattern that is possessed by modern lemurs (i.e., long femora).

The femoral head of notharctids is smaller and less spherical than in extant lemurs (Gregory, 1920; Rose & Walker, 1985). The femoral head in *Notharctus* is more flattened craniocaudally than it is in *Cantius* or *Smilodectes* (Covert, 1985). The femoral neck is short and the greater trochanter is equal to the height of the femoral head. These two features distinguish lemurs from notharctines (lemurs possess a higher greater trochanter). Notharctine hip anatomy suggests that adduction, flexion and medial rotation at this joint may have been more limited relative to that of living lemurs (Rose & Walker, 1985). Notharctine articular properties are similar to lemurs but they possess proportionately smaller humeral and femoral heads (Runestad, 1994).

At the proximal femur, the greater trochanter overhangs the femoral shaft anteriorly as it does in leaping prosimians (Walker, 1974*a*; Anemone, 1990; Dagosto, 1993). This feature is important in knee extension giving a better mechanical advantage to vastus lateralis (Jungers *et al.*, 1980). Posteriorly, a deep trochanteric fossa is present in *Notharctus* and this feature is also found in leaping primates like lemurs and indrids (Anemone, 1990). The lesser trochanter is plate-like and oriented more medially compared to lemurs (Dagosto, 1993). This orientation and flat attachment site suggests mechanical differences in iliopsoas function relative to hip flexion and thigh rotation. The third trochanter (attachment site for gluteus superficialis) is small and proximally placed, as it is in leaping mammals (Smith & Savage, 1956; Anemone, 1990).

The femoral shaft of *Notharctus* is round or barrel-like. Burr *et al.* (1989), Ruff & Runestad (1992) and Runestad (1994) have noted that femoral shaft shape and internal architecture reflect locomotor function. The femoral shaft of *Notharctus* suggests that forces were directed to resist anteroposterior forces, a normal situation for leaping primates. The cross-sectional properties of notharctine femora imply high axial rigidity indicating powerful but slower moving hindlimbs (Runestad, 1994).

The distal femur is deep anteroposteriorly and narrow mediolaterally. The patellar groove is narrow with a high rounded lateral rim that sits above the medial rim (Covert, 1985; Rose & Walker, 1985). The patellar groove is narrower than in lemurs. High knees with even higher lateral patellar rims imply forceful leaping (Anemone, 1990, 1993) and are common morphological features among extant leaping prosimians (Dagosto, 1993). Only specialized leapers elevate the patellar surface well above the shaft (Anemone, 1990). *Notharctus* and lemurs do not elevate this surface to the degree seen in tarsiers and galagos (Anemone, 1990).

The tibia in notharctids is slightly shorter compared to lemurs. The tibial condyles are smaller and tilted further backward. The tibial tuberosity (insertion for gracilis, sartorius and the hamstrings) is more distally positioned in notharctines than in lemurs (Covert, 1985; Rose & Walker, 1985). Among lemurs, a low insertion is found among the more quadrupedal forms (McArdle, 1981; Rose & Walker, 1985). The distal tibia is like that of lemurs in that the tibial malleolus is medially rotated and the joint surface is long rather than wide (Dagosto, 1985). The distal tibiofibular joint is unfused and the fibula is more bowed in notharctines relative to lemurs (Gregory, 1920).

Matthew (1915*a*) was the first to describe the foot bones of notharctids. The notharctid tarsal elements as well as the foot in general are broadly similar to those of lemurs. They both possess (1) nails; (2) a robust first metatarsal with a large peroneal tubercle; (3) similar calcaneal proportions; (4) highly mobile subtalar and transverse tarsal joints; and (5) moderate elongation of the navicular and cuneiforms (e.g., Gregory, 1920; Decker & Szalay, 1974; Dagosto, 1986; Szalay & Dagosto, 1988; Gebo *et al.*, 1991). The tarsals of notharctids resemble leaping prosimians with their high and short talar body, long talar neck, large posterior trochlear shelf, and elongated tarsals. Tarsals of five species of *Cantius* show increases and decreases in size over a several-million-years time period but no significant metric or qualitative changes in morphology in spite of the several speciation events that had occurred (Gebo *et al.*, 1991). However, it is misleading to compare notharctids too closely to lemurs since notharctid feet lack many of the lemurid and indrid specializations (see Gebo *et al.*, 1991). For example, notharctid feet are not folded like those of lemurs, their metatarsals are shorter, and the distal phalanges are much narrower. However, all notharctids, as do lemurs, show the characteristic sloping fibula facet, a more lateral position for the groove for flexor hallucis longus, and a navicular facet for the cuboid, which extends below the entocuneiform and mesocuneiform distal facets. These tarsal features link adapiform and lemuriform primates within Strepsirhini (Beard *et al.*, 1988). More recent discoveries of tali allocated to *Anchomomys* (Moyà-Solà & Köhler, 1993*a*) and to *Donrussellia* (M. Godinot, pers. comm.) show these same strepsirhine talar features.

For the toes, the metatarsals and phalanges are short and broad, although longer in *Smilodectes* and *Europolemur* (von Koenigswald, 1979; Covert, 1985). The first metatarsal has a saddle-shaped joint surface and a large peroneal tubercle (attachment site for peroneus longus). Both features are important aspects for grasping and hallucial opposability (Szalay & Dagosto, 1988). The range of motion for the first metatarsal is limited in notharctids relative to lemurs (Gebo *et al.*, 1991). The terminal phalanges are narrower and longer compared to lemurs (Gebo *et al.*, 1991). The narrower tips imply less expanded terminal pads for grasping. In contrast, the phalanx for the big toe is very broad and flat with a deep excavation proximally for flexor hallucis longus. No adapiform foot has been shown to possess a toilet claw like that of lemuriform primates (but see von Koenigswald, 1979).

Adapidae

Schlosser (1887) and Le Gros Clark (1959) originally suggested loris-like adaptations in the body parts of adapines. Dagosto's (1983) more detailed analysis cemented this interpretation of adapid adaptation. "*Adapis* is best characterized as an arboreally committed quadruped, ... an able and agile climber, capable of moving along relatively thin supports, but not relying on leaping ... *Adapis* was not capable of the extremes of limb mobility, intricate patterns of movement, or the remarkably strong grasp which make lorisine locomotion so unique" (Dagosto, 1983: 90). Thus, the body anatomy and locomotor movements for adapids are quite distinct from that of notharctid primates.

Adapids compare best with slow-climbing lorises and other non-leaping quadrupedal primates. On the basis of hand anatomy, Godinot (1991) has suggested a more monkey-like model for *Adapis* with an emphasis on above-branch quadrupedalism. However, cross-sectional studies on primate limb bones have suggested that adapine limbs are most similar to slow-climbing primates (Runestad, 1994). Adapines also have very short femora as do slow-climbing primates (Dagosto, 1983; Runestad, 1994). Two cranial features that further suggest an emphasis on climbing and quadrupedal abilities in adapids are the semicircular canals and orbital frontation. Spoor *et al.* (1998) showed that the semicircular canals of *Adapis* were more similar to those of slow-moving loris-like primates. Likewise, Lanèque's (1993) assessment of the low amount of frontation in adapine orbits implies a locomotor mode similar to pottos. She also noted that the great amount of convergence in adapine orbits "shows they probably did not have a galago-like locomotion" (Lanèque, 1993: 311).

New discoveries have forced a reappraisal of the fossil material attributed to *Adapis* (Godinot, 1991; Lanèque, 1992*b*). The species of *Adapis* (*A.* aff. *betillei*) from Escamps shows more extreme flexion at the elbow, more mobility at the hip, and more foot rotation than the species from

Rosieres (*A.* cf. *parisiensis*) (Godinot, 1991). Godinot concluded that "the Rosières species, with its restricted joint mobility and emphasis on parasagittal movements, as a predominately horizontal arboreal quadruped, in the sense of a large-branch walking and running form; and the Escamps species, with its increased joint mobility, as an accomplished climber, using more versatile movements for travel in a finer branch milieu" (Godinot, 1991: 400).

Forelimb

The fore and hindlimbs of *Adapis* appear to be more equal in length than the Notharctidae (Dagosto, 1983). Ratios calculated for humeral length to femoral length or humeral length to tibial length in *Adapis* are similar to lorises (Dagosto, 1983). Relative to *Adapis*, *Leptadapis* is known for its very robust limb bones (Dagosto, 1983). *Leptadapis*' humeri show great mechanical rigidity indicating very powerful forelimbs (Runestad, 1994). Adapine humeri have a broader and shallower bicipital groove, a thicker deltopectoral crest and a distally extensive teres major tuberosity relative to notharctids (Filhol, 1882; Dagosto, 1993). Adapine humeral heads are larger than those of notharctines (Runestad, 1994) while the deltopectoral crest and the brachialis flange are narrower. The large size of the brachialis and brachioradialis muscles implies powerful flexion of the forearm or slow flexion against resistance, an advantage for slow-climbing primates (Dagosto, 1983). *Adapis* also has a relatively short trochlea that is closely situated near the capitulum (Filhol, 1882; Dagosto, 1983). The zona conoidea is greatly reduced as it is in lorises. *Adapis* has a distinct lateral capitular tail or flange, a well-developed feature in *Nycticebus* and *Perodicticus* (Dagosto, 1983). The capitular flange is less developed in *Leptadapis*. Szalay & Dagosto (1980) surmise that the reduced trochlea and the capitular flange reflect a reduction in the importance of the ulna in forearm stabilization. *Leptadapis* is similar to *Adapis* but its humeri are very robust and they possess a sigmoid curvature. These features suggest frequent climbing in *Leptadapis*.

In the wrist, *Adapis* has "a more expanded radial facet on the lunate; lateral expansion of the centrale facet on the capitate; more proximally oriented facet for the triquetrum on the hamate; relatively flat rather than sellar-shaped articulation for the thumb and metacarpus; and broader ulna–pisiform contact (Godinot & Jouffroy, 1982; Beard & Godinot, 1988)" (cited in Dagosto, 1993: 205). These features imply greater ranges of carpal and midcarpal mobility in *Adapis* relative to notharctines (Beard & Godinot, 1988).

Adapis and *Leptadapis* have slightly shorter metacarpals with broader heads compared to lemurs (Schlosser, 1887). The proximal articular surfaces are more similar to lemurs than to lorises (Dagosto, 1993). The pollical–metacarpal joint is relatively flat compared to notharctines, indicating a smaller range of thumb movement (Godinot & Jouffroy, 1982). *Adapis* has relatively short digits (Godinot, 1992*a*) and the terminal phalanges are more dorsovolarly robust than in lemurs (Godinot, 1992*a*). Dagosto (1993: 209) notes that among adapids "the hands and the distal parts of the feet lack the extreme modifications characteristic of lorisines".

Hindlimb

Relative to notharctids, adapids have a shorter hindlimb (Dagosto, 1983). The femur, in particular, is very short (Runestad, 1994). *Adapis* has a more elongated femoral neck, a more spherical femoral head, a smaller greater trochanter that is less laterally extensive and a more distal position for the third trochanter (Dagosto, 1983). Smaller femoral head size in adapines, relative to notharctines, is unusual for modern slow-climbing primates (Runestad, 1994). The third trochanter is also very small in *Adapis*, as in lorisines (Walker, 1967*a*), but not in *Leptadapis* (Dagosto, 1993). The proximal femur of adapids most resembles that of *Perodicticus* (Dagosto, 1983). Adapine humeral to femoral shaft comparisons show higher than average values for axial rigidity, a characteristic of modern slow climbers (Runestad, 1994). Adapine femoral mechanics appear to be most similar to non-specialized primates (Runestad, 1994). Bacon & Godinot's (1998) analysis of several femoral and tibial remains from the Quercy localities in France suggest several morphological varieties among the species of *Adapis* and *Palaeolemur*. All of their varieties emphasize quadrupedalism and climbing in varying degrees.

The knee of adapids is more similar to lorises and anthropoids than to lemurs and notharctids. It is flatter and broader with a broad patellar groove relative to notharctids. The medial and lateral crests are low (Dagosto, 1983). The average height of the distal femur in *Adapis* lies between lorisines and lemurs (Dagosto, 1983). This anatomical region is not adapted for leaping. The broad joint surfaces and low height suggest capabilities for medial and lateral rotation, features important in climbing primates.

Adapis differs from lemurs and notharctids at the distal tibia. *Adapis* possesses a deeply excavated groove for tibialis posterior and flexor digitorum tibialis (Dagosto, 1993). These features are more similar to lorises than lemurs. *Leptadapis* is unlike *Adapis* in this region, being more similar to lemurs than to lorises.

In adapines, the tarsals are short, the talar trochlea is long, the talar neck is short and strongly angled medially ($\sim 40°$), the distal calcaneus is very short and the heel region is long (Decker & Szalay, 1974; Martin, 1979; Dagosto, 1983). In adapines, the medial rim of the talus curves well medially relative to the lateral rim, rather than being parallel as in notharctines. The talar body is low and the posterior trochlear shelf is also less developed in adapines compared to notharctids (Decker & Szalay, 1979; Dagosto, 1983). The posterior trochlear shelf is absent in *Adapis*. These adapine tarsal features resemble lorises, especially pottos, non-leaping prosimians. In contrast, the metatarsals of adapines are not similar to lorises and have articular surfaces like those of lemurs (Dagosto, 1993). The short load arm of the distal calcaneus also indicates quadrupedal versus a leaping form of locomotion in adapines.

Acknowledgements

I would like to thank Walter Hartwig for his invitation to participate in this volume. I also thank Chris Beard, Marc Godinot, Mark Mehrer and Marian Dagosto for their helpful comments. Lastly, I would like to thank and extend my appreciation to Kim Reed-Deemer for all of her hard work on the artwork in this chapter.

Primary References

The following list includes the publications that name the genera supported in this chapter. See also Szalay & Delson (1979), Fleagle (1999), and the references cited above.

Adapis

Cuvier, G. (1821). *Discours sur la théorie de la terre, servant d'introduction aux recherches sur les ossements fossiles*. Paris.

Adapoides

Beard, K. C., Qi, T., Dawson, M. R., Wang, B., & Li, C. (1994). A diverse new primate fauna from middle Eocene fissure-fillings in southeastern China. *Nature*, **368**, 604–609.

Aframonius

Simons, E. L., Rasmussen, D. T., & Gingerich, P. D. (1995). New cercamoniine adapid from Fayum, Egypt. *Journal of Human Evolution*, **29**, 577–589.

Agerinia

Crusafont-Pairo, M. & Golpe-Posse, J. M. (1973). Yacimientos del Eoceno prepirenaico (nuevas localidades del Cuisiense). *Acta Geologica Hispanica*, **8**(5), 145–147.

Anchomomys

Stehlin, H. G. (1916). Die Säugetiere des schweizerischen Eocäens: critischer Catalog der Materialen. VIIb. *Abhandlungen schweiznischen paläontologischen Gesellschaft*, **41**, 1299–1552.

Moyà-Soyà, S. & Köhler, M. (1993). Middle Bartonian Locality with *Anchomomys* (Adapidae, Primates) in the Spanish Pyrenees: Preliminary Report. *Folia Primatologica*, **60**, 158–163.

Barnesia

Thalmann, U. (1994). Die Primaten aus dem eozänen Geiseltal bei Halle/Saale (Deutschland). *Courier Forschungsinstitut Senckenberg*, **175**, 1–161.

Buxella

Godinot, M. (1988). Les primates adapidés de Bouxwiller (Eocène Moyen, Alsace) et leur apport à comprehension de la faune de Messel et à l'évolution des Anchomomyini. *Courier Forschungsinstitut Senckenberg*, **107**, 383–407.

Caenopithecus

Rütimeyer, L. (1862). Eocaene Säugetiere aus dem Gebiet des schweizerischen Jura. *Allgemeine schweizerische gesellschaft für die gesamten naturwisenschaften Denkschriften*, **19**, 1–98.

Cantius

Cope, E. D. (1875). Systematic catalogue of Vertebrata of the Eocene of New Mexico, collected in 1874. In *Geographical Expansion and Survey West of 100th Meridian*, ed. G. M. Wheeler, pp. 5–37. Washington, DC: Government Printing Office.

Simons, E. L. (1962*a*). A new Eocene primate genus, *Cantius*, and a revision of some allied European lemuroids. *Bulletin of the British Museum (Natural History) Geology Series*, **7**, 1–30.

Cercamonius

Stehlin, H. G. (1912). Die Säugetiere des schweizerschen Eocaens. Critischer Catalog der Materialen. VIIb. *Abhandlungen Schweizerischen Paläontologischen Gesellschaft*, **38**, 1165–1298.

Gingerich, P. D. (1975). A new genus of Adapidae (Mammalia, Primates) from the late Eocene of southern France, and its significance for the origin of higher primates. *Contributions from the Museum of Paleontology, University of Michigan*, **24**, 163–170.

Copelemur

Cope, E. D. (1877). Report upon the extinct Vertebrata obtained in New Mexico by parties of the expedition of 1874. Chapter XII. Fossils of the Eocene period. In *Geographical Surveys West of 100th Meridian*, ed. G. M. Wheeler, pp. 37–282. Washington, DC: Government Printing Office.

Gingerich, P. D. & Simons, E. L. (1977). Systematics, phylogeny and evolution of early Eocene Adapidae (Mammalia, Primates) in North America. *Contributions from the Museum of Paleontology, University of Michigan*, **24**, 245–279.

Cryptadapis

Godinot, M. (1984). Un nouveau genre témoignant de la diversité des Adapinés (Primates, Adapidae) à l'Éocène terminal. *Comptes rendus de l'Académie des sciences de Paris*, **299**, 1291–1296.

Djebelemur

Hartenberger, J. L. & Marandat, B. (1992). A new genus and species of an early Eocene primate from North Africa. *Human Evolution*, **7**, 9–16.

Donrussellia

Russell, D. E., Louis, P. & Savage, D. E. (1967). Primates of the

French early Eocene. *University of California Publications in the Geological Sciences*, **73**, 1–46.

Szalay, F. S. (1976). Systematics of the Omomyidae (Tarsiiformes, Primates): taxonomy, phylogeny, and adaptations. *Bulletin of the American Museum of Natural History*, **156**, 157–450.

Godinot, M. (1978). Un nouvel Adapidé (primate) de l'Éocène inférieur de Provence. *Comptes rendus de l'Académie des Sciences de Paris*, **286**, 1869–1872.

Europolemur

Weigelt, J. (1933). Neue Primaten aus der mitteleozänen (oberlutetischen) Braunkohle des Geiseltals. *Nova Acta Leopoldense*, **1**, 97–156.

Guangxilemur

Qi, T. & Beard, K. C. (1998). Late Eocene sivaladapid primate from Guangxi Zhuang Autonomous Region, People's Republic of China. *Journal of Human Evolution*, **35**, 211–220.

Hesperolemur

Gunnell, G. F. (1995). New notharctine (Primates, Adapiformes) skull from the Uintan (Middle Eocene) of San Diego County, California. *American Journal of Physical Anthropology*, **98**, 447–470.

Hoanghonius

Zdansky, O. (1930). Die alttertiären Säugetiere Chinas nebst stratigraphischen Bemerkungen. *Palaeontologica Sinica* (Series C), **6**, 1–87.

Indraloris

Lewis, G. E. (1933). Preliminary notice of a new genus of lemuroid from the Siwaliks. *American Journal of Science*, **26**, 134–138.

Leptadapis

Gervais, P. (1876). *Zoologie et paléontologie générale*, 2nd edn. Paris: Bertrand.

Lushius

Chow, M. (1961). A new tarsioid primate from the Lushi Eocene, Hunan. *Vertebrata PalAsiatica*, **5**, 1–5.

Mahgarita

Wilson, J. A. & Szalay, F. S. (1976). New adapid primate of European affinities from Texas. *Folia Primatologica*, **25**, 294–312.

Microadapis

Szalay, F. S. (1974). New genera of European Eocene adapid primates. *Folia Primatologica*, **22**, 116–133.

Notharctus

Leidy, J. (1870). Descriptions of *Palaeosyops paludosus*, *Microsus cuspidatus* and *Notharctus tenebrosus*. *Proceedings of the Academy of Natural Sciences, Philadelphia*, **22**, 111–114.

Gregory, W. K. (1920). On the structure and relations of *Notharctus*, an American Eocene primate. *Memoirs of the American Museum of Natural History*, new series, **3**(2), 49–243.

Omanodon

Gheerbrant, E., Thomas, H., Roger, J., Sen, S., & Al-Sulaimani, Z. (1993). Deux nouveaux primates dans l'Oligocène inférieur de Taqah (Sultanat d'Oman): premiers adapiformes (?Anchomomyini) de la Péninsula Arabique. *Palaeovertebrata*, **22**, 141–196.

Palaeolemur

Delfortrie, E. (1873). Un singe de la familie des Lémuriens. *Actes de la Société linnéenne de Bordeaux*, **24**, 87–95.

Panobius

Russell, D. E. & Gingerich, P. D. (1987). Nouveaux primates de l'Éocène du Pakistan. *Comptes rendus de l'Académie des sciences de Paris*, **304**, 209–214.

Pelycodus

Cope, E. D. (1875). Systematic catalogue of Vertebrata of the Eocene of New Mexico, collected in 1874. In *Geographical Expansion and Survey West of 100th Meridian*, ed. G. M. Wheeler, pp. 5–37. Washington, DC: Government Printing Office.

Gingerich, P. D. & Simons, E. L. (1977). Systematics, phylogeny and evolution of early Eocene Adapidae (Mammalia, Primates) in North America. *Contributions from the Museum of Paleontology, University of Michigan*, **24**, 245–279.

Periconodon

Stehlin, H. G. (1916). Die Säugetiere des schweizerischen Eocäens: critischer Catalog der Materialen. VIIb. *Abhandlungen Schweizerischen Paläontologischen Gesellschaft*, **41**, 1299–1552.

Pronycticebus

Grandidier, G. (1904). Un nouveau Lémurien fossile de France, le *Pronycticebus gaudryi*. *Bulletin du Muséum national d'histoire naturelle, Paris*, **10**, 9–13.

Protoadapis

Lemoine, V. (1878). Communication sur les ossements fossiles des terrains tertiaires inférieures des environs de Reims. *Société d'étude d'histoire naturelle de Reims*, 1–24.

Rencunius

Gingerich, P. D., Holroyd, P. A., & Ciochon, R. L. (1994). *Rencunius zhoui*, a new primate from the late middle Eocene of Henan, China, and a comparison with some early Anthropoidea. In *Anthropoid Origins*, eds. J. G. Fleagle & R. F. Kay, pp. 163–177. New York: Plenum Press.

Shizarodon

Gheerbrant, E., Thomas, H., Roger, J., Sen, S., & Al-Sulaimani, Z. (1993). Deux nouveaux primates dans l'Oligocène inférieur de Taqah (Sultanat d'Oman): premiers adapiformes (?Anchomomyini) de la Péninsula Arabique. *Palaeovertebrata*, **22**, 141–196.

Sinoadapis

Wu, R. & Pan, Y. (1985). A new adapid primate from the Lufeng Miocene, Yunnan Province. *Acta Anthropologica Sinica*, **4**, 1–6.

Pan, Y. (1988). Small fossil primates from Lufeng, a latest Miocene site in Yunnan Province, China. *Journal of Human Evolution*, **17**, 359–366.

Sivaladapis

Pilgrim, G. E. (1932). The fossil Carnivora of India. *Memoirs Geological Survey of India (Palaeontologica Indica)*, **18**, 1–232.

Gingerich, P. D. & Sahni, A. (1979). *Indraloris* and *Sivaladapis*: Miocene adapid primates from the Siwaliks of India and Pakistan. *Nature*, **279**, 415–416.

Smilodectes

Marsh, O. C. (1871). Notice of some fossil mammals from the Tertiary formation. *American Journal of Science*, **2**, 35–44, 120–127.

Wortman, J. L. (1903–1904). Studies of Eocene Mammalia in the Marsh Collection, Peabody Museum. *American Journal of Science*, **15**, 163–176, 399–414, 419–436; **16**, 345–368; **17**, 23–33, 133–140, 203–214.

Wadilemur

Simons, E. L. (1997). Discovery of the smallest Fayum Egyptian primates (Anchomomyini, Adapidae). *Proceedings of the National Academy of Sciences of the United States of America*, **94**, 180–184.

Wailekia

Ducrocq, S., Jaeger, J. J., Chaimanee, Y., & Suteethorn, V. (1995). New primate from the Palaeogene of Thailand, and the biogeographical origin of anthropoids. *Journal of Human Evolution*, **28**, 477–485.

5 | Tarsiiformes: Evolutionary history and adaptation

GREGG F. GUNNELL AND KENNETH D. ROSE

Introduction

Fossil tarsiiform primates are known from all three holarctic continents and may be represented in Africa as well. They range in time from the earliest Eocene to the earliest Miocene. They are very diverse and common in the North American early and middle Eocene, common but less diverse in the European middle and late Eocene, and present but neither diverse nor common in the early and middle Eocene of Asia. Only a few tarsiiform taxa survived beyond the end of the Eocene in North America and Europe and only *Tarsius* is known after the middle Eocene in Asia.

Tarsiiforms traditionally have been viewed as the "tarsier-like" primates of the early Cenozoic, as opposed to the "lemur-like" adapiforms that underwent a similar radiation during the same time on the same continents (see Gebo, this volume). More recently, it has become clear that most fossil tarsiiforms in general were less like living tarsiers than was once believed and that they filled a wider range of adaptive niches, with many forms being more anatomically galago-like.

Changing views of tarsiiform adaptation, along with ever-growing numbers of fossil remains, have led to a state of flux in higher-level taxonomic treatments of tarsiiform relationships (e.g., Szalay, 1976; Gingerich, 1976*a*, 1981*b*; Szalay & Delson, 1979; Bown & Rose, 1987; Beard, 1988*a*; Beard *et al.*, 1991; Beard & MacPhee, 1994; Rose, 1995*a*). In general, three different hypotheses have been put forward for the higher level placement of tarsiiforms: (1) Omomyiformes (all tarsiiforms except tarsiids) is recognized as a separate suborder within the order Primates, which also includes Adapiformes, Tarsiiformes, Lemuriformes and Simiiformes; (2) Omomyiformes is an infraorder within the suborder Prosimii that also includes Adapiformes, Lemuriformes and Tarsiiformes; or (3) Omomyidae is a family within the infraorder Tarsiiformes, included together with the infraorder Anthropoidea in the suborder Haplorhini. None of these arrangements is completely satisfactory given the state of knowledge of early Cenozoic fossil primates. For the purposes of this chapter we have chosen to deal with Tarsiiformes as a unified group including two families, Omomyidae and Tarsiidae, with the former family including three subfamilies, Microchoerinae, Anaptomorphinae and Omomyinae.

History of discovery and debate

Tarsiiformes have a long history, beginning with the discovery of *Microchoerus erinaceus* (Wood, 1844) by Searles Wood in 1843 at Hordwell in Hampshire, England. Wood (1844, 1846) believed that *Microchoerus* was related to suids, interpreting the taxon as that of a small pig. Leidy (1869) described the first omomyine, *Omomys carteri*, from the Bridger Formation in the Green River Basin, southwestern Wyoming. Leidy (1869) believed that *Omomys* was related to hedgehogs. The first anaptomorphine described was *Anaptomorphus aemulus*, named by Cope in 1872, also from the Bridger Formation in southwestern Wyoming. Cope (1872) compared *Anaptomorphus* with *Simia* (mis-reported as "*Limia*," but later corrected to *Simia* in a footnote in Cope, 1885) and *Homo*, apparently becoming the first person to recognize the possible primate affinities of omomyids. However, Marsh (1872) in a paper dated four days before Cope's 1872 contribution (see Gazin, 1958), noted similarities between *Notharctus* and extant lemurs and thus may have been the first to realize that primate relatives existed in the Eocene of North America. The holotype of *Necrolemur antiquus* Filhol (1873) is a well-preserved skull that proved the primate affinities of omomyids (Simons, 1961*a*).

Since 1872, two additional genera of microchoerines have been described from Europe (Stehlin, 1916) as have 14 genera of anaptomorphines and 19 genera of omomyines, all but three being exclusively from North America. Two other possible omomyid genera also are known from Asia, *Altanius* from Mongolia and *Kohatius* from Pakistan, while a third possible omomyid is known from Africa, *Altiatlasius* from Morocco.

Even with high diversity and abundance in the early and middle Eocene of North America, the interrelationships of omomyids are not well understood, nor are their possible relationships with later occurring primates. Szalay (1976) arranged anaptomorphine and omomyine taxa into several tribes and subtribes based on his interpretations of relationships within Omomyidae. These tribal units have remained relatively stable but taxa included within each tribe or subtribe have been a point of contention (Bown & Rose, 1987; Honey, 1990; Gunnell, 1995*a*; McKenna & Bell, 1997).

Hypotheses of higher-level relationships of omomyids revolve around their relationships with tarsiids (essentially extant tarsiers) and, by extension, anthropoid primates. Living primates can be divided into two suborders, either

Prosimii and Anthropoidea or Strepsirhini and Haplorhini, depending on the placement of tarsiers. Those who favor the prosimian–anthropoidean division (e.g., Gingerich, 1981*b*; Simons & Rasmussen, 1989) place tarsiers in the suborder Prosimii with lemurs and lorises citing the primitive dentition and neural organization of tarsiers as evidence that they are of a more primitive grade than are monkeys, apes and humans (Anthropoidea). However, there are few, if any, shared, derived character states that link tarsiers with lemurs and lorises (Martin, 1990; Shoshani *et al.*, 1996).

Those who favor the strepsirhine–haplorhine subdivision (e.g., Szalay, 1976; Szalay & Delson, 1979; Kay *et al.*, 1997; Ross *et al.*, 1998) cite soft anatomical features of the rhinarium as synapomorphies of tarsiers and anthropoids (Haplorhini) to the exclusion of lemurs and lorises (Strepsirhini). The haplorhine–strepsirhine dichotomy is difficult to trace in the fossil record, as these soft anatomical features are seldom preserved in sufficient detail. There is some evidence to suggest that both adapiforms and omomyids shared a strepsirhine nasal configuration (Rosenberger *et al.*, 1985; Beard, 1988*b*; Gebo, this volume) but it is possible that neither group did (Rasmussen & Nekaris, 1998).

Omomyids may have been strepsirhines but other osteological evidence favors a closer relationship with extant tarsiers (e.g., Simons, 1961*a*, 1961*b*; Szalay, 1975, 1976; Gingerich, 1981*b*; Beard *et al.*, 1991; Beard & MacPhee, 1994; Dagosto & Gebo, 1994). *Necrolemur* was felt to be distinctly tarsiiform (Stehlin, 1916) based on cranial anatomy. Hürzeler (1946, 1948) questioned this relationship but a thorough study by Simons (1961*a*) reaffirmed the tarsiiform affinities of *Necrolemur*. Beard *et al.* (1991) and Beard & MacPhee (1994) have described the cranial anatomy of the North American omomyine *Shoshonius* and favor a close relationship between washakiin omomyids and extant tarsiers, with microchoerines (or at least *Necrolemur*) as the sister taxon to that group. Postcranial evidence provides some weak support for a washakiin–tarsiid relationship as well, although omomyids in general appear more distantly related to tarsiers based on analyses of postcranial elements (Dagosto & Gebo, 1994; Dagosto *et al.*, 1999).

There is some merit to viewing omomyids as the sister group to adapiforms (Martin, 1990), since the earliest known representatives of each of these clades differ very little from one another (Rose & Bown, 1991; Rose, 1995*a*). The presence of what appear to be true tarsiids in the Eocene of Asia (Beard *et al.*, 1994; Beard, 1998*a*) lends some support to the idea of a very early split between tarsiids and adapiforms–omomyids, and it is possible that the divergence of a tarsiid–anthropoid lineage occurred prior to an adapiform–omomyid split. In this case, any similarities between omomyids and tarsiids would have to be viewed as homoplasies.

North American omomyids, unlike their adapiform counterparts, are highly diverse. It appears that omomyids were capable of adapting to a wide variety of habitats and micro-environments (Gunnell, 1997). The North American record of omomyids presents an interesting dichotomy with anaptomorphines being the most common omomyid group in the early Eocene and omomyines being much more common in the middle Eocene. A combination of climatic and paleoecological change along with competition between these two groups resulted in the distributions present in the early and middle Eocene of North America. Interestingly, recent work in the earliest middle Eocene has suggested that anaptomorphines are displaced rather than replaced by omomyines (Gunnell, 1997), with anaptomorphines being more common in distal and upland habitats while omomyines are more common in the much more frequently sampled proximal habitats.

Taxonomy

Systematic framework

Order Primates Linnaeus, 1758
- Suborder Prosimii Illiger, 1811
  - Infraorder Tarsiiformes Gregory, 1915
    - Family Omomyidae Trouessart, 1879
      - Subfamily Microchoerinae Lydekker, 1887
        - Genera *Microchoerus* Wood, 1846
          - *Necrolemur* Filhol, 1873
          - *Nannopithex* Stehlin, 1916
          - *Pseudoloris* Stehlin, 1916
      - Subfamily Anaptomorphinae Cope, 1883
        - Tribe Anaptomorphini
          - Genera *Anaptomorphus* Cope, 1872
            - *Tetonius* Matthew, 1915
            - *Absarokius* Matthew, 1915
            - *Teilhardina* Simpson, 1940
            - *Anemorhysis* Gazin, 1958
            - *Chlororhysis* Gazin, 1958
            - *Pseudotetonius* Bown, 1974
            - *Arapahovius* Savage & Waters, 1978
            - *Aycrossia* Bown, 1979
            - *Strigorhysis* Bown, 1979
            - *Gazinius* Bown, 1979
            - *Tatmanius* Bown & Rose, 1991
        - Tribe Trogolemurini
          - Genera *Trogolemur* Matthew, 1909
            - *Sphacorhysis* Gunnell, 1995
      - Subfamily Omomyinae Trouessart, 1879
        - Tribe Omomyini
          - Genera *Omomys* Leidy, 1869
            - *Steinius* Bown & Rose, 1984
            - *Chumashius* Stock, 1933
        - Tribe Rooneyini
          - Genera *Rooneyia* Wilson, 1966
        - Tribe Washakiini
          - Genera *Washakius* Leidy, 1873
            - *Shoshonius* Granger, 1910

Dyseolemur Stock, 1934
Loveina Simpson, 1940
Tribe Utahiini
Genera *Utahia* Gazin, 1958
Ourayia Gazin, 1958
Stockia Gazin, 1958
Asiomomys Wang & Li, 1990
Wyomomys Gunnell, 1995
Ageitodendron Gunnell, 1995
Chipetaia Rasmussen, 1996
Tribe Macrotarsiini
Genera *Macrotarsius* Clark, 1941
Hemiacodon Marsh, 1872
Yaquius Mason, 1990
Tribe Uintaniini
Genera *Uintanius* Matthew, 1915
Jemezius Beard, 1987
Family Tarsiidae
Genera *Tarsius* Storr, 1780
Xanthorhysis Beard, 1998
Problematic Taxa
Ekgmowechashala Macdonald, 1963
Altanius Dashzeveg & McKenna, 1977
Kohatius Russell & Gingerich, 1980
Afrotarsius Simons & Bown, 1985
Altiatlasius Sigé *et al.*, 1990

Family Omomyidae

Omomyidae, like adapiforms, can be recognized as primates by the presence of relatively low-crowned molars with broad basins, lower molars with reduced paraconids and an extended M_3 hypoconulid, and upper molars with postprotocingula (except in the most primitive taxa where this feature is often poorly developed or absent) and reduced stylar shelves. Additionally, omomyids and adapiforms, where known, also possess the following primate features: a fully ossified auditory bulla formed from the petrosal bone; front facing, enlarged orbits; a postorbital bar; a large braincase; a mobile elbow joint allowing extensive supination; an opposable hallux; and nails on all digits. Omomyids can be differentiated from adapiforms by possession of the following features: small body size; pointed, often enlarged central incisors; small, often non-overlapping canines with no evidence of canine sexual dimorphism; a tendency in many forms for the antemolar dentition to be mesiodistally compressed; ectotympanic formed into bony tube in most taxa; very large and front-facing orbits and a relatively shortened rostrum; elongate tarsal elements including the cuboid, navicular and cuneiforms; and distal tibia–fibula closely approximated, often fused into tibiofibula.

Subfamily Microchoerinae

All known microchoerines are from the Eocene and Oligocene of Europe. All microchoerines share enlarged, somewhat separated upper central incisors and upper lateral incisors placed directly posterior to the central pair. Lower central incisors are enlarged and canines and P_3 are slightly smaller than P_4. Upper canines are large and P^2 is reduced. Microchoerines, where known, have an upper dental formula of 2.1.3.3 and a lower dental formula of 2.1.2.3 (Szalay, 1976; Godinot *et al.*, 1992; but see Schmid, 1983; Hooker, 1986). Microchoerines resemble anaptomorphines and differ from omomyines in the presence of a postprotocingulum in several genera, but this structure appears to be of a different derivation from the postprotocingulum in anaptomorphines (Gingerich, 1981*b*).

In the following, the anatomical definition of each genus is based on the type species of that genus. Subsequent species in each genus are differentiated from the type and other recognized species but otherwise share the character states of the type species.

GENUS *Microchoerus* Wood, 1846 (Fig. 5.1A–B)
INCLUDED SPECIES *M. creechbarrowensis, M. edwardsi, M. erinaceus, M. ornatus, M. wardi*

TYPE SPECIES *Microchoerus erinaceus* Wood, 1846
TYPE SPECIMEN BMNH 25229, palate with left I^{1-2}, P^2–M^3 and right I^2, C^1, P^2–M^3, and left dentary P_4–M_3
AGE AND GEOGRAPHIC RANGE Late Eocene and early Oligocene, Europe
ANATOMICAL DEFINITION
The following anatomical description has been adapted from Hooker (1986): Cheek tooth enamel rugose and complex; upper molars quadrate with postprotocingulum well developed, making the hypocone nearly the size of the protocone on M^{1-2}; M^{1-2} with strong conules, metaconule doubled, and mesostyle present; M_{1-2} with discontinuous and displaced protocristid; M_3 hypoconulid relatively long and double-cusped; molar paraconid weak on M_1, absent on M_{2-3}; mandibular angular process enlarged, coronoid process relatively low. *Microchoerus* differs from *Necrolemur* in possessing much more rugose cheek tooth enamel, upper molar mesostyles, relatively well-developed hypoconulids on M_{1-2}, discontinuous M_{1-2} protocristids, stronger upper molar conules, and less reduced M_3, with a longer and broader hypoconulid.

SPECIES *Microchoerus edwardsi* (Filhol, 1880)
TYPE SPECIMEN MNHN Qu- 10946, left dentary I_1–C_1, P_3–M_3
AGE AND GEOGRAPHIC RANGE Late Eocene, England and France
ANATOMICAL DEFINITION
Microchoerus edwardsi shares an inflated M_3 lobe with *M. wardi* to the exclusion of all other *Microchoerus* species. Differs from *M. wardi* in being larger, in having a lower molar mesoconid always present, a double-lobed M_3 hypoconulid, and discontinuous lower molar postcristids.

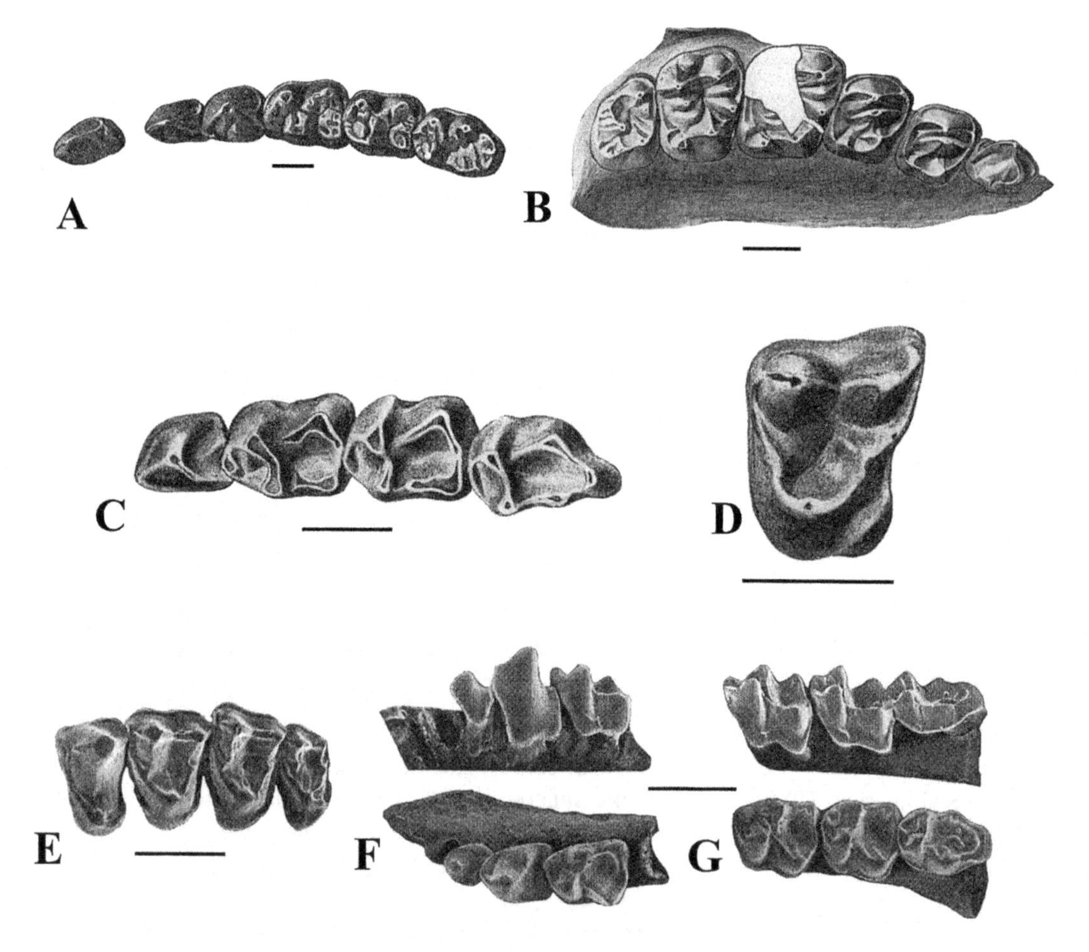

Fig. 5.1 (A) *Microchoerus erinaceus*, right dentary C–M_3 (BNM Db 507) in occlusal view. (B) *Microchoerus ornatus*, right maxilla C–M^3 (BNM Mt 552, holotype) in occlusal view. (C) *Pseudoloris parvulus*, right dentary P_4–M_3 (unnumbered specimen from Sosís) in occlusal view. (D) *Pseudoloris parvulus*, left M^1 (BNM St 1841) in occlusal view. (E) *Nannopithex zuccolae*, left maxilla P^4–M^3 (SLP 29-PE-450, holotype) in occlusal view. (F) *Nannopithex zuccolae*, left dentary P_3–M_1 (992-L) in lateral and occlusal views. (G) *Nannopithex zuccolae*, left dentary M_{1-3} (SLP 29-PE-850) in lateral and occlusal views. Scale bars represent 2 mm in A–B, E–G, 1 mm in C–D. Figures adapted from Stehlin (1916), Crusafont-Pairo (1967) and Godinot et al. (1992).

SPECIES *Microchoerus ornatus* Stehlin, 1916
TYPE SPECIMEN BNM Mt 552, right maxilla P^3–M^3
AGE AND GEOGRAPHIC RANGE Late Eocene, Switzerland and Spain
ANATOMICAL DEFINITION
Small size, similar to M. *edwardsi* but with more pronounced enamel crenulations. This species is only known from upper teeth – it may prove to be conspecific with M. *edwardsi* when lower teeth are known.

SPECIES *Microchoerus wardi* Hooker, 1986
TYPE SPECIMEN BMNH M37162, right M^1
AGE AND GEOGRAPHIC RANGE Middle to late Eocene, England
ANATOMICAL DEFINITION
Smaller than all other species of *Microchoerus* (approximately 15% smaller in tooth dimensions than any other known species of *Microchoerus*). Further differs from M. *creechbarrowensis* with which it co-occurs, in having P^{3-4} with more crenulated enamel (Hooker, 1986). Shares inflated hypoconulid lobe with M. *edwardsi*, but differs in the variable presence of a lower molar mesoconid (always present in M. *edwardsi*), lacking a double-lobed hypoconulid, and in having a continuous postcristid.

SPECIES *Microchoerus creechbarrowensis* Hooker, 1986
TYPE SPECIMEN BMNH M35732, right M^1
AGE AND GEOGRAPHIC RANGE Middle to late Eocene, England
ANATOMICAL DEFINITION
Morphologically similar to M. *erinaceus* but differs in being approximately 17% smaller in tooth dimensions. Both of these species differ from M. *wardi* and M. *edwardsi* in lacking an inflated M_3 hypoconulid lobe.

GENUS *Necrolemur* Filhol, 1873 (Fig. 5.2)
INCLUDED SPECIES *N. antiquus, N. zitteli*

TYPE SPECIES *Necrolemur antiquus* Filhol, 1873
TYPE SPECIMEN MNHN Qu 11013, skull and right dentary
AGE AND GEOGRAPHIC RANGE Middle to late Eocene, Europe
ANATOMICAL DEFINITION
Necrolemur skulls have enlarged orbits and relatively short rostra. Petromastoid regions are greatly inflated. Mandibular angles are short and broad. Like *Microchoerus*, and in contrast to *Nannopithex*, *Necrolemur* has relatively quadrate upper molars with well-developed hypocones. Lower third molars are reduced as in *Nannopithex*. P^{3-4} are less molariform and more triangular, as in *Nannopithex* and unlike *Microchoerus* where P^{3-4} are more quadrate and molariform. *Necrolemur* has a long, straight-shafted femur with a relatively deep distal end and distally fused tibia–fibula (Schlosser, 1907). The astragalus has a relatively high body with a long, narrow trochlea and a large posterior astragalar shelf (Godinot & Dagosto, 1983). These postcranial features suggest that *Necrolemur* was an active leaper, similar to extant *Galago*.

SPECIES *Necrolemur zitteli* Schlosser, 1887
TYPE SPECIMEN Munich No. 1879 XV 4, dentary with P_3–M_3
AGE AND GEOGRAPHIC RANGE Middle to late Eocene, France and Germany
ANATOMICAL DEFINITION
Differs from *Necrolemur antiquus* in being smaller, upper molars relatively shorter lingually with relatively smaller hypocones, P^4 with weaker postprotocrista, lower molars with relatively broader and shorter talonids, M_3 always smaller than M_2, and upper and lower cheek tooth enamel less complex.

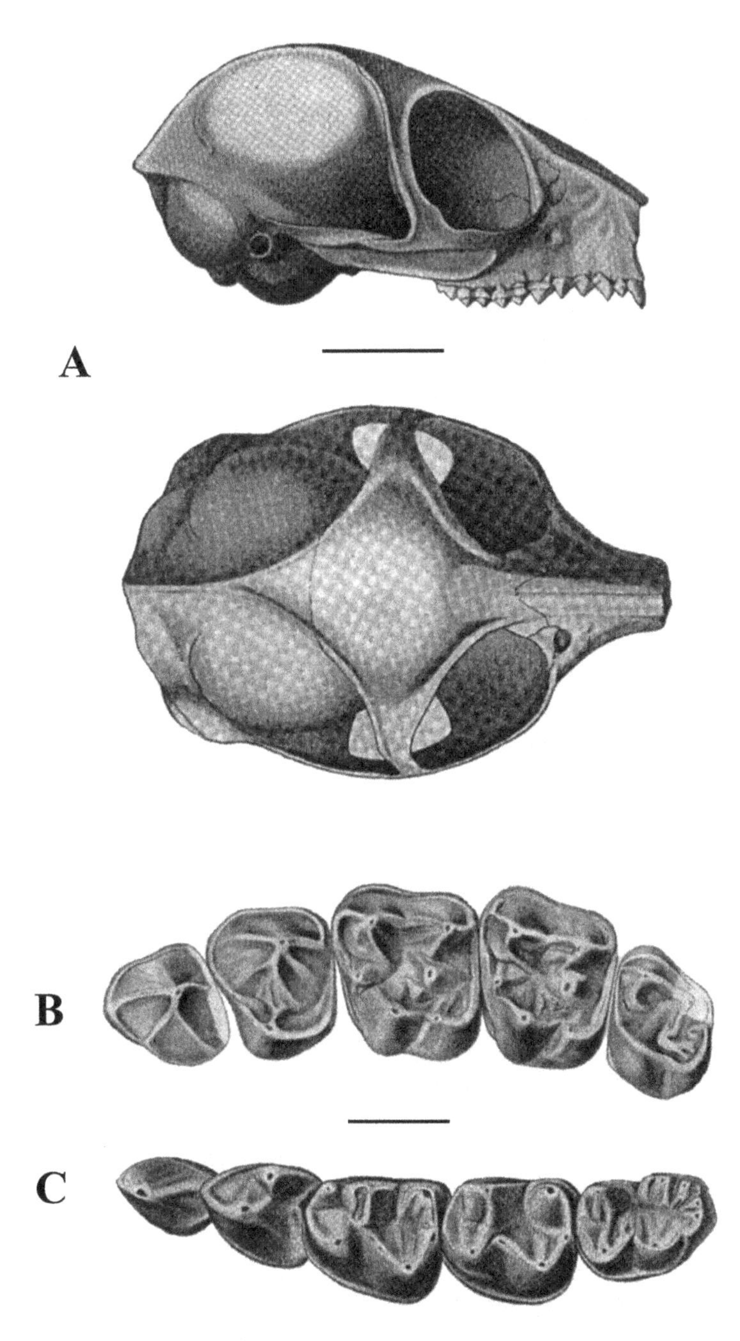

Fig. 5.2 *Necrolemur antiquus*. (A) Skull (BNM Qh 470 and Montauban 9) in lateral and dorsal views. (B) Left maxilla P^3–M^3 (Basel Qh 468) in occlusal view. (C) Left dentary P_3–M_3 (BNM Qh 445) in occlusal view. Scale bar in A represents 1 cm, scale bar in B–C represents 2 mm. Figures adapted from Stehlin (1916) and Hürzeler (1948).

GENUS *Nannopithex* Stehlin, 1916 (Fig. 5.1 E–G)
INCLUDED SPECIES *N. filholi, N. humilidens, N. quaylei, N. raabi, N. zuccolae*

TYPE SPECIES *Nannopithex filholi* Chantre & Gaillard, 1897
TYPE SPECIMEN FSL 4212, left dentary P_4–M_3
AGE AND GEOGRAPHIC RANGE Early and middle Eocene, Europe
ANATOMICAL DEFINITION
Nannopithex often has rugose cheek tooth enamel, but it is never as well developed or as complex as that exhibited by *Microchoerus*. Upper molars of *Nannopithex* are relatively triangular with small postprotocingulae, weak hypocones and relatively small conules. All upper molars lack mesostyles. Lower M_{1-2} with straight and continuous protocristids, entoconids lower than metaconids, and hypoconulids weak to absent. Lower third molars with variable hypoconulids, ranging from short and unicuspid to longer and double-cusped. All lower molars with paraconid, although this cusp often fuses with the metaconid on M_3. The mandibular angle is not greatly expanded, and the coronoid process is relatively high (Hooker, 1986).

SPECIES *Nannopithex raabi* Heller, 1930
TYPE SPECIMEN GMH CeI-4254, lectotype, right dentary C_1, P_3–M_3
AGE AND GEOGRAPHIC RANGE Middle Eocene, Germany

ANATOMICAL DEFINITION
Nannopithex raabi differs from *N. filholi* in having P^{3-4} with more distal protocones, P^2 longer, M^2 with less distinct postprotocingulum and M_3 with stronger hypoconulid and weaker enamel crenulation. This species probably is the senior synonym of *N. abderhaldeni* (Weigelt) and *N. barnesi* Thalmann (see Thalmann, 1994).

SPECIES *Nannopithex quaylei* Hooker, 1986
TYPE SPECIMEN BMNH M37145, left M^2
AGE AND GEOGRAPHIC RANGE Middle Eocene, England
ANATOMICAL DEFINITION
Nannopithex quaylei differs from all other *Nannopithex* species in being larger (as much as 30% larger in certain tooth dimensions; see Hooker, 1986).

SPECIES *Nannopithex zuccolae* Godinot *et al.*, 1992
TYPE SPECIMEN SLP 29-PE-450, left maxilla P^4–M^3
AGE AND GEOGRAPHIC RANGE Late early Eocene, France
ANATOMICAL DEFINITION
Nannopithex zuccolae differs from *N. filholi* and *N. raabi* in lacking upper molar hypocones, and in having small metaconules, variably present postprotocingula, and a more lingual and isolated paraconid on M_2; differs from *N. quaylei* in being much smaller, with a shorter and broader M_3 hypoconulid.

SPECIES *Nannopithex humilidens* Thalmann, 1994
TYPE SPECIMEN GMH IL-8, right dentary I_1–M_3
AGE AND GEOGRAPHIC RANGE Middle Eocene, Germany
ANATOMICAL DEFINITION
Nannopithex humilidens differs from *N. raabi*, the other recognized species of *Nannopithex* from Geiseltal, chiefly in being somewhat smaller with a relatively shorter P_4 (Thalmann, 1994).

GENUS *Pseudoloris* Stehlin, 1916 (Fig. 5.1C–D)
INCLUDED SPECIES *P. crusafonti, P. godinoti, P. isabenae, P. parvulus*

TYPE SPECIES *Pseudoloris parvulus* (Filhol, 1890)
TYPE SPECIMEN BNM Qh 476, left maxilla with P^4–M^3
AGE AND GEOGRAPHIC RANGE Late Eocene to Oligocene, Europe
ANATOMICAL DEFINITION
Pseudoloris is much smaller than any other microchoerine genus. The cheek teeth of *Pseudoloris* lack rugose enamel. Upper molars lack a postprotocingulum but have a strong postprotocrista that connects to a small metaconule. Paraconules are present and strong. Hypocones are cingular in origin, low, and much smaller than the protocone. All upper molars lack mesostyles. Lower premolars are long and relatively narrow. Lower M_{1-2} have straight and complete protocristids, entoconids the same height as metaconids, and tiny to absent hypoconulids. Lower third molars are not reduced and have a narrow, unicusped, but elongate hypoconulid. Lower molar paraconids are weak or absent but paracristids are well developed and separated from the metaconid by a relatively deep valley. The mandibular angle is not expanded and the coronoid process is low (Hooker, 1986).

SPECIES *Pseudoloris isabenae* Crusafont-Pairo, 1967
TYPE SPECIMEN Unnumbered specimen in Sabadell Museum (IPMC), right P_3–M_2
AGE AND GEOGRAPHIC RANGE Middle Eocene, Spain
ANATOMICAL DEFINITION
Pseudoloris isabenae differs from other species of *Pseudoloris* in being smaller.

SPECIES *Pseudoloris crusafonti* Louis & Sudre, 1975
TYPE SPECIMEN Collection of P. Louis, Gri. 382, right M^2
AGE AND GEOGRAPHIC RANGE Middle Eocene, France
ANATOMICAL DEFINITION
Pseudoloris crusafonti is larger than *P. isabenae* and *P. parvulus*. It further differs from *P. parvulus* in having a larger M_3 hypoconulid.

SPECIES *Pseudoloris godinoti* Köhler & Moyà-Solà, 1999
TYPE SPECIMEN IPMC 14041, left P^2–M^2
AGE AND GEOGRAPHIC RANGE Early Oligocene, Spain
ANATOMICAL DEFINITION
Pseudoloris godinoti differs from *P. isabenae* and *P. parvulus* in being larger. It is similar in size to *P. crusafonti* but differs from all other *Pseudoloris* species in having longer and narrower upper premolars and molars with mediolaterally compressed buccal cusps.

Subfamily Anaptomorphinae

Anaptomorphines differ from omomyines (Bown & Rose, 1987) in: usually having enlarged, often procumbent, central incisors (especially in more derived taxa); reduction in size and number of teeth between I_1 and P_4; P_4 often inflated and buccally distended; molar cusps less acute and more basally inflated; upper molar trigons more constricted with a postprotocingulum developed (absent in most omomyines); upper and lower third molars typically reduced. Anaptomorphines differ from microchoerines in: having I_2 less enlarged; P_2 primitively present; molar paraconids normally more lingually placed (especially on M_{2-3}); upper molars more transverse, lacking a continuous lingual cingulum, less distinct conules; upper and lower third molars typically reduced (also true of *Necrolemur*).

Tribe Anaptomorphini

GENUS *Anaptomorphus* Cope, 1872 (Fig. 5.3D)
INCLUDED SPECIES *A. aemulus, A. westi*

TYPE SPECIES *Anaptomorphus aemulus* Cope, 1872
TYPE SPECIMEN AMNH 5010, left dentary with P_4–M_2
AGE AND GEOGRAPHIC RANGE Middle Eocene, Wyoming
ANATOMICAL DEFINITION
Anaptomorphus has a lower dental formula of 2.1.2.3 (Szalay, 1976; Gunnell, 1995*a*). Lower teeth are relatively low with

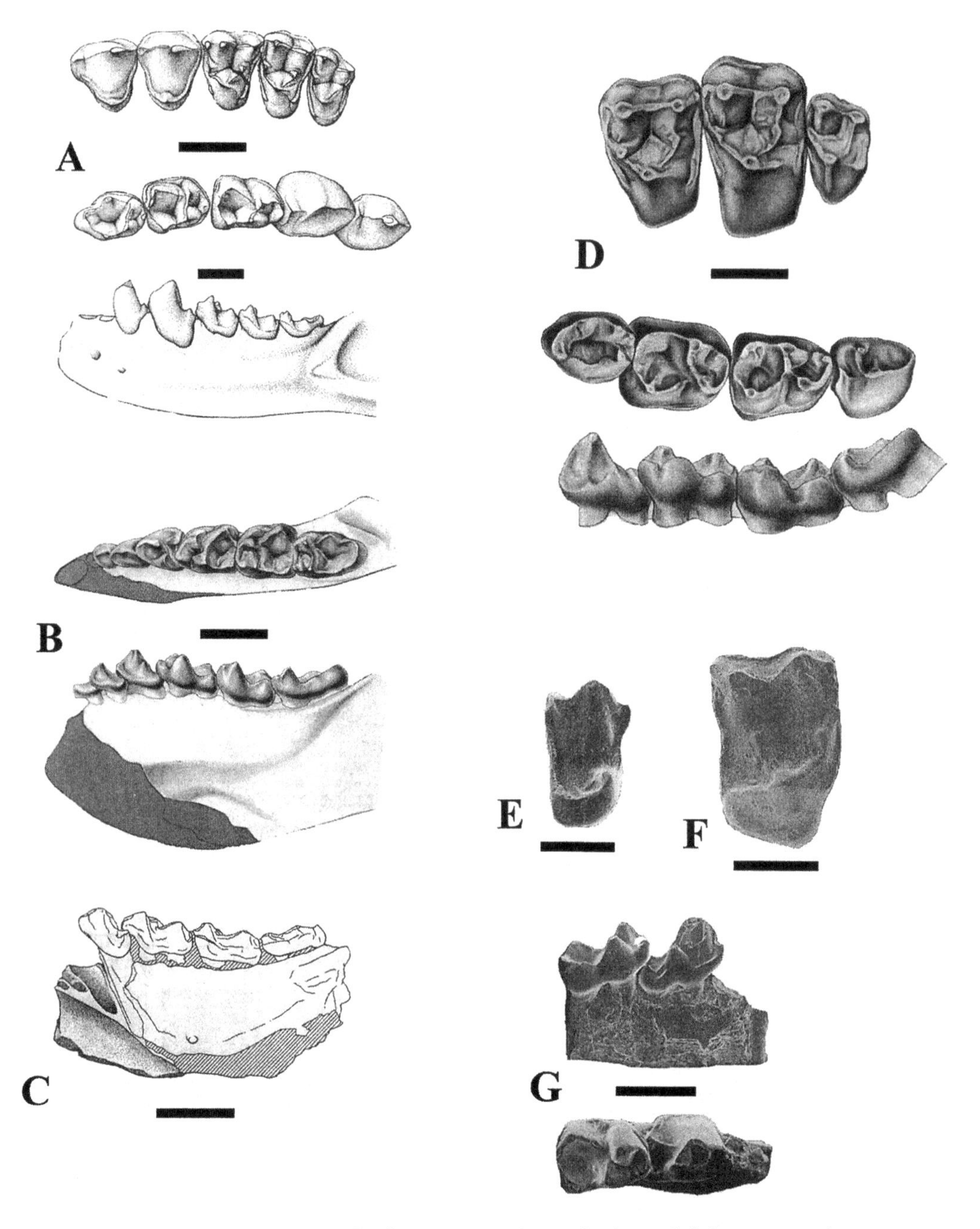

Fig. 5.3 (A) *Uintanius ameghini*, left maxilla P^3–M^3 (AMNH 13039) in occlusal view, left dentary P_3–M_3 (AMNH 12376, 55261, 56216) in occlusal and lateral views. (B) *Trogolemur myodes*, right dentary C–M_3 (AMNH 12599, holotype) in occlusal and medial views. (C) *Sphacorhysis burntforkensis*, left dentary P_4–M_3 (UM 30966, holotype) in lateral view. (D) *Anaptomorphus westi*, left maxilla M^{1-3} (FM 15661) in occlusal view, right dentary P_4–M_3 (FM 15041, M_{1-3}, holotype, FM 15684) in occlusal and medial views. (E–F) *Trogolemur myodes*, right P^4 (E, UM 31561), right M^1 (F, UM 31557) in occlusal view. (G) *Jemezius szalayi*, left dentary P_4–M_1 (CM 34843, holotype) in medial and occlusal views. Scale bars in A–D, G represent 2 mm and in E–F represent 1 mm. (A–F) Adapted from Szalay (1976) and Gunnell (1995*a*); (G) courtesy of K.C. Beard.

bulbous cusps and moderately developed crests. Paraconids are distinct on M_{1-2}, more weakly developed on M_3. Lower first and second molars approximately the same length with M_2 being somewhat broader. Lower and upper third molars are reduced, with M_3 having a weakly developed and short hypoconulid. Lower P_3 and P_4 are relatively simple, lacking paraconids and metaconids (P_4 sometimes with a weak enamel fold in position of metaconid). Both P_3 and P_4 have short, weakly developed talonids. Upper molars have distinct postprotocristae and are lingually extended (especially M^{1-2}). Paraconules are weak on M^{1-2}, metaconules are more strongly developed. Trigon basins are shallow and restricted on all upper molars. Despite being the type genus of the subfamily Anaptomorphinae,

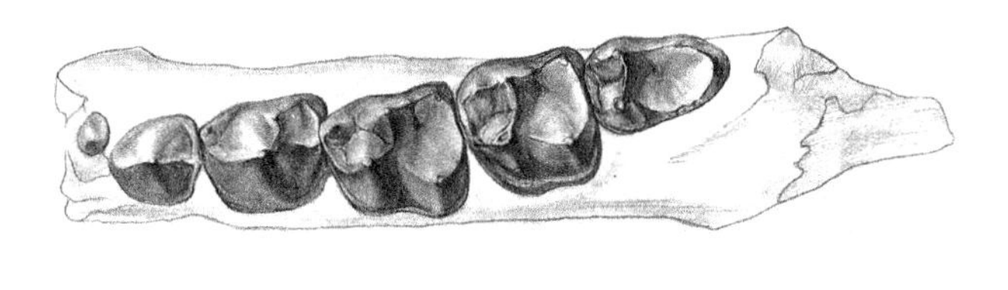

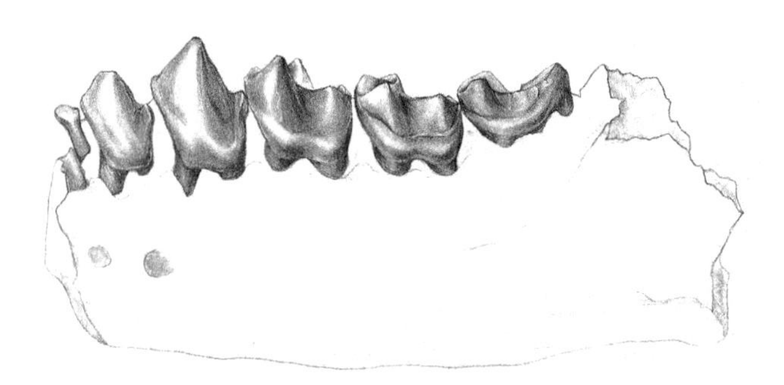

Fig. 5.4 *Tetonius matthewi*, left dentary P_2–M_3 (YPM 23031) in occlusal and lateral views. Scale in mm. Figures from Bown & Rose (1987).

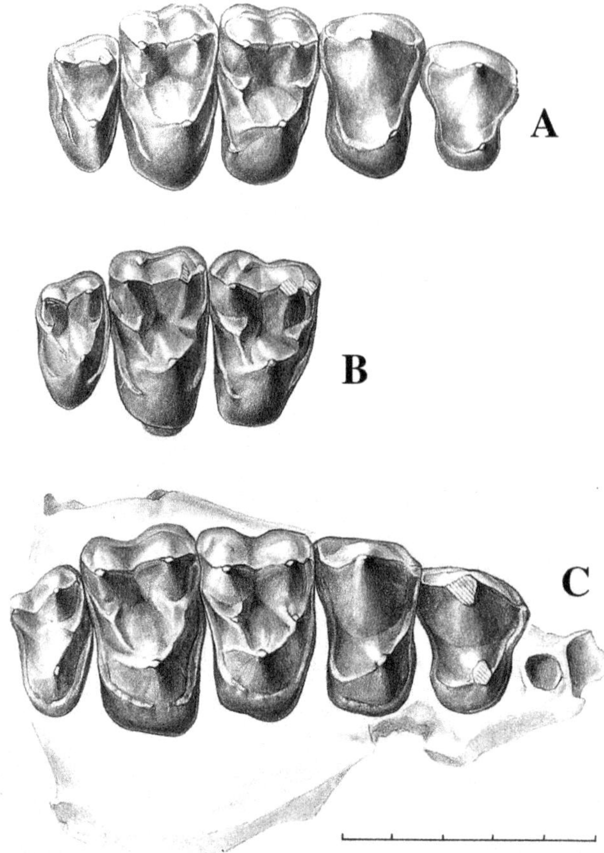

Fig. 5.5 *Pseudotetonius* and *Tetonius* upper dentitions. (A) *Pseudotetonius ambiguus*, right composite P^3–M^3 (USGS 3860, 9202). (B) *Tetonius* sp., right maxilla M^{1-3} (USGS 5992). (C) *Tetonius matthewi*, right maxilla P^3–M^3 (UM 76675), all in occlusal view. Scale in mm. Figures from Bown & Rose (1987).

Anaptomorphus is relatively poorly represented and much of its anatomy awaits discovery.

SPECIES *Anaptomorphus westi* Szalay, 1976
TYPE SPECIMEN FMNH 15041, right dentary with M_{1-3}
AGE AND GEOGRAPHIC RANGE Early middle Eocene, Wyoming
ANATOMICAL DEFINITION
Anaptomorphus westi differs from *A. aemulus* in being 30% to 40% larger in tooth dimensions (Szalay, 1976).

GENUS *Tetonius* Matthew, 1915 (Figs. 5.4, 5.5)
INCLUDED SPECIES *T. homunculus, T. matthewi, T. mckennai*

TYPE SPECIES *Tetonius homunculus* Cope, 1882
TYPE SPECIMEN AMNH 4194, crushed skull with left and right C–M^3
AGE AND GEOGRAPHIC RANGE Early Eocene, Wyoming and Colorado; an omomyid from the Bashi Formation in Mississippi may represent a species of *Tetonius* (Beard & Tabrum, 1990)
ANATOMICAL DEFINITION
Tetonius has a lower dental formula of 2.1.2–3.3 (Bown & Rose, 1987). The anterior teeth are not crowded together and P_3 is unreduced and double-rooted. The central incisor is always larger than I_2 and both incisors are relatively procumbent. *Tetonius* differs from *Teilhardina* in being larger and in having relatively broader cheek teeth, a taller P_4, and more reduced C and P_2. *Tetonius* differs from *Absarokius* in lacking a ventrobuccally distended (exodaenodont) P_4 and in having a larger, more procumbent I_1 and a relatively smaller P_2.

SPECIES *Tetonius mckennai* Bown & Rose, 1987
TYPE SPECIMEN UCMP 46192, left dentary with I_{1-2}, C, P_3–M_2
AGE AND GEOGRAPHIC RANGE Early Eocene of Colorado
ANATOMICAL DEFINITION
Tetonius mckennai is slightly larger than *Teilhardina americana*. It differs from all other species of *Tetonius* in being smaller and in having a narrower M_1.

SPECIES *Tetonius matthewi* Bown & Rose, 1987
TYPE SPECIMEN CM 12190, left dentary with I_1–M_3
AGE AND GEOGRAPHIC RANGE Early Eocene, Wyoming and Colorado
ANATOMICAL DEFINITION
Tetonius matthewi differs from *T. mckennai* in being larger and in having a reduced P_2 (sometimes absent in latest occurring samples of *T. matthewi*). It differs from later occurring *Tetonius* species in having a relatively unreduced and double-rooted P_3, and lacking crowding of antemolar teeth (Bown & Rose, 1987).

GENUS *Absarokius* Matthew, 1915 (Fig. 5.6)
INCLUDED SPECIES *A. abbotti, A. australis, A. gazini, A. nocerai, A. metoecus, A. witteri*

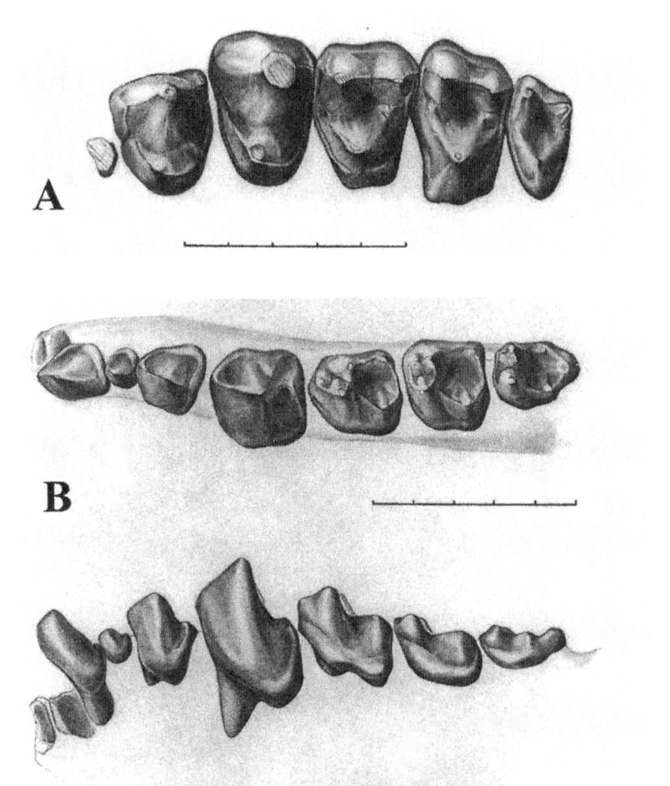

Fig. 5.6 *Absarokius abbotti*. (A) Left maxilla P^3–M^3 (YPM 18686) in occlusal view. (B) Left dentary C–M_3 (YPM 27791) in occlusal, lateral and medial views. Scale in mm. Figures from Bown & Rose (1987).

TYPE SPECIES *Absarokius abbotti* Loomis, 1906
TYPE SPECIMEN ACM 3479, right dentary P_3–M_3
AGE AND GEOGRAPHIC RANGE Early to early middle Eocene, Wyoming; late early Eocene, Colorado
ANATOMICAL DEFINITION
Absarokius has a lower dental formula of 2.1.2–3.3 and an upper dental formula of 2.1.3.3. *Absarokius* species have enlarged P_4s that are buccally distended and exodaenodont in more derived species. P^4s are enlarged and narrower and more quadrate than in *Tetonius* or *Pseudotetonius*, the two anaptomorphine taxa most similar to *Absarokius*. *Absarokius* also differs from these two genera in having relatively more reduced third molars. *Absarokius* further differs from *Pseudotetonius* in having a much smaller I_1 and a larger C and differs from *Tetonius* in having I_{1-2} much smaller and P_2 much larger (Bown & Rose, 1987).

SPECIES *Absarokius nocerai* Robinson, 1966
TYPE SPECIMEN AMNH 55215, left dentary with C–M_3
AGE AND GEOGRAPHIC RANGE Early middle Eocene, Colorado and Wyoming
ANATOMICAL DEFINITION
Absarokius nocerai differs from *A. metoecus* in having P_4 greatly enlarged; differs from *A. gazini* and *A. metoecus* in having relatively larger molars; differs from all species of *Absarokius* except *A. witteri* and *A. australis* in lacking P_2; differs from all species except *A. australis* in having single-rooted P_3; differs from *A. australis* in having distinct paraconids on M_{2-3} and in lacking a posteriorly enlarged talonid basin on M_3.

SPECIES *Absarokius witteri* Morris, 1954
TYPE SPECIMEN YPM PU 14972, left dentary with P_{3-4}, M_{2-3}
AGE AND GEOGRAPHIC RANGE Early middle Eocene, Wyoming
ANATOMICAL DEFINITION
Absarokius witteri differs from all other species of *Absarokius* in being larger and in having a relatively more hypertrophied P_4 that is basally very wide.

SPECIES *Absarokius metoecus* Bown & Rose, 1987
TYPE SPECIMEN USGS 492, right dentary with P_4–M_3
AGE AND GEOGRAPHIC RANGE Later early Eocene, Wyoming and Colorado
ANATOMICAL DEFINITION
Absarokius metoecus differs from all other *Absarokius* species in being smaller (except some specimens of *A. gazini*) and in having a relatively low-crowned, non-hypertrophied P_4.

SPECIES *Absarokius gazini* Bown & Rose, 1987
TYPE SPECIMEN UW 1644, right dentary with P_3–M_3
AGE AND GEOGRAPHIC RANGE Late early Eocene, Wyoming
ANATOMICAL DEFINITION
Absarokius gazini differs from all other species of *Absarokius* except *A. metoecus* in being smaller. Differs from *A. metoecus* in having a larger, more hypertrophied P_4.

SPECIES *Absarokius australis* Bown & Rose, 1987
TYPE SPECIMEN AMNH 55152, right dentary with P_3–M_3
AGE AND GEOGRAPHIC RANGE Earliest middle Eocene, Colorado
ANATOMICAL DEFINITION
Absarokius australis differs from all species of *Absarokius* except *A. nocerai* and *A. witteri* in lacking P_2 and in having P_3 very small relative to P_4; differs from *A. nocerai* and *A. witteri* in having P_4 less hypertrophied, paraconids smaller on M_{1-2} and absent on M_3, and M_3 with a broad hypoconulid lobe.

GENUS *Teilhardina* Simpson, 1940 (Figs. 5.7, 5.8)
INCLUDED SPECIES *T. americana, T. belgica, T. brandti, T. crassidens, T. demissa, T. tenuicula*

TYPE SPECIES *Teilhardina belgica* (Teilhard de Chardin, 1927)
TYPE SPECIMEN IRSNB 64, left dentary with P_3–M_3 (lectotype, Simpson, 1940)

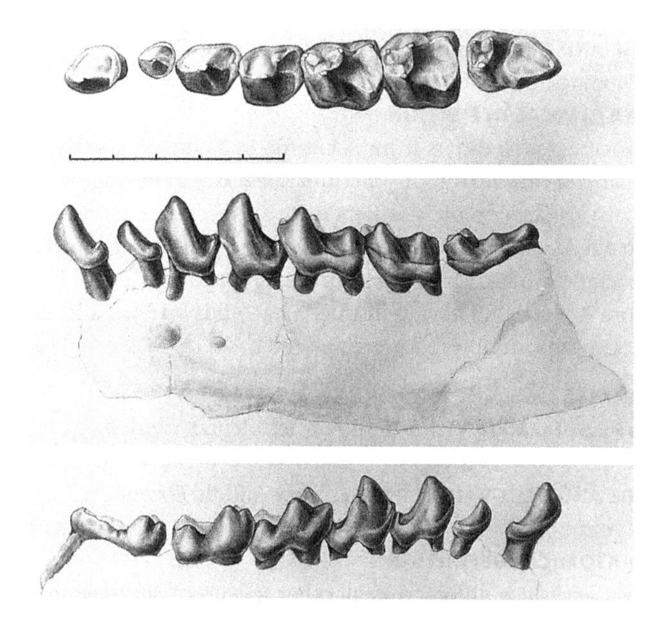

Fig. 5.7 *Teilhardina americana*. Left dentary C–M_3 (UW 6896, holotype) in occlusal, lateral and medial views. Scale in mm. Figures from Bown & Rose (1987).

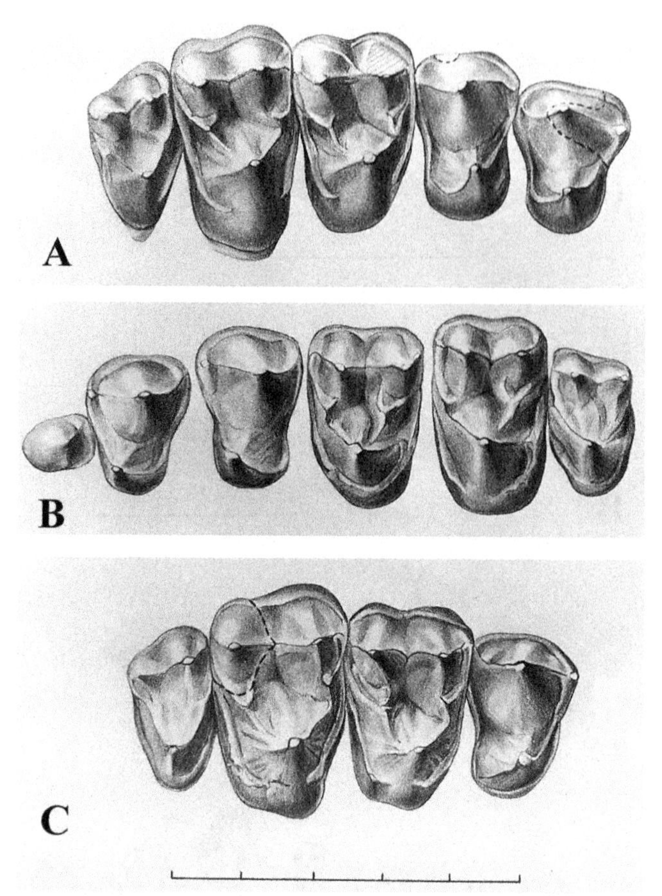

Fig. 5.8 *Teilhardina* upper dentitions. (A) *T. crassidens*, right maxilla P^3–M^3 (YPM 24626). (B) *Teilhardina* sp., left maxilla P^2–M^3 (UM 69783). (C) *T. americana*, right maxilla P^4–M^3 (UW 8871), all in occlusal view. Scale in mm. Figures from Bown & Rose (1987).

AGE AND GEOGRAPHIC RANGE Early Eocene, North America and Europe, possibly Asia

ANATOMICAL DEFINITION

Teilhardina is one of the most primitive omomyids known, rivaled only by *Steinius* and the possible omomyids *Altanius* and *Altiatlasius*. *Teilhardina* was relatively small, similar in size to *Anemorhysis*. The lower dental formula of *Teilhardina* is 2.1.3–4.3. Lower P_3 is simple and unreduced, P_4 may have small paraconid and metaconid cuspules and is not enlarged. *Teilhardina* differs from *Anemorhysis* in having a P_4 with a short, unbasined talonid and a more lingually directed cristid obliqua resulting in a deeper hypoflexid (Bown & Rose, 1987). *Teilhardina* is the only known anaptomorphine represented outside of North America with a single species, *T. belgica*, present in the early Ypresian of Europe. Recent informal reports indicate that *Teilhardina* may be present in Asia as well.

SPECIES *Teilhardina americana* Bown, 1976

TYPE SPECIMEN UW 6896, left dentary with C–M_3

AGE AND GEOGRAPHIC RANGE Early Eocene, Wyoming

ANATOMICAL DEFINITION

Teilhardina americana is very similar to *T. belgica*, differing only in being slightly larger, in having relatively broader lower cheek teeth, a slightly higher P_4 metaconid, and upper molars with a postprotocingulum consistently present.

SPECIES *Teilhardina crassidens* Bown & Rose, 1987

TYPE SPECIMEN UW 8959, left dentary with P_3–M_2

AGE AND GEOGRAPHIC RANGE Early Eocene, Wyoming

ANATOMICAL DEFINITION

Teilhardina crassidens differs from *T. belgica* and *T. americana* in having relatively broader cheek teeth with P_3 lower crowned and P_4 with higher and more distinct paraconid and metaconid; differs from *T. tenuicula* in having P_{3-4} higher crowned and M_1 trigonid more closed.

SPECIES *Teilhardina tenuicula* Jepsen, 1930

TYPE SPECIMEN YPM PU 13027, right maxilla with P^4–M^3

AGE AND GEOGRAPHIC RANGE Early Eocene, Wyoming

ANATOMICAL DEFINITION

Teilhardina tenuicula differs from *T. americana* and *T. crassidens* in being smaller and differs from all other species of *Teilhardina* (except *T. demissa*) in having relatively broad and low crowned P_{3-4} with relatively shorter and smaller talonids.

SPECIES *Teilhardina brandti* Gingerich, 1993

TYPE SPECIMEN UM 99031, right M_2

AGE AND GEOGRAPHIC RANGE Earliest early Eocene, Wyoming

ANATOMICAL DEFINITION

Teilhardina brandti differs from *T. belgica* in being larger, differs from all other North American species of *Teilhardina* in having M_2 relatively narrower with a weaker buccal cingulid. *Teilhardina brandti* is only represented by the single type tooth which differs little from *T. belgica* and *T. americana*. More complete material is required to determine the validity of *T. brandti*.

SPECIES *Teilhardina demissa* Rose, 1995
TYPE SPECIMEN USGS 25298, left dentary with C–M_3
AGE AND GEOGRAPHIC RANGE Later early Eocene, Wyoming
ANATOMICAL DEFINITION
Teilhardina demissa differs from *T. crassidens* in having less basally inflated lower molars; differs from all other *Teilhardina* species in having M_{1-2} enamel distended buccally, P_{3-4} lower crowned with talonids composed of high transverse ridge.

GENUS *Anemorhysis* Gazin, 1958 (Fig. 5.9)
INCLUDED SPECIES *A. natronensis, A. pattersoni, A. pearcei, A. savagei, A. sublettensis, A. wortmani*

TYPE SPECIES *Anemorhysis sublettensis* Gazin, 1952
TYPE SPECIMEN USNM 19205, left dentary with P_4–M_2
AGE AND GEOGRAPHIC RANGE Early Eocene, Wyoming, Utah, and North Dakota; earliest middle Eocene, Wyoming
ANATOMICAL DEFINITION
The main feature that distinguishes *Anemorhysis* from other similar anaptomorphines such as *Teilhardina* and *Tetonius* is the presence of a semimolariform P_4 (Bown & Rose, 1987). Lower fourth premolar has a relatively prominent paraconid and a high, well-developed metaconid. The talonid basin is well developed with a distinct hypoconid and a smaller entoconid. The cristid obliqua is short but distinct. Molar cusps are sharply defined and less basally inflated than in *Tetonius*. The postcristid of M_{1-2} is straight, not posteriorly convex as in *Teilhardina* and *Tetonius*.

SPECIES *Anemorhysis pearcei* Gazin, 1962
TYPE SPECIMEN USNM 22426, right dentary with P_3–M_2
AGE AND GEOGRAPHIC RANGE Early Eocene, Wyoming and North Dakota
ANATOMICAL DEFINITION
Anemorhysis pearcei differs from *A. wortmani, A. natronensis* and probably *A. pattersoni* in retaining a P_2. Differs from all other *Anemorhysis* species in having a weak P_4 cristid obliqua and a less well-developed talonid basin. *Anemorhysis pearcei* was originally described as the type species of a new genus, *Tetonoides* by Gazin (1962). Szalay (1976) synonymized *Anemorhysis* and *Tetonoides*, a practice supported by Bown & Rose (1984, 1987) but questioned by Gingerich (1981*b*) and Williams & Covert (1994). We follow Szalay (1976) in this paper.

SPECIES *Anemorhysis pattersoni* Bown & Rose, 1984
TYPE SPECIMEN USGS 476, left dentary with P_4–M_2
AGE AND GEOGRAPHIC RANGE Late early Eocene, Wyoming
ANATOMICAL DEFINITION
Anemorhysis pattersoni is larger than *A. sublettensis* and *A. pearcei*

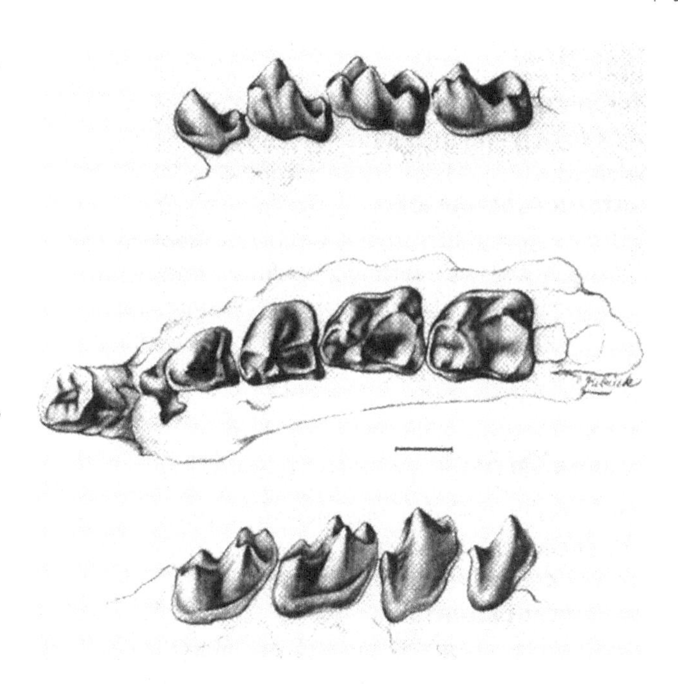

Fig. 5.9 *Anemorhysis wortmani*. Right dentary P_3–M_2 (USGS 6554, holotype) in medial, occlusal and lateral views. Scale bar represents 1 mm. Figures from Bown & Rose (1984).

and differs from all other *Anemorhysis* species in having a low and weak P_4 paraconid. P_2 is probably absent as in *A. wortmani* and *A. natronensis* but differing from other species of *Anemorhysis; A. pattersoni* differs from *A. natronensis* in having a shorter, more exodaenodont P_4 and a deeper horizontal ramus (Beard *et al.*, 1992*a*).

SPECIES *Anemorhysis wortmani* Bown & Rose, 1984
TYPE SPECIMEN USGS 6554, right dentary with P_3–M_2
AGE AND GEOGRAPHIC RANGE Middle early Eocene, Wyoming
ANATOMICAL DEFINITION
Anemorhysis wortmani lacks a P_2 as does *A. natronensis* and probably *A. pattersoni*; differs from *A. pearcei* and *A. pattersoni* in having a mesiodistally compressed, single-rooted P_3; differs from *A. pattersoni* and *A. natronensis* in having a stronger, higher P_4 paraconid and slightly narrower molar talonids; P_4 broader than in *A. pearcei* and *A. sublettensis* and relatively more distended buccally (Bown & Rose, 1987).

SPECIES *Anemorhysis natronensis* Beard *et al.*, 1992
TYPE SPECIMEN CM 41137, left dentary with I_2–M_2
AGE AND GEOGRAPHIC RANGE Earliest middle Eocene, Wyoming
ANATOMICAL DEFINITION
Anemorhysis natronensis differs from *A. wortmani* and *A. sublettensis* in having a weak, low paraconid on P_4; differs from *A. pattersoni* in having a relatively longer, less exodaenodont P_4, with a slightly better developed paraconid and in having a shallower horizontal ramus.

SPECIES *Anemorhysis savagei* Williams & Covert, 1994
TYPE SPECIMEN UCM 56410, right dentary with P_3–M_3
AGE AND GEOGRAPHIC RANGE Late early Eocene, Wyoming
ANATOMICAL DEFINITION
Anemorhysis savagei differs from *A. wortmani*, *A. natronensis* and probably *A. pattersoni* in retaining P_2; differs from all other *Anemorhysis* species in having a narrow P_4 talonid; differs from *A. sublettensis* in having a smaller, lower, crescentic P_4 paraconid.

GENUS *Chlororhysis* Gazin, 1958 (Fig. 5.10)
INCLUDED SPECIES *C. incomptus*, *C. knightensis*

TYPE SPECIES *Chlororhysis knightensis* Gazin, 1958
TYPE SPECIMEN USNM 21901, left dentary with C–P_4
AGE AND GEOGRAPHIC RANGE Early Eocene, Wyoming
ANATOMICAL DEFINITION
Chlororhysis is a very poorly known anaptomorphine represented by four specimens, two from the Green River Basin, and one each from the Bighorn and Washakie basins, all in Wyoming (Gazin, 1958, 1962; Bown & Rose, 1984, 1987). *Chlororhysis* is primitive in having a relatively unreduced lower canine and P_2, P_{3-4} relatively narrow and uninflated and P_4 with a tiny to absent metaconid and an unbasined talonid. Among known anaptomorphines, *Chlororhysis* seems most similar to *Loveina*, differing mostly in premolar proportions and in having a weak and low P_4 metaconid.

SPECIES *Chlororhysis incomptus* Bown & Rose, 1984
TYPE SPECIMEN YPM 24997, right dentary with P_{3-4}
AGE AND GEOGRAPHIC RANGE Early Eocene, Wyoming
ANATOMICAL DEFINITION
Chlororhysis incomptus differs from *C. knightensis* in having P_{3-4} narrower and less buccally distended, paracristids shorter and weaker, and lacking paraconids and metaconids.

GENUS *Pseudotetonius* Bown, 1974 (Figs. 5.5A, 5.11)
INCLUDED SPECIES *P. ambiguus*

TYPE SPECIES *Pseudotetonius ambiguus* Bown, 1974
TYPE SPECIMEN AMNH 15072, left dentary with P_3–M_2
AGE AND GEOGRAPHIC RANGE Early Eocene, Wyoming and Colorado
ANATOMICAL DEFINITION
Pseudotetonius is similar to *Tetonius* but differs in the following character states (Bown & Rose, 1987): P_3 reduced relative to P_4; P_3 with a single, compressed root (sometimes bilobed lingually); C and I_2 much smaller and more compressed; I_1 more robust; and antemolar portion of dentary shorter and more robust.

GENUS *Arapahovius* Savage & Waters, 1978 (Fig. 5.12)
INCLUDED SPECIES *A. advena*, *A. gazini*

TYPE SPECIES *Arapahovius gazini* Savage & Waters, 1978

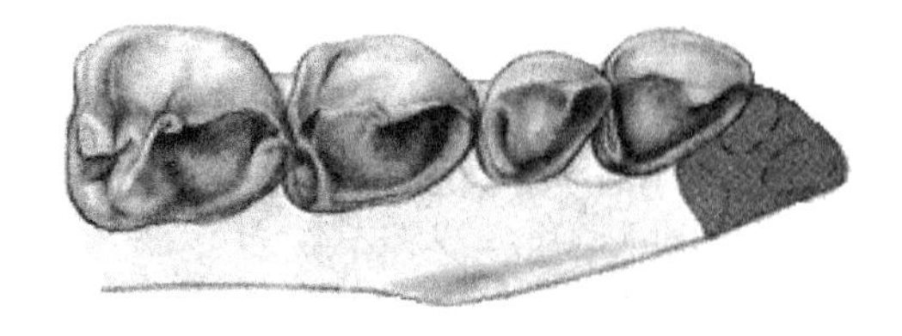

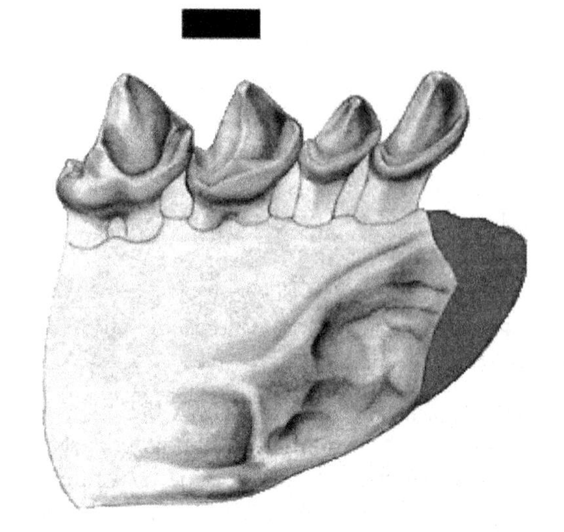

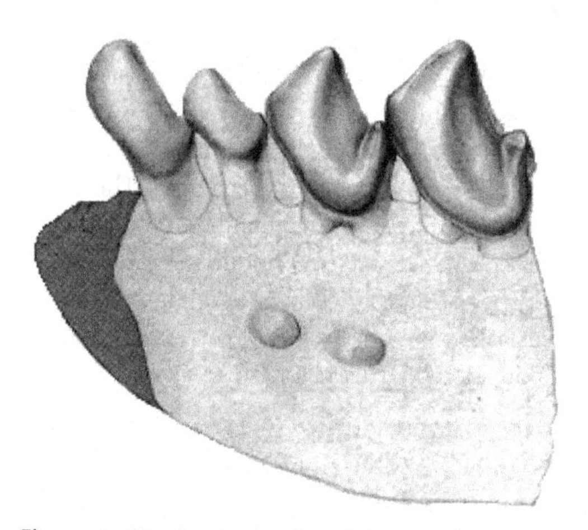

Fig. 5.10 *Chlororhysis knightensis*. Left dentary C–P_4 (USNM 21901, holotype) in occlusal, medial and lateral views. Scale bar represents 1 mm. Figure adapted from Szalay (1976).

TYPE SPECIMEN UCMP 100000, right maxilla with P^3–M^3
AGE AND GEOGRAPHIC RANGE Early Eocene, Wyoming
ANATOMICAL DEFINITION
Arapahovius has a lower dental formula of 2.1.3.3. I_1 is moderately enlarged and procumbent while I_2–C are small and P_2 is reduced and smaller than the C. Both P_{3-4} have well-developed trigonids with distinct paraconids and metaconids, better expressed on P_4. The talonid of P_3 is short while the talonid of P_4 is somewhat longer but not as well developed as in *Anemorhysis*. The most distinguishing feature of *Arapahovius* is the presence of rugose enamel on all upper and lower teeth (Savage & Waters, 1978). Some postcranial elements of *Arapahovius* are also available. The astragalus is of a typical omomyid design with a

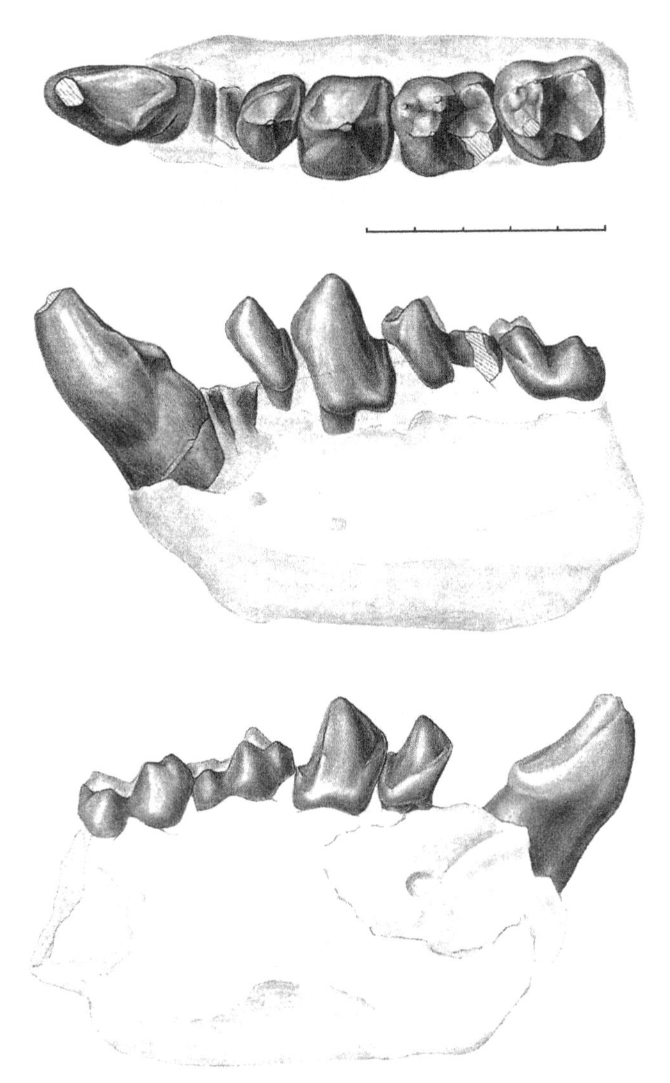

Fig. 5.11 *Pseudotetonius ambiguus*. Left dentary I_1, P_3–M_2 (MCZ 19010) in occlusal, lateral and medial views. Scale in mm. Figure from Bown & Rose (1987).

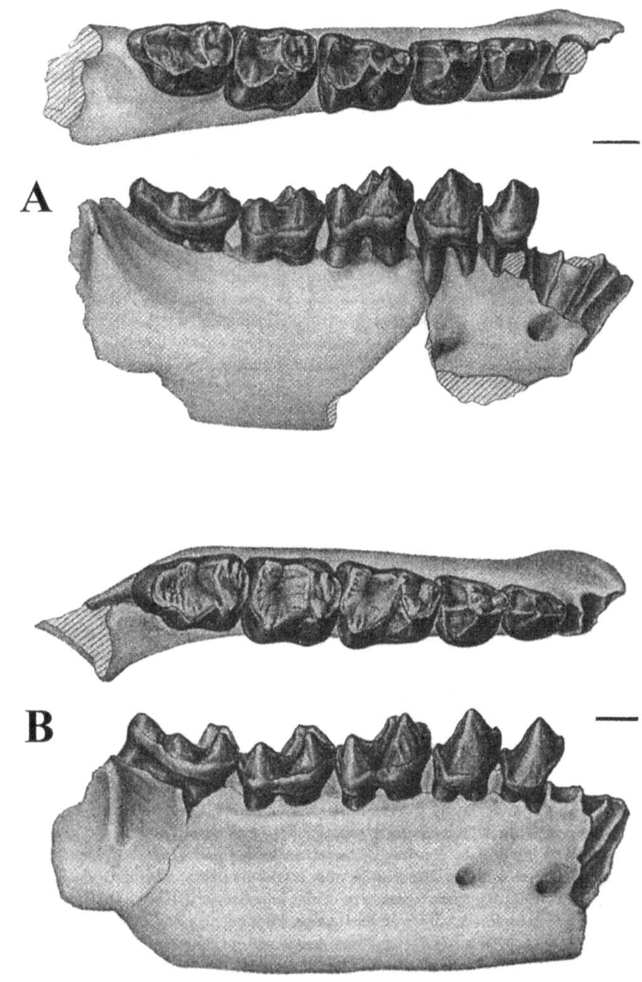

Fig. 5.12 *Arapahovius* lower dentitions. (A) *A. advena*, right dentary P_3–M_3 (USGS 25297) in occlusal and lateral views. (B) *A. gazini*, right dentary P_3–M_3 (UCMP 115668) in occlusal and lateral views. Scale bars represent 1 mm. Figures from Rose (1995*b*).

well-developed posterior shelf, an elongate neck and a continuous sustentacular–navicular facet. Other elements include an elongate calcaneum, navicular and cuboid. The elongation of all of these elements suggests that *Arapahovius* was an accomplished leaper. *Arapahovius* is well represented in the Washakie Basin of Wyoming but is almost unknown outside of this area, being otherwise represented by only a few specimens from the Bighorn Basin of Wyoming (Bown & Rose, 1991; Rose, 1995*b*).

SPECIES *Arapahovius advena* Bown & Rose, 1991
TYPE SPECIMEN USGS 21661, right maxilla with P^4–M^3
AGE AND GEOGRAPHIC RANGE Late early Eocene, Wyoming
ANATOMICAL DEFINITION
Arapahovius advena differs from *A. gazini* in being 15% to 20% smaller, in having more inflated lower molars, molar talonids relatively narrower, M_1 paraconid more buccal, M_{2-3} paraconids less appressed to metaconids, upper molars with lower postprotocristae, M^{1-2} stylar shelves larger, and trigon and talonid basins with less crenulated enamel (Bown & Rose, 1991; Rose, 1995*b*).

GENUS *Aycrossia* Bown, 1979 (Fig. 5.13)
INCLUDED SPECIES *A. lovei*

TYPE SPECIES *Aycrossia lovei* Bown, 1979
TYPE SPECIMEN USNM 250561, left maxilla with C–M^3
AGE AND GEOGRAPHIC RANGE Early middle Eocene, Wyoming
ANATOMICAL DEFINITION
Aycrossia has a lower dental formula of 2.1.3.3 and an upper dental formula of 2?.1.2.3 (Bown, 1979). P_3 is smaller relative to P_4, and P_4 is tall and premolariform, not enlarged or buccally distended. M_1 with a very large metaconid and a small but distinct parastylid located between the paraconid and metaconid. Lower molar talonids with weakly rugose

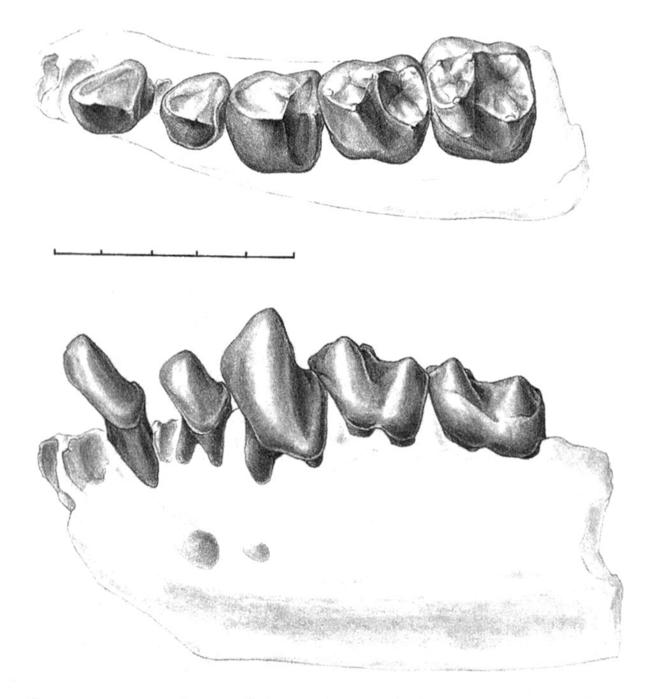

Fig. 5.13 *Aycrossia lovei*. Left dentary C, P_3–M_2 (USGS 2021) in occlusal and lateral views. Scale in mm. Figures from Bown & Rose (1987).

enamel. Upper molars are short and broad, lacking mesostyles, pericones and basal cingula and with slightly rugose enamel. Third molars relatively unreduced compared to second molars. *Aycrossia* is best represented from the Aycross Formation in the Owl Creek Mountains in the southern part of the Bighorn Basin. It may be represented in the Wind River Basin (Stucky, 1984) but this record has yet to be substantiated.

GENUS *Strigorhysis* Bown, 1979 (Fig. 5.14)
INCLUDED SPECIES *S. bridgerensis*, *S. huerfanensis*, *S. rugosus*

TYPE SPECIES *Strigorhysis bridgerensis* Bown, 1979
TYPE SPECIMEN USNM 250556, palate with left P^4–M^3 and right P^3–M^3
AGE AND GEOGRAPHIC RANGE Early middle Eocene, Wyoming and Colorado
ANATOMICAL DEFINITION
Strigorhysis has a lower dental formula of 2.1.2.3. The upper dental formula is probably the same but the presence of two incisors cannot be confirmed from available specimens. P_3 is relatively smaller than P_4, more so than in *Aycrossia*. P_4 is somewhat enlarged and buccally distended and lacks a paraconid. All upper and lower molars have rugose, complex enamel. Upper molars are lingually expanded and have strongly developed postprotocingula. *Strigorhysis*, like *Aycrossia*, is best represented in the Aycross Formation of the southern Bighorn Basin but is also present in the Willwood Formation in the central Bighorn Basin and in the Huerfano Formation in southern Colorado (Bown & Rose, 1987).

SPECIES *Strigorhysis rugosus* Bown, 1979

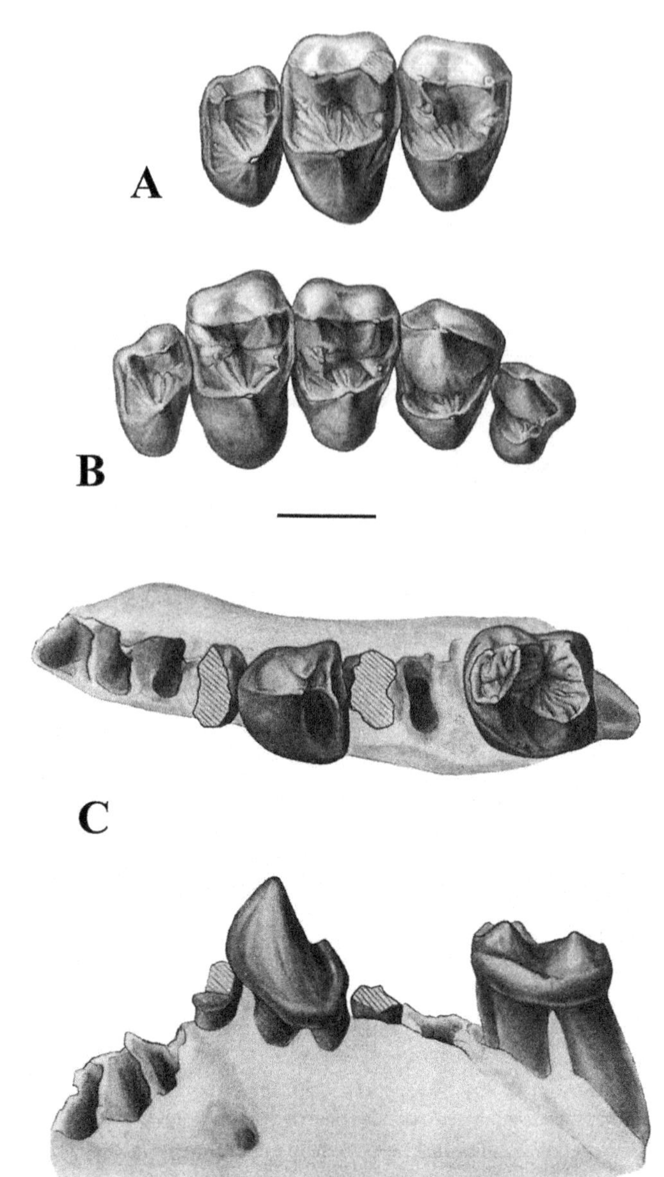

Fig. 5.14 *Strigorhysis*. (A) *S. rugosus*, right maxilla M^{1-3} (USNM 250553, holotype), (B) *S. bridgerensis*, right maxilla P^3–M^3 (USNM 250556, holotype), in occlusal views. (C) *S. bridgerensis*, left dentary P_4, M_2 (USNM 250559) in occlusal and lateral views. Scale bar represents 2 mm. Figures from Bown & Rose (1987).

TYPE SPECIMEN USNM 250553, right maxilla with M^{1-3}
AGE AND GEOGRAPHIC RANGE Early middle Eocene, Wyoming
ANATOMICAL DEFINITION
Strigorhysis rugosus differs from *S. bridgerensis* in having less distinct M^{1-2} conules and in lacking an M^2 metaconule; differs from both *S. bridgerensis* and *S. huerfanensis* in having more crenulated enamel and further differs from *S. huerfanensis* in being smaller.

SPECIES *Strigorhysis huerfanensis* Bown & Rose, 1987
TYPE SPECIMEN AMNH 55218, left dentary with P_4–M_3 and right maxilla with P^{3-4}, M^{2-3}

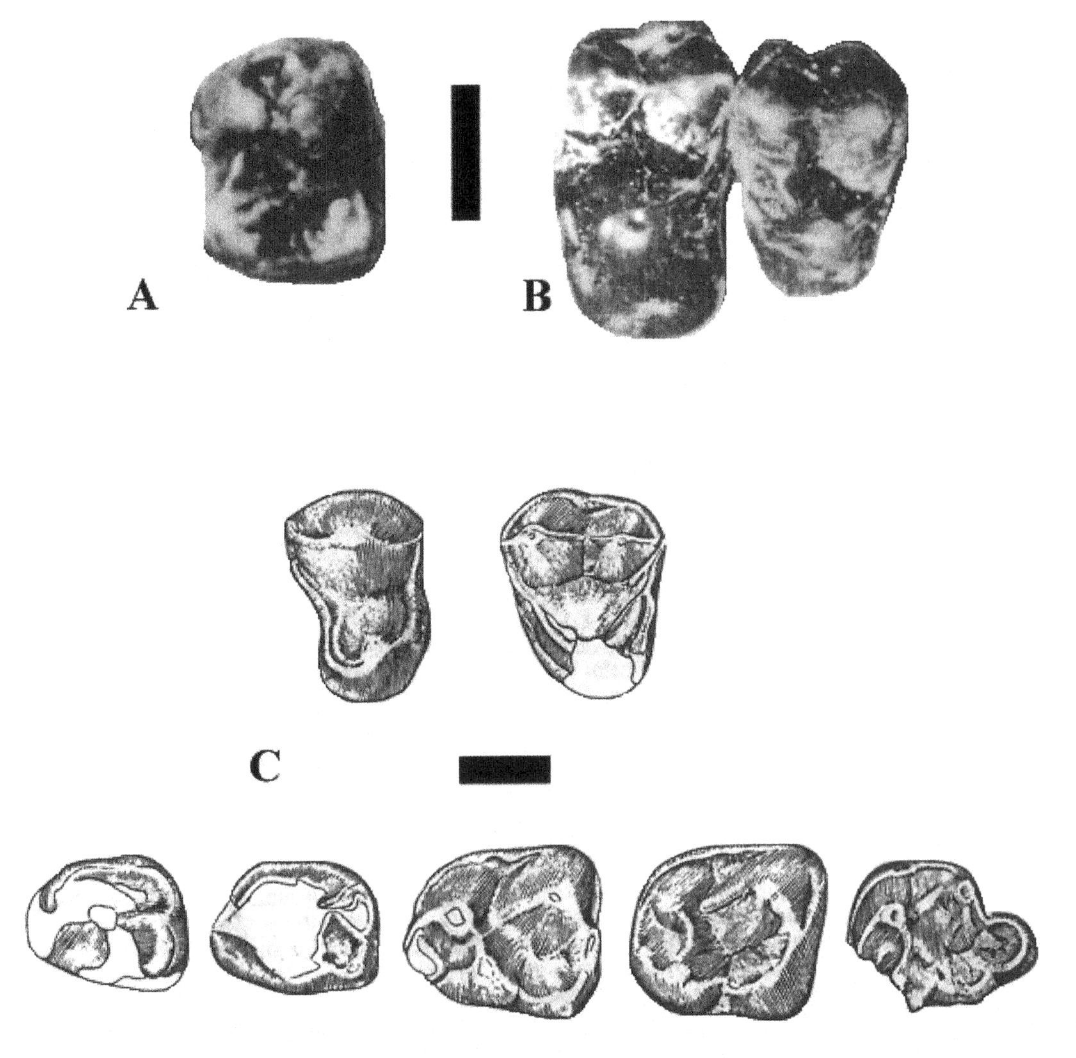

Fig. 5.15 (A) *Gazinius amplus*, left M_2 (USNM 250555). (B) *Gazinius amplus*, right maxilla M^{1-2} (USNM 250554, holotype). (C) *Yaquius travisi*, right P^4 (LACM 40240), left M^3 (LACM 40199), left P_3 (LACM 40001, reversed), right P_4 (LACM 40244), right M_1 (LACM 40202), right M_2 (LACM 40201, holotype) and right M_3 (LACM 40243), in occlusal views. Scale bars represent 2 mm. Figures adapted from Bown (1979) and Mason (1990).

AGE AND GEOGRAPHIC RANGE Earliest middle Eocene, Colorado

ANATOMICAL DEFINITION

Strigorhysis huerfanensis is larger than either of the other two known species of *Strigorhysis*, with M^3 larger relative to M^2; further differs from *S. rugosus* in having less crenulated molar basin enamel and from *S. bridgerensis* in having a taller P_4, a broader M_2 talonid, and a much more expanded M_3 hypoconulid lobe.

GENUS *Gazinius* Bown, 1979 (Fig. 5.15A–B)

INCLUDED SPECIES *G. amplus*, *G. bowni*

TYPE SPECIES *Gazinius amplus* Bown, 1979

TYPE SPECIMEN USNM 250554, right maxilla with M^{1-2}

AGE AND GEOGRAPHIC RANGE Early middle Eocene, Wyoming

ANATOMICAL DEFINITION

Gazinius is an extremely poorly known anaptomorphine represented by two specimens from the Aycross Formation and three specimens from the Bridger Formation in the southern Green River Basin. *Gazinius* differs from most other anaptomorphines in the following features (Bown, 1979): M^2 protocone centrally placed, massive and bulbous; M^2 lacking both conules and a postprotocingulum; M^2 greatly expanded lingually, more so than in *Strigorhysis*; M_2 trigonid compressed with low and bulbous cusps; distinct cristids (except for paracristid) absent on M_2; M_2 with short talonid, tall hypoconid, low entoconid and hypoconulid absent; the talonid basin of M_2 has weakly rugose enamel. *Gazinius amplus* is a large anaptomorphine but *G. bowni* is quite small (Bown, 1979; Gunnell, 1995*a*).

SPECIES *Gazinius bowni* Gunnell, 1995

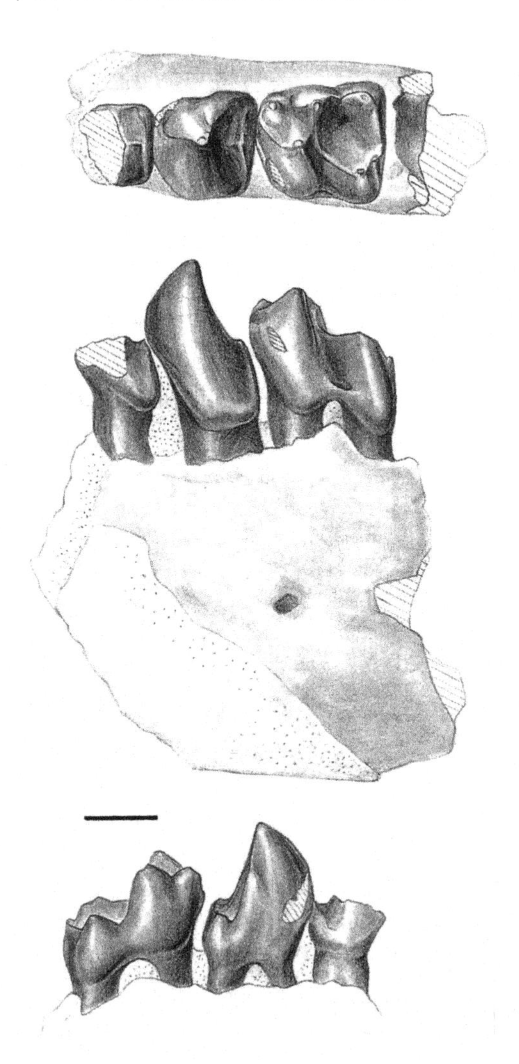

Fig. 5.16 *Tatmanius szalayi*. Left dentary P_3 (talonid), P_4–M_1 (USGS 21654, holotype) in occlusal, lateral and medial views. Scale bar represents 1 mm. Figures from Bown & Rose (1991).

TYPE SPECIMEN UM 31560, left M^2
AGE AND GEOGRAPHIC RANGE Middle Eocene, Wyoming
ANATOMICAL DEFINITION
Gazinius bowni differs from *G. amplus* in being 25% smaller, in having a well-developed postprotocingulum and postprotocrista, a weak paraconule, the protocone tallest molar cusp, and the trigon basin buccolingually restricted.

GENUS *Tatmanius* Bown & Rose, 1991 (Fig. 5.16)
INCLUDED SPECIES *T. szalayi*

TYPE SPECIES *Tatmanius szalayi* Bown & Rose, 1991
TYPE SPECIMEN USGS 21654, left dentary with P_3–M_1
AGE AND GEOGRAPHIC RANGE Early Eocene, Wyoming
ANATOMICAL DEFINITION
Tatmanius is represented by a single specimen from the Willwood Formation in the central Bighorn Basin, Wyoming (Bown & Rose, 1991). It is a relatively small anaptomorphine similar in size to *Anemorhysis*. It differs from all other anaptomorphines in possessing the following suite of character states (Bown & Rose, 1991): P_4 high-crowned and basally inflated but lacking a distinct paraconid or metaconid; P_4 root is bilobed being single-rooted on the buccal side and double-rooted on the lingual side (also characteristic of *Trogolemur*). The mandible of *Tatmanius* is very deep anteriorly indicating that it had a large I_1 as in *Trogolemur*, *Sphacorhysis* and *Pseudotetonius*. It probably descended from the *Tetonius–Pseudotetonius* lineage (Bown & Rose, 1991).

Tribe Trogolemurini

GENUS *Trogolemur* Matthew, 1909 (Fig. 5.3B, E–F)
INCLUDED SPECIES *T. amplior*, *T. fragilis*, *T. myodes*

TYPE SPECIES *Trogolemur myodes* Matthew, 1909
TYPE SPECIMEN AMNH 12599, right dentary with P_2–M_3
AGE AND GEOGRAPHIC RANGE Middle Eocene, Wyoming, Nevada (Emry, 1990) and Saskatchewan (Storer, 1990)
ANATOMICAL DEFINITION
Trogolemur has a lower dental formula of 2–1.1.2.3. *Trogolemur* is among the smallest known anaptomorphines, with lower first molar lengths averaging 1.8 mm. The lower central incisor is greatly enlarged and very procumbent, while I_2 (where present)–P_3 are very small with shallow roots. The anterior teeth posterior to the large I_1 are crowded together and overhang one another. The lower fourth premolar is small with a weak paraconid, a weak to absent metaconid, a protoconid that does not project above the toothrow and a very short talonid. The paraconid is distinct on M_1 but small to absent on M_{2-3}, where trigonids are anteroposteriorly compressed. The M_3 is relatively long and narrow with a well-developed, relatively broad hypoconulid. The P^4 has a distinct but low and anteriorly placed protocone and a high, centrally placed paracone. There is a relatively strong postprotocrista extending from the protocone that posteriorly borders a moderately developed trigon basin. Upper molars have prominent protocones, high, relatively narrow postprotocingula, weak paraconules and weak to absent metaconules. Mesostyles are absent on all molars. The mandible of *Trogolemur* is very deep anteriorly to accommodate the enlarged I_1 (Szalay, 1976; Emry, 1990; Gunnell, 1995*a*).

SPECIES *Trogolemur amplior* Beard *et al.*, 1992
TYPE SPECIMEN CM 40096, left dentary with M_{1-2}
AGE AND GEOGRAPHIC RANGE Early middle Eocene, Wyoming
ANATOMICAL DEFINITION
Trogolemur amplior is larger than *T. myodes* and *T. fragilis*. It further differs from *T. myodes* in having relatively wider and deeper molar talonid basins.

SPECIES *Trogolemur fragilis* Beard *et al.*, 1992
TYPE SPECIMEN CM 41152, right dentary with M_{2-3}

AGE AND GEOGRAPHIC RANGE Early middle Eocene, Wyoming
ANATOMICAL DEFINITION
Trogolemur fragilis differs from *T. myodes* in being slightly smaller and from *T. amplior* in being much smaller. Further differs from *T. myodes* in having M_3 with paraconid less lingual, a distinct valley separating the protoconid and metaconid, and a narrower talonid basin and hypoconulid lobe (Beard *et al.*, 1992*a*).

GENUS *Sphacorhysis* Gunnell, 1995 (Fig. 5.3C)
INCLUDED SPECIES *S. burntforkensis*

TYPE SPECIES *Sphacorhysis burntforkensis* Gunnell, 1995
TYPE SPECIMEN UM 30966, left dentary with P_4–M_3
AGE AND GEOGRAPHIC RANGE Early middle Eocene, Wyoming
ANATOMICAL DEFINITION
Sphacorhysis is represented by a single specimen from the upper Bridger Formation, southern Green River Basin, Wyoming (Gunnell, 1995*a*). It is the sister taxon of *Trogolemur*, differing from that taxon in the following ways: I_1 relatively less enlarged, I_2 and C smaller and less vertically implanted; P_4 with a very small metaconid and lacking a hypoconid or cristid obliqua; M_2 larger than M_1; lower molar talonids very shallow with moderately rugose enamel; M_3 trigonid very low relative to talonid; M_3 hypoconulid short and broad; and M_3 hypoconid massive and basally inflated, connected to the base of the protoconid by a short, straight cristid obliqua.

Subfamily Omomyinae

Omomyines differ from anaptomorphines (Szalay, 1976; Bown & Rose, 1987) in having: relatively small I_1s that are moderately procumbent to vertically implanted; teeth between I_1 and P_4 usually not as reduced in size or number; P_4 normally not inflated and buccally distended (*Uintanius* and *Jemezius* being notable exceptions); molar cusps more acute, less basally inflated, and positioned on the margins of the teeth; upper molar trigons more open with weaker postprotocingula; lower molar talonid basins relatively deep; upper and lower third molars not typically reduced. Omomyines differ from microchoerines in: less inflated mastoid regions (where known); I_{1-2} less enlarged; P_2 often retained; and molar cusps more acute, less basally inflated, and positioned on the margins of the teeth.

Tribe Omomyini

GENUS *Omomys* Leidy, 1869 (Fig. 5.17A)
INCLUDED SPECIES *O. carteri*, *O. lloydi*

TYPE SPECIES *Omomys carteri* Leidy, 1869
TYPE SPECIMEN ANS 10335, right dentary with P_{3-4}, M_2
AGE AND GEOGRAPHIC RANGE Middle Eocene, Wyoming,

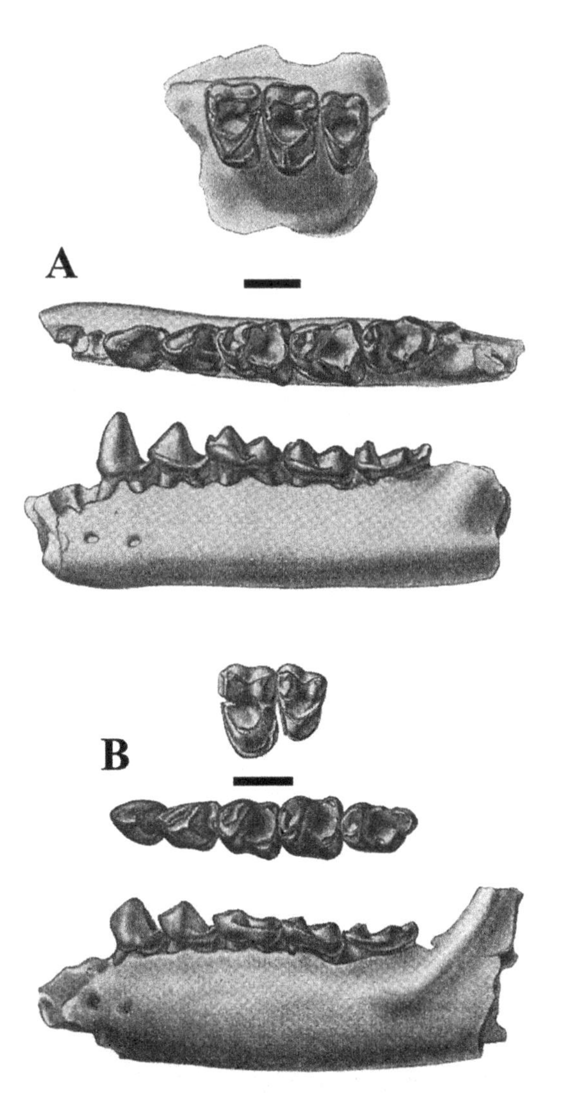

Fig. 5.17 Omomyini. (A) *Omomys carteri*, left maxilla M^{1-3} (YPM 11854) in occlusal view, left dentary P_3–M_3 (USNM 13289) in occlusal and lateral views. (B) *Chumashius balchi*, left M^{2-3} (LACM 1394) in occlusal view, left dentary P_3–M_3 (LACM 1391, holotype) in occlusal and lateral views. Scale bars represent 2 mm. Figures adapted from Gazin (1958).

Utah, Colorado and Texas; middle and late Eocene, Saskatchewan
ANATOMICAL DEFINITION
Omomys was the first omomyid and first primate described from North America (Leidy, 1869). Despite over 130 years of field work, *Omomys* was only known from dental remains until very recently (Rosenberger & Dagosto, 1992; Anemone *et al.*, 1997; Murphey *et al.*, 1998; Alexander & MacPhee, 1999*a*, 1999*b*; Anemone & Covert, 2000), when the first skull and postcranial remains were discovered. Dentally, *Omomys* differs from most other omomyines in having upper molars with lingually continuous cingula (although this characteristic is variable), well-developed pericones, distinct, cingular hypocones, and lacking postprotocingula; and in having a relatively tall P_3. Known postcranial remains include elongate tarsal elements as in *Arapahovius* and a long, slender, straight-shafted femur. The

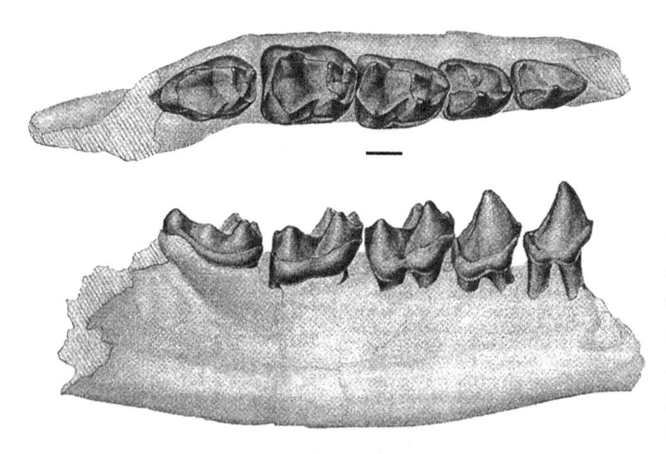

Fig. 5.18 *Steinius vespertinus*. Right dentary P_3–M_3 (USGS 25027) in occlusal and lateral views. Scale bar represents 1 mm. Figures from Bown & Rose (1991).

femur has a cylindrical head and an expansion of the articular surface onto the short, robust neck. The distal femur is anteroposteriorly deep. All of these indicate an arboreal animal with powerful leaping capabilities (Anemone *et al.*, 1997; Anemone & Covert, 2000). The one reported skull of *Omomys* has enlarged, but relatively laterally facing orbits and a more elongate rostrum than either *Shoshonius* or *Tetonius* (Alexander & MacPhee, 1999*b*). The enlarged orbits suggest that *Omomys*, like other known omomyids, was probably nocturnal.

Szalay (1976) synonymized *Omomys sheai* (Gazin, 1962) with *Utahia kayi*. *Omomys sheai* lacks the compressed molar trigonids of *Utahia* and retains distinctive paraconids on M_{2-3}. *Omomys sheai* is of the same size and similar morphology to *Loveina zephyri* and we believe that it can be assigned to that taxon (Stucky, 1984).

SPECIES *Omomys lloydi* Gazin, 1958
TYPE SPECIMEN CM 6417, left P_4–M_1
AGE AND GEOGRAPHIC RANGE Early middle Eocene, Utah
ANATOMICAL DEFINITION
Omomys lloydi differs from *O. carteri* in being somewhat smaller. No consistent morphological features separate these two species. *Omomys lloydi* is best represented at Powder Wash in Utah but is also known from South Pass in southwestern Wyoming.

GENUS *Steinius* Bown & Rose, 1984 (Fig. 5.18)
INCLUDED SPECIES *S. annectens, S. vespertinus*

TYPE SPECIES *Steinius vespertinus* Matthew, 1915
TYPE SPECIMEN AMNH 16835, left dentary with M_{1-3}
AGE AND GEOGRAPHIC RANGE Early Eocene, Wyoming
ANATOMICAL DEFINITION
Steinius is one of the most primitive omomyids known (Bown & Rose, 1984, 1987; Rose & Bown, 1991) even though it is not among the earliest occurring. The lower dental formula of *S. vespertinus* is 2.1.4.3 (Rose & Bown, 1991). The lower central incisor is only slightly larger than the canine and both are larger than I_2. Both P_1 and P_2 are small and single-rooted while P_{3-4} are simple and uninflated, P_3 being relatively tall and unreduced and P_4 having a small, low metaconid. Upper and lower molars are more omomyine-like in having rather acute cusps that are set on the margins of the teeth and lacking basal inflation. *Steinius* differs from *Omomys* (a taxon which is otherwise quite similar) in retaining P_1, in having a less projecting P_3, smaller molar paraconids with the M_1 paraconid less lingual and M_{2-3} paraconids more lingual than in *Omomys*. *Steinius* forms a plausible ancestral taxon for many later occurring omomyines.

Matthew (1915*a*) initially placed the type species questionably in *Omomys*, highlighting their similarity, but specimens found subsequently (Bown & Rose, 1984; Rose & Bown, 1991) show that "*Omomys?*" *vespertinus* differs from *Omomys* in retaining P_1 and a somewhat reduced M_3. *Steinius* appears to be linked to uintaniins as well through the Wasatchian genus *Jemezius*.

SPECIES *Steinius annectens* Bown & Rose, 1991
TYPE SPECIMEN USGS 28327, right dentary with P_4–M_2
AGE AND GEOGRAPHIC RANGE Late early Eocene, Wyoming
ANATOMICAL DEFINITION
Steinius annectens differs from *S. vespertinus* in being 20% to 30% larger and in having molars that are slightly basally inflated buccally.

GENUS *Chumashius* Stock, 1933 (Fig. 5.17B)
INCLUDED SPECIES *C. balchi*

TYPE SPECIES *Chumashius balchi* Stock, 1933
TYPE SPECIMEN LACM 1391, left dentary with P_3–M_3
AGE AND GEOGRAPHIC RANGE Middle Eocene, California; *Chumashius* may also be represented from the middle Eocene, Wyoming
ANATOMICAL DEFINITION
Chumashius is represented by only a small number of specimens, mostly from California. A few isolated teeth from the Badwater Creek area, Wind River Basin, Wyoming (Robinson, 1968) may represent *Chumashius* (Krishtalka, 1978; Lillegraven, 1980). *Chumashius* is very similar to *Omomys* differing only in lacking distinct pericones and hypocones on the upper molars and in having a relatively larger canine, and relatively lower P_{3-4} (Szalay, 1976). The relatively high variability of tooth morphology exhibited in *Omomys* suggests that *Chumashius* could easily be accommodated within that genus.

Tribe Washakiini

GENUS *Washakius* Leidy, 1873 (Fig. 5.19C)
INCLUDED SPECIES *W. insignis, W. izetti, W. laurae, W. woodringi*

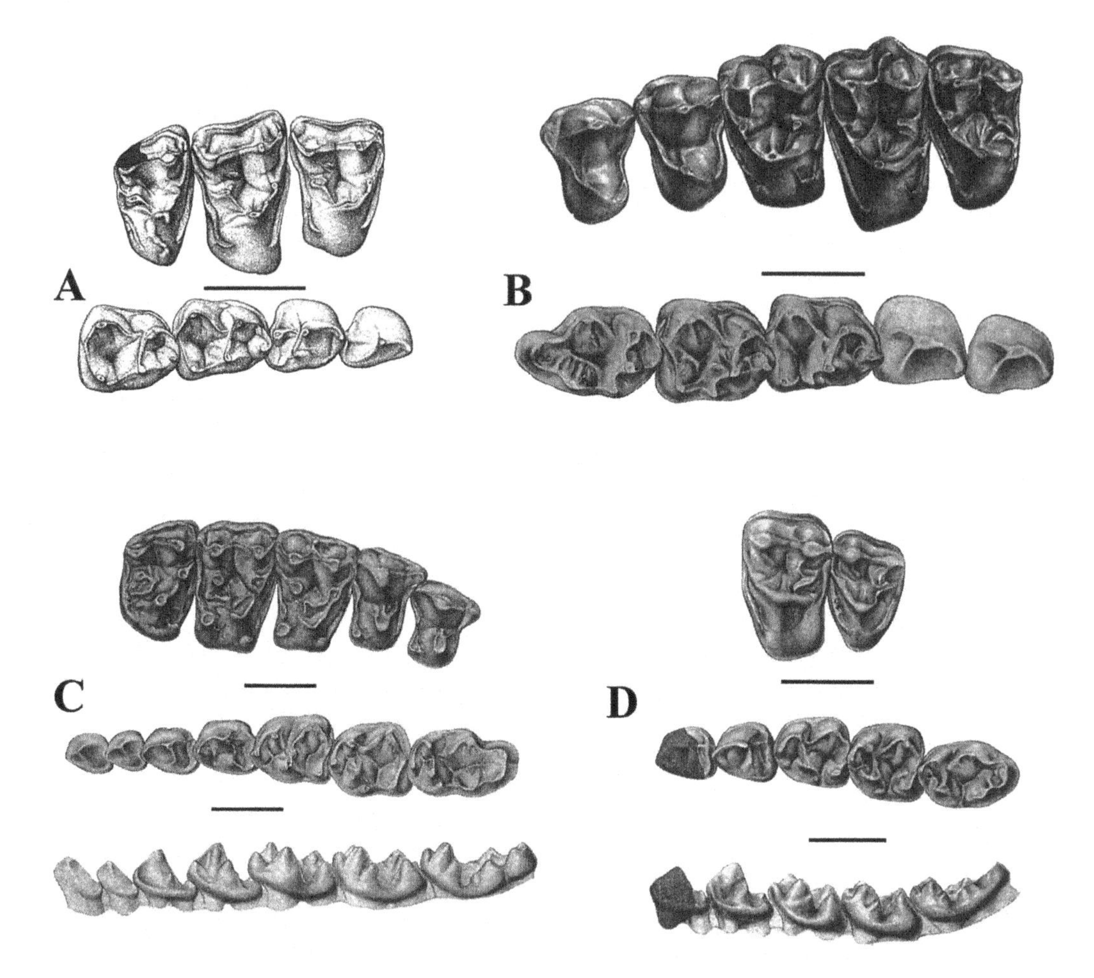

Fig. 5.19 Washakiini. (A) *Loveina zephyri*, right maxilla M^{1-3} (MCZ 3495), left dentary P_3–M_2 (AMNH 32517, P_3–M_1, holotype and 17554) in occlusal views. (B) *Shoshonius cooperi*, left maxilla P^3–M^3 (AMNH 14664, holotype), left dentary P_3–M_3 (CM 22105 and AMNH 14665) in occlusal views. (C) *Washakius insignis*, right maxilla P^3–M^3 (AMNH 55665, 55672, 56037, YPM 13236-1 and 15019) in occlusal view, right dentary C–M_3 (AMNH 12040, 55665, UW 1645) in occlusal and medial views. (D) *Dyseolemur pacificus*, left M^{2-3} (LACM 1520 and 1591) in occlusal view, right dentary P_3–M_3 (LACM 1395, holotype) in occlusal and medial views. Scale bars represent 2 mm. Figures adapted from Szalay (1976).

TYPE SPECIES *Washakius insignis* Leidy, 1873
TYPE SPECIMEN ANS unnumbered, right dentary with M_{2-3}
AGE AND GEOGRAPHIC RANGE Middle Eocene, Wyoming, Colorado and California
ANATOMICAL DEFINITION
Washakius is a geographically widespread and diverse omomyine taxon. Honey (1990) has rediagnosed washakiin omomyines and recognized *Washakius* based on the following character states: differs from *Loveina* in having M^{1-2} with second metaconule present, M^{1-2} metacone more buccally inflated, P_{3-4} less posteriorly distended, M_{1-3} talonids widely open lingually, and M_3 with a double-cusped hypoconulid; differs from *Shoshonius* in having: M^{1-3} without a strong mesostyle, M_2 paraconid more anterior with shorter paracristid and M_{1-3} talonid notch always present; differs from *Dyseolemur* in having: upper molars with two metaconules and more peripheral protocones, P_4 with a weaker buccal cingulid and a stronger paraconid and metaconid, M_{1-2} talonids relatively wider with more marginally placed cusps and M_3 with a double-cusped hypoconulid.

SPECIES *Washakius woodringi* Stock, 1938
TYPE SPECIMEN LACM 2233, right maxilla with M^{1-2}
AGE AND GEOGRAPHIC RANGE Late middle Eocene, California
ANATOMICAL DEFINITION
Washakius woodringi differs from *W. insignis* only in having more acute cusps and a better developed postprotocingulum; differs from *W. izetti* in having M^{1-2} with stronger hypocone and more distinct double metaconule, lower molars with stronger metastylids and P_4 distal cingulum cuspule stronger; differs from *W. laurae* in being smaller and in having better developed enamel crenulations.

SPECIES *Washakius izetti* Honey, 1990
TYPE SPECIMEN USGS 1522, left M_2
AGE AND GEOGRAPHIC RANGE Early middle Eocene, Colorado and Wyoming
ANATOMICAL DEFINITION
Washakius izetti differs from *W. insignis* in having P^4 with weak ectocingulum, parastyle, and protocone, and lacking a hypocone; upper molars with weak hypocone; lower molars with much weaker metastylids. *Washakius izetti* differs from *W. woodringi* in having M^{1-2} with weak hypocone and incipient double metaconule (much more strongly developed in *W. insignis* and *W. woodringi*); lower molars with weaker metastylids, P_4 distal cingulum cuspule weaker. *W. izetti* differs from *Washakius laurae* in having a more distinct P_4 paraconid and metaconid and a less distinct M_1 metastylid, and lower molars with more buccally placed paraconids.

SPECIES *Washakius laurae* Simpson, 1959
TYPE SPECIMEN AMNH 55672, left dentary with M_{1-3}
AGE AND GEOGRAPHIC RANGE Early middle Eocene, Wyoming
ANATOMICAL DEFINITION
Washakius laurae differs from *W. insignis* in having the P_4 metaconid high and closely appressed to protoconid, more lingually placed molar paraconids (trigonid closed), straight cristid obliquae, and in being somewhat smaller; differs from *W. woodringi* in being larger and in having less well-developed enamel crenulations; differs from *W. izetti* in having a less distinct P_4 paraconid and metaconid and a more distinct M_1 metastylid, and lower molars with more lingually placed paraconids. *Washakius laurae* was originally described as a possible species of *Shoshonius* by Simpson (1959). Szalay (1976) synonymized *W. laurae* with *W. insignis*. *Washakius laurae* was recognized as a distinct species of *Washakius* by Gunnell *et al.* (1992).

GENUS *Shoshonius* Granger, 1910 (Fig. 5.19B)
INCLUDED SPECIES *S. bowni*, *S. cooperi*

TYPE SPECIES *Shoshonius cooperi* Granger, 1910
TYPE SPECIMEN AMNH 14664, right maxilla with P^3–M^3
AGE AND GEOGRAPHIC RANGE Early to middle Eocene, Wyoming; early Eocene, Colorado
ANATOMICAL DEFINITION
Shoshonius, like *Omomys*, was until recently only represented by dental remains. Dentally, *Shoshonius* differs from other washakiins in the following ways (Honey, 1990): from *Loveina* in having upper molar mesostyles and distinct lower molar metastylids; from *Washakius* in having strong upper molar mesostyles, M_2 paraconid more lingual with longer paracristid, and lower molar talonid notch usually absent; from *Dyseolemur* in having upper molars with large mesostyles, M^2 hypocone less inflated, M_{1-3} paraconid more lingual with longer paracristid, M_{1-3} talonids relatively wider with more marginally placed cusps. Cranially (Beard *et al.*, 1991; Beard & MacPhee, 1994) *Shoshonius* has relatively large orbits, lacks postorbital closure, has a reduced rostrum, lacks a tubular ectotympanic, and has a posterior carotid foramen ventrolaterally placed on the auditory bulla. Postcranially (Dagosto *et al.*, 1999), *Shoshonius* resembles other known omomyids in having a relatively long femur with proximally placed second and third trochanters, a deep distal femur and elongate tarsal elements.

SPECIES *Shoshonius bowni* Honey, 1990
TYPE SPECIMEN USGS 2020, right maxilla with M^{1-3}
AGE AND GEOGRAPHIC RANGE Early middle Eocene, Wyoming
ANATOMICAL DEFINITION
Shoshonius bowni differs from *S. cooperi* in having a weaker metacone on P^3, weaker and less cuspate mesostyles on M^{1-2}, M^{1-2} paraconule and metaconule more inflated, taller, with better developed pre- and postmetaconule cristae, M^3 with larger paraconule, and M^{1-2} with more constricted trigon basins.

GENUS *Dyseolemur* Stock, 1934 (Fig. 5.19D)
INCLUDED SPECIES *D. pacificus*

TYPE SPECIES *Dyseolemur pacificus* Stock, 1934
TYPE SPECIMEN LACM 1395, right dentary with P_4–M_3
AGE AND GEOGRAPHIC RANGE Middle Eocene, California
ANATOMICAL DEFINITION
Dyseolemur differs from other washakiins in the following ways (Szalay, 1976): from *Shoshonius*, in lacking upper molar mesostyles and in having an expanded M^2 distolingual cingulum (defined as a "hypocone" by Szalay, 1976); from *Washakius*, in lacking a double upper molar metaconule, in having very buccally placed lower molar paraconids, and in having a relatively smaller M_3; differs from *Loveina* in possessing lower molar metastylids.

GENUS *Loveina* Simpson, 1940 (Fig. 5.19A)
INCLUDED SPECIES *L. minuta*, *L. wapitiensis*, *L. zephyri*

TYPE SPECIES *Loveina zephyri* Simpson, 1940
TYPE SPECIMEN AMNH 32517, left dentary with P_3–M_1
AGE AND GEOGRAPHIC RANGE Late early Eocene, Wyoming and Colorado
ANATOMICAL DEFINITION
Loveina is the most primitive member of the washakiin clade (Szalay, 1976; Honey, 1990). It differs from *Shoshonius*, *Dyseolemur*, and *Washakius* in lacking or having only incipient metastylids on lower molars (Gunnell *et al.*, 1992). *Loveina* further differs from *Washakius* in having upper molars that lack a double metaconule, hypocone and pericone. *Loveina* lacks a mesostyle, unlike *Shoshonius*. Bown & Rose (1984) moved *Notharctus minutus* Loomis, 1906 into *Loveina*. Stucky (1984) advocated moving both *Omomys minutus* and *O. sheai* to *Loveina*, a practice we have followed here except that we believe that *O. sheai* and *Loveina zephyri* represent the same taxon.

SPECIES *Loveina minuta* (Loomis, 1906)
TYPE SPECIMEN AC 3365, right dentary with M_{1-3}
AGE AND GEOGRAPHIC RANGE Late early Eocene, Wyoming
ANATOMICAL DEFINITION
Loveina minutus differs from *L. zephyri* and *L. wapitiensis* in being smaller, and in having a more closed M_2 trigonid, and more centrally placed hypoconulids on M_{1-2}.

SPECIES *Loveina wapitiensis* Gunnell *et al.*, 1992
TYPE SPECIMEN YPM PU 17317, right dentary with P_3–M_3
AGE AND GEOGRAPHIC RANGE Late early Eocene, Wyoming
ANATOMICAL DEFINITION
Loveina wapitiensis differs from *L. zephyri* in having P_3 less buccolingually inflated with a lower, relatively smaller paraconid and a less bulging metaconid; and P_4 with a less differentiated paraconid and metaconid and a less distinct metacristid; differs from *L. zephyri* and *L. minutus* in having M_{2-3} with tiny metastylids (absent in other species); further differs from *L. minutus* in being larger.

Tribe Utahiini

GENUS *Utahia* Gazin, 1958 (Fig. 5.20C)
INCLUDED SPECIES *U. kayi*

TYPE SPECIES *Utahia kayi* Gazin, 1958
TYPE SPECIMEN CM 6488, right dentary with P_3–M_3
AGE AND GEOGRAPHIC RANGE Late early Eocene, Wyoming; middle Eocene, Utah
ANATOMICAL DEFINITION
Utahia appears closely related to *Stockia* (Szalay, 1976) and perhaps slightly more distantly related to *Chipetaia* (Rasmussen, 1996). All three taxa have relatively anteroposteriorly constricted lower molar trigonids (M_{2-3} especially) and relatively enlarged talonid basins. *Utahia* differs from *Stockia* in having a relatively smaller P_4, higher molar trigonids and less crenulated molar enamel (Szalay, 1976); it differs from *Chipetaia* in having smaller and more anteriorly placed M_{1-2} entoconids, a more lingual cristid obliqua on M_{2-3}, and more lingual M_2 paraconids, in lacking a distinct P_4 metaconid, and in having less crenulated molar enamel (Rasmussen, 1996).

GENUS *Stockia* Gazin, 1958 (Fig. 5.20B)
INCLUDED SPECIES *S. powayensis*

TYPE SPECIES *Stockia powayensis* Gazin, 1958
TYPE SPECIMEN LACM 2235, right dentary with M_{1-3}
AGE AND GEOGRAPHIC RANGE Middle Eocene, California
ANATOMICAL DEFINITION
As noted for *Utahia*, *Stockia* is quite similar to that taxon and to *Chipetaia*. *Stockia* also shares many similarities with *Asiomomys* (Beard & Wang, 1991). *Stockia* differs from *Utahia* in having a relatively larger P_4, less elevated molar trigonids, and more crenulated molar enamel (Szalay, 1976). *Stockia* differs from *Chipetaia* in having less distinct M_{2-3} paraconids, less inflated metaconids, stronger premetacristids, more buccal cristid obliqua, and more closed talonid notches (Rasmussen, 1996). *Stockia* differs from *Asiomomys* in having relatively higher molar trigonids with protoconid and metaconid less distinctly separated, in having more distinct cusps and crests, and more crenulated molar enamel (Beard & Wang, 1991). Lillegraven (1980) suggested that *Stockia* was congeneric with *Omomys*, an idea rejected by Honey (1990) and Beard & Wang (1991). *Stockia*, *Utahia*, *Asiomomys* and *Chipetaia* appear relatively closely related, yet none of these taxa is well known and only *Chipetaia* is represented by any upper dental elements (Rasmussen, 1996). All of these taxa await much better material before certain taxonomic allocation will be possible.

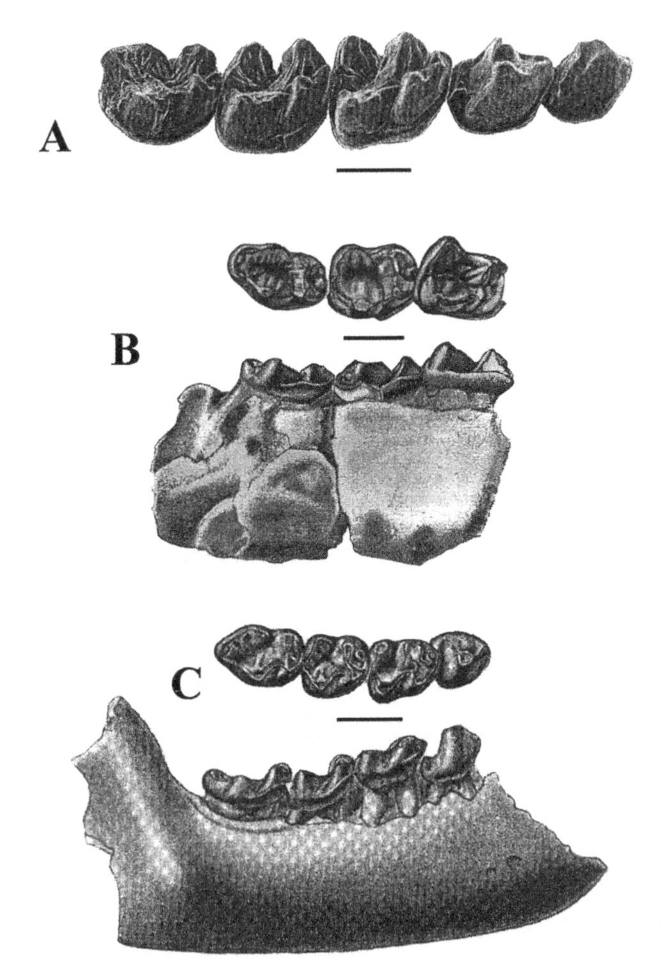

Fig. 5.20 Utahiina. (A) *Chipetaia lamporea*, right P_3–M_3 (CM 69800, M_{2-3}, holotype and 69804) in occlusal view. (B) *Stockia powayensis*, right dentary M_{1-3} (LACM 2234, holotype) in occlusal and lateral views. (C) *Utahia kayi*, right dentary P_4–M_3 (CM 6488, holotype) in occlusal and lateral views. Scale bars represent 2 mm. Figures adapted from Gazin (1958) and Rasmussen (1996).

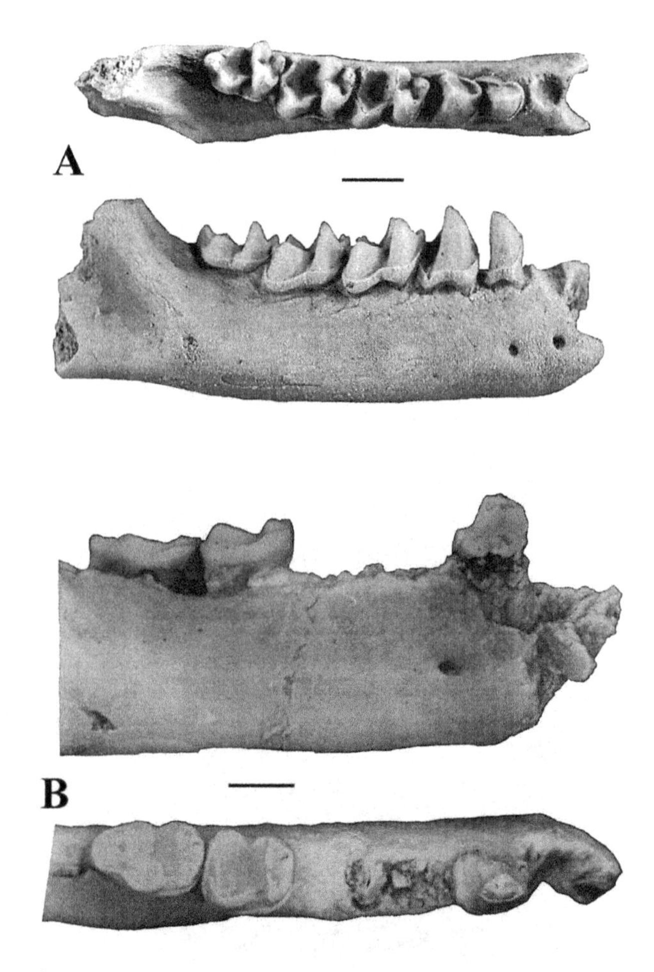

Fig. 5.21 (A) *Xanthorhysis tabrumi*, left dentary (reversed) P_3–M_3 (IVPP V12063, holotype) in occlusal and lateral views. (B) *Asiomomys changbaicus*, right dentary P_3, M_{2-3} (IVPP V8802, holotype) in lateral and occlusal views. Scale bars represent 2 mm. Figures courtesy of K.C. Beard.

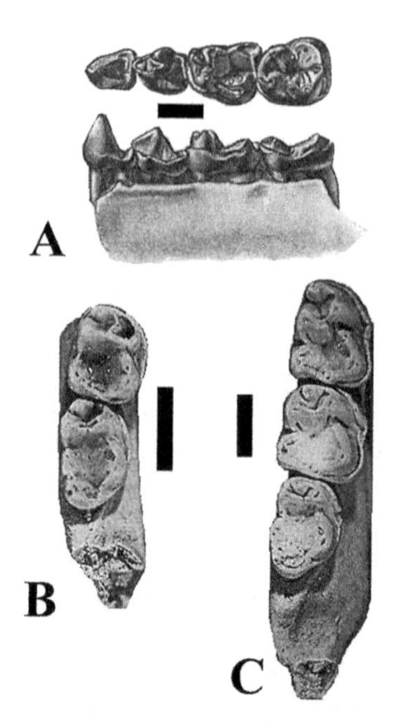

Fig. 5.22 Mytoniina. (A) *Ourayia uintensis*, left dentary P_3–M_2 (AMNH 1899, holotype) in occlusal and lateral views. (B) *Wyomomys bridgeri*, right dentary M_{2-3} (UM 98874, holotype) in occlusal view. (C) *Ageitodendron matthewi*, right dentary M_{1-3} (UM 30924, holotype) in occlusal view. Scale bars represent 2 mm. Figures adapted from Gazin (1958) and Gunnell (1995*a*).

GENUS *Chipetaia* Rasmussen, 1996 (Fig. 5.20A)
INCLUDED SPECIES *C. lamporea*

TYPE SPECIES *Chipetaia lamporea* Rasmussen, 1996
TYPE SPECIMEN CM 69800, right M_{2-3}
AGE AND GEOGRAPHIC RANGE Middle Eocene, Utah
ANATOMICAL DEFINITION
Chipetaia is most closely comparable with *Utahia* and *Stockia* among known omomyids (Rasmussen, 1996). *Chipetaia* differs from *Utahia* in: having relatively larger and more distally placed M_{1-2} entoconids; M_{2-3} with a more buccally placed cristid obliqua; molar paraconids more centrally placed; P_4 with a distinct metaconid and a broader distal shelf; and in having more rugose molar enamel. *Chipetaia* differs from *Stockia* in having M_{2-3} with: more distinct paraconids, more inflated metaconids, weaker premetacristid, more lingually placed cristid obliqua, and a more open talonid notch. *Chipetaia* also differs from other omomyids in having M_{2-3} with compressed trigonids and reduced paraconids.

GENUS *Asiomomys* Wang & Li, 1990 (Fig. 5.21B)
INCLUDED SPECIES *A. changbaicus*

TYPE SPECIES *Asiomomys changbaicus* Wang & Li, 1990
TYPE SPECIMEN IVPP V8802, right dentary with P_3, M_{2-3}
AGE AND GEOGRAPHIC RANGE Middle Eocene of China
ANATOMICAL DEFINITION
Asiomomys is one of two known omomyids from China and is represented by a single specimen. *Asiomomys* is very similar to *Stockia* (Beard & Wang, 1991), differing only in: having relatively less elevated molar trigonids with protoconid and metaconid more distinctly separated, less distinct cusps and crests, and weaker molar crenulation.

GENUS *Wyomomys* Gunnell, 1995 (Fig. 5.22B)
INCLUDED SPECIES *W. bridgeri*

TYPE SPECIES *Wyomomys bridgeri* Gunnell, 1995
TYPE SPECIMEN UM 98874, right dentary with M_{2-3}
AGE AND GEOGRAPHIC RANGE Middle Eocene, Wyoming
ANATOMICAL DEFINITION
Wyomomys is the most primitive known member of the Utahiini (= Ourayiini of Gunnell, 1995*a*). It is similar to *Omomys* but can be differentiated in the following ways: M_{2-3} trigonids compressed but with distinct paraconid separated from metaconid by sharply incised valley; paracristids long and relatively tall; hypoconids buccally inflated; M_2 with robust cristid obliqua and heavy, buccally extended ectocingulid; M_3 with relatively longer

entocristid. *Wyomomys* differs from *Ageitodendron* and *Ourayia* in being smaller, and in having M_{2-3} trigonids with distinct paraconid separated from metaconid by sharply incised valley, and M_3 as large or larger than M_2, and in lacking rugose talonid basin enamel and an inflected talonid notch.

GENUS *Ageitodendron* Gunnell, 1995 (Fig. 5.22C)
INCLUDED SPECIES *A. matthewi*

TYPE SPECIES *Ageitodendron matthewi* Gunnell, 1995
TYPE SPECIMEN UM 30924, right dentary with M_{1-3}
AGE AND GEOGRAPHIC RANGE Middle Eocene, Wyoming
ANATOMICAL DEFINITION
Ageitodendron is included in the subtribe Mytoniina with *Wyomomys* and *Ourayia* (Gunnell, 1995*a*). *Ageitodendron* differs from *Wyomomys* in having: M_{2-3} trigonids small, undifferentiated paraconids and a weakly twinned metaconid on M_2; M_3 relatively smaller than M_2; weakly rugose talonid basin enamel; an inflected talonid notch and a very robust ectocingulid that is notched basal to the hypoflexid; and in being approximately 25% larger. *Ageitodendron* differs from *Ourayia* in having: M_{1-2} with small but distinct hypoconulids; M_2 with weakly twinned metaconid; M_3 with relatively longer hypoconulid; molar talonid notch with very well-developed inflection of entocristid and postmetacristid; molar talonids with weaker development of rugose enamel; molar ectocingulids relatively less robust; and in being approximately 25% smaller.

GENUS *Ourayia* Gazin, 1958 (Fig. 5.22A)
INCLUDED SPECIES *O. hopsoni, O. uintensis*

TYPE SPECIES *Ourayia uintensis* Osborn, 1895
TYPE SPECIMEN AMNH 1899, left dentary with P_3–M_2
AGE AND GEOGRAPHIC RANGE Middle Eocene, Utah, Texas and California
ANATOMICAL DEFINITION
The taxonomic history of *Ourayia* is relatively long and confusing (Osborn, 1902; Wortman, 1903–1904; Gazin, 1958; Simons, 1961*b*; Robinson, 1968; Szalay, 1976; Krishtalka, 1978; Honey, 1990; Gunnell, 1995*a*; Rasmussen, 1996). Osborn originally named *Microsyops uintensis* (AMNH 1899) from the White River Pocket in Utah. Gazin (1958), recognizing the distinctiveness of this taxon, proposed a new genus, *Ourayia*, for this species.

We view *Ourayia* as the sister taxon to a clade formed by *Wyomomys* and *Ageitodendron* (Gunnell, 1995*a*). In turn the *Ourayia*, *Wyomomys*, *Ageitodendron* clade forms a sister-group relationship with a clade formed by *Utahia*, *Stockia* and *Chipetaia* (Rasmussen, 1996). The former clade (subtribe Mytoniina) can be differentiated from the latter clade (subtribe Utahiina) by the presence of: rounded and relatively shallow lower molar talonids; lower molars with very heavy, notched, buccally distended ectocingulids; hypoconid basally inflated, protoconid often basally inflated. Within mytoniinans, *Ourayia* differs from *Wyomomys* and *Ageitodendron* in being larger, having indistinct M_{2-3} paraconids, and having more rugose and complex molar enamel. It further differs from *Ageitodendron* in lacking M_{1-2} hypoconulids, and in having a talonid notch with a weaker inflection of the entocristid and postmetacristid, and a stronger ectocingulid.

SPECIES *Ourayia hopsoni* Robinson, 1968
TYPE SPECIMEN CM 12309, left M_2
AGE AND GEOGRAPHIC RANGE Late middle Eocene, Utah
ANATOMICAL DEFINITION
Ourayia hopsoni differs from *O. uintensis* in being smaller, and in having a centrally placed paraconid present on M_2, and a small and posteriorly placed metaconid on P_4. This species was originally placed in a new genus, *Mytonius*, by Robinson (1968). Szalay (1976) synonymized *Mytonius* with *Ourayia*, a practice we follow here, although unlike Szalay (1976), we believe that *O. hopsoni* is a valid species, distinct from *O. uintensis*.

Tribe Macrotarsiini

GENUS *Macrotarsius* Clark, 1941 (Fig. 5.23B–C)
INCLUDED SPECIES *M. jepseni, M. macrorhysis, M. montanus, M. roederi, M. siegerti*

TYPE SPECIES *Macrotarsius montanus* Clark, 1941
TYPE SPECIMEN CM 9592, right dentary with C, P_3–M_3
AGE AND GEOGRAPHIC RANGE Middle Eocene, Wyoming, Utah, Texas and California; late Eocene, Montana and Saskatchewan; middle Eocene, China
ANATOMICAL DEFINITION
Macrotarsius is one of the most geographically widespread omomyines known, being represented from deposits all along the Rocky Mountain corridor westward to California and across the Pacific into China (Clark, 1941; Robinson, 1968; Lillegraven, 1980; Kelly, 1990; Beard *et al.*, 1994). *Macrotarsius* is most closely related to *Hemiacodon* (Honey, 1990; Gunnell, 1995*a*) and can be differentiated from that taxon by the following character states: P^3 lingually broader with hypocone shelf better developed; upper molars with relatively weaker conules, moderate stylar shelves, and a well-developed mesostyle; upper molars more squared with lower, weaker hypocones and pericones; lower molar trigonids closed lingually and distally, and metaconid marginal and uninflated basally; and M_3 as broad as M_{1-2}.

SPECIES *Macrotarsius siegerti* Robinson, 1968
TYPE SPECIMEN CM 15122, right P_4
AGE AND GEOGRAPHIC RANGE Late middle Eocene, Wyoming
ANATOMICAL DEFINITION
Macrotarsius siegerti differs from *M. montanus* in having P_4 with an inflated paraconid and a large metaconid.

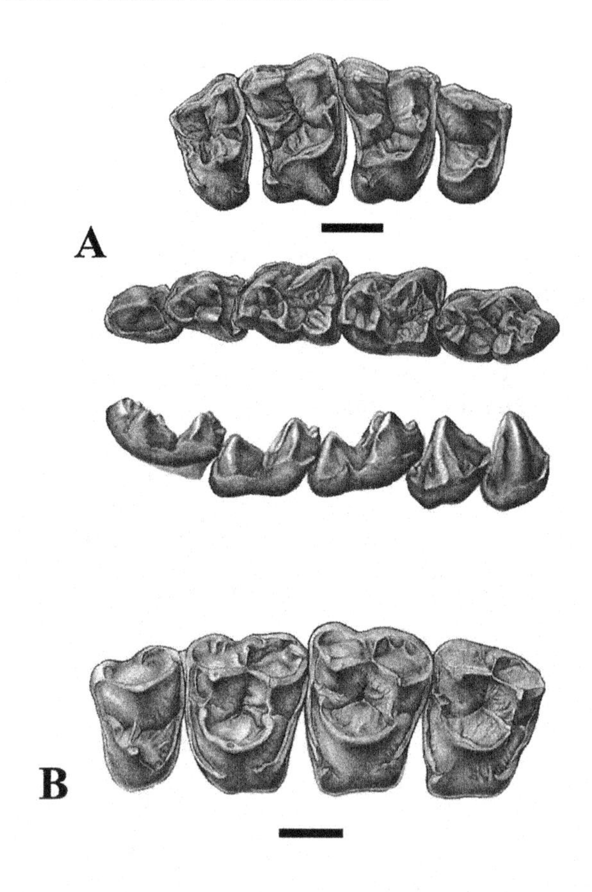

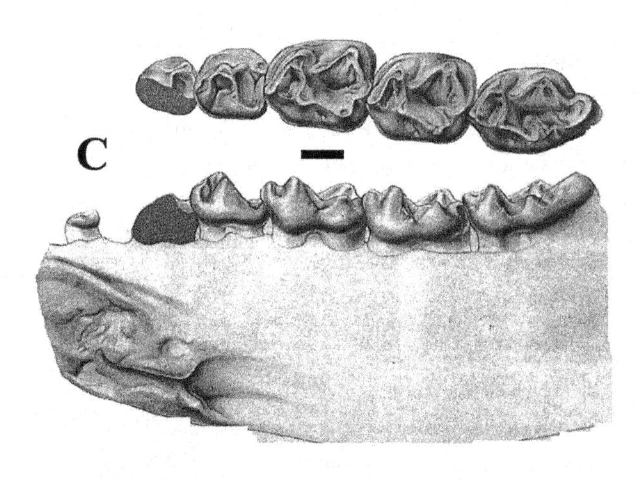

Fig. 5.23 Macrotarsiini. (A) *Hemiacodon gracilis*, right maxilla P^4–M^3 (AMNH 12030) in occlusal view (top), right dentary P_3–M_3 (AMNH 18991) in occlusal and lateral views (bottom). (B) *Macrotarsius siegerti*, left maxilla P^4–M^3 (CM 18646) in occlusal view. (C) *Macrotarsius montanus*, right dentary, canine, P_3–M_3 (CM 9592, holotype) in occlusal and medial views. Scale bars represent 2 mm. Figures adapted from Szalay (1976).

SPECIES *Macrotarsius jepseni* Robinson, 1968
TYPE SPECIMEN PU 16431, left and right dentaries and palate
AGE AND GEOGRAPHIC RANGE Late middle Eocene, Utah
ANATOMICAL DEFINITION
Macrotarsius jepseni differs from *M. siegerti* in having small mesostyles that are not connected to the centrocrista; differs from *M. siegerti* and *M. montanus* in having antemolar teeth more widely spaced, and lacking M_{1-3} exodaenodonty. Differs from all other *Macrotarsius* species in having small upper molar stylar shelves, a relatively longer and narrower P_4 and a diastema between P_{2-3}. The *M. jepseni* holotype and two other specimens were originally assigned to *Ourayia uintensis* by Simons (1961*b*). Robinson (1968) moved two of these specimens (the two that preserve teeth) to a new species of *Hemiacodon*, *H. jepseni*. Krishtalka (1978) assigned these specimens to *Macrotarsius jepseni*. Gunnell (1995*a*) reviewed all of the evidence and assigned one of the two specimens (YPM PU 11236) to *Ourayia uintensis*, leaving YPM PU 16431 as the only specimen representing *Macrotarsius jepseni*. The third original specimen (YPM PU 11288) is not referable to a specific taxon.

SPECIES *Macrotarsius roederi* Kelly, 1990
TYPE SPECIMEN LACM 128928, right dentary with P_2–M_3
AGE AND GEOGRAPHIC RANGE Late middle Eocene, California
ANATOMICAL DEFINITION
Macrotarsius roederi differs from *M. jepseni* in lacking a diastema between P_{2-3}, P_3 with a smaller paraconid, P_4 relatively larger, M_3 smaller and narrower; differs from *M. siegerti* and *M. montanus* in having premolars and molars more widely spaced, P_2 and P_4 relatively larger, and relatively smaller molars.

SPECIES *Macrotarsius macrorhysis* Beard *et al.*, 1994
TYPE SPECIMEN IVPP V11025, right P_4
AGE AND GEOGRAPHIC RANGE Middle Eocene, China
ANATOMICAL DEFINITION
Macrotarsius macrorhysis, represented by two isolated teeth from the Shanghuang fissure-fills in China, differs from other species of *Macrotarsius* in having a smaller, relatively narrower P_4 with a simpler talonid; further differs from *M. montanus* and *M. siegerti* in having a smaller M_1, and from *M. roederi* in lacking a complete M_1 ectocingulid.

GENUS *Hemiacodon* Marsh, 1872 (Fig. 5.23A)
INCLUDED SPECIES *H. casamissus*, *H. gracilis*

TYPE SPECIES *Hemiacodon gracilis* Marsh, 1872
TYPE SPECIMEN YPM 11806, right dentary with P_3–M_3
AGE AND GEOGRAPHIC RANGE Late early and middle Eocene, Wyoming and California (Lillegraven, 1980)
ANATOMICAL DEFINITION
Hemiacodon is one of the best-known omomyines dentally and is represented by some postcranial material as well. *Hemiacodon* is closely related to *Macrotarsius* (Gunnell, 1995*a*) and these two taxa can be distinguished from other omomyines by the following character states: sharply defined crests on all molars; short but well-developed talonid present on P_4 with distinct hypoconid and entoconid but cristid obliqua absent (*Macrotarsius* often has a

weak P_4 cristid obliqua); P_4 postvallid with distinct and sharply developed buccal and lingual ridges; M_{1-2} hypoconulids distinct and arcuate; molar talonids with rugose enamel (less complex and less well developed in *Macrotarsius*); P^3 with continuous basal cingulum and low, distinct hypocone shelf; upper molars with relatively mesiodistally narrow trigons and rugose enamel (less developed in *Macrotarsius*). *Hemiacodon* can be distinguished from *Macrotarsius* by the presence of very rugose enamel, distinct upper molar postprotocingula, large upper molar conules, more distinct hypocones, P_4 trigonid more open and better developed, and lower molar talonids that are relatively much wider than trigonids. Szalay (1976) suggested that *Hemiacodon* was more closely related to washakiins (including *Washakius*, *Loveina*, *Shoshonius* and *Dyseolemur*) than to *Macrotarsius*. Honey (1990) questioned this allocation and removed *Hemiacodon* from washakiins. Gunnell (1995*a*) argued for a close relationship between *Hemiacodon* and *Macrotarsius* based on the shared dental features outlined above.

SPECIES *Hemiacodon casamissus* Beard *et al.*, 1992
TYPE SPECIMEN CM 62035, left dentary M_{2-3}, right dentary M_{1-2}, right P_3
AGE AND GEOGRAPHIC RANGE Latest early Eocene, Wyoming
ANATOMICAL DEFINITION
Hemiacodon casamissus differs from *H. gracilis* in being smaller, in having P_3 lower crowned and lacking an entoconid, M_{1-2} with less crenulated enamel and stronger talonid notch. Gunnell (1995*a*) questioned the association of the dentaries with the isolated P_3, pointing out the wear differential between the heavily worn molars and the unworn premolar of *H. casamissus*. The validity of this species remains in doubt.

GENUS *Yaquius* Mason, 1990 (Fig. 5.15C)
INCLUDED SPECIES *Y. travisi*

TYPE SPECIES *Yaquius travisi* Mason, 1990
TYPE SPECIMEN LACM 40201, right M_2
AGE AND GEOGRAPHIC RANGE Middle Eocene, California
ANATOMICAL DEFINITION
Yaquius is represented by several isolated teeth from the middle Eocene of southern California. *Yaquius* is similar to *Macrotarsius* but differs from this taxon in the following ways (Mason, 1990): P_{3-4} buccal cingulids stronger; P_4 talonid basin less compressed anteroposteriorly; M_{1-3} protoconids and hypoconids more lingually placed, metaconids twinned, M_3 talonid relatively shorter and entoconid higher; M^3 conules tiny or absent. *Yaquius* is also similar to *Ourayia* but differs from this taxon in having broader P_3–M_3 buccal cingulids, more quadrate P_{3-4}, with P_4 being anteroposteriorly compressed, M_{2-3} talonid notch closed lingually by entocristid, M_{1-3} paraconids taller and metaconids twinned. McKenna & Bell (1997) included *Yaquius* within washakiins, but it appears that it is more likely a member of the macrotarsiin tribe as constituted by Krishtalka & Schwartz (1978).

Tribe Uintaniini

GENUS *Uintanius* Matthew, 1915 (Fig. 5.3A)
INCLUDED SPECIES *U. ameghini*, *U. rutherfurdi*

TYPE SPECIES *Uintanius ameghini* Wortman, 1904
TYPE SPECIMEN YPM 13241, left dentary with M_{2-3}
AGE AND GEOGRAPHIC RANGE Middle Eocene, Wyoming and Colorado
ANATOMICAL DEFINITION
Uintanius is a relatively rare taxon represented by only a few upper and lower dentitions (Szalay, 1976; Beard *et al.*, 1992; Gunnell, 1995*a*). It can be distinguished from all other omomyines by its very enlarged upper and lower third and fourth premolars (especially P^4/P_4). *Uintanius* differs from its probable sister taxon *Jemezius*, in having relatively smaller P^{3-4} protocones, stronger upper molar conules, a simpler P_4 talonid lacking a well-developed basin, lower molar paraconids more buccally placed, and in having a relatively smaller M_3.

SPECIES *Uintanius rutherfurdi* Robinson, 1966
TYPE SPECIMEN AMNH 55216, left dentary with P_3–M_2
AGE AND GEOGRAPHIC RANGE Early middle Eocene, Wyoming and Colorado
ANATOMICAL DEFINITION
Uintanius rutherfurdi differs from *U. ameghini* in having weaker and lower P_{3-4} paraconids (Beard *et al.*, 1992; Gunnell, 1995*a*).

GENUS *Jemezius* Beard, 1987 (Fig. 5.3G)
INCLUDED SPECIES *J. szalayi*

TYPE SPECIES *Jemezius szalayi* Beard, 1987
TYPE SPECIMEN CM 34843, left dentary with P_4–M_1
AGE AND GEOGRAPHIC RANGE Early Eocene, New Mexico
ANATOMICAL DEFINITION
Jemezius differs from its probable sister taxon *Uintanius*, in having relatively larger P^{3-4} protocones, weaker upper molar conules, a lower and more complex P_4 with a well-developed talonid basin, more lingually placed lower molar paraconids, and a relatively larger M_3 with a less compressed trigonid. *Jemezius* is only known from New Mexico but appears closely related to *Uintanius* and *Steinius* (Beard, 1987).

Tribe Rooneyini

GENUS *Rooneyia* Wilson, 1966 (Fig. 5.24A)
INCLUDED SPECIES *R. viejaensis*

TYPE SPECIES *Rooneyia viejaensis* Wilson, 1966
TYPE SPECIMEN UTBEG 40688-7, skull
AGE AND GEOGRAPHIC RANGE Latest Eocene, Texas

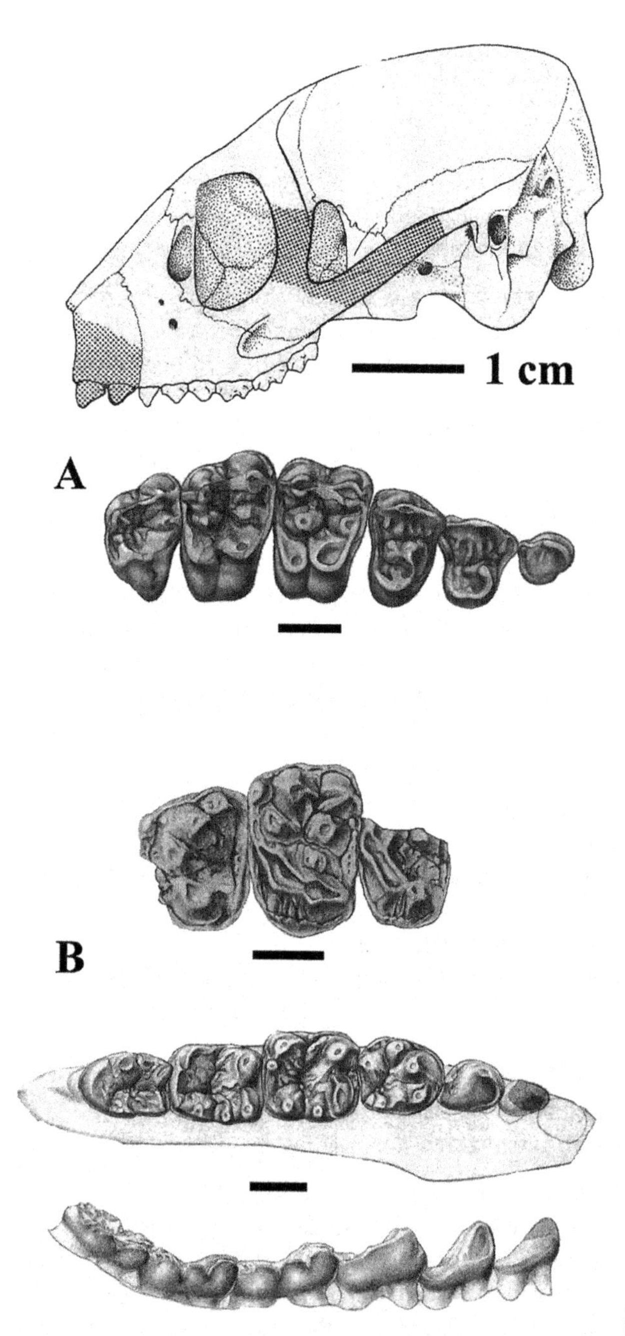

Fig. 5.24 (A) *Rooneyia viejaensis*, skull (UTBEG 40688-7, holotype) in lateral view, right P^2–M^3 (UTBEG 40688-7, holotype) in occlusal view. (B) *Ekgmowechashala philotau*, left P^4–M^2 (UCMP 128231) in occlusal view, left P_2–M_3 (SDSM 5550, P_3–M_2, holotype, SDSM 62104 and LACM 9207) in occlusal and medial views. Scale bars represent 2 mm (except for skull). Figures adapted from Szalay (1976) and Rose & Rensberger (1983).

ANATOMICAL DEFINITION

Rooneyia is represented only by the type skull (Wilson, 1966; Szalay, 1976). It differs from known omomyids in a number of features including: reduced and uninflated petromastoid region; fused frontal bones; squared orbits; upper molars low-crowned with well-developed conules and robust, true hypocones nearly as large as protocones. Other features of *Rooneyia* include the presence of a tubular ectotympanic, a lacrimal foramen opening anterior to the orbital margin, absence of postorbital closure, and a foramen magnum that opens posteriorly. The taxonomic position of *Rooneyia* is in question (Wilson, 1966; Simons, 1968; Szalay, 1976; Ross *et al.*, 1998) and further evidence is required to elucidate its phylogenetic relationships.

Family Tarsiidae

The fossil record of this family is very poor, consisting of only three Tertiary species from Asia, two of which are assigned to the extant genus *Tarsius*, and all of which are represented solely by teeth. The oldest uncontested tarsiers date from the middle Eocene of China. The five Recent species all belong to the genus *Tarsius* and are restricted to southeast Asia, particularly Indonesia and the Philippines (Groves, 1993).

At least three other taxa have sometimes been explicitly or implicitly considered tarsiids; the first two are here excluded from the family, while the third is included as a problematic taxon. Simons (1972; see also Rosenberger, 1985) argued that Microchoerinae should be assigned to Tarsiidae, based primarily on orbital size and structure, and basicranial anatomy. The current consensus, however, is that Microchoerinae belongs to the Omomyidae, as treated here, and that at least some of the similarities to tarsiers were achieved independently. Derived basicranial characters led Beard *et al.* (1991) to conclude that middle Eocene *Shoshonius* is the sister taxon of *Tarsius*, but most authorities consider it also to be an omomyid. Both of these taxa are known from well-preserved skulls, dentitions, and some postcrania. Finally, Simons & Bown (1985) allocated their new genus and species *Afrotarsius chatrathi* questionably to the Tarsiidae. Several subsequent workers suggested that certain characters of the single known dentary fragment (e.g., buccally shifted paraconid, shortened M_3) are derived traits shared with primitive anthropoids rather than tarsiids (e.g., Ginsburg & Mein, 1987; Kay & Williams, 1994). Van Valen (1994) went farther, tentatively grouping *Afrotarsius* with *Eosimias* in the Afrotarsiidae, which he considered to be the basal family of Anthropoidea. However, a fused tibiofibula, recently described from the same locality that produced the holotype jaw of *Afrotarsius*, is virtually identical to that of *Tarsius*. If it indeed pertains to that taxon, it supports assignment of *Afrotarsius* to the Tarsiidae (Rasmussen *et al.*, 1998).

GENUS *Tarsius* Storr, 1780 (Fig. 5.25A–C)
INCLUDED SPECIES *T. bancanus*, *T. dianae*, *T. eocaenus*, *T. pumilus*, *T. spectrum*, *T. syrichta*, *T. thailandica*

TYPE SPECIES *Tarsius syrichta* (Linnaeus, 1758)
TYPE SPECIMEN Presumably not specified
AGE AND GEOGRAPHIC RANGE Middle Eocene to Recent, eastern China and southeast Asia

ANATOMICAL DEFINITION

Tarsiers are among the smallest primates, with head and body lengths of about 85–160 mm and adult body weights of 80–165 g (Nowak, 1999). Recent species (except for the smaller *Tarsius pumilus*) do not differ significantly in head and body length or skull length, but they do show significant differences in tooth size, which increases from smallest to largest in the sequence: *T. pumilus*, *T. dianae* and *T. spectrum*, *T. syrichta*, *T. bancanus* (Musser & Dagosto, 1987; Niemitz *et al.*, 1991).

The most conspicuous feature of tarsiers is their enormous anteriorly facing eyes, each of which is larger than the brain. The orbits are incompletely closed posteriorly. The auditory bullae are large and elongate, and are continuous laterally with a short, tubular external auditory meatus. As in anthropoids, the promontorial branch of the internal carotid artery provides the principal blood supply to the brain. The mandible and upper jaws of tarsiers are V-shaped, and the mandibular symphysis is not fused. The dental formula is 2.1.3.3/1.1.3.3. The upper central incisors are enlarged and meet at the midline, in contrast to the condition in strepsirhines; the upper lateral and lower incisors are much smaller. The canines, though larger than the adjacent teeth, are small compared to those of anthropoids and most adapiforms. The premolars are smaller than the molars and relatively simple, each with a tall primary cusp. Tarsier molars are reminiscent of those of omomyids, having three distinct trigonid cusps and a broad, deep talonid basin. The hypoconulid is well defined only on M_3, where it forms an extended lobe. Upper molars of tarsiers are essentially tritubercular, with poorly developed conules. Tarsiers have long forearms, but overall forelimb length is short compared to their very long hindlimbs. The fibula is fused to the tibia for more than half of their length. The calcaneus and navicular are remarkably elongate in association with leaping habits; these specialized tarsal bones are reflected in the name *Tarsius*.

SPECIES *Tarsius thailandica* Ginsburg & Mein, 1987
TYPE SPECIMEN Specimen number and museum not specified (against Recommendation 73A of the ICZN, 1985, but not grounds for invalidating the species name), right M_2.
AGE AND GEOGRAPHIC RANGE Early Miocene, northwestern Thailand
ANATOMICAL DEFINITION

Tarsius thailandica is slightly smaller than *T. spectrum*, and resembles other tarsiers in having a very lingually placed paraconid and a deep talonid basin. It differs from *T. spectrum* in having a more mesial paraconid, but less mesial than in *T. bancanus* and *T. syrichta*. The protoconid and metaconid are of similar height and about the same level relative to the mesiodistal axis (Ginsburg & Mein, 1987).

SPECIES *Tarsius eocaenus* Beard *et al.*, 1994

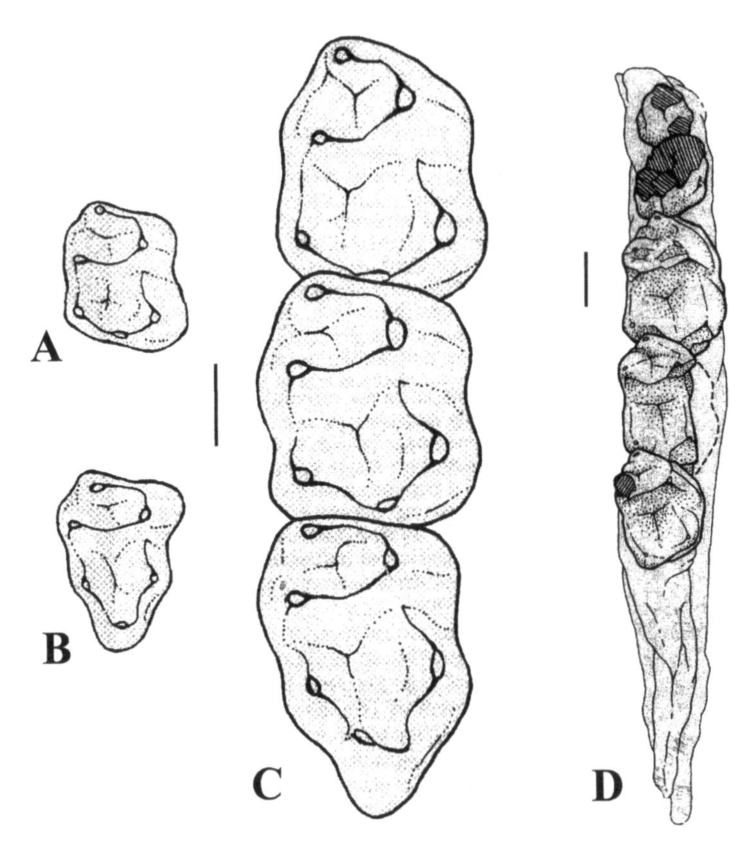

Fig. 5.25 (A) *Tarsius eocaenus*, right M_1 (IVPP V11030, holotype). (B) *Tarsius eocaenus*, right M_3 (IVPP V11027). (C) *Tarsius banacanus*, right M_{1-3} (USNM 300917). (D) *Afrotarsius chatrathi*, right dentary M_{1-3} (CGM 42830, holotype), in occlusal views. Scale bars represent 1 mm. Figures adapted from Beard *et al.* (1994) and Simons & Bown (1985).

TYPE SPECIMEN IVPP V11030, right M_1
AGE AND GEOGRAPHIC RANGE Middle Eocene, China
ANATOMICAL DEFINITION

This is the smallest known species of *Tarsius* (Beard *et al.*, 1994); linear dimensions of M_1 are about 60% of those in *T. bancanus* (the largest extant species) and about 80% of those in *T. pumilus* (smallest living species).

GENUS *Xanthorhysis* Beard, 1998 (Fig. 5.21A)
INCLUDED SPECIES *X. tabrumi* Beard, 1998

TYPE SPECIES *Xanthorhysis tabrumi* Beard, 1998
TYPE SPECIMEN IVPP V12063, left dentary with P_3–M_3
AGE AND GEOGRAPHIC RANGE Middle Eocene, China
ANATOMICAL DEFINITION

Based on molar size, *Xanthorhysis tabrumi* is smaller than *Tarsius syrichta* and *T. bancanus*. Its M_1 is about the same length as, but narrower than, that of *T. spectrum*. The lower dental formula is ?.1.3.3; P_2 had one root. *Xanthorhysis* differs from *Tarsius* in having relatively longer P_3 and P_4, with widely splayed roots, and a lower but more prominent metaconid on P_4. Its lower molars are relatively narrower and lower-crowned than in *Tarsius*, and have more distal entoconids (a resemblance to *Afrotarsius*). The molar paraconids are larger and more cuspidate than in *Afrotarsius*, however, and the hypoconulid lobe of M_3 is longer (as in *Tarsius*). Compared to omomyids, *Xanthorhysis* has a larger alveolus for the lower canine (as in *Tarsius*), and stronger, more trenchant paracristids and protocristids on P_{3-4} (Beard, 1998*a*).

Problematic taxa

GENUS *Ekgmowechashala* Macdonald, 1963 (Fig. 5.24B)
INCLUDED SPECIES *E. philotau*

TYPE SPECIES *Ekgmowechashala philotau* Macdonald, 1963
TYPE SPECIMEN SDSM 5550, left dentary with P_3–M_2
AGE AND GEOGRAPHIC RANGE Late Oligocene to early Miocene, South Dakota and Oregon
ANATOMICAL DEFINITION
Lower dental formula 2.1.3.3. *Ekgmowechashala* is an enigmatic omomyid from the latest Oligocene to earliest Miocene of Oregon and South Dakota. It differs from all other known omomyids in having low-crowned lower molars with very low trigonids, molars that are reduced in size progressively from M_1 to M_3, a flattened and molarized P_4 with closely joined and enlarged entoconid and metastylid and a large buccal cingulid, lower molars with complete absence of paraconids, weak to absent paracristids, and having large metastylids. P^4 has a very large paraconule situated between an enlarged, bulbous protocone and paracone (Rose & Rensberger, 1983). Enamel is heavily crenulated and all cusps are relatively low and rounded.

Rose & Rensberger (1983) described an upper dentition of *Ekgmowechashala* from Oregon and agreed with Szalay & Delson (1979) that its closest relative among omomyids was probably *Rooneyia*, although Rose & Rensberger (1983) did note some similarities between *Ekgmowechashala* and the microchoerine genera *Necrolemur* and *Microchoerus*. McKenna (1990) removed *Ekgmowechashala* from omomyids and placed it within the ?dermopteran family Plagiomenidae (see MacPhee *et al.*, 1989), in its own subfamily along with two new taxa, *Tarka* and *Tarkadectes*. No consensus yet exists as to the proper affinities of *Ekgmowechashala* but we have chosen to treat it as an omomyid in this paper.

GENUS *Kohatius* Russell & Gingerich, 1980 (Fig. 5.26C–D)
INCLUDED SPECIES *K. coppensi*

TYPE SPECIES *Kohatius coppensi* Russell & Gingerich, 1980
TYPE SPECIMEN GSP-UM 139, right M_1
AGE AND GEOGRAPHIC RANGE Middle Eocene, Pakistan
ANATOMICAL DEFINITION
The entire known sample of *Kohatius* consists of three complete teeth (right P_4, left P_4, right M_1) and two tooth fragments (Russell & Gingerich, 1980, 1987; Thewissen *et al.*, 1997). All of these specimens are from the Kuldana Formation, early middle Eocene of Pakistan. In nearly all morphological attributes, *Kohatius* is a very primitive omomyid. The lower fourth premolar (apparently the two known P_4s represent different species of *Kohatius*, see Thewissen *et al.*, 1997) has a relatively low protoconid, lacks a distinct paraconid but has a sharply defined paracristid, and has a low, rounded and weakly separated metaconid. The P_4 talonid is a short shelf with a weak and short cristid obliqua. One P_4 specimen (GSP-UM 212) is distolingually expanded and has a distinct metacristid that separates at the metaconid to continue down the postvallid to the cristid obliqua (Russell & Gingerich, 1987), while the other described P_4 (H-GSP 92166) lacks both of these features (Thewissen *et al.*, 1997). These teeth also differ in that GSP-UM 212 has a buccal cingulid while H-GSP 92166 does not. The lower first molar has distinct trigonid cusps with the paraconid placed mesiolingually and widely separated from the metaconid. The trigonid is buccolingually narrow. The cristid obliqua joins the postvallid near the lingual base of the protoconid and is not continuous with the buccal slope of the metaconid. The talonid basin is restricted and moderately deep with weak basal inflation of the hypoconid and entoconid. The talonid notch is closed by a short entocristid and there is a low shelf-like hypoconulid present. The relationships of *Kohatius* remain obscure and will not be fully understood until more complete specimens are available for study.

GENUS *Altanius* Dashzeveg & McKenna, 1977 (Fig. 5.26B)
INCLUDED SPECIES *A. orlovi*

TYPE SPECIES *Altanius orlovi* Dashzeveg & McKenna, 1977
TYPE SPECIMEN PSS 7/20-8, left dentary with P_4–M_3
AGE AND GEOGRAPHIC RANGE Early Eocene, Mongolia
ANATOMICAL DEFINITION
Altanius has a lower dental formula of 2.1.4.3 (Gingerich *et al.*, 1991) and retains many primitive dental features (Dashzeveg & McKenna, 1977; Rose & Krause, 1984; Gingerich *et al.*, 1991; Rose *et al.*, 1994). In the lower dental series the incisors appear relatively small and only slightly procumbent judging from alveoli. The canine alveolus is larger than those for I_1, I_2 and P_1, the latter tooth being single-rooted. P_2 is double-rooted and judging by alveolar dimensions, only slightly smaller than P_3. P_3 is a simple, trapezoidal tooth with a relatively tall protoconid and no paraconid or metaconid. P_4 is similar to P_3 except that it has a distinct, though small paraconid and metaconid and a slightly larger and more basined talonid. The lower molars have three cusped trigonids with the paraconid becoming progressively more appressed to the metaconid from M_1 to M_3. Hypoconids become relatively smaller from M_1 to M_3 while entoconids become larger. The M_3 hypoconulid forms a single-cusped posterior lobe. In the upper series the canine, P^1 and P^2 are separated by short diastemata, and the

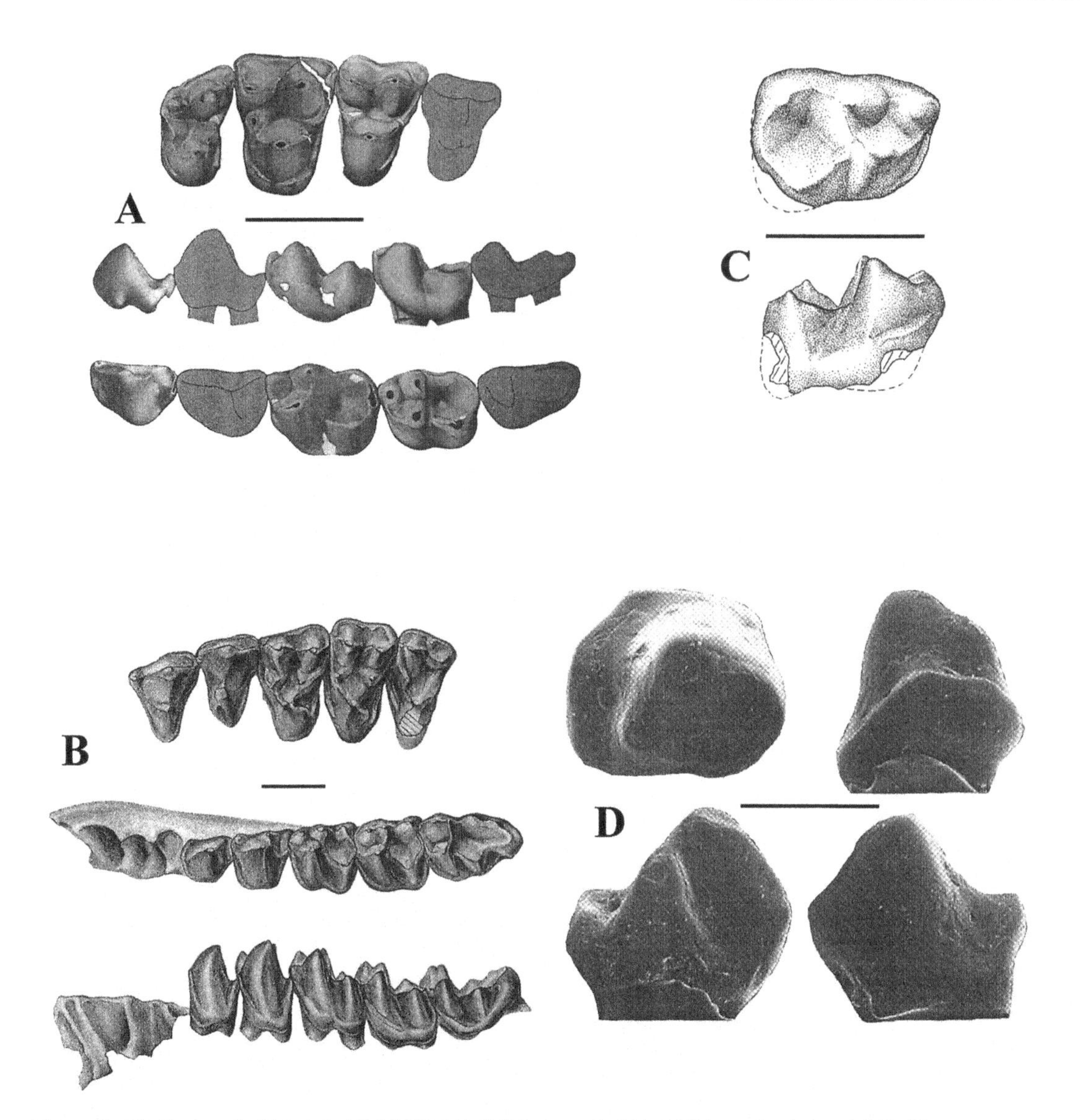

Fig. 5.26 (A) *Altiatlasius koulchii*, right M^{1-3} (THR 141 (M^2, holotype), 144, 145) in occlusal view, left P_3, M_{1-2} (THR 128, 135, 146) in lateral and occlusal views. (B) *Altanius orlovi*, left maxilla P^3–M^3 (PSS 20-61) in occlusal view, left dentary P_3–M_3 (PSS 20-58) in occlusal and lateral views. (C) *Kohatius coppensi*, right M_1 (GSP-UM 139, holotype) in occlusal and lateral views. (D) *Kohatius* sp., left P_4 (H-GSP 92166) in occlusal, distal, medial and lateral views. Scale bars in (A), (C–D) represent 2 mm, scale bar in (B) represents 1 mm. Figures adapted from Russell & Gingerich (1980), Sigé *et al.* (1990), Rose *et al.* (1994) and Thewissen *et al.* (1997).

canine alveolus is larger than the single-rooted P^1. P^{3-4} are triangular in outline and high-crowned. The upper molars have distinct proto-, para-, and metacones and well-developed conules (except for M^3 which has a reduced metacone and metaconule). M^1 has a short postprotocingulum that is not continuous with the postcingulum, while the postprotocingulum of M^2 nearly reaches the postcingulum. Pericones and hypocones, if present at all, are very rudimentary. The phylogenetic relationships of *Altanius* remain in doubt as it shares similarities with both primitive omomyids and primitive adapiforms (Gingerich *et al.*, 1991) as well as with certain plesiadapiforms (Dashzeveg & McKenna, 1977; Rose & Krause, 1984; Rose *et al.*, 1994). Unlike *Kohatius* which is poorly known morphologically, *Altanius* is represented by most of its upper and lower dentition but it retains so many primitive features that its taxonomic relationships remain obscure.

GENUS *Altiatlasius* Sigé *et al.*, 1990 (Fig. 5.26A)
INCLUDED SPECIES *A. koulchii*

TYPE SPECIES *Altiatlasius koulchii* Sigé *et al.*, 1990
TYPE SPECIMEN THR 141, left M^2
AGE AND GEOGRAPHIC RANGE Late Paleocene, Morocco
ANATOMICAL DEFINITION
Altiatlasius is known from only ten isolated teeth representing six tooth positions (Sigé *et al.*, 1990). M^2 is

larger than M^1 and both of these teeth lack hypocones, have moderate stylar shelves, and small conules. M^2 has a continuous lingual cingulum but M^1 lacks this feature. M^3 has a reduced metacone, an enlarged parastylar region with a small parastyle, lacks a hypocone and a continuous lingual cingulum, and is slightly smaller than M^1. Both M_1 and M_2 have distinct paraconids with M_2 having the paraconid slightly more appressed to the metaconid than M_1. The lower molar cristid obliqua is short and relatively straight and the entocristid is high, closing the talonid notch lingually. Hypoconulids are weakly formed, and postcristids are high and continuous with marginally placed hypoconids and entoconids. Entoconids are relatively mesially positioned. All tooth cusps are relatively blunt and rounded. Sigé *et al.* (1990) placed *Altiatlasius* in Omomyidae. Hooker *et al.* (1999) have suggested that *Altiatlasius* is a plesiadapiform, specifically a member of a new family of plesiadapiforms, Toliapinidae, otherwise only known from Europe. It is difficult to assess the affinities of *Altiatlasius* with so few remains available for study. It may be an early tarsiiform but, like *Altanius* and *Kohatius*, could also be a member of an, as yet, poorly known separate radiation of prosimian-like primates.

GENUS *Afrotarsius* Simons & Bown, 1985 (Fig. 5.25D)
INCLUDED SPECIES *A. chatrathi*

TYPE SPECIES *Afrotarsius chatrathi* Simons & Bown, 1985
TYPE SPECIMEN CGM 42830, right dentary with M_{1-3} and parts of P_{3-4}
AGE AND GEOGRAPHIC RANGE Early Oligocene, Egypt
ANATOMICAL DEFINITION
Afrotarsius chatrathi is slightly larger than *Tarsius spectrum* and slightly smaller than *T. syrichta* (based on M_1 length; Rasmussen *et al.*, 1998). Its molars resemble those of *Tarsius* (but differ from those of omomyids) in having high entocristids and buccolingually broad, shelf-like paraconids, which are set off from the metaconids and protoconids by a deep groove. *Afrotarsius* further differs from omomyids in having a more mesial metaconid, positioned lingual to the protoconid rather than distal to it. Compared to *Tarsius*, *Afrotarsius* has a more distal entoconid and, consequently, more widely separated metaconid and entoconid. Its molar paraconids (especially on M_2) are slightly buccally shifted relative to their position in *Tarsius*, and M_3 has a shorter talonid with a less expanded hypoconulid lobe (features that have led some to suggest that *Afrotarsius* is more closely allied with basal anthropoids, as noted above). The M_3 also lacks a distinct entoconid. In contrast to both *Tarsius* and omomyids, *Afrotarsius* has an indistinct talonid notch and $M_1 > M_2 > M_3$ (Simons & Bown, 1985). If the recently described tibiofibula is correctly attributed to *Afrotarsius*, it shares with *Tarsius* fusion of fibular and tibial shafts for at least the distal half of their length (Rasmussen *et al.*, 1998).

Afrotarsius has been linked tentatively with basal anthropoids (Ginsburg & Mein, 1987; Kay & Williams, 1994; Van Valen, 1994) and had its primate affinities questioned (Godinot, 1994).

Evolutionary history of Tarsiiformes

The earliest occurring definitive tarsiiforms are three species of *Teilhardina*, *T. belgica* (Europe) and *T. brandti* and *T. americana* (North America). The presence of the more primitive taxon *Steinius* in sediments stratigraphically younger than those in which *Teilhardina* occurs suggests that at least one ghost lineage of an even more primitive *Steinius*-like tarsiiform must have existed as early as or earlier than *Teilhardina* (Bown & Rose, 1984, 1987; Rose & Bown, 1991; Rose, 1995*a*).

Tarsiiforms radiated diversely through the early and middle Eocene in North America, with an ancestry that can be traced from *Teilhardina* or *Steinius* (Bown & Rose, 1987, 1991; Rose & Bown, 1991). Microchoerine ancestry is less clearly traced from *Teilhardina*. While they share many similarities with *Teilhardina* in cheek teeth, anterior dentitions are quite different suggesting that microchoerines may have been derived from an as yet unknown ancestry, perhaps from a lineage similar to *Tetonius*/*Pseudotetonius* (Hooker, 1986; Godinot *et al.*, 1992; Rose, 1995*a*). The interrelationships of microchoerine genera are not completely understood either. Hooker (1986) proposed a branching sequence wherein *Microchoerus* and *Necrolemur* are sister taxa with this clade being most closely related to *Nannopithex filholi*. In turn, the *Microchoerus*–*Necrolemur*–*N. filholi* clade is the sister taxon to a clade formed by *Pseudoloris* and two other species of *Nannopithex*. Hooker postulated that *Nannopithex raabi* (and by extension *N. zuccolae*) was the sister taxon to the rest of the microchoerine clade, rendering *Nannopithex* paraphyletic. This arrangement has not been rigorously tested and it remains to be seen just what the true interrelationships of microchoerines are.

Through the early Eocene in North America three separate clades of anaptomorphine omomyids can be recognized, one comprising *Teilhardina*, *Anemorhysis* and possibly *Arapahovius*, a second including *Tetonius* and *Pseudotetonius*, and a third consisting of *Absarokius* and probably *Strigorhysis*. All three of these lineages show evidence of anagenetic change through time as documented by dense samples in the Bighorn Basin of Wyoming (Bown & Rose, 1987). *Strigorhysis* and other later-occurring anaptomorphines such as *Anaptomorphus*, *Aycrossia* and *Gazinius* all are probably derived from the *Absarokius* clade.

The evolutionary history of middle Eocene omomyid tribes is not as well understood or as well documented as are early Eocene anaptomorphines but some relationships seem well supported. As noted above, middle Eocene anaptomorphines are likely derived from one of the lineages of *Absarokius* documented from the early Eocene (Bown & Rose, 1987). Trogolemurins seem distantly related to *Anemorhysis* (Beard *et al.*, 1992) but more closely related to *Arapahovius* or perhaps *Tatmanius* (Gunnell, 1995*a*). Omomyins may have been derived from a *Steinius*-like form (Bown & Rose, 1984; Honey, 1990; Rose & Bown, 1991; Gunnell, 1995*a*). Washakiins can be traced to latest early Eocene *Loveina* in Wasatchian biochron Wa-7 but their origins from that point are unclear. They too may be derived from a *Steinius*-like form as might uintaniins and utahiins. A *Teilhardina*-like form could have given rise to some of these lineages, which, if true, would require a reconstitution of subfamilies as these taxa would then be anaptomorphine, not omomyine.

Mytoniinans and utahiinans are especially rare in the early part of the middle Eocene and are never very abundant. Mytoniinan origins can be traced from an *Omomys*-like ancestry through *Wyomomys* and *Ageitodendron* to *Ourayia*. Utahiinans may also be derived from an *Omomys*-like ancestry but could also be more closely related to washakiins, which would then require a reconstitution of the tribe Utahiini. Macrotarsiins probably arose from a washakiin ancestry through *Hemiacodon* to *Macrotarsius* and *Yaquius*. Later-occurring North American taxa from the Uintan and Duchesnian of Utah, Texas, California and Canada can all trace their origins to middle Bridgerian omomyid tribes.

The latest known occurrences of North American possible omomyids are *Rooneyia* from the Chadronian (latest Eocene) of Texas and *Ekgmowechashala* from the Arikareean (latest Oligocene–earliest Miocene) of South Dakota and Oregon. Wilson (1966) originally considered *Rooneyia* to be a lemuriform while Simons (1968) believed that it might represent a catarrhine anthropoid. Szalay (1976) and Szalay & Delson (1979) placed *Rooneyia* in its own tribe (Rooneyiini) within omomyids while noting that it is similar in some features to South American platyrrhine anthropoids.

Adaptations

Dental adaptations

Tarsiiforms exhibit a wide range of dental morphologies. Many of the most primitive taxa (including *Teilhardina*, *Steinius*) have simple molars with well-developed, marginally placed cusps and little or no development of accessory cuspules. Upper molars are also simple with small stylar shelves, no stylar cusps, small conules, and weak or absent hypocones and postprotocingula. Upper and lower premolars are premolariform, pointed, and simple with only P_4 having a metaconid among lower premolars. Upper and lower molar basins are relatively shallow with smooth enamel. These primitive forms were almost certainly predominantly insectivorous (Strait, 1991).

There are several dental themes developed through various omomyid lineages. Certain lineages (most notably *Tetonius–Pseudotetonius* and possibly *Tatmanius* and *Trogolemur–Sphacorhysis*) are typified by dentaries that are progressively shortened and deepened anteriorly with development of a very large, pointed, procumbent central incisor. Anterior teeth between I_1 and P_4 are either lost or shortened, reduced, and crowded together with reduction in root number. Molars tend to be low-crowned with anteroposteriorly constricted trigonids (often with loss or reduction of the paraconid) and relatively broad, flattened talonid basins. This dental morphology, along with small body size, is suggestive of a predominantly insectivorous/faunivorous diet, perhaps supplemented by some fruits and seeds (Szalay, 1976; Strait, 1991).

A somewhat similar trend is seen in utahiinan omomyids. These taxa (including *Utahia*, *Stockia*, *Chipetaia* and *Asiomomys*) are all unknown from anterior dentitions (anterior to P_3) but have relatively low-crowned teeth with very anteroposteriorly constricted molar trigonids with reduced or absent paraconids on M_{2-3}. Talonid basins are flat and often exhibit very complex and crenulated enamel. These dental features indicate frugivory, perhaps with an emphasis on harder fruits and seeds (Szalay, 1976; Rasmussen, 1996). The microchoerine *Microchoerus* also has low-crowned teeth with complex enamel and probably shared a dietary specialization similar to utahiinans.

A different dental pattern is exhibited in washakiins. These taxa have sharply defined cusps with progressive proliferation of accessory cuspules including mesostyles, stylar cusps, metastylids, pericones and complex conules. These taxa retain paraconids on all lower molars and have more constricted, deeper talonid basins. Incisors tend to be small, not procumbent, and pointed. Szalay (1976) has suggested that some washakiins (in particular *Washakius*) may have been specialized herbivores, but small body size and cusp proliferation may better be interpreted as insectivorous/faunivorous adaptations (Strait, 1991, 1993).

Two omomyine groups, Macrotarsiini (*Hemiacodon*, *Macrotarsius*, *Yaquius*) and Mytoniina (*Wyomomys*, *Ageitodendron*, and *Ourayia*) are typified by increasing body size, but their dental morphology follows two different paths. In macrotarsiins, sharply defined crests and deep, constricted talon and talonid basins are developed. Cusps tend to be incorporated into crests and are not sharply defined. These morphological features suggest that macrotarsiins were predominantly folivorous. Mytoniinans developed very bulbous cusps that are basally inflated. Crests are relatively wide and short and teeth in general become flat and broad. Mytoniinans are interpreted as predominantly frugivorous.

Other omomyids such as uintaniins (*Uintanius* and *Jemezius*) and some *Absarokius* species, developed specialized premolars, especially P^4/P_4s that became tall, broad and exodaenodont. These animals may have specialized on hard fruits and seeds, or insects with hard exoskeletons, dietary items that required puncturing before ingestion (Szalay, 1976; Strait, 1991; Gunnell, 1995*a*).

Dental evidence suggests that tarsiiforms were capable of exploiting a wide variety of dietary niches including insectivory, omnivory, frugivory and folivory or combinations thereof. Omomyid diversity was often very high in the North American early and middle Eocene (as many as 10–12

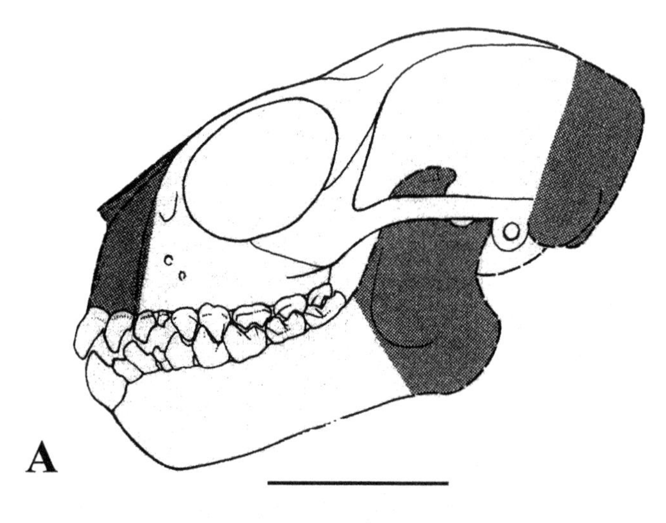

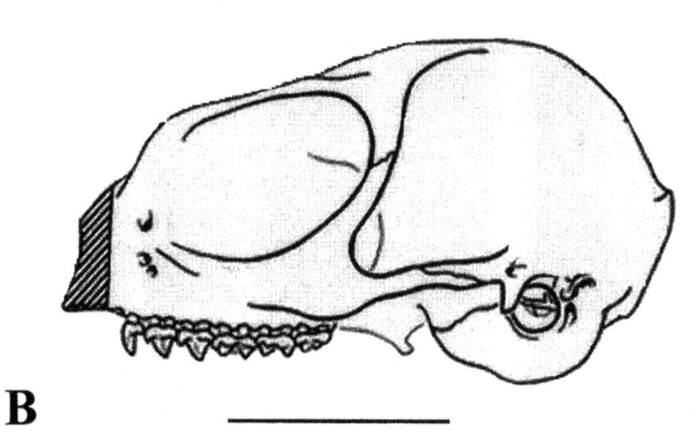

Fig. 5.27 (A) Reconstructed skull of *Tetonius homunculus* (AMNH 4194, holotype) in lateral view. (B) Reconstructed skull of *Shoshonius cooperi* (CM 31366, 31367, 60494) in lateral view. Scale bars represent 1 cm. Figures adapted from Szalay (1976) and Beard & MacPhee (1994).

co-existing species). The ability to subdivide habitats into specific dietary micro-habitats is one of the major contributing factors to high omomyid diversity.

Cranial adaptations

Only a few tarsiiform taxa are represented by cranial remains (Fig. 5.27), including the anaptomorphine *Tetonius* (Cope, 1882*b*; Szalay, 1976), the microchoerine *Necrolemur* (Simons, 1961*a*; Szalay & Delson, 1979; Gingerich, 1984*a*) and the omomyines *Omomys* (Alexander & MacPhee, 1999*a*, 1999*b*), *Shoshonius* (Beard *et al.*, 1991; Beard & MacPhee, 1994) and *Rooneyia* (Wilson, 1966; Szalay, 1975*b*, 1976). Some cranial features of the skull of *Shoshonius* have been used to argue for a close relationship between it and extant tarsiers (Beard *et al.*, 1991; Beard & MacPhee, 1994). These features include: a basioccipital that overlaps the posteromedial bullar wall; a ventrolaterally placed posterior carotid foramen (Beard *et al.*, 1991; later corrected by Beard & MacPhee, 1994, as also being high on the bulla, although these authors admit that none of their specimens preserve both the bulla and the posterior carotid foramen in an undistorted state); the presence of a stylomastoid foramen (Beard *et al.*, 1991; later corrected by Beard & MacPhee, 1994, as having been placed out of position by Beard *et al.*, 1991); narrow, peaked choanae and narrow central stem of basicranium; presence of a reduced snout, parotic fissure and suprameatal foramen. As Rose (1995*a*) noted, some of these features are also present in other omomyids and thus any resemblances with extant *Tarsius* could be convergent. Interestingly, *Smilodectes*, a notharctine adapiform with a relatively short face and rounded skull, also has a basioccipital that overlaps the posteromedial bullar wall (Gunnell, 1995*b*). While Beard & MacPhee (1994) dismiss this character state as "non-homologous", its presence in *Smilodectes* opens the question of how many of these possible shared derived characters of *Shoshonius* and *Tarsius* are simply the result of enlargement of orbits and a shortening of the face.

The overall pattern of cranial morphology including large, front-facing orbits and relatively short rostra indicates that omomyids were vision-dominated and nocturnal. Some may have used habitual vertical postures judging by the position of the foramen magnum underneath the skull (the foramen magnum in known omomyids is never as anteriorly positioned as in living tarsiers).

Postcranial adaptations

Several omomyids are represented by at least some postcranial elements (Gebo, 1988; Dagosto, 1993). Among these are the anaptomorphines *Teilhardina* (Szalay, 1976; Gebo, 1988), *Tetonius* (Szalay, 1976; Gebo, 1988; Rosenberger & Dagosto, 1992), *Absarokius* (Covert & Hamrick, 1993) and *Arapahovius* (Savage & Waters, 1978), the microchoerines *Microchoerus* (Schlosser, 1907; Schmid, 1979; Dagosto, 1993), *Necrolemur* (Schlosser, 1907; Godinot & Dagosto, 1983; Dagosto & Gebo, 1994) and *Nannopithex* (Weigelt, 1933), and the omomyines *Omomys* (Rosenberger & Dagosto, 1992; Anemone *et al.*, 1997; Hamrick, 1999; Anemone & Covert, 2000), *Hemiacodon* (Simpson, 1940; Szalay, 1976; Dagosto, 1985; Gebo, 1988), *Washakius* (Szalay, 1976; Gebo, 1988) and *Shoshonius* (Dagosto *et al.*, 1999).

While there is some variability within known omomyid postcranial elements, there is remarkable consistency as well (Fig. 5.28). Forelimb elements of omomyids are very poorly known, represented only by complete humeri of *Microchoerus* (Dagosto, 1993) and *Shoshonius* (Dagosto *et al.*, 1999), a proximal humerus of *Necrolemur* (Szalay, 1976; Dagosto, 1993), distal humeri of *Omomys* (Anemone *et al.*, 1997; Anemone & Covert, 2000) and *Hemiacodon* (Szalay, 1976; Dagosto, 1993) and some carpal elements of *Omomys* (Hamrick, 1999). *Shoshonius* has a relatively higher humerofemoral index (i.e. relatively shorter femur) than extant tarsiers, being more similar to galagos and cheirogaleids. Known

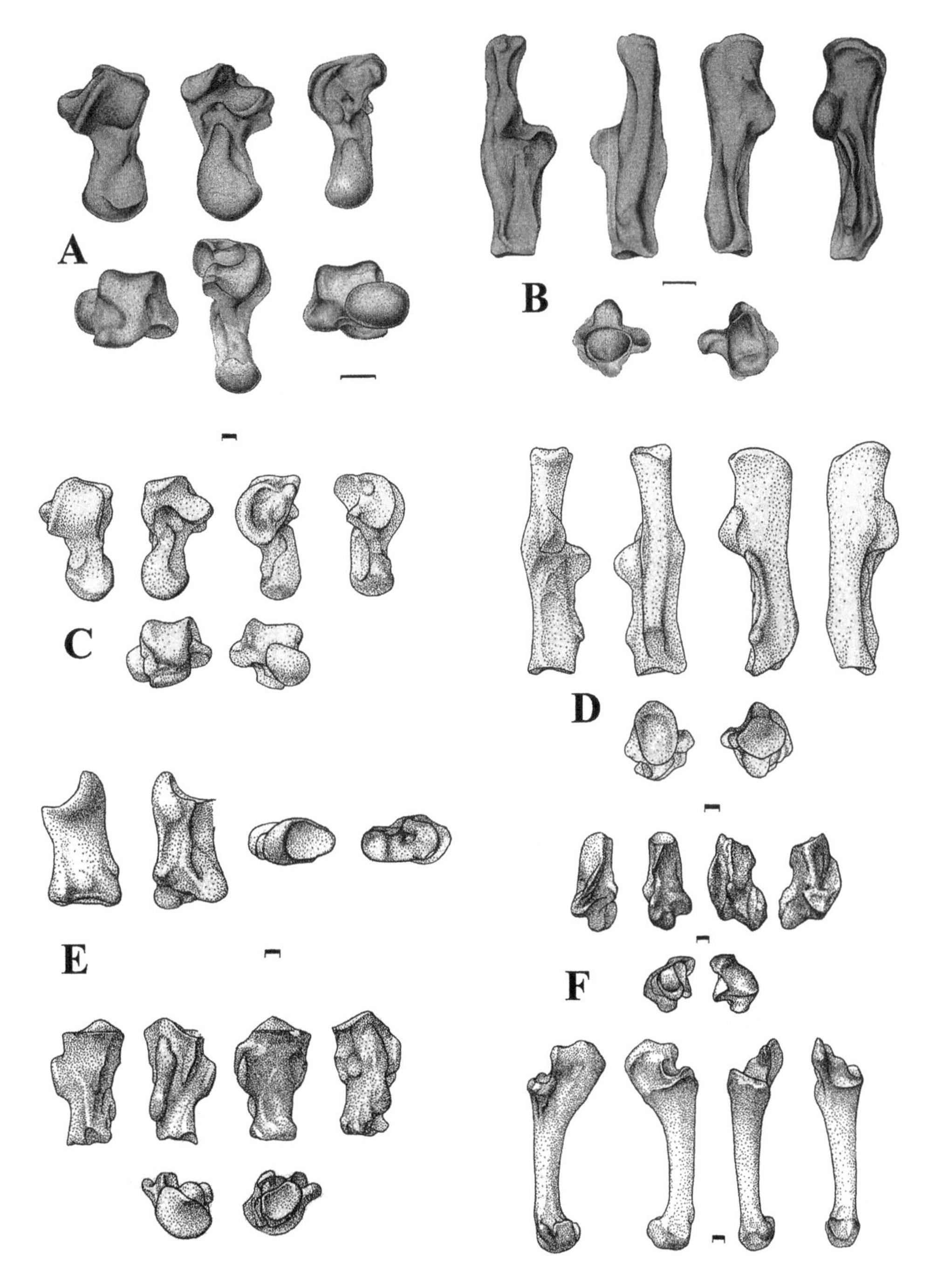

Fig. 5.28 Omomyid postcranial elements (all views described top to bottom, left to right). (A) *Teilhardina belgica*, right astragalus (IRNSB 4386) in dorsal, ventral, medial, proximal, lateral and distal views. (B) *Teilhardina belgica*, right calcaneum (IRNSB 61, 4385, 4390) in dorsal, ventral, lateral, medial, distal and proximal views. (C) *Hemiacodon gracilis*, right astragalus (based on several specimens) in dorsal, ventral, medial, lateral, proximal and distal views. (D) *Hemiacodon gracilis*, right calcaneum (based on several specimens) in dorsal, ventral, medial, lateral, proximal and distal views. (E) *Hemiacodon gracilis* (AMNH 12613), top row, right navicular in dorsal, ventral, proximal and distal views, bottom rows, right cuboid in dorsal, ventral, lateral, medial, proximal and distal views. (F) *Hemiacodon gracilis* (AMNH 12613), top rows, right entocuneiform in ventral, dorsal, lateral, medial, proximal and distal views, bottom row, right metatarsal I in ventral, dorsal, lateral and medial views. Scale bars represent 1 mm. Figures adapted from Szalay (1976).

omomyid hamates resemble those of extant monkeys, tarsiers and Tertiary adapiforms, differing only in having a less-well-developed hamulus (Hamrick, 1999). In general, omomyid carpals are consistent with powerful manual grasping during arboreal locomotion.

Hindlimb features of tarsiiforms (see Dagosto *et al.*, 1999 for summary) include a long, rod-like ilium, relatively long femur with semicylindrical head and an anteroposteriorly deep distal end, relatively long, bowed tibia that either has a long tibiofibular articulation (*Absarokius* and *Shoshonius*), a relatively shorter tibiofibular articulation (*Omomys*), or a distally fused tibiofibula (*Necrolemur*). Tarsiiform astragali have relatively long necks, a high astragalar body, a long and relatively narrow trochlea, and a well-developed posterior trochlear shelf (Gebo, 1988). The calcaneus is elongate with a well-developed cuboid pivot (except perhaps in *Shoshonius*, see Dagosto *et al.*, 1999). The cuboid, navicular and cuneiforms are also elongate and the first metatarsal has long and prominent peroneal tubercles. Like adapiforms, tarsiiforms had an opposable hallux. As in adapiforms, it appears that all omomyid digits terminated in nails, as there is no unequivocal evidence for the existence of toilet claws among known omomyids.

Judging from postcranial evidence, known omomyids were capable of powerful leaping but were not specialized vertical clingers and leapers like extant tarsiers. Omomyids were probably much more like extant cheirogaleids or galagos – active arboreal quadrupeds that included frequent leaping in their locomotor repertoire.

Acknowledgements

We thank Drs Jeremy Hooker (Natural History Museum, London) and Marc Godinot (Museum national d'histoire naturelle, Paris) for providing helpful information on microchoerines. Dr. K. Christopher Beard supplied Figs. 5.3G and 5.21. Ms. Kathleen M. Muldoon provided many useful comments on an earlier version of this chapter.

Primary References

The following list represents the publications naming the species recognized above for each genus. See also Szalay (1976), Szalay & Delson (1979), Bown & Rose (1984, 1987, 1991), Gunnell (1995*a*, 1995*b*), Fleagle (1999), and the references cited above.

Absarokius

Loomis, F. B. (1906). Wasatch and Wind River primates. *American Journal of Science*, **21**, 277–284.

Matthew, W. D. (1915*a*). A revision of the lower Eocene Wasatch and Wind River faunas. IV. Entelonychia, Primates, Insectivora (part). *Bulletin of the American Museum of Natural History*, **34**, 429–483.

Morris, W. J. (1954). An Eocene fauna from the Cathedral Bluffs Tongue of the Washakie Basin, Wyoming. *Journal of Paleontology*, **28**, 195–203.

Robinson, P. (1966). Fossil Mammalia of the Huerfano Formation, Eocene, of Colorado. *Peabody Museum of Natural History, Bulletin*, **21**, 1–95.

Bown, T. M. & Rose, K. D. (1987). Patterns of dental evolution in early Eocene anaptomorphine primates (Omomyidae) from the Bighorn Basin, Wyoming. *Journal of Paleontology*, **61**(**5** Supplement), 1–162.

Afrotarsius

Simons, E. L. & Bown, T. M. (1985). *Afrotarsius chatrathi*, new genus, new species: first tarsiiform primate (Tarsiidae?) from Africa. *Nature*, **313**, 475–477.

Ageitodendron

Gunnell, G. F. (1995*a*). Omomyid primates (Tarsiiformes) from the Bridger Formation, middle Eocene, southern Green River Basin, Wyoming. *Journal of Human Evolution*, **28**, 147–187.

Altanius

Dashzeveg, D. & McKenna, M. C. (1977). Tarsioid primate from the early Tertiary of the Mongolian People's Republic. *Acta Palaeontologica Polonica*, **22**, 119–137.

Gingerich, P. D., Dashzeveg, D., & Russell, D. E. (1991). Dentition and systematic relationships of *Altanius orlovi* (Mammalia, Primates) from the early Eocene, Mongolia. *Geobios*, **24**, 637–646.

Altiatlasius

Sigé, B., Jaeger, J. J., Sudre, J., & Vianey-Liaud, M. (1990). *Altiatlasius koulchii* n. gen. et sp., primate omomyidé du Paléocène supérieur du Maroc, et les origines des euprimates. *Palaeontographica Abteilung A*, **214**, 31–56.

Anaptomorphus

Cope, E. D. (1872). On a new vertebrate genus from the northern part of the Tertiary basin of Green River. *Proceedings of the American Philosophical Society*, **12**, 554.

Szalay, F. S. (1976). Systematics of the Omomyidae (Tarsiiformes, Primates): taxonomy, phylogeny, and adaptations. *Bulletin of the American Museum of Natural History*, **156**, 157–450.

Anemorhysis

Matthew, W. D. (1915*a*). A revision of the lower Eocene Wasatch and Wind River faunas. IV. Entelonychia, Primates, Insectivora (part). *Bulletin of the American Museum of Natural History*, **34**, 429–483.

Gazin, C. L. (1952). The lower Eocene Knight Formation of western Wyoming and its mammalian faunas. *Smithsonian Miscellaneous Collections*, **117**, 1–82.

Gazin, C. L. (1958). A review of the middle and upper Eocene Primates of North America. *Smithsonian Miscellaneous Collections*, **136**, 1–112.

Gazin, C. L. (1962). A further study of the lower Eocene mammalian faunas of southwestern Wyoming. *Smithsonian Miscellaneous Collections*, **144**, 1–98.

Bown, T. M. & Rose, K. D. (1984). Reassessment of some early Eocene Omomyidae, with description of a new genus and three new species. *Folia Primatologica*, **43**, 97–112.

Beard, K. C., Krishtalka, L., & Stucky, R. K. (1992). Revision of the Wind River faunas, Early Eocene, central Wyoming. XII. New species of omomyid primates (Mammalia: Primates: Omomyidae) and omomyid taxonomic composition across the Early–Middle Eocene boundary. *Annals of Carnegie Museum*, **61**, 39–62.

Williams, B. A. & Covert, H. H. (1994). New early Eocene anaptomorphine primate (Omomyidae) from the Washakie Basin, Wyoming, with comments on the phylogeny and paleobiology of anaptomorphines. *American Journal of Physical Anthropology*, **93**, 323–340.

Arapahovius

Savage, D. E. & Waters, B. T. (1978). A new omomyid primate from the Wasatch Formation of southern Wyoming. *Folia Primatologica*, **30**, 1–29.

Bown, T. M. & Rose, K. D. (1991). Evolutionary relationships of a new genus and three new species of omomyid primates (Willwood Formation, Lower Eocene, Bighorn Basin, Wyoming). *Journal of Human Evolution*, **20**, 465–480.

Asiomomys

Wang, B. & Li., C. (1990). First Paleogene mammalian fauna from northeast China. *Vertebrata PalAsiatica*, **28**, 165–205.

Aycrossia

Bown, T. M. (1979). New omomyid primates (Haplorhini, Tarsiiformes) from middle Eocene rocks of west-central Hot Springs County, Wyoming. *Folia Primatologica*, **31**, 48–73.

Chipetaia

Rasmussen, D. T. (1996). A new Middle Eocene omomyine primate from the Uinta Basin, Utah. *Journal of Human Evolution*, **31**, 75–87.

Chlororhysis

Gazin, C. L. (1958). A review of the middle and upper Eocene Primates of North America. *Smithsonian Miscellaneous Collections*, **136**, 1–112.

Bown, T. M. & Rose, K. D. (1984). Reassessment of some early Eocene Omomyidae, with description of a new genus and three new species. *Folia Primatologica*, **43**, 97–112.

Chumashius

Stock, C. (1933). An Eocene primate from California. *Proceedings of the National Academy of Sciences of the United States of America*, **19**, 954–959.

Dyseolemur

Stock, C. (1934). A second Eocene primate from California. *Proceedings of the National Academy of Sciences of the United States of America*, **20**, 150–154.

Ekgmowechashala

Macdonald, J. R. (1963). The Miocene faunas from the Wounded Knee area of western South Dakota. *Bulletin of the American Museum of Natural History*, **125**, 141–238.

Gazinius

Bown, T. M. (1979). New omomyid primates (Haplorhini, Tarsiiformes) from middle Eocene rocks of west-central Hot Springs County, Wyoming. *Folia Primatologica*, **31**, 48–73.

Gunnell, G. F. (1995*a*). Omomyid primates (Tarsiiformes) from the Bridger Formation, middle Eocene, southern Green River Basin, Wyoming. *Journal of Human Evolution*, **28**, 147–187.

Hemiacodon

Marsh, O. C. (1872). Preliminary description of new Tertiary mammals. I–IV. *American Journal of Science*, **4**, 22–28, 202–224.

Beard, K. C., Krishtalka, L., & Stucky, R. K. (1992). Revision of the Wind River faunas, Early Eocene, central Wyoming. XII. New species of omomyid primates (Mammalia: Primates: Omomyidae) and omomyid taxonomic composition across the Early–Middle Eocene boundary. *Annals of Carnegie Museum*, **61**, 39–62.

Jemezius

Beard, K. C. (1987). *Jemezius*, a new omomyid primate from the early Eocene, northwestern New Mexico. *Journal of Human Evolution*, **16**, 457–468.

Kohatius

Russell, D. E. & Gingerich, P. D. (1980). Un nouveau primate omomyide l'Éocène du Pakistan. *Comptes rendus de l'Académie des Sciences, Paris*, **291**, 621–624.

Thewissen, J. G. M., Hussain, S. T., & Arif, M. (1997). New *Kohatius* (Omomyidae) from the Eocene of Pakistan. *Journal of Human Evolution*, **32**, 473–477.

Loveina

Loomis, F. B. (1906). Wasatch and Wind River primates. *American Journal of Science*, **21**, 277–284.

Simpson, G. G. (1940). Studies of the earliest primates. *Bulletin of the American Museum of Natural History*, **77**, 185–212.

Bown, T. M. & Rose, K. D. (1984). Reassessment of some early Eocene Omomyidae, with description of a new genus and three new species. *Folia Primatologica*, **43**, 97–112.

Gunnell, G. F., Bartels, W. S., Gingerich, P. D., & Torres, V. (1992). Wapiti Valley faunas: early and middle Eocene fossil vertebrates from the North Fork of the Shoshone River, Park County, Wyoming. *Contributions from the Museum of Paleontology, University of Michigan*, **29**, 247–287.

Macrotarsius

Clark, J. (1941). An anaptomorphid primate from the Oligocene of Montana. *Journal of Paleontology*, **15**, 562–563.
Robinson, P. (1968). The paleontology and geology of the Badwater Creek area, central Wyoming. IV. Late Eocene primates from Badwater, Wyoming, with a discussion of material from Utah. *Annals of the Carnegie Museum*, **39**, 307–326.
Kelly, T. S. (1990). Biostratigraphy of Uintan and Duchesnean land mammal assemblages from the middle member of the Sespe Formation, Simi Valley, California. *Contributions in Science of the Natural History Museum of Los Angeles*, **419**, 1–42.
Beard, K. C., Qi, T., Dawson, M. R., Wang, B., & Li, C. (1994). A diverse new primate fauna from middle Eocene fissure-fillings in southeastern China. *Nature*, **368**, 604–609.

Microchoerus

Wood, S. (1844). Record of the discovery of an alligator with several new Mammalia in the freshwater strata at Hordwell. *Annals of the Magazine of Natural History, London*, **14**, 349–351.
Filhol, H. (1880). Note sur des mammifères fossiles nouveaux provenant des phosphorites du Quercy. *Bulletin de la Société philomathique, Paris*, **7**(4).
Stehlin, H. G. (1916). Die Säugetiere des schweizerischen Eocäns: critischer Catalog der Materialen. VIIb. *Abhandlungen schweizerischen paläontologischen Gesellschaft*, **41**, 1299–1552.
Hooker, J. J. (1986). Mammals from the Bartonian (middle/late Eocene) of the Hampshire Basin, southern England. *Bulletin of the British Museum of Natural History*, **39**, 191–478.

Nannopithex

Stehlin, H. G. (1916). Die Säugetiere des schweizerischen Eocäens. Critischer Catalog der Materialen. VIIb. *Abhandlungen Schweizerischen Palaontologischen Gesellschaft*, **41**, 1299–1552.
Heller, F. (1930). Die Säugetierfauna der mitteleozänen Braunkohle des Geiseltales bei Halle a. *Stuttgarter Jahrbuck Halleschen Verbandes*, **9**, 13–14.
Hooker, J. J. (1986). Mammals from the Bartonian (middle/late Eocene) of the Hampshire Basin, southern England. *Bulletin of the British Museum of Natural History*, **39**, 191–478.
Godinot, M., Russell, D. E., & Louis, P. (1992). Oldest known *Nannopithex* (Primates, Omomyiformes) from the early Eocene, France. *Folia Primatologica*, **58**, 32–40.
Thalmann, U. (1994). Die Primaten aus dem eozänen Geiseltal bei Halle/Saale (Deutschland). *Courier Forschungsinstitut Senckenberg*, **175**, 1–161.

Necrolemur

Filhol, H. (1873). Sur un nouveau genre de Lémurien fossile récemment découvert dans les gisements de phosphate de chaux du Quercy. *Comptes rendus de l'Académie des sciences de Paris*, **77**, 1111–1112.
Schlosser, M. (1887). Die Affen, Lemuren, Chiropteren, Insectivoren, Marsupialier, Creodonten, und Carnivoren des europaischen Tertiars und deren Beziehungen zu ihren lebenden und fossilen aussereuropaischen Verwandten. *Beitragezur paläontolog und Geologie Oesterreich–Ungarns und des Orients*, **6**, 1–227; **7**, 1–117.

Omomys

Leidy, J. (1869). Notice of some extinct vertebrates from Wyoming and Dakota. *Proceedings of the Academy of Natural Sciences, Philadelphia*, **21**, 63–67.
Gazin, C. L. (1958). A review of the middle and upper Eocene Primates of North America. *Smithsonian Miscellaneous Collections*, **136**, 1–112.

Ourayia

Osborn, H. F. (1895). Fossil mammals of the Uinta Basin. Expedition of 1894. *Bulletin of the American Museum of Natural History*, **7**, 71–105.
Gazin, C. L. (1958). A review of the middle and upper Eocene Primates of North America. *Smithsonian Miscellaneous Collections*, **136**, 1–112.
Simons, E. L. (1961*b*). The dentition of *Ourayia*: its bearing on relationships of omomyid primates. *Postilla*, **54**, 1–20.
Robinson, P. (1968). The paleontology and geology of the Badwater Creek area, central Wyoming. IV. Late Eocene primates from Badwater, Wyoming, with a discussion of material from Utah. *Annals of the Carnegie Museum*, **39**, 307–326.

Pseudoloris

Filhol, H. (1889–90). Description d'une nouvelle espèce de Lémurien fossile (*Necrolemur parvulus*). *Bulletin de la Société Philomathique, Paris*, **8**, 39–40.
Stehlin, H. G. (1916). Die Säugetiere des schweizerischen Eocäens. Critischer Catalog der Materialen. VIIb. *Abhandlungen schweizerischen Paläontologischen Gesellschaft*, **41**, 1299–1552.
Crusafont-Pairo, M. (1967). Sur quelques prosimiens de l'Eocène de la zone préaxiale pyrénaique et un essai provisoire de reclassification. *Colloques internationaux du Centre national de la recherche scientifique, Problèmes actuels de paléontologie*, **163**, 611–632.
Louis, P. & Sudre, J. (1975). Nouvelles données sur les primates de l'Éocène supérieur Européen. *Colloques internationaux du Centre national de la recherche scientifique*, **218**, 805–828.
Köhler, M. & Moyà-Solà, S. (1999). A finding of Oligocene primates on the European continent. *Proceedings of the National Academy of Sciences of the United States of America*, **96**, 14664–14667.

Pseudotetonius

Matthew, W. D. (1915*a*). A revision of the lower Eocene Wasatch and Wind River faunas. IV. Entelonychia, Primates, Insectivora (part). *Bulletin of the American Museum of Natural History*, **34**, 429–483.
Bown, T. M. (1974). Notes on some early Eocene anaptomorphine primates. *Contributions in Geology, University of Wyoming*, **13**, 19–26.

Rooneyia

Wilson, J. A. (1966). A new primate from the earliest Oligocene, west Texas, preliminary report. *Folia Primatologica*, **4**, 227–248.

Shoshonius

Granger, W. (1910). Tertiary faunal horizons in the Wind River Basin, Wyoming, with descriptions of new Eocene mammals. *Bulletin of the American Museum of Natural History*, **28**, 235–251.

Simpson, G. G. (1959). Primates. *Bulletin of the American Museum of Natural History*, **117**, 152–157.

Honey, J. G. (1990). New washakiin primates (Omomyidae) from the Eocene of Wyoming and Colorado, and comments on the evolution of the Washakiini. *Journal of Vertebrate Paleontology*, **10**, 206–221.

Sphacorhysis

Gunnell, G. F. (1995*a*). Omomyid primates (Tarsiiformes) from the Bridger Formation, middle Eocene, southern Green River Basin, Wyoming. *Journal of Human Evolution*, **28**, 147–187.

Steinius

Matthew, W. D. (1915*a*). A revision of the lower Eocene Wasatch and Wind River faunas. IV. Entelonychia, Primates, Insectivora (part). *Bulletin of the American Museum of Natural History*, **34**, 429–483.

Bown, T. M. & Rose, K. D. (1984). Reassessment of some early Eocene Omomyidae, with description of a new genus and three new species. *Folia Primatologica*, **43**, 97–112.

Bown, T. M. & Rose, K. D. (1991). Evolutionary relationships of a new genus and three new species of omomyid primates (Willwood Formation, Lower Eocene, Bighorn Basin, Wyoming). *Journal of Human Evolution*, **20**, 465–480.

Stockia

Gazin, C. L. (1958). A review of the middle and upper Eocene Primates of North America. *Smithsonian Miscellaneous Collections*, **136**, 1–112.

Strigorhysis

Bown, T. M. (1979). New omomyid primates (Haplorhini, Tarsiiformes) from middle Eocene rocks of west-central Hot Springs County, Wyoming. *Folia Primatologica*, **31**, 48–73.

Bown, T. M. & Rose, K. D. (1987). Patterns of dental evolution in early Eocene anaptomorphine primates (Omomyidae) from the Bighorn Basin, Wyoming. *Journal of Paleontology*, **61**(**5** Supplement), 1–162.

Tarsius

Ginsburg, L. & Mein, P. (1987). *Tarsius thailandica* nov. sp., premier Tarsiidae (Primates, Mammalia) fossile d'Asie. *Comptes rendus de l'Académie des sciences de Paris*, **304**, 1213–1215.

Musser, G. G. & Dagosto, M. (1987). The identity of *Tarsius pumilus*, a pygmy species endemic to the montane mossy forests of central Sulawesi. *American Museum Novitates*, **2867**, 1–53.

Niemitz, C., Nietsch, A., Warter, S., & Rumpler, Y. (1991). *Tarsius dianae*: a new primate species from central Sulawesi (Indonesia). *Folia Primatologica*, **56**, 105–116.

Beard, K. C., Qi, T., Dawson, M. R., Wang, B., & Li, C. (1994). A diverse new primate fauna from middle Eocene fissure-fillings in southeastern China. *Nature*, **368**, 604–609.

Tatmanius

Bown, T. M. & Rose, K. D. (1991). Evolutionary relationships of a new genus and three new species of omomyid primates (Willwood Formation, Lower Eocene, Bighorn Basin, Wyoming). *Journal of Human Evolution*, **20**, 465–480.

Teilhardina

Teilhard de Chardin, P. (1927). Les mammifères de l'Éocène inférieur de la Belgique. *Mémoires du Muséum royal d'histoire naturelle de Belgique*, **36**, 1–33.

Jepsen, G. L. (1930). New vertebrate fossils from the lower Eocene of the Bighorn Basin, Wyoming. *Proceedings of the American Philosophical Society*, **69**, 117–131.

Simpson, G. G. (1940). Studies of the earliest primates. *Bulletin of the American Museum of Natural History*, **77**, 185–212.

Bown, T. M. (1976). Affinities of *Teilhardina* (Primates, Omomyidae) with description of a new species from North America. *Folia Primatologica*, **25**, 62–72.

Bown, T. M. & Rose, K. D. (1987). Patterns of dental evolution in early Eocene anaptomorphine primates (Omomyidae) from the Bighorn Basin, Wyoming. *Journal of Paleontology*, **61**(**5** Supplement), 1–162.

Gingerich, P. D. (1993). Early Eocene *Teilhardina brandti*: oldest omomyid primate from North America. *Contributions from the Museum of Paleontology, University of Michigan*, **28**, 321–326.

Rose, K. D. (1995). Anterior dentition and relationships of the early Eocene omomyids *Arapahovius advena* and *Teilhardina demissa*, sp. nov. *Journal of Human Evolution*, **28**, 231–244.

Tetonius

Cope, E. D. (1882). Contributions to the history of the Vertebrata of the lower Eocene of Wyoming and New Mexico, made during 1881. I. The fauna of the Wasatch beds of the basin of the Bighorn River. II. The fauna of the *Catathlaeus* beds, or lowest Eocene, New Mexico. *Proceedings of the American Philosophical Society*, **20**, 139–197.

Matthew, W. D. (1915*a*). A revision of the lower Eocene Wasatch and Wind River faunas. IV. Entelonychia, Primates, Insectivora (part). *Bulletin of the American Museum of Natural History*, **34**, 429–483.

Bown, T. M. & Rose, K. D. (1987). Patterns of dental evolution in early Eocene anaptomorphine primates (Omomyidae) from the Bighorn Basin, Wyoming. *Journal of Paleontology*, **61**(**5** Supplement), 1–162.

Trogolemur

Matthew, W. D. (1909). The Carnivora and Insectivora of the Bridger Basin, middle Eocene. *Memoirs of the American Museum of Natural History*, **9**, 291–567.

Storer, J. E. (1990). Primates of the Lac Pelletier Lower Fauna (Eocene: Duchesnean), Saskatchewan. *Canadian Journal of Earth Sciences*, **27**, 520–524.

Beard, K. C., Krishtalka, L., & Stucky, R. K. (1992). Revision of the Wind River faunas, Early Eocene, central Wyoming. XII. New species of omomyid primates (Mammalia: Primates: Omomyidae) and omomyid taxonomic composition across the Early–Middle Eocene boundary. *Annals of Carnegie Museum*, **61**, 39–62.

Uintanius

Matthew, W. D. (1915*a*). A revision of the lower Eocene Wasatch and Wind River faunas. IV. Entelonychia, Primates, Insectivora (part). *Bulletin of the American Museum of Natural History*, **34**, 429–483.

Robinson, P. (1966). Fossil Mammalia of the Huerfano Formation, Eocene, of Colorado. *Peabody Museum of Natural History, Bulletin*, **21**, 1–95.

Beard, K. C., Krishtalka, L., & Stucky, R. K. (1992). Revision of the Wind River faunas, Early Eocene, central Wyoming. XII. New species of omomyid primates (Mammalia: Primates: Omomyidae) and omomyid taxonomic composition across the Early–Middle Eocene boundary. *Annals of Carnegie Museum*, **61**, 39–62.

Utahia

Gazin, C. L. (1958). A review of the middle and upper Eocene Primates of North America. *Smithsonian Miscellaneous Collections*, **136**, 1–112.

Washakius

Leidy, J. (1873). Contribution to the extinct vertebrate fauna of the western territories. *Reports of the United States Geological Survey of Territory*, Part 1.

Stock, C. (1938). A tarsiid primate and a mixodectid from the Poway Eocene, California. *Proceedings of the National Academy of Sciences of the United States of America*, **24**, 288–293.

Simpson, G. G. (1959). Primates. *Bulletin of the American Museum of Natural History*, **117**, 152–157.

Honey, J. G. (1990). New washakiin primates (Omomyidae) from the Eocene of Wyoming and Colorado, and comments on the evolution of the Washakiini. *Journal of Vertebrate Paleontology*, **10**, 206–221.

Gunnell, G. F., Bartels, W. S., Gingerich, P. D., & Torres, V. (1992). Wapiti Valley faunas: early and middle Eocene fossil vertebrates from the North Fork of the Shoshone River, Park County, Wyoming. *Contributions from the Museum of Paleontology, University of Michigan*, **29**, 247–287.

Wyomomys

Gunnell, G. F. (1995*a*). Omomyid primates (Tarsiiformes) from the Bridger Formation, middle Eocene, southern Green River Basin, Wyoming. *Journal of Human Evolution*, **28**, 147–187.

Xanthorhysis

Beard, K. C. (1998). A new genus of Tarsiidae (Mammalia: Primates) from the middle Eocene of Shanxi Province, China, with notes on the historical biogeography of tarsiers. *Bulletin of the Carnegie Museum of Natural History*, **34**, 260–277.

Yaquius

Mason, M. A. (1990). New fossil primate from the Uintan (Eocene) of southern California. *PaleoBios*, **13**, 1–7.

6 | Fossil lorisoids

ERICA M. PHILLIPS AND ALAN WALKER

Introduction

The prosimian superfamily Lorisoidea has a patchy fossil record extending as far back as the early Miocene. East African sites in Kenya and Uganda have yielded fossil evidence of lorisoids by 18 to 20 million years ago (Walker, 1978). If recent estimates of divergence dates for the strepsirhine primates (Yoder, 1997) are correct, then these lorisoids lived almost 40 million years after the origin of the superfamily. Cranial and dental features of some key fossils indicate that the African Miocene lorisoids also postdate the divergence of the families Lorisidae and Galagidae (Le Gros Clark & Thomas, 1952; Walker 1974*b*). At present, the early Miocene lorisids are represented by at least one genus, *Mioeuoticus*, while the galagids are represented by the genera *Progalago* and *Komba*. Unfortunately, the sparse and fragmentary nature of the fossil record has challenged our efforts to understand the early evolutionary stages of the families Lorisidae and Galagidae. Although lorisoid fossil material is fairly abundant at 18 to 20 Ma, their fossil record almost entirely disappears from the middle Miocene to the early Pliocene. The 14-million-year-old site of Fort Ternan and the 15-million-year-old site of Maboko Island in Kenya together have fossil evidence of three species, but each is represented by only one or two fossils (Walker, 1978; Shipman *et al.*, 1981; McCrossin, 1992*b*, 1999*b*). The Siwaliks of Pakistan have yielded fossils of a 9-Ma lorisid called *Nycticeboides*, as well as possibly additional taxa represented by isolated teeth (MacPhee & Jacobs, 1986; J.C. Barry, pers. comm.). By the middle Pliocene and early Pleistocene, lorisoid fossils are known again, but this time with fossils representing the earliest members of the living genus *Galago* from sites in East Africa.

History of discovery and debate

In 1931, Arthur Tindell Hopwood found the first lorisoid fossils, a fragmentary right mandibular ramus and left maxilla, at the site of Koru, Kenya. Unfortunately, these fossil fragments were not described until 36 years after their discovery. In 1933, G. Edward Lewis designated an isolated mandibular M_2 from the Siwaliks of India as the holotype for an ancestral lorisid called *Indraloris lulli* (Lewis, 1933). This specimen was later identified as an adapoid (Gingerich & Sahni, 1979; Qi & Beard, 1998). Thus, it was not until 1943 that Donald Gordon MacInnes provided the first published description of a true fossil lorisoid based upon a fragment of a left mandibular ramus with P_4 and M_2 (KNM-SO 379) from 20-million-year-old deposits at Songhor, Kenya. Dental similarities with living galagos led MacInnes to conclude that this fossil represented an ancestral galagid. The fragmentary mandible with two teeth became the holotype for a genus and species called *Progalago dorae* (MacInnes, 1943).

The holotype for *P. dorae* remained the total sum of the fossil lorisoid collection for nearly another decade. In 1952, Le Gros Clark and David Prys Thomas added to this meager hypodigm a large collection of lorisoid fossils found at Rusinga Island and Songhor during the 1947–1950 British–Kenya Miocene Expedition. Over a dozen lorisoid mandibular fragments, two partial maxillas and fragments of a basicranium associated with a natural endocranial cast were collected at these two early Miocene sites. With this new material, Le Gros Clark & Thomas (1952) expanded the hypodigm of *P. dorae*, as well as named two additional species, *Progalago robustus* and *Progalago minor*. The size of the mandible and associated dentition suggested that *P. dorae* was the largest of the three species, followed next by *P. robustus* and then by *P. minor*. All three species were believed to be more closely related to living galagos than to lorises, although *Progalago* had not yet achieved a fully modern galagid aspect. Particularly remarkable among this new collection of fossils was the partial skull KNM-RU 1940 and associated endocast. This specimen was represented by the posterior portions of the left and right parietals and temporals, most of the occipital bone, and an almost complete endocranial cast. Like the mandible and maxilla fossil fragments, this specimen resembled living galagos (Le Gros Clark & Thomas, 1952). Because the specimen was missing both the maxilla and mandible necessary for comparison with the hypodigms of the three *Progalago* species, Le Gros Clark & Thomas (1952) could only tentatively conclude that the specimen's large size aligned it with either *P. dorae* or *P. robustus*.

A much more complete lorisoid cranium, KNM-RU 2052, was found at Rusinga Island, Kenya in 1954 by Mary D. Leakey. The skull was nearly complete except for the lateral portions of the right and left orbits, the anterior maxillary dentition and the mandible. Two years after its discovery, Le Gros Clark published a detailed anatomical description of the Rusinga cranium (Le Gros Clark, 1956). The fossil's virtually unparalleled level of preservation provided a considerable amount of information regarding the dental, facial and cranial anatomy of the Miocene lorisoids. Despite the Rusinga cranium's excellent morphological detail, Le Gros Clark hesitated to draw any definite conclusions regarding the Rusinga cranium's phylogenetic relationship with living lorisoids. According to Le Gros Clark, the Rusinga cranium exhibited a unique mixture of galagid and Asian and African lorisid features (Le Gros Clark, 1956). The fossil cranium resembled extant galagids in the presence of a strong nuchal crest and the small size of the foramen lacerum in the basicranium. However, the fossil also shared with extant lorisids such features as an abbreviated snout and the absence of an inflated bulla (Le Gros Clark, 1956). As was the case with the partial cranium and endocast described in 1952, without a mandible, comparison with other Miocene lorisoid fossils could do little to clarify this ambiguous morphology (MacInnes, 1943; Le Gros Clark & Thomas, 1952). For convenience, Le Gros Clark temporarily assigned the Rusinga cranium to the genus *Progalago*, but deferred assigning the fossil to any particular species. Le Gros Clark predicted that future finds might later demonstrate that the Rusinga cranium represented a new species of *Progalago* or perhaps even a new, as yet unnamed, genus (Le Gros Clark, 1956).

In 1962, Louis S.B. Leakey named and described a new genus and species of fossil lorisoid he called *Mio-euoticus bishopi* Leakey 1962 (the hyphen in the genus name was later removed by Simpson (1967) in accordance with the Code of Zoological Nomenclature Article 32(c)(i)). The type specimen (NAP 1.3.6/58) was represented by the facial part of a skull and some associated maxillary dentition from the early Miocene site of Napak, Uganda. According to Leakey, similarities in the maxillary, palate and nasal regions of the Napak face aligned *Mioeuoticus bishopi* with living members of the galagid genus *Euoticus* (Leakey, 1962*a*). Others (Simpson, 1967; Walker, 1978) later challenged this conclusion, pointing out that the palate had been distorted from its original shape and did not particularly resemble extant galagids.

In 1967, George Gaylord Simpson presented a complete review of the Miocene lorisoid fossils collected to date and in the process made some fairly substantial revisions to the taxonomy presented by Le Gros Clark & Thomas in 1952. The wide size range included within *P. dorae* prompted Simpson to divide the hypodigm into two species, the original *P. dorae* as described by Le Gros Clark & Thomas (1952) and a smaller species Simpson called *Progalago songhorensis*. Furthermore, Simpson proposed that the morphological differences between *P. dorae* and the smaller *P. robustus* and *P. minor* were significant at the generic as well as the specific level, necessitating reassignment of the latter two species to a new genus he called *Komba*. Finally, he expressed reservations regarding the definition of the genus *Mioeuoticus*. Simpson felt that many of the dental and cranial features described by Leakey were inaccurate or ambiguous. Of those that remained, Simpson doubted their generic distinctiveness. Although Simpson ultimately retained *Mioeuoticus* as a distinct taxon, he suggested that the Napak face might be better still accommodated within *Progalago* (Simpson, 1967).

Unlike his predecessors, Simpson refused to speculate on the phylogenetic relationship between *Progalago*, *Komba* and *Mioeuoticus* and living lorisids and galagids (Simpson, 1967). So far, all three Miocene genera had been linked directly with the family Galagidae. However, Simpson argued that each of the existing taxa included specimens with a mixture of lorisid and galagid features, perhaps a condition not to be unexpected in the earliest lorisoids. In the absence of postcranial material that might have shed some light on the locomotor adaptations of the Miocene lorisoids, Simpson concluded that any family distinctions were meaningless (Simpson, 1967). Ironically, Simpson made an exception for one Miocene fossil lorisoid he believed to be clearly a lorisid, a new genus and species he named *Propotto leakeyi* based upon one left and two right mandibular rami and some associated premolars and molars. Walker (1969*a*) demonstrated that these specimens were actually fruit bats from the family Pteropidae.

In 1969, William W. Bishop discovered a left mandibular fragment from Napak that confirmed the presence of a toothcomb similar to that of living lorisoids by 19 Ma. Up until that time, all known lorisoid fossils were missing the anterior dentition and much of the associated alveolar process, leading to some speculation about the degree of procumbency of the lower incisors. Le Gros Clark & Thomas (1952) concluded that this feature was not expressed as strongly in the Miocene lorisoids as in modern lorisiforms. Simpson (1967) disagreed, finding no difference in the form of the anterior alveolar process in modern or fossil lorisoids. Bishop's partial mandible included an almost entirely intact alveolar process, confirming the presence of forwardly inclined lower incisors comparable to that of living lorisoids by the early Miocene (Walker, 1969*b*). Since then a lower comb-tooth complete with root and crown has been discovered at Koru. It is as procumbent as those in extant forms.

The first Miocene lorisoid postcranial remains were discovered within existing museum collections in 1970, but initially did little to clarify the phylogenetic relationship between fossil and living lorisoids. Alan Walker's (1970) description and analysis of 17 complete and fragmentary limb bones collected at Songhor, Kenya and Napak, Uganda suggested a galago-like vertical clinging and leaping locomotor pattern for all of the known Miocene lorisoid genera. Lacking direct association between the lorisoid cranial and dental fossils and any of the new postcranial material, Walker sorted the postcranial material by size, using modern species of *Galago* as a standard by which to judge species-level body size differences. Fragments of the humerus, femur, tibia and distal hallucial phalanx matched almost exactly the morphology seen in living galagos. According to Walker, the only significant difference between the skeletal morphology of extant galagos and these fossil specimens was the presence of a more distally shortened calcaneus. Even so, the fossil calcanei were still longer than seen in any living arboreal quadrupedal prosimian and so still implied a vertical clinging and leaping locomotor pattern. Walker's results had two possible implications: (1) all of the then known Miocene lorisoid fossils were galagids, with as yet no known Miocene lorisids, or (2) all Miocene lorisoids retained a primitive galago-like locomotion with the locomotor specializations of the family Lorisidae developing during the late Miocene or Pliocene (Walker, 1969, 1970). Walker favored the latter explanation and even suggested that the living lorisids' adaptations for slow climbing and suspensory quadrupedalism may have evolved on multiple occasions during the evolutionary history of the lorisids. In other words, the family Lorisidae could perhaps be better understood as a morphological taxon rather than as a phylogenetic one (Walker, 1969, 1970).

By 1974, the discovery of additional postcranial material as well as associated maxillary and mandibular dentitions prompted Walker to revise some of his earlier conclusions (Walker, 1974*b*, 1978). Specifically, Walker now believed that a family distinction within the early Miocene lorisoid material was possible, with the genus *Mioeuoticus* belonging to the family Lorisidae and the genera *Progalago* and *Komba* to the family Galagidae. According to Walker, all of the postcranial material discovered thus far was of fossil galagids, with no known material representing fossil lorisids. However, both the Rusinga cranium and Napak face exhibited at least six features suggesting a lorisid affinity: a strongly constructed cranium, distinctly raised temporal ridges, upward-facing orbits, wide internal nares, slight basicranial flexion, and a weakly inflated bulla and mastoid (Walker, 1974*b*, 1978). In contrast, the partial cranium with endocast shared a number of morphological features with living galagos, including a more lightly constructed cranium, weaker temporal ridges, greater basicranial flexion, and a more strongly inflated bulla. In light of these conclusions, Walker removed the Rusinga cranium from *Progalago* and designated it the type specimen for a new species within *Mioeuoticus* (Walker, 1974*b*). The name *Mioeuoticus shipmani* was later given to this taxon (Phillips & Walker, 2000). Based upon size, the partial cranium with endocast was one of either two galagid species, *K. robustus* or *P. songhorensis* (Walker, 1974*b*).

Fred Sigmond Szalay and C.C. Katz (1973) held a very different interpretation of the Rusinga cranium and the partial cranium with endocast. According to these authors, the partial cranium with endocast exhibited basicranial features and carotid circulation consistent with a phylogenetic position intermediate between hypothetical Oligocene and early Miocene-aged cheirogaleids and later Miocene and recent galagids. The Rusinga cranium shared the recent galagid condition, although essentially contemporaneous in age as the partial cranium and endocast. This interpretation of the fossil record supported the idea of a monophyletic lorisoid/cheirogalid group. Similarities in morphology, behavior and ecology between living cheirogalids and galagos prompted a number of researchers (Cartmill, 1975; Schwartz *et al.*, 1978) to conclude that the cheirogalids were more closely related to the lorisoids than to other lemurs. This was reflected in several primate taxonomies at the time (Schwartz *et al.*, 1978; Szalay & Delson, 1979; Schwartz & Tattersall, 1987) in which the family Cheirogaleidae was removed from Lemuroidea and placed within the superfamily Lorisoidea. Since then, molecular studies have overwhelmingly supported the existence of a monophyletic lemuroid clade (Dene *et al.*, 1976*a*, 1976*b*; Sarich & Cronin, 1976; Goodman *et al.*, 1994; Yoder, 1994, 1997; Yoder *et al.*, 1996), indicating that the similarities between the cheirogalids and galagos are either symplesiomorphic or the result of convergence.

In the 1980s, Daniel L. Gebo reexamined the lorisoid fossil postcranial material in two studies. The first focused on foot bones of *K. robustus*, *K. minor* and *P. songhorensis* (Gebo, 1986), while the second offered a more comprehensive discussion of the entire lorisoid postcranial anatomy (Gebo, 1989). Gebo confirmed Walker's (1970, 1974*b*) description of overall galago-like postcranium. However, Gebo also described a number of features of the calcaneus and talus in the foot that were more reminiscent of living lorises than galagos (Gebo 1986). Still other features of the distal humerus and proximal tibia gave these elements a more cheirogaleid-like appearance. The mosaic nature of the postcranial material implied a more generalized locomotor pattern than seen in either the living lorisids or galagids. Gebo concluded that the Miocene lorisoids' locomotion may be better approximated by the combination of quadrupedalism, climbing, leaping and suspension that characterizes living members of the family Cheirogaleidae. Although Gebo followed Walker (1974*b*, 1978) in allocating *Progalago* and *Komba* to the family Galagidae and *Mioeuoticus* to the family Lorisidae, he doubted that an unambiguous allocation could yet be made. If the postcranial resemblances between the Miocene genera and living galagos were primitive features, then they were of little help in allocation of Miocene fossil material to either the Lorisidae or Galagidae families.

The first lorisoid fossils found outside of Africa were described in 1986. A joint expedition by Yale University and the Geological Survey of Pakistan recovered several isolated teeth and a partial skeleton of a single individual (YGSP 8091) from middle Miocene localities in the Siwalik Group of northern Pakistan (Jacobs, 1981). The partial skeleton was represented by a fairly large portion of the maxilla and mandible, including nearly all of the dentition, as well as fragmentary material from the orbital region, right zygomatic, cranial vault, vertebral column, ribs, scapulas and forelimbs (MacPhee & Jacobs, 1986). Louis L. Jacobs designated this material as the type specimen for a new genus and species called *Nycticeboides simpsoni*. Features of the dentition and skeleton indicated that this was a true lorisid, complete with locomotor adaptations indicative of slow-climbing abilities (Jacobs, 1981). Several additional isolated teeth recovered from earlier sediments in the Siwaliks may represent additional taxa (MacPhee & Jacobs, 1986).

The discovery of *Nycticeboides simpsoni* gave researchers one of few glimpses of lorisoids after the early Miocene. With the exception of a single, as yet unattributed, maxillary fragment from the 14-million-year-old site of Fort Ternan, Kenya (Walker, 1978; Shipman *et al.*, 1981) and several fragmentary fossils from Pleistocene deposits at Olduvai Gorge described by Simpson (1965) as indistinguishable from living members of *Galago senegalensis*, all of the African lorisoid fossils came from a 2-million-year window between 18 and 20 million years ago. In the mid-1980s, discoveries of Plio-Pleistocene specimens provided new information about lorisoid evolution after the Miocene. In 1984, Henry (Hank) Wesselman described a partial maxilla and an isolated tooth from Omo, Ethiopia as *Galago howelli*, the first fossil species referred to an extant genus. Wesselman (1984) believed that *G. howelli* was most likely affiliated with the living species *G. crassicaudatus*. Additional isolated teeth were referred to *G. demidovii* and *G. senegalensis*. This was followed by the discovery of *G. sadimanensis* from Laetoli, Tanzania and Kapchebrit, Kenya (Walker, 1987). *Galago sadimanensis* possessed derived features of the mandibular symphysis and second premolar that indicated that it was not ancestral to living galagos (Walker, 1978).

In 1989, Brenda R. Benefit and Monte L. McCrossin announced the discovery of fossil lorisoid remains recovered from 14 to 16 Ma deposits at Maboko Island, Kenya. The right portion of a mandibular body and four associated teeth were made the holotype for a new species called *Komba winamensis* (McCrossin, 1992*b*). McCrossin recognized *K. winamensis* as a galagid, supporting Walker's earlier allocation of that genus to the family Galagidae (Benefit & McCrossin, 1989; McCrossin, 1992*b*). However, the absence of several derived galagid features led McCrossin to conclude that *Komba* represented a side branch in early galagid evolution. *Progalago*, on the other hand, shared several features with *Mioeuoticus* and living lorisids, including a distinct metaconid on the lower fourth premolar, a square upper second molar and a mandible that deepened posteriorly. McCrossin interpreted this morphology to mean *Progalago* was a lorisid. However, it is not clear that those features described by McCrossin actually preclude *Progalago* from the family Galagidae. As Rasmussen & Nekaris (1998) have pointed out, the morphology described by McCrossin of the lower fourth premolar and upper second molar is primitive. Additionally, not all specimens included within the *Progalago* hypodigm have a square second upper premolar. The partial maxilla KNM-SO 1361, which occludes well with the holotype for *Progalago dorae*, has an upper second molar with a distinct distal notch. Not all living lorisids have mandibles that deepen posteriorly. Both *Loris* and *Arctocebus* are like living galagos and *Komba* in that their mandibles maintain an equal depth along their entire length.

Unfortunately, there have been few new developments in the study of the lorisoid fossil record during the last decade. Although additional lorisoid fossils have been found, many of which certainly represent new fossil taxa, these have not yet been described. Isolated teeth from the Siwaliks of India may represent additional taxa besides *Nycticeboides simpsoni* (J.C. Barry, pers. comm.). McCrossin (1999*b*) has announced the discovery of another fossil galago from Maboko Island, Kenya. Initial accounts describe the specimens as representing a new genus and species smaller in body size than any other Miocene fossil taxon (McCrossin, 1999*b*).

Taxonomy

Systematic framework

As recommended by Schwartz *et al.* (1998), the taxonomy presented below uses the family names Lorisidae and Galagidae. The allocation of the Miocene lorisoid genera to these two families is based on taxonomies presented by Walker (1974*b*, 1978). Walker's (1978) classification has been followed by most researchers in broad outline but the reader should be aware that a few authors (e.g. McCrossin, 1992*b*; Rasmussen & Nekaris, 1998) have proposed alternate taxonomies for some genera. The present taxonomy differs from that of Walker (1978) in the addition of several new genera and species, including *Nycticeboides* from the late Miocene of Asia and *Galago* from the Plio-Pleistocene of East Africa.

Order Primates Linnaeus, 1758
- Suborder Prosimii Illiger, 1811
  - Infraorder Lemuriformes Gregory, 1915
    - Superfamily Lorisoidea Gray, 1821
      - Family Lorisidae Gray, 1821
        - Genus *Mioeuoticus* Leakey, 1962
          - *Mioeuoticus bishopi* Leakey, 1962
          - *Mioeuoticus shipmani* Phillips & Walker, 2000
        - Genus *Nycticeboides* Jacobs, 1981
          - *Nycticeboides simpsoni* Jacobs, 1981
      - Family Galagidae Gray, 1825
        - Genus *Progalago* MacInnes, 1943
          - *Progalago dorae* MacInnes, 1943
          - *Progalago songhorensis* Simpson, 1967
        - Genus *Komba* Simpson, 1967
          - *Komba robustus* (Le Gros Clark & Thomas, 1952)
          - *Komba minor* (Le Gros Clark & Thomas, 1952)
          - *Komba winamensis* McCrossin, 1992
        - Genus *Galago* Geoffroy, 1796
          - *Galago howelli* Wesselman, 1984
          - *Galago sadimanensis* Walker, 1987

Fig. 6.1 Inferior view of holotype of *Mioeuoticus bishopi*, NAP 1.3.6/58. Scale is 3 cm.

Family Lorisidae

GENUS *Mioeuoticus* Leakey, 1962
P^2 is a monocuspid tooth with two roots. P^4 is a bicuspid, nearly square, non-molariform tooth with a strong distolingual cingulum. Molars are bunodont with low crowns. M^1 and M^2 are relatively large, nearly square in outline, and possess a large hypocone. M^3 is triangular in outline due to a reduction in the hypocone. Dentition larger than in *Progalago*. Strong maxillary jugum. Large orbits with thin orbital floors. Strongly constructed cranium, distinctly raised temporal ridges, upwardly-facing orbits, weakly inflated auditory bulla and mastoid, wide internal nares and slight basicranial flexion.
INCLUDED SPECIES *M. bishopi*, *M. shipmani*

SPECIES *Mioeuoticus bishopi* Leakey, 1962 (Fig. 6.1)
TYPE SPECIMEN NAP.I.3.6/58, greater portion of the maxilla, infraorbital region and palate, right P^2 and P^4–M^2, left P^4–M^3
AGE AND GEOGRAPHIC RANGE Early Miocene, Napak, Uganda
ANATOMICAL DEFINITION
Mioeuoticus bishopi is similar in size to *Progalago songhorensis* and the living lorisid *Arctocebus calabarensis*. Unlike *M. shipmani*, P^2 is two-rooted while P^3 and P^4 are three-rooted. P^4 is parallel-sided, with nearly equal lingual and buccal widths and a strong lingual cingulum. M^2 is narrower buccolingually than in *M. shipmani* and has a more developed lingual cingulum (or possibly a less developed hypocone that has merged with the cingulum). M^3 is more oval in outline relative to *M. shipmani*. Single forward-facing infraorbital foramen far from orbital margin.

SPECIES *Mioeuoticus shipmani* Phillips & Walker, 2000 (Fig. 6.2)
TYPE SPECIMEN KNM-RU 2052, nearly complete cranium
AGE AND GEOGRAPHIC RANGE Early Miocene, Rusinga Island, Kenya
ANATOMICAL DEFINITION
Similar in size to the living lorisid *Perodicticus potto*. Unlike *M. bishopi*, both P^3 and P^4 are two-rooted. P^4 is wider mesiodistally on the buccal side than on the lingual side and has a cingulum limited to the distal corner. M^2 is wider buccolingually than in *M. bishopi* and has a groove that separates the hypocone and protocone. Molars exhibit strong buccal cingula but weak lingual cingula. M^3 has a more triangular outline than *M. bishopi*. Upper molars are larger than in *M. bishopi*. Single downward-facing infraorbital foramen near the orbital margin.

GENUS *Nycticeboides* Jacobs, 1981
INCLUDED SPECIES *N. simpsoni*

SPECIES *Nycticeboides simpsoni* Jacobs, 1981 (Fig. 6.3)
TYPE SPECIMEN YGSP 8091, partial skeleton
AGE AND GEOGRAPHIC RANGE Late Miocene, Siwalik Group, Pakistan
ANATOMICAL DEFINITION
Comparable in size to the living *Nycticebus coucang pygmaeus*. Possesses a robust vertical ramus, weakly molariform premolars and simple molars lacking the strong cingula that characterize *Mioeuoticus* and *Arctocebus*. Resembles *Nycticebus* in the shape of the anterior border of the ramus and the morphology of P_3 and P_4. Possesses weaker molar labial cingula and stronger styles on P^3 and P^4 than in *Nycticebus*. More molariform P^3 and stronger upper molar hypocones than in *Perodicticus*. Narrower P^4, more rounded molar cusps and stronger P_3 and P_4 cingula than in *Loris*.

Family Galagidae

GENUS *Progalago* MacInnes, 1943
Larger than species of *Komba*. Mandible deepens posteriorly, reaching greatest depth below M_3. Dentition more bunodont relative to *Komba*. Differs from living galagos in that P^4 is not molariform and molar hypocones are reduced. M_3 larger than recent galagids. Molar trigonids and talonids almost equal in height with trigonids narrowed mesiodistally. Prominent hypoconulid on M_3 and well-developed entoconid on lower molars. Lower molars possess distinct internal cingulum. Single mental foramen. Distal calcaneal elongation moderate like that of cheirogaleids.
INCLUDED SPECIES *P. dorae*, *P. songhorensis*

SPECIES *Progalago dorae* MacInnes, 1943 (Figs. 6.4, 6.5)
TYPE SPECIMEN KNM-SO 379, partial left mandibular corpus with P_4 and M_2
AGE AND GEOGRAPHIC RANGE Early Miocene, Koru, Songhor, and Mfangano, Kenya and Napak, Uganda
ANATOMICAL DEFINITION
Larger than *P. songhorensis*. Molar trigonids lack an external cingulum. Molar trigonids are shorter and talonids more

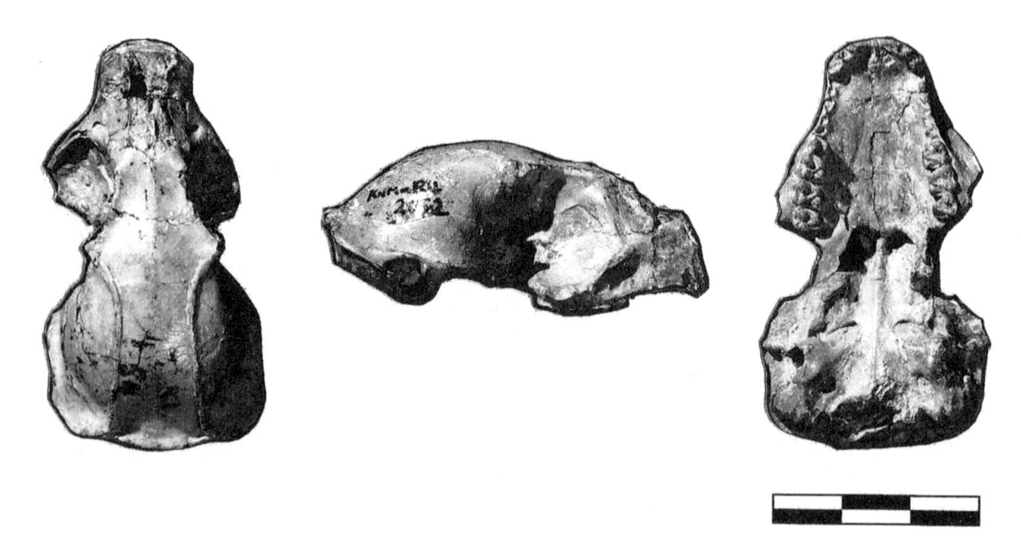

Fig. 6.2 Superior, right lateral, and inferior view of KNM-RU 2052. Originally assigned to *Progalago* (Le Gros Clark, 1956) but later designated the holotype of *Mioeuoticus shipmani* (Walker 1974*b*, 1978; Phillips & Walker, 2000). Scale is 3 cm.

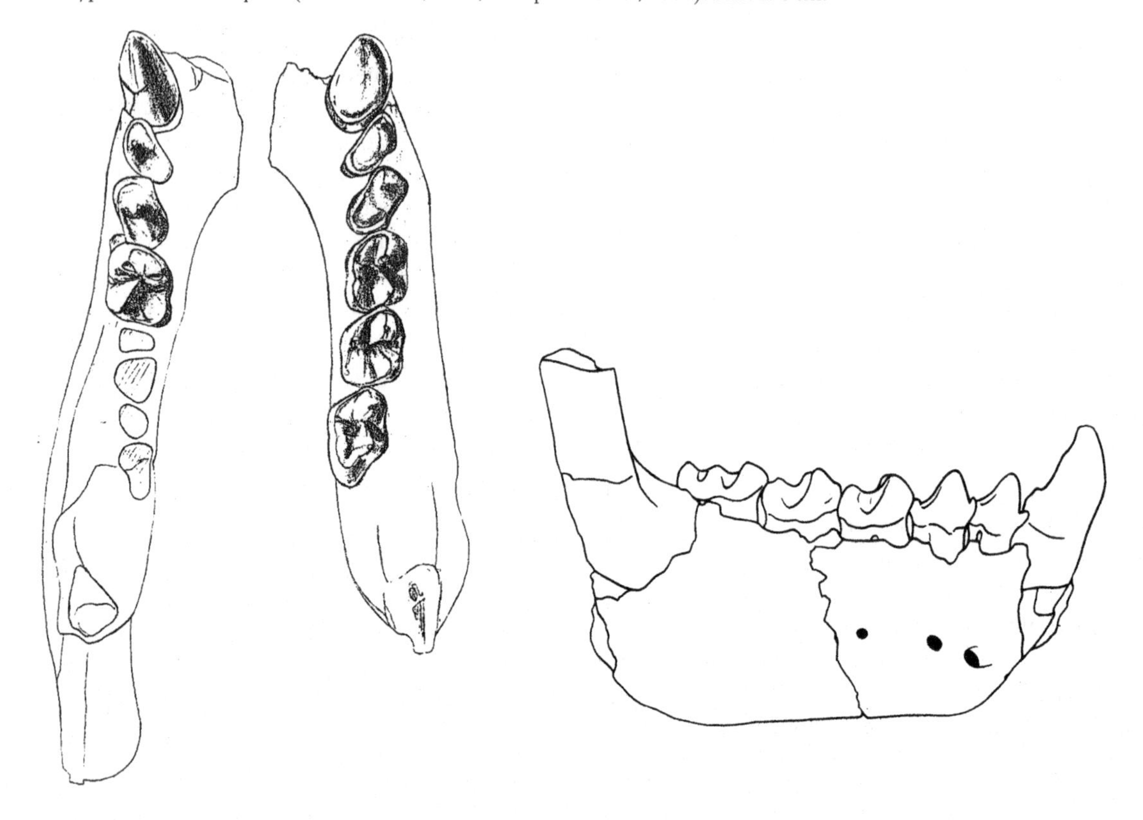

Fig. 6.3 *Nycticeboides simpsoni*. Occlusal and right lateral view of mandible of holotype of (YGSP 8091). Whole scale is 1 cm. Drawings courtesy of Dr. L.L. Jacobs.

expanded relative to *P. songhorensis*. M_2 more rounded than in *P. songhorensis*.

SPECIES *Progalago songhorensis* Simpson, 1967 (Figs. 6.6, 6.7, 6.8)

TYPE SPECIMEN KNM-SO 469, fragment of left mandibular ramus with M_2 and M_3

AGE AND GEOGRAPHIC RANGE Early Miocene, Songhor and Rusinga Island, Kenya

ANATOMICAL DEFINITION

Smaller than *P. dorae*. Molar trigonids are longer and talonids slightly less expanded than *P. dorae*. Molar trigonids exhibit weakly developed external cingulum. Angular M_2 relative to *P. dorae*. Upper teeth are more transversely widened than in *P. dorae*. Postcranially, *P. songhorensis* differs from *K. robustus* in that the dorsal calcaneal edge slopes upward. Similar to *K. minor* in that the height of the posterior calcaneal facet is

Fig. 6.4 *Progalago dorae*. Lateral and occlusal view of holotype of (KNM-SO 379), a left partial mandible with P_4 and M_2. Scale is 2 cm.

Fig. 6.5 *Progalago dorae*. Occlusal view of KNM-SO 1361, Left partial maxilla with P^3 to M^3. Scale is 2 cm.

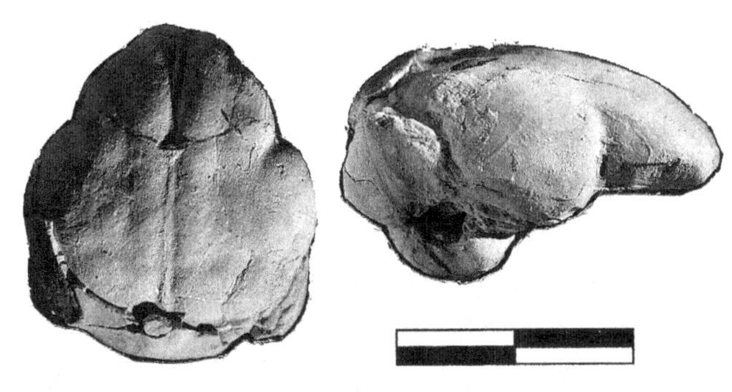

Fig. 6.6 *Progalago songhorensis* or *Komba robustus* (Walker, 1978). Superior and right lateral view of KNM-RU 1940, fragments of basicranium with associated endocast. Scale is 2 cm.

equal to the height of the dorsal heel process but differs by a shorter and more medially angled talar neck.

GENUS *Komba* Simpson, 1967

Smaller than species of *Progalago*. Mandible maintains an equal depth along entire tooth row. Dentition more acrodont than that of *Progalago*. Like *Progalago* (and unlike

Fig. 6.7 *Progalago songhorensis*. Left lateral view of KNM-RU 3420, left partial mandible with P_3 to M_2. Scale is 2 cm.

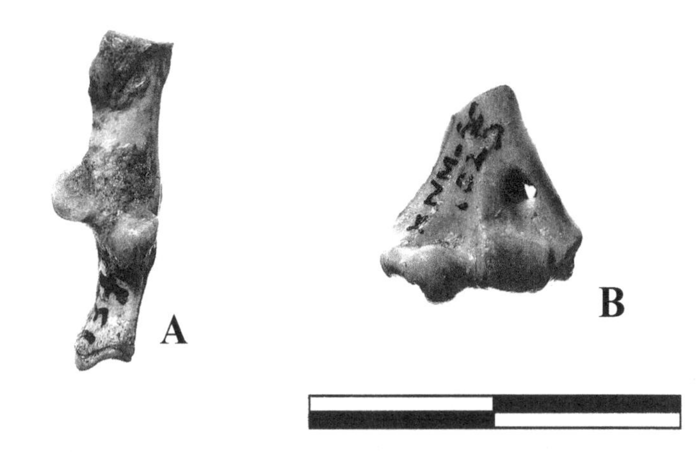

Fig. 6.8 *Progalago songhorensis*, superior view of KNM-SO 1371, right calcaneus. (B) *Komba robustus*, anterior view of KNM-SO 1025, right distal humerus. Scale is 2 cm.

living galagos) P^4 is not molariform and molar hypocones are reduced. P_4 has a narrower talonid relative to trigonid. Molar trigonids are taller than talonids. Distal calcaneal elongation similar to cheirogaleids and shorter than recent galagids.

INCLUDED SPECIES *K. minor*, *K. robustus*, *K. winamensis*

SPECIES *Komba robustus* Le Gros Clark & Thomas, 1952 (Figs. 6.6, 6.8B, 6.9A)

TYPE SPECIMEN KNM-SO 501, fragment of right mandibular ramus

AGE AND GEOGRAPHIC RANGE Early Miocene, Songhor, Rusinga Island, Koru and Mfangano, Kenya and Napak, Uganda

ANATOMICAL DEFINITION

Larger than *K. minor*. P^2 is two-rooted. P^3 and P^4 are three-rooted. P_4 slightly more molariform and M^3 less reduced than in *P. dorae*. Upper molars are more cuspidate and have more distinct molar hypocones than in *Progalago* spp. Lacks external cingula on lower molars. Widely separated, small, double mental foramina. Posterior calcaneal facet higher than dorsal heel process, downward sloping dorsal calcaneal edge, strong anteroposterior

Fig. 6.9 *Komba*. (A) *K. robustus*, right lateral and occlusal view of KNM-SO 1329, right partial mandible with P_3 to M_3; (B) *K. minor*, right lateral view of holotype of (KNM-SO 438), right partial mandible with M_1 to M_3. Scale is 2 cm.

Fig. 6.10 *Komba minor*. Anterior view of KNM-SO 1023, right distal femur. Scale is 2 cm.

curvature of the calcaneus along plantar surface, and strong medial curvature of the heel.

SPECIES *Komba minor* Le Gros Clark & Thomas, 1952 (Figs. 6.9B, 6.10)
TYPE SPECIMEN KNM-SO 438, right mandibular ramus
AGE AND GEOGRAPHIC RANGE Early Miocene, Songhor, Rusinga Island, Legetet and Chamtwara, Kenya and Napak and Moroto, Uganda
ANATOMICAL DEFINITION
Smaller than *K. robustus* but similar in size to the modern *Galago demidovii*. Small external and buccodistal cingulum on lower molars. P_4 shorter and wider with reduced talonid compared to *K. robustus*. Single mental foramen. Relatively long anterior calcaneal length, especially for small body size. Long and straight talar neck, slight anteroposterior curvature of the calcaneus along plantar surface, posterior calcaneal facet of equal height as dorsal heel process, moderately high talar body, small posterior trochlear shelf on talus, and parallel trochlear rims.

SPECIES *Komba winamensis* McCrossin, 1992
TYPE SPECIMEN KNM-MB 20200, right mandiblular fragment
AGE AND GEOGRAPHIC RANGE Middle Miocene, Maboko Island, Kenya
ANATOMICAL DEFINITION
Similar in size to *Galago crassicaudatus*. Molar cusps moderately low and blunt compared to *K. robustus* and *K. minor*. Reduced buccal cingulum on lower molars. Closely positioned, double mental foramina.

GENUS *Galago* Geoffroy, 1796
INCLUDED SPECIES *G. howelli*, *G. sadimanensis*

SPECIES *Galago howelli* Wesselman, 1984
TYPE SPECIMEN L.1-377, partial left maxillary corpus
AGE AND GEOGRAPHIC RANGE Upper Pliocene, Omo, Ethiopia
ANATOMICAL DEFINITION
Intermediate in size between the living species *G. crassicaudatus* and *G. alleni*. Resembles *G. crassicaudatus* in its robust maxilla, square P^4, the presence of a crest extending between the hypocone and metacone of M^1, and straight ventral mandibular edge. Resembles *G. alleni* in its shallow mandible and position of the single, large mental foramen below P^4. Maxillary and mandibular teeth are mesiodistally flattened.

SPECIES *Galago sadimanensis* Walker, 1987
TYPE SPECIMEN LAET 294, partial left mandibular corpus
AGE AND GEOGRAPHIC RANGE Pliocene, Laetoli, Tanzania and Kapchebrit, Kenya
ANATOMICAL DEFINITION
Comparable in size to the living *G. senegalensis*. Mandibular body is vertically deep near the symphyseal region, contributing to an inverted teardrop-shaped sagittal section. P_2 is shorter and more vertically oriented than those of *G. senegalensis*. P_3 and P_4 are nearly tetrahedral in shape, with P_3

lacking the mesial elongation and P_4 the double mesial cusps and enlarged basin found in all other *Galago* species. The molars are indistinguishable in form and within the size range of *G. senegalensis* but are smaller relative to the body of the mandible. Single mental foramen.

Evolution of the lorisoids

A detailed picture of lorisoid evolution is difficult to reconstruct given the small number of fossils and the geographically and temporally limited nature of the sites from which they come. The fossil record gives us brief glimpses of them at 20 to 18 Ma, 14 Ma, 9 Ma, and 4 to 2 Ma. Fortunately, genetic, morphological and biogeographic studies of living lorises and galagos have proven to be valuable sources of information regarding the evolutionary history of the Lorisoidea.

Living members of the families Galagidae and Lorisidae are easily distinguishable in appearance and behavior, particularly with respect to locomotion. Galagos are predominately vertical clingers and leapers while lorises are slow-climbing quadrupedalists. These differing locomotor styles are associated with distinctive postcranial anatomies (Hill, 1953). Galagos are characterized by slender bodies, large ears, long bushy tails, and elongated hindlimbs and feet. Galagos are capable of leaping distances as great as 14 times their body length (Jouffroy, 1974). In contrast, lorises are longer in body, have reduced ears and tails, and hindlimbs and forelimbs of equal length. Their hands, feet, wrists and ankles are specially adapted for increased mobility and stronger grasping abilities (Bearder, 1987; Fleagle, 1999). In both the hand and the foot, the first digit diverges at 180° from the others and the second digit is reduced to a stub, giving their hands a pincer-like appearance. The second digit of the foot is still equipped with a grooming claw. A unique circulation system in the limbs allows lorises to cling to branches for long periods of time without fatigue (Suckling *et al.*, 1969). Lorises move at a sloth-like speed and can travel through the forest canopy by meticulous placement of the hands along the terminal branches of two neighboring trees (Bearder, 1987; Sussman, 1999).

Despite their disparate appearance, molecular and karyological studies confirm that lorises and galagos form a monophyletic lorisoid clade separate from the lemurs (Sarich & Cronin, 1976; Dutrillaux, 1988; Yoder *et al.*, 1996). Within the superfamily Lorisoidea, the distinctive morphology of the living Lorisidae and Galagidae suggests that the two families form separate phylogenetic groups. With few exceptions, this has been the relationship depicted in nearly all contemporary primate phylogenies (e.g., Fleagle, 1999). The Lorisidae are usually further divided at the generic level into an African clade, including *Perodicticus*, *Arctocebus* and *Pseudopotto* (*sensu* Schwartz, 1996), and an Asian clade, composed of *Loris* and *Nycticebus*. However, some researchers (Simpson, 1967; Schwartz & Tattersall, 1987; Schwartz, 1992) prefer to divide the Lorisidae into two groups according to body type, with smaller, slender-bodied forms creating a clade separate from larger, more robust forms. This interpretation generates a different phylogeny since both Africa and Asia each support one slender loris (*Arctocebus* and *Loris*, respectively) and at least one robust loris (*Perodicticus*, *Pseudopotto* and *Nycticebus*). Evaluating the relative merit of these two models depends upon whether one weighs the geographic evidence as stronger than the morphological evidence or vice versa.

Unfortunately, attempts to resolve this conflict through molecular studies have failed. Indeed, their effect has been just the opposite, adding a new level of complexity to the issue. The initial genetic distance studies performed in the 1970s indicated that the Lorisoidea may not even be a true monophyletic clade. Results of early immunodiffusion studies (Dene *et al.*, 1976*a*, 1976*b*) indicated that the African lorises and galagos formed a clade separate from the Asian lorises. At least one microcomplement fixation study (Sarich & Cronin, 1976) supported a trichotomy, with the African and Asian lorises separated from one another as strongly as each was from the galagos. Although the exact branching relationships differ, all of these studies suggest that the Lorisoidea is not a natural phylogenetic group. If correct, the morphological and behavioral specializations of the Lorisidae must be the result of convergence rather than common ancestry. Unfortunately, these studies have not yet been adequately replicated during the past three decades. Nearly all of the more recent genetic distance studies have failed to include samples of both Asian and African lorises. Only Dutrillaux's (1988) study of primate chromosomal evolution and Yoder *et al.*'s (2000) recent analysis of mitochondrial and nuclear DNA have included members of both lorisid groups (both studies sampled *Perodicticus*, *Nycticebus* and *Loris*). Dutrillaux's results indicated that all of the lorisids formed a clade separate from the galagos (see also Martin, 1990; Yoder, 1997). Yoder's results suggested that the family was diphyletic, with the Asian lorisids grouping with the galagos, to the exclusion of *Perodicticus* (Yoder, 2000). However, in light of the long list of highly derived morphological characters of the lorisids, Yoder concluded that the genetic evidence considered alone was not an accurate indicator of lorisid phylogeny, though depending on weighting schemes, combined analyses of the morphological and molecular data also support lorisid monophyly. Until more comprehensive molecular studies are completed the most parsimonious relationship continues to be of a monophyletic lorisid group, probably composed of African and Asian clades.

Both the morphological and genetic data strongly support a natural galagid clade (Rasmussen & Nekaris, 1998), although higher branching relationships currently are unresolved. Early galagid taxonomies (Napier & Napier, 1967; Petter & Petter-Rousseaux, 1979) emphasized the apparent morphological homogenity among the galagos, with all species subsumed within the single genus *Galago*. In recent years, this practice has been challenged by the discovery of a surprisingly high degree of taxonomic diversity within the Galagidae (Yoder, 1997). Many investigators now recognize as many as three genera (e.g., Olson 1979, 1986) and more than a dozen species (e.g., Bearder *et al.*, 1995; Bearder, 1999). This number will almost certainly increase in the years to come as researchers learn more about the subtle morphological and behavioral features that differentiate closely related species (Bearder, 1999). Unfortunately, molecular phylogenetic studies have not been able to keep pace with the growing diversity in galagid taxa (Yoder, 1997; Rassmussen & Nekaris, 1998). Only a handful of genetic analyses have sampled more than one galago species, and of those few that have, the results are contradictory (Yoder, 1997). At least two studies (Crovella *et al.*, 1994*a*, 1994*b*; Masters *et al.*, 1994) have suggested that there may be a "greater" galago clade consisting of galagos of large body size (e.g., *G. garnetti*, *G. crassicaudatus*) distinct from a smaller body-sized "lesser" galago clade (e.g., *G. moholi*, *G. senegalensis*). However, other studies have described different relationships between these same species; where one found evidence of a close relationship between *G. senengalensis* and *G. crassicaudatus* to the exclusion of *G. demidoff* (Dene *et al.*,1976*a*, 1976*b*, 1980), another found no discernable branching pattern among them (Dutrillaux, 1988).

Recent estimates of divergence dates by molecular phylogenetic analyses suggest that the superfamily Lorisoidea diverged from the Lemuridae as early as the middle Paleocene (Sarich & Cronin, 1976; Yoder *et al.*, 1996; Yoder, 1997). The Lorisidae and Galagidae families appear to have split just a few million years after that event, probably during the early Eocene (Yoder, 1997). Unfortunately, the limited nature of the prosimian fossil record has made it impossible to confirm these molecular clock estimates (Yoder, 1997).

The Oligocene prosimian *Plesiopithecus teras* from the Fayum of Egypt has been described as the earliest known lorisiform (Simons & Rasmussen, 1994*a*). *Plesiopithecus teras* is represented by a nearly complete cranium and several partial mandibles (Simons, 1992; Simons & Rasmussen, 1994*a*). An isolated molar found several years earlier and thought to belong to an Eocene lorisoid was later attributed to this taxon (Simons *et al.*, 1986). Resemblances to living lorisoids, particularly with respect to cranial and facial structure, suggested an affinity to the tooth-comb strepsirhines (Simons & Rasmussen, 1994*a*). However, *P. teras* does not have a tooth-comb but has instead a highly derived form of anterior dentition, including enlarged and procumbent upper and lower canines (the lower tooth may be a lateral incisor). It is clearly not ancestral to either living lorisoids or lemuroids.

The oldest confirmed lorisoid fossils are only 20 million years old while the oldest fossil lemuroids are only a few thousand years old. These molecular dates are consistent, however, with Walker's (1974*b*, 1978) idea that there were separate lorisid and galagid lineages within the fossil record by the early Miocene. Perhaps not surprisingly, this distinction has been met with some criticism. This is due in large part to the fact that, as the fossil lorisoid taxonomy now stands, the African Miocene lorisids are represented by two fairly complete cranial fossils but as yet no definite postcranial material, while the galagids are represented by numerous postcranial fossils but only one partial cranium (Gebo, 1989; Rassmussen & Nekaris, 1998). Walker (1974*b*, 1978) has suggested that this disparity may be attributable to the smaller size and/or poorer preservation of more lightly constructed galago crania and/or the result of a higher proportion of galagos relative to lorises in Miocene-aged deposits, an ecological situation much like the one that exists today. However, Rasmussen & Nekaris (1998) have suggested that the Miocene lorisoids represented a basal stock characterized by a loris-like head atop a galago-like body, something akin to the condition seen in the monkey-bodied, ape-toothed early hominoid *Proconsul*. It would follow from their suggestion that the divergence between the two lorisoid families occurred some time after the early Miocene, but before the appearance of the definite loris *Nycticeboides* during the late Miocene, with galagos retaining the primitive postcranial condition and lorises retaining the primitive cranial condition. Certainly, we should not expect that the genetic and morphological divergence between species occurred simultaneously. The genetic divergence between species may begin long before these differences are expressed in the phenotype. But from the molecular clock estimates it seems most likely that by the Miocene, the two lorisoid families had diverged although the morphological differences between them probably were less extreme than between living galagos and lorises. A more complete fossil record is needed to detect the finer anatomical differences between these fossil species. We certainly expect that future fossils will necessitate revision of the current systematic framework. However, until they are found, the existing allocation of Miocene lorisoid genera to the families Lorisidae and Galagidae is the position we favor.

The discovery of the oldest lorisoid fossils in Africa has traditionally been interpreted as indicative of an African origin for the superfamily Lorisoidea. Since these same fossils are also believed to postdate the divergence between the Lorisidae and Galagidae families, this later event is generally presumed to have occurred within Africa as well (Walker, 1974*b*, 1978). However, MacPhee & Jacobs (1986) warn that the location of the oldest known fossil may not be sufficient evidence to establish a group's origin in that area. The extinct Eocene adapiforms may be possible candidates for the ancestors of living strepsirhines but none has a tooth-comb, the major derived feature of the Lemuriformes (Fleagle, 1999). The burgeoning Asian fossil record has raised the possibility that many early primate groups may have originated within that continent (Beard *et al.*, 1994, 1996; Beard, 1998*b*). How plausible is an Asian origin for

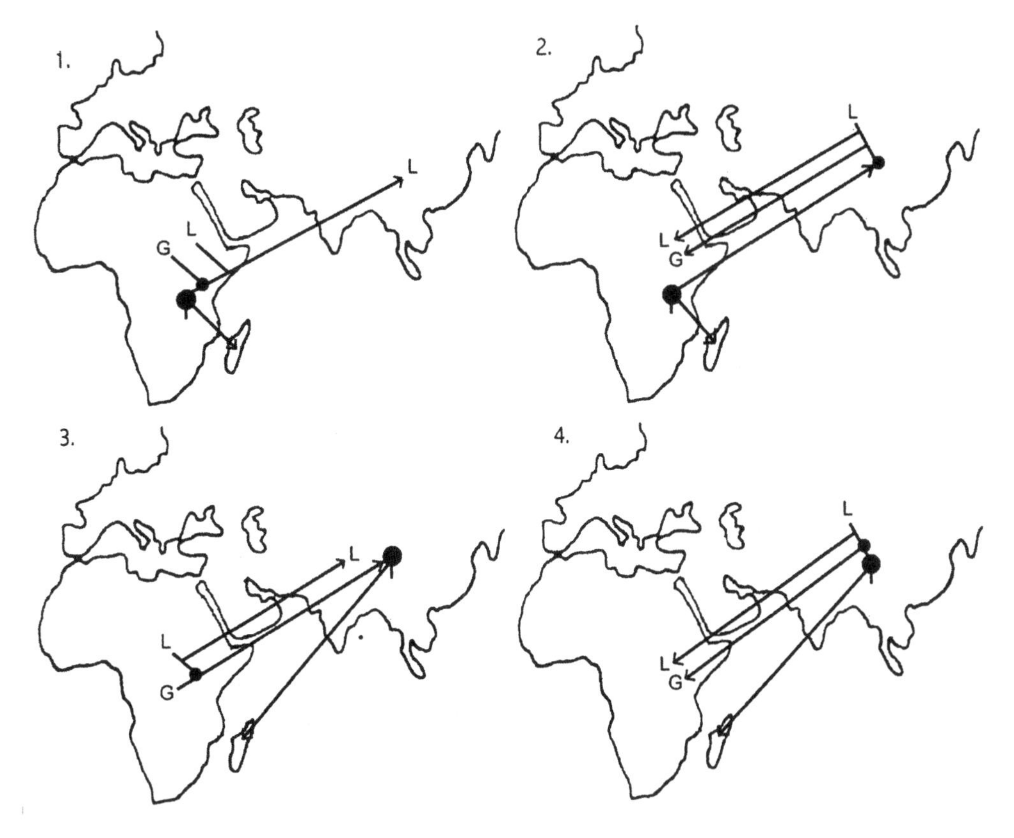

Fig. 6.11 Four models depicting the origins of the superfamily Lorisoidea (large dot) and families Lorisidae and Galagidae (small dot) in Africa and/or Asia. Arrows depict direction of dispersal. L = Lorisidae, G = Galagidae. Adapted from Yoder (1997).

the Lorisoidea superfamily and/or the Lorisidae and Galagidae families? There are four possible origin scenarios (Fig. 6.11):

1. Both the superfamily Lorisoidea and the families Lorisidae and Galagidae originated in Africa with a subsequent dispersal of a single lorisid group into Asia.
2. The superfamily Lorisoidea originated in Africa, migrated into Asia, and then diverged there into the families Lorisidae and Galagidae, with one lorisid group and all of the galagids subsequently migrating back into Africa.
3. The superfamily Lorisoidea originated in Asia, migrated into Africa, and then diverged there into the families Lorisidae and Galagidae, with one lorisid group migrating back into Asia.
4. The superfamily Lorisoidea and the families Lorisidae and Galagidae originated in Asia with one lorisid group and all of the galagids subsequently migrating into Africa.

Clearly, the two most complex scenarios, (2) and (3), which depict the origins of the Lorisoidea superfamily and families as occurring on separate continents, are the least likely of the four possibilities. Both of these models require multiple bidirectional migrations between Asia and Africa, most of which have failed to leave behind any living descendants or trace of a fossil record. All this would have had to happen within a time period of about 10 million years during the late Paleocene and early Eocene if Yoder's (1997) molecular clock estimates are correct.

The fourth scenario, which places both origin events within Asia, has been proposed by MacPhee & Jacobs (1986), the describers of the Asian fossil lorisid *Nycticeboides simpsoni*. The partial skeleton for which the taxon was named dates to the late Miocene, at approximately 9 Ma. Additional isolated lorisid teeth may extend the time range of the family in Asia to as early as 16 Ma but have not yet been described (J.C. Barry, pers. comm.). Since faunal exchange between South Asia and Africa is known to have occurred several times during the early Miocene, with possibly even earlier exchange events, MacPhee & Jacobs (1986) argue that there is nothing to preclude an Asian origin for the African fossil lorisoids. However, the earliest African lorisoids still predate *Nycticeboides simpsoni* by some 11 million years. Without stronger fossil evidence of earlier lorisoids, including galagids, in Asia, particularly of some nearer the origin date of the superfamily, the pattern of migrations described in this scenario seem less probable than the first scenario.

At the present time, an African origin for both the superfamily Lorisoidea and the families Lorisidae and Galagidae remains the most plausible of our four models. It is the simplest by far since it requires that only one lorisoid lineage disperse from Africa and includes a dispersal (by rafting or island-hopping during periods of low sea level) to Madagascar by ancestral lemurs that is far shorter than one from Asia. Although the fossil record of lorisoids in Africa is fragmentary and much younger than the estimated date of origin for the superfamily, it still remains the earliest fossil evidence of them.

Rasmussen & Nekaris (1998) have recently proposed an evolutionary model that attributes the early divergence of the lorisids and galagids to relatively minor differences in foraging strategies. Ancestors of living galagos may have gradually specialized in hunting active, fast-moving prey while the ancestors of living lorises specialized on slower-moving, but more toxic prey. From that initial difference, a "cascade" of morphological and behavioral adaptations followed (Rasmussen & Nekaris, 1998). Galagos developed a postcranial skeleton associated with rapid leaping and an increased ability to detect prey by sound. Lorises developed postcranial adaptations that allowed for more cryptic motion, an increased ability to detect prey by smell, and the ability to withstand high levels of toxins. When body size differences are accounted for, the thin, lightly-built skulls of galagos and the heavier, more strongly built skulls of lorises are also part of this same adaptive divergence. The better condition of cranial specimens (NAP 1.3.6/58 and KNM-RU 2052) attributed to *Mioeuoticus* compared to *Progalago* or *Komba* may simply be the result of differential preservation. It is for this reason that we think that the lorisid-like cranium of *Mioeuoticus* really belongs to a lorisid rather than a pre-divergence, basal lorisoid. Likewise, we think that the reason we have only one very damaged braincase (KNM-RU 1940) of the much more common *Progalago* or *Komba* is because they had lightly built, fragile skulls.

Very few postcranial fossils of lorisoids are associated with diagnostic teeth or cranial parts. Part of a pelvic bone is associated with a maxilla of *Progalago* (KNM-SO 1328) and the type of *Nycticeboides* includes a partial skeleton. All other assignments of postcranial bones are, therefore, guesses. In the case of the first published early Miocene bones (Walker, 1970), these guesses were based on regressions of the size of limb bones on the size of teeth in modern galagos. An early Miocene talus from Koru was assigned to *Mioeuoticus* by Gebo (1989) on the basis of its lorisid features, even though no teeth of this genus have been found there. In the case of the ischium associated with maxillary teeth of *Progalago*, the ischial tuberosity is set at right angles to the body as in galagos and not strongly recurved as in lorisids. It may be, of course, that the galago-like condition is primitive and the loris-like one derived, but this does suggest that *Progalago* had a galago-like hip joint and hints that galago-like femurs might be associated with it. However, without additional associations between postcrania and cranial and dental fossils, it is difficult to make definite determinations about the complete morphological complex of these extinct primates or be sure of the assignment of individual specimens to specific taxa.

It is perhaps as equally difficult to determine which, if any, of the African Miocene species are directly ancestral to living lorisids or galagids and which represent extinct lineages. Walker (1978) suggested that *Mioeuoticus* may be ancestral to living lorises while *Komba* may be ancestral to the living galagos. Species of *Progalago* may have represented an extinct branch of the family Galagidae. However, McCrossin (1992*b*) argued that both *Mioeuoticus* and *Progalago* were closely related to living lorises while *Komba* was an extinct galagid lineage. Rasmussen & Nekaris (1998) take the view that these fossils can not be assigned unambiguously to either family. Of course, these three viewpoints differ not only on the issue of ancestry but also on the family membership of these taxa. It is unlikely, then, without better and especially associated postcranial and cranial fossils that we will be able to resolve these issues. Further genetic studies will be of great benefit in helping understand the relationships between living lorisoids and, probably, in deciding the antiquity of lineage splitting. This may in turn constrain ideas about fossil taxa.

Acknowledgements

We would like to thank Dr. Walter Hartwig for inviting us to contribute this chapter, Dr. J. Barry, Dr. J. Kappelman, Dr. T. Olson and Dr. A. Yoder, for sharing their expertise with us, and Dr. L. Jacobs, Dr. F.C. Howell, Dr. H. Wesselman and Dr. T. White for lending us photographs and casts of specimens. We would also like to thank Dr. N. Vasey, L. Hlusko and H. Dunsworth for their helpful comments and discussion.

Primary references

Galago

Simpson, G. G. (1965). Mammalian fauna other than Bovidae. In *Olduvai Gorge 1951–1961*, ed. L. S. B. Leakey, pp. 15–16. Cambridge: Cambridge University Press.

Wesselman, H. B. (1984). The Omo micromammals: systematics and paleoecology of early man sites from Ethiopia. In *Contributions to Vertebrate Evolution*, vol. 7, eds. M. K. Hecht & F. S. Szalay, pp. 64–82. Basel: Karger.

Walker, A. C. (1987). Fossil Galaginae from Lateoli. In *Laetoli: A Pliocene Site in Northern Tanzania*, eds. M. D. Leakey & J. M. Harris, pp. 88–91. Oxford: Clarendon Press.

Komba

Simpson, G. G. (1967). The Tertiary lorisiform primates of Africa. *Bulletin of the Museum of Comparative Zoology*, **136**, 39–61.

McCrossin, M. L. (1992*b*). New species of bushbaby from the middle Miocene of Maboko Island, Kenya. *American Journal of Physical Anthropology*, **89**, 215–233.

Mioeuoticus

Le Gros-Clark, W. E. (1956). A Miocene lemuroid skull from East Africa. *Fossil Mammals of Africa*, No. 9. London: British Museum (Natural History).

Leakey, L. S. B. (1962). Primates. In *The Mammalian Fauna and Geomorphological Relations of the Napak Volcanics, Karamoja*, ed. W. W. Bishop, pp. 1–18. Entebbe: Recent Geological Survey of Uganda (1957–8).

Walker, A. C. (1974*b*). A review of the Miocene Lorisidae of East Africa. In *Prosimian Biology*, eds. R. D. Martin, G. A. Doyle, & A. C. Walker, pp. 435–447. London: Duckworth.

Phillips, E. M. & Walker, A. C. (2000). A new species of fossil lorisid from the Miocene of East Africa. *Primates*, **41**, 367–372.

Nycticeboides

Jacobs, L. L. (1981). Miocene lorisid primates from the Pakistan Siwaliks. *Nature*, **289**, 585–587.

MacPhee, R. D. E. & Jacobs, L. L. (1986). *Nycticeboides simpsoni* and the morphology, adaptations, and relationship of Miocene Siwalik Lorisidae. *Contributions to Geology, University of Wyoming, Special Paper* **3**, 131–161.

Progalago

MacInnes, D. G. (1943). Notes on the East African Miocene Primates. *Journal of the East Africa and Uganda Natural History Society*, **17**, 141–181.

Le Gros-Clark, W. E. & Thomas, D. P. (1952). The Miocene Lemuroids of East Africa. *Fossil Mammals of Africa*, No. 5. London, British Museum of Natural History.

Simpson, G. G. (1967). The Tertiary lorisiform primates of Africa. *Bulletin of the Museum of Comparative Zoology*, **136**, 39–61.

7 | Quaternary fossil lemurs

LAURIE R. GODFREY AND WILLIAM L. JUNGERS

Introduction

Evolving in isolation, the lemurs of Madagascar represent one of the order Primates' most extraordinary adaptive radiations. They belong to a unique island fauna characterized, as is typical, by high endemicity and taxonomic imbalance. No camels, elephants, giraffes, gazelles, felids, canids or other "typical" African continental big mammals ever reached the Great Red Island. "Typical" African birds (including woodpeckers, hornbills and trogons), reptiles, amphibians (pipids and bufonids) and fishes (elephant fish, minnows and carps) are nowhere to be found (Krause *et al.*, 1997). Of artiodactyl ungulates, only hippopotami reached Madagascar, and no perissodactyl did. In the not-too-distant past, giant tortoises, dwarf hippopotami (Steunes, 1989), an obscure insect-eating ungulate called *Plesiorycteropus* (Lamberton, 1946; MacPhee, 1994), a large-bodied primitive carnivoran called *Cryptoprocta spelea* (Lamberton, 1939*c*), and baboon-to-gorilla-sized lemurs roamed Madagascar alongside the animals that still survive (see Goodman & Patterson, 1997; Garbutt, 1999). There were half-ton elephantbirds (Lamberton, 1934*d*; Amadon, 1947), and eagles large enough to snatch animals heavier than adult sifakas or indris (Goodman 1994*a*, 1994*b*; Goodman & Rakotozafy, 1995; Karpanty & Goodman, 1999; T. Rasmussen, pers. comm.). The 30-odd still-extant lemur species include the smallest primate and the largest living prosimian. All of the island's *extinct* lemurs were larger than the largest of the living lemurs, and adaptively diverse – they have been compared to baboons, patas monkeys, sloths, koalas, cave bears, pigs, tapirs, orangutans and gorillas.

History of discovery and debate

The primate fossil record of Madagascar is geologically shallow. The oldest radiocarbon date thus far obtained on a specimen of extinct lemur is 26 000 BP (Simons *et al.*, 1995*a*). Most Quaternary terrestrial taxa are believed to have descended from ancestors who arrived via rafting (or swimming) between the end of the Mesozoic and the late Cenozoic Era (Krause *et al.*, 1997; Yoder, 2000). The ancestor of Malagasy lemurs has been estimated to have colonized Madagascar by the lower mid-Eocene (Yoder *et al.*, 1996; Yoder, 1997, 2000) if not before (Masters *et al.*, 1995). But Tertiary terrestrial deposits are almost non-existent in Madagascar, and Madagascan lemurs prior to the late Pleistocene are unknown. This implies that the vast majority of the record of the history of lemurs on Madagascar has been lost.

Subfossil remains of at least 17 species of extinct lemurs are recognized from the Holocene and late Pleistocene deposits of Madagascar. Fourteen were named as a result of excavations conducted in southern, western and central Madagascar during the early part of the twentieth century (Tattersall, 1982). After a hiatus of about a half century (see Ekblom, 1951; Mahé, 1965; Walker, 1967*a*), paleontological field work began anew in the early 1980s (e.g., MacPhee *et al.*, 1984; Godfrey *et al.*, 1990, 1997*b*; Simons *et al.*, 1990, 1992, 1995*b*). Ontogenetic series are now available for some taxa (Ravosa & Simons, 1994; Godfrey *et al.*, 2001). Associated skeletal remains of single individuals include rarely found elements such as carpal bones, the baculum, and distal phalanges (Simons *et al.*, 1992). Virtually whole hands and feet are now known for some species (MacPhee *et al.*, 1984; Wunderlich *et al.*, 1996; Jungers *et al.*, 1998; Hamrick *et al.*, 2000). Standard long-bone indices can be calculated with some confidence, and body masses estimated from long-bone circumferences (Table 7.1). Preserved fecal pellets of an *Archaeolemur* have enabled researchers to dissect the foods it ate (Burney *et al.*, 1997). New research tools (such as electron microscopy and DNA amplification via polymerase chain reaction) have added to our ability to probe the life-ways and phylogenetic relationships of Madagascar's extinct lemurs (Godfrey *et al.*, 1997*a*; Yoder *et al.*, 1999; Yoder, 2000; Jungers *et al.*, 2001).

Radiocarbon dates confirm that several of the giant lemurs (*Archaeolemur*, *Megaladapis*, *Palaeopropithecus*) survived into the past millennium, along with gigantic flightless birds, hippopotami, and other subfossil fauna (Mahé & Sourdat, 1972; Dewar, 1984, 1997; MacPhee & Burney, 1991; Simons *et al.*, 1995*a*; Burney *et al.*, 1997; Simons, 1997*b*). A few may have succumbed only very recently. In the seventeenth century, French naturalist/explorer Étienne de Flacourt (1658) described an animal called the "tretretretre" living in southeast Madagascar. The Malagasy accounts might well describe *Palaeopropithecus*. By 400 BP, except for dispersed patches or strips, highland forests had almost entirely disappeared. Nevertheless, dwindling populations of giant lemurs might have endured in coastal regions less affected by cutting and burning. A specimen of *Palaeopropithecus ingens* recently was

Table 7.1. Estimated body mass and long bone indices for subfossil taxa (sources in parentheses)

Taxon	Estimated body mass[a] (kg.)	Humerofemoral index[b]	Intermembral index[b]	Brachial index[b]	Crural index[b]
Mesopropithecus globiceps	9	90 (s)	97 (s)	101 (s)	85 (s)
Mesopropithecus pithecoides	10	91 (s)	99 (s)	101 (s)	85 (s)
Mesopropithecus dolichobrachion	11	104 (s)	113 (s)	104 (s)	88 (s)
Babakotia radofilai	16	115 (s)	118 (s)	100 (s)	93 (s)
Palaeopropithecus ingens	45	146 (v)	135 (v)	89 (v)	105 (v)
		149 (j1)	138 (j2)	87 (j2)	102 (j2)
Palaeopropithecus maximus	52	152 (v)	145 (v)	99 (v, j2)	110 (v, j2)
		150 (j1)	144 (j2)		
Pachylemur jullyi	13	82 (v)	94 (v)	111 (v)	84 (v)
		81 (j1)			
Pachylemur insignis	10	85 (v)	97 (v)	113 (v)	86 (v)
		86 (j1)			
Daubentonia robusta	13	78 (v)	82 (v)	108 (v)	98 (v)
		80 (j1)	85 (j2)	101 (j2)	99 (j2)
Archaeolemur edwardsi	24	85 (v)	92 (v)	107 (v)	86 (v)
		82 (j1)			
Archaeolemur majori	14	86 (v)	92 (v)	100 (v)	94 (v)
		85 (j1)			
Hadropithecus stenognathus	27	103 (g)	—	84 (g)	—
Megaladapis edwardsi	75	107 (v)	119 (v)	90 (v)	73 (v, j2)
		110 (j1)	120 (j2)	87 (j2)	
Megaladapis grandidieri	63	114 (v)	117 (v)	90 (v)	84 (v)
		113 (j1)	115 (j2)	88 (j2)	85 (j2)
Megaladapis madagascariensis	38	112 (v, j1)	118 (v)	94 (v, j2)	83 (v)
			114 (j2)		90 (j2)

[a] All body mass estimates from j1 are rounded to the nearest kg.
[b] In all cases including indices, 0.5 is rounded down.
j1 = Jungers *et al.* (2001), j2 = Jungers (1980); g = Godfrey *et al.* (1997*b*); s = Simons *et al.* (1995*b*); v = Vuillaume-Randriamanantena (1982).

radiocarbon dated at 510 ± 80 BP (Simons, 1997*b*). Confidence limits on this date include the historical period (Burney & Ramilisonina, 1998; Burney, 1999). Eyewitness accounts of a very large, pronograde lemur with a dark coat and white patches on the forehead and under the chin were recorded in 1995 by Burney & Ramilisonina (1998). The several people who independently described it were knowledgeable of local fauna and flora, and they insisted that the animal was neither a sifaka nor an indri. Their description best fits *Archaeolemur*.

The first formal "research" discoveries of extinct lemurs were those of Alfred Grandidier, in 1865 at Ambolisatra (including the distal humerus of *Palaeopropithecus* and a tibia of a *Propithecus*; see Filhol, 1895; Vuillaume-Randriamanantena, 1990; Tattersall *et al.*, 1992). But giant extinct lemurs were not formally described until the 1890s. The first to be described was a peculiar lemur with a long and narrow skull – *Megaladapis madagascariensis* (Forsyth-Major, 1893, 1894). Its teeth resembled those of considerably smaller-bodied sportive lemurs, but it had laterally directed orbits and downturned nasals – entirely uncharacteristic for a primate.

Numerous extinct primate species names were erected in the next several years based on subfossil specimens collected in central, southeastern and southwestern Madagascar. Almost as quickly, many were sunk. Some specimens were distributed to museums in Europe while others were kept in Madagascar. Field associations, if they existed, were often lost. Unsurprisingly, postcrania belonging to different genera of extinct lemurs (especially those of *Megaladapis*, *Palaeopropithecus* and *Hadropithecus*) were confounded. Some giant lemur bones were assigned to non-primates, and, even into the 1950s, bones belonging to non-primates were attributed to lemurs (see MacPhee & Raholimavo, 1988).

The monkey lemurs

Soon after Filhol (1895) had named and described *Archaeolemur majori* on the basis of postcrania from Belo-sur-mer in western Madagascar, Forsyth-Major (1896) described fragmentary dental and facial specimens that he believed belonged to an extinct Malagasy monkey, *Nesopithecus roberti*. This animal had a decidedly flat facial profile and bilophodont molars resembling some cebids and cercopithecids. Forsyth-Major (1896: 436) was convinced that an animal with such teeth and facial profile must also have postorbital closure, and other telltale anthropoid features. He classified it as a member of a new monkey family, the Nesopithecidae. Ironically, Forsyth-Major (1893) previously had described a partial (faceless) cranium and, convinced of *its* lemur affinities, named it *Globilemur* (Forsyth-Major, 1897). Thus, in the late 1890s, specimens actually belonging to *Archaeolemur* were ascribed to both lemurs and monkeys. Additional "monkeys" and synonymous "lemurs" were chaotically named in the mid-1890s (see Filhol, 1985; Grandidier, 1899*b*; Forsyth-Major, 1900*a*; Lorenz von Liburnau, 1900*a*).

References to Madagascar's extinct monkeys (Earle, 1897; Trouessart, 1897; Lorenz von Liburnau, 1899) were short-lived. The discovery of an almost perfect cranium of *Archaeolemur* from Andrahomana (in southeastern Madagascar) forced Forsyth-Major (1900*c*) to acknowledge the synonymy and lemur status of *Globilemur* and *Nesopithecus*. With synonymy largely sorted out in the early 1900s (Forsyth-Major, 1900*c*; Grandidier, 1902*b*; Standing, 1908), two species remained: the more gracile *A. majori* and the more robust *A. edwardsi*, both named by Filhol (1895).

Still Forsyth-Major and his colleague, Herbert F. Standing, were reluctant to abandon a special relationship for *Archaeolemur* to monkeys and apes. Forsyth-Major believed that this creature might be related to the stock from which anthropoids evolved. Standing (1908) argued that *Archaeolemur*'s combined lemur- and monkey-likenesses proved that it (and other Malagasy lemurs) had *devolved* from a primitive simian stock, and any Eocene "lemur"-likenesses were signs of the degeneration ("retrogression to a lower and more primitive type", Standing, 1908: 161) that had anticipated its extinction.

Guillaume Grandidier (1905) championed the view that later appeared so obvious: *Archaeolemur* was a genuine lemur with anthropoid convergences. *Archaeolemur*'s monkey-image, however, has endured. Carleton (1936: 284) characterized *Archaeolemur* as having "begun to experiment with brachiation but [having] failed to reach the level attained in higher Primates." Le Gros Clark (1945), Jean-Jacques Piveteau (1950) and William Charles Osman Hill (1953) described its brain as advanced. As some of its presumed craniofacial (and particularly neurological) convergences to monkeys were scrutinized critically (e.g., Smith, 1908; Radinsky, 1970; Tattersall, 1973), postcranial "macaque-" or "baboon-" likenesses rose to the forefront (Lamberton, 1938, 1939*b*; Jouffroy, 1963; Walker, 1967*a*, 1974*a*; Tattersall, 1973).

Ludwig R. Lorenz von Liburnau (1899) assigned the nomen *Hadropithecus stenognathus* to a mandible from Andrahomana. Shortly thereafter (1900*a*), he ascribed several forelimb and axial fragments to *Hadropithecus*. There were no hindlimb bones in the lot. Similarities between the crania of *Hadropithecus* and *Archaeolemur* were immediately recognized (Forsyth-Major, 1900*c*; Lorenz von Liburnau, 1900*a*, 1902; Grandidier, 1902*b*, 1905; see Tattersall 1973, for a review), and Lorenz von Liburnau (1902; also Grandidier, 1905) posited a close relationship to extant indrids. Standing (1908) assigned the two genera to their own subfamily within the family Indridae. Szalay & Delson (1979) gave this group independent familial status. Saban (1956, 1963) described similarities of the ear region to indrids, and G. Elliot Smith (1908) and Leonard Radinsky (1970) noted similarities in the morphology of their brains.

In 1938, Charles Lamberton wrote a monograph on *Hadropithecus*. By this time, additional forelimb and craniodental specimens of *Hadropithecus* but still no hindlimb bones had been recovered. Lamberton tried to remedy the situation. He assigned to *Hadropithecus* one femur, two tibiae and two fibulae from Tsirave (west) and one tibia and fibula from Anavoha (south). No definitive associations with forelimb bones or crania of *Hadropithecus* existed for these hindlimb specimens, and Lamberton admitted uncertainty in making these allocations.

The papionin model for *Archaeolemur* and *Hadropithecus* was rooted in Lamberton's (e.g., 1938) comparisons of *Archaeolemur* and *Hadropithecus* to macaques. Françoise Jouffroy (1963) also noted postcranial similarities of *Archaeolemur* to baboons and macaques, but stressed its greater robusticity. Alan Walker (1967*a*, 1974*a*), Clifford Jolly (1970*a*) and Ian Tattersall (1973, 1982) popularized the baboon model for both *Archaeolemur* and *Hadropithecus*, emphasizing their similarities to the most terrestrial papionins, and this notion persists in secondary sources (Mittermeier *et al.*, 1994; Nowak, 1999). Tattersall (1973, 1982) focused on the highly derived dental proportions and cranial architecture. *Archaeolemur* has enormous, vertically implanted median upper incisors that contact one another at the midline and form a monkey-like bite with the highly modified lower incisors. The premolars form a continuous shearing blade that wears, along with the incisors, considerably faster than the molars. Tattersall took the relative bunodonty of the molars, combined with the fast-wearing incisal/premolar shearing device, to suggest that *Archaeolemur* was predominantly frugivorous, feeding (as do baboons and mangabeys) on a mixed diet of herbs and hard-skinned fruit. He rejected folivory on the grounds that committed leaf-eaters have molars with greater relief and sharper cusps than more frugivorous taxa. Recent studies continue to inform dietary reconstructions of *Archaeolemur*. Taken together, the dental morphology, molar microwear, molar microstructure and fecal pellet data (Burney *et al.*, 1997; G. Schwartz, M. Teaford, and G. Semprebon, pers. comm.) suggest that *Archaeolemur* was a generalist capable of exploiting a variety of hard-to-process foods.

Laurie Godfrey (1977, 1988) argued that a baboon analogy is too simplistic for *Archaeolemur*. In its general proportions, *Archaeolemur* is more macaque- than baboon-like, and some parts of the anatomy, such as the scapula and pelvis, are not at all baboon-like. *Archaeolemur* was a heavily-built animal with relatively short limbs (in comparison to its head and body length) and a wide trunk. The cheiridia are now well known (Jungers *et al.*, 1998, 2001; Hamrick *et al.*, 2000); *Archaeolemur* had far shorter metapodials (both absolutely and relatively) than baboons; its phalanges were also shorter but more curved than those of baboons (Jungers *et al.*, 1997). Hamrick *et al.* (2000) describe the carpus of *Archaeolemur* as most similar to *Cercopithecus mitis*, an arboreal pronograde cercopithecine. Nevertheless, the cheiridia exhibit numerous specializations for terrestriality, including a reduced, somewhat adducted hallux, extreme digital reduction, and dorsal extensions of the metapodial distal articular surfaces. The heels of both the hand (pisiform) and foot (calcaneus) are robust.

Hadropithecus postcrania show even more specializations for terrestrial locomotion. *Hadropithecus* has been likened specifically to patas monkeys (Walker 1967*a*) and gelada baboons (Jolly, 1970*a*; Tattersall, 1973, 1982). Neither analogy is very good, but the evidence is compelling that *Hadropithecus* spent much time on the ground.

The absence of strong crests or rugosities for muscle insertion on the humerus caused Lamberton to conclude that *Hadropithecus* was less "adept" at quadrupedal climbing than *Archaeolemur*, and therefore less arboreal. Walker (1967*a*, 1974*a*) agreed (though for different reasons) that *Hadropithecus* was more terrestrial than *Archaeolemur*. In effect, Walker saw *Hadropithecus* as Madagascar's patas monkey. Jolly (1970*a*) disliked the *Erythrocebus* analogy (e.g., it was difficult to reconcile *Hadropithecus*' very straight humerus with the strongly bowed humerus of the patas monkey). Jolly substituted another terrestrial model – *Theropithecus*. With such a model, Jolly (1970*a*: 624) insisted, *Hadropithecus*' long forelimbs made sense: "Since the combination of a humerofemoral index above a hundred with extreme terrestrial adaptation is apparently unique, among primates, to Pleistocene representatives of *Theropithecus*, the implication is clear.... The functional implication of his long arms is that *Hadropithecus* foraged in the sitting position, reaching forward to gather grass-blades with his hands, in the same way as *Theropithecus gelada*."

Hadropithecus's cranial architecture and dentition seemed to bolster Jolly's argument. Like *Theropithecus*, *Hadropithecus* has relatively small, vertically implanted anterior teeth and enlarged, thick-enameled, flat-wearing molars. The mandibular corpus is extremely robust. Jolly (1970*a*) and Tattersall (1973) defended a graminivorous and granivorous diet for *Hadropithecus*. Molar microwear suggests that *Hadropithecus* was indeed a seed predator (M. Teaford, pers. comm.), but not a *Theropithecus*-like grazer (G. Semprebon, pers. comm.).

But even if Jolly's dietary model has partially stood the test of time, the vision of *Hadropithecus* as a terrestrial cursor with long, slender limbs has not. Godfrey *et al.* (1997*b*) revised hind-limb attributions for the genus and argued that Lamberton (1938) had incorrectly allocated to *Hadropithecus* hindlimb specimens belonging to *Archaeolemur majori*. Godfrey *et al.* (1997*b*) instead ascribed to *Hadropithecus* a suite of robust archaeolemurid hindlimb bones from exactly those subfossil sites that have yielded cranial and forelimb specimens of *Hadropithecus*. In 1901, Grandidier had assigned one of them (a femur) incorrectly to *Palaeopropithecus ingens*. Carleton (1936) recognized its morphology as archaeolemurid; she ascribed it to "*Bradylemur robustus*". But the synonymy of *cranial* specimens of *Bradylemur* and *Archaeolemur* had already been established, and most researchers (e.g., Tattersall, 1982; Godfrey, 1988) simply ignored Carleton's "*Bradylemur*" postcranial attributions. When Vuillaume-Randriamanantena (1982) rediscovered Lamberton's robust archaeolemurid femora in the collections of the Académie Malgache, she sought to resurrect *Bradylemur*, recognizing (see also Lamberton, 1947) that these specimens could not belong to *Archaeolemur*.

The revised humerofemoral index is well within the range of quadrupedal mammals. However, the new attributions make *Hadropithecus* hardly more gracile than *Archaeolemur*! *Hadropithecus* was largely terrestrial, but probably also a cautious climber. It weighed between 25 and 30 kg. Like *Archaeolemur*, it differed from most other giant extinct lemurs in lacking suspensory adaptations.

The sloth lemurs

The Palaeopropithecidae are called sloth lemurs in recognition of the remarkable postcranial convergences exhibited by some genera, in particular *Palaeopropithecus*, to arboreal sloths. Cranially they resemble indrids. Four genera are currently recognized as belonging in this family: *Palaeo-propithecus* (the type genus); the truly gigantic *Archaeoindris*; *Mesopropithecus*; and the recently discovered *Babakotia*.

The distal humerus of *Palaeopropithecus ingens* that Alfred Grandidier found at Ambolisatra in 1865 was not described until 1895 when it was formally named *Thaumastolemur grandidieri* (Filhol, 1895). The nomen *Palaeopropithecus ingens* was assigned four years later to a large mandibular fragment with sifaka-like teeth (G. Grandidier, 1899*b*). In 1902, A. Grandidier's distal humerus was synonymized incorrectly with *Megaladapis madagascariensis* (G. Grandidier, 1902*a*). Nine decades later *Thaumastolemur*'s taxonomic priority over *Palaeopropithecus* was recognized (Vuillaume-Randriamanantena, 1990), and then quickly suppressed by the International Commission for Zoological Nomenclature, for lack of use (see Tattersall *et al.*, 1992).

Standing (1903) named two more species of *Palaeopropithecus* (*P. maximus* and *P. raybaudii*) on the basis of one cranium each from Ampasambazimba (central Madagascar). Shortly thereafter, with 13 additional crania available, he synonymized them (Standing, 1908). Whether *P. ingens* and *P. maximus* truly represent different species has been questioned (Walker, 1967*a*; Tattersall, 1973). With the recent discovery of a complete skeleton of a new species of *Palaeo-propithecus* from the northwest (MacPhee *et al.*, 1984; W.L. Jungers *et al.*, unpublished data) and the very recent discovery of new specimens of *P. ingens* from the southwest (Ankilitelo, Ankomaka), further investigation is warranted.

The major difficulty researchers faced during the early twentieth century lay in figuring out associations of crania and postcrania. Forelimb specimens belonging to *Palaeopropithecus* were allocated to *Megaladapis* and vice versa (e.g., Grandidier, 1905; Standing, 1908). A femur of *Palaeopropithecus ingens* was at first attributed to a gigantic sloth ("*Bradytherium*", see Grandidier, 1901), while *Palaeopropithecus ingens* was given the femur of a robust archaeolemurid (Grandidier, 1902*a*, 1905); see above, on *Hadropithecus*.

Enthusiasm for Grandidier's giant fossil sloth waned as ongoing field excavations yielded many more "sloth" femora but not a single sloth cranium. Standing (1908) was convinced that these new femora belonged to a giant lemur. Initially, he failed to associate them with *Palaeopropithecus*, as he believed, following Grandidier, that *Palaeopropithecus* had an *Archaeolemur*-like femur. His subsequent publications make the correct hindlimb attributions (see Standing, 1910, 1913). Faulty forelimb attributions were corrected later by Carleton (1936) and Lamberton (1947). With Lamberton's (1934*c*) discoveries of apparently associated cranial and postcranial specimens of *Megaladapis*, it became obvious that the elongated forelimb bones that Grandidier had ascribed to *Megaladapis* must belong instead to *Palaeopropithecus*. The new postcranial attributions for *Palaeo-propithecus* were confirmed when an associated skeleton was discovered at Anjohibe near Mahajanga (MacPhee *et al.*, 1984). *Palaeopropithecus* did indeed have an elongated forelimb and short hindlimb.

Carleton (1936) was first to argue forcefully that *Palaeopropithecus* was arboreal-slothlike in its positional behavior. Standing (1903, 1909) had earlier reconstructed *Palaeopropithecus* as aquatic, a scenario embraced by some paleontologists well into the mid-twentieth century (Sera, 1935, 1938, 1950; Hill, 1953; for critique see Lamberton, 1957). Two realistic competing models for *Palaeopropithecus* arose later in the twentieth century – the orang model and sloth model. Lamberton (1947) had compared *Palaeopropithecus* to both sloths and orangutans, but clearly preferred the orang model, as did Walker (1967*a*, 1974*a*) and Tattersall (1982). Carleton (1936) championed the sloth model. The knee in particular convinced Tardieu & Jouffroy (1979) that *Palaeopropithecus* moved via quadrupedal suspension. As these authors pointed out, whereas the elongated forelimb might support an orang model, the hindlimb morphology emphatically could not. William Jungers (1980) and Vuillaume-Randriamanantena (1982) favored the sloth model on the basis of hindlimb and cheiridial morphology, postcranial proportions and long-bone allometries.

Recent research on bone geometry (Demes & Jungers, 1993), joint surface area ratios (Godfrey *et al.*, 1995), vertebral morphology (Shapiro *et al.*, 1994) and cheiridial morphology (Jungers *et al.*, 1997) has further elucidated sloth likenesses of *Palaeopropithecus*. Numerous postcranial adaptations point to sloth lemur commitment to slow, vertical climbing and suspensory modes of locomotion (Jungers *et al.*, 1991; Simons *et al.*, 1992). Additional support for the sloth model comes from the skeletal anatomy of new species of sloth lemurs, including *Mesopropithecus dolichobrachion* (Simons *et al.*, 1995*b*), *Babakotia radofilai* (Godfrey *et al.*, 1990; Jungers *et al.*, 1991; Simons *et al.*, 1992), and the *Palaeopropithecus* of the northwest (W.L. Jungers *et al.*, unpublished data).

The spectrum of postcranial variation within the sloth lemurs is much like that described by William Straus and George Wislocki (1932) for lorises and sloths. In effect, *Mesopropithecus dolichobrachion* and *Babakotia radofilai* are members of a morphocline showing a range in specialization for quadrupedal suspension from the most quadrupedal (and loris-like) *Mesopropithecus pithecoides* and *M. globiceps* to the most arboreal slothlike *Palaeopropithecus* spp. *Palaeopropithecus* was the most specialized of the sloth lemurs, with exceptionally long forelimbs and short hindlimbs, and extremely curved manual and pedal proximal phalanges. Its hook-like hands and feet (with reduced thumbs and halluces, and very reduced hindfeet) are entirely unsuited for weight-bearing in terrestrial locomotion. As a group, the palaeopropithecids were undoubtedly climber/hangers, sharing with indrids adaptations for hang-feeding (Gebo, 1987*b*), but not for leaping.

Archaeoindris fontoynontii was established by Standing (1909) on basis of three specimens: a mandible with complete dentition, a left maxillary fragment and a right partial maxilla. It remains one of the most poorly known of all subfossil lemur species. Lamberton (1934*a*) discovered the single known cranium with associated mandible. There are only six known postcranial specimens: a damaged humerus of an adult, a nearly complete adult femur and four bones of an immature individual (a damaged humerus, a damaged ulna and a pair of femora lacking proximal and distal epiphyses); see Vuillaume-Randriamanantena (1988). All known specimens are from Ampasambazimba. By all measures, whether dental or postcranial, *Archaeoindris* was the largest-bodied lemur – about the size of a large adult male gorilla.

Misattributions confounded early interpretations of the lifeways of *Archaeoindris*. Lamberton (1934*a*) had ascribed incorrectly an undamaged adult tibia and two fibulae that actually belonged to *Megaladapis grandidieri*. In effect, Lamberton's reconstruction of *Archaeoindris* was based on a few bones of an immature individual, and a few bones of an adult *Megaladapis*. Carleton (1936) rejected Lamberton's attributions for the tibia and two fibulae; these corrections were followed by more from Walker (1967*a*, 1974*a*) and later confirmed by Jungers (1977). In 1988, Vuillaume-Randriamanantena summarized all known elements of the postcranial skeleton of *Archaeoindris*, and provided a lucid description of their consistent similarity to *Palaeopropithecus*.

Both Standing (1910, 1913) and Lamberton (1934*a*) reconstructed *Archaeoindris* as similar in positional behavior to *Megaladapis* (i.e., a slow-moving arborealist). Piveteau (1961) and Walker (1974*a*) accepted and repeated Lamberton's view, and considered *Archaeoindris* a *Megaladapis*-like, arboreal climber and clinger. Recent secondary sources (Mittermeier *et al.*, 1994; Nowak, 1999) have perpetuated these *Megaladapis*-like inferences.

Lamberton (1934*a*) also noted some postcranial similarities of *Archaeoindris* to *Palaeopropithecus*, and he tentatively suggested a ground sloth model for *Archaeoindris*. Unencumbered by misattributions, Jungers (1980) embraced the ground sloth model more fully. Specializations for suspension similar in degree to *Palaeopropithecus* are unknown, and Jungers felt it likely, given the sheer size of *Archaeoindris* (approximately 200 kg), that this species spent most of its time on the ground. At the same time, hip joint functional morphology suggests higher mobility than might be expected for a fully terrestrial creature. The humerus was considerably longer than the femur, and the humerus and ulna of the immature individual were each considerably longer than the femur (Vuillaume-Randriamanantena, 1988). Given the species' gorilla-like bulk, it would prove surprising indeed to discover hand and foot bones as derived for hang-feeding as those seen in *Palaeopropithecus*. Without that information, however, one can only speculate that *Archaeoindris* was a capable but deliberate, scansorial browser, and that it also frequented the ground to feed and travel.

Standing (1905) named *Mesopropithecus pithecoides* on the basis of four crania from Ampasambazimba. Lamberton (1936*a*) gave the name *Neopropithecus globiceps* to a single cranium from Tsirave (southwest) and *N. platyfrons* to two from the extreme south (Anavoha). He considered *Neopropithecus* distinct and intermediate between *Mesopropithecus* and *Propithecus*. Tattersall (1971) sank *N. platyfrons* into *N. globiceps*, and further synonymized *Neopropithecus* and *Meso-propithecus*. Tattersall & Schwartz (1974) considered *Meso-propithecus* the sister taxon specifically to *Propithecus*. They argued that *Mesopropithecus* and *Propithecus* share reduced third molars and particularly reduced premolars (relative to molar length). However, these features do not distinguish *Meso-propithecus* and *Propithecus* from *Palaeopropithecus* and *Archaeoindris*, which also have reduced (indeed very reduced) mandibular and maxillary third molars, and relatively short premolars. Most other traits that ally *Mesopropithecus* with *Propithecus* are clearly plesiomorphic for the indrid–palaeopropithecid clade. For example, unlike *Palaeopropithecus* and *Archaeoindris*, *Mesopropithecus* has a conventional tooth-comb (though with only four teeth) and a typical inflated bulla. *Mesopropithecus* also lacks some features of the cheek teeth that distinguish *Palaeopropithecus* and *Archaeoindris* (e.g., the elongated, buccolingually compressed first and second molars). Its forelimb is similar to that of indrids (though relatively larger), but its hindlimb is not at all like that of *Propithecus*.

Primarily on the basis of the postcrania, Godfrey (1986*b*, 1988) defended a closer phylogenetic relationship of *Mesopropithecus* to *Palaeopropithecus*. She also maintained, however, that features of the crania support this inference. Postorbital constriction is considerably more marked in *Mesopropithecus* than in any indrid (despite its having a larger brain and shorter skull than the largest extant indrid, *Indri indri*). As in *Palaeopropithecus* and *Archaeoindris*, the interorbital breadth is relatively small, the cheek tooth enamel is wrinkled and the zygoma are robust. Also as in *Palaeopropithecus*, the orbits of *Mesopropithecus* are small (in absolute and not merely relative size).

Paucity of postcranial specimens and poor cranial/postcranial associations hampered early reconstructions of *Mesopropithecus* positional behavior. Carleton (1936) assumed that the postcrania of *Mesopropithecus* were similar to *Propithecus* and assigned to *M. pithecoides* a suite of hindlimb bones that actually belonged to *Propithecus diadema*. Lamberton (1948*b*) made the correct hindlimb attributions, having at his disposal some associated cranial and postcranial specimens from the southwest. He recognized that *Mesopropithecus* lacked structural adaptations for leaping. Walker, in contrast, interpreted *Mesopropithecus* as a "modifed" and somewhat sedentary vertical clinger and leaper (Walker, 1967*a*: 329). Walker (1967*a*) also thought it resembled *Megaladapis* and *Archaeoindris* (on a far reduced scale). He later displayed more caution, stating only that the few bones available "resemble small *Megaladapis* bones" and that "until more evidence is available it seems wisest to leave open the question of presumed locomotor habits" (Walker, 1974*a*: 376). In contrast, Vuillaume-Randriamanantena (1982) saw loris-like features in the appendicular anatomy of *Mesopropithecus*. Godfrey (1988) drew similar conclusions and likened *Mesopropithecus* to *Palaeopropithecus*. New discoveries at Ankarana (*Babakotia radofilai* and *Mesopropithecus dolichobrachion*) provided a remarkable series of morphological intermediates between *M. globiceps/pithecoides* and *Palaeopropithecus* spp. *Babakotia* and *Mesopropithecus* appear to have been arboreal quadrupeds, adapted for slow climbing and, to a lesser extent, suspension.

Babakotia radofilai was first discovered in 1988 at Antsiroandoha, a cave in the Ankarana Massif, northern Madagascar (Godfrey *et al.*, 1990). Specimens of a dozen or so individuals soon were found, including a rather spectacular skull and skeleton (Simons *et al.*, 1992). With an intermembral index of about 118, *Babakotia* was morphologically intermediate between *Mesopropithecus* and *Palaeopropithecus*. Its skull has very elongated premolars and its cheek teeth have crenulated enamel; its forelimbs are long, hindlimbs short, and hands and feet were long and adapted for suspension. Many features of its axial and appendicular skeleton ally it with *Palaeopropithecus* (Jungers *et al.*, 1991; Simons *et al.*, 1992).

The third species of *Mesopropithecus* (*M. dolichobrachion* – the long-armed *Mesopropithecus*) was discovered at Ankarana and named in 1995 (Simons *et al.*, 1995*b*). The skeleton included previously unknown elements, such as a partial innominate, fibula and ulna. The hindfoot is reduced, but not as reduced as in *Babakotia* or especially *Palaeopropithecus*. *Mesopropithecus dolichobrachion* is larger than *M. pithecoides* (and bears a sagittal crest, as might be expected). It differed from both *M. pithecoides* and *M. globiceps* in having postcranial morphology and proportions somewhat closer to those of *Babakotia* (Simons *et al.*, 1995*b*). *Mesopropithecus* is the least specialized of the sloth lemurs, with an intermembral index ranging from about 97 (*M. pithecoides* and *M. globiceps*) to 113 (*M. dolicho-brachion*).

Oddly incorrect information about *Mesopropithecus* is still being perpetuated in the secondary literature. Contra Jenkins (1987, following Lamberton), *M. pithecoides* and *M. globiceps* do not differ in the degree of postorbital constriction. Contra Nowak (1999, again following Lamberton), *M. globiceps* is hardly more similar to *Propithecus* than is *M. pithecoides*. The notion that the two species of *Mesopropithecus* have very different crural indices (Jenkins 1987) is apparently grounded in an arithmetic error. Jenkins (1987) reports crural indices of 84.9 for *M. globiceps* and 98.3 for *M. pithecoides*, based on Walker (1967*a*). We obtained crural indices of 85.2 and 85.5 for the two taxa. The two have virtually identical humerofemoral indices, brachial indices and intermembral indices (Simons *et al.*, 1995*b*). The forelimbs of both species are about the same length as the hindlimbs. Finally, the notion that *Mesopropithecus* resembles *Megaladapis* is still current in the literature (Nowak, 1999), but any such resemblances are superficial at best.

Sloth lemurs do have indrid-like dentitions (Tattersall & Schwartz, 1974), but their molar shearing quotients are higher (Jungers *et al.*, 2001), perhaps suggesting higher degrees of folivory. On the basis of the robusticity of the mandibular corpus and symphysis, Matthew J. Ravosa (1991) similarly argued that *Palaeopropithecus* and *Archaeoindris* were strongly folivorous. These larger taxa have long, oblique and very robust mandibular symphyses that fuse early. It is also noteworthy that all palaeopropithecid species for which immature individuals are known can be shown to have had unusually accelerated dental development, in apparent preparation for early processing of fibrous foodstuffs (Schwartz *et al.*, 2000; Godfrey *et al.*, 2001).

The koala lemurs

As was seen for other taxa, the early taxonomic history of the genus *Megaladapis* documents a period of naming many new taxa followed by judicious lumping and pruning. Considerable confusion was created early on by separate diagnoses of dental, cranial and postcranial specimens, by nearly simultaneous discoveries of subfossils by expeditions from different countries, as well as by differential treatment of mature and immature individuals.

Forsyth-Major prefaced his verbal report to the Royal Society of London on June 15, 1893 with a passing reference to "a strange gigantic Lemuroid skull (*Megaladapis madagascariensis* Maj.)" that J.T. Last had discovered in the marsh deposits of Ambolisatra in the southwest. In the published technical account he drew pointed analogies to the Australian koala bear, *Phascolarctos* (Forsyth-Major, 1894). (This characterization has taken hold, so much so that Nowak (1999) recently dubbed the three currently recognized species of *Megaladapis* the "koala lemurs".)

Shortly thereafter, Filhol (1895) proposed the nomen "*Dinolemur grevei*" for a nearly complete humerus, a distal femur fragment and complete calcaneus discovered at Belo-sur-mer by M. Grevé. Primarily on the basis of its size, Filhol suggested that the humerus might prove to be associated with a species of *Megaladapis*. Jungers (1976) doubted this attribution, but Godfrey *et al.* (1997*b*) make a strong metric case that it probably does belong to *M. madagascariensis*. Filhol had also named "*Thaumastolemur grandidieri*" in the same article on the basis of a distal humerus. Simons's (1972) proposal that "*Thaumastolemur*" is a junior synonym of *Megaladapis* perpetuated an erroneous synonomy made much earlier by Grandidier (1902*a*) with respect to this humerus, and can be rejected on both metric and morphological criteria (Jungers, 1976; Vuillaume-Randriamanantena, 1990).

Forsyth-Major (1897) described the brain endocast of the specimen of *M. madagascariensis* mentioned above, and also proposed (Forsyth-Major, 1900*b*) a new species, "*M. insignis*" on the basis of upper and lower teeth that were considerbly larger than those from *M. madagascariensis*. Unbeknownst at the same time in France, G. Grandidier (1899*a*) erected a new genus and species "*Peloriadapis edwardsi*" from a fragmentary lower third molar. Shortly thereafter (Grandidier, 1899*b*), he added a maxillary fragment with second and third molars to the hypodigm of "*Peloriadapis*". "*Peloriadapis*" is now used as a subgeneric nomen for part of the genus *Megaladapis* (Vuillaume-Randriamanantena *et al.*, 1992), but "*edwardsi*" was given priority for the species designation – hence *Megaladapis* (*Peloriadapis*) *edwardsi*, the largest of the species in this genus.

The year 1900 was especially confusing. Grandidier (1900) attributed several postcranial elements to *M. madagascariensis* and noted a skull but did not attribute it to species. Almost simultaneously, Forsyth-Major (1900*b*) expanded his diagnosis of dental remains of "*M. insignis*". Working in Vienna, Lorenz von Liburnau was preparing and publishing on subfossils collected by F. Sikora from Andrahomana (southeastern Madagascar). He proposed two more species, "*M. brachycephalus*" and "*M. dubius*" (Lorenz von Liburnau, 1900*a*), for a cranium and postcranial fragments, respectively, and misdiagnosed an immature skull as "*Mesoadapis destructus*". In his review article on "our present knowledge of extinct primates from Madagascar", Forsyth-Major (1900*c*) formally sank all three of these new taxa, along with a third genus proposed by Lorenz von Liburnau (1900*b*), "*Paleolemur*", into "*M. insignis*". He drew attention to the apparent loss of upper incisors in adults, and he pointed out the similarity of this condition to that seen in "*Lepidolemur*" (= *Lepilemur*). Although Forsyth-Major noted that the diaphyseal flattening of *Megaladapis* femora recalled that seen in African lorids, he had doubts about its climbing capability and hinted for the first time that an aquatic adaptation was possible.

The task of clarifying and consolidating the chaotic taxomony of *Megaladapis* fell to Grandidier (1902*a*). He accepted Forsyth-Major's revision of Lorenz von Liburnau's allocations. Arguing for priority, he also sank "*M. insignis*" into *M. edwardsi*, and properly reallocated previously confused postcrania. At the same time, he described a manual phalanx of *M. edwardsi* from Andrahomana, the curvature of which suggested to him a modest degree of arborealism not unlike the gorilla; this inference was reinforced by his precocious conclusion, based on incomplete long bones, that the upper limb of *Megaladapis* was longer than the lower.

In a report on excavations on the central plateau at Ampasambazimba, Standing (1903) announced a new species, *M. grandidieri*, from among a copious collection of exciting new fossils. Apparently working in isolation without the benefit of Grandidier's revisions, Antony Jully & Standing (1904) contended that this region of Madagascar had been home to three different species of *Megaladapis* ("*M. insignis*", *M. grandidieri* and *M. madagascariensis*). Standing (1908) later correctly recognized *M. grandidieri* as the only species from this region.

In the interim, Lorenz von Liburnau (1905) published his excellent descriptions of *Megaladapis* fossils from Andrahomana in the southeast. He accepted Grandidier's (1905) suggestion that cranial variation there did not rise to the level of specific differences. Although he could not assign new postcrania to specific skulls, it was clear that these were the first truly associated postcrania of the genus. It is a pity that this work was not fully appreciated by some contemporaries (e.g., Grandidier, 1905) and later workers (e.g., Standing, 1908), who continued to draw faulty functional inferences in part from incorrect allocations of skulls, teeth and postcrania. In step with Grandidier, Lorenz von Liburnau also reconstructed *Megaladapis* as an arboreal animal, but noted similarities in skeletal robusticity to cave bears. From the long and flexed nasal bones and the absence of permanent upper incisors, he reconstructed a fleshy nasal and labial apparatus to help in plucking leaves and fruits, and viewed the broad ilia as evidence for an expanded gut needed to digest foliage.

Standing's (1908) monographic treatment of the subfossil material from Ampasambazimba included quite accurate, concise and very useful descriptions and metric comparisons of *Megaladapis* skulls and teeth. Unfortunately, the section on the appendicular skeleton was fraught with inaccuracy and misattribution, along with some fairly fanciful functional scenarios. Even with the highly derived postcrania of *Palaeopropithecus* in hand as his *M. grandidieri*, Standing believed that *Megaladapis* was both terrestrial and arboreal, repeating the large-bodied ape analogies of Grandidier. Ironically, the functional deductions from the material that really did appertain to *Megaladapis* led Standing to suggest an aquatic niche for *Palaeopropithecus* (see above).

Lamberton's (1934*c*) monograph described and figured many postcranial elements for the first time and included an important synthesis of information known to that date on *Megaladapis*. He published additional elements sporadically afterwards (Lamberton, 1939*b*, 1946), but one of his notes merits special attention (Lamberton, 1936*b*). During his 1936 campaign in the southwest, a complete cranium of *M. edwardsi* was recovered from Beavoha in association with "les principaux os du squelette groupés à coté" ("the principal bones of the skeleton grouped beside", Lamberton, 1936*b*: 7). This association has since been lost in the collection when it was moved from the Académie Malgache to the University of Antananarivo, but it did vindicate Lorenz von Liburnau's much earlier associations and prove that Standing was wrong. Lamberton (1934*c*) initially believed that *Megaladapis* was an adroit tree-climber, which Carleton (1936) corroborated from her observations on the distal humerus. Lamberton later reversed his position after reporting on new tali (Lamberton, 1939*b*) and what he believed to be first metatarsals of *M. edwardsi* (Lamberton, 1948*b*), which are now known to be the first metacarpals of *M. madagascariensis* (Wunderlich *et al.*, 1996). A locomotor reconstruction emerged with a greater emphasis on terrestrial quadrupdedalism than on climbing, perhaps not unlike some gorillas. Piveteau (1957, 1961) adopted and promoted the gorilla analogy. Wed now to a more terrestrial reconstruction, Lamberton (1939*b*) tentatively suggested that *Megaladapis* may have rooted in the soil for tubers and bulbs. This feeding strategy was repeated later in the French literature as a virtual certainty (Genet-Varcin, 1963; Mahé, 1976).

Erich Thenius's (1953) analysis of the dentition of *Megaladapis* once again drew upon pointed analogies to *Lepilemur* (e.g., loss of upper incisors), and he concluded (Thenius, 1970) unequivocally that *Megaladapis* was a folivore. Tattersall (1975) reinvoked the folivorous koala as the best available analogy for the feeding apparatus (also see Hofer, 1953; Wall, 1997). Analyses of molar microwear and the especially well-developed lower molar shearing crests (summarized in Jungers *et al.*, 2001) corroborate the specialized folivore analogy, but dietary differences among the recognized species are possible.

The long digits and moderately curved phalanges of *Megaladapis* strongly colored Helmuth Zapfe's (1963) "Lebensbild", wherein he favored the reconstruction of an arboreal climber despite the apparently bulky body build. Holger Preuschoft's (1971) theoretical analysis of digital stresses led to the conclusion that *Megaladapis* was a specialized climber. Walker's (1967*a*, 1974*a*) wide-ranging analysis once again used a koala analogy, but the strong-grasping hands and feet indicated that all species probably engaged in a highly modified form of vertical climbing and clinging (with very reduced leaping abilities compared to extant lemurs). Jungers (1976, 1977) subsequently incorporated Walker's koala model into Cartmill's (1974*a*) elegant biomechanical model for climbing mammals lacking claws. Limb-bone allometries were interpreted as further evidence for competence in climbing at large body size (Jungers, 1978, 1980), but anatomical limits to the postcranial koala model were also noted. Recent recovery of nearly complete and relatively enormous hand and foot skeletons of *Megaladapis* (Wunderlich *et al.*, 1996; Hamrick *et al.*, 2000) confirm powerful arboreal grasping capabilities (see Godfrey *et al.*, 1997*a*; Jungers *et al.*, 2001, for reviews). We do not doubt that *Megaladapis* also frequented the ground, if only to travel between trees, but the primary impression is that of a large-bodied, deliberate folivore that relied on pincer-like grips when arboreal.

Tore Ekblom (1951) described a variety of "new" subfossils (the Ljungqvist Collection in Uppsala), including remains of *Megaladapis*, but these had been collected much earlier by a Swedish school teacher in 1928–9. Expeditions led by Elwyn Simons (Simons *et al.*, 1990) and Dewar recovered material that, along with that of Mahé (1965), has been referred to *Megaladapis* cf. *grandidieri/madagascariensis* in a reconsideration of the ecogeographic variation within the genus (Vuillaume-Randriamanantena *et al.*, 1992). We suspect that the exisiting tripartite classification of species may be overly simplified. It is clear that *Megaladapis* does not comprise three anatomically identical, merely scaled versions of one another. Rather, two distinct morphs can be recognized, one comprising the wide-ranging subgenus *Megaladapis* (for *M. madagascariensis*, *M. grandidieri* and the northern subfossils) and a second, the subgenus *Peloriadapis*, for the very large-bodied *M. edwardsi* from the south and southwest.

Impressed by the well-known similarities in the skulls and jaws of *Lepilemur* and *Megaladapis*, Tattersall (1982; also see Tattersall & Schwartz, 1974) placed both in the family Lepilemuridae with subfamilies Lepilemurinae and Megaladapinae. Jungers (1980), Mittermeier *et al.* (1994) and Nowak (1999) also lumped them into the same family, either in Lepilemuridae or Megaladapidae. Preliminary phylogenetic analysis of ancient DNA has cautioned against inferring an especially close relationship between these two genera (Yoder *et al.*, 1999; but see Montagnon *et al.*, 2001). We have opted to follow Hill (1953), Szalay & Delson (1979) and Jenkins (1987) in separating the two into their own families, Lepilemuridae and Megaladapidae (see below).

Close cousins of extant lemurs

Whereas most of the extinct lemurs are only distantly related to (different) extant lemur families, a few species appear to be sister taxa to particular extant species. They belong to two genera, *Daubentonia* and *Pachylemur*. *Daubentonia robusta* is without doubt the sister taxon to *D. madagascariensis*. *Pachylemur insignis* and *P. jullyi* are lemurids apparently closely connected to *Varecia* (Crovella *et al.*, 1994*b*).

The first-discovered specimens of the giant aye-aye were a pair of pierced incisors from Lamboharana in southwest Madagascar (see MacPhee and Raholimavo, 1988). Grandidier (1905, 1929) recognized their similarity to living aye-ayes but, despite their larger size, did not designate a new species. Discovery of postcranial specimens at two more sites in southwest Madagascar (Anavoha and Tsirave) prompted Lamberton (1934*b*) to name a new species. The Tsirave skeleton had all of its long bones, bits of scapula, the complete innominate and bones of the cheiridia, including the specialized phalanges of the third digit. Very recently, additional incisors belonging to a giant aye-aye were found at Ankilitelo (a deep sinkhole pit in southwest Madagascar), and a partial tibia found long ago at Ampasambazimba in central Madagascar was located in the collections of the University of Antananarivo (Godfrey *et al.*, 1999). Thus, it appears that *Daubentonia robusta* ranged not merely through the southwest but also into the Central Highlands.

The giant aye-aye was remarkably similar to the living aye-aye, but it was much stockier and larger at roughly 10 kg, and it had different limb proportions (Lamberton, 1934*b*; Jungers, 1980; Simons, 1994*b*; Godfrey *et al.*, 1997*a*). Despite the fact that no skull is known, incisal and manual morphology prove that the giant aye-aye used the same feeding strategy. Like its extant congener, the giant aye-aye almost certainly specialized on "structurally defended" resources (see Sterling, 1994*b*, on living aye-ayes). It probably consumed nuts, seeds, fruit and limited amounts of fauna (including some insect larvae), and used its hypertrophied incisors to gnaw into very hard objects. Simons (1994*b*) has speculated that its diet may have also included giant burrowing crickets and termites, but woodpeckers are an inappropriate analogy for both aye-aye species. The ramy nut dominates the diet of *D. madagascariensis* (Garbutt, 1999), and both living and extinct species are better compared to arboreal rodents, especially squirrels (Jungers *et al.*, 2001, and references therein).

Upon its initial discovery, *Pachylemur* was recognized as a large-bodied lemurid. Filhol named *Lemur insignis* in 1895 on the basis of a humerus. The very similar "*Paleochirogalus jullyi*" (Grandidier, 1899*b*) from Antsirabe in central Madagascar was soon afterward placed in the genus *Lemur* by Forsyth-Major (1900*c*) and Standing (1904). Lamberton (1948*a*) proposed that a separate subgenus, *Pachylemur*, of the genus *Lemur*, be erected for the extinct forms. Since then, *Pachylemur* has been afforded full separate genus status (Jouffroy, 1963; Tardieu & Jouffroy, 1979), or, alternatively, the extinct species have been referred to the genus *Varecia* (see Walker, 1974*a*). Szalay & Delson (1979) synonymized the two species, but more research is needed to confirm this given the variation in their bones.

Compared to lemurids, *Pachylemur* limbs are shorter and more robust relative to the vertebral column, and the proportions are different. The humerofemoral index is high, the brachial index is high and the crural index is low (Jungers, 1980). At about 10 kg, *Pachylemur* was more than twice the size of *Varecia*. The forelimb was nearly as long as the hindlimb. Carleton (1936) characterized the extinct "*Lemur*" as "bent-limbed", clumsy and slow. Long-bone proportions and robusticity convinced Jouffroy (1960, 1963) that *Pachylemur* was less agile than extant lemurids. The hip joint morphology and the relative size of the femoral head suggested to Godfrey (Godfrey, 1988; Godfrey *et al.*, 1995) that hindlimb suspension may have been more important to *Pachylemur* in feeding and other activities than it is even in *Varecia* (Meldrum *et al.*, 1997). Tardieu & Jouffroy (1979), on the basis of knee morphology, reconstructed *Pachylemur* as a slow quadrupedal climber that definitively did not leap. *Pachylemur* also differs from *Varecia* in having absolutely shorter vertebral bodies (despite its much larger body size) and far weaker anticliny of its neural spines, implying less bounding and a greater emphasis on slow climbing (Ravololonarivo, 1990). Interlimb robusticities corroborate the same conclusion; leapers tend to have humeri that are far more robust than their femora, and this disparity does not exist for *Pachylemur*. Ravololonarivo's (1990) cladistic analysis of the vertebral column of lemurids gave *Pachylemur* sister taxon status to all other lemurid genera.

Those who favor placing *Pachylemur* in the genus *Varecia* generally do so on the basis of its strikingly *Varecia*-like dentition (Seligsohn & Szalay, 1974). Crovella *et al.* (1994*a*) concluded from ancient DNA that *Pachylemur* and *Varecia* are indeed sister taxa. It thus appears that the postcranial distinctions of *Pachylemur* are uniquely derived, and we believe these apomorphies merit generic distinction. Most evidence suggests that *Pachylemur* was a cautious arboreal quadruped. Claims for greater terrestriality in *Pachylemur* (e.g., Walker, 1974*a*; Mittermeier *et al.*, 1994) are based on its greater robusticity and limb proportions (e.g., fore and hindlimbs of roughly equal length), but overall support for significant amounts of terrestriality is weak at best. We conclude that the functional morphology of the limb bones signals arboreality unequivocally, especially compared to archaeolemurids.

Taxonomy

Systematic framework[1]

Order Primates Linnaeus, 1758
- Suborder Strepsirhini É. Geoffroy Saint-Hilaire, 1812
  - Infraorder Lemuriformes Gregory, 1915
    - Family Archaeolemuridae G. Grandidier 1905
      - Genus *Archaeolemur* Filhol, 1895
        - *Archaeolemur majori* Filhol, 1895
        - *Archaeolemur edwardsi* Filhol, 1895
      - Genus *Hadropithecus* Lorenz von Liburnau, 1899
        - *Hadropithecus stenognathus* Lorenz von Liburnau, 1899
    - Family Palaeopropithecidae Tattersall, 1973
      - Genus *Mesopropithecus* Standing, 1905
        - *Mesopropithecus globiceps* Lamberton, 1936
        - *Mesopropithecus pithecoides* Standing, 1905
        - *Mesopropithecus dolichobrachion* Simons *et al.*, 1995
      - Genus *Babakotia* Godfrey *et al.*, 1990
        - *Babakotia radofilai* Godfrey *et al.*, 1990
      - Genus *Palaeopropithecus* G. Grandidier, 1899
        - *Palaeopropithecus ingens* G. Grandidier, 1899
        - *Palaeopropithecus maximus* Standing, 1903
        - [*Palaeopropithecus* sp. nov.]
      - Genus *Archaeoindris* Standing, 1909
        - *Archaeoindris fontoynontii* Standing, 1909
    - Family Megaladapidae Forsyth-Major, 1894
      - Genus *Megaladapis* Forsyth-Major, 1894
        - *Megaladapis edwardsi* G. Grandidier, 1899
        - *Megaladapis grandidieri* Standing, 1903
        - *Megaladapis madagascariensis* Forsyth-Major, 1894
    - Family Daubentoniidae Gray, 1863
      - Genus *Daubentonia* É. Geoffroy Saint-Hilaire, 1795
        - *Daubentonia robusta* Lamberton, 1934
    - Family Lemuridae Gray, 1821
      - Genus *Pachylemur* Lamberton, 1948
        - *Pachylemur insignis* Filhol, 1895
        - *Pachylemur jullyi* G. Grandidier, 1899

Family Archaeolemuridae

GENUS *Archaeolemur* Filhol, 1895 (Figs. 7.1–7.3)
Dental formula 2.1.3.3/2.0.3.3. Mandible with fused symphysis. Central incisors enormous and spatulate. The lateral upper incisors are relatively small. Dental formula as in lemurids except the modified "tooth-comb" has four instead of six teeth. These teeth are long and slender, and obliquely (but not horizontally) implanted; their tips wear flat. *Archaeolemur* is atypical in having hypertrophied central upper incisors that contact one another at the midline; it thus lacks the typical strepsirhine interincisal gap. The premolar series is modified into a continuous shearing blade. P^4 is molariform, its posterior portion is buccolingually expanded, and it bears a distinct protocone. The upper canine is very broad and low-crowned; P_2 is caniniform and very robust. The molars are buccolingually expanded; the first two exhibit classic bilophodonty. Third

[1] See Tattersall & Schwartz (1974) for more complete dental descriptions of most taxa; see Table 7.1 for complete list of standard long bone studies for all taxa.

molars reduced but may exhibit incipient bilophodonty. Primitive configuration of bulla, i.e., there is an inflated petrosal bulla with tympanic ring intrabullar and free. The carotid foramen is located on the posterior wall of the bulla, again as in other lemurs. Limb bones are relatively short, straight and robust, although the femora of the smallest-bodied species may exhibit slight anteroposterior bowing. Hands and feet relatively short. Greater tubercle projects above humeral head. Deep olecranon fossa with strong retroflexion of the medial epicondyle. Free os centrale, large pisiform, reduced pollex. Enormous apical tufts on distal phalanges; no evidence for grooming claw. Greater trochanter projects above femoral head. Calcaneus, cuboid and fifth metatarsal sport large tuberosities. Reduced hallux. Metapodial heads II–V have dorsal extensions of the articular surface. Proximal phalanges relatively straight. Broad pelvic girdle.
INCLUDED SPECIES *A. edwardsi*, *A. majori*

SPECIES *Archaeolemur majori* Filhol, 1895
TYPE SPECIMEN From Belo-sur-mer (west Madagascar, and) presumably located in the collection of the Muséum National d'Histoire Naturelle, Paris, a humerus and superior portions of a radius and ulna
AGE AND GEOGRAPHIC RANGE Late Quaternary, southern and western Madagascar; possibly central and northern Madagascar
ANATOMICAL DEFINITION
Archaeolemur majori is smaller in body size at approximately 14 kg than *A. edwardsi*; intermembral index 92; skull length averages 128 mm. Sagittal and nuchal crests often absent. Facial profile not as steep as in *A. edwardsi*.

SPECIES *Archaeolemur edwardsi* Filhol, 1895
TYPE SPECIMEN 1906–17 (Institut de Paléontologie), in the collection of the Muséum National d'Histoire Naturelle, Paris, two lower jaws and "several postcrania"
AGE AND GEOGRAPHIC RANGE Late Quaternary, Central Madagascar; possibly western, northern, and southeastern Madagascar; Ampasambazimba, Ampoza–Ankazoabo, Belo-sur-mer, Masinandraina, Morarano-Betafo, Sambaina, Vakinankaratra (marshes of Sirabe, Central Madagascar); possibly also Ambolisatra, Amparihingidro, Andrahomana, Anjohibe, Anjohikely, Ankarana, Bungo-Tsimanindroa, Mt. des Français
ANATOMICAL DEFINITION
Larger of the two species at ~ 24 kg; intermembral index 92. Skull length averages 147 mm. Sagittal and nuchal crests common. Facial profile steeper than in *A. majori*.

GENUS *Hadropithecus* Lorenz von Liburnau, 1899 (Figs. 7.3, 7.4)
Body mass estimated at approximately 27 kg; skull length is approximately 141 mm. Intermembral index cannot be estimated, but humerofemoral index (approximately 103) is considerably higher than in *Archaeolemur*, and brachial index (approximately 84) is considerably lower. Face is short, facial profile steep, mandible fairly deep and very robust. Zygomatic arch and postorbital bar well developed, robust. Neurocranium relatively broad and short. Anterior

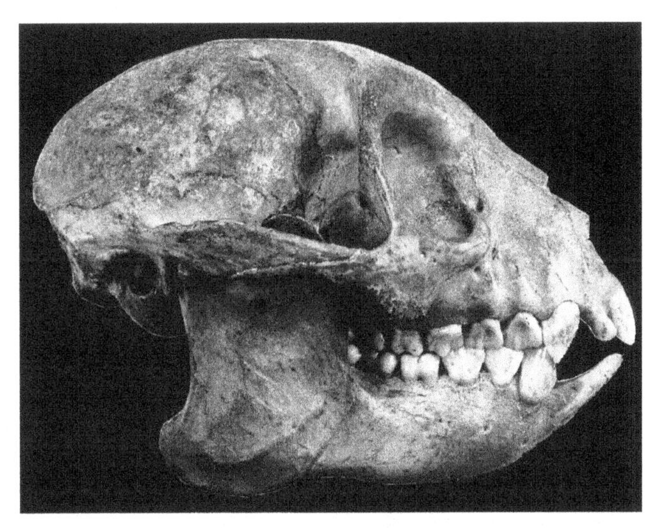

Fig. 7.1 *Archaeolemur majori*. Skull of from Andrahomana (M7374 in the Natural History Museum of London). Courtesy of Ian Tattersall.

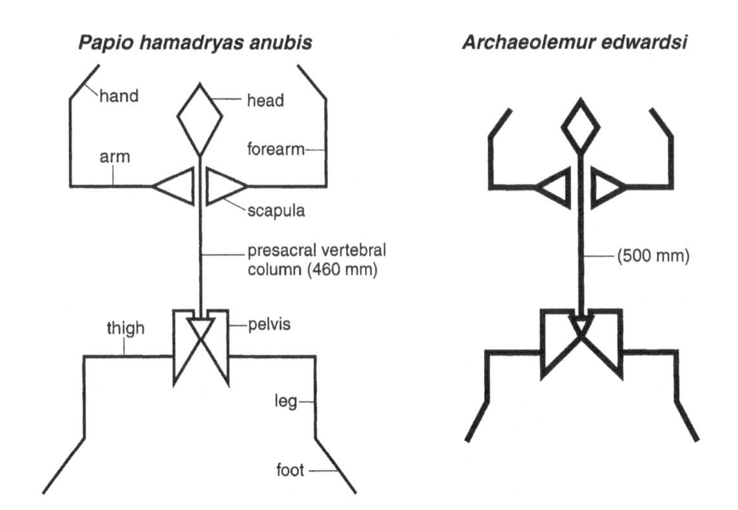

Fig. 7.2 Schematic of body proportions in an adult male *Papio hamadryas anubis* and *Archaeolemur edwardsi* (the latter based primarily on a skeleton from Anjohikely, in the northwest of Madagascar). Traits measured to derive this comparison were: skull length, bizygomatic breadth, presacral vertebral column length, sacrum length, scapula length (parallel to the scapular spine), scapula breadth (perpendicular to the scapular spine), innominate length, bi-iliac breadth, iliac breadth, humerus length, radius length, total hand length, femur length, tibia length and total foot length. Drawings are scaled such that the presacral vertebral columns are made the same length; the actual presacral length is greater in *Archaeolemur*. Both animals would have been similar in body mass (approximately 25 kg). The bolder lines for *Archaeolemur* denote its greater skeletal robusticity. For example, humeral and femoral robusticity values (midshaft transverse × 100 / maximum length) average 7.0 (SD = 0.6) and 6.5 (SD = 0.4) respectively for 19 *Papio hamadryas*. They are 10.6 (SD = 0.5) for 11 humeri and 10.6 (SD = 0.4) for 9 femora belonging to the robust *Archaeolemur* from the north.

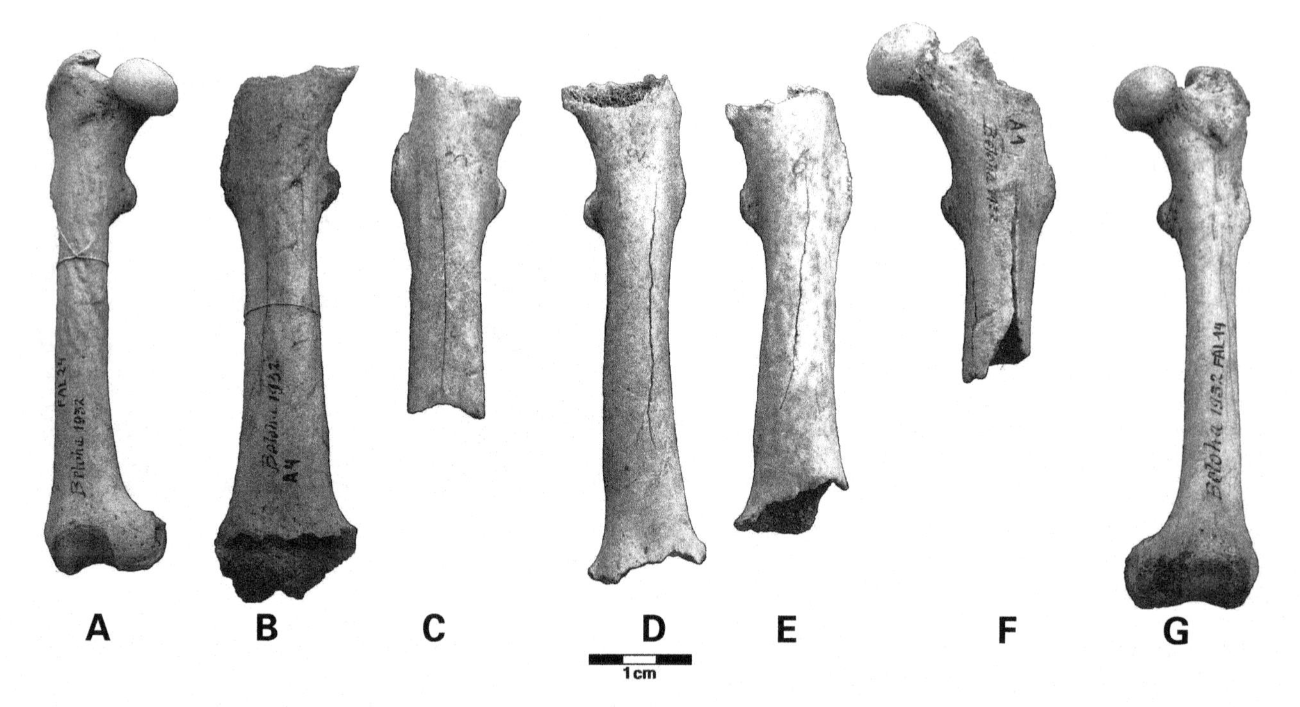

Fig. 7.3 Comparison of femora of *Archaeolemur majori* (A, G) and *Hadropithecus stenognathus* (B–F) from Anavoha (south).

teeth diminutive, lower incisors quite orthally implanted. Basicranium anteroposteriorly compressed. Mandibular symphysis fuses early and is quite orthal. Premolar series modified, like that of *Archaeolemur*, into continuous shearing blade, but anterior premolars are reduced (as is the upper canine and all incisors). All upper premolars have protocone developed to some extent, P^4 is broader than M^1 and completely molariform; enamel very thick. Bullae inflated as in *Archaeolemur*, tympanic ring free. Relatively short, straight and robust limb bones that recall those of *Archaeolemur* in many respects. Epiphyseal morphologies of long bones also similar in many respects to *Archaeolemur*.
INCLUDED SPECIES *H. stenognathus*

SPECIES *Hadropithecus stenognathus* Lorenz von Liburnau, 1899 (Fig. 7.4)
TYPE SPECIMEN VNH 1934. IV 1/1, mandible
AGE AND GEOGRAPHIC RANGE Late Quaternary; southern, western and central Madagascar; "Ambovombe", Ampasambazimba, Ampoza–Ankazoabo, Anavoha, Andrahomana, Belo-sur-mer, "Morondava", Tsirave
ANATOMICAL DEFINITION
As for genus.

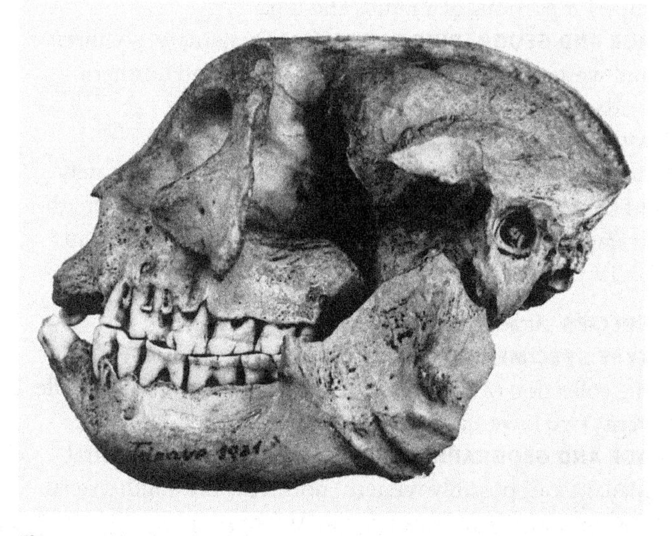

Fig. 7.4 *Hadropithecus stenognathus*. Skull of uncatalogued Académie Malgache display specimen from Tsirave. Courtesy of Ian Tattersall.

Family Palaeopropithecidae

GENUS *Mesopropithecus* Standing, 1905
Midsized indroid, ca. 10 kg. Sagittal crest may be present or temporal lines are anteriorly confluent; nuchal ridge is confluent with posterior root of zygoma; orbits small and more convergent than in *Propithecus*; postorbital constriction marked; postorbital bar more robust; muzzle is wide and anteriorly squared; zygoma are robust and cranially convex; facial angle is steeper than in *Propithecus*. Typical inflated auditory bulla with intrabullar ectotympanic ring. Dental formula (2.1.2.3/2.0.2.3), dental morphology similar to other palaeopropithecids and indrids. Short upper and lower premolars, moderately buccolingually constricted M^3. Possesses conventional tooth-comb of indrid type, with four teeth. Forelimb relatively conservative (indrid-like); hindlimb and axial skeleton more specialized for suspension (more like *Palaeopropithecus* and *Babakotia*). Fore and hindlimbs approximately equal in length.
INCLUDED SPECIES *M. dolichobrachion*, *M. globiceps*, *M. pithecoides*

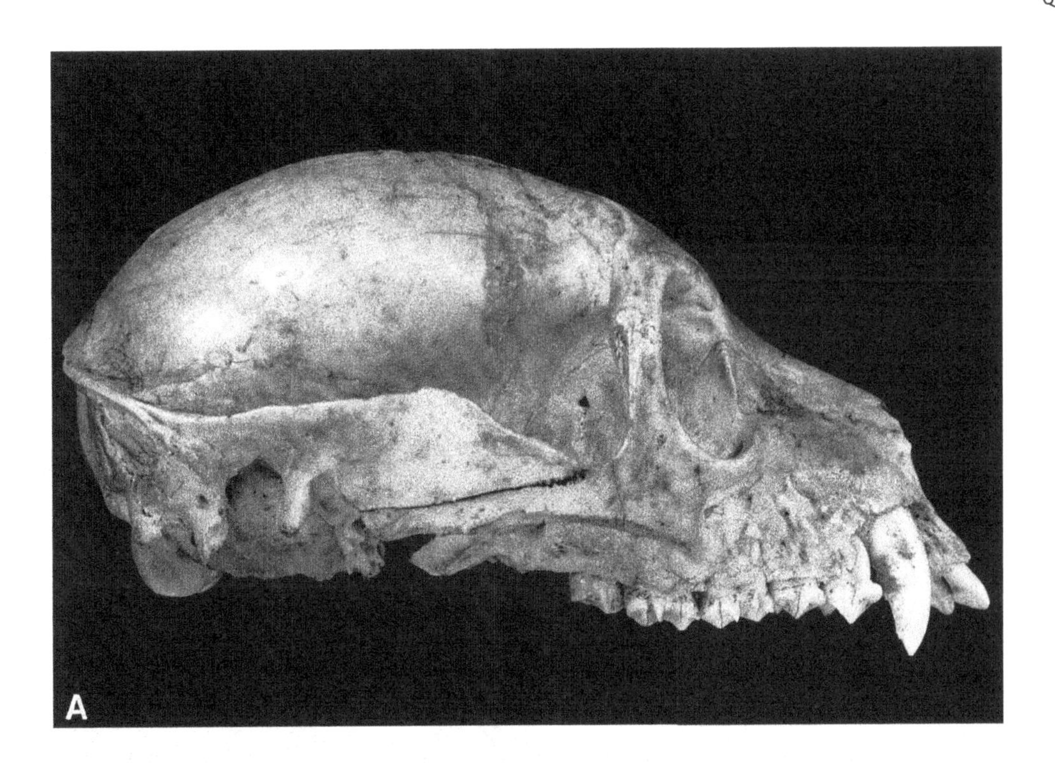

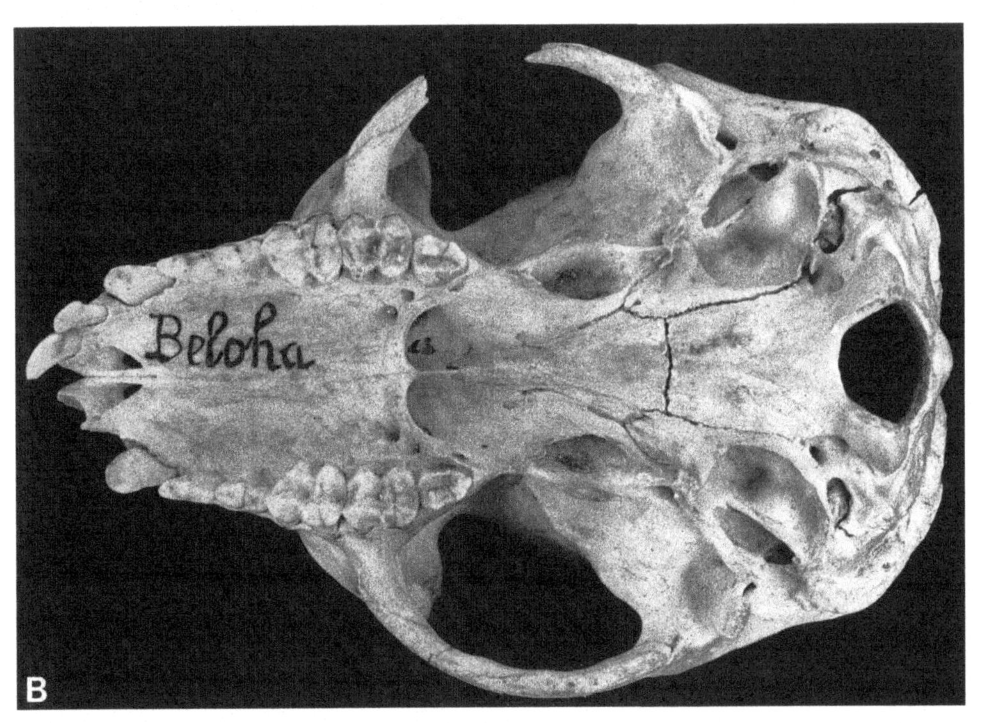

Fig. 7.5 *Mesopropithecus globiceps*. The skull in (A) lateral and (B) basal view is NpH3 from Beloha Anavoha, in the collection of the University of Antananarivo and originally called *Neopropithecus platyfrons*.

SPECIES *Mesopropithecus globiceps* Lamberton, 1936 (Fig. 7.5)
TYPE SPECIMEN Specimen No. 211011 (Tattersall's (1971) reference number, following Mahé (1976), in the collection of the University of Antananarivo, cranium
AGE AND GEOGRAPHIC RANGE Late Quaternary, southern and southwestern Madagascar, including Anavoha, Ankazoabo-Grotte, Belo-sur-mer, Manombo-Toliara, Taolambiby, Tsiandroina, and Tsirave

ANATOMICAL DEFINITION
At approximately 9 kg, slightly smaller than M. *pithecoides*. Intermembral index 97. Skull length averages 94 mm. Skull is more gracile and snout narrows anteriorly.

SPECIES *Mesopropithecus pithecoides* Standing, 1905
TYPE SPECIMEN The known crania of *Mesopropithecus* from Ampasambazimba include: MpH1 (Mahé's 221023, cast

available at the Natural History Museum, London), MpH2 (Mahé's 221022, missing; cast available at the Natural History Museum, London), Mahé's 221021 (immature, missing; cast available at the Natural History Museum, London), an uncatalogued specimen at the Laboratoire d'Anatomie Comparée, Paris; and M9915 (Natural History Museum, London)

AGE AND GEOGRAPHIC RANGE Late Quaternary, Central Madagascar: Ampasambazimba, Antsirabe

ANATOMICAL DEFINITION
Approximately 10 kg; intermembral index 99; skull length averages 98 mm. Well-developed sagittal and nuchal cresting; massive zygomatic arches, muzzle broader anteriorly.

SPECIES *Mesopropithecus dolichobrachion* Simons et al., 1995 (Figs. 7.6, 7.7)

TYPE SPECIMEN UA-LPV 9100 at the University of Antananarivo, associated cranium, mandible, vertebrae, and hand and foot bones

AGE AND GEOGRAPHIC RANGE Late Quaternary, northern Madagascar: Ankarana

ANATOMICAL DEFINITION
Largest of the three species at approximately 11 kg. Both humerofemoral (approximately 104) and intermembral (approximately 113) indices relatively high (hence its specific nomen). Skull length averages 102 mm. Sagittal and nuchal crests present, postorbital constriction marked, muzzle wide and squared anteriorly. Dentition very like that of congeners. Humerus substantially longer and more robust, and unique in exceeding the length of the femur. Indrid-like carpus but strongly curved proximal phalanges. Moderately reduced spines of lumbar vertebrae.

GENUS *Babakotia* Godfrey et al., 1990
Adult dental formula (2.1.2.3/2.0.2.3) as in indrids and other palaeopropithecids, dental morphology similar in many respects to these taxa, but with relatively greater mesiodistal elongation of the premolars. Cheek teeth with crenulated enamel. Possesses conventional tooth-comb of the indrid type, with four teeth. Extensive shearing crests on cheek teeth. Cranium recalls that of *Indri indri*, but more postorbital constriction and mandible is much more robust. No orbital tori or circumorbital protuberances; postorbital bar is robust. Inflated auditory bulla with intrabullar ring-like ectotympanic. Body mass estimated at approximately 16 kg, intermembral index 118; skull length averages 114 mm. Moderate degree of spinous process reduction in thoracolumbar region. Incipient ischial spine, reduced anterior inferior iliac spine. Long pubis with some degree of superoinferior flattening. Globular femoral head somewhat cranially directed (but not to extent seen in *Palaeopropithecus* and *Archaeoindris*). Reduced tibial malleolus. Some reduction in relative length of pollex and hallux. Long and curved proximal phalanges. Grooming claw present. Calcaneus is reduced in size.

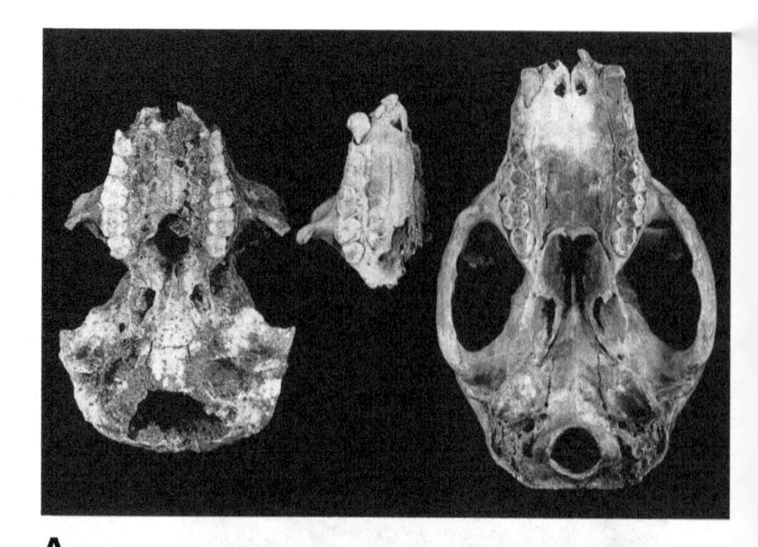

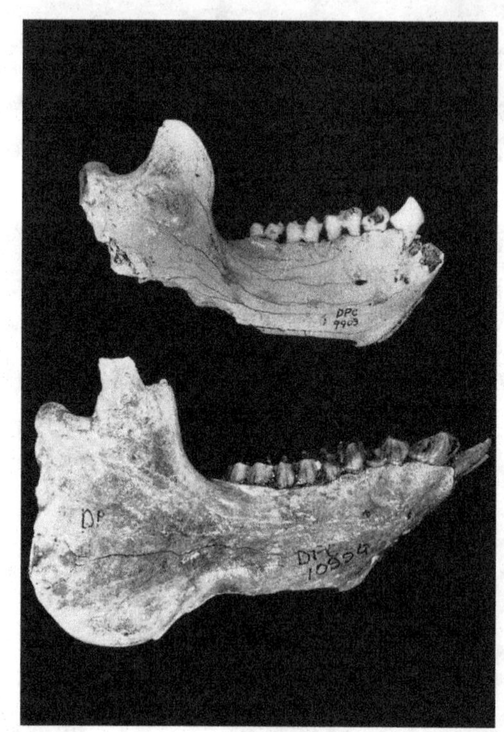

Fig. 7.6 Occlusal views of (A) the maxillary dentitions of *Mesopropithecus dolichobrachion* (left, DPC 11755 and middle, DPC 9903) and *Babakotia radofilai* (right, DPC 10994), and (B) mandibular dentitions of *Mesopropithecus dolichobrachion* (top, DPC 9903) and *Babakotia radofilai* (bottom, DPC 10994). *Mesopropithecus* and *Babakotia* have teeth that resemble those of living indrids.

INCLUDED SPECIES *B. radofilai*

SPECIES *Babakotia radofilai* Godfrey et al., 1990 (Figs. 7.6–7.8)

TYPE SPECIMEN UA-LPV 8713 in the collection of the University of Antananarivo, left upper jaw fragment, right

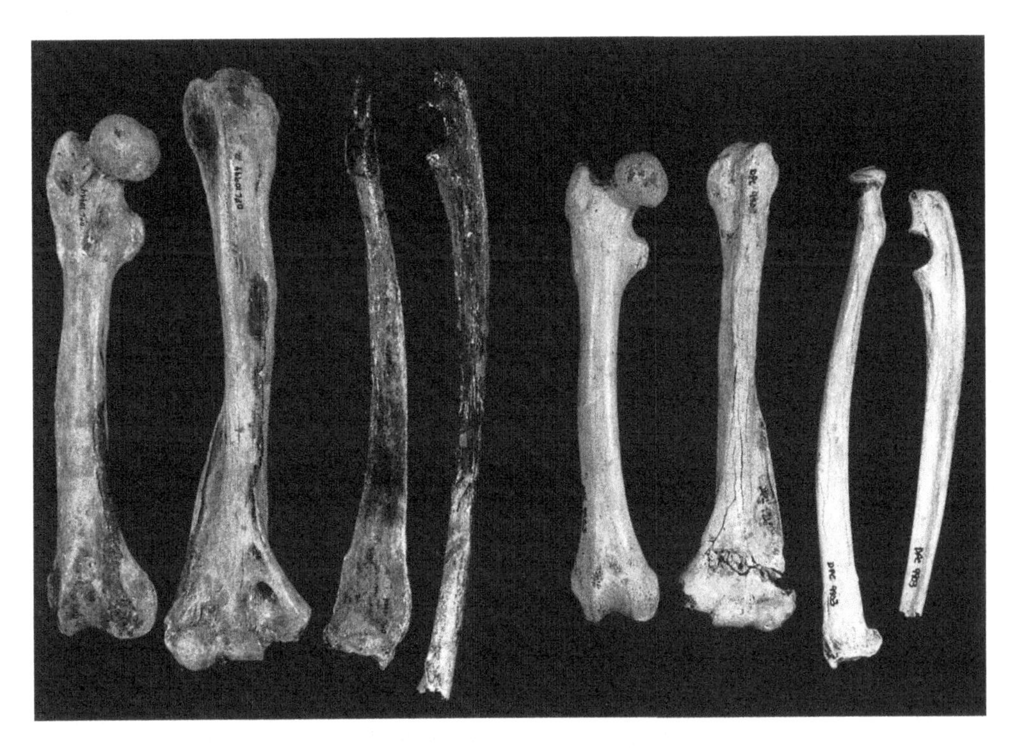

Fig. 7.7 Comparison of the postcrania of *Babakotia radofilai* (left, DPC 10994) and *Mesopropithecus dolichobrachion* (right, DPC 9903) femur, humerus, radius and ulna, both from the Ankarana Massif.

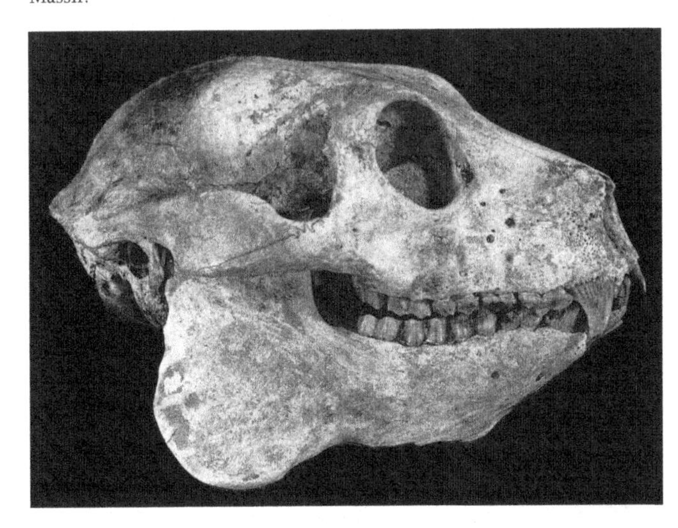

Fig. 7.8 *Babakotia radofilai*. Skull of DPC 10994 in lateral view, from the Cave of the Lone Barefoot Stranger, Ankarana (northern Madagascar).

maxilla, apparently associated humeral and femoral midshafts

AGE AND GEOGRAPHIC RANGE Late Quaternary, northern and northwestern Madagascar: Ankarana, Anjohibe

ANATOMICAL DEFINITION

As for genus.

GENUS *Palaeopropithecus* G. Grandidier, 1899

The dorsal portions of the premaxillae (as well as, to a far lesser extent, the lateral terminus of the nasals) are inflated, bulbous. Orbits small, orbital margin raised to form a bony rim; petrosal elongated to form a tube, and bulla not inflated; brain small, frontal region depressed; postorbital constriction strong; facial retroflexion strong; sagittal crest often present; mandible deep (particularly in gonial region) but mandibular corpus thin; dental rows nearly parallel. Large paraoccipital processes. Dental formula as in extant indrids: there are only two pairs of premolars in the adult dentition, and four procumbent but stubby teeth in the anterior portion of the mandible. A diastema is present between the lower anterior and posterior premolars, and the 4-toothed "tooth-comb" is highly modified and reduced. The cheek teeth resemble those of the indrids (especially *Propithecus*), but the first and second molars are more buccolingually compressed and mesiodistally elongated, and the third molars are smaller in relative size. The molar cusps (especially the hypocones) are relatively low. Intermembral index (approximately 144) exceeds all living primates except orangutans. The humerus is long and robust. The femur is short and anteroposteriorly flattened, with a shallow patellar groove, reduced greater trochanter, high collodiaphyseal angle and huge femoral head lacking a fovea capitis. Ankle with very reduced (essentially absent) medial and lateral malleoli. The hands and feet bear long, strongly curved metapodials and phalanges with deep flexor grooves; notched metacarpo- and metatarsophalangeal joints. The hindfoot is very reduced, and the talar head articulates uniquely with both the navicular and cuboid; ball-like proximal surface of talus. The vertebral neural spines are flattened throughout

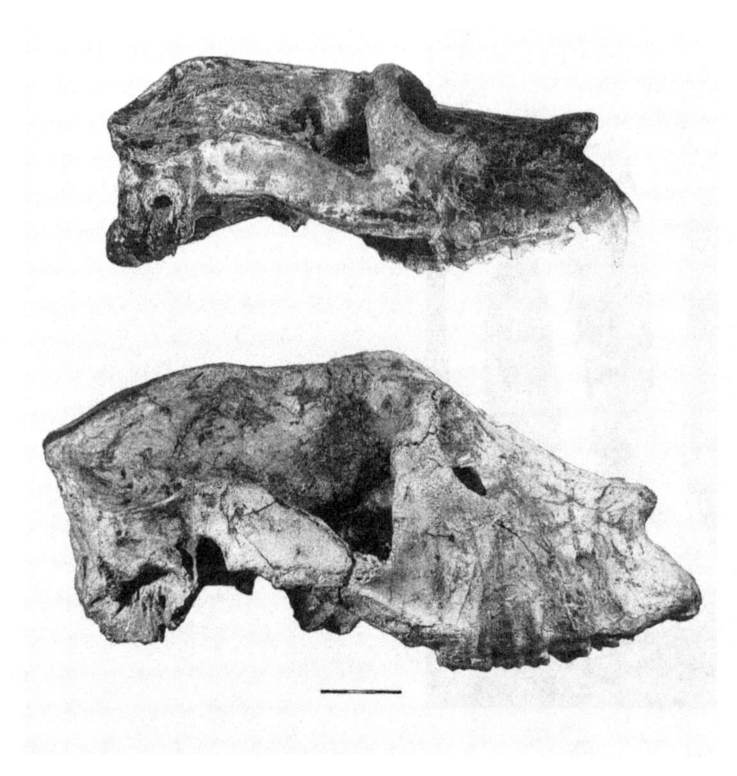

Fig. 7.9 Comparison of the skulls of *Palaeopropithecus maximus* and *Archaeoindris fontoynontii*. Note the differences in the facial profile, the position of the orbits and orbital rimming, and the elevation of the nasals. Both specimens are uncatalogued and on display at the Académie Malgache. Courtesy of Ian Tattersall.

the entire thoracosacral vertebral column. Transverse processes of thoracic and lumbar vertebrae arise from the vertebral arch. Pubis is long and flattened superoinferiorly. Ischial spine present; anterior inferior iliac spine rudimentary.
INCLUDED SPECIES *P. ingens*, *P. maximus*, [*P.* sp. nov.]

SPECIES *Palaeopropithecus maximus* Standing, 1903 (Fig. 7.9)
TYPE SPECIMEN Current whereabouts and specimen number unknown, skull from Ampasambazimba
AGE AND GEOGRAPHIC RANGE Late Quaternary, central, possibly northern Madagascar
ANATOMICAL DEFINITION
The largest species of the genus, about twice the size of the new species from the northwest. Shares with *P. ingens* extreme reduction and shelf-like appearance of the hypocone on M^1 and M^2 (note: a small but distinct hypocone cusp is manifested on the first and second molars of the new species from the northwest). Body mass estimated at 52 kg; intermembral index 144. Skull length averages 191 mm.

SPECIES *Palaeopropithecus ingens* G. Grandidier, 1899
TYPE SPECIMEN Figured by Grandidier (1899*b*), current whereabouts and specimen number unknown (Muséum National d'Histoire Naturelle, Paris?), portion of right mandible containing posterior premolar and first two molars, from Belo-sur-mer
AGE AND GEOGRAPHIC RANGE Late Quaternary, southern and western Madagascar; Ambolisatra, Ampoza-Ankazoabo, Anavoha, Andranovato, Ankazoabo-Grotte, Ankilitelo, Ankomaka, Beavoha, Belo-sur-mer, Betioky-Toliara, Itampolobe, "Lower Menarandra", Manombo-Toliara, Taolambiby, Tsiandroina, Tsivonohy
ANATOMICAL DEFINITION
Body mass estimated at 45 kg; intermembral index 138; single undamaged skull has length of 184 mm. Similar to *P. maximus* in most respects, but slightly smaller overall.

SPECIES *Palaeopropithecus* sp. nov.
AGE AND GEOGRAPHIC RANGE Late Quaternary, northwestern Madagascar, perhaps west: Amparihingidro, Anjohibe; perhaps Ampoza-Ankazoabo
ANATOMICAL DEFINITION
(Description in preparation.) Smallest and most gracile of the *Palaeopropithecus* species.

GENUS *Archaeoindris* Standing, 1909 (Fig. 7.9)
Archaeoindris was much larger and more robust than *Palaeopropithecus*. Body mass estimated at approximately 200 kg, intermembral index estimated well over 100, but probably lower than that of *Palaeopropithecus*. Femur is short but massive, femoral head is huge, collodiaphyseal angle high, and greater trochanter reduced, as in *Palaeopropithecus*. Length of single known skull is 269 mm. Adult dental formula (2.1.2.3/2.0.2.3) is identical, and dental morphology similar, to other palaeopropithecids and indrids. No tooth comb. Small diastema between P_2 and P_4. Rectangular palate. Crenulated enamel on cheek teeth. Paired protuberances over the nasal aperture. Marked postorbital closure; orbits less dorsally oriented than in *Palaeopropithecus*. Sagittal and nuchal crests present. Tubular external auditory meatus is probably petrosal in origin; deflated auditory bulla.
INCLUDED SPECIES *A. fontoynontii*

SPECIES *Archaeoindris fontoynontii* Standing, 1909
TYPE SPECIMEN AM-6239 (maxillae) and AM-6237 (mandible), in the collection of the University of Antananarivo
AGE AND GEOGRAPHIC RANGE Late Quaternary, central Madagascar: Ampasambazimba
ANATOMICAL DEFINITION
As for genus.

Family Megaladapidae

GENUS *Megaladapis* Forsyth-Major, 1894
Adult dental formula 0.1.3.3/2.1.3.3. Mandibular symphysis is fused in adults. Typical strepsirhine tooth-comb present (six teeth). No permanent upper incisors. Diastemata present between upper canine and first

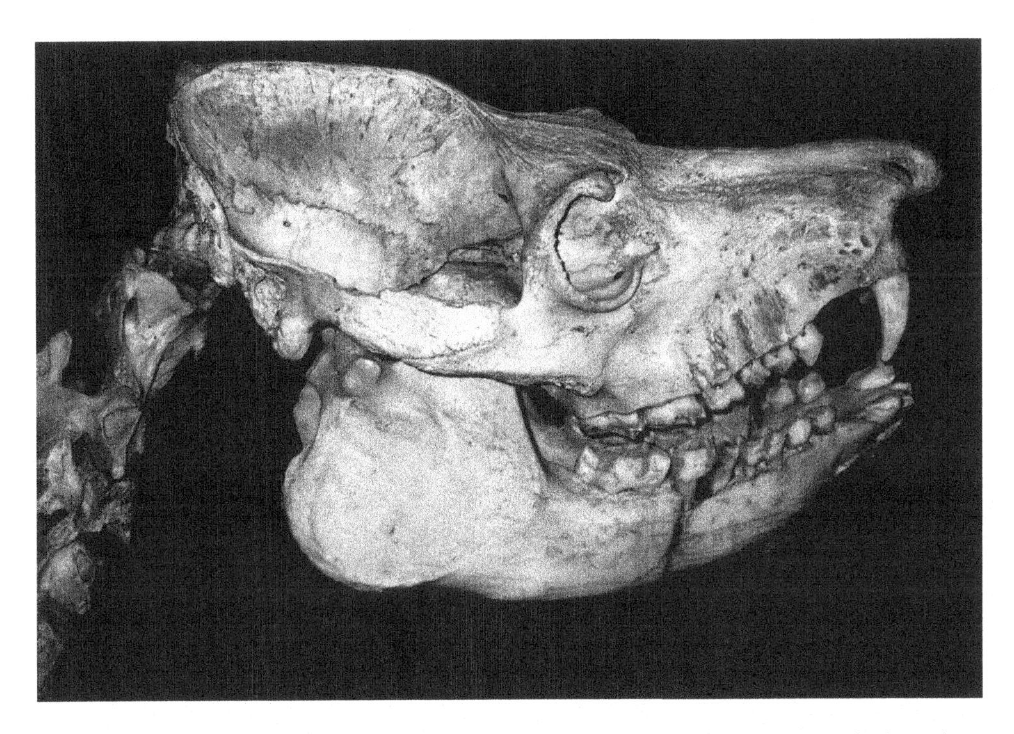

Fig. 7.10 *Megaladapis edwardsi*. The skull is part of a composite reconstruction of the skeleton on display at the Académie Malgache. Courtesy of Russell Mittermeier and Stephen Nash.

premolar and between caniniform lower premolar and P_3. Molar size increases from M1 to M3. Posterior extension of the mandibular condyle's articular surface (and reciprocal expansion of the postglenoid process). Skull is narrow, elongate and decorated with sagittal and nuchal crests. Strong postorbital constriction and large frontal sinus. Nuchal plane is vertical and occipital condyles face posteriorly. Long paraoccipital processes. Orbits are relatively divergent and encircled by bony tori. Facial axis is retroflexed (airorhynchy). Nasals long, projecting beyond prosthion, and flexed downwards above the nasal aperture. Very long olfactory tracts. Relatively small optic foramina. Auditory bulla is not inflated. The tubular external auditory meatus is petrosal in origin. Humerofemoral and intermembral indices are greater than 100. Both hands and feet relatively enormous; long divergent pollex and hallux. Moderately curved proximal phalanges. Very robust limb bones that are also short relative to body size. Spinous processes of thoracolumbar vertebrae reduced (but not to extent seen in *Palaeopropithecus*). Transverse processes arise from the vertebral arch in the thoracolumbar region.

INCLUDED SPECIES *M. edwardsi*, *M. grandidieri*, *M. madagascariensis*, [*M.* sp., cf. *grandidieri*/*madagascariensis*]

SPECIES *Megaladapis (Peloriadapis) edwardsi* Grandidier, 1899 (Fig. 7.10)

TYPE SPECIMEN Figured by Grandidier (1899*a*), current exact whereabouts and specimen number unknown, fragmentary lower third molar from Ambolisatra (west)

AGE AND GEOGRAPHIC RANGE Late Quaternary, southern and southwestern Madagascar: Ambolisatra, Ampanihy, Ampoza-Ankazoabo, Anavoha, Andrahomana, Andranovato, Ankomaka, Beavoha, Betioky-Toliara, Itampolobe, Lamboharana, Mitoho, Taolambiby, Tsiandroina

ANATOMICAL DEFINITION

Large body size (approximately 75 kg). Intermembral index approximately 120. Cranial length averages 296 mm. Absolutely short diastemata. Extremely large molars (e.g., mesiodistal length of M^1 is 18.8 mm on average). Relatively straight radial diaphysis. Extremely varus knee joint. Dominance of medial condyle of proximal tibia and very lateral projection of tibial tuberosity. Relatively small tubercle on fifth metatarsal. Flattened surface of talar trochlea and malleolar facets. Low crural index.

SPECIES *Megaladapis (Megaladapis) madagascariensis* Forsyth-Major, 1894

TYPE SPECIMEN M4848, housed at the Natural History Museum (London), cranium with mandible

AGE AND GEOGRAPHIC RANGE Late Quaternary, southern and southwestern Madagascar: Ambararata-Mahabo, Ambolisatra, Ampoza-Ankazoabo, Anavoha, Andrahomana, Ankilitelo, Beavoha, Belo-sur-mer, Bemafandry, Itampolobe, Taolambiby, Tsiandroina, Tsirave, Tsivonohy

ANATOMICAL DEFINITION

Smallest of the three species, at approximately 38 kg. Intermembral index approximately 114. Skull length averages 245 mm. Mean length of M^1 is 14.0 mm. Humeral head exhibits greater longitudinal curvature. Olecranon fossa is deeper. Broad distal humerus with projecting medial epicondyle and broad brachialis flange. Radial diaphysis quite curved. Relatively large lesser trochanter. Prominent lateral tubercle of fifth metatarsal. Large posterior calcaneus with medially projecting tuberosity.

SPECIES *Megaladapis (Megaladapis) grandidieri* Standing, 1903
TYPE SPECIMEN M9916, Natural History Museum (London), anterior portion of skull
AGE AND GEOGRAPHIC RANGE Late Quaternary, from localities on the high plateau of central Madagascar, including Ampasambazimba, Antsirabe, "Itasy," Morarano-Betafo
ANATOMICAL DEFINITION
Intermediate in body size but closer to *M. edwardsi* at approximately 63 kg, but individual elements may be very variable. Intermembral index approximately 115. Skull length estimated at 289 mm; M^1 length is 15.4 mm. Absolutely and relatively large diastemata. Larger but morphologically similar to *M. madagascariensis*.

SPECIES *Megaladapis* sp., cf. *grandidieri/madagascariensis* (provisional)
AGE AND GEOGRAPHIC RANGE Late Quaternary, northwestern and extreme northern Madagascar, including Amparihingidro, Anjohibe, Ankarana, Mt. des Français
ANATOMICAL DEFINITION
Postcranially intermediate in size between *M. madagascariensis* and *M. grandidieri*, and very similar anatomically to both. Variant from Anjohibe has particularly small teeth.

Family Daubentoniidae

GENUS *Daubentonia* E. Geoffroy Saint-Hilaire, 1795
Incisors are hypertrophied and curved, chisel-like, with enamel on the anterior surface only. Both upper and lower incisors are laterally compressed and open-rooted; the mesial enamel and distal dentine create a sharp cutting edge through differential wear. A long diastema separates these teeth from the reduced cheek teeth. Canines are absent except in the deciduous dentition. Cheek teeth are flattened and exhibit indistinct, rounded cusps; there is a single peg-like upper premolar. Adult dental formula: 1.0.1.3/1.0.0.3. Postcrania exhibit a number of distinctive features: for example, the brachialis flange is enormous and wing-like. The forelimb is short and robust in comparison to the hindlimb. The femoral head is relatively small; the ilia are narrow and rod-like. *Daubentonia* possesses a suite of appendicular and especially manual adaptations that facilitate the manual extraction (through bored holes) of nuts, insects, insect larvae and other foodstuffs.
INCLUDED SPECIES *D. madagascariensis, D. robusta*

SPECIES *Daubentonia robusta* Lamberton, 1934 (Fig. 7.11)
TYPE SPECIMEN Postcranial skeleton from Tsirave (southwest), in the collections of the University of Antananarivo [individual bones were assigned distinct specimen numbers]

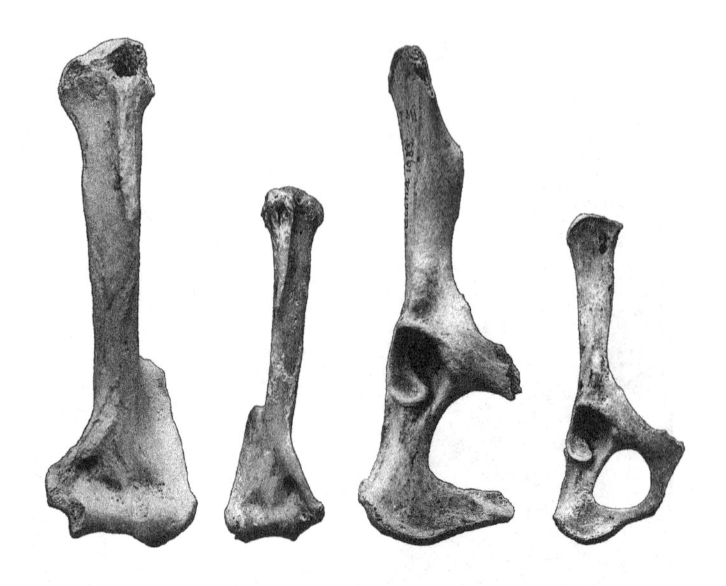

Fig. 7.11 Comparison of humeri (left) and pelves (right) of *Daubentonia robusta* from Anavoha (Antananarivo Collection) and *D. madagascariensis*.

AGE AND GEOGRAPHIC RANGE Late Quaternary, southwestern to central Madagascar: Ampasambazimba?, Anavoha, Ankilitelo, Lamboharana, Tsirave
ANATOMICAL DEFINITION
The extinct *Daubentonia* possessed hypertrophied incisors (virtually identical to those of *D. madagascariensis* but larger), unusually thin, filiform phalanges on manual digit III, and an elongated 3 metacarpal. The postcranial skeleton was very similar in morphology to that of its extant congener but more robust. Limbs were short in comparison to body mass, and humerofemoral index was higher than in congener. Intermembral index approximately 85. Estimated mass (approximately 13 kg) is roughly five times that of living congener.

Family Lemuridae

GENUS *Pachylemur* Lamberton, 1946
Adult dental formula (2.1.3.3/2.1.3.3), dental morphology similar in most respects to *Varecia*. Like *Varecia*, *Pachylemur* is distinguished from *Lemur* and *Eulemur* by a suite of dental traits (elongate talonid basins, the absence of entoconids on the lower molars, the protocone fold on the first upper molar, the anterior expansion of the lingual cingulum of the first and second upper molars, and so on). Sagittal and nuchal crests present. Orbits more frontally oriented than in *Varecia*. Skull relatively broader than that of *Varecia*, jaws more massive, molar teeth considerably larger. Intermembral index higher than in *Varecia* (the fore and hindlimbs more equal in length). Greater tubercle and greater trochanter project just proximal to humeral and femoral heads, respectively. About three to four times larger in body size than *Varecia*, and far more robust. Short

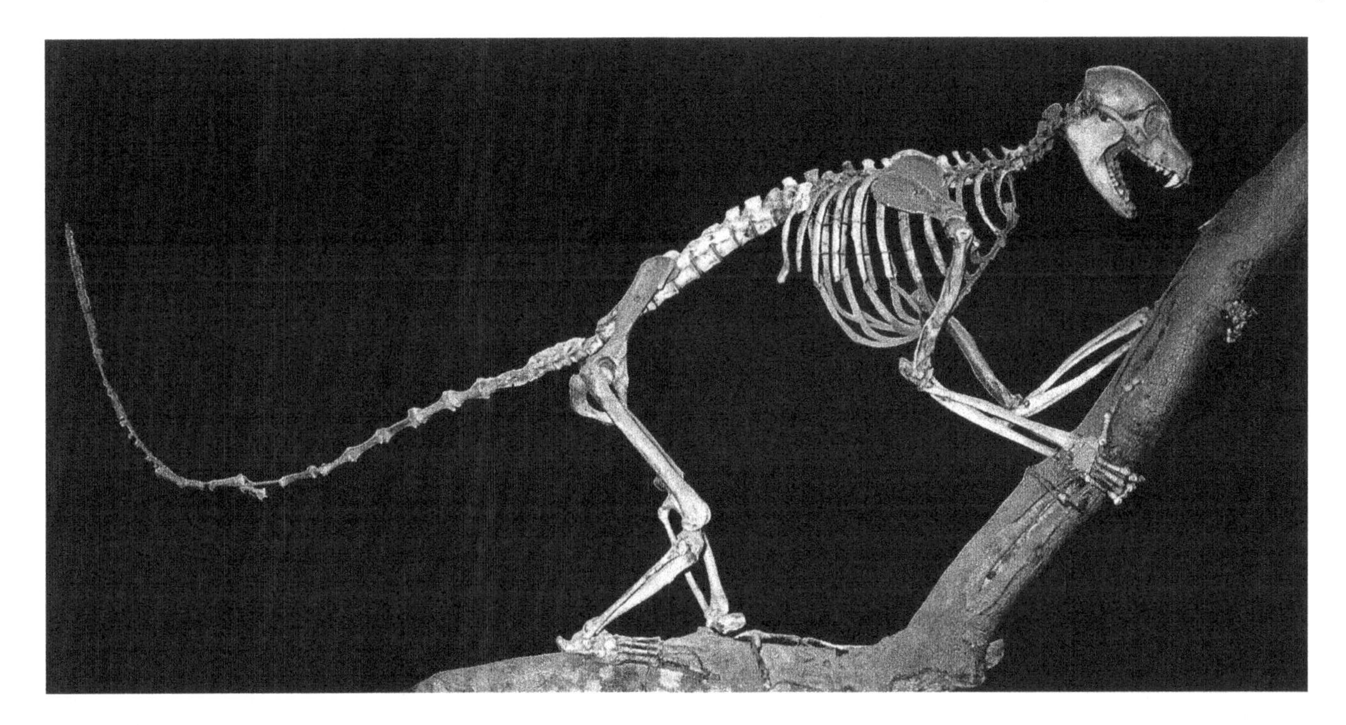

Fig. 7.12 *Pachylemur insignis*. Reconstructed composite skeleton on display at the Académie Malgache. Courtesy of Russell Mittermeier and Stephen Nash.

lumbar vertebral bodies; lumbar spinous processes somewhat reduced and exhibit less anticliny.
INCLUDED SPECIES *P. insignis*, *P. jullyi*

SPECIES *Pachylemur insignis* Filhol, 1895 (Fig. 7.12)
TYPE SPECIMEN A humerus from Belo-sur-mer (west), presumably in the collections of the Muséum National d'Histoire Naturelle, Paris [holotype unfigured, exact whereabouts and specimen number unknown]
AGE AND GEOGRAPHIC RANGE Late Quaternary, southern and southwestern Madagascar, perhaps northwest: Ambararata-Mahabo, Ambolisatra, Anavoha, Andrahomana, Ampoza-Ankazoabo, Belo-sur-mer, Bemafandry, Lamboharana, Manombo-Toliara, Taolambiby, Tsiandroina, and Tsirave; perhaps Amparihingidro?
ANATOMICAL DEFINITION
Smaller of the two species at approximately 10 kg. Intermembral index approximately 97. Skull length averages 117 mm. Superior temporal lines generally do not meet at midline.

SPECIES *Pachylemur jullyi* Lamberton, 1948
TYPE SPECIMEN Isolated upper molars from Antsirabe (center), presumably in the collections of the Muséum National d'Histoire Naturelle, Paris [exact whereabouts and specimen number unknown]
AGE AND GEOGRAPHIC RANGE Late Quaternary, high plateau of central Madagascar, possibly north: Ampasambazimba, Antsirabe, Morarano-Betafo; possibly Ankarana
ANATOMICAL DEFINITION
Larger of the two species at approximately 13 kg; intermembral index approximately 94. Average skull length is 125 mm. Sagittal and nuchal crests common. Humerofemoral, intermembral, brachial and crural indices all slightly higher than in *P. insignis*.

Evolution and extinction of giant lemurs

The extinct lemurs of Madagascar have left a remarkable fossil record that has enabled researchers to probe their skeletal anatomies in some detail. Still, some of the most basic questions concerning the evolutionary history of Malagasy lemurs remain unresolved. Higher taxonomic relationships among the eight recognized lemur families are still poorly understood, perhaps because speciation was explosive early in the evolutionary history of lemurs on Madagascar (Fleagle & Reed, 1999), or because the Malagasy primates represent an old radiation. Our record of intrageneric geographic variation and the past distributions of species has changed dramatically even over the past two decades, and new questions concerning temporal changes in species' geographic ranges and community structure have emerged. Finally, perhaps because there is good evidence for *both* recent climate change and human habitat disturbance, the question of what exactly precipitated the late Quaternary extinction of a full third of Madagascar's primate species continues to engender sharp debate (see Goodman & Patterson, 1997; Burney, 1999).

With regard to higher taxonomic relationships, the Palaeopropithecidae and Indridae appear to comprise a clade, with numerous shared derived morphological and developmental similarities in the dentitions (Tattersall &

Schwartz, 1974; Tattersall, 1982; Godfrey *et al.*, 2001). On the basis of similarities in both the dentition and the mandibular condyle, Tattersall (1975, 1982) and Wall (1997) have supported close ties between the Megaladapidae and Lepilemuridae. The Archaeolemuridae have been interpreted morphologically as distant relatives of a palaeopropithecid–indrid clade (see also Maier, 1980), but here the evidence is less compelling. Ancient DNA extracted from subfossil bones (Crovella *et al.*, 1994*b*; Yoder, 2000) has confirmed a close relationship of *Palaeopropithecus* to indrids (Yoder *et al.*, 1999) but no special relationship between *Megaladapis* and *Lepilemur* (but see Montagnon *et al.*, 2001), or between the Archaeolemuridae and members of the palaeopropithecid–indrid clade.

One open question is the relationship of the Daubentoniidae to other lemurs. Historically, *Daubentonia* has been considered a squirrel, a marsupial, a tarsier, a bushbaby, or a member of a unique suborder within the order Primates (see review by Sterling, 1994*a*). Standing (1908) included *Daubentonia* (along with the archaeolemurids and palaeopropithecids) in the indrid family. A close relationship to extant indrids has since been embraced by Schwartz (1974, 1975) and Tattersall & Schwartz (1974) on the basis of dental morphology and development. Groves (1974*b*) and Oxnard (1981, 1984) defended a distant relationship for *Daubentonia* to other prosimians. Features that were presumed to separate *Daubentonia* from other primates included (1) claws on all digits except the hallux (Groves 1974*b*) and (2) retention of three pairs of deciduous incisors (Luckett & Maier, 1986). However, Soligo & Müller (1999) showed that *Daubentonia madagascariensis* has a typical strepsirhine toilet claw on its second pedal digit and nails (which do not differ in histological structure from those of other primates) on all other digits. Ankel-Simons (1996) has shown, contra Luckett & Maier (1986), that *Daubentonia* does not possess three deciduous upper and lower incisors. Instead, newborn aye-ayes exhibit advanced calcification of the single permanent replacement as well as *single* milk incisors, and well-developed crowns of the deciduous canines.

Recent genetic studies (e.g., Adkins & Honeycutt, 1994; DelPero *et al.*, 1995, 2001; Yoder *et al.*, 1996; Yoder, 1997) have failed to support a close relationship of the Daubentoniidae and Indridae. Using karyotype fission theory, Kolnicki (1999) generated an elegantly simple interpretation of lemur chromosomal evolution. Kolnicki suggests that *Lepilemur* retains the primitive karyotype and may be the sister to all other Malagasy lemurs (including *Daubentonia*). No special relationship between the Daubentoniidae and the Indridae was suggested. Using fusion theory, Dutrillaux (1988) interpreted the same chromosomal data as suggesting a sister taxon relationship of *Daubentonia* to Malagasy lemurs – a position that Yoder *et al.* (1996) support on the basis of DNA sequence data; see also Yoder (1997). Adkins & Honeycutt (1994) situate aye-ayes at the base of the entire strepsirhine clade, on the basis of sequence data for the mitochondrial cytochrome oxidase II gene. On balance, we believe the evidence places them at or near the base of a monophyletic Malagasy clade of strepsirhines.

New paleontological discoveries, particularly in the north and northwest, have contributed significantly to our database on the geographic variation of extinct genera. It is evident, for example, that there were two species of *Archaeolemur* inhabiting the extreme north (C. Simons, 1997; Godfrey *et al.*, 1999) and northwest (Gommery *et al.*, 1998). These populations (particularly the more abundant robust form) differ somewhat from *Archaeolemur* in other regions (Godfrey *et al.*, 1999; Rasoloharijaona, 1999). An analysis of island-wide geographic variation in *Megaladapis* (Vuillaume-Randriamanantena *et al.*, 1992) has demonstrated the existence of two clades: M. (*Peloriadapis*) *edwardsi* and M. (*Megaladapis*) *grandidieri*/*madagascariensis*. While it is clear that the populations of *Megaladapis* in the extreme north and northwest belong to the M. *grandidieri*/*madagascariensis* group, they do not fit comfortably in either M. *grandidieri* or M. *madagascariensis*. *Palaeopropithecus* is extremely variable in the west, and two species may be represented at sites such as Ampoza-Ankazoabo. Several pairs of recognized species of extinct lemurs are very similar to one another, differing primarily in body size. This is true of *Mesopropithecus globiceps* and M. *pithecoides*, *Palaeopropithecus ingens* and P. *maximus*, and *Pachylemur insignis* and P. *jullyi*. Their variation warrants reanalysis in light of new materials belonging to each genus. New analyses of island-wide patterns of geographic variation are in order.

One focus of current research is primate community niche characteristics and the evolution of primate community structure (Fleagle & Reed, 1996; Ganzhorn, 1997, 1998; Fleagle *et al.*, 1999; Ganzhorn *et al.*, 1999). Using body size and lifeway reconstructions, along with first-hand knowledge of the size and habits of the extant species whose bones have been found in the subfossil deposits, one can document the ecological niche space ("ecospace") occupied by species at different subfossil sites. Godfrey *et al.* (1997*a*) used multivariate tools to analyze changes in the ecospace occupied by past and present primate communities in different regions of Madagascar. Madagascan primate communities did not simply shrink; they changed dramatically in ecological character. The lemur fauna of Madagascar today cannot be regarded as ecologically intact. Hardest hit by the extinctions were the large-bodied, slow-climbing seed predators and folivores.

Specimen rosters at the different subfossil sites trace geographic changes in primate species composition and the contraction of the geographic ranges of individual species (Vuillaume-Randriamanantena *et al.*, 1985; Jungers *et al.*, 1995; Godfrey *et al.*, 1999). Specimens of up to 20 primate species are represented at some subfossil sites; this probably underrepresents the species total living in the forests of the recent past. Given the short temporal range of many subfossil deposits, we can reasonably infer very high degrees of sympatry. No forest in Madagascar today has more than 13 primate species (Ganzhorn *et al.*, 1999).

Several hypotheses have been put forth to explain the extinctions of the giant lemurs (e.g., the overkill or "Blitzkrieg" hunting hypothesis, the natural change or aridification hypothesis, the human habitat destruction hypothesis, the human domesticant competition hypothesis, the human–climate synergy hypothesis, and the hypervirulent disease hypothesis; see Battistini and Vérin, 1966; Walker, 1967*b*; Dewar, 1984, 1997; Burney, 1997, 1999; Goodman & Patterson, 1997; Tattersall, 1999). Not all are mutually exclusive. Given the "dreadful syncopation" of human arrival on Madagascar (about 2000 years ago) and megafaunal disappearance (MacPhee & Marx, 1997), it is impossible to conceive of humans as blameless. There was, however, a prolonged window of human and megafaunal overlap (at least 1500 years), and it seems that gradual habitat degradation and increasing forest fragmentation might well have triggered progressive local extirpation and the ultimate extinction of the most vulnerable species. Three main arguments have been proffered against forest degradation and destruction playing a major role. It is claimed that: (1) many of the giant lemurs were terrestrial and thus not vulnerable to forest degradation and fragmentation; (2) the disappearance of megafauna from undisturbed habitats (e.g., the pristine eastern rainforest) suggests that something other than deforestation precipitated their extinction; and (3) forests could not have been critical to the survival of the megafauna, because much of Madagascar was not forested when humans arrived.

These counterarguments are not as compelling as they might at first seem. The lesson of paleontology is clear: from their skeletal and dental anatomy, we know that most of the larger-bodied lemurs of Madagascar were at least partly arboreal or scansorial, and that the primary food resources of most of these species were leaves and seeds (Godfrey *et al.*, 1997*a*; Jungers *et al.*, 1997, 2001). Even members of the family Archaeolemuridae, the group with the clearest terrestrial adaptations, had numerous skeletal adaptations for climbing (Godfrey *et al.*, 1997*a*; Hamrick *et al.*, 2000). Even the most terrestrial of living lemurs, *Lemur catta*, is forest-dependent. Finally, the species that disappeared entirely (mainly large-bodied arboreal folivores and seed predators) are exactly those species that might be *expected* to be most vulnerable to forest disturbance and fragmentation, given the area needs and low reproductive resilience of large-bodied species.

The geography of species loss is equally telling. It is the regions that are the most degraded – not the most arid – that have suffered the greatest losses of primate species. To be sure, there are many subfossil sites in what are today very dry habitats. The driest portion of Madagascar, the southwest, suffered a natural period of aridification just prior to the arrival of humans on the island (Mahé & Sourdat, 1972; Burney, 1993; Goodman & Rakotozafy, 1997). Subfossil sites in the arid southwest have yielded numerous specimens of arboreal primates. Some site names (such as Ankilitelo, the "home of three kily trees", where there are none today) bespeak of a richer floral environment in the recent past. But the Central Highlands, not the south, has suffered the greatest losses of primate species.

Table 7.2 lists the primate species that are known or believed (on the basis of skeletal remains) to have lived at selected subfossil sites in diverse geographic regions, along with those species that still survive at neighboring Special Reserves. Of the 20 primate species whose remains have been (at least tentatively and conservatively) identified at Ampasambazimba (in central Madagascar), only four (20%) survive at the nearby Special Reserve, Ambohitantely. Six (46.2%) of the 13 primate species represented at Ankilitelo and Ankomaka Caves (in southwest Madagascar) survive today at Beza Mahafaly. Nine (47.4%) of the 19 species whose remains have been found in the caves of Ankarana (extreme north) survive today in the surrounding forests. Finally, six (approximately 54.5%) of the 10 or 11 species in the subfossil fauna at Anjohibe (northwest Madagascar) survive at the nearby Ankarafantsika Strict Nature Reserve. Clearly, the Central Highlands have lost the most species.

Gade (1996) documents the loss of evergreen forest in the Central Highlands. He shows that the Central Highlands was, at least in places, densely forested until AD 1600, with lingering forested patches continuing well into the nineteenth and twentieth centuries. Today, small native highland forests are isolated by huge expanses of anthropogenic savanna. Deliberately set fires were the main cause of deforestation in this region. This region is certainly the most degraded of any in Madagascar. Gade (1996) maintains that the average annual precpitation in the Central Highlands today is quite sufficient to support evergreen forests, but recolonization by endemic forest vegetation has been hampered by soil erosion and the invasion of highly fire-resistant exotic grasses.

There are no known subfossil primate sites in eastern Madagascar; thus, there is no way to calculate the percentage of primate taxa that was lost there in the recent past. However, there is strong ethnohistorical and archaeological evidence that areas of eastern rainforest were more disturbed in the past than today (Goodman *et al.*, 1997; Rakotoarisoa 1997). Furthermore, displacement by humans and hunting or trapping can have had major impact on (particularly the large-bodied) primates of the eastern rainforests. Hunting continues to threaten the still-extant lemurs of the eastern forests. Sophisticated lemur traps are set regularly by people living in this region today (M. Irwin, pers. comm.).

In summary, both the anatomical and the geographic evidence support the synergistic (Burney, 1999) and differential susceptibility (Godfrey *et al.*, 1997*a*, 1999) argument that habitat disturbance and resource exploitation by humans

Table 7.2. Comparison of past and present primate faunas in different regions of Madagascar

Region and sites compared	Primate specimens at subfossil site	Primate taxa living today at neighboring protected area	Percentage survived
Central Highlands	*Microcebus* sp.	*Microcebus rufus*	20.0
Ampasambazimba vs.	*Cheirogaleus major*	*Cheirogaleus major*	
Ambohitantely Special Reserve	*Eulemur fulvus*	*Eulemur fulvus*	
	Eulemur mongoz	*Avahi laniger*	
	Varecia variegata		
	Hapalemur simus		
	Lepilemur microdon/mustelinus		
	Indri indri		
	Propithecus diadema		
	Propithecus verreauxi?		
	Avahi laniger		
	Pachylemur jullyi		
	Megaladapis grandidieri		
	Hadropithecus stenognathus		
	Archaeolemur edwardsi		
	Archaeolemur majori?		
	Palaeopropithecus maximus		
	Mesopropithecus pithecoides		
	Archaeoindris fontoynontii		
	Daubentonia robusta?		
Southwest	*Microcebus griseorufus*	*Microcebus griseorufus*	46.2
Ankilitelo-Ankomaka vs. Beza	*Microcebus murinus*	*Microcebus murinus*	
Mahafaly Special Reserve	*Cheirogaleus medius*	*Cheirogaleus medius*	
	Lepilemur leucopus	*Lepilemur leucopus*	
	Lemur catta	*Lemur catta*	
	Eulemur fulvus	*Propithecus verreauxi*	
	Propithecus verreauxi		
	Pachylemur sp.		
	Megaladapis madagascariensis		
	Megaladapis edwardsi		
	Archaeolemur majori		
	Palaeopropithecus ingens		
	Daubentonia robusta		
Northeast	*Microcebus* sp.	*Microcebus* sp.	47.4
Ankarana vs. Ankarana Special	*Cheirogaleus* sp.	*Cheirogaleus medius*	
Reserve	*Lepilemur* sp.	*Lepilemur septentrionalis*	
	Hapalemur griseus	*Hapalemur griseus*	
	Hapalemur simus	*Eulemur fulvus*	
	Eulemur fulvus	*Eulemur coronatus*	
	Eulemur coronatus	*Avahi occidentalis*	
	Avahi sp.	*Propithecus diadema*	
	Propithecus diadema	*Daubentonia madagascariensis*	
	Propithecus tattersalli	(*Phaner furcifer*)	
	Indri indri		
	Daubentonia madagascariensis		
	Pachylemur sp. cf. *jullyi*		
	Archaeolemur sp. cf. *edwardsi*		
	Archaeolemur sp. cf. *majori*		
	Megaladapis sp. cf. *grandidieri*		
	Palaeopropithecus sp. cf. *maximus/ingens*		
	Babakotia radofilai		
	Mesopropithecus dolichobrachion		

Table 7.2. (*cont.*)

Region and sites compared	Primate specimens at subfossil site	Primate taxa living today at neighboring protected area	Percentage survived
Northwest Anjohibe–Anjohikely vs. Ankarafantsika Strict Nature Reserve	*Microcebus* sp. *Cheirogaleus medius* *Lepilemur edwardsi* *Eulemur fulvus* *Eulemur mongoz* *Hapalemur simus* *Propithecus verreauxi* *Megaladapis madagascariensis* *Archaeolemur* sp. cf. *edwardsi* *Palaeopropithecus* sp. nov. *Babakotia radofilai*	*Microcebus* spp. (several) *Cheirogaleus medius* *Lepilemur edwardsi* *Eulemur fulvus* *Eulemur mongoz* *Propithecus verreauxi* (*Avahi occidentalis*)	54.5

(via burning, displacement and hunting) were major factors contributing to the extinction and local extirpation of some of the world's most remarkable primate species. We conclude that the destruction of Madagascar's natural tree cover contributed significantly to the disappearance of the giant lemurs of the Great Red Island.

Acknowledgements

We wish to thank Walter Hartwig for his invitation to write this chapter and for his continuing encouragement. Special thanks go to our wonderful colleagues for help and inspiration, in Madagascar and elsewhere: Elwyn Simons, Prithijit Chatrath and Don DeBlieux at the Duke University Primate Center; Berthe Rakotosamimanana, Gisèle Randria and many students at the University of Madagascar in Antananarivo, Benjamin Randriamihaja and his staff at MICET in Antananarivo, David Burney at Fordham University, Peter Andrews and Paula Jenkins at the Natural History Museum in London, Brigitte Senut and Françoise Jouffroy at the Muséum National d'Histoire Naturelle in Paris, Ian Tattersall and Ross MacPhee at the American Museum of Natural History, Steve Goodman at the Field Museum of Natural History, Maria Rutzmoser at the Museum of Comparative Zoology in Cambridge, Patricia Wright at the ICTE in Stony Brook, David Krause at Stony Brook, Mike Sutherland at the University of Massachusetts in Amherst, Martine Vuillaume-Randriamanantena in Paris, Alan Walker at Pennsylvania State University, and Tab Rasmussen at Washington University in St. Louis. Thanks also to Brigitte Demes, Mark Hamrick, Helen James, Mitch Irwin, Stephen King, Pierre Lemelin, Robert Paine, Andrew Petto, Kathy Rafferty, Jonah Ratsimbazafy, Jean Claude Razafimahaimodison, Brian Richmond, Karen Samonds, Gary Schwartz, Liza Shapiro, Cornelia Simons, Mark Teaford, Christine Wall, Frank Williams, Trevor Worthy and Roshna Wunderlich, for advice or collaboration on various aspects of the research reported here, including some unpublished work. We are also grateful for the opportunity to work in Madagascar and wish to acknowledge financial support from the following sources: the National Science Foundation, the Margot Marsh Fund, the Boise Fund, the School of Medicine at Stony Brook, and the University of Massachusetts at Amherst. We are forever indebted to Paul Godfrey and Leslie Jungers for years of patience and understanding as we pursued our obsession with primate evolution in Madagascar. Thanks also to Ian Tattersall, Russell Mittermeier and Stephen Nash for supplying some of the figures, and to Luci Betti-Nash for helping to prepare all the figures. Misaotra betsaka Madagasikara!

Primary References

Archaeoindris

Standing, H. F. (1909, for the year 1908). Subfossiles provenant des fouilles d'Ampasambazimba. *Bulletin de l'Académie malgache*, **6**, 9–11.

Standing, H. F. (1910, for the year 1909). Note sur les ossements subfossiles provenant des fouilles d'Ampasambazimba. *Bulletin de l'Académie Malgache*, **7**, 61–64.

Lamberton, C. (1934*a*). Contribution à la connaissance de la faune subfossile de Madagascar: Lémuriens et Ratites. *L'Archaeoindris fontoynonti* Stand. *Mémoires de l'Académie malgache*, **17**, 9–39 plus plates.

Archaeolemur

Forsyth-Major, C. I. (1893). Verbal report on an exhibition of a specimen of a subfossil Lemuroid skull from Madagascar. *Proceedings of the Zoological Society of London*, **36**, 532–535.

Filhol, H. (1895). Observations concernant les mammifères contemporains des *Aepyornis* à Madagascar. *Bulletin du Muséum national d'histoire naturelle, Paris*, **1**, 12–14.

Forsyth-Major, C. I. (1896). Preliminary notice on fossil monkeys from Madagascar. *Geological Magazine, New Series, Decade 4*, **3**, 433–436.

Forsyth-Major, C. I. (1897). On the brains of two sub-fossil Malagasy lemuroids. *Proceedings of the Royal Society of London*, **62**, 46–50.

Grandidier, G. (1899*b*). Description d'ossements de lémuriens disparus. *Bulletin du Muséum national d'histoire naturelle, Paris*, **5**, 344–348.

Forsyth-Major, C. I. (1900*a*). Exhibition of, and remarks upon, specimens of two subfossil mammals from Madagascar. *Proceedings of the Zoological Society of London* (for 1899), 988–989.

Lorenz von Liburnau, L. (1900*a*). Über einige Reste ausgestorbener Primaten von Madagaskar. *Denkschriften kaiserlichen Akademie der Wissenschaften in Wien*, **70**, 1–15.

Hamrick, M. W., Simons, E. L., & Jungers, W. L. (2000). New wrist bones of the Malagasy giant subfossil lemurs. *Journal of Human Evolution*, **38**, 635–650.

Babakotia

Godfrey, L. R., Simons, E. L., Chatrath, P. S., & Rakotosamimanana, B. (1990). A new fossil lemur (*Babakotia*, Primates) from northern Madagascar. *Comptes rendus de l'Académie des sciences, Paris*, **310**, 81–87.

Jungers, W. L., Godfrey, L. R., Simons, E. L., Chatrath, P. S., & Rakotosamimanana, B. (1991). Phylogenetic and functional affinities of *Babakotia radofilai*, a new fossil lemur from Madagascar. *Proceedings of the National Academy of Sciences of the United States of America*, **88**, 9082–9086.

Simons, E. L., Godfrey, L. R., Jungers, W. L., Chatrath, P. S., & Rakotosamimanana, B. (1992). A new giant subfossil lemur *Babakotia* and the evolution of the sloth lemurs. *Folia Primatologica*, **58**, 190–196.

Daubentonia

Grandidier, G. (1905). Recherches sur les lémuriens disparus et en particulier sur ceux qui vivaient à Madagascar. *Nouvelles Archives du Muséum national d'histoire naturelle, Paris*, **7**, 1–142.

Grandidier, G. (1929, for the year 1928). Une variété du *Cheiromys madagascariensis* actuel et un nouveau *Cheiromys* subfossile. *Bulletin de l'Académie malgache* (nouvelle série), **11**, 101–107.

Lamberton, C. (1934*b*). Contribution à la connaissance de la faune subfossile de Madagascar: Lémuriens et Ratites. *Chiromys robustus* sp. nov. Lamb. *Mémoires de l'Académie malgache*, **17**, 40–46 plus plates.

Hadropithecus

Lorenz von Liburnau, L. (1899). Über einen fossilen Anthropoid·en von Madagaskar. *Anzeiger der kaiserlichen Akademie der Wissenschaften in Wien*, **37**, 8–9.

Lorenz von Liburnau, L. (1900*a*). Über einige Reste ausgestorbener Primaten von Madagaskar. *Denkschriften kaiserlichen Akademie Wissenschaften in Wien*, **70**, 1–15.

Lorenz von Liburnau, L. (1902). Über *Hadropithecus stenognathus* Lz. Nebst bemerkungen zu einigen anderen austestorbenen Primaten von Madagascar. *Denkschriften kaiserlichen Akademie der Wissenschaften in Wien*, **72**, 243–254.

Grandidier, G. (1902*a*). Observations sur les lémuriens disparus de Madagascar. Collections Alluaud, Gaubert, Grandidier. *Bulletin du Muséum national d'histoire naturelle, Paris*, **8**, 497–505.

Lamberton, C. (1938, for the year 1937). Contribution à la connaissance de la faune subfossile de Madagascar. Note III. Les Hadropithèques. *Bulletin de l'Académie malgache* (nouvelle série), **20**, 127–170 plus plates.

Godfrey, L. R., Jungers, W. L., Wunderlich, R. E., & Richmond, B. G. (1997*b*). Reappraisal of the postcranium of *Hadropithecus* (Primates, Indroidea). *American Journal of Physical Anthropology*, **103**, 529–556.

Megaladapis

Forsyth-Major, C. I. (1893). Verbal report on an exhibition of a specimen of a subfossil lemuroid skull from Madagascar. *Proceedings of the Zoological Society of London*, **36**, 532–535.

Forsyth-Major, C. I. (1894). On *Megaladapis madagascariensis*, an extinct gigantic lemuroid from Madagascar, with remarks on the associated fauna, and on its geologic age. *Philosophical Transactions of the Royal Society of London, Series B*, **185**, 15–38.

Filhol, H. (1895). Observations concernant les mammifères contemporains des *Aepyornis* à Madagascar. *Bulletin du Muséum national d'histoire naturelle, Paris*, **1**, 12–14.

Forsyth-Major, C. I. (1897). On the brains of two subfossil Malagasy lemuroids. *Proceedings of the Royal Society of London*, **62**, 46–50.

Grandidier, G. (1899*a*). Description d'ossements de lémuriens disparus. *Bulletin du Muséum national d'histoire naturelle, Paris*, **5**, 272–276.

Grandidier, G. (1899*b*). Description d'ossements de lémuriens disparus. *Bulletin du Muséum national d'histoire naturelle, Paris*, **5**, 344–348.

Grandidier, G. (1900). Sur les lémuriens subfossiles de Madagascar. *Comptes rendus de l'Académie des Sciences, Paris*, **130**, 1482–1485.

Forsyth-Major, C. I. (1900*b*). Extinct mammalia from Madagascar. 1. *Megaladapis insignis*. sp.n. *Philosophical Transactions of the Royal Society, Series B*, **193**, 47–50.

Standing, H. F. (1903). Rapport sur des ossements subfossiles provenant d'Ampasambazimba. *Bulletin de l'Académie malgache*, **2**, 227–235.

Ekblom, T. (1951). Studien uber subfossile Lemuren von Madagaskar. *Bulletin of the Geological Institute of Uppsala*, **34**, 123–190.

Wunderlich, R. E., Simons, E. L., & Jungers, W. L. (1996). New pedal remains of *Megaladapis* and their functional significance. *American Journal of Physical Anthropology*, **100**, 115–138.

Mesopropithecus

Standing, H. F. (1905). Rapport sur des ossements sub-fossiles provenant d'Ampasambazimba. *Bulletin de l'Académie malgache*, **4**, 95–100.

Lamberton, C. (1936*a*). Nouveaux lémuriens fossiles du groupe des Propithèques et l'intérêt de leur découverte. *Bulletin du Muséum national d'histoire naturelle, Paris*, **8**, 370–373.

Simons, E. L., Godfrey, L. R., Jungers, W. L., Chatrath, P. S., & Ravaoarisoa, J. (1995*b*). A new species of *Mesopropithecus* (Primates, Palaeopropithecidae) from Northern Madagascar. *International Journal of Primatology*, **16**, 653–682.

Pachylemur

Filhol, H. (1895). Observations concernant les mammifères contemporains des *Aepyornis* à Madagascar. *Bulletin du Muséum national d'histoire naturelle, Paris*, **1**, 12–14.

Grandidier, G. (1899*b*). Description d'ossements de lémuriens disparus. *Bulletin du Muséum national d'histoire naturelle, Paris*, **5**, 344–348.

Lamberton, C. (1948*a*, for the year 1946). Contribution à la connaissance de la faune subfossile de Madagascar. Note XVII. Les Pachylemurs. *Bulletin de l'Academie Malgache* (nouvelle série), **27**, 7–22 plus plates.

Palaeopropithecus

Filhol, H. (1895). Observations concernant les mammifères contemporains des *Aepyornis* à Madagascar. *Bulletin du Muséum national d'histoire naturelle, Paris*, **1**, 12–14.

Grandidier, G. (1899*b*). Description d'ossements de lémuriens disparus. *Bulletin du Muséum national d'histoire naturelle, Paris*, **5**, 344–348.

Grandidier, G. (1901). Un nouvel édenté subfossile de Madagascar. *Bulletin du Muséum national d'histoire naturelle, Paris*, **7**, 54–56.

Standing, H. F. (1903). Rapport sur des ossements sub-fossiles provenant d'Ampasambazimba. *Bulletin de l'Académie malgache*, **2**, 227–235.

MacPhee, R. D. E., Simons, E. L., Wells, N. A., & Vuillaume-Randriamanantena, M. (1984). Team finds giant lemur skeleton. *Geotimes*, **29**, 10–11.

Jungers, W. L., Simons, E. L., Godfrey, L. R., Chatrath, P. S., & Rakotosamimanana, B. (in prep.). A new sloth lemur (Palaeopropithecidae) from Northwest Madagascar.

The origin and diversification of anthropoid primates

8 | The origin and diversification of anthropoid primates: Introduction

MARIAN DAGOSTO

Introduction

The anthropoid or "man-like" primates include the monkeys, apes, and man. The living members of this clade are divided into two major groups, the Platyrrhini (New World monkeys) and the Catarrhini (Old World monkeys, apes, and man). Earlier zoologists entertained the notion that the anthropoid "grade" may have been attained independently by these two groups (i.e., Anthropoidea is not monophyletic; Matthew, 1915*b*; Simpson, 1945), but this pre-continental drift idea is now rejected. Current morphological and molecular evidence strongly supports anthropoid monophyly (Delson & Rosenberger, 1980). However, the means of deployment of anthropoids from the Old to the New World is still a matter of controversy (Ciochon & Chiarelli, 1980*a*).

Fossils of anthropoid primates have long been known. Most finds fit comfortably within the firmly established platyrrhine and catarrhine clades, e.g., the late Oligocene–Pleistocene South American primates, and the Miocene–Pleistocene monkeys, hominoids and hominids. The taxonomic status of many Eocene and Oligocene anthropoids, however, is more problematic.

Early anthropoids

In 1907, Richard Markgraf began collecting fossils in the Fayum deposits of Egypt. These early collections were vastly increased by the work of Elwyn L. Simons in the 1970s to the present (reviewed by Simons, 1995*b*). The Fayum fossils from the upper part of the Jebel Qatrani formation date to the early–mid-Oligocene, and thus filled the temporal gap between Eocene prosimian primates and the Miocene anthropoids. They were also obviously more anatomically primitive than Miocene or extant anthropoids, giving us our first clues to the time and place of anthropoid origins.

It soon became apparent that the Fayum primates could be divided into two groups, the Parapithecidae and the Propliopithecidae. In earlier work, it was common to link the parapithecids to Old World monkeys and the propliopithecids to apes, but more recent studies have concluded that neither group is a member of these more modern clades. Propliopithecids are now viewed as primitive catarrhines, and parapithecids either as basal catarrhines, or as the sister group of platyrrhines+catarrhines (Fleagle & Kay, 1987; Harrison, 1987; Kay & Williams, 1994; Kay *et al.*, 1997*b*). Another Fayum family, oligopithecids (*Oligopithecus* and *Catopithecus*), are considered catarrhines by some (e.g., Rasmussen & Simons, 1992; Simons & Rasmussen, 1994*b*), but as basal anthropoids by others (e.g., Kay *et al.*, 1997*b*). Anthropoids similar to those of the Fayum have also been discovered on the Arabian Peninsula (Thomas *et al.*, 1988, 1991).

In the 1980s, additional work in the Fayum uncovered an even earlier deposit, L-41, which dates from the late Eocene. These deposits have yielded an array of even more primitive anthropoids and other primate taxa whose affinities to later Fayum groups and other primates are still being worked out (see Simons & Rasmussen, 1994*b* for a review).

The Fayum primates provided strong evidence for the hypothesis that the anthropoid lineage arose in Africa sometime before the late Eocene. Within the last decade, however, increased knowledge of the fossil record from North Africa and Asia has provided some intriguing new evidence that the anthropoid lineage may be even older than suggested by the Fayum deposits, and that Asia must also be considered as a possible area of origin.

Expeditions led by French scientists in North Africa have discovered an array of interesting primate fossils, many of which predate the Fayum sequence (except perhaps for the L-41 locality). These fossils are unfortunately rare and very incomplete, leading to various interpretations of their phylogenetic affinties. *Biretia* (late Eocene, Algeria) known from a single lower molar, has been considered a catarrhine by de Bonis *et al.* (1988*b*), but a parapithecid by Rasmussen & Simons (1992) and Kay & Williams (1994). *Djebelemur* (early Eocene, Tunisia), *Omanodon* and *Shizarodon* (early Oligocene, Oman) were considered adapids by the original describers (Hartenberger & Marandat, 1992; Gheerbrant *et al.*, 1993) but as anthropoids by Godinot (1994; Godinot & Mahboubi, 1994).

In 1992, Marc Godinot and Mohamed Mahboubi described *Algeripithecus* from the early Eocene of Algeria. Despite its minute size (estimated body weight 150–300 g) its teeth show remarkable anatomical similarities to those of Fayum anthropoids. In particular, the molars are surprisingly bunodont for such small animals. In 1994 another anthropoid, *Tabelia*, from the same locality was described (Godinot

& Mahboubi, 1994). Given the similarities between these two early Eocene genera and the later Fayum anthropoids, Godinot proposes a long independent history for the anthropoid lineage on the Afro-Arabian continent (Godinot, 1994). Supporting this view, *Altiatlasius* (late Paleocene, Morocco; Sigé *et al.*, 1990) is viewed as a potentially basal anthropoid by Godinot (1994). Kay (Kay *et al.*, 1997*b*; Kay & Williams, 1994), and Simons (Simons & Rasmussen, 1994*b*) however, are not convinced that these taxa are anthropoids.

Similar kinds of discoveries from Asia have also caused some to question both the age and site of origin of the anthropoid lineage. In 1994 K. Christopher Beard and colleagues described *Eosimias* in deposits dating from the middle Eocene (Beard *et al.*, 1994, 1996; Gebo *et al.*, 2000*a*). Although very primitive in many respects, it nevertheless exhibits derived features of the teeth and postcranium which link it to anthropoids. Some workers, however, question the relationship of *Eosimias* to anthropoids (Godinot, 1994; Simons & Rasmussen, 1994*b*).

First discovered in the 1930s, *Pondaungia* and *Amphipithecus* are from the late middle Eocene of Burma. These fragmentary fossils have always been enigmatic. Opinions as to affinities varied, some workers viewing them as adapids, and some as anthropoids (see Ciochon & Holroyd, 1994 for a review). The recent recovery of more complete material and new taxa both from Burma and Thailand (e.g., Ciochon *et al.*, 1985; Chaimanee *et al.*, 1997; Ducrocq, 1998, 1999; Jaeger *et al.*, 1999) has helped to corroborate the anthropoid status of some of these forms, although the status of *Wailekia* is still controversial. It has also documented a long history of anthropoid evolution throughout the Eocene of Asia. These Asian primates demonstrate that Anthropoidea may not be an endemic African radiation of primates, but indicates significant and previously unsuspected faunal exchange between Asia, Arabia and North Africa during the Eocene. This is supported by evidence from other mammalian groups (Holroyd & Maas, 1994; Ducrocq *et al.*, 1995; Ducrocq, 1999). Whether anthropoids first arose in Africa and spread to Asia or vice versa is a matter of speculation (Gunnell & Miller, 2001), although most authors still favor an African origin.

Anthropoid origins and higher primate phylogeny: Which if any prosimian group gave rise to/is the sister group of Anthropoidea?

Zoologists have recognized the division between the anthropoid primates and the more "primitive" prosimian primates (lemurs, tarsiers) since the time of St. George Jackson Mivart (1864). The subordinal taxa Anthropoidea and Prosimii are commonly used in classification to acknowledge this separation. More recently, this taxonomy has been considered a gradistic rather than a phylogenetic view of primate relationships. It is now widely (but not universally) accepted that some/one of these prosimians (or their fossil relatives) must be more closely related to or form the stem group of the anthropoids. Which prosimian group fills this role is a matter of intense debate and several competing hypotheses still have strong proponents. Each hypothesis has evidence (potential homologies) in its favor, but each also has conflicting evidence to explain away (Fig. 8.1).

Hypothesis 1

Anthropoids descended from Adapiformes, a lemur-like radiation of early Eocene primates (Gidley, 1923; Gingerich & Schoeninger, 1977; Gingerich, 1980*b*; Rasmussen, 1986, 1990*b*, 1994; Rasmussen & Simons, 1988; Franzen, 1994*b*). This is based on similarities of the upper molars of some adapids and early anthropoids, and the perception of a more primitive, generalized nature of adapiform anterior dentition, skull and postcranium which is argued to provide a better model for an anthropoid ancestor than does *Tarsius* or any omomyid (see Table 8.1 for list of characters). In earlier versions of this hypothesis, tarsiers and omomyids were seen as more closely related to Plesiadapiformes (Gingerich, 1976*a*; Schwartz *et al.*, 1978; Krishtalka & Schwartz, 1978), but this idea has been rejected, even by its original proponents. Instead, *Tarsius* is seen as nested within the Omomyidae (Rasmussen, 1994) or within Lorisiformes (Schwartz & Tattersall, 1987). In either case, *Tarsius* (and omomyids) is argued not to share a close relationship with anthropoids. Characters shared by *Tarsius* and anthropoids, omomyids and anthropoids, or adapids and lemurs must be interpreted as homoplasies or primitive retentions.

Hypothesis 2

Anthropoids descended from Omomyidae, a tarsier-like radiation of early Eocene primates (Szalay, 1975*a*; Rosenberger & Szalay, 1980; Hoffstetter, 1982; Szalay *et al.*, 1987). This hypothesis is based on shared-derived soft anatomical and molecular features shared by *Tarsius* and anthropoids (*Tarsius* is assumed to be nested within the omomyid clade), and features of the skull and basicranium shared by omomyids, *Tarsius* and anthropoids (see Table 8.2 for list of characters). Adapids share derived features with lemurs (Rosenberger & Szalay, 1980; Beard *et al.*, 1988), and thus cannot be more closely related to anthropoids than they are to lemurs. Characters shared by adapids and anthropoids must be interpreted as homoplasies or primitive retentions, as are characters shared exclusively by *Tarsius* and anthropoids, but not characteristic of omomyids.

Hypothesis 3

Anthropoids share a more recent common ancestry with *Tarsius* than with any other extant or fossil primate (Cartmill & Kay, 1978; Cartmill, 1980; Cartmill *et al.*, 1981; Godinot, 1994; Kay & Williams, 1994; Ross, 1994). This hypothesis is

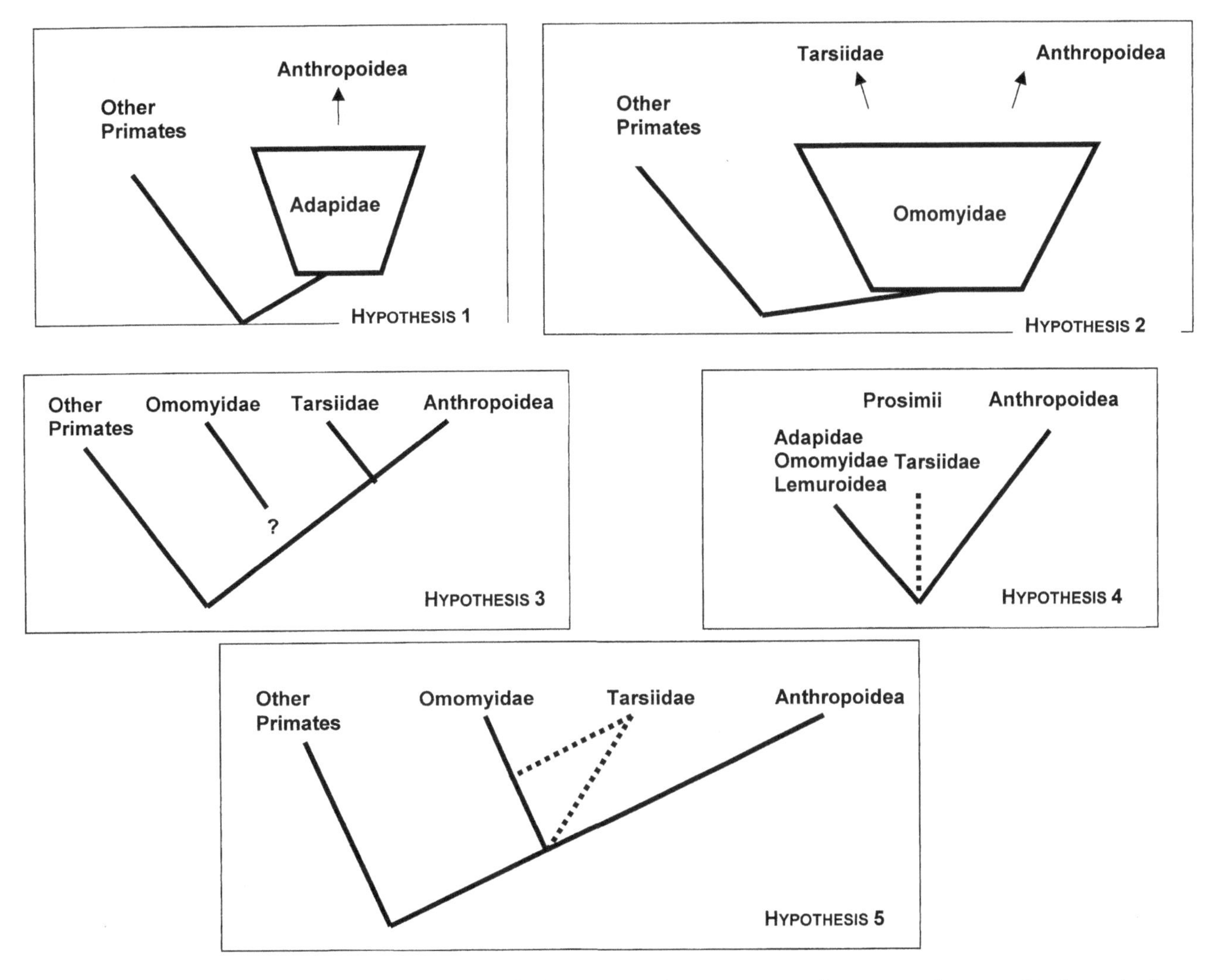

Fig. 8.1 Hypotheses of anthropoid origins and relationship of anthropoids to other primates. Hypothesis 1 posits the origin of Anthropoidea from within Adapidae. Hypothesis 2 postulates separate origin of *Tarsius* and Anthropoidea from within Omomyidae. Hypothesis 3 posits a sister-group relationship between *Tarsius* and Anthropoidea. Hypothesis 4 proposes an early separation between the prosimians and Anthropoidea; no prosimian is ancestral to anthropoids. Hypothesis 5 also suggests an early origin for Anthropoidea, but from a haplorhine ancestor. See text and Tables 8.1–8.3 for more details.

based on shared-derived soft anatomical, molecular, and cranial features shared by *Tarsius* and anthropoids to the exclusion of all other primates including all known omomyids (see Table 8.3 for list of characters). Characters shared by adapids and anthropoids must be interpreted as homoplasies or primitive retentions, as must be any characters shared exclusively by *Tarsius* and any omomyid (e.g., *Necrolemur*, *Shoshonius*).

Hypothesis 4

Anthropoids are an independent lineage of primates with no close relationship to any known prosimian group (Ford, 1988, 1994); i.e, the basal division within Primates is between Anthropoidea and Prosimii. This hypothesis is derived from parsimony analyses of postcranial evidence, which indicate that prosimians (adapids, omomyids, *Tarsius*, lemurs) form a clade; none of them group most closely with anthropoids. An independent line of evidence is the long isolated evolutionary history for anthropoids implied by the fossil evidence from both Africa and Asia (Godinot & Mahboubi, 1992; Beard *et al.*, 1994, 1996; Godinot, 1994). Anthropoids are presumed to retain a primitive morphology; any characters shared with any prosimian group (characters from hypotheses 1–3) must be homoplasies. An alternate version of this "third group" hypothesis could postulate an unresolved (and possibly real) trichotomy among Strepsirhini (adapids plus lemurs), Tarsiiformes (*Tarsius* plus omomyids) and anthropoids.

Table 8.1. Features supporting hypothesis 1, that anthropoids are descended from adapids

Potential shared-derived features of adapids and anthropoids	Comment	Homology, similarity or polarity disputed by investigators
Small, vertical, spatulate lower incisors (Gidley, 1923; Gingerich, 1980*b*)	Primitive retention	Gregory, 1915; Kay, 1980; Rosenberger & Szalay, 1980
I_1 smaller than I_2 (Gingerich, 1980*b*)	Homoplasy	Delson & Rosenberger, 1980
Large body size > 500 g (Gingerich, 1980*b*)	Homoplasy	
Fused mandibular symphysis; deep corpora (Gidley, 1923; Gingerich, 1980*b*)	Homoplasy	Kay, 1980; Ravosa & Hylander, 1994
Interlocking canine occlusion (Gingerich, 1980*b*)	Homoplasy	Delson & Rosenberger, 1980
Projecting canines (Gidley, 1923)	Primitive retention	Gregory, 1915
Canine sexual dimorphism (Gingerich, 1980*b*)		
Canine–premolar hone (Gingerich, 1980*b*)	Homoplasy; developed independently in different anthropoids	Kay, 1980
Free ectotympanic (Gingerich, 1980*b*)	Not found in anthropoids	Delson & Rosenberger, 1980
Calcaneum–navicular short (Gingerich, 1980*b*)	Primitive retention	Dagosto, 1988
Tibia fibula unfused (Gingerich, 1980*b*)	Primitive retention	Dagosto, 1988
Molar form quadrate (presence of hypocone) (Gingerich, 1980*b*)	Homoplasy	Hoffstetter, 1974
Lose nannopithex fold on upper molars (Kay, 1980)	Homoplasy	Kay, 1980
Loss of middle premolar of five (Schwartz *et al.*, 1978)	Five premolars is not characteristic of *Tarsius* and omomyids	Cartmill & Kay, 1978
Dental features shared by *Oligopithecus* and *Hoanghonius* (Gingerich, 1977*a*)	Homoplasy	Godinot, 1994; Kay & Williams, 1994
Dental features shared by *Protoadapis* and *Oligopithecus* (Rasmussen & Simons, 1988)	Homoplasy	Godinot, 1994; Kay & Williams, 1994
Cranial features shared by *Mahgarita* and anthropoids (Rasmussen, 1990*b*)	Homoplasy or lack of similarity	Beard & MacPhee, 1994; Ross, 1994

Hypothesis 5

Anthropoids are descended from an early as yet unknown or unrecognized haplorhine radiation of fossil primates; dubbed the "pre-tarsioid" hypothesis (Rasmussen, 1994). This is very similar to hypothesis 2, as it supports a monophyletic Haplorhini, but recognizes that *Tarsius* and many known omomyids are too derived either cranially, dentally or postcranially to be direct ancestors of anthropoids. It postulates that the common haplorhine ancestor is likely to be very omomyid-like in some ways, but not necessarily a member of the omomyid clade as currently constituted. The anthropoid ancestor (and possibly the tarsiid ancestor) probably did not diverge much from this ancestor either. It may not, therefore, be possible to resolve a trichotomy between *Tarsius*, anthropoids and omomyids; i.e., perhaps Omomyidae is not a paraphyletic clade giving rise to *Tarsius* and/or anthropoids (as implied by hypothesis 2), but a real monophyletic clade with no living descendants. An alternate version of this hypothesis proposes that while *Tarsius* may have arisen from Omomyidae, anthropoids did not, but are likely the sister of this clade (e.g., Beard *et al.*, 1991). Like hypothesis 4, hypothesis 5 also implies a long independent history for Anthropoidea, but with its earliest members being much more primitive than late Eocene or extant anthropoids. New fossil evidence of anthropoids from North Africa and China provide support for this view (Godinot & Mahboubi, 1992; Beard *et al.*, 1994, 1996; Godinot, 1994). Characters shared by adapids and anthropoids must be homoplasies or primitive retentions, as are characters shared exclusively by *Tarsius* and anthropoids, but not characteristic of omomyids.

Later deployment of anthropoids

During the Eocene and most of the Oligocene anthropoid primates were confined to Africa, Arabia and Asia. The first occurrence of an anthropoid outside this area is of *Branisella*, a very late Oligocene primate from Salla, Bolivia. Most workers recognize *Branisella* as a platyrrhine, but some consider it primitive enough to be outside this clade (Hoffstetter, 1969; Hershkovitz, 1977). Nevertheless, this fossil marks the first appearance of primates in South America. An adaptive

Table 8.2. Features supporting hypothesis 2, that anthropoids (and *Tarsius*) are descended from omomyids. In addition to these features, all the soft anatomy/molecular features outlined in Table 8.3 (features shared by *Tarsius* and anthropoids) are assumed to be found in omomyids as well (except those identified as exclusive tarsier–anthropoid synapomorphies, e.g., postorbital septum)

	Comment	Homology, similarity or polarity disputed by investigators
Potential shared-derived features of anthropoids, omomyids and Tarsius		
Promontory canal (and artery) is larger in haplorhines than strepsirhines; stapedial canal (and artery) is reduced (Gregory, 1915; Szalay, 1975*a*)	Variable within each fossil group	MacPhee & Cartmill, 1986; Rasmussen, 1990*b*
Medial rather than lateral entrance of internal carotid artery (Szalay, 1975*a*)	Polarity uncertain	MacPhee & Cartmill, 1986
Increasingly extrabullar (phaneric) position of ectotympanic	Ectotympanic is still (primitively) intrabullar in omomyids	MacPhee & Cartmill, 1986; Ross, 1994
Reduced facial skull length (Rosenberger & Szalay, 1980)	Homoplasy	Ross, 1994
Apical interorbital septum (Haines, 1950; Cave, 1967; Cartmill, 1975)	Homoplasy – also in some lorisiformes; not typical of some anthropoids	Simons & Rasmussen, 1989
Interorbital constriction is below olfactory tract	Possibly primitive for euprimates	Ross, 1994
Hypotympanic sinus is anteromedial to promontorium (Rosenberger & Szalay, 1980)	Present in all primates; omomyid sinus not homologous to anterior accessory cavity of *Tarsius* and anthropoids	MacPhee & Cartmill, 1986; Simons & Rasmussen, 1989
Carotid canal well separated from fenestra cochlea (Ross, 1994)		
Humeral trochlea medially downturned (Szalay & Dagosto, 1980)	?Polarity	Dagosto & Gebo, 1994
Progressive loss of paraconid and absence of premetacristid (Kay, 1980)	Does not rule out more primitive adapids (i.e., *Donrussellia*)	
Reduction of lower third molars (Kay, 1980)		
Features shared by omomyids and anthropoids (Wortman, 1903–04)	Primitive retentions	Kay, 1980
Features shared by *Washakius* and anthropoids (Gazin, 1958)	Primitive retentions	Kay, 1980
Features shared by *Omomys* and anthropoids (Simons, 1961*b*)	Primitive retentions	Kay, 1980
Features shared by *Chumashius* and anthropoids (Kay, 1980)	Homoplasy	Kay & Williams, 1994
Shared derived features of omomyids and Tarsius		
Features shared by *Necrolemur* and *Tarsius* (Rosenberger, 1985)	Homoplasy or lack of similarity	Ross, 1994
Features shared by *Shoshonius* and *Tarsius* (Beard *et al.*, 1991)	Homoplasy or lack of similarity	Ross, 1994

radiation of Platyrrhini in South and Central America took place during the Miocene, Pleistocene and Recent.

The time of origin, place of origin and mode of deployment of the Platyrrhini remains a controversial topic. Although most primatologists agree that the platyrrhine ancestor must be found in the Old World, probably in Africa, no platyrrhine has been identified there. Neither has an unambiguous platyrrhine ancestor, although Hoffstetter (1977) thought that *Parapithecus* might fill this role, and more recently Simons (1997*c*) and Takai *et al.* (2000*a*) have suggested a relationship between the Fayum taxon *Proteopithecus* and the South American anthropoid *Branisella*. Fleagle & Kay (1987) and Harrison (1987) have suggested that parapithecids may be primitive anthropoids that predate the platyrrhine–catarrhine split, which would give a date of origin for the platyrrhine lineage possibly not earlier than late Eocene. On the other hand, if, as Godinot (1994) suggests, *Tabelia*, *Algeripithecus*, *Oligopithecus* or parapithecids are not basal

Table 8.3. Features supporting hypothesis 3, that anthropoids and *Tarsius* are sister groups

Potential shared-derived featured of *Tarsius* and anthropoids	Comment	Homology, similarity or polarity disputed by investigators
Fetal membranes and placentation (Hubrecht, 1908; Luckett, 1975)		
Eccentric implantation	Homoplasy; correlated characters	Schwartz & Tattersall, 1987
Primordial amniotic cavity present		
Yolk sac small, free reduced later		
Choriovitelline placenta absent		
Allantoic vesicle small, rudimentary		
Chorioallantoic placenta discoid, hemochorial		
Eyes		
Lack of tapetum lucidum (Wolin & Massopust, 1970)		
Presence of central fovea	Presence may be variable	Castenholz, 1984
Face		
Naked rhinarium; unfused nasal process (Pocock, 1918; Cave, 1967)	Possible homoplasy	Hofer, 1979
Skull		
Apical interorbital septum (Cave, 1967; Cartmill, 1975)	Homoplasy – also in some lorisiformes	Simons & Rasmussen, 1989
Postorbital septum (Pocock, 1918; Cartmill & Kay, 1978; Cartmill, 1980)	Homoplasy	Simons & Russell, 1960; Cachel, 1979; Rosenberger & Szalay, 1980; Simons & Rasmussen, 1989
Alisphenoid contribution to postorbital septum		
Perbullar carotid artery pathway (Cartmill *et al.*, 1981)	Homoplasy	Simons & Rasmussen, 1989
Vertical internal carotid entry	Not typical of platyrrhines	Rosenberger & Szalay, 1980
Presence of accessory anterior cavity of middle ear (= hypotympanic sinus) (MacPhee & Cartmill, 1986)	Present in all primates Present in omomyids	Simons & Rasmussen, 1989 Rosenberger & Szalay, 1980
Intrabullar transverse septum separating tympanic cavity from hypotympanic sinus (Cartmill & Kay, 1978)	Morphology not similar in platyrrhines and *Tarsius*	Rosenberger & Szalay, 1980; Simons & Rasmussen, 1989
Small aditus connecting the tympanic and anterior accessory cavity (Cartmill & Kay, 1978)	Morphology not similar in *Tarsius* and anthropoids	Simons & Rasmussen, 1989
Ectotympanic fused to lateral bulllar wall (= phaneric bulla; = extrabullar tympanic ring; = loss of annular bridge; = loss of subtympanic recess) (Cartmill & Kay, 1978)	Homoplasy	Rosenberger & Szalay, 1980
Extreme reduction of stapedial artery and canal	Homoplasy	Rosenberger & Szalay, 1980
Anastomosis between meningeal branches of stapedial and maxillary arteries to supply dura (Bugge, 1974; Cartmill, 1980)		
Teeth		
Short cristid obliqua (Kay & Williams, 1994)		
Shallow hypoflexids (Kay & Williams, 1994)		
Molecular		
Alpha and beta hemoglobins (Koop *et al.*, 1989*a*)		
Alpha crystallin (de Jong & Goodman, 1988)		
Involucrin repeats (Djian & Green, 1991)		
Alu repeats (Zietkiewicz *et al.*, 1999)		
DNA–DNA hybridization (Bonner *et al.*, 1980)		
Vitamin A synthesis (Pollock, 1987)		

anthropoids but more closely related to catarrhines than platyrrhines, the split between the platyrrhine and catarrhine lineages must be at least pre-middle Eocene.

There is no record of primates in South America before the late Oligocene. This of course raises the question of how and when they got there. Three hypotheses are proposed to explain the presence of primates in South America.

1. Origin of platyrrhines (or platyrrhine ancestors) in Africa: migration from West Africa to South America by crossing the Atlantic.
2. Origin of platyrrhines (or platyrrhine ancestors) in North America: migration from North America to South America by crossing the Caribbean Sea.
3. Origin of platyrrhines (or platyrrhine ancestors) in Old World (Africa or Asia): migration from Asia to North America (via the Bering Strait), then to South America by crossing the Caribbean Sea.

The first hypothesis is the current consensus (Ciochon & Chiarelli, 1980*a*), and is supported by the presence of the earliest (and possibly pre-platyrrhine) anthropoids in North Africa, the presence of the sister group of platyrrhines in Africa, the co-distribution of South American caviomorph and African phiomorph rodents, reconstructions of oceanic circulation patterns, and the possible presence of mid-oceanic volcanic islands in the mid-Atlantic during the Eocene. The drawbacks are the lack of known platyrrhines in Africa, and the probable great distance of sea that needs to be crossed (Hartwig, 1994). Promoters of the second hypothesis cite the presence of suitable (tarsiiform or adapid) platyrrhine ancestors in North America during the Eocene and the presence of volcanic island arcs between North and South America. The problems with this hypothesis are the lack of anthropoids in North America, the probable oceanic circulation patterns in the primitive Caribbean Sea which do not favor north-to-south rafting, and the great distance between the two continents during the early Tertiary. The third hypothesis is a variant of the second. Supporting it are the presence of primitive anthropoids in Asia, and the extensive faunal exchange between Asia and North America during the Eocene. The drawbacks are the same as the second hypothesis, as this route involves crossing not only the Bering Strait, but also the Caribbean Sea. A fourth possibility, that of a Gondwandan route through Antarctica, has been reviewed and rejected by Houle (1999).

In the Eocene, Oligocene and early Miocene, catarrhine primates were confined to the African continent. (Two Asian taxa may be early Miocene, but the dates are uncertain.) In the early Miocene, the first true cercopithecoids (*Victoriapithecus* and *Prohylobates*) and "apes" (proconsulids; some include these taxa in the superfamily Hominoidea; some reserve this term for the extant apes) are recognized. The "apes" underwent an impressive adaptive radiation during the Miocene (see this volume and Ciochon & Corruccini, 1983; Begun *et al.*, 1997*b*). The early Miocene proconsulids and oreopithecids are known primarily from sites in East Africa, but by the Middle Miocene, "apes" invaded Eurasia and radiated extensively there. They began to decline in the late Miocene and all of the Eurasian apes have become extinct except for the orangutan, gibbon and humans. Likewise, in contrast to the large number of species present during the early and middle Miocene, African ape taxa are represented today by only the chimpanzee, gorilla and man. The phylogenetic relationships among the fossil taxa and between fossil and living taxa are still largely unsettled.

The Old World monkeys were only modestly diversified during the Miocene, but radiated extensively during the latest Miocene and Pleistocene. Baboons, magabeys and guenons stayed on the African continent. Two groups, the macaques and the colobine monkeys, extended their ranges into Eurasia, and are still represented by numerous species there today.

The fossil record and anthropoid origins

Although the time and place of anthropoid origins has been informed by the fossil record, recovery of more fossils has not really helped resolve the question of the source group for anthropoids. There are several reasons for this. (1) Fossil anthropoids of the early Oligocene and late Eocene are already derived enough not to provide indisputably strong evidence of a transition from any particular prosimian group (or lack of such a transition). No known North American or European adapids or omomyids from the Eocene are convincingly intermediate forms bridging the gap between prosimians and anthropoids. Early and middle Eocene anthropoids (e.g., *Algeripithecus, Eosimias*) are different enough from North American and European adapids and omomyids to suggest that none of the known omomyid or adapid species are directly ancestral to anthropoids. (2) As demonstrated in Tables 8.1–8.3, the available evidence is interpreted very differently by investigators. The homology, polarity and even similarity of features is disputed. Some seemingly critical features (for example, molecules and traits related to reproductive biology) are, of course, unavailable for fossil taxa. These two facts imply the possibility that (3) the split between anthropoids and other primates may have taken place much earlier than previously thought, and that this is yet another example of the "long branch problem" in phylogeny reconstruction (Felsenstein, 1978). A long time of separation between branches has two effects. First, it is likely that the earliest members of the anthropoid lineage are morphologically more primitive than anyone expects, and therefore finding the few characters that distinguish them from early adapids, omomyids or tarsiers may be extremely difficult. "Classic" anthropoid features (e.g., postorbital closure) may have appeared much later in the evolutionary history of this lineage. Second, the long time of separation between the anthropoid and other primate lineages allows the possibility of considerable homoplasy between them, complicating phylogenetic analyses. The best way of solving such problems is to discover new taxa which "break up" the long branches (Graybeal, 1998). It is most likely that fossil taxa will fill this role best, since they have the greatest chance of preserving combinations of character states not seen in living taxa and thus revealing the homo-plasy of disputed characters (Donoghue *et al.*, 1989). The ancestor–descendant and even sister group relationship between Anthropoidea and other primates will probably remain a controversial topic until such suitable fossil taxa are discovered.

9 | Basal anthropoids

K. CHRISTOPHER BEARD

Introduction

Living anthropoids differ from other primates in such profound ways that many aspects of anthropoid origins remain unresolved. By the late Eocene (~36 Ma), a series of dramatic evolutionary changes had already transformed some unknown but fairly primitive group of haplorhine primates into animals that were equipped with virtually the full arsenal of anthropoid anatomy. Several lines of evidence suggest that the anthropoid clade is much older than many previous models implied. Fossils of basal anthropoids are now well documented from sites dating to the middle Eocene to early Oligocene. Longitudinally, they have been recovered as far east as the central coastline of China and as far west as the border region between Morocco and Algeria. Although many basal anthropoid taxa are known from only fragmentary fossils, significant aspects of anatomy have been described for certain key taxa, including *Eosimias*, *Proteopithecus* and *Apidium*. Because of their apparently basal phylogenetic position and their unique combination of primitive and derived character states, Asian eosimiids provide crucial information for reconstructing anthropoid origins.

History of discovery and debate

The history of scientific debates regarding the origin of anthropoid primates is older than the discovery of fossils that are directly germane to this issue. By the late 1800s, paleontological exploration in Europe and the United States had revealed an astonishing variety of Eocene primates, taxa that now are regarded as members of the adapiform and tarsiiform radiations. Given the stable Earth paradigm that then prevailed in the geosciences, as well as the lingering influence of *scala naturae* models in evolutionary biology, it is not surprising that earlier workers believed that living anthropoids of the Old and New Worlds reached the anthropoid "grade" by separate evolutionary pathways. For example, such influential North American paleontologists as William Diller Matthew (1915*b*), G.G. Simpson (1945) and Bryan Patterson (1954) touted the likelihood that playrrhines and catarrhines evolved independently from different groups of Eocene prosimians. Obviously, if platyrrhines and catarrhines do not share a close common ancestry, it would be futile to attempt to find fossils documenting the stem lineage of all living anthropoids. Demonstrating that Anthropoidea is a natural group of organisms descended from a single common ancestor is one of the great triumphs of primate evolutionary biology during the twentieth century.

The first discovery of a fossil anthropoid was not recognized initially as such. Rather, in his terse description of *Apidium phiomense*, Osborn (1908) compared it with small, bunodont artiodactyls such as *Cebochoerus* as well as with primates. A few years later, the description of additional fossils from the Fayum region of Egypt, including the virtually complete lower dentition of *Parapithecus fraasi*, allowed Max Schlosser (1911) to refer both *Apidium* and *Parapithecus* to the Primates. At the same time, Schlosser (1911) described the first propliopithecid fossils from the Fayum (see Rasmussen, this volume). These discoveries of parapithecid and propliopithecid anthropoids in early Oligocene strata of the Fayum region of Egypt established a baseline for the antiquity of anthropoids in the Old World.

The discovery of additional basal anthropoids in Myanmar (Burma) soon challenged Africa's role as the possible ancestral homeland for the anthropoid clade. Guy Ellock Pilgrim (1927) described the first of these Burmese Eocene primates as *Pondaungia cotteri*, based on fragmentary upper and lower jaws thought to belong to a single individual. Duplicating Osborn's (1908) earlier concerns with respect to *Apidium*, Pilgrim compared *Pondaungia* and bunodont artiodactyls extensively before concluding that *Pondaungia* most likely represented an early anthropoid primate. Edwin Harris Colbert (1937) described a second Burmese Eocene primate as *Amphipithecus mogaungensis*, based on a single lower jaw preserving P_3–M_1 and roots or alveoli for part of the anterior dentition. Emphasizing the length of the lower premolar roots, their distinctive crown morphology, and the great depth and robusticity of the mandible, Colbert advocated anthropoid affinities for *Amphipithecus*. Colbert favored catarrhine affinities for *Amphipithecus*, despite its three premolar dental formula. He also dismissed the possibility that *Amphipithecus* and *Pondaungia* are closely related. Although the phylogenetic affinities of these Burmese primates remain controversial (see Szalay & Delson, 1979), they were acknowledged from the time of their initial description to be significantly older than the Fayum parapithecids and pro-pliopithecids described by Osborn (1908) and Schlosser (1911).

For the next two decades, no additional fossils bearing on the problem of anthropoid origins were collected or described. The diphyletic origin of New and Old World anthropoids continued to be advocated by leading students of fossil primates (e.g., Gazin, 1958).

The study of anthropoid origins was reinvigorated by Elwyn Simons, who located and described some previously unrecognized primate specimens from the Fayum region in the collections of the American Museum of Natural History. The description by Simons (1959) of an isolated primate frontal bone from the Fayum, which is now thought to pertain to *Apidium*, revealed that small Fayum anthropoids already possessed a relatively complete postorbital septum, orbits apparently smaller than those of nocturnal primates, and reduced olfactory lobes. Based on this success, and cognizant of the potential significance of securing additional fossil primate specimens from that region, Simons initiated a series of paleontological expeditions to the Fayum in 1961 (Simons, 1995*b*). To date, no other series of paleontological expeditions has had greater impact on the study of anthropoid origins.

The early Fayum expeditions greatly enhanced our knowledge of the anatomy and taxonomic diversity of parapithecids, which are among the most abundant small mammals in the upper part of the Fayum's Jebel Qatrani Formation. Simons (1962*b*, 1974) described new species of *Apidium* and *Parapithecus* and repeatedly explored their possible phylogenetic relationships. At first, it was unclear that *Apidium* and *Parapithecus* were specially related to one another, and Simons (1960) cited *Apidium* as a possible close relative of the late Miocene Italian catarrhine *Oreopithecus*. Subsequently discovered specimens of both *Apidium* and *Parapithecus* demonstrated that they are closely related (e.g., Simons, 1967*c*), and both have been classified as members of the Parapithecidae ever since. The broader affinities of parapithecids with respect to other anthropoids remain controversial, but at that time they were suggested to be closely related to cercopithecids (Simons, 1974).

Political difficulties interrupted field work in the Fayum for nearly a decade (1967–77). Among the more important advances made during this interval was Glenn Conroy's (1976*b*) analysis of Fayum primate postcranial elements, most of which were attributed to *Apidium*. These specimens suggested that the locomotor behavior of *Apidium* would have emphasized arboreal quadrupedalism and leaping, similar to generalized South American monkeys. Following the resumption of active field work in the Fayum in 1977, additional *Apidium* discoveries indicated relatively long ischia and extensive tibiofibular apposition, characters suggesting an even greater emphasis on leaping (Fleagle & Simons, 1979, 1983, 1995).

Contributions to an edited volume on the phylogenetic origin and paleobiogeographical history of early platyrrhines (Ciochon & Chiarelli, 1980*a*) highlighted the conflicting models of anthropoid origins prevalent then. One model, based largely on similarities in dental morphology, posited that early anthropoids evolved from Eocene adapids (Gingerich, 1980*b*). A second model, consistent with neontological evidence favoring the monophyly of anthropoids and *Tarsius* among living primates, proposed that anthropoids evolved from Eocene omomyids (Rosenberger & Szalay, 1980). A third model, built upon similarities in the ear region and postorbital septum in *Tarsius* and anthropoids, viewed these living primates as sharing more recent common ancestry with each other than with either of the two major groups of Eocene prosimians (Cartmill, 1980; Cartmill *et al.*, 1981). Although these three competing models implied conflicting phylogenetic tree topologies and mutually exclusive reconstructions of character evolution, they agreed in accepting the monophyly of New World and Old World anthropoids. This represented a significant advance over persistent claims that anthropoids in the New and Old Worlds may have evolved in parallel from earlier "prosimian" ancestors (Simons, 1976). With the notable exception of publications by Hoffstetter (1977, 1980), most models of anthropoid origins in vogue at this time agreed that anthropoids originated relatively late in the Paleogene, probably sometime near the Eocene–Oligocene boundary (Delson & Rosenberger, 1980; Gingerich, 1980*b*).

After a hiatus of more than four decades, additional basal anthropoid fossils from Myanmar were discovered and described by Ba Maw *et al.* (1979) and Ciochon *et al.* (1985). Although these relatively fragmentary specimens only modestly improved knowledge of *Pondaungia* and *Amphipithecus*, they rekindled the paleobiogeogaphic controversy regarding anthropoid origins.

In the meantime more primates were found in the Fayum. Simons & Kay (1983) described the first of these as *Qatrania wingi*, a small parapithecid characterized by retention of a small paraconid on M_1 and extremely cuspidate lower molars, with little development of molar shearing crests. They subsequently described a second species, *Q. fleaglei*, along with additional material of *Q. wingi* (Simons & Kay, 1988). Lower molars of *Qatrania* lack development of wear facet X, a feature previously thought to unite parapithecids with undoubted catarrhines (Kay, 1977). The novel dental anatomy shown by *Qatrania* and the additional postcranial data then available for *Apidium* provoked a more comprehensive analysis of parapithecid affinities. Fleagle & Kay (1987) concluded that parapithecids occupied a more basal position on the anthropoid tree than most earlier workers had supposed. Rather than viewing parapithecids as being specially related to cercopithecids (Simons, 1974), catarrhines (Szalay & Delson, 1979) or platyrrhines (Hoffstetter, 1980), Fleagle & Kay (1987) proposed that they were probably the sister group of a clade including both catarrhines and platyrrhines.

Among the more controversial yet poorly documented primate taxa discovered in the Fayum is *Afrotarsius chatrathi*, based on a lower jaw fragment originally described as a tarsiiform primate possibly related to Tarsiidae (Simons & Bown, 1985). The tarsiid or tarsiiform affinities of *Afrotarsius* have been disputed by most subsequent workers, who advocate anthropoid affinities for *Afrotarsius* instead (Fleagle & Kay, 1987; Ginsburg & Mein, 1987; Kay & Williams, 1994; van Valen, 1994; Beard, 1998*a*; Ross *et al.*, 1998; but see

Rasmussen *et al.*, 1998; Gunnell & Rose, this volume).

Since the late 1980s collaborative teams of French and Algerian paleontologists have recovered fragmentary but extremely important fossil anthropoids in the Eocene of Algeria. These demonstrate that basal anthropoids inhabited northern Africa substantially prior to the interval documented in the Fayum. Louis de Bonis *et al.* (1988*b*) described the first of these as *Biretia piveteaui*, based on an isolated lower molar from Bir el Ater, in northeastern Algeria. *Biretia* was originally classified as a small, primitive member of the Parapithecoidea, and subsequent workers have noted similarities to the small parapithecid *Qatrania wingi* (Kay & Williams, 1994). Godinot & Mahboubi (1992, 1994) described two additional basal anthropoid taxa (*Algeripithecus minutus* and *Tabelia hammadae*) from a second site called Glib Zegdou, which is apparently older than either Bir el Ater or the Fayum sequence (Godinot, 1994). Both of these taxa are known only from isolated teeth; however, the advanced dental anatomy implies that they are nested well within the anthropoid radiation (Godinot, 1994).

The Algerian Eocene anthropoids were described at a critical juncture in the debate over anthropoid origins. New skulls of the early Eocene omomyid primate *Shoshonius cooperi* had just been described, and their surprisingly tarsier-like anatomy suggested that the tarsier–anthropoid dichotomy had already occurred prior to 50 Ma (Beard *et al.*, 1991). The great antiquity of the Algerian anthropoids therefore corroborated the hypothesis that anthropoids are an ancient primate clade, a view previously endorsed by a small minority (Hoffstetter, 1977).

A new fossiliferous locality in the lower part of the Jebel Qatrani Formation, dubbed L-41, began to yield fossil primates in the fall of 1987 (Simons, 1989, 1992, 1997*b*; Miller and Simons, 1997). Among other primates described to date from L-41, three taxa qualify as basal anthropoids – *Proteopithecus sylviae*, *Serapia eocaena* and *Arsinoea kallimos*. The discovery of this diverse radiation of anthropoid primates in the lower part of the formation constrained hypotheses about anthropoid origins in two main ways. First, the 35.5–36 Ma age of the L-41 site (Kappelman, 1992) reinforced the emerging consensus that the anthropoid clade diverged from all other primates relatively early in the Paleogene. Second, primitive anatomical characters shown by some of the basal anthropoid taxa from L-41, including unfused mandibular symphyses, small body size and the retention of prominent paraconids on the lower molars, implied that anthropoids acquired their large suite of derived anatomical features mosaically, rather than as a package.

At this time Simons also considered the relationships of Fayum basal anthropoids to other primates. Relying heavily on similarities in dental morphology shared by the oligopithecid *Catopithecus* and cercamoniine adapiforms, Simons (1989, 1990, 1995*c*) advocated an adapiform ancestry for anthropoids. However, comparisons between cercamoniine adapiforms and non-oligopithecid anthropoids from L-41, especially *Proteopithecus*, did not support close relationship between anthropoids and adapiforms (Miller & Simons, 1997).

A team of scientists from the Carnegie Museum of Natural History and the Institute of Vertebrate Paleontology and Paleoanthropology (Chinese Academy of Sciences) began exploring various early Cenozoic deposits in the People's Republic of China in early 1992. Initially, these expeditions focused on exploration of middle Eocene fissure-fillings near the village of Shanghuang, in southern Jiangsu Province. Screen-washing these fossiliferous fissure-fillings yielded an extremely diverse primate fauna. Among the Shanghuang primate taxa described to date, one species – *Eosimias sinensis* – has been interpreted as a very basal member of the anthropoid radiation, lying outside the clade including oligopithecids, parapithecids, platyrrhines and catarrhines (Beard *et al.*, 1994). A new family (Eosimiidae) was erected to accommodate its distinctive combination of primitive and derived dental characters and its apparently basal phylogenetic position within Anthropoidea.

Initially, *Eosimias sinensis* was documented only by incomplete lower dentitions, which showed that it was a small primate (approximately 67–137 g) that retained an unfused mandibular symphysis and a lower dental formula of 2.1.3.3. Alveoli of the missing incisor crowns indicated that the lower incisors were implanted nearly vertically in a dorsoventrally robust symphysis, a key difference from Eocene adapiforms and tarsiiforms. Relative to each other, the lower anterior teeth of *Eosimias* showed a very anthropoid-like pattern of size relationships and root configurations. In contrast, the lower molar occlusal pattern differed rather dramatically from that of other basal anthropoids, and indeed from that of other Eocene primates. Beard *et al.* (1994) interpreted *Eosimias* as further evidence that the anthropoid clade was established during the early part of the Paleogene. They also emphasized that the east Asian distribution of *Eosimias* once again raised the specter of an Asian origin for anthropoids. However, the hypothesis that *Eosimias* is a basal anthropoid proved to be highly controversial, and it was dismissed by several other specialists on early anthropoids (Godinot, 1994; Godinot & Mahboubi, 1994; Simons & Rasmussen, 1994*b*; Simons, 1995*c*).

Subsequent discoveries have augmented significantly our understanding of the anatomy, taxonomic diversity and geographic distribution of eosimiids. The anatomy of an eosimiid petrosal bone described by MacPhee *et al.* (1995*a*) differs appreciably from that of living and fossil anthropoids, but is similar to Eocene tarsiiform primates such as *Necrolemur*. MacPhee *et al.* (1995*b*) interpreted it as further evidence that anthropoids acquired their suite of diagnostic attributes mosaically, and that eosimiids lie outside the clade including oligopithecids, parapithecids, platyrrhines and catarrhines. Based on their collaborative field work in the late middle Eocene Heti Formation in central China's Yuanqu Basin, Beard *et al.* (1996) described the first complete lower dentition of an eosimiid as the new species *E. centennicus*. The dental anatomy of *E. centennicus* confirmed the anthropoid-like anterior lower dentition that had been predicted on the basis of fragmentary specimens of *E. sinensis* by Beard *et al.* (1994). Tong (1997) described additional isolated teeth of *E. centennicus*, including the first known upper teeth of this species.

The first eosimiid fossil to be found outside the People's Republic of China was described from the late middle Eocene of Myanmar as *Bahinia pondaungensis* by Jaeger *et al.* (1999). Larger and somewhat more derived than *Eosimias*, *Bahinia* was interpreted as corroborating the anthropoid affinities of eosimiids, particularly in terms of upper molar anatomy (Jaeger *et al.*, 1999). Most recently, Gebo *et al.* (2000*a*) described the first eosimiid postcranial elements. The tarsals of *Eosimias* exhibit both primitive (haplorhine) and derived (anthropoid) characters. This combination of characters supports phylogenetic reconstructions that place *Eosimias* at the base of the anthropoid radiation. Remarkably small tarsals from Shanghuang that are morphologically very similar to those of *Eosimias* were described by Gebo *et al.* (2000*b*).

Renewed field work in Myanmar during the late 1990s resulted in the discovery of more complete specimens of *Pondaungia* and *Amphipithecus* (Jaeger *et al.*, 1998*b*; Chaimanee *et al.*, 2000*a*; Takai *et al.*, 2000*b*). At roughly the same time, a new but closely related primate, *Siamopithecus eocaenus*, was recovered from late Eocene lignites in southern Thailand (Chaimanee *et al.*, 1997, 2000*b*). These southeast Asian primates resemble one another in many details of dental morphology, and it seems likely that they belong to the same clade, the family Amphipithecidae (Chaimanee *et al.*, 2000*a*). Because *Siamopithecus* retains eosimiid-like P_{3-4} that are not mesiodistally compacted as they are in *Amphipithecus* and *Pondaungia*, the two Burmese genera appear to be sister taxa within Amphipithecidae (Chaimanee *et al.*, 2000*b*). Despite a large difference in size, eosimiids and amphipithecids share a relatively small P_2, in contrast to platyrrhines, parapithecids and proteopithecids. Like eosimiids, amphipithecids are currently unknown outside of Asia.

Taxonomy

Systematic framework

Order Primates Linnaeus, 1758
- Suborder Haplorhini Pocock, 1918
  - Infraorder Anthropoidea Mivart, 1864
    - Family *incertae sedis*
      - Genus *Algeripithecus* Godinot & Mahboubi, 1992
        - *Algeripithecus minutus* Godinot & Mahboubi, 1992
      - Genus *Arsinoea* Simons, 1992
        - *Arsinoea kallimos* Simons, 1992
      - Genus *Tabelia* Godinot & Mahboubi, 1994
        - *Tabelia hammadae* Godinot & Mahboubi, 1994
    - Family Afrotarsiidae Ginsburg & Mein, 1987
      - Genus *Afrotarsius* Simons & Bown, 1985
        - *Afrotarsius chatrathi* Simons & Bown, 1985
    - Family Eosimiidae Beard *et al.*, 1994
      - Genus *Eosimias* Beard *et al.*, 1994
        - *Eosimias sinensis* Beard *et al.*, 1994
        - *Eosimias centennicus* Beard *et al.*, 1996
      - Genus *Bahinia* Jaeger *et al.*, 1999
        - *Bahinia pondaungensis* Jaeger *et al.*, 1999
    - Family Amphipithecidae Godinot, 1994
      - Genus *Pondaungia* Pilgrim, 1927
        - *Pondaungia cotteri* Pilgrim, 1927
      - Genus *Amphipithecus* Colbert, 1937
        - *Amphipithecus mogaungensis* Colbert, 1937
      - Genus *Siamopithecus* Chaimanee *et al.*, 1997
        - *Siamopithecus eocaenus* Chaimanee *et al.*, 1997
    - Family Proteopithecidae Simons, 1997
      - Genus *Proteopithecus* Simons, 1989
        - *Proteopithecus sylviae* Simons, 1989
      - Genus *Serapia* Simons, 1992
        - *Serapia eocaena* Simons, 1992
    - Family Parapithecidae Schlosser, 1911
      - Genus *Apidium* Osborn, 1908
        - *Apidium phiomense* Osborn, 1908
        - *Apidium moustafai* Simons, 1962
        - *Apidium bowni* Simons, 1995
      - Genus *Parapithecus* Schlosser, 1910
        - *Parapithecus fraasi* Schlosser, 1910
        - *Parapithecus grangeri* Simons, 1974
      - Genus *Qatrania* Simons & Kay, 1983
        - *Qatrania wingi* Simons & Kay, 1983
        - *Qatrania fleaglei* Simons & Kay, 1988
      - Genus *Biretia* Bonis *et al.*, 1988
        - *Biretia piveteaui* Bonis *et al.*, 1988

Family *incertae sedis*

GENUS *Algeripithecus* Godinot & Mahboubi, 1992
INCLUDED SPECIES *A. minutus*

SPECIES *Algeripithecus minutus* Godinot & Mahboubi, 1992 (Fig. 9.1A–D)
TYPE SPECIMEN GZC 1, an isolated left M^2
AGE AND GEOGRAPHIC RANGE Late early or middle Eocene, Glib Formation, Glib Zegdou, Algeria
ANATOMICAL DEFINITION
A small, yet relatively bunodont anthropoid. M^2 bears a strong hypocone, an incipient pericone and a complete lingual cingulum. Paracone and metacone on M^2 are situated internally, without steep labial margins. P_3 is simply constructed, with a relatively strong lingual cingulid and a trigonid composed of a single cusp. There is no indication that P_3 would have been either exodaenodont or obliquely implanted in the dentary. M_3 lacks a paraconid, and bears a mesiodistally reduced hypoconulid lobe.

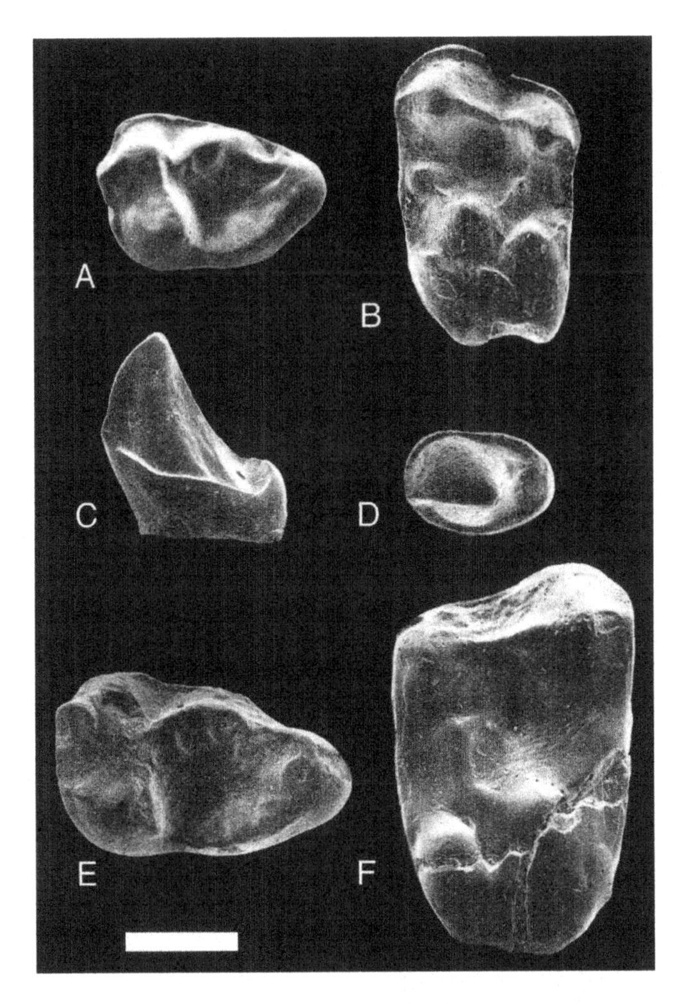

Fig. 9.1 Isolated teeth of basal anthropoids from Glib Zegdou, Algeria. (A) *Algeripithecus minutus* (GZC 2), right M_3 in occlusal view. (B) *Algeripithecus minutus* (GZC 1), holotype left M^2 in occlusal view. (C, D) *Algeripithecus minutus* (GZC 11), right P_3 in lingual (C) and occlusal (D) views. (E) *Tabelia hammadae* (GZC 5), holotype right M_3 in occlusal view. (F) *Tabelia hammadae* (GZC 4), right M^2 in occlusal view. Adapted from Godinot & Mahboubi (1994). Scale bar = 1 mm.

GENUS *Arsinoea* Simons, 1992
INCLUDED SPECIES *A. kallimos*

SPECIES *Arsinoea kallimos* Simons, 1992 (Fig. 9.2A)
TYPE SPECIMEN CGM 42310, a left dentary fragment preserving the crowns of I_2–M_3 and the base of the crown of I_1
AGE AND GEOGRAPHIC RANGE Late Eocene, site L-41, lower part of the Jebel Qatrani Formation, Fayum Province, Egypt
ANATOMICAL DEFINITION
A small anthropoid, probably intermediate in size between *Algeripithecus minutus* and *Proteopithecus sylviae*. The lower dental formula is 2.1.3.3. I_1 is smaller than I_2, and both lower incisors are oriented relatively vertically in the symphysis. I_2 is high-crowned and semispatulate. C_1 in the holotype is only slightly taller than the adjacent I_2 and P_2, suggesting that the holotype was female. C_1 is stoutly constructed and broad in basal circumference, with a strong lingual cingulid and a distinct distal cuspule. P_2 is single-rooted, yet its crown is mesiodistally elongated. All lower premolars possess strongly developed lingual cingulids and elevated hypoconids. P_{3-4} are moderately exodaenodont, obliquely oriented in the toothrow, and bear distinct metaconids that are situated distolingual to their respective protoconids. Paraconids are lacking on all lower premolars, although each possesses a distinct swelling of enamel near the junction of the preprotocristid and the lingual cingulid. Lower molar protoconids are taller and more voluminous than their corresponding metaconids. A distinct paraconid is retained on M_{1-2}. Lower molars bear twinned hypoconulid and entoconid cusps, the latter of which is relatively mesial in position, near the lingual base of the trigonid. M_3 possesses a highly reduced hypoconulid lobe.

GENUS *Tabelia* Godinot & Mahboubi, 1994
INCLUDED SPECIES *T. hammadae*

SPECIES *Tabelia hammadae* Godinot & Mahboubi, 1994 (Fig. 9.1E, F)
TYPE SPECIMEN GZC 5, an isolated right M_3
AGE AND GEOGRAPHIC RANGE Late early or middle Eocene, Glib Formation, Glib Zegdou, Algeria
ANATOMICAL DEFINITION
A small anthropoid, slightly larger than *Algeripithecus*. M_3 lacks a paraconid but possesses a strong premetacristid and a less reduced hypoconulid lobe than that of *Algeripithecus*. M^2 bears a relatively larger trigon basin than occurs in *Algeripithecus*. Its hypocone is weaker than that of *Algeripithecus*, and there is no pericone.

Family Afrotarsiidae

GENUS *Afrotarsius* Simons & Bown, 1985
INCLUDED SPECIES *A. chatrathi*

SPECIES *Afrotarsius chatrathi* Simons & Bown, 1985 (Fig. 9.3)
TYPE SPECIMEN CGM 42830, a left dentary fragment preserving the bases of the crowns of P_{3-4} and the crowns of M_{1-3}
AGE AND GEOGRAPHIC RANGE Early Oligocene, Quarry M, Jebel Qatrani Formation, Fayum Province, Egypt
ANATOMICAL DEFINITION
A small anthropoid, roughly similar in size to *Tabelia*. Judging from the preserved bases of the premolar crowns, P_3 seems to be significantly smaller than P_4. Both P_3 and P_4 were probably oriented somewhat obliquely with respect to the molars (with the mesial root being positioned farther labially than the distal root). Strong, trenchant paraconids are retained on all lower molars. Lower molar protoconids appear to be substantially larger than the corresponding metaconids. Lower molar paraconids and metaconids do

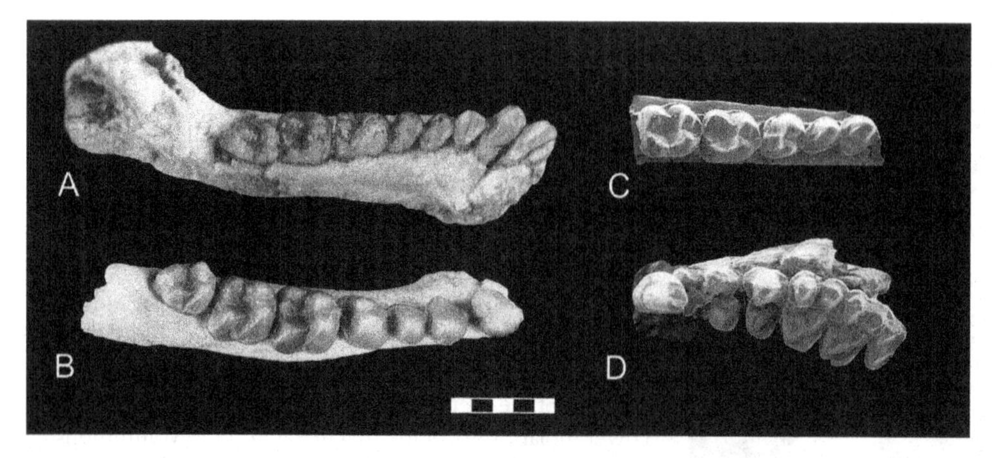

Fig. 9.2 Basal anthropoid primates from the Fayum region of Egypt. (A) *Arsinoea kallimos*, holotype left dentary bearing crowns of I_2–M_3 (CGM 42310) in occlusal view. (B) *Serapia eocaena*, holotype right dentary bearing crowns of C_1–M_3 (CGM 42286) in occlusal view. Reproduced by permission of Elwyn Simons. (C) *Proteopithecus sylviae*, left dentary bearing crowns of P_2–M_2 (CGM 42209) in occlusal view. (D) *Proteopithecus sylviae*, left maxilla bearing crowns of C^1–M^3 (DPC 15518) in occlusal view. Reproduced from Miller, E.R. & Simons, E.L. (1997). Dentition of *Proteopithecus sylviae*, an archaic anthropoid from the Fayum, Egypt. *Proceedings of the National Academy of Sciences of the United States of America*, **94**, 13760–13764. © 1997 National Academy of Sciences, USA, reproduced by permission. Scale bar = 5 mm.

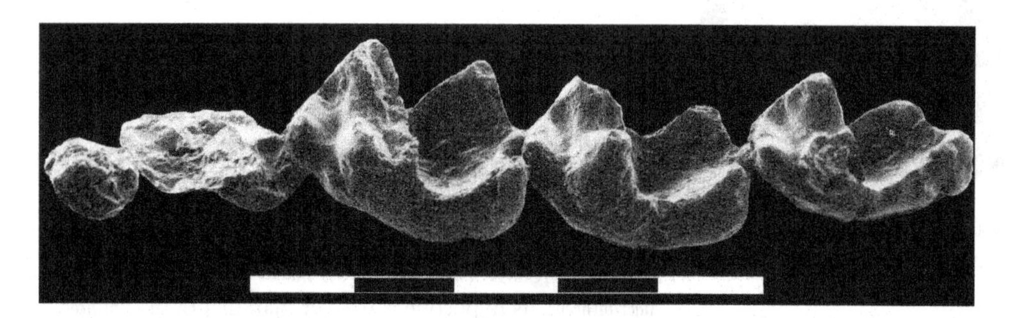

Fig. 9.3 *Afrotarsius chatrathi*. Holotype right dentary bearing crowns of M_{1-3} and bases of crowns of P_{3-4} (CGM 42830) in oblique occlusal/lingual view. Reproduced from Kay, R.F. & Williams, B.A. (1994). Dental evidence for anthropoid origins. In *Anthropoid Origins*, eds. J.G. Fleagle & R.F. Kay, pp. 361–445. New York: Plenum Press. © 1994 Kluwer Academic/Plenum Press, reproduced by permission. Scale bar = 5 mm.

not become increasingly connate posteriorly, so that the trigonids of M_{1-3} are nearly identical. On M_{1-2} the entoconid is situated distally, far from the lingual base of the trigonid. M_3 bears a reduced talonid, primarily due to the abbreviated nature of the hypoconulid lobe. On none of the lower molars are the entoconid and hypoconulid "twinned".

Family Eosimiidae

GENUS *Eosimias* Beard *et al.*, 1994

Given its small size, the dentary of *Eosimias* is deep and robust, especially in the symphyseal region. The lower dental formula is 2.1.3.3. Lower incisors are nearly vertically implanted, and I_1 is smaller than I_2. The lower canine is stout, tall and recurved. P_2 is small, simply constructed and single-rooted. P_{3-4} are moderately exodaenodont and obliquely oriented in the toothrow, so that the mesial root is situated farther labially than the distal root. All lower molars retain prominent paraconids, but these cusps are more inflated than is the case in *Afrotarsius*. The entoconid on M_{1-3} is relatively mesial in position, being near the lingual base of the trigonid. Other lower molar characters, including widely spaced paraconid and metaconid on M_{2-3}, lack of "twinned" entoconid and hypoconulid, and a reduced talonid/hypoconulid lobe on M_3, resemble conditions in *Afrotarsius*. Upper molars are tritubercular, with strong pre- and postprotocristae and no postprotocingulum. Upper molar conules are absent.

INCLUDED SPECIES *E. centennicus*, *E. sinensis*

SPECIES *Eosimias sinensis* Beard *et al.*, 1994 (Fig. 9.4A, C, E)

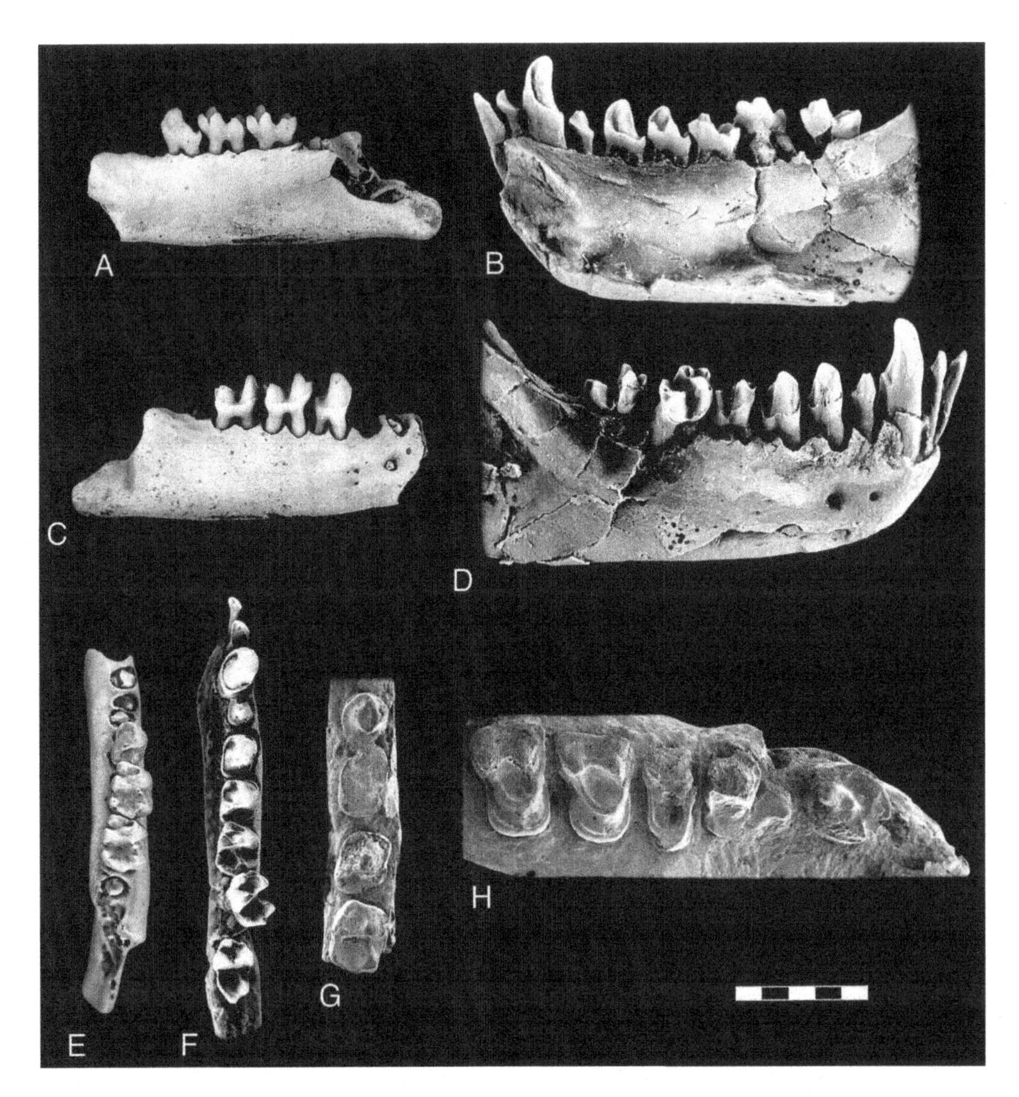

Fig. 9.4 Eosimiid primates from China and Myanmar. (A,C,E) *Eosimias sinensis*, holotype right dentary bearing crowns of P_4–M_2 (IVPP V10591) in lingual (A), buccal (C) and occlusal (E) views. Reproduced from Beard, K.C., Qi, T., Dawson, M.R., Wang, B. & Li, C. (1994). A diverse new primate fauna from middle Eocene fissure-fillings in southeastern China. *Nature*, **368**, 604–609. © 1994 Macmillan Magazines Ltd, reproduced by permission. (B,D,F) *Eosimias centennicus*, holotype right dentary bearing complete dentition (IVPP V11000) in lingual (B), buccal (D) and occlusal (F) views. Reproduced from Beard, K.C., Tong, Y., Dawson, M.R., Wang, J. & Huang, X. (1996). Earliest complete dentition of an anthropoid primate from the late middle Eocene of Shanxi Province, China. *Science*, **272**, 82–85. © 1996 American Association for the Advancement of Science, reproduced by permission. (G–H) *Bahinia pondaungensis*, holotype lower (G) and upper (H) dentition (NMMP 15, 16) in occlusal view. Courtesy of Stephan Ducrocq. Scale bar = 5 mm.

TYPE SPECIMEN IVPP V10591, a right dentary fragment preserving the alveoli for C_1–P_3 and M_3 and the crowns of P_4–M_2

AGE AND GEOGRAPHIC RANGE Middle Eocene, Shanghuang Fissure B, Liyang County, southern Jiangsu Province, People's Republic of China

ANATOMICAL DEFINITION

Eosimias sinensis is slightly smaller than *E. centennicus*. Morphology of P_4 is apparently more primitive in *E. sinensis* than in *E. centennicus*, in that there is no distinct paraconid and the metaconid is weaker and situated lower on the crown.

SPECIES *Eosimias centennicus* Beard *et al.*, 1996 (Fig. 9.4B, D, F)

TYPE SPECIMEN IVPP V11000, associated left and right dentaries of a single individual preserving all lower tooth crowns except left I_{1-2}

AGE AND GEOGRAPHIC RANGE Late middle Eocene, Locality 1 (also known as "River Section" locality or Tuqiaogou), Zhaili Member, Heti Formation, Yuanqu Basin, southern Shanxi Province, People's Republic of China

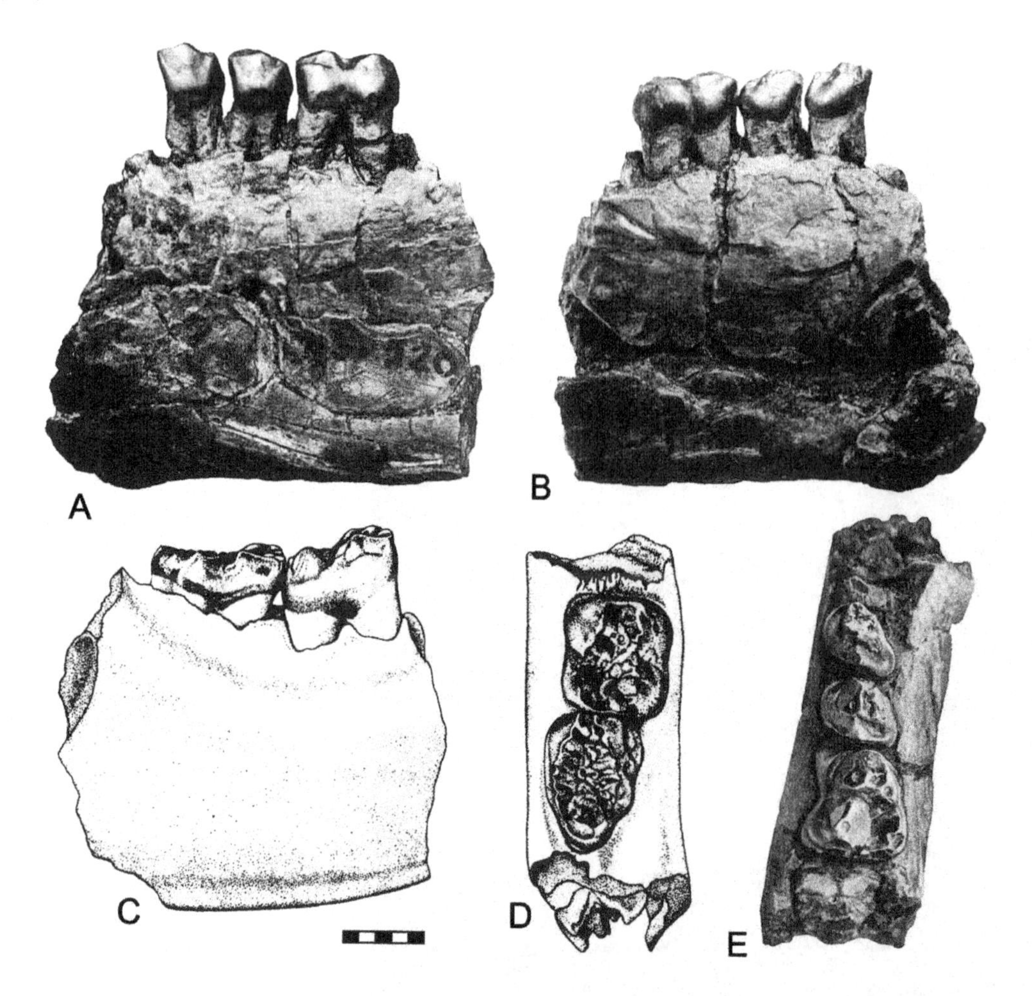

Fig. 9.5 Amphipithecid primates from Myanmar. Holotype left dentary of *Amphipithecus mogaungensis* bearing crowns of P_3–M_1 (AMNH 32520) in buccal (A), lingual (B) and occlusal (E) views. Right dentary of *Pondaungia cotteri* bearing crowns of M_{2-3} (UCMP 120377) in buccal (C) and occlusal (D) views. (A–C) reproduced from Ciochon, R. L. & Holroyd, P.A. (1994). The Asian origin of Anthropoidea revisited. In *Anthropoid Origins*, eds. J.G. Fleagle & R.F. Kay, pp. 143–162. New York: Plenum Press. © 1994 Kluwer Academic/Plenum Press, reproduced by permission. Scale bar = 5 mm.

ANATOMICAL DEFINITION
Eosimias centennicus is slightly larger than *E. sinensis*. Its P_4 is more nearly molariform in having a distinct paraconid and a stronger metaconid that occurs higher on the crown.

GENUS *Bahinia* Jaeger *et al.*, 1999
INCLUDED SPECIES *B. pondaungensis*

SPECIES *Bahinia pondaungensis* Jaeger *et al.*, 1999 (Fig. 9.4G, H)
TYPE SPECIMEN NMMP 14–16, associated maxillary and dentary fragments of a single individual
AGE AND GEOGRAPHIC RANGE Late middle Eocene, Yashe Kyitchaung locality, near Bahin village, Pondaung Formation, Myanmar
ANATOMICAL DEFINITION
Bahinia is the largest known eosimiid primate. P^3 lacks a distinct protocone. Upper molars are tritubercular, with occlusal outlines less extended transversely than in *Eosimias*. Upper molars bear complete lingual cingula and highly trenchant pre- and postprotocristae. P_2 is relatively larger than in *Eosimias*.

Family Amphipithecidae

GENUS *Pondaungia* Pilgrim, 1927
INCLUDED SPECIES *P. cotteri*

SPECIES *Pondaungia cotteri* Pilgrim, 1927 (Fig. 9.5C, D)
TYPE SPECIMEN GSI D201-203, associated maxillary and dentary fragments of a single individual; left maxilla fragment preserves M^{1-2}, left dentary fragment preserves M_{2-3} and right dentary fragment preserves M_3
AGE AND GEOGRAPHIC RANGE Late middle Eocene, Pondaung Formation, Myanmar

ANATOMICAL DEFINITION

The dentary of *Pondaungia* is robustly constructed. The lower dental formula is almost certainly 2.1.3.3. Lower incisor crowns remain unknown, but these teeth are small and vertically oriented in the unfused symphysis. The lower canine, known only from a broken root, is larger than either the incisors or the premolars. The three lower premolars are reduced in mesiodistal length relative to the molars. P_2 is single-rooted and simple in shape. P_{3-4} are more nearly molariform and oriented obliquely with respect to the remainder of the toothrow. Upper and lower cheek teeth are distinctive in being strongly bunodont and in having highly crenulated enamel. On M^{1-2} the protocone is linked to a prominent distolingual cusp (?hypocone) by a mesiodistally oriented crest. On both M_2 and M_3 the paraconid is highly reduced, and the entoconid is situated near the lingual base of the trigonid.

GENUS *Amphipithecus* Colbert, 1937
INCLUDED SPECIES *A. mogaungensis*

SPECIES *Amphipithecus mogaungensis* Colbert, 1937 (Fig. 9.5A, B, E)
TYPE SPECIMEN AMNH 32520, a left dentary fragment preserving the roots of C_1 and P_2 and the crowns of P_3–M_1
AGE AND GEOGRAPHIC RANGE Late middle Eocene, Pondaung Formation, Myanmar
ANATOMICAL DEFINITION
Amphipithecus is somewhat smaller than *Pondaungia*, but with a similarly robust dentary. P_{3-4} are less reduced relative to the lower molars compared to *Pondaungia*, but lower premolar morphology is otherwise very similar. The lower molars show less enamel crenulation than do those of *Pondaungia*. M_3 in *Amphipithecus* is reduced in size relative to M_{1-2}.

GENUS *Siamopithecus* Chaimanee et al., 1997
INCLUDED SPECIES *S. eocaenus*

SPECIES *Siamopithecus eocaenus* Chaimanee et al., 1997 (Fig. 9.6)
TYPE SPECIMEN TF 3635, a right maxillary fragment preserving the crowns of P^3–M^3 (only the lingual half of M^3 is preserved)
AGE AND GEOGRAPHIC RANGE Late Eocene, main lignite seam exposed in Krabi mine, peninsular Thailand
ANATOMICAL DEFINITION
Siamopithecus is a large amphipithecid, roughly similar in size to *Pondaungia*. Its dentary is robust and deep, even for an amphipithecid. The lower dental formula is almost certainly 2.1.3.3, although the lower incisors remain unknown. The symphysis is unfused but robustly constructed and not at all procumbent. The lower canine is large, tall and caniniform, with a strong lingual cingulid. Judging from its diminutive, single root, the unknown crown of P_2 is relatively small, as in eosimiids. P_{3-4} are relatively larger than in other amphipithecids, and their crown morphology strongly resembles that of eosimiids. Both P_3 and P_4 are oriented obliquely with respect to the rest of the toothrow, with the mesial root being positioned farther labially than the distal root. P_3 is morphologically simple, bearing only a single trigonid cusp, while P_4 is more nearly molariform, bearing a metaconid both inferior and distal to the protoconid, as in eosimiids. The lower molars are strongly bunodont and lack paraconids, but their entoconids are relatively mesial in position, a further similarity to eosimiids. Upper molars are robust yet bunodont, with the apices of major cusps located internally because of the shallow slopes formed by the labial and lingual margins of these teeth. M^1 bears a nearly complete crista obliqua uniting the protocone with the metacone. As in *Pondaungia*, M^{1-2} each possess a relatively well-developed distolingual cusp (?hypocone) that is united with the protocone by a crest.

Family Proteopithecidae

GENUS *Proteopithecus* Simons, 1989
INCLUDED SPECIES *P. sylviae*

SPECIES *Proteopithecus sylviae* Simons, 1989 (Figs. 9.2C, D, 9.7C)
TYPE SPECIMEN CGM 41886, a left maxillary fragment preserving the crowns of M^{1-3} (lingual part of M^1 broken) and associated left P^2
AGE AND GEOGRAPHIC RANGE Late Eocene, Quarry L-41, lower part of the Jebel Qatrani Formation, Fayum Province, Egypt.
ANATOMICAL DEFINITION
Proteopithecus is a small anthropoid approximately similar in size to the eosimiid *Bahinia pondaungensis*. There are three premolars in each jaw quadrant. In at least some specimens, C^1 is relatively large and bears a distinct mesial groove. Based on small samples, canines are probably sexually dimorphic. P^2 is small relative to P^{3-4}. Upper molars are simple in structure, with weak or absent conules, strong pre- and postprotocristae, and no postprotocingulum. Crestiform hypocones are present on M^{1-2}. P_2 is larger relative to P_{3-4} than would be expected based on the reduced P^2. P_{3-4} are obliquely oriented, with the mesial root situated farther labially than the distal root, and exhibit only a minor degree of exodaenodonty. The metaconid on P_4 is strong, situated directly lingual to the protoconid, and both of these cusps are united by a strong transverse crest. M_1 retains a paraconid, and all lower molars bear twinned hypoconulid and entoconid cusps.

Recently described postcranial specimens (Simons & Seiffert, 1999) include a femur and tibia. They resemble small platyrrhines in general aspect and reinforce the primitive identity of the genus.

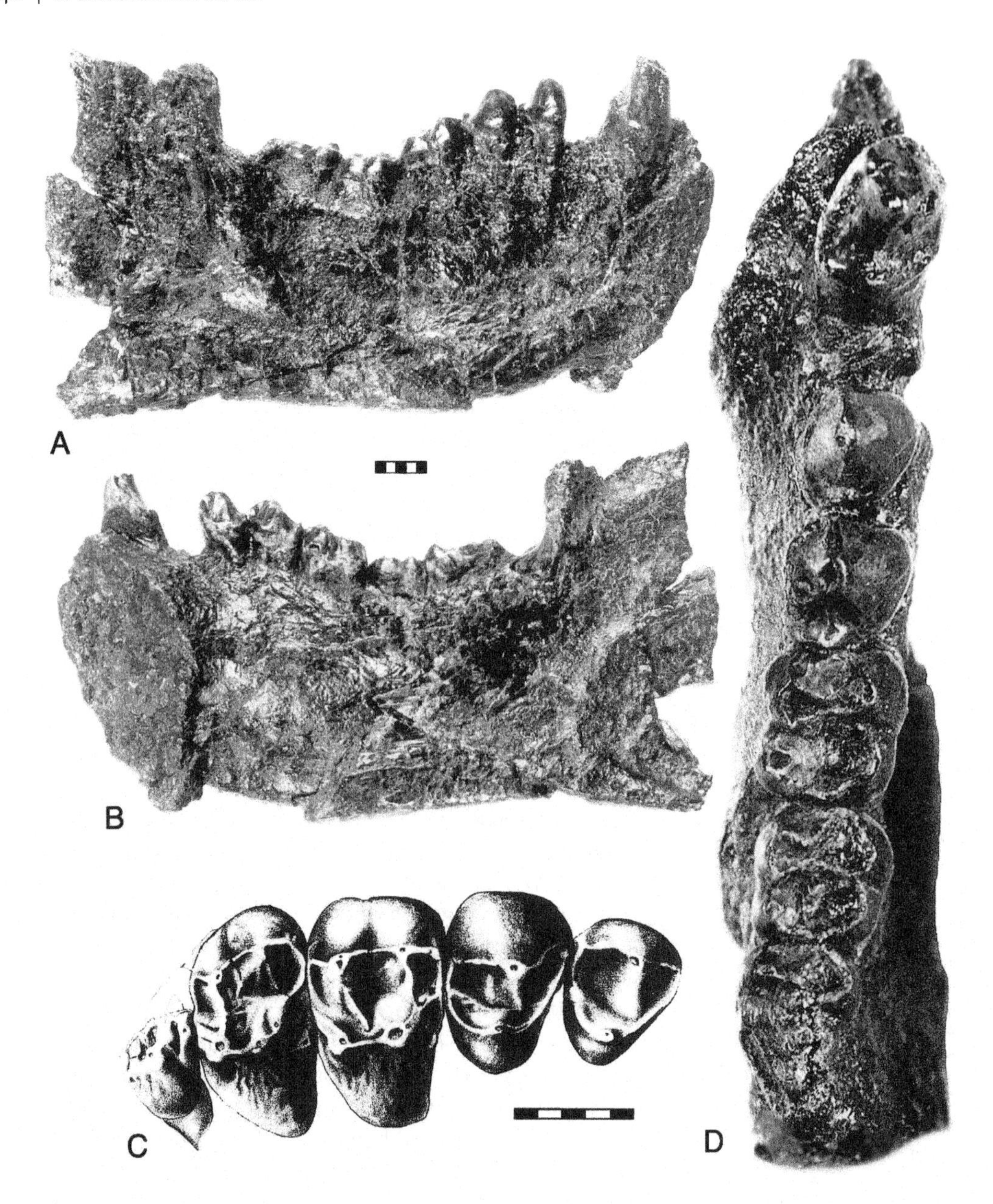

Fig. 9.6 *Siamopithecus eocaenus*. Right dentary bearing crowns of C_1, P_3–M_3 (TF 7624) in buccal (A), lingual (B) and occlusal (D) views, adapted from Chaimanee *et al.* (2000*b*). (C) Right maxilla bearing P^3–M^3 (TF 3635) in occlusal view. Reproduced from Chaimanee, Y., Suteethorn, V., Jaeger, J.-J. & Ducrocq, S. (1997). A new late Eocene anthropoid from Thailand. *Nature*, **385**, 429–431. © 1997 Macmillan Magazines Ltd, reproduced by permission. Scale bars = 5 mm. Upper scale bar is for (A, B); lower scale bar is for (C, D).

GENUS *Serapia* Simons, 1992
INCLUDED SPECIES *S. eocaena*

SPECIES *Serapia eocaena* Simons, 1992 (Fig. 9.2B)
TYPE SPECIMEN CGM 42286, a right dentary fragment preserving the crowns of C_1–M_3
AGE AND GEOGRAPHIC RANGE Late Eocene, Quarry L-41, lower part of the Jebel Qatrani Formation, Fayum Province, Egypt
ANATOMICAL DEFINITION
Serapia is larger than *Proteopithecus*, but the lower dentitions are remarkably similar in these two genera. *Serapia* differs from *Proteopithecus* chiefly in having more bunodont cheek teeth. Compared to *Proteopithecus*, *Serapia* possesses a more rudimentary talonid on P_4 with no distinct entoconid. M_{2-3} of *Serapia* each bear a small, neomorphic crest mesial to the entoconid, which encloses a tiny fovea that occurs on the mesiolingual margin of the talonid. Lower molar paraconids are diminutive and are situated mesiobuccal to the metaconid rather than directly mesial to it.

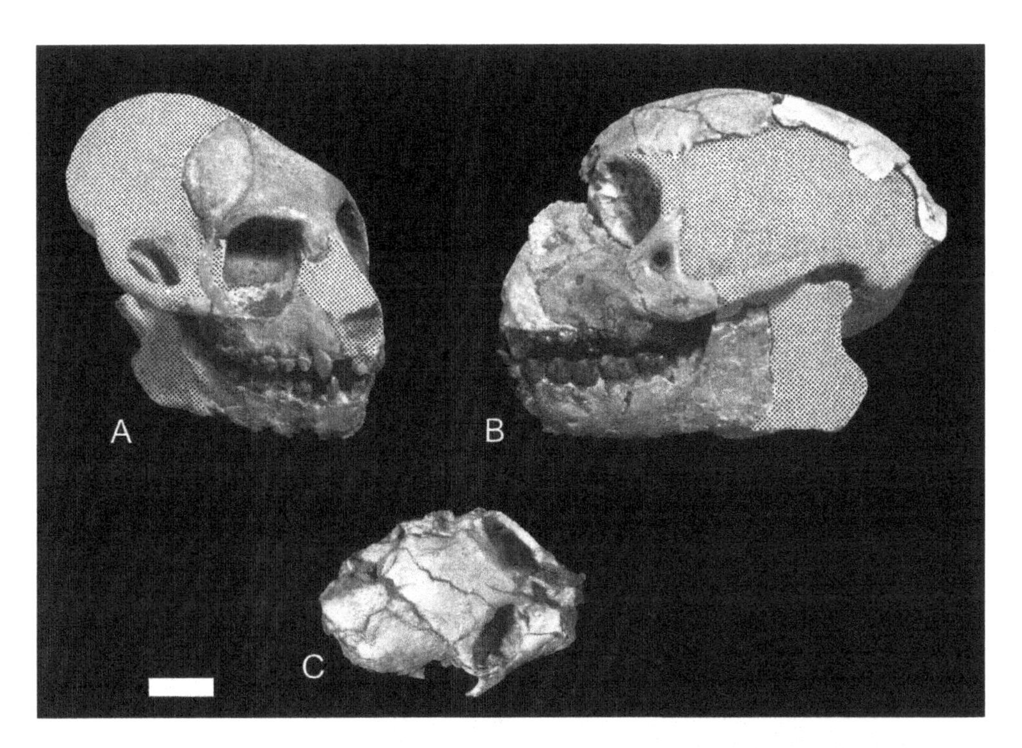

Fig. 9.7 Partial skulls of basal anthropoid primates from the Fayum region of Egypt. (A) *Apidium phiomense*, composite reconstruction of the skull, based on AMNH 14556, CGM 26929 and YPM 21018. (B) *Parapithecus grangeri*, composite reconstruction of the skull, based on DPC 2385, DPC 6641, DPC 1098 and DPC 2807. Courtesy of Elwyn Simons. (C) *Proteopithecus sylviae*, partial skull (CGM 42214). Reproduced from Simons, E.L. (1997). Preliminary description of the cranium of *Proteopithecus sylviae*, an Egyptian late Eocene anthropoidean primate. *Proceedings of the National Academy of Sciences of the United States of America*, **94**, 14970–14975. © 1997 National Academy of Sciences, USA, reproduced by permission. Scale bar = 1 cm.

Family Parapithecidae

GENUS *Apidium* Osborn, 1908
Partial skulls and skull fragments show that postorbital closure is relatively complete in *Apidium*, the metopic suture fuses early in ontogeny, the olfactory bulbs are relatively voluminous for an anthropoid of its size, and the zygomaticofacial foramen is large. Dental formula is 2.1.3.3. All cheek teeth bear inflated cusps and are generally bunodont. Upper premolars bear strong paraconules. Upper molars bear well-developed lingual cingula with prominent pericones and hypocones. P_2 is smaller relative to P_{3-4} than in *Parapithecus*. P_{3-4} are obliquely oriented with respect to the toothrow as is typical of basal anthropoids. Both P_3 and P_4 bear metaconid cusps that are located inferior and slightly distal to their respective protoconids. There is no transverse crest uniting the metaconid and protoconid on P_{3-4}. The talonids of P_{3-4} are notable in having a massively inflated and elevated hypoconid/cristid obliqua that merges with the metaconid mesially. Lower molars lack paraconids but possess a neomorphic cusp (centroconid or mesoconid) near the mesial termination of the cristid obliqua. The hypoconulid cusp is pronounced and relatively central in position (rather than being "twinned" with the entoconid) on all lower molars.

INCLUDED SPECIES *A. bowni*, *A. moustafai*, *A. phiomense*

SPECIES *Apidium phiomense* Osborn, 1908 (Figs. 9.7A, 9.8A,D)
TYPE SPECIMEN AMNH 13370, a left dentary fragment preserving the crowns of P_4–M_3
AGE AND GEOGRAPHIC RANGE Early Oligocene, upper part of the Jebel Qatrani Formation, Fayum Province, Egypt
ANATOMICAL DEFINITION
Apidium phiomense is larger than either *A. moustafai* or *A. bowni*. *Apidium phiomense* also possesses a relatively longer M_3 than do the other two species of the genus.

SPECIES *Apidium moustafai* Simons, 1962
TYPE SPECIMEN CGM 26919, a left dentary fragment preserving the crowns of P_3–M_1
AGE AND GEOGRAPHIC RANGE Earliest Oligocene, middle part of the Jebel Qatrani Formation, Fayum Province, Egypt
ANATOMICAL DEFINITION
Apidium moustafai is smaller than *A. phiomense*, but larger than *A. bowni*. M_3 is relatively shorter mesiodistally than is the case in *A. phiomense*.

SPECIES *Apidium bowni* Simons, 1995
TYPE SPECIMEN CGM 42199, a right dentary fragment preserving the crowns of P_3–M_3 and the alveoli for the anterior dentition

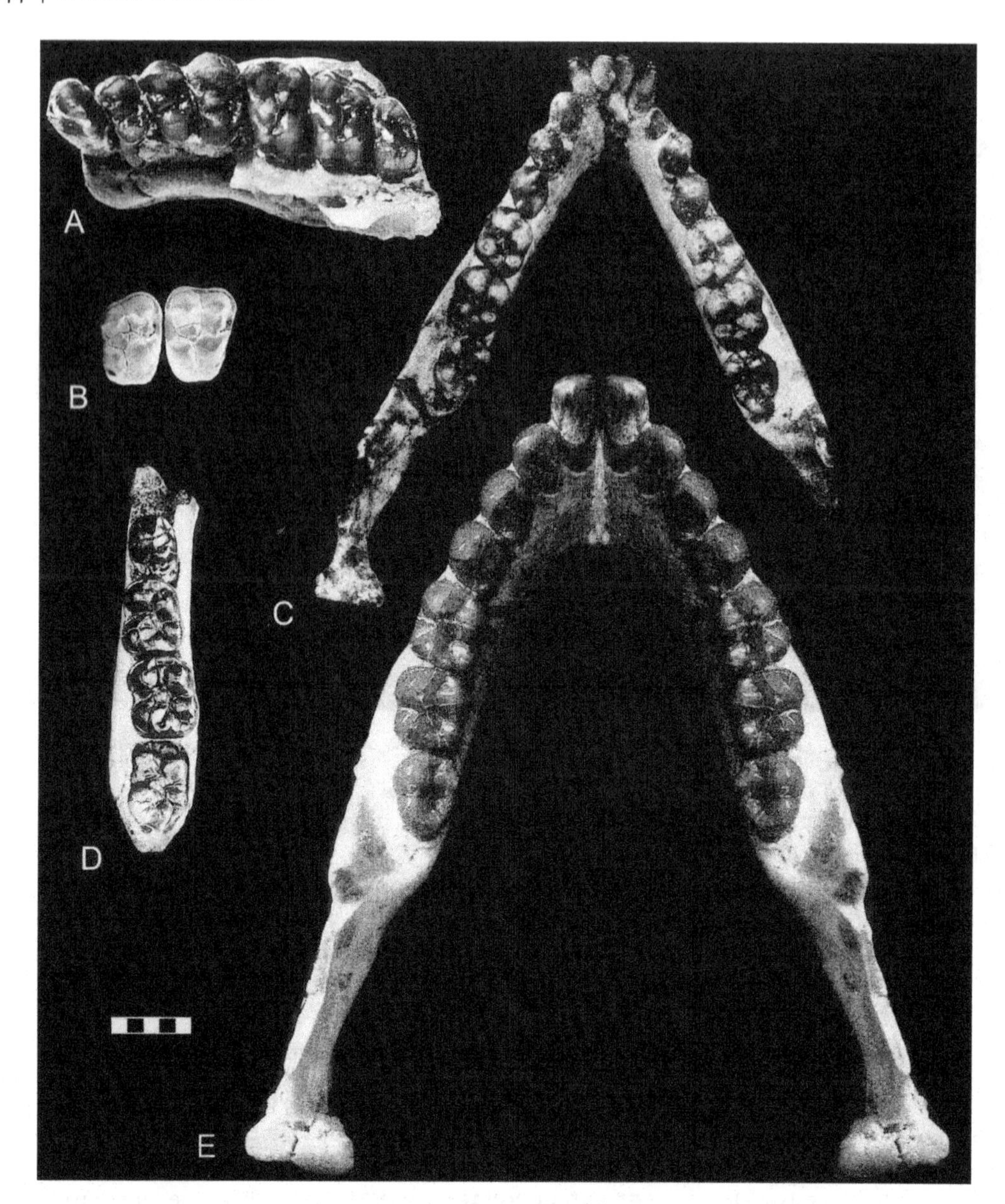

Fig. 9.8 Parapithecid primates from the Fayum region of Egypt. (A) Left maxillary fragment of *Apidium phiomense* bearing crowns of C^1–M^3 in occlusal view. Reproduced by permission of Elwyn Simons. (B) Right M^{1-2} of *Parapithecus grangeri* (DPC 1123) in occlusal view. Reproduced from Kay, R.F. & Williams, B.A. (1994). Dental evidence for anthropoid origins. In *Anthropoid Origins*, eds. J.G. Fleagle & R.F. Kay, pp. 361–445. New York: Plenum Press. © 1994 Kluwer Academic/Plenum Press, reproduced by permission. (C) Holotype mandible of *Parapithecus fraasi* bearing complete lower dentition aside from right P_2 (SNM 12639a) in occlusal view. Reproduced from Simons, E.L. & Rasmussen, D.T. (1991). The generic classification of Fayum Anthropoidea. *International Journal of Primatology*, **12**, 163–178. © 1991 Kluwer Academic/Plenum Press, reproduced by permission. (D) Holotype left dentary of *Apidium phiomense* bearing crowns of P_4–M_3 (AMNH 13370) in occlusal view. Reproduced from Simons, E.L. (1995*a*). Egyptian Oligocene primates: a review. *Yearbook of Physical Anthropology*, **38**, 199–238. © 1995 Wiley–Liss, Inc., a Subsidiary of John Wiley & Sons, Inc., reproduced by permission. (E) Complete lower dentition of *Parapithecus grangeri* in occlusal view, based on mirror-imaging of complete left dentition. Reproduced from Simons, E.L. & Rasmussen, D.T. (1991). The generic classification of Fayum Anthropoidea. *International Journal of Primatology*, **12**, 163–178. © 1991 Kluwer Academic/Plenum Press, reproduced by permission. Scale bar = 5 mm.

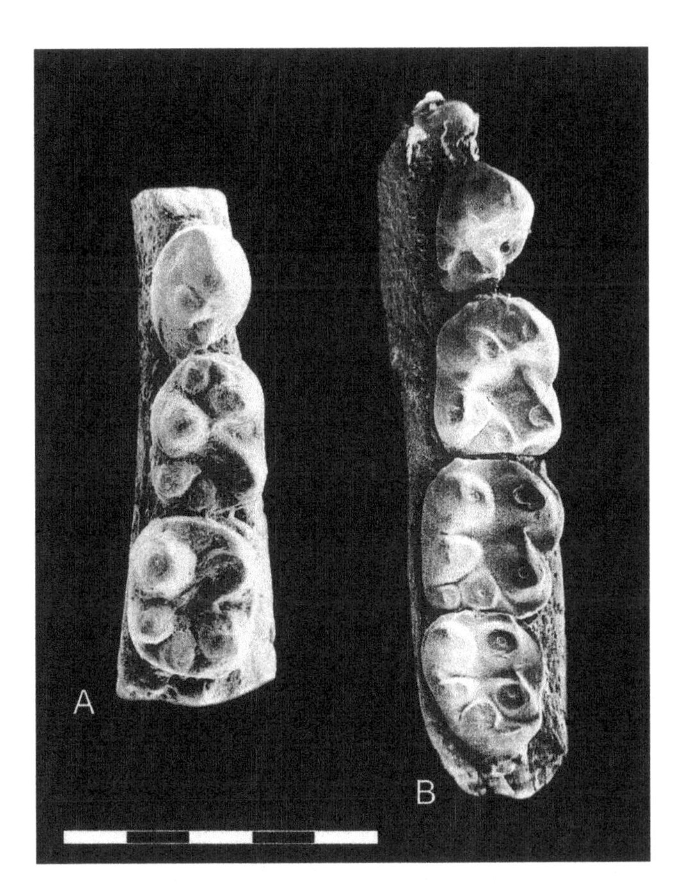

Fig. 9.9 *Qatrania*. (A) *Q. fleaglei*, holotype right dentary bearing crowns of P_4–M_2 (CGM 41850) in occlusal view. (B) *Q. wingi*, right dentary bearing crowns of P_4–M_3 (DPC 6125) in occlusal view. Reproduced from Simons, E.L. & Kay, R.F. (1988). New material of *Qatrania* from Egypt with comments on the phylogenetic position of the Parapithecidae (Primates, Anthropoidea). *American Journal of Primatology*, **15**, 337–347. © 1995 Wiley–Liss, Inc., a Subsidiary of John Wiley & Sons, Inc., reproduced by permission. Scale bar = 5 mm.

AGE AND GEOGRAPHIC RANGE Earliest Oligocene, middle part of the Jebel Qatrani Formation, Fayum Province, Egypt

ANATOMICAL DEFINITION

Apidium bowni is the smallest known species of *Apidium*.

GENUS *Parapithecus* Schlosser, 1910

Fragmentary cranial material shows that *Parapithecus* possesses a fused metopic suture and a large zygomaticofacial foramen, as does its relative *Apidium*. The face of *Parapithecus* seems to have been foreshortened relative to that of *Apidium*, a feature that is probably correlated with reduction or loss of incisors. Cheek teeth are generally similar to those of *Apidium*, although in *Parapithecus* the lower molars lack centroconids and the upper molars show reduced lingual cingula and lack pericones. *Parapithecus* also possesses a relatively larger P_2 than does *Apidium*, and the elevated talonids of P_{3-4} are more cuspidate, with little or no development of the cristid obliqua.

INCLUDED SPECIES *P. fraasi*, *P. grangeri*

SPECIES *Parapithecus fraasi* Schlosser, 1910 (Fig. 9.8C)

TYPE SPECIMEN SNM 12639a, nearly complete left and right dentaries of a single individual

AGE AND GEOGRAPHIC RANGE Early Oligocene, Jebel Qatrani Formation, Fayum Province, Egypt (exact stratigraphic provenance is unknown)

ANATOMICAL DEFINITION

Parapithecus fraasi is smaller than *P. grangeri*, possesses a relatively larger M_3, and retains a pair of deciduous lower incisors beyond the time of eruption of the permanent lower canines.

SPECIES *Parapithecus grangeri* Simons, 1974 (Figs. 9.7B, 9.8B, E)

TYPE SPECIMEN CGM 26912, left dentary preserving P_3–M_3

AGE AND GEOGRAPHIC PROVENANCE Early Oligocene, upper part of the Jebel Qatrani Formation, Fayum Province, Egypt

ANATOMICAL DEFINITION

Parapithecus grangeri is larger than *P. fraasi*. *Parapithecus grangeri* entirely lacks lower incisors.

GENUS *Qatrania* Simons & Kay, 1983

Qatrania is a small parapithecid anthropoid characterized by extremely cuspidate molars that lack virtually any development of molar crests. The lower dental formula is ?.1.3.3. Based on preserved alveoli for P_3 and the crown of P_4, these teeth are obliquely oriented in the toothrow, with the mesial root being positioned farther labially than the distal root. P_4 is moderately exodaenodont. The metaconid on P_4 is small and located distolingual and inferior to the protoconid; there is no strong transverse crest uniting the protoconid and metaconid. The hypoconid/cristid obliqua of P_4 is strongly elevated, as is the case in *Apidium*, *Parapithecus* and, to a lesser extent, *Arsinoea*. A distinct paraconid is retained on M_1, but this cusp is indistinct or absent on M_{2-3}. As in *Apidium* and *Parapithecus*, the lower molars bear strong hypoconulids, but these cusps are more nearly central in position than is the case in proteopithecids and oligopithecids, in which the molar hypoconulids and entoconids are twinned. The hypoconulid lobe of M_3 is reduced to a single small cusp.

INCLUDED SPECIES *Q. fleaglei*, *Q. wingi*

SPECIES *Qatrania wingi* Simons & Kay, 1983 (Fig. 9.9B)

TYPE SPECIMEN CGM 40240, a right dentary fragment preserving the distal alveolus for P_3 and the crowns of P_4–M_3

AGE AND GEOGRAPHIC RANGE Late Eocene, Quarry E, lower part of the Jebel Qatrani Formation, Fayum Province, Egypt

ANATOMICAL DEFINITION

Qatrania wingi is smaller than *Q. fleaglei*. Its M_1 bears a relatively weak paraconid.

SPECIES *Qatrania fleaglei* Simons & Kay, 1988 (Fig. 9.9A)
TYPE SPECIMEN CGM 41850, a right dentary fragment preserving roots or alveoli for I_2–P_3 and M_3 and the crowns of P_4–M_2
AGE AND GEOGRAPHIC RANGE Early Oligocene, Quarry M, upper part of the Jebel Qatrani Formation, Fayum Province, Egypt
ANATOMICAL DEFINITION
Qatrania fleaglei is larger than *Q. wingi*. It bears a relatively prominent paraconid on M_1.

GENUS *Biretia* Bonis et al., 1988
INCLUDED SPECIES *B. piveteaui*

SPECIES *Biretia piveteaui* Bonis et al., 1988
TYPE SPECIMEN BRT 17-84, Université Pierre-et-Marie-Curie, Paris, France, an isolated right M_1
AGE AND GEOGRAPHIC RANGE Late Eocene, Bir el Ater locality, south of the Nementcha Mountains, northeastern Algeria
ANATOMICAL DEFINITION
Biretia piveteaui is similar in size to *Qatrania wingi*. The only tooth described for this taxon to date resembles lower molars of parapithecids in lacking a distinct paraconid and in having a prominent hypoconulid that is centrally located on the distal margin of the talonid. Distinctive crests connect the three talonid cusps.

Evolution of basal anthropoids

Less than a decade ago, many specialists on early anthropoids agreed on several key points. First, anthropoids were thought to have originated relatively late in the Paleogene, probably sometime near the Eocene–Oligocene boundary (Gingerich, 1980*b*, 1993*b*; Rasmussen & Simons, 1992; Rasmussen, 1994). Second, anthropoids were thought to have originated in Africa (Fleagle & Kay, 1987; Godinot & Mahboubi, 1992; Holroyd & Maas, 1994; Simons, 1995*b*). Finally, morphological distinctions between basal anthropoids and their Eocene prosimian relatives were thought to be so subtle that only relatively complete skulls showing such diagnostic traits as postorbital closure could demonstrate the anthropoid affinities of a controversial fossil (Rasmussen, 1994; Simons et al., 1994; Simons & Rasmussen, 1994*b*; Simons, 1995*c*). It now appears that all of these ideas regarding early anthropoid evolution are wrong. Furthermore, while a consensus regarding anthropoid origins is still lacking, one previously influential hypothesis – that anthropoids evolved from Eocene adapiforms – is now widely rejected.

Diverse and abundant evidence supports an early origin for the anthropoid clade. Paleontological data indicate that anthropoids must have diverged from all other primates by at least the beginning of the Eocene (Beard & MacPhee, 1994). Basal anthropoids are now known from middle Eocene sites ranging from China to Algeria (Godinot & Mahboubi, 1992; Beard et al., 1994). The taxonomic diversity of these middle Eocene anthropoids and their geographic range across two continents imply an earlier interval of evolutionary history that remains very poorly documented. Intriguing in this regard is *Altiatlasius koulchii*, a primate from the late Paleocene of Morocco that has been interpreted as a possible close relative of anthropoids (Sigé et al. 1990; Godinot, 1994). However, because *Altiatlasius* is known only from a small sample of isolated teeth and a single fragmentary jaw bearing one molar, its affinities will remain controversial until more complete material is recovered.

The stratigraphic ranges of possible sister groups for Anthropoidea also imply an early origin for the anthropoid clade. Detailed anatomical studies of adapiforms and omomyids suggest that by the early Eocene these taxa were already basal members of the strepsirhine and tarsiiform clades, respectively (Dagosto, 1985, 1988; Rosenberger, 1985; Gebo, 1986, 1988; Covert, 1988; Beard & Godinot, 1988; Beard et al., 1988, 1991; Covert & Williams, 1994). If so, the anthropoid clade must also have been established by this early date, because the sister group of Anthropoidea must be Tarsiiformes, Strepsirhini, or a Tarsiiformes + Strepsirhini clade (Beard & MacPhee, 1994).

Neontological data further corroborate an ancient origin for the anthropoid clade. Molecular data routinely show that Anthropoidea is among the most robustly supported clades in Primates (Koop et al., 1989*b*; Andrews et al., 1998), which accords with a relatively long interval of common descent for all living anthropoids after their divergence from other primates. Morphological data derived from living forms show similarly robust support for the monophyly of Anthropoidea (Ross et al., 1998).

An ancient origin for anthropoids has significant implications for reconstructing higher-level primate phylogeny. Previous hypotheses positing origin from some middle or late Eocene prosimian stock (either adapiform or omomyid) can no longer be sustained. These hypotheses were never robustly supported, and they were often founded upon poorly known fossils. For example, most of the allegedly transitional taxa that were once purported to link anthropoids with adapiforms (Gingerich, 1977*b*, 1980*b*; Rasmussen & Simons, 1988; Rasmussen, 1990b, 1994; Gingerich et al., 1994), including *Mahgarita*, *Oligopithecus*, *Pondaungia*, *Amphipithecus*, *Hoanghonius* and *Rencunius*, have subsequently been shown to be irrelevant to this issue. In the case of *Pondaungia* and *Amphipithecus*, recently discovered specimens have demonstrated that they are nested well within Anthropoidea, and are not transitional between adapiforms and anthropoids (Jaeger et al., 1998*b*; Chaimanee et al., 2000*a*). In the case of *Oligopithecus*, the discovery of relatively complete material of the closely related genus *Catopithecus* has likewise removed any doubt about its supposedly transitional status (Simons, 1990, 1995*c*). Both *Hoanghonius* and *Rencunius* are now regarded as Eocene members of the Sivaladapidae, an adapiform clade that persisted into the late Miocene in southern

Asia that has never been regarded as being closely related to anthropoids (Qi & Beard, 1998). Finally, critical reappraisal of the cranial anatomy of *Mahgarita* reveals that this primate is a fully conventional adapiform, rather than a link between adapiforms and anthropoids (Ross, 1994).

With the collapse of the adapiform–anthropoid hypothesis only a few options remain for linking anthropoids to the rest of the primate evolutionary tree. One possibility, often favored by phylogenetic analyses of postcranial characters, is that the anthropoid lineage was the first major primate clade to diverge from other primates in the early Cenozoic (Ford, 1988, 1994; Godinot, 1991, 1992*b*). If accurate, this model would vindicate the traditional prosimian/anthropoid duality within primates on strictly phylogenetic grounds. However, a serious weakness of this model is that it reverses the traditional polarities of many postcranial character states in basal primates, from prosimian-like conditions that imply an emphasis on leaping to more anthropoid-like conditions that imply a greater emphasis on arboreal quadrupedalism (Dagosto, 1990; Dagosto & Gebo, 1994). Empirical evidence from eosimiid tarsals, which are in many ways morphologically intermediate between those of omomyids and anthropoids, suggests that traditional interpretations of primate postcranial character polarities are correct, and that anthropoids are therefore derived from ancestors that retained many omomyid-like traits (Gebo *et al.*, 2000*a*).

Having eliminated both the adapiform–anthropoid and the prosimian/anthropoid models of anthropoid origins, only hypotheses consistent with the monophyly of Haplorhini remain viable. Haplorhine monophyly is supported by a wide variety of neontological and paleontological data (Martin, 1990; Beard & MacPhee, 1994; Ross *et al.*, 1998; Gebo *et al.*, 2000*b*). Currently, there are two major variants of this hypothesis. The first variant advocates a fundamental dichotomy within Haplorhini between Anthropoidea and Tarsiiformes (Hoffstetter, 1977; Beard *et al.*, 1991; Beard & MacPhee, 1994), and is based primarily on derived features of cranial anatomy shared by omomyids, microchoerids and tarsiers (Rosenberger, 1985; Beard & MacPhee, 1994). The second variant holds that anthropoids and tarsiers share a more recent common ancestor with each other than either group does with omomyids and/or microchoerids (Cartmill, 1980; Cartmill *et al.*, 1981; Ross, 1994; Ross *et al.*, 1998). The strict tarsier/anthropoid clade is based primarily on similarities in the morphology of the auditory region and the postorbital septum. However, several researchers remain unconvinced that these similarities are actually homologous (Simons & Rasmussen, 1989; Beard & MacPhee, 1994; MacPhee *et al.*, 1995*a*). A strict tarsier/anthropoid clade is also difficult to reconcile with available postcranial data (Dagosto *et al.*, 1999). As such, current data accord best with the hypothesis that the sister group of Anthropoidea is Tarsiiformes, the two clades having been separate since the beginning of the Eocene.

A consensus regarding the phylogenetic relationships among basal anthropoids themselves is also lacking (Ross *et al.*, 1998). Eosimiids certainly appear to occupy an extremely basal phylogenetic position with respect to other living and fossil anthropoids (Beard *et al.*, 1994, 1996; MacPhee *et al.*, 1995*a*; Ross *et al.*, 1998; Gebo *et al.*, 2000*a*). Eosimiids are plausibly the sister group of all other living and fossil anthropoids described to date. The only African forms that could potentially be in such a basal phylogenetic position among anthropoids are *Afrotarsius* and the problematic *Altiatlasius*. Indeed, some workers regard *Afrotarsius* as possibly being derived from eosimiids (Ross *et al.*, 1998). The second major Asian radiation of basal anthropoids, the Amphipithecidae, also appears to occupy a very basal phylogenetic position. Possibly, eosimiids and amphipithecids together comprise a basal, Asian clade of early anthropoids that left no living descendants. Alternatively, they may form successive sister groups to the clade that includes the more advanced anthropoid taxa: the parapithecids, proteopithecids, oligopithecids, platyrrhines and catarrhines.

Although the African fossil record of the critical early Cenozoic is scant, an Asian origin for the anthropoid clade is the simplest biogeographic hypothesis given available data. Undoubted tarsiiforms, the likely sister group of anthropoids, have never been recovered from Africa, whereas both tarsiid and omomyid primates are known from Asia (Beard *et al.*, 1994; Beard, 1998*b*). Similarly, two of the most basal anthropoid clades, Eosimiidae and Amphipithecidae, are so far restricted to Asia. Additional fossils from the early Cenozoic (particularly the early Eocene) will be needed in order to rule out either Asia or Africa as the ancestral homeland for all anthropoids.

It now seems inescapable that the vast morphological, biochemical and behavioral differences that distinguish modern anthropoids from prosimians evolved in mosaic fashion over millions of years, rather than as a package. For example, eosimiids evolved highly diagnostic anthropoid features of dental and jaw anatomy well before their mandibular symphyses fused (Beard *et al.*, 1996). These same basal anthropoids retained certain omomyid-like features in their tarsal bones, which in other respects were very anthropoid-like (Gebo *et al.*, 2000*b*). Although skull fragments preserving the postorbital region have yet to be recovered for eosimiids, they probably retained a very microchoerid-like ear region (MacPhee *et al.*, 1995*a*). Even the basal anthropoids from locality L-41 in the Fayum retain certain primitive features, such as unfused mandibular symphyses and lower molar paraconids, that their modern relatives lack (Simons, 1989, 1992; Simons & Rasmussen, 1994*b*).

Documenting mosaic evolution among early anthropoids is hardly surprising, partly because this mode of character evolution has long been assumed (Gingerich, 1980*b*; Rosenberger & Szalay, 1980). However, this does not necessarily imply that complete skulls are required to tell early anthropoids apart from their prosimian relatives (Rasmussen, 1994; Simons & Rasmussen, 1994*b*; Simons *et al.*, 1994;

Simons, 1995*b*). We still do not know when such distinctive cranial features as postorbital closure and anthropoid-like ear regions evolved. Possibly, these highly diagnostic traits evolved relatively recently in anthropoid phylogeny, in a manner analogous to the relatively late evolution of increased cranial capacity in hominids. The whole issue is moot, however, because it remains a relatively trivial task to distinguish eosimiids and other basal anthropoids from their prosimian relatives on dental criteria alone. While more nearly complete fossils of eosimiids and other basal anthropoids will be required to determine the sequence in which key anthropoid characters evolved, they are not required merely to recognize their anthropoid affinities.

Acknowledgements

Many of my ideas on early anthropoid phylogeny could only have been generated through access to specimens derived from collaborative field work in China. I therefore extend my deepest appreciation to the many colleagues who have participated in these collaborative expeditions through the years, including M. R. Dawson, D. L. Gebo, M. Godinot, Guo Jianwei, Huang Xueshi, J. R. Kappelman, R. D. E. MacPhee, Qi Tao, A. R. Tabrum, Tong Yongsheng, Wang Banyue, Wang Jingwen, and others. I also thank J.-J. Jaeger and S. Ducrocq for providing images of important basal anthropoid taxa from southeast Asia. Many colleagues, especially M. Dagosto, J.G. Fleagle, D. L. Gebo, M. Godinot and R. F. Kay, have discussed issues relating to anthropoid origins with me through the years, and I thank them all. M. Klingler expertly rendered the figures. Financial support was provided by grants from the L. S. B. Leakey Foundation and the National Science Foundation (SBR 9615557).

Primary references

Afrotarsius

Simons, E. L. & Bown, T. M. (1985). *Afrotarsius chatrathi*, new genus, new species: first tarsiiform primate (?Tarsiidae) from Africa. *Nature*, **313**, 475–477.

Algeripithecus

Godinot, M. & Mahboubi, M. (1992). Earliest known simian primate found in Algeria. *Nature*, **357**, 324–326.

Godinot, M. & Mahboubi, M. (1994). Les petits primates simiiformes de Glib Zegdou (Éocène inférieur à moyen d'Algérie). *Comptes rendus de l'Académie des sciences, Paris*, **319**, 357–364.

Amphipithecus

Colbert, E. H. (1937). A new primate from the upper Eocene Pondaung Formation of Burma. *American Museum Novitates*, **951**, 1–18.

Ciochon, R. L., Savage, D. E., Tint, T., & Ba Maw (1985). Anthropoid origins in Asia? New discovery of *Amphipithecus* from the Eocene of Burma. *Science*, **229**, 756–759.

Jaeger, J. J., Soe, A. N., Aung, A. K., Benammi, M., Chaimanee, Y., Ducrocq, R. M., Tun, T., Thein, T., & Ducrocq, S. (1998). New Myanmar middle Eocene anthropoids: an Asian origin for catarrhines? *Comptes rendus de l'Académie des sciences, Paris*, **321**, 953–959.

Apidium

Osborn, H. F. (1908). New fossil mammals from the Fayum Oligocene, Egypt. *Bulletin of the American Museum of Natural History*, **24**, 265–272.

Simons, E. L. (1962). Two new primate species from the African Oligocene. *Postilla*, **64**, 1–12.

Simons, E. L. (1995*a*). Crania of *Apidium*: primitive anthropoidean (Primates, Parapithecidae) from the Egyptian Oligocene. *American Museum Novitates*, **3124**, 1–10.

Fleagle, J. G. & Simons, E. L. (1995). Limb skeleton and locomotor adaptations of *Apidium phiomense*, an Oligocene anthropoid from Egypt. *American Journal of Physical Anthropology*, **97**, 235–289.

Arsinoea

Simons, E. L. (1992). Diversity in the early Tertiary anthropoidean radiation in Africa. *Proceedings of the National Academy of Sciences of the United States of America*, **89**, 10743–10747.

Bahinia

Jaeger, J. J., Thein, T., Benammi, M., Chaimanee, Y., Soe, A.N., Lwin, T., Tun, T., Wai, S., & Ducrocq, S. (1999). A new primate from the middle Eocene of Myanmar and the Asian early origin of anthropoids. *Science*, **286**, 528–530.

Biretia

Bonis, L. de, Jaeger, J. J., Coiffat, B., & Coiffat, P. E. (1988). Découverte du plus ancien primate catarrhinen connu dans l'Éocène supérieur d'Afrique du Nord. *Comptes rendus de l'Académie des sciences de Paris*, **306**, 929–934.

Eosimias

Beard, K. C., Qi, T., Dawson, M. R., Wang, B., & Li, C. (1994). A diverse new primate fauna from middle Eocene fissure-fillings in southeastern China. *Nature*, **368**, 604–609.

Beard, K. C., Tong, Y., Dawson, M. R., Wang, J., & Huang, X. (1996). Earliest complete dentition of an anthropoid primate from the late middle Eocene of Shanxi Province, China. *Science*, **272**, 82–85.

Tong, Y. (1997). Middle Eocene small mammals from Liguanqiao Basin of Henan Province and Yuanqu Basin of Shanxi Province, central China. *Palaeontologica Sinica* (Ser. C), **26**, 1–256.

Gebo, D. L., Dagosto, M., Beard, K. C., Qi, T., & Wang, J. (2000*b*). The oldest known anthropoid postcranial fossils and the early evolution of higher primates. *Nature*, **404**, 276–278.

Parapithecus

Schlosser, M. (1911). Beiträge zur Kenntnis der oligozänen Land-

säugetiere aus dem Fayum (Ägypten). *Beiträge zur Paläontologie und Geologie Österreich–Ungarns und des Orients*, **24**, 51–167.
Simons, E. L. (1974). *Parapithecus grangeri* (Parapithecidae, Old World higher primates): new species from the Oligocene of Egypt and the initial differentiation of Cercopithecoidea. *Postilla*, **166**, 1–12.
Simons, E. L. (1986). *Parapithecus grangeri* of the African Oligocene: an archaic catarrhine without lower incisors. *Journal of Human Evolution*, **15**, 205–213.

Pondaungia

Pilgrim, G. E. (1927). A *Sivapithecus* palate and other primate fossils from India. *Memoirs of the Geological Survey of India (Palaeontologica Indica)*, **14**, 1–26.
Ba Maw, Ciochon, R. L., & Savage, D. E. (1979). Late Eocene of Burma yields earliest anthropoid primate, *Pondaungia cotteri*. *Nature*, **282**, 65–67.
Jaeger, J. J., Soe, A. N., Aung, A. K., Benammi, M., Chaimanee, Y., Ducrocq, R. M., Tun, T., Thein, T., & Ducrocq, S. (1998). New Myanmar middle Eocene anthropoids. An Asian origin for catarrhines? *Comptes rendus de l'Académie des sciences, Paris*, **321**, 953–959.
Chaimanee, Y., Thein, T., Ducrocq, S., Soe, A. N., Benammi, M., Tun, T., Lwin, T., Wai, S., & Jaeger, J. J. (2000*a*). A lower jaw of *Pondaungia cotteri* from the late middle Eocene Pondaung Formation (Myanmar) confirms its anthropoid status. *Proceedings of the National Academy of Sciences of the United States of America*, **97**, 4102–4105.
Takai, M., Shigehara, N., Tsubamoto, T., Egi, N., Aung, A. K., Thein, T., Soe, A. N., & Tun, S. T. (2000). The latest middle Eocene primate fauna in Pondaung area, Myanmar. *Asian Paleoprimatology*, **1**, 7–28.

Proteopithecus

Simons, E. L. (1989). Description of two genera and species of late Eocene Anthropoidea from Egypt. *Proceedings of the National Academy of Sciences of the United States of America*, **86**, 9956–9960.
Miller, E. R. & Simons, E. L. (1997). Dentition of *Proteopithecus sylviae*, an archaic anthropoid from the Fayum, Egypt. *Proceedings of the National Academy of Sciences of the United States of America*, **94**, 13760–13764.
Simons, E. L. (1997). Preliminary description of the cranium of *Proteopithecus sylviae*, an Egyptian late Eocene anthropoidean primate. *Proceedings of the National Academy of Sciences of the United States of America*, **94**, 14970–14975.
Simons, E. L. & Seiffert, E. R. (1999). A partial skeleton of *Proteopithecus sylviae* (Primates, Anthropoidea): first associated dental and postcranial remains of an Eocene anthropoidean. *Comptes rendus de l'Académie des sciences, Paris*, **329**, 921–927.

Qatrania

Simons, E. L. & Kay, R. F. (1983). *Qatrania*, new basal anthropoid primate from the Fayum, Oligocene of Egypt. *Nature*, **304**, 624–626.
Simons, E. L. & Kay, R. F. (1988). New material of *Qatrania* from Egypt with comments on the phylogenetic position of the Parapithecidae (Primates, Anthropoidea). *American Journal of Primatology*, **15**, 337–347.

Serapia

Simons, E. L. (1992). Diversity in the early Tertiary anthropoidean radiation in Africa. *Proceedings of the National Academy of Sciences of the United States of America*, **89**, 10743–10747.

Siamopithecus

Chaimanee, Y., Suteethorn, V., Jaeger, J. J., & Ducrocq, S. (1997). A new late Eocene anthropoid from Thailand. *Nature*, **385**, 429–431.
Chaimanee, Y., Khansubha, S., & Jaeger, J. J. (2000*b*). A new lower jaw of *Siamopithecus eocaenus* from the late Eocene of Thailand. *Comptes rendus de l'Académie des sciences, Paris*, **323**, 235–241.

Tabelia

Godinot, M. & Mahboubi, M. (1994). Les petits primates simiiformes de Glib Zegdou (Êocène inférieur à moyen d'Algérie). *Comptes rendus de l'Académie des sciences, Paris*, **319**, 357–364.

10 | Platyrrhine paleontology and systematics: The paradigm shifts

ALFRED L. ROSENBERGER

Prologue

The risk I take in this essay is that of a critic. Ultimately, my intention is to shed more light on platyrrhine evolution and on problem areas where we may have gone astray because of method. Accomplishing this without being critical is difficult.

While debates about primate phylogeny are real, and scientifically healthy, a philosophical context veils the circumstances. First, the posture of our contemporary systematic literature is to advocate positions rather than elucidate hypotheses by conjecture and refutation. Debates about phylogeny are often entwined in a web of taxa and traits which *requires* an exegesis if it is to be properly understood. Second, with automated tree-building, and a limited capacity to objectively select the "truth" from among many potential tree-solutions, we are flooded with hypotheses that are methodologically immunized from rejection, so they are presented as viable models for the sake of consistency. Peer review and editorial direction can easily change the *status quo* and move us toward a more constructive dialogue.

My sense is that "knowing" the phylogeny of platyrrhines is within our grasp for several interconnected reasons. First, the empirical evidence reveals a surprising number of long-lived lineages (Rosenberger, 1979*b*; Delson & Rosenberger, 1984), which should make easier the job of reconstructing history. Second, from an analytical perspective, the living forms are known to comprise a number of high-level clades that are morphologically coherent, distinctive and derived in pattern, whose behavioral ecology is also fairly well known. To stereotype them: pitheciins are dentally bizarre seed-eaters; atelines are postcranially modified climbers; callitrichines are dentally specialized, small-bodied claw-clinging locomotors; cebines are large-headed, predaceous, frugivorous omnivores. Thus, we are dealing for the most part with what we might call a "shallow phylogeny", a radiation where solid knowledge of the living can be extended to the past in order to maximize our interpretive capacity.

A third reason why we should have confidence in the interpretation of fossil platyrrhines comes from the force of genetic evidence. Schneider & Rosenberger (1996) have stressed the satisfying congruence in the results of morphological and molecular studies of platyrrhine cladistics. While I would not go so far as Fleagle (2000) did in endorsing our position, the independent corroboration of many facets of the cebid–atelid cladistic model (Rosenberger, 1981*b*, 1984, 1992) more than confirms the branching sequence for the living forms. It validates many of the characters used to generate the tree, and these are eminently applicable to fossils.

Nearly the entire literature on platyrrhine higher phylogeny and systematics over the past 20 years is slanted toward cladistics. Rare is the paper that eschews PAUP (Phylogenetic Analysis Under Parsimony: Swofford, 1993), but some of the best analyses of fossils (e.g., Meldrum & Kay, 1997*a*, on *Nuciruptor*) prove not to need algorithms at all. I believe there are some systematic problems that can benefit from a judicious use of numerical cladistics. But these methods are not easily applied to the broad sweep of morphological characters we are used to dealing with in primate systematics. The promise that large data sets and parsimony algorithms would bring greater objectivity to systematics has not been realized, for that pivotal series of decisions upon which all else is based – character selection – is by definition a subjective, idiosyncratic process, often rooted in experience and training.

Cladistics in a strict sense narrowly defines "phylogeny" as a branching sequence, as recency of common ancestry, or a network of collateral relationships. Among fossil New World monkeys there is an excellent opportunity to find examples of true phyletic evolution, ancestral–descendant relationships. Their importance to the story of platyrrhine evolution is being misread if ancestors and descendants are simply labeled sister taxa without further inquiry. Several likely generic lineages have already been identified (Delson & Rosenberger, 1984; Setoguchi & Rosenberger, 1987; Rosenberger, 1979*b*). Indeed, the preponderance of long-lived lineages seems to be a high-level evolutionary pattern among platyrrhines. The fossils themselves beckon a broader set of questions, and methods suitable to a more inclusive phylogenetic enterprise. Neither cladistic analysis nor molecular systematics can help us retrieve the entire story. In my view, the non-automated approach to morphology and character analysis that pays particular attention to homology, polarity, character weighting, functional morphology, behavior, etc., is a superior methodology. It stands up well against parsimony routines based on any form of data. In fact, the broad confirmation emerging from the genetic evidence (simple characters for which I think parsimony algorithms work well) of the most important cladistic hypotheses emerging from morphology (e.g., Schneider & Rosenberger, 1996) should be viewed as a scientific triumph for both approaches.

A turning point

Philip Hershkovitz's *Living New World Monkeys (Platyrrhini)* (1977) is sometimes described as the most important reference work on platyrrhine systematics of the last 200 years. With more than 1000 pages and over 2500 references, the book is famous for its ultra-encyclopedic account of nomenclature and place-name geography. For a work of such extraordinary dimensions by a man of stunning ability, ambition and purpose, Hershkovitz's book (1977) was oddly self-limiting. It hardly dealt with fossils. By 1977, eight fossil genera were described, two recently named by Hershkovitz himself, *Stirtonia* in 1970 and *Tremacebus* in 1974. In his opening, Hershkovitz stated that there were then no known callitrichine fossils, and he may have set aside the non-callitrichine fossils for Volume 2, or even a third volume. But the short shrift he gave paleontology, in a book of this scope, symbolizes a turning-point. It marks the end of an era when platyrrhine systematics could confidently advance without being fully informed by the fossil record.

In even more dramatic fashion, Hershkovitz (1977) epitomized the last gasp of non-synthetic, gradistic thinking (Rosenberger, 1980, 1981*b*). The sheer volume of information he assembled on callitrichines and other platyrrhines precipitated a crisis in theory, because there was none. In his effort to touch upon all things platyrrhine, ranging from Hershkovitz's own morphological forte to reproductive physiology, ontogeny, behavior, disease, parasites, nutrition, feeding habits, locomotion, longevity, social organization, mating strategies, growth, cognition, and more, the book was virtually born an intellectual white elephant – gigantic, obvious, going nowhere and anxious to be spoken for. It was hardly likely that Hershkovitz's model of platyrrhine evolution, summarized in an opening paragraph, could *explain* patterns embedded in such a huge body of information.

An integrating idea and a robust method were needed to mold key pieces of the data into a testable evolutionary model. This occurred as other advances were taking place in the late 1960s and 1970s, and included major refinements in systematic methodology, growing knowledge of the form and function of modern and early primate morphology, exploding awareness of primate behavior and ecology, and evolutionary models that explained evolutionary differentiation and diversity. The intellectual mix revived a profound challenge to Hershkovitz's (1977) central organizing hypothesis, that callitrichines were primitive. The factual bases for this observation were all part of the book, but they were misinterpreted (apparently). So, it is a coincidence brought about by time rather than epiphany which has this masterwork punctuating the first major reformation in the scientific study of the New World monkeys.

A century-and-half of fossils

The history of platyrrhine paleontology was a quiet one until the late 1960s and 1970s. The early history of platyrrhine paleontology was dominated by one fossil species, *Homunculus patagonicus*, based on a small collection of craniodental and postcranial specimens (Ameghino, 1891*a*, 1891*b*). Although subfossils were already known from the Lagoa Santa caverns of Brazil for several decades (Lund, 1838), only two other significant discoveries were made in the entire continent of South America in the first half of the twentieth century. Both *Tremacebus harringtoni* and *Dolichocebus gaimanensis* were overinterpreted as congeners of *Homunculus* (Rusconi, 1933; Bordas, 1942) until they were recognized as generically distinct (Kraglievich, 1951; Hershkovitz, 1974). With so little known, non-specialists such as Simpson and Gregory, whose brilliance contributed much to contemporary thinking on primates, also saw platyrrhine evolution through the lens of *Homunculus*.

The breakthrough of the twentieth century was Ruben Arthur Stirton's discovery of many vertebrate fossils at the middle Miocene site of La Venta (Stirton, 1951; Stirton & Savage, 1951; Kay *et al.*, 1997*a*). Given the nature of this material, even the first analysis was hardly hampered by the *Homunculus* specter. Stirton and Donald Elvin Savage's major primate discoveries were of taxa remarkably similar to living forms, *Neosaimiri fieldsi* and *Cebupithecia sarmientoi*.

At about the same time, *Xenothrix mcgregori* was released from 30 years of anonymity in a scrap-drawer of unallocated bones and diagnosed a platyrrhine (Williams & Koopman, 1952). The extraordinary discovery of extinct New World monkeys from Jamaica opened up a new geographical dimension in the evolutionary history of primates, and a chapter that grows more and more interesting with new fossils (MacPhee, 1996; see MacPhee & Horovitz, this volume). By the 1950s, after over a century of platyrrhine paleontology, the few fossils still had no appreciable impact on platyrrhine classification, systematics, or historical reconstruction.

The modern era of platyrrhine paleontology and systematics arose in the years bracketed by the publication in 1969 of the Oligocene *Branisella boliviana*, by Robert Hoffstetter, a seasoned paleontologist who knew the big questions, and by the appearance of Hershkovitz's book, in 1977. *Branisella*, more than any fossil found before, hinted at something different, early South American primates that might bridge the morphological gap between platyrrhines and Fayum catarrhines, or early anthropoids, or even North American omomyids. *Branisella* seemed relevant to platyrrhine origins, and the new biogeography based on continental drift and plate tectonics.

The rate of discovery of fossils increased dramatically in the 1980s and 1990s, with increasing interest, more participants, new ideas and a wealth of related research. The symbolic importance here was that fossil platyrrhines had finally become a field of study in their own right (Fleagle & Rosenberger, 1990). The scientific importance was that the foundation data was being multiplied at an unprecedented rate. Something of a critical mass was achieved in the early 1990s, when the number of fossil genera eclipsed the tally of living genera.

The paradigms shift

The first critical insight after Hershkovitz (1977) was that callitrichines were a sister group to a living platyrrhine clade. That clade could be found by tracing back apomorphies that callitrichines shared at least one link up or down a platyrrhine cladogram. This notion complemented the finding that non-marmosets could not be held together as a monophyletic group, which freed up both cebine genera, *Cebus* and *Saimiri* – the non-marmosets most like callitrichines, with short faces, reduced rear teeth, shallow jaws and gracile zygomatic arches – as possible relatives. Together, these ideas clinched the need for a wholesale revision in thinking. The cebid–atelid model began to replace the cebid–callitrichid schema (Rosenberger, 1981*b*). New ingredients fomented this revolution in the form of character analyses.

Rosenberger (1979*b*, 1981*b*; see also Szalay & Delson, 1979) and Susan Marie Ford (1986*a*, 1986*b*) were the first to present explicit, modernistic studies of platyrrhine interrelationships based on the morphocline polarity of characters. Fossils provided critical temporal evidence in character analyses and, secondarily, as heuristic checks on the relationships between groups whose modern relatives were morphologically divergent, such as *Cebus* and *Saimiri*, and *Aotus* and *Callicebus*. A feeding–foraging model was also proposed to explain the diversity of platyrrhines (Rosenberger, 1980, 1992) as an array of lineages inhabiting different adaptive zones and further differentiating along different adaptive modalities.

To better evaluate the correspondence between these hypotheses I developed a simple method to calculate a correlation coefficient that reflects congruency of the summary cladograms. The approach was inspired by methods used to test the match between cladisitc hypotheses and stratigraphic evidence from the fossils record (see Benton, 1998). Nodes of the cladograms (Fig. 10.1) were systematically numbered to reflect their branching order, thus ascribing a numerical designation for each monophyletic group. Trichotomies were treated the same as a dichotomous branching. Pair-wise comparisons of each of the genera, which for the 16 living platyrrhines amount to 120 entries, then defined the last common ancestors they shared. For example, in the Rosenberger tree: *Cebuella*:*Callithrix* = 7, *Cebuella*:*Leontopithecus* = 6 ... *Cebuella*:*Cacajao* = 1 ... *Cebuella*:*Alouatta* = 1, etc. The resulting table is a summary of all the cladistic relationships depicted in the tree. Matrices were developed for six studies and correlation coefficients were computed to compare their correspondence.

The results are instructive (Table 10.1). The Schneider (2000) and Horovitz (1999) cladograms are virtually identical, as expected, with a coefficient of 0.997. This is a welcome result, in part, because Horovitz used Schneider's data on four nuclear genes, to which she added mtDNA data and a large number of morphological characters analyzed simultaneously. (This part of the Horovitz study involved the 16 living genera only, not fossils; see below.) The close congruence between the two Ford studies (1986*b*), with a correlation coefficient of 0.842, compares two of the several possible permutations of her base tree, which presented two alternative cladistic positions for *Saimiri* and *Ateles*. This value is essentially the same as the correlations (0.849, 0.842) between Rosenberger's tree (1984, 1992), based on morphology, and the results of Schneider and of Horovitz. Yielding coefficients between 0.284 and 0.484, the least congruous of all the trees is Kay's (1990), which is radically different in both high-level branches and lower, genus-to-genus nodes.

I believe an important lesson that we can draw from this relates to characters. A common aspect of the projects of Ford (1986*b*), Horovitz (1999) and Rosenberger (1981*b*, 1992) is that they combined data from different systems. Kay's study (1990), on the other hand, used dental characters. An earlier project by Ford (1986*a*), which yielded different results from her synthetic studies, was also based on a several joint complexes of a single system, the postcranium. Both these efforts produced many characters, but the results of neither have been well replicated.

Lessons from a character tree

Differences of opinion regarding the phylogenetic relationships of several fossils relate to contrasting views on the valence of single-system characters. That is what underscores the debate between Kay (1990) and Rosenberger *et al.* (1990) over *Mohanamico* and *Aotus dindensis*. A bias toward dentition seems also to have influenced Horovitz's (1999) study of all the platyrrhine fossils, which turned into a "character tree" instead of a cladogeny (Fig. 10.2). She showed that the cladistic relationships of the extant genera can be retrieved reasonably well with molecular and morphological data when the latter is composed of craniodental characters. However, the same data set and algorithms did a rather poor job of allocating fossils to clades. The reason for this may be that the dental characters dominated the matrix and drove the results.

The anomalous cladistic linkages of *Neosaimiri*/*Tremacebus* and *Dolichocebus*/*Soriacebus ameghinorum* are instructive (Table

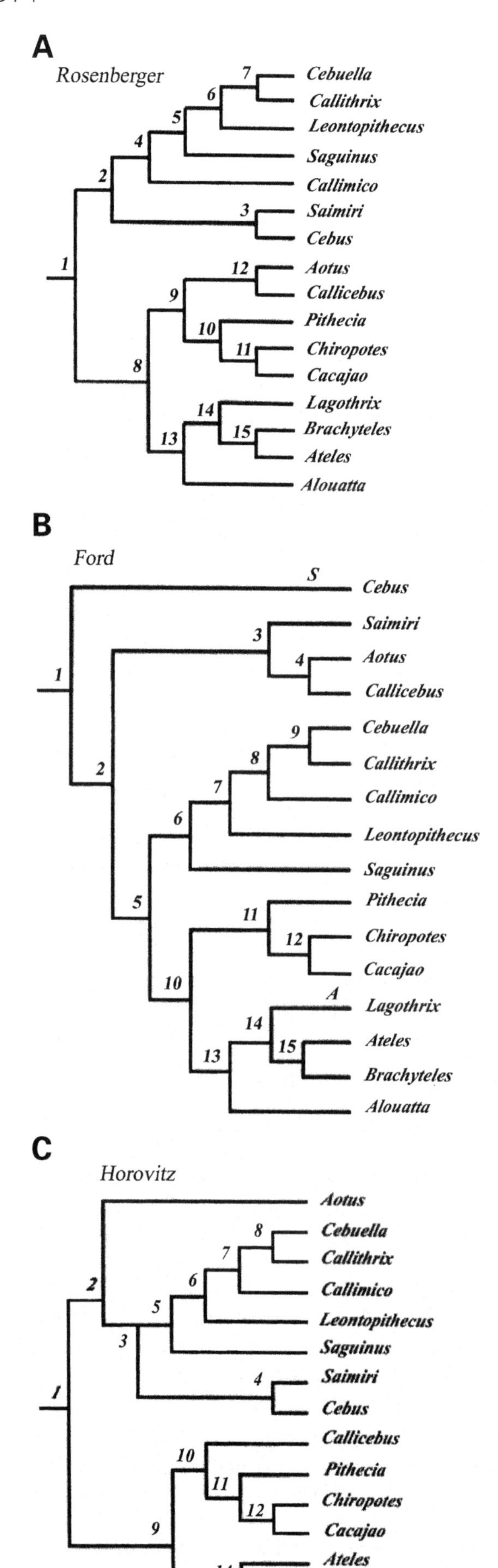

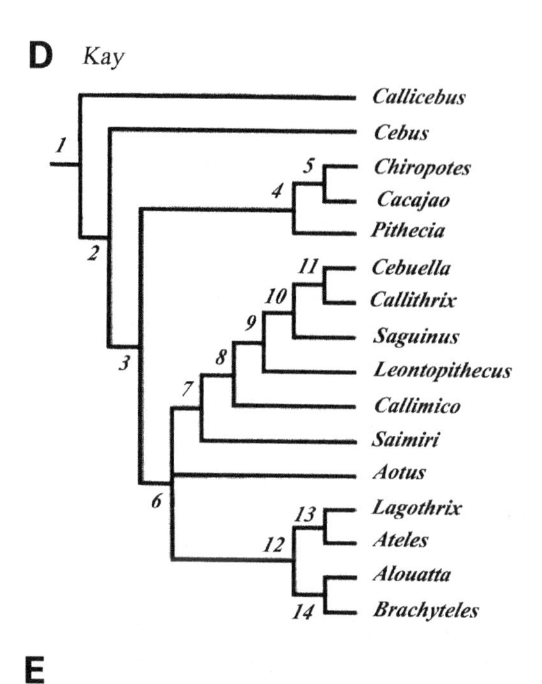

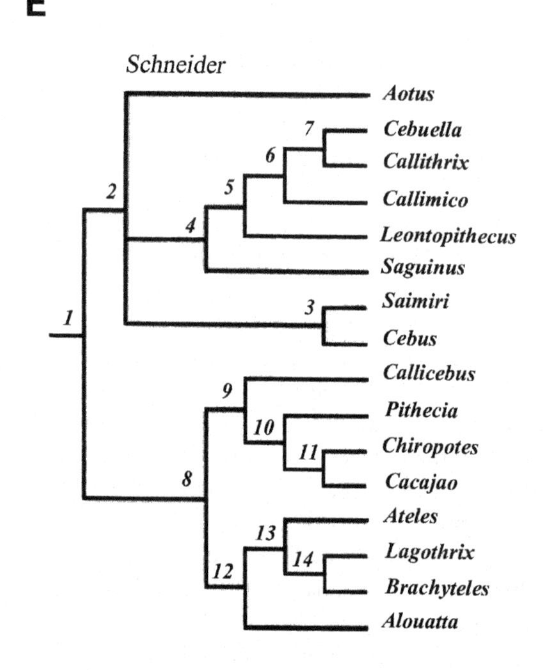

10.1 Cladograms of major morphological and molecular studies of platyrrhine interrelationships. Numbers at each node were used to calculate a correlation coefficient between trees as a measure of their correspondence. See text. The S and A in Ford's tree refer to alternative positions of *Saimiri* and *Ateles*, respectively.

Table 10.1. Correlation coefficients measuring the correspondence of cladograms

	Kay	Ford A	Ford B	Schneider	Horovitz
Rosenberger	0.279	0.662	0.634	0.849	0.842
Kay		0.484	0.406	0.288	0.284
Ford A			0.843	0.748	0.732
Ford B				0.757	0.744
Schneider					0.997

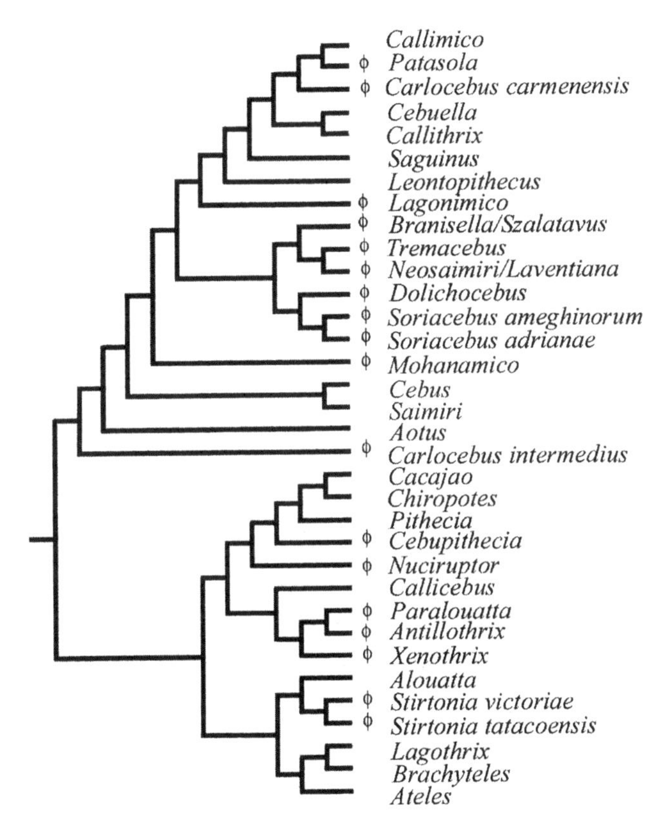

Fig. 10.2 Horovitz's (1999) cladogram of living and fossil (ϕ) platyrrhines, using her genus-level taxonomy.

10.2, Fig. 10.2). Fifty-five of the 86 characters (64%) used in Horovitz's (1999) analysis were dental features, so at the outset it would appear likely that teeth would weigh heavily in the analysis. For the *Neosaimiri/Tremacebus* link, it is evident that premolar characters would be dominating followed by the molars, because of the high frequency of these features in *Neosaimiri*, in contrast with only one premolar trait and two molar states that could be scored in *Tremacebus*. *Neosaimiri*, in fact, shows a disproportionately high frequency of derived premolar features in this study. For the *Dolichocebus/S. ameghinorum* pairing, there was much greater equality in the distribution of "informative" features. Here the question is more a matter of the believability of results rather than an expectation of inherent data bias (see below). But what is also interesting is that there is substantial fossil cranial

Table 10.2. Distribution of derived dental character states from Horovitz's (1999) matrix

	Incisors	Canines	Premolars	Molars
Neosaimiri	0	3	14	10
Tremacebus	0	0	1	2
Dolichocebus	0	3	4	5
Soriacebus ameghinorum	2	3	5	7
Total Study Group (n = 41)	7	9	20	21

material for one of the taxa in each of these sets that was thoroughly overshadowed by the quantity of dental evidence representing the other form.

Tremacebus is known from the type skull and a questionably allocated, damaged jaw from another locality, and there is no skull known for *Neosaimiri*. Since there was little anatomical overlap between these taxa in the input matrix (Table 10.2), there could not be much that would support this cladistic result directly. The orientation of the M_1 cristid obliqua proved to be the only derived feature sustaining this node. One obvious question is: Why should we have confidence in this result when the cranium of *Tremacebus* presents far more compelling data (e.g., Fleagle & Rosenberger, 1983)?

There are 29 potential cranial characters in the matrix that might drive the placement of *Dolichocebus* but none aligns it with *Tremacebus*, even as both supposedly belong to the same monophyletic group. *Dolichocebus* and *Tremacebus* present two of the best-known crania among the fossil platyrrhines, and cranial characters have been shown to be informative regarding New World monkey relationships. All have apparently been swamped by the dental data. Why? Perhaps it is the nature of the cranial traits selected. Three are listed as shared derived in *Dolichocebus*, the shape of the ectotympanic, cranial capacity and infraorbital foramen position. None could be scored for *Soriacebus*. Nor are any of the features proposed by Rosenberger (1979b) as shared derived features linking *Dolichocebus* with *Saimiri* and other cebines listed in the Horovitz data set.

I believe these results are methodological artifacts – a character tree not based on homology – rather than a reconstruction of relationships. This is not necessarily an objectionable result, so long at it corresponds with other evidence or presents a heuristically valuable perspective. Horovitz (1999) proposes a clade of: (((*Soriacebus*, *Dolichocebus*) ((*Neosaimiri*, *Tremacebus*) *Branisella*))). Nothing in the literature seems to support this. The single derived feature that holds this branch together is a P^4 hypocone. Putting this into other terms, it suggests that from a full range of craniodental features found distributed in this adaptive radiation of at least five genera, a premolar cusp is the only one that can explain differentiation. Nowhere among extant platyrrhines is a

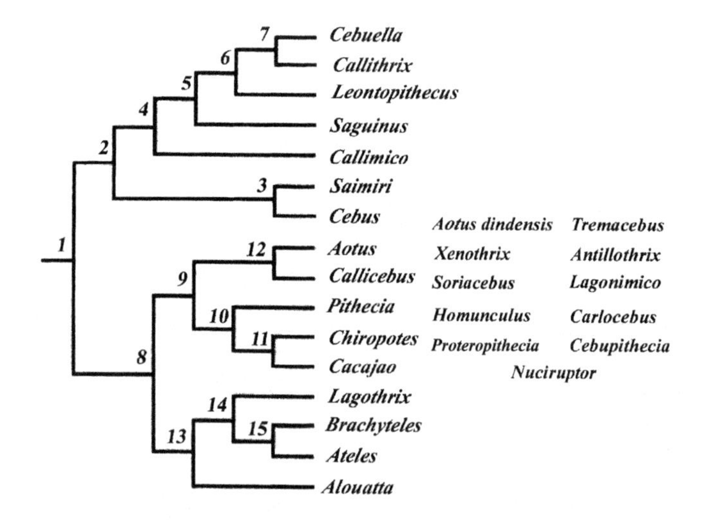

Fig. 10.3 Rosenberger's cladogram of living platyrrhine genera with a list of fossil taxa assigned to pitheciines. *Aotus dindensis* is an extinct species of *Aotus*.

clade of similar diversity supported by such a narrow anatomical basis.

Lessons from the pitheciines

Pitheciines are a major fraction of the platyrrhine radiation (Fig. 10.3). Kay and colleagues (Kay, 1990; Meldrum & Kay, 1997*a*; Kay *et al.*, 1998*a*) recognize three fossil genera, and I recognize eleven, plus five living – saki-uakaris, owl and titi monkeys. All of these fossils, in one way or another, share craniodental and mandibular characters that are typically found only in pitheciins or pitheciins and the *Callicebus–Aotus* clade among the living. The inclusion of *Callicebus* in this group seems to be accepted by all at this point, as suggested by the cladograms illustrated above (Fig. 10.1), but the implications of this for character analysis have yet to be broadly appreciated. The status of *Aotus* is also an interesting aspect of the pitheciine story.

The key to interpreting pitheciines comes from two directions. One is that the dental morphology of pitheciins is virtually self-polarizing, meaning it is so different from other platyrrhines that all workers agree the flat-basined cheek teeth, stylate incisors and pyramidal canines are homologous and derived among New World monkeys. The other clues come from molecules (e.g., Horovitz, 1999; von Dornum & Ruvolo, 1999; Schneider, 2000). They are unanimous in supporting saki-uakari monophyly and an immediate linkage of *Callicebus* as their sister taxon. By extension, the original morphological argument (Rosenberger, 1981*b*, 1984, 1992) which linked *Callicebus* and *Aotus* with saki-uakaris as pitheciines is confirmed. This was based on a dental and mandibular transformation series shared in step-wise fashion by the five genera, and cranial features also shared by *Callicebus* and *Aotus*.

Several numerical cladistic studies have confirmed the affiliation of *Callicebus* and saki-uakaris. Horovitz and colleagues used morphological as well as genetic evidence, in various combinations (e.g., Horovitz & Meyer, 1995; Horovitz *et al.*, 1998; Horovitz, 1999; Horovitz & MacPhee, 1999). Kay (1990), also using a subset of the Rosenberger (1979*b*) dental characters as well as his own, arrived at a very different conclusion regarding *Callicebus* (Fig. 10.1). Neither of these studies included features of the mandible which distinguish all pitheciines. I suspect that is part of the reason why *Aotus*, and for Kay *Callicebus*, failed to group with them. The clustering of *Aotus* with cebines and callitrichines by algorithms (e.g., Horovitz, 1999) based largely on dentition is not surprising, for there are numerous phenetic similarities shared by them in cheek tooth morphology (see also Rosenberger *et al.*, 1990).

The sheer anatomical oddness of saki-uakaris would seem to predict that some type of intermediacy in form and adaptive context would eventually be discovered, probably involving taxa that occupied a transitional adaptive zone and exercised characters preadaptive to the derived saki-uakari patterns. Kinzey (1992) and Rosenberger (1992) have argued from morphologic and behavioral evidence that *Callicebus* and *Aotus* fit this prediction, but there are fossils that fit this picture even better.

With the concept of pitheciines broadened, one is faced with another paleontological dilemma, to select an analytical approach that balances the wealth of data available for the living with the dearth of evidence presented by the fossils. Results so far do not support the implied contention that more anatomical data points can get around a lack of well-understood, taxon-defining features. Although Horovitz (1999) scored a great many dental traits, the mandibular and cranial data did not achieve parity.

Pitheciines show that character weighting is a powerful tool that should not be ignored. The recipes for polarity determination, e.g., parsimony, in-group – out-groups distribution, temporal precedence, ontogeny, function, etc., are valuable perspectives but there are many circumstances when a polarity hypothesis is robust even when ancestral conditions are moot. This is the case for pitheciines. It follows that homology decisions, which I believe are a fundamental precondition to investigating morphocline polarity, can be equally well guided by realizing the power of autapomorphous traits. Any of the stereotypical features of saki-uakaris that are shared with any platyrrhine genus is more likely than not to be homologous. The likelihood of homology increases as unique shared similarities span anatomical/adaptive systems to form a pattern, e.g., from molars, to incisors, to mandibles, to heads, to feet. Some of the present difficulties with parsimony studies of platyrrhine fossils is that this strong, coherent body of data is subdivided into minutia in order to extract individual character states which together form a long list rather than an integrated working unit, thus causing them to lose phylogenetic signal.

Lessons from the record: *Xenothrix*

Xenothrix, long known from a single mandible with only two molar teeth, has been an especially vexatious taxon (Williams & Koopman, 1952; Hershkovitz, 1970; Rosenbeger, 1977; Rosenberger *et al.*, 1990; MacPhee & Fleagle, 1991; MacPhee, 1996; Horovitz, 1999; Horovitz & MacPhee, 1999). The first studies could not place it unambiguously (Williams & Koopman, 1952; Hershkovitz, 1970). As my ideas of platyrrhine phylogeny developed, I became more convinced that *Xenothrix* is a Jamaican pitheciine, possibly most closely related to *Callicebus* among the living. Recently, with new material in hand, Ross D. E. MacPhee and colleagues proposed that *Xenothrix* is part of a monophyletic Antillean clade, Xenotrichidae, whose sister taxon is *Callicebus* (Fig. 10.2). The proposed clade is comprised of *Xenothrix*, *Paralouatta* and *Antillothrix*.

After studying the new material I come to a different conclusion, briefly outlined here. I believe *Xenothrix* is a Jamaican owl monkey most closely related to *Aotus* and *Tremacebus*, which I believe are sister taxa to the *Callicebus* lineage. Among the new mandibles are a few important morphological details that confirm earlier studies and extend a set of derived similarities to include *Aotus*. For example, more is known of the gonial region, which was even more highly expanded (in at least some individuals) than could have been guessed from the type specimen, which was a young animal as MacPhee (1996) noted. In part, this is because the mandibular corpus is extraordinarily deep below the molar region – again, more than the type demonstrates. The ramus of the jaw is anteroposteriorly short, and probably had a significant coronoid process with a deep sigmoid notch. Alveoli in the several jaws also confirm a consistently small lower canine, unlikely to be gender-related.

These features are all consistent with the idea that *Xenothrix* is a pitheciine closely resembling *Callicebus* and *Aotus*. The small canines and deep jaws are probably synapomorphies shared with *Callicebus* and possibly with *Aotus* as well. Two high-weight, derived features link *Xenothrix* and *Aotus*, one dental and one cranial. Regarding the dentition, the first upper incisor alveolus is greatly enlarged in the fossil, relative to the I^2 socket. This is paralleled by a relatively large interalveolar distance separating right and left I^1s. I interpret the morphology as an indication of a greatly broadened I^1 crown, which is a novelty of *Aotus* (Rosenberger, 1992) among the extant platyrrhines.

Of even more importance, the orbit of *Xenothrix* is enlarged, like *Aotus* and *Tremacebus*. This is evident in two ways: (1) The lower anterior rim of the right orbit is preserved in a wide arc that indicates a large orbital diameter. (2) The posterior floor of the orbit preserves intact on both sides evidence of the inferior orbital fissure, an opening between the orbital and temporal fossae related to eyeball hypertrophy. This fissure is typically closed in *Callicebus*, where the orbit is tightly sealed from the temporal fossa. In all important respects, the morphology of this region, including the shape of the maxillary tuberosities, conforms with *Aotus*. One way that *Xenothrix* differs from *Aotus* in orbit-related features is the size of the maxillary sinus, which is secondarily reduced in living *Aotus* and in the fossil *A. dindensis* (Setoguchi & Rosenberger, 1987). The deep face of *Tremacebus*, however, probably reflects a large maxillary sinus, more like *Xenothrix* perhaps.

In the final analysis by Horovitz (1999: Table 2B), three synapomorphies linked *Xenothrix*, *Antillothrix* and *Paralouatta*: nasal fossa width, C_1/P_4 alveolus size, and the M_1 bulging protoconid. How much confidence should we attribute to them? For nasal fossa width, generally speaking, no data could be collected on half the taxa (20/41) in the matrix and the other half were deemed to have the primitive state. For the canine/premolar ratio, homoplastic derived states occur in *Tarsius* and *Carlocebus carmenensis*. This alone would not be a cause for alarm, even though one immediately wonders about homologies. However, *C. carmenensis* falls out as a callitrichine in this analysis (linked with *Callimico* and the fossil *Patasola*) and its congener, *Carlocebus intermedius*, is linked with *Soricacebus* and *Dolichocebus* (see above), leaving me to wonder even more about the signal potential of *Carlocebus* characters, which are all dental.

The features that I worked with in the past (Rosenberger,1977; Rosenberger *et al.*, 1990) helped place *Xenothrix* within a clade for the first time. The analyses of Antillean primates by MacPhee (1995) and MacPhee *et al.* (1995) supported the linkage of *Xenothrix* with *Callicebus* based on small, probably non-projecting canines, occlusal anatomy and the expanded gonial region. With discovery of the face, I now prefer to weight more heavily a new set of characters crucial to the adaptive zone occupied by a taxon closely related to *Callicebus* (see Rosenberger, 1992). Enlarged orbits and eyeballs are the fundamental adaptive breakthrough of owl monkeys, as far as we know. A second character that implies the same phylogenetic interpretation, the inferred size of I^1, may be linked with how a taxon exploits an adaptive zone. In the *Aotus* lineage these involve harvesting adaptations, i.e., the uniquely enlarged central incisors of *Aotus* (Rosenberger, 1992).

Lessons from *Xenothrix*, *Soriacebus* and *Lagonimico*

There is a broader context in which a rethinking of *Xenothrix* has importance for platyrrhine systematics. Studies of *Xenothrix* (Rosenberger, 1977; MacPhee, 1996; Horovitz, 1999) tend to agree that it is a pitheciine on account of its mandibular morphology, inferences about the anterior teeth and premolars, and a few details of occlusal morphology. The conundrum has been this: The odd, two-molar dental formula and elongate molar shapes bear little resemblance to any living pitheciine and, until recently, to all fossils. The reasonable default position has been that the molars are autapomorphous, thus irrelevant to cladogeny and to morphotype reconstruction. However, labeling traits as autapomorphies must always be a provisional statement. Discovery of a new fossil sharing that particular trait forces a revision of its polarity status. This is the pitheciine story.

Extending the above analysis to include non-Antillean fossils brings us a step closer to resolving the analytical challenges of the *Xenothrix* dental formula. The Miocene fossil *Soriacebus*, from Argentina, offers new perspective. Its phylogenetic position is a matter of debate (e.g, Rosenberger, 1992; Fleagle *et al.*, 1997*b*; Meldrum & Kay, 1997*a*; Kay *et al.* 1998*a*; Horovitz, 1999). My view is that *Soriacebus* is a pitheciine, based on a suite of traits of the anterior dentition and mandible that are indisputably pitheciine-like, and a second tier of premolar characters that also align the genus with high confidence among pitheciines. Molar morphology is the most serious source of disagreement (see Fleagle *et al.*, 1987, 1997*b*; Kay, 1990). *Soriacebus* lower molars are long and narrow and otherwise unremarkable, in contrast to saki-uakaris, the modern-looking *Cebupithecia* and *Proteropithecia*, and also of *Nuciruptor*. Another feature to emphasize here is that the upper molars of *Soriacebus* are unusual for an atelid in having a small hypocone, which is a sharp contrast to the traditional way we see the pitheciine pattern, as a quadrate large-hypocone tooth. The long lower molars and reduced upper molar morphology of *Xenothrix* and *Soriacebus* are unexpected similarities, and rather unique.

A second genus, *Lagonimico*, provides yet another clue. It, too, presents an interesting character combination (Kay, 1994), with posteriorly deep jaws, small third molars and upper molars lacking any real trace of a hypocone. Using a parsimony algorithm, Kay interpreted *Lagonimico* as a *Callicebus*-sized callitrichine, emphasizing loss of the hypocone among other features. I believe *Lagonimico* is a pitheciine, in part because of its diagnostically deep mandible and flaring gonial region, and the shape of its ascending ramus. The latter uniquely resembles *Callicebus* and *Aotus* in being tall and narrow anteroposteriorly, with a sloping anterior border and a small sigmoid notch. Related features are also seen in *Soriacebus* and *Xenothrix*. The single troubling feature of *Lagonimico* is the tricuspid upper molar pattern.

All these facets come together in a simple, parsimonious hypothesis that binds together *Xenothrix*, *Soriacebus* and *Lagonimico* as pitheciines. The notion is that there is another dimension to the pitheciine radiation that has only surfaced now. These fossils may reflect the "deep phylogeny" of pitheciines, even if *Xenothrix* (if not the others) is a closer relative of *Aotus* and *Tremacebus*. In some aspects, the three may reflect the morphological antecedents of the morphotype we were able to distill from comparing extant pitheciines, their fossil isomorphs and information from atelines, the pitheciine outgroup. In another respect, they indicate a new evolutionary pattern in which some pitheciines (*Aotus* and *Xenothrix*) parallel *Callimico* and callitrichins, with a transformation series involving reduction and loss of the third molar.

No splendid isolation

Extinct platyrrhines ranged more widely than the present boundaries of the tropical and subtropical forests of South and Central America, proving that the zoogeographic history of New World monkeys is likely to be interesting and complex. Patagonia has yielded about eight platyrrhine genera, none of which are congeneric with the larger Miocene fauna of Colombia. The relationships of some of these taxa are disputed, but I will focus on three. Based on cranial anatomy, I have argued (Rosenberger, 1979*a*; Fleagle & Rosenberger, 1983) that *Dolichocebus* and *Tremacebus* are early representatives of modern lineages, possibly sister taxa or even ancestors of living genera. I think *Dolichocebus* is closely related to *Saimiri* and *Tremacebus* to *Aotus*.

Kay (1990) and Fleagle *et al.* (1997*b*: 482) presented different views of the relationships of these taxa, and they proposed another way of looking at the collection of Patagonian platyrrhines, summarized as follows:

> Miocene Argentine fossil platyrrhines cannot be readily allied with the commonly recognized clades of living platyrrhines. Rather, they are generally more primitive, each possessing a mosaic of similarities to extant platyrrhine taxa that is incongruent with many current phyletic reconstructions based solely on the anatomy of living New World monkeys.

This is one of several ideas presented by Fleagle, Kay and colleagues about the fossils, yet it echoes the persistent theme that there was a distinct platyrrhine radiation in the south (Kay, 1990; Kay *et al.*, 1998*b*). A generous reading might take the meaning of this statement as: None of the Patagonian taxa are cladistically related to extant forms or their close fossil relatives. A more restrictive reading might be: The Patagonian fossils are a monophyletic group outside the crown group of platyrrhines.

Both hypotheses require a high degree of parallel or convergent evolution, implying that: (1) an *Aotus*-like (and *Xenothrix*-like), nocturnal genus or lineage, exemplified by *Tremacebus*, developed analogously enlarged orbits; (2) a *Saimiri*-like interorbital fenestra (or an annexed, *Cebus*- and *Saimiri*-like conformation of the orbit's medial walls if the fenestrated condition should prove to be misinterpreted) evolved twice among platyrrhines, once in *Dolichocebus*; and, (3) a pitheciin-like dental complex involving incisors and canines, related to hard-fruit eating, evolved independently in *Soriacebus*, which also presents other pitheciine characters of inadequately studied functional significance, such as a deep, thin mandible. It stretches the imagination to think so much parallelism could occur among platyrrhines *within* platyrrhines.

A similar case of mistaken monophyly involves the Greater Antillean primate fauna. MacPhee and colleagues (e.g., MacPhee, 1996; Horovitz & MacPhee, 1999) have argued explicitly that the three Antillean primates, *Xenothrix*, *Antillothrix* and *Paralouatta*, are monophyletic, the descendants of a single ancestral population. As discussed above, I believe *Xenothrix* is closely related to *Aotus*. I also have confidence in the original assessment of *Paralouatta* as a howler relative (Rivero & Arredondo, 1991), based on a comprehensive series of derived cranial features seen nowhere else but in *Alouatta*, in spite of differences in dental anatomy. The latter are likely to reflect the "deep phylogeny" of atelines, just as *Soriacebus* morphology, for one, reflects the "deep phylogeny" of pitheciines. Here, again, we must weigh the likelihood that two regionally grouped taxa sharing unique morphological patterns with other adaptively specialized platyrrhines living elsewhere are anything but their cousins. Rather than a monospecific origin for Antillean primates, I think the evidence indicates at least two subfamilies belonging to a community of primates got into the Caribbean. Since the crossing distance is also likely to have been small, more than one colonization is not an improbable scenario. In fact, it is likely that there was more than one emigration out of South America and into Central America. An early dispersal produced the community that eventually arrived in the Caribbean. Another introduced the modern Central American forms after the isthmus arose. Their closest relatives are not the Antillean monkeys but the South American species.

Epilogue

The paradigms that dominated this field have shifted in concept and method, promoted partly by a steadily increasing fossil record. With a broad concurrence by morphological and genetic studies, the cebid–atelid model of phylogeny and classification has gained support as a central organizational theme, with callitrichines, pitheciines (with qualifications) and atelines universally recognized as derived, monophyletic subtaxa. The evidence for cebine monophyly is also increasing. Knowledge of the diversity of the pitheciine lineage has greatly increased, owing to a redefinition of their taxonomic composition based on phylogenetic concepts and the discovery of several new fossil genera that pertain to this group.

Approaches to the study of platyrrhine systematics have drifted toward a reliance on parsimony-based numerical cladistic methods but without improving the state of the underlying comparative morphology, which would probably make these methods more valuable. Teeth, the most plentiful of fossils, may still prevail as the favored source of cladistic information – what Fred Szalay calls the "Tyranny of the Teeth" – but the evidence produced thus far shows that the dentition alone is an unreliable phylogenetic mentor, especially as the homologous parts of fossils are unevenly represented in the record.

It is prudent not to assume that areas now outside the center of gravity of tropical and subtropical South America were major, independent theatres of platyrrhine evolution. There is no evidence for the scenarios which hold that fossil Patagonian primates, and Antillean primates, are separate monophyletic groups. Each area has at least two genera that belong to two separate families, cladistically. To play devil's advocate in this regard one must, at the very least, directly account for the strong suite of derived morphological features that bind together *Paralouatta* and *Alouatta*, and *Tremacebus* and *Aotus*.

Clearly, the systematics of New World monkeys has emerged from a long dormant state dominated by a scarcity of fossils and a shortage of ideas. The future is promising, for the past 20 years has also proven that fossils can be found in taxonomic abundance. Our methods will become better also, particularly as new technologies improve data input by bringing a new level of clarity to comparative morphology. All these trends are already established. To secure its intellectual future, however, we need to encourage more students to take the path of systematics, particularly in the countries of South and Central America where the living species, and the extinct, are home.

Acknowledgements

For asking me to write an essay for this volume rather than a "paper", I thank Walter Hartwig. Robert Costello contributed to the development of the arguments. Thanks to Richard Kay and Ross MacPhee for making it possible to see fossils under their care, and to a host of colleagues and museums for continuing to facilitate study of collections.

11 | Early platyrrhines of southern South America

JOHN G. FLEAGLE AND MARCELO F. TEJEDOR

Introduction

The fossil record of New World monkeys in southern South America extends from the Bolivian Altiplano at Salla, 15° below the equator, to numerous surfside localities in southernmost Argentina barely north of Tierra del Fuego. Within this vast area are 15 more or less distinct sites that have yielded close to 300 platyrrhine fossils, placed in eight genera and ten species. The vast majority of these are in the southern part of Argentina, known colloquially as Patagonia. The absolute ages of South American land mammal ages, and the individual localities yielding fossil primates, have been a source of continuous debate and successive upward revisions during the past three decades. By the most recent dates, the primate-bearing deposits from southern South America seem to lie within a 10-million-year span from approximately 25 Ma to 15 Ma. These include the earliest fossil New World monkeys yet recovered. They probably lie near the base of several extant lineages.

History of discovery and debate

The earliest discovery and description of fossil New World monkeys (and indeed most other fossil mammals) in southern South America was due to the extraordinary efforts of the Ameghino brothers of Argentina (e.g., Simpson, 1980). Carlos Ameghino (1865–1936), the younger brother, made an astonishing 16 collecting expeditions to Patagonia between 1887 and 1903. His collections were described and interpreted by his older brother Florentino, an extraordinary scientist who described many hundreds of fossil mammals and whose collected works fill a very large bookshelf.

Homunculus patagonicus, the first fossil platyrrhine recovered from southern South America, was described by Florentino Ameghino on August 1, 1891. It was based on a mandibular specimen that had been collected by Carlos earlier that year from Santacrucian (early Miocene) deposits near the town of Rio Gallegos along the banks of the river of the same name (Ameghino, 1891*a*). The holotype specimen, a right mandibular corpus preserving I_2 through M_1 and the roots of M_2, is apparently lost, but casts exist.

Months later, Alcides Mercerat (1891) described "*Ecphantodon ceboides*" which Ameghino promptly synonymized with *Homunculus patagonicus*. In the same paper, Ameghino (1891*b*) named another fossil monkey, "*Anthropops perfectus*", based on a mandibular symphysis preserving a right P_4 and alveoli or roots for the other premolars, canines and incisors on both sides. In 1893, Ameghino described *Pithecus australis*, based on a mandible with a single first lower molar. This specimen was figured in 1899. Most subsequent researchers (Bluntschli, 1913; Hershkovitz, 1970; Rosenberger, 1979*b*; Szalay & Delson, 1979; Fleagle, 1988, 1999; Tejedor, in press; but see Stirton, 1951) have recognized these taxa as junior synonyms of *Homunculus*. Ameghino also named numerous other "primate" taxa based on specimens that have subsequently proved to belong to other orders of mammals. Finally, in a reverse twist, Ameghino named a species of caenolestoid marsupial *Stylotherium grandis* based on a tooth that subsequently proved to be the dP_4 of *Homunculus*, presumably *H. patagonicus* (Hershkovitz, 1981).

The hypodigm of *Homunculus* was augmented substantially in the years following its description with a complete, but heavily worn, lower jaw and many parts of a skeleton from the famous site of Corriguen Aike near the Rio Coyle. A partial face was also found 100 meters away from the jaw and partial skeleton. It is not clear exactly when the face was found. It has often been assumed that the face and partial skeleton are from the same individual but there seems no good reason for this assumption. Corriguen Aike is a very rich fossil locality and 100 m is a substantial distance for a single individual to be dispersed.

The material collected and described by the Ameghinos was reviewed by the Swiss anatomist Hans Bluntschli in 1913 (reprinted by Scott in 1928) and again in a much longer paper in 1931. In the later paper, Bluntschli described specimens in detail and summarized the scattered literature on fossil platyrrhines in the works of Florentino Ameghino. Bluntschli recognized that many of the taxa Ameghino described as primates were, in fact, members of other mammalian groups, and he synonymized Ameghino's taxa *Anthropops perfectus* and *Pitheculus australis* with *Homunculus patagonicus*. In addition, Bluntschli described the lower jaw from Corriguen Aike as a new species, *Homunculus ameghinoi*, which subsequent researchers have not recognized.

In the 1930s and 1940s, two additional fossil platyrrhines were discovered in Patagonia. Both were isolated crania from deposits of the Colhuehuapian Land Mammal Age in Chubut Province of Argentina. In 1933, Carlos Rusconi reported on a primate cranium discovered at Sacanana in northern Chubut by Tomas Harrington. Rusconi (1933) named the specimen *Homunculus harringtoni* and subsequently (1935*b*) provided an extensive description of the new fossil and a comparison with other Patagonian primates known at the time. He attributed the Sacanana locality to the Colhuehuapian (Colpodense beds) Land Mammal Age, which he assigned to the lower Oligocene.

In 1942, the Argentine paleontologist Alejandro F. Bordas reported the discovery of an edentulous primate skull from the well-known Colhuehuapian deposits of Gaiman on the banks of the Chubut River. In his brief description of the fossil, Bordas (1942) attributed it to *Homunculites pristinus*, a Colhuehuapian taxon that Ameghino had identified as an ancestor to the Santacrucian *Homunculus*, but others had recognized properly as a marsupial (e.g., Bluntschli, 1931). The Gaiman cranium was subsequently described in greater detail by Jorge Lucas Kraglievich (1951) who assigned it to the new species *Dolichocebus gaimanensis*. Kraglievich identified roots of only two molars in the *Dolichocebus* cranium and thus argued that this taxon was related to callitrichids. Interestingly, his paper also includes a discussion regarding the geographic origin of platyrrhines that summarizes the strengths and weaknesses of an African origin for platyrrhines in much the same terms as modern discussions of this topic (e.g., Hartwig, 1994; Fleagle, 1999).

In the course of his study of new fossil platyrrhines from La Venta, Stirton (1951; Stirton & Savage, 1951) reviewed the fossil monkeys from Patagonia, and assigned one taxon from the Miocene of Colombia to the genus *Homunculus*. He argued that this species, *H. tatacoensis*, supported a close phylogenetic relationship between *Homunculus* and the living howling monkeys, *Alouatta* (see Hartwig & Meldrum, this volume).

The study of Patagonian primates, and platyrrhine fossils in general, lay fallow throughout most of the 1950s and 1960s. Interest in platyrrhine fossils picked up tremendously at the end of the 1960s and throughout the 1970s because of the extensive work by Hershkovitz in Chicago and Hoffstetter in Paris. In 1969, Hoffstetter described the oldest known platyrrhine, *Branisella boliviana*, based on a maxillary fragment recovered by the prominent Bolivian paleontologist Leonardo Branisa from Deseadan deposits at Salla, Bolivia. The Deseadan was thought to be early Oligocene. In his description of *Branisella*, and in a series of subsequent papers, Hoffstetter (along with the rodent specialist Rene Lavocat), challenged the currently accepted view of a North American origin for platyrrhines, and, by implication, a parallel evolution of anthropoids in the Old World and the New World. Rather, they argued that platyrrhines were more closely related to Old World catarrhines than to any prosimians and that platyrrhines (along with caviamorph rodents) must have rafted to South America from Africa during the early Cenozoic (e.g., Hoffstetter, 1972, 1974, 1977, 1980, 1982). Hoffstetter and Lavocat's theory of an African origin for platyrrhines generated a tremendous renaissance of interest in platyrrhine evolution as well as in anthropoid origins that has yet to subside (e.g., Ciochon & Chiarelli, 1980*a*; Rose & Fleagle, 1981; Fleagle & Kay, 1994; Kay *et al.*, 1997*a*).

In 1970, Hershkovitz began a series of studies reviewing fossil platyrrhines. He provided the correct dental formula of *Dolichocebus* (2.1.3.3); he separated the La Venta taxon from *Homunculus* and placed it in a new genus, *Stirtonia*; and he argued that *Homunculus* was not related to *Alouatta* as Stirton had believed. In a subsequent paper, Hershkovitz (1974) reanalyzed the skull from Sacanana that had been described by Rusconi as *Homunculus harringtoni* and renamed it *Tremacebus harringtoni*. The name was based on Hershkovitz's interpretation that it retained a large opening (trema) in the posterior orbit wall thus showing that postorbital closure was attained independently in New World platyrrhines and in Old World catarrhines. In 1981 he described some newly discovered or newly recognized platyrrhine fossils from Patagonia – a lower jaw from Colhuehuapian deposits at the Gran Barranca in Chubut (e.g., Simpson, 1980) and the aforementioned deciduous tooth of *Homunculus* that Ameghino mistook for a marsupial. Hershkovitz returned to this latter topic in 1987 and described a more complete specimen of *Homunculus* with both deciduous and permanent molars in place. Throughout these works and in his massive 1977 treatise on New World monkeys, Hershkovitz remained steadfast in his views that most of the known fossil platyrrhines had nothing to do with any specific extant taxa.

A very different view of platyrrhine evolution, and the significance of fossil platyrrhines from Patagonia (and elsewhere), is that of Alfred Rosenberger, whose theories of platyrrhine evolution first appeared in the late 1970s and have slowly come to dominate the field. In Rosenberger's view the radiation of living platyrrhines reflects the results of an ancient split into two very distinct clades occupying different adaptive zones (e.g., Rosenberger, 1979*b*, 1980, 1985, 1992). One clade is the generally smaller, more insectivorous cebids (including *Cebus*, *Saimiri*, and the callitrichines *Callithrix*, *Cebuella*, *Callimico*, *Leontopithecus* and *Saguinus*). The other clade, the atelids, includes the larger, often more folivorous taxa, made up of the seed-eating pitheciines (*Pithecia*, *Chiropotes*, and *Cacajao* as well as *Callicebus* and possibly *Aotus*) and the very large suspensory atelines (*Alouatta*, *Lagothrix*, *Ateles* and *Brachyteles*). Fossil platyrrhines, especially those from Patagonia, have played a critical role in support of the antiquity of this major dichotomy. Thus, beginning in 1979, Rosenberger identified the Colhuehuapian *Dolichocebus* as an early direct ancestor of the squirrel monkey, *Saimiri*, based on its possession of an interorbital fenestra and numerous other cranial characteristics (Rosenberger, 1979*a*, 1982; but see Hershkovitz, 1982); the Colhuehuapian *Tremacebus* was allied with the *Aotus* lineage; and the Santacrucian *Homunculus* was allied with *Callicebus* (Rosenberger, 1979*b*, 1980, 1985; Fleagle & Rosenberger, 1983).

Beginning in 1982 expeditions from the State University of New York, the Museo Argentino de Ciencias Naturales in Buenos Aires, and more recently the Centro Nacional

Patagonia in Puerto Madryn, have recovered additional fossils from the Colhuehuapian sites that yielded *Dolichocebus* and *Tremacebus*, more remains of *Homunculus* from the younger Santacrucian deposits, and an extensive fauna of fossil platyrrhines from the Pinturas deposits in western Santa Cruz Province. Thus, in 1983, Fleagle and Bown described several isolated teeth and a talus from Gaiman that they provisionally attributed to *Dolichocebus* (Fleagle & Bown, 1983; see also Reeser, 1984; Fleagle & Kay, 1989) and a lower jaw from Sacanana (Fleagle *et al.*, 1987). Four new taxa have been described from the early Santacrucian deposits of the Pinturas Formation in western Santa Cruz Province: *Soriacebus ameghinorum* (Fleagle *et al.*, 1987); *S. adrianae*, *Carlocebus carmenensis* and *C. intermedius* (Fleagle, 1990). In addition, numerous limb bones have been described for several of these taxa (Anapol & Fleagle, 1988; Meldrum, 1990; Stevens & Fleagle, 1998).

The Deseadan deposits near Salla, Bolivia that yielded the remains of *Branisella* have likewise been the focus of nearly continuous geological and paleontological research during the past two decades. In 1981, Rosenberger described a mandible of *Branisella* from a collection of Salla fossils at Princeton University. Paleontological and geological studies at Salla by a team from the University of Florida yielded two more specimens of *Branisella* (Wolff, 1984) that were subsequently designated a new taxon, *Szalatavus* (Rosenberger *et al.*, 1991*b*). More recently, expeditions from Duke University and Kyoto University (Japan) in conjunction with the Bolivian National Museum have recovered still more dental remains of *Branisella* (Kay & Williams, 1995; Takai & Anaya, 1996; Takai *et al.*, 2000*a*).

Beginning in the mid-1980s and continuing to the present, Adan Tauber of the University of Cordoba has conducted detailed biostratigraphic research in the southernmost deposits of the Santa Cruz Formation and has recovered new material of *Homunculus*, including a partial cranium (Tauber, 1991). More recently, expeditions from the Museo de la Plata (Argentina) and Duke University have conducted paleontological research in many parts of Patagonia and have recovered additional platyrrhine fossils. These include a new taxon, *Proteropithecia neuquenensis* from Neuquen (Kay *et al.*, 1998*a*, 1999*a*), and additional undescribed primates from Gaiman, and from the Gran Barranca. From deposits of Colhuehuapian age in the Andes of Chile, Flynn *et al.* (1995) have described a new platyrrhine taxon, *Chilecebus carrascoensis*, based on a nearly complete skull that is still being prepared and described.

Analysis has lagged behind description in studies of the recently discovered fossil platyrrhines from southern South America so our understanding of the significance of these remains is far from resolved. Nevertheless, for many taxa there are competing hypotheses and there have been few recent attempts at synthesis. For example, *Soriacebus* has been identified commonly as a basal pitheciine, and *Carlocebus* and *Homunculus* have been generally considered to show similarities to *Callicebus* (see Fleagle, 1990; Rosenberger, 1992; Tejedor, 1998*a*, in press). However, others have argued that all Patagonian primates are a geographically isolated radiation with no phylogenetic relationship to extant platyrrhines. In recent years, studies in molecular systematics appear to have largely resolved the major aspects of platyrrhine phylogeny, and most current analyses of fossil platyrrhines are aimed at synthesizing evidence from the fossil record with the results of molecular systematics (e.g., Schneider & Rosenberger, 1996; Fleagle & Reed, 1999; Horovitz, 1999; Brenoe, 2000).

Taxonomy

Systematic framework

Order Primates Linnaeus, 1758
- Infraorder Platyrrhini É. Geoffroy Saint-Hilaire, 1812
  - Family Atelidae Gray, 1825
    - Subfamily Pitheciinae Mivart, 1865
      - Tribe Pitheciini Gray, 1849
        - Genus *Soriacebus* Fleagle *et al.*, 1987
          - *Soriacebus ameghinorum* Fleagle *et al.*, 1987
          - *Soriacebus adrianae* Fleagle, 1990
        - Genus *Proteropithecia* Kay *et al.*, 1999
          - *Proteropithecia neuquenensis* Kay *et al.*, 1998
      - Tribe Homunculini (includes *Callicebus*) Ameghino, 1894
        - Genus *Homunculus* Ameghino, 1891
          - *Homunculus patagonicus* Ameghino, 1891
        - Genus *Carlocebus* Fleagle, 1990
          - *Carlocebus carmenensis* Fleagle, 1990
          - *Carlocebus intermedius* Fleagle, 1990
  - Family Cebidae Bonaparte, 1831
    - Subfamily Cebinae Bonaparte, 1831
      - Tribe Saimiriini Miller, 1912
        - Genus *Dolichocebus* Kraglievich, 1951
          - *Dolichocebus gaimanensis* Kraglievich, 1951
      - Tribe *incertae sedis*
        - Genus *Chilecebus* Flynn *et al.*, 1995
          - *Chilecebus carrascoensis* Flynn *et al.*, 1995
    - Subfamily Aotinae Elliot, 1913
      - Tribe *incertae sedis*
        - Genus *Tremacebus* Hershkovitz, 1974
          - *Tremacebus harringtoni* Rusconi, 1933
  - Family *incertae sedis*
        - Genus *Branisella* Hoffstetter, 1969
          - *Branisella boliviana* Hoffstetter, 1969

Family Atelidae

Subfamily Pitheciinae

Tribe Pitheciini

GENUS *Soriacebus* Fleagle *et al.*, 1987
INCLUDED SPECIES *S. adrianae*, *S. ameghinorum*

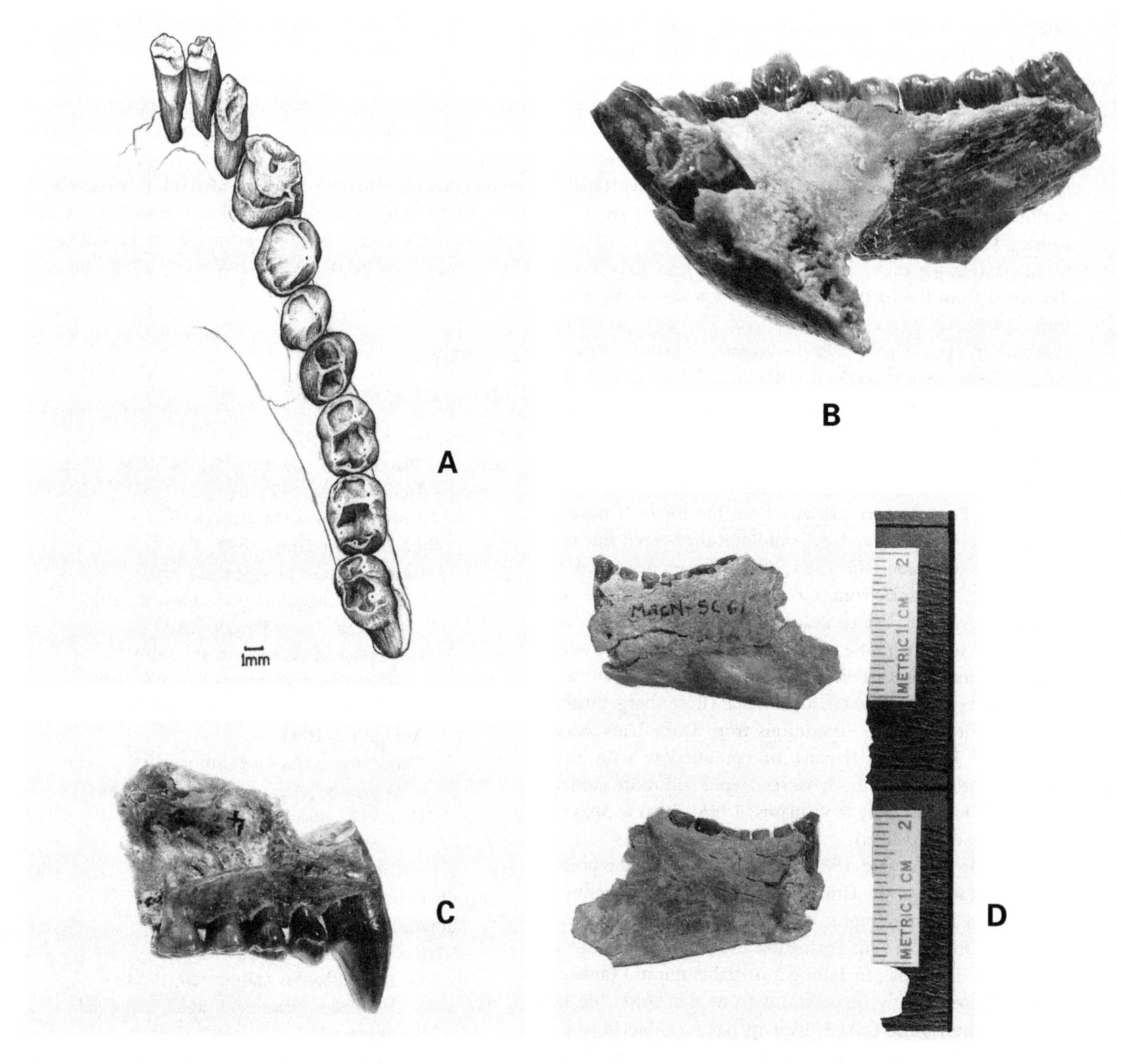

Fig. 11.1 *Soriacebus ameghinorum*. (A) MACN-SC 2, holotype mandible; reconstruction based on MACN-SC 2 and MACN-SC 33; (B) lateral view of MACN-SC 2 + 33; (C) lingual view of maxillary fragment MACN-SC 4; (D) buccal view of mandibular fragment MACN-SC 61. (B) and (C) approximately 2.5× natural size.

SPECIES *Soriacebus ameghinorum* Fleagle *et al.*, 1987 (Fig. 11.1)
TYPE SPECIMEN MACN-SC 2, mandibular ramus preserving right P_2–M_3
AGE AND GEOGRAPHIC RANGE Late early Miocene (early Santacrucian Land Mammal Age), approximately 18 Ma (Bown & Fleagle, 1993; Fleagle *et al.*, 1995), Santa Cruz and Chubut provinces, Argentina
ANATOMICAL DEFINITION
Soriacebus ameghinorum is a saki-sized (1800 g) primate known primarily from dental and gnathic remains and a few pedal elements. The lower jaw has relatively narrow and tall procumbent incisors, a relatively large canine that is oval in cross-section, a large unicuspid P_2; P_3 is smaller than P_4 with a poorly developed trigonid that is open lingually and a much smaller metaconid than protoconid; P_4 has a well-developed trigonid that is closed lingually with protoconid and metaconid subequal in size. The lower molars of *S. ameghinorum* are characterized by a relatively long trigonid and an oblique protocristid. There is often a remnant of a tiny hypoconulid on M_{1-2}. $M_1 = M_2 > M_3$. The individual cusps of the lower premolars and molars of *S. ameghinorum* are poorly developed and connected by low crests. The mandible has a robust symphysis and deepens posteriorly as in pitheciines.

Fig. 11.2 *Soriacebus adrianae*. Holotype mandible (MACN-SC 59) and referred specimens. Scale = 1 cm.

The upper canine of *S. ameghinorum* is a laterally compressed, blade-like tooth with a deep mesial groove and a wear facet of the distolingual surface where it sheared against the buccal surface of P_2. P^2 is a unicuspid, slightly asymmetrical tooth with a prominent lingual cingulum. P^3 and P^4 are larger than P^2. Each has a large paracone, small subequal protocone and hypocone, and a small lingual cingulum. P^{2-4} each have three roots. The upper molars of *S. ameghinorum* are broad and triangular in outline. Paracone is slightly larger than the metacone. There is a large protocone that occupies most of the lingual half of the tooth. The hypocone is small, slightly lingual and lies on the prominent lingual cingulum. $M^1 > M^2 > M^3$. The maxilla is relatively deep between the inferior orbital and alveolar margins. An isolated talus has been attributed to this species (Meldrum, 1990).

SPECIES *Soriacebus adrianae* Fleagle, 1990 (Fig. 11.2)
TYPE SPECIMEN MACN-SC 59, a mandibular fragment preserving the right P_{3-4}, the roots of the right canine and P_2, and alveoli for incisors
AGE AND GEOGRAPHIC RANGE Upper part of the Pinturas Formation, dated to the late early Miocene (early Santacrucian Land mammal Age), slightly older than 16.5 Ma (Bown & Fleagle, 1993; Fleagle *et al.*, 1995); *S. adrianae* comes from younger deposits than its congener *S. ameghinorum*
ANATOMICAL DEFINITION
Like *S. ameghinorum*, *S. adrianae* has tall, narrow, procumbent lower incisors, a large canine that is oval in cross-section, a large unicuspid P_2, a smaller P_3 and P_4. In *S. adrianae*, the premolars are more similar is size than in *S. ameghinorum*. The lower molars are relatively broader and the trigonid relatively shorter and the protocristid more transverse compared with *S. ameghinorum*. As in *S. ameghinorum* premolars and molars are high-crowned with low cusp relief. The shearing coefficient is extremely low (Fleagle *et al.*, 1997).

M_3 was probably very small, based on alveolar size.

The upper dentition is poorly known. As in *S. ameghinorum*, the upper canine is laterally compressed with a mesial grove and a distal shearing facet. P^{2-3} are broad with a prominent paracone and smaller protocone. Upper molars are only known from isolated teeth. They have a smaller hypocone than in *S. ameghinorum*. As in *S. ameghinorum*, the maxilla is relatively deep between the orbital and alveolar margins.

Subfamily Pitheciinae

Tribe Pitheciini

GENUS *Proteropithecia* Kay *et al.*, 1999
INCLUDED SPECIES *P. neuquenensis*

SPECIES *Proteropithecia neuquenensis* Kay *et al.*, 1998
TYPE SPECIMEN MLP 91-IX-1-125, an isolated right M_1 or M_2
AGE AND GEOGRAPHIC RANGE Colon Curan Formation, just above an ignimbrite dated at 15.71 ± 0.07 Ma, Neuquen Province, Argentina
ANATOMICAL DEFINITION
Proteropithecia neuquenensis is known from only 15 isolated teeth from two localities documenting I_1, C_1, P_2 or P_3, M_1 and/or M_2, C^1 and $P^?$. The lower incisors are mesiodistally compressed and tall as in pitheciines. The worn canine is oval in cross-section. The lower molars are relatively high-crowned with a well-developed trigonid and talonid with low cusp relief, crenulated enamel and a shallow ectoflexid. They are relatively broader than the molar teeth of *Soriacebus*. The upper canine is rounded in cross-section with a distinct mesial groove and a distal shearing facet. Overall, the isolated teeth of *Proteropithecia* more closely resemble extant and middle Miocene pitheciines than do any of the Santacrucian taxa.

The only postcranial element is a talus that shows overall similarities to *Callicebus*.

Subfamily Pitheciinae

Tribe Homunculini

GENUS *Homunculus* Ameghino, 1891
INCLUDED SPECIES *H. patagonicus*

SPECIES *Homunculus patagonicus* Ameghino, 1891 (Fig. 11.3)
TYPE SPECIMEN MACN-A 12498, a right mandibular fragment with a complete symphysis from the banks of Rio Gallegos (Ameghino, 1891*a*). This specimen has probably been lost. Tejedor (in press) has designated MACN-A 5757, a mandible, as a neotype. It is associated with several postcranial elements, including a complete right femur (MACN-A 5758), a complete right radius (MACN-A 5760), the shaft of an ulna (MACN-A 5759) and the distal end of a humerus (MACN-A 5761, missing).

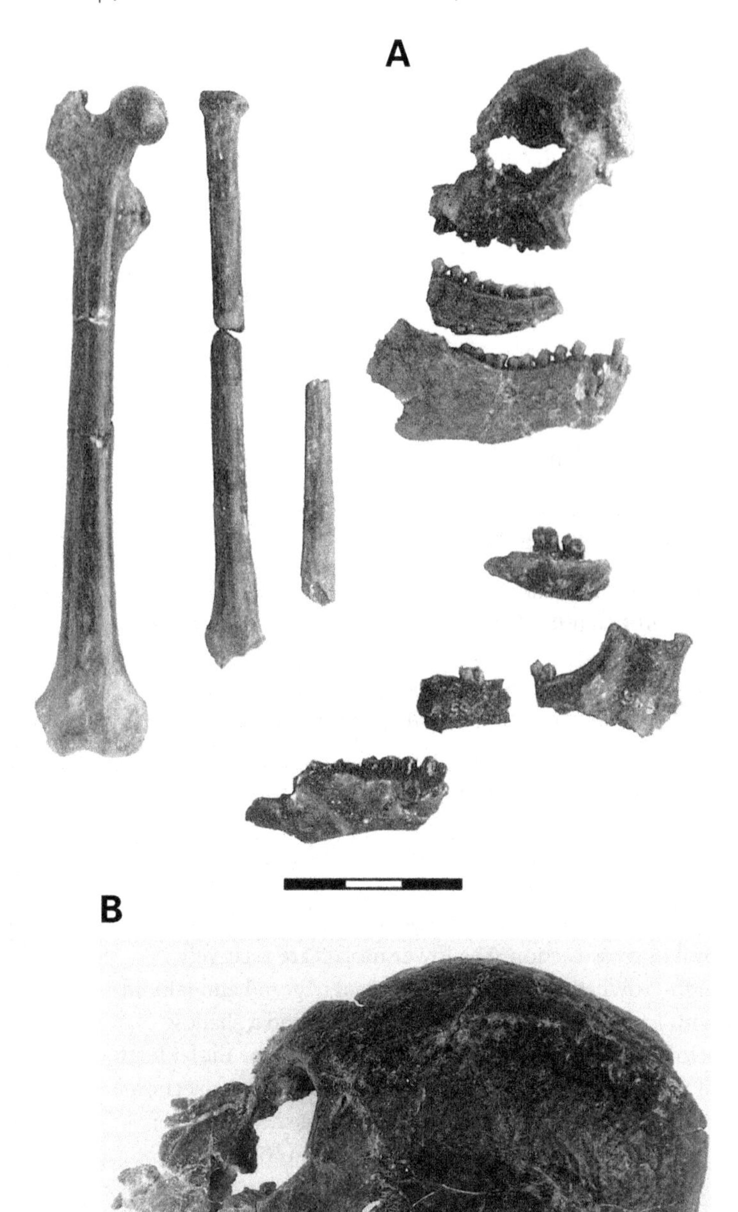

Fig. 11.3 *Homunculus patagonicus*. (A) Various specimens including holotype mandible (MACN 12498, middle right), right femur (MACN-A 5758), right radius (MACN-A 5760) and partial face (upper right). (B) CORD-PZ 1.130, cranium of *Homunculus patagonicus*. Scales = 1 cm. (A) Composite artwork by Luci Betti-Nash. (B) Photograph courtesy of Adan Tauber.

AGE AND GEOGRAPHIC RANGE Late early Miocene–early middle Miocene Santa Cruz Formation (Santacrucian Land Mammal Age) dated to 16.5–16.0 Ma (Bown & Fleagle, 1993; Fleagle *et al.*, 1995), southernmost Santa Cruz Province, Argentina

ANATOMICAL DEFINITION

The lower incisors, canines and premolars of *H. patagonicus* are known only from broken or worn specimens and roots. Lower incisors are small, narrow and vertically implanted. The lower canines of *H. patagonicus* are all relatively small, slightly recurved and oval in cross-section. It is a relatively small tooth with a large protoconid. The distal surface of the lower canine and the mesial surface of P_2 show extensive facets for occlusion with projecting upper canine. P_3 is a semimolariform tooth with subequal protoconid and metaconid, a small lingually open trigonid and a small talonid basin. P_4 is similar in size and morphology to P_3 with a larger trigonid and broader talonid basin with a small entoconid. The lower molars of *H. patagonicus* are characterized by a trigonid that is higher than the talonid. The trigonid on M_1 is enclosed by a series of crests; on M_2 it is open lingually. Protoconid, metaconid, hypoconid and entoconid are prominent and individual cusps are connected by strong crests so that *H. patagonicus* has a moderately high shearing coefficient. There is a well-developed ectoflexid on all lower molars. All lower molars are similar in length. The mandible deepens slightly distally and is very similar in shape to the mandible of *Aotus*.

The upper central incisor is broad and spatulate. The upper canine is laterally compressed and relatively short with a robust root. P^3 and P^4 are oval teeth with a large paracone, a smaller protocone and a tiny hypocone distal to the protocone. The upper molars of *H. patagonicus* are quadrate teeth with a prominent trigon and moderate-sized hypocone, and an expanded distal basis. There is a moderately developed lingual cingulum and often a pericone lingual to the protocone.

Homunculus has a prominent rostrum, relatively deep face, moderate-sized orbit and a narrow interorbital pillar. The braincase is globular with distinct temporal lines.

The femur of *Homunculus* has a straight shaft, a head that extends proximally to the level of the greater trochanter, a very large lesser trochanter, and relatively deep distal condyles. Rosenberger (1979*b*) has argued that the femur is too large to belong to the same taxon as the referred craniodental specimens. The radius has a relatively straight shaft and a rounded head.

GENUS *Carlocebus* Fleagle, 1990

INCLUDED SPECIES *C. carmenensis*, *C. intermedius*

SPECIES *Carlocebus carmenensis* Fleagle, 1990 (Fig. 11.4)

TYPE SPECIMEN MACN-SC 266, a right mandibular fragment preserving P_4–M_2 and alveoli for P_{2-3} and M_3

AGE AND GEOGRAPHIC RANGE Early Miocene (early Santacrucian Land Mammal Age), Pinturas Formation; the oldest fossiliferous deposits of this formation are older than 17.5 Ma and the youngest are older than 16.5 Ma (Bown & Fleagle, 1993; Fleagle *et al.*, 1995)

ANATOMICAL DEFINITION

Carlocebus carmenensis was a medium-sized monkey with an estimated body mass of 2500 g. It is known from dental,

Fig. 11.4 *Carlocebus carmenensis*. (A) Holotype mandible (MACN-SC 266); (B–D) composite maxillary tooth row; (E) lateral view of holotype mandible (bottom) and referred maxillary fragment (top); (F) MACN-SC 101, proximal ulna referred to *Carlocebus*. Scales = 1 cm. (F) Photograph courtesy of Fred Anapol.

cranial and postcranial remains (Anapol & Fleagle, 1988; Fleagle, 1990; Meldrum, 1990; Fleagle & Rae, 1992). Lower incisors are known primarily from roots. They are smaller and less procumbent than the lower incisors of *Soriacebus*. P_2 is a relatively simple tooth with a large protoconid. P_3 is broader than P_2. The protoconid is somewhat larger than the metaconid. P_4 is a semimolariform tooth with subequal protoconid and metaconid, an enclosed trigonid basin and a broad talonid basin. The molars of *Carlocebus* are relatively broader than those of *Homunculus* with more bulbous cusps and less prominent crests. There is also greater flare of the base of the molars in *Carlocebus*. The mandible is robust and deepens posteriorly as in *Callicebus*.

P^{3-4} of *Carlocebus* are ovoid with a large paracone, a large protocone and an expanded distal basin with a distinct hypocone, similar to the premolars of *Callicebus*. The upper molars are quadrate teeth with more bulbous cusps than those of *Homunculus*, a larger hypocone and a more prominent lingual cingulum. There is often a small parastyle and metastyle. The metacone and hypocone are relatively smaller in M^2 than M^1 and are absent in M^3.

The few cranial remains of *Carlocebus* show a broad palate, a relatively deep face, and moderate-sized orbits, indicating diurnal habits.

Several limb elements have been described for *Carlocebus*, including parts of a scapula, an ulna, a femur, a tibia and several tali (Anapol & Fleagle, 1988; Meldrum, 1990; Stevens & Fleagle, 1998). The limb elements group with a wide range of extant platyrrhines, including *Homunculus*, and pitheciines. All suggest quadrupedal and climbing habits rather than leaping.

SPECIES *Carlocebus intermedius* Fleagle, 1990 (Fig. 11.5)
TYPE SPECIMEN MACN-SC 3 a left mandibular ramus with C–M_2
AGE AND GEOGRAPHIC RANGE Late early Miocene (early Santacrucian Land Mammal Age), approximately 18 Ma (Bown & Fleagle, 1993; Fleagle *et al.*, 1995), in the northwestern part of Santa Cruz Province, Argentina
ANATOMICAL DEFINITION
Carlocebus intermedius is similar to *C. carmenensis* in dental morphology, but smaller.

Family Cebidae

Subfamily Cebinae

Tribe Saimiriini

GENUS *Dolichocebus* Kraglievich, 1951
INCLUDED SPECIES *Dolichocebus gaimanensis*

SPECIES *Dolichocebus gaimanensis* Kraglievich, 1951 (Fig. 11.6)
TYPE SPECIMEN MACN 14128, an edentulous and distorted cranium

Fig. 11.5 *Carlocebus intermedius*. Holotype mandible (MACN-SC 3), lingual view. Approximately 2.5× natural size.

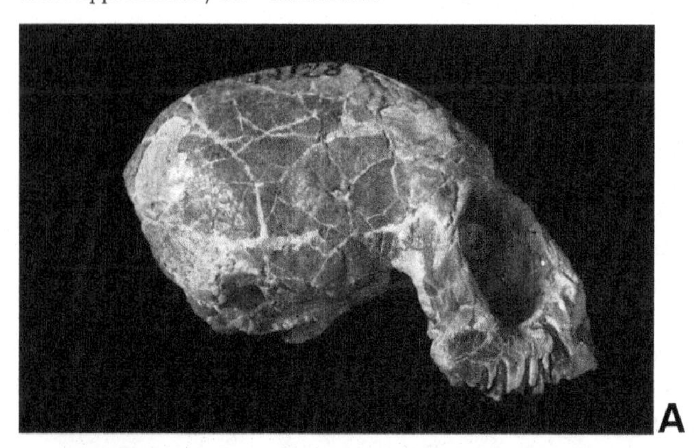

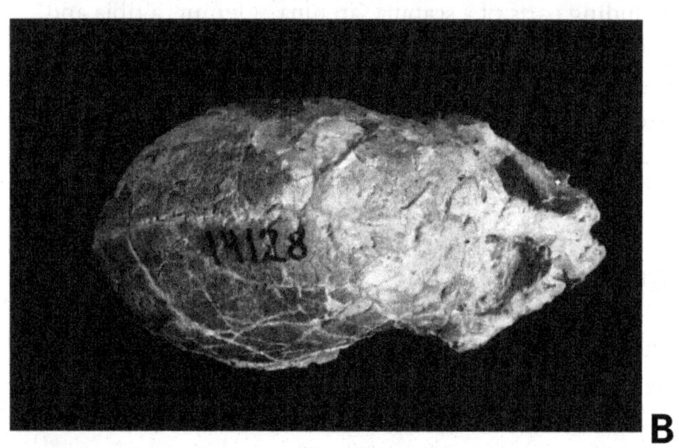

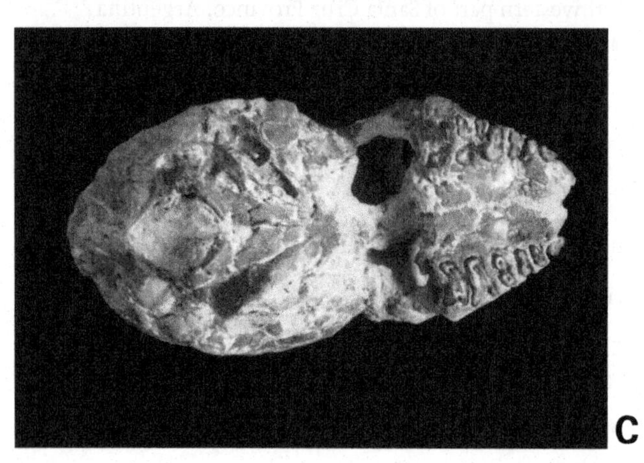

Fig. 11.6 *Dolichocebus gaimanensis*. Holotype cranium (MACN 14128) in (A) lateral, (B) superior and (C) inferior views. Natural size.

AGE AND GEOGRAPHIC RANGE Early Miocene (Colhuehuapian Land Mammal Age) dated at greater than 20.5 Ma (Kay *et al.*, 1999*b*), Gaiman, Chubut Province, Argentina

ANATOMICAL DEFINITION

The cranium of *Dolichocebus* indicates a medium-sized platyrrhine with an estimated body mass of 2700 g. The cranium shows moderate-sized orbits indicating diurnal habits, a narrow interorbital region, a relatively narrow braincase (independent of distortion) and relatively broad molar and premolar teeth, based on the size and shape of the roots. The most distinctive feature of the skull that has yielded controversial interpretations is the interorbital fenestra connecting the right and left orbital cavities. Rosenberger (1979*a*) has argued that this is a natural opening that indicates a phylogenetic link between *Dolichocebus* and *Saimiri*. Hershkovitz (1982) has argued that the hole is the result of breakage. Isolated teeth from the same site that have been allocated to *Dolichocebus* include an upper molar with a broad lingual cingulum and a large hypocone (Fleagle & Bown, 1983).

Subfamily Cebinae

Tribe *incertae sedis*

GENUS *Chilecebus* Flynn *et al.*, 1995
INCLUDED SPECIES *C. carrascoensis*

SPECIES *Chilecebus carrascoensis* Flynn *et al.*, 1995 (Fig. 11.7)
TYPE SPECIMEN SGOPV 3213, a nearly complete slightly deformed skull
AGE AND GEOGRAPHIC RANGE Early Miocene (Colhue-huapian Land Mammal Age), 20.09 ± 0.27 Ma, Abanico Formation, 100 km south-southeast of Santiago, central Chile

ANATOMICAL DEFINITION

Chilecebus was a small monkey with an estimated body mass of just over 1000 g. The skull has a rounded braincase with no cresting and relatively small orbits indicating diurnal habits. The palate shows a posteriorly diverging dental arcade. The upper dentition shows relatively broad spatulate central incisors, small canines, a unicuspid P^2, and very broad P^3 and P^4 with a large paracone and a small protocone. M^{1-2} are quadrate with well developed hypocones but no metaconule. M^3 is ovoid and lacks either a metacone or a hypocone. P^3–M^3 show a broad stylar shelf and small stylar crests. A detailed description of the specimen is in progress. However, the large, broad premolars suggest cebine affinities.

Subfamily Aotinae

Tribe *incertae sedis*

GENUS *Tremacebus* Hershkovitz, 1974
INCLUDED SPECIES *T. harringtoni*

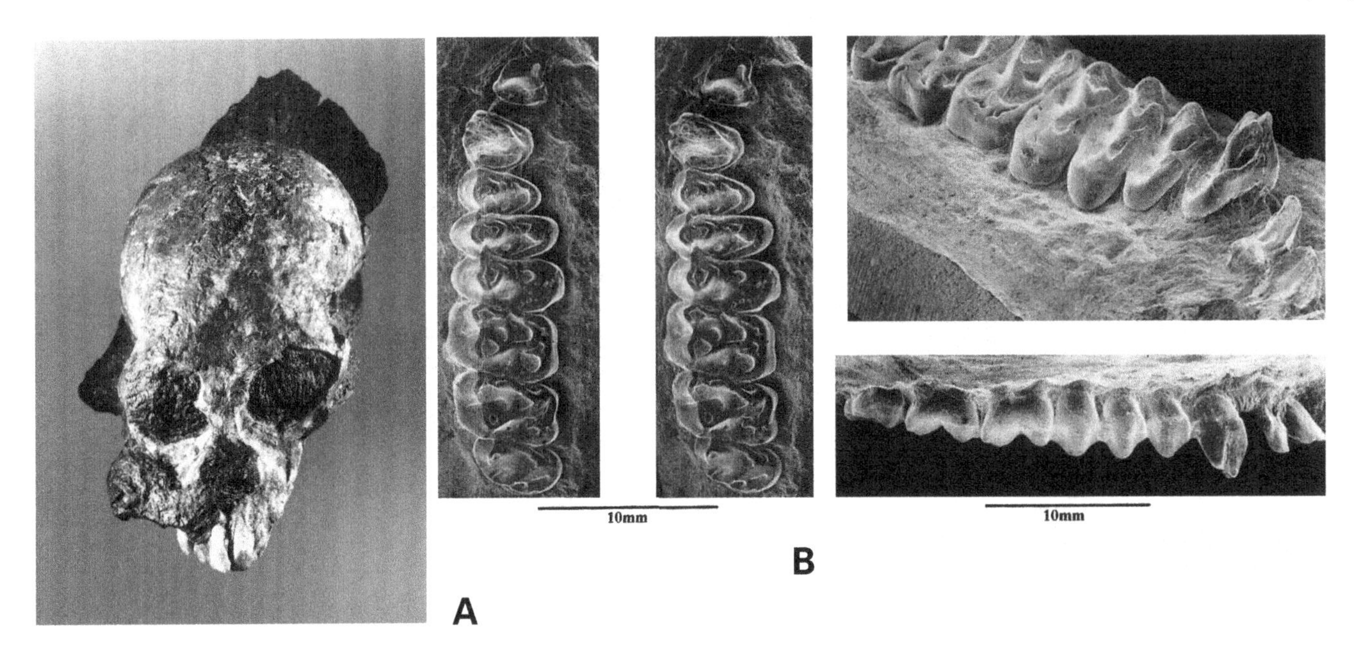

Fig. 11.7 *Chilecebus carrascoensis*. Holotype cranium (SGOPV 3213): (A) superior view; (B) occlusal (left), lingual (upper right) and buccal (lower right) views of the dentition. (A) Photograph by John Weinstein, © The Field Museum; (B) images courtesy of John Flynn.

SPECIES *Tremacebus harringtoni* Rusconi, 1933 (Fig. 11.8)
TYPE SPECIMEN FLM 619, a relatively complete and undistorted cranium with broken molar teeth
AGE AND GEOGRAPHIC RANGE Early Miocene (Colhuehuapian Land Mammal Age), Sacanana, Chubut Province, Argentina
ANATOMICAL DEFINITION
The cranium of *Tremacebus harringtoni* has a broad snout, a wide interorbital region and relatively large orbits, suggesting perhaps nocturnal habits and perhaps a phylogenetic relationship with *Aotus*. The nuchal region is relatively vertical for a platyrrhine. The fragmentary upper molars are quadrate with a large hypocone and a small pericone on the lingual cingulum.

Family *incertae sedis*

GENUS *Branisella* Hoffstetter, 1969 (includes *Szalatavus* Rosenberger et al., 1991)
INCLUDES SPECIES *B. boliviana*

SPECIES *Branisella boliviana* Hoffstetter, 1969 (Fig. 11.9)
TYPE SPECIMEN MNHN-Bol-V 3460, left maxillary fragment
AGE AND GEOGRAPHICAL RANGE Deseadan Land Mammal Age 25-26 Ma (MacFadden, 1990; Kay et al., 1999b), Salla, Bolivia
ANATOMICAL DEFINITION
Branisella is known only from dental and gnathic specimens collected by many expeditions over more than three decades (Takai et al., 2000a). Based on molar dimensions, *Branisella* was a medium-sized platyrrhine with an estimated body mass of roughly 1000 g. Based on the roots, the lower incisors appear to be small and somewhat procumbent. The lower canines are laterally compressed. P_2 is a simple, unicuspid tooth with a strong lingual cingulum. P_3 and P_4 are both bicuspid teeth with protoconid, metaconid and a distinct trigonid. In P_4 the cusps are subequal: in P_3 the metaconid is relatively smaller. The lower molars have four bulbous cusps: protoconid, metaconid, hypoconid, entoconid, and sometimes a small hypoconulid on M_2. The lower molars are relatively high-crowned, a feature that has been interpreted as evidence for terrestrial habits (Takai et al., 2000a). The mandible is robust but not particularly deep.

The upper canine is a projecting laterally compressed tooth with mesial groove and sharpening facet distally; P^2 has a relatively small, single (fused) root and reportedly is a tiny unicuspid tooth (Kay & Williams, 1995; Takai et al., 2000a). P^{3-4} are broader, bicuspid teeth with a strong lingual cingulum. M^{1-2} of *Branisella* have a prominent trigon and a moderate-sized hypocone on the strong lingual cingulum. The upper molars seem to show considerable variability in overall shape with some being more quadrate and others more triangular (Rosenberger et al., 1991b; Takai & Anaya, 1996). The distal cusps are reduced or absent on M^3.

The large sample of dental material collected from the Salla locality in recent years demonstrates that the remains originally described as a new taxon "*Szalatavus attricuspis*" (Rosenberger et al., 1991b) are part of a continuous range of variation within what is best interpreted as a single taxon, *Branisella boliviana* (Takai & Anaya, 1996; Takai et al., 2000a).

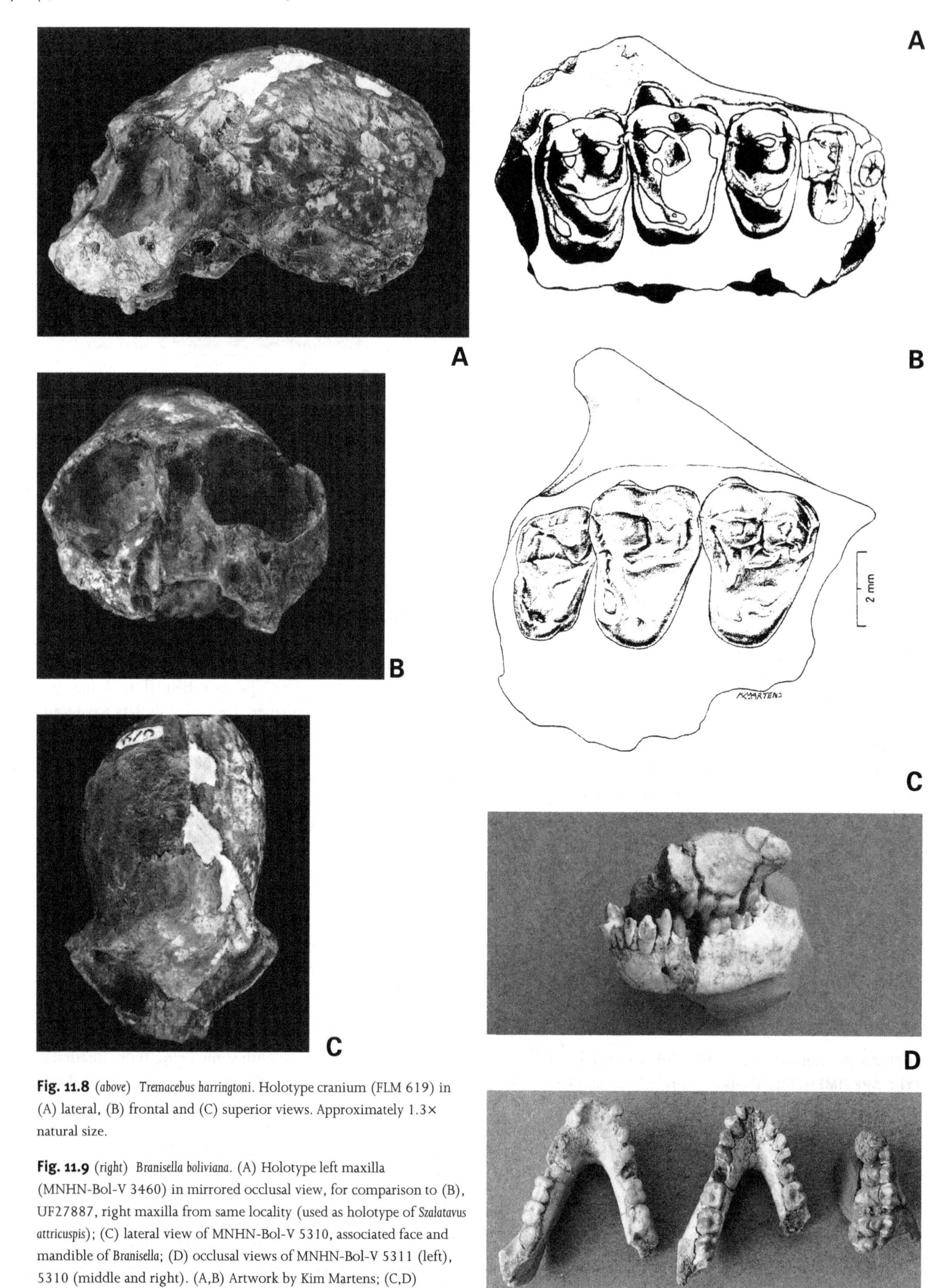

Fig. 11.8 (*above*) *Tremacebus harringtoni*. Holotype cranium (FLM 619) in (A) lateral, (B) frontal and (C) superior views. Approximately 1.3× natural size.

Fig. 11.9 (*right*) *Branisella boliviana*. (A) Holotype left maxilla (MNHN-Bol-V 3460) in mirrored occlusal view, for comparison to (B), UF27887, right maxilla from same locality (used as holotype of *Szalatavus attricuspis*); (C) lateral view of MNHN-Bol-V 5310, associated face and mandible of *Branisella*; (D) occlusal views of MNHN-Bol-V 5311 (left), 5310 (middle and right). (A,B) Artwork by Kim Martens; (C,D) courtesy of Masanaru Takai.

Evolution of fossil platyrrhines from southern South America

Fossil platyrrhines from southern South America have been described, discussed and debated for over a century now. In contrast with most aspects of primate evolution, there is very little debate over the alpha taxonomy. However, there is a tremendous diversity of opinion over the phylogenetic relationships of the fossil taxa to living clades of platyrrhines.

The fossil primates from southern South America document a critical time in platyrrhine evolution. They include the oldest fossil monkeys in the New World, and they occupy the time period between 25 and 15 million years ago in which virtually all extant lineages diverged according to calculations from molecular systematics (e.g., Porter *et al.*, 1997). Thus, for most of the fossil primates from this region, the major debate is whether these extinct taxa can be placed correctly in a living lineage.

Branisella boliviana provides the first documentation of higher primates in the New World. The absence of primates from older deposits is not obviously a sampling problem as there is an abundant fossil record from deposits up to 10 million years earlier. The fact that *Branisella* is roughly 10 million years younger than the earliest undoubted anthropoids in the Old World has been put forth as evidence in support of the argument that New World monkeys are an immigrant group that somehow rafted to South America from Africa (e.g., Hoffstetter, 1980; Fleagle, 1999).

The most striking aspect of the paleobiology of *Branisella* is the recent suggestion (Kay & Williams, 1995; Takai *et al.*, 2000*a*) that the molars of *Branisella* are more high-crowned than those of any other living or fossil platyrrhine and show greatest similarity in this feature to terrestrial Old World monkeys. This reconstruction of its locomotion based on teeth accords with interpretations of the paleoenvironment of Salla, which MacFadden (1990) has reconstructed as a semiarid environment based on the associated fauna and soil characteristics.

Although *Branisella* is the oldest platyrrhine, few authorities have argued that it is particularly close to a platyrrhine morphotype, except in possessing a full complement of molars and in having generalized quadritubercular upper molars (e.g., Rose & Fleagle, 1981; Miller & Simons, 1997). Thus, until most recently, *Branisella* has been generally interpreted as an early platyrrhine with no clear position relative to extant lineages. Most recently, however, Horovitz (1999) argued on the basis of a total evidence analysis of living and fossil platyrrhines that *Branisella*, and most other fossil platyrrhines, are best placed in a clade that is the sister taxon of callitrichines. Similarly, Takai *et al.* (2000*a*) have argued that *Branisella*, and *Proteopithecus* from the late Eocene of Egypt, might be uniquely related to callitrichines. This interpretation would place the divergence of callitrichines nearly 20 million years earlier than most current estimates and also place the diversification of platyrrhine lineages in Africa. It seems unlikely that the few features shared by *Branisella*, *Proteopithecus* and callitrichines warrant such a drastic reinterpretation of the timing of platyrrhine origins and diversification.

Dolichocebus and *Tremacebus* are both known primarily from cranial remains and have been subject to extremely divergent views. Rosenberger (1979*a*, 1992) has argued extensively that *Dolichocebus* is an early member of the lineage leading to *Saimiri* based on the interorbital foramen and other features of cranial structure. Others have questioned the value of the fenestra (Hershkovitz, 1982) and any relationship between *Dolichocebus* and any extant platyrrhines. However, an isolated talus from the Gaiman locality that yielded *Dolichocebus* was found to be most similar to *Saimiri*, *Cebus*, *Aotus* or *Callicebus* (Reeser, 1984), and a cladistic analysis of isolated teeth from Gaiman found a similarity to *Saimiri* one of the most parsimonious alternatives (Fleagle & Kay, 1989). Pending analysis of additional remains from this site, *Saimiri* affinities seem the most strongly supported.

The large postorbital opening that led Hershkovitz (1974) to identify *Tremacebus* as a primitive platyrrhine unrelated to any later taxa is almost certainly an artifact of breakage as the openings on the right and left side are totally different in shape. The relatively large orbit size suggests possible nocturnal or crepuscular habits and also affinities with the owl monkey. Again, in the absence of more material and more detailed analysis this is the most reasonable suggestion.

The skull of *Chilecebus* has only been described preliminarily and the authors of this genus mainly emphasized its distinctiveness as a new taxon. However, the extremely broad premolars of *Chilecebus* certainly suggest affinities with cebines (*Saimiri* and *Cebus*) among extant platyrrhines and the teeth are also similar to those of *Dolichocebus*, which is roughly the same age.

The Pinturas Formation in the pre-cordillera of northwestern Santa Cruz Province, Argentina, appears to have sampled a forested habit in an area of moderate local topographic relief (Bown & Fleagle, 1993). The associated fauna includes water fowl, passerine birds, aderitine marsupials, browsing ungulates and arboreal sloths with less abundance of the open-country edentates and ratite birds commonly found in most other Miocene sites in Argentina. This habitat accords well with the great abundance and diversity of fossil platyrrhines in the Pinturas deposits compared with other Argentine localities. Four species and two genera have been described so far, although there is evidence of additional unnamed taxa among the primate remains (Fleagle, 1990; Tejedor, 1995*a*, 1995*b*). Of the two genera that have been described so far, *Soriacebus* has been the source of the greatest debate. *Soriacebus* has large procumbent incisors, large anterior premolars, long relatively flat lower molar teeth and a very deep mandible. It was most probably some type of hard-seed predator like modern pitheciines. On the basis of these and other features, many authorities have identified *Soriacebus* as an early pitheciine. At the same time, there are many ways in which *Soriacebus* differs from extant pitheciines and many young fossils that are assigned to the Pitheciinae, such as the long rather than square lower molars, the triangular rather than square upper molars, and the oval, relatively narrow canine. These features have led Kay (1990) and Meldrum & Kay (1997*a*) to question the pitheciine affinities of *Soriacebus*. However, most of the purported pitheciine features of *Soriacebus* such as the procumbent incisors, large P_2, and deep mandible are also found in younger more derived fossil pitheciines suggesting that *Soriacebus* is just closer to the base of the radiation. Whether the unusual features of *Soriacebus* are primitive for this clade or unique features of this genus is unclear. This interpretation also concords with the greater similarity between the younger *Soriacebus adrianae* and other, younger, fossil pitheciines such as *Proteropithecia*, discussed below.

Carlocebus, from the same deposits as *Soriacebus*, has been the subject of less discussion. Although it shares a number of features with *Soriacebus* such as hypocones on the upper premolars, it has very different dental proportions with smaller incisors and anterior premolars compared to relatively large molars. *Carlocebus* was a medium-sized, diurnal monkey. The shearing crests suggest that *Carlocebus* was predominantly frugivorous (Fleagle *et al.*, 1997*b*). Limb bones assigned to this species suggest an arboreal quadruped with possible incipient suspensory abilities (Anapol & Fleagle, 1988). The molar and premolar teeth of *Carlocebus* are strikingly similar to those of *Callicebus* suggesting possible phyletic affinities with titi monkeys. Alternatively these similarities may well be primitive features for the atelid clade.

Homunculus occupied a very different habitat than *Soriacebus* and *Carlocebus*, based on the associated fauna and sediments (Bown & Fleagle, 1993). There is evidence of trees based on root casts, but the fauna includes more taxa with high-crowned dentition and terrestrial birds. *Homunculus* was a medium-sized diurnal monkey with greater development of shearing crests than *Carlocebus* or *Soriacebus* suggesting a greater component of either leaves or insects in its diet. Its dentition is most similar to that of *Aotus* or *Callicebus*. Likewise, the limbs (especially the femur) of *Homunculus* are most similar to those of *Aotus*, *Callicebus* or callitrichines, suggesting a quadrupedal, leaping monkey (e.g., Ford, 1988).

Most authorities have noted the dental and gnathic similarities between *Homunculus* and *Aotus* and *Callicebus*; however, as Tejedor (1997, 1998*b*) has emphasized, these features are likely to be primitive for platyrrhines. Tauber (1991) found a cranial endocast that showed similarities to pitheciines.

In contrast with the younger fossil platyrrhines from Colombia, the early Miocene fossil platyrrhines from southern Argentina do not show unequivocal affinities with extant subfamilies (or genera), except, perhaps, for *Proteropithecia* (Kay *et al.*, 1998*a*). Rather they show a novel mix of possible derived similarities to various extant subfamilies combined with other features not found in the extant clades. Many workers (e.g., Rosenberger, Tejedor) have interpreted this character distribution as evidence that many of these taxa lie near the base of modern clades (e.g., *Soriacebus*, *Carlocebus* and *Homunculus* with pitheciines, including *Callicebus*; *Dolichocebus* with cebines), before the total suite of features characterizing the modern clades had accumulated. This interpretation accords well with molecular estimates of the timing of platyrrhine evolution (e.g., Porter *et al.*, 1997). There are, however, two alternative views. Meldrum (1993; Meldrum & Kay, 1997*b*) has repeatedly argued that the Patagonian primates (except for *Proteropithecia*) are an early, geographically isolated radiation of primates that may have had no direct relationship with any extant platyrrhine subfamilies. In another strikingly different view, Horovitz (1999) found that all of the Patagonian primates were most closely allied with callitrichines. Thus far there has been no attempt to address these contrasting views.

Primary References

Branisella

Hoffstetter, R. (1969). Un primate de l'Oligocène inférieur sud-americain: *Branisella boliviana* gen. et. sp. nov. *Comptes rendus de l'Académie des Sciences de Paris*, **269**, 434–437.

Rosenberger, A. L. (1981). A mandible of *Branisella boliviana* (Platyrrhini, Primates) from the Oligocene of South America. *International Journal of Primatology*, **2**, 1–7.

Wolff, R. (1984). New fossil specimens of *Branisella boliviana* from the Early Oligocene of Salla, Bolivia. *Journal of Vertebrate Paleontology*, **4**, 570–574.

Rosenberger, A. L., Hartwig, W. C., & Wolff, R. G. (1991). *Szalatavus attricuspis*, an early platyrrhine primate. *Folia Primatologica*, **56**, 225–233.

Takai, M. & Anaya, F. (1996). New specimens of the oldest fossil platyrrhine, *Branisella boliviana*, from Salla, Bolivia. *American Journal of Physical Anthropology*, **99**, 301–317.

Takai, M., Anaya, F., Shigehara, N., & Setoguchi, T. (2000*a*). New fossil materials of the earliest New World Monkey, *Branisella boliviana*, and the problem of platyrrhine origins. *American Journal of Physical Anthropology*, **111**, 263–282.

Carlocebus

Anapol, F. & Fleagle, J. G. (1988). Fossil platyrrhine forelimb bones from the early Miocene of Argentina. *American Journal of Physical Anthropology*, **76**, 417–428.

Fleagle, J. G. (1990). New fossil platyrrhines from the Pinturas Formation, southern Argentina. *Journal of Human Evolution*, **19**, 61–85.

Meldrum, D. J. (1990). New fossil platyrrhine tali from the early Miocene of Argentina. *American Journal of Physical Anthropology*, **83**, 403–418.

Fleagle, J. G. & Rae, T. C. (1992). Primate cranial remains from the Pinturas Formation, Argentina. *American Journal of Physical Anthropology*, Supplement **14**, 75–76.

Stevens, N. J. & Fleagle, J. G. (1998). A new fossil platyrrhine femur from the Pinturas Formation of Argentina. *American Journal of Physical Anthropology*, Supplement **26**, 209.

Chilecebus

Flynn, J. J., Wyss, A. R., Charrier, R., & Swisher, C. C. (1995). An early Miocene anthropoid skull from the Chilean Andes. *Nature* **373**, 603–607.

Dolichocebus

Bordas, A. F. (1942). Anotaciones sobre un "Cebidae" fosil de Patagonia. *Physis*, **19**, 265–269.

Kraglievich, J. L. (1951). Contribuciones al conocimiento de los primates fosiles da la Patagonia I. Diagnosis previa de un nuevo primarte fosil del Oligocene superior (Colhuehuapiano) de Gaiman, Chubut. *Comunicaciones, Instituto nacional de investigación de las ciencias naturales*, **2**, 57–82.

Fleagle, J. G. & Bown, T. M. (1983). New primate fossils from late Oligocene (Colhuehuapian) localities of Chubut, Province, Argentina. *Folia Primatologica*, **41**, 240–266.

Reeser, L. A. (1984). Morphological affinities of new fossil talus of *Dolichocebus gaimanensis*. *American Journal of Physical Anthropology*, **63**, 206–207.

Homunculus

Ameghino, F. (1891*a*). Los monos fosiles del Eoceno de la Republica Argentina. *Revista argentina de historia natural*, **1**, 383–397.

Ameghino, F. (1891*b*). Nuevos restos de mamiferos fosiles descubiertos por C. Ameghino en el Eoceno inferior de la Patagonia austral. Especies nuevas, adiciones y correcciones. *Revista argentina de historia natural*, **1**, 289–328.

Bluntschli, H. (1931). *Homunculus patagonicus* und die ihm zugereihten Fossilfunde aus den Santa-Cruz-Schichten Patagoniens: eine morphologische Rivision an Hand der Originalstucke in der Sammlung Ameghino zu La Plata. *Genenbaurs morphologisches Jahrbuch* **67**, 811–892.

Hershkovitz, P. (1981). Comparative anatomy of platyrrhine mandibular cheek teeth dpm4, pm4, m1, with particular reference to those of *Homunculus* (Cebidae), and comments on platyrrhine origins. *Folia Primatologica*, **35**, 179–217.

Hershkovitz, P. (1984). More on *Homunculus* dpm4 and M1 and comparisons with *Alouatta* and *Stirtonia* (Primates, Platyrrhini, Cebidae). *American Journal of Primatology*, **7**, 261–283.

Fleagle, J. G., Buckley, G. A., & Schloeder, M. E. (1988). New fossil primates from Monte Observación, Santa Cruz Formation (lower Miocene), Santa Cruz Province, Argentina. *Journal of Vertebrate Paleontology*, **8**, 14A.

Tauber, A. (1991). *Homunculus patagonicus* Ameghino, 1891 (Primates, Cebidae), Mioceno medio, de la costa Atlantica austral, Provincia de Santa Cruz, Republica Argentina. *Academia nacional de ciencias (Cordoba, Argentina), Miscelanea*, **82**, 1–32.

Tejedor, M. (in press). New material of *Homunculus patagonicus* (Primates, Platyrrhini) from the Santacrucian of Argentina. *American Journal of Physical Anthropology*.

Proteropithecia

Kay, R. F., Johnson, D., & Meldrum, D. J. (1998). A new pitheciin primate from the middle Miocene of Argentina. *American Journal of Primatology*, **45**, 317–336.

Kay, R. F., Johnson, D., & Meldrum, D. J. (1999). Corrigendum. *American Journal of Primatology*, **47**, 347.

Soriacebus

Fleagle, J. G., Powers, D. W., Conroy, G. C., & Watters, J. P. (1987). New fossil platyrrhines from Santa Cruz Province, Argentina. *Folia Primatologica*, **48**, 65–77.

Fleagle, J. G. (1990). New fossil platyrrhines from the Pinturas Formation, southern Argentina. *Journal of Human Evolution*, **19**, 61–85.

Meldrum, D. J. (1990). New fossil platyrrhine tali from the early Miocene of Argentina. *American Journal of Physical Anthropology*, **83**, 403–418.

Tremacebus

Rusconi, C. (1933). Nuevos restos de monos fosiles del terciario antiguo de la Patagonia. *Anales ciencias argentina*, **116**, 286–289.

Rusconi, C. (1935). Los especies de primates del oligoceno de Patagonia (gen. *Homunculus*). *Revista argentina de paleontologia y antropologia ameghinia*, **1**, 39–126.

Hershkovitz, P. (1974). A new genus of late Oligocene monkey (Cebidae, Platyrrhini) with notes on postorbital closure and platyrrhine evolution. *Folia Primatologica*, **21**, 1–35.

12 | Miocene platyrrhines of the northern Neotropics

WALTER CARL HARTWIG AND D. JEFFREY MELDRUM

Introduction

The fossil record of New World monkeys in the northern Neotropics is confined to La Venta, a small region of the Magdalena River valley in the Cordillera Central of Colombia. Nonetheless, the variety of taxa from this locality is comparable to the Fayum and is the most diverse assemblage of extinct platyrrhines yet discovered. Dated to 10–14 Ma (Flynn *et al.*, 1997), La Venta also constitutes the limit of the known fossil record of New World monkeys prior to the terminal Pleistocene.

Similarity with living species is the most obvious trend among these fossils. Unlike fossil primates of the Miocene from elsewhere, the platyrrhines of La Venta can be classified into modern subfamilies with little debate. Authorities disagree on the taxonomic expression of this similarity, but agree in general on the modern aspect of the fossils (Fleagle, 1999).

The existence of so many different and modern-looking species indicates that platyrrhine biodiversity has a deep and extensive evolutionary history. Compared to the ambiguous relationships of earlier platyrrhine fossils from southern South America and the unusually derived appearance of most Pleistocene–Holocene species, La Venta remains the most direct reflection of the ancestry of living New World monkeys. The La Venta vertebrate fauna is rich enough, in fact, to allow a rare view of the paleoecology of an extinct primate radiation. Kay & Madden (1997) conclude that the mammalian community of the monkey-bearing deposits best matches the profile of a forested community in the modern Neotropics.

History of discovery and debate

The history of primate discoveries at La Venta began with survey activity in the 1940s and peaked in the late 1980s and early 1990s. For the full history of paleontology in the La Venta badlands please read the excellent review by Madden *et al.* (1997). The fossil record here first came to light during the paleontological explorations of Ruben Arthur Stirton of the University of California, Berkeley, and José Royo y Gomez of the National Geological Service of Colombia in the mid-1940s. During several years of collecting by the UC–Berkeley team a primate partial skeleton and two fragmentary primate mandibles were recovered.

The fossil material clearly constituted three different taxa. The partial skeleton, named *Cebupithecia sarmientoi*, was recognized to be like living saki monkeys (Stirton & Savage, 1951), and the smaller partial mandible, named *Neosaimiri fieldsi*, was recognized to resemble living squirrel monkeys (Stirton, 1951). The larger partial mandible showed less obvious similarities to living taxa and was attributed by Stirton (1951) to a new species (*tatacoensis*) of the ambiguous extinct primate genus *Homunculus*.

No more fossil platyrrhines from this part of South America would be discovered until the early 1980s. In the meantime, with one exception the initial taxonomic assessments of Stirton (1951) and Stirton & Savage (1951) became accepted, if unexamined, in texts and syntheses (Patterson & Pascual, 1972; Simons, 1972, 1976; Orlosky & Swindler, 1975; Paula-Couto, 1979; Szalay & Delson, 1979; Rose & Fleagle, 1981). The exception was *Homunculus tatacoensis*, which Hershkovitz (1970) reassessed to be distinctly similar to living howler monkeys. He pulled the specimen out of *Homunculus* and proposed the current genus name, *Stirtonia*.

Szalay & Delson (1979) formalized the affinity between the La Venta genera and living taxa by classifying them within modern higher categories: *Stirtonia* within Tribe Alouattini, *Neosaimiri* within Subfamily Cebinae, and *Cebupithecia* within Tribe Pitheciini. This assignment, together with dissertation research by Frank James Orlosky (1973) and Rosenberger (1979*b*), neutralized any question of affinity between *Stirtonia* and *Homunculus* and dismissed arguments by Hershkovitz (1970) that *Cebupithecia* and the pitheciines were not closely related. Subsequent discoveries at La Venta have expanded the breadth of platyrrhine diversity there but in large measure have affirmed the trend of modernity established by the Berkeley field seasons.

In the late 1970s a team of Japanese researchers based at Kyoto University began a long-term field project in the La Venta deposits. They reported discovery of additional *Stirtonia* teeth (Setoguchi, 1980; Setoguchi *et al.*, 1981), and would continue to survey La Venta for monkey fossils for several more years. In the meantime Miocene platyrrhines were included in broader analyses of New World monkey evolution (Ciochon & Chiarelli, 1980*a*; Delson & Rosenberger, 1984). Morphological similarities between the La Venta primates and living radiations were emphasized by Rosenberger (1980, 1981*b*, 1984), who established a collaboration with the Kyoto University project.

In 1980 a brief expedition to La Venta by a Field Museum of Natural History team recovered a primate partial mandible, which was first described and named to a new genus (*Mohanamico*) in 1986 (Luchterhand *et al.*, 1986). Although this field survey was incidental compared to the efforts of the Kyoto University and Duke University projects (see below), the publication of *Mohanamico* initiated a debate between the two university-based research teams that has influenced interpretation of more recent discoveries.

In 1983 the Kyoto University team recovered several isolated teeth and fragmentary primate fossils. Material originally named *Kondous laventicus* (Setoguchi, 1985; Setoguchi & Rosenberger, 1985*b*) is now considered to be part of *Stirtonia*. A small isolated dentine core of a lower molar tooth became the type specimen of *Micodon*, nominated by Setoguchi & Rosenberger (1985*a*) as evidence of the Miocene emergence of marmosets. A partial mandible and maxillary fragment discovered in 1986 so closely resembled living owl monkeys to Setoguchi & Rosenberger (1987) that they nominated it to a species of the same genus – *Aotus dindensis*. The *Aotus dindensis* mandible resembled *Mohanamico*, however, and a vigorous taxonomic debate ensued (see below).

Shortly after the Kyoto project began Richard Frederick Kay and colleagues at Duke University began a long-term project at La Venta that has recovered several new platyrrhine fossils and initiated a new period of interest in New World monkey evolutionary history. Between the efforts of the Kyoto and Duke projects nine new species have been recovered at La Venta, bringing the total to 12 for this region dated to between 10 and 14 million years ago. The first new species published by the Duke project was named *Stirtonia victoriae* and consisted of subadult right and left partial maxillae, an abraded adult right maxilla and a small frontal fragment (Kay *et al.*, 1987). This represented the first indication of something beyond a monospecific genus at La Venta. The Duke project also recovered postcranial material, including two partial humeri in 1986 referred ultimately to the existing taxa *Cebupithecia* and *Neosaimiri* (Meldrum *et al.*, 1990), and a partial skeleton referred to *Cebupithecia* (Meldrum & Kay, 1997*b*). Together with a detailed analysis of the *Cebupithecia* axial skeleton (Meldrum & Lemelin, 1991), these discoveries firmly established the unity of the taxon and its phylogenetic relationship to living pitheciines. In the meantime both the Kyoto and Duke University projects were refining the chronology of the La Venta sediments (Hayashida, 1984; Takemura & Danhara, 1986).

The impact of the 1980s fieldwork is reflected in a special issue of the *Journal of Human Evolution* entitled "The platyrrhine fossil record" and edited by John G. Fleagle and Rosenberger (1990). Papers presented in that symposium volume include a detailed analysis of the *Cebupithecia* postcranium (Ford, 1990*a*) supporting its affinities with modern *Pithecia* (see also Davis, 1987). More controversial were two papers debating the differences between *Mohanamico* and *Aotus dindensis* and initiating a decade of exchanges about phylogenies and systematics of the La Venta fossils. By 1990 La Venta was established as a Miocene snapshot of platyrrhine diversity (e.g., Fleagle, 1988), even if the exact number of different taxa was unresolved.

Until this point the phylogenetic arguments advanced by Rosenberger and colleagues and Kay and colleagues had proceeded independently of one another because of the almost simultaneous publication of *Mohanamico* and *Aotus dindensis*. The original description of *Mohanamico* placed it as a primitive pitheciine with affinities to *Callimico*, thus implying a possible monophyletic relationship between callitrichines and pitheciines (Luchterhand *et al.*, 1986). Rosenberger *et al.* (1990) supported the relationship between *Mohanamico* and callitrichines on the basis of shared features in the canine–premolar honing complex. They also distanced it from the pitheciine clade by dismissing the synapomorphic characters described by Luchterhand *et al.* (1986) and Kay (1990). Kay (1990) argued, however, that features shared between *Mohanamico* and *Callimico* are either present in other New World monkey groups or are outweighed by other derived features of the same teeth that unite *Mohanamico* with pitheciines. Kay also disputed the similarities between *Aotus dindensis* and living owl monkeys and noted instead that *Aotus dindensis* shared many proportional dental measurements and features with *Mohanamico*. Kay (1990) advocated lumping *Aotus dindensis* into *Mohanamico* and maintaining the genus as a primitive pitheciine. Rosenberger *et al.* (1990) maintained that the occlusal design in the two fossils was significantly different and reiterated that the maxillary fragment of *Aotus dindensis* was consistent with that region in living owl monkeys. They maintained separate generic status, as well as phylogenetic placement, for *Mohanamico* and *Aotus dindensis*.

This debate reflects not only subjective interpretation of qualitative features, but even more fundamentally the methodological differences that prevail in many interpretations of the primate fossil record. In the specific case of La Venta primates, Rosenberger and colleagues have emphasized the diversity of the fossils and any apparent links between them and living taxa. Kay and colleagues, on the other hand, have emphasized a more cladistic approach to the same fossils and are less persuaded by their similarities to modern species. This difference in approaches crystallized in the debate over *Mohanamico*/*Aotus dindensis* and would be amplified by later discoveries at La Venta that, fortuitously perhaps, reinforced each position.

The 1988 Kyoto field season recovered two primate tali and a partial mandible. Gebo *et al.* (1990) allocated one talus to cf. *Aotus dindensis*, and the other to a new taxon based also on the partial mandible. Setoguchi *et al.* (1990) previewed this new taxon as closely related to *Neosaimiri*, and Rosenberger *et al.* (1991*b*) formalized it as *Laventiana annectens*. They described it to be similar in many ways to *Neosaimiri*, differing in molar proportions and the presence of an apparently autapomorphic molar crown feature, a postentoconid notch. At the same time Rosenberger *et al.* (1991*a*) sunk *Neosaimiri* into the living genus *Saimiri* on the grounds that no significant features of the preserved anatomy in *Neosaimiri* were outside of the range of the living genus.

The 1989 and 1990 Kyoto field seasons, however, would recover specimens inconsistent with this analysis. Masanaru Takai (1994) reported more than 200 dental specimens and allocated them all to *Neosaimiri* based on provenance and morphology. The additional specimens clarified enough differences between *Neosaimiri* and *Saimiri* to convince Takai (1994) of the generic integrity of *Neosaimiri*. The molar series displayed a range of expression of the postentoconid notch morphology from *Neosaimiri*-like to *Laventiana*-like, sometimes within the same jaw fragment. This led Takai (1994) to subsume *Laventiana annectens* into the hypodigm of *Neosaimiri*. Kay & Meldrum (1997) later separated the *Laventiana* mandible out as a second species of *Neosaimiri* (*N. annectens*).

Coincident with the annual surveys by Kyoto University, the Duke University expedition continued to expand the survey region around La Venta. Four new genera of extinct platyrrhines have been recovered as a result of this effort (Kay, 1994; Kay & Meldrum, 1997; Meldrum & Kay, 1997a; Kay *et al.*, 1998*a*). The first to be published was a badly crushed cranium aptly named *Lagonimico* (pancake monkey) and referred to the callitrichines (Kay, 1994) based on visible occlusal morphology and a very reduced third molar.

In 1997 the Duke University project announced two new genera, both based on mandibular specimens. Meldrum & Kay (1997*a*) referred one, *Nuciruptor rubricae*, to the pitheciines and Kay & Meldrum (1997) referred the other, *Patasola magdalenae*, broadly to the callitrichines. The incisor–canine–premolar battery of *Nuciruptor* indicated to Meldrum & Kay (1997*a*) that a primitive pitheciine morphological pattern was well expressed at La Venta, perhaps to the exclusion of *Mohanamico* if it includes the *A. dindensis* hypodigm. Moreover, the combination of traits in *Nuciruptor* that placed it as a sister group to *Cebupithecia* and living pitheciines suggested that other fossil material attributed to *Cebupithecia* (as reported by Setoguchi *et al.*, 1987; Meldrum & Kay, 1997*b*) may in fact be part of *Nuciruptor*.

Patasola is represented by a subadult lower jaw and is in the size range of the living callitrichines. Molar wear facet morphology indicates a hypocone on upper molars, and *Patasola* does retain a third molar (although it is partially encrypted in the fossil). For these and other reasons Kay & Meldrum (1997) interpreted *Patasola* to be the sister taxon to the two-molared marmosets and tamarins. Meldrum & Kay (1997*b*) also reviewed the postcranial fossils known at that time from La Venta. In addition to the new material referred by them to *Cebupithecia* (see above), they reported a distal tibia whose morphology is consistent with the *Laventiana* talus (Gebo *et al.*, 1990).

The *Patasola* and postcranial descriptions are chapters in the second major synthetic edited work to feature La Venta primates. *Vertebrate Paleontology in the Neotropics* (1997; edited by Kay, Madden, Cifelli, & Flynn) comprehensively reviews the paleontological data of La Venta and assesses, for the first time, the paleobiological and paleoecological implications of the fauna. A summary chapter of the primate data (Fleagle *et al.*, 1997*b*) promotes a general theme of taxonomic diversity among the platyrrhines while emphasizing that recent discoveries are not as modern in aspect as the older discoveries appear to be.

The most recent new material reported from La Venta is a large collection of postcranial material referred to *Neosaimiri* (Nakatsukasa *et al.*, 1997). Specimens include parts of the humerus, ulna, femur, tibia, talus and calcaneus. The overall morphology of the new material is consistent with a close functional and phylogenetic relationship between *Neosaimiri* and living squirrel monkeys.

La Venta accounts for the entire discussion of the Miocene fossil record of platyrrhines in the northern Neotropics, but fossil platyrrhine teeth have been found at one other locality. The late Miocene fauna recovered from the western Brazil site of Rio Acre (Frailey, 1986) contains a lower molar allocated to *Stirtonia* sp. (Kay & Frailey, 1993), and a substantially sized upper molar referred to the cebine subfamily (see Fleagle *et al.*, 1997*b*).

The first edition of Fleagle's influential text *Primate Adaptation and Evolution* was published prior to the flood of new information from La Venta. In the second edition of this book (Fleagle, 1999) he offers the most objective assessment of the La Venta primates currently available. To manage this he cautiously recognizes the distinctiveness of *Aotus dindensis* and *Laventiana annectens* and affirms the modern aspect of many La Venta specimens. At the same time he notes the diversity of taxa that have been considered pitheciine or callitrichine, or both, and affirms the difficulty of recognizing stem taxa of these radiations.

For 30 years prior to the 1980s only three taxa of ancient platyrrhines were known from the northern Neotropics. The pace of discovery in the last 20 years has nearly quadrupled this diversity and created a cloud of taxonomic interpretation. The number of monotypic genera at La Venta is high, and monotypic genera are the most unstable units in phylogenetic systematics. On the other hand, to the extent that these monotypic genera accord with extant clades they may survive future systematic revisions. Moreover, the potential for discovering new taxa is high. The modern subfamilies of New World monkeys are not equally represented at La Venta. The atelins, *Cebus* and *Callicebus*, which are widely distributed today, are poorly sampled or absent at La Venta.

Taxonomy

Systematic framework

To put this part of the fossil record into taxonomic perspective, the following classification combines La Venta primate taxa with the 16 living genera of New World monkeys. At issue in many previous phylogenies of the living genera was the placement of the owl monkey, genus *Aotus*, and the titi monkey, genus *Callicebus* (see Ford, 1990*a*; Kay, 1990; Rosenberger, 1992). Because of disagreements among morphologists, the classification used here follows a general consensus platyrrhine phylogeny derived from several recent molecular studies (Canavez et al., 1999; Horovitz, 1999; Porter et al., 1999; von Dornum & Ruvolo, 1999). Current molecular studies support morphological arguments (Rosenberger, 1992) that group *Callicebus* with the pitheciines. They also consistently place *Aotus* as an outgroup to cebines and callitrichines within the family Cebidae.

A classification of four to five subfamilies of living New World monkeys with some debate about family level grouping is widely promoted (Fleagle, 1999). The following classification lists all taxa supported by Fleagle (1999), the validity of some of which is currently debated (see above). Assignment of La Venta taxa within the subfamilies is widely debated (e.g., Rosenberger et al., 1990, 1991*a* vs. Kay & Meldrum, 1997; Takai, 1994).

Order Primates Linnaeus, 1758
- Infraorder Platyrrhini É. Geoffroy Saint-Hilaire, 1812
  - Family Atelidae Gray, 1825
    - Subfamily Atelinae Gray, 1825
      - Tribe Atelini Gray, 1825
        - Subtribe Atelina
          - Genus *Ateles* É. Geoffroy Saint-Hilaire, 1806
        - Subtribe Lagotrichina Gray, 1825
          - Genus *Lagothrix* É. Geoffroy Saint-Hilaire, 1812
          - Genus *Brachyteles* Spix, 1823
      - Tribe Alouattini Elliott, 1904
        - Genus *Alouatta* Lacépède, 1799
        - Genus *Stirtonia* Hershkovitz, 1970
          - *Stirtonia tatacoensis* Stirton, 1951
          - *Stirtonia victoriae* Kay et al., 1987
    - Subfamily Pitheciinae Mivart, 1865
      - Tribe Pitheciini Gray, 1849
        - Genus *Pithecia* Desmarest, 1820
        - Genus *Chiropotes* Lesson, 1840
        - Genus *Cacajao* Lesson, 1840
        - Genus *Cebupithecia* Stirton & Savage, 1951
          - *Cebupithecia sarmientoi* Stirton & Savage, 1951
        - Genus *Nuciruptor* Meldrum & Kay, 1997
          - *Nuciruptor rubricae* Meldrum & Kay, 1997
      - Tribe Callicebini Thomas, 1903
        - Genus *Callicebus* Thomas, 1903
  - Family Cebidae Bonaparte, 1831
    - Subfamily Cebinae Bonaparte, 1831
      - Genus *Cebus* Erxleben, 1777
      - Genus *Saimiri* Voigt, 1831
      - Genus *Neosaimiri* Stirton, 1951
        - *Neosaimiri fieldsi* Stirton, 1951
      - Genus *Laventiana* Rosenberger et al., 1991
        - *Laventiana annectens* Rosenberger et al., 1991
    - Subfamily Aotinae Elliot, 1913
      - Genus *Aotus* Illiger, 1811
        - *Aotus dindensis* Setoguchi & Rosenberger, 1987
    - Subfamily Callitrichinae Thomas, 1903
      - Tribe Callitrichini
        - Genus *Callithrix* Erxleben, 1777
        - Genus *Cebuella* Gray, 1866
      - Tribe Saguini
        - Genus *Saguinus* Hoffmannsegg, 1807
        - Genus *Leontopithecus* Lesson, 1840
      - Tribe Callimiconi Thomas, 1913
        - Genus *Callimico* Miranda Ribeiro, 1912
        - Genus *Mohanamico* Luchterhand et al., 1986
          - *Mohanamico hershkovitzi* Luchterhand et al., 1986
      - Tribe *incertae sedis*
        - Genus *Patasola* Kay & Meldrum, 1997
          - *Patasola magdalenae* Kay & Meldrum, 1997
        - Genus *Lagonimico* Kay, 1994
          - *Lagonimico conclutatus* Kay, 1994
        - Genus *Micodon* Setoguchi & Rosenberger, 1985
          - *Micodon kiotensis* Setoguchi & Rosenberger, 1985

Family Atelidae

Subfamily Atelinae

Tribe Alouattini

GENUS *Stirtonia* Hershkovitz, 1970
This genus is based on a partial mandible described by Stirton in 1951 and referred by him to the extinct genus *Homunculus*. Hershkovitz (1970) demonstrated differences between Stirton's mandible and the *Homunculus* collection in incisor morphology, molar morphology, and overall size. Hershkovitz (1970) proposed a new genus name for the partial mandible, *Stirtonia*, which now includes some referred specimens and a second species.

Features of the genus include relatively hypertrophied molars that display a prominent buccal cingulum and crista obliqua. P_4 is molariform and the mandibular margin diverges markedly posteriorly. In general aspect the specimens are very similar to but not as specialized as the same elements in the living howler monkeys (*Alouatta*). Unlike other La Venta genera, *Stirtonia* may extend beyond the locality to tropical Brazil based on an isolated tooth recovered in Rio Acre and assigned to *Stirtonia* sp. (Kay & Frailey, 1993).

INCLUDED SPECIES *S. tatacoensis*, *S. victoriae*

SPECIES *Stirtonia tatacoensis* Stirton, 1951 (Fig. 12.1)
TYPE SPECIMEN UCMP 38989, partial mandible
AGE AND GEOGRAPHIC RANGE 12.6–13.7 Ma, La Venta, Colombia
ANATOMICAL DEFINITION
This species is based on Stirton's (1951) partial mandible,

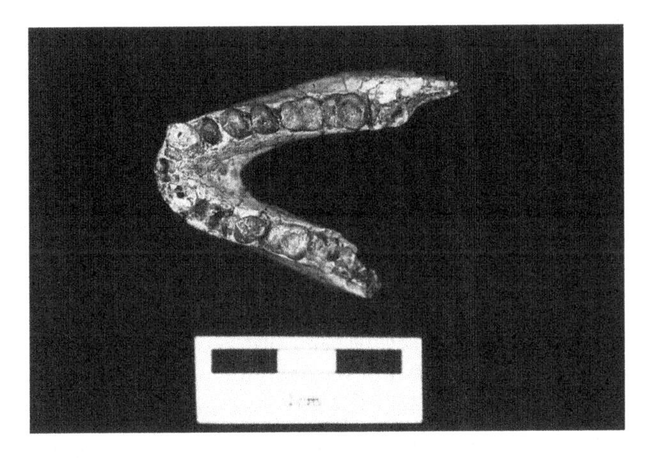

Fig. 12.1 *Stirtonia tatacoensis*. Holotype (UCMP 38989).

several isolated teeth and a maxillary fragment described by Setoguchi (1985). Collectively the material displays a relatively large molar battery and a mandibular corpus that is divergent posteriorly. The symphysis region is well buttressed as in *Alouatta*. It differs from *Alouatta* by having relatively constant mandibular depth, a trigon that opens lingually and in the subequal proportions of the trigonid and talonid on P_4. Smaller size distinguishes it from *S. victoriae*.

SPECIES *Stirtonia victoriae* Kay *et al.*, 1987
TYPE SPECIMEN DU-IGM 85-400, partial right maxilla and premaxilla; DU-IGM 86-534, partial left palate with associated left frontal fragment
AGE AND GEOGRAPHIC RANGE 12.6–13.7 Ma, La Venta, Colombia
ANATOMICAL DEFINITION
This species is also known from dental remains. The molars show greater development of stylar shelves and cusps, and the premolars are three-rooted. Kay *et al.* (1987) describe the preserved facial skeletal anatomy as similar to *Alouatta seniculus* in nearly all respects. Some of these features include a small premaxilla, inflated maxillary sinuses, and a stout and flaring zygomatic arch.

Subfamily Pitheciinae

Tribe Pitheciini

GENUS *Cebupithecia* Stirton & Savage, 1951
INCLUDED SPECIES *C. sarmientoi*

SPECIES *Cebupithecia sarmientoi* Stirton & Savage, 1951 (Fig. 12.2)
TYPE SPECIMEN UCMP 38762, partial skeleton
AGE AND GEOGRAPHIC RANGE 12.1–12.5 Ma, La Venta, Colombia
ANATOMICAL DEFINITION
Cebupithecia is represented by one of the most complete

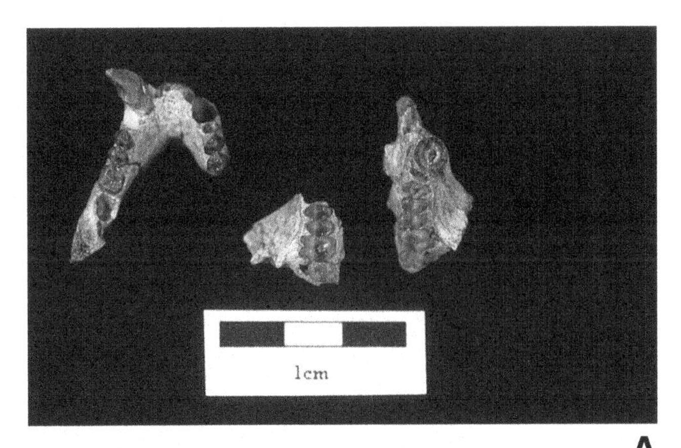

A

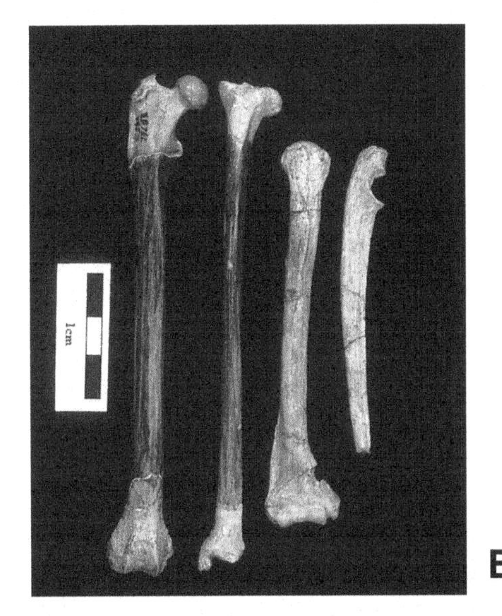

B

C

Fig. 12.2 *Cebupithecia sarmientoi*. Holotype (UCMP 38762): (A) mandible (left), maxillary fragments (center, right); (B) representative postcrania of the holotype; from left to right – femur, tibia, humerus, ulna; (C) composite postcranial skeleton of the holotype.

skeletons of a fossil New World primate. The dentition bears numerous synapomorphies to the modern pitheciins, especially in the incisor–canine complex. The postcanine dentition is less specialized than in the modern forms, but the molars are very bunodont with shallow hypoflexids and

A

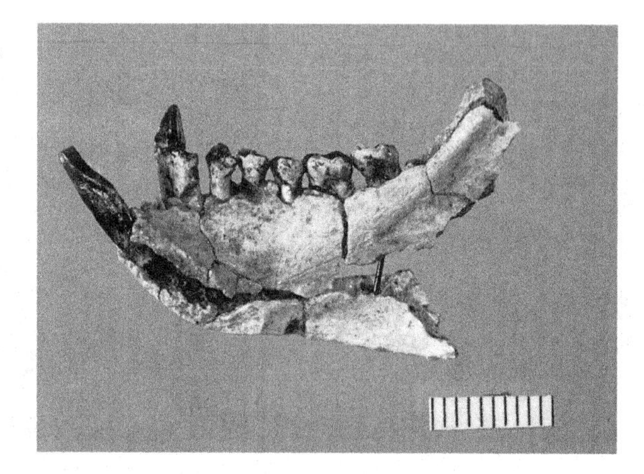

B

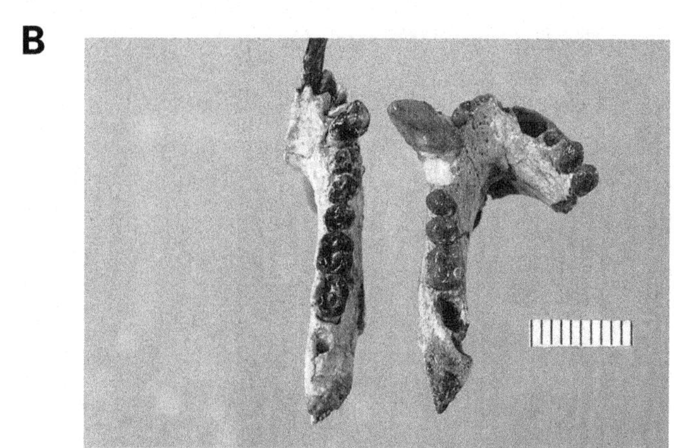

Fig. 12.3 *Nuciruptor rubricae*. (A) Holotype mandible (IGM 251074). (B) Occlusal view of *Nuciruptor* (left) and *Cebupithecia* (right).

raised talonids. The third molar is reduced in size to a greater extent than even the modern pitheciins.

The holotype preserves much of the postcranial skeleton, which bears numerous phenetic resemblences to *Pithecia* and yet lacks many derived features that characterize modern pitheciins, especially those associated with suspensory postures. Instead arboreal quadrupedal running and leaping are indicated as the dominant locomotor modes.

GENUS *Nuciruptor* Meldrum & Kay, 1997
INCLUDED SPECIES *N. rubricae*

SPECIES *Nuciruptor rubricae* Meldrum & Kay, 1997 (Fig. 12.3)
TYPE SPECIMEN IGM 251074, partial mandible
AGE AND GEOGRAPHIC RANGE 12.1–12.5 Ma, La Venta, Colombia
ANATOMICAL DEFINITION
Nuciruptor is a *Pithecia*-sized primate represented by a right mandibular corpus and well-preserved dentition. The corpus deepens posteriorly and the dentition exhibits trends evident in the modern pitheciin clade, including styliform incisors and bunodont molars. However these are not expressed to the same degree in that the incisors are shorter and less procumbent, the canine although projecting lacks a lingual crest, the premolars are less enlarged and the molars have more cusp relief. *Nuciruptor* appears to confirm the prediction of Kinzey (1992) that adaptations of the anterior dentition for sclerocarpy would precede modifications to the postcanine dentition.

A talus was recovered from the same locality as the type mandible of *Nuciruptor* (Ford *et al.*, 1991). It measures approximately 14.0 mm in length, suggesting a monkey intermediate in size to *Callicebus* and *Pithecia* (Meldrum, 1990). It shares a number of similarities with the talus of *Cebupithecia*, but has a somewhat narrower head and neck, a slightly more rounded lateral trochlear crest. The correlation of talar length and M_1 length leaves room for a possible allocation to this species.

Family Cebidae

Subfamily Cebinae

GENUS *Neosaimiri* Stirton, 1951
INCLUDED SPECIES *N. fieldsi*

SPECIES *Neosaimiri fieldsi* Stirton, 1951 (Fig. 12.4)
TYPE SPECIMEN UCMP 39205, partial mandible
AGE AND GEOGRAPHIC RANGE 12.1–12.5 Ma, La Venta, Colombia
ANATOMICAL DEFINITION
The holotype partial mandible very closely resembles the same element of the living squirrel monkey, genus *Saimiri*. Molars of the holotype present moderate development of the buccal cingulid. Several additional fragmentary jaws and gnathic remains described by Takai (1994) indicate a relatively parabolic mandibular arcade shape, and relatively small incisors and mesiodistally longer molars compared to living *Saimiri*. Upper teeth show moderately developed lingula cingula on I^{1-2}, sexually dimorphic canines, a cingulum-derived hypocone on P^4, and relatively small third molars.

A distal humerus described by Meldrum *et al.* (1990) also closely resembles *Saimiri* and presents a relatively deep radial fossa separated from the coronoid fossa by a prominent ridge, and a relatively deep olecranon fossa accentuated by retroflexion of the medial epicondyle and prominence of the dorsolateral margin of the trochlea. More postcrania attributed to *Neosaimiri* (Nakatsukasa *et al.*, 1997) include a proximal ulna with a relatively long olecranon process, thus matching the fossa of the 1990 distal humerus. The proximal femur resembles both living squirrel monkeys and the tamarins (genus *Saguinus*), while the distal femur, distal tibia, calcaneus and four tali share traits with these taxa and *Aotus* (Fig. 12.5).

GENUS *Laventiana* Rosenberger *et al.*, 1991
INCLUDED SPECIES *L. annectens*

A

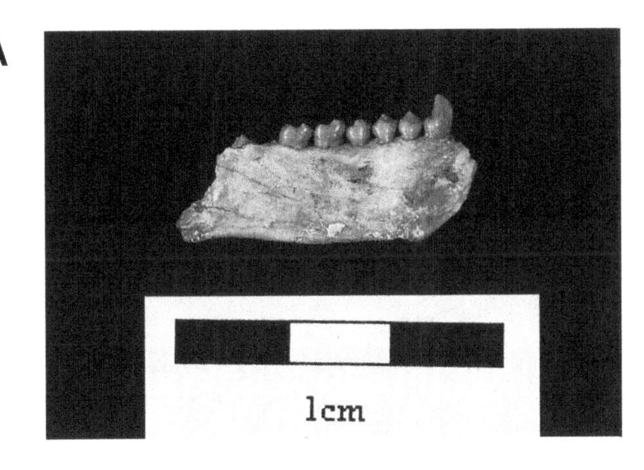

B

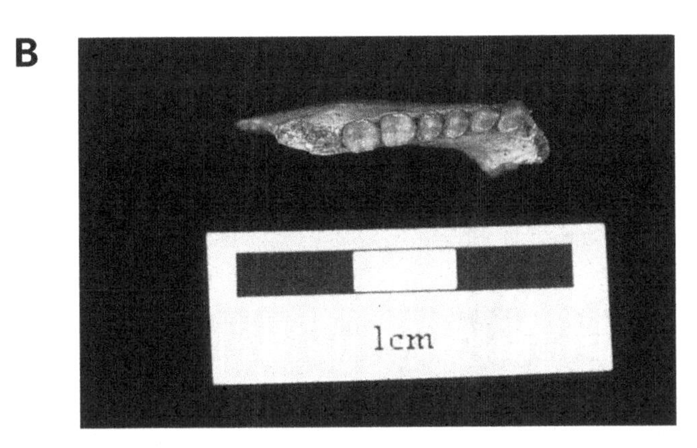

C

D

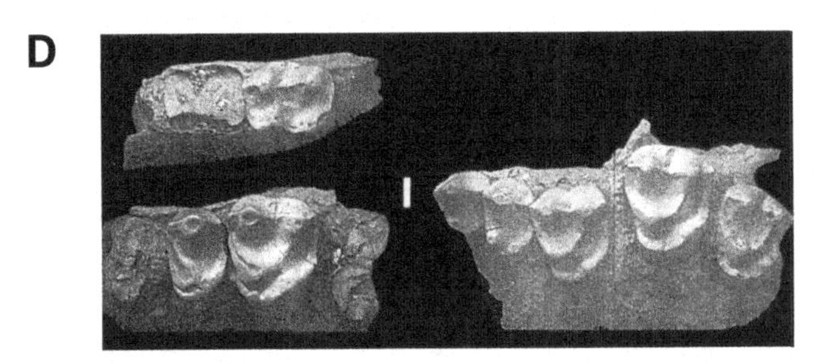

Fig. 12.4 *Neosaimiri fieldsi*. (A–C). Holotype (UCMP 39205); (D) IGM-KU 90028 (above) and 90005 (below), the first examples of upper dentition in *N. fieldsi*. (D) Courtesy of Masanaru Takai.

SPECIES *Laventiana annectens* Rosenberger *et al.*, 1991 (Fig. 12.6)
TYPE SPECIMEN IGM-KU 8801a, partial mandible; IGM-KU 8803, right talus (referred to as IGM-KU 8801b in Rosenberger *et al.*, 1991c)
AGE AND GEOGRAPHIC RANGE 10–12.5 Ma, La Venta, Colombia
ANATOMICAL DEFINITION
The *Laventiana* mandible is *Saimiri*-sized and similar to squirrel monkeys and to *Neosaimiri fieldsi* in many ways. The mandible presents acutely cusped postcanine crowns, a relatively broad P_2, and a shallow mandibular corpus (7.6 mm) that does not deepen posteriorly. A distinct postentoconid notch is present on M_{1-2}, but the autapomorphic nature of this trait relative to living species is disputed (Takai, 1994; Schmidt & Meldrum, 2000). The relatively long and narrow molars also display incipient buccal cingulids.

A talus was associated with the type mandile of *Laventiana* (Gebo *et al.*, 1990) is very similar to extant *Saimiri*, but lacks a dorsal tibial stop. A distal tibia fragment recovered from the same site accords in size and morphology (Meldrum & Kay, 1997*b*). Similarities to *Saimiri* include marks of a well-developed syndesmosis of the distal tibiofibular joint. It differs from *Saimiri*, and from most callitrichines and small-bodied platyrrhines in that the articular surface of the tibial trochlea does not extend onto the anterior surface of the tibial shaft. These and subsequent postcranials have been subsumed within *Neosaimiri* by Meldrum & Kay (1997*b*) and Nakatsukasa *et al.* (1997).

Subfamily Aotinae

GENUS *Aotus* Illiger, 1811
INCLUDED SPECIES *A. dindensis*

SPECIES *Aotus dindensis* Setoguchi & Rosenberger, 1987 (Fig. 12.7)
TYPE SPECIMEN IGM-KU 8601, partial mandible and left maxillary fragment
AGE AND GEOGRAPHIC RANGE 12.1–12.5 Ma, La Venta, Colombia
ANATOMICAL DEFINITION
As the *Aotus* designation implies, this partial mandible very closely resembles the same element in living owl monkeys. It differs by having relatively small incisors, smaller P_{3-4} metaconids and less elevated premolar trigonids. In addition, the molar trigonids are slightly smaller and their metaconids less prominent. Mandibular depth increases posteriorly, to 10.00 mm below M_2. The maxillary fragment preserves enough of the lower orbital margin to suggest that *A. dindensis* was nocturnal.

A small isolated talus was recovered within the Monkey Unit and was referred to cf. *Aotus dindensis* on the basis of a suite of characters including a moderately short talar neck,

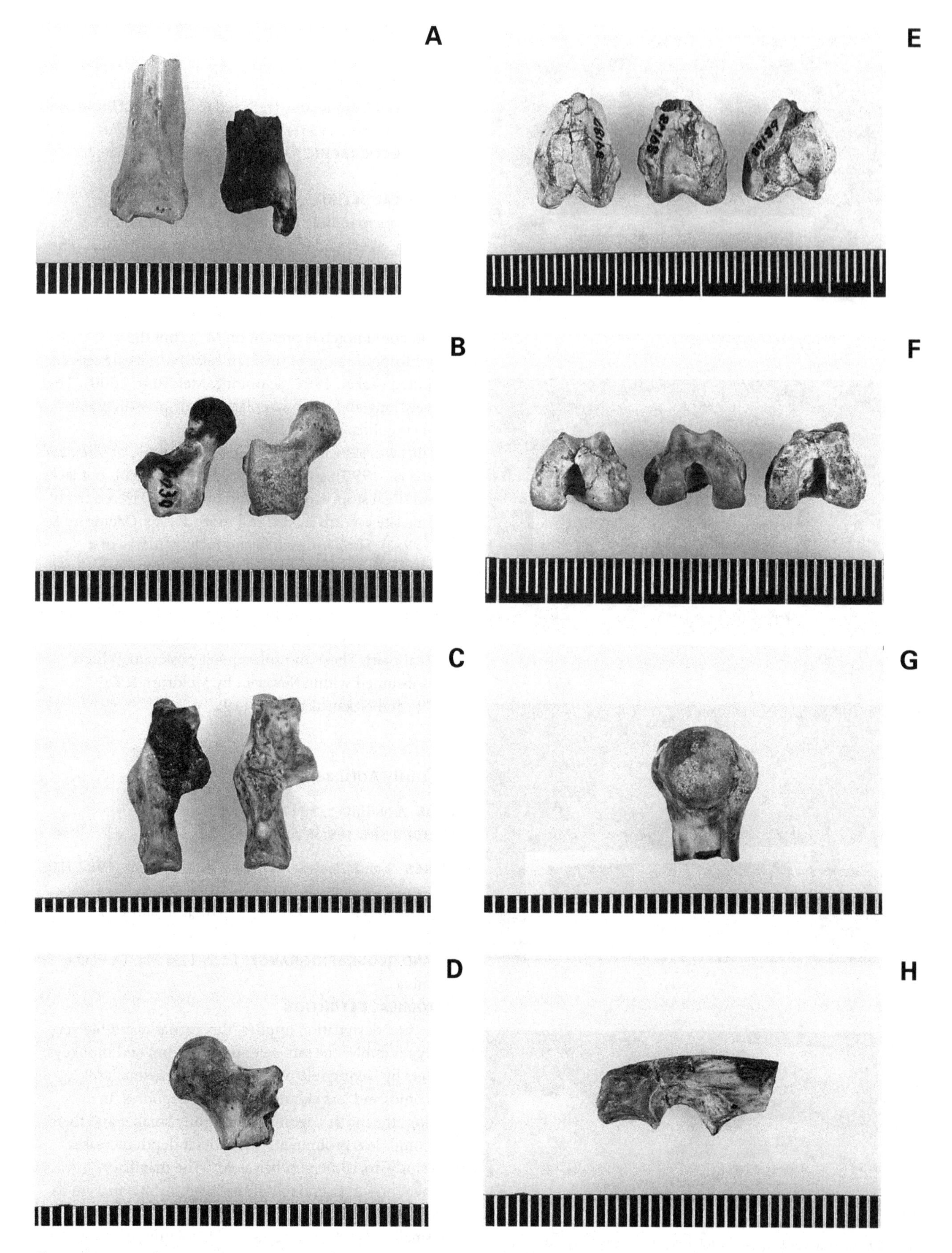

Fig. 12.5 *Neosaimiri fieldsi*. (A) Distal tibia IGM-KU 89194, 89195; (B) tali IGM-KU 89030, 89031; (C) calcanei IGM-KU 89201, 89202; (D) proximal femur IGM-KU 89181; (E) distal femora (dorsal view) IGM-KU 89187, 89188, 89189; (F) distal femora (distal view) IGM-KU 89187, 89188, 89189; (G) proximal humerus IGM-KU 89171; (H) proximal ulna IGM-KU 89174. Scale in mm. All photographs courtesy of Masato Nakatsukasa.

A

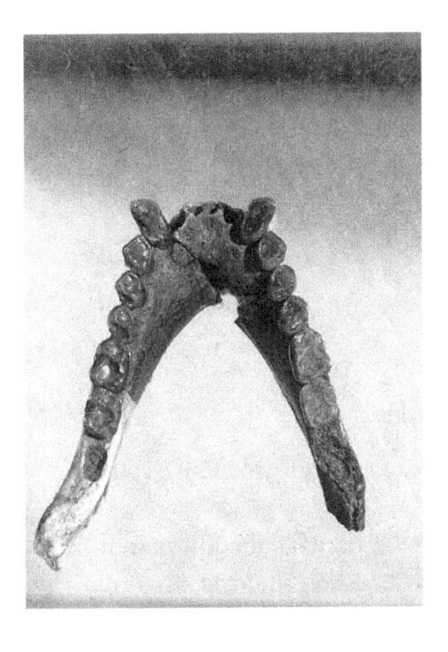

B

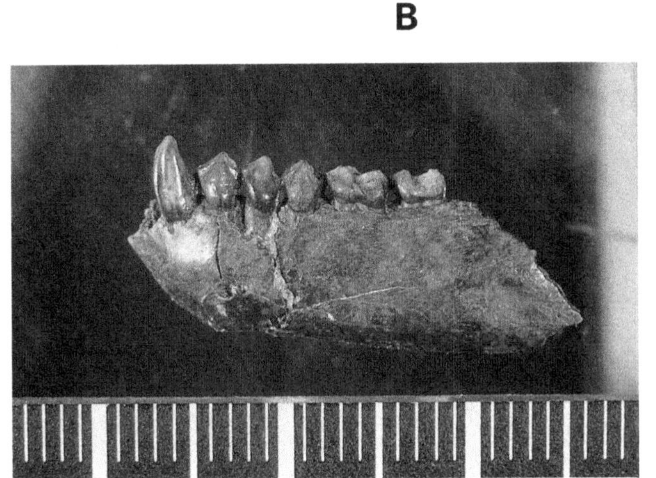

Fig. 12.6 *Laventiana annectens*. Holotype (IGM-KU 8801a) in (A) occlusal and (B) lateral views. Courtesy of Masanaru Takai.

wide talar head, moderately high talar body, long and narrow talar body, and medium to large medial protuberance, sometimes faceted, which characterize *Aotus* and *Callicebus* (Gebo *et al.*, 1990). On the other hand, Gebo *et al.* (1990) note that the talus clearly differs from extant *Aotus* in having a more square-shaped (relatively wide and short) talar body. Indeed, the proportions and configuration of the talar head and trochlea are very similar to those reported for callitrichines (Meldrum, 1990). Given the controversies surrounding this taxon, the attribution of this isolated talus remains in question.

Subfamily Callitrichinae

Tribe Callimiconi

GENUS *Mohanamico* Luchterhand *et al.*, 1986
INCLUDED SPECIES *M. hershkovitzi*

SPECIES *Mohanimico hershkovitzi* Luchterhand *et al.*, 1986 (Fig. 12.8)

A

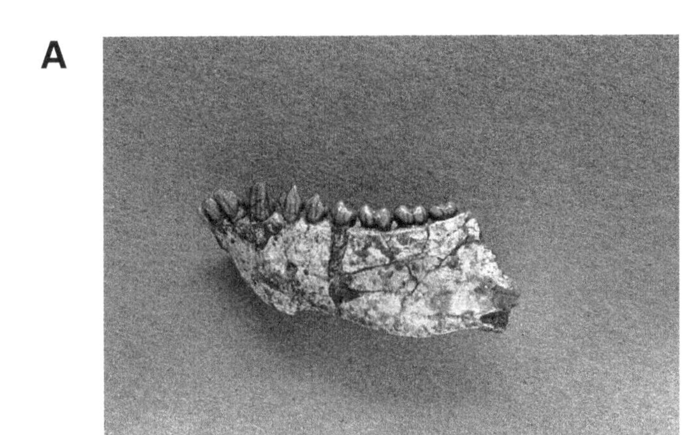

B

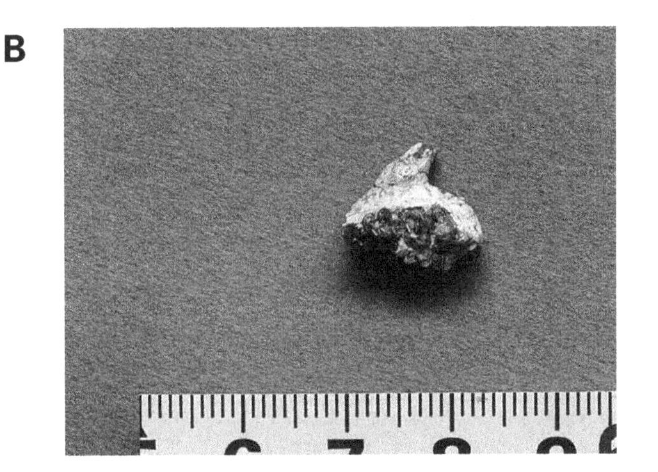

C

Fig. 12.7 *Aotus dindensis*. Holotype (IGM-KU 8601), mandible (A) and maxillary fragment (B); (C) occlusal view of *A. dindensis* holotype (left) and a referred mandible fragment (right) also from La Venta. All photographs courtesy of Masanaru Takai.

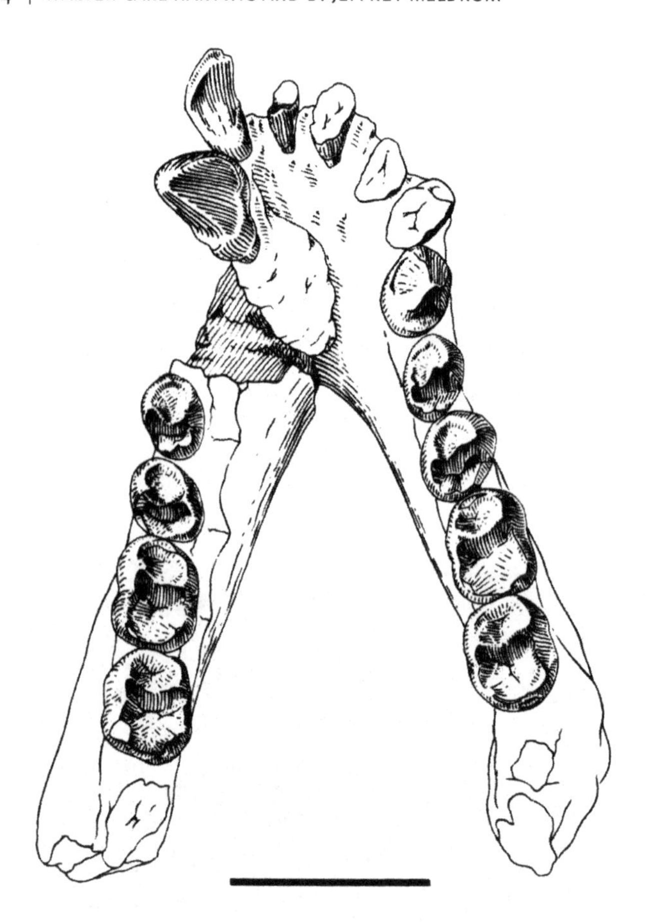

Fig. 12.8 *Mohanamico herskovitzi*. Holotype (IGM 181500). Scale bar = 1 cm. Adapted from Luchterhand *et al.*, 1986.

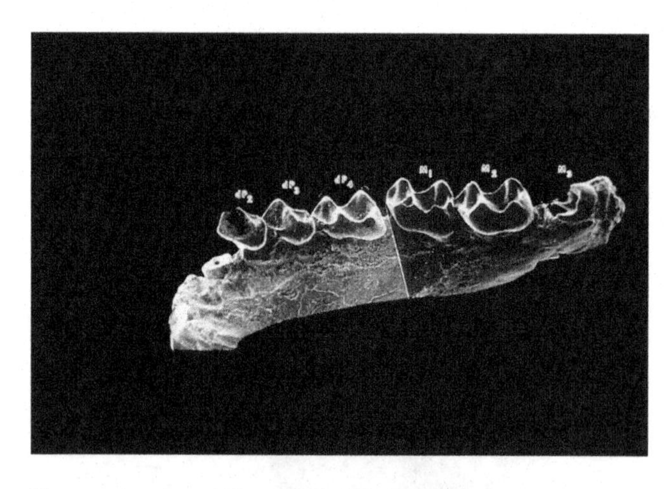

Fig. 12.9 *Patasola magdalenae*. Holotype (IGM 184332).

TYPE SPECIMEN IGM 181500, partial mandible
AGE AND GEOGRAPHIC RANGE 10–12.5 Ma, La Venta, Colombia
ANATOMICAL DEFINITION
Mohanamico is an *Aotus*-sized partial mandible. The incisors are high-crowned and semistyliform, canine tall and sharp with a lingual crest. P_2 is large and projects above the crowns of P_3 and P_4. The molars are of moderate crown relief, with M_1 and M_2 nearly equal in size. They lack hypoconulids and postentoconid sulci. Mandibular depth is 9.9 mm and does not increase posteriorly.

Tribe *incertae sedis*

GENUS *Patasola* Kay & Meldrum, 1997
INCLUDED SPECIES *P. magdalenae*

SPECIES *Patasola magdalenae* Kay & Meldrum, 1997 (Fig. 12.9)
TYPE SPECIMEN IGM 184332, partial subadult mandible
AGE AND GEOGRAPHIC RANGE 12.5–13.7 Ma, La Venta, Colombia
ANATOMICAL DEFINITION
Patasola is a small *Leontopithecus*-sized callitrichid represented by a shallow mandibular corpus retaining much of the deciduous dentition. It was also possible to extract the crypts of two of the permanent premolars from the fractured mandible. Permanent M_{1-2} protoconids much taller than metaconids. The cheek teeth are very narrow with high crown relief, in details most similar to the callitrichines. However, *Patasola* retains a small third molar, which is relatively larger than the third molar in *Callimico*. The preserved aspects of the mandibular corpus are relatively shallow (6 mm) and deepen very slightly posteriorly.

GENUS *Lagonimico* Kay, 1994
INCLUDED SPECIES *L. conclutatus*

SPECIES *Lagonimico conclutatus* Kay, 1994 (Fig. 12.10)
TYPE SPECIMEN IGM 184531, flattened skull
AGE AND GEOGRAPHIC RANGE 13.5 Ma, La Venta, Colombia
ANATOMICAL DEFINITION
Lagonimico is represented by a nearly complete but badly crushed skull and mandible. Larger than *Callimico* or any other callitrichine, yet the well-preserved maxillary and mandibular dentition is very similar to the extant callitrichines, with elongate slender incisors, simple premolars, and lack of molar hypocones, but retaining small third molars. The mandible deepens posteriorly more than in extant callitrichines. The shearing coefficient falls around values for species with diets of fruits and gums. Reconstruction of the crushed orbits indicates a diurnal habit.

GENUS *Micodon* Setoguchi & Rosenberger, 1985
INCLUDED SPECIES *M. kiotensis*

SPECIES *Micodon kiotensis* Setoguchi & Rosenberger, 1985 (Fig. 12.11)
TYPE SPECIMEN IGM-KU 8401, left M^1

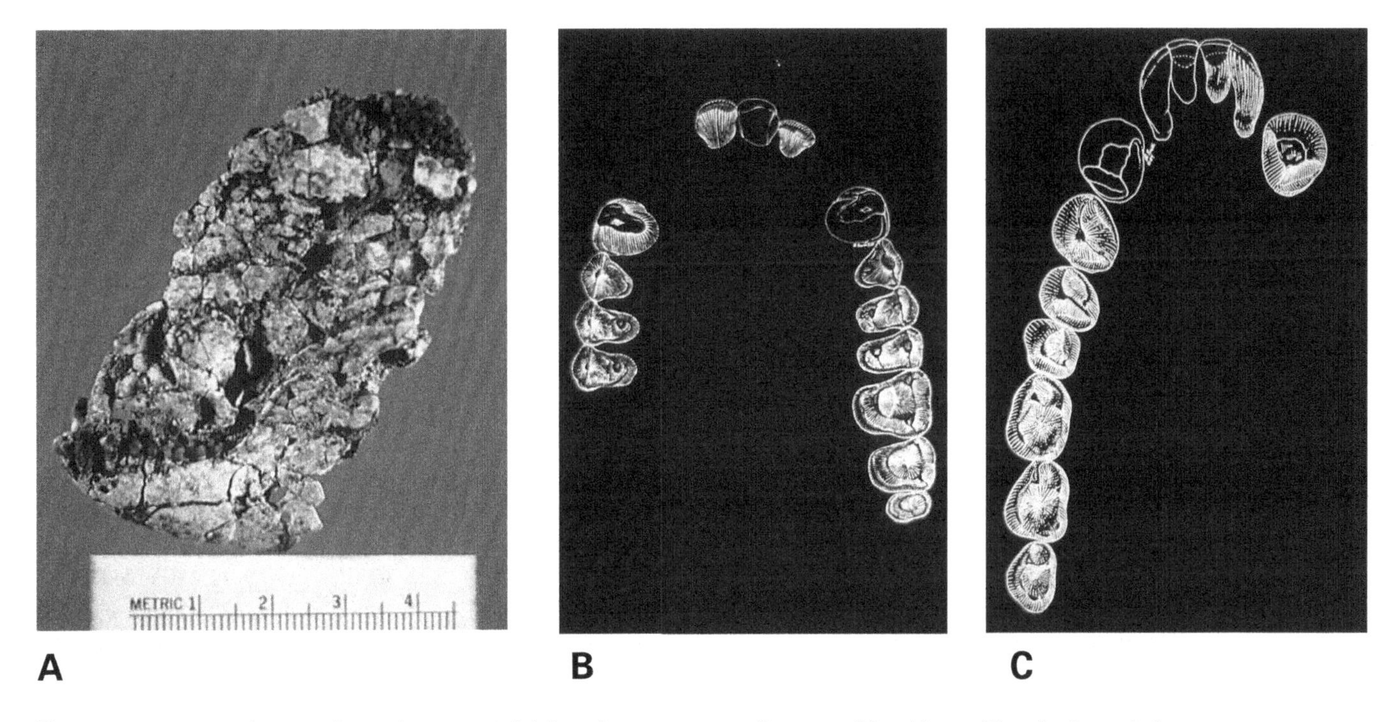

Fig. 12.10 *Lagonimico conclutatus*. Holotype (IGM 184531) (A), with reconstruction of its upper (B) and lower (C) occlusal morphology.

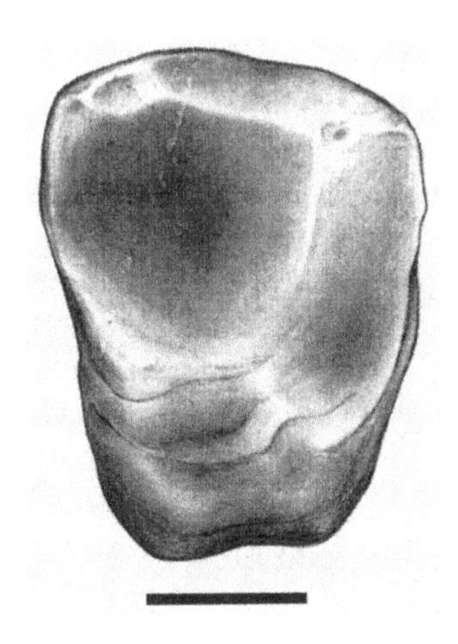

Fig. 12.11 *Micodon kiotensis*. Holotype (IGM-KU 8401). Scale = 1 mm.

AGE AND GEOGRAPHIC RANGE 12.2–12.5 MA, La Venta, Colombia

ANATOMICAL DEFINITION

This taxon is based on a single dentine core interpreted originally to be from an upper first molar. Little can be determined from this fossil. The original diagnosis describes an open, triangular trigon basin, low occlusal relief, large talon basin, a weakly developed lingual cingulum at the base of the protocone and a relatively large hypocone. No other fossils from La Venta have been attributed to this genus since its discovery.

Evolution of New World monkeys in the Miocene northern Neotropics

At the present time more than 10 genera of Miocene platyrrhines are recognized from the single Colombian region of La Venta. Some of these genera are so similar to living New World monkey taxa that they are classified (by some authorities) as congeneric. Other genera are not as instantly recognizable, but all of the fossils can be considered platyrrhines of modern aspect (Fleagle, 1999).

Of the modern clades the pitheciines are well represented at La Venta (*Cebupithecia*, *Nuciruptor*, and by some allocations *Aotus dindensis*). The atelines, especially outside of *Alouatta*, are poorly represented despite having a wide geographic distribution today. Relatives of the smaller cebine (*Saimiri*) abound, but to date no fossils closely related to the generalized, widespread and "successful" radiation of the larger cebine (the capuchin monkey, genus *Cebus*) have been found.

The information available from La Venta suggests several taxa within a broad cebine–callitrichine adaptive zone. The *Neosaimiri* and *Laventiana* material is overtly similar to modern squirrel monkeys, while *Mohanamico*, *Lagonimico*, *Patasola* and *Micodon*(?) display a variety of gnathic features in between the distinctiveness of living callitrichines and the affinities of the *Saimiri–Neosaimiri–Laventiana* cohort. Rosenberger's (e.g., 1992) long-held belief in a close relationship between cebines and callitrichines has so far been supported by the La Venta community of taxa. Fleagle (1999) noted that several fossil taxa from La Venta have been proposed as fossil evidence of the callitrichine radiation, none of which are convincing beyond a reasonable doubt and each of which implies a different evolutionary origin for the group. This critical mass and diversity of callitrichine-like forms likely reflect the dynamic transitions from Miocene lineages into the modern lineages seen today.

Postcranial fossils from La Venta are rare, but they reveal at least as much about the adaptive diversity of Miocene platyrrhines as the cranial evidence. Beginning with the thesis research of Ford (1980*b*), evolutionary changes in the platyrrhine postcranium have been studied coincident with advances in molecular studies (e.g., Horovitz 1999) and an increasing inventory of fossils (Ford 1988, 1990*a*; Meldrum 1990; Meldrum & Lemelin, 1991; Meldrum & Kay, 1997*b*; Nakatsukasa *et al.*, 1997). Ford (1988, 1990*a*) reconstructs a hypothetical ancestral morphotype for platyrrhines that postcranially resembles an *Aotus* monkey (approximately 1.0 kg) with a generalized locomotor repertoire dominated by arboreal quadrupedalism and limited leaping and suspensory behavior. If this hypothetical ancestor is accurate, to what extent have middle Miocene subfamilial postcranial morphotypes differentiated from it?

Only two groups are well represented by fossil postcrania, the subfamily Pitheciinae and the Saimiriini tribe. The extant pitheciines are larger than the hypothetical ancestor, weighing on average between 2.0 and 3.5 kg (Fleagle, 1988). *Cebupithecia* was within the lower limits of this range, with an estimated body weight of 2.2 kg (Fleagle, 1999). The extant pitheciines display more climbing and suspensory behaviors than the hypothetical ancestor, as reflected in their proportionately longer limbs. The relative length of the forelimb (humerus + radius / trunk) ranges between 70 and 96 for species of *Chiropotes* and *Pithecia* (Hershkovitz, 1985, 1987). Using an estimate of trunk length (cervical – sacral vertebrae inclusive) of 250 mm (Meldrum & Lemelin, 1991), *Cebupithecia* had forelimbs relatively shorter than the extant pitheciines, with an estimated value of 65. This falls within the range of values for *Aotus*, *Callicebus* and *Saimiri* (52–73; Hershkovitz, 1990). The distal femur of the extant pitheciines is relatively broad and shallow with a wide shallow patellar groove, very similar to the condition in the extant atelines. Likewise, the talar trochlea is low with a rounded lateral margin approaching the condition found in the ateline talus (Meldrum, 1990). These and other features of the pitheciine limbs that resemble the ateline condition are generally lacking in the postcranium of *Cebupithecia*. The distal femur of *Cebupithecia* is very narrow and deep and the talar trochlea is moderately high with a sharply crested lateral margin and more closely resembles such forms as *Aotus*, *Callicebus*, or some callitrichines (Ciochon & Corruccini, 1975; Davis, 1988). Therefore, while *Cebupithecia* had evolved the body size and dental morphology characteristic of modern pitheciines, it still retained many aspects of the hypothetical ancestral postcranial skeleton.

By comparison less is known concerning the postcranial anatomy of *Neosaimiri*. The fragmentary elements that are known are nearly identical in size and morphology to those of the extant squirrel monkey, *Saimiri*. *Saimiri* differs from the hypothetical platyrrhine ancestral morphotype in several ways. Squirrel monkeys are moderately smaller, with mean body weights of 850 g for males and 680 g for females (Hartwig, 1996; Fleagle, 1999). *Neosaimiri* had an estimated body weight of 840 g (Fleagle, 1999). *Saimiri* has relatively longer hindlimbs than the hypothetical ancestor, and leaps with greater frequency. Although nothing can be said about the limb proportions of *Neosaimiri*, postcranial morphology clearly indicates adaptations for frequent leaping. They may represent part of a radiation of platyrrhines, including the Callitrichinae and Callimiconinae, which underwent reduction in body size to exploit a primarily insectivorous/frugivorous diet (Rosenberger, 1992). Reduction in body size may have selected for leaping adaptations to negotiate the relatively larger gaps in the canopy as perceived by a smaller primate.

By comparing corresponding elements of the postcranium of *Cebupithecia* and *Neosaimiri*, some aspects of the ecological adaptive diversity present in the primate community of the Colombian Miocene can be inferred. Their modern counterparts, *Pithecia* and *Saimiri*, occur sympatrically over much of their respective ranges. *Pithecia* has a rather broad range of forest habitat tolerance. It frequents the understory and lower canopy strata and is frequently encountered on vertical supports. Locomotion is primarily by leaping and bounding. Few observations are available for postures during feeding. It is almost exclusively a frugivore/seed-predator. The morphology of the distal humerus of *Cebupithecia* suggests that it also adopted frequent clinging postures on vertical supports (Meldrum *et al.*, 1990). The morphology of the hindlimb also suggests frequent leaping, although after a fashion somewhat different than *Pithecia* (Fleagle & Meldrum, 1988). The dental and mandibular structure of *Cebupithecia* strongly implies that it had progressed toward a seed-eating diet as well. This adaptation is indicated especially by the broad flat molars, large chisel-like canines (Kay, 1990) and procumbent, mesiodistally compressed lower incisors of *Nuciruptor*.

Saimiri is also environmentally tolerant within the context of forest habitats, and also occupies the understory and lower canopy strata. It moves primarily by quadrupedal walking and running, interspersed with frequent leaps. Its diet differs from *Pithecia* in that it consists largely of insects and some fruit. By comparison, the distal humerus of *Neosaimiri* also suggests arboreal quadrupedalism, and the extensive syndesmosis of the distal tibiofibular joint suggests frequent leaping. The dentition of *Neosaimiri* also indicates a primarily insectivorous/frugivorous diet, as especially illustrated by the well-developed molar shearing crests (Fleagle *et al.*, 1997*b*). Given the similarities between the fossil taxa and their modern "counterparts" it seems reasonable to infer that the sympatric *Cebupithecia* and *Neosaimiri* had partitioned the forest understory in a manner comparable to living *Pithecia* and *Saimiri*.

The numerous primitive features of the postcranial skeleton of *Cebupithecia* demonstrate that the distinctive modern features of the dentition and postcranium may not have evolved in synchrony. Therefore, it is impossible to predict confidently the postcranial adaptations of fossil platyrrhines known only from dental remains. For example, the platyrrhine subfamily Alouattinae is represented in the Miocene of Colombia by *Stirtonia*, which displays many similarities to the modern genus *Alouatta* and had a comparably large body size (Kay *et al.*, 1987). But only the discovery of postcranial fossils of *Stirtonia* will determine whether the distinctive climbing and suspensory adaptations associated with the foraging strategy of the large-bodied alouattines had emerged at this point in their evolutionary history.

Three concentrated surveys at La Venta by researchers at UC–Berkeley, Kyoto University and Duke University have recovered critical evidence of the adaptive radiation of New World monkeys. The fossil record at La Venta is far from exhausted, however. Moreover, vast areas of Tertiary South America remain undersurveyed or unsurveyed. Future research certainly will help to resolve the issues apparent today and will present a fuller picture of the evolution of platyrrhine primates in the Neotropics.

Primary References

Aotus

Setoguchi, T. & Rosenberger, A. L. (1987). A fossil owl monkey from La Venta, Colombia. *Nature*, **326**, 692–694.

Cebupithecia

Stirton, R. A. (1951). Ceboid monkeys from the Miocene of Colombia. *Bulletin of the University of California Publications in the Geological Sciences*, **28**, 315–356.

Stirton, R. A. & Savage, D. E. (1951). A new monkey from the La Venta late Miocene of Colombia. *Compilación de los estudios geologicos oficiales en Colombia*, **7**, 345–346.

Setoguchi, T., Takai, M., Villarroel, A. C., Shigehara, N., & Rosenberger, A. L. (1988). New specimen of *Cebupithecia* from La Venta, Miocene of Colombia, South America. *Kyoto University Special Publications*, **1988**, 7–9.

Meldrum, D. J., Fleagle, J. G., & Kay, R. F. (1990). Partial humeri of two Miocene Colombian primates. *American Journal of Physical Anthropology*, **81**, 413–422.

Meldrum, D. J. & Lemelin, P. (1991). Axial skeleton of *Cebupithecia sarmientoi* (Pitheciinae, Platyrrhini) from the middle Miocene of La Venta, Colombia. *American Journal of Primatology*, **25**, 69–90.

Meldrum, D. J. & Kay, R. F. (1997*b*). Postcranial skeleton of Laventan platyrrhines. In *Vertebrate Paleontology in the Neotropics: The Miocene Fauna of La Venta, Colombia*, eds. R. F. Kay, R. H. Madden, R. L. Cifelli, & J. J. Flynn, pp. 459–472. Washington, DC: Smithsonian Institution Press.

Lagonimico

Kay, R. F. (1994). "Giant" tamarin from the Miocene of Colombia. *American Journal of Physical Anthropology*, **95**, 333–353.

Laventiana

Gebo, D. L., Dagosto, M., Rosenberger, A. L., & Setoguchi, T. (1990). New platyrrhine tali from La Venta, Colombia. *Journal of Human Evolution*, **19**, 737–746.

Setoguchi, T., Takai, M., & Shigehara, N. (1990). A new ceboid primate, closely related to *Neosaimiri*, found in the Upper Red Bed in the La Venta badlands, middle Miocene of Colombia, South America. *Kyoto University Overseas Research Reports on New World Monkeys*, **7**, 9–14.

Rosenberger, A. L., Setoguchi, T., & Hartwig, W. C. (1991). *Laventiana annectens*, new genus and species: fossil evidence for the origins of callitrichine monkeys. *Proceedings of the National Academy of Sciences of the United States of America*, **88**, 2137–2140.

Micodon

Setoguchi, T. & Rosenberger, A.L. (1985). Miocene marmosets: first evidence. *International Journal of Primatology*, **6**, 615–625.

Mohanamico

Luchterhand, K., Kay, R. F., & Madden, R. H. (1986). *Mohanamico hershkovitzi*, gen. et sp. nov., un primate du Miocène moyen d'Amérique du Sud. *Comptes rendus de l'Académie des sciences de Paris*, **303**, 1753–1758.

Neosaimiri

Stirton, R. A. (1951). Ceboid monkeys from the Miocene of Colombia. *Bulletin of the University of California Publications in the Geological Sciences*, **28**, 315–356.

Meldrum, D. J., Fleagle, J. G., & Kay, R. F. (1990). Partial humeri of two Miocene Colombian primates. *American Journal of Physical Anthropology*, **81**, 413–422.

Takai, M. (1994). New specimens of *Neosaimiri fieldsi* from La Venta, Colombia: a middle Miocene ancestor of the living squirrel monkeys. *Journal of Human Evolution*, **27**, 329–360.

Nakatsukasa, M., Takai, M., & Setoguchi, T. (1997). Functional morphology of the postcranium and locomotor behavior of *Neosaimiri fieldsi*, a *Saimiri*-like middle Miocene platyrrhine. *American Journal of Physical Anthropology*, **102**, 515–544.

Nuciruptor

Meldrum, D. J. & Kay, R. F. (1997). *Nuciruptor rubricae*, a new pitheciin seed predator from the Miocene of Colombia. *American Journal of Physical Anthropology*, **102**, 407–427.

Patasola

Kay, R. F. & Meldrum, D. J. (1997). A new small platyrrhine and the phyletic position of Callitrichinae. In *Vertebrate Paleontology in the Neotropics: The Miocene Fauna of La Venta, Colombia*, eds. R. F. Kay, R. H. Madden, R. L. Cifelli, & J. J. Flynn, pp. 435–458. Washington, DC: Smithsonian Institution Press.

Stirtonia

Stirton, R. A. (1951). Ceboid monkeys from the Miocene of Colombia. *Bulletin of the University of California Publications in the*

Geological Sciences, **28**, 315–356.

Hershkovitz, P. (1970). Notes on Tertiary platyrrhine monkeys and description of a new genus from the late Miocene of Colombia. *Folia Primatologica*, **12**, 1–37.

Setoguchi, T. (1980). Discovery of a fossil primate from the Miocene of Colombia. *Monkey*, **24**, 64–69.

Setoguchi, T., Watanabe, T., & Mouri, T. (1981). The upper dentition of *Stirtonia* (Ceboidea, Primates) from the Miocene of Colombia, South America, and the origins of the posterointernal cusp of upper molars of howler monkeys. *Kyoto University Overseas Research Reports on New World Monkeys*, **3**, 51–60.

Setoguchi, T. & Rosenberger, A. L. (1985). Some new ceboid primates from the La Venta, Colombia, South America. *Memorias VI Congreso latinoamericano de geologica*, **1**, 187–198.

Setoguchi, T. (1985). *Kondous laventicus*, a new ceboid primate from the Miocene of La Venta, Colombia, South America. *Folia Primatologica*, **44**, 96–101.

Kay, R. F., Madden, R. H., Plavcan, J. M., Cifelli, R. L., & Diaz, J. G. (1987). *Stirtonia victoriae*, a new species of Miocene Colombian primate. *Journal of Human Evolution*, **16**, 173–196.

Kay, R. F. & Frailey, C. D. (1993). Large fossil platyrrhines from the Rio Acre local fauna, late Miocene, western Amazonia. *Journal of Human Evolution*, **25**, 319–327.

13 | Extinct Quaternary platyrrhines of the Greater Antilles and Brazil

ROSS D. E. MACPHEE AND INES HOROVITZ

Platyrrhine remains are occasionally found in Quaternary paleontological and archaeological contexts in the northern Neotropics (e.g., Olsen, 1982); in virtually all cases, such remains pertain to extant species only. The known exceptions are limited to two geographical areas, both of which have yielded extinct platyrrhines of late Quaternary age: the Greater Antilles and south-central Brazil. From the standpoint of extinction studies it makes sense to discuss these late-surviving monkeys in one place, as they constitute the only verified instances of Neotropical primate losses during end-Pleistocene or post-Pleistocene times (MacPhee & Flemming, 1999; see Godfrey & Jungers, this volume). However, as the insular and mainland species are not otherwise closely related, they are discussed under separate headings in this chapter.

EXTINCT QUATERNARY PLATYRRHINES OF THE GREATER ANTILLES

Introduction

In this chapter "Antillean monkeys" are to be understood as a monophyletic group of platyrrhines that were, as far as is now known, exclusively endemic to the western Greater Antilles (MacPhee, 1996; Horovitz & MacPhee, 1999). For obvious reasons, this common name should not be understood to cover either non-platyrrhines (e.g., introduced green monkeys of St. Kitts, Barbados, and some other Lesser Antillean islands (Denham, 1987)) or extant platyrrhines that have simply extended their range onto certain Caribbean shelf islands during the late Quaternary (e.g., Trinidad and Isla Margarita).

The Antillean monkeys are of special interest for three distinct reasons. First, the evidence now available strongly indicates that these monkeys, although markedly different from one another skeletally and dentally, are closely related *inter se* and to the extant titi monkey (*Callicebus*) of mainland South America. Previously, they were thought to represent separate offshoots from different mainland lineages. Second, Antillean monkeys display a variety of features that are rare or absent in platyrrhines from the mainland. Presumably, these features appeared in response to selective regimes characteristic of islands (Whittaker, 1998). Third, the Antillean monkeys died out within recent times – very recent times in the case of the Jamaican monkey *Xenothrix* (c. ?1700). Whether these losses were due to human impacts or other forces remains unresolved. This also applies to the case of extinctions on the mainland of the two "megafaunal" genera, *Protopithecus* and *Caipora*.

History of discovery and debate

The fact that endemic monkeys once existed in the West Indies was not established satisfactorily until the middle of the twentieth century, although speculation on this matter began much earlier. Which among the earliest relevant finds should be considered "first", however, is a matter for debate. There are three candidates: (1) the handful of isolated teeth comprising the holotype of *Montaneia anthropomorpha* (Ameghino, 1910), said by their discoverer, Dr. Luis Montané, to have been discovered in 1888 in a pre-Columbian context in a cave near Sancti Spíritus, Cuba; (2) the jaw and postcranial bones found by Harold Anthony in 1920 at Long Mile Cave, Jamaica; and (3) the "?cercopithecine" distal tibia found by Gerrit Miller in 1928 within an archaeological midden on a cay in Bahia de Samaná, Hispaniola.

The claim of "*Montaneia*" (sometimes referred to as *Ateles anthropomorphus*) can be considered vacated: the teeth are those of a modern individual of *Ateles fusciceps*, as both morphology and radiocarbon dating have shown conclusively (see Miller, 1916; MacPhee & Rivero de la Calle, 1996). It is difficult to be certain, but it seems likely that this discovery was a deliberate hoax (although not necessarily on the part of Montané himself; MacPhee, 1996; but see Arredondo & Varona, 1983). The right half of the "*Montaneia*" dentition is in the collections of the Facultad de Biología, Universidad de La Habana; the left half is in the Museo Argentino de Ciencias Naturales, Buenos Aires (M. F. Tejedor, pers. comm.).

The Long Mile jaw undeniably would have been "first" had it been described and published expeditiously. Unfortunately, Anthony did not recognize its distinctiveness, apparently believing it to be an exotic introduced during European times. The jaw remained unremarked in a museum drawer until Ernest Williams & Karl F. Koopman (1952) demonstrated convincingly that it represented a heretofore unknown species, which they named *Xenothrix mcgregori*. Platyrrhine postcranial remains were also recovered by Anthony at Long Mile; these were not finally described until 1991 by MacPhee & Fleagle.

This leaves Miller's (1929) find, which has bibliographical pride of place because he immediately committed to print a description of the Samaná distal tibia (USNMM 254682). However, Miller was not at all sure what he had found. He could not refer the Samaná specimen to any known species of New or Old World monkey, but could not believe that it was an endemic primate otherwise unknown to science. In the end, he cited some vague similarities to cercopithecines as a basis for concluding that perhaps it belonged to a monkey recently imported from Africa.

And this is how matters stood until mid-century, when Williams & Koopman (1952) placed beyond doubt the fact that at least one kind of endemic monkey had existed, very late in the Quaternary, on at least one island of the Greater Antilles. In their paper they were chiefly concerned with demonstrating that the Jamaican monkey could not be considered a member of any known mainland genus. They were rather agnostic about its affinities, claiming in the end only that it was clearly "cebid" (which, in their taxonomy, was coextensive with "platyrrhine"). Rosenberger's (1977) paper should be consulted for a detailed study of the anatomy of the type jaw.

No new allocation for the Samaná tibia was proposed until 1980, when Ford (1980*a*; see also Ford 1986*a*) concluded that the specimen might actually be callitrichid. One impediment to accepting this conclusion is that the tibia is in the size range of a *Cebus* monkey (average body mass, 2.5 kg), while the largest living callitrichid weighs only a few hundred grams (Jungers, 1985). Perhaps, Ford (1986*a*) reasoned, the Dominican monkey was a member of a callitrichid lineage that had evolved toward gigantism after lengthy isolation in Hispaniola.

In the meantime, another platyrrhine fossil had come to light in Hispaniola: a partial maxilla recovered by Renato Rímoli and coworkers from late Holocene ($>$ 3800 rcyrbp) sediments at Cueva de Berna in southeastern Hispaniola. This specimen – the first new Antillean monkey fossil to come to light in nearly half a century – was an important find because it included teeth (P^4–M^2) and was clearly distinct at the species level from any platyrrhine then known. However, concluding that the morphology of the Berna specimen indicated a close relationship to extant squirrel monkeys, Rímoli (1977) named it *Saimiri bernensis*. Other commentators questioned whether this was a plausible allocation (e.g., Rosenberger, 1978): although it could be said that the Berna fossil evinced some resemblances to squirrel monkeys, overall agreement was not especially close. Moreover, the teeth were much larger than those of any extant squirrel monkey species, approaching in size those of a capuchin (*Cebus*).

In the early 1980s, MacPhee & Woods (1982) described yet another monkey fossil from Hispaniola – a partial ramus preserving M_1 only, from the locality of Trou Wòch Sa Wo in southern Haiti (for locality information see MacPhee *et al.*, 2000). These authors raised questions about the validity of associating the Hispaniolan monkey finds either with squirrel monkeys or, in reference to Ford's (1980*a*) "giant tamarin" hypothesis, with callitrichids. They preferred to leave all Hispaniolan platyrrhine fossils outside the limits of both *Saimiri* and *Cebus*, while recognizing that they probably represented a species closely related to the latter genus.

Later in the same decade, concepts of the platyrrhine radiation in the Greater Antilles were extended considerably by discoveries in western Cuba. In 1988, members of the Grupo "Pedro A. Borrás" of the Sociedad Espeleológica de Cuba recovered a remarkably complete skull and other bones belonging to a large and hitherto unknown monkey. Rivero & Arredondo (1991), who described the skull, concluded that it represented a taxon with phylogenetic ties to living howler monkeys (*Alouatta* and near relatives).

Rivero & Arredondo (1991) were not able to identify many dental characters in their evaluation of the type skull of *Paralouatta* because the teeth are excessively worn. As a result, their basis for assigning *Paralouatta* to a position close to living alouattines was largely predicated on general aspect and nondental cranial characters. For example, in both *Alouatta* and *Paralouatta* the face is flaring, the neurocranium is hafted on the splanchnocranium in a similar manner, and the cranial vault is low. At the same time, there are some marked differences. Of particular interest is the fact that the canine is tiny in *Paralouatta*, whereas the canines of *Alouatta* and other atelines are relatively large in both sexes (although proportionately smaller in females). Also, the orbits are distinctively different from those of any atelid, being proportionately almost as large as those of *Aotus*.

Subsequent collecting expeditions by the American Museum of Natural History, Museo Nacional de Historia Natural (La Habana) and Grupo Borrás have led to the recovery of a number of additional elements of *P. varonai* (Jaimez Salgado *et al.*, 1992). Unfortunately, no new sites have been identified in recent years, despite the fact that the region abounds in caves (many with superbly well-preserved remains of late Quaternary Cuban vertebrates). The hypodigm of *P. varonai* now includes a number of isolated teeth, a beautifully preserved jaw, and several limb bones (Horovitz & MacPhee, 1999).

With the discovery and analysis of extra material of *P. varonai* during the mid-1990s, the systematic histories of the different endemic primates of the Greater Antilles began to converge. MacPhee *et al.* (1995*b*) and Horovitz & MacPhee (1999) found that *Paralouatta* does not group among atelines next to *Alouatta*, but instead occupies a position within a distinct clade that includes other Antillean monkeys and the mainland taxon *Callicebus*. In particular, MacPhee *et al.* (1995*b*) argued that a subtle combination of dental features found in "*Saimiri*" *bernensis* was seen in only one other species, the Cuban primate *Paralouatta varonai*. Further, parsimony analysis indicated that *Paralouatta varonai* and "*Saimiri*" *bernensis* should be grouped as phyletic sisters, related more generally and at a higher level to atelines. No indication of a close relationship with true *Saimiri* was supported by this study. With no reason to continue the fiction that the Berna specimen belonged in (or was even closely related to) *Saimiri*, MacPhee *et al.* (1995*b*) fashioned a replacement name, *Antillothrix bernensis*.

Although the distinctiveness of *Antillothrix* as based on the Berna type specimen is well established, several other monkey bones from Hispaniola remain to be discussed. Miller's (1929) distal tibia from Bahia de Samaná is in the correct size range to belong in the *Antillothrix* hypodigm, as are a number of still-undescribed monkey teeth, skull parts and postcranial bones found in Haitian caves and sinkholes by Charles A. Woods and associates in the early 1980s (see MacPhee *et al.*, 2000). According to Ford (1990*b*), these specimens probably also belong to the same species as the Berna maxilla. In particular, Ford (1990*b*: 242) noted that recovered canines are "small", a possible indication of similarity to other Antillean monkeys (although the partial canine alveolus in the Sa Wo ramus seems rather large).

However, things may be somewhat more complicated than this. Hispaniola is environmentally the most diverse of the Greater Antilles, and, for this reason, speciation has reached remarkable proportions in some extant vertebrate taxa (e.g., *Eleutherodactylus*, *Anolis*). If there were one place in the West Indies that might have supported more than one monkey species, it should be this island. Interestingly, Ford (1990*b*: 247) has stated that Woods' Haitian material includes a femur similar to the one recovered at Coco Ree – so similar that "it appears likely that these two specimens represent the same species." Such a degree of similarity is startling given that these two islands do not share any other endemic land-mammal conspecifics. In our view, the only way these two islands could share the same monkey species would be if human agency was involved in their dispersal, which seems unlikely. A better alternative would be to assume that different species are represented on the two islands, and that any similarities exhibited by the femora are simply primitive. In any case, the evidence at present is just not good enough to reach any settled conclusions about the nature of this "second" Hispaniolan primate. Jamaica presents a similar paleontological puzzle (see below).

In 1991, MacPhee & Fleagle discussed some highly distinctive postcranial remains from Long Mile Cave that Anthony (n. d.) had collected in 1920 along with the type jaw of *Xenothrix*. The humerus (AMNHM 259906) has a straight but very robust shaft, slightly raised deltopectoral crest (as we now know from new Jackson's Bay material), and a rather ovoid head that is only slightly elevated above the level of the tuberosities. The femur is similarly robust, with a markedly constricted distal end, like that of a sloth. The os coxae and tibiae are not well enough preserved to make useful functional inferences, although the os coxae in particular is more primate-like than anything else. In the aggregate, the referred postcranials indicate that *Xenothrix* was a non-leaping, non-swinging, slow-moving arborealist, probably unlike any living platyrrhine in locomotor behavior but perhaps somewhat convergent on lorises, sloths, and the kinkajou, *Potos flavus*.

Several important new finds of *Xenothrix mcgregori* have been made in recent years by teams from the American Museum of Natural History and McKenna College (Claremont) working in caves in the Jackson's Bay area, southern Jamaica. The most important new specimen is a partial skull, a preliminary description of which has been given by Horovitz & MacPhee (1999). Other remains include a partial maxilla, two mandibular fragments and several long bones. This material is now in the process of being described, but some significant details can be noted here. The new material confirms Williams & Koopman's (1952) original observation that the molars in *Xenothrix* were reduced to two. Although this apomorphy is encountered in callitrichines, tooth structure is otherwise so different as to make it highly improbable that dental reduction is homologous in *Xenothrix* and marmosets. Another remarkable feature of the cheek teeth is the relatively large size of their occlusal surfaces, which were highly bunodont (very low cusp relief, lacking pronounced shearing crests). The teeth in the holotype show almost no evidence of wear (?young animal), whereas both occlusal and interproximal wear in the new jaws is extreme. Enamel thickness is fairly substantial, since "lakes" of dentine are not exposed on the occlusal surface despite the degree of enamel wear. Although none of these specimens has a canine preserved, it is obvious from the size of the alveolus that this tooth could not have been large and projecting, and that a diastema was small or absent.

Intriguingly, two other clearly primate-like femora have been found at Coco Ree (UF 40097) in St. Catherine and Sheep Pen (UF 58350) in Trelawny (Ford & Morgan, 1986, 1988; Ford, 1990*b*; see also Hershkovitz, 1988). Although these specimens are obviously of some antiquity, exactly how old they are is somewhat controversial, especially in the case of the Coco Ree specimen (Goodfriend & Mitterer, 1987). The femora are similar in size and in most details of their morphology, although neither is well preserved (see MacPhee & Fleagle, 1991: Figure 2). Since the Coco Ree and Sheep Pen femora are distinctly primate-like, whereas that from Long Mile is quite unusual, it would be reasonable to conclude that the former actually represents *Xenothrix*, while the latter belongs to another, non-primate taxon. The trouble with this conclusion is that there is no conceivable candidate other than *Xenothrix* for the Long Mile femur (MacPhee & Fleagle, 1991). We are thus left with contemplating the real possibility that Jamaica harbored an otherwise unknown Quaternary mammal that was itself either platyrrhine or massively convergent thereon. Elucidation of this point is eagerly awaited.

Finally, there is definitely another platyrrhine in the Cuban record, although this species is known only from an astragalus from the early Middle Miocene locality of Domo

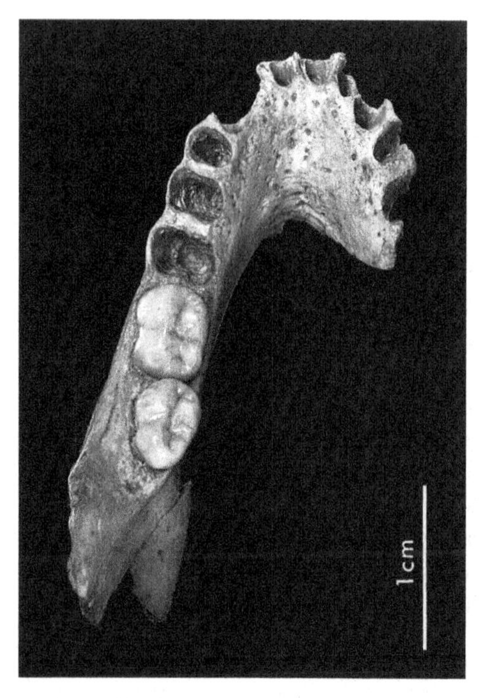

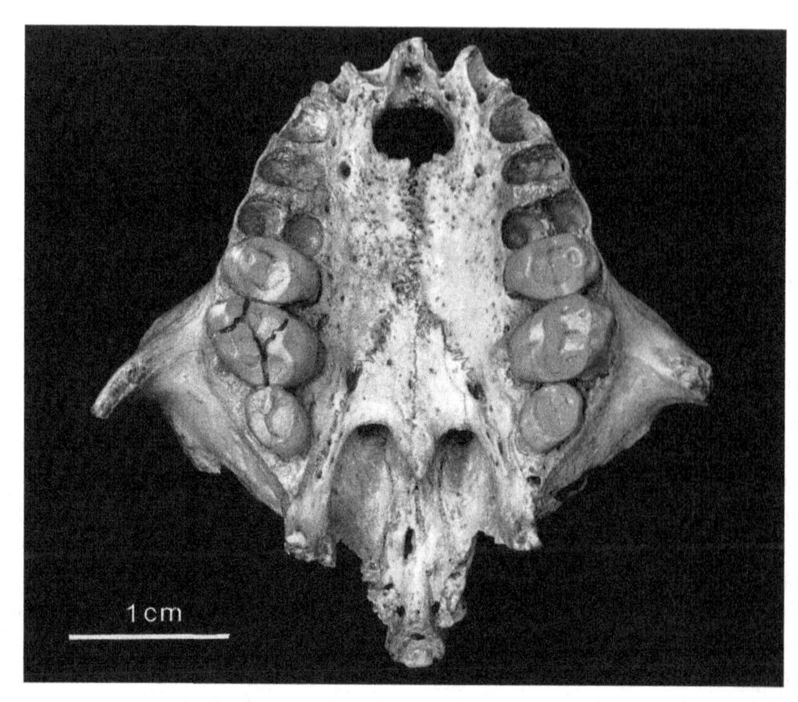

Fig. 13.1 *Xenothrix mcgregori*. (A) Holotype mandible of AMNHM 148198 (Long Mile Cave, Jamaica), left mandibular ramus, occlusal view; (B) partial skull (AMNHM 268006) (Lloyd's Cave, Jamaica), occlusal view.

de Zaza, near Sancti Spíritus. The Domo de Zaza astragalus is reasonably well preserved, as large as that of an extant *Ateles* and rather primitive. Perhaps unsurprisingly, it is remarkably similar to an incomplete astragalus referred to *Paralouatta* (see MacPhee & Iturralde-Vinent, 1995*a*). The Domo de Zaza specimen is important for several reasons. First, discovery of a mid-Cenozoic platyrrhine together with a significant range of other terrestrial taxa confirms that this part of the Greater Antilles must have been able to support land vertebrates for a considerable period before the Middle Miocene (MacPhee & Iturralde Vinent, 1994, 1995*b*; Iturralde-Vinent & MacPhee, 1999). Second, not only is the Zaza astragalus the oldest platyrrhine fossil from the West Indies, at 17–18 Ma it is also the earliest from the northern part of the continental Neotropics. Together with the somewhat later (and far richer) La Venta fauna (Madden *et al.*, 1997), it attests that platyrrhines were already present and well differentiated in northern South America early in the Neogene.

Taxonomy

Systematic framework

Our framework follows Horovitz (1997, 1999) and Horovitz & MacPhee (1999, in press), in which the three Antillean genera are grouped in one monophyletic entity (currently unnamed), the sister group of which is extant *Callicebus*. These taxa are subsumed in Tribe Callicebini, Subfamily Pitheciinae. At present this is the only robust phylogenetic hypothesis of Antillean monkey relationships, although several alternative relationships are proposed in the older literature.

Order Primates Linnaeus, 1758
  Infraorder Platyrrhini É. Geoffroy Saint-Hilaire, 1812
    Family Atelidae Gray, 1825
      Subfamily Pitheciinae (Callicebini) Mivart, 1865
        Genus *Xenothrix* Williams & Koopman, 1952
          Xenothrix mcgregori Williams & Koopman, 1952
        Genus *Antillothrix* MacPhee *et al.*, 1995
          Antillothrix (= "*Saimiri*") *bernensis* Rímoli, 1977
        Genus *Paralouatta* Rivero & Arredondo, 1991
          Paralouatta varonai Rivero & Arredondo, 1991

Family Atelidae

Subfamily Pitheciinae (Callicebini)

GENUS *Xenothrix* Williams & Koopman, 1952
INCLUDED SPECIES *X. mcgregori*

SPECIES *Xenothrix mcgregori* Williams & Koopman, 1952 (Fig. 13.1)
TYPE SPECIMEN AMNH 148198, partial mandible with two teeth
AGE AND GEOGRAPHIC RANGE Quaternary, ?surviving to c. AD 1700, insular Neotropics, Jamaica only

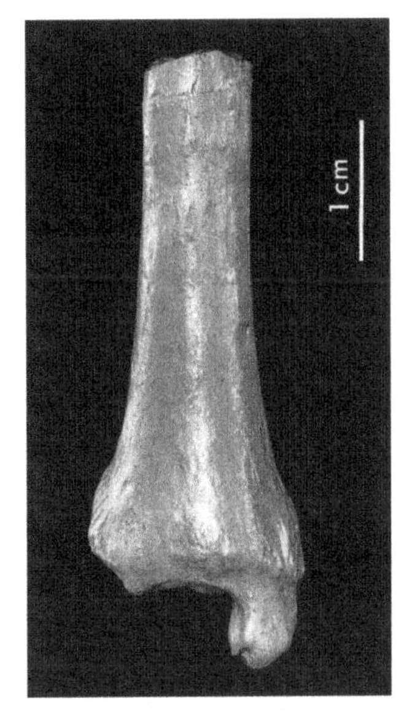

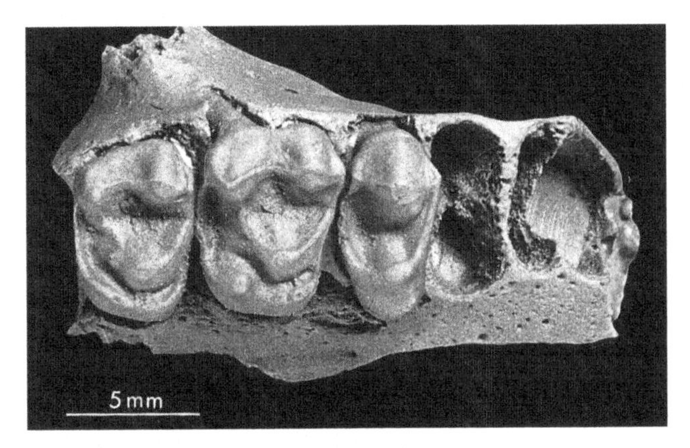

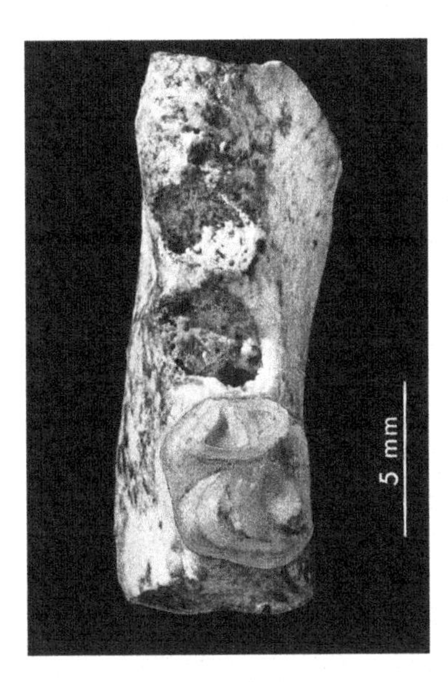

Fig. 13.2 *Antillothrix bernensis*. (A) Distal tibia of USNMM 254682 (Bahia de Samaná, Dominican Republic), anterior view; (B) holotype right maxilla (CENDIA 1, Cueva de Berna, Dominican Republic), occlusal view; (C) mandibular fragment with M_1 (right; reversed) of UF 28038 (Trou Wòch Sa Wo, Haiti), occlusal view.

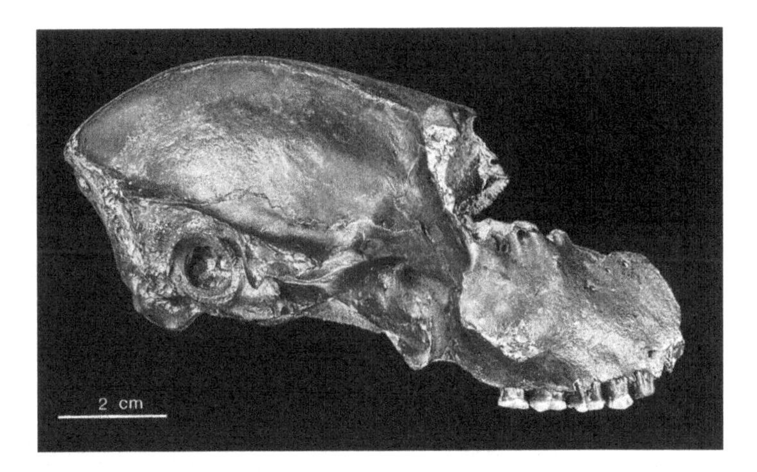

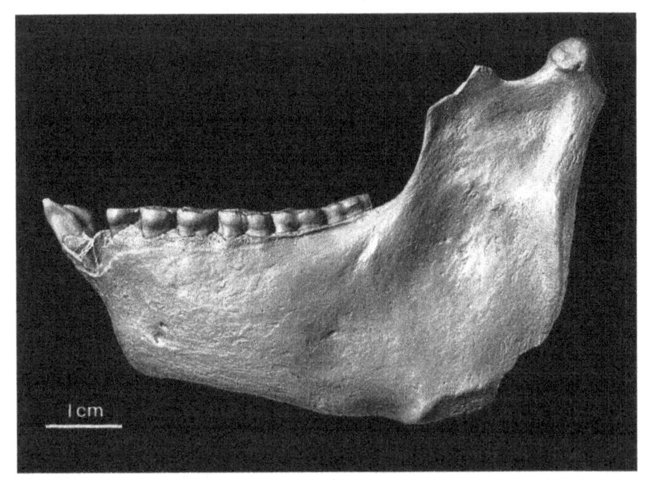

Fig. 13.3 *Paralouatta varonai*. (A) Holotype skull (MNHNH V 194, Cueva de Mono Fósil, Cuba), right lateral view; (B) mandible (MNHNH V 195, Cueva Alta, Cuba), left lateral view.

ANATOMICAL DEFINITION
Medium-sized monkey; only two molars in both jaws; M_1 with pronounced paracone; canines much reduced; femur distal epiphysis ventrodorsally flattened.

GENUS *Antillothrix* MacPhee *et al.*, 1995
INCLUDED SPECIES *A. bernensis*

SPECIES *Antillothrix* (= "*Saimiri*") *bernensis* Rímoli, 1977 (Fig. 13.2)
TYPE SPECIMEN CENDIA 1, partial maxilla including P^4–M^2
AGE AND GEOGRAPHIC RANGE Quaternary, surviving to late Holocene ("last occurrence" ^{14}C date 3850 ± 150 rcybrp), insular Neotropics, Hispaniola only
ANATOMICAL DEFINITION
Medium-sized species; molars resemble those of *Paralouatta* in general construction, but are considerably smaller;

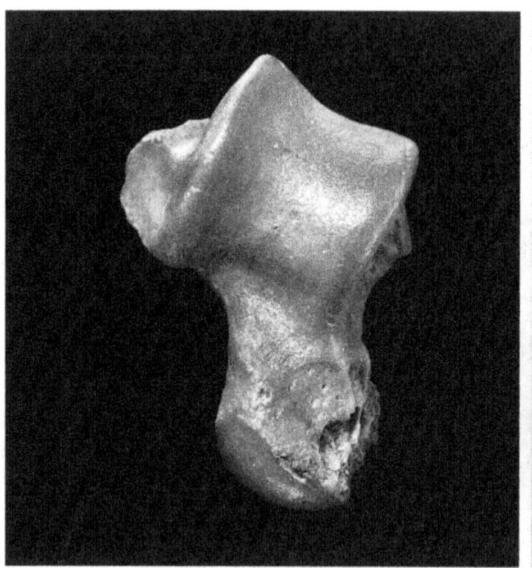

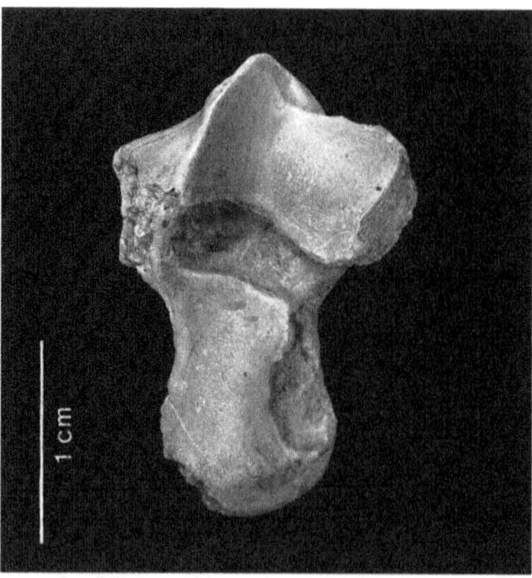

Fig. 13.4 Right astragalus of a platyrrhine monkey (MNHNH P 3059) from early Miocene locality of Domo de Zaza, Cuba, in dorsal (left) and ventral (right) views.

canine ?secondarily enlarged (as judged by width of partial alveolus).

GENUS *Paralouatta* Rivero & Arredondo, 1991
INCLUDED SPECIES *P. varonai*

SPECIES *Paralouatta varonai* Rivero & Arredondo, 1991 (Fig. 13.3)
TYPE SPECIMEN MNHNH V 194, cranium with damaged face and cheek tooth dentition
AGE AND GEOGRAPHIC RANGE Quaternary, perhaps early rather than late? Limited to insular Neotropics, Cuba only
ANATOMICAL DEFINITION
Large, long cranium with relatively large orbits; canines much reduced and integrated with incisors; mandible has small "chin"; gonial region not enlarged; humerus with large entepicondylar foramen.

Evolution of Antillean monkeys

Precladistic interpretations of the Antillean species tended either to allocate them to extant platyrrhine genera or families, or to leave them essentially *incertae sedis* within Platyrrhini. However, none of these earlier studies analyzed all Antillean species simultaneously. In recent years, several cladistic analyses have investigated their affiliations within the context of the entire platyrrhine radiation. The results of these investigations, which utilize new craniodental material that has become available in the last few years (MacPhee *et al.*, 1995*b*; Horovitz & MacPhee, 1996, 1999, in press; Horovitz, 1997, 1999), require conclusions that differ considerably from the traditional views.

The systematics of *Xenothrix mcgregori* have been elusive, owing in part to this species' unparallelled combination of derived characters. Although Williams & Koopman (1952) did not undertake a higher level systematic analysis of *Xenothrix*, they did notice similarities to several other platyrrhines. These included possession of a deep mandible (shared with *Callicebus*), reduced dental formula (shared with callitrichins), bunodont crowns (shared with *Cebus*), reduction of talonid basin and greater anterior width of M_1 and M_2 (shared with *Saimiri*), presence of hypoconulids (variably present in *Ateles*), and so on. In his review of the evidence almost a quarter-century later, Rosenberger (1977) noted several characters indicating a close relationship among *Xenothrix*, *Callicebus* and *Pithecia* (e.g., offset entoconid and post-talonid extension, low cusp relief, short cristid obliqua, and differentiated postprotocristid). Rosenberger (1977) also noted that, at a more nested level, *Xenothrix* shares with *Callicebus* a small canine socket, premolar alveoli that broaden posteriorly, a parabolic dental arcade, and a jaw that strongly deepens posteriorly. The similarities of *Xenothrix* and *Cebus* (e.g., marked bunodonty), detected by Williams & Koopman (1952), are notable but fewer (Rosenberger *et al.*, 1990). Ford (1986*a*; 1990*b*) pointed out that *Xenothrix* might be nested among callithrichines based on the lack of a third molar. Other modifications in the dentition of *Xenothrix* could be the result of a change of diet resulting from increased body size. However, as noted above, cladistic analyses based on the new material of *Xenothrix* reveal that the Jamaican primate is the sister group of *Paralouatta* and *Antillothrix* (Horovitz, 1999; Horovitz & MacPhee, 1999).

Paralouatta varonai was originally considered to be closely related to *Alouatta* (Rivero & Arredondo, 1991), a conclusion echoed in its generic name. Later discoveries, however, have painted a different picture of the Cuban monkey's relationships. Cladistic analyses by MacPhee *et al.* (1995*b*), Horovitz & MacPhee (1999) and Horovitz (1999) indicated that *Paralouatta* is not closely related to *Alouatta*, but is instead a close affine of *Antillothrix bernensis*.

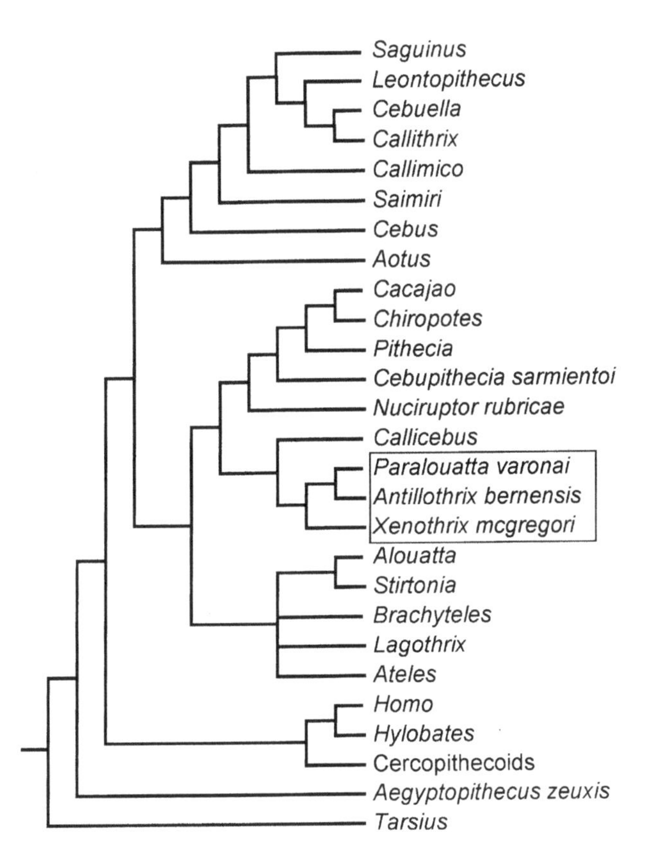

Fig. 13.5 Phylogenetic relationships of Antillean monkeys. After Horovitz & MacPhee (in press).

Antillothrix bernensis was originally considered to be similar to squirrel monkeys because it exhibits a relatively large molar trigon, a separate and offset hypocone (especially on M^2), and a small M^2 relative to M^1. MacPhee *et al.* (1995*b*) found no evidence of affinity with *Saimiri*: rather, *A. bernensis* appeared on the cladogram as the sister group of *Paralouatta*, these two taxa in turn being the sister group of *Callicebus*. This result was later confirmed with more extensive analyses of platyrrhine systematics, which included *Xenothrix* as well (Horovitz, 1999; Horovitz & MacPhee, 1999).

Figure 13.5 shows the strict consensus of the three most-parsimonious trees recovered by Horovitz & MacPhee (1999; see also Horovitz, 1999). These point to a sister group relationship between *Paralouatta* and *Antillothrix*, followed in the hierarchy by *Xenothrix*, and at a more inclusive level by *Callicebus*. The sister group of this clade is constituted as follows: (*Nuciruptor* (*Cebupithecia* (*Pithecia* (*Cacajao*, *Chiropotes*)))). Unambiguous characters supporting the *Callicebus*/Antillean monkey clade are: two prominences present on the lateral wall of the promontorium, derived from the condition of a flat surface or a single prominence; the ventral border of the zygomatic arch extends below the plane of the alveolar border, derived from a condition in which the zygomatic arch occupies a higher position on the face; the mandibular canine root is highly compressed, derived from a more rounded condition; and finally, the alveolus for the maxillary canine is smaller than that of P^4, derived from the reverse condition.

The Antillean clade itself was supported by three unambiguously placed characters (Horovitz & MacPhee, 1999): the nasal fossa is wider than the palate (the latter measured at the level of the M^1), derived from a narrower condition; the alveolus of the mandibular canine is buccolingually smaller than that of P_4, derived from the reverse condition; and the mandibular M_1 protoconid has a bulging buccal surface, derived from absence of this feature.

The dyad consisting of *Paralouatta* + *Antillothrix* was supported by six unambiguous characters (Horovitz & MacPhee, 1999): M_1 oblique cristid intersects the protolophid lingual to the protoconid, derived from a position directly distal to this cusp; P^4 lingual cingulum projects mesially, derived from a cingulum that projects directly lingually; P^4 subequal to M^1 in buccolingual dimension, derived from a smaller P^4; M^1 possesses a distinct pericone, derived from absence of this cusp; M^1 postmetacrista slopes distobuccally, derived from a slope directed distally or distolingually; and M^1 hypocone is located lingually with respect to protocone, derived from a condition in which this pair of cusps is aligned mesiodistally.

The chief biogeographical implication of the discovery that Greater Antillean monkeys form a monophyletic group is that it is now parsimonious to assume that only one primate colonization took place from the South American mainland, as opposed to the several unrelated events invoked or implied by previous studies. The minimum date for this colonization event is early Miocene, since this is the age of the Zaza astragalus. However, colonization could have taken place well before this, since Paleogene land-mammal fossils have been recovered from Puerto Rico (MacPhee & Iturralde-Vinent, 1995*b*) and Jamaica (Domning *et al.*, 1997). The Cenozoic biogeography of the Caribbean region, especially as it relates to the history of the terrestrial mammal fauna, is treated exhaustively by Iturralde-Vinent & MacPhee (1999).

EXTINCT QUATERNARY PLATYRRHINES OF BRAZIL

Introduction

Considering the diversity of continental platyrrhines recorded for the early Neogene, the negligible late Neogene record is surprising. The only platyrrhine species that have been recorded so far from late Quaternary localities in the continental Neotropics are still extant – with two exceptions, *Protopithecus brasiliensis* and *Caipora bambuiorum*, from the caves of Minas Gerais and Bahia states in east-central Brazil.

History of discovery and debate

It is of some historical interest that the first primate fossils to be correctly recognized and named as such were found in

July, 1836, by the Danish naturalist Peter Wilhelm Lund at Lapo de Periperi, a cave system in Lagoa Santa, Brazil. Lund based the species *Protopithecus brasiliensis* on these finds (a left proximal femur and a right distal humerus) in a letter dated November 16, 1837 (published as Lund (1838), and subsequently republished by other European journals (Lund, 1839, 1840*a*, 1840*b*)).

Several decades later, Paul Gervais and Florentino Ameghino (Gervais & Ameghino, 1880) erected a new species, *Protopithecus bonariensis*, for some isolated incisors recovered near Buenos Aires. Inasmuch as these specimens were not described, illustrated, or catalogued, it is impossible to say at this juncture what their pertinency (if any) may have been to Lund's monkey. A few years later, Herluf Winge (1895) insured that *Protopithecus brasiliensis* would fall into obscurity by sinking it into *Eriodes* (now *Brachyteles*) as *E. protopithecus*. He took this step because he felt that, although the limbs of Lund's monkey were clearly that of a very large platyrrhine, their morphology was similar enough to that of the woolly spider monkey to place them in the same genus.

Hill (1962) performed essentially the same maneuver 70 years later. Grossly underestimating the size of *P. brasiliensis*, he transferred it to *Brachyteles* (as *B. brasiliensis*), on the ground that the apparently slight differences in size between the fossil species and living *B. arachnoides* did not justify a generic distinction. A few later commentators made mention of *Protopithecus brasiliensis* (Aguirre, 1971; Paula-Couto, 1979; Pinto da Silveira, 1985), but the question of its distinctiveness was not reopened until the 1990s, by Rosenberger (see Hartwig, 1995*b*).

Hartwig (1995*b*) pointed out that the difference in size between Lund's monkey and the woolly spider monkey, together with certain discrete features, fully justified making a genus-level distinction between them. While agreeing that Lund's monkey was certainly ateline, he noted that the evidence did not support a close relationship between *Protopithecus brasiliensis* and *Brachyteles arachnoides* to the exclusion of other atelines as Winge (1895) had argued. Hartwig (1995*a*) also showed that Lund's specimens came from an animal whose body size (approximately 25 kg) was 40% larger than that of any living platyrrhine species.

In 1993, Castór Cartelle announced the discovery of two nearly complete skeletons, found by Grupo Bambui de Pesquisas Espeleologicas in 1992 in Toca da Boa Vista, a large cave in Bahia, Brazil. Hartwig & Cartelle (1996) attributed one of the skeletons, which included the skull and mandible with a partial dentition, to *Protopithecus brasiliensis*. The other skeleton was that of a hitherto unknown large-bodied ateline (approximately 20 kg), which Cartelle & Hartwig (1996*a*) named *Caipora bambuiorum*. The genus name means "dweller in the wood" and is a reference to Lund's (1840*b*) report of an Amerindian story concerning a large monkey that inhabited the Brazilian interior highlands.

Taxonomy

Systematic framework

Allocation follows Hartwig & Cartelle (1996) and Cartelle & Hartwig (1996*a*).

Order Primates Linnaeus, 1758
- Infraorder Platyrrhini É. Geoffroy Saint-Hilaire, 1812
  - Family Atelidae Gray, 1825
    - Subfamily Atelinae Gray, 1825
      - Tribe *incertae sedis*
        - Genus *Protopithecus* Lund, 1838
          - *Protopithecus brasiliensis* Lund, 1838
      - Tribe Atelini
        - Genus *Caipora* Cartelle & Hartwig, 1996
          - *Caipora bambuiorum* Cartelle & Hartwig, 1996

Family Atelidae

Subfamily Atelinae

Tribe *incertae sedis*

GENUS *Protopithecus* Lund, 1838
INCLUDED SPECIES *P. brasiliensis*

SPECIES *Protopithecus brasiliensis* Lund, 1838 (Figs. 13.6, 13.7)
TYPE SPECIMEN UZM 82, left proximal femur, and UZM 3530, right distal humerus
AGE AND GEOGRAPHIC RANGE Quaternary, late Pleistocene or possibly early Holocene, South America, Brazil only
ANATOMICAL DEFINITION
Large cranium with compound temporonuchal crest, flat caudally directed nuchal plane, rounded cusps on upper and lower molars, postcranium extremely robust, long upper limbs, reduced olecranon, strong muscle markings for brachial flexion and leg extension (Fig. 13.7).

Tribe Atelini

GENUS *Caipora* Cartelle & Hartwig, 1996
INCLUDED SPECIES *C. bambuiorum*

SPECIES *Caipora bambuiorum* Cartelle & Hartwig, 1996 (Fig. 13.8)
TYPE SPECIMEN IGC-UFMG 05, a nearly complete skeleton
AGE AND GEOGRAPHIC RANGE Quaternary, late Pleistocene or possibly early Holocene, South America, Brazil only
ANATOMICAL DEFINITION
Large, skull widest high on the parietals, upper and lower molars are quadrate and bunodont, lower third molar set in the mandibular corpus, extremely long upper limbs, reduced olecranon, metacarpals and metatarsals subequal in length, relatively robust caudal vertebrae.

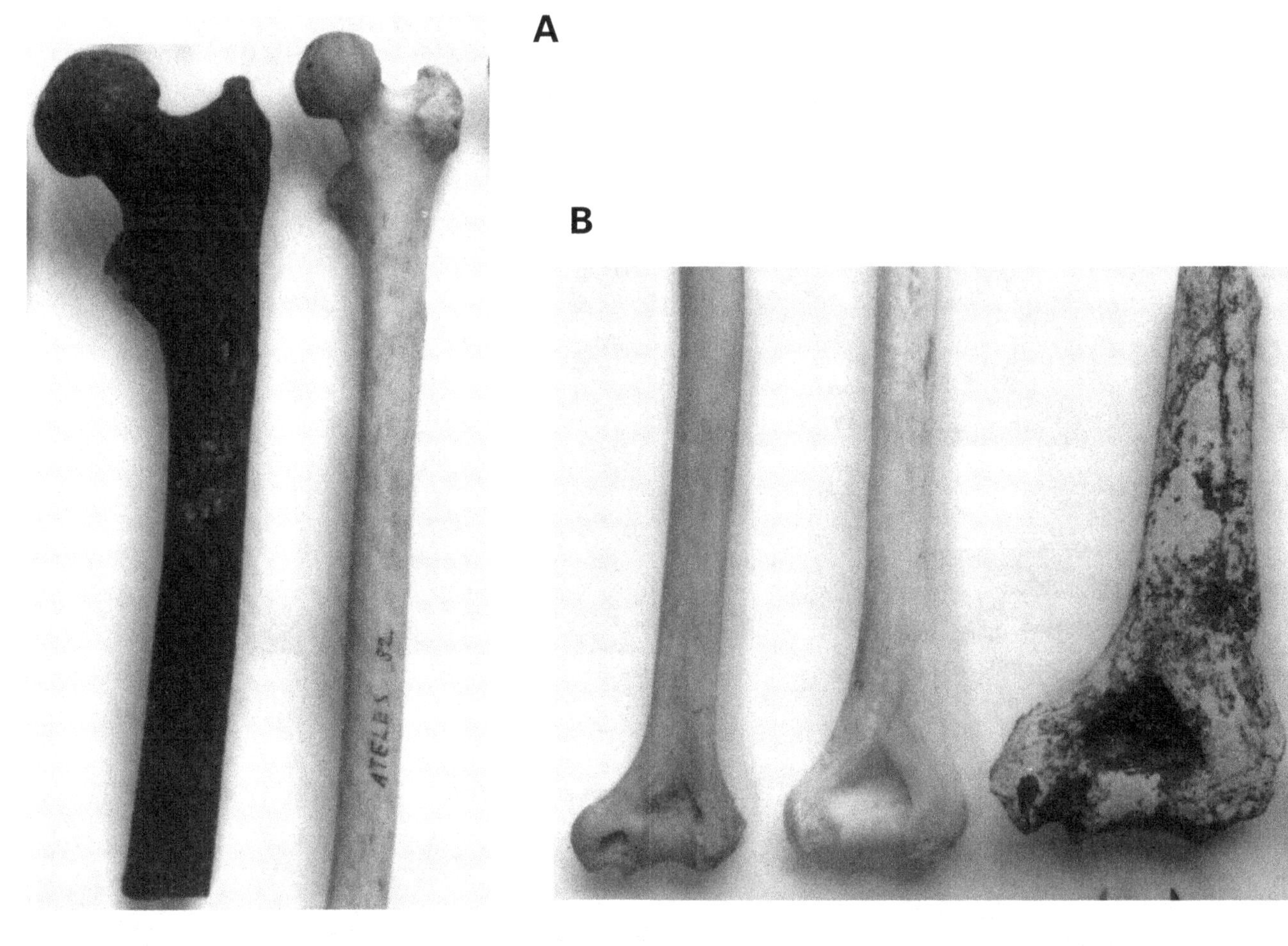

Fig. 13.6 *Protopithecus brasiliensis*. (A) Holotype femur (UZM 1623, left) compared to *Brachyteles arachnoides* (bone label "*Ateles*" reflects former classification of *Brachyteles* as a subgenus of *Ateles*), anterior view; (B) UZM 3530 (right), holotype humerus compared to *Alouatta* (left) and *Ateles* (center), posterior view.

Evolution of extinct Brazilian atelines

Protopithecus brasiliensis displays a suite of derived characters shared with *Alouatta*, *Ateles*, *Brachyteles* and *Lagothrix*, and on this basis it has been classified as an ateline by Hartwig & Cartelle (1996). All atelines are characterized by large body size, but *Protopithecus* shares a suite of derived postcranial characters with a more restricted group, the atelins (*Ateles*, *Brachyteles* and *Lagothrix*). Relevant features include a high intermembral index, reduced olecranon process of the ulna and articular morphology typical of rotational mobility at the glenohumeral joint.

These features have been interpreted as modifications for a substantial brachiating component in locomotion (Hartwig & Cartelle, 1996). The postcranium of *Protopithecus* is very robust, in contrast with that of *Ateles* and *Lagothrix*. On the other hand, the skull morphology of *Protopithecus* most resembles that of *Alouatta* in regard to the presence of a compound temporonuchal crest, a flat caudally directed nuchal plane, and an extended basicranium, all of which could be related with the presence of an enlarged hyoid bone as seen in *Alouatta*. The dentition of *Protopithecus* lacks specializations for folivory present in *Alouatta*, such as molar shearing crests and reduced incisors.

This combination of cranial and postcranial characters is unparalleled among living atelines and poses a dilemma for a finer resolution of the phylogenetic relationships of *Protopithecus* (Hartwig & Cartelle, 1996). If *Protopithecus* is a close relative of the alouattins, either its postcranial anatomy must be derivedly convergent on that of atelins, or the typical atelin postcranial anatomy must have been present in the common ancestor of all atelines (and reversed in *Alouatta*). If *Protopithecus* is the sister group of atelins, then its cranial morphology might be either convergent on that of *Alouatta* or it was present in the common ancestor of all atelines (and reversed in atelins). In either case, the discovery of additional remains of *Protopithecus* changes interpretations of the evolution of atelines based more or less exclusively on living taxa.

Caipora bambuiorum has been classified as an atelin; within this group its resemblance is greatest to *Ateles*, especially in regard to dental and mandibular morphology (Cartelle & Hartwig, 1996*a*). The occlusal morphology is bunodont, and dental length and breadth are not significantly greater than those in large individuals of *Ateles*. Although the individ

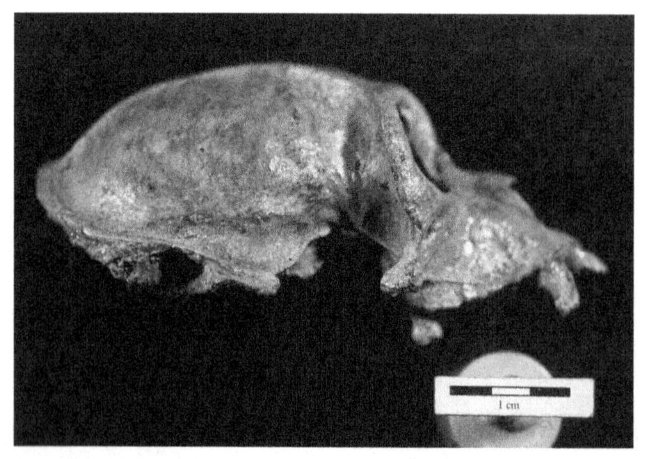

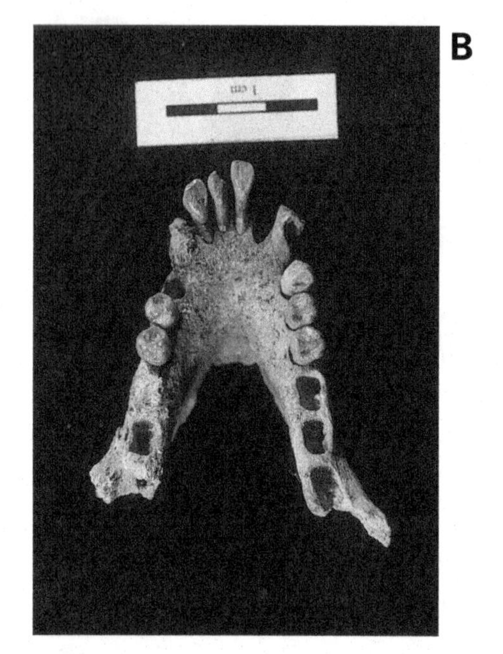

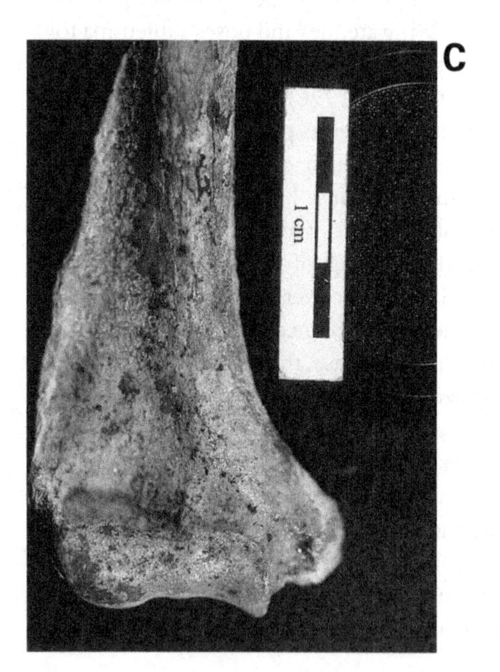

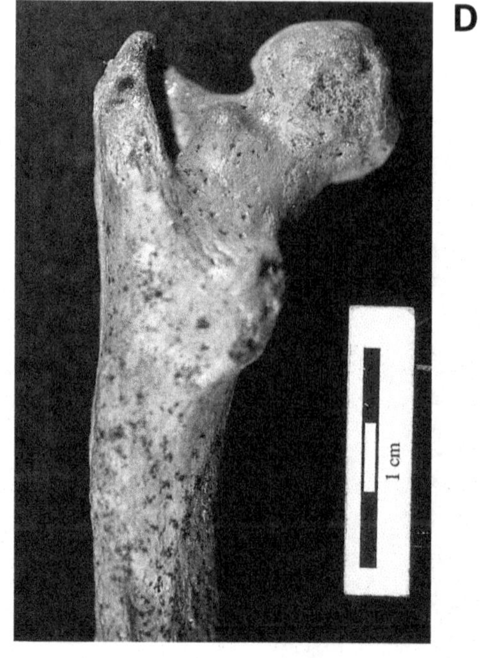

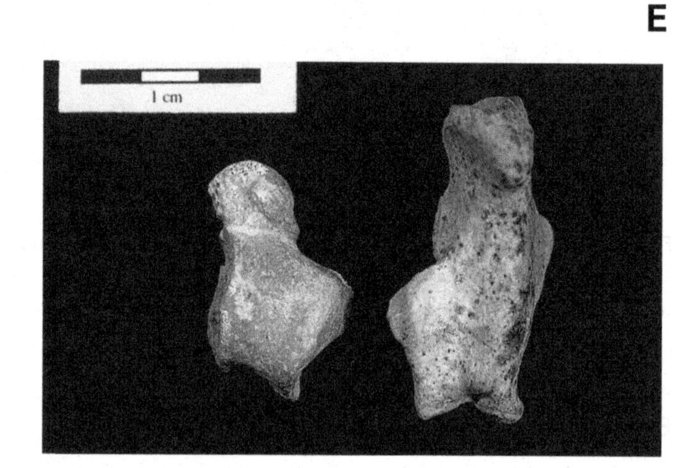

Fig. 13.7 *Protopithecus brasiliensis*. Aspects of holotype skeleton (IGC-UFMG 06): (A) cranium, correct canine reconstruction; (B) mandible, occlusal view; (C) distal right humerus, anterior view; (D) proximal left femur, posterior view; (E) astragalus (left) in dorsal view and calcaneus (right) in ventral view.

Fig. 13.8 (opposite) *Caipora bambuiorum*. Aspects of holotype skeleton (IGC-UFMG 05), and comparisons of *Caipora* and *Protopithecus*: (A) skull; (B) palate, occlusal view; (C) articulated left foot; (D) proximal ulnae, anterior view, of *Caipora* (left), *Protopithecus* (right); (E) acetabular view of os coxae of *Protopithecus* (left), *Caipora* (right); (F) distal humeri, posterior view, of, from left to right: *Ateles*, *Caipora*, *Protopithecus*, *Caipora*; (G) anterior view of crania of *Protopithecus* (left, incorrect canine reconstruction), *Caipora* (right); (H) occlusal view of mandibles of *Caipora* (left), *Protopithecus* (right).

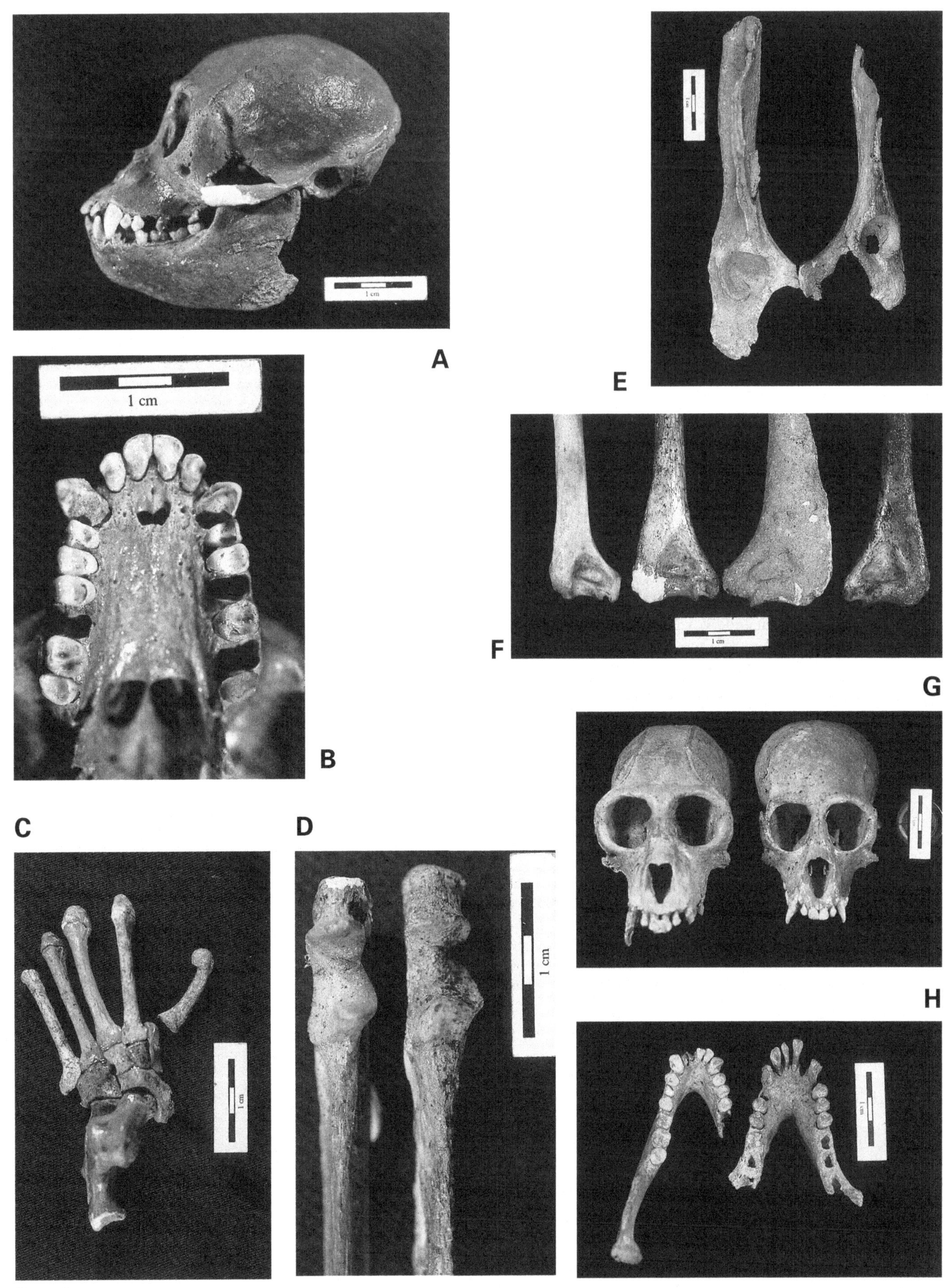

Fig. 13.8

more robust than any living platyrrhine. It shares several derived postcranial characters with atelins, which are brachiating monkeys, such as extremely long upper limbs, spherical humeral head, short ulnar olecranon process, radial articular facet flush with the shaft of the ulna, metacarpals and metatarsals of subequal length, and robust proximal caudal vertebrae.

No radiometric dates are available for these specimens, although they were found – as were Lund's – in apparent association with extinct as well as still-extant late Quaternary species (Cartelle & Hartwig, 1996*b*; Hartwig & Cartelle, 1996). This is of interest because *Protopithecus* and *Caipora* can be considered as "megafaunal" in comparison to their relatives. The fact that they – but no other Quaternary monkey species – possibly were lost during the terminal Pleistocene New World extinction event fits in well with the pattern of extinction seen in many other mammalian groups (MacPhee & Marx, 1997; MacPhee & Flemming, 1999). Most (but not all) of these species were in the megafaunal size category defined by Martin (1984). The cause of these losses remains elusive, although overhunting, climate change, and emergent infectious diseases have all been suggested.

Acknowledgements

The authors thank Walter Hartwig for inviting us to contribute to this important endeavor and for supplying the photographs of the Brazilian specimens presented in this paper. We also thank Lorraine Meeker (photographs) and Clare Flemming (editing) for their usual excellent work.

Primary References

Antillothrix

Miller, G. S. (1929). Mammals eaten by Indians, owls, and Spaniards in the coast region of the Dominican Republic. *Smithsonian Miscellaneous Collections*, **82**(5), 1–16.

Rímoli, R. (1977). Una nueva especie de mono (Cebidae: Saimirinae: *Saimiri*) de La Hispaniola. *Cuadernos de CENDIA, Universidad Autónoma de Santo Domingo*, **242**, 5–14.

MacPhee, R. D. E. & Woods, C. A. (1982). A new fossil cebine from Hispaniola. *American Journal of Physical Anthropology*, **58**, 419–436.

MacPhee, R. D. E., Horovitz, I., Arredondo, O., & Jiménez Vasquez, O. (1995). A new genus for the extinct Hispaniolan monkey *Saimiri bernensis* (Rímoli, 1977), with notes on its systematic position. *American Museum Novitates*, **3134**, 1–21.

Caipora

Cartelle, C. & Hartwig, W. C. (1996). A new extinct primate among the Pleistocene megafauna of Bahia, Brazil. *Proceedings of the National Academy of Sciences of the United States of America*, **93**, 6405–6409.

Paralouatta

Rivero de la Calle, M. & Arredondo, O. (1991). *Paralouatta varonai*, a new Quaternary platyrrhine from Cuba. *Journal of Human Evolution*, **21**, 1–11.

Horovitz, I. & MacPhee, R. D. E. (1999). The Quaternary Cuban platyrrhine *Paralouatta varonai* and the origin of Antillean monkeys. *Journal of Human Evolution*, **36**, 33–68.

Protopithecus

Lund, P. W. (1838). Blik paa Brasiliens dyreverden for sidste jordomvaeltning. *Det Kongelige Danske Videnskabernes Selskabs Naturvidenskabelige og Mathematiske Afhandlinger*, **8**, 61–144.

Hartwig, W. C. & Cartelle, C. (1996). A complete skeleton of the giant South American primate *Protopithecus*. *Nature*, **381**, 307–311.

Xenothrix

Williams, E. E. & Koopman, K. F. (1952). West Indian fossil monkeys. *American Museum Novitates*, **1546**, 1–16.

MacPhee, R. D. E. & Fleagle, J. G. (1991). Postcranial remains of *Xenothrix mcgregori* (Primates, Xenotrichidae) and other late Quaternary mammals from Long Mile Cave, Jamaica. *Bulletin of the American Museum of Natural History*, **206**, 287–321.

Unattributed

Ford, S. M. & Morgan, G. S. (1986). A new ceboid femur from the Late Pleistocene of Jamaica. *Journal of Vertebrate Paleontology*, **6**, 281–9.

MacPhee, R. D. E. & Iturralde-Vinent, M. A. (1995*a*). Earliest monkey from Greater Antilles. *Journal of Human Evolution*, **28**, 197–200.

The fossil record of early catarrhines and Old World monkeys

14 | Early catarrhines of the African Eocene and Oligocene

DAVID TAB RASMUSSEN

Introduction

One of the best-known fossil primates of the early Tertiary is *Aegyptopithecus zeuxis*, an early catarrhine represented by several fossil crania, much of the limb skeleton, and many dozens of jaws and teeth. The name "*zeuxis*" is derived from the Greek word for yoke, meaning to join or connect, which appropriately conveys that *Aegyptopithecus* is one of paleontology's not-so-missing links – a mosaic of primitive and specialized characters linking the earliest anthropoids of the Eocene to modern apes and monkeys. The most notable catarrhine features of *Aegyptopithecus* are its extremely ape-like dental structure and a 2.1.2.3 dental formula; the primitive traits include a low, long cranial vault containing a very small brain, and a primitively constructed snout. *Aegyptopithecus* is related to several other early catarrhine genera that, in some cases, show even more primitive features (such as an unfused mandibular symphysis). The primitiveness of these anthropoids has made their catarrhine status controversial; some authorities believe only a few of the taxa discussed in this chapter can stake a legitimate claim as true catarrhines. Others have concluded that early catarrhines were indeed very primitive, and that some of the specialized traits we would define as "anthropoid" based on today's living primates actually evolved in parallel among basal catarrhines, platyrrhines and other primitive anthropoids.

All of the primate taxa discussed in this chapter occur in the late Eocene and early Oligocene (30–35 Ma) of the combined Afro-Arabian continent including today's Arabian Peninsula. Africa was an isolated continent during most of the early Tertiary, floating separately from its Gondwanaland siblings of South America, Antarctica, Australia, Madagascar and India, and yet still attached until the Oligocene to the Arabian Peninsula. The isolated Afro-Arabian continent served as the evolutionary stage for many key events in mammalian evolution, among them the origin and diversification of anthropoid primates, despite reasonable views to the contrary discussed elsewhere in this book.

The earliest putative catarrhines of Africa date to the latest Eocene and early Oligocene of the Fayum, Egypt, and the Oligocene of Oman and Angola (a single isolated canine). The catarrhines from Oman and Egypt occur in vertebrate faunas that contain other primate taxonomic groups.

The sedimentary deposits of the Fayum that contain primate fossils are all part of the Jebel Qatrani Formation, which is made up of over 340 m of rock deposited primarily as point bars and overbank deposits in a lowland riverine environment. The age of the Fayum has been assessed by analysis of depositional unconformities, faunal comparisons and paleomagnetic correlations, all of which indicate that the lowest parts of the Jebel Qatrani Formation are late Eocene, while the upper parts are early Oligocene (Kappelman *et al.*, 1992; Rasmussen *et al.*, 1992). The depositional environment of the Jebel Qatrani Formation consisted of the wet, muddy edges of rivers, bayous and overbank swamps of humid, lowland tropical forests bordering a complex network of distributary river channels close to the southern border of the tropical Tethys Sea. The forests included rich riverine strips, or broader lowland tracts, of gigantic tropical trees flush with primates and brightly colored, canopy-dwelling birds (Bown *et al.*, 1982; Rasmussen *et al.* 1987). The terrestrial habitats are harder to interpret because they were dominated by divergent groups extinct today, such as herbivorous arsinoitheres, pliohyracid hyracoids, anthracotheres and meat-eating creodonts (Gagnon, 1997). Fossils from the Fayum come from dozens of localities initially lettered alphabetically in order of discovery, and once the alphabet was exhausted, numbered sequentially following the letter "L" (e.g., quarries A, B, C, I, M, quarry L-41).

The Omani catarrhines come from a littoral marine deposit of the Ashawq Formation in the coastal strip of Dhofar (Thomas *et al.*, 1989). Well over 1000 isolated mammalian teeth have been found, including parapithecid and probable adapoid primates, but among the primates, the catarrhines are the most diverse group, especially at a site called Taqah. The Taqah sediments are interpreted to have accumulated in a shallow coastal swamp periodically inundated by fluctuating sea levels (Roger *et al.*, 1993; Thomas *et al.*, 1999). Because of the intercalation of marine and terrestrial deposits within the Ashawq Formation their age can be assessed faunally using nummulites; the species present is typical of the early Oligocene. This age has been supported by paleomagnetic assessments dated to about 33 million years ago.

History of discovery and debate

The first catarrhines reported from the Fayum were found in 1907 by an amateur German collector living in Egypt named Richard Markgraf, who delivered his specimens to the German paleontologist Max Schlosser. Markgraf's collecting efforts were inspired by previous discoveries in the

Fayum made by the British zoologist Charles William Andrews (1906) and the subsequent collecting crew sent to Egypt by the American paleontologist Henry Fairfield Osborn. Osborn (1908) identified the first primate from the Fayum (the parapithecid *Apidium*), but it was Schlosser (1910, 1911) who described the first two catarrhine specimens. One of these was a beautiful lower jaw of what was clearly an ape-like primate, which Schlosser named *Propliopithecus haeckeli*. Schlosser identified this specimen as a catarrhine based on its 2.1.2.3 dental formula and its gibbon-like dentition. As suggested by the name, the specimen closely resembled lower jaws of Miocene *Pliopithecus*. At the time, there was no reason to believe that a fossil so ape-like was not indeed an advanced ape. While early phylogeneticists had no problem understanding the relevant importance of primitive and specialized features, there was no solid foundation of empirical knowledge available at that time to believe ape-like morphology might indeed be primitive for extant catarrhines. Ernst Haeckel, namesake of the Fayum fossil, had produced a hypothetical phylogeny of primates that differed little from preconceptions of a *scala naturae*, in which humans resided at the apex of an evolutionary tree and apes (and by inference, everything about ape morphology) occupied a position just two ladder steps down from that of modern humans (just below Haeckel's famous missing link "*Pithecanthropus*"). On the sole basis of the material at hand, Schlosser and many subsequent workers were justified in interpreting *Propliopithecus* as a small, early ape – it would be more than a half-century before cranial and postcranial finds demonstrated that *Propliopithecus* was considerably less ape-like than would be judged from its dentition alone.

The other early catarrhine described by Schlosser was based on a mandibular fragment containing only M_{1-2}, which Schlosser (1910, 1911) bestowed with the name *Moeripithecus markgrafi*. As has been pointed out many times over the last century (e.g., Simons, 1995*b*), these two teeth alone do not provide good basis for cogent taxonomic assessments. The molars had slightly sharper crests than the bunodont teeth of *Propliopithecus*, which led some researchers to consider that *Moeripithecus* might be a cercopithecoid (Abel, 1931), while others considered its affinities to be with *Propliopithecus* (Kälin, 1961). Later researchers, frustrated that such an incomplete specimen had been made the type of a new genus and species, typically have sunk *Moeripithecus markgrafi* into the genus *Propliopithecus* (Simons, 1967*b*; Kay *et al.*, 1981), even though it clearly "differs significantly from other species [of the genus *Propliopithecus*] in the occlusal morphology of the first and second lower molars" (Simons *et al.*, 1987). The possible significance of the differences finally became evident when older, more primitive catarrhines were recovered from the Fayum, and as new catarrhines possibly attributable to *Moeripithecus* were later found in Oman (Thomas *et al.*, 1991).

For half a century (1910–60), the Fayum primates were subjects of much discussion, but no one committed to finding more material until American paleontologist and anthropologist Elwyn Simons began searching in 1961 (Simons, 1995*b*). His expeditions spanning 40 years have been wildly successful in recovering a tremendous database of fossil catarrhines and other primates. As a fitting testament to the serendipity of field paleontology, the first primate specimen that Simons's field crew recovered from the Fayum was of an utterly different kind of catarrhine than those found early in the century. It proved to be an extremely primitive anthropoid whose special relationship to catarrhines was questioned at the time and which remains controversial up to the present. This primate, given the name *Oligopithecus savagei*, was small and cresty-toothed, bearing a distinctive prosimian appearance. The molar trigonids were raised and the cusps were uninflated, and yet it had a catarrhine dental formula of 2.1.2.3. The crestiness, like that of *Moeripithecus markgrafi* before it, led some to suggest a relationship to the bilophodont cercopithecoids while others considered it an adapoid prosimian (Szalay, 1970; Gingerich, 1977*a*). Unfortunately, despite daunting man-hours spent screening sediment at the site where the initial specimen was found, no additional material came to light for more than 20 years.

While the Simons expeditions continued to be stymied in their efforts to find additional material of *Oligopithecus*, they found a spectacular abundance of true catarrhines in the upper part of the stratigraphic section. Two years after finding the type jaw of *Oligopithecus*, quarry I in the Fayum's upper sequence was discovered, which would become one of the Fayum's greatest fossil localities. Among abundant material of the parapithecid *Apidium*, Simons also found jaws of a larger catarrhine whose existence had been suggested to him already by an edentulous jaw fragment in the American Museum of Natural History that had been collected decades earlier during the Osborn expeditions (Simons, 1995*b*). This siamang-sized catarrhine differed from *Propliopithecus* in its larger size, more robust mandible, and in diagnostic features of the molar teeth, and Simons (1965) named it *Aegyptopithecus zeuxis*. Quarry I also yielded a mandible of quite distinct morphology initially named *Aeolopithecus chirobates* (Simons, 1965); the most distinctive anatomical attributes of the type specimen soon were shown to be due to corrosion of the dental enamel; and so the species was transferred into Schlosser's genus *Propliopithecus* (Simons, 1967*b*; Kay *et al.*, 1981).

While the taxonomic details of the increasing diversity of early catarrhines were being worked out, a spectacular find was made in 1966 by Fayum researcher Grant Meyer while surface collecting on quarry M; he "noticed orbits and brow ridges protruding from the sand" which proved to be the uppermost exposure of an entire cranium of *Aegyptopithecus* (Simons, 1995*b*). This find immediately impacted paleoprimatology because of its mosaic of primitive and specialized features. The specimen was undoubtedly anthropoid as judged from its closed orbits, single frontal bone and distinctly ape-like upper dentition. At the same time, the specimen bore a low, long braincase proportionately smaller than those of living anthropoids, it lacked an ectotympanic tube, and had a protruding snout that in details resembled some Eocene primates rather than long-nosed anthropoids like baboons. Because of the indisputable ape-like features of the dentition and cranium, Simons coined a term for the Fayum early catarrhines: "dawn apes".

The mosaic of primitive and specialized features in *Aegyptopithecus* was further substantiated when postcranial remains discovered during the 1960s finally were analyzed. The first bone to be examined was an ulna, which indicated *Aegyptopithecus* was an arboreal quadruped lacking specialized features always seen in extant apes (Fleagle *et al.*, 1975). A hallucal metatarsal showed *Aegyptopithecus* had a typical anthropoid grasping foot, but it also retained a scar for the primitive prehallux, lost in living cercopithecoids and great apes (Conroy, 1976*a*). Additional material continued to support the early interpretations that *Aegyptopithecus* (and *Propliopithecus*) were generalized arboreal quadrupeds lacking special hominoid or even catarrhine features (Conroy, 1976*a*; Fleagle & Simons, 1978*a*, 1982*a*, 1982*b*; Gebo & Simons, 1987; Gebo, 1993; Ankel-Simons *et al.*, 1998). One of the most important results to come out of the study of early catarrhines is that they show a primitive morphology that would not have been predicted by a parsimony analysis of character states found among extant catarrhines, and indeed, in many important ways the Fayum catarrhines are notably prosimian-like.

The locomotor interpretations of the new limb bones along with functional interpretations of the teeth of early catarrhines (Kay & Simons, 1980; Fleagle & Kay, 1985) allowed researchers to reconstruct what these primates had been like in life. *Aegyptopithecus* and *Propliopithecus* were interpreted as arboreal, quadrupedal frugivores that lived in polygynous social groups (as interpreted from patterns of canine dimorphism; Fleagle *et al.*, 1980). From the growing body of data Cachel (1979*b*, 1981) developed the first specific evolutionary models of anthropoid origins. At the time, it seemed that diagnostic anthropoid characters might have been acquired in response to a dietary shift and increase in body size associated with hypothesized deteriorating climates at the Eocene–Oligocene boundary.

At about the same time, paleontological investigations of other mammals, plants and birds from the Fayum provided the basis for reconstructing the habitat as warm, wet tropical swamp forests and riparian strips of rainforest cut by meandering rivers with open sand bars (Bown *et al.*, 1982; Olson & Rasmussen, 1986). Unlike in the northern continents, where the transition from the Eocene to the Oligocene was interpreted as a dramatic loss of wet, tropical rainforest habitats, it appeared that the early Oligocene of the Fayum still harbored classic, tropical primate habitat. Later studies failed to identify a dramatic faunal or environmental shift across the Eocene–Oligocene boundary in the Fayum, and by

Table 14.1. Stratigraphic distribution and age of Fayum catarrhines

Quarry	Paleomagnetism	Age (Ma)	Species
I, M	C13 normal	33.1–33.4	*Aegyptopithecus zeuxis* *Propliopithecus chirobates*
V, G	C13 reversed	33.8–34.0	*Propliopithecus ankeli* ? *P. haeckeli*
	– Major unconformity possibly encompassing the Eocene–Oligocene boundary, 34.0 Ma –		
E	C13 reversed	34.0–35.1	*Oligopithecus savagei*
L-41	C15 reversed	35.6–35.9	*Catopithecus browni*

Source: Kappelman *et al.* (1992).

inference, possibly in other parts of Africa as well (Rasmussen *et al.*, 1992; Gagnon, 1997). By then, Eocene anthropoids had been discovered (Godinot, 1994; Simons *et al.*, 1994), and so the early models of anthropoid or catarrhine origins associated with an Eocene–Oligocene climatic shift were abandoned (but see Gingerich, 1993*b*).

One of the important breakthroughs in the Simons expeditions, in contrast to the work early in the century, was that a concerted effort was made to sample and record fossils within a stratigraphic framework (Table 14.1). Sedimentological work and geological sampling to obtain an absolute age of the deposits were initiated in the 1960s and reached their apex in the 1980s with the publication of major works on sedimentation and age (Bown & Kraus, 1988; Kappelman *et al.*, 1992). The fossil catarrhines collected by the Simons expedition came from two distinct stratigraphic zones, initially informally called the Lower and Upper Fossil Wood Zone (after the abundance of dark-stained fossil wood in the deposits) and later formalized as the lower and upper sequences of the Jebel Qatrani Formation (Bown & Kraus, 1988). *Oligopithecus* occurred in the lower sequence while *Aegyptopithecus zeuxis* and *Propliopithecus chirobates* were found only high in the section. The material found early in the century – *P. markgrafi* and *P. haeckeli* – was of unknown stratigraphic provenience. The Simons crews found a few specimens of *Propliopithecus* low in the upper sequence that seemed to be allocatable to *P. haeckeli*. Then in 1985 another dawn ape was recovered from a quarry low in the upper sequence (quarry V). This species, named *Propliopithecus ankeli*, was interpreted as a large-bodied member of the genus with an emphasis on enlarged anterior dentition, including a uniquely broad incisor series and short, robust canines (Simons *et al.*, 1987). Among extant primates the dental battery functionally resembled the South American capuchin monkey, *Cebus*.

This was the last novel taxonomic discovery of catarrhines in the upper part of the Jebel Qatrani Formation before a wholesale shift of collecting efforts downwards in the stratigraphic column. In the late 1980s, renewed efforts were put into recovering additional specimens of *Oligopithecus* from the lower sequence. Several isolated teeth of this enigmatic primate were found at the same quarry yielding the type specimen in 1961 (quarry E), including for the first time upper and anterior teeth that helped clarify its dental morphology (Rasmussen & Simons, 1988). The new material was interpreted as indicating that *Oligopithecus* was indeed a basal catarrhine, not a cercopithecoid or adapid, but one which may have evolved from an adapid ancestry. Also in the middle 1980s, American geologist and paleontologist Thomas M. Bown explored additional areas within the Jebel Qatrani Formation as part of a stratigraphic project. In 1984 Bown made what would prove to be one of the most important discoveries ever in the Fayum: he encountered very low in the stratigraphic section a distinctive deposit of greenish-gray mudstones full of fossil mammals. This locality, designated quarry L-41, eventually would yield an impressive late Eocene fauna, including what may be the world's earliest catarrhine primate, *Catopithecus browni*.

At about the same time, French and Omani researchers working on the Ashawq Formation in Oman found new mammal-bearing localities of early Oligocene age. Two important sites bearing terrestrial mammals (Thaytiniti and Taqah) were found in 1986 by French geologist and paleontologist Jean Roger (Thomas *et al.*, 1988, 1989; Roger *et al.*, 1993). Taqah is particularly rich in primates, yielding about 10 species, six of which are reportedly catarrhines (four oligopithecines and two propliopithecines: Thomas *et al.*, 1999). In the Fayum, no more than two species of catarrhines are known to coexist at a single stratigraphic level. Among the important discoveries in Oman was the recognition of a species similar to the holotype of *Moeripithecus markgrafi*, which led to the resurrection of that genus name (Thomas *et al.*, 1991). The Oman material referred to *M. markgrafi* shows similarities both to Fayum *Moeripithecus* and to species of *Propliopithecus*. The most common primate from Taqah is *Oligopithecus rogeri*, known by 120 isolated teeth, and interpreted as insectivorous (Gheerbrant *et al.*, 1995). Although the great majority of the finds in Oman have been isolated teeth, they are important for illuminating catarrhine diversity in Afro-Arabia during the early Oligocene.

In addition to the Omani discoveries, an isolated canine from a long-neglected Oligocene fauna in Cabinda, Angola, which had been misinterpreted as a Miocene assemblage, was identified as belonging to *Propliopithecus* (Pickford, 1986c). Together, the Omani discoveries and the Angolan tooth imply that catarrhines must have had a broad Afro-Arabian distribution during the early Tertiary, although a rarity of sites obscures our view of the radiation. All in all, the 1980s has to qualify as a spectacular decade in terms of the number of dramatic new field discoveries of early anthropoids and catarrhines.

Back in the Fayum, work at quarry L-41 yielded remains of an early and very primitive catarrhine, *Catopithecus browni* (Simons, 1989, 1990). The L-41 fossil assemblage was remarkable for its preservation not only of jaws and teeth but also crushed crania and postcrania. *Catopithecus* soon rivaled *Aegyptopithecus* as the best-known catarrhine of the early Tertiary. The cranium of *Catopithecus* was shown to have complete postorbital closure, a relatively tiny braincase by anthropoid standards, and a more heavily constructed lower face than those of comparably-sized small platyrrhines. Immediately, several important debates emerged concerning *Catopithecus* and quarry L-41, one of which concerned its age. When *Catopithecus* was first discovered, credible Eocene anthropoids had not yet been reported from anywhere in the world, but several lines of evidence suggested that quarry L-41 may be late Eocene (Kappelman *et al.*, 1992; Rasmussen *et al.*, 1992). Others concluded that the entire Jebel Qatrani Formation was Oligocene (Gingerich, 1993*b*) or Eocene (van Couvering & Harris, 1991). New taxa recovered from quarry L-41 since those publications tend to support a late Eocene age for that site (e.g., Simons *et al.*, 1995).

The very primitive morphology of *Catopithecus* has also generated controversy, with some researchers rejecting *Catopithecus* as a catarrhine based on parsimony analyses (Ross *et al.*, 1998), while others favor catarrhine status based primarily on the morphological continuity evident among early Afro-Arabian catarrhines (Gebo *et al.*, 1994; Simons & Rasmussen, 1996; Seiffert *et al.*, 2000). The primitive but distinctly non-tarsioid attributes of *Catopithecus* and *Oligopithecus* have led some to conclude that anthropoids did not come from tarsioid primates (Rasmussen & Simons, 1988; Simons & Rasmussen, 1996), while others who favor tarsioid origins have rejected *Catopithecus* as a model for primitive anthropoids and instead have turned their attention to an Asian tarsioid, *Eosimias* (Kay *et al.*, 1997*b*; Ross *et al.*, 1998). At the time of this writing, the relationships of *Catopithecus* and any interpretations drawn from them remain controversial, both with respect to debates about the earlier prosimian origin of anthropoids as well as the possible affiliation of *Catopithecus* to later true catarrhines.

Ironically, debates about higher level taxonomy often increase as the density of the fossil record improves. Higher-level taxa were initially defined and diagnosed as such because the species that happened to survive to the present often clump into distinct groups, each easily defined by a substantial assemblage of shared specialized traits and well separated from its nearest relatives. But the gaps between extant forms are only an illusion generated by extinctions, and the fossil record fills in the gaps. One by one, supposedly diagnostic traits are peeled away from the assemblage so that the recognition of a higher taxon comes to rely on fewer and fewer traits, thus becoming less and less reliable. The cranium of *Aegyptopithecus* showed, for example, that the tubular ectotympanic is not a diagnostic catarrhine trait, and *Catopithecus* demonstrated that mandibular fusion is not a diagnostic anthropoid trait. Our assessment of *Catopithecus* as an anthropoid or as a catarrhine now depends on a controversial fraction of the characters that were initially used to define these clades. As the fossil record improves paleontologists must inevitably place less value on assemblages of shared derived features found within living clades and rely more on evolutionary patterns that are discernible in the stratigraphic and geographic contexts of the morphological finds. For an excellent review of the problem, see Bown & Rose (1987).

Taxonomy

Systematic framework

Once it became evident that the Fayum "dawn apes" actually preceded the evolutionary divergence between extant hominoids and cercopithecoids, cladistic *classifiers* ceased calling them apes or hominoids. While nearly all researchers agreed on the basic phylogenetic issues, there was disagreement about nomenclatural technicalities. Grouping Fayum catarrhines plus modern apes (to the exclusion of cercopithecoids) is perfectly monophyletic (all species share a common ancestor) but in a paraphyletic, not holophyletic sense (the difference being that the latter requires *all* descendants of the ancestor to be recognized within the taxon). The main advantage of trying to use a strict holophyletic taxonomy rather than a monophyletic but paraphyletic one is that we will not confuse cladistic classifiers. The disadvantage is that for each new taxon that is discovered splitting off between early Tertiary catarrhines and Neogene forms *we must erect a new higher-level taxon*. Obviously, such a system is unwieldy and genealogically misleading. A valuable classification of related animals distributed deeply in time must artfully blend undoubted paraphyletic taxa and occasionally useful but often hypothetical holophyletic taxa. I adopt a system that defines the two extant catarrhine superfamilies (Hominoidea, Cercopithecoidea) on the basis of their split from a common ancestor. All catarrhines currently interpreted to lie outside the true hominoid–cercopithecoid clade can be conveniently placed within a third (probably paraphyletic) superfamily, Propliopithecoidea.

A more substantive debate concerns how *Catopithecus* and *Oligopithecus* relate to other catarrhines. Because of its extremely primitive nature, some researchers have suggested that *Catopithecus* may be an outgroup to both catarrhines plus some other basal anthropoids. Based on a parsimony analysis of a very large number of characters all equally weighted, Ross *et al.* (1998) placed *Catopithecus* and *Oligopithecus* in their own family separate from Catarrhini. Of course, mega-matrix parsimony-based algorithms cannot *test* the catarrhine status of *Catopithecus* because they enforce parsimony (minimize parallelism) from the start; thus the "answer" is predetermined by the method. The essence of the hypothesis that *Catopithecus* is a catarrhine is that extensive parallelism has indeed occurred among early anthropoid lineages, and a more appropriate test is to examine patterns of morphological continuity and discontinuity among the Afro-Arabian primates stratigraphically. The challenge to researchers who believe evolutionary parallelism is common is to figure out which characters are misleading and which few are reliable. For example, many dental details associated with trigonid reduction and development of bunodonty can be seen to evolve piecemeal through the Fayum stratigraphic column, from *Catopithecus* to *Oligopithecus*, *Propliopithecus* and *Aegyptopithecus* (Simons & Rasmussen, 1996: 282). For those who believe coincidence in space and continuity in time are as scientifically important as morphological parsimony, this leads to an interpretation that parallel development of these dental traits occurred in the two-premolared ancestry of catarrhines and the three-premolared ancestry of parapithecids (and probably a third time in platyrrhines). The dental formula, on the other hand, shows greater evolutionary conservatism in time and space, and therefore, given the current evidence, deserves greater weight.

In the past I have used a single family to encompass all Fayum primates with a 2.1.2.3 dental formula, in large part because there are possible evolutionarily transitional forms between *Catopithecus* and *Aegyptopithecus*. However, in order to highlight the substantive debate about the phylogenetic position of *Catopithecus* and *Oligopithecus*, I will here place them in their own family (but still retain the family as a member of Catarrhini).

Order Primates Linnaeus, 1758
- Infraorder Catarrhini É. Geoffroy Saint-Hilaire, 1812
  - Superfamily Propliopithecoidea Simons, 1965
    - Family Oligopithecidae Kay & Williams, 1994
      - Genus *Catopithecus* Simons, 1989
        - *Catopithecus browni* Simons, 1989
      - Genus *Oligopithecus* Simons, 1962
        - *Oligopithecus savagei* Simons, 1962
        - *Oligopithecus rogeri* Gheerbrant *et al.*, 1995
    - Family Propliopithecidae Straus, 1961
      - Genus *Moeripithecus* Schlosser, 1910
        - *Moeripithecus markgrafi* Schlosser, 1910
      - Genus *Propliopithecus* Schlosser, 1910
        - *Propliopithecus haeckeli* Schlosser, 1910
        - *Propliopithecus ankeli* Simons *et al.*, 1987
        - *Propliopithecus chirobates* Simons, 1965
      - Genus *Aegyptopithecus* Simons, 1965
        - *Aegyptopithecus zeuxis* Simons, 1965

Infraorder Catarrhini

Superfamily Propliopithecoidea

Family Oligopithecidae

GENUS *Catopithecus* Simons, 1989
INCLUDED SPECIES *C. browni*

SPECIES *Catopithecus browni*
TYPE SPECIMEN CGM 41885, right mandible with P_3–M_3
AGE AND GEOGRAPHIC RANGE Lower sequence of the Jebel

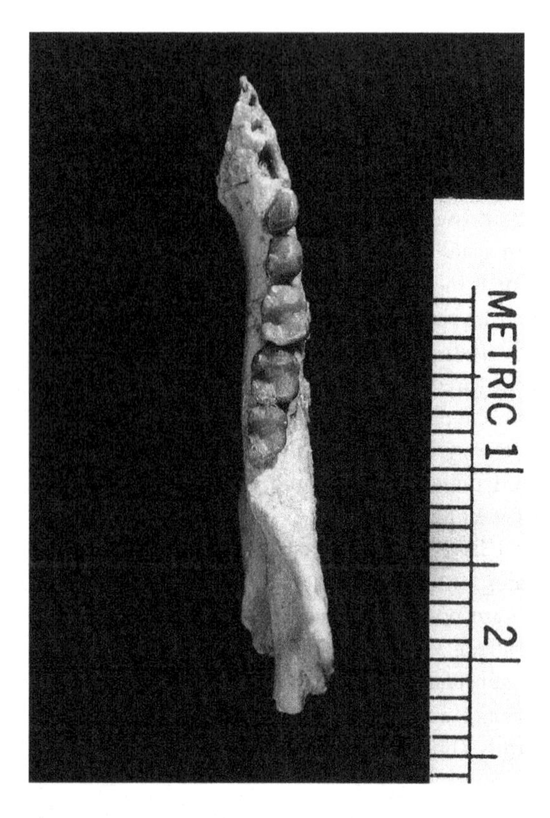

Fig. 14.1 *Catopithecus browni*. A mandible fragment (CGM 41885) showing the unfused mandibular symphysis, the sockets for I_{1-2} and canine and the crowns of P_3–M_3. From Simons and Rasmussen (1994*b*).

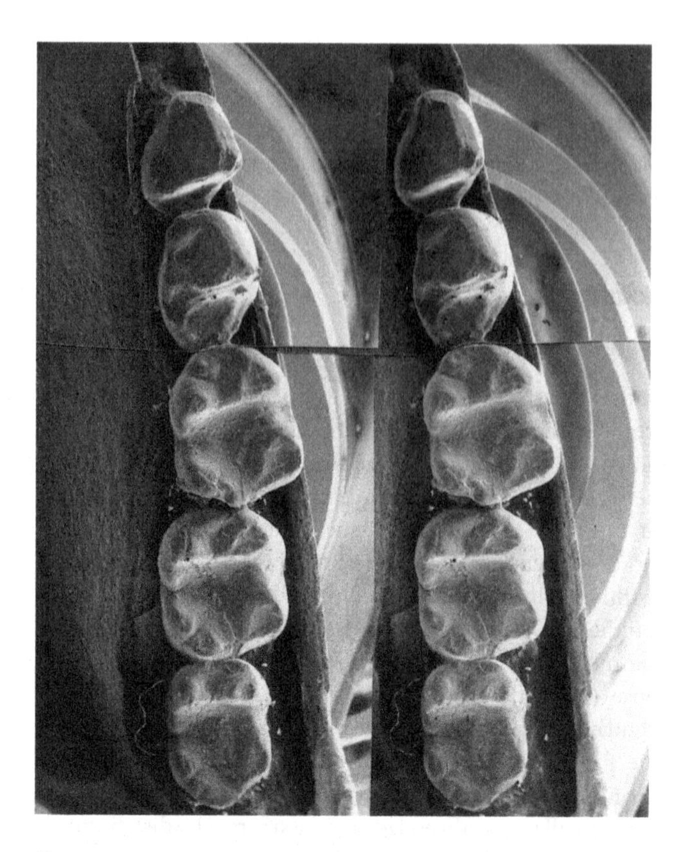

Fig. 14.2 *Catopithecus browni*. Scanning electron stereopair images of the lower dentition of CGM 41885. Note on M_{1-2} the well-defined trigonids and the entoconid–hypoconulid pairing.

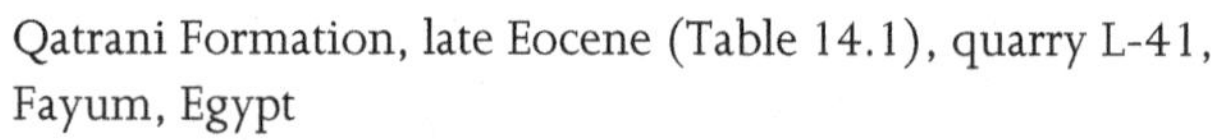

Qatrani Formation, late Eocene (Table 14.1), quarry L-41, Fayum, Egypt

ANATOMICAL DEFINITION

Small-bodied primitive catarrhines with an unfused mandibular symphysis (Fig. 14.1). The lower molars retain sharply-crested, raised trigonids, including slight paraconids on M_1 and M_2, and the upper molars have only moderate development of hypocones (Simons, 1989; Simons & Rasmussen, 1996) (Figs. 14.2, 14.3). The upper and lower molar series are characterized by sequentially reduced size of the posterior (distal) teeth. The hypoconulid is distinct, bulbous, displaced lingually and pressed closely against the entoconid ("twinned entoconid–hypoconulid"). The canines are projecting and sexually dimorphic (Fig. 14.4). The upper M^{1-2} are simple triangular teeth with crested paracone, metacone and protocone and just a small hypocone riding on the distolingual cingulum (Fig. 14.5). The M^3 is reduced in size and has a weak metacone and no hypocone. The incisors are spatulate (Simons, 1995*c*) (Fig. 14.6). Like other catarrhines, *Catopithecus* has complete postorbital closure, a 2.1.2.3 dental formula, and a honing facet between the upper canine and the lower P_3.

Since the type specimen was named, additional lower jaws have been found along with several whole or partial crania, and specimens of the humerus and femur (Simons & Rasmussen, 1996; Seiffert *et al.*, 2000). Cranially, *Catopithecus* is characterized by a relatively deep and projecting face for an anthropoid its size (Simons, 1990; Simons & Rasmussen, 1996). The ascending process of the premaxilla, which contributes to the front of the snout, is much broader than that of any living primate, anthropoid or prosimian. The orbits are small, indicating diurnality, and the interorbital region is broad. The postorbital plate closing the socket (which along with that of quarry mate

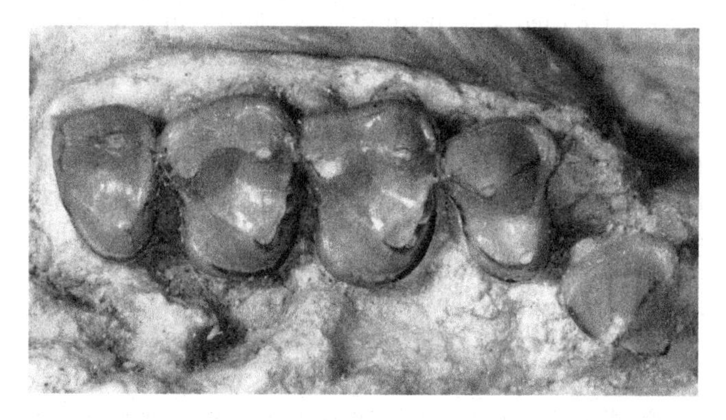

Fig. 14.3 *Catopithecus browni*. Upper right dentition (P^3–M^3) of DPC 8701 showing the triangular molar structure with very slight hypocone development.

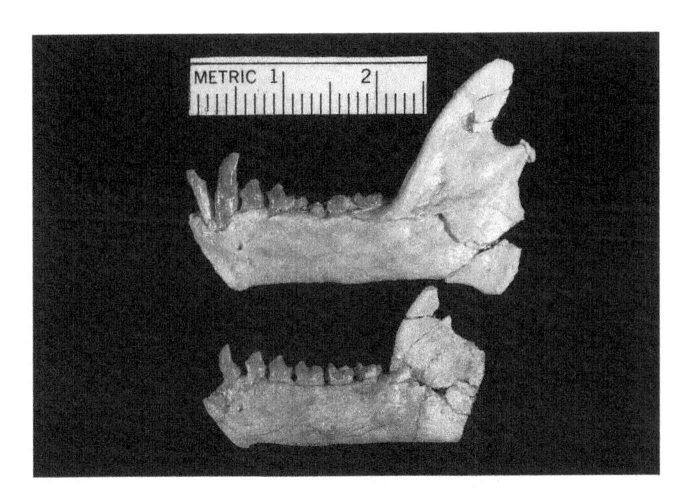

Fig. 14.4 *Catopithecus browni*. Two left mandibles showing sexual dimorphism in canine size.

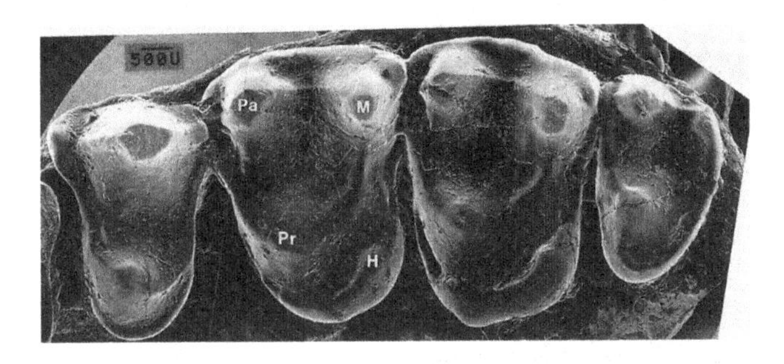

Fig. 14.5 *Catopithecus browni*. Scanning electron micrograph of the left maxillary P^4–M^3 of DPC 11594. H, hypocone; M, metacone; Pa, paracone; Pr, protocone. From Simons and Rasmussen (1996).

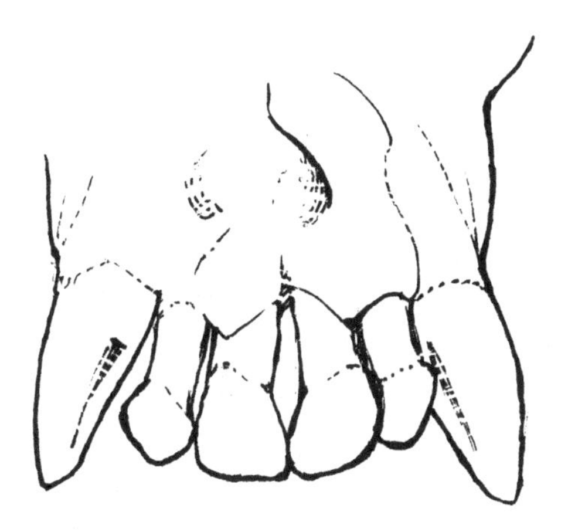

Fig. 14.6 *Catopithecus browni*. Reconstructed upper anterior dentition showing large canines and spatulate incisors. From Simons and Rasmussen (1994); illustration by E. L. Simons.

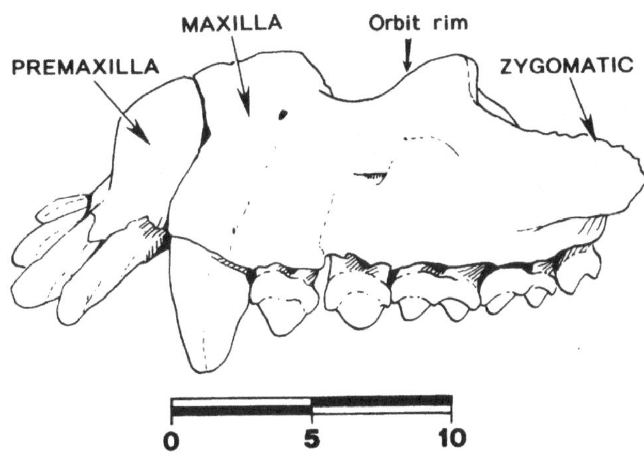

Fig. 14.7 *Catopithecus browni*. Left side of the face of DPC 11594 showing long and deep maxilla and the broad ascending wing of the premaxilla; the apparent prognathic orientation of the incisors is due to postmortem distortion. From Simons and Rasmussen (1996).

Proteopithecus is the earliest one known in the fossil record) consists primarily of an expansion of zygomatic bone. The cranial vault is small, indicating a brain notably smaller than those of living anthropoids. Temporal lines converge toward the midline, forming a sagittal crest along the posterior half of the vault. All fully adult specimens have a fused metopic suture. Nuchal cresting is notable. The ectotympanic has a platyrrhine-like annular shape, rather than a tubular one (Figs. 14.7–14.10). All known crania of *Catopithecus* are badly crushed.

Postcranially, *Catopithecus* is also very primitive (Gebo *et al.*, 1994). The proximal femur, for example, has a distinct, prosimian-like third trochanter (Fig. 14.11). Initially, a very prosimian-like distal humerus was allocated to *Catopithecus* (Gebo *et al.*, 1994) but now, with the discovery of fossil prosimians at L-41, and the analysis of new humeri attributable to *Catopithecus*, it appears that *Catopithecus* has a less primitive humerus than first reported. Seiffert *et al.* (2000) described specific resemblances between the humerus of *Catopithecus* and propliopithecines. Behaviorally, *Catopithecus* is interpreted to be a generalized arboreal quadruped.

In addition to the Fayum species, *C. browni*, there is a distinct, unnamed species of *Catopithecus* in the slightly younger deposits (basal Oligocene) of Taqah, Oman. It is currently known on the basis of unassociated isolated teeth (Thomas *et al.*, 1999).

GENUS *Oligopithecus* Simons, 1962
Resembles *Catopithecus* in having primitive, cresty molars and slightly raised trigonids, but differs from *Catopithecus* in its relatively larger M_2; the absence of a paraconid on M_2; a larger metaconid on P_4 with a better developed talonid basin; a proportionately longer P_3 with more elaborate honing facet; and smaller hypocones on the upper molars (Figs. 14.12, 14.13).
INCLUDED SPECIES *O. rogeri*, *O. savagei*

SPECIES *Oligopithecus savagei* Simons, 1962
TYPE SPECIMEN Cairo Geological Museum No. 18000, left mandible with C–M_2

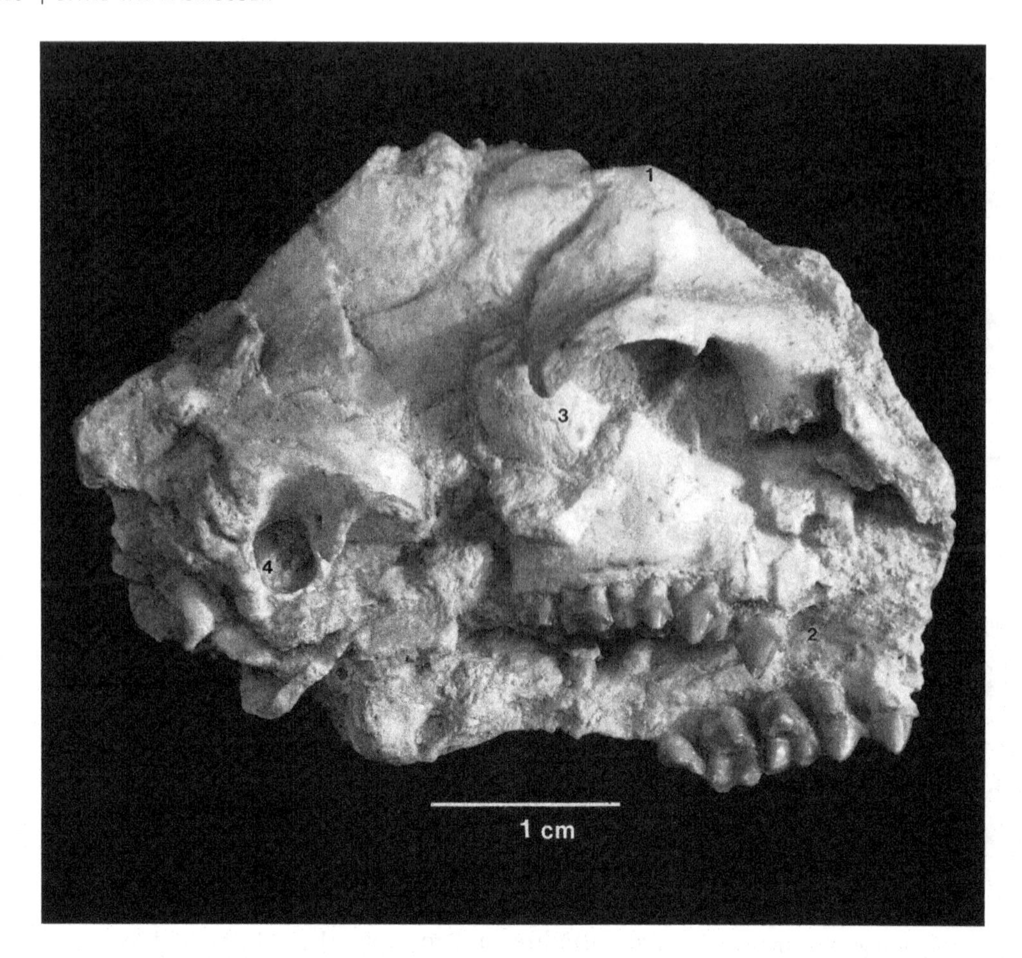

Fig. 14.8 *Catopithecus browni*. Crushed cranium of DPC 8701.

AGE AND GEOGRAPHIC RANGE Lower sequence of the Jebel Qatrani Formation, latest Eocene (Table 14.1), known only from quarry E, Fayum, Egypt

ANATOMICAL DEFINITION

Differs from *O. rogeri* in its smaller size, the smaller, less robust premolars, especially P_3; the less molariform P_4 with a proportionately shorter talonid; the shorter, broader lower molars. Like *Catopithecus*, this species has a prosimian-like aspect to its dentition, with sharp crests and somewhat raised trigonids, although unlike *Catopithecus*, the paraconid has been lost on M_2 (Simons, 1962*b*; Rasmussen & Simons, 1988; Gheerbrant *et al.*, 1995). The species is known only from the holotype and a few isolated teeth, including an upper molar, premolar and incisor.

SPECIES *Oligopithecus rogeri* Gheerbrant *et al.*, 1995

TYPE SPECIMEN TQ 313 (provisionally curated at the Muséum National d'Histoire Naturelle de Paris), lower right molar

AGE AND GEOGRAPHIC RANGE Known only from Taqah, Dhofar, Oman, from the basal Oligocene Ashawq Formation. The precise relationship between Taqah and localities in the Fayum section remains unknown. Based on the described primates from Taqah (a species of *Oligopithecus* more derived than the one in the Fayum, and a species of propliopithecid more primitive than those found in the Fayum) a position stratigraphically intermediate between quarry E and quarry V seems to be a good hypothetical approximation (Table 14.1). The reported paleomagnetic correlations place the Taqah material slightly higher than this (Kappelman *et al.*, 1992; Roger *et al.*, 1993).

ANATOMICAL DEFINITION

This species is represented by 120 isolated teeth representing all the postcanine upper and lower tooth positions, and thus it is better known than the type of the genus, *O. savagei*. The Oman species is larger by 5%–15% in linear dimensions and has distinctly narrow and elongated talonids on P_4 and the lower molars (Figs. 14.14, 14.15). The P_3 is very large and robust, especially in males. The upper molar crowns are lower and the hypocone is reduced when compared to the few upper specimens known for *O. savagei* (Gheerbrant *et al.*, 1995) (Figs. 14.15, 14.16). There are a few fragmentary limb bones from Taqah that may belong to this species, based on size (Senut & Thomas, 1994). Gheerbrant *et al.* (1995) pointed out that *O. savagei* and *O. rogeri* probably represent separate lineages rather than different time samples along a single evolving lineage.

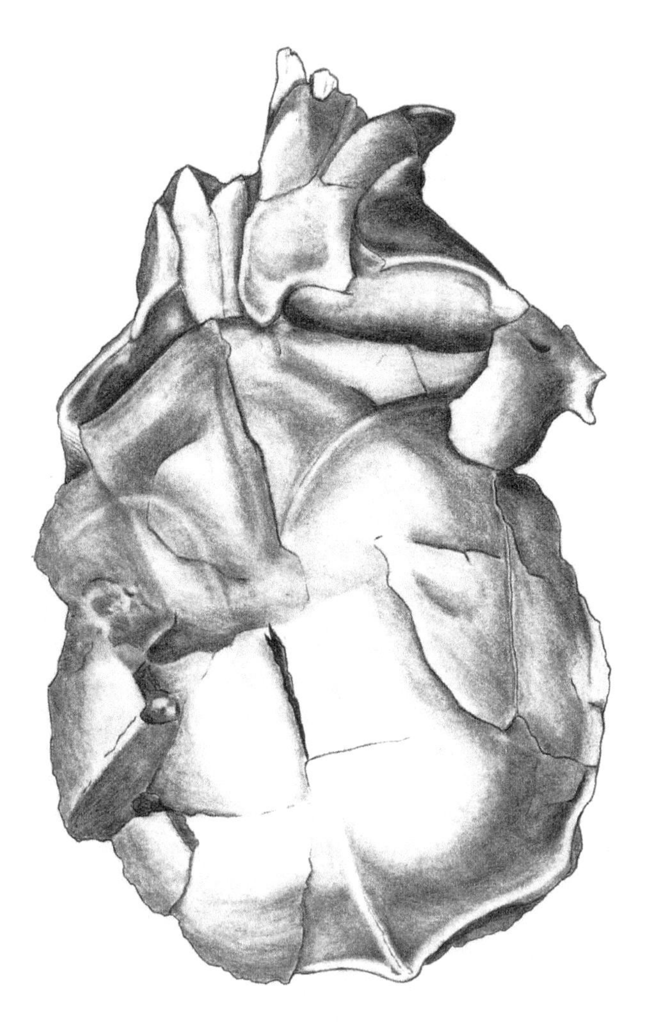

Fig. 14.9 *Catopithecus browni*. Crushed cranium of DPC 11388 viewed from above. From Simons and Rasmussen (1996); illustration by J. A. Rehg.

Family Propliopithecidae

GENUS *Moeripithecus* Schlosser, 1910
INCLUDED SPECIES *M. markgrafi*

SPECIES *Moeripithecus markgrafi* Schlosser, 1910
TYPE SPECIMEN SNM 12639G, Stuttgart, right mandible with M_{1-2}
AGE AND GEOGRAPHIC RANGE Type specimen from an unknown locality within the Jebel Qatrani Formation; referred material from Oman is from the locality of Taqah in the basal Oligocene Ashawq Formation (30–35 Ma)
ANATOMICAL DEFINITION
Diagnosis of taxon is difficult because the holotype (the only known specimen from the Fayum) has but two molars, and the referred material from Oman is remote geographically and differs somewhat from the type. Based strictly on the type, the key anatomical features of *M. markgrafi* are a twinned entoconid–hypoconulid pair and crests sharply defined (as in oligopithecids) in combination with reduced trigonid height and loss of molar paraconids (specialized traits found in other propliopithecids). The distinctive sharpness of the crests may be due partly to the fact that their teeth are unworn; however, young specimens of other propliopithecids are more bundont even without wear. The buccal margin of the crown is greatly expanded and the cusps are placed fairly close together (described by Kay *et al.* (1981), as having a narrow occlusal surface) compared to species of *Propliopithecus*. The referred material from Oman resembles the type of *M. markgrafi* in buccal inflation and cusp sharpness, but resembles species of *Propliopithecus* in having greater entoconid–hypoconid separation, the sulcus better developed than in the type (Thomas *et al.*, 1991). The referred lower P_3 is very broad; the referred upper molars have a small hypocone and a small but distinct pericone. The Oman teeth are larger than those in the type specimen of *M. markgrafi*. It is important to note that the Oman material referred to *Moeripithecus* probably represents a species distinct from the Fayum type. Simons (1995*b*) warned that the diagnosis of *M. markgrafi* (which he considered a species of *Propliopithecus*) should not be "extended" by the Oman material because this "changes the original concept of the genus that started out being inadequate in the first place." While several authors have been frustrated by the incompleteness of the type specimen, they have all recognized it as being morphologically distinct from other propliopithecids. I recognize the genus here because despite the poor type, it is easily recognizable morphologically, and because the researchers working on the Oman material have raised important issues concerning the taxon. The dental structure of the Fayum type specimen and the material from Oman suggests a primate that is the more primitive ancestor or sister taxon to a combined clade of *Propliopithecus* and *Aegyptopithecus*.

GENUS *Propliopithecus* Schlosser, 1910
Medium-sized to large propliopithecid that differs from *Moeripithecus* in having more bunodont molar cusps, greater separation between the entoconid and hypoconulid with variable development of a distal fovea between them, and the consequent central positioning of the hypoconulid. These primates are sexually dimorphic in body size and canine size (like other early catarrhines). The species of the genus have variably expanded antemolar teeth ranging from broad and robust premolars and canines in *P. ankeli* to gracile and narrow ones in *P. haeckeli*. A distinctive feature of the genus is the relatively steep-sided lower molars (or in other words, the molars have cusps placed marginally) which lack the extreme buccal inflation of *Moeripithecus* and the overall crown inflation of *Aegyptopithecus* (especially on M_2). Upper molars are quadrate, with hypocones nearly enlarged to the size of the principal three cusps (unlike the smaller hypocones in the oligopithecids and in the Oman material referred to *Moeripithecus*).

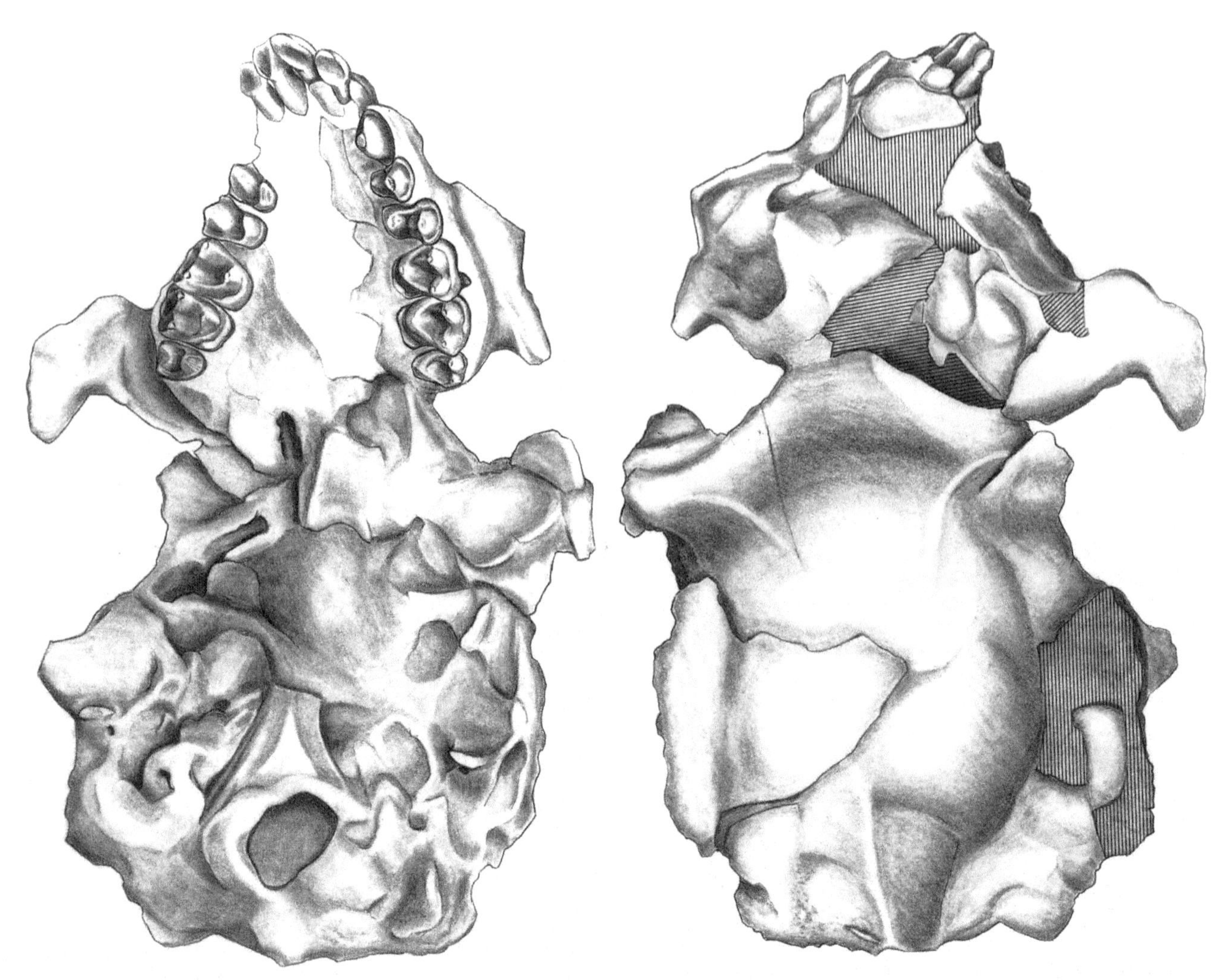

Fig. 14.10 *Catopithecus browni*. A badly crushed and distorted cranium of DPC 11594 with a fairly well-preserved upper dentition. From Simons & Rasmussen (1996); illustration by J. A. Rehg.

INCLUDED SPECIES *P. ankeli, P. chirobates, P. haeckeli*

SPECIES *Propliopithecus ankeli* Simons *et al.*, 1987
TYPE SPECIMEN CGM No. 42847, maxilla with P^3–M^3
AGE AND GEOGRAPHIC RANGE Lower part of the upper sequence of the Jebel Qatrani Formation, probably early Oligocene (Table 14.1), known only from quarry V, Fayum, Egypt
ANATOMICAL DEFINITION
This is a large species of the genus with hypertrophied antemolar teeth (Figs. 14.17, 14.18). The canines are large and robust; the P_3 is large, very broad buccolingually, and oval in shape. The molars resemble those of other species of *Propliopithecus* in size and morphology, with bunodont cusps and centrally positioned hypoconulids. The upper molars have a large hypocone and no pericone (unlike the Oman material referred to *Moeripithecus* but equated with this species by Thomas *et al.*, 1991).

SPECIES *Propliopithecus haeckeli* Schlosser, 1910
TYPE SPECIMEN SNM 12638, Stuttgart, mandible with canine to M_3
AGE AND GEOGRAPHIC RANGE Lower part of the upper sequence of the Jebel Qatrani Formation, probably early Oligocene (Table 14.1); type from an unknown locality within the Jebel Qatrani Formation, tentatively referred material from quarry G, Fayum, Egypt
ANATOMICAL DEFINITION
A small species of *Propliopithecus* with relatively small, rounded premolars and a slender canine, pronounced buccal cingulum on the premolars and molars, and a relatively deep mandible (Fig. 14.19). The slender canine suggests the type specimen may be of a female while the deep mandible suggests a male. The combination in a single specimen is unique. A few isolated and fragmentary teeth from quarry G have been attributed to this species, mainly on the basis of their relatively small size (Kay *et al.*, 1981).

SPECIES *Propliopithecus chirobates* Simons, 1965
TYPE SPECIMEN CGM No. 26923, mandible with right and left canine to M_3 and the sockets for the incisors

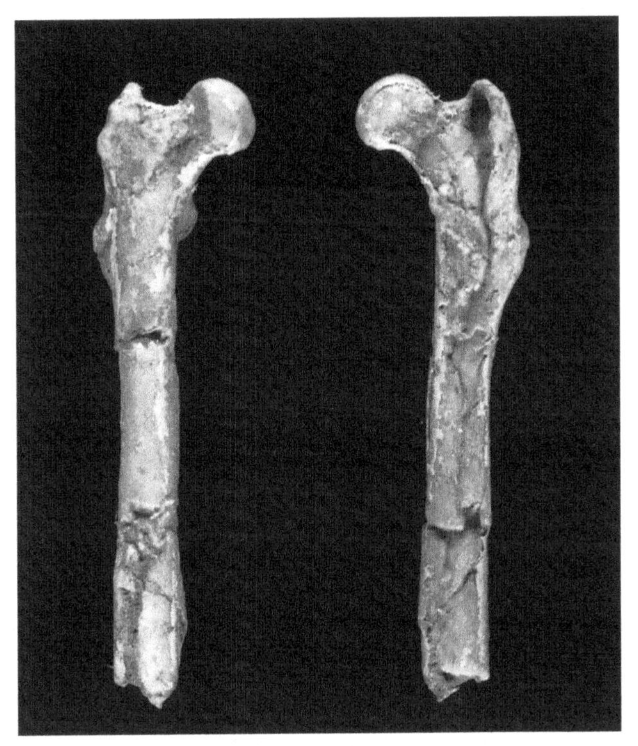

Fig. 14.11 *Catopithecus browni*. Proximal femur (DPC 8256) in anterior (left) and posterior (right) views. From Gebo *et al.* (1994).

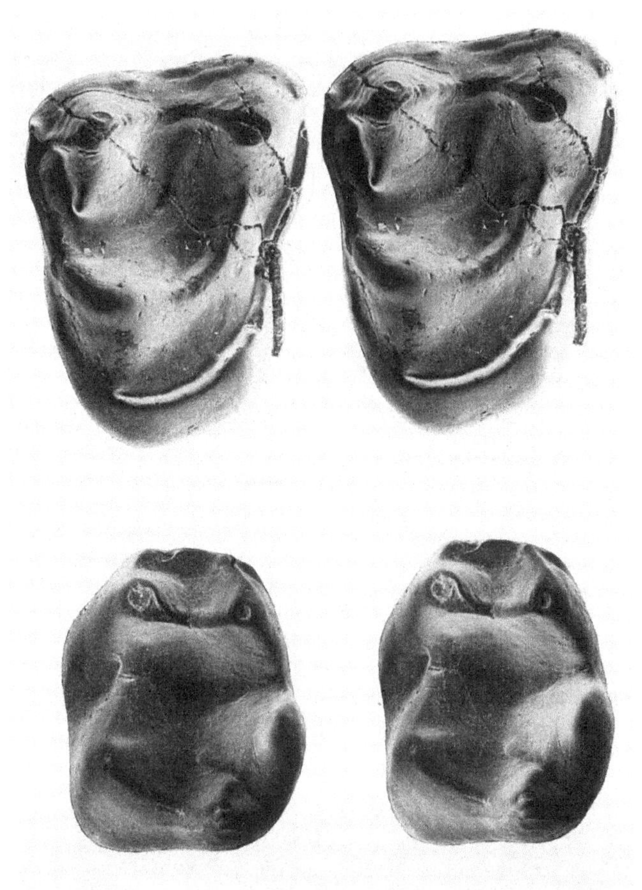

Fig. 14.13 *Oligopithecus savagei*. Scanning electron stereopairs of an upper M^1 (DPC 6020, top) and a lower M_2 (DPC 6330, bottom). From Rasmussen and Simons (1988).

Fig. 14.12 *Oligopithecus savagei*. Mandible (CGM 18000) in occlusal and lateral views.

AGE AND GEOGRAPHIC RANGE Upper sequence of the Jebel Qatrani Formation, early Oligocene (Table 14.1), several quarries at Fayum, Egypt, notably quarries I and M

ANATOMICAL DEFINITION

Propliopithecus chirobates is the best-known species of the genus. It is a medium-sized propliopithecid (slightly larger than *P. haeckeli*, smaller than *P. ankeli* and *Aegyptopithecus zeuxis*; Figs. 14.18C, 14.20) which further differs from *P. haeckeli* in having a less developed cingulum on the lower premolars and molars, a more oval P_3, and proportionally shallower jaw. It is characterized by steep-sided bunodont lower molars and upper molars with large, inflated hypocones. The canines are large, projecting and sexually dimorphic. The incisors are spatulate. Overall, the dentition of this species is very ape-like (Kay *et al.*, 1981). A cranium is not known for the species. Postcranially, *P. chirobates* is smaller and slightly more gracile than is *Aegyptopithecus zeuxis* (Gebo, 1993).

The canine of an early catarrhine resembling *Propliopithecus* has been reported from Angola, providing the only direct fossil evidence to date that early catarrhines must have had a broad sub-Saharan distribution (Pickford, 1986*b*) (Fig. 14.21).

GENUS *Aegyptopithecus* Simons, 1965
INCLUDED SPECIES *A. zeuxis*

SPECIES *Aegyptopithecus zeuxis* Simons, 1965
TYPE SPECIMEN CGM No. 26901, left mandible with P_4–M_2

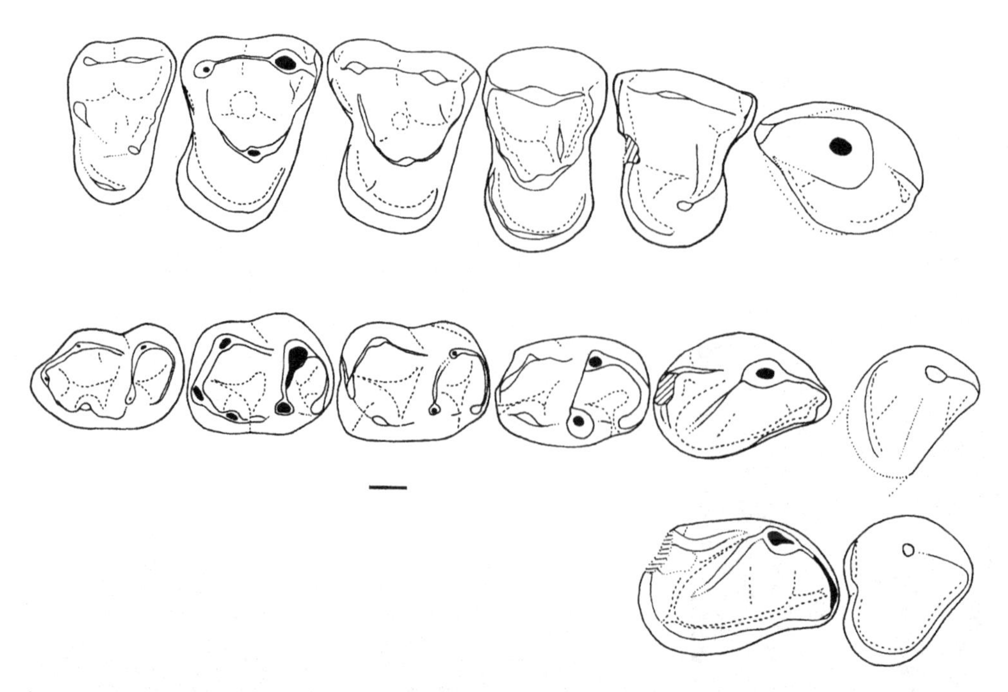

Fig. 14.14 *Oligopithecus rogeri*. Composite upper and lower dentition showing the smaller female form and larger male form of the lower canine and P_3. Scale bar = 1 mm. Courtesy of Emmanuel Gheerbrant.

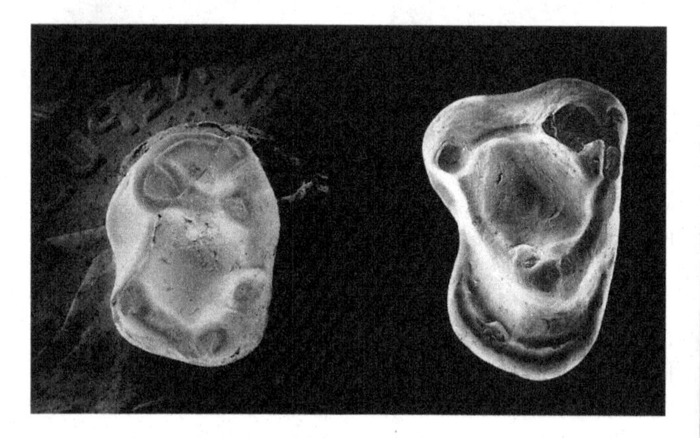

Fig. 14.15 *Oligopithecus rogeri*. Scanning electron micrographs of a lower M_2 (left, TQ 312) and an upper M^2 (right, TQ 221). From Gheerbrant et al. (1995).

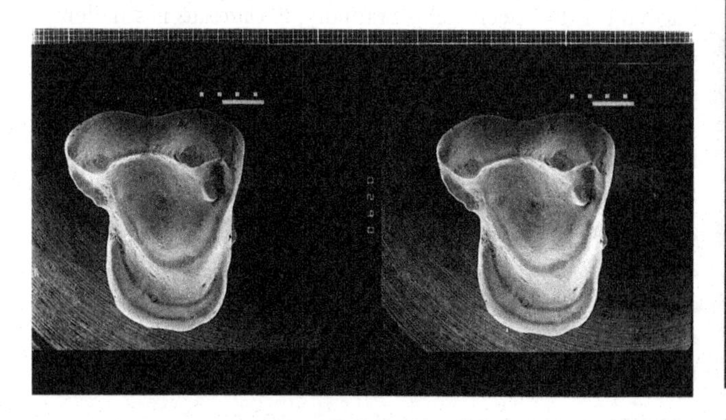

Fig. 14.16 *Oligopithecus rogeri*. Scanning electron stereopair of M^1 (TQ 290). Courtesy of Emmanuel Gheerbrant.

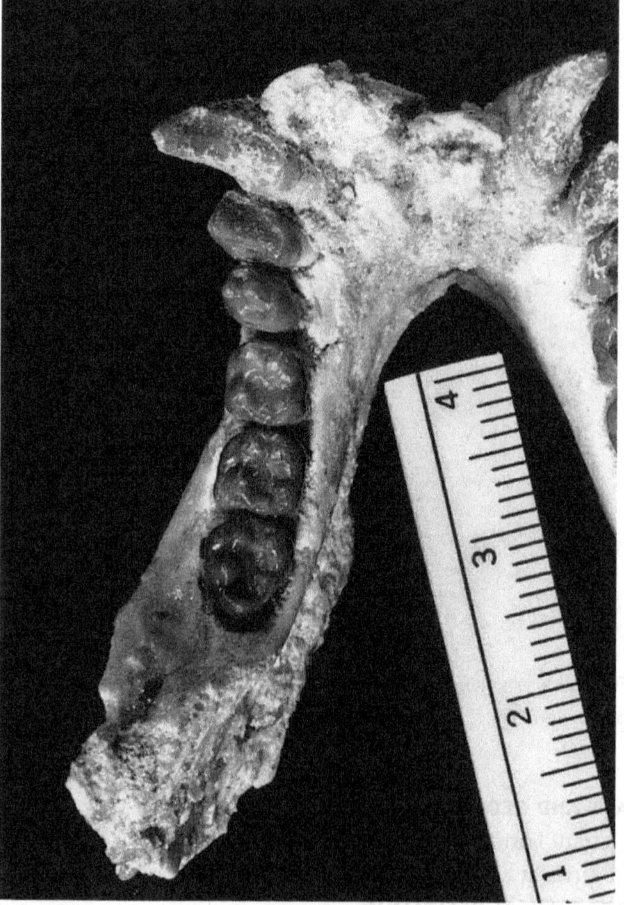

Fig. 14.17 *Propliopithecus ankeli*. The lower left dentition of DPC 5392 showing the approximately equal size and shape of M_1 and M_2, the broad premolars, and the robust canine. Scale in cm.

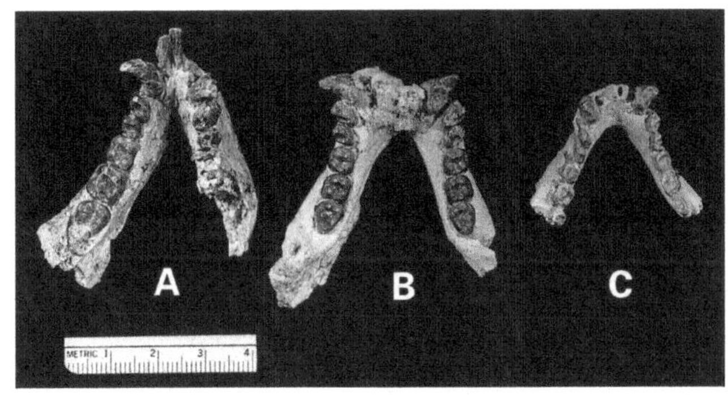

Fig. 14.18 Lower jaws of (A) male *Aegyptopithecus zeuxis* (DPC 1112), (B) *Propliopithecus ankeli* (DPC 5392) and (C) badly corroded type specimen of *Propliopithecus chirobates* (CGM 26923). From Simons *et al.* (1987).

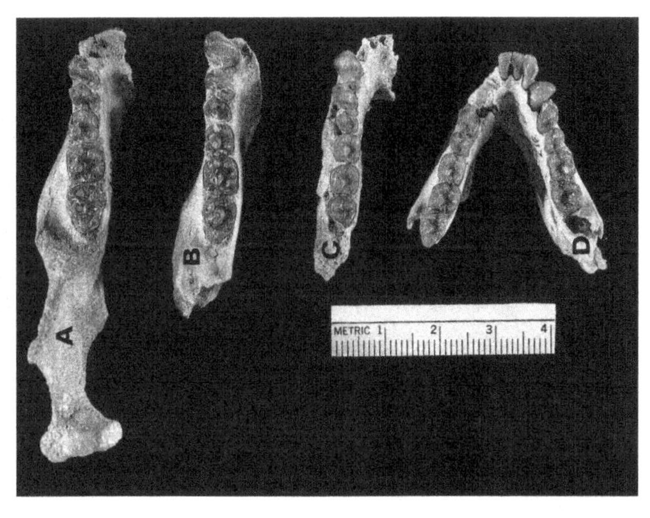

Fig. 14.20 Lower jaws of *Aegyptopithecus zeuxis* (A,B) and *Propliopithecus chirobates* (C,D). From Simons and Rasmussen (1991).

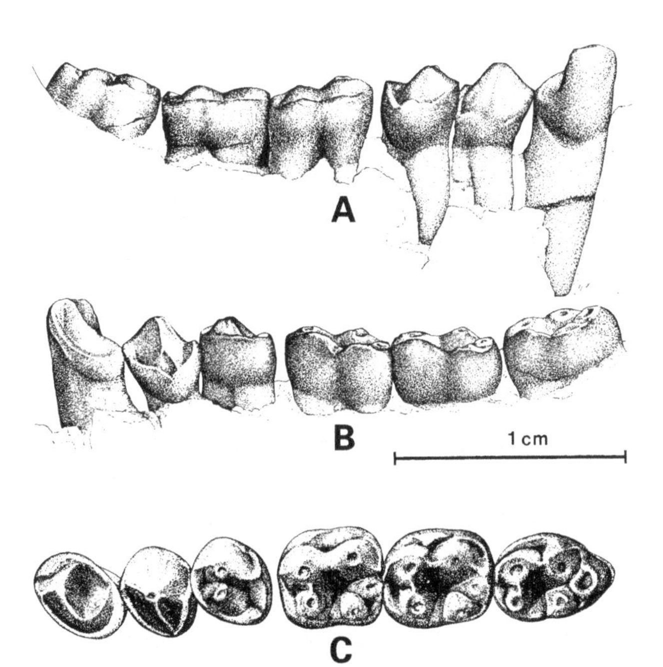

Fig. 14.19 *Propliopithecus haeckeli*. Lower right dentition (C–M_3) of SNM 12638 in (A) lateral, (B) lingual, and (C) occlusal views. From Simons (1995).

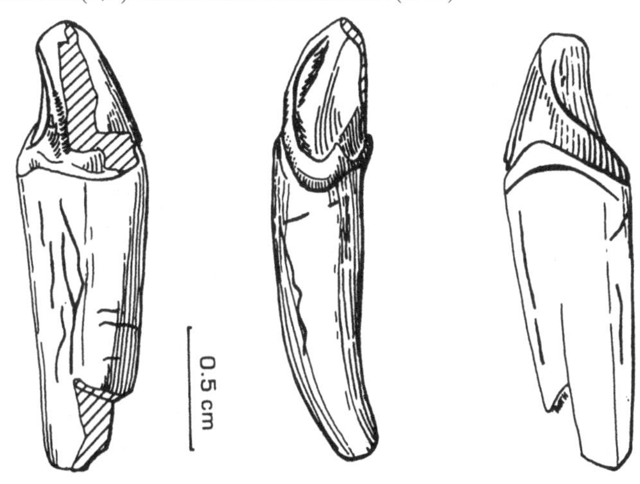

Fig. 14.21 Isolated canine from Malembe, Angola, resembling that of *Propliopithecus*, in three views. From Pickford (1986c).

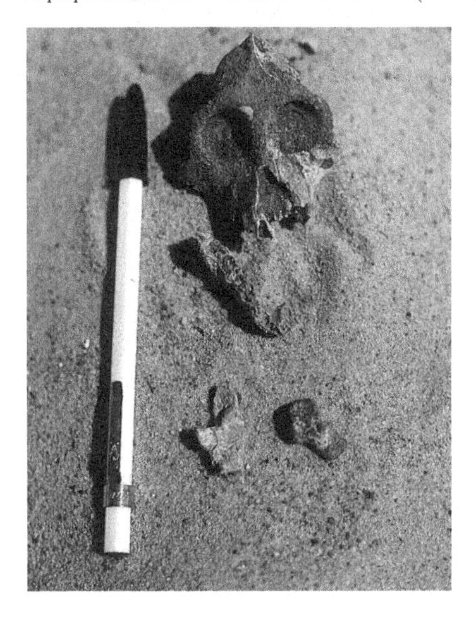

Fig. 14.22 *Aegyptopithecus zeuxis* lying in the sands of Fayum quarry M on the day of discovery, including a facial cranium, a juvenile mandible, a calcaneus and an astragalus, with a pen for scale.

AGE AND GEOGRAPHIC RANGE Upper sequence of the Jebel Qatrani Formation, early Oligocene (Table 14.1), several quarries at Fayum, Egypt, notably quarries I and M (Fig. 14.22)

ANATOMICAL DEFINITION

Large and robust propliopithecids (about the size of gibbons and howling monkeys) with greatly inflated molars (Simons, 1965; Kay *et al.*, 1981; Simons & Rasmussen, 1991). The second molar is substantially larger than the first, with an inflated crown (rather than steep-sided, as seen in *Propliopithecus*: Figs. 14.18, 14.20). The third molar is long with a large hypoconulid, yielding a sequence for mesiodistal length of $M_3 > M_2 > M_1$. The upper molars are characterized by four nearly equal-sized cusps. The hypocone is large, bulbous and arises from a

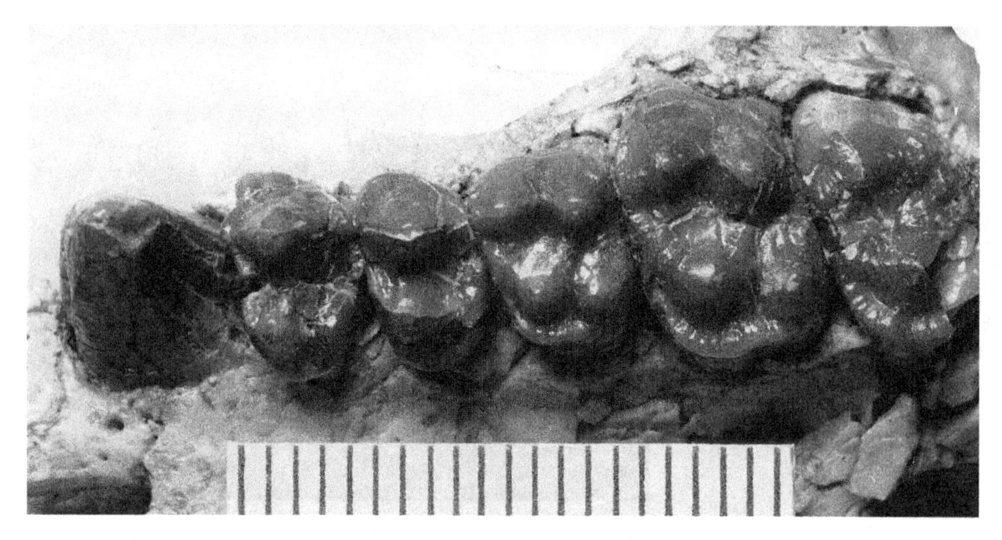

Fig. 14.23 *Aegyptopithecus zeuxis*. Maxillary dentition. Scale in mm.

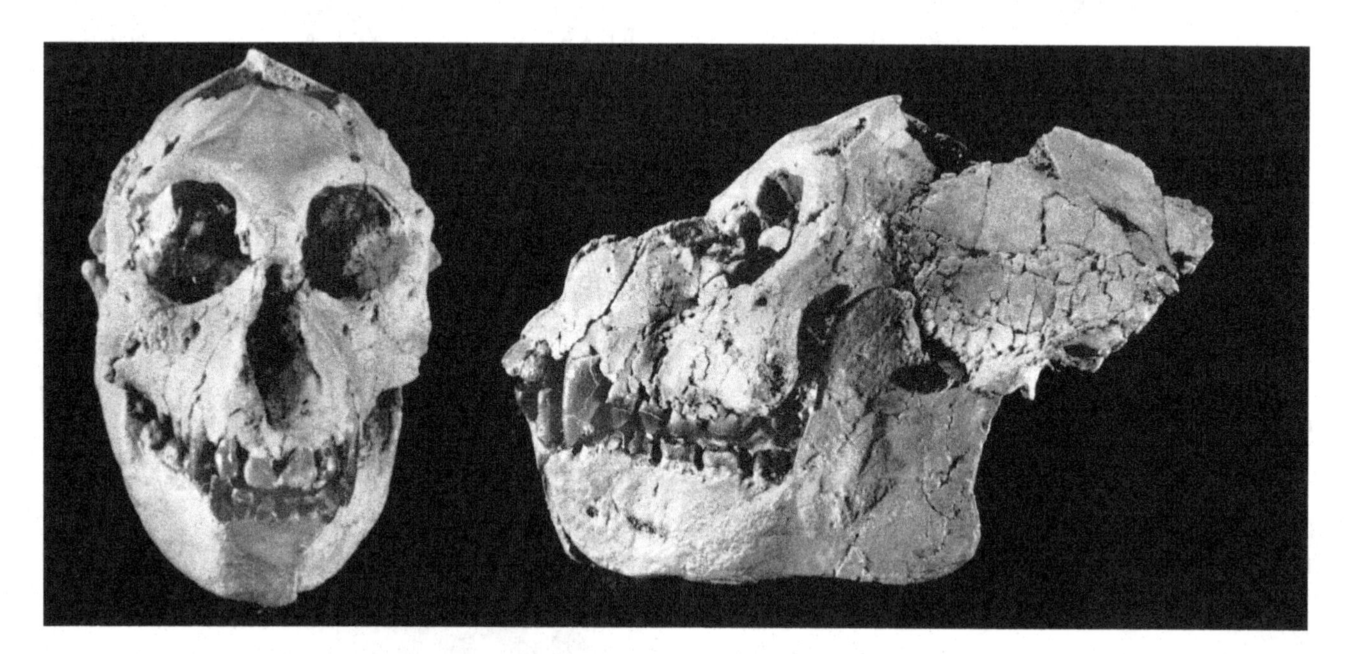

Fig. 14.24 *Aegyptopithecus zeuxis*. The 1966 skull in front and side view.

broad, beaded cingular shelf (Fig. 14.23). Upper premolars are bicuspid like those of other early catarrhines. Like other propliopithecoids, *A. zeuxis* is sexually dimorphic. The antemolar teeth are relatively small in diameter and gracile compared to those of *P. ankeli* (Simons *et al.*, 1987).

Several cranial specimens of *Aegyptopithecus* are known (Simons, 1967*b*, 1987, 1993). The facial skeleton is robust (Figs. 14.24–14.26). The ascending process of the premaxilla is large, both in height and length. The interorbital distance is broad with two intervening plates of bone separating the eyes, rather than a single bony interorbital septum (Fig. 14.27). The zygomatic is deep. Temporal lines converge above the orbits in males forming a prominent, robust sagittal crest (Figs. 14.25–14.27). The braincase is low and long, and the olfactory fossa is moderately developed (Figs. 14.24, 14.27). The basicranium is broad, and nuchal cresting is exaggerated. There is an annular, rather than tubular, ectotympanic. The posterior carotid foramen enters the auditory bulla fairly far back (as in *Alouatta*), there is a large promontory canal and the anteromedial cavity is at least moderately trabeculated (Simons & Rasmussen, 1989).

The brain of *Aegyptopithecus* has frontal lobes smaller than those of any extant anthropoid, but relatively larger olfactory bulbs than extant forms (Radinsky, 1973; Simons, 1993). The brain shape is low and elongated, and the surface complexity seems slight compared to later catarrhines.

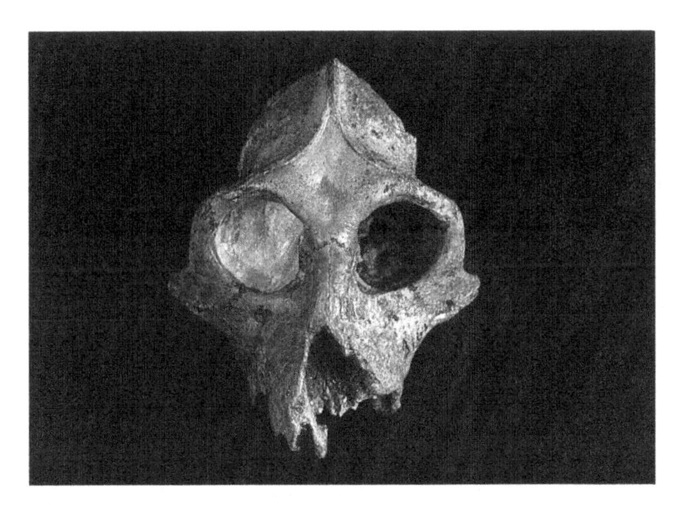

Fig. 14.25 *Aegyptopithecus zeuxis*. A facial cranium of a young male (DPC 3161) in front view (the right orbit is full of wax for the purpose of casting).

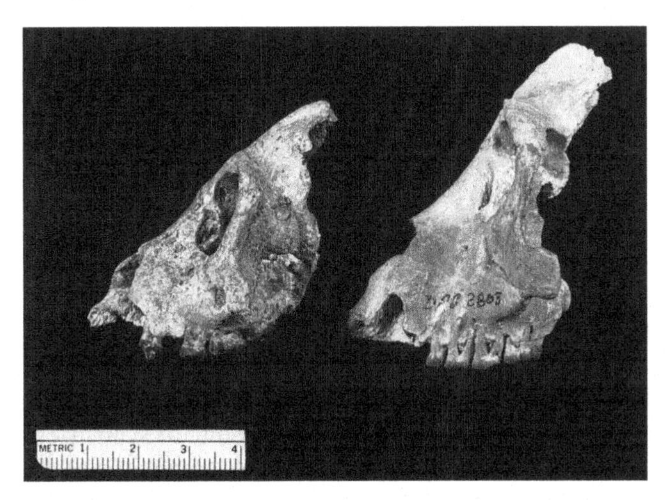

Fig. 14.26 *Aegyptopithecus zeuxis*. Two facial crania (DPC 3161, young male on left; DPC 2803, old male on right) in side view, showing the distinctive angled facial profile.

Functional aspects of the limb skeleton of *A. zeuxis* have been compared to ateline platyrrhines, especially the relatively robust genus, *Alouatta*. The humerus is relatively short and robust, with large muscle crests, particularly the deltopectoral crest (Fig. 14.28). The humeral head is not as spherical as in extant apes. The distal humerus has a prominent medial epicondyle, a patent entepicondylar foramen, a small capitular tail, and a shallow olecranon fossa. These characters resemble the condition in some platyrrhines and prosimians, rather than extant catarrhines (Gebo, 1993). The articular surfaces of the elbow are catarrhine-like and indicate nontranslatory movement (Rose, 1988; Gebo, 1993). The ulna is similarly robust with a large coronoid process. Other bones that have been described include the talus, calcaneus, and some metatarsals and phalanges. The first metatarsal has a prehallux facet, a primitive trait absent in extant catarrhines (Conroy, 1976*a*).

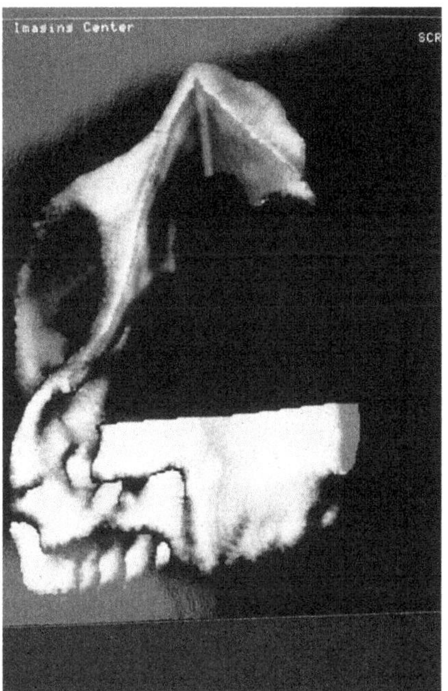

A

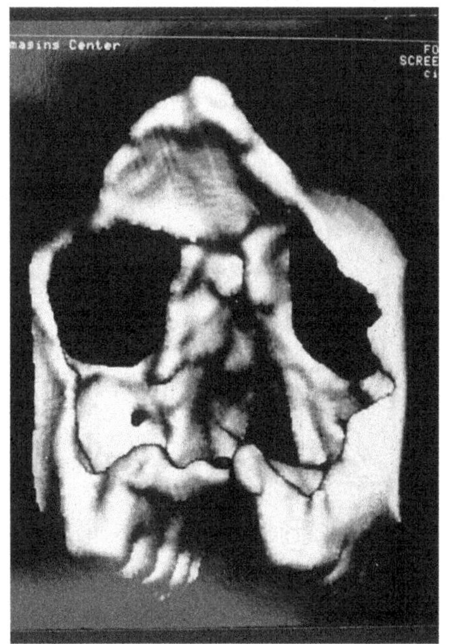

B

Fig. 14.27 CT scans of the face of *Aegyptopithecus*: (A) anterolateral view with orbital region removed to show thick frontal region; (B) posterolateral view showing interorbital structures, which are damaged posteriorly. Images courtesy of Suellen Gauld and Dan Valentino.

Evolution of early catarrhines

The known early catarrhines of Afro-Arabia present a remarkable evolutionary transition from primitive *Catopithecus* to relatively ape-like *Aegyptopithecus*. For several morphological systems (e.g., lower molar structure) one can model an evolutionary sequence of character states by using the fol-

Table 14.2. Lower molar changes among propliopithecoids, to be read as an approximate evolutionary sequence from most primitive (bottom) to most derived (top)

Taxon	Palaeomagnetic chronology[a]	Dental traits
Aegyptopithecus, Propliopithecus	C13 normal	honing edge of P_3 lengthened distal fovea becomes larger molar hypoconulid shifts centrally acute crests lost entirely
Moeripithecus	C13 reversed	hypocone size becomes large molar trigonid reduced in height wear facet X appears M_1 paraconid lost distal fovea attains moderate size increasing molar bunodonty
Oligopithecus	C13 reversed	P_4 trigonid closed M_2 paraconid lost slight distal fovea appears increasing molar bunodonty
Catopithecus	C15 reversed	catarrhine dental formula prosimian-like molar features

[a]Paleomagnetic data from Thomas *et al.* (1989) and Kappelman *et al.* (1992).

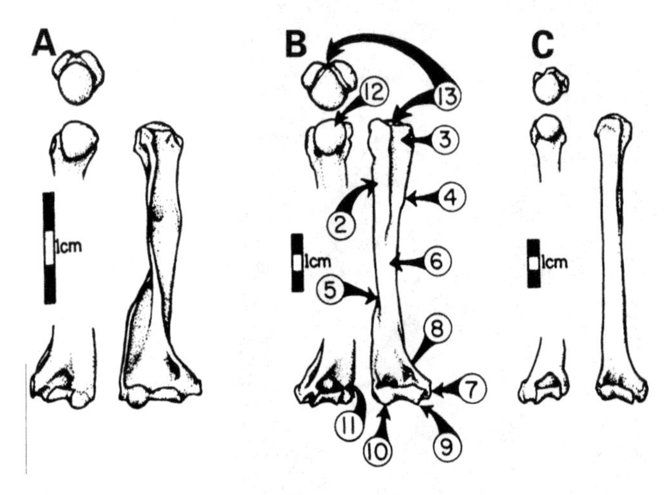

Fig. 14.28 Humeri of (A) *Notharctus*, (B) *Aegyptopithecus* (based on two specimens, DPC 1275 and CGM 40855) and (C) *Alouatta*, in a classic comparison made by Fleagle & Simons (1982) showing that the early catarrhine humerus is intermediate in many respects between that of an Eocene prosimian and a modern anthropoid. The numbers indicate key features analyzed in detail by Fleagle & Simons (1982): 2, deltoid plane; 3, bicipital groove; 4, tuberosity for teres major; 5, supinator crest; 6, shaft below the deltopectoral ridge; 7, entocondyle; 8, entepicondylar foramen; 9, trochlea; 10, capitulum; 11, olecranon fossa; 12, head; 13, greater and lesser tuberosities.

lowing taxa in order: *Catopithecus*, *Oligopithecus*, *Moeripithecus*, *Propliopithecus*, *Aegyptopithecus*. While the differences between the two end-taxa appear to be vast, the morphological distances between any two adjacent taxa in the continuum are fairly slight (Table 14.2). The morphological gradient corresponds somewhat (but not perfectly) to a chronological sequence. While we cannot be sure of the precise cladistic fit of certain known species, the sample of taxa that we have from just two regions – Oman and Egypt – provides us with a surprisingly clear model of the evolution of early catarrhines. Early forms retained primitive features also shared with other basal anthropoids, such as unfused mandibular symphyses, raised trigonids, small hypocones and sharp molar crests. In the catarrhine lineage through time the symphysis fused, trigonid height reduced, the paraconid was lost, the hypocone enlarged, and molar bunodonty increased.

The greatest morphological leap in the hypothetical sequence is between the oligopithecids and the more primitive propliopithecids. Gheerbrant *et al.* (1995) noted that *Oligopithecus* may show secondary acquisition of dental characters related to insectivory, with sharp crests, reduced hypocones and narrow talonids that seem to deviate from the more primitive condition seen in older *Catopithecus* and other basal anthropoids. At the same time, Gheerbrant *et al.* (1995) pointed out that the enlarged, robust P_3, and the dimorphism evident in P_3 and the canine, closely resemble the Oman material referred to *Moeripithecus*. In addition to these characteristics, the cresty and narrow occlusal surfaces of *Moeripithecus* with their twinned entoconid–hypoconulid pair seem easily derivable from the condition seen in *Oligopithecus*. An evolutionary transition between derived oligopithecids and primitive propliopithecids requires little more than increased body size and the concomitant shift to greater frugivory.

While parismony algorithms assess simplicity among unweighted matrices of discrete morphological characters, they do not take into account issues of geographic distribution and temporal sequences. It seems wildly improbable to me that *Oligopithecus* and *Moeripithecus*, two taxa with the shared morphological attributes listed above (a unique combination of traits among all known primates), which live together in the same region at the same time, should be unrelated because an older relative of *Oligopithecus* (*Catopithecus*) has an unfused mandibular symphysis and a presumably younger relative of *Moeripithecus* (*Aegyptopithecus*) is considered to be a true catarrhine. Separating *Catopithecus* and *Oligopithecus* from catarrhines and placing them in a distinct anthropoid clade (Kay *et al.*, 1997*b*; Ross *et al.*, 1998) breaks the morphological and chronological continuum shown among these forms through the late Eocene and Oligocene of northern Afro-Arabia (Seiffert *et al.*, 2000).

In constrast to the controversial connection between oligopithecids and propliopithecids, there seems to be general consensus that *Aegyptopithecus zeuxis* is positioned near the common ancestry of later catarrhines (Conroy, 1990; Fleagle, 1999). The phylogenetic and morphological implications have been examined by numerous authors (e.g., Simons, 1965; Fleagle, 1983; Fleagle & Kay, 1983; Harrison, 1987; Benefit & McCrossin, 1991; Leakey *et al.*, 1991; Ankel-Simons *et al.*, 1998). There is an enormous gap in the fossil record following the Fayum's upper sequence quarries, stretching from about 32 to 22 million years. Current simple views of the evolutionary transition from *A. zeuxis* to early Miocene apes and monkeys will undoubtedly become more complex should we luck into fossils filling this gap. A major lesson from the Fayum (and from other areas rich in fossil primates) is that more data in the form of more fossils can often wreak havoc with simple expectations generated from study of a more limited sample of primates.

Acknowledgements

I thank Prithijit S. Chatrath of Duke University for his expert and tireless efforts over a quarter of a century in managing the field and laboratory operations of the Fayum project, Elwyn L. Simons for inviting me to participate in the research on the Fayum's birds and mammals, Emmanuel Gheerbrant for his review of an earlier version of this manuscript, and Walter Hartwig for his expert advice and editorial assistance. Illustrations of the Omani primates were graciously provided by Gheerbrant, while many of the illustrations of the Fayum primates were provided by Elwyn Simons and other researchers associated with the Fayum project.

Primary References

Aegyptopithecus

Simons, E. L. (1961*d*). An anthropoid mandible from the Oligocene Fayum beds of Egypt. *American Museum Novitates*, **2051**, 1–20.

Simons, E. L. (1965). New fossil apes from Egypt and the initial differentiation of Hominoidea. *Nature*, **205**, 135–139.

Simons, E. L. (1967*b*). The earliest apes. *Scientific American*, **217**, 28–35.

Fleagle, J. G., Simons, E. L., & Conroy, G. C. (1975). Ape limb bone from the Oligocene of Egypt. *Science*, **189**, 135–137.

Conroy, G. C. (1976*b*). Primate postcranial remains from the Oligocene of Egypt. *Contributions in Primatology*, **8**, 1–134.

Fleagle, J. G. & Simons, E. L. (1978*a*). Humeral morphology of the earliest apes. *Nature*, **276**, 705–707.

Kay, R. F., Fleagle, J. G. & Simons, E. L. (1981). A revision of the Oligocene apes from the Fayum Province, Egypt. *American Journal of Physical Anthropology*, **55**, 293–322.

Gebo, D. L. & Simons, E. L. (1987). Morphology and locomotor adaptations of the foot in early Oligocene anthropoids. *American Journal of Physical Anthropology*, **74**, 83–101.

Simons, E. L. (1987). New faces of *Aegyptopithecus* from the Oligocene of Egypt. *Journal of Human Evolution*, **16**, 273–290.

Hamrick, M. W., Meldrum, D. J., & Simons, E. L. (1995). Anthropoid phalanges from the Oligocene of Egypt. *Journal of Human Evolution*, **28**, 121–145.

Ankel-Simons, F., Fleagle, J. G., & Chatrath, P. S. (1998). Femoral anatomy of *Aegyptopithecus zeuxis*, an early Oligocene anthropoid. *American Journal of Physical Anthropology*, **106**, 413–424.

Catopithecus

Simons, E. L. (1989). Description of two genera and species of late Eocene Anthropoidea from Egypt. *Proceedings of the National Academy of Sciences of the United States of America*, **86**, 9956–9960.

Simons, E. L. (1990). Discovery of the oldest known anthropoidean skull from the Paleogene of Egypt. *Science*, **247**, 1567–1569.

Gebo, D. L., Simons, E. L., Rasmussen, D. T., & Dagosto, M. (1994). Eocene anthropoid postcrania from the Fayum, Egypt. In *Anthropoid Origins*, eds. J. G. Fleagle & R. F. Kay, pp. 203–233. New York: Plenum Press.

Simons, E. L. (1995*c*). Skulls and anterior teeth of *Catopithecus* (Primates: Anthropoidea) from the Eocene and anthropoid origins. *Science*, **268**, 1885–1888.

Simons, E. L. & Rasmussen, D. T. (1996). Skull of *Catopithecus browni*, an early Tertiary catarrhine. *American Journal of Physical Anthropology*, **100**, 261–292.

Seiffert, E. R., Simons, E. L., & Fleagle, J. G. (2000). Anthropoid humeri from the late Eocene of Egypt. *Proceedings of the National Academy of Sciences of the United States of America*, **97**, 10062–10067.

Moeripithecus

Schlosser, M. (1910). Über einige fossile Säugetiere aus dem Oligocän von Ägypten. *Zoologischer Anzeiger*, **35**, 500–508.

Schlosser, M. (1911). Beiträge zur Kenntnis der oligozänen Landsäugetiere aus dem Fayum (Ägypten). *Beiträge zur Paläontologie und Geologie Österreich–Ungarns und des Orients*, **24**, 51–167.

Thomas, H., Roger, J., Sen, S., & Al-Sulaimani, Z. (1991). The discovery of *Moeripithecus markgrafi* Schlosser (Propliopithecidae, Anthropoidea, Primates) in the Ashawq Formation (early Oligocene of Dhofar Province, Sultanate of Oman). *Journal of Human Evolution*, **20**, 33–49.

Oligopithecus

Simons, E. L. (1962*b*). Two new primate species from the African Oligocene. *Postilla*, **64**, 1–12.

Rasmussen, D. T. & Simons, E. L. (1988). New specimens of *Oligopithecus savagei*, early Oligocene primate from the Fayum, Egypt. *Folia Primatologica*, **51**, 182–208.

Thomas, H., Roger, J., Sen, S., & Al-Sulaimani, Z. (1988). Découverte des plus anciens "anthropoïdes" du continent arabo-africain et d'un primate tarsiiforme dans l'Oligocène du Sultanat d'Oman. *Comptes rendus de l'Académie des sciences de Paris*, **306**, 823–829.

Senut, B. & Thomas, H. (1994). First discoveries of anthropoid postcranial remains from Taqah (early Oligocene, Sultanate of Oman). In *Current Primatology*, eds. B. Thierry, J. R. Anderson, J. J. Roeder, & N. Herrenschmidt, pp. 255–260. Strasbourg: Université Louis Pasteur.

Gheerbrant, E., Thomas, H., Sen, S., & Al-Sulaimani, Z. (1995). Nouveau primate Oligopithecinae (Simiiformes) de l'Oligocène inférieur de Taqah, Sultanat d'Oman. *Comptes rendus de l'Académie des sciences de Paris*, **321**, 425–432.

Propliopithecus

Schlosser, M. (1910). Über einige fossile Säugetiere aus dem Oligocän von Ägypten. *Zoologischer Anzeiger*, **35**, 500–508.

Schlosser, M. (1911). Beiträge zur Kenntnis der Oligozänen Landsäugetiere aus dem Fayum (Ägypten). *Beiträge zur Paläontologie und Geologie Österreich–Ungarns und des Orients*, **24**, 51–167.

Simons, E. L. (1965). New fossil apes from Egypt and the initial differentiation of Hominoidea. *Nature*, **205**, 135–139.

Kay, R. F., Fleagle, J. G. & Simons, E. L. (1981). A revision of the Oligocene apes from the Fayum Province, Egypt. *American Journal of Physical Anthropology*, **55**, 293–322.

Fleagle, J. G. & Simons, E. L. (1982). Skeletal remains of *Propliopithecus chirobates* from the Egyptian Oligocene. *Folia Primatologica*, **39**, 161–177.

Simons, E. L., Rasmussen, D. T., & Gebo, D. L. (1987). A new species of *Propliopithecus* from the Fayum, Egypt. *American Journal of Physical Anthropology*, **73**, 139–148.

15 | The Pliopithecoidea

DAVID R. BEGUN

Introduction

A primitive catarrhine group with no known descendants, pliopithecoids were a diverse and fascinating group of primates ranging from southwestern France to China, from about 17 to 7 million years ago. Equally impressive is the range of morphological diversity within the Pliopithecoidea. Pliopithecoids were highly diverse and successful, varying in size from approximately 3 to 20 kg, and spanning dietary adaptations from generalized frugivory to highly specialized folivory. Smaller taxa were probably more monkey-like in their positional behavior, moving along the tops of branches, while the larger forms appear to have been more suspensory. The fossil record of the Pliopithecoidea displays widespread homoplasy, or parallel evolution, and adaptive radiation. Pliopithecoids resemble South American monkeys in diversity and adaptation and, indeed, are an example of the "Splendid Isolation" phenomenon typified by the faunas of Australia and South America (Simpson, 1980). In contrast, the evolutionary relations remain unclear, both within the Pliopithecoidea and between it and other anthropoids.

History of discovery and debate

Pliopithecoids were among the first fossil primates to be discovered and described. The famous femur from Eppelsheim and the famous lower jaw from Sansan set the stage for both the catarrhine fossil record and the sciences of paleoprimatology and paleoanthropology. However, discovery of fossil catarrhines more closely resembling great apes and humans shifted attention away from the Pliopithecoidea. After discovery and recognition of *Pliopithecus antiquus* only a few specimens from several localities, and a large collection of mostly isolated teeth from one (Göriach, Austria), were discovered in the next 100 years. Overall they attracted relatively little attention from the scientific community.

Paul Gervais (1849*a*) nominated *Pliopithecus* from discoveries first announced by Édouard Amand Isidore Hippolyte Lartet in 1837 and referred by Henri-Marie Ducrotay de Blainville (1839, 1840) to *Pithecus antiquus*. Several years later, a new and slightly larger species of *Pliopithecus*, *P. platyodon*, was described from collections in Switzerland by Biedermann (1863). A much larger collection of jaws and isolated teeth, first described by Hofmann (1863) from Göriach, was attributed to *Pliopithecus antiquus*, but eventually moved into *P. platyodon* (Hürzeler, 1954*a*; Harrison *et al.*, 1991). More specimens were recovered from Sansan and another locality in France, La Grive St. Alban (Depéret, 1887). Isolated teeth were found in various localities in the Loire valley of central France (Gervais, 1867; Lecointre, 1912), in Germany near Augsburg and in Bavaria (Roger, 1898; Schlosser, 1900) and in Poland (Wegner, 1913). These highly fragmentary partial dentitions or isolated teeth expanded the known geographic range of *Pliopithecus*, but offered little insight into the nature of *Pliopithecus*. Researchers assumed with little doubt, based on the simple and primitive morphology of the teeth and their small size, that this taxon was directly ancestral to hylobatids.

During the middle of the twentieth century the pace of discovery of, and scientific interest in, pliopithecoid fossils, continued to lag behind that of the hominoids. New isolated teeth from Switzerland (Stehlin, 1914) and an older specimen from Děvínská Nová Ves, known since the end of the previous century but only described in Glässner (1931), preceded the exceptional review monograph by Johannes Hürzeler (1954*a*). He referred to recent discoveries to be described elsewhere, including the remains of several skeletons of a pliopithecoid from the Děvínská Nová Ves fissures (Zapfe, 1952; Zapfe & Hürzeler, 1957). Eventual publication of these specimens would jump-start research and interest in this group.

The spectacular discoveries made by Helmuth Zapfe in the fissures of Děvínská Nová Ves culminated in what many consider the best monograph ever written on a fossil primate – Zapfe's monumental treatise on *Pliopithecus* (*Epipliopithecus*) *vindobonensis* (Zapfe, 1960). These specimens include the first postcranial fossils described for a pliopithecoid, although Lartet (1837*b*) noted some in his announcement. Major portions of three individuals are preserved at Děvínská Nová Ves, including a well-preserved skull, associated limb bones, vertebrae, scapulae and an ilium, which indicated extreme primitiveness. While he stressed resemblances to gibbons, Zapfe (1960) noted numerous similarities to platyrrhines, and even suggested that had they not been found in association, the humerus and the ear region of the temporal bone would scarcely have been recognized as anthropoid (Zapfe, 1958).

Along with more isolated specimens found at this time, a new type of pliopithecoid was discovered in 1959 and named *Plesiopliopithecus* (Zapfe, 1961; Bergounioux & Crouzel, 1965), which several authors have recognized as belonging to a distinct subfamily, the Crouzeliinae (Ginsburg & Mein, 1980; Andrews *et al.*, 1996). In fact, most of the more recently discovered pliopithecoids have been placed in the Crouzeliinae by these authors, though the justification for this is by no means clear (see below). This includes large samples of jaws, teeth and postcrania from Rudabánya, Hungary and Lufeng, Yunnan Province, China (Kretzoi, 1975; Pan, 1988). These latter specimens are also among the latest surviving members of the Pliopithecoidea, and dentally the most peculiar. Interestingly, these most recently discovered specimens may clear up one of the oldest controversies of paleoanthropology, the affinities of the famous femur from Eppelsheim (see below). Also discovered and described in this most recent phase of research are the oldest and most primitive pliopithecoids, *Dionysopithecus* and *Platodontopithecus*, from the early Miocene of China (Li, 1978; Gu & Lin, 1983), as well as a more advanced form, *Pliopithecus zhanxiangi* (Harrison *et al.*, 1991).

Beginning with Cuvier and his apparent dismissal of the Eppelsheim femur, through the great breakthrough by Lartet, and up to the present, pliopithecoids have been in and out of the limelight. Hürzeler and Zapfe did much to revive interest in this group, as did the discovery of "gibbon-like" fossils from east Africa (Le Gros Clark & Thomas, 1951; Ferembach, 1958; Fleagle, 1975; Andrews & Simons, 1977). However, in the final analysis, pliopithecoids tell us more about the dynamics of macroevolution than about the evolutionary history of gibbons, which remains shrouded in mystery.

Taxonomy

Systematic framework

The Pliopithecoidea

Because they are extinct, distant relatives of living catarrhines, a brief description of their defining features, assuming there are some, is appropriate before embarking on their systematics. Beyond the fact that they have only two premolars per quadrant in their adult dentitions, few characters define them as catarrhines and none offers convincing evidence that they are more closely related to modern catarrhines than are the earliest members of this group, the propliopithecoids of the Eocene and Oligocene of Egypt. In their recent review of the pliopithecoids, Harrison & Gu (1999) cite only three characters linking pliopithecoids to Old World monkeys and apes, and these are relatively unimpressive. They suggest that upper molars are narrower than those of *Propliopithecus* or *Aegyptopithecus*, lower molars are broader than *Propliopithecus* or *Aegyptopithecus*, and that the tubular ectotympanic is partially ossified. However, neither of the molar features is consistent in all pliopithecoids, and is also variable in propliopithecoids and platyrrhines. Worse still, the ectotympanic character is a presumed intermediate morphology and not an actual synapomorphy.

Most authors have assumed a unilinear direction in the evolutionary transformation (morphocline) of the catarrhine ectotympanic (e.g., Szalay, 1975*a*; Szalay & Delson, 1979; Andrews *et al.*, 1996). The scenario is as follows: the ectotympanic resembles a short bony tube fused to the outer surface of the auditory bulla in New World monkeys; it becomes an elongated tube, forming the canal of the outer ear (external auditory meatus). This occurs, it is assumed, through the intermediate step of a partially ossified tube, as seen in *Epipliopithecus vindobonensis* and superficially resembling some very young modern catarrhines. Direct evidence for this is lacking, however, and it is just as probable that the ectotympanic morphology of *Epipliopithecus vindobonensis* is either unique to that species or to the Pliopithecoidea, and independent of the evolution of an ectotympanic tube in Old World monkeys and apes, or primitive for Anthropoidea. In fact, among anthropoids the external auditory meatus in *Epipliopithecus* resembles those of *Tremacebus harringtoni*, a primitive platyrrhine from the late Oligocene of Argentina (Hershkovitz, 1974), and *Aegyptopithecus*, from the Oligocene of Egypt (Szalay & Delson, 1979; Fleagle & Kay, 1983). These forms also resemble much more primitive primates or primate relatives such as *Ignacius* and *Shoshonius* (Kay *et al.*, 1992; Beard & MacPhee, 1994). While the inferior or ventral portions of the ectotympanic tube may be slightly more ossified in *Epipliopithecus*, convergence in the evolution of ectotympanics in fossil and living prosimians, plesiadapiforms and tarsiers is well noted (e.g., Szalay, 1975*a*; MacPhee, 1977, 1981; MacPhee & Cartmill, 1986; Kay *et al.*, 1992; Beard & MacPhee, 1994) and there is no reason to think that anthropoids were immune to such phenomena.

In many ways pliopithecoid dental morphology is more similar to platyrrhines than to propliopithecoids. Pliopithecoids tend to have narrow lower incisors, occasionally waisted, or constricted at the junction of the crown and the root (the cervix), and this morphology is also found in a number of platyrrhines. In addition, one of the few defining traits that seems to be consistently present in pliopithecoids is a P_3 with a tall crown, roughly triangular in outline, with a comparatively short, vertically oriented mesiobuccal face. The mesiobuccal face of the anterior premolar is not expanded to accommodate the upper canine, which in catarrhines is honed or sharpened by this structure (the sectorial premolar) (Andrews, 1978*a*; Harrison & Gu, 1999).

In most pliopithecoids the protoconid and metaconid are not transversely aligned, but slightly offset such that the protoconid is mesial to the metaconid. In some cases the fovea mesial to these cusps is also expanded and bears a small mesial cusp that may be homologous to the paraconid. These features are never found in even the most primitive hominoid or Old World monkey, but they are common in platyrrhines and many adapids and living prosimians (Begun, 1989*b*). Some pliopithecoids appear to have had shorter faces and larger brains than propliopithecoids, and these features are shared with catarrhines (Fleagle & Kay, 1983), but also with many platyrrhines. Ford (1994) found no unambiguous derived postcranial character shared between

Epipliopithecus and catarrhines. Unlike any catarrhine, *Epipliopithecus vindobonensis* retains such primitive features as an entepicondylar foramen at the distal end of the humerus, a hinge-like carpometacarpal joint of the thumb, a rectangular as opposed to squared posterior articular facet of the calcaneus, a posterior position of the anterior articular facet of the calcaneus, and possibly a prehallux bone in the foot (Zapfe, 1960; Fleagle, 1983; Ford, 1994; Andrews *et al.*, 1996).

The point of this brief review of the Pliopithecoidea is to dispel the impression that they are little apes (Szalay & Delson, 1979; Fleagle, 1988) and focus attention on their primitiveness, as Zapfe (1958) noted long ago. All researchers who have analyzed pliopithecoids in some detail agree that they are not hominoids, and that they pre-date the division of Old World monkeys and apes. Pliopithecoids may be related to living catarrhines, may be an independent descendant lineage of the propliopithecoids, or may even be the sister clade to living anthropoids. More interesting, perhaps, is their diversity and success as the first modern-looking anthropoid to evolve and radiate in Eurasia. In many ways they mirror the evolutionary history of the New World monkeys, having found their way onto a land mass devoid of anthropoids.

The following taxonomy cautiously recognizes the Pliopithecoidea as primitive catarrhines, based only on their dental formula. The superfamily is subdivided into two families, which differs from the classification of Harrison & Gu (1999), who only recognize subfamilial distinctions. Within the two families few synapomorphies unite the included taxa, and there remains the distinct possibility that they, particularly the Pliopithecidae, will be found to be paraphyletic.

Order Primates Linnaeus, 1758
- Infraorder Catarrhini É. Geoffroy Saint-Hilaire, 1812
  - Superfamily Pliopithecoidea Zapfe, 1960
    - Family Pliopithecidae Zapfe, 1960
      - Subfamily Dionysopithecinae
        - Genus *Dionysopithecus* Li, 1978
          - *Dionysopithecus shuangouensis* Li, 1978
          - *Dionysopithecus orientalis* Suteethorn *et al.*, 1990
        - Genus *Platodontopithecus* Li, 1978
          - *Platodontopithecus jianghuaiensis* Li, 1978
      - Subfamily Pliopithecinae Zapfe, 1960
        - Genus *Pliopithecus* Gervais, 1849
          - *Pliopithecus piveteaui* Hürzeler, 1954
          - *Pliopithecus antiquus* Gervais, 1849
          - *Pliopithecus platyodon* Biedermann, 1863
          - *Pliopithecus zhanxiangi* Harrison *et al.*, 1991
          - *Pliopithecus* sp.
        - Genus *Epipliopithecus* Zapfe & Hürzeler, 1957
          - *Epipliopithecus vindobonensis* Zapfe & Hürzeler, 1957
        - Genus *Egarapithecus* Moyà-Solà *et al.*, 2001
          - *Egarapithecus narcisoi* Moyà-Solà *et al.*, 2001
    - Family Crouzeliidae
      - Subfamily Crouzeliinae
        - Genus *Plesiopliopithecus* Zapfe, 1961
          - *Plesiopliopithecus lockeri* Zapfe, 1961
          - *Plesiopliopithecus auscitanensis* Bergounioux & Crouzel, 1965
          - *Plesiopliopithecus rhodanica* Ginsburg & Mein, 1980
          - *Plesiopliopithecus priensis* Welcomme *et al.*, 1991
        - Genus *Anapithecus* Kretzoi, 1975
          - *Anapithecus hernyaki* Kretzoi, 1975
        - Genus *Laccopithecus* Wu & Pan, 1984
          - *Laccopithecus robustus* Wu & Pan, 1984
    - Family *incertae sedis*
      - Genus *Paidopithex* Pohlig, 1895
        - *Paidopithex rhenanus* Pohlig, 1895

Superfamily Pliopithecoidea

Family Pliopithecidae

Subfamily Dionysopithecinae

GENUS *Dionysopithecus* Li, 1978
A genus of small primitive catarrhine approximating the size of gibbons. Of the two species recognized here one is known only from a single lower molar. Thus, the genus is defined essentially by the morphology of the better-known species, *Dionysopithecus shuangouensis*. Four isolated teeth from Pakistan are also referred to *Dionysopithecus* but no species is defined, so these teeth are not described here (Bernor *et al.*, 1988).
INCLUDED SPECIES *D. orientalis, D. shuangouensis*

SPECIES *Dionysopithecus shuangouensis* Li, 1978
TYPE SPECIMEN IVPP V5597 (from Songlinzhuang, Sihong County, Jiangsu Province, People's Republic of China), a left maxillary fragment with M^{1-3}.
AGE AND GEOGRAPHIC RANGE The earliest species of the genus, known from the Xiacaowan Formation at the type locality of Songlinzhuang, dated by faunal associations to between 17 and 18 Ma; also recognized from the Zhenji locality, thought to be of similar age (Harrison & Gu, 1999; Qiu *et al.*, 1999).
ANATOMICAL DEFINITION
Dionysopithecus shuangouensis is known only from isolated teeth. It has a broad I^1 with a pronounced lingual cingulum and female upper canines that are triangular in horizontal cross-section. Upper premolars are narrow while upper molars are comparatively broad with well-developed lingual cingula and moderately developed buccal cingula. M^3 has strongly reduced distal cusps. Incisors are tall-crowned, narrow and waisted. P_3 is vertical and lacks the crown flare of a structurally sectorial P_3 (see above). P_4 has a lingual cusp (metaconid) that is lower in cusp height

than the buccal cusp (protoconid), a feature not found in other currently recognized pliopithecoids, but possibly present in one specimen from Kenya (see below). The lower molars are long and narrow, often preserving the paraconid, a mesial cusp otherwise only found in prosimians and the most primitive anthropoids. The other mesial cusps are not aligned transversely, as in most other catarrhines, but, common to pliopithecoids, the buccal cusp (protoconid) is more anterior or mesial than the lingual cusp (metaconid). The cristid obliquid is obliquely oriented, again as in most pliopithecoids and more primitive anthropoids, and unlike most other catarrhines. Finally, a pliopithecine triangle, one of the few defining characters of the Pliopithecoidea, is present. This feature consists of a subtle set of ridges on the buccal side of the crown defining a small triangular shaped pit between the protocone and hypocone (Hürzeler, 1954*a*). *Dionysopithecus shuangouensis* (and *Platodontopithecus jianghuaiensis*) are distinguished from other pliopithecids in having relatively rounded molar cusps and moderately developed upper molar buccal cingula, narrow M^1 with a distinctively convex lingual edge, and small M^3 with reduced lingual cusps (Ginsburg & Mein, 1980; Harrison & Gu, 1999).

SPECIES *Dionysopithecus orientalis* Suteethorn *et al.*, 1990
TYPE SPECIMEN TF 2451, an M_1
AGE AND GEOGRAPHIC RANGE Known only from a single tooth dated by faunal association to between 16 and 17 Ma (Ducrocq *et al.*, 1994; Qiu *et al.*, 1999), from Ban San Klang, northern Thailand
ANATOMICAL DEFINITION
Though originally referred to the east African genus *Dendropithecus*, *Dionysopithecus orientalis* is referred to *Dionysopithecus* in Harrison & Gu (1999) based on strong similarities to the type species, *D. shuangouensis*. These include size and basic morphological attributes of the occlusal surface typical of pliopithecoids. For example, *D. orientalis* has mesial cusps that are offset, such that the buccal cusp is more mesial than the lingual cusp, as in most pliopithecoids, but unlike *Dendropithecus* and hominoids. It should be noted however, that while the morphology of the Ban San Klang molar is clearly pliopithecoid, its distinctiveness from the type species of *Dionysopithecus* remains to be proven. Harrison & Gu (1999) are cautious in recognizing a separate species, and this is wise given the known range of variation in molar morphology among pliopithecoids.

GENUS *Platodontopithecus* Li, 1978
INCLUDED SPECIES *P. jianghuaiensis*

SPECIES *Platodontopithecus jianghuaiensis* Li, 1978
TYPE SPECIMEN PA 870, currently in the collections of the IVPP, Beijing, PRC. a left M^3
AGE AND GEOGRAPHIC RANGE Known from the Xiacaowan Formation at the type locality of Songlinzhuang, Sihong County, Jiangsu Province, PRC, dated by faunal associations to between 17 and 18 Ma; also recognized from the Zhenji locality, thought to be of similar age (Harrison & Gu, 1999; Qiu *et al.*, 1999)
ANATOMICAL DEFINITION
Platodontopithecus jianghuaiensis is known only from isolated teeth, which are considerably larger than those of *D. shuangouensis*, being somewhat larger than siamang teeth. Harrison & Gu (1999) estimate the body mass at about 15 kg. Presumed male upper canines are tall and bilaterally compressed. The upper premolars are broader than in *Dionysopithecus*, and the lower P_4 has subequal mesial cusps unlike *Dionysopithecus*. The molars are like *Dionysopithecus* but slightly narrower, with higher cusps and crests, including a better developed pliopithecine triangle.

Subfamily Pliopithecinae

GENUS *Pliopithecus* Gervais, 1849
A genus of small primitive catarrhine approximating the size range of hylobatids. Pliopithecines share with dionysopithecines a suite of dental characters that are almost all primitive for anthropoids. These include incisors that are tall-crowned, narrow and waisted, spatulate but labiolingually flat upper central incisors, narrow, pointed and asymmetrical upper lateral incisors, tall, broad P_3 crown lacking a truly sectorial morphology, long and narrow lower molars, often preserving a paraconid, mesial cusps that are aligned obliquely, an obliquely oriented cristid obliquid, and a pliopithecine triangle. In most species of *Pliopithecus* the P_4 and lower molars tend to be long and narrow with large anterior or mesial pits (fovea) and well-developed buccal cingula. Though most species have teeth close in size to *Hylobates*, the mandibles tend to be more massive. Upper premolars and molars tend to be broad and short. The premolars have heteromorphic cusps, the buccal ones always the more prominent. The upper molars usually have well-developed, shelf-like lingual cingula, and commonly buccal cingula or stylar shelves. Upper molars commonly lack a distal transverse ridge between the hypocone and metacone, but have a ridge connecting the hypocone to the protocone or the crista obliqua (this is true of crouzelines as well, in contrast to the opinion of Andrews *et al.* (1996)).
INCLUDED SPECIES *P. antiquus*, *P. piveteaui*, *P. platyodon*, *P. zhanxiangi*

SPECIES *Pliothecus antiquus* Gervais, 1849 (Fig. 15.1)
TYPE SPECIMEN From the E. Lartet collection from Sansan at the Muséum National d'Histoire Naturelle, Paris, a mandible lacking only the rami and portions of the right canine and left I_2 crowns

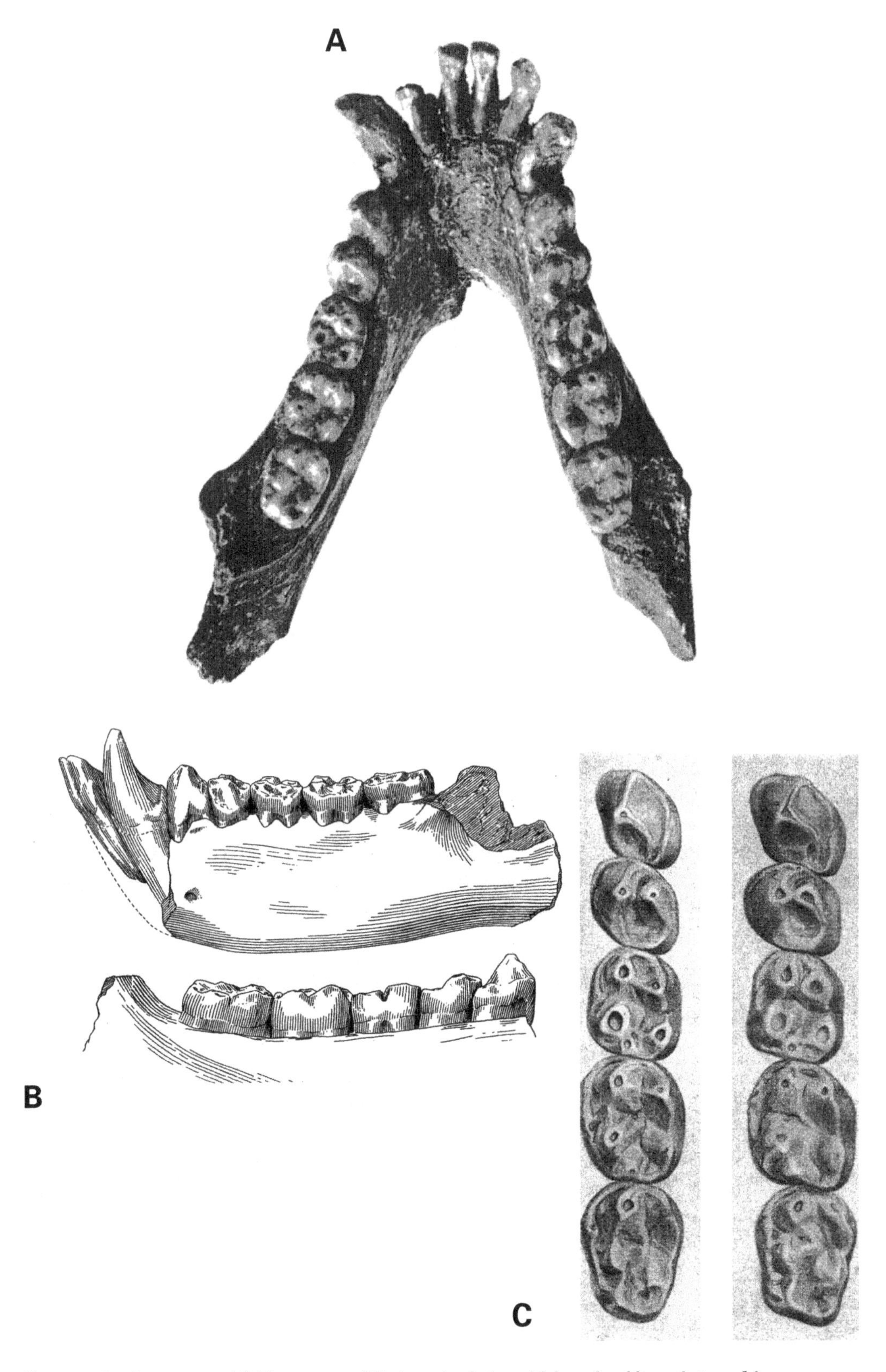

Fig. 15.1 *Pliopithecus antiquus*. (A) The type mandible in occlusal view; (B) buccal and lingual view of the type; (C) occlusal drawings of the type (both appear to be from the left side, but in fact the row on the left is a photographically reversed image of the right side dentition, for ease of comparison with the left side and with other dentitions). Adapted from Simons (1972) and Hürzeler (1954*a*).

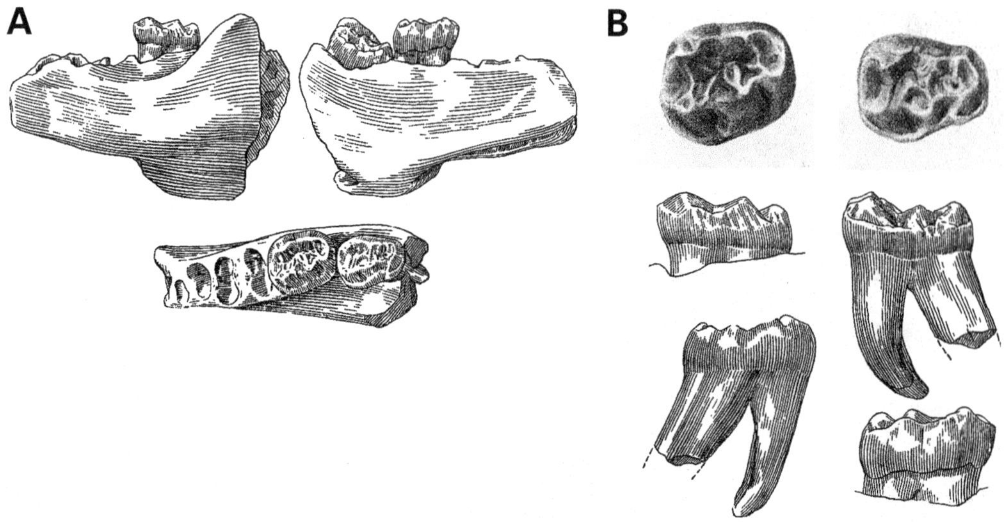

Fig. 15.2 *Pliopithecus piveteaui*. (A) Three views of the type and only specimen, a right mandible photographically reversed here; (B) studies of the M_2 (left) and M_3 (right). Adapted from Hürzeler (1954*a*).

AGE AND GEOGRAPHIC RANGE *Pliopithecus antiquus* is definitively identified only at Sansan and La Grive, both in France and dated to MN 6 (about 15 Ma). Hürzeler (1954*a*) was of the view that the species is only known from Sansan, but most subsequent authors also include the La Grive specimen, which comes from older sediments than the *Dryopithecus* teeth from the same site (Ginsburg, 1975, 1986). *Pliopithecus antiquus* may also be known from other MN 6 localities in Germany (Diessen am Ammersee, Stätzling, Ziemetshausen, Gallenboch) and Switzerland (Kreuzlingen and Rümikon), though these are isolated teeth and their species attribution is uncertain. Similarly, *P. antiquus* has been tentatively identified at later localities in Poland (MN 7 of Opole, Poland; MN 8 of Castel de Barbera, Spain and Przeworno II, Poland; MN 9 of Doué-la-fontaine and Meigné-le-vicomte, France), though again only on the basis of isolated teeth. In a number of cases however, authors note similarities to the sample from Göriach referred previously to *P. antiquus* but here, following Andrews *et al.* (1996), referred to *P. platyodon*. In addition, one of these samples, from Castel de Barbera, is recognized here as a distinct species (see below). Conservatively, we can conclude that *P. antiquus* is an MN 6 taxon from France that may have persisted into later periods (MN 6–9) in more central areas of Europe.

ANATOMICAL DEFINITION

Pliopithecus antiquus is dentally among the smallest species of *Pliopithecus*, though there is extensive overlap among the European species (Fig. 15.1). In addition to size, *P. antiquus* can be distinguished from some other species only by a number of subtle dental characters. To avoid repetition, these are listed in the anatomical definitions of the other species.

SPECIES *Pliopithecus piveteaui* Hürzeler, 1954 (Fig. 15.2)

TYPE SPECIMEN In the Lecointre collections at la Chapelle-Blanche, Manthelan (Indre et Loire), a right mandibular fragment with M_{2-3} and alveoli for the roots of the P_4 and M_1

AGE AND GEOGRAPHIC RANGE Dated to MN 5 (16–17 Ma), only from the Loire valley of France (Faluns de Touraine, Anjou, Pontevoy-Thenay, Manthelan) (Ginsburg & Mein, 1980; Ginsburg, 1986).

ANATOMICAL DEFINITION

The combination of subtle morphological differences, more primitive morphology, geography and greater age suggest that *P. piveteaui* is a distinct species, as originally recognized by Hürzeler (1954*a*) and more recently by Ginsburg (1975, 1986) and Ginsburg & Mein (1980). The type specimen is unusual for pliopithecines in having a very small M_3 in relation to M_2. The teeth are small but within the range of *P. antiquus*. They are considerably smaller than in other most species, but close to those of the small species from Castel de Barbera (see below). Both teeth narrow distally, a feature common in M_3 but unusual in M_2. The M_3 of the type has an even more tapered morphology than is typical for the genus. The M_3 has a very reduced entoconid (the distal lingual cusp) and the M_2 has a smooth and flared, or bulging buccal surface lacking the buccal cingulum typical of *P. antiquus*. Finally, the cusps on both molars are more bilaterally compressed, the crests that connect them more strongly defined, and the fovea and basins that separate them are larger, all compared to *P. antiquus*. These latter features are found in a number of other pliopithecoids, including *P. platyodon*, *P.* sp. from Spain and in most crouzeliines, and thus may be primitive for the superfamily. The right P_4 is broad and also has a more bulging buccal surface than in *P. antiquus*, with a larger talonid basin, a lower protoconid and a more strongly developed hypoconid (Ginsburg, 1975). Ginsburg (1975) also described additional lower molars that resemble the type. Finally, two upper teeth are known, a P^4 and M^3, which cannot be compared directly to *P. antiquus* from Sansan. Compared to an isolated P^4 from Poland that may belong to *P. antiquus* or *P. platyodon* (Kowalski & Zapfe, 1974) the *P. piveteaui* P^4 is smaller, and relatively broader or shorter, with a relatively larger protocone, a pronounced lingual cingulum, and a much shorter talon. Both upper teeth more closely resemble *P. platyodon* P^4 specimens from Göriach, though again they are much smaller (Fig. 15.4).

A number of researchers have recently suggested that *P. piveteaui* is indistinguishable from *P. antiquus* (Harrison *et al.*, 1991; Andrews *et al.*, 1996). Andrews *et al.* (1996) consider the relative size of the M_3 and its unusual morphology, which distinguishes *P. piveteaui* from *P. antiquus*, to be unreliable given known ranges of variability in this tooth. They do not comment on other aspects of the dental morphology of *P. piveteaui*. In light of the number of differences from *P. antiquus* and the consistency of those differences in the larger samples described by Ginsburg (1975), *P. piveteaui* is recognized here as a separate species. Its apparently primitive morphology may be an important hint to understanding some aspects of the evolutionary history of the Pliopithecoidea (Ginsburg & Mein, 1980, and see below).

SPECIES *Pliopithecus platyodon* Biedermann, 1863 (Fig. 15.3)
TYPE SPECIMEN In the collections of the Museum of Winterthur, Zurich, a damaged female maxilla with heavily worn dentition
AGE AND GEOGRAPHIC RANGE *Pliopithecus platyodon* from Elgg (near Zurich) is considered to be MN 5 in age, based on biostratigraphic correlations (Ginsburg, 1986); however, the bulk of the sample currently attributed to this taxon is from the MN 6 locality of Göriach, Austria
ANATOMICAL DEFINITION
Pliopithecus platyodon, based on the sample from Göriach, is dentally larger on average than *P. antiquus*. According to Andrews *et al.* (1996), *P. platyodon* has a relatively broader P_3 and slightly longer, more rectangular lower molars that increase in size from M_1 to M_3 more than in *P. antiquus*. However, these characters are very variable in the Göriach sample. It is safe to say that *P. antiquus* and *P. platyodon* are very similar and essentially differ only in size (Fig. 15.3).

Hürzeler (1954) and Zapfe (1960) both suggested that the Elgg and Göriach samples may represent the same species, but were reluctant to attribute the Göriach material to *P. platyodon* due to the poorly preserved occlusal morphology of the type. Harrison *et al.* (1991) formally combined the two samples. Andrews *et al.* (1996) cite a number of differences between the Göriach and Sansan samples as evidence of their taxonomic distinction (see below). Of course, it remains unclear if Elgg and Göriach are really the same taxon or if another species of *Pliopithecus*, *P. goeriachensis* (Sera, 1917) should be recognized. Here I follow Harrison *et al.* (1991). As noted above, a number of other specimens usually attributed to *P. antiquus* may in fact belong to *P. platyodon* (Hürzeler, 1954).

SPECIES *Pliopithecus zhanxiangi* Harrison *et al.*, 1991
TYPE SPECIMEN BPV-1021 (found at Maerzuizigou (BN 87021), Tongxin County, People's Republic of China, and in the collections of the IVPP, Beijing), a damaged female cranium
AGE AND GEOGRAPHIC RANGE Considered to be contemporary with Sansan (MN 6), based on faunal similarities (Harrison *et al.*, 1991; Qiu *et al.*, 1999), China
ANATOMICAL DEFINITION
Pliopithecus zhanxiangi is the largest species of the genus. Lower molars, which bear pliopithecine triangles, increase markedly in size from M_1 to M_3. They commonly show secondary wrinkling of the occlusal surface. The mandible is robust with a large extramolar sulcus. The upper canine is thick and relatively low-crowned, and the upper premolars and molars are very broad with strong buccal cingula (Harrison *et al.*, 1991). On the buccal sides of the $M_{2\text{-}3}$ there is a depression or notch that gives the tooth a waisted appearance.

Unlike other species of the genus, *P. zhanxiangi* is also known from a partial cranium, distinguished from *E. vindobonensis* primarily by size. The anterior palate is somewhat more complete in *P. zhanxiangi*, revealing large, broad incisive foramina, similar to those of hylobatids and other non-hominid primates. The inferior orbital fissure is large, a feature also more typical of non-catarrhine primates (e.g., Hershkovitz, 1974). *Pliopithecus zhanxiangi* shares with *E. vindobonensis* a short face with narrow premaxilla, broad incisive foramina, narrow, oval-shaped nasal apertures, broad orbits with mildly projecting rims, low cheek bones (zygoma), and restricted maxillary sinuses.

GENUS *Epipliopithecus* Zapfe & Hürzeler, 1957
INCLUDED SPECIES *E. vindobonensis*

SPECIES *Epipliopithecus vindobonensis* Zapfe & Hürzeler, 1957 (Figs. 15.4–15.6; see also Fig. 20.1)
TYPE SPECIMEN In the collections of the Naturhistorische Museum, Vienna, Individual III from the Děvínská Nová Ves fissures (Harrison *et al.* (1991) describe a palate as part of the holotype, but this is from Individual II (Zapfe, 1960); their Figure 9 with a view of the type (Individual III) is correct), portions of a mandible, maxilla, cranial fragments, vertebrae, a left clavicle, left humerus, distal left ulna, carpals, metacarpals and phalanges
AGE AND GEOGRAPHIC RANGE Based on faunal comparisons, generally considered to be lower MN 6 or upper MN 5 in age, about 15 to 15.5 Ma (Zapfe, 1958; Ginsburg, 1986; Rögl, 1999), eastern Europe
ANATOMICAL DEFINITION
Epipliopithecus vindobonensis was originally named as a subgenus of *Pliopithecus*, *P. (Epipliopithecus) vindobonensis* (Zapfe &

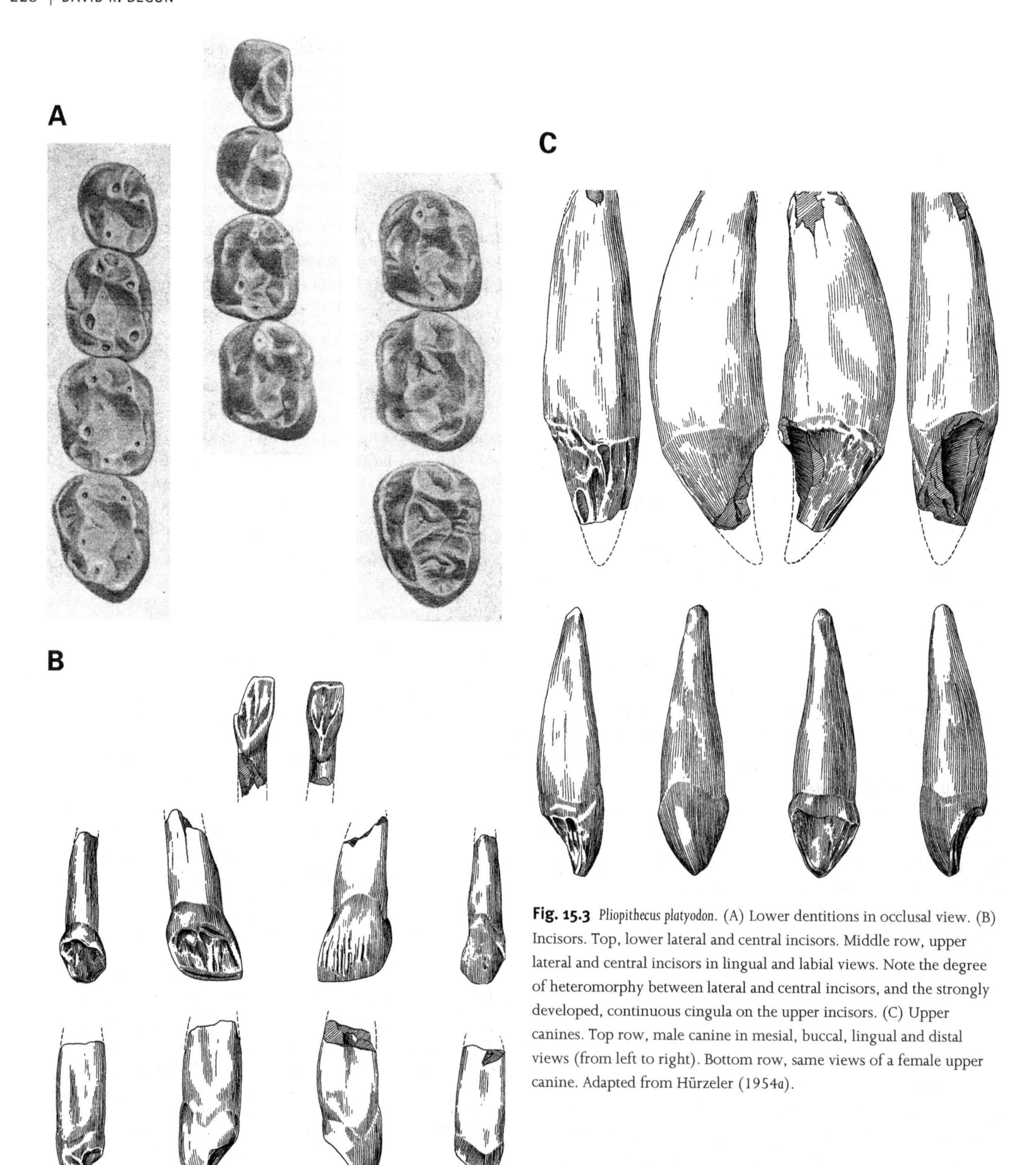

Fig. 15.3 *Pliopithecus platyodon*. (A) Lower dentitions in occlusal view. (B) Incisors. Top, lower lateral and central incisors. Middle row, upper lateral and central incisors in lingual and labial views. Note the degree of heteromorphy between lateral and central incisors, and the strongly developed, continuous cingula on the upper incisors. (C) Upper canines. Top row, male canine in mesial, buccal, lingual and distal views (from left to right). Bottom row, same views of a female upper canine. Adapted from Hürzeler (1954*a*).

Hürzeler, 1957). In addition to the impressive type specimen, *E. vindobonensis* is known from two more partial skeletons and a number of isolated remains, all from the same fissure deposit. Dentally *E. vindobonensis* can be distinguished from *Pliopithecus* by a number of features. These include differences from *P. antiquus* in overall larger dental size, higher-crowned lower incisors, upper central incisor broad with a notched lingual cingulum, P_4 and lower molars slightly narrower, indistinct or missing pliopithecine triangle, slightly broader upper molars, small trigone basin on molars, less well-developed buccal cingulum on upper molars, and greater size differences between molars. It is due to this more impressive suite of dental differences, particularly the absence of a

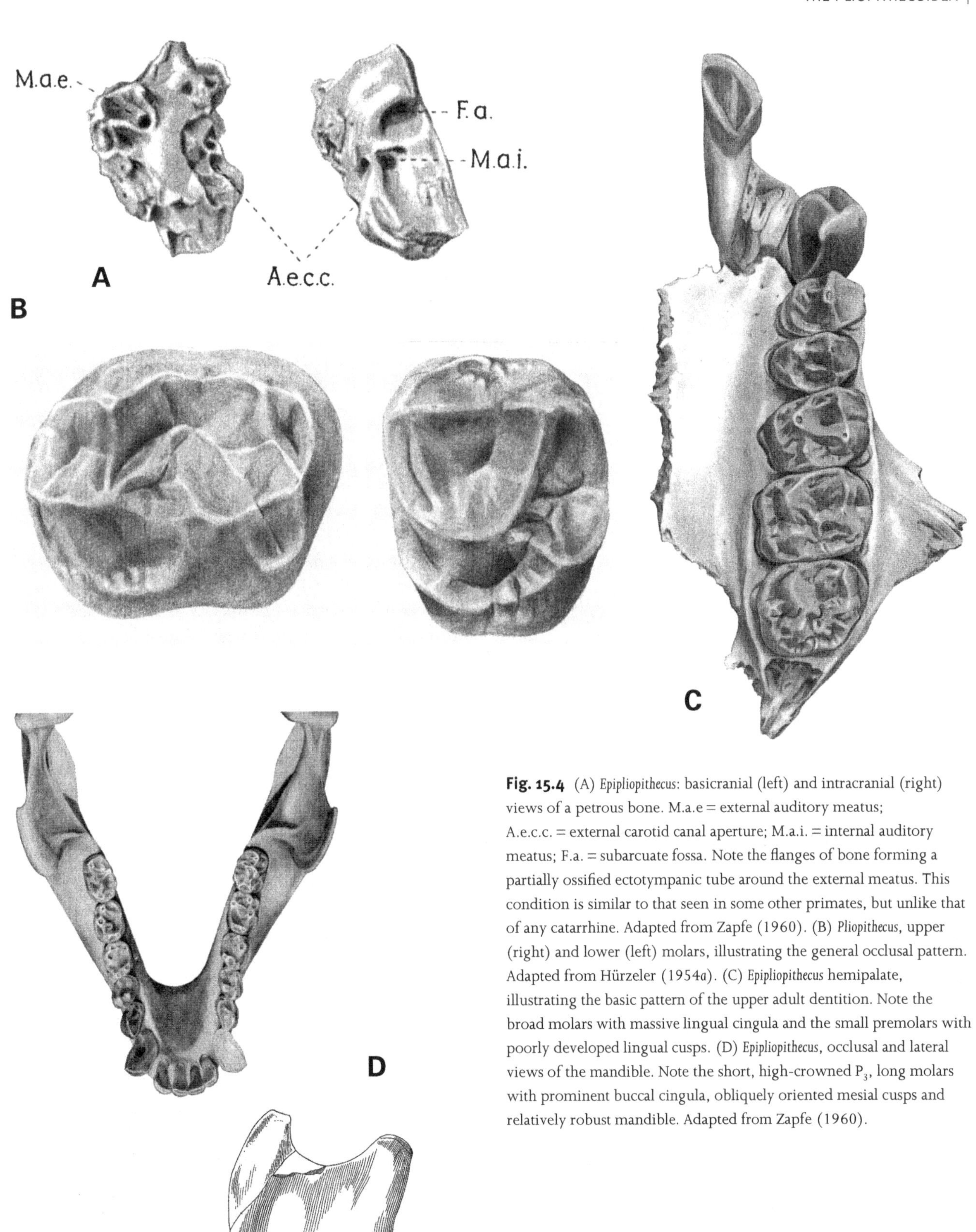

Fig. 15.4 (A) *Epipliopithecus*: basicranial (left) and intracranial (right) views of a petrous bone. M.a.e = external auditory meatus; A.e.c.c. = external carotid canal aperture; M.a.i. = internal auditory meatus; F.a. = subarcuate fossa. Note the flanges of bone forming a partially ossified ectotympanic tube around the external meatus. This condition is similar to that seen in some other primates, but unlike that of any catarrhine. Adapted from Zapfe (1960). (B) *Pliopithecus*, upper (right) and lower (left) molars, illustrating the general occlusal pattern. Adapted from Hürzeler (1954*a*). (C) *Epipliopithecus* hemipalate, illustrating the basic pattern of the upper adult dentition. Note the broad molars with massive lingual cingula and the small premolars with poorly developed lingual cusps. (D) *Epipliopithecus*, occlusal and lateral views of the mandible. Note the short, high-crowned P_3, long molars with prominent buccal cingula, obliquely oriented mesial cusps and relatively robust mandible. Adapted from Zapfe (1960).

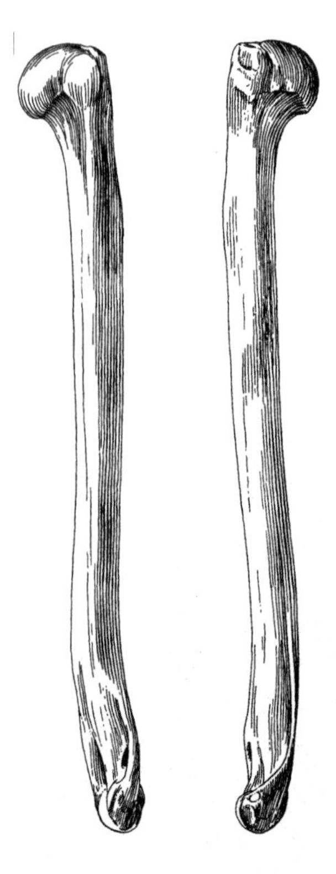

Fig. 15.5 *Epipliopithecus vindobonensis*. Medial (left) and lateral (right) views of the humerus illustrating the primitive nature of this taxon. Note the entepicondylar foramen. Adapted from Zapfe (1960).

pliopithecine triangle and the distinctive morphology of the incisors that *E. vindobonensis* is recognized here as a separate genus from *Pliopithecus*.

Epipliopithecus vindobonensis has a fairly large and globular braincase with well-developed temporal lines (in one specimen meeting in the midline to form a low sagittal crest), a relatively projecting snout (though less than in *Aegyptopithecus*), orbits slightly laterally deviated with a prominent, projecting glabellar region, supraorbital costae that do not meet in the midline to form a torus, depressed frontal trigon, relatively short but vertical frontal squama, broad interorbital space, prominent lacrimal crest obscuring the lacrimal fossa from anterior view, petrous bones with a large subarcuate fossa, and an incompletely ossified ectotympanic tube. The mandibles are long and have robust corpora and broad rami with prominent, flared gonial angles. Postcranially *E. vindobonensis* most closely resembles long-limbed New World monkeys but also relatively leggy Old World monkeys. In brachial index it is more like suspensory New World monkeys and prosimians, but hindlimb overall length and the crural index are closer to Old World monkeys (Zapfe, 1960). However, the crural index of *E. vindobonensis* is also very close to gibbons and chimpanzees, and within the human range (Zapfe, 1960). Unlike apes, however, the forelimb was slightly shorter than the hindlimb, being most comparable to howling monkeys but also baboons (Zapfe, 1960). The trunk was long and slender, and probably had seven lumbar vertebrae, a long sacrum and possibly a tail (Zapfe, 1960; Ankel, 1965). *Epipliopithecus vindobonensis* had comparatively long hands and feet, and long, curved fingers, and was mostly likely an agile climber. This is also suggested by the morphology of the joint surfaces of the limb bones (Fig. 15.6).

GENUS *Egarapithecus* Moyà-Solà *et al.*, 2001
INCLUDED SPECIES *E. narcisoi*

SPECIES *Egarapithecus narcisoi* Moyà-Solà *et al.*, 2001
TYPE SPECIMEN IPS 2943, a fragmentary mandible
AGE AND GEOGRAPHIC RANGE Torrent de Febulines, in the Vallés Penedés basin of Catalonia near Barcelona, is biostratigraphically dated to MN 10 and is placed in Chron C4An of the Geomagnetic Polarity Time Scale, indicating an age of about 9 Ma (Moyà-Solà *et al.*, 2001).
ANATOMICAL DEFINITION

The type specimen and an associated palatal fragment with a right P^3 have been mentioned in the literature by Golpe-Posse (1982) and Andrews *et al.* (1996), and recently named to a new genus by Moyà-Solà *et al.* (2001). *Egarapithecus* has a very distinctive occlusal morphology that surely merits a genus-level distinction. The lower teeth are narrow and elongated with sharply defined occlusal crests and large talonid basins. The P_3 has a strong metaconid and the P_4 an exceptionally elongated talonid with well-formed distal cusps. The molars have small mesial fovea and larger distal fovea, opposite the condition in *Anapithecus*. The M_3 is exceptionally long and narrow. The mandible is extremely deep relative to transverse breadth compared to *Epipliopithecus*, *Anapithecus* and *Pliopithecus*. Moyà-Solà *et al.* (2001) suggest that the small canines in the symphyseal fragment are unerupted, despite the wear on the M_3 (canines almost always erupt before the M_3 in catarrhines). However, unerupted canines lack root apical closure, while the canines in the *Egarapithecus* type have completed roots. The fact that the canines in *Egarapithecus* remain embedded in the mandible probably results from a pathology, another possibility suggested by Moyà-Solà *et al.* (2001). The incisors (based on the exposed roots) and canines are very small, even for a female, but it is not clear to what extent this is normal for the genus. The palatal fragment preserves the distal surface of the canine alveolus indicating the presence of a large upper canine, probably of a male. Thus it is unlikely to have come from the same individual as the type. The P^3 is also unusual in being rectangular with roughly equal mesial and distal moieties, unlike most other pliopithecoids.

As noted by Andrews *et al.* (1996) and Moyà-Solà *et al.* (2001), *Egarapithecus* has some crouzeliine similarities, and indeed these authors assign this taxon to the Crouzeliinae. These similarities mainly involve lower dental elongation and sharply developed occlusal crests, which, while more strongly developed in crouzeliines, are nevertheless present

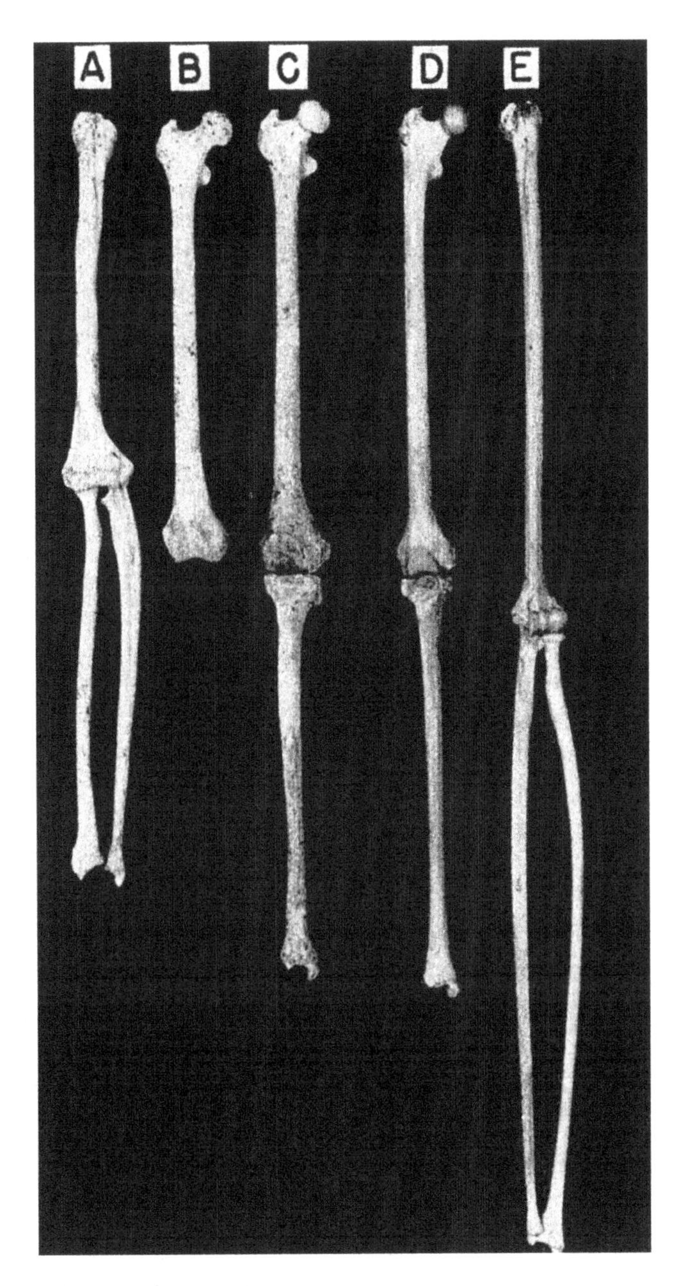

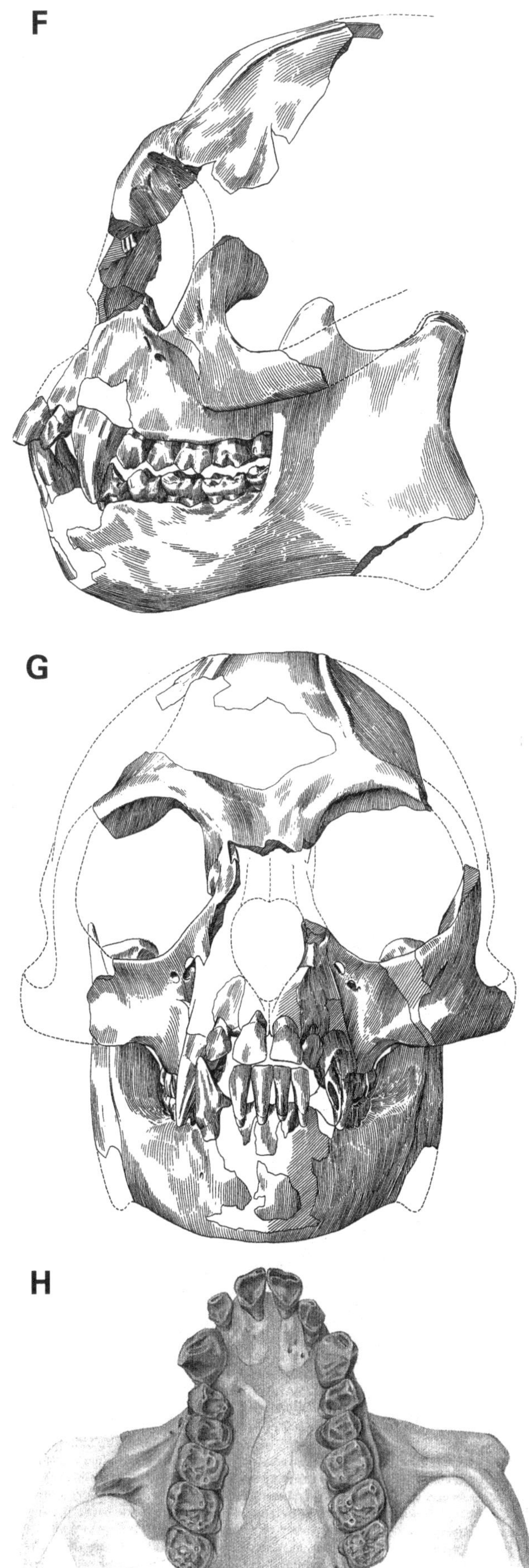

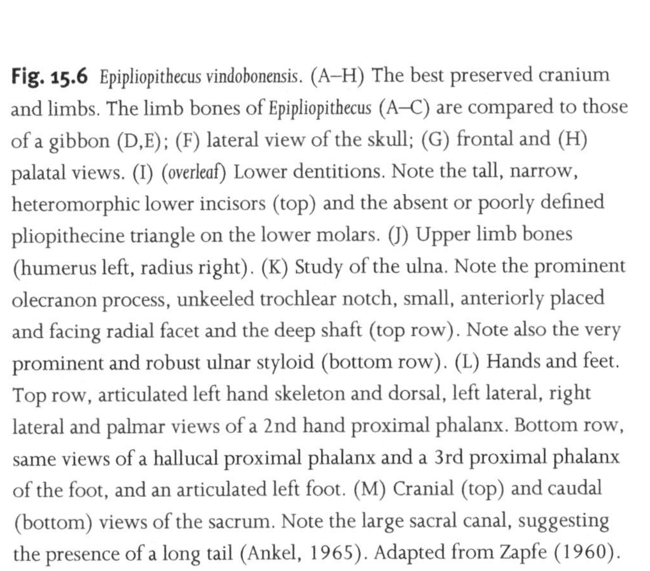

Fig. 15.6 *Epipliopithecus vindobonensis*. (A–H) The best preserved cranium and limbs. The limb bones of *Epipliopithecus* (A–C) are compared to those of a gibbon (D,E); (F) lateral view of the skull; (G) frontal and (H) palatal views. (I) (*overleaf*) Lower dentitions. Note the tall, narrow, heteromorphic lower incisors (top) and the absent or poorly defined pliopithecine triangle on the lower molars. (J) Upper limb bones (humerus left, radius right). (K) Study of the ulna. Note the prominent olecranon process, unkeeled trochlear notch, small, anteriorly placed and facing radial facet and the deep shaft (top row). Note also the very prominent and robust ulnar styloid (bottom row). (L) Hands and feet. Top row, articulated left hand skeleton and dorsal, left lateral, right lateral and palmar views of a 2nd hand proximal phalanx. Bottom row, same views of a hallucal proximal phalanx and a 3rd proximal phalanx of the foot, and an articulated left foot. (M) Cranial (top) and caudal (bottom) views of the sacrum. Note the large sacral canal, suggesting the presence of a long tail (Ankel, 1965). Adapted from Zapfe (1960).

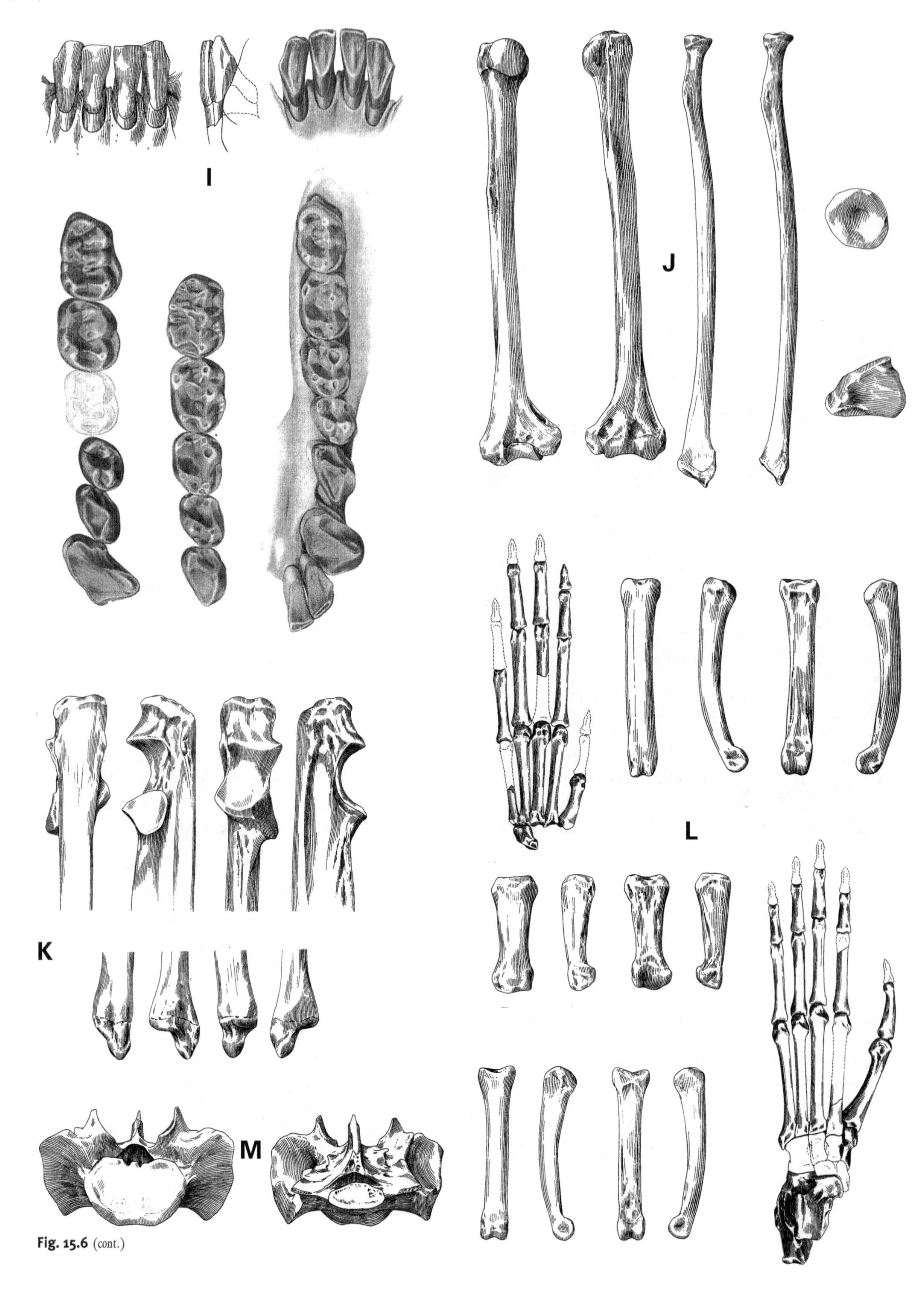

Fig. 15.6 (cont.)

in some pliopithecids, including those from the nearby and somewhat older locality of Castel de Barbera (see below). However, *Egarapithecus* lacks key features, such as large mesial basins, and has unique characters, such as very long molars and premolars, P_4 with large distal cusps, very obliquely oriented oblique crests and very small lower anterior teeth, suggesting that this taxon evolved independently from the crouzeliines, the view adopted here. It may be a descendant of the Castel de Barbera taxon, which has some similar characteristics but in a less well-developed state.

New pliopithecine

At least one new taxon belonging to this subfamily is probably present in Spain, but given the scope of this work it is not formally named here. A new species of *Pliopithecus* is represented by specimens from two other localities in the Vallés Penedés basin. The most informative specimen is an associated upper and lower dentition of a small catarrhine from the MN 8 locality of Castel de Barbera. Recognition of a new but unnamed species here contrasts with the view of Andrews *et al.* (1996), who consider these teeth to belong to *P. antiquus*. However, the Castel de Barbera specimens are smaller on average than *P. antiquus*, and are morphologically distinctive. They have bilaterally compressed cusps with well-developed crests and large occlusal basins. In some respects these are similarities with *P. piveteaui*, but they are more strongly developed in the Spanish taxon. In addition, a right dP_4 is known from Can Feliu, also considered to be of MN 8 age (Ginsburg, 1986). Andrews *et al.* (1996) consider this specimen to be a crouzeliine based on the fact that in size it is more consistent with *Egarapithecus*, which they consider to be a crouzeliine. As noted above, *Egarapithecus* lacks diagnostic crouzeliine characters. In addition, the Can Feliu specimen is morphologically closer to homologous *Pliopithecus* teeth from Göriach than to crouzeliine specimens from Rudabánya. It is long compared to breadth and has a narrower talonid, a more mesial protoconid and a distinctive trigonid with a long basin divided by a transverse ridge ending mesiolingually at a small paraconid. It is probably from a larger individual of the same taxon as the Castel de Barbera dentitions. An isolated male canine from Castel de Barbera (IPS 1823), considerably larger than that from the associated dentitions, may also represent this taxon.

Family Crouzeliidae

Subfamily Crouzeliinae

Ginsburg & Mein (1980) defined Crouzeliinae based on the type genus *Crouzelia*. Subsequent authors (e.g., Andrews *et al.*, 1996) have recognized that *Crouzelia* cannot be distinguished from *Plesiopliopithecus*, but the suprageneric taxon remains valid. Here it is elevated to a family based on the numerous differences from the Pliopithecidae.

Crouzeliids in general differ from pliopithecids in having sharper, more bilaterally compressed cusps more displaced toward the margins of the crowns, resulting in larger, relatively deep occlusal basins (except the distal basin, which is restricted and lingually offset) (Ginsburg & Mein, 1980; Begun, 1989b; Andrews *et al.*, 1996). These traits are similar to those in pliopithecines from Spain, but more strongly expressed. Crouzeliines also differ in having elongated molars and premolars with sharp, well-developed crests, particularly between the trigonids and talonids and along the crown margins.

GENUS *Plesiopliopithecus* Zapfe, 1961
Plesiopliopithecus is a small primitive catarrhine, most of the species of which are, on average, smaller in dental size than *Pliopithecus*. Like *Epipliopithecus*, *Plesiopliopithecus* was originally recognized as a subgenus of *Pliopithecus* by Zapfe (1961). I follow Ginsburg & Mein (1980) in elevating *Plesiopliopithecus* to genus status here. *Plesiopliopithecus* species are only known from lower teeth. They are distinguished from other crouzeliines in being much smaller and in having very reduced hypoconulids. In addition to their crouzeliine characters, they retain typical features of the pliopithecoids including a variably expressed pliopithecine triangle, obliquely oriented oblique crest and a protoconid that is mesial to the metaconid.
INCLUDED SPECIES *P. auscitanensis, P. lockeri, P. priensis, P. rhodanica*

SPECIES *Plesiopliopithecus lockeri* Zapfe, 1961 (Fig. 15.7)
TYPE SPECIMEN In the collections of the Naturhistorische Museum, Vienna, a left mandibular fragment from Trimmelkam, Austria
AGE AND GEOGRAPHIC RANGE Trimmelkam is considered to date to MN 6, but the fauna from the site is poor (Ginsburg, 1986); the only other taxon definitively identified at the site, *Palaeomeryx eminens*, is generally considered to be an MN 7/8 taxon (Gentry *et al.*, 1999)
ANATOMICAL DEFINITION
Like all the species of *Plesiopliopithecus*, *P. lockeri* is known from one individual. The P_3 is long and oval with a small distal fovea, compared to the other crouzeliines *Anapithecus* and *Laccopithecus*. The P_4 is like other crouzeliines in being elongated with a particularly large talonid surrounded by tall, sharp crests (Zapfe, 1961; Ginsburg & Mein, 1980). Like *P. auscitanensis* the P_4 has a distinct entoconid. Like *P. auscitanensis* and *Laccopithecus*, but unlike *Anapithecus*, the M_1 is broader distally due to the presence of a large, buccally displaced hypoconid. The lower incisors are tall-crowned and narrow. The I_1 is flared or wider at the occlusal edge than at the cervix and the I_2 is asymmetrical (Fig. 15.7).

SPECIES *Plesiopliopithecus auscitanensis* Bergounioux & Crouzel, 1965 (Fig. 15.8C, D)
TYPE SPECIMEN Sa 999 (MNHN), a left mandibular fragment with P_4–M_1

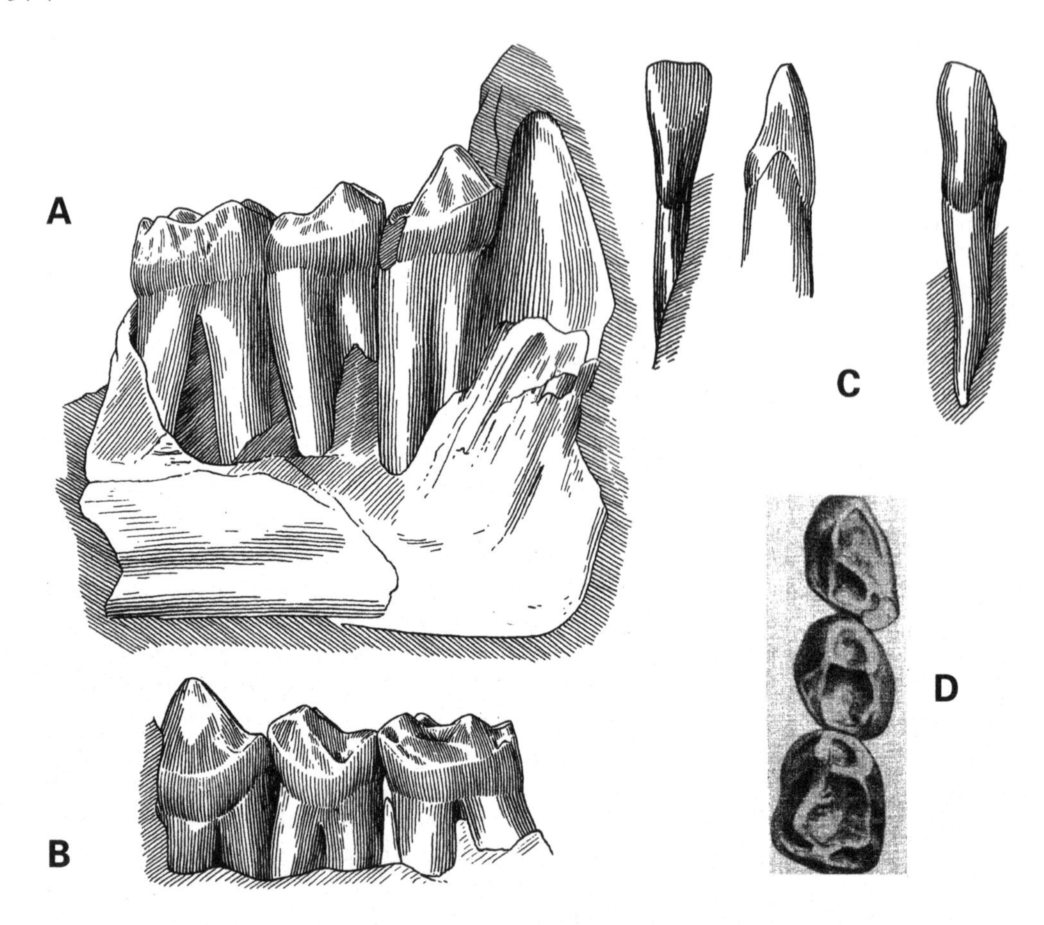

Fig. 15.7 *Plesiopliopithecus lockeri*. (A) Lingual and (B) buccal views of P_3 to M_1 and the impression in the matrix of the canine. (C) Lower incisors and (D) an occlusal view of the postcanine dentition. Note the oblique cristid obliquid between the protoconid and hypoconid and the broad, distally flared talonid basin, and reduced hypoconulid. Adapted from Zapfe (1961).

AGE AND GEOGRAPHIC RANGE Sansan, France is in MN 6 (by definition, since it is the reference locality for this zone); it is considered to date to about the middle of this zone, roughly 14.5 Ma

ANATOMICAL DEFINITION

Plesiopliopithecus auscitanensis is very difficult to distinguish from *P. lockeri*. According to Ginsburg & Mein (1980) it has a somewhat smaller P_4 talonid and an M_1 with compressed mesial cusps and a reduced hypoconulid. According to Andrews *et al.* (1996) it has a less well-developed M_1 buccal cingulum and a less well-defined distal fovea, lacking the ridge that separates this basin from the talonid in *P. lockeri*. Unfortunately, these apparently distinct morphologies are found together in larger single species samples of pliopithecoids, such as *Anapithecus hernyaki* and *Pliopithecus platyodon*. *Plesiopliopithecus auscitanensis* is dentally smaller than *P. lockeri*, but the only specimen of *P. lockeri* is a male, which can be expected to have been in the upper end of the range of variation in dental size. Reluctantly, two separate species are recognized here (Fig. 15.8).

SPECIES *Plesiopliopithecus rhodanica* Ginsburg & Mein, 1980 (Fig. 15.8A, B)

TYPE SPECIMEN FSL 65626, collections of the Faculté des Science, Université Claude-Bernard, Lyon, an M_2

AGE AND GEOGRAPHIC RANGE La Grive Saint Alban, Isère, France, has primate-bearing sediments of differing ages; Fissure L7 is dated to MN 7 (Ginsburg, 1986). Many authors currently combine MN 7 and MN 8, so that the ages of the crouzeliine and hominid from La Grive may in fact be quite close. This would be an interesting combination of primates, already known from a number of localities (Salmendingen, Austria, Rudabánya, Hungary and Lufeng, China)

ANATOMICAL DEFINITION

Plesiopliopithecus rhodanica is even more difficult to distinguish because it is only known from one tooth. Differences include smaller size, slightly narrower crown, elongated crests, reduced buccal cingulum, smaller mesial fovea and very small hypoconulid (Andrews *et al.*, 1996), but again, a similar range of variation is easy to match in larger samples of other pliopithecoids. Ginsburg & Mein (1980) consider these characters, which tend to distinguish pliopithecids from crouzeliines, better developed in *P. rhodanica* (apart from size). Given this evidence of an evolutionary change, and given the temporal separation of the samples, this

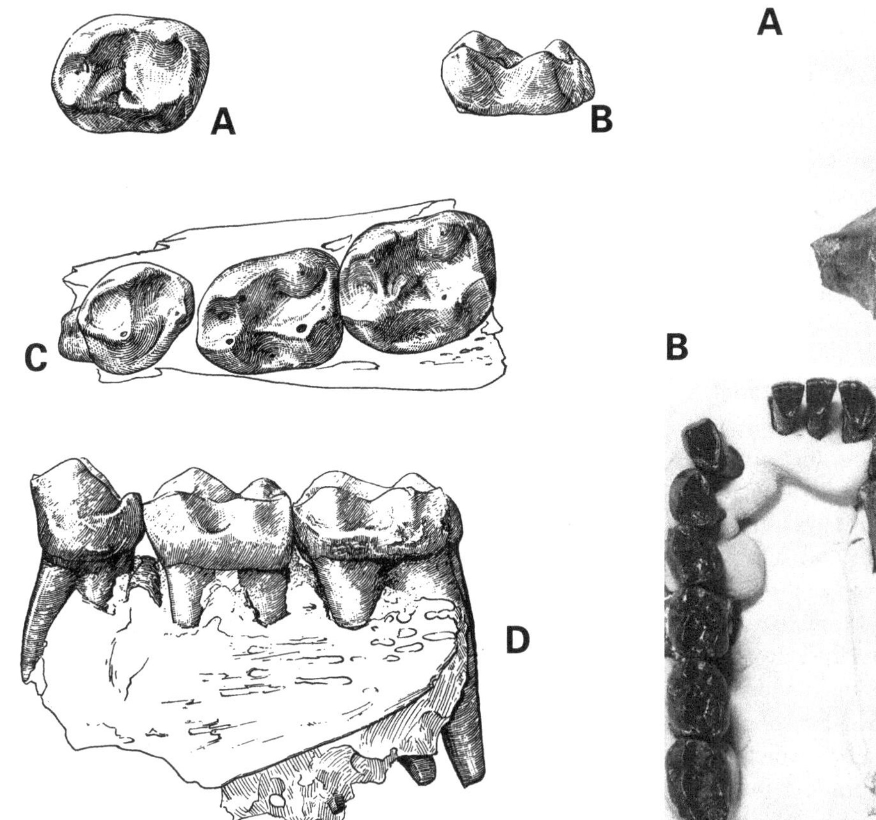

Fig. 15.8 *Plesiopliopithecus*. (A,B) *P. rhodanica* in occlusal (A) and buccal (B) views. (C,D) *P. auscitanensis* in occlusal (C) and buccal (D) views. Adapted from Ginsburg and Mein (1980).

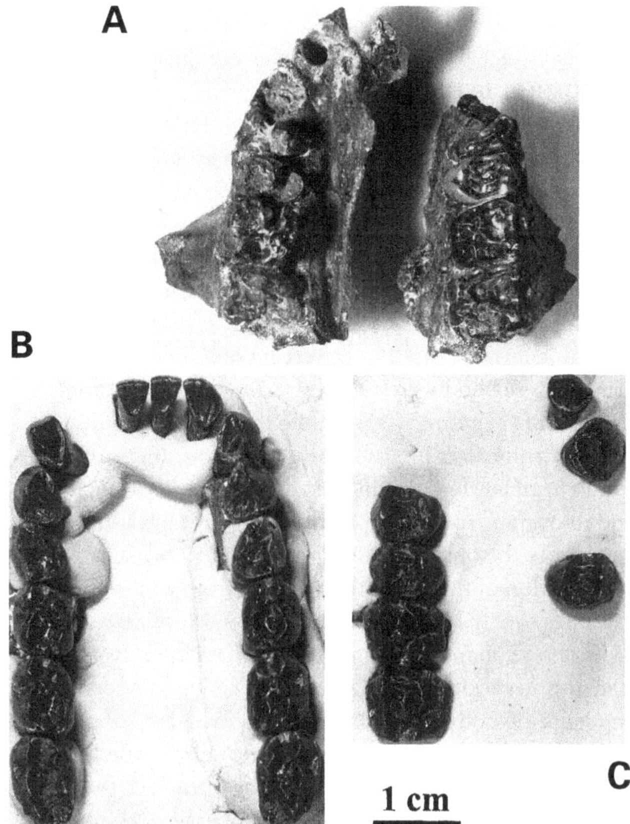

Fig. 15.10 *Anapithecus hernyaki*. (A) Palatal view from a partial cranium of a female with a heavily worn and damaged dentition, and, bottom, well-preserved lower (B) and upper (C) dentitions of a subadult female.

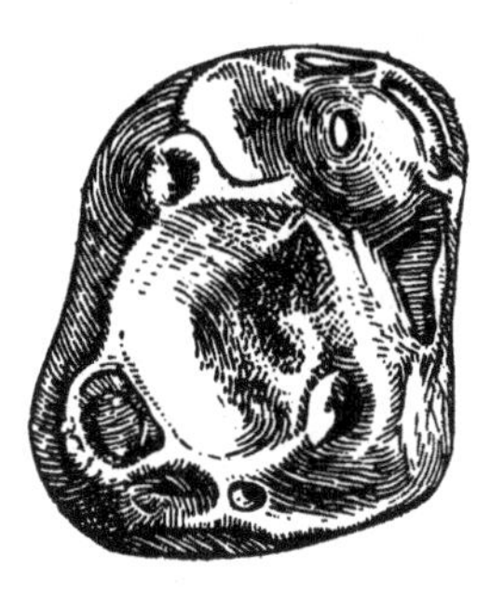

Fig. 15.9 *Plesiopliopithecus priensis*. Occlusal view.

species is recognized here as distinct as well (Fig. 15.8).

SPECIES *Plesiopliopithecus priensis* Welcomme *et al.*, 1991 (Fig. 15.9)
TYPE SPECIMEN Specimen in the collections of the Muséum Nationale d'Histoire Naturelle, Paris, a right mandibular fragment
AGE AND GEOGRAPHIC RANGE Priay (Ain, France) contains a good micromammal assemblage that unambiguously dates the locality to the upper part of MN 9, or about 9.5 Ma, considerably younger than other species of the genus (Welcomme *et al.*, 1991)

ANATOMICAL DEFINITION
Plesiopliopithecus priensis can be distinguished by its much larger size, but it retains the typical morphology of the genus with its broad talonid and reduced hypoconulid. The buccal cingulum is also more strongly developed than is typical for *Anapithecus*. This specimen has been attributed to *Pliopithecus* in Welcomme *et al.* (1991) and Andrews *et al.* (1996), but it is clearly distinct as described in general for crouzeliines and more specifically for *Plesiopliopithecus* (see above).

GENUS *Anapithecus* Kretzoi, 1975
INCLUDED SPECIES *A. hernyaki*

SPECIES *Anapithecus hernyaki* Kretzoi, 1975 (Fig. 15.10)
TYPE SPECIMEN RUD 9, in the collections of the Geological Museum of Hungary but currently stored in the National Museum of Hungary, a mandibular fragment
AGE AND GEOGRAPHIC RANGE *Anapithecus hernyaki* is known from several localities at Rudabánya, Hungary, the ages of which are essentially contemporaneous at MN 9. Unlike other pliopithecoids, *Anapithecus* appears to have had a broad distribution. It is identified at Salmendingen and Götzendorf, in the Vienna basin of Austria. Götzendorf is considered to be slightly younger than Rudabánya although

still in MN 9, while Salmendingen is considered to be MN 10 (Rögl *et al.*, 1993; and see Chapter 20, this volume).

ANATOMICAL DEFINITION

Anapithecus hernyaki is medium-sized primitive catarrhine that is larger than all other pliopithecoids on average. It was originally recognized as a subgenus of *Pliopithecus*. Ginsburg & Mein (1980) elevated it to a distinct genus assigned to the crouzeliines. A distinctive system of crests between the mesial cusps and the M_1 hypoconid unambiguously identifies *Anapithecus*. These crests form a Y, with the vertical component represented by the cristid obliquid and the arms represented by crests going to the protoconid and metaconid (Begun, 1989*b*). While this looks superficially like catarrhine deciduous molar morphology there are fundamental differences in crest development, orientation and in overall crown morphology (Begun, 1991). In contrast to the opinion of Andrews *et al.* (1996), a careful examination of the original specimens from all three localities clearly reveals their morphological similarities.

Anapithecus hernyaki is dentally somewhat larger than a siamang and probably weighed about 15 kg. The specimens suggest little body-mass sexual dimorphism but substantial canine dimorphism. Lower incisors are tall-crowned but also relatively robust transversely (long), and lack the "waisting" of other pliopithecoids. Upper central incisors are broad and low-crowned with marked lingual cingula. Upper lateral incisors are very distinctive, being much smaller than the centrals, pointed, symmetrical and relatively flat labiolingually, resembling miniature upper female canines. Lower premolars and molars are long with very large mesial fovea and talonids and small, lingually displaced distal fovea. The M_3 is especially long and tapered distally. These characters are more strongly developed in *A. hernyaki* than in other crouzeliines. In addition, lower molars have typical crouzeliine and pliopithecoid characters including mesially placed protoconids, obliquely oriented oblique crests, bilaterally compressed, marginalized cusps and prominent occlusal crests. Many specimens preserve either remnants or well-developed pliopithecine triangles. Upper molars and premolars are broad with large basins as well. Premolars have substantial cusp heteromorphy, the buccal cusps being taller than the lingual ones, and a distinctive, hexagonal shape (L. Kordos, pers. comm.). The upper molars have strong lingual and buccal cingula, relatively large talons, and well-developed ridges connecting the hypocone to the protocone. Though poorly preserved it is clear that the mandible was transversely robust, as in many pliopithecoids.

One cranial specimen is broadly similar to *Pliopithecus zhanxiangi* and *Epipliopithecus vindobonensis* but much larger. Shared characters include short faces with short premaxilla, fenestrated palates, narrow, oval shaped nasal apertures, broad orbits with projecting rims, low cheek bones (zygoma), and restricted maxillary sinuses (Kordos & Begun, 2000). Like *E. vindobonensis*, *Anapithecus hernyaki* has a fairly large and globular neurocranium. The orbits also face slightly laterally and are surrounded laterally by prominent, projecting supraorbital costae and infraorbital rims, and the frontal bone has a depressed frontal trigon and a relatively short but vertical frontal squama, and broad interorbital space. Reconstruction suggests that the interorbital region was relatively somewhat narrower than in *E. vindobonensis*, and that the medial ends of the supraorbital costae dipped down toward glabella, as in *E. vindobonensis* and hylobatids. Compared to *E. vindobonensis* the temporal lines are less well-developed and the snout relatively less projecting. The orbits are more elongated, the root of the zygomatic on the maxilla is higher and positioned more anteriorly, the postorbital breadth is relatively greater, and the frontal is shorter and more vertical (Kordos & Begun, 2000). Compared to *Pliopithecus zhanxiangi* and *Epipliopithecus vindobonensis* the anterior palate is broad. The few postcranial fragments include phalanges and some foot bones that are also broadly similar to *Epipliopithecus vindobonensis*, but with features that suggest more suspensory postures (e.g., more strongly curved phalanges) (Begun, 1988*a*, 1993*a*).

GENUS *Laccopithecus* Wu & Pan, 1984
INCLUDED SPECIES *L. robustus*

SPECIES *Laccopithecus robustus* Wu & Pan, 1984 (Fig. 15.11)
TYPE SPECIMEN PA 880, a nearly complete female lower dentition and fragmentary mandible, and PA 876, two halves of a badly damaged maxilla. Wu & Pan (1984) interpreted these specimens to be parts of a single individual, but PA 876 is clearly a male, based on canine size and morphology (Pan *et al.*, 1989). Technically these specimens are thus syntypes rather than a holotype, because they are different specimens and different individuals. In view of the fact that PA 880 has a better preserved dentition, was figured first in the original publication, has been figured elsewhere in a higher-quality image (Pan *et al.*, 1989), and is more directly comparable to most other type specimens of pliopithecoids, it should be designated as the lectotype. All specimens are from Lufeng County, Yunnan Province, and are in the collections of the Institute of Vertebrate Paleontology and Paleoanthropology, Beijing, PRC.
AGE AND GEOGRAPHIC RANGE Lufeng appears to be one of the latest occurrences of non-cercopithecid primates in Eurasia, and has been correlated to NMU 10, the Chinese mammal unit equivalent to MN 11–12 of Europe (Qiu, 1990; Qiu & Storch, 1990; Qiu *et al.*, 1999), about 8 Ma; only *Oreopithecus* persists beyond this time in Eurasia

ANATOMICAL DEFINITION

Laccopithecus robustus is known from a rich sample of about 90 specimens, including a partial cranium with a well-preserved face and palate, as well as a number of more fragmentary jaws, associated dentitions, isolated teeth and a proximal phalanx (Pan, 1998). *Laccopithecus robustus* is a

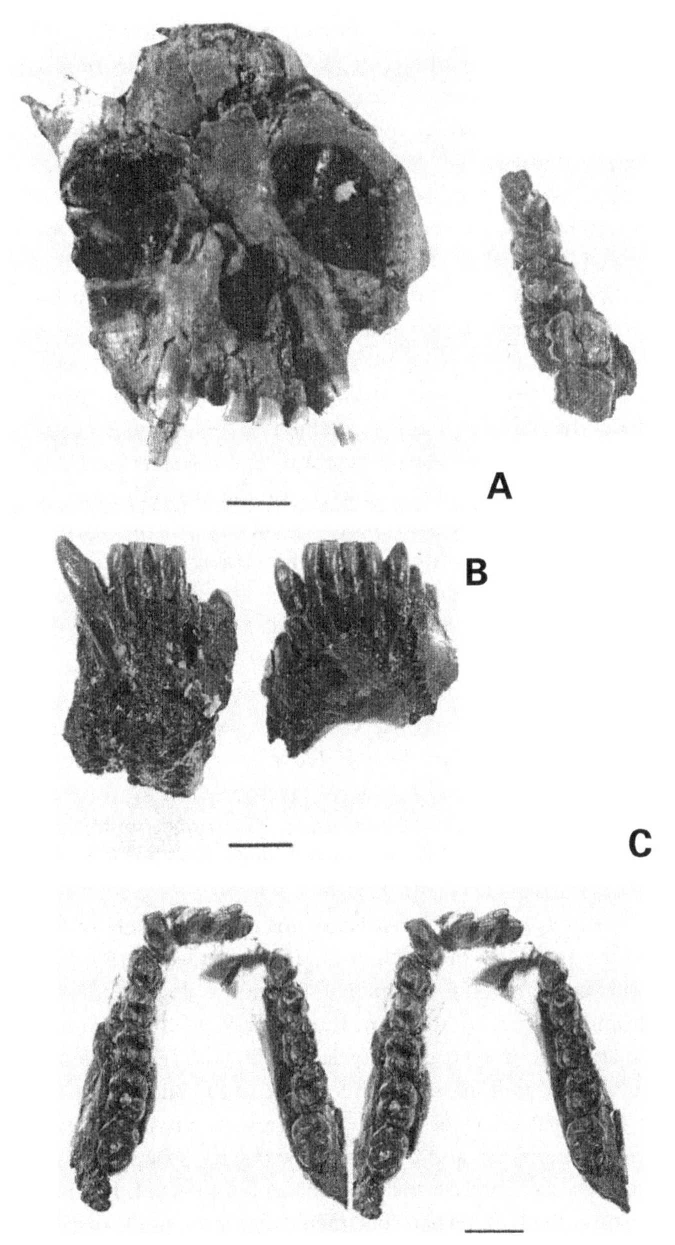

Fig. 15.11 *Laccopithecus robustus*. (A) The best-preserved cranium and an occlusal view of an upper tooth row; (B) anterior views of a male (left) and female (right) mandible illustrating the degree of sexual dimorphism; (C) stereophotograph of the lower dentition (type specimen). Adapted from Pan *et al.* (1989).

medium-sized primitive catarrhine, in most dimensions dentally smaller than *Anapithecus*. Females are close in molar size to siamangs, whereas males are somewhat larger. It has a number of typical crouzeliine characters such as large occlusal basins, compressed, marginalized cusps, and obliquely oriented oblique crests. Like *Anapithecus*, *L. robustus* has a fairly broad P_3 but like *Plesiopliopithecus* the molars and premolars are less elongated than in *Anapithecus*, the occlusal basins are less expanded, especially the mesial fovea, and the hypoconulid is rather small, though not with the degree of reduction seen in *Plesiopliopithecus*. There are no pliopithecine triangles on any of the lower molars and the buccal cingula are minimally developed. The lower canines are more massive than in *Anapithecus* while the lower incisors more closely resemble *P. lockeri* in that they are mesiodistally shorter. Two damaged mandibular symphyses suggest that the anterior part of the mandible was robust with well-developed transverse tori while the posterior portion appears to have been relatively gracile in transverse dimensions. Upper incisors lack the strong size differential of *Anapithecus*, the I^1 being a very short tooth barely longer than the I^2. Both upper incisors are labiolingually robust. The male upper canine is very large and strongly bilaterally compressed. All three upper anterior teeth in fact closely resemble their homologues in hylobatids, though this is not the case for the lower incisors, which are lower-crowned and broader in hylobatids. Another important distinction from hylobatids is the presence of sexual dimorphism in canine morphology, as in nearly all other anthropoids. The upper premolars are longer or less broad than in *Anapithecus*, and lack the degree of cusp heteromorphy of this taxon. The upper molars are also less broad than in *Anapithecus* with minimally developed lingual cingula and no stylar shelves. The talons tend to be smaller and the cusps more rounded.

The cranium of *L. robustus* is badly damaged, as is the case for most specimens from Lufeng. The palate is narrow anteriorly, as in *Epipliopithecus* and it appears to have been fenestrated. As in *Epipliopithecus*, *Anapithecus* and *Hylobates*, the nasal aperture is tall but small and narrow overall, with a narrow base and an apex that reaches above the lower level of the orbits. As in *Epipliopithecus* the orbits appear to have been more squared, or less elongated than in *Anapithecus*, though they are distorted. Relative to orbital dimensions the interorbital space is similar to that of *Epipliopithecus*. The root of the zygomatic is placed fairly high on the maxilla, and the zygomatic bone itself is robust and separated from the body of the maxilla by a prominent malar notch, all more like *Anapithecus* than *Epipliopithecus*, while the orbital rims are less prominent than in *Anapithecus*, more like the condition in *Epipliopithecus* (Fig. 15.11). Finally, the single proximal phalanx of *Laccopithecus* is long and curved, with strong muscle markings suggestive of suspensory positional behavior (Meldrum & Pan, 1988).

Family *incertae sedis*

GENUS *Paidopithex* Pohlig, 1895
INCLUDED SPECIES *P. rhenanus*

SPECIES *Paidopithex rhenanus* Pohlig, 1895 (Fig. 15.12; see also Fig. 20.1)
TYPE SPECIMEN The Eppelsheim femur, a nearly complete specimen first described in Kaup (1861), in the collections of the Hessisches Landesmuseum, Darmstadt, Germany
AGE AND GEOGRAPHIC RANGE Known only from

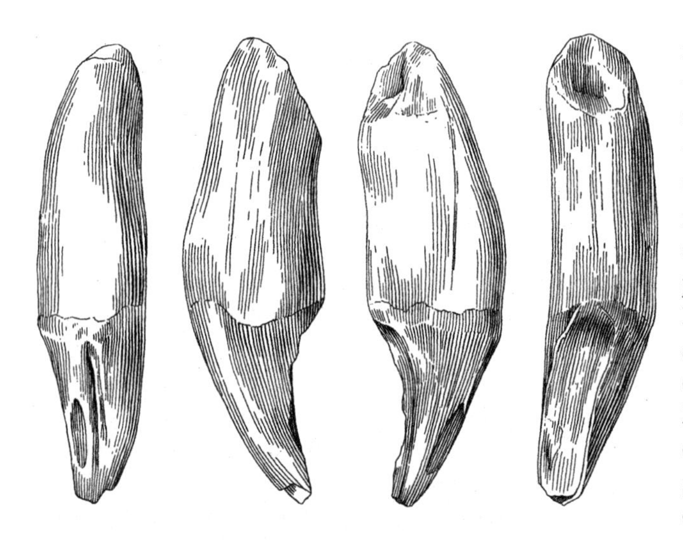

Fig. 15.12 *Paidopithex rhenanus*. Upper male canine in mesial, buccal, lingual and distal views. Adapted from Hürzeler (1954).

Eppelsheim, dated securely to MN 9 based on its rich associated fauna (Mein, 1986)

ANATOMICAL DEFINITION

Paidopithex rhenanus is a large primitive catarrhine known from the type femur and an upper male canine (Fig. 15.12). Both specimens resemble *Epipliopithecus* but are much larger. The femur routinely is attributed to *Dryopithecus*, but is different from any fossil or living hominoid (Begun, 1992*a*) as confirmed by the recent discovery of femoral fragments of *Dryopithecus* from Spain (Moyà-Solà & Köhler, 1996). Recently recovered femora from Rudabánya possibly attributable to *Anapithecus* based mostly on size, are very different in morphology from *Paidopithex rhenanus* (Kordos & Begun, 1999). *Paidopithex rhenanus* can be distinguished morphologically from hominoids but in terms of size it is consistent with a number of pliopithecoids. Here it is considered to be pliopithecoid but left unassigned beyond the superfamily level.

The femur of *Paidopithex rhenanus* is long and gracile with a comparatively short neck with a lower neck–shaft angle than in hominoids generally, including *Dryopithecus*, and in comparison to the Rudabánya femora (Kordos & Begun, 1999). The shaft is straight and the condyles are shallow and separated by a broad patella groove. The canine is long and dagger-like, closely resembling the canines of *Epipliopithecus* and *Laccopithecus*. The root and cervix are similar in dimensions to the single upper male canine specimen of *Anapithecus*, which unfortunately does not preserve much of the crown. It is smaller and less robust than *Dryopithecus* upper male canines (Fig. 15.12).

Evolution of European Miocene catarrhines

What if anything is a pliopithecoid?

Distinguishing among many of the species of pliopithecoids is difficult due to their exceptionally primitive and conservative dental morphology. It is possible that the dionysopithecines are broadly ancestral to all other pliopithecoids, though more fossils are required to establish the affinities of these primitive Asian catarrhines. Within the Pliopithecinae it appears that *Pliopithecus platyodon* and *P. antiquus* are most closely related, differing essentially only in size. *Pliopithecus piveteaui* is smaller and more primitive than the other European species, and may be their ancestor. Technically this should require a new genus name for the older sample to distinguish it from the more closely related descendants. *Pliopithecus zhanxiangi* and *Epipliopithecus vindobonensis* are more distinctive, and it may be justified to recognize separate genera for both of these taxa. Because *E. vindobonensis* lacks one of the few diagnostic characters of the pliopithecines, a pliopithecine triangle on the lower molars, it is recognized here as a distinct genus. In fact, *E. vindobonensis* is no more morphologically like *Pliopithecus* than are some specimens attributed tentatively to *Limnopithecus*, such as KNM-FT 20 from Fort Ternan, Kenya (Andrews, 1978*a*).

Within the Crouzeliinae a morphological gradient is apparent from the more conservative *Plesiopliopithecus lockeri* to the more derived *Anapithecus hernyaki*. *Laccopithecus* does not fit clearly with this trend and may have diverged early in the evolutionary history of the crouzeliines. The dionysopithecines are most similar to early Miocene African taxa (Bernor *et al.*, 1988; Harrison & Gu, 1999).

The early Miocene of Africa seems the most likely origin for the pliopithecoids (Thomas, 1985; Made, 1999). *Limnopithecus* shares the distinctive P_3 morphology with the Pliopithecoidea. In addition, the lower P_4 of the specimen tentatively identified as *Limnopithecus* from Fort Ternan, Kenya has a metaconid that is much lower than the buccal cusp, as in *Dionysopithecus* (see above). This however is not true of other *Limnopithecus*, and it may well be that the specimen from Fort Ternan is not *Limnopithecus* but a pliopithecoid. If this is the case, the Fort Ternan specimen may represent the dispersal of a pliopithecoid into Africa, since Fort Ternan is considerably younger than much of the record of pliopithecoids in Europe and China (Pickford, 1986*a*). A number of Fort Ternan taxa, including rodents, ruminants and carnivores, are also thought to be Eurasian in origin (Tong & Jaeger, 1993; Gentry & Heizmann, 1996; Werdelin & Solounias, 1996; Made, 1999). At Wadi Moghara in Egypt, a humerus is known with a relatively modern elbow joint, as in proconsulids and more modern hominoids, and an entepicondylar foramen, reminiscent of prosimians, many New World monkeys, Oligocene catarrhines and, of course, *Epipliopithecus* (Simons, 1994*a*). This early Miocene specimen may also be associated with the ancestry of the Pliopithecoidea. At any rate, these fossil samples suggest a complex connection between the Miocene primate faunas of Asia and Africa. It may be, as suggested by Harrison *et al.* (1991) that the earliest pliopithecoids lived in East Asia. They seem to diversify however in Europe. If the taxonomy proposed here is correct, then it appears that on two separate occasions the pliopithecoids sent representatives to Asia, once in the form of the pliopithecine *P. zhanxiangi* and another time the crouzeliine *Laccopithecus*. The evolutionary relations within the Pliopithecoidea are so poorly defined that it is possible that the Asian taxa evolved independently from the European taxa. However, given the evidence of migration of other mammalian genera between Asia and Europe during the middle and late Miocene (Flynn *et al.*, 1986; Bernor *et al.*, 1988; Qiu, 1990; Pickford, 1993*b*; Qiu & Qiu, 1995; Qiu *et al.*, 1999; Made, 1999), a similarly complex pattern of biogeography in the Pliopithecoidea is certainly possible.

In conclusion, the systematics and evolutionary history of the Pliopithecoidea are about as simple as those of the Platyrrhini. This is no coincidence. The histories of both groups are remarkably similar. Both emerge from a primitive ancestor on a new land mass devoid of anthropoids, and both diverge into their respective major groups very shortly after their first appearance. Both undergo evolutionary changes that seem to be emergent or structurally inevitable in anthropoids: molarization, encephalization, reduction of the snout, limb gracilization and the development of suspensory positional behavior. Pliopithecoids were unable to maintain their splendid isolation, having been joined first by hominoids and then by cercopithecoids during the Miocene. Like the hominoids, they were unable to adapt to changing climates. It may be this more than anything else that led to the extinction of both groups at practically the same moment at the end of the Miocene.

Primary References

Anapithecus

Kretzoi, M. (1975). New ramapithecines and *Pliopithecus* from the lower Pliocene of Rudabánya in north-eastern Hungary. *Nature*, **257**, 578–581.

Begun, D. R. (1988). Catarrhine phalanges from the Late Miocene (Vallesian) of Rudabánya, Hungary. *Journal of Human Evolution*, **17**, 413–438.

Begun, D. R. (1993*a*). New catarrhine phalanges from Rudabánya (northeastern Hungary) and the problem of parallelism and convergence in hominoid postcranial morphology. *Journal of Human Evolution*, **24**, 373–402.

Kordos, L. & Begun, D. R. (1999). Femora of *Anapithecus* from Rudabánya. *American Journal of Physical Anthropology*, Supplement **28**, 173.

Kordos, L. & Begun, D. R. (2000). Four catarrhine crania from Rudabánya. *American Journal of Physical Anthropology*, Supplement **30**, 193.

Kordos, L. & Begun, D. R. (2001). Primates from Rudabánya: allocation of specimens to individuals, sex and age categories. *Journal of Human Evolution*, **40**, 17–39.

Dionysopithecus

Li, C. (1978). A Miocene gibbon-like primate from Shihhung, Kiangsu Province. *Vertebrata PalAsiatica*, **16**, 187–192.

Bernor, R. L., Flynn, L. J., Harrison, T., Hussain, S. T., & Kelley, J. (1988). *Dionysopithecus* from Southern Pakistan and the biochronology and biogeography of early Eurasian catarrhines. *Journal of Human Evolution*, **17**, 339–358.

Suteethorn, V., Buffetaut, E., Buffetaut-Tong, H., Ducrocq, S., Helmcke-Ingavat, R., Jaeger, J. J., & Jongkanjanasoontorn, Y. (1990). A hominoid locality in the Middle Miocene of Thailand. *Comptes rendus de l'Académie des sciences de Paris*, **311**, 1449–1454.

Egarapithecus

Moyà-Solà, S., Köhler, M., & Alba, D. M. (2001). *Egarapithecus narcisoi*, a new genus of Pliopithecidae (Primates, Catarrhini) from the late Miocene of Spain. *American Journal of Physical Anthropology*, **114**, 312–324.

Epipliopithecus

Glässner, M. F. (1931). Neue Zähne von Menschenaffen aus dem Miozän des wiener Beckens. *Annalen des Naturhistorischens Museums in Wien*, **46**, 15–27.

Zapfe, H. (1952). Die *Pliopithecus*-Funde aus der Spaltenfüllung von Neudorf an der March (CRS). Vienna: *Verhandlungen der geologischen Bundesanstalt*, Sonderheft C, 126–130.

Zapfe, H. & Hürzeler, J. (1957). Die Fauna der miozänen Spaltenfüllung von Neudorf a. d. March (CSR). *Sitz. -Ber. Österr. Akad. Wiss. math.-nat.. Kl. Abt. I*, **166**, 113–123.

Zapfe, H. (1958). The skeleton of *Pliopithecus* (*Epipliopithecus*) *vindobonensis* Zapfe & Hürzeler. *American Journal of Physical Anthropology*, **16**, 441–455.

Zapfe, H. (1960). Die primatenfunde aus der miozänen Spaltenfüllung von Neudorf an der March (Dévinská Nová Ves), Tschechoslovakei. Mit anhang: Der Primatenfund aus dem Miozän von Klein Hadersdorf in Niederosterreich. *Abhandlungen der schweizerische paläontologischen Gesellschaft* **78**, 1–293.

Zapfe, H. (1961). Ein Primatenfund aus der miozänen Molasse von Oberösterreich. *Zeitschrift für Morphologie und Anthropologie*, **51**, 247–267.

Andrews, P. J. & Simons, E. (1977). A new Miocene gibbon-like genus, *Dendropithecus* (Hominoidea, Primates) with distinctive postcranial adaptations: its significance to origin of Hylobatidae. *Folia Primatologica*, **28**, 161–169.

Laccopithecus

Wu, R. & Pan, Y. (1984). A late Miocene gibbon-like primate from Lufeng, Yunnan Province. *Acta Anthropologica Sinica*, **3**, 185–194.

Meldrum, D. J. & Pan, Y. (1988). Manual proximal phalanx of *Laccopithecus robustus* from the latest Miocene site of Lufeng. *Journal of Human Evolution*, **17**, 719–731.

Pan, Y. (1988). Small fossil primates from Lufeng, a latest Miocene site in Yunnan Province, China. *Journal of Human Evolution*, **17**, 359–366.

Pan, Y. R., Waddle, D. M., & Fleagle, J. G. (1989). Sexual dimorphism in *Laccopithecus robustus*, a late Miocene hominoid from China. *American Journal of Physical Anthropology*, **79**, 137–158.

Pan, Y. (1998). Middle–small-bodied apes from Neogene in China and their significance. *Acta Anthropologica Sinica*, **17**, 283–292.

Paidopithex

Kaup, J. J. (1861). *Beitrage zur naheren Kenntniss der urweltlichen Säugethiere*, vol. 5. Darmstadt and Leipzig.

Platodontopithecus

Gu, Y. & Lin, Y. (1983). First discovery of *Dryopithecus* in East China. *Acta Anthropologica Sinica*, **2**, 305–314.

Plesiopliopithecus

Zapfe, H. (1961). Ein Primatenfund aus der miozänen Molasse von Oberösterreich. *Zeitschrift für Morphologie und Anthropologie*, **51**, 247–267.

Bergounioux, F. M. & Crouzel, F. (1965). Les Pliopithèques de France. *Annales de paléontologie*, **51**, 45–65.

Ginsburg, L. & Mein, P. (1980). *Crouzelia rhondanica*, nouvelle espèce de primate catarrhinien, et essai sur la position systématique de Pliopithécidae. *Bulletin du Muséum national d'histoire naturelle, Paris*, **4**, 57–85.

Welcomme, J. L., Aguelar, J. P., & Ginsburg, L. (1991). Découverte d'un nouveau Pliopithèque (Primates, Mammalia) associé à des rongeurs dans les sables du Miocène supérieur de Priay (Ain, France) et remarques sur la paléogéographie de la Bresse au Vallésien. *Comptes rendus de l'Académie des sciences de Paris*, **313**, 723–729.

Pliopithecus

Lartet, E. (1837). Note sur les ossements fossiles des terrains tertiaires de Simorre, de Sansan, etc., dans le département du Gers, et sur la découverte récente d'une máchoire de singe fossile. *Annales des sciences naturelles*, **13**, 116–123.

Gervais, P. (1849*a*). Note sur une nouvelle espèce de singe fossile. *Comptes rendus de l'Académie des sciences de Paris*, **XXVIII**, 699–700, 789.

Gervais, P. (1849*b*). *Zoologie et paléontologie françaises*, 1st edn. Paris: Bertrand.

Gervais, P. (1859). *Zoologie et paléontologie françaises*, 2nd edn, 2 vols., Texte et Atlas. Paris: Bertrand.

Biedermann, W. G. A. (1863). *Petrefacten aus der Umgegend von Winterthur*, Vol. 2, Die Braunkohlen von Elgg. Winterthur: Bleuler-Hausheer.

Hofmann, A. (1863). Die Fauna von Göriach. *Abhandlungen der kaiserlich-königlichen geologischen Reichsanstalt Wien*, **15**, 1–87.

Gervais, P. (1867). *Zoologie et paléontologie générales*. Paris: Arthus.

Depéret, C. (1887). Sur la faune vertébrés miocène de La Grive-Saint Alban (Isère). *Archives du Muséum d'histoire naturelle de Lyon*, **5**.

Roger, O. (1898). Wirbeltierreste aus dem Dinotheriensande der bayerisch-schwäbischen Hochebene. *Bericht des Naturwissenschaftlichen Vereins für Schwaben und Neuburg in Augsburg*, **33**, 1–46.

Schlosser, M. (1900). Die neueste Litteratur über die ausgestorbenen Anthropomorphen. *Zoologischer Anzeiger*, **23**, 289–301.

Lecointre, G. (1912). *Les Faluns de Touraine*. Tours: Société d'agriculture, science, arts et belles-lettres de Départment Indre-et-Loire.

Wegner, R. N. (1913). Teriär und umgelagerte Kreide bei Oppeln (Oberschlesien). *Palaeontographica*, **60**, 175–274.

Stehlin, H. G. (1914). Übersicht über die Säugetiere der schweizerischen Molasseformation, ihre Fundorte und ihre stratigraphische Verbreitung. *Verhandlungen der naturforschenden Gesellschaft, Basel*, **25**.

Sera, G. L. (1917). La testimianza dei fossili di anthropomorfi per la questione dell'origine dell'uomo. *Attidella Società italiana di scienze naturali*, **56**, 1–156.

Hürzeler, J. (1954). Contribution à l'odontologie et à la phylogénèse du genre *Pliopithecus* Gervais. *Annales de paléontologie*, **40**, 1–63.

Bergounioux, F. M. & Crouzel, F. (1965). Les Pliopithèques de France. *Annales de Paléontologie*, **51**, 45–65.

Ginsburg, L. (1975). Les Pliopithèques des faluns helvétiens de la Touraine et de l'Anjou. *Colloques Internationaux du Centre national de la recherche scientifique*, **218**, 877–886.

Harrison, T., Delson, E., & Jian, G. (1991). A new species of *Pliopithecus* from the middle Miocene of China and its implications for early catarrhine zoogeography. *Journal of Human Evolution*, **21**, 329–362.

Harrison, T. & Gu, Y. (1999). Taxonomy and phylogenetic relationships of early Miocene catarrhines from Sihong, China. *Journal of Human Evolution*, **37**, 225–277.

Pliopithecinae gen. et sp. indet.

Golpe Posse, J. M. (1982). Un pliopitécido persistente en el Vallesience Medio-Superior de los alrededores de Terrassa (cuenca del Vallés, España) y problemas de su adaptación. *Boletín di geología y minería*, **93**, 287–296.

16 | The Victoriapithecidae, Cercopithecoidea

BRENDA R. BENEFIT AND MONTE L. MCCROSSIN

Introduction

Victoriapithecidae is the geologically oldest family of Old World monkey, pre-dating and representing the sister taxon to the Cercopithecidae which includes both fossil and extant colobine and cercopithecine monkeys (von Koenigswald, 1969; Benefit, 1987, 1993). Its phylogenetic position is indicated by its unusual craniodental morphology which is intermediate between that of the last common ancestor shared by cercopithecoids and hominoids and modern monkeys. Currently, this family includes two genera, *Victoriapithecus* and *Prohylobates* (Leakey, 1985). The Victoriapithecidae differ greatly from the arboreal, folivorous and colobine-like ancestral morphotype that was once proposed for cercopithecoid monkeys. Instead, known limbs, teeth and crania of this 19–12.5-million-year-old African family of monkeys indicate that they were as terrestrial as vervet monkeys, among the most frugivorous of cercopithecoids with a diet of hard fruits and/or seeds, and had a somewhat macaque-like skull with a long and low cranial vault and moderately long snout (von Koenigswald 1969; Benefit & Pickford 1986; Benefit 1987, 1993, 1994, 1999*a*; Harrison 1989*c*; McCrossin & Benefit, 1994, 1998; Benefit & McCrossin, 1997*a*). Estimated body size based on postcrania indicates that *Victoriapithecus* weighed 3–5 kg and is therefore one of the smallest terrestrially adapted anthropoids (Harrison, 1989*c*; Zambon *et al.*, 1999).

History of discovery and debate

Technically, victoriapithecid fossils were first discovered in early to middle Miocene North African deposits at the site of Wadi Moghara. However, the three partial mandibles with weathered and worn P_4–M_3 of *Prohylobates tandyi* Fourtau 1918 originally were described as hominoid and not recognized as cercopithecoid until 1950 (Le Gros Clark & Leakey, 1950). Their significance in cercopithecoid evolution was not fully discussed until 1969 (Simons, 1969). A second species from Gebel Zelten, *P. simonsi*, was added to the genus in 1979 based on a single mandible with M_2–M_3 (Delson, 1979). Simons (1969) and Delson (1979) consider the lower molars of these species to be incompletely bilophodont. Although Leakey (1985) pointed out that *Prohylobates* molars are too worn and weathered to be certain of this, an M_1 recently discovered in a partial mandible appears to lack a distal lophid between entoconid and hypoconid, providing some support for the hypothesis (Simons, 1994*a*). Because of its history, paucity of fossil evidence and poor preservation, *Prohylobates* has played a less important role in deciphering cercopithecoid origins than do victoriapithecid fossils belonging to the genus *Victoriapithecus* from eastern Africa.

Victoriapithecus remains were first recovered by Archdeacon Walter Edwin Owen and Donald Gordon MacInnes in 1933 and 1934 on Maboko Island in western Kenya, at a locality now known as Maboko Main. The strata in which the monkeys are found (Beds 3 and 5) are radiometrically dated as older than 14.7 Ma and biostratigraphically as younger than 16 Ma (Le Gros Clark & Leakey, 1951; Leakey, 1967*b*; Feibel & Brown, 1991). In 1943 MacInnes attributed a mandible (KNM-MB 1) and three isolated upper molars from Maboko to the colobine genus *Mesopithecus* best known from the late Miocene site of Pikermi in Greece (Zapfe, 1991). In 1947 Leakey and MacInnes quarried an additional 37 cercopithecoid craniodental remains and 11 postcranial remains from a large southeast trending trench ("Leakey's Trench") cutting into a low hill at Maboko Main (Le Gros Clark & Leakey, 1951). The material was not described until 1969, when Gustav Heinrich Ralph von Koenigswald attributed the collection to two species of a new genus *Victoriapithecus*. He defined *Victoriapithecus macinnesi* as having smaller and more posteriorly constricted upper molars, more weakly defined and/or absent upper molar crista obliqua, and a shorter and simpler M_3 hypoconulid than *Victoriapithecus leakeyi*. The two species were interpreted as pre-dating the split between colobine and cercopithecine monkeys, and as intermediate in morphology between modern monkeys and apes because their upper molars retained primitive crista obliqua not found among modern cercopithecoids (von Koenigswald, 1969). The genus was therefore placed in a distinct subfamily Victoriapithecinae von Koenigswald 1969.

Just four years after von Koenigwald described *Victoriapithecus*, Eric Delson (1973, 1975*a*, 1975*c*; Simons & Delson, 1978; Szalay & Delson 1979) revised the descriptions and hypodigms (although not explicitly given) of *V. leakeyi* and *V. macinnesi* within the context of broader papers concerning cercopithecoid evolution. Although he never published a detailed study of the material, Delson claimed (Szalay & Delson, 1979: 422) that: "'*Victoriapithecus*' *leakeyi* is thus seen as a dentally more conservative, somewhat cercopithecine-like species, with narrow teeth, long M_3, and retained crista obliqua on upper molars (M^3 uncertain)."

Only two upper molars (an M^1 and M^2) with crista obliqua were placed into "*V*". *leakeyi* and all others without the crest into *V. macinnesi*. The cercopithecine affinity of "*V*". *leakeyi* was further supported by placing all lower molars which Delson perceived to be longer relative to width, and to have longer trigonid and talonid basins than others into that species, whereas all other molars were attributed to *V. macinnesi*. *Victoriapithecus macinnesi* was described as having squarish lower molars that combine ancestrally shallow notches and bulging cingulum remnants with apparently derived colobine-like short trigonids, conforming to Delson's predicted mosaic dentition for ancestral colobines (Delson 1973, 1975*a*, 1975*c*; Simons & Delson, 1978; Szalay & Delson, 1979). The two known distal humeri of *Victoriapithecus* were perceived to differ in the degree to which they indicated arboreal versus terrestrial adaptations, with the more arboreal specimen KNM-MB 3 being placed in the supposed colobine species *V. macinnesi* (Delson 1973, 1975*a*, 1975*c*; Simons & Delson, 1978; Szalay & Delson, 1979). Based on these revisions, Delson proposed that the split between colobine and cercopithecine monkeys "had indeed become distinct by early in the middle Miocene" (Szalay & Delson, 1979: 425) and was well documented at Maboko Island, making Victoriapithecinae an invalid taxon.

Field work resumed at Maboko in 1973, when David Pilbeam and Peter Andrews collected 28 isolated cercopithecoid permanent teeth (Andrews *et al.*, 1981*b*). However, it was not until the 1980s that field work by Martin Pickford led to the discovery of 282 dentognathic and 52 postcranial remains of *Victoriapithecus*, providing sufficient evidence to help resolve phylogenetic questions (Benefit & Pickford, 1986; Senut, 1986; Harrison, 1987, 1989*c*; Benefit, 1993, 1994). Pickford recovered the new fossils by wet-screening the backfill to the south of Leakey's Trench during three field seasons between 1982 and 1984 (Pickford, 1986*a*). Among the new specimens were the first partial maxillae with associated upper molars of *Victoriapithecus*. They provided an important reference from which to assign isolated molars to element.

As the pre-1985 *Victoriapithecus* dentition and postcrania were analyzed, it became evident that only a single taxon was preserved in the deposits. With upper molars correctly assigned to element, differences perceived between *V. macinnesi* and *V. leakeyi* by von Koenigswald (1969) clearly were attributable to misidentification of second and third molars, the former being more quadrate in shape than the latter. It became obvious that *Victoriapithecus* crista obliqua are differentially expressed across the molar row, being more frequent on M^1 (91%) than M^2 (71%) and on M^2 than M^3 (33%) (Benefit & Pickford, 1986; Benefit, 1987, 1993). Since quadrangular upper molars are also those that usually have crista obliqua, *V. leakeyi* corresponds to the morphology of M^1 and M^2, while *V. macinnesi* corresponds to the M^3. The evidence suggests that presence or absence of crista obliqua cannot be used to sort specimens into two species. Colobine-like and cercopithecine-like lower molars similarly were found on associated molars in mandibles (Benefit, 1987, 1993). Levels of metric variation for a variety of detailed measurements of molar morphology and shape, including the lengths of trigonids and talonids, indicate that only a single cercopithecoid species, *V. macinnesi* (based on page priority), is present in deposits at Maboko. In fact, coefficients of variation for *Victoriapithecus* I^1–M^2 measurements seldom exceeded that observed for populations of wild-trapped Kenyan vervet monkeys (Benefit, 1987, 1993). Upper and lower third molar variation is high within the sample, but similar to that found in species in which M^3 sexual dimorphism is high, including *Macaca mulatta*, *Cercocebus albigena*, *Papio cynocephalus*, *Cercopithecus cephus*, *Cercopithecus mitis* and *Nasalis larvatus*.

Harrison (1989*c*) found that the range of morphological variation within the postcranial sample did not exceed that observed for seven out of 14 anthropoid species he examined. On this basis, he also concluded that only a single species, *V. macinnesi*, is preserved at Maboko. While it is evident that some individual *V. macinnesi* were more arboreal than others, the same degree of variation is found within many anthropoid species including *Cercopithecus ascanius*. Senut (1986) has argued instead that two congeneric species of *Victoriapithecus*, one more arboreal than the other, are preserved at Maboko.

As recognized by von Koenigswald (1969), the dentition of *V. macinnesi* differs from that of extant Old World monkeys. *Victoriapithecus* teeth are characterized by a mosaic of traits that are distinctly colobine-like (short M^{1-2} mesial shelf length and upper premolars wider relative to length than those of cercopithecines) and cercopithecine-like (I^1 with curved distal margin, tall P_4 metaconid, lower molar mesial width subequal to distal width, and molars with high flare, close molar cusp proximity relative to crown length and short central basins) (Benefit, 1993). Comparison to hominoids and outgroups which predate the cercopithecoid–hominoid divergence (parapithecids, pro-pliopithecids and pliopithecids), indicates that features shared exclusively between *Victoriapithecus* and one extant subfamily are either primitive or phylogenetically equivocal. In comparison to *Victoriapithecus*, the Colobinae and Cercopithecinae share complete loss of the crista obliqua and development of a new crest between the hypocone and metacone to form consistently bilophodont upper molars, loss of M_1–M_2 hypoconulids and a realignment of the long axis of P_4 such that it is parallel rather than oblique to the molar row (von Koenigswald, 1969; Leakey, 1985; Benefit & Pickford, 1986; Benefit, 1987, 1993). Based on its dentition, *Victoriapithecus* is interpreted as representing a distinct family Victoriapithecidae from its sister taxon the Cercopithecidae (von Koenigswald, 1969; Leakey, 1985; Benefit & Pickford, 1986; Benefit, 1987, 1993).

More recently, excavation and wet-screening in the two fossiliferous strata near Leakey's Trench at Maboko Main were carried out by Brenda Benefit and Monte L. McCrossin during eight field seasons between 1987 and 1997. The excavations led to the collection of 840 *in situ* *Victoriapithecus* specimens from Bed 3, and 1939 specimens from Bed 5, with numerous others collected from surface exposures (Benefit, 1999*a*). Within each unit *Victoriapithecus* is the most common mammal. The new collection of specimens includes a complete cranium and several craniofacial fragments (Benefit & McCrossin, 1991, 1993*a*, 1997*a*), complete male and female mandibles (Benefit, 1999*b*), the full set of permanent and deciduous dentition (Benefit, 1993, 1994), complete humeri, radii and femora, and all elements of the postcrania except the trapezoid, os centrale and portions of the axial skeleton (Strasser, 1997; McCrossin *et al.*, 1998). Consequently, *Victoriapithecus* is one of the more completely documented fossil primates.

Cranial and deciduous dentition morphology confirms interpretation of *Victoriapithecus* as representing a family distinct from Cercopithecidae. The skull differs from that of extant cercopithecoids but resembles the basal catarrhine *Aegyptopithecus* and many Miocene hominoid genera in having supraorbital costae, a trigon formed in part by the anterior convergence of temporalis lines, orbits that are taller than wide, and a deep malar region of the zygomatic (Benefit & McCrossin, 1991, 1993*a*, 1997*a*). *Victoriapithecus* deciduous teeth differ from those of Cercopithecidae and share with hominoids and other non-cercopithecoid catarrhines: upper deciduous premolars that are wide relative to length; presence of crista obliqua on 87% of dP^4s and hypoconulids on 51% of dP_4s, lack of distal lophid between entoconid and hypoconid on 30% of dP_4s, absence of transverse distal lophs between the hypocone and metacone of dP^3–dP^4 (Benefit, 1994). Uniquely among catarrhines, the dP^3 is truly intermediate between that of apes and monkeys, with an obliquely oriented metacone crest that terminates in the trigon basin and intersects no other cusps. The dP^4 is unique in having a prehypocrista that intersects with the crista obliqua and a small cuspule that occurs at the point of intersection.

As progress was being made regarding *Victoriapithecus* from Maboko, seven cercopithecoid partial mandibles and maxillae as well as eight isolated teeth and a pisiform of an unnamed species of *Prohylobates* were being recovered from the 17-Ma early Miocene site of Buluk in northern Kenya (Leakey, 1985). If the generic attribution is correct, this sample of *Prohylobates* fossils includes the first upper molars and unworn lower molars known for the genus. Like *Victoriapithecus*, *Prohylobates* sp. from Buluk differs from colobine and cercopithecine monkeys in having M^{1-2} that retain crista obliqua and lack hypometacone crests, an M^3 that approaches the bilophodont pattern, P_4 long axis oriented oblique to that of the molar row, and completely bilophodont M_{1-2} which retain hypoconulids (Leakey, 1985). Given that *Prohylobates* shares several traits with *Victoriapithecus* to the exclusion of both Colobinae and Cercopithecinae, Leakey (1985) assigned the former genus to Victoriapithecidae. These species are maintained in *Prohylobates* unless further fossil evidence proves otherwise. Until more is known about the anatomy of the three species, description of the skull, postcranial anatomy and deciduous dentition of Victoriapithecidae is based largely on *Victoriapithecus macinnesi*.

Taxonomy

Systematic framework

Order Primates Linnaeus, 1758
- Infraorder Catarrhini É. Geoffroy Saint-Hilaire, 1812
  - Superfamily Cercopithecoidea Gray, 1821
    - Family Victoriapithecidae von Koenigswald, 1969
      - Genus *Victoriapithecus* von Koenigswald, 1969
        - *Victoriapithecus macinnesi* von Koenigswald, 1969
      - Genus *Prohylobates* Fourtau, 1918
        - *Prohylobates tandyi* Fourtau, 1918
        - *Prohylobates simonsi* Delson, 1979
        - *Prohylobates* sp. Leakey, 1985

Family Victoriapithecidae

Victoriapithecidae is an extinct family of Old World monkeys which shares the following derived features with other Cercopithecoidea: absence of a maxillary sinus; high mandibular genial pit; upper male canines with a deep sulcus continuing onto the root; upper molars that are nearly as long as wide and which have an expanded hypocone that directly opposes and is nearly as large as the metacone; tall and sexually dimorphic extension of enamel onto the mesiobuccal root of P_3; high frequency of bilophodont lower molars and deciduous premolars; constriction of molar width between mesial and distal aspects of the crown; strong medial trochlear keel on the distal humerus; and ischial callosities (von Koenigswald, 1969; Simons, 1969; Strasser & Delson, 1987; Benefit, 1987; Harrison, 1987, 1989*c*; Benefit & McCrossin, 1991, 1993*b*; McCrossin & Benefit, 1992).

The skull of Victoriapithecidae differs from cercopithecids in having: a lower neuro-cranium relative to length and width (unknown for Miocene apes); more airorhynchous hafting of the neurocranium and face with less flexed basicranium and longer midcranial region as measured from postglenoid process to M^3 (unknown for Miocene apes); steep and linear facial profile; frontal trigon formed in part by the anterior convergence of temporal lines and presence of supraorbital costae; taller than wide orbits; orbits and zygomatic that are angled dorsally relative to the Frankfurt plane, so that the top of the orbit is posterior to its inferior margin instead of perpendicular to the anatomical plane; deep malar region of the zygomatic; and shallow palate (Benefit & McCrossin, 1991, 1993a, 1997). In these cranial features victoriapithecids closely resemble the basal catarrhine *Aegyptopithecus* (Simons, 1987) and many Miocene hominoid genera, especially *Afropithecus* (Leakey & Leakey, 1986*a*; Leakey *et al.*, 1991; Leakey & Walker, 1997), *Sivapithecus* (Pilbeam, 1982), and to some extent *Dryopithecus* (Moyà-Solà & Köhler, 1993*b*). Additional cranial differences between victoriapithecids and modern cercopithecids include a mandibular coronoid process that extends well above the condyle (shared with *Aegyptopithecus* and some cercopithecoids), a longer than wide neuro-cranium (shared with the fossil colobine *Libypithecus*), and pronounced sagittal and nuchal crests (shared with *Libypithecus* and fossil *Theropithecus*). These features may have been primitive for cercopithecoids, but the limited distribution of the former two traits, and potential homoplasy of the latter, make this hypothesis difficult to test.

Dentally, victoriapithecid teeth differ from those of cercopithecids in the variable occurrence of crista obliqua and distal lophs on dP^4–M^3; upper and lower M^2 and M^3 crowns that are highly flared inferiorly with occasional bulging of the crown below the base of the median buccal cleft on M_2 and M_3, and medial lingual cleft on M^2 and M^3; molars with lower crown relief and lower shearing potential relative to length; upper molars and deciduous premolars that are wider than long; dP^3 metacone set mesial to a very small hypocone; an obliquely oriented metacone crest terminating in the trigon basin; dP^3 trigon continuous with the distal shelf; obliquely oriented P_4, such that its distal heel is set lingual to the anterior fovea and the lingual side of the crown is oriented oblique to that of the molars; variable presence of dP_4–M_2 hypoconulid and dP_4–M_1 distal lophid; unworn height of the M_3 entoconid lower relative to that of the metaconid than is observed among extant monkeys; and M_1 wider relative to length and hence more square than those of cercopithecids. Victoriapithecids share only two apparently derived dental features exclusively with Cercopithecinae (curved distal margins of I^1 and dI^1) and only one seemingly derived trait with Colobinae (high dI^2 crown height) (Benefit, 1993, 1994).

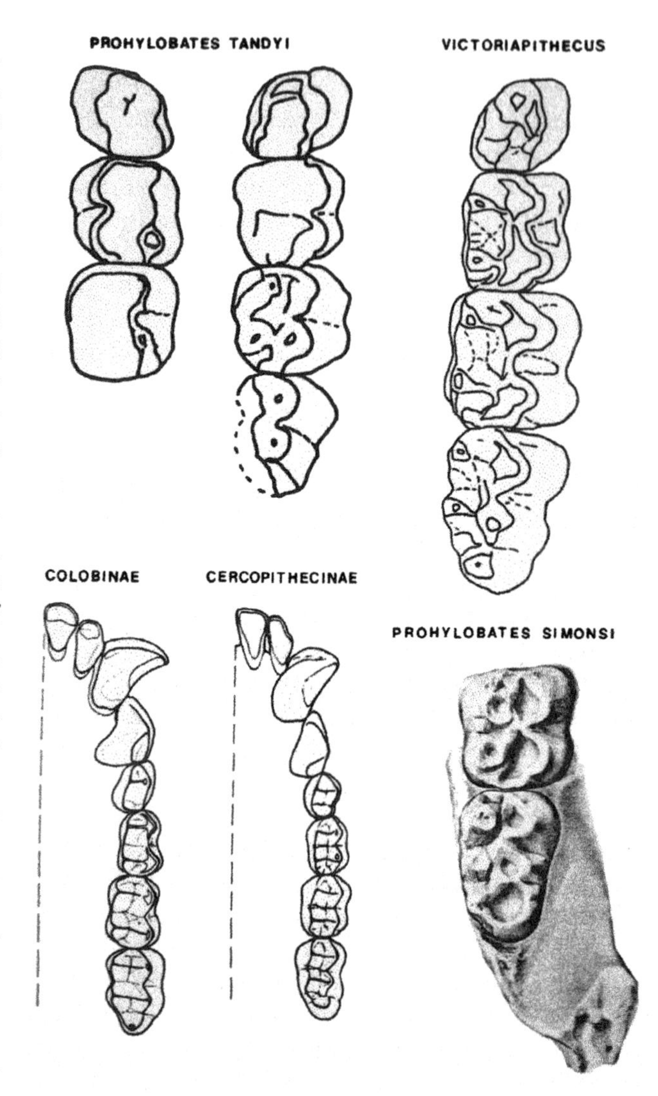

Fig. 16.1 *Prohylobates tandyi* and *Victoriapithecus macinnesi*. P_4–M_3s compared with those of extant monkeys.

In other aspects of their morphology, victoriapithecids incorporate a mosaic of several cercopithecine-like and some colobine-like features. Victoriapithecids are more cercopithecine-like in having a long and low cranial vault; posteriorly inclined orientation of the temporal muscles; low and relatively narrow nasal aperture which does not restrict the size of the I^1 roots; narrow interorbital septum; long and narrow nasal bones; moderate snout length; long and anteriorly tapering premaxilla; heteromorphic incisors; I^1 that is mesiodistally long, with a curved distal margin, long and curved root, and is procumbently implanted anterior to I^2; molar cusps that are more closely approximated in the mesiodistal and buccolingual direction resulting in greater molar flare, lower cusp relief, and shorter shear crests than those of extant cercopithecines; mesial molar width equal to distal width; humeral greater tubercle extending above the head; commonly retroflected humeral entepicondyle and ulnar olecranon process; short and stout phalanges; restricted range of big toe abduction (based on the morphology of the entocuneiform); small acetabulum and long ischial body (Harrison, 1989*c*; Benefit & McCrossin, 1991, 1993*a*, 1997*a*; McCrossin & Benefit, 1992; Benefit, 1993, 1994, 1999*a*, 2000). Features shared by victoriapithecids and colobine monkeys include a wide palate; anteroposteriorly wide and vertically oriented mandibular ramus; taller dI^2 crown; and greater dP_4 crown relief (Benefit & McCrossin, 1991, 1993*a*, 1997*a*; Benefit, 1993, 1994, 1999*a*, 1999*b*, 2000).

GENUS *Victoriapithecus*

Victoriapithecus is a genus of Victoriapithecidae distinguished from *Prohylobates* (Fig. 16.1) by the presence of a cercopithecine-like inferior transverse torus on most specimens; smaller P^4 relative to molar size; P_3 metaconid; relative molar proportions such that in general $M^1 < M^2 > M^3$ and $M_1 < M_2 < M_3$ in size; squarish M_1 that is smaller relative to M_2 and M_3 than that of other

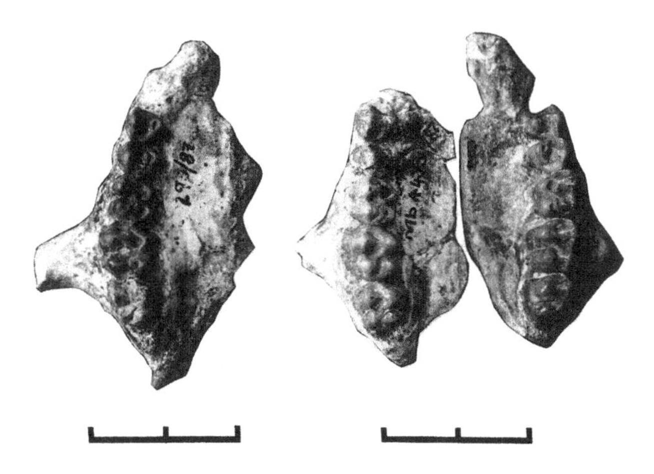

Fig. 16.2 *Victoriapithecus macinnesi*. Maxillae KNM-MB 18995 (left) and KNM-MB 18994 (right) showing crista obliqua on all upper molars and the relative size and shape of M^{1-3}.

cercopithecoids; more elongated M_2 than those of *P. tandyi* and *P. simonsi*; lower M_2–M_3 crown height below the median lingual notch; and for M_{1-2}: invariably complete distal lophid, hypoconulid positioned closer to the buccal side of the crown, shorter mesial shelf length relative to crown length, and low metaconid height relative to crown length.

SPECIES *Victoriapithecus macinnesi* von Koenigswald, 1969 (Figs. 16.2–16.18)
TYPE SPECIMEN KNM-MB 1, partial mandible
AGE AND GEOGRAPHIC RANGE 19–12.5 Ma; Napak V Uganda, 19 Ma (Bishop, 1964; Pilbeam & Walker, 1968; Pickford *et al.*, 1986*a*); Loperot, 17.5–15.8 Ma (Baker *et al.*, 1971); Maboko Formation, Kenya, 14.8–16 Ma; Ombo, middle Miocene (Pickford, 1981); Kipsaramon, middle Miocene (Hill, 1995); Nachola, middle Miocene (Ishida *et*

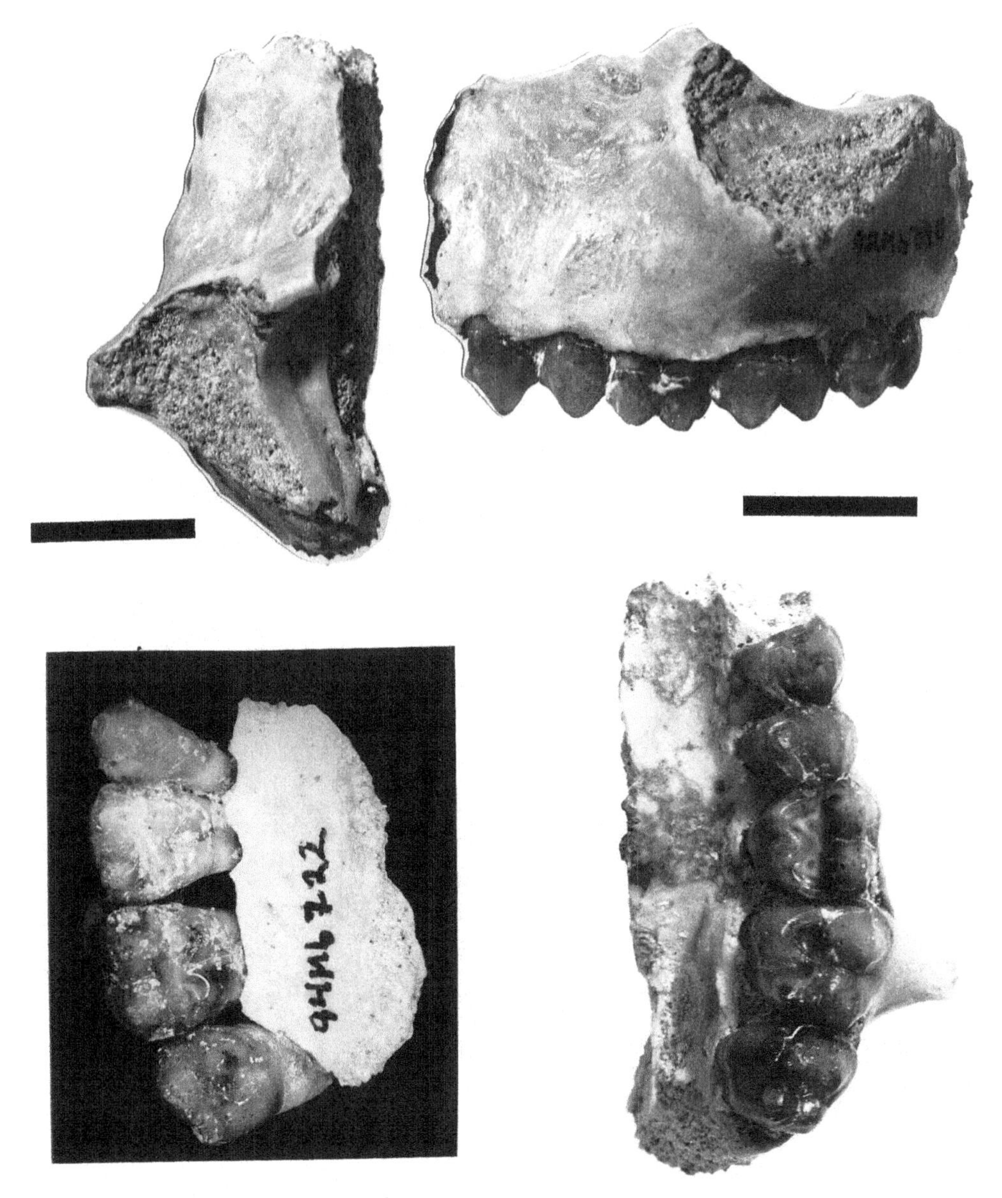

Fig. 16.3 *Victoriapithecus macinnesi*. Left maxilla 92 Mb 395 from superior (upper left), lateral (upper right); and palatal view (lower right): and right maxilla 94 Mb 722 (lower left). Note that M^3s of both specimens lack crista obliqua.

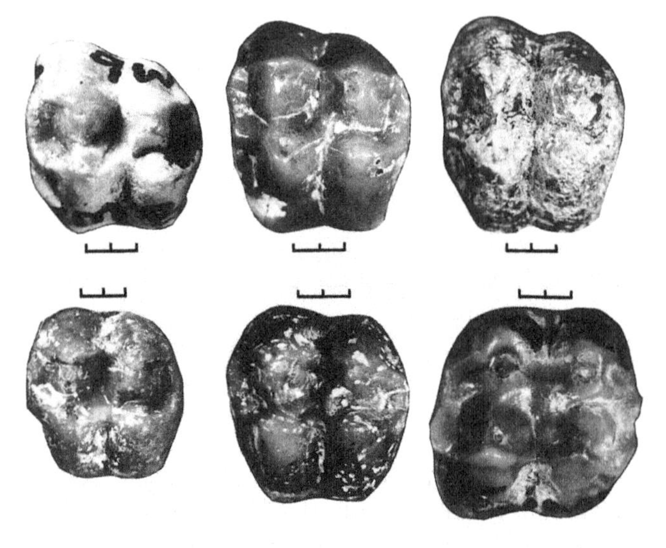

Fig. 16.4 *Victoriapithecus macinnesi*. Left M^2s of KNM-MB 11741 (upper left), KNM-MB 319 (upper center), 11887 (lower left) and 18724 (lower center) showing variation in crista obliqua development; left M^2 of KNM-MB 14308 (upper right) with neither a crista obliqua nor a distal loph; and right M^2 of KNM-MB 320 (lower right) showing a distal trigon wall formed by crista obliqua merging smoothly with the floor of the basin. For all teeth, mesial is to the left.

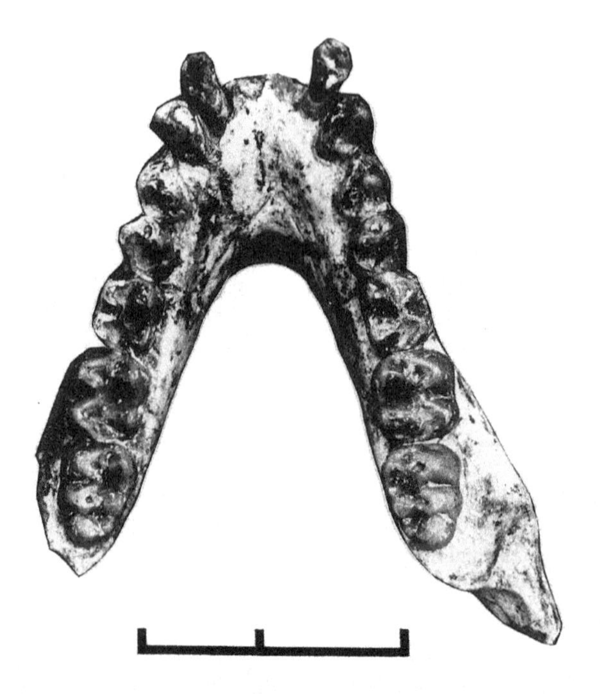

Fig. 16.5 *Victoriapithecus macinnesi*. Female mandible KNM-MB 18993 showing the oblique orientation of the long axis of P_4 and the relative size and shape of lower molars.

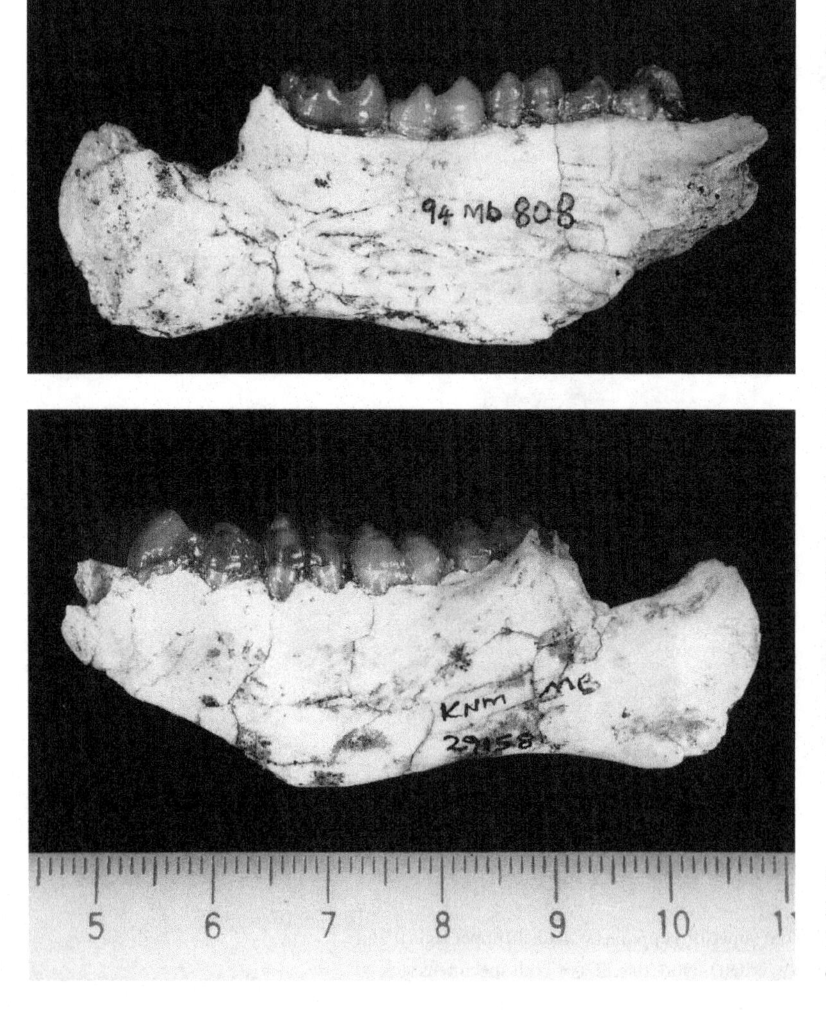

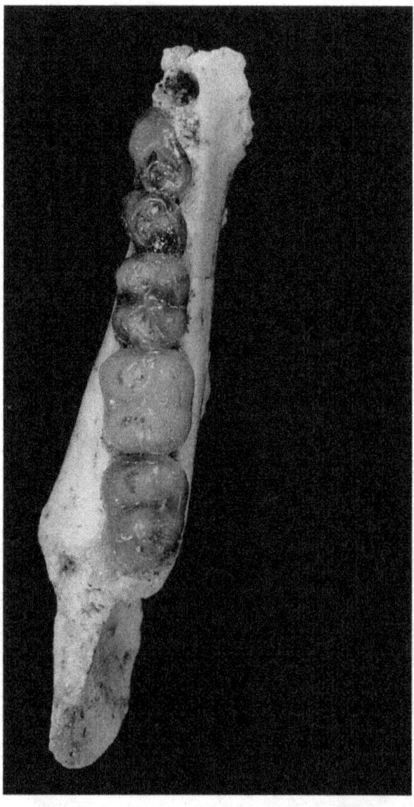

Fig. 16.6 *Victoriapithecus macinnesi*. Mandible KNM-MB 29158 from lingual (upper left), buccal (lower left) and occlusal views. Scale in cm.

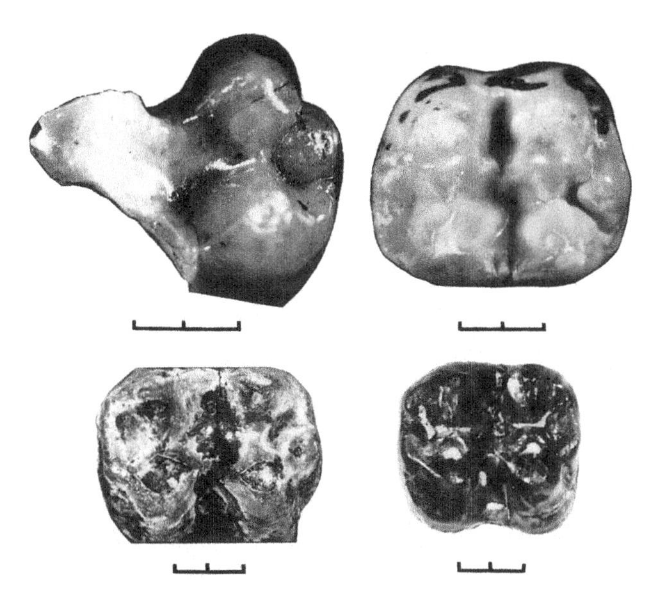

Fig. 16.7 *Victoriapithecus macinnesi*. Occlusal views (mesial to left) of some lower first molars with hypoconulids. Upper row: unworn right M_1s of KNM-MB 310 (left) and KNM-MB 324 (right). Lower row: worn left M_1s of KNM-MB 11674 (left) and KNM-MB 20765 (right).

al., 1984); Ngorora site BPRP 38 (Nakali) in the Tugen Hills Sequence, Kenya, 12.5 Ma (Hill, 1995); and probably Ongoliba, Zaïre, middle Miocene (Hooijer, 1963)

ANATOMICAL DEFINITION

As for genus.

GENUS *Prohylobates* Fourtau, 1918

Prohylobates differs from *Victoriapithecus* in having lower molars with taller metaconids, taller crown height below the median lingual notch, longer mesial shelves relative to crown length, and more centrally positioned M_{1-2} hypoconulids (Benefit 1993).

INCLUDED SPECIES *P. simonsi*, *P. tandyi*, *Prohylobates* sp. nov.

SPECIES *Prohylobates tandyi* Fourtau, 1918 (Fig. 16.1)

TYPE SPECIMEN CGM 30936, right mandible fragment

AGE AND GEOGRAPHIC RANGE Middle Miocene deposits, Wadi Moghara, Egypt

ANATOMICAL DEFINITION

Prohylobates tandyi differs from other species of *Prohylobates* in having a large P_4 relative to molar size, and a very small M_3 relative to M_2 and equal to M_1 in size, a condition that is highly unusual among cercopithecoids which retain hypoconulids.

SPECIES *Prohylobates simonsi* Delson, 1979

AGE AND GEOGRAPHIC RANGE Middle Miocene deposits, Gebel Zelten, Libya

TYPE SPECIMEN AMNH 17768, partial mandible

ANATOMICAL DEFINITION

Prohylobates simonsi differs from other species of *Prohylobates* in having an M_2 that is as large as M_3.

SPECIES *Prohylobates* sp. nov. Leakey, 1985

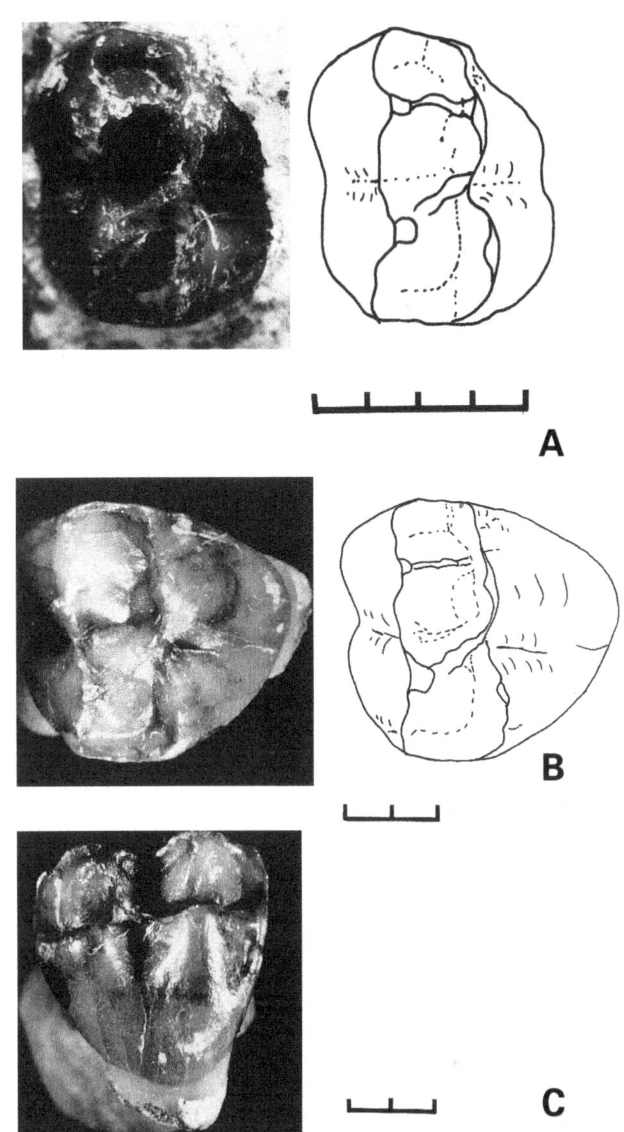

Fig. 16.8 *Victoriapithecus macinnesi*. Right dP^4s: (A) occlusal view of KNM-MB 18742 (mesial oriented toward top); (B) occlusal view of KNM-MB 21047 (mesial oriented toward top): (C) lingual view of KNM-MB 21047 (mesial oriented toward right). Scale in mm.

AGE AND GEOGRAPHIC RANGE 17 Ma, Buluk, Kenya

ANATOMICAL DEFINITION

Differs from other species of *Prohylobates* in having a larger M_1 relative to other molars, and from *Victoriapithecus* in having a larger P^4 and M^3 relative to other molars (Leakey, 1985; Benefit, 1993; Fleagle *et al.*, 1997*a*).

Evolution of Victoriapithecidae

The victoriapithecids provide important information from which to understand the cercopithecoid–hominoid split which took place approximately 25 million years ago (Fig. 16.19). Of these two catarrhine superfamilies, the

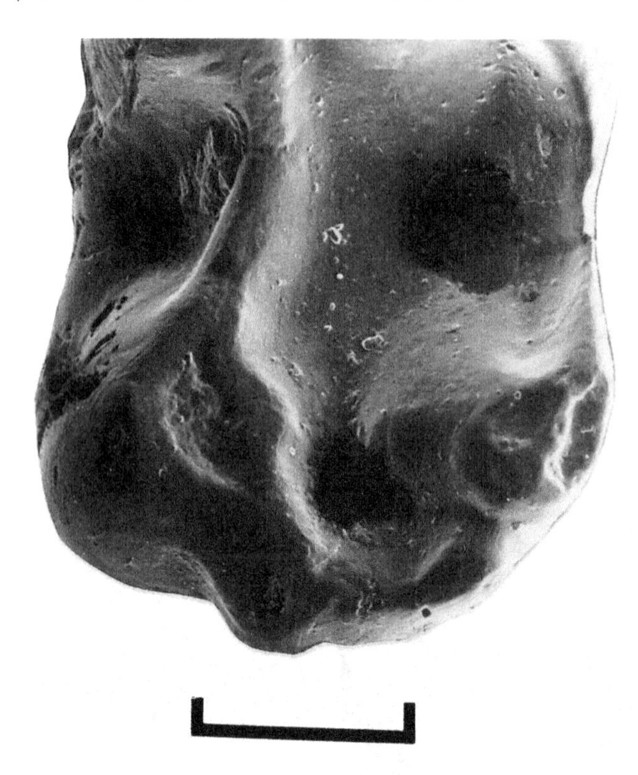

Fig. 16.9 *Victoriapithecus macinnesi*. Distal half of left dP_4 of KNM-MB 21021 (mesial oriented toward top) showing lack of a distinct distal lophid. Scale bar = 1 mm.

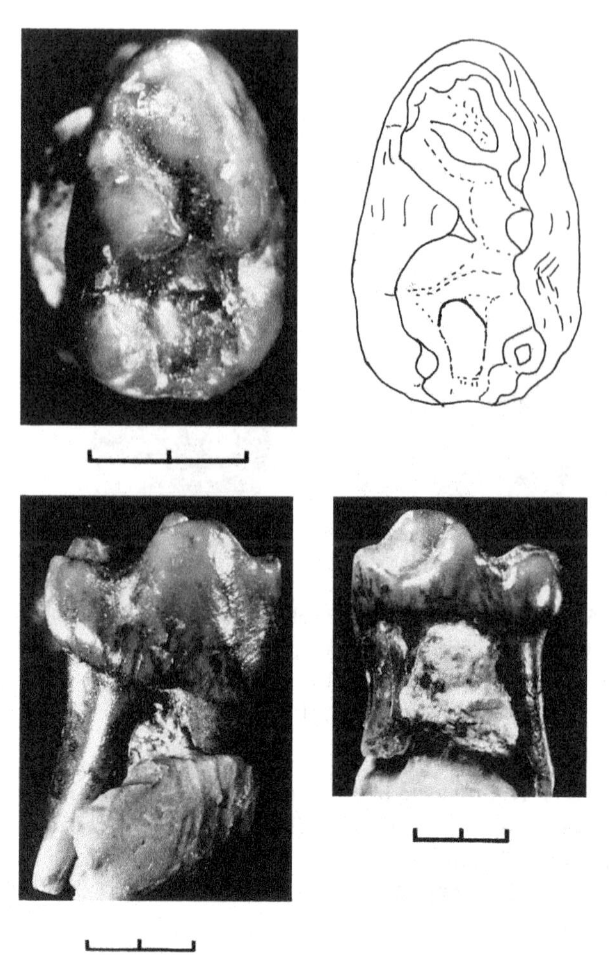

Fig. 16.11 *Victoriapithecus macinnesi*. Right dP_3 of KNM-MB 20839 from occlusal view (top) with mesial oriented at top, labial view (lower left) with mesial oriented toward right, and lingual view (lower right) with mesial oriented toward left. Scales in mm.

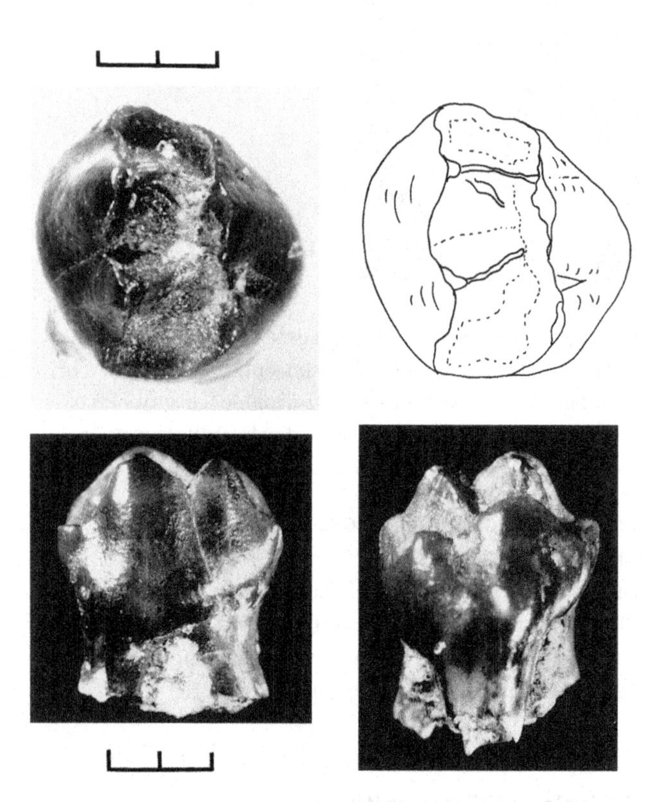

Fig. 16.10 *Victoriapithecus macinnesi*. Right dP^3 of KNM-MB 18809 from occlusal view (top) with mesial oriented at top, labial view (lower left) with mesial oriented toward left, and lingual view (lower right) with mesial oriented toward right.

cercopithecoids display highly derived dental and postcranial morphologies and adaptations. Aside from lacking an entepicondylar foramen on the distal humerus (a catarrhine feature shared with cercopithecoids), Miocene ape postcrania are sometimes so similar to those of generalized arboreal quadrupedal platyrrhines and propliopithecids, that their hominoid status has been questioned (Harrison, 1987; Rose *et al.*, 1992; McCrossin & Benefit, 1994; Benefit & McCrossin, 1995). Victoriapithecids and cercopithecids differ from these generalized arboreal quadrupeds in having a narrow distal humerus, well-defined medial trochlear keel, posteriorly rather than medially oriented medial epicondyle, well-defined ischial callosities, and restricted hip and ankle joints among other features (Napier & Davis, 1959; Jolly, 1967; Rose, 1983; Harrison, 1989c; McCrossin & Benefit, 1992, 1994). The more limited elbow, hip and ankle joint flexibility of Old World monkeys provided them with enhanced running abilities, which are particularly useful on the ground. In addition, the short and stout phalanges and proximal humerus morphology indicate that early monkeys were as terrestrial as modern vervets (von Koenigswald,

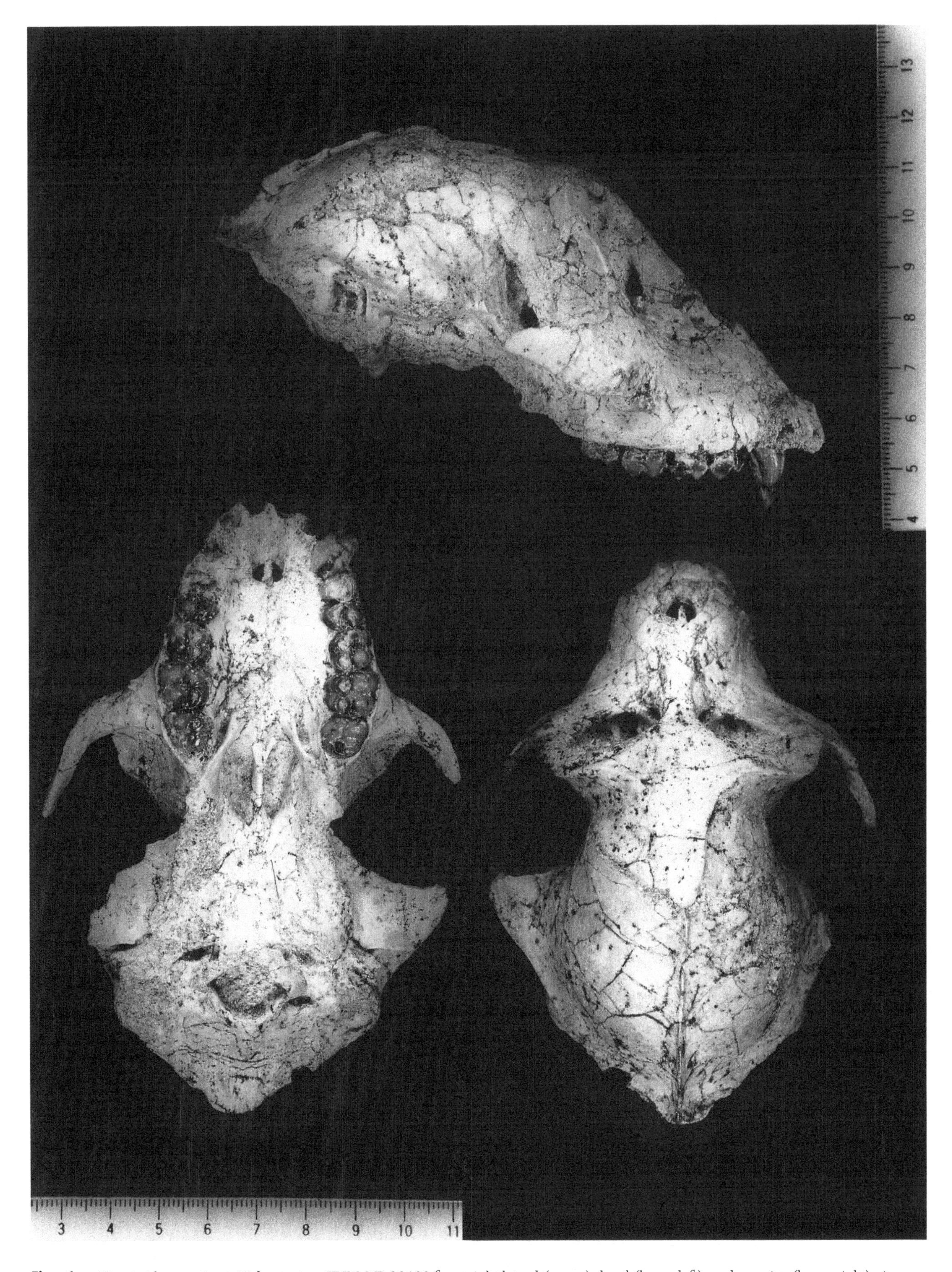

Fig. 16.12 *Victoriapithecus macinnesi*. Male cranium KNM-MB 29100 from right lateral (upper), basal (lower left), and superior (lower right) views.

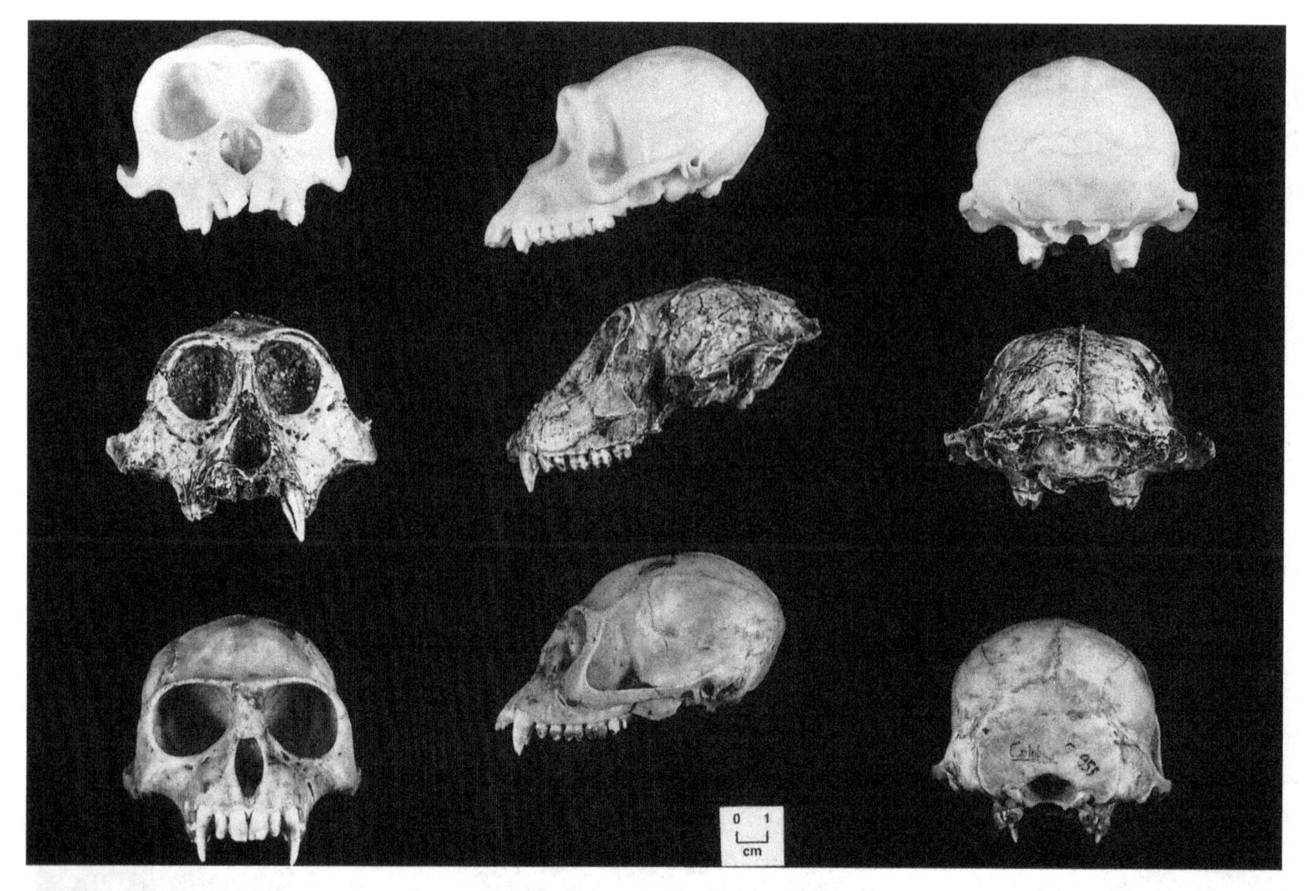

Fig. 16.13 *Victoriapithecus macinnesi*. Male cranium KNM-MB 29100 (center), compared with female *Macaca fascicularis* (above) and male *Colobus guereza* (below), facial view (left), left lateral view (center) and posterior view (right).

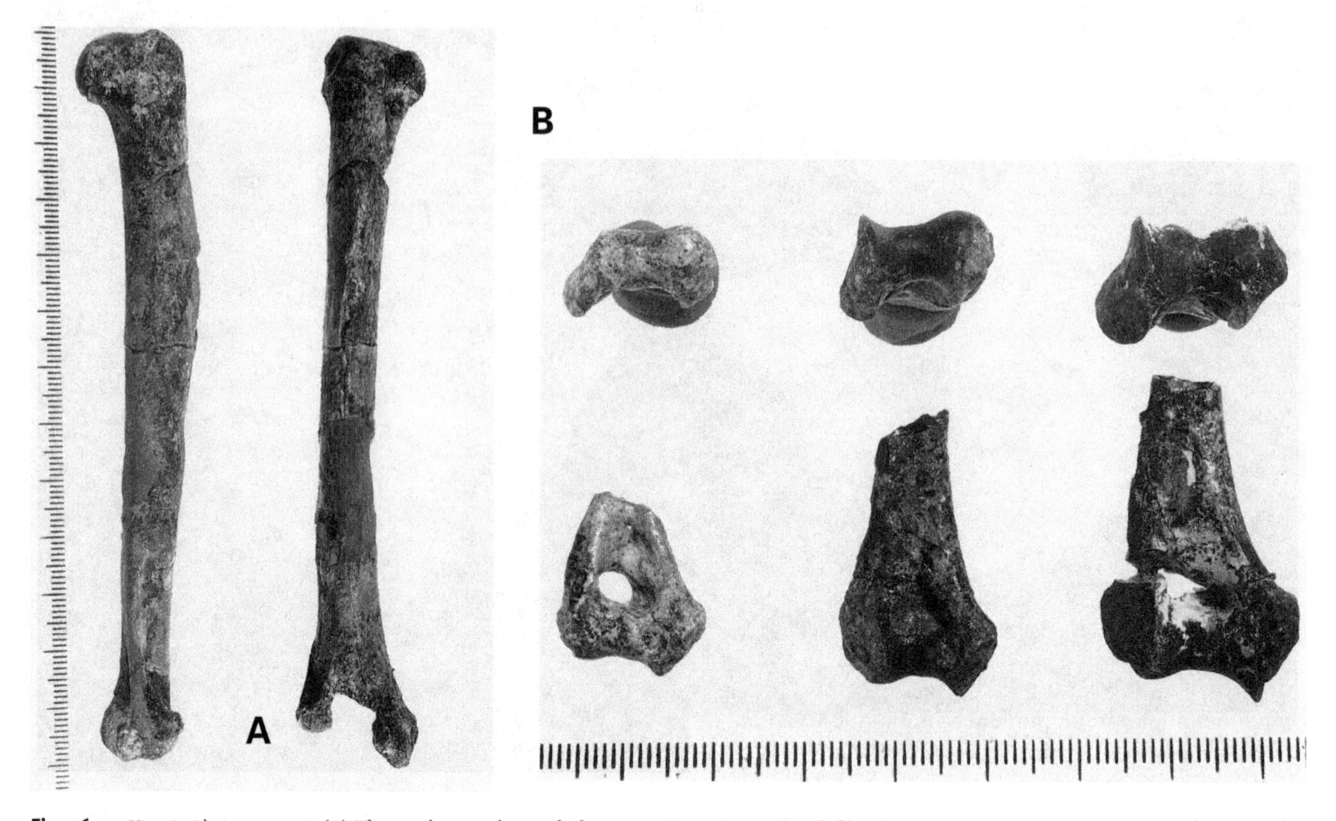

Fig. 16.14 *Victoriapithecus macinnesi*. (A) The nearly complete right humerus KNM-MB 21809 (left) in lateral and anterior views shows the greater tubercle extending above the articular surfaces as in semiterrestrial monkeys. (B) Distal and posterior views of distal humeri KNM-MB 3 (left column), KNM-MB 19 (center column) and 87 Mb 3947 (right column) show the strong medial trochlear keel characteristic of cercopithecoids and the posterior orientation of the medial epicondyle characteristic of semiterrestrial monkeys. Scales in mm.

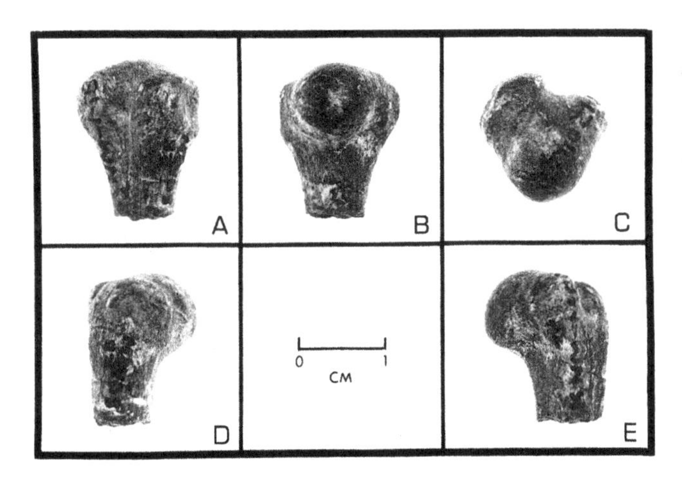

Fig. 16.15 *Victoriapithecus macinnesi*. Isolated left proximal humerus (KNM-MB 12044) in anterior (A), posterior (B), proximal (C), lateral (D) and medial (E) views.

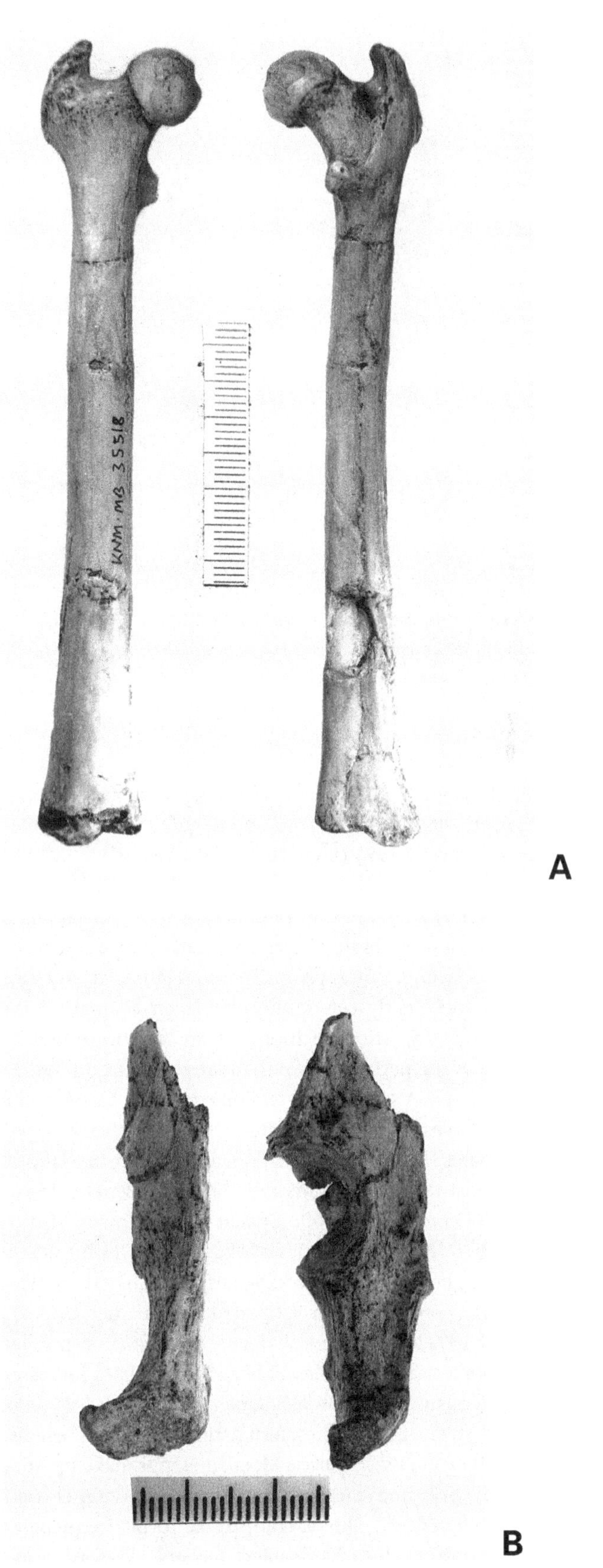

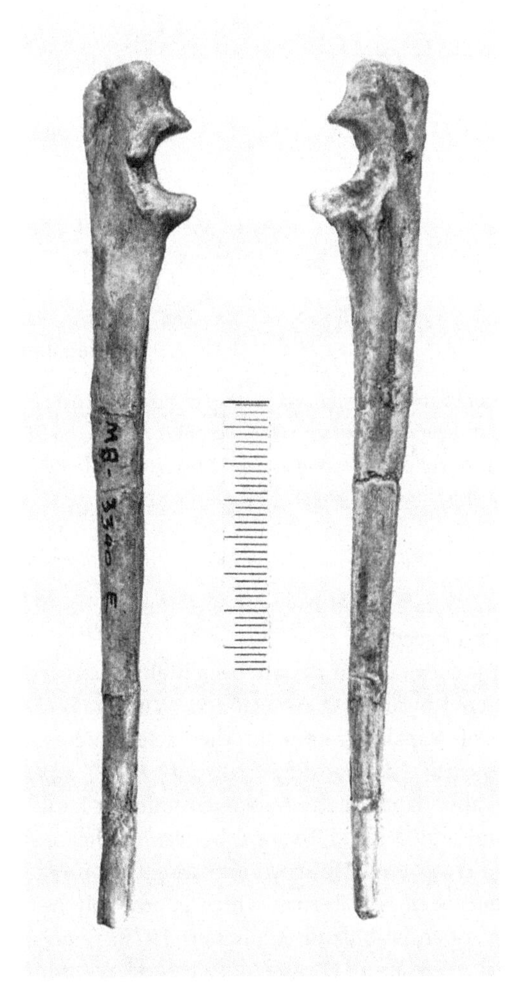

Fig. 16.16 *Victoriapithecus macinnesi*. Nearly complete left ulna from medial (left) and lateral (right) views.

Fig. 16.17 *Victoriapithecus macinnesi*. (A) Nearly complete femur KNM-MB 35518 from anterior (left) and posterior (right) views; (B) left ischium KNM-MB 25329 from dorsal (left) and lateral (right) views. Scales in mm.

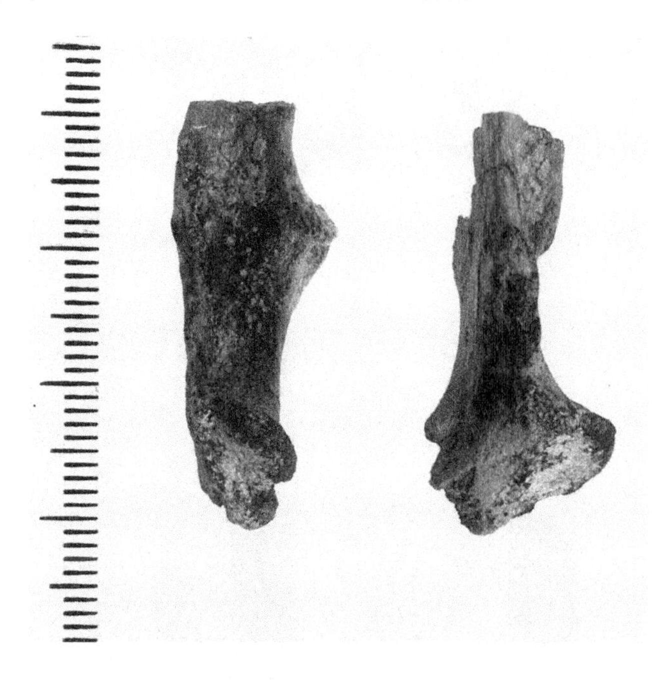

Fig. 16.18 *Victoriapithecus macinnesi*. Ischium KNM-MB 20229 in lateral (left) and posterior (right) views showing its length and presence of ischial callosities. Scale in cm.

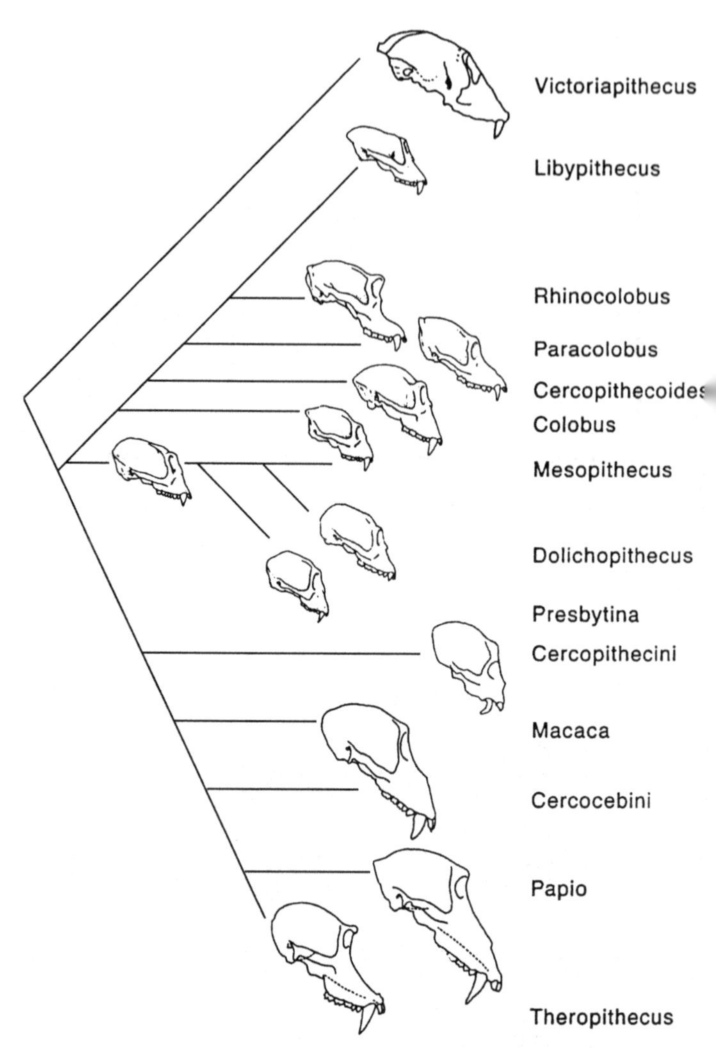

Fig. 16.19 Phylogenetic tree showing relationship between selected Miocene, Plio-Pleistocene and modern cercopithecoids based on cranial morphology. Drawings of colobine crania adapted from Delson (1994), and of cercopithecine crania from Vogel (1968).

1969; Delson, 1975*a*; Senut, 1986; Harrison, 1989*c*; McCrossin, 1995; McCrossin & Benefit, 1992, 1994; McCrossin *et al.*, 1998) (Figs. 16.14–16.18).

Dentally, Miocene apes are much more conservative than victoriapithecids. While the bilophodonty of *Victoriapithecus* and *Prohylobates* is highly derived relative to molars of pliopithecids and other primitive catarrhines, it did not evolve in these early cercopithecoids as an adaptation for folivory. Instead, Miocene monkeys are as committed to frugivory as, or perhaps more than, any extant Old World monkey or ape. A series of regression equations based on 11 dental variables strongly correlated with diet in extant species estimates the annual diet of *Victoriapithecus* from Maboko to have consisted of 79% fruits and 7% leaves (Benefit, 1986, 1987, 1999*a*, 1999*c*, 2000; Benefit & McCrossin, 1990; McCrossin & Benefit, 1994). *Prohylobates* from Buluk has a slightly higher shearing coefficient (−4.6) than *Victoriapithecus* (−6.9) and an estimated annual diet of 74% fruits and 13% leaves. The low degree of lingual cusp relief on *Prohylobates tandyi* and *P. simonsi* from North Africa indicates that they consumed about 84% fruits annually and were as highly frugivorous as their eastern African relatives (Benefit, 1987, 1993). On *Victoriapithecus* M_2 enamel, pits make up 46% of phase II (grinding) facet microwear features (Palmer *et al.*, 1998, 1999, 2000). This is comparable to pit frequencies found among modern hard-object feeders (Teaford *et al.*, 1996). In addition, striations on *Victoriapithecus* non-occlusal molar enamel (Ungar & Teaford, 1996), and of bowl-shaped depressions on the worn molar cusp tips of the victoriapithecines (a wear pattern found only among extreme frugivores), corroborate reconstruction of its diet as consisting of hard fruits or seeds.

The earliest monkeys appear to have been highly restricted in the range of environments they occupied. With the possible exception of Napak V, none of the victoriapithecid localities is reconstructed as rainforest (Bishop, 1968). Apes are diverse and abundant at early Miocene rainforest localities such as Songhor, Koru and Rusinga, but cercopithecoids are conspicuously absent. Monkeys are absent from the middle Miocene site of Fort Ternan which is probably best reconstructed as open woodland (Churcher, 1970; Gentry, 1970; Evans *et al.*, 1981), but also seems to have elements of savanna (Shipman *et al.*, 1981; Retallack, 1992) and forest (Pickford, 1983; Cerling *et al.*, 1991).

Of the diverse range of Miocene ape species, few coexisted with *Victoriapithecus* and *Prohylobates*. The only site at which victoriapithecids might occur with *Proconsul* (specifically *P. major*) is Napak V (Pilbeam & Walker, 1968). At Buluk (17

Ma) *Prohylobates* occurs with *Afropithecus* (Leakey, 1985; Leakey & Leakey, 1986*a*), and *Victoriapithecus* is found with *Kenyapithecus* at Maboko Beds 3 and 5, Kipsaramon and Nachola (von Koenigswald, 1969; Ishida *et al.*, 1984; Pickford, 1987; Benefit, 1993; Hill, 1995). Like victoriapithecids, *Afropithecus* and *Kenyapithecus* probably fed upon hard fruits and seeds, using their anterior teeth to prepare tough foods (McCrossin & Benefit, 1993*a*, 1993*b*, 1994, 1997; Benefit & McCrossin, 1995; Leakey & Walker, 1997). Microwear on *Kenyapithecus* molars clearly falls out with modern hard object feeders in percentage of pits and pit width. However, *Kenyapithecus* has much wider pits (18.2 μm) than *Victoriapithecus* (13.3 μm), as well as a higher percentage of pits (61% versus 46%), and may have had a more specialized diet (Palmer *et al.*, 1998, 1999, 2000). Whereas *Afropithecus* was apparently an arboreal quadruped (Leakey *et al.*, 1988*a*; Rose, 1993), *Kenyapithecus* was the first, and so far only, Miocene ape recognized to have a semiterrestrial pattern of locomotion, similar to that of *Mandrillus* and *Cercocebus* (McCrossin, 1994*a*, 1994*b*, 1995, 1996; McCrossin & Benefit, 1994; Benefit & McCrossin, 1995).

Hard-object feeding in combination with semiterrestriality seems to have allowed *Kenyapithecus* and *Victoriapithecus* to cope with the changing vegetation of the eastern African middle Miocene by exploiting a disturbed mosaic of seasonally flooded and wooded environments where food sources were abundant closer to the ground and seasonally unpredictable (Benefit, 1987; McCrossin, 1994*a*; McCrossin & Benefit, 1997). It would also have permitted them to exploit fallen fruits (McCrossin, 1994*a*), a resource that is crucial to allowing some semiterrestrial primates, such as the ring-tailed lemur, to survive the dry season (Sussman, 1977). Ecologically, *Victoriapithecus* strongly resembles cercopithecine lineages that replaced them, and appears to have been preadapted to the increasingly more seasonal climate and open habitats of the Plio-Pleistocene. Why they did not disperse into Eurasia during the middle Miocene, and why they went extinct approximately 12.5 million years ago, remain a mystery.

Primary References

Prohylobates

Simons, E. L. (1969). Miocene monkey (*Prohylobates*) from northern Egypt. *Nature*, **223**, 687–689.

Delson, E. (1979). *Prohylobates* (Primates) from the Early Miocene of Libya: a new species and its implications for cercopithecid origins. *Geobios*, **12**, 725–733.

Leakey, M. G. (1985). Early Miocene cercopithecids from Buluk, Northern Kenya. *Folia Primatologica*, **44**, 1–14.

Victoriapithecus

Koenigswald, G. H. R. von (1969). Miocene Cercopithecoidea and Oreopithecoidea from the Miocene of East Africa. In *Fossil Vertebrates of Africa*, vol. 1, ed. L. S. B. Leakey, pp. 39–51. London: Academic Press.

Senut, B. (1986). Nouvelles découvertes de restes postcraniens de primates (Hominoidea et Cercopithecoidea) sur le site Maboko au Kenya occidental. *Comptes rendus de l'Académie des science, de Paris*, **303**, 1359–1362.

Benefit, B. R. (1987). The molar morphology, natural history, and phylogenetic position of the middle Miocene monkey *Victoriapithecus*. PhD dissertation, New York University.

Harrison, T. (1989*c*). New postcranial remains of *Victoriapithecus* from the middle Miocene of Kenya. *Journal of Human Evolution*, **18**, 3–54.

Benefit, B. R. & McCrossin, M. L. (1991). Ancestral facial morphology of Old World higher primates. *Proceedings of the National Academy of Sciences of the United States of America*, **88**, 5267–5271.

Benefit, B. R. (1993). The permanent dentition and phylogenetic position of *Victoriapithecus* from Maboko Island, Kenya. *Journal of Human Evolution*, **25**, 83–172.

Benefit, B. R. & McCrossin, M. L. (1993). The facial anatomy of *Victoriapithecus* and its relevance to the ancestral cranial morphology of Old World monkeys and apes. *American Journal of Physical Anthropology*, **92**, 329–370.

Benefit, B. R. (1994). Phylogenetic, paleodemographic and taphonomic implications of *Victoriapithecus* deciduous teeth from Maboko, Kenya. *American Journal of Physical Anthropology*, **95**, 277–331.

Benefit, B. R. & McCrossin, M. L. (1997). Earliest known Old World monkey skull. *Nature*, **388**, 368–371.

17 | Fossil Old World monkeys: The late Neogene radiation

NINA G. JABLONSKI

Introduction

By the early and middle Miocene, the fossil record shows that the Old World monkeys were well established, but apparently not taxonomically diverse. Beginning in the late Miocene, the variety and geographical distribution of cercopithecoid taxa increased dramatically, peaking in the late Pliocene and early Pleistocene. The recency of most of the Old World monkey radiation denies the long, independent evolutionary branches necessary to clearly demarcate subgroups. This has created difficulties in interpreting the evolutionary relationships within the superfamily from fossil evidence. For the Old World monkeys, skeletal and dental morphology is conservative. Observed variations often are unique to species and therefore not taxonomically informative. The wealth of Old World monkey fossils has told us more about the niches and adaptations of cercopithecoid species, and about ancient primate communities, than it has about evolutionary relationships among species.

History of discovery and debate

The 1830s were years of tremendous activity and active exploration in many areas of natural history, and especially the nascent fields of geology and paleontology. The first fossil primate of any kind to be collected and reported in the scientific literature was, in fact, that of an Old World monkey. This specimen, the maxilla of a large macaque-like monkey, was recovered from Siwalik deposits during the construction of the Doab Canal, near the town of Madhopur, on the border of the states of Himachal Pradesh and Jammu and Kashmir in northwestern India. Two British army engineers working in India, William E. Baker and Sir Henry Marion Durand, reported the fossil, now generally classified as *Procynocephalus subhimalayanus*, in 1836. Their paper does not include the naming of the fossil, but rather is occupied with speculation as to why the remains of fossil monkeys were so rare as compared to those of other animals. Soon thereafter, an astragalus of a fossil Old World monkey from the Siwaliks came to light and was reported rather more colorfully by Major Proby Thomas Cautley and Dr. Hugh Falconer of the Indian Civil Service in Bengal (Cautley, 1837; Falconer & Cautley, 1837; Cautley & Falconer, 1840). These authors carefully compared the morphology and size of their find to the modern temple langur, *Semnopithecus entellus*, and concluded that the ancient monkey "agrees so closely in size and general form with the astragalus of the Entellus, that it probably belonged to the same sub-genus: still the points of difference are sufficient to leave no doubt, that the fossil must be assigned to a distinct species" (Cautley & Falconer, 1840). It is salutary to reflect on the fact that, over 150 years later, our insights into the classification of this find have progressed no further. Cautley & Falconer's report is remarkable in other respects, especially their observations on what we now refer to as taphonomic and paleoecological contexts of their ancient primate:

> If we refer to the remote epochs when the climate was suitable, and when genera now associated with the Monkeys were abundant, it is easy to conceive that the latter might have existed in numbers, without their remains being entombed. It requires in all instances many unconnected circumstances for the preservation of organic bodies, and their subsequent disclosure. Amongst the most important of these are the habits and organization of the animals themselves... A flood might suffocate in their dens, over a large tract of country, the burrowing tribes; and might sweep from under the feet of the monkey hundreds of its herbivorous and predaceous fellow-tenants of the forest, and bury them in the near shingle or far-distant estuary, or drown them and deposit them in the stagnant swamp – while he would remain secure. The tree on which he was perched might totter, and yield to the undermining current, and he still escape and feed on his wonted fruits, undisturbed by the destruction around. When the debt of nature comes to be paid, his carcase falls to the ground, and immediately becomes the prey of the numerous predaceous scavengers of torrid regions, the Hyæna, the Chacal, and the Wolf. So speedily does this occur, that in India, where Monkeys occupy, in large societies, mango groves around villages, unmolested and cherished by man, the traces of casualties among them are so rarely seen, that the simple Hindoo believes that they bury their dead by night. (Cautley & Falconer, 1840: 499–500)

Discoveries of Old World monkey fossils, mostly from Europe, were reported steadily for the balance of the nineteenth and beginning of the twentieth centuries. The first description of the extinct Eurasian colobid *Mesopithecus pentelicus* appeared in 1839 (Wagner, 1839), followed by those of fossil macaques by Sir Richard Owen (Owen, 1845) and Gervais (Gervais, 1859), and of the colobid *Dolichopithecus ruscinensis* by Charles Jean Julien Depéret in 1889 (Depéret, 1889). In these years, controversies in the literature were almost unheard of because there was little basis for alternate interpretations. The discovery of fossil monkeys in areas where they did not occur today – *Mesopithecus* in Greece and *Dolichopithecus* in France, for instance – posed bigger problems of interpretation, but did not invite extensive controversy. As pointed out by Fleagle & Hartwig (1997), the nineteenth century saw few real syntheses of the primate fossil record, because so few fossils and so little information about them were available. Schlosser's (1887) review of fossil primates within his treatise on fossil mammals is an exception and a landmark work in its detail and establishment of the first phylogenetic tree of fossil primates.

In the early twentieth century, discovery of Old World monkey fossils often occurred in the course of searching for other ancient materials, notably human fossils or artifacts. Although most paleoanthropologists think of early twentieth-century expeditions to Africa in this context, these efforts were matched at the same time in China. It was widely held that the Central Asian Plain could have been the Garden of Eden that nurtured the human species (Jia & Huang, 1990) and thus warranted extensive survey for traces of ancient humans and extinct civilizations. Working for the Geological Survey of China, a Swedish mining geologist, Johan Gunnar Andersson, secured support for paleontological and archaeological explorations in northern China. As a result, by the late 1910s and 1920s, significant discoveries of ancient macaques and their close relatives were made at Mianchi, Henan Province and at the Paleolithic cave site of Zhoukoudian, near Beijing (Yang, 1999). Andersson's teams at various times included Otto Zdansky and Pierre Teilhard de Chardin, who both contributed to the description of cercopithecoid fossil material (Zdansky, 1928; Teilhard de Chardin, 1938). Andersson also established the practice of gathering experts from different disciplines to research and excavate paleontological and archaeological sites (Yang, 1999), a tradition we now take completely for granted.

The Japanese invasion of China in 1937 brought an end to this era of exploration, but new initiatives were begun again in 1949. Diligent searching for Tertiary and Quaternary mammalian fossils in China since 1949 has brought many significant finds to light, but relatively few informative Old World monkey fossils, virtually all of which are from Pleistocene deposits (Pan & Jablonski, 1987; Jablonski, 1993*b*; Jablonski *et al*., 2000). The majority represent the near relatives of cercopithecoids living in China today, most being macaques or closely related forms; a much smaller fraction represents *Rhinopithecus* and other colobids. Elsewhere in eastern Asia, the history of recovery of Old World monkey fossils and the nature of the fossil record are similar. Small collections of fossil macaques have been made in China, Korea and Japan in the twentieth century, but generally these finds have not been reported or studied thoroughly. Large collections of fragmentary cercopithecoid fossils from Pleistocene and Holocene caves in Indonesia were described by Dirk Albert Hooijer (Hooijer, 1962*a*, 1962*d*), but have not inspired great interest because they appear to represent minor variants of the modern local *Macaca*, *Trachypithecus* and *Presbytis* species. Fragmentary finds of latest Miocene colobids from western and southern Asia, namely *Mesopithecus pentelicus* from Afghanistan and *Presbytis sivalensis* from northern India and Pakistan, have added to our knowledge of cercopithecoid evolution. However, the relatively sparse fossil record of Asia has shed little light on the major events in Old World monkey evolution.

Africa, and particularly sub-Saharan Africa, has been quite another story. The largest numbers and greatest diversity of Pliocene and Pleistocene cercopithecoid fossils have come from Africa, beginning in the early twentieth century. Among the first cercopithecoid fossils to be recovered in sub-Saharan Africa were the remains of a "large baboon" by Dr. Felix Oswald at Kanjera, Kenya. These remains, named *Simopithecus oswaldi* (Andrews, 1916), represented the first of a large series of the same species to be recovered from the site, and would come to be recognized as the most widespread radiation of any non-human primate, that of the genus *Theropithecus*.

The discovery of *Australopithecus* remains in the limestone breccias of many South African caves in the 1920s and 1930s was accompanied by the discovery of an unparalleled diversity of Old World monkey fossils. The site of Sterkfontein alone yielded four species, and Makapansgat five (Freedman, 1970). In what can best be described as the Victorian scientific tradition, descriptions of new species of Old World monkeys from Africa proceeded with a minimum of comparison and debate through to the early 1950s. As late as 1952, new species such as *Brachygnathopithecus peppercorni* (now considered to be *Theropithecus* (*Theropithecus*) *darti*) were being described (Kitching, 1952), without reference to finds of anatomically similar fossils described in previous decades. This style of science, as applied to Old World monkey fossils at least, came to an end with Leonard Freedman's landmark 1957 study of "The fossil Cercopithecoidea of South Africa". Freedman (1957) revised the classification of several species and proposed two new species. His reliance on morphometric as well as morphological comparisons between fossil and living cercopithecoid species represented a unique approach and serves as a model for analyses to the present day. His review politely noted the incorrect observations, incomplete species descriptions and dubious interpretations of other investigators. Regarding Robert Broom's original description of *Parapapio major*, Freedman commented "Nowhere in his short description is there any clue as to what Broom had in mind" (Freedman, 1957: 220). The breadth and thoroughness of Freedman's comparative approach set the tone for the modern study of Old World monkey systematic paleontology.

In the late 1920s, European paleontologists began long-term surveys for ancient humans in Ethiopia, Kenya and

Tanzania. Louis Leakey's discoveries of cercopithecoid fossils from Olduvai Gorge in Tanzania from the early 1930s onward are particularly relevant (e.g., Leakey & Whitworth, 1958). His son, Richard, and his daughter-in-law, Meave, took on this work in the 1960s, and through their efforts several new genera and species of extinct Old World monkeys have been described, including *Paracolobus chemeroni*, *P. mutiwa*, *Cercopithecoides kimeui*, *Theropithecus* (formerly *Papio*) *baringensis* and *Microcolobus tugenensis* (e.g., Leakey, 1969; Leakey, 1982). Without the recovery and basic description of the literally hundreds of new fossils of Old World monkeys in the last 30 years in Africa by teams led by – in alphabetical order – Camille Arambourg, Yves Coppens, Andrew Hill, F. Clark Howell, Donald Johanson, Meave Leakey, Richard Leakey, Tim White and others, our knowledge of cercopithecoid evolution would be a shadow of what it is today. Despite almost 170 years of collecting efforts, several regions, especially northern Africa and central Asia, still have not been explored fully for late Neogene vertebrates.

By the 1960s and increasingly in the 1970s, individuals who had not been involved in the discovery of cercopithecoid fossils nonetheless shed light on their taxonomic status and evolutionary relationships. John and Prue Napier brought Old World monkeys to the attention of anthropologists, mammalogists and paleontologists with a symposium in 1969 on this then neglected group of primates. This effort culminated in the publication in 1970 of their volume *Old World Monkeys: Evolution, Systematics, and Behavior*, in which several young scientists (e.g., Colin Groves, Clifford Jolly and Elwyn Simons) made some of their first important contributions to cercopithecoid systematics and evolution. The classification of the Old World monkeys published in that volume (Thorington & Groves, 1970) was virtually the only one widely used until the mid-1990s, and continues to serve as a starting-point for all other beta taxonomic works on the group.

In 1972 Clifford Jolly published his influential monograph "The classification and natural history of *Theropithecus* (*Simopithecus*) (Andrews, 1916), Baboons of the African Plio-Pleistocene". This was the first work in which the fossils comprising a single lineage were examined from taxonomic, functional anatomical and ecological perspectives. Jolly's wide-ranging and penetrating intellect suffuses this work. His monograph signified that the study of fossil monkeys had long gone past the point of merely describing new species. In doing so, Jolly set another standard against which subsequent works should be judged. In this and other related works on the evolution of *Theropithecus*, Jolly explored the anatomical similarities and ecological interactions of theropith and hominid primates. His 1970 paper "The seed-eaters: a new model of hominid differentiation based on a baboon analogy" (Jolly, 1970*b*) was the first to invoke a non-human primate as a model organism in the study of the adaptive radiation of Plio-Pleistocene hominids. It is in part because of "The seed-eaters" that *Theropithecus* is a more popular and celebrated fossil Old World monkey genus than most, and has been the only largely extinct cercopithecoid genus to have become the subject of its own scholarly edited volume (Jablonski, 1993*a*).

Most of the wide-ranging and synthetic interpretations of the cercopithecoid fossil record that are now used as starting points for debates are those of Eric Delson, who has devoted his nearly 30-year career to the study of Old World monkey fossils. His PhD dissertation provided the most comprehensive review of the history of discovery and interpretation of fossil Old World monkeys up to 1970 (Delson, 1973). His most widely cited early papers (e.g., Delson, 1975*a*, 1975*c*, 1980) and the section on cercopithecoid evolution in the book *Evolutionary History of the Primates* (Szalay & Delson, 1979), provided syntheses of the history of Old World monkeys that have not yet been equaled. He was among the first in vertebrate paleontology to supplement anatomically based discussions of classification with information on the geochronological, paleobiogeographical and paleoenvironmental contexts of the fossils under discussion.

Three main factors have driven debates on the Old World monkey fossil record. The first, alluded to above, is the recency of the cercopithecoid radiation itself, the lack of morphological differentiation within and between major lineages, and the commonness of convergent evolution leading to homoplasy. Character state exhaustion, when character change more often yields homoplasy than novelties, probably has been produced by a combination of intrinsic constraints and similar selective agents. Because morphological character spaces for fossil taxa are exhaustible, these taxa often can only be distinguished by combinations of character states, not by uniquely novel shared-derived characters (Wagner, 2000). This concept is inimical to many practitioners of cladistics within paleoprimatology. The frequency of homoplasy in the Cercopithecoidea has contributed to spirited exchanges, in particular, over the evolutionary relationships and taxonomy of long-muzzled fossil taxa. Questions persist as to which fossils should be assigned to *Papio* and *Theropithecus* (see, e.g., Eck & Jablonski, 1984; Delson & Dean, 1993).

The second has to do with problems created by the simple absence of fossils representing putative ancestral forms of many cercopithecoid lineages. Many living species groups, including some of the most diverse, are represented by tiny or non-existent fossil records. Efforts to utilize the few fossils available or the morphology of living species alone to infer the evolutionary relationships of these groups and to clarify their taxonomy have been inconclusive (e.g., Brandon-Jones, 1995). Conversely, the fossil record has proven richest for many genera that have the fewest or no living representatives (e.g., *Theropithecus*, *Parapapio* and *Mesopithecus*).

The third and most recent factor has been the introduction of a new source of phylogenetic information for the Cercopithecoidea in the form of mitochondrial and nuclear DNA sequences from modern species. Some results of molecular analyses have been at variance with results of morphological analyses, and have inspired reanalyses of problematical morphological questions such as possible diphyly of the mangabeys (Fleagle & McGraw, 1999). Molecular data have also led to important attempts to reconcile or integrate molecular evidence with the fossil record and the biogeographic scenarios based on it (e.g., Cronin & Meikle, 1982; Stewart & Disotell, 1998; Zhang & Ryder, 1998). For those groups with complex recent histories, such as the macaques, molecular data (e.g., Melnick & Kidd, 1985; Hoelzer & Melnick, 1996) are proving critical for interpreting recent genus history.

Taxonomy

Systematic framework

Old World monkeys comprise three well-characterized clades, one accommodating the primitive, extinct genera *Victoriapithecus* and *Prohylobates*, one for the leaf-eating monkeys with complex stomachs, and one for the cheek-pouch-bearing monkeys. The last two groups are usually distinguished at the subfamilial level. Hill (1966) and Groves (1989, 2000) have advocated distinction at the family level, because this would provide more taxonomic flexibility within each to recognize fine degrees of relationship among genera. This arrangement is adopted here for this reason and also, as Groves (1989, 2000) has pointed out, the highest taxonomic level of living taxa should be at the same rank as, not lower than, a plesiomorphic fossil group, in this case the Victoriapithecidae.

Although many suggested classification schemes of the Old World monkeys have been published since Simpson's (1945) classification of mammals (e.g., Thorington & Groves, 1970; Simons, 1972; Szalay & Delson, 1979; Napier, 1981, 1985), the only recent revision of the group was published by Groves (Groves, 1989). As a result of this, classification has been unstable, and non-taxonomists within primatology have tended to view it with consternation and confusion. Old World monkey lower-level classifications recently have been reconfigured to reflect phyletic hypotheses generated by morphological and molecular studies. Laudable as many of these studies have been, most have examined interspecific relationships and the ramifications of these for classifications at the genus level. Only a few recent studies (Delson, 1993; Brandon-Jones, 1995) have ventured formal taxonomic revisions. The classification presented below incorporates the positions published by several workers, including Groves (2000), Szalay & Delson (1979), Strasser & Delson (1987), Jablonski & Peng (1993) and others. Because the phylogeny of the colobids and the position of many fossil taxa have proven particularly difficult to resolve, the classification presented here recognizes only one tribe, Colobini, with the remainder of the group relegated to Family Colobidae Blyth, 1875, *incertae sedis* pending further analysis. To as great an extent as possible, absolute ages for fossil specimens have been taken from up-to-date reviews of the subject (Behrensmeyer *et al.*, 1997; Gundling & Hill, 2000).

- Order Primates Linnaeus, 1758
  - Infraorder Catarrhini É. Geoffroy Saint-Hilaire, 1812
    - Superfamily Cercopithecoidea Gray, 1821
      - Family Colobidae Blyth, 1875
        - Subfamily Colobinae Jerdon, 1867 (1825)
          - Tribe Colobini Blyth, 1875
            - Genus *Colobus* Illiger, 1811
              - *Colobus guereza* Rüppell, 1835
              - *Colobus* sp. indet.
            - Genus *Microcolobus* Benefit & Pickford, 1986
              - *Microcolobus tugenensis* Benefit & Pickford, 1986
            - Genus *Rhinocolobus* M.G. Leakey, 1982
              - *Rhinocolobus turkanaensis* M. G. Leakey, 1982
      - Family Colobidae Blyth, 1875, *incertae sedis*
        - Genus *Mesopithecus* Wagner, 1839
          - *Mesopithecus pentelicus* Wagner, 1839
          - *Mesopithecus monspessulanus* Gervais, 1849
          - *Mesopithecus* sp. indet.
        - Genus *Pygathrix* É. Geoffroy Saint-Hilaire, 1812
          - *Pygathrix* cf. *nemaeus* Linnaeus, 1771
        - Genus *Rhinopithecus* É. Geoffroy Saint-Hilaire, 1812
          - Subgenus *Rhinopithecus* É. Geoffroy Saint-Hilaire, 1812
          - *R.* (*Rhinopithecus*) *roxellana* Milne Edwards, 1870
          - *R.* (*Rhinopithecus*) *roxellana tingianus* Matthew & Granger, 1923
          - *R.* (*Rhinopithecus*) *lantianensis* Hu & Qi, 1978
          - *R.* (*Rhinopithecus*) sp. indet.
        - Genus *Dolichopithecus* Depéret, 1889
          - *Dolichopithecus ruscinensis* Depéret, 1889
        - Genus *Libypithecus* Stromer, 1913
          - *Libypithecus markgrafi* Stromer, 1913
        - Genus *Presbytis* Eschscholtz, 1821
          - *Presbytis sivalensis* Lydekker, 1878
          - *Presbytis comata* Desmarest, 1822
          - *Presbytis* sp. indet.
        - Genus *Semnopithecus* Desmarest, 1822
          - *Semnopithecus entellus* Dufresne, 1797
          - *Semnopithecus palaeindicus* Lydekker, 1884
        - Genus *Parapresbytis* Kalmykov & Maschenko, 1992
          - *Parapresbytis eohanuman* Borissoglebskaya, 1981
        - Genus *Trachypithecus* Reichenbach, 1862
          - *Trachypithecus auratus* É. Geoffroy Saint-Hilaire, 1812
          - *T. auratus robustus* Hooijer, 1962 (new combination)

T. auratus sangiranensis Jablonski & Tyler, 1999
Trachypithecus cf. *phayrei* Blyth, 1847
Trachypithecus sp. indet.
Genus *Cercopithecoides* Mollett, 1947
Cercopithecoides williamsi Mollett, 1947
Cercopithecoides kimeui M. G. Leakey, 1982
Genus *Paracolobus* R. E. F. Leakey, 1969
Paracolobus chemeroni R. E. F. Leakey, 1969
Paracolobus mutiwa M. G. Leakey, 1982
Paracolobus sp. indet.
Family Colobidae gen. et. sp. indet.
Family Cercopithecidae Gray, 1821
Subfamily Cercopithecinae Gray, 1821
Tribe Cercopithecini Gray, 1821
Genus *Cercopithecus* Brunnich 1772
Cercopithecus sp. indet.
Tribe Papionini Burnett, 1828
Subtribe Macacina Owen, 1843
Genus *Macaca* Lacépède, 1799
Macaca sylvanus Linnaeus, 1758
Macaca sylvanus cf. *sylvanus* Linnaeus, 1758
Macaca sylvanus florentina Cocchi, 1872
Macaca sylvanus prisca Gervais, 1859
Macaca sylvanus majori Azzaroli, 1946
Macaca sylvanus subsp. indet.
Macaca libyca Stromer, 1920
Macaca cf. *nemestrina* Linnaeus, 1766
Macaca fascicularis Raffles, 1821
Macaca anderssoni Schlosser, 1924
Macaca jiangchuanensis Pan *et al.*, 1992
Macaca arctoides I. Geoffroy Saint-Hilaire, 1830
Macaca arctoides subfossilis Jouffroy, 1959 (new combination)
Macaca mulatta Zimmerman, 1780
Macaca cf. *thibetana* Milne Edwards, 1870
Macaca cf. *fuscata* Blyth, 1875
Macaca sp. indet.
Genus *Procynocephalus* Schlosser, 1924
Procynocephalus wimani Schlosser, 1924
Procynocephalus subhimalayanus von Meyer, 1848
Genus *Paradolichopithecus* Necrasov et al., 1961
Paradolichopithecus arvernensis Depéret, 1929
Subtribe Papionina Burnett, 1928
Genus *Parapapio* Jones, 1937
Parapapio broomi Jones, 1937
Parapapio jonesi Broom, 1940
Parapapio whitei Broom, 1940
Parapapio antiquus Haughton, 1925
Parapapio ado Hopwood, 1936
Genus *Dinopithecus* Broom, 1937
Dinopithecus ingens Broom, 1937
Genus *Gorgopithecus* Broom & Robinson, 1949
Gorgopithecus major Broom, 1940
Genus *Cercocebus* É. Geoffroy Saint-Hilaire, 1812
Cercocebus sp. indet.
Genus *Lophocebus* Palmer, 1903
Lophocebus sp. indet.
Genus *Theropithecus* I. Geoffroy Saint-Hilaire, 1843
Subgenus *Theropithecus* Delson, 1993
T. (*Theropithecus*) *darti* Broom & Jensen, 1946
T. (*Theropithecus*) *oswaldi* Andrews, 1916
T. (*Theropithecus*) *oswaldi oswaldi* Andrews, 1916
T. (*Theropithecus*) *oswaldi leakeyi* Hopwood, 1934
T. (*Theropithecus*) *oswaldi delsoni* Gupta & Sahni, 1981
Subgenus *Omopithecus* Delson, 1993
T. (*Omopithecus*) *baringensis* R. E. F. Leakey, 1969
T. (*Omopithecus*) *quadratirostris* Iwamoto, 1982 (new combination)
T. (*Omopithecus*) *brumpti* Arambourg, 1947
Genus *Papio* Müller, 1773
Papio hamadryas Linnaeus, 1758
Papio hamadryas robinsoni Freedman, 1957
Papio hamadryas izodi Gear, 1926

Superfamily Cercopithecoidea

Family Colobidae

Subfamily Colobinae

Tribe Colobini

GENUS *Colobus* Illiger, 1811
Comparable in size and morphology with living species of *Colobus*, especially *C. polykomos*.
INCLUDED SPECIES *C. guereza*, *Colobus* sp. indet.

SPECIES *Colobus guereza* Rüppell, 1835
REFERRED SPECIMEN YPM 19063, nearly complete fossil cranium, originally designated as *C. polykomos abyssinicus* (Simons, 1967*a*).
AGE AND GEOGRAPHIC RANGE Presumed Pleistocene age, near Wad (Wadi) Medani, central Sudan
ANATOMICAL DEFINITION
The cranium shows a short face, with short nasal bones, as in modern *C. polykomos*. The supraorbital ridges are fairly well marked, and the postorbital constriction is pronounced. In most features, very similar to *C. polykomos abyssinicus*, except for the buccally bowed cheek toothrows and relatively small teeth, which more closely resemble *C. polykomos polykomos*.

SPECIES *Colobus* sp. indet.
REFERRED SPECIMENS Mandibular fragment from Kanam East (M 15911) (Harrison & Harris, 1996); right M_3 (F8-14), lower molar fragment (F8-15) (Eck, 1976*a*, 1976*b*), three isolated teeth and three associated teeth (P997-15) from the Omo (Leakey, 1987); mandible fragment (KN 41'88) from the Western Rift (Senut, 1994); "prolific" remains, from Middle Awash (Kalb *et al.*, 1982*a*, 1982*b*); identified but undescribed remains from Hadar (Johanson *et al.*, 1982*b*).
AGE AND GEOGRAPHIC RANGE Plio-Pleistocene, Kanam East, Kenya; Plio-Pleistocene, Omo Group, Hadar, and Middle Awash deposits, Ethiopia; Late Pleistocene, Kazinga, Western Rift, Uganda
ANATOMICAL DEFINITION
All specimens described as very similar in size and morphology to living black-and-white colobus monkeys, but not identifiable at the species level. The Kanam East mandible conforms closely to the general morphology seen in *Colobus*, except in unique details of P_4 (Harrison & Harris, 1996). The isolated molars from the Omo Valley are considerably smaller than modern *C. polykomos*. A well-preserved mandible from lower Pliocene Kuseralee Member of the Adu-Asa Formation in the Middle Awash is comparable in size to the largest living colobid monkeys. The Kazinga mandible morphology and size fall within the range of variability of *Piliocolobus badius* and *Colobus guereza* (Senut, 1994).

GENUS *Microcolobus* Benefit & Pickford, 1986
Slightly smaller than *Procolobus verus* and *Trachypithecus obscurus*; the mandibular symphysis lacks an inferior transverse torus below the genioglossal fossa and a median symphyseal foramen. P_3 possesses a well-developed heel, and, unlike *Mesopithecus*, lacks a mesiolingual sulcus. The P_4 metaconid and protoconid are of almost equal heights. The molar cusps are tall and anteriorly directed, and the lingual notches are deep. The height of the M_1 and M_2 cusps above the base of the lingual notch is lower than in many colobids (Benefit & Pickford, 1986).
INCLUDED SPECIES *M. tugenensis*

SPECIES *Microcolobus tugenensis* Benefit & Pickford, 1986
TYPE SPECIMEN KNM-BN 1740, an almost complete mandible
AGE AND GEOGRAPHIC RANGE 11-10 Ma, Ngeringerowa, Kenya
ANATOMICAL DEFINITION
Same as for the genus. The type mandible is slightly deeper below M_3 than below M_1, as in *Colobus*, *Rhinopithecus*, *Nasalis* and *Mesopithecus*. The species is dentally most similar to females of *Mesopithecus pentelici*, with moderately high molar cusps angled slightly forward, and well developed mesial and distal shelves.

GENUS *Rhinocolobus* M.G. Leakey, 1982
Significantly larger than living *Colobus* species, but with smaller teeth than *Paracolobus chemeroni* (see Delson *et al.*, 2000). Its long braincase and muzzle distinguish it from other colobids, except *Libypithecus* and *Nasalis*. Interorbital width is narrow, postorbital constriction marked, orbits relatively small and postglenoid process relatively large. Mandibular body is deep and slender, the gonial region expanded and the median symphyseal foramen present. Canines and P_3 are highly sexually dimorphic. *Rhinocolobus* is distinguished from *Libypithecus* by its larger overall size, larger and more rounded calvaria, longer muzzle, and lack of a large sagittal crest. It is distinguished from *Nasalis* by its short nasal bones, wide malar region, presence of a postglabellar sulcus, and nuchal crests (Leakey, 1982). Its long muzzle and lateral facial profile resemble *Theropithecus gelada*. The long calvaria with a marked postorbital constriction is not usually seen in colobids. The dentition is typically colobid but the P^3 bears a protocone, in contrast to *Colobus*. Postcrania are described as indistinguishable from extant colobids.
INCLUDED SPECIES *R. turkanaensis*

SPECIES *Rhinocolobus turkanaensis* M. G. Leakey, 1982 (Fig. 17.1)
TYPE SPECIMEN Omo 75 1969-1012, a partial cranium
REFERRED SPECIMENS The hypodigm includes a female skull (KNM-ER 1485); associated cranial and postcranial fragments from a male individual (KNM-ER 1542); complete and fragmentary female mandibles (Omo 57.4 189; C68 Omo 75s; KNM-ER 1520); male complete and fragmentary mandibles (L627-238; L412-1; KNM-ER 5406); unsexed mandibles and mandibular fragments (Omo 57.4/72 50; Omo 47 68-2164; Omo 29 406; KNM-ER 3862); a left maxillary fragment (KNM-ER 146); and over 60 isolated teeth (Leakey, 1982). Further specimens include several mandibular fragments and isolated teeth from the Omo Valley, Ethiopia (Leakey, 1987).
AGE AND GEOGRAPHIC RANGE 3.4–1.8 Ma, Omo Valley, Ethiopia, and East Turkana, Kenya
ANATOMICAL DEFINITION
Same as for the genus.

Subfamily Colobinae *incertae sedis*

GENUS *Mesopithecus* Wagner, 1839
Colobids of small to medium size, with pronounced sexual dimorphism in the skull, dentition and postcranium. Muzzle is relatively short and upright, with long nasals. The upper face resembles extant *Pygathrix* with squarish orbits, moderate interorbital distance and average supraorbital torus. Temporal lines in males meet posterior to bregma (fully mature or older individuals only). Ascending ramus is upright and the gonial region is expanded posteriorly.

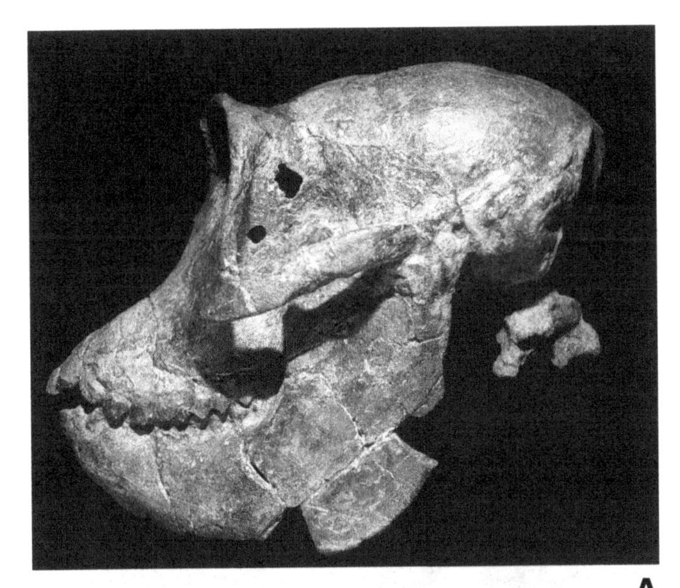

Fig. 17.1 *Rhinocolobus turkanaensis*. (A) KNM-ER 1485 female skull, left lateral view; (B) same, frontal view. Copyright National Museums of Kenya.

The P^3 is bicuspid. The molars are typically colobid, showing deep relief and a buccally directed talonid groove. The stout quadrate to trapezoidal shape of the molar crowns resembles Asian colobids such as *Semnopithecus entellus*, *Presbytis comata* and *Nasalis larvatus*, but differs from *Colobus* spp. M_3 hypoconulid size is variable. Long bone robustness and forelimb morphology indicate a terrestrial adaptation (Delson, 1973). The extremity indices and body proportions suggest an unspecialized, semiterrestrial monkey.

INCLUDED SPECIES *M. pentelicus*, *M. monspessulanus*, *Mesopithecus* sp. indet.

SPECIES *Mesopithecus pentelicus* Wagner, 1839 (Fig. 17.2)
TYPE SPECIMEN Munich AS II 11, a partial maxilla
AGE AND GEOGRAPHIC RANGE Late Miocene to Late Pliocene, southern and central Europe between 40° and 50° N and 0° to 30° E, Iran, and Afghanistan
ANATOMICAL DEFINITION
Same as for the genus.
OTHER SPECIMENS This species is represented by a large number of specimens, the majority of which derive from the type locality of Pikermi (Attiki, Greece). The most complete cranial remains include a male skull with mandible (Vienna A 4714), a female skull with mandible (Athens No. 8), a male cranium (Paris Pik 035), a male facial skeleton (London M 8947), a female facial skeleton (Munich AS II 7), a male mandible (Paris Pik 034), and two unnumbered female mandibles (from collections in Vienna and the University of Turin) (Mottura & Ardito, 1992). Many of these specimens are illustrated in Zapfe (Zapfe, 1991). The species is also represented from deposits of the lower Axios Valley of Macedonia (Greece) by a female cranium with both cheek toothrows preserved (LGPUT VTK-56) and a maxilla with M^{1-3} (MNHN unnumbered) (de Bonis *et al.*, 1997). The eastern extremity of the distribution of *M. pentelicus* is represented by finds from Molayan, Afghanistan, the most complete of which is a partial mandible of a young adult female (MOL 001) (Heintz *et al.*, 1981). The species is also known from Maragheh in northwestern Iran (Campbell *et al.*, 1980), Wissberg in Germany (Tobien, 1986), Baltavar in Hungary and Casino in Italy (Andrews *et al.*, 1996).

SPECIES *Mesopithecus monspessulanus* Gervais, 1849
TYPE SPECIMEN three molars and a possibly associated canine (unnumbered) (Gervais, 1849*b*).
REFERRED SPECIMENS The more complete specimens referred to the species are a male partial mandible with C–M_3 from Villafranca d'Asti (unnumbered) (Gentili *et al.*, 1998), and a male partial mandible (DIT 22) and female partial mandible (DTK 235) from Macedonia (Andrews *et al.*, 1996). An isolated upper molar crown from Suffolk, England (BM (NH) A M9171), extends the distribution of the genus significantly northward (Napier, 1985).
AGE AND GEOGRAPHIC RANGE Pliocene, France and England through Romania and Greece
ANATOMICAL DEFINITION
Distinguished from *M. pentelicus* by smaller size, narrower molars and an elbow joint that reflects a less terrestrial adaptation.

SPECIES *Mesopithecus* sp. indet.
REFERRED SPECIMENS IGF-7425V and IGF 7505V, right mandibular fragments and M_{2-3} from Baccinello, Italy; lost specimens from Gravitelli, Italy (Andrews *et al.*, 1996; Rook, 1999).
AGE AND GEOGRAPHIC RANGE Late Miocene of Italy
ANATOMICAL DEFINITION
Same as for the genus. Insufficient information to diagnose at the species level.

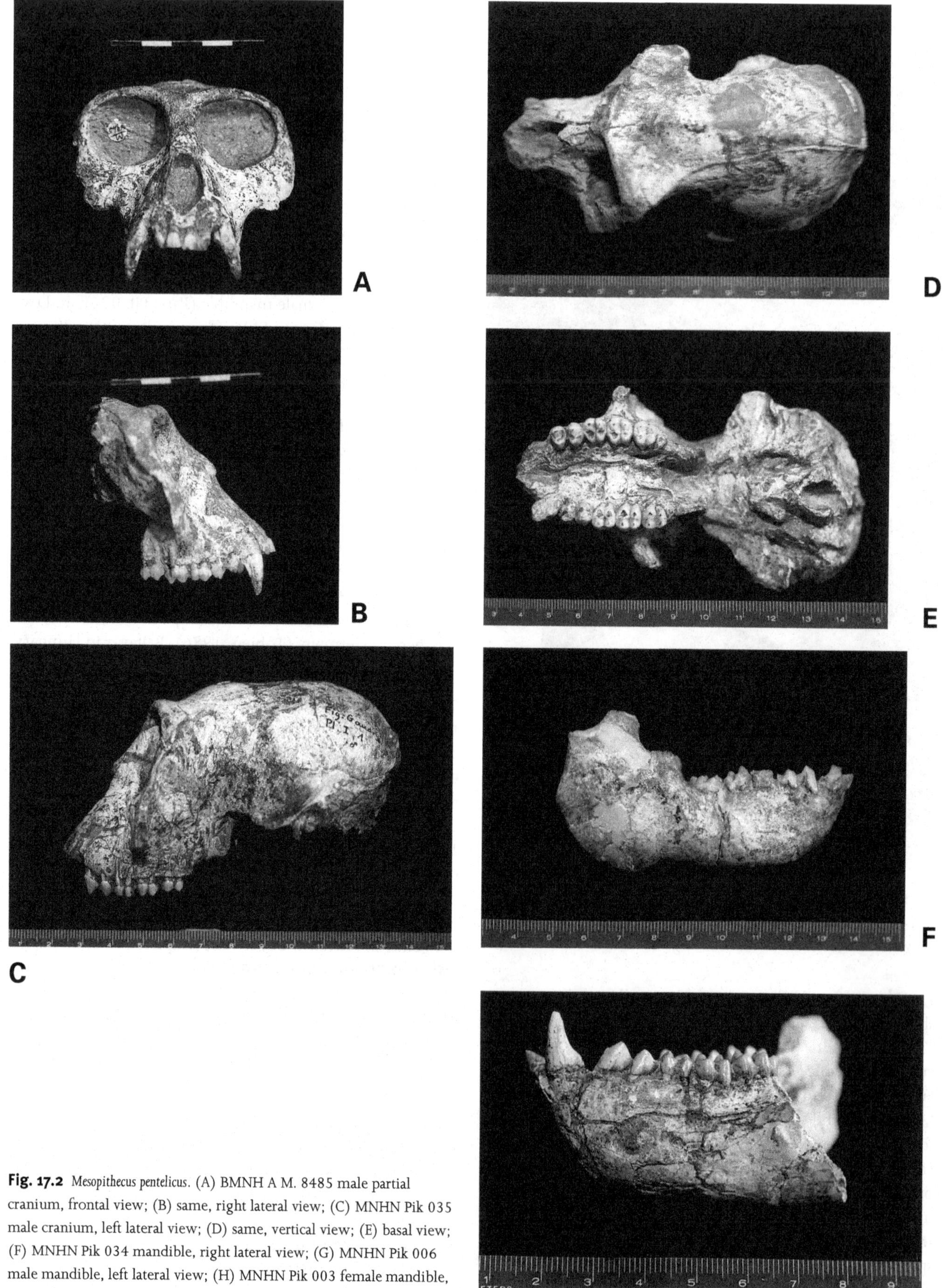

Fig. 17.2 *Mesopithecus pentelicus*. (A) BMNH A M. 8485 male partial cranium, frontal view; (B) same, right lateral view; (C) MNHN Pik 035 male cranium, left lateral view; (D) same, vertical view; (E) basal view; (F) MNHN Pik 034 mandible, right lateral view; (G) MNHN Pik 006 male mandible, left lateral view; (H) MNHN Pik 003 female mandible, left lateral view; (I) MNHN Pik 335 male? and Pik 1727 female humeri. Scales in cm.

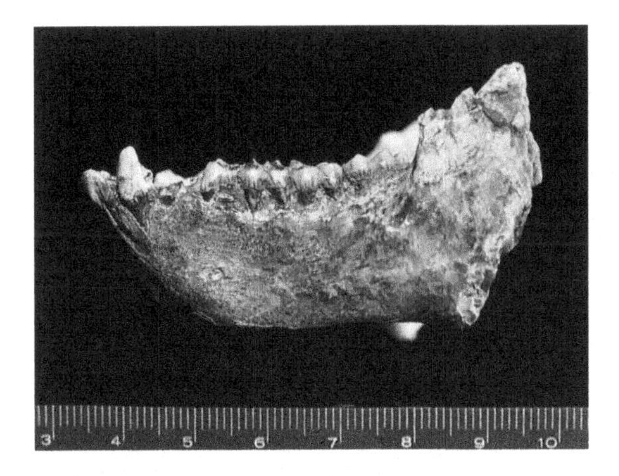

H

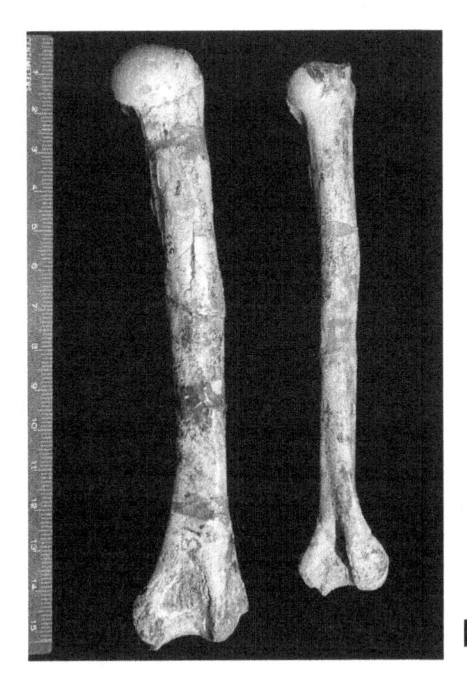

I

Fig. 17.2 (cont.)

GENUS *Pygathrix* É. Geoffroy Saint-Hilaire, 1812
A colobid of moderate size, with a relatively short muzzle and broad interorbital region. Molar teeth are typically colobid, with steep sides and high relief. Extremity indices are similar to *Nasalis* and *Macaca*, but limb bone robustness is only moderate. Distinguished from *Rhinopithecus* by its narrower face (including relatively narrower interorbital region), nasal bones of moderate length, and simple internasal and nasomaxillary sutures.
INCLUDED SPECIES *P.* cf. *nemaeus*

SPECIES *Pygathrix* cf. *nemaeus* Linnaeus, 1771
REFERRED SPECIMENS LPS231, LPX193 and LPS191, isolated molars from the caves at Luoding, Guangdong, China (Gu *et al.*, 1996)
AGE AND GEOGRAPHIC RANGE Pleistocene, southeastern China
ANATOMICAL DEFINITION
The small sample of isolated teeth is very similar in size and morphology to those of living *Pygathrix nemaeus*. Particularly striking are the long M_3 crowns with steep, unflared buccal and lingual surfaces.

GENUS *Rhinopithecus* É. Geoffroy Saint-Hilaire, 1812
Distinguished from *Pygathrix* and other Asian colobids by its larger size, more robust facial skeleton including strong, shelf-like supraorbital torus, markedly concave facial profile, extremely broad interorbital region, abbreviated nasal bones, and complex internasal and naxomaxillary sutures. Postcranial proportions are similar to *Nasalis* and *Macaca*, and long bones are generally robust.
SUBGENUS *Rhinopithecus* É. Geoffroy Saint-Hilaire, 1812
Distinguished from subgenus *Presbytiscus* by extreme nasal bone shortening and the greater puffiness of the buccal and lingual molar faces. Long bone, metacarpus and metatarsus robustness resembles *Mesopithecus pentelicus*.
INCLUDED SPECIES AND SUBSPECIES *R.* (*Rhinopithecus*) *lantianensis*, *R.* (*R.*) *roxellana*, *R* (*R.*) *roxellana tingianus*, *Rhinopithecus* sp. indet.

SPECIES *Rhinopithecus* (*Rhinopithecus*) *roxellana* Milne Edwards, 1870
REFERRED SPECIMEN IVP V9543, cranium (Gu & Hu, 1991)
AGE AND GEOGRAPHIC RANGE Middle Pleistocene, Henan, east–central China
ANATOMICAL DEFINITION
Metrically and morphologically indistinguishable from modern *R.* (*R.*) *roxellana* (Jablonski, 1993*b*).

SUBSPECIES *Rhinopithecus* (*Rhinopithecus*) *roxellena tingianus* Matthew & Granger, 1923 (Fig. 17.3A–C)
TYPE SPECIMEN AMNH 18466, a subadult cranium
REFERRED SPECIMENS Hypodigm includes a left maxilla with P^4–M^2 (AMNH 18467), a left maxilla with P^4–M^1 (AMNH 21788), a right maxilla with P^4–M^3 (AMNH 18468), a right maxilla with P^3–M^2 (AMNH 18746) and a left mandibular ramus with P_4–M_3 (AMNH 18469).
AGE AND GEOGRAPHIC RANGE Probably Late Pleistocene, Sichuan, central China
ANATOMICAL DEFINITION
Subadult morphology has prevented definitive assignment to species (Jablonski, 1993*b*; Jablonski & Peng, 1993). The specimen is similar to subadult crania of all species of *R.* (*Rhinopithecus*) to which it has been compared (Fig. 17.3). The dentitions of the nonholotype specimens cannot be distinguished metrically or morphologically from those of extant *R.* (*R.*) *roxellana* (Colbert & Hooijer, 1953).

SPECIES *Rhinopithecus* (*Rhinopithecus*) *lantianensis* Hu & Qi, 1978 (Fig. 17.3D, E)
TYPE SPECIMEN IVPP V2934.1, a damaged mandible
REFERRED SPECIMENS Hypodigm includes a right male maxillary fragment with P^3–M^3 (IVPP V2934.2), a left male maxillary fragment with C–M^2 (IVPP V2934.3) and an

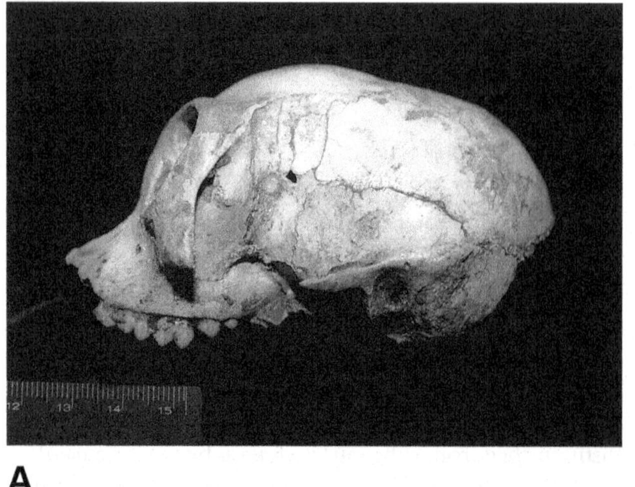

A

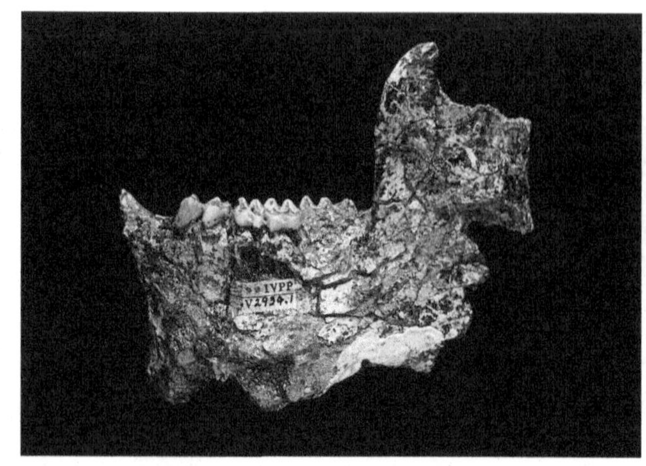

D

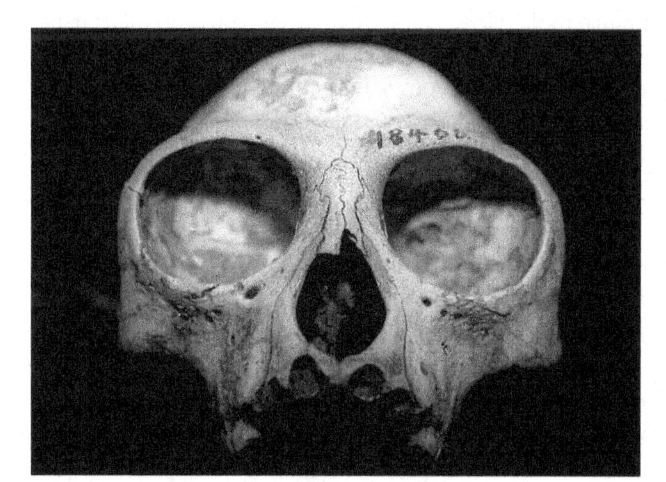

B

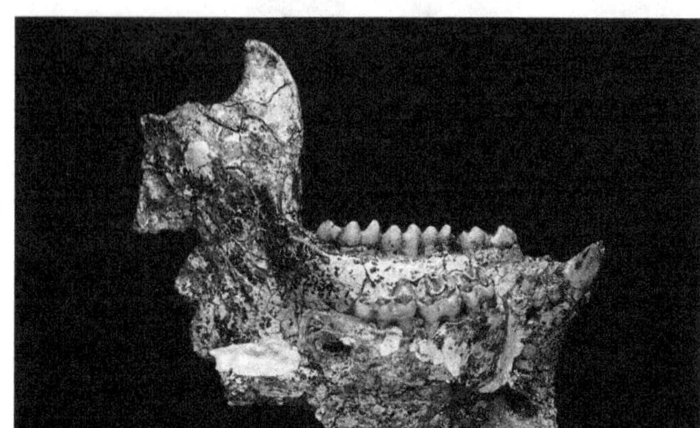

E

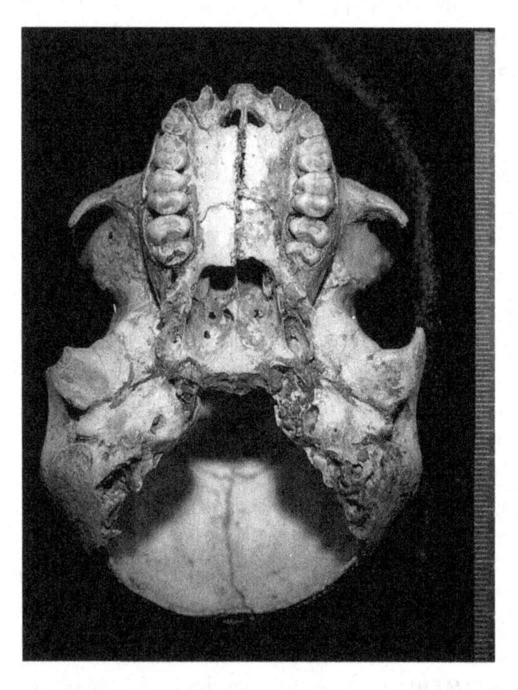

C

Fig. 17.3 *Rhinopithecus* (*Rhinopithecus*) specimens. (A) *R.* (*R.*) *roxellana tingianus*, AMNH 18466 (holotype), juvenile cranium, left lateral view; (B) same, frontal view; (C) basal view; (D) *R.* (*R.*) *lantianensis*, IVPP V.29341 (holotype), left lateral view; (E) same, right lateral view tilted to show the deeply incised lingual notches of molars characteristic of colobids.

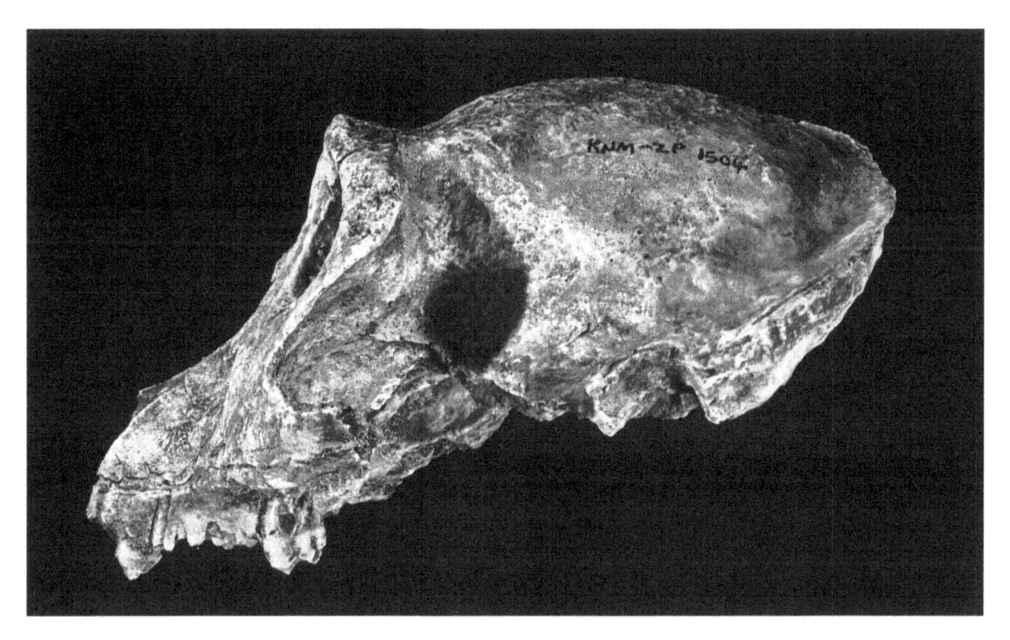

Fig. 17.4 *Libypithecus markgrafi*. Cast of BSM 1914 II 1 male cranium (holotype), left lateral view. Copyright National Museums of Kenya.

isolated right M^1 (IVPP V2934.4). A partial male mandible from Yunxian, Hubei has been referred to the species but is unpublished (T. Li & D. A. Etler, unpublished data).
AGE AND GEOGRAPHIC RANGE Late Early Pleistocene, Shaanxi and Hubei, central China
ANATOMICAL DEFINITION
Distinguished by its large size and relatively thin mandible.

SPECIES *Rhinopithecus (Rhinopithecus)* sp. indet.
REFERRED SPECIMENS Isolated premolars (LPS266, LPX152) and molars (LPS223, LPS249, LPS254, LPS256, LPS258 and LPS263, LPX206, LPX156, and LPX166) from the caves at Luoding, Guangdong, China (Gu *et al.*, 1996)
AGE AND GEOGRAPHIC RANGE Pleistocene, southeastern China
ANATOMICAL DEFINITION
The molars are typical of those of modern *Rhinopithecus* from China, with high relief but remarkably puffy sides, resembling some species of *Macaca*.

GENUS *Dolichopithecus* Depéret, 1889
A colobid of moderately large size with a rather long face, relatively narrow interorbital distance, considerable sexual dimorphism, and postcrania reflecting a highly terrestrial, quadrupedal habitus. Its mandibular angle is somewhat enlarged and the mandibular corpus is of even depth below the cheek teeth. Distinguished from *Mesopithecus* by its more extreme adaptation to terrestrial life and its larger cranial and body size.
INCLUDES SPECIES *D. ruscinensis*

SPECIES *Dolichopithecus ruscinensis* Depéret, 1889
TYPE SPECIMEN MNHN-P PER 001, a partial cranium (lectotype designated by Delson (1973))
REFERRED SPECIMENS The species is represented by relatively complete but crushed specimens from the type locality of Perpignan, France, including female skulls (FSL 41327, FSL 4309), a male skull (MNHN-P PER 002), male mandibles (FSL 40906, MNHN-P PER 003, ML Pp1) and a large collection of jaw fragments and isolated teeth (Delson, 1973). A nearly complete, probably male ulna from Pestzentlorinc, Hungary (Andrews *et al.*, 1996) and portions of a female mandible from Megalo Emvolon, Macedonia (Greece) (Koufos *et al.*, 1991) have also been recovered.
AGE AND GEOGRAPHIC RANGE Pliocene; western and central Europe into Russia, Macedonia, Greece
ANATOMICAL DEFINITION
Same as for the genus.

GENUS *Libypithecus* Stromer, 1913
A colobid of medium size, with a cranium comparable in size and shape to *Piliocolobus badius*. Cranium characterized by a projecting face, narrow interorbital region, long nasal bones, and a typical colobid dentition featuring moderately small incisors and molars that increase in size from anterior to posterior. Marked sagittal and nuchal crests increase in height toward inion.
INCLUDES SPECIES *L. markgrafi*

SPECIES *Libypithecus markgrafi* Stromer, 1913 (Fig. 17.4)
TYPE SPECIMEN BSM 1914 II 1, cranium
AGE AND GEOGRAPHIC RANGE Late Miocene, northern Egypt
ANATOMICAL DEFINITION
Same as for the genus. The anterior dental arch of the type specimen was deformed during life as a result of breakage

of the crown of the right canine. Damage to the midface has precluded accurate reconstruction of the angular relationships between the face and basicranium.

GENUS *Presbytis* Eschscholtz, 1821

Small, short-faced colobids with rounded crania, weakly developed supraorbital tori, and reduced third molars, limited dental and somatic sexual dimorphism, and postcrania adapted to arboreal climbing and jumping.

INCLUDES SPECIES *P. comata*, *P. sivalensis*, *Presbytis* sp. indet.

SPECIES *Presbytis sivalensis* Lydekker, 1878

TYPE SPECIMEN GSI D 2, left maxilla with M^3

REFERRED SPECIMENS The most complete known specimen is a cranium and attached mandible (GSI K16/49); also includes a left maxillary fragment with M^3 (GSI D 2), a right maxillary fragment with M^{1-3} (GSI D 2a), a right maxillary fragment with P^4, dP^4 and M^1 (GSI D 120), a left maxillary fragment with dP^3 and dP^4 (GSI D 121), a right mandibular corpus fragment with M_1 and M_2 (GSI D 184), a left mandibular corpus fragment with M_3 (YPM 19135), and left and right mandibular corpus fragments with P_4–M_3 and P_4–M_1 and M_3' respectively (GSP 14043).

AGE AND GEOGRAPHIC RANGE Early Pliocene, Punjab, Pakistan

ANATOMICAL DEFINITION

Represented by mostly fragmentary gnathic and dental specimens collected over many decades in the Potwar Plateau of Pakistan. Originally assigned to *Macaca* and *Cercopithecus*, but now widely agreed to be a single colobid species best accommodated in *Presbytis* mostly because of its small size (Simons, 1970; Szalay & Delson, 1979; Barry, 1987). Jaw and tooth morphology strongly recalls *Mesopithecus*, except much smaller.

SPECIES *Presbytis comata* Desmarest, 1822

REFERRED SPECIMENS Coll. Dub. 3780, partial palate

AGE AND GEOGRAPHIC RANGE Middle Pleistocene, Java

ANATOMICAL DEFINITION

The referred specimen is indistinguishable from modern representatives of the same species (Hooijer, 1962*c*).

SPECIES *Presbytis* sp. indet.

REFERRED SPECIMENS Several unnumbered upper and lower jaws from Niah Cave, Sarawak, Malaysia (Hooijer, 1962*b*)

AGE AND GEOGRAPHIC RANGE Latest Pleistocene and early Holocene of eastern Malaysia (Sarawak)

ANATOMICAL DEFINITION

Morphologically identical, but metrically slightly larger than the living *Presbytis* species of Borneo; insufficient anatomical information is available to make a more specific diagnosis (Hooijer, 1962*b*).

GENUS *Semnopithecus* Desmarest, 1822

Medium to large-size colobids distinguished from *Presbytis* and *Trachypithecus* by more robust skulls with especially prominent, shelf-like supraorbital tori, straight-sided molars with deep relief, and postcrania indicating a more exclusively terrestrial adaptation.

INCLUDES SPECIES *S. entellus*, *S. palaeindicus*

SPECIES *Semnopithecus entellus* Dufresne, 1797 (Fig. 17.5D)

REFERRED SPECIMENS Associated and isolated teeth from the Billa Surgam Caves, Kurnool, Andhra Pradesh, India

AGE AND GEOGRAPHIC RANGE Pleistocene, India

ANATOMICAL DEFINITION

Originally described as *S. priamus*, the collection of isolated teeth from Kurnool includes a group of associated teeth from an adult male (lower canine, P_{3-4}, M_1, and M_2) (BMNH 2963a), two isolated lower canines, one male (BMNH 2963b), and one female (BMNH 2963c) and a deciduous lower premolar (BMNH 2963d.) The specimens are very similar in size and morphology to those of living *S. entellus* in Andhra Pradesh today.

SPECIES *Semnopithecus palaeindicus* Lydekker, 1884 (Fig. 17.5)

TYPE SPECIMEN BMNH a 15710, lectotype designated as a right partial mandible

REFERRED SPECIMENS Recognized from two specimens in addition to the lectotype, a right mandibular fragment with M_3 (BMNH 15711) and a right talus (BMNH M1539) (Napier, 1985). Delson's (1975*b*, 1975*c*, 1980) assignment of them to ?*Macaca palaeindica* is rejected because premolar, molar and talar morphology closely resembles large *S. entellus* (Fig. 17.5D).

AGE AND GEOGRAPHIC RANGE Pliocene to earliest Pleistocene, "Siwalik Hills", northwestern India or Pakistan

ANATOMICAL DEFINITION

Comparable in size and morphology to the largest living subspecies of *S. entellus*.

GENUS *Parapresbytis* Kalmykov & Maschenko, 1992

A large colobid displaying a mosaic of cranial and dental features seen in *Semnopithecus*, *Nasalis*, *Pygathrix* and *Rhinopithecus*. Most similar to *Semnopithecus* and *Rhinopithecus* in its broad interorbital distance, shelf-like supraorbital torus, pronounced ophyronic groove, molar morphology, molar dimensions, and postcranial robustness; distinct from them and other colobids in subnasal and incisor morphology, and a suite of mandibular characteristics.

INCLUDED SPECIES *P. eohanuman*

SPECIES *Parapresbytis eohanuman* Borissoglebskaya, 1981

TYPE SPECIMEN PEN 3381-235, partial mandible

REFERRED SPECIMENS Hypodigm includes a distorted fragment of a mandible with I_1 (PIN 3381-286), a left distal humerus (PIN 3381-210) and a nearly complete right ulna (PIN 3381-211). More recently referred specimens include a partial premaxilla with both central incisors and left I^2 (GIN 987/878[2]) a fragment of left premaxilla and maxilla with I^1–P^3 (GIN 987/878[1]), fragmentary right maxilla with P^3–M^3 (GIN 987/445), and a crushed fragment of calvarial roof preserving the supraorbital torus and temporal lines (GIN 987/ 493[1]) (Kalmykov & Maschenko, 1992, 1995).

AGE AND GEOGRAPHIC RANGE Pliocene, Mongolia

ANATOMICAL DEFINITION

Same as for the genus.

GENUS *Trachypithecus* Reichenbach, 1862

Colobids of small to large size, distinguished by pronounced but narrow supraorbital torus, pronounced postorbital constriction, prominent muzzle, elongated palate, broad interorbital regions, pronounced sexual dimorphism in the canine–premolar complex, relatively large mandible with a high ascending ramus, and extremity indices typical of a generalized quadruped.

INCLUDED SPECIES *T. auratus robustus*, *T. auratus sangiranensis*, *T. auratus* subsp. indet, *T.* cf. *phayrei*, *Trachypithecus* sp. indet.

SPECIES *Trachypithecus auratus* É. Geoffroy Saint Hiliare, 1812

A medium-sized colobid distinguished from *T. cristatus* of the Indonesian Archipelago by a dental arcade longer relative to the calvaria and longer associated mandibular measurements (Weitzel *et al.*, 1988).

SUBSPECIES *Trachypithecus auratus robustus* Hooijer, 1962 (comb. nov.)

TYPE SPECIMEN Coll. Dub. 3778, subadult cranium (Hooijer, 1962*c*)

AGE AND GEOGRAPHIC RANGE Middle Pleistocene, Java

ANATOMICAL DEFINITION

Same as for the genus. Distinguished from extant *T. auratus* by larger size (Hooijer, 1962*c*).

SUBSPECIES *Trachypithecus auratus sangiranensis* Jablonski & Tyler, 1999 (Fig. 17.6)

TYPE SPECIMEN Jtr-1993.05-SNJ, complete right and partial left maxillae

AGE AND GEOGRAPHIC RANGE Latest Pliocene, Java

ANATOMICAL DEFINITION

Same as for the genus. Distinguished from extant *T. auratus* by shorter height of the buccal cusps of the upper molars, the puffiness of the sides of the molars, small foveae near the bases of the median buccal clefts, and shallower and broader molar trigon basins. The last characteristic also distinguishes the subspecies from *T. a. robustus* (Jablonski & Tyler, 1999).

SUBSPECIES *Trachypithecus auratus* subsp. indet.

REFERRED SPECIMENS Unnumbered upper and lower jaws from Niah Cave, Sarawak, Malaysia (Hooijer, 1962*b*); right palate from Bangle, Java (Coll. Dub. 3781) (Hooijer, 1962*c*); left fragmentary maxilla from Soember Kepoeh, Java (Coll. Dub. 2779); and a left partial mandible Sibrambang Cave, Sumatra (Coll. Dub. 11688) (Hooijer, 1962*c*)

AGE AND GEOGRAPHIC RANGE Latest Pleistocene and early Holocene, Sarawak, Java and Sumatra

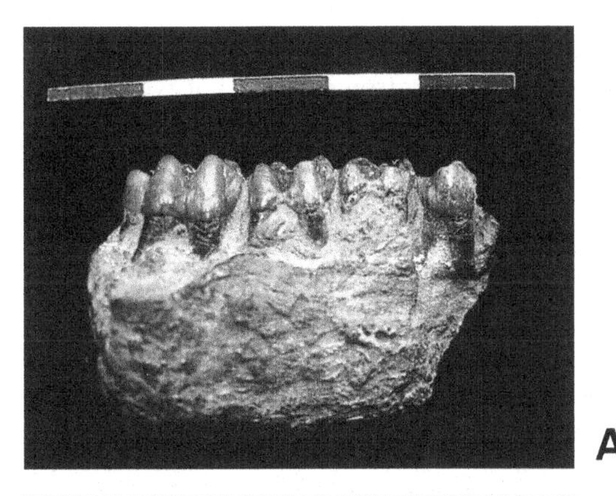

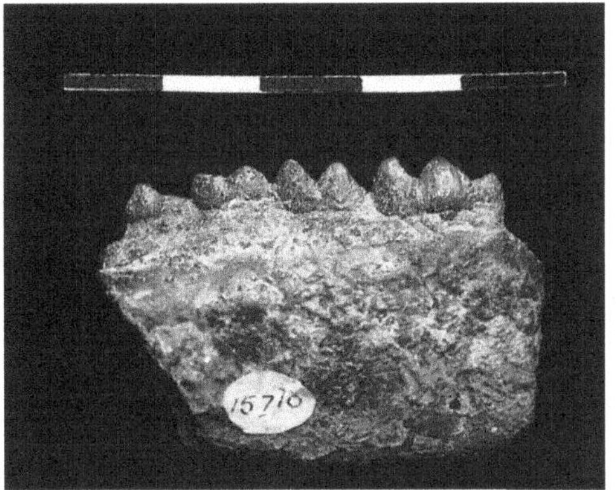

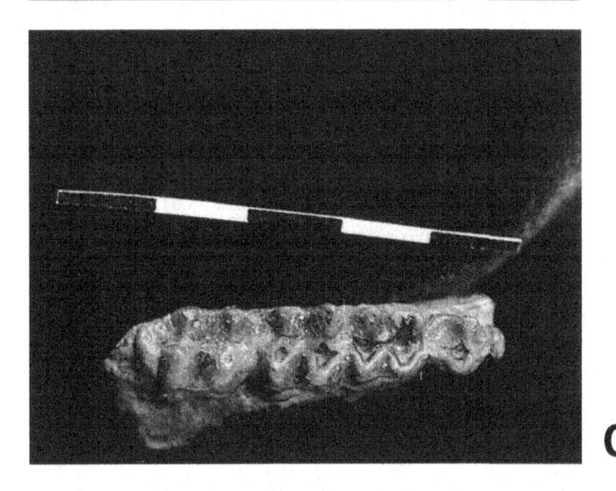

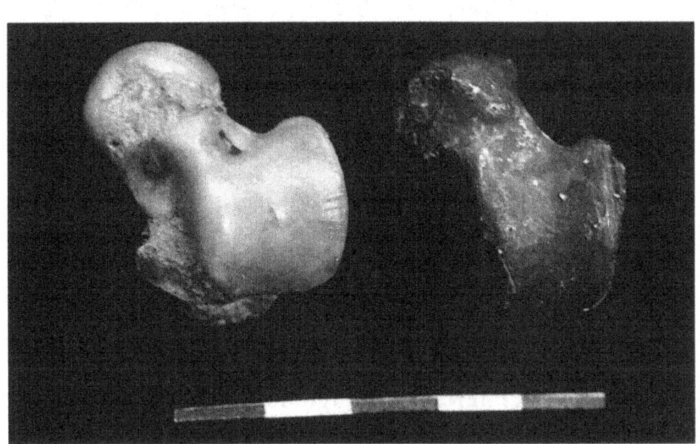

Fig. 17.5 *Semnopithecus palaeindicus*. (A) BMNH 15710 partial mandible, right buccal view; (B) same, lingual view; (C) same, occlusal view; (D) cast of BMNH 1539 talus (right) compared with *Semnopithecus entellus* BMNH 10256 (left). Scales in cm.

Fig. 17.6 *Trachypithecus auratus sangiranensis*. I.T.B. 001, renumbered Jtr-1993.05-SNJ (holotype), male partial maxilla, basal view. Photograph by Dong Lin.

ANATOMICAL DEFINITION

Same as for the species. Javan specimens are metrically within the limits for or slightly larger than *T. a. pyrrhus*, but these and other specimens are not sufficiently well known to permit diagnosis at the subspecies level.

SPECIES *Trachypithecus* cf. *phayrei* Blyth, 1847
REFERRED SPECIMENS YV 1071 and YV 1073, a distorted partial face and a mandibular fragment (Pan *et al.*, 1992)
AGE AND GEOGRAPHIC RANGE Early Holocene, southwestern China
ANATOMICAL DEFINITION
In size and morphology, near the ranges of variation represented by the *Trachypithecus* species in Yunnan today, *T. françoisi* and *T. phayrei*, but more similar to the latter (Pan *et al.*, 1992; Jablonski *et al.*, 1994).

SPECIES *Trachypithecus* sp. indet.
REFERRED SPECIMENS Isolated premolars (LPS258, LPX195, LPS110, LPX227 and LPX228) and molars (LPS251, LPS253, LPX205, LPS173, LPS175, LPX198, LPX199, LPX170, LPX200, LPX202 and LPX224) from the caves at Luoding, Guangdong, China
AGE AND GEOGRAPHIC RANGE Pleistocene, southeastern China
ANATOMICAL DEFINITION
Nearly identical to living *Trachypithecus* species in southern China (*T. françoisi* and *T. phayrei*), to which they were originally referred (Gu *et al.*, 1996).

GENUS *Cercopithecoides* Mollett, 1947
A fairly large extinct colobid, with a short, relatively narrow muzzle, and a large and rounded calvaria. Wide face and large orbits, narrow frontal process of zygoma, wide interorbital region, moderately long nasals, small nasal aperture. Slight postorbital constriction, thick supraorbital torus. Males exhibit relatively long and narrow crania, females shorter crania with more globular calvariae. The mandible exhibits a shallow body with a marked lateral ridge and a flat anterior surface with a median symphyseal foramen and unexpanded gonial region. Ascending ramus is low and oriented obliquely relative to the occlusal plane. Teeth are small, especially P^3, M^3 and M_3, but exhibit a typical colobid structure. Molars exhibit very high cusps and large central foveae. P^3 lacks a protocone. Canines and P_3 are sexually dimorphic. Distinguished from all other colobids by its low, shallow mandible with an obliquely oriented ascending ramus. Distinguished from *Rhinocolobus*, *Libypithecus* and *Nasalis* by its short, rounded calvaria and short muzzle; differs from Asian colobids *Paracolobus* and *Rhinocolobus* in the absence of a P^3 protocone. Postcranial elements similar to more terrestrial cercopithecoids.
INCLUDES SPECIES *C. kimeui*, *C. williamsi*

SPECIES *Cercopithecoides williamsi* Mollett, 1947 (Fig. 17.7A–D)
TYPE SPECIMEN AD. 1326/3, damaged cranium; and M. 2038, mandible
REFERRED SPECIMENS Represented by numerous fossils from South Africa, including an associated cranium and mandible (Bolt's Farm 56784), nearly complete crania (STS 394 A, M 2999, M3055), an almost complete calvaria (KB 122 [= KA 195]), a male muzzle (M 3000), maxillary fragments (STS 350 and M 666), and several mandibles or mandibular fragments (M 2990, M 2989, M 2987, LW 7/48, STS 394B) including one juvenile (KB 3108) (Freedman, 1957, 1960, 1965; Maier, 1970; Freedman & Brain, 1972; Keyser, 1991). Also represented by several specimens from east of Lake Turkana, including associated cranial and postcranial fragments (KNM-ER 4420), a partial cranium (KNM-ER 6000) and partial mandibles (KNM-ER 2124, KNM-ER 2133 and KNM-ER 3852) (Leakey, 1982).
AGE AND GEOGRAPHIC RANGE Plio-Pleistocene, South Africa and East Turkana, Kenya
ANATOMICAL DEFINITION
Same as for the genus.

SPECIES *Cercopithecoides kimeui* M. G. Leakey, 1982 (Fig. 17.7E)
TYPE SPECIMEN KNM 068/6514, partial cranium
REFERRED SPECIMENS Hypodigm includes two female crania (KNM-ER 398 and KNM-ER 991) (Fig. 17.7E), two sets of associated cranial and postcranial fragments (KNM-ER 3065 and KNM-ER 3069), maxillary fragments (KNM-ER 3015 and KNM-ER 3838) and mandibular fragments (KNM-ER 6005, KNM-ER 1529, KNM-ER 976 and KNM-ER 879).

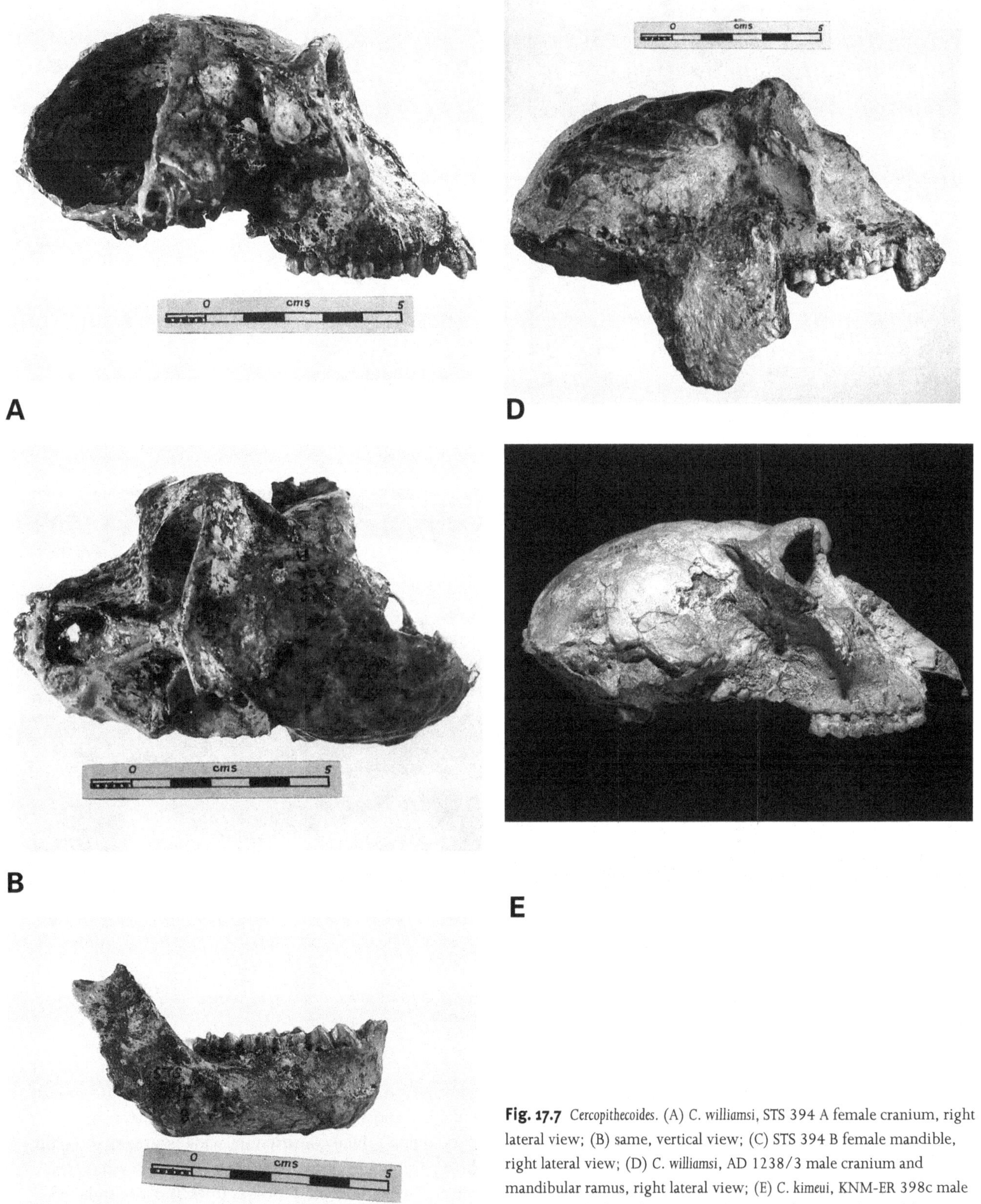

Fig. 17.7 *Cercopithecoides*. (A) *C. williamsi*, STS 394 A female cranium, right lateral view; (B) same, vertical view; (C) STS 394 B female mandible, right lateral view; (D) *C. williamsi*, AD 1238/3 male cranium and mandibular ramus, right lateral view; (E) *C. kimeui*, KNM-ER 398c male cranium, right lateral view. A–D Courtesy of Len Freedman; (E) Copyright National Museums of Kenya.

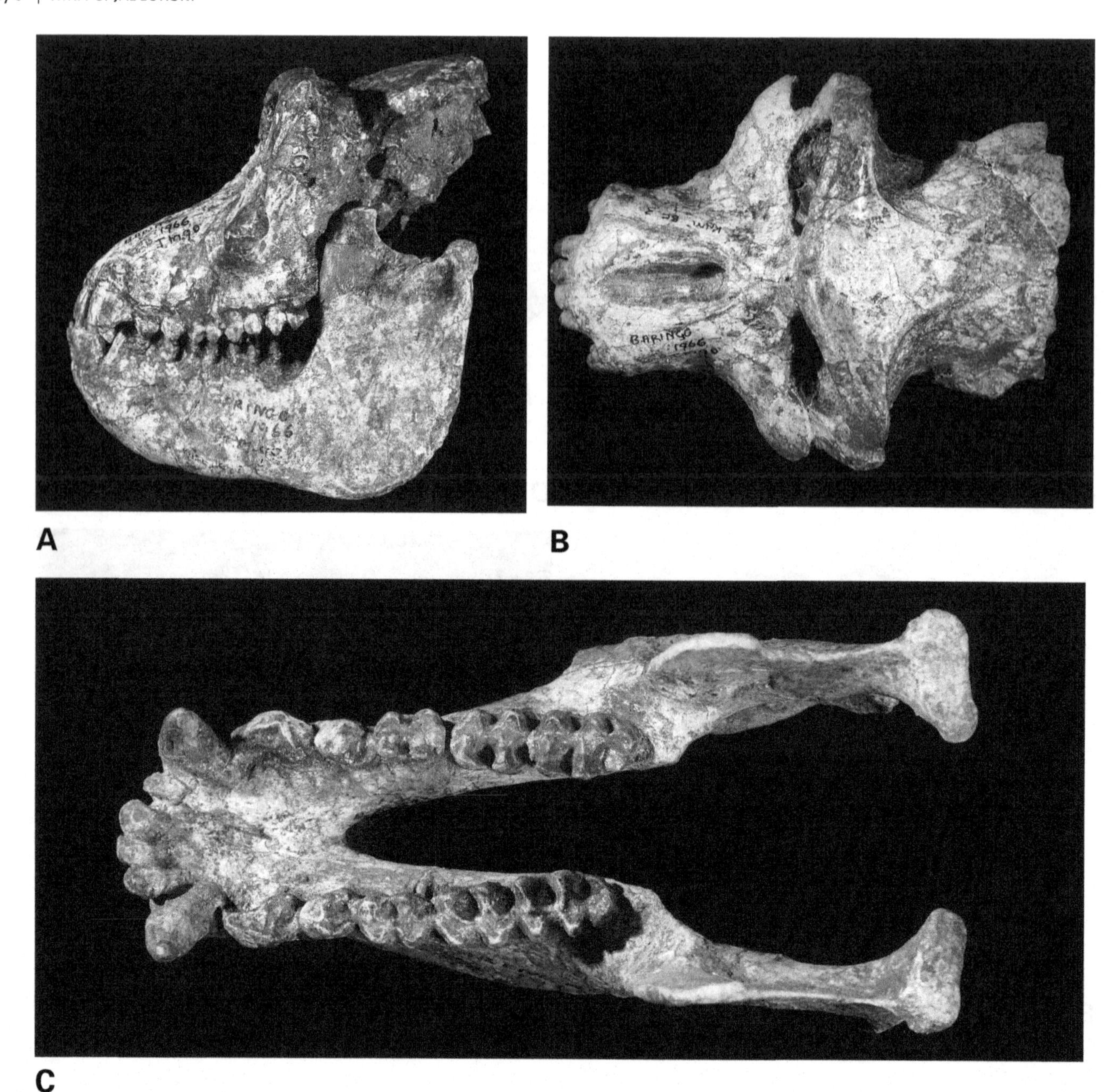

Fig. 17.8 *Paracolobus chemeroni*. (A) KNM-BC 3a and b male skull, left lateral view; (B) KNM-BC 3a male cranium, vertical view; (C) KNM BC 5b male mandible, occlusal view. Copyright National Museums of Kenya.

AGE AND GEOGRAPHIC RANGE Approximately 2.0–1.88 Ma, Olduvai Gorge, Tanzania and East Turkana, Kenya

ANATOMICAL DEFINITION

Larger than *C. williamsi*, with a more robust mandible and upper molars that are wide, low-cusped and flared toward the cervix.

GENUS *Paracolobus* R. E. F. Leakey, 1969

A large monkey exhibiting characteristics more similar to *Colobus* than to *Papio*, *Theropithecus* or other large Pleistocene Cercopithecoidea (Fig. 17.8). According to the original (Leakey, 1969) and revised (Leakey, 1982) diagnoses, the cranium is moderately long and possesses a broad muzzle, wide face and a wide frontal process of the zygoma. Marked postorbital constriction, wide interorbital region, short nasal bones, long nasal aperture, and sagittal crest minor. The supraorbital torus is thick and a postglabellar sulcus is present. The ascending ramus of the mandible is tall and vertically oriented. Mandibular corpus is uniformly slender and deep below the premolars and molars, but deepens posteriorly to form an expanded gonion. Most postcranial features resemble arboreal colobids, but some are intermediate between the colobid and terrestrial cercopithecid conditions. As in *Colobus*, the costal surface of

the scapula is concave and bears a marked keel on the lateral margin. The deltoid tuberosity of the humerus is poorly developed and the trochlea is shallow. The olecranon process of the ulna is inclined anteriorly.
INCLUDED SPECIES *P. chemeroni*, *P. mutiwa*, cf. *Paracolobus* sp. indet.

SPECIES *Paracolobus chemeroni* R. E. F. Leakey, 1969 (Fig. 17.8)
TYPE SPECIMEN KNM-BC 3, a cranium missing the posterior calvaria, a complete mandible and much of the postcranial skeleton
AGE AND GEOGRAPHIC RANGE Pliocene, Chemeron, Baringo Basin, Kenya
ANATOMICAL DEFINITION
Same as for the genus.

SPECIES *Paracolobus mutiwa* M. G. Leakey, 1982
TYPE SPECIMEN KNM-ER 3843, partial cranium
REFERRED SPECIMENS Hypodigm includes a mandibular fragment from East Turkana (KNM-ER 125), maxillary fragments from the Omo (Omo 56 Sup. 68 2162, and L9-14A), mandibular fragments from the Omo (Omo 56 Sup 68 2160, L131-40, L35-59, Omo 29 68 1404, Omo 29 1968 1402, and L390-3) and 46 isolated teeth (Leakey, 1982). Further specimens include several gnathic fragments and isolated teeth from the Omo Valley (Leakey, 1987), and an associated maxilla, mandible and assorted postcrania of a male from West Turkana (KNM-WT 16287) (Harris *et al.*, 1988*a*).
AGE AND GEOGRAPHIC RANGE Approximately 3.36–1.88 Ma, East and West Turkana, Kenya, and the Omo Valley, Ethiopia
ANATOMICAL DEFINITION
High muzzle, maxillary fossae, narrower interorbital region, less sharply converging temporal lines, larger teeth and a wide frontal process of the zygoma differ from *P. chemeroni*. Expanded gonial region in males.

SPECIES cf. *Paracolobus* sp. indet.
REFERRED SPECIMENS Mandibular, maxillary and isolated teeth specimens from Laetoli, Tanzania (Leakey, 1982; Leakey & Delson, 1987); isolated molars and a distal humerus (NK 322'88) from the Western Rift (Senut, 1994); and, possibly, three fragments including an isolated M_3 from Makapansgat (Eisenhart, 1974)
AGE AND GEOGRAPHIC RANGE Approximately 3.8–3.5 Ma, Laetoli, Tanzania and Western Rift, Uganda; approximately 3.0–2.5 Ma, Makapansgat, South Africa
ANATOMICAL DEFINITION
Distinguished from *P. chemeroni* and *P. mutiwa* by smaller teeth. Tooth and femur morphology indicate affinities with *Paracolobus* (rather than *Rhinocolobus*) despite the smaller size.

FAMILY Colobidae gen. et sp. indet.
REFERRED SPECIMENS This heterogeneous collection includes a large colobid from apparent late Pliocene horizons of Nakatsu, Japan, tentatively referred to ?*Dolichopithecus eohanuman* by Delson (1994), a small series of dental specimens assigned to ?*Colobus flandrini* by Delson (1973), a single isolated upper molar (with field number 1951-17-21) from late Neogene Lusso Beds of Upper Semliki Valley, Zaïre (Boaz *et al.*, 1992), a set of 17 isolated teeth recovered from Members B, C and G of the Shungura Formation of the Omo Group, Ethiopia (Leakey, 1987) and a right distal humerus (13P15A) from Sahabi, Libya (Meikle, 1987).
AGE AND GEOGRAPHIC RANGE Pliocene, Japan, Zaïre, Libya
ANATOMICAL DEFINITION
This diverse group of specimens is united only by the apparent colobid qualities of their respective pieces of anatomy identified by their describers. The only clearly colobid feature of the Nakatsu cranium is its relatively great interorbital breadth.

Family Cercopithecidae

Tribe Cercopithecini

GENUS *Cercopithecus* Brunnich, 1772
Small-bodied cercopithecids with small, globular crania, small, low-crowned and puffy-sided cheek teeth, reduced third molars, and relatively little dental and somatic sexual dimorphism.
INCLUDED SPECIES *Cercopithecus* sp. indet.

SPECIES *Cercopithecus* sp. indet.
REFERRED SPECIMENS Although several small cercopithecids possibly assignable to *Cercopithecus* have been recovered from East African sites, only two sets of fossils have been fully described and conclusively diagnosed as belonging to this genus (see Szalay & Delson, 1979). The larger of these sets derives from the Omo Valley, Ethiopia, and comprises two mandibular fragments from a male individual with left I_1–C, P_4–M_3 and P_3–M_1 (L621-4a and b, respectively), a mandibular fragment with an M_1 only of a possible female (Omo 18-'69-532), a right maxillary fragment (P994-8b), four isolated upper molars, three deciduous premolars and one left upper canine (Eck & Howell, 1972; Eck & Jablonski, 1987). The smaller set comprises a left fragmentary mandible of an immature individual with the roots of dP_4 and the unerupted crown of M_1 from Kanam East (M 15923) (Harrison & Harris, 1996).
AGE AND GEOGRAPHIC RANGE Late Pliocene, Ethiopia
ANATOMICAL DEFINITION
Size and morphology of the referred specimens fall mostly within the range of *C. nictitans*. The narrowness of the distal lophids of the M_2 and M_3 in L621-4a and b mirrors the pattern seen in modern *Cercopithecus*.

Tribe Papionini
Subtribe Macacina

GENUS *Macaca* Lacépède, 1799
Cercopithecids displaying narrow interorbital regions, generally rounded muzzles of moderate length, molars with flared sides and relatively short crowns, and moderate to great sexual dimorphism in the canine–premolar complex and postcranium. Interspecific variation in craniodental anatomy, especially in the size and form of the muzzle and supraorbital torus, as well as postcranial anatomy, is considerable. Distinguished from other papionins by the absence, in general, of suborbital fossae and maxillary ridges, and a lack of extreme lateral flare of the molars.
INCLUDED SPECIES AND SUBSPECIES *M. anderssoni, M. fascicularis, M. jiangchuanensis, M. libyca, M. nemestrina, M. sylvanus, M. sylvanus* cf. *sylvanus, M. sylvanus florentina, M. sylvanus majori, M. sylvanus prisca, M. sylvanus* subsp. indet., *Macaca* sp. indet.

SPECIES *Macaca sylvanus* Linnaeus, 1758
A large species distinguished by strong supraorbital torus, deep facial skeleton, relatively small incisors and molars with relatively little lateral flare, relatively longer shearing blades, high cusps and larger crushing basins. This species now comprises nearly all known fossil macaques of the circum-Mediterranean originally designated as separate species.

SUBSPECIES *Macaca sylvanus* cf. *sylvanus* Linnaeus, 1758
REFERRED SPECIMENS Unnumbered isolated molars from Aïn Brimba, Tunisia
AGE AND GEOGRAPHIC RANGE Pleistocene, Tunisia
ANATOMICAL DEFINITION
Relatively brachyodont molars with deep foveae, within the morphological and metric range represented by the living Barbary macaque.

SUBSPECIES *Macaca sylvanus florentina* Cocchi, 1872
TYPE SPECIMEN IGF 10034, partial mandible from Upper Val d'Arno, Italy
AGE AND GEOGRAPHIC RANGE Late Pliocene, southern and central Europe
ANATOMICAL DEFINITION
A geographically widespread and anatomically variable subspecies represented by several mandibles, gnathic fragments, isolated teeth and a few partial long bones from sites in Italy, Spain, Germany, the Netherlands and Croatia (summarized by Delson, 1980; Ardito & Mottura, 1987; Basilici *et al.*, 1991; Rook *et al.*, 1996). Some mandibular corpora possess fossae. Closely comparable in size and morphology to *M. s. sylvanus*; distinguished by its occurrence in middle and late Villafranchian (late Pliocene) horizons of mostly southern Europe.

SUBSPECIES *Macaca sylvanus prisca* Gervais, 1859
TYPE SPECIMEN Partial mandible from Montpellier, France
REFERRED SPECIMENS A poorly known subspecies thought to represent the earliest occurrence of *Macaca* in Europe; it is represented by several isolated teeth and partial jaws from Italy, France, Hungary, Spain and Germany (Delson, 1980).
AGE AND GEOGRAPHIC RANGE Early to middle Pliocene, southern and central Europe
ANATOMICAL DEFINITION
Slightly smaller than extant *M. sylvanus*; distinguished from *M. s. florentina* by smaller size and older provenience.

SUBSPECIES *Macaca sylvanus majori* Azzaroli, 1946
TYPE SPECIMEN unnumbered partial mandible
REFERRED SPECIMENS Known from over 100 specimens, mostly cranial fragments and isolated teeth, including a palate and frontal fragment (BMNHA M15842) and a right M_3 (BMNH 11713). The type series comprises nine unnumbered specimens in addition to the holotype, including an adult mandible, a partial facial skeleton, fragments of a juvenile mandible and palate, two femora and a metatarsal (Azzaroli, 1946).
AGE AND GEOGRAPHIC RANGE Latest Pleistocene, Sardinia
ANATOMICAL DEFINITION
An apparent "island dwarf" form of *Macaca sylvanus, M. s. majori* is about 5%–10% smaller dentally and possesses cheek teeth slightly puffier than in the living Barbary macaque.

SPECIES *Macaca sylvanus* subsp. indet.
REFERRED SPECIMENS Included here are several geographically disjunct but anatomically similar fossils resembling the extant Barbary macaque. These include a right M^2 (BMNH 1892) from Essex and four isolated teeth (including a left M^2 (Hoxne 5219) and a left M^1 (Hoxne 13474)) from Suffolk, England (Napier, 1981; Singer *et al.*, 1982); fragmentary cranial and postcranial remains from Villafranca d'Asti, Italy, including a left maxilla (V. J. 88), and 20 associated limb bones (V. J. 130) (Rook *et al.*, 2001); a series of fragmentary phalanges and a patella fragment from Deutsch-Alternburg, Austria (Fladerer, 1987); and a left P_3 from near Graz, Austria (Fladerer, 1991).
AGE AND GEOGRAPHIC RANGE Pleistocene, Great Britain and Europe
ANATOMICAL DEFINITION
The assemblages represented in this mixed group are united only by their metric and morphological similarities to the living Barbary macaque.

SPECIES *Macaca libyca* Stromer, 1920
TYPE SPECIMEN Unnumbered partial mandible
AGE AND GEOGRAPHIC RANGE Late Miocene, Wadi Natrun, Egypt
ANATOMICAL DEFINITION
Poorly known species apparently lacking some of the cranial distinctions of later Asian and Mediterranean

macaque lineages, including a fossa of the mandibular corpus; teeth in the size range of modern *M. sylvanus*.

SPECIES *Macaca nemestrina* Linnaeus, 1766
REFERRED SPECIMENS Over 100 isolated teeth from three caves in the Padang highlands of Sumatra and six isolated teeth from the Niah Caves of northern Sarawak (Hooijer, 1962*b*, 1962*c*); an associated left maxilla (with I^2–M^3) and mandible with complete dentition (G.M.B.K. 102) (Aimi, 1981).
AGE AND GEOGRAPHIC RANGE Holocene, Sumatra, Java and Sarawak
ANATOMICAL DEFINITION
Large species with deep face, broad muzzle, heavy supraorbital torus continuous with a broad and robust zygoma, broad premolars, cheek teeth with inflated sides and relatively closely approximated cusp apices, and an abbreviated tail of less than 20 caudal vertebrae. The measurements of the teeth referred to *M. nemestrina* by Hooijer (1962*c*) fall within the range of modern *M. n. nemestrina*.

SPECIES *Macaca fascicularis* Raffles, 1821
REFERRED SPECIMENS A large number of isolated teeth and small gnathic fragments from west Malaysia and Sarawak, and from the Indonesian islands of Java (see Fooden (1995) for a complete summary)
AGE AND GEOGRAPHIC RANGE Latest Pleistocene and Holocene, Greater and Lesser Sunda Islands
ANATOMICAL DEFINITION
A small to moderate-sized macaque with a lightly built cranium, a small and laterally rounded supraorbital torus, relatively shallow facial skeleton, temporal lines that meet in a modest sagittal crest in some older males, large anterior teeth relative to size of cheek teeth, and long tails that decrease in length with increasing latitude. Most of the referred specimens are subfossils and many represent the remains of human meals. Their molars are distinguished from the often sympatric *M. nemestrina* by virtue of their smaller size.

SPECIES *Macaca anderssoni* Schlosser, 1924 (Fig. 17.9)
TYPE SPECIMEN PMU M3651, a nearly complete facial skeleton of an adult male with a complete upper dentition (Schlosser, 1925)
REFERRED SPECIMENS *Macaca anderssoni* is represented by a large number of mostly fragmentary gnathic remains. This includes a large series of remains from Locality 1 at Zhoukoudian, near Beijing, including a nearly complete left maxilla with P^3–M^3 (IVPP 1817), a fragmentary mandible of a female with P_3–M_3 (IVPP 1820), a nearly complete mandibular corpus of a female with a complete dentition (IVPP 1824) and an incomplete humerus (Young & Pei, 1933). Fragmentary remains from other Zhoukoudian localities, as well as from localities in Shaanxi and Sichuan Provinces are also recognized (see Pan & Jablonski, 1987; Fooden, 1990 for summaries).

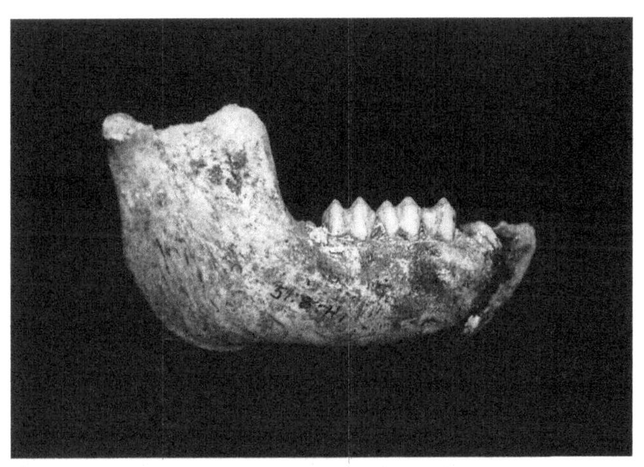

A

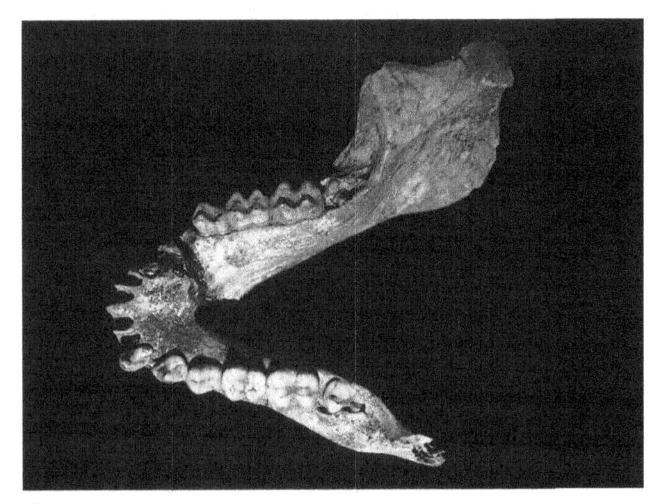

B

Fig. 17.9 *Macaca anderssoni*. (A) Zhoukoudian 51 : 8 : HI female mandible, right buccal view; (B) same, occlusal view.

AGE AND GEOGRAPHIC RANGE Early to middle Pleistocene, north–central China
ANATOMICAL DEFINITION
This species is taken here to include a very large number of Middle Pleistocene fossil macaques originally assigned to *M. robustus* Young, 1934. The teeth of the *M. anderssoni* and *M. robustus* samples cannot be distinguished metrically or morphologically and are, therefore, tentatively judged here to be conspecific, following Szalay & Delson (1979). The prominent elevation of the nasals in the skull of *M. robustus* (e.g., AMNH 39044) is not shared with *M. anderssoni*, however, as noted by Fooden (1990) and may ultimately warrant a recognition of the two samples as separate species.

SPECIES *Macaca jiangchuanensis* Pan *et al.*, 1992
TYPE SPECIMEN YV 3000, a mandibular corpus of a male with P_3–M_3 (Pan *et al.*, 1992)
AGE AND GEOGRAPHIC RANGE Early Pleistocene, Yunnan, southwestern China

ANATOMICAL DEFINITION
A macaque morphologically similar to but larger than extant *M. arctoides*; more robust and dentally larger than *M. anderssoni*, but dentally smaller that *Procynocephalus wimani*. Recognized only from the type specimen, the most remarkable aspect of this species is its deep and robust mandibular corpus, with prominent sublingual and mylohyoid lines. The cheek teeth bear thick enamel and in morphology are very similar to those of extant *M. arctoides* (Jablonski, 1993*b*; Jablonski *et al.*, 1994).

SPECIES *Macaca arctoides* I. Geoffroy Saint-Hilaire, 1803
REFERRED SPECIMEN NBV 00101.1-2, a broken and distorted partial cranium of a probable female with right I^2–M^2 and left I^2–M^3 (Li *et al.*, 1982)
AGE AND GEOGRAPHIC RANGE Late Pleistocene, Jiangsu, China
ANATOMICAL DEFINITION
A large macaque with a relatively short, rounded muzzle, broad and laterally projecting frontal processes of the zygomata, robust zygomatic arches, very wide bizygomatic breadth, laterally rounded supraorbital torus, temporal lines that in adult males meet anteriorly to form a sagittal crest, a mandible that deepens slightly from anterior to posterior, premolar and molar teeth with relatively thick enamel, and a tail with generally 10 or fewer caudal vertebrae. The allocation of this rather battered specimen to *Macaca arctoides* is made on the basis of its tooth dimensions, which compare favorably to those of the extant form (Li *et al.*, 1982).

SUBSPECIES *Macaca arctoides subfossilis* Jouffroy, 1959 (comb. nov.)
TYPE SPECIMEN (LPV-P PV.F1, a nearly complete cranium of a male with right P^3–M^3 and left P^3–M^3) (Jouffroy, 1959)
AGE AND GEOGRAPHIC RANGE Late Pleistocene, northern Vietnam
ANATOMICAL DEFINITION
A subspecies slightly smaller than extant *M. arctoides*, but resembling the extant species in the anterior inclination of the frontal process of the zygoma and the relatively great breadth of the zygoma. This subspecies is recognized by the type specimen only, which resembles extant male *M. arctoides* except in cranial and dental size, where the subfossil is at or near the lower limit of the range of variation of the modern form.

SPECIES *Macaca mulatta* Zimmerman, 1780
REFERRED SPECIMENS A small series of isolated teeth and fragmentary jaws from eastern and southern China, Vietnam and India, mostly summarized by Fooden (2000), but also including specimens from Luoding, Guangdong, China (Gu *et al.*, 1996)
AGE AND GEOGRAPHIC RANGE Pleistocene and Holocene, eastern and southern China, Vietnam and India
ANATOMICAL DEFINITION
A small and lightly built species with a relatively short, rounded muzzle, rounded orbits, a thin and laterally rounded supraorbital torus, lightly built zygomatic arches, relatively bunodont molars with increased lateral flare. Distinguished from *M. fascicularis* only in the somewhat less marked protrusion of the nasal rostrum and in the possession of significantly shorter tails that vary little in length with respect to latitude. Species-level diagnosis of fossil remains has been conducted mostly on the basis of the relatively small size of the teeth and the occurrence of the fossils in areas where the extant rhesus macaque occurs today; it should be noted, however, that the size range represented by these teeth overlaps almost completely with that of *M. fascicularis*.

SPECIES *Macaca* cf. *thibetana* Milne Edwards, 1870
REFERRED SPECIMENS A series of isolated teeth, mostly molars (including LPX 97, LPX 85, LPX 115, LPX 136, LPX 137, LPS 246 and LPS 248) from the caves at Luoding, Guangdong, China (Gu *et al.*, 1996)
AGE AND GEOGRAPHIC RANGE Pleistocene, southeastern China
ANATOMICAL DEFINITION
A large macaque distinguished from *M. arctoides* and *M. assamensis* by slightly less anteriorly protruding frontal processes of the zygomata, a squared muzzle cross-section, maxillary fossae in some males, elevated nasal bones and molars with relatively taller, less inflated crowns. This collection of isolated teeth is referred to *M. thibetana* on the basis of strong similarities between the fossils and the living species in tooth crown shape and size.

SPECIES *Macaca* cf. *fuscata* Blyth, 1875
REFERRED SPECIMENS Three specimens from Japan, being an almost complete adult cranium from the Shikoku District (unnumbered) (Iwamoto, 1975), an isolated upper canine from northern Japan and an isolated molar from western Japan (Iwamoto & Hasegawa, 1972)
AGE AND GEOGRAPHIC RANGE Pleistocene, Japan
ANATOMICAL DEFINITION
A macaque of moderate size, distinguished by long, elevated nasal bones, a supraorbital torus that is enlarged medially and indented over glabella in many individuals, and a curved temporal process of the zygoma that dips inferiorly, giving the zygomatic arch a slightly sigmoid shape. Similarities in the form and orientation of the frontal process of the zygoma between *M. fuscata* and *M. anderssoni* render specific allocation of the Shikoku cranium uncertain (Iwamoto, 1975); unfortunately, the highly diagnostic temporal process of the zygoma is absent in the specimen. The isolated teeth, on the other hand, are very similar to those of the extant Japanese macaque (Iwamoto & Hasegawa, 1972).

SPECIES *Macaca* sp. indet.
REFERRED SPECIMENS A large assortment of fragmentary remains, mostly isolated teeth and jaws, from the length and breadth of the genus's distribution, including Great Britain (Gibbard, 1994), Spain (Penela, 1983; Moyà-Solà *et al.*, 1992), Germany (Thenius, 1965; Fladerer, 1989), the Ukraine (Tesakov & Maschenko, 1992) and China (Pan & Jablonski, 1987; Jablonski *et al.*, 1994). Among the more complete and intriguing specimens placed here are apparently late Miocene mandibular fragments from Spain (Köhler *et al.*, 2000), and a Pliocene right mandibular fragment from Sahabi, Libya (Meikle, 1987).
AGE AND GEOGRAPHIC RANGE Late Miocene through Pleistocene, Europe and north Africa
ANATOMICAL DEFINITION
This highly variable assemblage of materials is united only by its possession of "primitive papionin" characteristics associated with the highly variable array of modern macaques, but especially reminiscent of *M. sylvanus*.

GENUS *Procynocephalus* Schlosser, 1924
Large, macaque-like monkeys with evenly sloping muzzles without fossae, broad palates, molars with low, rounded cusps, and baboon-like long bones featuring characteristics indicating a terrestrial habitus. The genus is distinguished from *Macaca* by its larger size.
INCLUDED SPECIES *P. subhimalayanus*, *P. wimani*

SPECIES *Procynocephalus wimani* Schlosser, 1924
TYPE SPECIMEN A female mandible (unnumbered) from Xin'an County, Henan, China (Schlosser, 1924)
REFERRED SPECIMENS The fragmentary teeth and partial postcranial bones from Locality 12 of Zhoukoudian (Teilhard de Chardin, 1938) represent the most significant assemblage for this species. The species has also been reported as being present at Yushe, Shaanxi, and in Qingxing County, Hebei (Young & Pei, 1933; Teilhard de Chardin, 1938).
AGE AND GEOGRAPHIC RANGE Late Pliocene, north–central and northeastern China
ANATOMICAL DEFINITION
Same as for the genus. The proportions of the distal humerus and the retroflexion of the olecranon indicate that the species was highly terrestrial (Jolly, 1967).

SPECIES *Procynocephalus subhimalayanus* von Meyer, 1848 (Fig. 17.10)
TYPE SPECIMEN BMNH 37157, fragment of right maxilla of a female with damaged canine and P^3, and P^4–M^3 (Meyer, 1848)
REFERRED SPECIMENS The more recently described *Procynocephalus pinjorii* Verma, 1969 is also here referred to this species, following Szalay & Delson (1979). *Cynocephalus falconeri* Lydekker, 1886, considered by Jolly (1967) to be a macaque, is also referred here to *P. subhimalayanus*, following Szalay & Delson (1979). Pending a review of the original materials, the large fossil cercopithecids from northern India are probably all best referred to the single species, *P. subhimalayanus*, in light of their gnathic and dental similarities to one another and to the type specimen of *P. subhimalayanus*.
AGE AND GEOGRAPHIC RANGE Pleistocene, northwestern India
ANATOMICAL DEFINITION
Same as for the genus. The type specimen, originally classified as *Cynocephalus subhimalayanus*, has long been recognized as a large macaque-like cercopithecid best classified as *Procynocephalus* rather than *Papio* (Jolly, 1967).

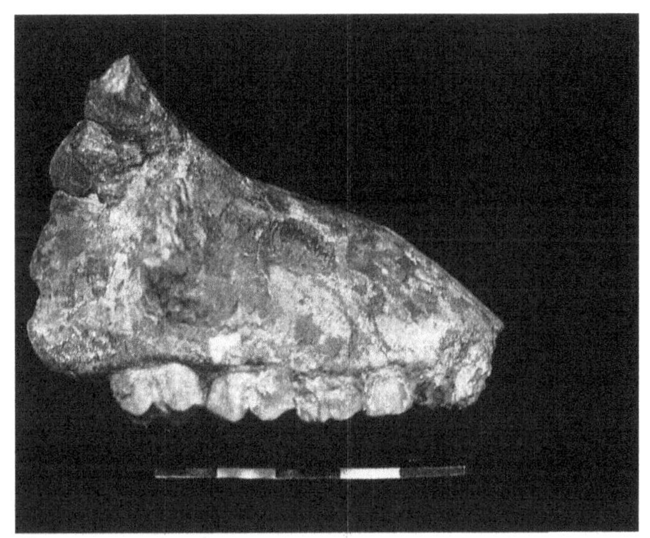
A

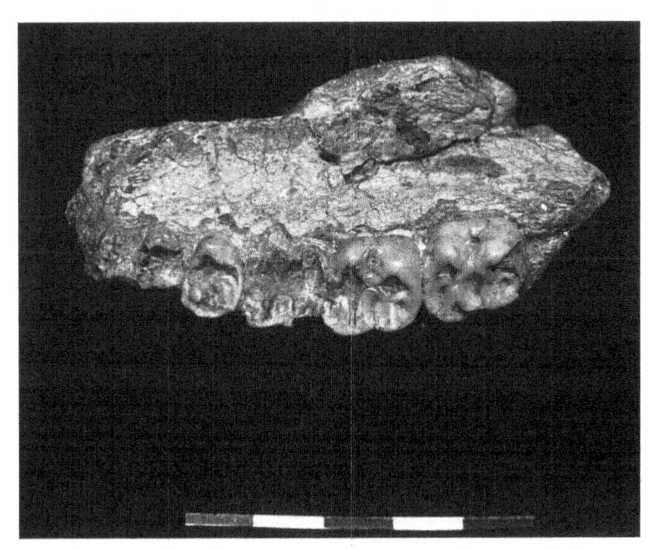
B

Fig. 17.10 *Procynocephalus subhimalayanus*. (A) BMNH 37157 (holotype) maxilla, right lateral view; (B) same, occlusal view.

GENUS *Paradolichopithecus* Necrasov *et al.*, 1961
A large cercopithecid exhibiting strongly macaque-like

cranial and dental features such as an anteriorly tapering muzzle of moderate length and squarish cross-section, absent or very shallow facial and mandibular fossae, a slightly posteriorly inclined mandibular ascending ramus, a shallow mandibular corpus that deepens slightly anteriorly, thick-enameled, low-crowned molars, and moderate canine–premolar dimorphism. This suite of characters is combined with a highly terrestrially adapted, baboon-like postcranial skeleton. The history of its classification is tortuous (for details, see Szalay & Delson, 1979). Originally described as a colobid related to *Dolichopithecus ruscinensis* (Depéret, 1929), the type specimen from France went on to be compared with a partial face of a large monkey from Romania by Necrasov and colleagues (Necrasov *et al.*, 1961). The latter authors maintained that the French and Romanian fossils were probably colobid with some cercopithecid characteristics, but further research by Delson (Szalay & Delson, 1979) clearly demonstrated the unambiguous cercopithecid affinities of both sets of remains. Simons (1970) noted that *Paradolichopithecus* might be synonymized appropriately with *Procynocephalus* given strong similarities in the dental and gnathic remains available for comparison. The two genera are retained here, following Szalay & Delson (1979), because the similarities invoked to unite *Paradolichopithecus* and *Procynocephalus* could be as easily invoked, in error, to unite other large, dentally conservative cercopithecid genera that are easily distinguished from one another in cranial form. The genus is distinguished from *Macaca* by its larger size, and cannot be readily distinguished from *Procynocephalus*. Were further remains of *Procynocephalus* to be recovered in China or India, the discussion about synonymy of *Paradolichopithecus* and *Procynocephalus* clearly would need to be reopened.
INCLUDED SPECIES *P. arvernensis*

SPECIES *Paradolichopithecus arvernensis* Depéret, 1929
TYPE SPECIMEN cranium and mandible of a female (unnumbered) (Depéret, 1929)
REFERRED SPECIMENS A large series of fossils attributable to *P. arvernensis* from Romania is now recognized (Szalay & Delson, 1979; E. Delson, pers. comm.) and ongoing work by Delson and colleagues promises to significantly illuminate the biology of this species. Referred here are specimens described by Trofimov (1977) as *P. suskhini* from Tajikistan.
AGE AND GEOGRAPHIC RANGE Later Pliocene, France, Romania and Tajikistan
ANATOMICAL DEFINITION
Same as for the genus.

Subtribe Papionina

GENUS *Parapapio* Jones, 1937
A papionin of moderate size distinguished by a suite of cranial features, including a facial skeleton that displays a generally straight or only slightly concave profile from nasion to rhinion, a lightly built supraorbital torus that lacks a prominent glabella or an ophyronic groove behind it, weak maxillary ridges and poorly excavated or absent maxillary and mandibular fossae, weak temporal lines and slender zygomatic arches, a wide and relatively long muzzle, and molars and premolars of the typical papionin type displaying moderate amounts of lateral flare (Fig. 17.11). Except for moderate sexual dimorphism in the canine–premolar complex, dental sexual dimorphism is slight. Much of the definition of species is based on absolute dental dimensions. Extensive discussion in the literature over the validity of the number of recognized species (Eisenhart, 1974; Freedman, 1976; Szalay & Delson, 1979) has not resulted in significant changes in specimen allocation.
INCLUDED SPECIES *P. ado, P. antiquus, P. broomi, P. jonesi, P. whitei, Parapapio* sp. indet.

SPECIES *Parapapio broomi* Jones, 1937 (Fig. 17.11A, B)
TYPE SPECIMEN STS 564, a badly damaged male cranium lacking teeth
REFERRED SPECIMENS The species is recognized from Makapansgat and Bolt's Farm, but the largest series of specimens derives from Sterkfontein. The most complete specimens of the species include a male cranium (M202, 1326/2), a cranium of an immature female (MP 224), a fragmentary cranium (M3065) of a male with a possibly associated mandible (M3067), a female cranium (STS 254A), a nearly complete mandible (STS 363) and a partial mandible (STS 562) (Freedman, 1957, 1976; Maier, 1970).
AGE AND GEOGRAPHIC RANGE Plio-Pleistocene, South Africa
ANATOMICAL DEFINITION
Distinguished from other *Parapapio* species by its large cranium, short, broad palate, straight nasal profile and deeply set orbits framed by a slender, non-protruding supraorbital torus. Some later examples of the species show elongated muzzles (Eisenhart, 1974). In dental size, *P. broomi* is intermediate between *P. jonesi* and *P. whitei*.

SPECIES *Parapapio jonesi* Broom, 1940 (Fig. 17.11C)
TYPE SPECIMEN STS 565, a damaged female cranium with a nearly complete dentition
REFERRED SPECIMENS Over 20 specimens from Sterkfontein, and small numbers of specimens from Swartkrans, Kromdraai, Taung and, probably, Makapansgat.

Fig. 17.11 (opposite) *Parapapio*. (A) *P. broomi*, STS 254 A female cranium, right lateral view; (B) *P. broomi*, M 202 male cranium, right lateral view; (C) *P. jonesi*, STS 565 (holotype) cranium, left lateral view; (D) *P. whitei*, STS 563 female mandible (holotype), right buccal view; (E) same, occlusal view; (F) *P. antiquus*, Tvl. 639 cranium, vertical view; (G) *P. antiquus*, CT 5364 male cranium, left lateral view; (H) *P. ado*, BMNH 14940 (holotype) partial mandible, right buccal view. (A–G) Photographs courtesy of Len Freedman.

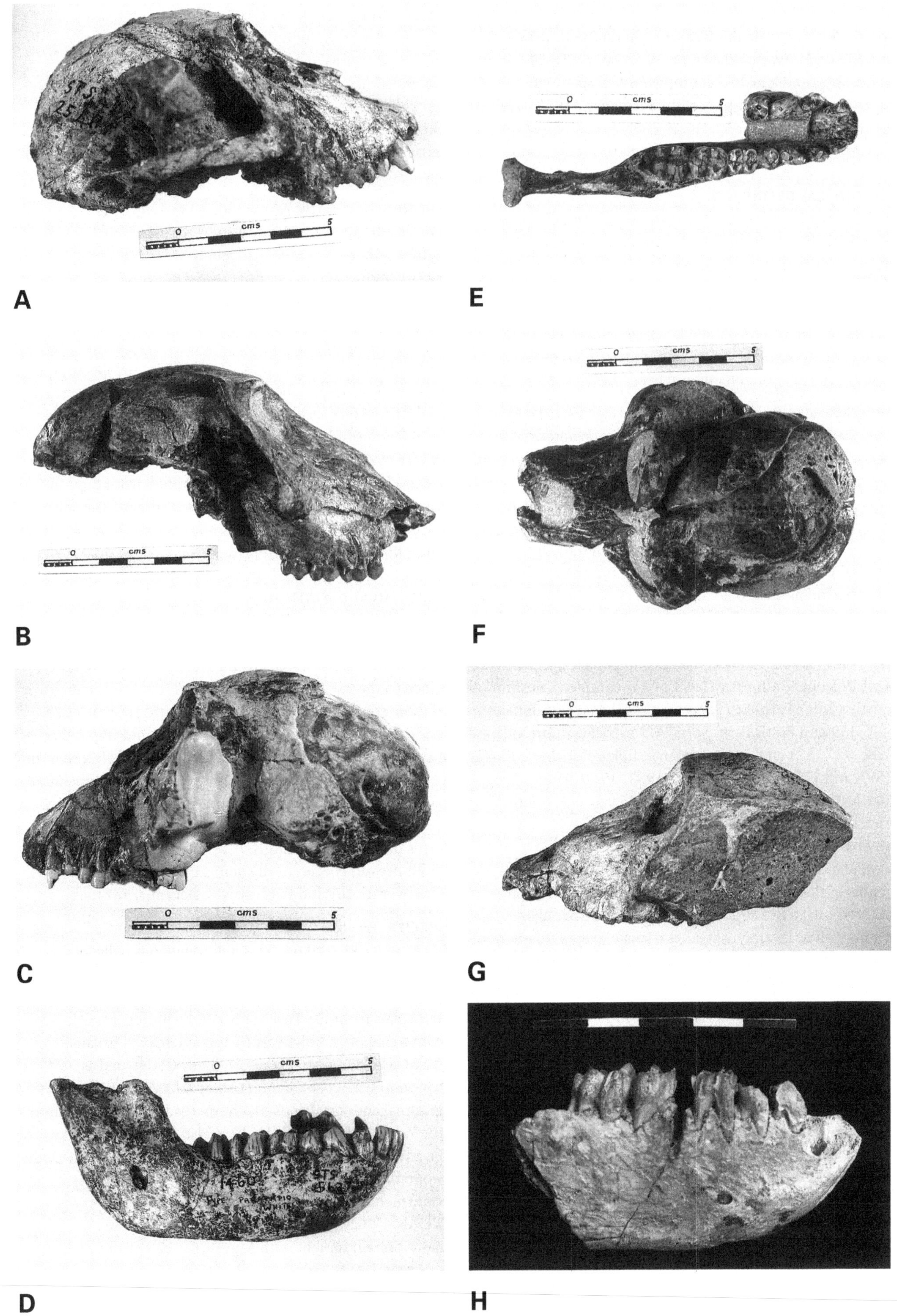
A
B
C
D
E
F
G
H
0
cms
5

The best (and least ambiguous) representatives of this species are two male muzzles from Sterkfontein (STS 250 and STS 367), two partial crania of females (STS 547 and STS 372A), and a male cranium from Makapansgat (M2961) previously recorded by Freedman (1957) as *P. broomi* (Maier, 1970).

AGE AND GEOGRAPHIC RANGE Plio-Pleistocene, South Africa

ANATOMICAL DEFINITION

A species of *Parapapio* distinguished by its relatively short muzzle, as demonstrated by the presence of nasal bones that slope inferiorly at a steep angle. It also appears to be smaller in overall size relative to *P. broomi* and *P. whitei*. According to Eisenhart (1974), the specimens attributed to *P. jonesi* represent a metrically heterogeneous collection, with some specimens warranting placement in other *Parapapio* species or even other genera.

SPECIES *Parapapio whitei* Broom, 1940 (Fig. 17.11D, E)

TYPE SPECIMEN STS 563, the right half and anterior part of the left of a mandible of a female with I_1–M_3 present on the right side and I_1–P_4 on the left

REFERRED SPECIMENS Referred specimens have been recovered from the South African sites of Sterkfontein, Taung, Makapansgat and Bolt's Farm, with the largest sample deriving from Sterkfontein. Significantly, however, two specimens attributed to *Parapapio* cf. *P. whitei* have been recorded from West Turkana, Kenya (Harris *et al.*, 1988). The more complete of these is a partial left mandible with M_{1-3} (WT 16751). Among the most complete specimens of *P. whitei* from South Africa are a nearly complete cranium of a male (Bolt's Farm 56785), two partial crania of females (STS 259 and Bolt's Farm 56652), a partial cranium of a male (M2962) and two partial mandibles of males (STS 389A and STS 533) (Freedman, 1957, 1960, 1965).

AGE AND GEOGRAPHIC RANGE Plio-Pleistocene, South Africa and 3.36–2.35 Ma, Kenya

ANATOMICAL DEFINITION

In dental dimensions, *P. whitei* is the largest species of its genus. Well-defined maxillary ridges, moderate steepness of the nasals, and a high, relatively short muzzle also distinguish it. Distinguished from smaller representatives of *Papio* by the absence of clear dental sexual dimorphism (apart from that in the canine–premolar complex).

SPECIES *Parapapio antiquus* Haughton, 1925 (Fig. 17.11F, G)

TYPE SPECIMEN CT 5364, a cranium of a probable male, lacking teeth and the posterior portion of the calvaria (Haughton, 1925)

REFERRED SPECIMENS Known only from Taung, the species is represented by a modest series of fossils, the most complete of which are four male crania (TP 9, CT 5364, CT 5356 and T 10) and three female crania (Tvl 639,T 17 and M3078) (Freedman, 1957, 1976; Maier, 1971).

AGE AND GEOGRAPHIC RANGE Plio-Pleistocene, South Africa

ANATOMICAL DEFINITION

The most morphologically distinctive of the *Parapapio* species, *P. antiquus* exhibits prominent maxillary ridges, excavated maxillary fossae, elevated nasals that are straight in profile from glabella to the rhinion, a relatively long cheek toothrow, and significantly reduced M^3s in the female. The muzzle dorsum is considerably flatter than in *P. broomi*.

SPECIES *Parapapio ado* Hopwood, 1936 (Fig. 17.11H)

TYPE SPECIMEN BMNH 14940, right mandibular corpus with weathered P_3–M_3 and alveoli of I_1–C (Hopwood, 1936)

REFERRED SPECIMENS The most complete of the Laetoli *Parapapio* fossils is a mandible of a young adult female (LAET 1209) and a partial mandible of a female with left P_3–M_3 and right C (root) and P_3–M_2 (LAET 223). A collection of hand bones and one foot bone associated with an isolated P_4 (LAET 3904) is also referred here. The species has also been recognized from a well-preserved mandible of a male, WT 16752, from West Turkana (Harris *et al.*, 1988*a*). The species has also been tentatively identified from deposits at Allia Bay, on the eastern shore of Lake Turkana, Kenya (Leakey *et al.*, 1995*b*).

AGE AND GEOGRAPHIC RANGE Pliocene, Tanzania and Kenya

ANATOMICAL DEFINITION

Originally described as a species of *Cercocebus*, Delson (Szalay & Delson, 1979) has referred the type specimen to *Parapapio*. A large number of fossils from Laetoli have now been added to the list of referred specimens for the species. Most of these specimens are fragmentary gnathic remains that lack the diagnostic features of the muzzle and upper face most strongly diagnostic of *Parapapio*; they are referred to the genus because the available anatomy is that of a generalized papionin within the range of variation seen in *Parapapio* species from southern Africa (Leakey & Delson, 1987). In terms of dental dimensions, *P. ado* may differ from the South African species of the genus by its possession of narrower molars and P_3 (Leakey & Delson, 1987).

SPECIES *Parapapio* sp. indet.

REFERRED SPECIMENS This heterogeneous collection includes fossils originally described as *Papio serengetensis* from Laetoli, Tanzania (Dietrich, 1942) and a large series of specimens from the Turkana Basin, including a specimen from Kanapoi (KNM-KP 286) and a damaged isolated M_3 (KNM-KP 287) from Ekora originally designated as Papionini gen. et sp. indet. (Leakey & Leakey, 1976). The proximal and distal ends of a humerus from the Albertine Rift Valley of Uganda–Zaïre have also been referred to as cf. *Parapapio* (Senut, 1994). The occurrence of cf. *Parapapio* in the geographic corridor between east and south Africa is confirmed by the record of isolated teeth from Botswana (KOAN 1'90a–KOAN 1'90e) (Senut, 1996) and a mandibular fragment with M_2 (HCRP-128) from Malawi

(Bromage & Schrenk, 1986). The occurrence of the genus early in the Pliocene of South Africa is attested by the retrieval of isolated premolars (left P^4 SAM-L 20660a and left P_4 SAM-L 20660b) from Langebaanweg (Grine & Hendey, 1981). Large collections of fragmentary gnathic remains and isolated teeth referable to *Parapapio* have also been recovered from Lothagam, Allia Bay, and east of Lake Turkana.

AGE AND GEOGRAPHIC RANGE Pliocene, Turkana Basin, Kenya, Uganda–Zaïre, Malawi and South Africa

ANATOMICAL DEFINITION

Small papionins distinguished by the absence of specializations of the facial skeleton and dentition characteristic of *Papio* or *Theropithecus*. The fragmentary postcranial material assigned to the genus has been so on the basis of its size and generalized papionin morphology. The postcranium of *Parapapio* is poorly documented. This series comprises a large number of specimens spanning two million years, which vary greatly in size and morphology. It is likely that the series includes more than one species and that it includes material referable to existing *Parapapio* species as well as new species. A comprehensive examination of this material, including a comparison with known South African specimens, is clearly needed.

GENUS *Cercocebus* É. Geoffroy Saint-Hilaire, 1812

Large monkeys distinguished cranially from *Cercopithecus* and *Macaca* by the presence of suborbital fossae and large incisors relative to molar size; distinguished from *Lophocebus* by the presence of relatively shallow (as opposed to deeply excavated) maxillary fossae and larger, more laterally flared molars.

INCLUDED SPECIES *Cercocebus* sp. indet.

SPECIES *Cercocebus* sp. indet.

REFERRED SPECIMENS Fragmentary remains from geographically disjunct sub-Saharan sites, including a partial cranium from Makapansgat (M3057/8/9, M218), fragmentary jaws from Kromdraai (Eisenhart, 1974), a large series of mandibles (e.g., KNM-ER 594, KNM-ER 568, KNM-ER 604 and KNM-ER 822), one associated mandible and maxilla (KNM-ER 590) and isolated teeth from East Turkana (Leakey & Leakey, 1976), and a mandibular fragment from Olduvai (OLD/55 108) (Jolly, 1965).

AGE AND GEOGRAPHIC RANGE Late Pliocene and Early Pleistocene, South Africa, Kenya and Tanzania

ANATOMICAL DEFINITION

The fragmentary nature of most of the material assigned to *Cercocebus* sp. indet. inspires caution. Most of the referred materials have shown anatomical tendencies toward *Cercocebus* in molar form and size, and the ratio of incisor size to the length of the cheek toothrow, but have lacked the parts of the anatomy – the midfacial skeleton – most diagnostic of the genus.

GENUS *Lophocebus* Palmer, 1903

Large monkeys distinguished cranially from *Cercopithecus* and *Macaca* by the presence of suborbital fossae and large incisors relative to molar size; distinguished from *Cercocebus* by the presence of deeply excavated maxillary fossae, small, *Cercopithecus*-like molars with little lateral flare, and robust, convexly curved incisor teeth.

INCLUDED SPECIES *Lophocebus* sp. indet.

REFERRED SPECIMENS M 15922, associated right I^1, right I^2, and right M_2, all heavily worn, and M 18800, a fragment of right temporal bone, from Kanam East (Harrison & Harris, 1996)

AGE AND GEOGRAPHIC RANGE Plio-Pleistocene, Kanam East, Kenya

ANATOMICAL DEFINITION

The general size of these remains indicates that they derive from a medium-size papionin. The robustness and mesiodistal convexity of the incisors clearly indicates that they are from *Lophocebus* rather than *Macaca*, *Papio* or *Cercocebus*. The associated lower molar also shows similarities to *Lophocebus* in its shape, relative degree of buccal flare and overall morphology (Harrison & Harris, 1996).

GENUS *Dinopithecus* Broom, 1937

A very large and sexually dimorphic papionin with a large and rugged skull. The presence of strong masticatory and nuchal muscles is inferred by the robustness of the zygomatic arches, the strength of the temporal lines in both sexes which, in males, join to form a sagittal crest, and strong nuchal crests. The supraorbital torus is robust and the interorbital region is broad. The muzzle, known only for females, lacks maxillary fossae. The genus is distinguished by large, broad molar teeth that often show accessory cuspules and short P_3s with large anterior foveae in males (Freedman, 1957).

INCLUDED SPECIES *D. ingens*

SPECIES *Dinopithecus ingens* Broom, 1937 (Fig. 17.12)

TYPE SPECIMEN SB 7, most of the left and part of the right corpus of a mandible of a male with some damaged teeth (Broom, 1937) (Fig. 17.12C)

REFERRED SPECIMENS Unequivocal remains of the species are known from two sites only, the type locality of Skurweberg (one specimen) and Swartkrans (over 30 specimens). Among the collection from Swartkrans are three nearly complete crania of females (SK 553, SK 600 and SK 603), a fragmentary cranium of a male (SK 599), a partial mandible of a male (SK 401), a partial muzzle of a female (SK 574) and the cranium of a juvenile (SK 554) (Freedman, 1957). Freedman (1957) noted morphological similarities of the large (and worn) molar teeth of *D. ingens* to those of *Gorgopithecus major*. A series of specimens from Leba, Angola considered possibly referable to *Dinopithecus* (Simons & Delson, 1978), have been shown to belong to an early, large representative of *Theropithecus* (Jablonski, 1994).

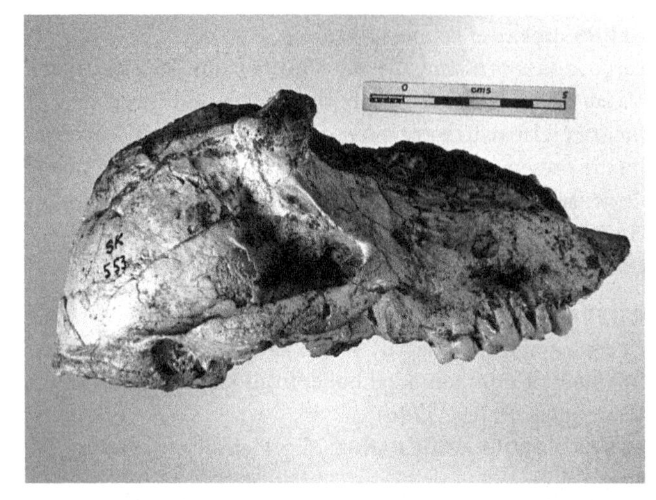

A

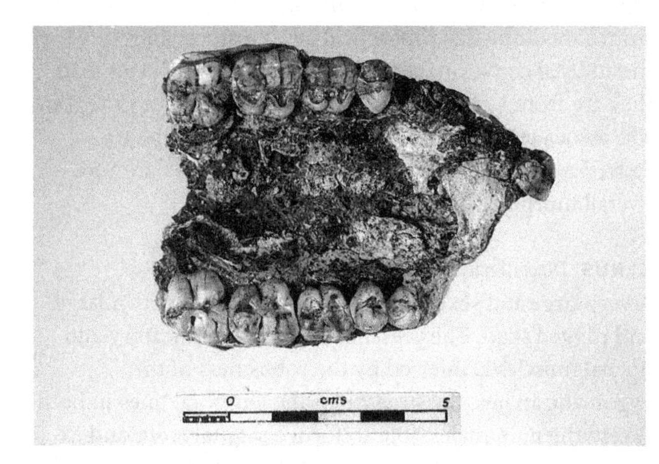

B

C

Fig. 17.12 *Dinopithecus ingens.* (A) SK553 female cranium, right lateral view; (B) SK 604 female palate, occlusal view; (C) SB 7 male partial mandible, occlusal view. Photographs courtesy of Len Freedman.

AGE AND GEOGRAPHIC RANGE Pliocene, South Africa
ANATOMICAL DEFINITION
Same as for the genus. *Dinopithecus ingens* represents one of the largest cercopithecoids known, exceeded in overall size only by the largest representatives of *Theropithecus oswaldi* of the Late Pleistocene.

GENUS *Gorgopithecus* Broom & Robinson, 1946
A very large papionin with a relatively short, high and narrow muzzle, deep maxillary fossae and long calvaria. In the male the orbits are laterally elongated and the interorbital breadth is great. The teeth show great similarities to those of *Dinopithecus ingens*; dental sexual dimorphism in all elements except the canine–premolar complex, is slight (Freedman, 1957). The facial skeleton of this species is highly distinctive, with its maxillae dropping steeply inferiorly from the sides of the nasal bones and nasal aperture to the alveolar margin. The short nasals are oriented almost horizontally to the posterior edge of the nasal aperture; the muzzle drops steeply from there to the anterior margin of the aperture on the premaxilla (Freedman, 1957). The zygomatic arch in both sexes is heavily built and the bizygomatic breadth is great.
INCLUDED SPECIES *G. major*

SPECIES *Gorgopithecus major* Broom, 1940 (Fig. 17.13)
TYPE SPECIMEN KA 193, two linked teeth, M^2 and M^3, considerably worn, from Kromdraai, South Africa (Broom & Jensen, 1946)
REFERRED SPECIMENS This species is best known from the crushed male cranium that is still imbedded in matrix and that is missing the right I^2 and M^3 and the left I^2, M^{2-3} (KA 192); a partial female cranium with C–M^3 present on the right side and C–P^4 present on the left (KS 153) and two badly damaged mandibles (KA 150 and KA 152) are also recognized.
AGE AND GEOGRAPHIC RANGE Pleistocene, Kromdraai, South Africa
ANATOMICAL DEFINITION
Same as for the genus.

GENUS *Theropithecus* I. Geoffroy Saint-Hilaire, 1843
Large, heavily built and highly sexually dimorphic papionins distinguished by a suite of craniodental characteristics related to the prolonged chewing of tough and abrasive vegetation and by postcranial features reflecting adaptations to extreme terrestriality and the dexterous manipulation of objects in the hand. The main craniodental specializations of *Theropithecus* are its high- and columnar-cusped cheek teeth with deep foveae and pronounced infoldings of thick enamel, arrangement of the mandibular teeth in an anteroposteriorly convex curve (reversed Curve of Spee) in most, a deep mandibular with a robust and vertically oriented symphysis, a high and generally vertically oriented mandibular ascending ramus, a

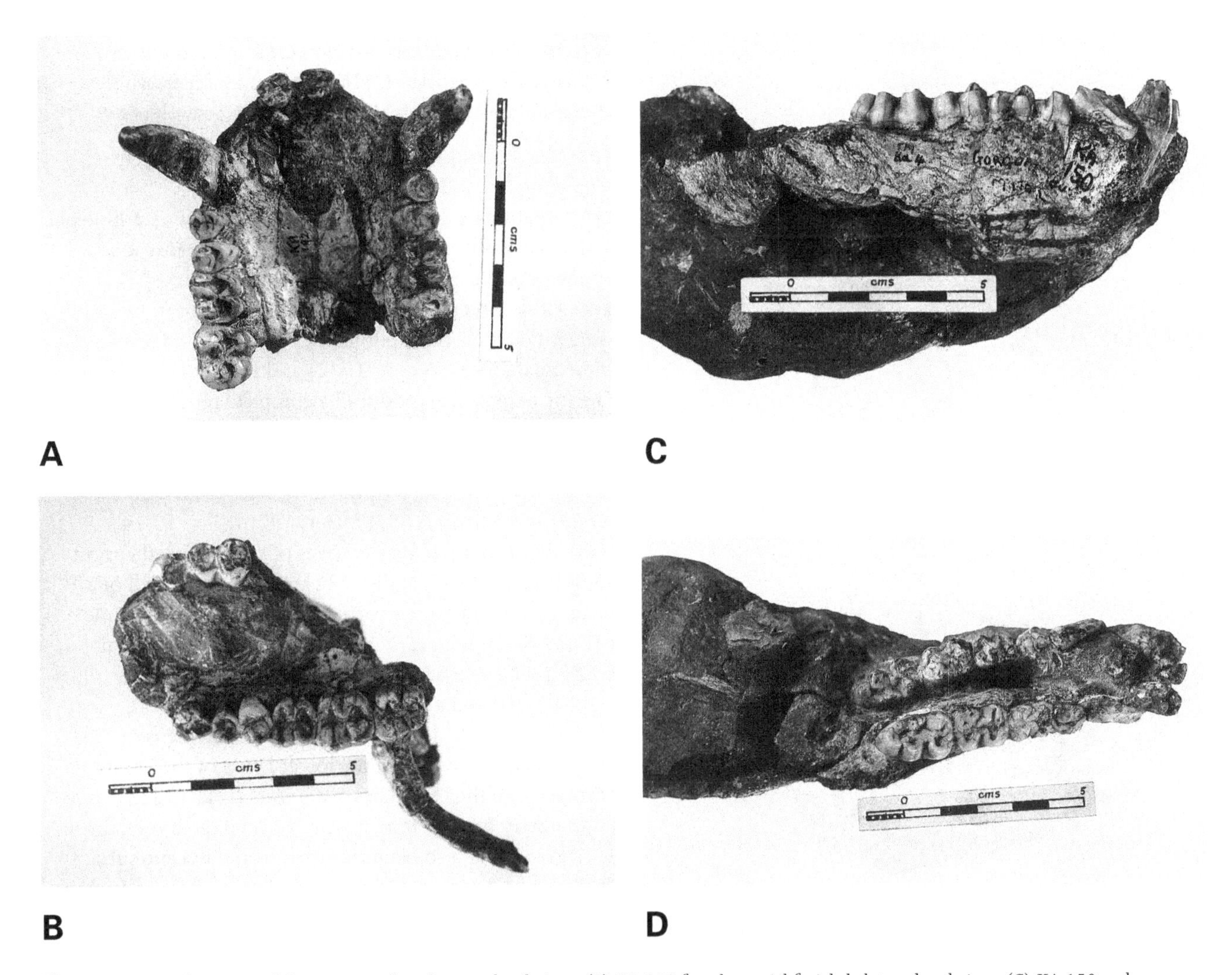

Fig. 17.13 *Gorgopithecus major.* (A) KA 192 male palate, occlusal view; (B) KA 153 female partial facial skeleton, basal view; (C) KA 150 male mandible, right buccal view; (D) same, occlusal view. Photographs courtesy of Len Freedman.

deep posterior maxilla, a strong postorbital constriction, and an anteriorly set temporalis musculature that in males produces a sagittal crest at or anterior to the bregma. Postcranial characteristics of the genus include the retroflexed ulnar olecranon process seen in other highly terrestrial papionins, but unique to *Theropithecus* is the high opposability index produced by its elongated pollex and abbreviated index finger. Sexual dimorphism in the canine–premolar complex and in somatic dimensions is considerable. The subgeneric distinction between *Theropithecus* (*Theropithecus*) and *Theropithecus* (*Simopithecus*), proposed by Jolly (1972), that distinguished between living and fossil representatives of the genus, respectively, is not followed here, in favor of the subgeneric definitions advanced by Delson (1993).

SUBGENUS *Theropithecus* (*Theropithecus*) I. Geoffroy Saint-Hilaire, 1843
Distinguished from *Theropithecus* (*Omopithecus*) by tendencies toward reduction of muzzle length, weak development of maxillary ridges and corresponding rounding of the muzzle cross-section, reduction of incisors and molarization of premolars. Many of these characteristics can be related to an adaptation for ingestion and prolonged chewing of large amounts of fibrous and abrasive vegetation. This subgenus includes the living representative of *Theropithecus*, the gelada, as well as the fossil species with abbreviated muzzles. All workers would agree that the fossils to be included in this subgenus form a geographically and temporally extensive array, and that morphological differentiation between fossil populations is often subtle. Some workers have advocated the lumping of these populations into a single species, *T. oswaldi* (Dechow & Singer, 1984), with designation of three slightly morphologically differentiated "chrono-subspecies": *T. oswaldi darti*, *T. oswaldi oswaldi* and *T. oswaldi leakeyi* (Leakey, 1993). Here, the integrity of *T. darti* as a separate species is retained, following Eck & Jablonski (1987), but the two other subspecific names are accepted, as well as that of *T. oswaldi delsoni*, the only Asian form of the genus formally recognized, following Delson (1993).

A

B

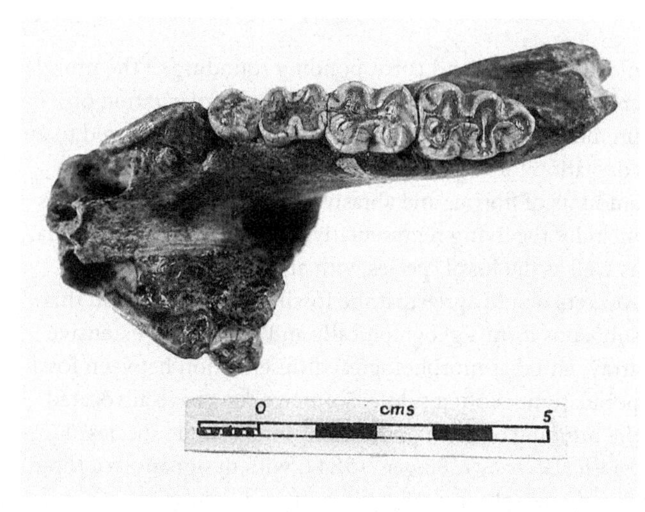

C

Fig. 17.14 *Theropithecus (Theropithecus) darti*. (A) M 636 (left) M 672 (upper right) and M 669 (lower right), occlusal views; (B) M 626 male mandible, right buccal view; (C) same, occlusal view. Photographs courtesy of Len Freedman.

INCLUDED SPECIES AND SUBSPECIES *T. (Theropithecus) darti, T. (Theropithecus) oswaldi, T. (Theropithecus) oswaldi oswaldi, T. (Theropithecus) oswaldi delsoni, T. (Theropithecus) oswaldi leakeyi*

SPECIES *Theropithecus (Theropithecus) darti* Broom & Jensen, 1946 (Fig. 17.14)
TYPE SPECIMEN M 201, 1326/1, a damaged mandible of a male with badly damaged teeth, from Makapansgat Limeworks (Broom & Jensen, 1946)
REFERRED SPECIMENS The most significant samples of this species that have been published derive from Makapansgat (Freedman, 1957; Maier, 1972) and Hadar (Eck, 1993). The former site has produced a complete female skull (M3073), fragmentary maxillae of females (M669, M672 and M636), a partial mandible of a male (BPI M626) and mandibular fragments from two females (M672 and M669) (Freedman, 1957; Maier, 1972) (Fig. 17.14). Hadar has produced a large series of specimens, the most complete of which are the only known partial skull of a male (AL205-1a–1c), a partial skull of a female juvenile (AL185-5a–5c), a nearly complete cranium of a female (AL321-12), a neurocranium of a young adult male (AM187-10) and a partial mandible of a female (AL196-3a) (Eck, 1993). Fragmentary specimens of *T. (T.) darti* are also known from the lower levels of the Shungura Formation of the Omo Group deposits (Eck, 1987). Some fragmentary gnathic remains and isolated teeth referable to *T. (T.) darti* have also been retrieved from Pliocene sites of the Turkana Basin, including Koobi Fora and West Turkana.
AGE AND GEOGRAPHIC RANGE Pliocene, South Africa and East Africa, including Ethiopia, Kenya and Tanzania
ANATOMICAL DEFINITION
Similar to *Theropithecus (Theropithecus) oswaldi*, *T. (T.) darti* is distinguished by its smaller overall size, elevated nasal bones (in most specimens), a narrow, ovoid piriform aperture, broadly bowed and smoothly curved zygomatic arches that create very large infratemporal fossae, and a deeply excavated anteroinferior angle of the triangular depression of the ramus of the mandible.

SPECIES *Theropithecus (Theropithecus) oswaldi* Andrews, 1916
TYPE SPECIMEN Because a holotype was not designated in Andrews's original description, two following syntypes were named retrospectively: BMNH 11539, a cranium of a female with complete dentition, and BMNH 11537, maxillae of a female, both from Kanjera, Kenya
AGE AND GEOGRAPHIC RANGE Late Pliocene and Pleistocene, Spain, sub-Saharan Africa and northern India
ANATOMICAL DEFINITION
Same as for the genus and subgenus. This long-lived and widespread species exhibits considerable variation in time and space. The species is distinguished from other papionins and, particularly, from *Theropithecus darti* by its larger size, robust postglenoid processes and greater relative reduction of incisors and canines. The recognition of

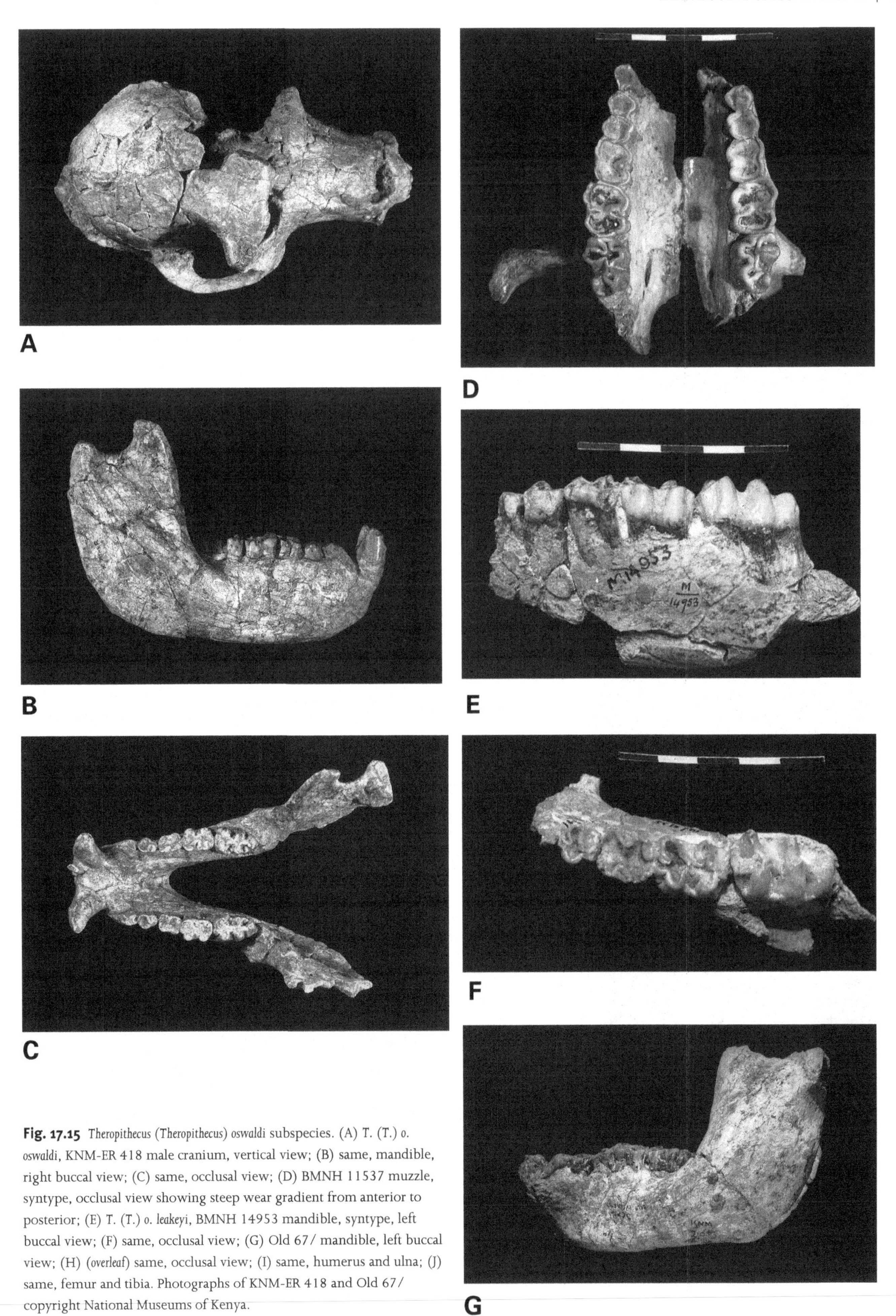

Fig. 17.15 *Theropithecus (Theropithecus) oswaldi* subspecies. (A) *T.* (*T.*) *o. oswaldi*, KNM-ER 418 male cranium, vertical view; (B) same, mandible, right buccal view; (C) same, occlusal view; (D) BMNH 11537 muzzle, syntype, occlusal view showing steep wear gradient from anterior to posterior; (E) *T.* (*T.*) *o. leakeyi*, BMNH 14953 mandible, syntype, left buccal view; (F) same, occlusal view; (G) Old 67/ mandible, left buccal view; (H) (*overleaf*) same, occlusal view; (I) same, humerus and ulna; (J) same, femur and tibia. Photographs of KNM-ER 418 and Old 67/ copyright National Museums of Kenya.

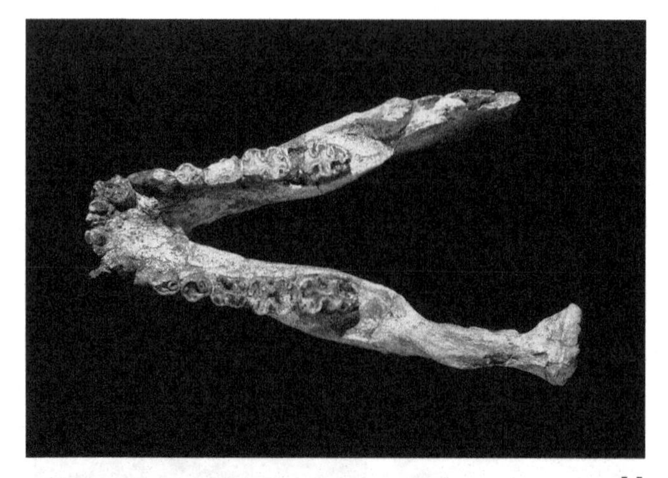

H

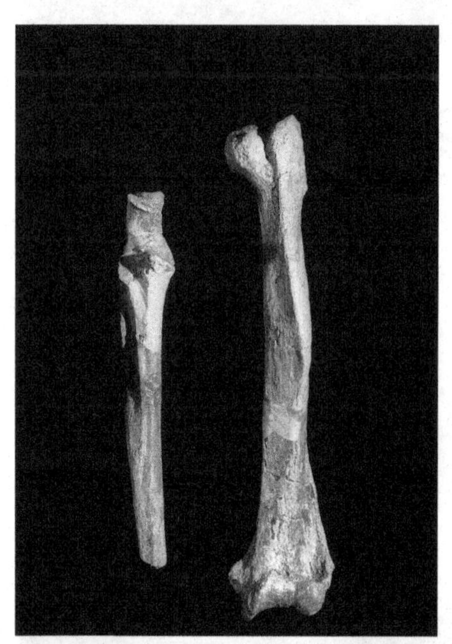

I

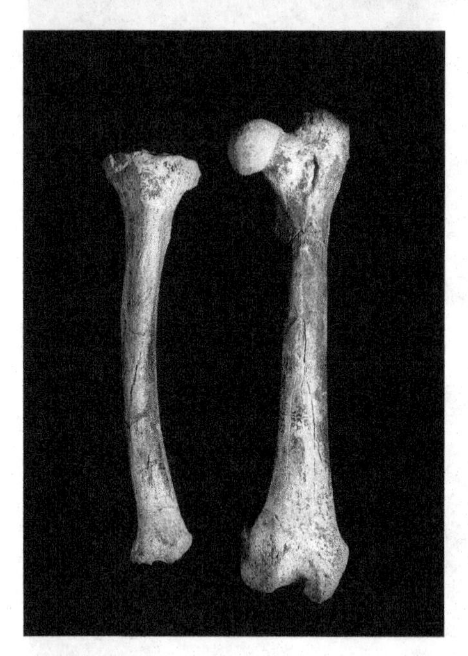

J

Fig. 17.15 (cont.)

Theropithecus (*Theropithecus*) *oswaldi oswaldi* and *Theropithecus* (*Theropithecus*) *oswaldi leakeyi* recognizes M. G. Leakey's (1993) and Delson's (1993) distinction of two time-successive subspecies, a smaller earlier one and a larger, later one, respectively. It should be noted, however, that the fossil populations included in these respective subspecies have not been formally identified. Therefore, specimens have been allocated here to the two subspecies according to the spirit, rather than the letter, of their distinction.

SUBSPECIES *Theropithecus* (*Theropithecus*) *oswaldi oswaldi* Andrews, 1916 (Fig. 17.15A–D)
TYPE SPECIMEN Same as for *Theropithecus* (*Theropithecus*) *oswaldi* Andrews, 1916
REFERRED SPECIMENS This subspecies comprises a very large sample of cranial materials, including several complete crania and mandibles, and some postcranial remains (Fig. 17.15A–D). Napier (1981) provides comprehensive listings of most of the known specimens for Kanjera, Kenya, and Delson *et al.* (1993) for other sites. Among the most complete specimens now recognized are several from the type locality of Kanjera, including a complete cranium of a female (M 14936), distal humeri (M18789, M18721), a complete ulna (M 11546), and a complete radius (M 1154). Other complete or nearly complete specimens include a large series from East Turkana: two male crania (KNM-ER 1531 and KNM-ER 18925), two female crania (KNM-ER 180 and KNM-ER 971), two male mandibles (KNM-ER 865 and KNM-ER 18925) and one female mandible (KNM-ER 567). At the geographical extremities of the subspecies's distribution are a female cranium from Swartkrans (SK 561) and an isolated lower first or second molar from Spain (Gibert *et al.*, 1995).
AGE AND GEOGRAPHIC RANGE Late Pliocene and Early Pleistocene, Murcia (Spain) and sub-Saharan Africa, specifically, Kanjera and East and West Turkana (Kenya), Olduvai (Bed I and Lower Bed II) and Peninj (Tanzania), Omo Valley (Ethiopia), Kaiso (Uganda), Swartkrans and Hopefield (South Africa)
ANATOMICAL DEFINITION
Distinguished from *Theropithecus* (*Theropithecus*) *oswaldi leakeyi* by the smaller size of its cheek teeth and the slightly larger size of its incisors and canines.

SUBSPECIES *Theropithecus* (*Theropithecus*) *oswaldi leakeyi* Hopwood, 1934 (Fig. 17.15E–J)
TYPE SPECIMEN BMNH 14680, left corpus of a mandible of a juvenile male from Olduvai Gorge, Tanzania (Hopwood, 1934)
REFERRED SPECIMENS The subspecies is recognized from a large number of cranial and postcranial specimens, some of the best examples coming from the upper portions of the Olduvai sequence: a partial skeleton (Old 1472, 57, the famed type specimen of the giant "*Simopithecus jonathani*"), a nearly complete mandible of a male (O67/5603) and a muzzle with partial left zygomatic arch of a male (OLD/69,

S.133) (Leakey & Leakey, 1973). Ternifine has produced well-preserved portions of a mandible of a male (MNHN-P TER 1702, 1703 and 1815) (Delson & Hoffstetter, 1993). Many specimens of this subspecies have not yet been described.

AGE AND GEOGRAPHIC RANGE Late early to late Pleistocene, northern Africa: Ternifine (Algeria) and Thomas Quarries (Morocco), and East Africa: Olorgesailie and Kapthurin (Kenya), upper Bed II, Bed III, Bed IV and Masek Beds, Olduvai Gorge (Tanzania), Middle Awash (Ethiopia) and Kaiso (Uganda)

ANATOMICAL DEFINITION

This subgenus is distinguished from T. (T.) *o. oswaldi* by its greater body size, slightly greater cheek tooth dimensions, further molarization of the premolars and more pronounced reduction of the anterior dentition. The incisors of T. (T.) *o. leakeyi* are quite tiny, and the canines are significantly reduced in height, but not in basal dimensions. Compared to all other forms of *Theropithecus*, the P_3 of this subspecies retains the least honing function. This subspecies comprises the largest cercopithecoids known, with late representatives from Olduvai estimated to weigh in excess of 60 kg (Leakey, 1993).

SUBSPECIES *Theropithecus (Theropithecus) oswaldi delsoni* Gupta & Sahni, 1981

TYPE SPECIMEN A small fragment of maxilla bearing M^{2-3} from Mirzapur, Punjab, India (unnumbered) (Gupta & Sahni, 1981)

AGE AND GEOGRAPHIC RANGE Later early or middle Pleistocene, northern India

ANATOMICAL DEFINITION

Described originally as the distinct species *Theropithecus delsoni* (Gupta & Sahni, 1981), this sole Asian representative of the genus is now considered a subspecies of T. *oswaldi* because its teeth fall metrically within the range of T. *o. leakeyi* (Delson, 1993). Delson, to recognize its geographical more than its morphological distinctiveness, has preserved the subspecific nomen.

SUBGENUS *Theropithecus (Omopithecus)* Delson, 1993

Distinguished from *Theropithecus (Theropithecus)* by its elongated muzzle with a relatively flat dorsum and well-developed maxillary ridges, a large and anteriorly expanded zygoma, a robust zygomatic arch that is triangular in cross-section, a robust mandibular symphysis bearing sinusoidal mental ridges, and features of the postcranial skeleton that reflect an adaptation for elbow stability but shoulder flexibility (Delson, 1993).

INCLUDED SPECIES AND SUBSPECIES T. *(Omopithecus) baringensis*, T. *(Omopithecus) brumpti*, T. *(Omopithecus) quadratirostris*

SPECIES T. *(Omopithecus) baringensis* R. Leakey, 1969 (Fig. 17.16)

TYPE SPECIMEN KNM-BC 2, an associated partial cranium and mandible from Chemeron

REFERRED SPECIMENS The species is recognized in the literature by a further mandibular specimen from Chemeron (KNM-BC 1647), a mandibular specimen from East Turkana (KNM-ER 3038) (Leakey, 1993) and a series of cranial remains from the Angolan site of Leba including a complete juvenile cranium (TCH 38 '90), a partial cranium of a male (TCH 25 '90) and a partial mandible of a male (CAN 30 '90) (Jablonski, 1994) (Fig. 17.16). Other remains referable to this species are discussed and illustrated as *Papio quadratirostris* by Delson & Dean (1993). More remains of the species from sites along the eastern and western shores of Lake Turkana have been tentatively identified and are awaiting study (N. Jablonski & M. G. Leakey, unpublished data).

AGE AND GEOGRAPHIC RANGE Pliocene, Omo and Turkana Basins (Ethiopia and Kenya), Angola

ANATOMICAL DEFINITION

Same as for the subgenus. *Theropithecus (Omopithecus) baringensis* shows the most conservative facial morphology of the subgenus, with its broad muzzle lacking well-defined maxillary ridges and the zygomatic arches showing anterior thickening, not flare. The molars are somewhat lower-crowned and exhibit less pinching of the cusps than in T. (O.) *brumpti*.

SPECIES *Theropithecus (Omopithecus) quadratirostris* Iwamoto, 1982 (comb. nov.)

TYPE SPECIMEN A nearly complete cranium from the Omo Valley, Ethiopia (unnumbered) (Iwamoto, 1982)

REFERRED SPECIMENS The species is only recognized with certainty from the type specimen; specimens allocated by Delson & Dean (1993) to *Papio quadratirostris* are here assigned to T. (O.) *baringensis*.

AGE AND GEOGRAPHIC RANGE Pliocene, Turkana Basin (Ethiopia and Kenya)

ANATOMICAL DEFINITION

Same as for the subgenus. Characters of the cranium that are somewhat intermediate between T. (O.) *baringensis* and T. (O.) *brumpti* distinguish this species. The muzzle is long and broad and bears rounded maxillary ridges; the zygomatic arches are robust, triangular in cross-section at the zygomaticomaxillary suture, and thickened anteriorly; the frontal process of the zygoma is broad, lending a great depth to the malar region as in other *Theropithecus* species. The species is distinguished from T. (O.) *baringensis* by its elongated neurocranium, slightly more posteriorly oriented temporalis musculature, and a union of the temporal lines just posterior to bregma to form a sagittal crest in males.

SPECIES *Theropithecus (Omopithecus) brumpti* Arambourg, 1947 (Fig. 17.17)

TYPE SPECIMEN The syntype of the species comprises three specimens: Omo 001, a left maxillary fragment with M^2 and unerupted M^3, Omo 002, a right mandibular fragment with M_{1-2}, and Omo 003, a left mandibular fragment with a worn M_3; the lectotype was subsequently

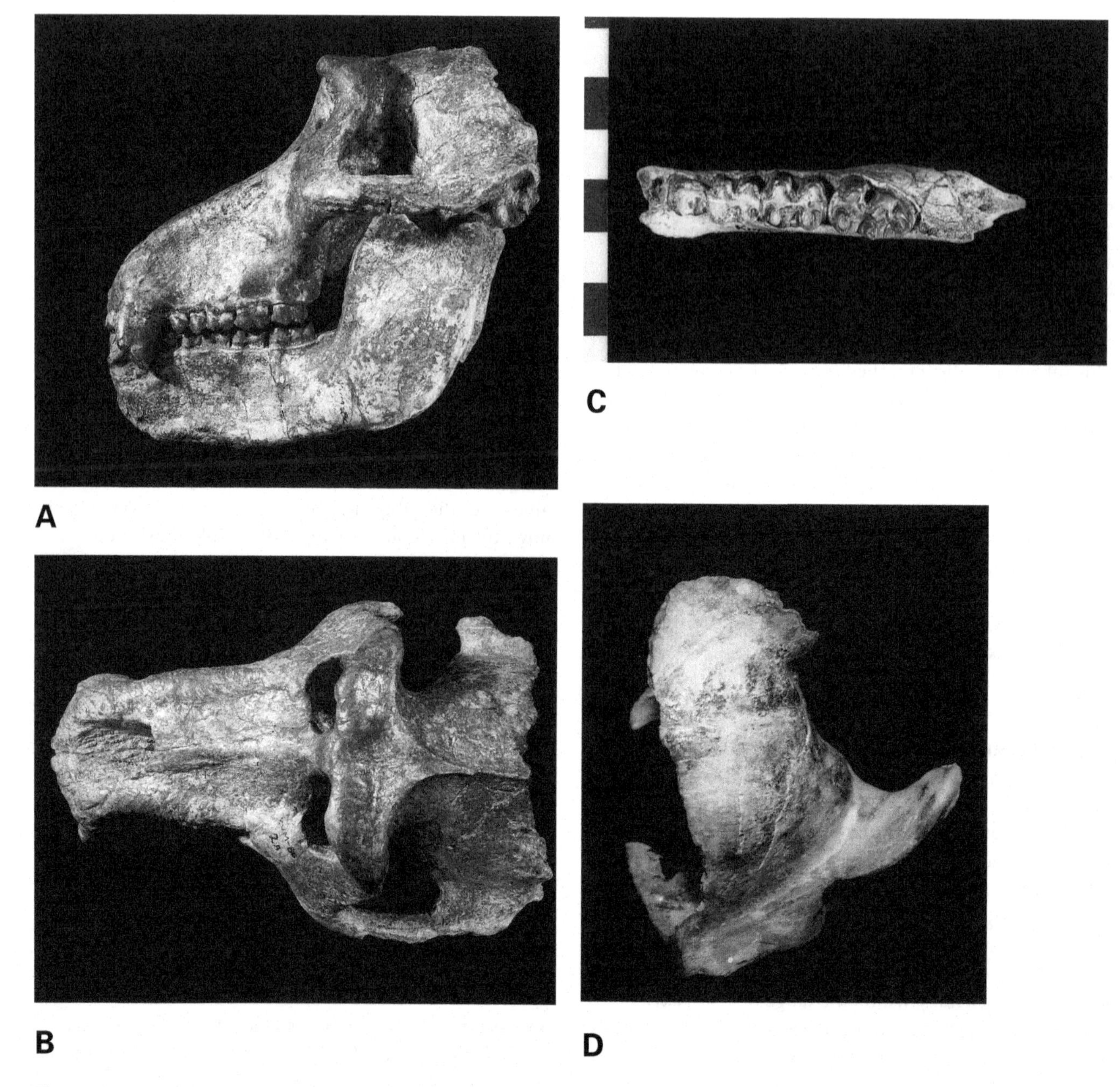

Fig. 17.16 *Theropithecus (Omopithecus) baringensis*. (A) KNM-BC 2 a, b (holotype) skull, left lateral view; (B) same, vertical view; (C) KNM 1647-A partial mandible, occlusal view; (D) TCH 25 '90 male partial cranium, vertical view. Photographs of KNM-BC 2 a, b and KNM 1647-A copyright National Museums of Kenya.

designated as Omo 001 (Eck & Howell, 1972)

REFERRED SPECIMENS Several good examples of this species have been retrieved from deposits of the Omo and Turkana Basins, including two nearly complete crania of adult males (L345-287 and KNM-ER 16828), partial crania of males (L338y-2257, L345-3, and L32-154), partial crania of females (L32-155, young adult, and L122-34) and nearly complete mandibles of males (L576-8, KNM-ER 2015) (Eck & Jablonski, 1987; Leakey, 1993).

AGE AND GEOGRAPHIC RANGE Pliocene, Turkana Basin (Ethiopia and Kenya)

ANATOMICAL DEFINITION

Same as for the subgenus, with exaggeration of some facial features, specifically, those of the maxillae and zygomatic arch. *Theropithecus (Omopithecus) brumpti* is distinguished from other species of the subgenus by its very high-crowned, straight-sided and columnar-cusped molars, more deeply excavated fossae of the mandibular corpus, more pronounced maxillary ridges and by the flaring zygomatic arches present in both sexes the underside of which accommodated the attachment of a greatly enlarged superficial masseter muscle (Eck & Jablonski, 1987). The shape of the maxillary ridges and the shape and orientation of the zygomaticomaxillary flare vary considerably between individuals of the same sex.

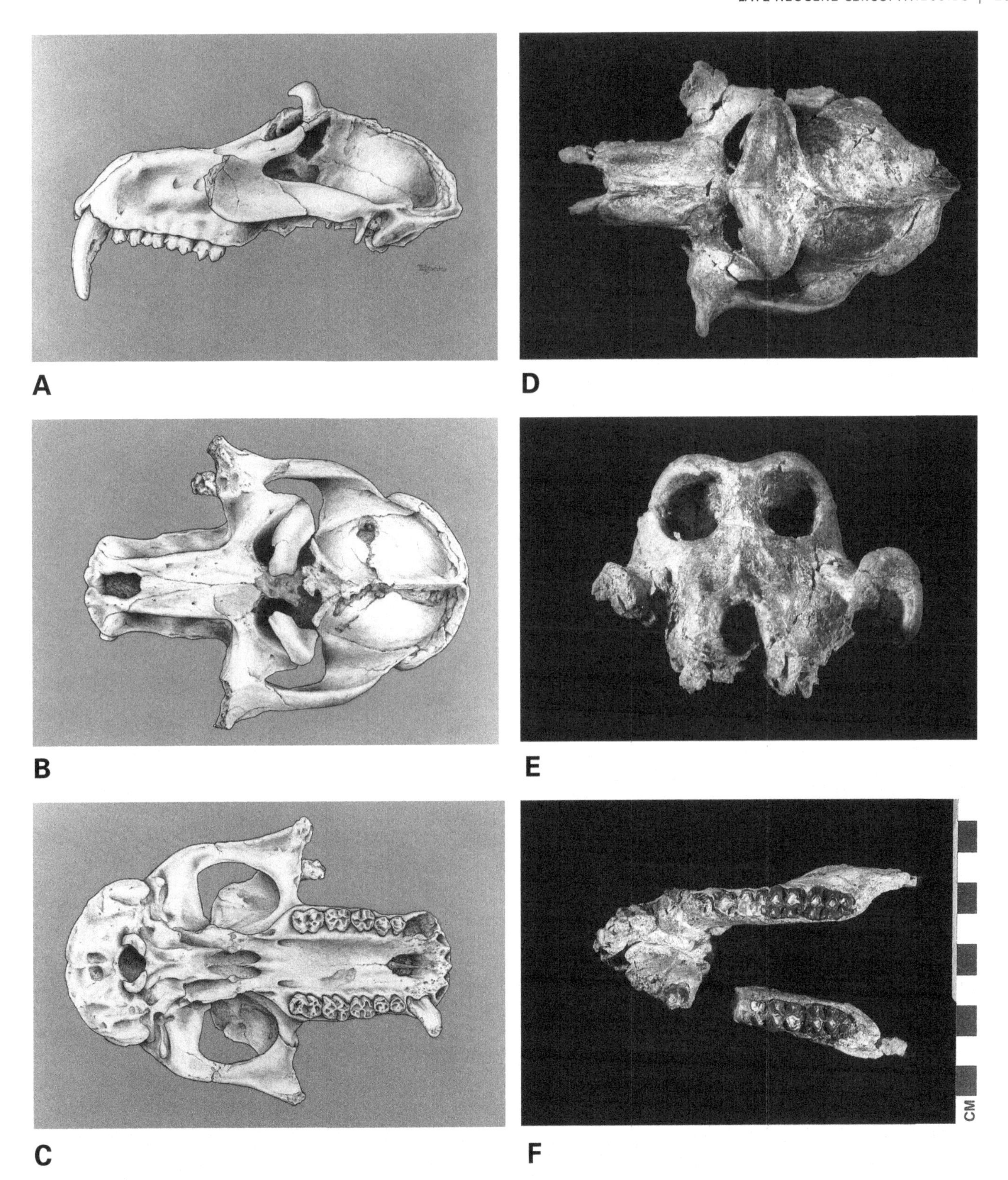

Fig. 17.17 *Theropithecus (Omopithecus) brumpti*. (A) L345-287 male cranium, left lateral view; (B) same, vertical view; (C) same, basal view; (D) KNM-WT 16828 male cranium, vertical view; (E) same, frontal view; (F) KNM-ER 3038 female mandible, occlusal view. (A–C) Illustrations by Peter Gaede; (D–F) photographs copyright National Museums of Kenya.

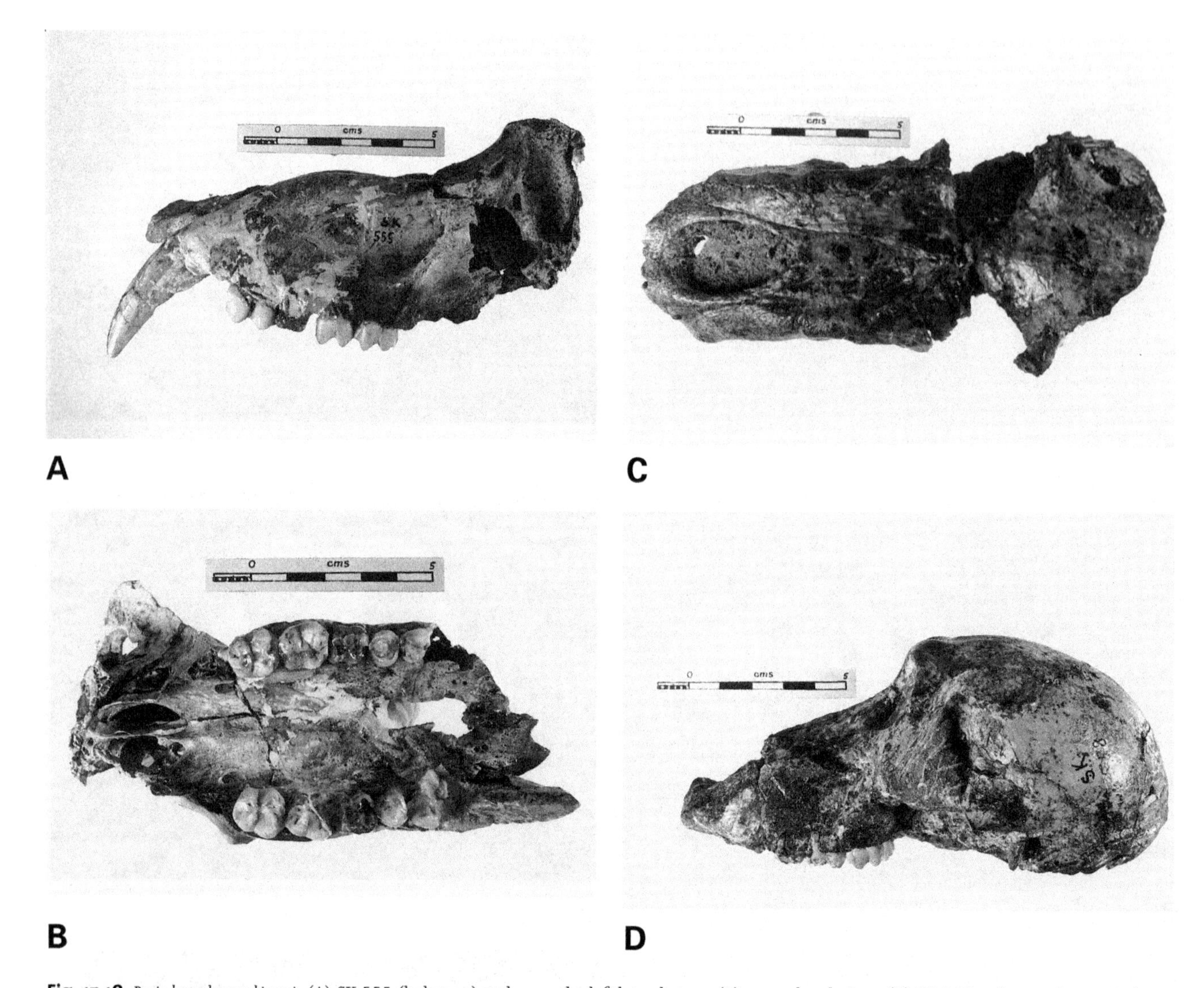

Fig. 17.18 *Papio hamadryas robinsoni*. (A) SK 555 (holotype) male muzzle, left lateral view; (B) same, basal view; (C) SK 560 male muzzle, vertical view; (D) SK 588 female cranium, left lateral view. Photographs courtesy of Len Freedman.

GENUS *Theropithecus* subgen. et spec. indet.
REFERRED SPECIMEN Two lower molars from locality Senga 5, Upper Semliki Valley, Zaïre (Boaz *et al.*, 1992).
AGE AND GEOGRAPHIC RANGE Pliocene, Zaïre
ANATOMICAL DEFINITION
These teeth are reported as being equivalent in size to fossils of *T.* (*Theropithecus*) *o. oswaldi* and *T.* (*Omopithecus*) *brumpti* from the Omo Group Deposits, Shungura Formation Member 6 (Boaz *et al.*, 1992).

GENUS *Papio* Müller, 1773
A long-muzzled cercopithecid of moderate to large body size with extreme levels of canine–premolar and somatic sexual dimorphism, and long bones exhibiting clear adaptations to terrestriality (Fig. 17.18). Exhibiting a wide range of body size that grades into that of *Parapapio* at its lower end, *Papio* specimens and species have often been and continue to be confused with *Parapapio*. *Papio* can be distinguished from *Parapapio*, however, by its pronounced concavity between glabella and inion, an angle between the plane of the orbits and the slope of the muzzle, a heavy supraorbital torus with a prominent glabella, generally deep facial fossae, and prominent temporal lines in males that fuse in older, larger individuals to create a posterior sagittal crest. Classification of extinct and extant species within *Papio* has been unstable. The assignment of all living forms of the genus into several subspecies within the single species *P. hamadryas*, following Groves (1989), is widely but not universally advocated. A consolidation into *Papio izodi* of three small species of *Papio* recognized from South Africa (*P. izodi*, *P. wellsi* and *P. angusticeps*) was advocated by Szalay & Delson (1979) and is tentatively followed here. The retention of individual species names for these forms by several workers (e.g., McKee, 1993; McKee & Keyser, 1994) must, however, be noted.
INCLUDED SPECIES AND SUBSPECIES *P. izodi*, *P. hamadryas robinsoni*, *P. hamadryas ursinus*, *P. hamadryas* subsp. indet., cf. *Papio* sp. indet.

SPECIES *Papio hamadryas* Linnaeus, 1758

SUBSPECIES *Papio hamadryas ursinus* Kerr, 1792
Distinguished from other subspecies by an extremely long and narrow muzzle, with sharp parallel maxillary ridges
REFERRED SPECIMENS A partial cranium of a male from "Pretoria", the Transvaal, originally described as *Papio spelaeus* (unnumbered) (Broom, 1936*b*)
AGE AND GEOGRAPHIC RANGE Pleistocene, Transvaal, South Africa
ANATOMICAL DEFINITION
Distinguished from living *Papio hamadryas ursinus* by its much larger teeth and larger maxillary ridges (Broom, 1936*b*; Freedman, 1976).

SUBSPECIES *Papio hamadryas robinsoni* Freedman, 1957 (Fig. 17.18)
TYPE SPECIMEN SK 555, a well-preserved muzzle of an adult male
REFERRED SPECIMENS Referred materials include over 100 specimens from Swartkrans, and smaller numbers of specimens from Kromdraai, Cooper's, Swartkrans II, Gladysvale, Bolt's Farm and Skurveberg. Among the most complete specimens is another well-preserved muzzle of a male (SK 560), a damaged muzzle of a male (Bolt's Farm 56795), a nearly complete but somewhat distorted cranium of a female (SK 588), two partial crania of female (SK 557 and Bolt's Farm 56767), a well-preserved muzzle and palate of a female (SK 562) and a nearly complete mandible of a female (SK 407) (Freedman, 1957, 1965).
AGE AND GEOGRAPHIC RANGE Pleistocene, South Africa
ANATOMICAL DEFINITION
A large species distinguished by the expression of a very flat muzzle dorsum and nasal bones that lie slightly below the level of the maxillary ridges instead of rising progressively above them posteriorly as in *P. h. ursinus*. Freedman (1957) notes that *P. h. robinsoni* is distinguished by "cramping" of the molar row and overlapping of the lower canine and P_3 in several specimens.

SUBSPECIES *Papio hamadryas* subsp. indet.
REFERRED SPECIMENS Juvenile cranium from Olduvai Gorge, Tanzania (unnumbered) (Remane, 1925); a series of mostly fragmentary cranial specimens and isolated teeth from the Omo Valley, Ethiopia, including a partial skull (L185-6), a maxillary fragment with P^3–M^3 (L4-13) and a left mandibular fragment with left P_3–M_2 (L310-1a) (Eck, 1976*a*; Eck, 1976*b*).
AGE AND GEOGRAPHIC RANGE Pleistocene, East Africa
ANATOMICAL DEFINITION
The specimen from Olduvai described by Remane remains one of the few unambiguous specimens of fossil *Papio* described from East Africa. Although the juvenile status of the cranium prevents more detailed assignment of the fossil, the concave profile of the orbital region and sloping muzzle clearly identify the specimen as *Papio* as opposed to *Parapapio* from the same region. The relatively simple, bulbous form of the molars of the specimen clearly distinguishes it from *Theropithecus*. The remains from the Omo Valley are similar in size and shape to modern *Papio*, but their specific affinities are unclear.

SPECIES *Papio izodi* Gear, 1926
TYPE SPECIMEN AD 992, the lectotype designated by Jones (1937), is a badly damaged cranium of a probable female from Taung, South Africa
REFERRED SPECIMENS The more complete specimens of this papionin include a female muzzle with attached mandible (AD 946) and a damaged mandible (CT 5357) from Taung (Freedman, 1957), a nearly complete cranium of a female from Kromdraai (KA 194), a cranium of a young adult of indeterminate sex from Sterkfontein (STS 262) (McKee, 1993) and a large series of over 80 craniodental remains from Haasgat Cave (McKee & Keyser, 1994).
AGE AND GEOGRAPHIC RANGE Pleistocene, South Africa
ANATOMICAL DEFINITION
A small species of *Papio* distinguished by its steep anteorbital drop, weaker maxillary ridges and a well-developed supraorbital torus defined by a relatively deep ophyronic groove in adult specimens. Comprising specimens originally assigned to *Papio wellsi* and *Papio angusticeps* in addition to those originally referred to *Papio izodi*, this species has a long temporal span and has been reported from a broad range of sites throughout South Africa, including Sterkfontein, Kromdraai, Cooper's Minnaar's Cave, and Haasgat Cave (Freedman, 1957, 1970; McKee, 1993; McKee & Keyser, 1994). Specimens allocated here are dentally indistinguishable from *Parapapio* species, but are rendered distinct cranially by the concavity and steep drop of the orbital profile.

SPECIES cf. *Papio* sp. indet.
REFERRED SPECIMENS A distal humerus (LAET 4925) and right dP^4 (LAET 4765) from Laetoli, Tanzania (Leakey & Delson, 1987); specimens "of a papionin close in size to modern representatives of the genus *Papio*" have also been recorded from high in the Koobi Fora succession of Lake Turkana (Harris *et al.*, 1988*a*).
AGE AND GEOGRAPHIC RANGE Pliocene, Tanzania, Kenya
ANATOMICAL DEFINITION
The two specimens from Laetoli were tentatively referred to *Papio* sp. because they were clearly remains of a large papionin, but were not strongly similar to *Theropithecus*. The deciduous premolar shows affinities to *Dinopithecus ingens*, but the absence of comparative postcranial material for *D. ingens* precludes drawing conclusions about the humerus (Leakey & Delson, 1987).

FAMILY Cercopithecidae gen. et sp. indet.
REFERRED SPECIMENS As of this writing, an important set of cercopithecid remains has been recovered from Aramis,

Ethiopia and is being described (S. Frost, pers. comm.). These fossils, when published, will help illuminate the early history of the African Papionini.

Evolution

Old World monkeys are the most taxonomically and ecologically diverse, as well as the most geographically widespread, group of non-human primates. Their environments range from tropical forests with relatively little environmental seasonality and modest annual floristic changes to the some of the most highly seasonal forests with marked annual cycles of photoperiod, temperature and vegetative productivity. Their range of diets enjoyed covers everything from mixed diets of succulent fruits, shoots and seeds to more restricted diets of hanging lichen or grass parts. What accounts for this extraordinary adaptability? What are its origins? And, what light can the fossil record shed on it?

The cercopithecoid fossil record presents a different picture of the taxonomic diversity and range of adaptations of the superfamily (Table 17.1). Only 12 of 19 living genera are also present in the fossil record, and only six of these (*Macaca, Papio, Theropithecus, Colobus, Rhinopithecus* and *Trachypithecus*) are represented substantially. Perhaps most striking is that the fossil record is far richer in terrestrial genera. A total of 13 genera of cercopithecoids can be considered mostly terrestrial; of these, nine – including all except one of the terrestrial colobids – are extinct. Twelve cercopithecoid genera can be considered mostly arboreal; eight of these are present in both the modern and extinct monkey faunas, and four are extant genera only. Some of these trends almost certainly are attributable to taphonomic factors or selective species destruction. It is unlikely, though, that the great differences between the arrays of fossil and living cercopithecoids are due only to taphonomic biases. An overview of environmental contexts helps to shed light on this apparent conundrum.

The earliest Old World monkeys, the Victoriapithecidae, showed strong trends toward the modern cercopithecoid condition in their molar teeth, but they retained features of the ancestral catarrhine morphology in other aspects of their cranial anatomy, in particular in their moderately long, anteriorly tapering snouts, a narrow interorbital region, a moderately deep cheek region relative to facial height and procumbent upper incisors (Benefit & McCrossin, 1991, this volume). Between the early middle and very late Miocene, Old World monkeys are absent from the fossil record entirely, probably due to the near absence of the entire African fossil record during that period (Jablonski & Kelley, 1997). By 7–9 million years ago cercopithecoids are once again detected in the fossil record as two clearly differentiated stocks – colobids and cercopithecids – that were able to adapt well to increasingly seasonal and drier environments evolving in Eurasia and Africa in the late Miocene (see Behrens-meyer *et al.*, 1992; Janis, 1993). The extensive climatic variation resulted in significant changes in the size of the Antarctic Ice Sheet that, in turn, led to an appreciable drop in global sea levels. This phenomenon, coupled with the closure of the Straits of Gibraltar due to tectonic impingement of the African Plate against Europe, resulted in the

Table 17.1. Living cercopithecoid genera compared with those of the Late Tertiary and Quaternary, based on the taxonomy presented in this chapter for fossil groups and that of Groves (1997) for extant forms

Genus	Extant	Extinct	Arboreal [A], terrestrial [T] or both [A/T][a]
Cercopithecidae			
Allenopithecus	•		A
Cercocebus	•	•	A
Cercopithecus	•	•	A
Chlorocebus	•		A/T
Dinopithecus		•	T?
Erythrocebus	•		T
Gorgopithecus		•	?
Lophocebus	•	•	A
Macaca	•	•	A/T
Mandrillus	•		T
Miopithecus	•		A
Papio	•	•	T
Paradolichopithecus		•	T
Parapapio		•	T?
Procynocephalus		•	T?
Theropithecus	•	•	T
Colobidae			
Cercopithecoides		•	T?
Colobus	•	•	A
Dolichopithecus		•	T
Libypithecus		•	?
Mesopithecus		•	T
Microcolobus		•	?
Nasalis	•		A
Paracolobus		•	T
Parapresbytis		•	T?
Presbytis	•	•	A
Procolobus	•		A
Pygathrix	•	•	A
Rhinocolobus		•	?
Rhinopithecus	•	•	A
Semnopithecus	•	•	T
Trachypithecus	•	•	A
Totals	19	25	

[a]Classification of habitus (arboreal [A] versus terrestrial [T] or both [A/T]) is based on available information on postcranial morphology for living and fossil genera, and the generally agreed patterns of behavior based on observations of living genera.

Messinian Salinity Crisis and the eventual isolation and desiccation of the Mediterranean Sea (Kingston & Hill, 1999). The resulting increasing aridity in low latitudes influenced African and Arabian floras significantly (Kingston and Hill, 1999), including the character and distribution of circum-Mediterranean cercopithecoid faunas. In Africa, intense tectonic action along the axis of the Western Rift beginning 3 Ma increased the severity of the rainshadow effect and created even drier environments in which some cercopithecoids (especially large *Theropithecus* and, later, *Papio*, species) enjoyed considerable success (Pickford, 1993*a*). Whereas expansion and diversification of many primate groups (e.g., lorises, tarsiers, New World monkeys and apes) have been limited by their inability to colonize relatively dry, seasonal and non-forested environments, this ability has proven the strong evolutionary gambit of many Cercopithecoidea.

The late Miocene cercopithecoid radiation was concentrated in the circum-Mediterranean and was dominated initially by terrestrial colobids. At this time, eastern Europe was characterized by considerable habitat variation, including ecotones between forest, open vegetation and marshy areas (Behrensmeyer *et al.*, 1992). Western Asia and Arabia were characterized by more open-country environments and became the center of what has been referred to as the open-country woodland chronofauna, which is said to have had a diversity greater than that of modern-day savanna communities (Bernor, 1983, 1984). The disappearance, about 10 Ma, of thin-enameled hominoids associated with the moist and closed woodlands of western Europe was followed by the radiation of cercopithecoids and hominoids with thick occlusal enamel into the drier and more open environments of Eurasia (Behrensmeyer *et al.*, 1992). It is easy to see why terrestrial colobids such as *Mesopithecus* and, later, *Dolichopithecus*, did so well under such conditions. They had locomotor systems that allowed them to forage efficiently on the ground in open environments, and both arboreally and terrestrially in mixed environments. Their digestive systems were well adapted to the chewing and chemical breakdown of many types of vegetation. Equipped with foregut fermentation, these pioneering monkeys actually thrived on leaves, seeds and other types of "low-quality" vegetation that hominoids generally avoided.

The geographically extensive fossil record of *Mesopithecus*, especially *M. pentelicus*, attests to its widescale penetration of latest Miocene Eurasian environments (Delson, 1994). The many morphological similarities of *Mesopithecus* with the "odd-nosed colobine" genera, especially *Pygathrix*, suggest a particularly close phyletic relationship between them (Jablonski, 1998), but the presence of apparent shared-derived attributes of cranial and dental morphology between *Mesopithecus* and the late Miocene Siwalik colobid *Presbytis sivalensis* suggest possible sister-group or ancestor–descendant relationships between them (E. Delson, pers. comm.; N. Jablonski, unpublished data). Populations of *Mesopithecus* may have given rise to or are the sister taxa of other colobids in Asia as well. This leaves open the possibility that *Presbytis* is a dwarfed sister taxon of the common ancestor of itself and *Mesopithecus*, or is a dwarfed descendant of *Mesopithecus* (N. Jablonski, unpublished data). The latter case would raise the possibility that *Mesopithecus* as it is now understood in the fossil record is paraphyletic.

For Old World monkeys, the next most important event of the terminal Miocene was the circum-Mediterranean radiation of the genus *Macaca*. The similarities in craniodental morphology between *Macaca* and *Parapapio*, especially their sharing of a facial profile not marked by a steep anteorbital drop and the absence in both of maxillary and mandibular corpus fossae, strongly suggest a sister-group relationship of the two, and their probable origin from a common ancestor in eastern or northern Africa. The details of the earliest episode of the macaque radiation are unclear, but it is signaled by the appearance, approximately 6 Ma, of *Macaca libyca* in northern Egypt, a species that may predate the bifurcation of the macaques into Asian and Mediterranean lineages (Delson, 1980). Interpretation of macaque history from fossil evidence has proven extremely difficult because of the lack of phylogenetically meaningful craniodental variation within the group. Constructing fossil species and subspecies, therefore, has been an uncertain enterprise and determining evolutionary relationships between extinct and extant forms highly problematical. The one firm statement that can be made on the basis of the fossil evidence, that is reinforced by now copious molecular data, is that *Macaca sylvanus* is the sister clade to all the Asian macaques (Hayasaka *et al.*, 1996; Morales & Melnick, 1998).

The success of macaques in the Pliocene environments of Europe and Asia owes much to the fact that they blend hominoid and colobid functional and ecological characteristics. In their craniodental and postcranial anatomy, macaques functionally resemble some of the last Miocene thick-enameled hominoids – able to forage well on the ground and in the trees, and capable of exploiting a wide variety of plant foods. Because they lacked a specialization for foregut fermentation, their diets could include ripe fruits, which are shunned by colobids because of the disruptive effect of simple sugars on their gut flora. Macaques, like *Papio* baboons, can exploit resources more intensively in a given area than either colobids or hominoids (Andrews, 1981*a*; Wrangham *et al.*, 1998). Like colobids and other cercopithecoids, however, macaques differ significantly from hominoids in many life-history and reproductive parameters. Their shorter gestation times, shorter weaning periods, shorter interbirth intervals and higher intrinsic rates of increase of population made it possible for them to reproduce successfully in highly seasonal environments not tolerated by hominoids (Jablonski *et al.*, 2000). For macaques, the combination of dietary eclecticism and a typically cercopithecoid life-history pattern appears to have been the key both to their initial success in the Pliocene environments of Eurasia, and in surviving, at least in some areas, the extreme climatic fluctuations of the Pleistocene.

As was the case with *Mesopithecus*, it is likely that some geographically disparate species of stem *Macaca* or its Mediterranean or Asian lineages gave rise to forms so morphologically distinct as to be considered separate genera by most students. *Paradolichopithecus* and *Procynocephalus* are basically large, highly terrestrial Pliocene macaques. Whether or not they are distinct genera has implications for classification: If *Paradolichopithecus* and *Procynocephalus* are shown to be one biological entity, the latter name has priority. If they are distinct entities that originated from separate stocks of stem, Mediterranean or Asian macaques, then *Macaca* as a genus would be rendered paraphyletic. Groves (1989) has reached this conclusion for a different reason, that is, that morphological evidence suggests that the lineages leading to the Mediterranean macaques, the Asian macaques and Allen's swamp monkey (*Allenopithecus*), diverged from a papionin stem more or less simultaneously. Molecular data, by contrast, currently support monophyly of the genus because the branch separating all the macaques from other cercopithecoids is long and well defined (Morales & Melnick, 1998).

The fossil record sheds little light on macaque Plio-Pleistocene evolution. Far more significant insights are being derived from molecular studies, especially from the analysis of patterns of variation in mitochondrial DNA (e.g., Melnick *et al.*, 1993). In broadest outline the fossil record demonstrates that macaques were widespread in the early and middle Pleistocene in Europe and eastern Asia, that they were far less widespread by the end of the Pleistocene, and that their distribution rebounded significantly through the early Holocene (Pan & Jablonski, 1987; Jablonski *et al.*, 2000). Molecular studies have shown clearly that Asian macaques underwent rapid speciation, and that speciation occurred repeatedly from highly polymorphic ancestral populations (Hayasaka *et al.*, 1996; Hoelzer *et al.*, 1992; Melnick *et al.*, 1993). The most widespread of today's macaque species, *Macaca fascicularis* and *M. mulatta*, appear to have established their geographic dominance in mainland Asia during the Holocene, as the result of successful recolonization from southeast Asian refugia after the last glacial maximum (Eudey, 1980; Melnick & Kidd, 1985; Jablonski *et al.*, 2000).

By the late Pliocene, the modern pattern of ecological roles filled by cercopithecoids had, in large part, been established. In Asia, the most widespread species appear to have been macaques and their close relatives, who were mostly ecological generalists. Colobids, with a few notable exceptions, by then had evolved into more geographically circumscribed and specialized niches, nearly all of which were arboreal and folivorous. This was accompanied, at least in one lineage (*Trachypithecus*), by the evolution of a further refinement of gut anatomy and physiology, gastrocolic fermentation, that allowed them to concentrate more exclusively on leaves and retain a body size of under 10 kg (Caton, 1999). In another lineage, *Rhinopithecus*, a specialization for the eating of hanging lichens in high-altitude coniferous forests evolved in some species, accompanied by modifications of molar crown shape, enamel thickness and gnathic morphology converging on those of macaques and other papionins (Jablonski, 1993c). An important exception to this trend toward stenotopy in the Asian colobids was *Semnopithecus*, which maintained the Miocene colobid tradition of a generalized, arboreal and terrestrial niche from the Pliocene to the present day.

For the Plio-Pleistocene, the cercopithecoid fossil record of Africa provides a rich and clear picture of cladogenesis, extinctions and adaptations. Colobid evolution in Africa has taken a different course from that in Asia. Firstly, the diversity of African colobid species today is less than it is in Asia. This is probably because, in Africa, species of *Cercopithecus* occupy niches comparable to those occupied by smaller *Presbytis* species in southeast Asia. In addition, all the terrestrial African colobids became extinct in the Pleistocene, presumably because of competition with large papionins. During the Plio-Pleistocene, however, the most widespread of the fossil colobids of Africa, *Cercopithecoides*, appears to have filled a niche similar to those of the Eurasian terrestrial colobids of the Mio-Pliocene. Preliminary observations of its locomotor skeleton by Birchette (quoted in Leakey, 1982) indicate that the genus has a postcranium very much like that of a terrestrial cercopithecid, especially in features of its humeral morphology. Dental evidence further supports this analogy. In *Cercopithecoides kimeui*, at least, the molar teeth are papionin-like, with low crowns, lingual flare and a wide cervix (Leakey, 1982). The extreme dental wear observed in several specimens of this species suggests that it may have subsisted on an abrasive diet.

Rhinocolobus is thought to have been an arboreal folivore similar to modern *Colobus* in its dietary and locomotor habits, whereas *Paracolobus* was probably more terrestrial (Birchette quoted in Leakey, 1982). Leakey (1982) speculates that although the Plio-Pleistocene colobids of sub-Saharan Africa may have achieved initial evolutionary success by occupying largely terrestrial niches, their dental heritage of thin-enameled molars may have rendered them less successful in competing with cercopithecids in drier, more open environments. This hypothesis is strongly supported by the near disappearance about 2 Ma of *Cercopithecoides* and *Paracolobus*, against the backdrop of the strong rise of *Theropithecus (Theropithecus) oswaldi* and the *Theropithecus (Omopithecus)* lineage in the same environments. The more arboreal *Colobus* remained, albeit as an apparently less common element of the primate paleofauna, through the Pleistocene to the present day.

Papionins dominate the Plio-Pleistocene African cercopithecid fauna. Cercopithecins are represented by a small number of fragmentary remains that attest to their presence but not the nature of their adaptation. The near absence of this tribe from the fossil record belies its long and complex history. Molecular estimates place the divergence of the

Cercopithecini from the Papionini at 9–10 Ma (Page *et al.*, 1999), the splitting of *Cercopithecus–Chlorocebus* from *Erythrocebus* at approximately 8 Ma, and the divergence of *Chlorocebus* from *Erythrocebus* at about 5 Ma (Page *et al.*, 1999). Today, African monkeys are considered to be primarily arboreal forest-dwellers, on the basis of the great diversity of arboreal guenons inhabiting the forests across central Africa (Leakey, 1988). Because many cercopithecins are small (body mass < 10 kg; Delson *et al.*, 2000) and live in moist forested environments, however, preservational bias probably explains their absence from the fossil record.

When it comes to fossil evidence, the papionins of Africa are by far the most dominant elements of the late Neogene cercopithecoid fauna. Because their radiation appears to have been concentrated in relatively xeric environments, their preservation in the fossil record has been inordinately enhanced. And the story that they reveal of their own evolution is vivid and complex. Papionins are frequently recovered in the Plio-Pleistocene fossil faunas of East and South Africa, but the genera and species were not those common today (Leakey, 1988). The Pliocene and Pleistocene cercopithecoid faunas were dominated by *Parapapio* (especially in South Africa) and by *Theropithecus* (especially in East Africa). Their morphology reveals that they were well adapted, in different ways, for harvesting and eating a wide variety of vegetation found in forest fringe environments, woodlands and savannas.

Paleontological evidence suggests that the African papionin radiation arose from *Parapapio* or a very similar form, essentially the sub-Saharan equivalent of *Macaca* in Asia. The fossil evidence of the genus is plentiful, but – as in the case of fossil *Macaca* – does not represent critical time periods and does not provide enough information to permit reconstruction of hypotheses of phylogeny and deployment. Fossil and molecular evidence suggests that the macaque and papionin lineages diverged at 7–8 Ma (Disotell *et al.*, 1992; Page *et al.*, 1999), probably in northern Africa. Judging from craniodental morphology, the ancestral papionin was a primitive species of *Parapapio* or its immediate ancestor. The nature of its radiation in the latest Miocene and earliest Pliocene is unknown, but can be reconstructed tentatively thanks to copious molecular data and recent cercopithecid finds from the Pliocene site of Aramis in Ethiopia (S. Frost, pers. comm.).

The lineages comprising *Mandrillus* and *Cercocebus* on the one hand and *Theropithecus*, *Papio* and *Lophocebus* on the other appear to have diverged from a primitive papionin ancestor 6–8 Ma (Disotell *et al.*, 1992; Page *et al.*, 1999). This scenario reflects the now strong consensus of morphological and molecular data that point to the diphyletic origin of baboons and mandrills and the diphyly of mangabeys (Disotell *et al.*, 1992; Fleagle & McGraw, 1999; Page *et al.*, 1999). The divergence of the *Lophocebus–Theropithecus* clade from *Papio* appears to have occurred 4 Ma, followed more or less immediately by the *Theropithecus–Papio* split (Page *et al.*, 1999). The earliest morphological evidence of *Parapapio* in the interval from 3.5 to 3 Ma provides no conclusive evidence as to its affinities with either the *Lophocebus–Theropithecus* or *Papio* clade. Rather, the primitive morphology of East African *Parapapio* neither conflicts with the molecular-based hypotheses nor precludes it having given rise to the more derived papionins.

The major radiation of *Parapapio* itself occurred in South Africa, in the latest Pliocene and Pleistocene. In a comprehensive study of molar morphology in African fossil baboons, Benefit & McCrossin (1990) concluded that South African *Parapapio* species subsisted largely on diets of herbaceous materials harvested in savanna situations, and that they supplemented their diets with grasses during periods when fruits were seasonally unavailable, in a manner similar to that of extant *Papio*. It is suggested here that *Dinopithecus ingens* and *Gorgopithecus major*, poorly known large papionins recognized by most workers as being restricted to southern Africa, are large-bodied, thick-enameled descendants of South African *Parapapio*.

The African papionin genus represented by the largest and most geographically widespread array of fossils is *Theropithecus*. Species of this genus appear to present a more mixed picture of ecological and dietary adaptations than does *Parapapio*. This is probably due to the fact that the genus is relatively common in the fossil record of many East African sites, but is most diverse in the fossil faunas of the Turkana Basin, where a variety of forest, woodland and savanna environments have been recognized (Harris *et al.*, 1988*a*). Some of the extinct species of *Theropithecus* are thought to have included more fruits and leaves in their diet than do extant geladas (Benefit & McCrossin, 1990). The earliest species (*T. (Theropithecus) darti*, *T. (Omopithecus) quadratirostris* and *T. (Omopithecus) brumpti*) appear to have been largely folivorous and are geologically older than the grass-eating species (*T. (Theropithecus) oswaldi* and *T. (Theropithecus) gelada*) (Benefit & McCrossin, 1990). This is borne out especially by the teeth of *T. (Omopithecus) brumpti*, which possess the longer shearing blades and thinner enamel common to folivorous cercopithecoids (Benefit & McCrossin, 1990). Further research is necessary to reconcile this evidence with other features of the species' cranial morphology that suggest a possible emphasis on large-object feeding requiring a wide gape (Eck & Jablonski, 1987).

Dental evidence suggests that *Theropithecus* origins are associated with the beginnings of folivory in a large, forest-dwelling papionin (Benefit & McCrossin, 1990). The specialization for grass-eating that characterizes the genus did not arise until after the environmental desiccation associated with the rise of the Ruwenzoris and the creation of the Western Rift (Pickford, 1993*a*; but see Behrensmeyer *et al.*, 1997). It is significant, however, that the postcranial skeleton indicates that the genus has been adapted, from its beginning, to the dexterous manual harvesting of food items because of its high opposability index (Jablonski, 1986; Krentz, 1993). This ability would have been handy in the harvesting of grass parts (as geladas do today) and other foods such as young leaves or fruits, and may have given Plio-Pleistocene *Theropithecus* an adaptive advantage over cercopithecoids competing for the same resources.

The reasons for the decline and eventual extinction of *Parapapio* and *Theropithecus* (except for the relict species *T. (Theropithecus) gelada*) are still not well understood. The ecological roles filled by *Parapapio* and *Theropithecus* species in the Plio-Pleistocene are now filled almost entirely by subspecies of *Papio hamadryas*. In addition to good explanations for the extinction of large-bodied *Theropithecus* based on energetics (Lee & Foley, 1993) and social dynamics (Dunbar, 1993), their decline and the rise of *Papio* were related to their respective failure and success at living in sympatry with hunting hominids. The butchered remains of young *Theropithecus oswaldi leakeyi* at Olorgesailie (Isaac, 1977) testify to the fact that ancient theropiths fell prey to human hunters, possibly because their dependence on permanent water sources made them easy targets (Dunbar, 1993). The combination of guile and an ability to survive for longer periods of time without water might have provided *Papio* baboons with a critical advantage under these circumstances. Evidence that other, now widely distributed lineages of African mammals were colonized from South African refugia following the recent cold periods of the last glaciations may also have also influenced the fate of the genus, as these baboons were already well established there in the Early Pleistocene (Arctander *et al.*, 1999; Hewitt, 2000).

The fossil record attests to the fact that the Cercopithecoidea was extremely successful as a mammalian radiation, especially in the drier, seasonal environments of post-Miocene Eurasia and Africa. Its successful occupation of terrestrial, in addition to arboreal, niches is the group's ecological hallmark. Having said this, the membership in those terrestrial niches has changed over time. The few generalist species that now populate them – *Semnopithecus entellus* and *Macaca* species in Asia and *Papio hamadryas* subspecies in Africa – are those that can live in close proximity to humans and in many cases exploit human resources. For the Old World monkeys of the late Pleistocene and Holocene, it has been in the arboreal niches of southeast Asian and central African forests that the most diversification has occurred. In these environments, bipedal humans – at least until recently – have had little impact.

Acknowledgements

My love of and interest in Old World monkeys owes much to the inspiration of my PhD advisor, Gerald Eck, who wisely advised me 25 years ago that fossil monkeys were more plentiful than other fossil primates and tended to be studied by nice people. This work also benefited greatly from the input of many colleagues and friends in more recent years. Meave Leakey has generously permitted me to study and photograph many original fossil specimens at the National Museums of Kenya since the mid-1980s, and has openly shared her ideas of cercopithecoid evolution with me. Peter Andrews generously made fossil collections at the Natural History Museum (London) available to me, and access to those at the Muséum National d'Histoire Naturelle (Paris) was made possible by Brigitte Senut, and at the American Museum of Natural History (New York) by Richard Tedford, Michael Novacek and Mark Norell. I am also very grateful to Eric Delson, for permitting me to study his extensive personal collection of casts of Old World monkeys, as well as several undescribed original fossil specimens. Regular exchanges about Old World monkey evolution with him over the last 20 years have been extremely useful, even when we did not see eye to eye. He and his student Steve Frost are also thanked for sharing their new insights about the classification of *Theropithecus* fossils with me in recent months. In keeping with his usual spirit of intellectual generosity, Len Freedman gave me his notes and original photographs of South African cercopithecoids when he retired. I am tremendously grateful to him for allowing me to draw upon these materials in this work. In the physical production of this chapter, I am grateful to Patty Shea-Diner for extensive assistance in gathering references, Maria Tom for compiling the bibliography, Dinah Crawford for assistance with references and in compiling the fossil locality database, Peter Gaede for creating the illustrations of *Theropithecus (Omopithecus) brumpti* and Dong Lin for specimen photography and printing of photographs. Finally, I thank my husband, George Chaplin, for many hours of stimulating discussions about Old World monkey evolution over many years, and for his consistent and strong support of my research and career.

Primary References

FAMILY COLOBIDAE

Cercopithecoides

Mollett, O. (1947). Fossil mammals from the Makapan Valley, Potgietersrust. I. Primates. *South African Journal of Science*, **43**, 295–303.

Freedman, L. (1957). The fossil Cercopithecoidea of South Africa. *Annals of the Transvaal Museum*, **23**, 8–262.

Freedman, L. (1960). Some new cercopithecoid specimens from Makapansgat, South Africa. *Palaeontologica Africana*, **7**, 7–45.

Freedman, L. (1965). Fossil and subfossil primates from the limestone deposits at Taung, Bolt's Farm and Witkrans, South Africa. *Palaeontologica Africana* **9**, 19–48.

Maier, W. (1970). New fossil Cercopithecoidea from the lower Pleistocene cave deposits of the Makapansgat Limeworks, South Africa. *Palaeontologica Africana*, **13**, 69–108.

Freedman, L. & Brain, C. K. (1972). Fossil cercopithecoid remains from the Kromdraai australopithecine site (Mammalia: Primates). *Annals of the Transvaal Museum*, **28**, 1–16.

Leakey, M. G. (1982). Extinct large colobines from the Plio-Pleistocene of Africa. *American Journal of Physical Anthropology*, **58**, 153–172.

Keyser, A. W. (1991). The palaeontology of Haasgat: a preliminary account. *Palaeontologica Africana*, **28**, 29–33.

Colobus

Simons, E. L. (1967). A fossil *Colobus* skull from the Sudan (Primates, Cercopithecidae). *Postilla*, **111**, 1–12.

Eck, G. G. (1976*b*). Cercopithecoidea from Omo Group deposits. In *Earliest Man and Environments in the Lake Rudolf Basin*, eds.Y. Coppens, F. C. Howell, G. L. Isaac & R. E. F. Leakey, pp. 332–344. Chicago, IL: University of Chicago Press.

Johanson, D. C., Taieb, M., & Coppens, Y. (1982). Pliocene hominids from the Hadar Formation, Ethiopia (1973–1977): stratigraphic, chronologic, and paleoenvironmental contexts, with notes on hominid morphology and systematics. *American Journal of Physical Anthropology*, **57**, 373–402.

Kalb, J. E., Jolly, C. J., Mebrate, A., Tebedge, S., Smart, C., Oswald, E. B., Cramer, D., Whitehead, P., Wood, C. B., Conroy, G. C., Adefris, T., Sperling, L., & Kana, B. (1982*a*). Fossil mammals and artifacts from the Middle Awash Valley, Ethiopia. *Nature*, **298**, 25–29.

Kalb, J. E., Jolly, C. J., Tebedge, S., Mebrate, A., Smart, C., Oswald, E. B., Whitehead, P. F., Wood, C. B., Adefris, T., & Rawn–Schatzinger, V. (1982*b*). Vertebrate faunas from the Awash Group, Middle Awash Valley, Afar, Ethiopia. *Journal of Vertebrate Paleontology*, **2**, 237–258.

Leakey, M. G. (1987). Colobinae (Mammalia, Primates) from the Omo Valley, Ethiopia: Cercopithecidae de la formation de Shungura. *Faunes de Plio-Pleistocene, Valleé Omo*, **3**, 147–169.

Senut, B. (1994). Cercopithecoidea néogènes et quaternaires du Rift Occidental (Ouganda). In *Geology and Palaeobiology of the Albertine Rift Valley, Uganda–Zaïre*, Vol. 2, *Palaeobiology*, pp. 195–205. Orléans, France: CIFEG.

Harrison, T. & Harris, E. E. (1996). Plio-Pleistocene cercopithecids from Kanam, western Kenya. *Journal of Human Evolution*, **30**, 539–561.

Dolichopithecus

Delson, E. (1973). Fossil colobine monkeys of the circum-Mediterranean region and the evolutionary history of the Cercopithecidae (Primates, Mammalia). PhD dissertation, Columbia University.

Koufos, G. D., Syrides, G. E., & Koliadiou, K. K. (1991). A Pliocene primate from Macedonia (Greece). *Human Evolution*, **2**, 283–294.

Libypithecus

Stromer, E. (1913). Mitteilungen über Wirbeltierreste aus dem Mittelpliocän des Natrontales (Ägypten). *Zeitschrift deutsche geologische Gesellschaft*, **65**, 350–372.

Mesopithecus

Wagner, A. (1839). Fossile Überreste von einem Affenschadel und anderen Säugtieren aus Griechenland. *Gelehrte Anziegen Bayerisches Akademie Wissenschaften*, **38**, 306–311.

Gervais, P. (1849). *Zoologie et paléontologie françaises*, 1st edn. Paris: Bertrand.

Delson, E. (1973). Fossil colobine monkeys of the circum-Mediterranean region and the evolutionary history of the Cercopithecidae (Primates, Mammalia). PhD dissertation, Columbia University.

Campbell, B. G., Amini, M. H., Bernor, R. L., Dickinson, W., Drake, R., Morris, R., Van Couvering, J. A., & Van Couvering, J. A. H. (1980). Maragheh: a classical late Miocene vertebrate locality in northwestern Iran. *Nature*, **287**, 837–841.

Heintz, E., Brunet, M., & Battail, B. (1981). A cercopithecid primate from the late Miocene of Molayan, Afghanistan, with remarks on *Mesopithecus*. *International Journal of Primatology*, **2**, 273–284.

Tobien, H. (1986). An early Upper Miocene (Vallesian) *Mesopithecus* molar from Rheinhessen. *Primate Report*, **14**, 49.

Zapfe, H. (1991). *Mesopithecus pentelicus* Wagner aus dem Turolien von Pikermi bei Athen, Odontologie und Osteologie (Eine Dokumentation). Vienna: Ferdinand Berger & Sohne.

Mottura, A. & Ardito, G. (1992). Observations on the Turin specimen of *Mesopithecus pentelici* (Wagner, 1839). *Human Evolution*, **7**, 67–73.

Bonis, L. de, Bourvrain, G., Geraads, D., & Koufos, G. D. (1997). New material of *Mesopithecus* (Mammalia, Cercopithecidae) from the late Miocene of Macedonia, Greece. *Neues Jahrbuch für Mineralogie, Geologie und Paläontologie* 1997, **5**, 255–265.

Gentili, S., Mottura, A., & Rook, L. (1998). The Italian fossil primate record: recent finds and their geological context. *Geobios*, **31**, 675–686.

Rook, L. (1999). Late Turolian *Mesopithecus* (Mammalia, Primates, Colobinae) from Italy. *Journal of Human Evolution*, **36**, 535–547.

Microcolobus

Benefit, B. R. & Pickford, M. (1986). Miocene fossil cercopithecoids from Kenya. *American Journal of Physical Anthropology*, **69**, 441–464.

Paracolobus

Leakey, R. E. F. (1969). New Cercopithecidae from the Chemeron Beds of Lake Baringo, Kenya. *Fossil Vertebrates of Africa*, **1**, 53–69.

Leakey, M. G. (1982). Extinct large colobines from the Plio-Pleistocene of Africa. *American Journal of Physical Anthropology*, **58**, 153–172.

Leakey, M. G. (1987). Colobinae (Mammalia, Primates) from the Omo Valley, Ethiopia: Cercopithecidae de la formation de Shungura. *Faunes de Plio-Pleistocene, Valleé Omo*, **3**, 147–169.

Leakey, M. G. & Delson, E. (1987). Fossil Cercopithecidae from the Laetolil Beds. In *Laetoli: A Pliocene Site in Northern Tanzania*, eds. M. D. Leakey & J. M. Harris, pp. 91–107. Oxford: Clarendon Press.

Harris, J. M., Brown, F. H., & Leakey, M. G. (1988). Stratigraphy and paleontology of Pliocene and Pleistocene localities west of Lake Turkana, Kenya. *Contributions in Science*, **399**, 1–128.

Senut, B. (1994). Cercopithecoidea néogènes et quaternaires du Rift Occidental (Ouganda). In *Geology and Palaeobiology of the Albertine Rift Valley, Uganda–Zaïre*, Vol. 2, *Palaeobiology*, 195–205. Orléans, France: CIFEG.

Parapresbytis

Borissoglebskaya, M. B. (1981). New species of monkey (Mammalia, Primates) from the Pliocene of N. Mongolia. *Iskopaemye pozvonochnye Mongolii*, **15**, 95–108.

Kalmykov, N. P. & Maschenko, E. N. (1992). The most northern representative of early Pliocene Cercopithecidae from Asia. *Paleontologicheskifi Zhurnal*, **4**, 136–138.

Kalmykov, N. P. & Maschenko, E. N. (1995). *Parapresbytis eohanuman* cercopithecid monkey (Primates, Cercopithecidae) from Pliocene of Baikal. *Voprosy antropologii*, **88**, 91–116.

Presbytis

Lydekker, R. (1878). Notices of Siwalik Mammals. *Records of the Geological Survey of India*, **11**, 64–104.
Hooijer, D. A. (1962*b*). Prehistoric bone: the gibbons and monkeys of Niah Great Cave. *The Sarawak Museum Journal*, **11**, 428–449.
Hooijer, D. A. (1962*c*). Quaternary langurs and macaques from the Malay Archipelago. *Zoologische Verhandelingen*, **55**, 3–64.
Simons, E. L. (1970). The deployment and history of Old World Monkeys (Cercopithecidae, Primates). In *Old World Monkeys: Evolution, Systematics, and Behavior*, eds. J. R. Napier & P. H. Napier, pp. 99–137. New York: Academic Press.
Barry, J. C. (1987). The history and chronology of Siwalik cercopithecids. *Human Evolution*, **2**, 47–58.

Pygathrix

Gu, Y., Huang, W., Chen, D., Guo, X., & Jablonski, N. G. (1996). Pleistocene fossil primates from Luoding, Guangdong. *Vertebrata PalAsiatica*, **34**, 235–250.

Rhinocolobus

Leakey, M. G. (1982). Extinct large colobines from the Plio-Pleistocene of Africa. *American Journal of Physical Anthropology*, **58**, 153–172.
Leakey, M. G. (1987). Colobinae (Mammalia, Primates) from the Omo Valley, Ethiopia: Cercopithecidae de la formation de Shungura. *Faunes de Plio-Pleistocene, Valleé Omo*, **3**, 147–169.

Rhinopithecus

Matthew, W. D. & Granger, W. (1923). New fossil mammals from the Pliocene of Sze-Chuan, China. *Bulletin of the American Museum of Natural History*, **48**, 563–598.
Colbert, E. H. & Hooijer, D. A. (1953). Pleistocene Mammals from the limestone fissures of Szeschwan, China. *Bulletin of the American Museum of Natural History*, **102**, 1–134.
Hu, C. & Qi, T. (1978). Gongwangling Pleistocene mammalian fauna of Lantian, Shaanxi. *Chinese Paleontology*, **155**, 1–64.
Gu, Y. & Hu, C (1991) A fossil cranium of *Rhinopithecus* found in Xinan, Henan Province. *Vertebrata PalAsiatica*, **29**, 55–58.
Gu, Y., Huang, W., Chen, D., Guo, X., & Jablonski, N. G. (1996). Pleistocene fossil primates from Luoding, Guangdong. *Vertebrata PalAsiatica*, **34**, 235–250.

Semnopithecus

Lydekker, R. (1884). Indian Tertiary and post-Tertiary Vertebrata: rodents, ruminants and synopsis of Mammalia. *Memoirs of the Geological Survey of India (Palaeontologica Indica)*, **3**, 105–134.

Trachypithecus

Hooijer, D. A. (1962*b*). Prehistoric bone: the gibbons and monkeys of Niah Great Cave. *The Sarawak Museum Journal*, **11**, 428–449.
Hooijer, D. A. (1962*c*). Quaternary langurs and macaques from the Malay Archipelago. *Zoologische Verhandelingen*, **55**, 3–64.
Pan, Y. R., Peng, Y. Z., Zhang, X. Y., & Pan, R. L. (1992). Cercopithecid fossils discovered in Yunnan and their stratigraphical significance. *Acta Anthropologica Sinica*, **11**, 303–311.
Jablonski, N. G., Pan, Y. R., & Zhang, X. Y. (1994). New cercopithecid fossils from Yunnan Province, People's Republic of China. In *Current Primatology*, eds. B. Thierry, J. R. Anderson, J. J. Roeder & N. Herrenschmidt, pp. 303–311. Strasbourg: Université Louis Pasteur.
Gu, Y., Huang, W., Chen, D., Guo, X., & Jablonski, N. G. (1996). Pleistocene fossil primates from Luoding, Guangdong. *Vertebrata PalAsiatica*, **34**, 235–250.
Jablonski, N. G. & Tyler, D. E. (1999). *Trachypithecus auratus sangiranensis*, a new fossil monkey from Sangiran, Central Java, Indonesia. *International Journal of Primatology*, **20**, 319–326.

Colobidae gen. et sp. indet.

Leakey, M. G. (1987). Colobinae (Mammalia, Primates) from the Omo Valley, Ethiopia: Cercopithecidae de la formation de Shungura. *Faunes de Plio-Pleistocene, Valleé Omo*, **3**, 147–169.
Meikle, W. E. (1987). Fossil Cercopithecidae from the Sahabi Formation. In *Neogene Paleontology and Geology of Sahabi*, eds. N. T. Boaz, A. El-Arnauti, A. W. Gaziry, J. de Heinzelin, & D.D. Boaz, pp. 119–127. New York: Alan R. Liss.
Boaz, N. T., Bernor, R. L., Brooks, A. S., Cooke, H. B. S., De Heinzelin, J., Dechamps, R., Delson, E., Gentry, A. W., Harris, J. W. K., Meylan, P., Pavlakis, P. P., Sanders, W. J., Stewart, K. M., Verniers, J., Williamson, P. G., & Winkler, A. J. (1992). A new evaluation of the significance of the Late Neogene Lusso Beds, Upper Semliki Valley, Zaïre. *Journal of Human Evolution*, **22**, 505–517.

FAMILY CERCOPITHECIDAE

Cercocebus

Leakey, M. G. & Leakey, R. E. F. (1976). Further Cercopithecinae (Mammalia, Primates) from the Plio/Pleistocene of East Africa. *Fossil Vertebrates of Africa*, **4**, 121–146.

Cercopithecus

Eck, G. G. & Howell, F. C. (1972). New fossil *Cercopithecus* material from the Lower Omo Basin, Ethiopia. *Folia Primatologica*, **18**, 325–355.
Harrison, T. & Harris, E. E. (1996). Plio-Pleistocene cercopithecids from Kanam, western Kenya. *Journal of Human Evolution*, **30**, 539–561.

Dinopithecus

Broom, R. (1937). On some new Pleistocene mammals from limestone caves of the Transvaal. *South African Journal of Science*, **33**, 750–768.
Freedman, L. (1957). The fossil Cercopithecoidea of South Africa. *Annals of the Transvaal Museum*, **23**, 8–262.

Gorgopithecus

Broom, R. & Jensen, J. S. (1946). A new fossil baboon from the

caves at Potgietersrust. *Annals of the Transvaal Museum*, **20**, 337–340.
Freedman, L. (1957). The fossil Cercopithecoidea of South Africa. *Annals of the Transvaal Museum*, **23**, 8–262.

Lophocebus

Harrison, T. & Harris, E. E. (1996). Plio-Pleistocene cercopithecids from Kanam, western Kenya. *Journal of Human Evolution*, **30**, 539–561.

Macaca

Gervais, P. (1859). *Zoologie et paléontologie françaises*, 2nd edn, 2 vols., Texte et Atlas. Paris: Bertrand.
Cocchi, I. (1872). Su di due scimmie fossili italiane. *Bollettino del reale Comitato geologico d'Italia*, **3**, 59–71.
Schlosser, M. (1925). Fossil primates from China. *Palaeontologica Sinica* (C), **1**, 1–16.
Young, C. C. & P'Ei, W. C. (1933). On the fissure deposits of Chinghsinghsien with remarks on the Cenozoic geology of the same area. *Bulletin of the Geological Society of China*, **13**, 63–71.
Azzaroli, A. (1946). La scimmia fossile della Sardegna. *Rivista di scienze preistoriche*, **1**, 68–76.
Jouffroy, F. K. (1959). Un crâne subfossile de macaque du Pleistocène du Viet Nam. *Bulletin du Muséum d'histoire naturelle, Paris*, **31**, 309–316.
Hooijer, D. A. (1962*b*). Prehistoric bone: the gibbons and monkeys of Niah Great Cave. *The Sarawak Museum Journal*, **11**, 428–449.
Hooijer, D. A. (1962*c*). Quaternary langurs and macaques from the Malay Archipelago. *Zoologische Verhandelingen*, **55**, 3–64.
Thenius, E. (1965). Ein Primaten-Rest aus dem Altpleistozän von Voigtstedt in Thuringen. *Paläontologische Abhandlungen*, **2**, 681–686.
Iwamoto, M. & Hasegawa, Y. (1972). Two macaque fossil teeth from the Japanese Pleistocene. *Primates*, **13**, 77–81.
Iwamoto, M. (1975). On a skull of a fossil macaque from the Shikimizu Limestone Quarry in the Shikoku District, Japan. *Primates*, **16**, 83–94.
Aimi, M. (1981). Fossil *Macaca nemestrina* (Linnaeus, 1766) from Java, Indonesia. *Primates*, **22**, 409–413.
Li, W., Zhang, Z., Gu, Y., Luin, Y., & Yan, F. (1982). A fauna from Lianhua Cave, Dantu, Jiangsu. *Acta Anthropologica Sinica*, **1**, 169–179.
Singer, R., Wolff, R. G., Gladfelter, B. G., & Wymer, J. J. (1982). Pleistocene *Macaca* from Hoxne, Suffolk, England. *Folia Primatologica*, **37**, 141–152.
Penela, A. M. (1983). Presence of the genus *Macaca* in the Pleistocene site of the Solana del Zamborino (Fonelas, Granada, Spain). Preliminary study. *Boletí de la real Sociedad española de historia natural, Sección Geológica*, **81**, 187–195.
Ardito, G. & Mottura, A. (1987). An overview of the geographic and chronologic distribution of west European cercopithecoids. *Human Evolution*, **2**, 29–45.
Fladerer, F. A. (1987). *Macaca* (Cercopithecidae, Primates) in the lower Pleistocene from Deutsch-Altenberg, lower Austria. *Beitrage zur Paläontologie von Osterreich*, **13**, 1–24.
Meikle, W. E. (1987). Fossil Cercopithecidae from the Sahabi Formation. In *Neogene Paleontology and Geology of Sahabi*, eds. N. T. Boaz, A. El-Arnauti, A. W. Gaziry, J. de Heinzelin, & D. D. Boaz, pp. 119–127. New York: Alan R. Liss.
Fladerer, F. A. (1989). Hohlenschutz und Eiszeitforschung Erstnachweis von Affen (Gattung *Macaca*) im Jungpleistozän Mitteleuropas. *Mitteilungen des naturwissenschaftlichen Vereines für Steiemark*, **119**, 23–26.
Fooden, J. (1990). The bear macaque, *Macaca arctoides*: a systematic review. *Journal of Human Evolution*, **19**, 607–686.
Basilici, G., Faraone, A. G., & Gentili, S. (1991). Un nuovo reperto di *Macaca* nelle brecce ossifere pleistoceniche di Monte Peglia (Terni, Italia centrale). *Bollettino della Società paleontologica italiana*, **30**, 251–254.
Fladerer, F. A. (1991). Der erste Fund von *Macaca* (Cercopithecidae, Primates) im Jungpleistozän von Mitteleuropa. *Zeitschrift für Säugetierkunde*, **56**, 272–283.
Moyà-Solà, S., Moyà, J. P., & Köhler, M. (1992). Primates catarrinos (Mammalia) del Neogeno de la peninsula Iberica. *Paleontologia i Evolució*, **23**, 41–45.
Pan, Y. R., Peng, Y. Z., Zhang, X. Y., & Pan, R. L. (1992). Cercopithecid fossils discovered in Yunnan and their stratigraphical significance. *Acta Anthropologica Sinica*, **11**, 303–311.
Tesakov, A. S. & Maschenko, E. N. (1992). The first reliable finding of a macaque (Cercopithecidae, Primates) in the Pliocene of Ukraine. *Paleontologicheskiaei Zhurnal*, **4**, 47–52.
Gibbard, P. L. (1994). *Pleistocene History of the Lower Thames Valley*. Cambridge: Cambridge University Press.
Jablonski, N. G., Pan, Y. R., & Zhang, X. Y. (1994). New cercopithecid fossils from Yunnan Province, People's Republic of China. In *Current Primatology*, eds. B. Thierry, J. R. Anderson, J. J. Roeder, & N. Herrenschmidt, pp. 303–311. Strasbourg: Université Louis Pasteur.
Fooden, J. (1995). Systematic review of southeast Asian longtail macaques, *Macaca fascicularis* (Raffles, [1821]). *Fieldiana*, **81**, 1–205.
Gu, Y., Huang, W., Chen, D., Guo, X., & Jablonski, N. G. (1996). Pleistocene fossil primates from Luoding, Guangdong. *Vertebrata PalAsiatica*, **34**, 235–250.
Rook, L., Harrison, T., & Engesser, B. (1996). The taxonomic status and biochronological implications of new finds of *Oreopithecus* from Baccinello (Tuscany, Italy). *Journal of Human Evolution*, **30**, 3–27.
Fooden, J. (2000). Systematic review of the rhesus macaque, *Macaca mulatta* (Zimmermann, 1780). *Fieldiana*, **96**, 1–180.
Köhler, M., Moyà-Solà, S., & Alba, D. M. (2000). *Macaca* (Primates, Cercopithecidae) from the Late Miocene of Spain. *Journal of Human Evolution*, **38**, 447–452.

Papio

Remane, A. (1925). Der fossile Pavian (*Papio* Sp.) von Oldoway nebst Bemerkungen über die Gattung *Simopithcus* C.W. Andrews. *Wissenschaftliche Ergebnisse der Oldoway Expedition*, **2**, 83–90.
Gear, J. H. S. (1926). A preliminary account of the baboon remains from Taungs. *South African Journal of Science*, **23**, 731–747.
Broom, R. (1936). A new fossil baboon from the Transvaal. *Annals of the Transvaal Museum*, **18**, 393–396.
Freedman, L. (1957). The fossil Cercopithecoidea of South Africa. *Annals of the Transvaal Museum*, **23**, 8–262.
Freedman, L. (1965). Fossil and subfossil primates from the limestone deposits at Taung, Bolt's Farm and Witkrans, South Africa. *Palaeontologica Africana*, **9**, 19–48.
Eck, G. G. (1976*b*). Cercopithecoidea from Omo Group Deposits. In *Earliest Man and Environments in the Lake Rudolf Basin*, eds. Y. Coppens, F. C. Howell, G. L. Isaac, & R. E. F. Leakey, pp. 332–344. Chicago, IL: University of Chicago Press.
Freedman, L. (1976). South African fossil Cercopithecoidea: a

re-assessment including a description of new material from Makapansgat, Sterkfontein and Taung. *Journal of Human Evolution*, **5**, 297–315.

Leakey, M. G. & Delson, E. (1987). Fossil Cercopithecidae from the Laetolil Beds. In *Laetoli: A Pliocene Site in Northern Tanzania*, eds. M. D. Leakey & J. M. Harris, pp. 91–107. Oxford: Clarendon Press.

Harris, J. M., Brown, F. H., & Leakey, M. G. (1988). Stratigraphy and paleontology of Pliocene and Pleistocene localities west of Lake Turkana, Kenya. *Contributions in Science*, **399**, 1–128.

McKee, J. K. (1993). Taxonomic and evolutionary affinities of *Papio izodi* fossils from Taung and Sterkfontein. *Palaeontologica Africana*, **30**, 43–49.

McKee, J. K. & Keyser, A. W. (1994). Craniodental remains of *Papio angusticeps* from the Haasgat Cave Site, South Africa. *International Journal of Primatology*, **15**, 823–841.

Paradolichopithecus

Depéret, C. (1929). Nouveau singe du Pliocène supérieur de Sénèze (Hte-Loire). *Travaux du Laboratoire de géologie de la Faculté des sciences de Lyon*, **12**, 5–12.

Necrasov, O., Samson, P., & Radulesco, C. (1961). Sur un nouveau singe catarrhinien fossile, découvert dans un nid fossilifère d'Olténie (R.P.R.). *Analele stiintifice ale Universitatii "Al. I. Cuza" din Iasi*, **7**, 401–416.

Trofimov, B. A. (1977). Primate *Paradolichopithecus pushkini* sp. nov. from Upper Pliocene of the Pamirs Piedmont. *Journal of the Paleontological Society of India*, **20**, 26–32.

Parapapio

Haughton, R. H. (1925). A note on the occurrence of a species of baboon in limestone deposits near Taungs. *Transactions of the Royal Society of South Africa*, **12**, 68.

Hopwood, A. T. (1936). New and little-known fossil mammals from the Pleistocene of Kenya Colony and Tanganyika Territory. *Annals and Magazine of Natural History (Series 10)*, **17**, 636–641.

Jones, T. R. (1937). A new fossil primate from Sterkfontein, Krugersdorp, Transvaal. *South African Journal of Science*, **33**, 709–728.

Broom, R. (1940). The South African Pleistocene cercopithecid apes. *Annals of the Transvaal Museum*, **20**, 89–100.

Dietrich, W. O. (1942). Altestquartiare Säugetiere aus der sudlichen Serengeti, Deutsch-Ostafrika. *Palaeontographica (A)*, **94**, 43–133.

Freedman, L. (1957). The fossil Cercopithecoidea of South Africa. *Annals of the Transvaal Museum*, **23**, 8–262.

Freedman, L. (1960). Some new cercopithecoid specimens from Makapansgat, South Africa. *Palaeontologica Africana*, **7**, 7–45.

Freedman, L. (1965). Fossil and subfossil primates from the limestone deposits at Taung, Bolt's Farm and Witkrans, South Africa. *Palaeontologica Africana*, **9**, 19–48.

Maier, W. (1970). New fossil Cercopithecoidea from the lower Pleistocene cave deposits of the Makapansgat Limeworks, South Africa. *Palaeontologica Africana*, **13**, 69–108.

Maier, W. (1971). Two new skulls of *Parapapio antiquus* from Taung and a suggested phylogenetic arrangement of the genus *Parapapio*. *Annals of the South African Museum*, **59**, 1–16.

Freedman, L. (1976). South African fossil Cercopithecoidea: a re-assessment including a description of new material from Makapansgat, Sterkfontein and Taung. *Journal of Human Evolution*, **5**, 297–315.

Leakey, M. G. & Leakey, R. E. F. (1976). Further Cercopithecinae (Mammalia, Primates) from the Plio/Pleistocene of East Africa. *Fossil Vertebrates of Africa*, **4**, 121–146.

Grine, F. E. & Hendey, Q. B. (1981). Earliest primate remains from South Africa. *South African Journal of Science*, **77**, 374–376.

Bromage, T. G. & Schrenk, F. (1986). A cercopithecoid tooth from the Pliocene of Malawi. *Journal of Human Evolution*, **15**, 497–500.

Leakey, M. G. & Delson, E. (1987). Fossil Cercopithecidae from the Laetolil Beds. In *Laetoli: A Pliocene Site in Northern Tanzania*, eds. M. D. Leakey & J. M. Harris, pp. 91–107. Oxford: Clarendon Press.

Harris, J. M., Brown, F. H., & Leakey, M. G. (1988). Stratigraphy and paleontology of Pliocene and Pleistocene localities west of Lake Turkana, Kenya. *Contributions in Science*, **399**, 1–128.

Senut, B. (1994). Cercopithecoidea néogènes et quaternaires du Rift Occidental (Ouganda). In *Geology and Palaeobiology of the Albertine Rift Valley, Uganda–Zaïre*, vol. 2, *Palaeobiology*, pp. 195–205. Orléans, France: CIFEG.

Leakey, M. G., Feibel, C. S., McDougall, I., & Walker, A. C. (1995). New four-million-year-old hominid species from Kanapoi and Allia Bay, Kenya. *Nature*, **376**, 565–571.

Senut, B. (1996). Plio-Pleistocene Cercopithecoidea from the Koanaka Hils (Ngamiland, Botswana). *Comptes rendus de l'Académie des sciences de Paris*, **322**, 423–428.

Procynocephalus

Meyer, H. von (1848). In *Index Palaeontologicus*, ed. H. G. Bronn. Stuttgart: E. Schweizerbart.

Schlosser, M. (1924). Fossil primates from China. *Palaeontologica Sinica*, **1**, 1–16.

Young, C. C. & P'Ei, W. C. (1933). On the fissure deposits of Chinghsinghsien with remarks on the Cenozoic geology of the same area. *Bulletin of the Geological Society of China*, **13**, 63–71.

Teilhard de Chardin, P. (1938) Fossils from Locality 12, Choukoutien. *Palaeontologica Sinica*, **5**, 1–50.

Theropithecus

Andrews, C. W. (1916). Note on a new baboon (*Simopithecus oswaldi*, gen. et sp. nov.) from the Pliocene of British East Africa. *The Annals and Magazine of Natural History, Including Zoology, Botany, and Geology*, **18**, 410–419.

Hopwood, A. T. (1934). New fossil mammals from Olduvai, Tanganyika Territory. *Annals and Magazine of Natural History (Series 10)*, **14**, 546–550.

Broom, R. & Jensen, J. S. (1946). A new fossil baboon from the caves at Potgietersrust. *Annals of the Transvaal Museum*, **20**, 337–340.

Arambourg, C. (1947). *Mission scientifique à l'Omo*, vol. 1, *Géologie, anthropologie*. Paris: Muséum national d'histoire naturelle.

Freedman, L. (1957). The fossil Cercopithecoidea of South Africa. *Annals of the Transvaal Museum*, **23**, 8–262.

Leakey, R. E. F. (1969). New Cercopithecidae from the Chemeron Beds of Lake Baringo, Kenya. *Fossil Vertebrates of Africa*, **1**, 53–69.

Maier, W. (1972). The first complete skull of *Simopithecus darti* from Makapansgat, South Africa, and its systematic position. *Journal of Human Evolution*, **1**, 395–405.

Leakey, M. G. & Leakey, R. E. F. (1973). Further evidence of *Simopithecus* (Mammalia, Primates) from Olduvai and Olorgesailie. *Fossil Vertebrates of Africa*, **3**, 101–120.

Gupta, V. J. & Sahni, A. (1981). *Theropithecus delsoni*, a new cercopithecine species from the Upper Siwaliks of India. *Bulletin of the Indian Geological Association*, **14**, 69–71.

Iwamoto, M. (1982). A fossil baboon skull from the Lower Omo Basin, Southwest Ethiopia. *Primates*, **23**, 533–541.

Dechow, P. C. & Singer, R. (1984). Additional fossil *Theropithecus* from Hopefield, South Africa: a comparison with other African sites and a reevaluation of its taxonomic status. *American Journal of Physical Anthropology*, **63**, 405–435.

Eck, G. G. (1987). *Theropithecus oswaldi* from the Shungura Formation, Lower Omo Basin, Southwestern Ethiopia. In *Les Faunes Plio-Pleistocènes de la Vallée de l'Omo (Éthiopie)*, eds. G. G. Eck, N. G. Jablonski, & M. Leakey, pp. 123–140. Paris: Centre national de la recherche scientifique.

Boaz, N. T., Bernor, R. L., Brooks, A. S., Cooke, H. B. S., De Heinzelin, J., Dechamps, R., Delson, E., Gentry, A. W., Harris, J. W. K., Meylan, P., Pavlakis, P. P., Sanders, W. J., Stewart, K. M., Verniers, J., Williamson, P. G., & Winkler, A. J. (1992). A new evaluation of the significance of the Late Neogene Lusso Beds, Upper Semliki Valley, Zaïre. *Journal of Human Evolution*, **22**, 505–517.

Delson, E. (1993). *Theropithecus* fossils from Africa and India and the taxonomy of the genus. In Theropithecus: *The Rise and Fall of a Primate Genus*, ed. N. G. Jablonski, pp. 157–189. Cambridge: Cambridge University Press.

Delson, E., Eck, G. G., Leakey, M. G., & Jablonski, N. G. (1993). Appendix I: A partial catalogue of fossil remains of *Theropithecus*. In Theropithecus: *The Rise and Fall of a Primate Genus*, ed. N. G. Jablonski, pp. 499–525. Cambridge: Cambridge University Press.

Delson, E. & Hoffstetter, R. (1993). *Theropithecus* from Ternifine, Algeria. In Theropithecus: *The Rise and Fall of a Primate Genus*, ed. N. G. Jablonski, pp. 191–208. Cambridge: Cambridge University Press.

Eck, G. G. (1993). *Theropithecus darti* from the Hadar Formation, Ethiopia. In Theropithecus: *The Rise and Fall of a Primate Genus*, ed. N. G. Jablonski, pp. 15–83. Cambridge: Cambridge University Press.

Leakey, M. G. (1993). Evolution of *Theropithecus* in the Turkana Basin. In Theropithecus: *The Rise and Fall of a Primate Genus*, ed. N. G. Jablonski, pp. 85–123. Cambridge: Cambridge University Press.

Jablonski, N. G. (1994). New fossil cercopithecid remains from the Humpata Plateau, southern Angola. *American Journal of Physical Anthropology*, **94**, 435–464.

Gibert, J., Ribot, F., Gibert, L., Leakey, M., Arribas, A., & Martinez, B. (1995). Presence of the cercopithecid genus *Theropithecus* in Cueva Victoria (Murcia, Spain). *Journal of Human Evolution*, **28**, 487–493.

The fossil record of hominoid primates

18 | Perspectives on the Miocene Hominoidea

DAVID R. PILBEAM

Introduction

The boundaries of the Miocene are arbitrary here in that hominoids (or at least non-cercopithecoid catarrhines that might be cladistically hominoids) are probably present in the latest Oligocene of Africa while most subtropical hominoids of Euro-Asia disappear during the middle of the late Miocene rather than its very end. Would we pay these catarrhines such attention were they not related to us? I think we probably would because the Miocene ape record, unsatisfactory though it still is in many ways, fascinates because of its diversity, ambiguity and difference. As each species slowly emerges from obscurity with recovery of new material, we are often finding that our earlier expectations about basic biological adaptations and evolutionary relationships are not met (Pilbeam, 1997).

One of the editorial requirements for these introductory "perspective" chapters is that they should provide a commentary by explaining "what is known, what can be known, and what will always be speculative about the fossil record". The first part is the responsibility of the following chapters. The second is more challenging. We can expand the record by continuing to search known sites and areas. We particularly need fossils from those equatorial regions in which living apes are found – they are effectively unrepresented in our current samples.

As to what will be forever speculative, this is the most difficult to address. One way to think about this would be to look back at earlier reviews of the Miocene apes, and note which "predictions", if they can be so characterized, strike us now as reasonable and which as unreasonable. I suspect that such an exercise would confirm the wisdom of my decision to pursue the point no further!

Let me begin this essay with the review I co-authored 15 years ago (Kelley & Pilbeam, 1986), just before discovery of important new material that has clarified some taxonomic issues. To pick just two examples, consider the recognition that more "good" genera are represented in the record than had been suspected, and that some Miocene apes were more dimorphic in certain features than any living species (Kelley & Xu, 1991; Kelley, 1995).

Despite the considerable expansion of rigorous "cladistic" approaches to phylogenetic analysis ("rigor" being associated with an ever-growing list of morphological characters and easy access to personal-computer-based phylogenetic programs), there is hardly more consensus now than earlier about evolutionary relationships among Miocene apes or their relationships to living hominoids.

This reflects primarily two factors. Firstly, the still less than adequate nature of the fossil record; only *Proconsul* and *Oreopithecus* have most of their skeletal parts preserved while all other Miocene apes are less well sampled, sometimes considerably so. Secondly, and in my opinion more importantly, there is no adequate agreement on the nature of characters; that is, on how to describe, or in the current jargon "atomize" complex anatomy in ways that will be phylogenetically informative. I do not anticipate that this will change greatly, and accordingly I believe that it is the evolutionary relationships of the Miocene apes, along with their center(s) of radiation, that are most likely to remain "speculative".

Perhaps the most important introductory point to make about these Miocene apes is that they are mostly unlike living forms, and not just in largely lacking autapomorphous features of the various extant lineages. Many also appear to lack a number of features shared by all living (especially large) apes, and therefore expected in ancestral morphotypes. One example of many would be the morphology of the metacarpal–phalangeal epiphyses (Susman, 1979; Rose, 1986, 1997) which is common to all living apes and is found among Miocene apes only in *Oreopithecus* (Moyà-Solà *et al.*, 1999). Likewise, the postcranials of most Miocene apes differ rather considerably from those of all the living great apes (Rose, 1997; Ward, 1997). Hence arguments now focus on the extent to which living-ape postcranial similarities are homoplasies rather than homologies. This, like so many other aspects of any current debate in paleoprimatology and paleoanthropology, is a return to a much older argument.

Living hominoids, briefly

Distribution, diet and ecology

Before interpreting the ape fossil record, we need to consider briefly some basic features of the living apes. Living apes, and particularly the great apes, mature, grow and reproduce more slowly and live longer than other primates (Kelley, 1997). They are almost entirely restricted to tropical forest (see, for example, Dolhinow & Fuentes, 1999). We know very little of their distribution over the past few million years. Fossil orangutans are known from the Pleistocene of southern China and southeast Asia (Szalay & Delson, 1979), and Pleistocene chimpanzees may have been recorded from northern Tanzania (Lönnberg, 1936), but they are conspicuous by their absence from the abundant later Neogene (mostly non-forest) African fossil record, and it looks as though they have always been restricted largely to tropical rainforest.

Living apes are ripe-fruit-eaters; in the case of chimpanzees and orangutans their obsession with and dependence upon ripe fruit is now well documented (Leighton, 1993; Conklin-Brittain *et al.*, 1998; Knott, 1998; Wrangham *et al.*, 1998), and hylobatids have similar needs (Bartlett, 1999). Gorillas are able to subsist on a diet higher in more fibrous plant food, but when given the chance they are ripe-fruit-eaters (Remis, 1997). If we think of foods as falling into two categories, "preferred" and "fallback" foods (Wrangham *et al.*, 1998), we can say that ape preferred foods are ripe fruits, while fallback foods vary, being principally leaves for hylobatids, mainly terrestrial piths for the African apes, and a range of less nutritious plant parts including cambium for orangutans.

Many living-ape morphologies can be interpreted as feeding adaptations to a strongly preferred diet of ripe fruit (e.g., Schultz, 1968; Kay & Ungar, 1997). These features are related to current ape positional behaviors – suspension, arm-swinging and vertical climbing – and presumably also reflect past behaviors. They are adaptations to moving into and within a small-branch milieu and doing so effectively at a range of body sizes. Given the clearly "arboreal" nature of many features of the African ape torso and forelimb, and hindlimb, it is reasonable to interpret the "terrestrial" features as an "added" and more recently evolved positional component (Gebo, 1996).

Living-hominoid phylogenetic relationships: genetic and phenotypic

The phylogenetic relationships of extant Hominoidea are now resolved, based on abundant genetic evidence (Caccone & Powell, 1989; Bailey *et al.*, 1991; Ruvolo, 1994, 1997; Horai *et al.*, 1995; Takahata, 1995; Goodman *et al.*, 1998; Kaessman *et al.*, 1999; Shen *et al.*, 2000). Chimpanzees and humans are closest relatives, with gorillas, orangutans and gibbons successively more distant. Ruvolo (1994) further noted that in any phylogenetic analysis it was essential to explain similarities that are highly unlikely to be synapomorphies. For example, the involucrin gene tree supports chimpanzee–gorilla, but involucrin shows evidence of gene conversion, which makes it an inappropriate estimator.

This chimpanzee–human monophyly is also supported by the DNA–DNA hybridization experiments of Caccone & Powell (1989). As noted by several workers (Sibley & Ahlquist, 1984; Sarich *et al.*, 1989) DNA–DNA hybridization is in principle an excellent way of estimating relationships. Although the technique is phenetic, comparisons of all single-copy DNA of two species yield distance estimates that are for many groups markedly clock-like as determined using the rate test. Although there has been some controversy over some data sets (Sarich *et al.*, 1989), the Caccone & Powell (1989) data are concordant with mtDNA (Horai *et al.*, 1995) and nuclear DNA (Ruvolo, 1997; Takahata, 1995) analyses in showing a clear chimpanzee-human relationship and a non-trivial internode back to the divergence of the gorilla lineage.

Recently Deinard *et al.* (1998) reported a data set apparently supporting chimpanzee-gorilla, but based on non-sequence data. But until we have a clear understanding of the genetic sequences involved this is little different from using any other phenotypic characters. Probably these similarities are symplesiomorphic. Whatever these similarities reflect genetically, they do not show that a multiply supported, sequence-based human–chimpanzee clade is incorrect.

Although the phylogenetic relationships of the living hominoids are, I believe, now established, the debates based on phenotypic characters are informative, for they tell a cautionary tale. By the 1950s the predominant view was that the great apes were monophyletic, with hominids anciently diverged (perhaps as long ago as 20 Ma), splitting after hylobatids and before a pongid radiation. Then Morris Goodman began to publish a series of important papers in which he used comparative immunology to establish phylogenetic relationships (e.g., Goodman, 1962). Of revolutionary importance was the recognition that humans and African apes formed a monophyletic group to the exclusion of the orangutan, and by the 1960s many paleontologists had incorporated this pattern into their hominoid phyogenies (Simpson, 1963). Hominids, in the form of *Ramapithecus*, were still thought to be anciently derived, though perhaps no older than 15 Ma (Pilbeam, 1996, 1997). Also in the early 1960s equally revolutionary comparative immunological work placed the hominid divergence from the African apes at 4 or 5 Ma rather than 15 to 20 Ma (Sarich & Wilson, 1967).

Through the 1960s, indeed into the 1990s, morphologists continued to argue about living-hominoid relationships. The 1980s was a particularly interesting decade. Kluge (1983) supported a great-ape clade, Schwartz (1984) proposed that humans and orangutans were sister taxa, Andrews & Martin (1987*a*) supported a closer chimpanzee–gorilla relationship, and Groves (1986) concluded that chimpanzees and humans were closer. More recent and more comprehensive analyses of phenotypic characters (Groves & Paterson, 1991; Shoshani *et al.*, 1996; Begun *et al.*, 1997*b*) show that results vary somewhat based on choice of characters, body parts and outgroups. Collard & Wood (2000) used hard-tissue characters from hominid phylogenetic studies to examine living-hominoid relationships, and failed to retrieve the genetically based most probable tree. The same characters failed to retreive the preferred genetic tree for papionins, within which phenetic and sequence-based trees were already incongruent (Disotell, 1994, 1996; Harris & Disotell, 1998).

Three other phenotypic studies are worth mentioning here. Hartwig-Scherer (1993) examined hominoid skeletal growth allometrically. Confirming the great similarity of the African apes to each other in growth patterns, she concluded that chimpanzees and gorillas are mostly (of course not entirely) scaled versions of "the same" animal (see also Shea, 1984, 1985). But she further argued that this allometric similarity implies that the African apes are closest relatives, to the exclusion of humans. Allometric patterns are here essentially used then as characters, and I would argue that they reveal homologous but symplesiomorphous similarity.

Two other studies support a different result. Braga's (1995) analysis of qualitative (non-metric) cranial features frequently used in systematic research yielded a human–bonobo relationship, with chimpanzees next most closely related followed by gorillas and orangutans. Gorilla and orangutan subspecies showed deep separations, and overall the network from which the most parsimonious tree was generated was congruent with that generated by mitochodrial gene sequence data (Ruvolo, 1997; Gagneux *et al.*, 1999). Finally, Gibbs (Gibbs, 1999; Gibbs *et al.*, 2000) analyzed hominoid soft-tissue anatomy (mainly muscles, blood vessels, nerves) and recovered a tree strongly supporting chimpanzee–human monophyly. Significant for this analysis is that the data used were not collected with the intention of being part of any phylogenetic analysis.

Can we then make any single summary statement about hominoid phylogeny using phenotypic characters? One is struck, first of all, by the disagreements. These arise not because of different analytical techniques but primarily because different characters are being used by different analysts. In a perceptive but rarely cited paper, Sarich (1993: 111) neatly characterized the problem:

> There are always going to be apparently conflicting signals that support the picture you have already decided upon intuitively (and which in fact may be correct), even though another worker, possibly equally experienced, can recognize and choose another signal, equally explicit, that supports quite another picture.

The problem is that there are no objective rules or criteria for defining phenotypic characters. As Sarich (1993: 111) further notes:

> there is no objective, unbiased way of randomly choosing and, perhaps even more important, counting characters of anatomy. It is at this point that one has to emphasize that one has to know one's organisms very well indeed to know what characters are phylogenetically relevant and which are not.

Sarich (1993: 112) continues:

> But how do you choose when, as will almost always be the case, different phylogenies can be supported by different character sets? Parsimony will not do, I am afraid, because that would assume the characters were randomly chosen – or perhaps chosen because of their phylogenetic utility. But how do you judge such utility without already having come to some conclusions as to the phylogeny involved? I cannot see how you can, and therefore conclude that one cannot legitimate the statistical testing of various phylogenetic hypotheses derived from anatomical data using cladistic reasoning.

And of course these points apply equally if not more depressingly to the fossils.

An increasing amount of attention is being paid to improving the definition of morphological characters (Atchley & Hall, 1991; Pilbeam, 1996; Lieberman, 1999; Lovejoy *et al.*, 1999; McCollum, 1999) to improve the "parsing" or "atomizing" of complex three-dimensional morphologies with the goal of better revealing genealogical relationships. Working on groups with well resolved molecular phylogenies will greatly assist our morphological interpretations. We can expect progress in recognizing homoplasy a posteriori once relationships are well resolved. For example, the papionin trees based on genetic comparisons made it clear that the cranial resemblances of *Papio* and *Mandrillus* were likely to be homoplasies (Disotell, 1996; Harris & Disotell, 1998). Careful subsequent morphological analysis has shown this to be the case, as well as enabling the identification of probable morphological synapomorphies of *Mandrillus* and *Cercocebus* (Fleagle & McGraw, 1999).

But determining polarity is at least as big a problem.

Living-hominoid phylogenetic relationships: divergence times

As is now widely recognized, genetic and paleontological data can be used together to estimate divergence ages. This is because certain kinds of genetic data can provide us with trees with reasonably accurate relative branch lengths. These would be data sets for which the rate test would indicate either relatively unvarying rates of change along all or most lineages, or data sets in which "local branch length" estimates can be made (Bailey *et al.*, 1991). The best tree proportions are likely to be those derived from data sets in which a significant fraction of the nuclear genome is sampled. DNA–DNA hybridization data sets are excellent (Sibley & Ahlquist, 1984; O'Brien *et al.*, 1985; Caccone & Powell, 1989; Sarich *et al.*, 1989) because they reflect assays of the entire single-copy nuclear genome, as are data sets that involve a large number of independent nuclear genes or proteins (Dene *et al.*, 1976*a*; Takahata, 1995; Ruvolo, 1997; Kumar & Hedges, 1998). Slowly evolving genes such as the globins (Bailey *et al.*, 1991) or albumins and transferrins (Sarich & Cronin, 1976) are also valuable. More rapidly evolving genes, such as those in the mitochondrial genome, have to be approached more cautiously.

Paleontological data are used to calibrate one or more nodes of the "relative" tree, enabling us to estimate ages for other branching points. Ultimately all inferred ages have to be compatible with plausible interpretations of the fossil record, but how to define and recognize "plausibility" is of course a significant problem. Martin makes the important point (1993, 1998) that paleontological data inevitably provide minimum ages and, he argues, frequently greatly underestimate true divergence ages.

Before addressing the hominoid record directly, we need to consider it in broader perspective as part of the overall mammalian record. One useful way for us to focus the discussion is to ask ourselves whether, given a broad consensus on the age of diversification of placental mammal orders, the crown catarrhines radiated around 25 Ma (the clear majority consensus) or around 50 Ma. The current consensus among vertebrate paleontologists places the radiation of most placental orders as occuring around 90 to 110 Ma (Novacek, 1992*b*; Archibald, 1996; Shoshani & McKenna, 1998; Flynn *et al.*, 1999; Foote *et al.*, 1999; Rowe, 1999). We can then look at various genetic data sets and predict ages for more recent splitting events, including the Hominoidea.

DNA–DNA hybridization data sets for simians (Benveniste & Todaro, 1976; O'Brien *et al.*, 1985; Caccone & Powell, 1989; Sibley *et al.*, 1990) and for some other groups (for example of Carnivora: Wayne *et al.*, 1989) are clock-like enough to yield trees with reliable relative branch lengths. There is evidence in some but not all data sets for rate heterogeneity among non-simian primates (Bonner *et al.*, 1980; Bailey *et al.*, 1991; C. C. Sibley & J. E. Ahquist, pers. comm.), but it is still possible to estimate branching times for primates using a *c.* 100 Ma date for the main placental radiation. With this as a calibration point, the strepsirhine–haplorhine divergence can be dated to around 65 to 75 Ma, the New World–Old World simian divergence to around 40 Ma, and the hominoid–cercopithecoid divergence to around 25 Ma (Dene *et al.*, 1976*b*; Sarich & Cronin, 1976; Koop *et al.*, 1989*a*, 1989*b*; Bailey *et al.*, 1991).

A much earlier data set (Dene *et al.*, 1976*a*) used immunological assays of a large number of proteins to estimate genetic distances between taxa. Setting the radiation of most placentals at 100 Ma, the New World–Old World simian split is 53 Ma and the cercopithecoid–hominoid 25 Ma. Transferrin and albumin immunological data (Sarich & Cronin, 1976) calibrated at 90–100 Ma for placentals and 125 Ma for the marsupial–placental split yield New World–Old World simian and cercopithecoid–hominoid divergences of 35 Ma and 20 Ma, respectively.

A large amount of sequence data for the globin gene complex (Bailey *et al.*, 1991; Goodman *et al.*, 1998; Koop *et al.*, 1989*a*, 1989*b*) shows that, given a basal placental radiation centered around 100 Ma, the primate radiation as a whole would have to fall in the latest Cretaceous or early Cenozoic. Given the pattern of genetic differences among primates, and using a calibration of 25 Ma for cercopithecoid–hominoid divergence, the New World–Old World split is 35 to 40 Ma.

If we turn now to the fossil record, we can ask which paleontological data sets are useful for calibration. For the primates, because the early and middle Miocene African fossil record is dense it is possible to make some plausible inferences about the cercopithecoid/hominoid and colobine/cercopithecine divergences. For the former pair, apomorphic features of hominoids are present by 20 Ma (Gebo *et al.*, 1997*b*; MacLatchy & Pilbeam, 1999) and for cercopithecoids by 18 Ma (Pilbeam & Walker, 1968; Benefit & McCrossin, 1997*a*; Benefit, 1999*c*). Although the record between around 22 and 30 Ma is poor, no evidence for divergence is found in the abundant material from the Oligocene. Cercopithecoid material is plentiful between 18 and 14 Ma with no evidence that colobines and cercopithecines had diverged; this divergence had occurred by 8 Ma (Benefit, 1993, 1994, 1999*a*). Genetic data would predict, based on a cercopithecoid–hominoid split of 25 Ma, a colobine–cercopithecine divergence of around 14 to18 Ma (Benveniste & Todaro, 1976; Todaro, 1980).

Since there is general agreement across a number of genetic systems that the catarrhine radiation begins around 25 Ma (although I believe this may be an overestimate; Pilbeam, 1996), we can then ask about divergence times for the crown hominoids. Predicted DNA–DNA hybridization based branching times are: gibbons and siamangs 17 Ma; orangutans 12 Ma; gorillas 9 Ma; and chimpanzees and bonobos 6 Ma. Analyses based on the DNA–DNA hybridization (Caccone & Powell, 1989), total mitochondrial DNA sequence (Horai *et al.*, 1995), and 7–13 nuclear gene sequence data sets (Takahata, 1995), which are likely to give the best estimates, show that the time between the divergence of gorillas and the splitting of humans and chimpanzees/bonobos is a substantial fraction of the time since the *Pan–Homo* split, on the order of 35% to almost 60%.

Note that shifting the cercopithecoid–hominoid calibration point older to 30 Ma or younger to 20 Ma, which spread is likely to contain the true age, would make the hybridization-based chimpanzee–human estimated split vary only between 4.5 Ma and 7 Ma. From a paleontologist's perspective this is a quite narrow window. Recently, de Heinzelin *et al.* (1999) noted that the excellent Middle Awash exposures in Ethiopia document hominids spanning the past 6 Ma: we await with interest publication of more complete descriptions of *Ardipithecus ramidus ramidus* (White *et al.*, 1994), and of the recently described 5.2–5.8 Ma sample of *A. ramidus kadabba* (Haile-Selassie, 2001).

Where does this leave us with Martin's (1993, 1998: 41) caveat that "It now seems likely that hominids originated at an earlier date than has commonly been assumed in recent discussion..."? The dates for the primate radiation and non-edentate mammals based on genetically derived tree proportions are entirely compatible with the fossil record. As more genetic systems are analyzed and understood, we can expect improvement in the shapes of these "relative" genetic trees. But I believe that at least for hominoids it will be difficult to justify major departures from these age estimates. The chimpanzee–human split cannot be much older than about 7 Ma without requiring all other branch times to become unreasonably old. The genetic approach in fact enables us to see a way out of "Martin's dilemma" – the concern that because only a fraction of the fossil record is sampled, branch points will inevitably be underestimated (as shown graphically, literally and figuratively, in Martin, 1993: 226 Fig. 2). If the inferred tree is framed by genetic data, carefully tuned in strategic places to paleontological data, there is no necessary reason to seriously misinterpret the totality of the record.

Reconstructing morphotypes using extant species

What were the various hominoid ancestral stages like? Because I believe the Miocene ape record to be too fragmentary and ambiguous of interpretation I propose first reviewing this issue from the perspective of the living taxa. But before continuing, let me quote again from Bob Martin's recent review (1998):

> There is a common tendency to assume that it has now been established beyond reasonable doubt that chimpanzees . . . are closer to humans than gorillas. Because of this, many authors simply take chimpanzees as models for . . . the early ancestry of hominids. This, however, is unjustifiable on two counts. Firstly, no modern species should ever be taken as a direct model for an ancestral condition . . . Secondly, despite the fact that most molecular studies indicate a closer relationship between chimpanzees and humans . . . the evidence is by no means conclusive.

I trust I have demonstrated that the evidence of a closer relationship is indeed conclusive and that this phylogenetic issue is now settled. But what about the chimpanzee–human morphotype? Is it really so unreasonable to describe it as "chimp-like" enough to be placed in *Pan* (Washburn, 1968; Sarich, 1971; Zihlman, 1979; Pilbeam, 1996)? Is it really the case that "no modern species should ever be taken as a direct model for the ancestral condition"?

To clarify, I would not claim that the chimpanzee–human ancestor looked exactly like a living *Pan troglodytes* or *paniscus*. But I think it likely that it would be recognized as a species of *Pan*. Is this "conservatism" improbable from a broader comparative perspective? Surely not. One can cite many examples of speciose mammalian genera which are now known, based on genetic data, to have considerable time depth. For example, based on genetic comparisons (Goodman *et al.*, 1998; Todaro, 1980; Zehr, 1999) species of *Hylobates* share a probable latest Miocene or earliest Pliocene common ancestor, as do those of *Macaca* (Hayasaka *et al.*, 1996). These ancestors would surely be assigned either to *Hylobates* or *Macaca*. The common ancestor of *Cercopithecus* species must have lived no later than the mid Pliocene (Ruvolo, 1988), and probably earlier, and would have been a *Cercopithecus*. The eastern and western gorilla subspecies shared a *Gorilla* ancestor that is at least as ancient as that of chimpanzees and bonobos, and the same applies to Bornean and Sumatran orangutans (Gagneux *et al.*, 1999). For two decades it was considered quite reasonable to include *Pan* and *Gorilla* in the same genus (e.g., Simpson, 1963; Szalay & Delson, 1979), and it was assumed that their ancestor would have been *Pan*-like if *Pan*-sized. In principle there is no reason to exclude a chimpanzee-like common ancestor for chimpanzees and humans.

Additional support comes from the aforementioned allometric studies of chimps and gorillas (Shea, 1984; Hartwig-Scherer, 1993), which show very clearly for a wide range of cranial and postcranial features that these are scaled taxa: substantial evidence supports the view that their ancestor would have been, if chimpanzee-sized, chimpanzee-like. And their common ancestor is likely to have resembled the chimpanzee–human morphotype. As Groves (1988) has demonstrated, based on his preferred phenotypic characters, *Pan* shows less change than other hominoid lineages.

What about the ancestor of the living large hominoids, or of all extant hominoids? The former question becomes one of the extent to which the shared features of chimpanzees and orangutans represent homologies, and the extent in turn to which shared features among chimpanzees, orangutans and gibbons are also homologies. Until recently, as I noted a few years ago (Pilbeam, 1996), most of us accepted the idea that many of the postcranial similarities of the crown hominoids were shared-derived homologies (Schultz, 1968; Larson, 1998). But largely because of relatively recent Miocene ape discoveries, this assumption should now be questioned (Begun & Kordos, 1997; Larson, 1998). Recognizing that recently discovered Miocene ape postcranial fossils are often surprisingly unlike what might have been the expected morphotypes, it is quite legitimate to raise the possibility that some substantial fraction of living-ape postcranial resemblances are homoplasies (e.g., Begun, 1993*a*).

In an important recent study Larson (1998) reviewed the distribution in anthropoids of 34 postcranial features that had been proposed to typify living Hominoidea. She concluded that many of them were variable within Hominoidea and often overlapped with non-hominoid anthropoids, particularly *Ateles*. She inferred from this that considerable parallelism and convergence in the hominoid postcranium was distinctly possible (see also Erikson, 1963). However, other analyses which combine characters in various ways have reached different conclusions. Thus the multivariate analyses of shoulder and arm dimensions by Ashton, Oxnard, and colleagues (Ashton *et al.*, 1965, 1976; Oxnard, 1977) group hominoids together, especially large hominoids, with *Ateles* more or less close to hominoids. A separate set of multivariate analyses by Corruccini & Ciochon (1976, 1978) involving the forelimb and shoulder joint confirmed the homogeneity of hominoids and their clear distinction from non-hominoids, but an intermediate position for *Ateles* and *Brachyteles*. Corruccini & Ciochon also added (1978: 542) that "the pattern shared among hominoids might also have partially arisen in parallel, especially in the case of gibbons". Finally, a cladistic character analysis of Larson's data set using gap-coding (N. Young, unpublished data) shows that hominoids form a monophyletic group, separate from a "monkey" group in which Old and most New World monkeys are sister taxa, with *Ateles* forming a third clade. The fossil-oriented (in the sense of determining character selection) postcranial analyses of Ward (1997) and Rose (1997) also support at least large hominoid monophyly. Thus, whether characters are combined in multivariate or cladistic analyses, the data patterning suggests quite strongly that the postcranial similarities of hominoids, and particularly the large apes, are mostly synapomorphies.

As far as the hominoid cranial morphotype is concerned, Shea (1985) has made the case that both the African ape and orangutan patterns are each likely to be derived relative to their common ancestor, a view with which I concur. Biegert (1957) showed that hylobatids are intermediate between the two large-ape patterns in craniofacial hafting and degrees of relative klinorynchy and airorynchy. It is possible that the crown hominoid morphotype was most like the larger hylobatids. Of interest here is that, with the exception of *Sivapithecus*, which resembles the orangutan, Miocene hominoids differ facially and in the anterior cranium from all living large hominoids.

Fossil hominoids: some brief notes

The late Oligocene and Miocene time-span is more than three times as long as the Pliocene and Pleistocene together. For its first half or so, from 25 Ma until around 15 Ma, the hominoid (or non-cercopithecoid catarrhine) record is confined to a small area of eastern Africa (Kenya and Uganda). Close to ten genera are known, and a couple of species (of *Proconsul*) are as well represented as *Australopithecus afarensis* or *africanus*. Most of the others are more poorly sampled. This surely underestimates the geographical range of apes because tropical forest in west and central Africa is not sampled. In contrast, for another third or so of the interval, from 15 Ma to 8 Ma, at least a dozen genera and a few more than a dozen hominoid species are distributed throughout the Old World land mass from Namibia to Spain to China. But only one of these (*Oreopithecus*) is well sampled as far as body-part representation is concerned. African and Asian regions that are now and were then tropical forest are effectively unsampled.

As far as taxonomy is concerned, there is little disagreement on the species-level assignments. Previous disagreements have been largely settled about the number of ape species represented in *Proconsul* and at sites in Spain, Hungary, Turkey, Indo-Pakistan and Lufeng. A high degree of at least tooth-size dimorphism is clearly represented in some cases, perhaps higher than seen in extant apes (Kelley & Xu, 1991; Kelley, 1995). With the dissolution of a too-generous *Sivapithecus* (Kelley & Pilbeam, 1985) we can now recognize a genuinely high level of between-regional generic diversity, and outside Africa there seems minimal within-regional generic or even specific diversity. Above the genus level (subtribe to family) there is less agreement; the rise of cladistic approaches to classification means that generic membership of suprageneric categories varies with the phylogeny inferred, and until there is more agreement on phylogenetic relationships it is as well to remain agnostic about family-group classifications.

Perhaps the most contentious current issues in the study of Miocene apes concern genus-level phylogenetics. With the exception of the near-universal acceptance of *Sivapithecus* and *Pongo* as sister taxa, there is quite widespread disagreement on the relationships of other genera. Almost all phylogenetic analyses infer that most middle and late Miocene hominoids are crown large hominoids (see Begun *et al.*, 1997*a*). *Proconsul*, *Afropithecus* and *Kenyapithecus* are generally inferred to be either sister to all hominoids or to all large hominoids (as surely would be the recently described *Equatorius*). An important unresolved issue is whether or not the early Miocene genus *Proconsul* is cladistically a hominoid (Harrison & Rook, 1997; Walker, 1997). In some analyses, *Ouranopithecus* and *Dryopithecus* are sisters to the African hominoid clade, in others to the orangutan clade (Begun, 1993*b*; Moyà-Solà & Köhler, 1995). Given that with few exceptions postcrania of these taxa do not resemble hominoids (Rose, 1997; Ward, 1997), this implies considerable postcranial homoplasy. For particular data sets (Shoshani *et al.*, 1996; Begun *et al.*, 1997*b*), in addition to choice of characters, the choice of both body part (cranial, dental, postcranial, forelimb or hindlimb characters) and included taxa can influence the results (N. Young, unpublished data). Young's analyses show that the inclusion of some Miocene apes with crown hominoids (albeit in varying combinations depending on an author's preferred characters) is driven by cranial–dental traits; but when postcranial characters alone are used all Miocene apes fall outside crown large hominoids (compare Begun *et al.*, 1997*b*; Rose, 1997; Ward, 1997).

Andrews & Bernor (1999) reviewed the different cladograms proposed by various authors, along with a consensus or concordance cladogram. One of the consensus cladograms shows living large hominoids as a monophyletic group, but with most Miocene apes as sisters with *Hylobates* more distant still. Forced to choose, I would support this.

These analyses raise two major issues. First, the question of how postcranial similarities between the orang and chimpanzee–gorilla evolved. I assume that postcranially gorillas can be treated as scaled chimpanzees, so the question we need to address is the extent to which chimpanzee–orangutan postcranial similarities are convergent, as implied by most current character-based phylogenetic analyses of

Miocene apes. Just how likely is this? Begun *et al.* (1997*b*) suggested that postcranial convergence is likely, driven largely by the more abundant and more phylogenetically informative cranial–dental characters used in the analysis. But a closer look at the postcranial characters used in the analysis shows that, unlike the data sets of, for example, Gibbs (Gibbs, 1999; Gibbs *et al.*, 2000), Larson (1998) and Shoshani *et al.* (1996), the characters are overwhelmingly uninformative for inferring generic relationships. This is because the character states are predominantly of the form: non-hominoid/hominoid. In fact, because they are quite unvarying, they make it appear as though the living large hominoids are a monophyletic group when analyzed separately, which might be taken to imply that the postcranial similarities of living apes are mostly homologous! It is more likely the case that these characters are inadequate for this particular kind of analysis, and another look at the postcranium would be desirable.

This brings us to the second major issue. These disagreements over Miocene ape phylogenetic relationships are no different from those we noted for extant hominoids when the analyses had to rely only on phenotypic characters, especially hard-tissue features. This forces us once again to confront the realities of the flaws in current morphologically based cladistic analyses of these forms, this time without the safety net of a possible genetic solution to the ambiguities.

If phylogenetic analysis has its frustrations, a more fruitful and positive aspect of the study of Miocene hominoids involves functional and behavioral reconstruction. What were these apes like as animals? Given the fact that Miocene apes differ quite markedly from living hominoids throughout the skeleton, they emerge as rather different animals in many ways.

Because of the abundance of teeth and jaws it is not surprising that there have been a number of attempts at dietary reconstruction. Most species are reconstructed as having diets falling more or less within the range of living apes (Ungar & Kay, 1995) although some species, for example *Ouranopithecus macedoniensis* and *Oreopithecus bambolii*, fall outside that range, reflecting differences in tooth design and molar microwear. Those Miocene apes that do appear on the basis of one or two features to have diets similar to those of living species nonetheless present some interesting problems. Three examples will illustrate the point: *Sivapithecus*, *Griphopithecus* and *Dryopithecus*.

The molar microwear of *Sivapithecus* from Pakistan resembles that of *Pan troglodytes* (Teaford & Walker, 1984). However, their tooth designs differ quite radically; for example, *Sivapithecus* has considerably thicker enamel. Incisor microwear resembles *Cercocebus* (Kelley, 1986*a*). Mandibular robusticity is also markedly greater than in chimpanzees or orangs (Ward & Brown, 1986). The environment of *Sivapithecus* was probably not the kind of equatorial tropical rainforest in which orangutans live; associated faunas suggest more subtropical woodland (Andrews *et al.*, 1997; but see Scott *et al.*, 1999), so available foods may have differed from those of *Pongo* or *Pan*. More work is needed.

Griphopithecus is another intriguing case. The site of Paşalar in Turkey preserves abundant fauna representing a "snapshot" of the middle Miocene. Various lines of evidence suggest a wooded habitat but not tropical forest (Andrews, 1990; Viranta & Andrews, 1995). Although clearly generically different from *Sivapithecus*, *Griphopithecus* teeth are quite similar in design. Incisor microwear resembles *Sivapithecus* (Kelley, 1986*a*), while molar microwear is heavier (King *et al.*, 1999), resembling *Pongo* rather than *Pan*, although occlusal design differs from the orangutan (Andrews & Martin, 1991). Mandibles are more robust than *Pongo*. Isotopic analysis (carbon and oxygen) of a single ape tooth suggests feeding and drinking in both open and closed portions of the habitat (Quade *et al.*, 1995). A diet unlike that of living apes is again suggested.

The third example, *Dryopithecus*, raises similar issues. The habitat in Spain and Hungary was forested, but forests probably structurally unlike equatorial forests (Andrews *et al.*, 1997). Molar occlusal morphology and mandibular design are more like those of chimpanzees than either *Sivapithecus* or *Griphopithecus*; enamel is thinner than either and closer to *Pan* (Ungar, 1996). Incisor wear is modest while molar microwear is heavier and more like *Pongo* (Kelley, 1986*a*; Ungar, 1996). Incisors are tiny compared to those of either chimpanzees or orangutans (Moyà-Solà & Köhler, 1995). A different diet is again suggested. However, unlike the previous two fossil taxa, *Dryopithecus* appears to have included a significant suspensory component in its positional repertoire (Moyà-Solà & Köhler, 1996).

In summary, considering what can be inferred about habitats and from mandibular and tooth morphology, tooth wear and isotopes, it is likely that all three taxa were "frugivores" of some kind, but different from either of the living large ape frugivores. They probably did not subsist on, because they did not have access to, a preferred diet largely of ripe fruits.

Miocene hominoid positional behavior is inferred from increasing amounts of postcranial material. Most taxa differ from living apes in that they appear to have been pronograde arboreal quadrupeds (Rose, 1997; Ward, 1997). There are three likely exceptions to this. Given possible large body size and perhaps a more open habitat, *Ouranopithecus* may have been at least partly terrestrial (de Bonis & Koufos, 1997); only the discovery and description of postcranial material would confirm this. *Dryopithecus*, as noted earlier, had a suspensory component to its positional repertoire, but also probably a component of arboreal (above-branch) quadrupedalism (Moyà-Solà & Köhler, 1996). *Oreopithecus* has been interpreted as a suspensory ape (Harrison & Rook, 1997), a hominoid exhibiting significant bipedalism (Köhler & Moyà-Solà, 1997; Moyà-Solà *et al.*, 1999; Rook *et al.*, 1999), and a sloth-like suspensor (Wunderlich *et al.*, 1999).

The predominant pattern for middle and late Miocene

apes of mainly above-branch arboreal quadrupedalism raises some interesting issues. At the inferred body sizes (smaller morphs 20 kg, largest perhaps 90 kg) locomotion must have been both very slow and very deliberate, which in turn raises questions about foraging and movement between food sources. Adaptations rather different from those of living apes are again suggested.

There are two possible reactions to the disjunct geographic distribution of living and extinct hominoids. One, that the absence of fossils from living catchment areas is a real phenomenon, reflecting the fact that living apes originated outside the tropics – for example in Turkey (Begun & Güleç, 1998; Stewart & Disotell, 1998). Given what we know of habitat structure of middle and late Miocene times in that part of the world, this would imply that original ape adaptations were not to tropical forest and a strongly ripe-fruit-dominated diet.

A second possibility is that the absence of tropical equatorial ape fossils, particularly late Miocene and Pliocene, is a taphonomic artifact and that until (or if) they are found it will not be possible to improve our currently inadequate phylogenetic picture of the hominoid radiation(s). My own view is that the African and Asian tropics have been home to apes with the feeding adaptations of the living species at least since the middle Miocene, while the majority of non-tropical Miocene apes are adaptively and phylogenetically somewhat different.

The place of origin, whether central–west Africa, east Asia, or somewhere between, must also remain unclear until we have a fuller sampling of Neogene sediments. The proposal (Begun & Güleç, 1998; Stewart & Disotell, 1998) that hominoids originated in "Asia" (northern Arabia/southern Turkey) is certainly possible, but is driven by the inadequacies of the fossil record and, as importantly, is too accepting of the quality of characters used in phylogenetic analyses of an anatomically and geographically very patchy Miocene ape fossil record.

In summary, returning to the "forever speculative" we began with, what do we need to make progress and what might we reasonably expect to find and accomplish? First, what are prospects for future and more complete fossils, particularly the postcranial material we so desperately need? Clearly, this is impossible to predict, but collecting in currently un- or undersampled areas in the (equatorial) tropics is of great importance. West and central Africa are likely to remain long shots because of the apparent scarcity of exposed sediments of the right age or appropriate constitution. Southern China, Indonesia, Indo-China and Malaysia are all critically important target areas, and appropriate Neogene fossils, particularly from the last 10 Ma, would help resolve the extent to which the postcranial similarities of orangutans and chimpanzees are or are not likely to reflect common ancestry, and this in turn will do much to improve our understanding of Miocene ape phylogenetic relationships (and particularly that of the tantalizingly ambiguous *Sivapithecus*).

I expect that increased concern with "characters" and their definition will continue and that we shall see much more explicit descriptions of characters and rationales for their selection (for example, see Lieberman, 1999). In particular I anticipate further analysis of postcranials in search of more and "better" characters for phylogenetic analysis. But I would expect that different scholars will continue to pursue their favored characters and ignore or dismiss those of others, so discourse on phylogenetics will continue to resemble ships passing in the night. Given this, I expect that more progress will occur with behavioral and functional reconstructions, along with improved understanding of paleohabitats and paleoecology, rather than with our understanding of phylogenetic relationships.

Acknowledgements

I would like to thank John Barry, Walter Hartwig, Daniel Lieberman and Nathan Young for useful comments.

19 | Late Oligocene to middle Miocene catarrhines from Afro-Arabia

TERRY HARRISON

Introduction

The fossil anthropoid primates from the late Oligocene to early Miocene of Afro-Arabia are part of a major radiation of later Tertiary catarrhines of modern aspect that are critical for understanding the initial stages of hominoid evolution (Harrison, 1987, 1988, 1993; Harrison & Gu, 1999). The fossil record for this group is primarily known from early Miocene (approximately 16–23 Ma) localities in East Africa. Additional finds have been recovered from the late Oligocene of Kenya and the middle Miocene (approximately 10–16 Ma) of East Africa and Saudi Arabia. Twenty-one species are currently recognized, with additional species almost certainly to be added to the taxonomic roster in the next few years. This undoubtedly represents only a limited sampling of the actual diversity of catarrhines that lived in Afro-Arabia during the Miocene. Contemporary early Miocene localities in western and northern Kenya have entirely different catarrhine communities, which suggests extensive regional differentiation. In addition to taxonomic diversity, early Miocene catarrhines exhibit a wide range of adaptive diversity, especially in body size and inferred dietary behavior.

From a phylogenetic perspective, the fossil anthropoids from East Africa are stem catarrhines and stem hominoids[1] (Andrews, 1985, 1992; Andrews & Martin, 1987*a*; Harrison, 1987, 1988, 1993; Groves, 1989; Rose *et al.*, 1992; Fleagle, 1999; Harrison & Gu, 1999). Current debate centers on which taxa should be recognized as hominoids and on determining the nature of their relationships to extant apes. They occupy a narrow evolutionary grade close to the initial radiation of all recent catarrhines as evident in the general similarity of their craniodental and postcranial anatomy (see Harrison, 1987, 1988, 1993). The early Miocene catarrhines are undoubtedly critically important for documenting the evolutionary stages in the divergence and initial diversification of hominoids.

History of discovery and debate

Fossil catarrhine primates from the Miocene of East Africa were first collected by H. L. Gordon in 1927 at Koru in western Kenya, and the area was revisited by Edward Wayland, Arthur Tindell Hopwood and E. Nielsson between 1928 and 1932. These fossils were described by Hopwood (1933*a*, 1933*b*) as belonging to three new genera and species – *Proconsul africanus*, *Xenopithecus koruensis* and *Limnopithecus legetet*. *Proconsul africanus* was described as a medium-sized hominoid, probably closely related to the extant chimpanzee, while *Xenopithecus* was regarded as a somewhat specialized and "aberrant" anthropoid primate (Hopwood, 1933*b*: 445). *Limnopithecus legetet* was considered to be closely related to *Propliopithecus*, *Prohylobates* and *Pliopithecus*, and possibly to the extant gibbons. However, Hopwood argued that because of the derived morphology of its deciduous teeth, *Limnopithecus* was a more advanced hominoid (preferring to recognize it in his diagnosis as a "Gibbon-like Simiidae": Hopwood, 1933*b*: 438).

Between 1931 and 1942 a large number of fossil catarrhines was recovered from the early Miocene localities of Songhor, Rusinga Island and Mfangano Island, Kenya, principally by Donald MacInnes and L. S. B. Leakey, in conjunction with members of the East African Archaeological Expedition (MacInnes, 1943). Several mandibular specimens from Songhor formed the basis for the description of a new species of *Limnopithecus*, *L. evansi* (MacInnes, 1943). A palate and lower face of *Proconsul* recovered from Rusinga Island in 1932 was described by MacInnes (1943) as belonging to *P. africanus*, but was subsequently made the type specimen of *P. nyanzae* (Le Gros Clark & Leakey, 1950).

The British–Kenya Miocene Expedition, organized by L. S. B. Leakey and W. E. Le Gros Clark between 1947 and 1956, recovered further fossil primates from Rusinga, Mfangano, Songhor, Koru, Mteitei Valley, Maboko and Karungu. Le Gros Clark & Leakey (1950) described several new species, including *Proconsul nyanzae*, *P. major* and *Limnopithecus macinnesi*, and published a classic monograph describing these collections the following year (Le Gros Clark & Leakey, 1951). The most important finds were from Rusinga Island, and these included: a complete mandible of *Proconsul nyanzae* from R1A found in 1942 (Leakey, 1943; MacInnes, 1943); a partially complete skull of *P. africanus* (now *P. heseloni*) found at R106 by

[1] In this chapter: Hominoidea (hominoids) = Hylobatidae + Hominidae; Hominidae = Ponginae + Hominae; Ponginae = *Pongo*; Homininae = African apes + Hominini (humans).

Mary Leakey in 1948 (Le Gros Clark & Leakey, 1951); associated postcranial and craniodental remains of at least four individuals of *Limnopithecus macinnesi* (now *Dendropithecus macinnesi*) from R3A recovered by Louis Leakey in 1948 (Le Gros Clark, 1950; Le Gros Clark & Thomas, 1951); and a partial skeleton of *P. africanus* (now *P. heseloni*) discovered by Thomas Whitworth in 1951 at R114 (Napier & Davis, 1959; Davis & Napier, 1963). Le Gros Clark & Leakey (1951) formally sunk *Limnopithecus evansi* into *L. legetet*, considering the morphological differences between the two taxa as insufficient to merit a specific separation. They also regarded *Xenopithecus koruensis* as a junior synonym of *P. africanus*, a view largely adopted by subsequent workers (e.g., Andrews, 1978*a*; Szalay & Delson, 1979; Kelley & Pilbeam, 1986; Walker *et al.*, 1993; but see Madden, 1980; Pickford, 1986*c*). The collections made on Maboko Island in 1951 also included an isolated M_3, described by von Koenigswald (1969) as the holotype of *Mabokopithecus clarki*. He considered this new taxon as having affinities with *Oreopithecus bambolii* from the late Miocene of Italy, and consequently placed *Mabokopithecus* in the superfamily Oreopithecoidea. However, Szalay & Delson (1979) and Andrews (1981*a*) questioned whether *Mabokopithecus* was even a primate.

In a series of important papers published in the late 1940s and early 1950s, L. S. B. Leakey, W. E. Le Gros Clark and their colleagues (Leakey, 1946; Le Gros Clark, 1949, 1950, 1952; Le Gros Clark & Leakey, 1950, 1951; Le Gros Clark & Thomas, 1951) argued that the East African Miocene small and large catarrhines were early representatives of the gibbon and great ape lineages respectively (see Harrison, 1988). *Proconsul* was considered an ideal common ancestor for great apes, while *Limnopithecus* was regarded as a potential ancestor for modern gibbons. This phylogenetic and taxonomic scheme was further formalized and elaborated during the 1960s and early 1970s (Leakey, 1963; Frisch, 1965; Simons & Pilbeam, 1965; Pilbeam, 1969, 1972; Simons, 1972; Simons & Fleagle, 1973). During this period most authorities considered *Proconsul* as a subgenus of *Dryopithecus*, and as a member of the Pongidae (Simons & Pilbeam, 1965; Pilbeam, 1969, 1972; Simons, 1972). However, Leakey (1963) recognized that *Proconsul* possessed a suite of unique characteristics, and preferred to include the genus in a separate family within the Hominoidea, the Proconsulidae. Camille Arambourg (1943*a*, 1943*b*, 1948, 1952) suggested that *Limnopithecus* represented a possible phylogenetic precursor of Plio-Pleistocene hominins, and placed the genus within the Australopithecidae. Denise Ferembach (1953, 1954, 1958) believed the hylobatid affinities of *Limnopithecus* were based on primitive hominoid characteristics, and that the genus was more closely related to extant great apes. Adolph Remane (1965), followed by Colin Groves (1972, 1974*a*) and Ciochon & Corruccini (1977), considered *Limnopithecus* a stem hominoid or primitive catarrhine that could be grouped with *Propliopithecus* and *Pliopithecus* in a separate family, the Pliopithecidae.

Between 1959 and 1965 Louis Leakey turned his attentions to the middle Miocene locality of Fort Ternan, and was responsible for recovering the majority of the fossil catarrhines known from this locality (Leakey, 1962*a*, 1967*a*, 1968; Andrews & Walker, 1976; Harrison, 1986*b*, 1992). The collection included mandibular fragments and isolated teeth of a small catarrhine primate (Leakey, 1968; Andrews & Walker, 1976; Andrews, 1978*a*) that have subsequently been referred to *Simiolus* (Harrison, 1992). During the early 1960s, Leakey also collected extensively at Songhor, and found a relatively complete palate, later designated the holotype of *Rangwapithecus gordoni* (Andrews, 1974).

During the late 1960s and early 1970s, field research in western Kenya focused primarily on better documentation of geological and chronological context (Evernden *et al.*, 1965; Bishop *et al.*, 1969; Van Couvering & Miller, 1969; Van Couvering, 1972; Van Couvering & Van Couvering, 1976). Meanwhile, William W. Bishop and colleagues initiated active field work in eastern Uganda, with fossil primates being discovered at Napak, Bukwa and Moroto (Allbrook & Bishop, 1963; Bishop, 1964; Pilbeam & Walker, 1968; Walker, 1968; Walker & Rose, 1968; Pilbeam, 1969). Large catarrhine specimens recovered from Moroto between 1961 and 1965 were referred to *P. major* (Allbrook & Bishop, 1963; Bishop, 1964; Pilbeam, 1969), although this taxonomic attribution was questioned (Leakey, 1963, 1970; Martin, 1981; Pickford, 1982; Kelley & Pilbeam, 1986; see Pickford *et al.*, 1999). A palate and lower face of a small catarrhine primate recovered from Napak (Bishop, 1964; Fleagle, 1975) later became the holotype of a new genus and species, *Micropithecus clarki* (Fleagle & Simons, 1978*b*).

In the early 1970s, Peter Andrews, David Pilbeam and Alan Walker led a series of expeditions to many of the major fossil localities in western Kenya (Andrews & Van Couvering, 1975; Andrews, 1981*b*; Andrews *et al.*, 1981; Shipman et al., 1981). The new collections from these expeditions allowed Andrews (1970, 1973, 1974, 1978*a*; Andrews & Simons, 1977) to revise completely the taxonomy of the fossil catarrhines from the early Miocene of East Africa. Andrews (1974) recognized two species of *Dryopithecus*, which he included in a new subgenus, *Rangwapithecus*, to distinguish them from species of *Dryopithecus* (*Proconsul*). These new species, *D.* (*Rangwapithecus*) *vancouveringi* and *D.* (*Rangwapithecus*) *gordoni*, were based mainly on newly discovered specimens from Songhor, Rusinga Island and Maboko Island. Andrews (1974) also suggested that *Limnopithecus legetet* was more closely related to *Dryopithecus* (*Proconsul*), and therefore a member of the Pongidae, whereas *L. macinnesi* should be retained in the Hylobatidae. This phylogenetic and taxonomic distinction was formally recognized by the creation of a separate genus, *Dendropithecus*, for "*L.*" *macinnesi* (Andrews & Simons, 1977; Andrews 1978*a*). Later, Andrews (1980) amended this taxonomic and phylogenetic scheme, by removing *Dendropithecus* from the Hylobatidae, and placing it in a basal catarrhine family, the Pliopithecidae, following Groves (1972).

Between 1975 and 1980 Martin Pickford directed the Western Kenya Project and discovered a large number of fossil primates from Koru, Songhor, Mteitei Valley and Meswa Bridge (Pickford & Andrews, 1981). These fossils were described by Harrison (1981, 1982), Martin (1981), and Andrews *et al.* (1981), and formed the basis for a taxonomic revision of the smaller fossil catarrhines (Harrison, 1982, 1988). Harrison (1988) formally resurrected *Limnopithecus evansi* as a valid taxon, and recognized a new genus and species, *Kalepithecus songhorensis* to accommodate Andrews's (1978*a*) subspecies of *D. macinnesi* from Songhor. Harrison's (1982, 1987) cladistic analysis confirmed earlier suggestions that *Dendropithecus* was a primitive sister group to extant catarrhines, but more derived than both propliopithecids and pliopithecids. Moreover, Harrison (1982, 1987, 1988, 1993) concluded that the early Miocene East African catarrhines in general, including *Proconsul*, represent a fairly narrow grade, situated close to the initial radiation of all recent catarrhines, and he classified them together in the Proconsulidae. Subsequent authors have broadly adopted this scheme with various modifications (e.g., Andrews, 1985, 1992; Fleagle, 1986, 1999; Feldesman, 1986; Andrews & Martin, 1987*a*; Pilbeam, 1996, 1997). Most prefer to regard *Proconsul* (and the other large fossil East African catarrhines) as basal hominoids (e.g., Andrews, 1985, 1992), while Harrison (1982, 1987, 1988, 1993) advocates recognizing *Proconsul* as a stem catarrhine, along with other proconsulids.

Pickford's expedition to the middle Miocene locality of Maboko Island, Kenya in 1982 and 1983 recovered more than 100 small catarrhines. Among these were specimens belonging to a new genus and species, *Nyanzapithecus pickfordi* (Harrison, 1986*b*), characterized by a distinctive suite of dental specializations. Harrison (1986*b*) also transferred the smaller species of *Rangwapithecus*, *R. vancouveringi*, to this genus, and emended its spelling (to *Nyanzapithecus vancouveringorum*). *Nyanzapithecus* was considered closely related to *Rangwapithecus* and *Mabokopithecus*, with which it shares dental features (Harrison, 1986*b*; Kunimatsu, 1992*a*, 1992*b*, 1997). It was also initially considered to be an early relative of *Oreopithecus bambolii*, and included in the Oreopithecidae (Harrison, 1986*b*, 1987). However, Harrison (Andrews *et al.*, 1996; Harrison & Rook, 1997; Harrison, 1999) now prefers instead to include *Nyanzapithecus* in the Proconsulidae. Kunimatsu (1997) recently described an additional species of *Nyanzapithecus*, *N. harrisoni*, from the middle Miocene locality of Nachola in Kenya.

A recent find of a nearly complete mandible containing an M_3 that closely matches the morphology of the holotype of *Mabokopithecus clarki* (Benefit *et al.*, 1998) indicates that *Nyanzapithecus* is possibly a junior synonym of *Mabokopithecus*. Since *Mabokopithecus* von Koenigswald, 1969 has priority over *Nyanzapithecus* Harrison, 1986, all three species of *Nyanzapithecus* may need to be transferred to the former genus. Harrison (1989*a*) also recognized a new species of *Micropithecus*, *Mi. leakeyorum*, from Maboko Island. More recently, Benefit (1991) and Gitau & Benefit (1995) have reported the discovery on Maboko Island of additional craniodental specimens of this species, and have suggested that it should be transferred to the genus *Simiolus*.

During the mid-1970s and early 1980s a series of papers was published re-evaluating the functional and phylogenetic implication of the postcranium of *Proconsul*, especially the forelimb (see Fleagle, 1983; Rose, 1983; McHenry & Corruccini, 1983; Senut, 1989). Renewed interest in Rusinga and Mfangano Islands during this period led to the discovery (or rediscovery in museum collections) of important new finds of *Proconsul*, including a lower leg and foot of *P. nyanzae* from Rusinga (KNM-RU 5872), a partial skeleton of *P. nyanzae* from Mfangano (KNM-MW 13412); a palate and lower face of a male individual of *P. heseloni* (KNM-RU 16000); additional parts of the *P. heseloni* skeleton (KNM-RU 2036) and the 1948 skull (KNM-RU 7290), and the remarkable discovery of at least nine partial skeletons of *P. heseloni* and *P. nyanzae* ranging in age from infant to adult at the Kaswanga Primate Site on Rusinga Island (Walker & Pickford, 1983; Walker *et al.*, 1983, 1985; Beard *et al.*, 1986; Walker & Teaford, 1988, 1989; Teaford *et al.*, 1988; Ward, 1993; Ward et al., 1993, 1995; Begun *et al.*, 1994).

During the 1980s and early 1990s, following the earlier studies by Leonard Greenfield (1972, 1973), a major debate arose about species recognition and sexual dimorphism in *Proconsul*, which centered around the number of species represented on Rusinga Island and the allocation of individual specimens to these species (Bosler, 1981; Martin, 1981; Kelley, 1986*b*, 1995; Kelley & Pilbeam, 1986; Pickford, 1986*b*; Teaford *et al.*, 1988, 1993; Ruff *et al.*, 1989; Cameron, 1991, 1992; Walker *et al.*, 1993). A consensus emerged that there were two species – *P. nyanzae* and a smaller species that could be distinguished at the species level from *P. africanus* from Koru and Songhor, leading to recognition of a new species, *P. heseloni* Walker *et al.*, 1993.

Between 1983 and 1987, Meave Leakey, Richard Leakey and Alan Walker directed expeditions to late Oligocene and early Miocene localities (i.e., Lothidok, Moruorot, Buluk, Kalodirr) in the Lothidok range on the western side of Lake Turkana in northern Kenya (Leakey & Walker, 1985*a*; Leakey & Leakey, 1986*a*, 1986*b*, 1987; Leakey *et al.*, 1988*a*, 1988*b*; Leakey & Walker, 1997). Fossil catarrhines had been collected in this region several decades earlier by the University of California African Expedition in 1948 and by Louis Leakey in 1959, but no concerted effort to explore the region had previously been attempted. Specimens of a large catarrhine primate recovered from Lothidok by the University of California team led to the recognition of a new species, *Proconsul (Xenopithecus) hamiltoni* (Madden, 1980). With additional material recovered in 1986, this late Oligocene species was recognized as a distinct genus, *Kamoyapithecus* (Leakey *et al.*, 1995*a*).

The well-preserved material from Kalodirr includes

several skulls and partial skeletons that have provided the basis for three new genera and species: *Simiolus enjiessi*, a primitive small catarrhine, apparently closely related to *Dendropithecus macinnesi*; *Turkanapithecus kalakolensis*, considered to be a primitive hominoid with its own distinctive suite of characteristics; and *Afropithecus turkanensis*, regarded as a specialized large hominoid (Leakey & Leakey, 1986*a*, 1986*b*, 1987; Leakey *et al.*, 1988*a*, 1988*b*; Rose *et al.*, 1992). Additional finds of *Afropithecus* were made by William Anyonge in 1986–7 at the locality of Locherangan in northern Kenya (Anyonge, 1991). *Afropithecus* has been linked to several different taxa: *Heliopithecus* from Ad Dabtiyah, Saudi Arabia, *Morotopithecus* from Moroto, Uganda, *Equatorius* from Maboko and Kipsaramon, Kenya and *Otavipithecus* from Otavi, Namibia (Andrews & Martin 1987*a*, 1987*b*; Andrews, 1992; S. Ward *et al.*, 1999; Singleton, 2000). In fact, some authors have even suggested that *Heliopithecus leakeyi* may be congeneric with *Afropithecus turkanensis* (Andrews & Martin, 1987*a*, 1987*b*; Andrews, 1992). Andrews (1992) included *Afropithecus*, *Heliopithecus*, *Otavipithecus* and *Equatorius* within the tribe Afropithecini (Subfamily Dryopithecinae; Family Hominidae) as distinct from *Proconsul*, *Rangwapithecus*, *Nyanzapithecus*, *Limnopithecus* and *Kamoyapithecus* (Family Proconsulidae).

Renewed interest in the Ugandan Miocene sites in the mid-1980s and early 1990s led to the discovery of additional catarrhine primates from Moroto and Napak (Pickford *et al.*, 1986*b*, 1999; Gebo *et al.*, 1997*b*; Gommery *et al.*, 1998; MacLatchy & Pilbeam, 1999; MacLatchy *et al.*, 2000). The age of Moroto was found to be older than 20.6 Ma based on new radiometric dates (Gebo *et al.*, 1997*b*). In addition, the new finds prompted a re-evaluation of the relationship of the Moroto catarrhine, and eventually led to the recognition of a new genus and species, *Morotopithecus bishopi* (Gebo *et al.*, 1997*b*). Based on important derived hominid characteristics of the scapula and vertebrae in *Morotopithecus*, not seen in more primitive catarrhines such as *Proconsul*, Gebo *et al.* (1997*b*) identified it as the sister taxon of crown hominoids or hominids. Some of the new material from Napak (Gommery *et al.*, 1998), initially assigned to *P. major*, has been used to distinguish the species as a separate taxon, *Ugandapithecus major* (Senut *et al.*, 2000) from the other *Proconsul* species. This distinction is not supported in this chapter.

Taxonomy

Systematic framework

The following scheme removes all of the early Miocene Afro-Arabian catarrhines from the Hominoidea, and recognizes them as a paraphyletic grouping of stem catarrhines. The exception is *Morotopithecus*, provisionally retained in the Hominoidea on the basis of its postcranial synapomorphies. Three superfamilies are recognized: Proconsuloidea, Dendropithecoidea and Hominoidea. The Proconsuloidea contains a single family, the Proconsulidae, divided into three subfamilies – Proconsulinae, Afropithecinae and Nyanzapithecinae. It may even be desirable to elevate these subfamilies to family rank at some later date. The Afropithecinae may also include *Otavipithecus*, *Equatorius* and *Nacholapithecus* (discussed in Ward & Duren, this volume). Recognition of the Nyanzapithecinae as a separate clade follows Harrison (1999). *Kalepithecus*, *Limnopithecus* and *Kamoyapithecus* are not well enough known to classify them with any confidence, so they are left as *incertae sedis*. It is probable, however, that they are members of the Proconsuloidea or the Dendropithecoidea. Given the primitive morphology of the upper molars of *Kamoyapithecus*, this taxon may represent the sister taxon of all other catarrhines from the Miocene and later (see below), including the Proconsuloidea and Dendropithecoidea.

Order Primates Linnaeus, 1758
- Infraorder Catarrhini É. Geoffroy Saint-Hilaire, 1812
  - Superfamily Proconsuloidea Leakey, 1963
    - Family Proconsulidae Leakey, 1963
      - Subfamily Proconsulinae Leakey, 1963
        - Genus *Proconsul* Hopwood, 1933
          - *Proconsul africanus* Hopwood, 1933
          - *Proconsul nyanzae* Le Gros Clark & Leakey, 1950
          - *Proconsul major* Le Gros Clark & Leakey, 1950
          - *Proconsul heseloni* Walker *et al.*, 1993
      - Subfamily Afropithecinae Andrews, 1992 (new rank)
        - Genus *Afropithecus* Leakey & Leakey, 1986
          - *Afropithecus turkanensis* Leakey & Leakey, 1986
        - Genus *Heliopithecus* Andrews & Martin, 1987
          - *Heliopithecus leakeyi* Andrews & Martin, 1987
      - Subfamily Nyanzapithecinae (new Subfamily group name)
        - Genus *Nyanzapithecus* Harrison, 1986
          - *Nyanzapithecus vancouveringorum* (Andrews, 1974)
          - *Nyanzapithecus pickfordi* Harrison, 1986
          - *Nyanzapithecus harrisoni* Kunimatsu, 1997
        - Genus *Mabokopithecus* von Koenigswald, 1969
          - *Mabokopithecus clarki* von Koenigswald, 1969
        - Genus *Rangwapithecus* Andrews, 1974
          - *Rangwapithecus gordoni* Andrews, 1974
        - Genus *Turkanapithecus* Leakey & Leakey, 1986
          - *Turkanapithecus kalakolensis* Leakey & Leakey, 1986
  - Superfamily Dendropithecoidea (new Superfamily group name)
    - Family Dendropithecidae (new Family group name)
      - Genus *Dendropithecus* Andrews & Simons, 1977
        - *Dendropithecus macinnesi* (Le Gros Clark & Leakey, 1950)
      - Genus *Micropithecus* Fleagle & Simons, 1978
        - *Micropithecus clarki* Fleagle & Simons, 1978
        - *Micropithecus leakeyorum* Harrison, 1989
      - Genus *Simiolus* Leakey & Leakey, 1987

Simiolus enjiessi Leakey & Leakey, 1987
Superfamily Hominoidea
Family *incertae sedis*
Genus *Morotopithecus* Gebo *et al.*, 1997
Morotopithecus bishopi Gebo *et al.*, 1997
Superfamily *incertae sedis*
Family *incertae sedis*
Genus *Limnopithecus* Hopwood, 1933
Limnopithecus legetet Hopwood, 1933
Limnopithecus evansi MacInnes, 1943
Genus *Kalepithecus* Harrison, 1988
Kalepithecus songhorensis (Andrews, 1978)
Genus *Kamoyapithecus* Leakey *et al.*, 1995
Kamoyapithecus hamiltoni (Madden, 1980)

Superfamily Proconsuloidea

Family Proconsulidae

Subfamily Proconsulinae

GENUS *Proconsul* Hopwood, 1933
Face moderately short and broad. Premaxilla relatively short. Incisive fossa forms a large aperture with wide communication into the nasal passage. Diastema between upper canine and lateral incisor relatively large in males, but small in females. Nasal aperture relatively broad, rhomboidal in shape, narrowing inferiorly between the central incisor roots, and mediolaterally widest just below mid-height. Nasoalveolar clivus short. Nasals long and narrow, more or less parallel-sided with superior expansion, and supported laterally by premaxillary alae. Premaxilla makes contact with the nasal bones, thereby excluding the maxilla from the pyriform aperture margin (*contra* Andrews, 1978*a*). Prominent facial pillar or canine jugum accommodates the root of the upper canine. Canine fossa shallow. Single large infraorbital foramen. Palate long, rectangular and relatively shallow. Massive and rugose tuberosity located posterior to the alveolar process of M^3. Maxillary sinus extensive and penetrates anteriorly at least as far as the premolars, and laterally into the anterior root of the zygomatic arch. Zygomatic arch originates relatively low on the face, and sweeps superiorly and posteriorly. Malar tuberosity well developed. Articular fossa gutter-like, bordered anteriorly by a well-developed eminence and posteriorly by a prominent postglenoid process. External auditory meatus forms a short and broad tube. Well-developed mastoid process. Nuchal plane short and steeply angled, with a strongly developed external occipital protuberance located high on the neurocranium. Subarcuate fossa large. Frontal process of the zygomatic perforated by multiple zygomaticofacial foramina situated slightly above the horizontal level of the inferior margin of the orbits. Superior orbital margin is sharp, bordered by low indistinct supraorbital costae, and a slightly swollen glabella. Extensive frontal sinus. Lacrimal duct located within the orbit. Interorbital region relatively broad. Orbits subrectangular, slightly broader than high, with a distinct angulation of the superolateral margin. Frontal bone anteroposteriorly short. Neurocranium relatively large, with slight postorbital constriction. Superior temporal lines are strongly marked and converge posteriorly, but do not meet to form a sagittal crest, at least in female individuals. Mandibular symphysis with superior transverse torus moderate to well developed, and inferior transverse torus generally weaker or entirely absent. Corpus shallows slightly posteriorly. Single mental foramen located below the lower premolars (Le Gros Clark & Leakey, 1951; Napier & Davis, 1959; Davis & Napier, 1963; Corruccini & Henderson, 1978; Whybrow & Andrews, 1978; McHenry *et al.*, 1980; Walker & Pickford, 1983; Walker et al., 1983; T. Harrison, unpublished data).

Upper incisors procumbently inclined. I^1 relatively narrow and high-crowned, and much larger than I^2 (the mesiodistal length of I^2 is on average 70% that of I^1). Lower incisors narrow. Upper canines relatively stout, moderately bilaterally compressed, with a single mesial groove. Canines strongly sexually dimorphic (with upper canines of females having an average cross-sectional area only 60% that of males, at least in *P. heseloni*). Upper premolars mesiodistally short and broad, with a marked height differential between the buccal and lingual cusps, especially on P^3, and a weak lingual cingulum variably present on P^4. P_3 moderately sectorial, with relatively long mesiobuccal face. P_4 usually broader than long, with a weak buccal cingulum. Upper molars rectangular to rhomboidal in shape, and buccolingually broader than long. Cusps and crest moderately elevated, and occlusal basins generally well defined. Protoconule usually conspicuous. Lingual cingulum broad, and commonly beaded. Buccal cingulum variably developed. Hypocone large, with poorly developed crests linking it to the protocone or crista obliqua. M^3 subequal in size to M^2 or slightly larger, with variable regression of the distal cusps. Lower molars relatively long and narrow, with simple "crystalline" cusps, and few secondary wrinkles. Buccal cingulum variably developed. Mesial and distal foveae generally well defined. $M_1 < M_2 < M_3$. Enamel on molars relatively thick (Andrews, 1978*a*; Walker *et al.*, 1983, 1993; Andrews & Martin, 1991; Walker, 1997; Beynon *et al.*, 1998; T. Harrison, unpublished data).

INCLUDED SPECIES *P. africanus*, *P. heseloni*, *P. major*, *P. nyanzae*

SPECIES *Proconsul africanus* Hopwood, 1933 (Fig. 19.1)
TYPE SPECIMEN BMNH M 14084, left maxilla with C–M^3
AGE AND GEOGRAPHIC RANGE Early Miocene (approximately 19–20 Ma), Kenya (Bishop *et al.*, 1969; Pickford & Andrews, 1981; Pickford, 1981, 1983, 1986*c*; Harrison, 1988)
ANATOMICAL DEFINITION
A medium-sized catarrhine, intermediate in dental size between *Hylobates syndactylus* and *Pan troglodytes*. It is comparable in size to *Proconsul heseloni*, although the teeth

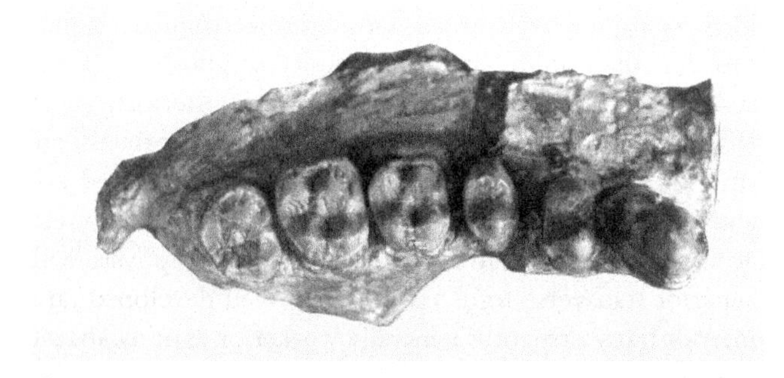

A

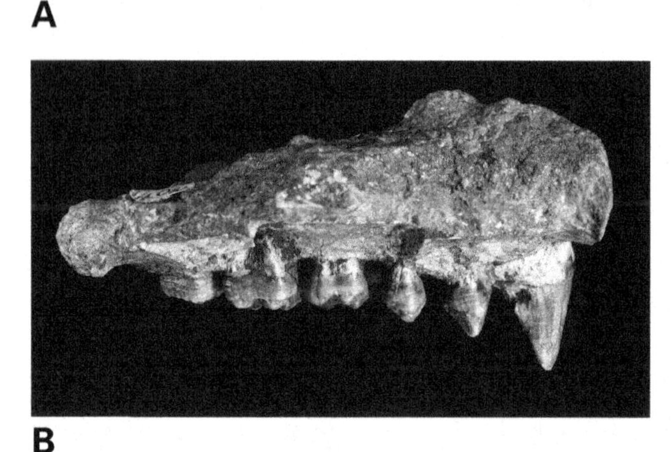

B

Fig. 19.1 *Proconsul africanus*. BMNH 14084 (holotype), left maxilla with C–M^3: (A) occlusal view; (B) medial view. Photographs courtesy of Peter Andrews.

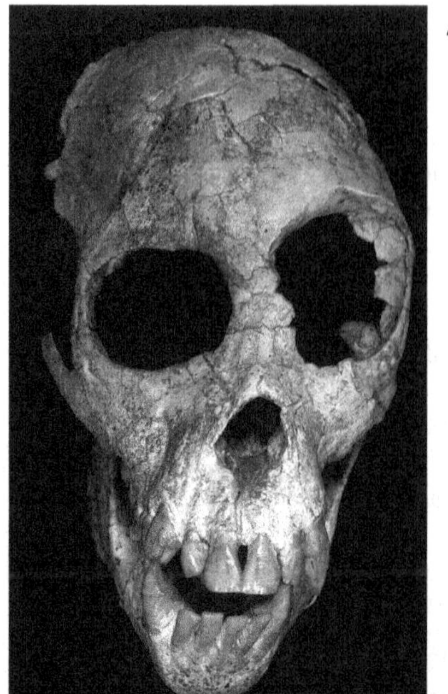

A

B

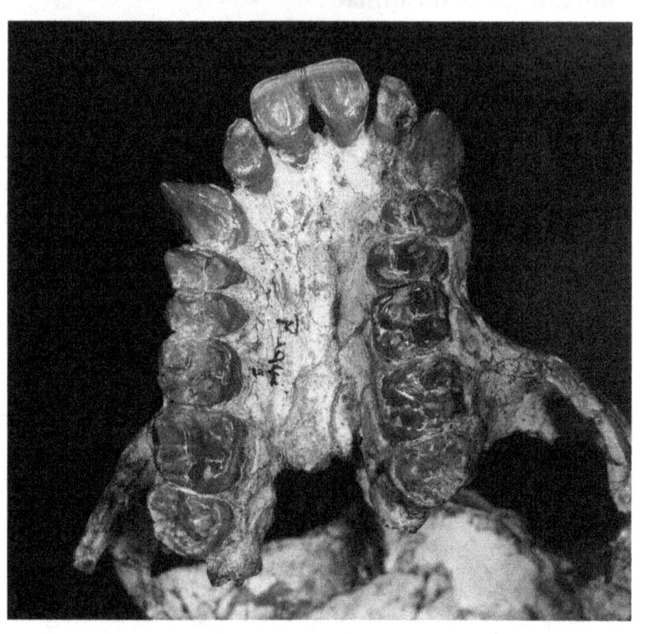

Fig. 19.2 *Proconsul heseloni*. KNM-RU 7290, partial skull: (A) facial view; (B) maxilla, occlusal view. Photographs courtesy of Peter Andrews.

tend to be slightly smaller. Differs from *P. heseloni* in the following characteristics: lower canines more bilaterally compressed; upper premolars narrower, with greater height differential between the buccal and lingual cusps; buccal and lingual cingula and occlusal crests better developed on upper molars; hypocone and protocone subequal in size (hypocone smaller in *P. heseloni*); M^1 relatively narrower; $M^1 < M^3 < M^2$; lower molars with better-developed buccal cingulum; M_3 hypoconulid and hypoconid subequal in size (hypoconulid larger in *P. heseloni*), entoconid relatively larger and more distally placed relative to the hypoconid, distal fovea broader with better-developed crest linking the entoconid and hypoconulid, broader less triangular talonid with more reduced distal cusps; mandibular symphysis with massive superior transverse torus only (the inferior transverse torus is variably developed in *P. heseloni*, although it is generally smaller than the superior transverse torus); mandibular corpus tends to be deeper, and shallows posteriorly more strongly (Andrews, 1978*a*; Walker *et al.*, 1993; T. Harrison, unpublished data).

SPECIES *Proconsul heseloni* Walker *et al.*, 1993 (Fig. 19.2)
TYPE SPECIMEN KNM-RU 2036, partial skull and a large portion of the postcranial skeleton of a subadult female individual
AGE AND GEOGRAPHIC RANGE Early Miocene (approximately 17.0–18.5 Ma), Kenya (Drake *et al.*, 1988)

ANATOMICAL DEFINITION
A medium-sized catarrhine, similar in dental size to *P. africanus*, and somewhat larger than the extant siamang (*Hylobates syndactylus*). Regressions of postcranial dimensions indicate that female individuals of *P. heseloni* weighed around 9–11 kg. Given the sexual dimorphism in gnathic material it is likely that males may have averaged approximately 20 kg, and that overall it was similar in body size to *Nasalis larvatus*, the largest arboreal Old World monkey.

The skull of *P. heseloni* has been described in some detail previously, and provides the primary basis for the description of the genus (Le Gros Clark, 1950; Le Gros

Clark & Leakey, 1951; Napier & Davis, 1959; Davis & Napier, 1963; Corruccini & Henderson, 1978; McHenry *et al.*, 1980; Walker & Pickford, 1983; Walker et al., 1983; Teaford *et al.*, 1988). Walker *et al.* (1983) have estimated the cranial capacity of the type specimen at 167.3 cm^3, and have inferred from this that *Proconsul heseloni* was more encephalized than modern Old World monkeys of comparable body size. However, using foramen magnum area as a surrogate, Manser & Harrison (1999) have predicted a brain size of only 130.3 cm^3, close to the mean for degree of encephalization among anthropoid primates. The cortical sulcal pattern in *P. heseloni* is generally similar to that found in extant platyrrhine monkeys (Falk, 1983).

The postcranial skeleton of *Proconsul heseloni* is well known, being represented by several partial skeletons from Rusinga Island (i.e., KNM-RU 2036, RU 7290, and at least seven individuals from the Kaswanga Primate Site) (Le Gros Clark & Leakey, 1951; Napier & Davis, 1959; Walker & Pickford, 1983; Walker *et al.*, 1985; Walker & Teaford, 1989). The postcranium corresponds closely to the primitive catarrhine morphotype (Harrison, 1982, 1987, 1993), although a few traits have been inferred to be synapomorphies linking *Proconsul* with extant hominoids (Walker & Pickford, 1983; Rose, 1988, 1992, 1993*b*, 1997; Ward *et al.*, 1991, 1993; Walker *et al.*, 1993; Ward, 1993, 1997, 1998; Kelley, 1997; Walker, 1997). Scapula resembles those of colobines and non-suspensory platyrrhines (Rose, 1993*a*, 1997). Thorax is relatively long and mediolaterally narrow (Ward, 1993, 1997; Ward et al., 1993). Lumbar vertebrae with long centrum, small cranial and caudal surface areas relative to estimated body weight (Sanders & Bodenbender, 1994; Harrison & Sanders, 1999), and moderately well-developed ventral keel. Sacrum relatively narrow, with small sacroiliac joint (Rose, 1993*b*). It has been inferred from a purported partial sacrum from the Kaswanga Primate Site that *P. heseloni* did not have a tail (Ward et al., 1991; Ward, 1997), but this has been contested (Harrison, 1998; C. Ward *et al.*, 1999). Ilium narrow and monkey-like. Ischial tuberosities lacking, implying sitting and sleeping behaviors similar to platyrrhines, rather than to extant catarrhines (Rose, 1993*b*; Harrison & Sanders, 1999). Estimated intermembral index of 88 (Walker & Pickford, 1983; T. Harrison, unpublished data). Estimated brachial and crural indices of 96 and 92 respectively (Walker & Pickford, 1983; T. Harrison, unpublished data). Limb bones relatively robust (Ruff *et al.*, 1989). Humeral head lacks medial torsion, and proximal shaft retroflexed. Distal humerus with distinct lateral keel, inflated capitulum, and narrow and recessed zona conoidea (Rose, 1993*a*, 1997). Entepicondylar foramen and dorsal epitrochlear fossa absent (Harrison, 1987). Radial head ovoid, with beveled margin. Olecranon process of ulna well developed. Ulnar styloid process articulates directly with the pisiform and triquetral (Napier & Davis, 1959; Harrison, 1982, 1987; Beard *et al.*, 1986). Os centrale unfused. Hand relatively long, with hand length index of 35 (similar to arboreal cercopithecines) (Walker *et al.*, 1993). Pollex relatively long, with a mobile trapezium–1st metacarpal joint (Rafferty, 1990; Rose, 1992, 1993*a*). High femoral neck angle (Walker, 1997). Distal end of femur relatively broad, with medial condyle only slightly larger than lateral condyle. Fibula stout. Foot closely resembles those of arboreal non-hominoid primates (Harrison, 1982; Langdon, 1986; Rose, 1993*b*; Strasser, 1993). Manual and pedal phalanges quite stout and slightly curved (Begun *et al.*, 1994). Hallux well-developed with a powerful grasping capability. Overall, the postcranium indicates that *P. heseloni* was an arboreal, quadrupedal catarrhine, most similar in its locomotor repertoire to arboreal cercopithecids, particularly colobines, and to the larger platyrrhines (Rose, 1983, 1993*b*, 1994; Walker & Pickford, 1983; Walker, 1997).

SPECIES *Proconsul major* Le Gros Clark & Leakey, 1950
TYPE SPECIMEN BMNH M 16648, right mandibular corpus with P_4–M_3
AGE AND GEOGRAPHIC RANGE Early Miocene (approximately 19–20 Ma), Kenya and Uganda (Bishop *et al.*, 1969; Pickford, 1981, 1983, 1986*b*; Pickford & Andrews, 1981; Pickford *et al.*, 1986*a*; Harrison, 1988)
ANATOMICAL DEFINITION
A large catarrhine, similar to or slightly larger in dental size to *Pongo pygmaeus*. Estimates based on postcranial dimensions indicate a body weight of 60–90 kg (Harrison, 1982; Rafferty *et al.*, 1995; Gommery *et al.*, 1998), comparable to female lowland gorillas or male orangutans (Smith & Jungers, 1997). *Proconsul major*, as its name implies, is the largest species of *Proconsul*, with average dental dimensions almost 20% larger than those of *P. nyanzae*. *Proconsul major* is distinguished from other species of *Proconsul* in the following combination of features: massive superior transverse torus with no inferior torus; upper and lower incisors relatively broader; upper canines less bilaterally compressed and more tusk-like; upper and lower canines with distinctive sinusoidal curvature of distal crest and a blade-like tip; upper premolars narrower, with cusps more similar in height; lower molar proportions similar to those in *P. heseloni*, and differs from *P. nyanzae* in having a less pronounced size differential between M_1 and M_2; M_3 variable in size, but tends to be larger relative to M_2; lower molars have a stronger buccal cingulum than in *P. nyanzae* and *P. heseloni*; M^3 relatively large as in *P. nyanzae* (Andrews, 1978*a*; Martin, 1981).

A few postcranials referable to *P. major* have been recovered from Koru, Songhor and Napak (MacInnes, 1943; Le Gros Clark & Leakey, 1951; Preuschoft, 1973; Harrison, 1982; Conroy & Rose, 1983; Langdon, 1986; Rafferty *et al.*, 1995; Gommery *et al.*, 1998). These include humeral shaft fragments, metapodials and phalanges, a navicular, an associated talus and partial calcaneus, and femoral and tibial fragments. Despite their much greater

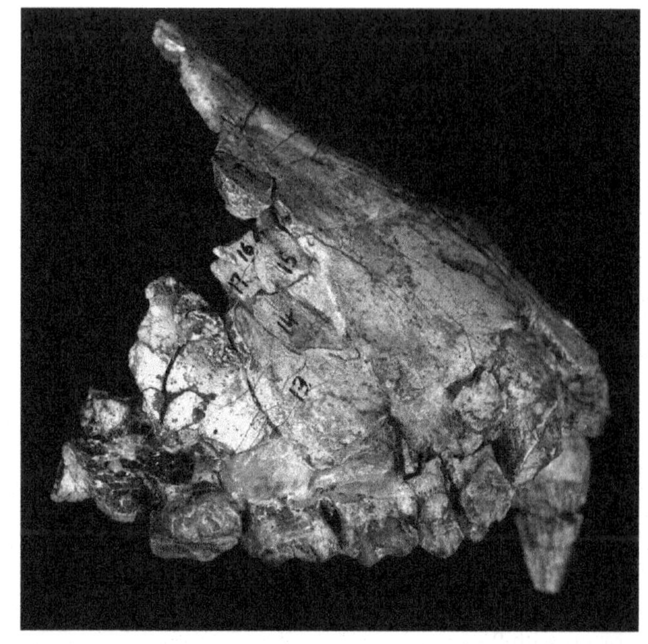

A

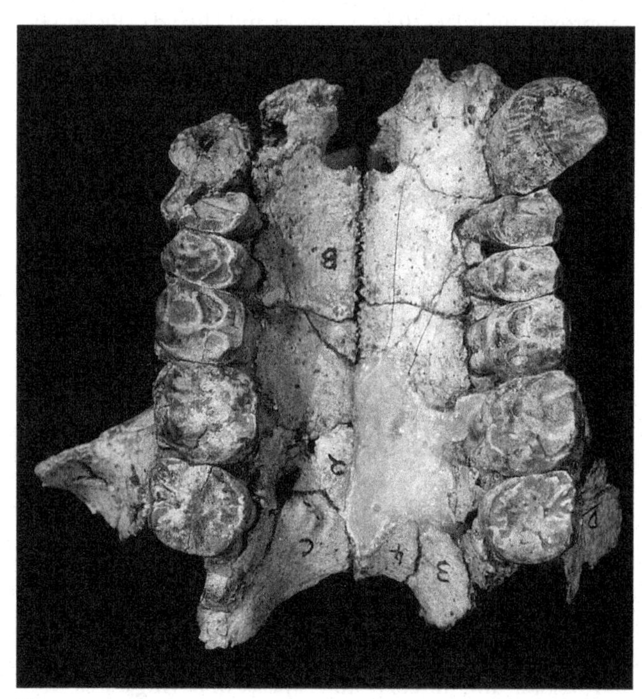

B

Fig. 19.3 *Proconsul nyanzae*. BMNH 16647 (holotype), lower face and palate: (A) lateral view; (B) occlusal view. Photographs courtesy of Peter Andrews.

size than corresponding elements of *P. heseloni* and *P. nyanzae*, they are very similar in morphology. However, minor distinctions do, perhaps, indicate subtle differences in locomotor and positional behavior (Harrison, 1982), and support inferences based on the craniodental evidence that *P. major* may be phyletically more distantly related than are *P. heseloni* and *P. nyanzae* to each other. Nengo & Rae (1992) have described a fragmentary distal ulna of *P. major* as being much more hominoid-like than that of *P. heseloni*, but the attribution of this specimen is uncertain (Rose, 1997; Walker, 1997).

SPECIES *Proconsul nyanzae* Le Gros Clark & Leakey, 1950 (Fig. 19.3)
TYPE SPECIMEN BMNH M 16647, lower face and palate with left C–M^3 and right P^3–M^3
AGE AND GEOGRAPHIC RANGE Early Miocene (approximately 17.0–18.5 Ma), Kenya (Drake *et al.*, 1988)
ANATOMICAL DEFINITION
A large catarrhine, intermediate in dental size between *P. heseloni* and *P. major*. Estimates based on postcranial dimensions indicate a body-weight range of 20–50 kg (Ruff *et al.*, 1989; Rafferty *et al.*, 1995), with males and females probably averaging 28 kg and 40 kg respectively, slightly smaller than *Pan troglodytes schweinfurthii* (Smith & Jungers, 1997).

Proconsul nyanzae is morphologically closely similar to *P. heseloni*, and differs primarily in its larger size (although the two species may have overlapped in body weight). However, a number of dental features do serve to distinguish *P. nyanzae* from its smaller contemporary. These include: upper and lower canines less bilaterally compressed; P^3 with greater height differential between buccal and lingual cusps; lower molars relatively broader; greater size differential between M_1 and M_2 (the area of M_2 is 61% greater than M_1 in *P. nyanzae*, compared with only 34% in *P. heseloni*); M_3 has a larger entoconid connected to the hypoconulid by a well-developed crest; size differential between M^1 and M^2 greater (the area of M^2 is 54% greater than M^1, compared with only 33% in *P. heseloni*); upper M^1 and M^2 relatively narrower, with hypocone subequal in size to metacone, lingual cingulum better developed, and distal transverse crest more pronounced; M^3 relatively larger; greater degree of secondary wrinkling on the occlusal surface of upper and lower molars. In addition, there is no inferior transverse torus on the mandible.

The postcranium of *Proconsul nyanzae* is represented by a partial skeleton from Mfangano Island (Ward *et al.*, 1993), two partial skeletons from the Kaswanga Primate Site, and a number of associated and isolated postcranial elements from Rusinga Island (Le Gros Clark & Leakey, 1951; Le Gros Clark, 1952; Preuschoft, 1973; Harrison, 1982). Despite the size difference, *Proconsul nyanzae* is remarkably similar in its postcranial morphology to *P. heseloni*, which, in conjunction with the craniodental similarities, supports a sister-group relationship between these species.

Subfamily Afropithecinae

GENUS *Afropithecus* Leakey & Leakey, 1986
INCLUDED SPECIES *A. turkanensis*

SPECIES *Afropithecus turkanensis* Leakey & Leakey, 1986 (Fig. 19.4)

B

Fig. 19.4 *Afropithecus turkanensis*. KNM-WK 16999 (holotype), partial cranium: (A) frontal view; (B) occlusal view. Photographs courtesy National Museums of Kenya.

TYPE SPECIMEN KNM-WK 16999, face and frontal region of cranium with complete upper dentition

AGE AND GEOGRAPHIC RANGE Early Miocene (approximately 17–18 Ma), Kenya (Leakey & Walker, 1985*b*; McDougall & Watkins, 1985; Leakey & Leakey, 1986*a*; Watkins, 1989; Anyonge, 1991; Boschetto *et al.*, 1992)

ANATOMICAL DEFINITION

A large catarrhine, comparable in dental size to *Proconsul nyanzae* and *Pan troglodytes*. Postcranials provide an estimated body weight of 35 kg (Leakey & Walker, 1997), which confirms an overall size similar to that of extant chimpanzees (with a range of 20–50 kg). Cranium (based on the holotype, a male individual) with the following characteristics: long, broad and domed muzzle; palate shallow, long and narrow, and with tooth rows parallel-sided or converging slightly posteriorly; incisive foramen with large paired openings; large diastema between C and I^2; premaxilla narrow but anteriorly protruding, with broad contact superiorly with the nasals; steeply inclined frontal; strong postorbital constriction; temporal lines strongly marked, and converge in the midline far anteriorly to form a frontal trigon; a frontal sinus is possibly present in the glabellar region; supraorbital costae are slender; small supraorbital notch at the medial angle of the orbital margin; interorbital distance is great; nasals long and narrow, with midline keeling; pyriform aperture only slightly higher than broad, and oval in shape; nasoalveolar clivus relatively deep; canine jugum prominent, with shallow canine fossa; distinct maxillary fossa on the snout just below and anterior to the orbit; double infraorbital foramina; anterior root of the zygomatic arch deep, superiorly sloping and attaches relatively low on the face; maxillary sinus extensive, extending from the tip of the canine root to beyond M^3; orbit broader than high, and asymmetrical in shape; orbital process of frontal narrow; lacrimal fossa extends onto the face just anterior to the margin of the orbit. Mandible has a very deep corpus, with a distinct mandibular fossa, and a single mental foramen. Ascending ramus set at an oblique angle to the corpus. Symphyseal region characterized by the lack of development of a superior transverse torus, and a steeply sloping subincisive planum.

Upper incisors strongly procumbent, and angled obliquely towards the midline; I^1 relatively broad, and much larger than I^2 (the mesiodistal breadth of I^2 is only 65.5% that of I^1); lower incisors broad, particularly I_2; upper canine (in males) is broad and tusk-like, with an almost circular basal cross-section and a deep mesial groove; lower canine stout, bilaterally compressed and relatively low-crowned; P^3 larger than P^4; upper premolars broad, with only moderate difference in height between buccal and lingual cusps, and lacking a lingual cingulum; P_3 relatively large, narrow and sectorial; P_4 generally broader than long; upper premolars and molars have marked lingual and buccal basal flare; upper molars relatively narrow, with small mesial fovea, moderate development of lingual cingulum, and large hypocone (subequal in size to protocone); $M^1 < M^2 < M^3$; lower molars relatively broad; $M_1 < M_2 < M_3$; enamel of cheek teeth very thick with heavy wrinkling (Leakey & Leakey, 1986*a*; Leakey *et al.*, 1988*a*; Leakey & Walker, 1997).

A few postcranial bones of *Afropithecus* are known from Kalodirr and Buluk, including two fragmentary distal humeri, a proximal ulna, a fibula, carpals, tarsals and metapodials, and more than a dozen phalanges (Leakey &

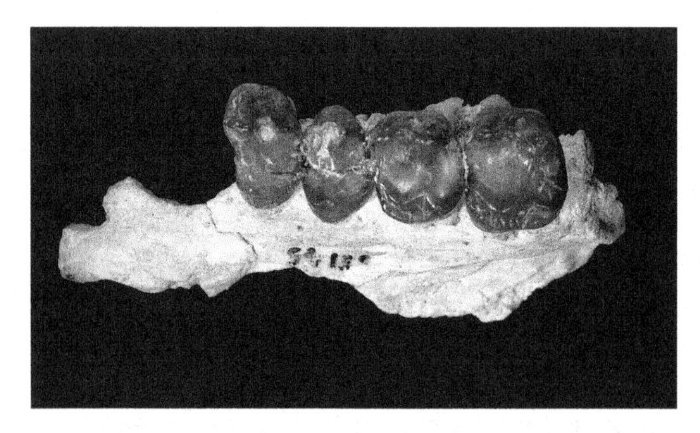

Fig. 19.5 *Heliopithecus leakeyi*. BMNH 35145 (holotype), left maxilla with P^3–M^2, occlusal view. Photograph courtesy of Peter Andrews.

Leakey, 1986*a*; Leakey *et al.*, 1988*a*; Leakey & Walker, 1997). These are all similar in size and morphology to those of *P. nyanzae* (Leakey *et al.*, 1988*a*; Rose, 1997; Ward, 1997, 1998).

GENUS *Heliopithecus* Andrews & Martin, 1987
INCLUDED SPECIES *H. leakeyi*

SPECIES *Heliopithecus leakeyi* Andrews & Martin, 1987 (Fig. 19.5)
TYPE SPECIMEN BMNH M 35145, left maxilla with P^3–M^2
AGE AND GEOGRAPHIC RANGE Early Middle Miocene, Saudi Arabia (Andrews *et al.*, 1978; Andrews & Martin, 1987*b*)
ANATOMICAL DEFINITION
A large catarrhine, slightly smaller in dental size than *Pan troglodytes*. It is intermediate in size between *Proconsul heseloni* and *P. nyanzae*, and somewhat smaller than *Afropithecus turkanensis*, which some authorities recognize as congeneric (Andrews & Martin, 1987*b*; Andrews, 1992). The genus and species is based on a maxillary fragment (the holotype) and four isolated teeth (i.e., P^4, M^3, dC^1, dP^4), so knowledge of its anatomy is rather limited (Andrews et al., 1978; Andrews & Martin, 1987*b*). The main characteristics of the species are as follows: palate relatively shallow and narrow, with parallel toothrows; sizable diastema present between C and I^2 (at least in males); upper premolars large in relation to molars; P^3 larger than P^4; P^3 with marked difference in height between the buccal and lingual cusps; P^4 with lingual cingulum; upper cheek teeth relatively low-crowned with voluminous cusps and quite thick enamel; upper molars slightly broader than long, with moderate development of the lingual cingulum, and a small buccal cingulum. *Heliopithecus* differs from *Proconsul* and resembles *Afropithecus* (as well as *Morotopithecus* and *Equatorius*) in the following derived characters: upper premolars relatively large, and upper molars narrower with reduced development of the lingual cingulum and more bunodont cusps. Andrews *et al.* (1987) and Andrews & Martin (1987*b*) have suggested that *Heliopithecus* may be congeneric with *Afropithecus* (which has priority) and with the large catarrhine from Moroto (subsequently named *Morotopithecus*), a view echoed by other workers (Leakey *et al.*, 1988*a*). However, *Heliopithecus* is distinguished from *Afropithecus* in having relatively broader cheek teeth, a greater differential between the heights of the buccal and lingual cusps on P^3, presence of a lingual cingulum on P^4, upper molars with a relatively smaller hypocone and a better-developed lingual cingulum. It differs from *Morotopithecus* in having a narrower palate, narrower upper cheek teeth, P^3 with less marked height differential between the buccal and lingual cusps, a better-developed lingual cingulum on P^4 and upper molars. Although more detailed comparisons between *Heliopithecus*, *Afropithecus* and *Morotopithecus* are needed, these differences indicate that *Heliopithecus* is more primitive than the other two taxa, and generically distinct.

Subfamily Nyanzapithecinae

GENUS *Nyanzapithecus* Harrison, 1986
Nyanzapithecus is a small to medium-sized catarrhine close to *Macaca mulatta* and *Hylobates syndactylus* in dental size, and intermediate in size between *Proconsul heseloni* and *Dendropithecus macinnesi*. Judging from the dental remains it is reasonable to infer that *N. vancouveringorum* and *N. pickfordi* had average body weights of 11 kg and 8 kg for males and females respectively, while *N. harrisoni* was probably slightly smaller. *Nyanzapithecus* is distinguished from other proconsulids by the following combination of dental features: I^1 broad and spatulate, relatively low-crowned, and stoutly constructed; I^2 broad, moderately low-crowned and robust, and approaching I^1 in size (the mesiodistal length of I^2 is 89% that of I^1 in *N. pickfordi*, compared with 73% in *P. heseloni*); lower incisors broad and moderately high-crowned; P^3 structurally similar to P^4; upper premolars ovoid in occlusal outline and relatively long and narrow, with elevated and inflated cusps of similar height, poorly developed occlusal crests, and an inflated lingual cingulum, at least on P^4; P_3 long and slender, with only slight extension of enamel onto the buccal aspect of the mesial root; P_4 long and narrow with high cusps, and mesial fovea much more elevated than the distal basin; upper molars long and narrow, with low, rounded and voluminous cusps, buccally displaced protocone, restricted trigon basin and foveae, well-developed lingual cingulum, low and rounded occlusal crests; $M^1 < M^2 \leqslant M^3$; lower molars very long and narrow, with low, rounded and inflated cusps, short and rounded crests, long and narrow talonid basin, restricted mesial and distal foveae, poorly developed buccal cingulum, and deep lingual notch; $M_1 < M_2 < M_3$; dP^4 longer than broad, with voluminous cusps and relatively restricted occlusal foveae (Harrison, 1986*b*; Kunimatsu, 1992*a*, 1992*b*, 1997).

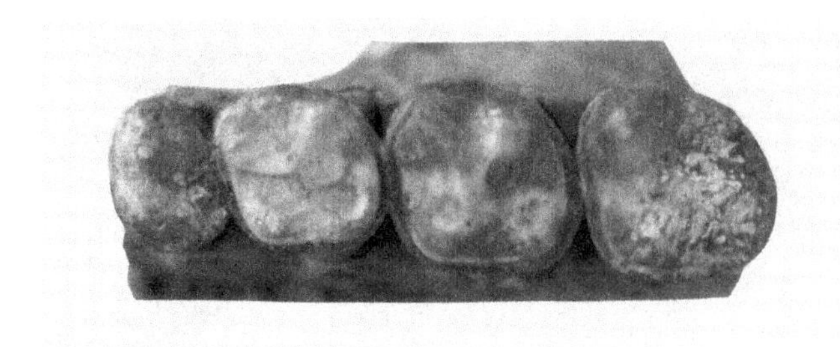

A

B

Fig. 19.6 *Nyanzapithecus vancouveringorum*. (A) KNM-RU 2058 (holotype), left maxilla with P^4–M^3, occlusal view; (B) KNM-RU 1855, mandible with right P_4–M_3 and left M_{1-3}.

Nyanzapithecus is poorly known cranially and postcranially. The fragmentary maxillae and premaxilla of *N. vancouveringorum* and *N. pickfordi* indicate that the genus has a relatively short face, low and broad nasal aperture, and robust premaxilla (Harrison, 1986*b*). There are no associated postcranials, but McCrossin (1992*a*) provisionally attributed a proximal humerus from the middle Miocene locality of Maboko Island to *N. pickfordi*. In addition, a proximal humerus (KNM-RU 17376) from Rusinga Island previously attributed by Gebo *et al.* (1988) to *D. macinnesi* or *P. heseloni*, is probably best assigned to *N. vancouveringorum* on the basis of size. These two specimens are comparable in morphology and, like the postcranials of other proconsulids, correspond closely to the inferred primitive catarrhine morphotype (Gebo *et al.*, 1988; McCrossin, 1992*a*).

INCLUDED SPECIES *N. harrisoni*, *N. pickfordi*, *N. vancouveringorum*

SPECIES *Nyanzapithecus vancouveringorum* (Andrews, 1974) (Fig. 19.6)

TYPE SPECIMEN KNM-RU 2058, fragment of a left maxilla with P^4–M^3

AGE AND GEOGRAPHIC RANGE Early Miocene (approximately 17–18.5 Ma), Kenya (Drake *et al.*, 1988)

ANATOMICAL DEFINITION

Until recently, this species was represented by only seven craniodental specimens, but Kelley (1986*b*) and Harrison (1999) have recognized additional specimens previously attributed to *P. heseloni*. It differs from other species of *Nyanzapithecus* in the following features: upper premolars both have well-developed lingual cingulum; upper molars and dP^4 only slightly longer than broad, and generally rectangular to square in occlusal outline; upper molar cusps moderately inflated, and encroach only partially into the occlusal basins; the trigon basin and mesial and distal foveae restricted, but well defined; hypocone connected to the protocone by a short crest; lingual cingulum on the upper molars well developed, both lingually and mesially; lower molars moderately long and narrow, with reduced mesial and distal foveae, relatively expansive talonid basin, and moderately inflated cusps (Harrison, 1986*b*).

OTHER SPECIMENS

Andrews (1978*a*) and Harrison (1986*b*) previously referred a worn and eroded upper molar from Songhor to this species, but Kunimatsu (1992*a*) has suggested that the specimen may be an M^1 of *Rangwapithecus gordoni*.

SPECIES *Nyanzapithecus pickfordi* Harrison, 1986

TYPE SPECIMEN KNM-MB 11645, left M^1

AGE AND GEOGRAPHIC RANGE Middle Miocene (approximately 15–16 Ma), Kenya (Pickford, 1981, 1983, 1986*b*; Feibel & Brown, 1991)

ANATOMICAL DEFINITION

A species of *Nyanzapithecus* distinguished from *N. vancouveringorum* by the following characteristics: P^3 lacking a lingual cingulum; upper molars higher-crowned, much longer than broad, tending to taper distally and become waisted midway along their length, with inflated cusps that crowd the occlusal basins and restrict the mesial and distal foveae; hypocone connected by a crest to the crista obliqua, but with no direct connection to the protocone; lingual cingulum particularly well developed mesially, but reduced lingually; lower molars longer and narrower with very inflated cusps and extremely restricted occlusal basins (Harrison, 1986*b*).

OTHER SPECIMENS When originally described, the hypodigm of this species consisted entirely of isolated teeth, except for a fragmentary premaxilla with I^{1-2} (Harrison, 1986*b*). Recent excavations at Maboko Island have led to the recovery of a large sample of additional specimens, including a nearly complete mandible of a subadult female individual and a proximal humerus (McCrossin, 1992*a*; Benefit & McCrossin, 1997*b*; Gitau *et al.*, 1998).

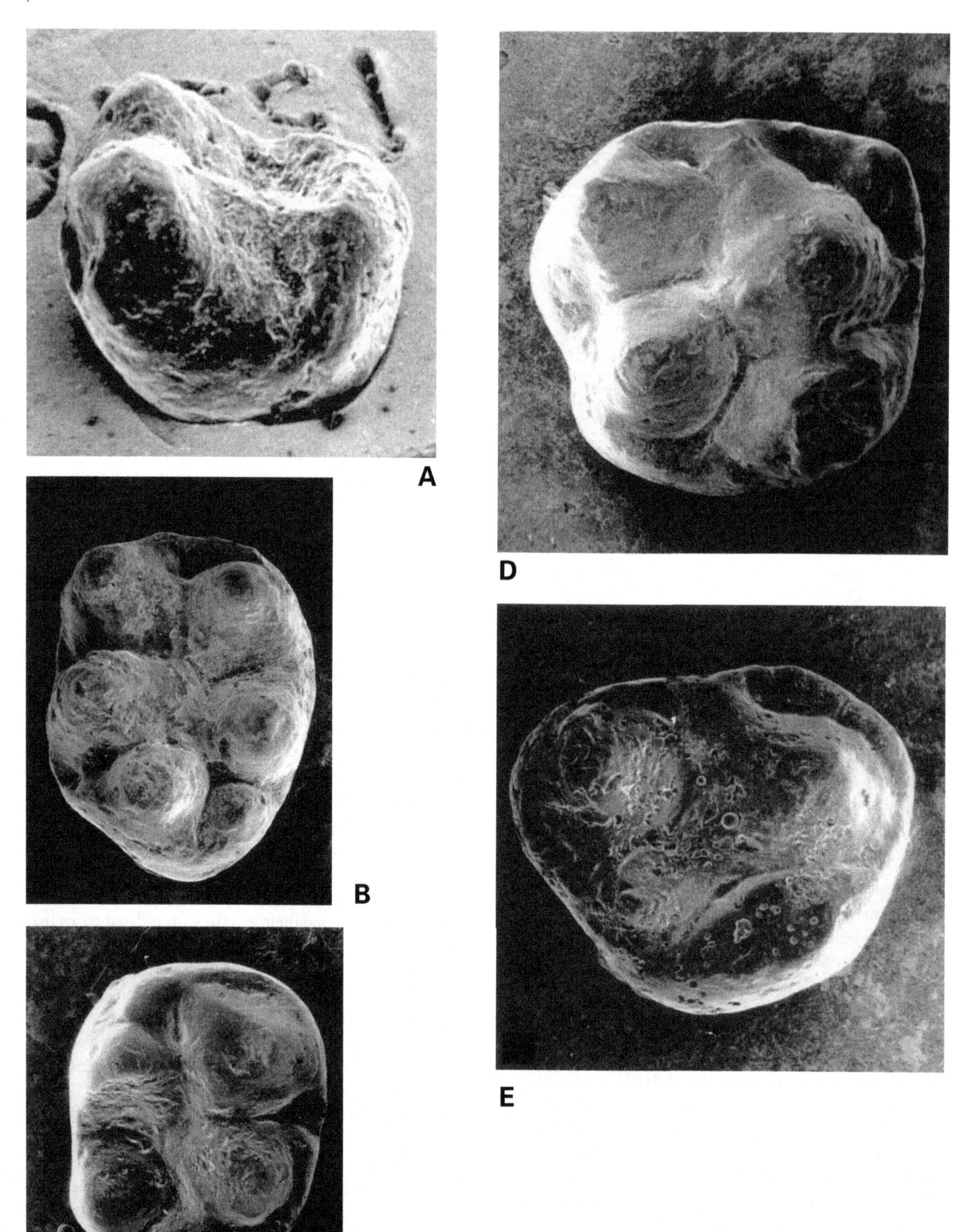

Fig. 19.7 *Nyanzapithecus harrisoni*. (A) KNM-BG 15318, left P_4, occluso-buccal view; (B) KNM-BG 15235, left M_2, occlusal view; (C) KNM-BG 15227, right M_3, occlusal view; (D) KNM-BG 15237, right M^2, occlusal view; (E) KNM-BG 15344, right M^3, occlusal view. Photographs courtesy of Y. Kunimatsu.

SPECIES *Nyanzapithecus harrisoni* Kunimatsu, 1997 (Fig. 19.7)
TYPE SPECIMEN KNM-BG 15237, right M^2
AGE AND GEOGRAPHIC RANGE Middle Miocene (approximately 13–15 Ma), Kenya (Kunimatsu, 1992*a*, 1992*b*, 1997)
ANATOMICAL DEFINITION
A species of *Nyanzapithecus* distinguished from *N. vancouveringorum* and *N. pickfordi* by smaller size. It is distinguished from *N. vancouveringorum* in having: upper molars higher crowned; molar cusps higher and more inflated; occlusal basins and foveae more restricted; upper molars that tend to taper distally; more distinct lingual cingulum; M^3 crown shorter; P_4 more elongated; lower molars relatively short (especially M_3); mandible more gracile (Kunimatsu, 1997). It differs from *N. pickfordi* in having: upper and lower molars less elongated; lingual cingulum on upper molars with less well-developed mesial portion and better developed lingual part; hypocone linked to the protocone directly, rather than to the crista obliqua; less waisted upper molars; P_3 with weak but continuous buccal cingulum; hypoconulid on M_3 tends to be located more medially (Kunimatsu, 1997). The three species of *Nyanzapithecus* can be arranged in a phyletic series of increasing specialization from the early Miocene *N. vancouveringorum* through *N. harrisoni* to *N. pickfordi* in the middle Miocene (Harrison, 1986*b*; Kunimatsu, 1997).

GENUS *Mabokopithecus* von Koenigswald, 1969
INCLUDED SPECIES *Ma. clarki*

SPECIES *Mabokopithecus clarki* von Koenigswald, 1969
TYPE SPECIMEN KNM-MB 76, isolated M_3
AGE AND GEOGRAPHIC RANGE Middle Miocene (approximately 15–16 Ma), Kenya (Pickford, 1981, 1986*b*; Feibel & Brown, 1991)
ANATOMICAL DEFINITION
A small to medium-sized catarrhine comparable in dental size to *Nyanzapithecus pickfordi*. Until recently, this species was known only from two isolated M_3s (von Koenigswald, 1969; Harrison, 1986*b*), so comparisons with other species were extremely limited. The M_3 of *Mabokopithecus clarki* is characterized by the following features: crown long and narrow, with a distinctive curvature; cusps high, conical, and voluminous; protoconid and metaconid large, well developed and transversely aligned, and separated by a deep median groove which communicates with an elongated mesial fovea; hypoconid lingually displaced, so that the buccal cusps are not aligned, leading to the distinctive concavity along the buccal margin of the crown; low rounded crests descend from the metaconid and protoconid into the talonid basin and converge with a similar crest from the hypoconid at a distinct mesoconid (present in the holotype only); hypoconulid very large; entoconid and hypoconulid linked by a prominent crest, which defines a pit-like distal fovea; subsidiary tubercle located between the metaconid and entoconid; and buccal cingulum narrow and irregular.

Recent excavations at Maboko Island have yielded a nearly complete mandible of a female individual containing an M_3 that provides a close match with the holotype of *Ma. clarki* (Benefit *et al.*, 1998). The rest of the dentition is extremely similar in morphology to *Nyanzapithecus pickfordi*, and they should be included together in the same genus. Since *Mabokopithecus* von Koenigswald, 1969 has priority over *Nyanzapithecus* Harrison, 1986, all three species of *Nyanzapithecus* should eventually be transferred to the former genus once the new specimens from Maboko Island have been formally described. Benefit *et al.* (1998) consider *Ma. clarki* to be specifically distinct from "N." *pickfordi* because it differs in having: a more strongly developed planum alveolare; a more robust mandibular corpus; a vertical ramus; less caniniform I_2; higher lingual cingulum and presence of a small metaconid on P_3; more strongly developed buccal cingulum around the protoconid of the lower molars; M_2 and M_3 slightly larger; and M_2 hypoconulid more centrally positioned and sharing a continuous wear facet with the entoconid. These differences may prove to be taxonomically significant, but from my own brief comparison of the specimens, I favor including all of the nyanzapithecine specimens from Maboko in a single species, *Mabokopithecus clarki*.

GENUS *Rangwapithecus* Andrews, 1974
INCLUDED SPECIES *R. gordoni*

SPECIES *Rangwapithecus gordoni* Andrews, 1974 (Fig. 19.8)
TYPE SPECIMEN KNM-SO 700, premaxilla and maxilla with right and left C–M^3
AGE AND GEOGRAPHIC RANGE Early Miocene (approximately 19–20 Ma), Kenya (Bishop *et al.*, 1969; Pickford & Andrews, 1981; Pickford, 1983, 1986*b*; Harrison, 1988)
ANATOMICAL DEFINITION
A medium-sized catarrhine similar in dental size to *Proconsul africanus* and *P. heseloni*. Differs from *Proconsul* in the following respects: upper and lower incisors high-crowned and relatively narrow; upper incisors moderately procumbent; upper canines strongly bilaterally compressed, with a blade-like distal crest; upper premolars and molars relatively elongated; upper premolars with more ovoid occlusal outline, buccal and lingual cusps more similar in height, inflated lingual cingulum on both P^3 and P^4; $P^3 < P^4$; upper molars with a strong lingual cingulum, enlarged hypocone, and rhomboidal arrangement of cusps and occlusal outline; molars with low cusps, well-developed crests, wrinkled occlusal surface, and strong wear differential; $M^1 < M^2 < M^3$; lower canine high-crowned and bilaterally compressed in males; canines strongly sexually dimorphic; P_3 elongated and bilaterally compressed; P_4 and lower molars long and narrow; lower

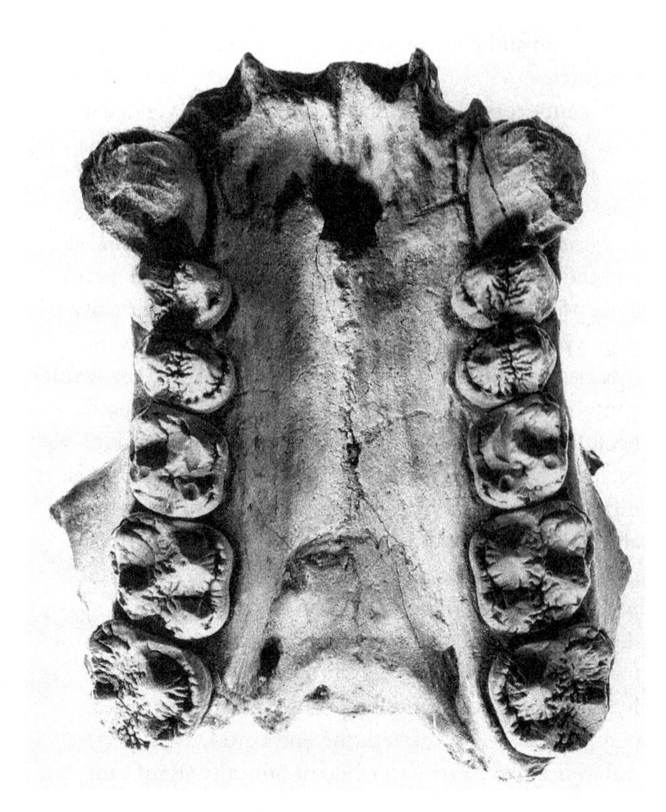

Fig. 19.8 *Rangwapithecus gordoni*. KNM-RU 700, lower face and palate, occlusal view. Photograph courtesy of Peter Andrews.

molars with buccal cingulum represented by deep foveae between the buccal cusps; $M_1 < M_2 < M_3$, with marked size increase along the series (with an $M_1 : M_2 : M_3$ area ratio of 72 : 100 : 127, compared with that in *P. heseloni* of 75 : 100 : 116) (Andrews, 1978*a*; Nengo & Rae, 1992). Several mandibular specimens recently recovered from Songhor (Hill & Odhiambo, 1987; Nengo & Rae, 1992) document more fully the lower jaw and dentition, but these still await detailed description. Cranial and mandibular specimens exhibit the following features: premaxilla relatively short; diastema small; palate long and narrow, and broadens posteriorly; maxillary sinus deeply excavated between the roots of the upper molars; nasal aperture probably relatively broad; anterior root of the zygomatic arch positioned low on the face above M^{1-2}; mandibular corpus deep with a strongly developed superior transverse torus (Andrews, 1978*a*). A few isolated postcranial elements are provisionally assigned to this species (Preuschoft, 1973; Harrison, 1982; Langdon, 1986). They are similar in morphology to those of *Proconsul heseloni*, and indicate a pronograde, arboreal primate.

GENUS *Turkanapithecus* Leakey & Leakey, 1986
INCLUDED SPECIES *T. kalakolensis*

SPECIES *Turkanapithecus kalakolensis* Leakey & Leakey, 1986 (Fig. 19.9)

A

B

C

Fig. 19.9 *Turkanapithecus kalakolensis*. KNM-WK 16950 (holotype), partial skull: (A) cranium, facial view; (B) palate, occlusal view; (C) mandible, occlusal view. Photographs courtesy of the National Museums of Kenya.

TYPE SPECIMEN KNM-WK 16950A & B, a partial cranium with right C–P^3, M^{1-3} and left P^3–M^3, and a mandible with left M_2 and right M_3
AGE AND GEOGRAPHIC RANGE Early Miocene (16.6–17.7 Ma), Kenya (Boschetto *et al.*, 1992)
ANATOMICAL DEFINITION
A medium-sized catarrhine in which male specimens are comparable in cranial and postcranial size to female specimens of *Proconsul heseloni* or male specimens of *Colobus guereza* (with an average body weight of approximately 10 kg: Smith & Jungers, 1997). Similar in dental size to *Nyanzapithecus vancouveringorum* and *N. pickfordi*.

The main characteristics of the skull are as follows: face relatively short, with broad and domed snout; large incisive fossa; narrow palate, with toothrows that converge posteriorly; premaxillary suture makes contact with the nasals; nasals are relatively broad and expand inferiorly and superiorly; nasal aperture broad and ovoid in shape, but narrows between the central incisors in the midline; very broad interorbital region; orbits subcircular in outline, with shallow supraorbital notch; slightly thickened supraorbital tori, with depressed glabella region; possibly a large frontal sinus is present; single large infraorbital foramen, located just below the orbital margin; lacrimal fossa located just anterior to the margin of the orbit; well-developed facial pillars or canine juga to accommodate the roots of the canines; canine fossa indistinct; extensive maxillary sinus; anterior root of zygomatic arch situated low on the face; zygomatic arches relatively deep and widely flaring, with a slight upward sweep; postorbital constriction marked, with large temporal fossae; multiple zygomaticofacial foramina all located above the inferior margin of the orbit; frontal process of the malar mediolaterally narrow, with a rugose anterior face; temporal lines strongly marked and converge posteriorly, possibly resulting in a sagittal crest posteriorly, at least in males; cranial capacity relatively small (estimated from the area of the foramen magnum to be only 84.3 cm^3, significantly less encephalized than *P. heseloni*: Manser & Harrison, 1999); glenoid articular surface saddle-shaped, with a strong mediolateral concavity, and no distinct eminence; postglenoid process well developed; inferior orbital fissure large; nuchal plane relatively short, and bordered posteriorly by a heavy nuchal crest; mandibular symphysis with weakly developed superior transverse torus and indistinct inferior transverse torus; intercanine distance relatively narrow; corpus shallow and relatively slender, with a constant depth below the molars; ramus anteroposteriorly long, and superoinferiorly low, with its anterior margin sloping posteriorly at an angle of approximately 120° to the alveolar plane; pronounced inferior expansion of the angular region of the ramus; articular condyle knob-like; multiple mental foramina located below the premolars (Leakey & Leakey, 1986*b*; Leakey *et al.*, 1988*b*; T. Harrison, unpublished data).

Dental characteristics include: upper canine large and strongly bilaterally compressed in males, with a deep mesial groove and a flange-like distal margin; upper premolars relatively narrow, with buccal cusp not much higher than the lingual cusp; lingual cingulum present on both P^3 and P^4; $P^4 < P^3$; upper molars with elongated rhomboidal-shaped crowns that narrow distally; trigon narrow and dominated by a voluminous protocone; hypocone closely appressed to protocone and linked to it (or the crista obliqua) by a distinct crest; buccal cingulum moderately well developed, with variable development of accessory cuspule; M^2 and M^3 with distinct paraconule at the termination of the preparacrista; lingual cingulum very broad and continues mesially around the protocone to form a distinct mesial ledge; prominent secondary conule on the mesiolingual margin of the cingulum; $M^1 < M^3 < M^2$; lower molars long and narrow, with only slight development of a buccal cingulum between the main cusps; $M_1 < M_2 < M_3$; M_3 only slightly larger than M_2 (the area is only 5.2% larger, compared with 15.9% in *P. heseloni* and 23.1% in *P. nyanzae*) (Leakey & Leakey, 1986*b*; Leakey *et al.*, 1988b; T. Harrison, unpublished data).

Postcranial specimens from Kalodirr (Leakey & Leakey, 1986*b*; Leakey *et al.*, 1988*b*; Rose, 1993*b*, 1994; Ward, 1997) are generally similar to those of *Proconsul*, but do differ in the following key features: proximal ulna with moderately long and proximally directed olecranon process and narrow sigmoid notch; distal ulna with long styloid process that articulates with carpus; oval radial head with a moderately developed lateral lip, and beveled margin; radial neck relatively long; metacarpals and phalanges comparable to those of *Proconsul*; trapezium–1st metacarpal joint mobile; femur less robust than in *Proconsul*, with high bicondylar angle; distal articular surfaces of femur less expanded mediolaterally than in *Proconsul* (Leakey *et al.*, 1988*b*; Rose, 1993*b*, 1994; Ward, 1997). *Turkanapithecus* was an arboreal quadrupedal primate generally similar in its locomotor repertoire to *Proconsul*, but possibly with enhanced climbing capabilities (Rose, 1993*b*).

Superfamily Dendropithecoidea

Family Dendropithecidae

GENUS *Dendropithecus* Andrews & Simons, 1977
INCLUDED SPECIES *D. macinnesi*

SPECIES *Dendropithecus macinnesi* (Le Gros Clark & Leakey, 1950) (Fig. 19.10)
TYPE SPECIMEN BMNH M 16650, almost complete, but badly crushed mandible
AGE AND GEOGRAPHIC RANGE Early Miocene (approximately 17–20 Ma), Kenya and Uganda (Bishop *et al.*, 1969; Harrison, 1981, 1982, 1988; Pickford, 1981, 1983, 1986*b*; Pickford & Andrews, 1981; Pickford *et al.*, 1986*a*; Drake *et al.*, 1988)

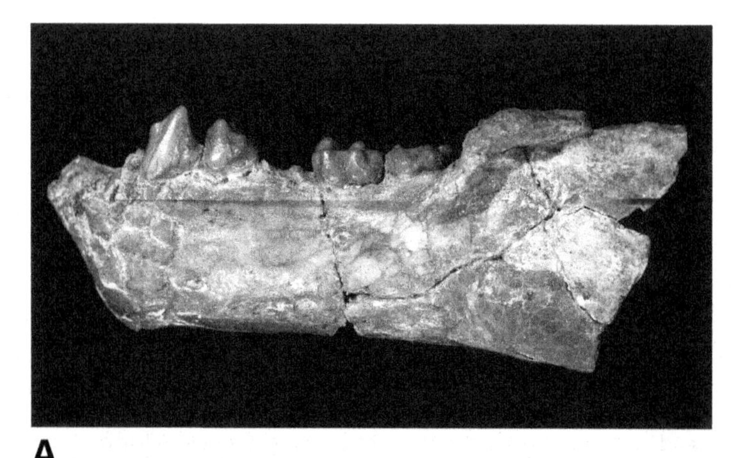

A

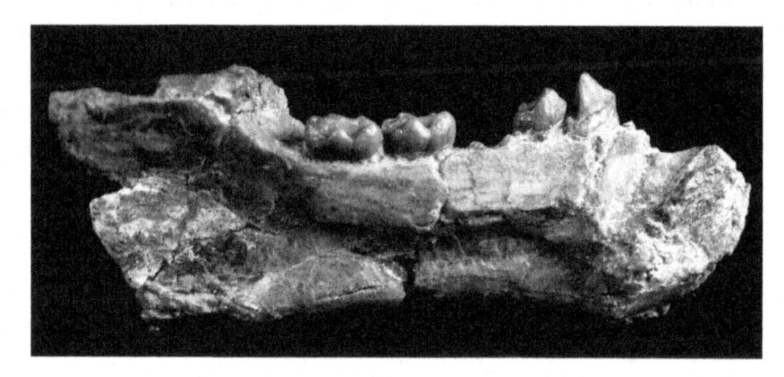

B

Fig. 19.10 *Dendropithecus macinnesi*. BMNH 16650 (holotype), left mandibular fragment with P_{3-4} and M_{2-3}: (A) lateral view; (B) medial view. Photographs courtesy of Peter Andrews.

ANATOMICAL DEFINITION

A small to medium-sized catarrhine, approximating to *Macaca mulatta* or *Hylobates syndactylus* in dental and postcranial size, and slightly smaller than the contemporary *N. vancouveringorum*. Body weights of approximately 9 kg and 5–6 kg for males and females respectively are reasonable estimates judging from comparisons of the postcranial specimens and the degree of sexual dimorphism exhibited by the gnathic remains. The main craniodental characteristics are as follows: incisors high-crowned and mesiodistally relatively narrow; I_2 asymmetrical in shape with a convex distal margin; canines strongly sexually dimorphic in size and morphology; canines high-crowned and bilaterally compressed in males, lower-crowned and less compressed in females; upper canine in males with double mesial groove (single in other early Miocene catarrhines); upper premolars broad, with strongly projecting buccal cusps; P_3 sectorial, with a high and bilaterally compressed crown, and a long mesiobuccal honing face for occlusion with the upper canine; upper molars relatively broad and rectangular in occlusal outline, and characterized by high and voluminous cusps, well-developed crests, well-defined mesial and distal foveae and trigon basin, and a broad shelf-like lingual cingulum; $M^1 < M^3 < M^2$; lower molars long and quite broad, with high conical cusps, sharp and well-developed occlusal crests, broad and transverse mesial fovea, large, well-defined and slightly obliquely oriented distal fovea, broad and deep talonid basin, and moderately well-developed buccal cingulum; $M_1 < M_2 < M_3$, with marked increase in size from M_1 to M_3; incisors relatively small in comparison to the molars; palate long and narrow, and perforated anteriorly by a pair of large incisive foramina; nasal aperture narrow and tapers inferiorly between the roots of the upper central incisors to approach quite close to the alveolar margin of the premaxilla; maxillary sinus extensive, penetrating anteriorly as far as the canine root, posteriorly into the tuberosity of the alveolar process, and laterally into the anterior root of the zygomatic arch; mandibular corpus low and robust, especially below M_3; symphysis relatively robust, and buttressed by moderately well-developed superior and inferior transverse tori (Le Gros Clark & Leakey, 1951; Le Gros Clark & Thomas, 1951; Andrews & Simons, 1977; Andrews, 1978*a*; Harrison, 1981, 1982, 1988).

Dendropithecus macinnesi is known from several partial skeletons from Rusinga Island (R3a, Hiwegi Formation), comprising at least four individuals (Le Gros Clark & Thomas, 1951; Ferembach, 1958; Harrison, 1982; Fleagle, 1983; Rose, 1993*a*). The key characteristics are as follows: long and slender limb bones; proximal humerus lacks torsion; humeral shaft slightly retroflexed proximally; distal humerus lacks an entepicondylar foramen, but exhibits a large dorsal epitrochlear fossa on the posterior aspect of the medial epicondyle (a primitive primate feature typically absent in modern catarrhines: Harrison, 1982; Fleagle, 1983; Rose, 1993*a*); distal articulation of humerus with globular capitulum, broad spool-shaped trochlea, and low lateral trochlear keel; proximal ulna with large and proximally directed olecranon process; radius with oval head and relatively long neck; tarsals, metapodials, and phalanges generally similar to those of *Proconsul*. The combined postcranial evidence suggests that *D. macinnesi* was an active, arboreal quadrupedal primate, capable of powerful climbing activities, and at least some degree of forelimb suspension. Comparisons with extant primates suggest that *D. macinnesi* may have been most similar in its morphology and functional capabilities to the larger New World monkeys (Le Gros Clark & Thomas, 1951; Harrison, 1982; Fleagle, 1983; Rose, 1983, 1993*b*).

GENUS *Micropithecus* Fleagle & Simons, 1978

Small catarrhine primates similar in dental size to *Cercopithecus cephus* (in which males and females weigh approximately 4.3 and 2.9 kg respectively: Smith & Jungers, 1997). The key characteristics are as follows: upper incisors broad and relatively high-crowned; upper lateral incisors relatively bilaterally symmetrical; canines high-crowned, moderately bilaterally compressed, and exhibiting a marked degree of sexual dimorphism; upper premolars long and

narrow, with well-developed transverse crests; P_3 moderately to strongly sectorial, with long and bilaterally compressed crown; P_4 small, ovoid to circular in outline and generally longer than broad; upper molars relatively small with a high breadth–length index, more lingually positioned hypocone relative to protocone, breadth of trigon only slightly greater than its length, distal fovea and cingulum expansive, and development of lingual cingulum poor to moderate; lower molars ovoid in shape, with low rounded and poorly developed occlusal ridges, and a slightly oblique mesial fovea; anterior dentition large relative to the size of the cheek teeth; posterior molars relatively small; face very short; premaxilla small, with short nasoalveolar clivus, and probably not making contact with the nasals; nasal aperture quite broad, and narrows inferiorly between the roots of the central incisors; orbits relatively large, and inferior margin overlaps with the nasal aperture; broad interorbital region; anterior root of the zygomatic arch originates above M^2, very close to the alveolar margin of the maxilla; inferior orbital fissure large; maxillary sinus extensive; palate broad and shallow; large incisive foramina; sulcal pattern on endocranial surface of frontal similar to that of *Proconsul* (Radinsky, 1975); mandible moderately high and gracile, with a low superior transverse torus and a small inferior transverse torus (Pilbeam & Walker, 1968; Fleagle, 1975; Fleagle & Simons, 1978; Harrison, 1981, 1982, 1988, 1989*a*).
INCLUDED SPECIES *Mi. clarki*, *Mi. leakeyorum*

SPECIES *Micropithecus clarki* Fleagle & Simons, 1978 (Fig. 19.11)
TYPE SPECIMEN UMP 64-02, palate and lower face
AGE AND GEOGRAPHIC RANGE Early Miocene (approximately 19–20 Ma), Kenya and Uganda (Bishop *et al.*, 1969; Pickford & Andrews, 1981; Pickford, 1983, 1986*b*; Pickford *et al.*, 1986; Harrison, 1988)
ANATOMICAL DEFINITION
A species distinguished from *Mi. leakeyorum* by the following features: P_3 strongly sectorial, with long and bilaterally compressed crown; P_4 and lower molars relatively broader, with a less pronounced buccal cingulum and poorly defined mesial and distal fovea; M_3 much smaller than M_2, and exhibits a marked regression of the cusps and crests distally; $M_3 \leqslant M_1 < M_2$; upper molars slightly narrower, with narrower trigon and smaller hypocone; M^3 relatively larger with better developed cusps in its distal moiety; $M^3 < M^1 < M^2$. A few isolated postcranial bones from Koru provisionally referred to this species (Harrison, 1982) are morphologically consistent with *D. macinnesi*.

SPECIES *Micropithecus leakeyorum* Harrison, 1989
TYPE SPECIMEN KNM-MB 11660, left mandibular fragment with P_3–M_2
AGE AND GEOGRAPHIC RANGE Middle Miocene (approximately 15–16 Ma), Kenya (Andrews *et al.*, 1981; Pickford, 1981, 1983, 1986*b*; Feibel & Brown, 1991)

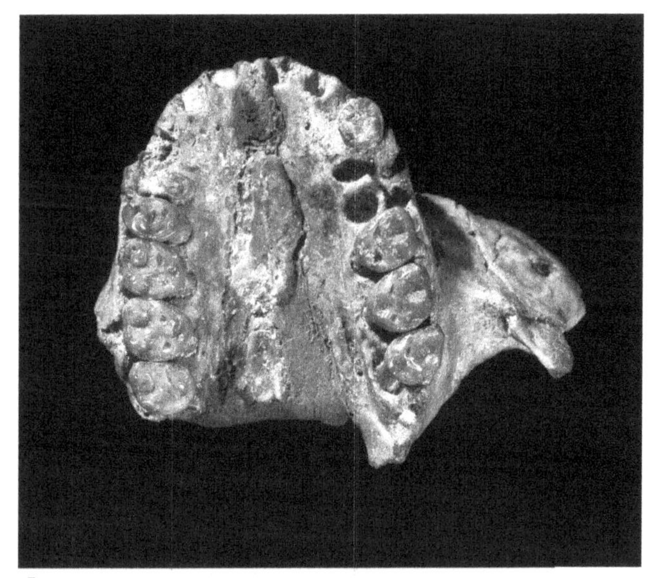
A

B

Fig. 19.11 *Micropithecus*. (A) *Mi. clarki*, UMP 64-02 (holotype), palate and lower face, occlusal view (photograph courtesy of John G. Fleagle); (B) *Mi. leakeyorum*, KNM-CA 380, mandible, occlusal view.

ANATOMICAL DEFINITION
A species distinguished from *Mi. clarki* by the following features: P_3 more bilaterally compressed, with only moderate development of a honing face mesially for occlusal contact with the upper canine; P_4 relatively longer and narrower; lower molars relatively narrower, with a more pronounced buccal cingulum and better-defined mesial and distal foveae; M_3 subequal to or slightly larger in occlusal area than M_2, and no indication on M_3 of marked regression of the cusps and occlusal crests distally; $M_1 < M_2 \leqslant M_3$; upper molars slightly broader, with a shorter and more restricted trigon and a larger hypocone (concordant with the larger and better-defined mesial and distal foveae in the lower molars); M^3 relatively larger with

better-developed cusps in its distal moiety; $M^1 < M^3 < M^2$ (Harrison, 1989*a*).

Benefit (1991) and Gitau & Benefit (1995) have reported the discovery of additional craniodental specimens from Maboko Island that may be referable to this species, and have suggested that *Mi. leakeyorum* should be transferred to the genus *Simiolus* (see also Fleagle, 1999). Gitau & Benefit (1995) have indicated that a facial fragment of a male individual from Maboko Island differs from the type specimen of *Mi. clarki* (a female individual) in having a deeper lower face and a different orientation of the anterior root of the zygomatic arch, but these differences could well be related to sexual dimorphism. There are general similarities in the lower cheek teeth of *Mi. leakeyorum* and *S. enjiessi*, but the following features distinguish *Mi. leakeyorum*: cheek teeth smaller in size (18% smaller on average); P_3 has a shorter mesiobuccal face and a less obliquely aligned crown relatively to the long axis of the molar row; P_4 lower-crowned; lower molars relatively narrower, with smaller mesial and distal foveae, a relatively longer mesial fovea, a more transversely aligned distal fovea, a hypoconid and hypoconulid that are more closely twinned, and a more strongly developed buccal cingulum; M_3 subequal in size to M_2 or only slightly larger (the mean area of M_3 is 4.3% larger in *Mi. leakeyorum*, whereas it is 20.9% larger in *S. enjiessi*); buccal cusps on M_3 are linearly arranged, with the hypoconulid placed more buccally; upper M^2 and M^3 are relatively broader (the length–breadth index of M^2 is 83.1–89.8, compared with 97.1 in *S. enjiessi*); size differential between M^1 and M^2 much less marked (the area of M^2 is only 16.4% larger in *Mi. leakeyorum*, compared with 26.9% in *S. enjiessi*), and lacks the peculiar shape difference between M^1 and M^2 typical of *S. enjiessi*; lingual margin of the upper molars more convex, giving the cingulum a C-shaped configuration (it is L-shaped in *S. enjiessi*), mesial fovea relatively narrower, crest linking the hypocone and metacone less well developed, lingual cingulum does not continue around the hypocone; superior transverse torus of the mandible may have been more strongly developed than the inferior transverse torus (equally developed in *S. enjiessi*).

The morphological difference between *Mi. leakeyorum* and *S. enjiessi* is of sufficient magnitude to justify including the two species in separate genera. As noted by Benefit (1991), the similarities between *Mi. leakeyorum* and *S. enjiessi* are explainable as adaptations for a more folivorous diet (see also Harrison, 1989*a*), and are potentially functionally convergent. Given the closer similarities of *Mi. leakeyorum* to *Mi. clarki*, I prefer to retain both species in *Micropithecus*, although it is conceivable that the former taxon may belong to a new genus.

GENUS *Simiolus* Leakey & Leakey, 1987
INCLUDED SPECIES *S. enjiessi*, *Simiolus* sp. nov.

SPECIES *Simiolus enjiessi* Leakey & Leakey, 1987 (Fig. 19.12)

TYPE SPECIMEN KNM-WK 16960, left mandibular and maxillary fragments
AGE AND GEOGRAPHIC RANGE Early Miocene (16.6-17.7 Ma), Kenya (Boschetto *et al.*, 1992)
ANATOMICAL DEFINITION
A small catarrhine, slightly larger in dental size than *Limnopithecus legetet*. It is comparable in dental and postcranial dimensions to *Cercopithecus mitis*, in which males and females weigh approximately 6 kg and 4 kg respectively (Rose *et al.*, 1992; Smith & Jungers, 1997). Characteristic features include: face relatively short with orbits positioned far anteriorly; incisive foramen large; mandibular symphysis with superior transverse torus subequal to or larger than the inferior transverse torus; incisors relatively narrow and high-crowned; canines buccolingually compressed; P_3 high-crowned and strongly sectorial, with a relatively long mesiobuccal face; P^3 almost triangular in occlusal outline, with pronounced degree of extension of enamel onto the buccal root; P_4 long and narrow; P^4 with limited lingual or buccal flare of the cusps, and a well-developed shelf-like lingual cingulum; lower molars relatively long and narrow, with poorly developed buccal cingulum, and well-defined, slightly oblique distal fovea; upper molars have a large talon basin and a well-developed lingual cingulum that continues around the hypocone; M^1 much smaller than M^2, and differs in shape, being much shorter and more rectangular; M^2 and M^3 relatively elongated mesiodistally and subequal in size; M^3 relatively large without marked regression of the distal cusps; upper molars with a distinct crest linking the hypocone and metacone; molars have elevated cusps and sharp occlusal crests (Leakey & Leakey, 1987).

Postcranial features include: humeral shaft slender and relatively straight, with only slight anteflexion; entepicondylar foramen absent, but distinct dorsal epitrochlear fossa; distal articulation of humerus resembles platyrrhines, with modest lateral trochlear keel; femur has a relatively small head, with a high neck angle, a distinct tubercle on the neck; talus comparable in morphology to other dendropithecids and proconsulids, and resembles quadrupedal arboreal non-hominoid anthropoids (Harrison, 1982; Rose *et al.*, 1992; Rose, 1993*b*); metacarpals and phalanges indicate a narrow hand with good flexion-grasping capabilities in an arboreal milieu (Rose *et al.*, 1992; Rose, 1993*a*). *Simiolus enjiessi* and *D. macinnesi* share a common morphological and functional–behavioral pattern (Rose et al., 1992; Rose, 1993*b*, 1997).

SPECIES *Simiolus* sp. nov. (Fig. 19.12D, E)
A second species of *Simiolus* is known from Fort Ternan, but has not been diagnosed formally (Andrews & Walker, 1976; Harrison, 1992). *Simiolus* sp. from Fort Ternan is distinguished from *S. enjiessi* by the following features: I_2 slightly broader, with a more distinctly angular distal

A

C

B

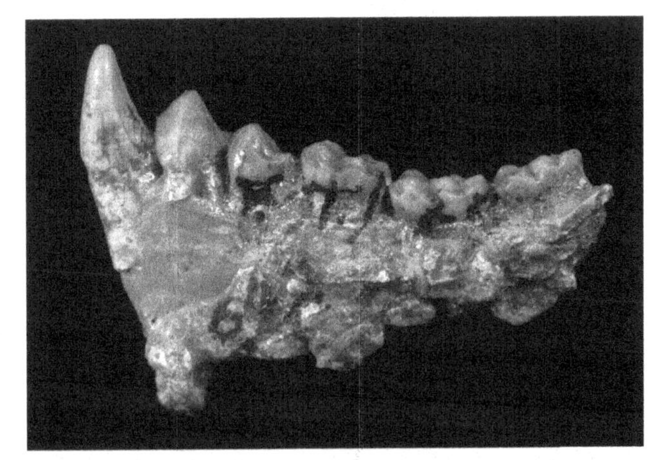

D

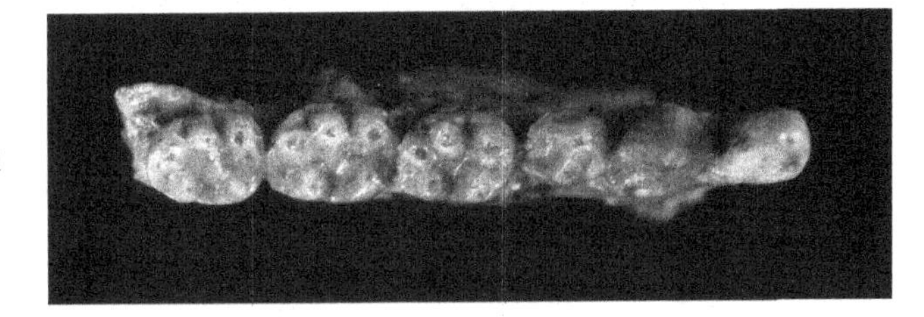

E

Fig. 19.12 *Simiolus.* (A–C) *S. enjiessi*: (A) KNM-WK 16960 (holotype), left premaxilla/maxilla with $C–P^3$ and M^{1-3}, and right M^{1-3}, occlusal view; (B) KNM-WK 16960 (holotype), left mandible with $I_1–M_3$, lateral view; (C) KNM-WK 16960 (holotype), left mandible with $I_1–M_3$, occlusal view; (D,E) *Simiolus* sp.: (D) KNM-FT 20, left mandibular fragment with $C–M_3$, lateral view; (E) KNM-FT 20, occlusal view. (A–C) Photographs courtesy of National Museums of Kenya.

margin; lower canine slightly lower-crowned and more slender; P_4 has less well-developed buccal cingulum; M_2 has more transversely aligned protocristid and a broader distal fovea; M_3 smaller than M_2; M^3 mesiodistally shorter and smaller in size. Most of the differences are minor, and could plausibly be due to intraspecific variation. However, the reduction of the third molars in the Fort Ternan material may provide adequate grounds for recognizing a species distinction (Harrison, 1982, 1992). Given that the samples from Kalodirr and Fort Ternan are rather limited, Harrison (1992) has opted to await the recovery of additional material before establishing a new species.

Superfamily Hominoidea

Family *incertae sedis*

GENUS *Morotopithecus* Gebo *et al.*, 1997
INCLUDED SPECIES *M. bishopi*

SPECIES *Morotopithecus bishopi* Gebo *et al.*, 1997 (Fig. 19.13)
TYPE SPECIMEN UMP 62-11, palate
AGE AND GEOGRAPHIC RANGE Early (or middle?) Miocene of Moroto (Moroto I and II), eastern Uganda. Recently published $^{40}Ar/^{39}Ar$ dates indicate an age for both sites older than 20.6 Ma (Gebo *et al.*, 1997*b*). However, Pickford (1998) and Pickford *et al.* (1999) suggest a middle Miocene age (approximately 15 Ma) based on faunal correlation

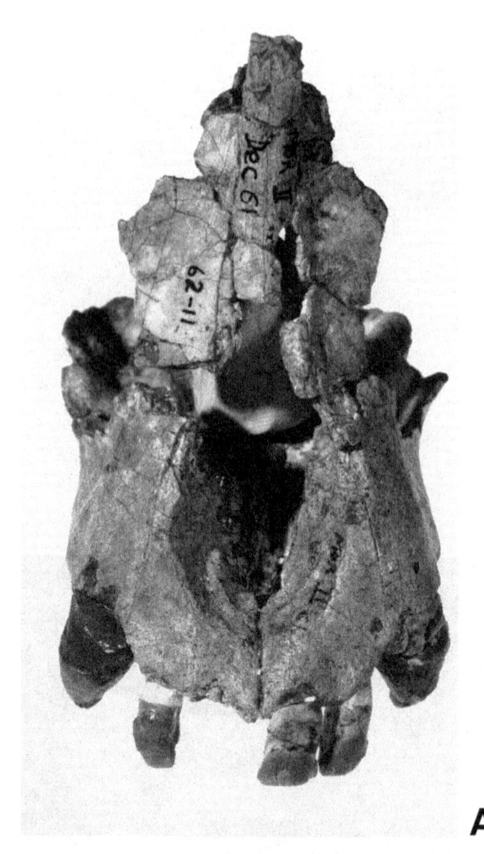

Fig. 19.13 *Morotopithecus bishopi*. UMP 62-11 (holotype), palate and lower face: (A) facial view; (B) lateral view; (C) occlusal view. Photographs courtesy of David Pilbeam.

consistent with earlier K–Ar dates (Bishop *et al.*, 1969) on an overlying lava at Moroto II (14.3 ± 0.3 Ma).

ANATOMICAL DEFINITION

A large catarrhine primate comparable in dental size to *Proconsul major*, but possibly smaller in body size. Estimates based on isolated postcranials indicate a body weight range of approximately 36–54 kg (Sanders & Bodenbender, 1994; Gebo *et al.*, 1997*b*).

The species is known only from the holotype, two isolated upper canines and a P_3, two mandibular fragments (possibly associated with the holotype), three vertebral fragments, parts of the right and left femora, a scapular fragment, and a distal end of a proximal phalanx (Allbrook & Bishop, 1963; Pilbeam, 1969; Gebo *et al.*, 1997*b*; Pickford *et al.*, 1999). *Morotopithecus* is characterized by the following cranial features: palate broad, especially anteriorly; premaxilla relatively short; interpremaxillary suture patent; large incisive fossa; face relatively long, high and narrow; massive canine jugum and shallow maxillary fossa; diastema large (at least in male individuals); relatively narrow interorbital region; broad pyriform aperture, being widest at approximately mid-height; nasals long and narrow, and expand inferiorly; nasoalveolar clivus relatively short; frontal sinus present; maxillary sinus extensive (Allbrook & Bishop, 1963; Pilbeam, 1969; Andrews, 1978*a*). The dentition is characterized by having: upper incisors relatively narrow and procumbent; I^2 with mesiodistal length almost 80% of that of I^1; upper canine stout, with almost circular basal cross-section in males, and blade-like tips as in *P. major*; strong sexual dimorphism in canine size (linear dimensions of the single known female canine are only 56% that of the male average, exceeding the degree of sexual dimorphism in extant *Gorilla*); upper premolars broad, and relatively large in relation to M^1; upper molars similar to those of *P. major*, with bunodont cusps, wrinkled enamel, and moderately well-developed beaded lingual cingulum; $M^1 < M^3 \leqslant M^2$ (Pilbeam, 1969; Andrews, 1978*a*). The lower dentition is not preserved well enough for detailed comparison, except for an isolated P_3, which has a relatively long honing face. The lumbar vertebrae are derived in the direction of modern hominoids in having robust pedicles, lack of anapophyses, reduced ventral keeling, caudally inclined spinous process, and dorsally oriented transverse process arising from the pedicle (Walker & Rose, 1968; Ward, 1993; Sanders & Bodenbender, 1994; MacLatchy *et al.*, 2000). The glenoid articular surface of the scapula is rounded and expanded superiorly, as in hominoids (MacLatchy & Pilbeam, 1999; MacLatchy *et al.*, 2000), although several authors have argued that this scapula fragment is not diagnostic, and may not even belong to a primate (Pickford *et al.*, 1999; Johnson *et al.*, 2000). The femoral and phalangeal fragments are similar in morphology to those of *Proconsul*, and imply an arboreal habit (Gebo *et al.*, 1997b; MacLatchy & Pilbeam, 1999; MacLatchy *et al.*, 2000).

Superfamily *incertae sedis*

GENUS *Limnopithecus* Hopwood, 1933
Small catarrhine primates intermediate in dental size between *Simiolus* and *Micropithecus* (probably averaging ⩾ 5 kg). *Limnopithecus* is represented mainly by isolated teeth and jaw fragments, the key characteristics of which are as follows: lower face short, with a shallow nasoalveolar clivus of the premaxilla; nasal aperture elliptical in shape, and very narrow; orbits situated low on the face, and positioned relatively far forward, with the anterior margin situated just above the apex of the upper canine root; maxilla has relatively inflated maxillary sinus that extends into the tuberosity of the alveolar process posterior to M^3 and laterally into the anterior root of the zygomatic arch; mandible is shallow and gracile; superior transverse torus prominent, while the inferior transverse torus is poorly developed (*L. evansi*) or possibly absent (*L. legetet*); incisors comparatively small in relation to the size of the molars; I_2 bilaterally asymmetrical in shape; upper premolars relatively broad with buccal and lingual cusps of approximately equal height; upper molars with high conical cusps, protocone relatively small, trigon only slightly broader than long, and lingual cingulum well developed; $M^1 < M^3 \leqslant M^2$; M^3 only slightly reduced in size compared with M^2; lower molars ovoid to rectangular in occlusal outline, with a broad talonid basin, and well-developed buccal cingulum; $M_1 < M_2 < M_3$; (Harrison, 1981, 1982, 1988). A few isolated postcranial bones referred to *Limnopithecus* on the basis of size and frequency of occurrence at sites are morphologically similar to the corresponding elements in *Dendropithecus macinnesi* (Harrison, 1982).
INCLUDED SPECIES *L. evansi*, *L. legetet*

SPECIES *Limnopithecus legetet* Hopwood, 1933 (Fig. 19.14A, B)
TYPE SPECIMEN BMNH M 14079, right mandibular fragment with M_{1-2}

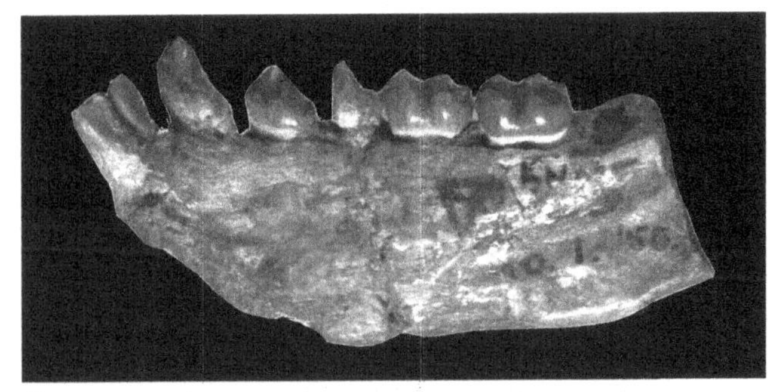

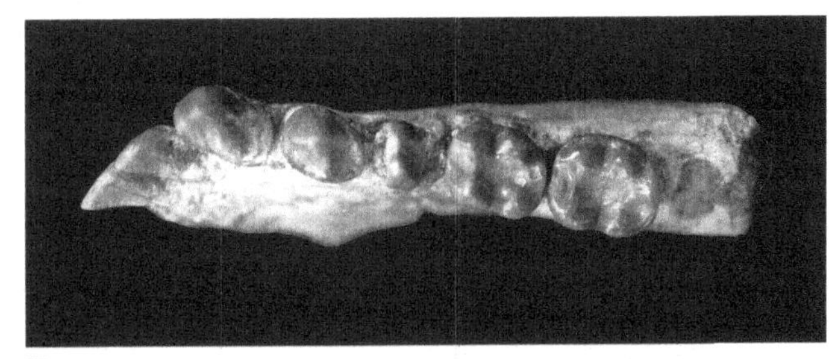

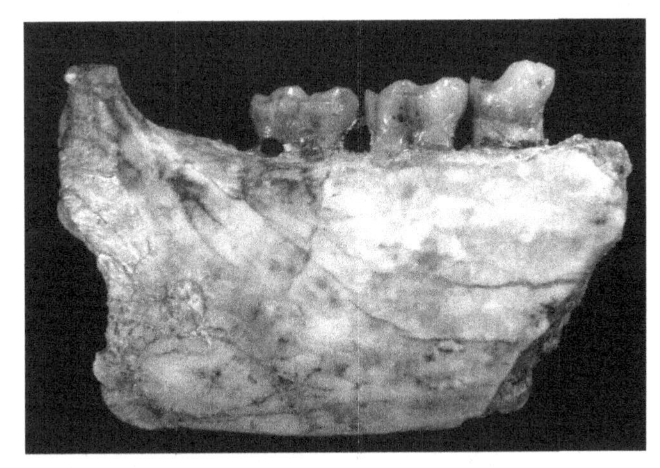

Fig. 19.14 *Limnopithecus*. (A, B) *L. legetet*, KNM-KO 8, right mandibular fragment with I_1–M_2: (A) medial view; (B) occlusal view. (C,D) *L. evansi*, KNM-SO 385 (holotype), right mandibular fragment with P_4–M_2: (C) lateral view; (D) occlusal view.

AGE AND GEOGRAPHIC RANGE Early Miocene (approximately 17–22 Ma), Kenya and Uganda (Walker, 1968; Bishop *et al.*, 1969; Andrews & Pickford, 1981; Harrison, 1981, 1982, 1988; Pickford, 1981, 1983, 1986*b*)

ANATOMICAL DEFINITION

Upper and lower incisors broad and low-crowned; canines relatively small; P_3 ovoid to almost circular in shape with a short mesiobuccal face (*L. legetet* may have been unique among early Miocene East African catarrhines in having limited development of the C^1–P_3 honing complex); P_4 broad and ovoid to circular in occlusal outline; upper premolars and molars moderately long and broad, with well-defined occlusal crests; lower molars broad and rectangular, with high and sharp cusps and occlusal crests, large and well-defined talonid basin and mesial and distal foveae; distal fovea broad and slightly obliquely oriented in relation to the transverse axis of the crown in M_1 and M_2, and very obliquely oriented in M_3; M_3 relatively large, with the entoconid situated transversely opposite the hypoconid (Harrison, 1981, 1982, 1988).

SPECIES *Limnopithecus evansi* MacInnes, 1943 (Fig. 19.14C, D)

TYPE SPECIMEN KNM-SO 385, right mandibular fragment

AGE AND GEOGRAPHIC RANGE Early Miocene (approximately 19–20 Ma) Kenya and possibly Uganda (Bishop *et al.*, 1969; Andrews *et al.*, 1981; Pickford, 1981, 1983, 1986*b*; Harrison, 1981, 1982, 1988)

ANATOMICAL DEFINITION

Limnopithecus evansi is distinguished from the type species by the following morphological characteristics: incisors narrower and relatively higher-crowned; canines somewhat larger; P_3 is narrower with a moderately developed sectorial face on the mesiobuccal aspect of the crown (more typical of other early Miocene catarrhines); P_4 long and narrow, with a large mesial fovea; upper premolars and molars relatively broader, with less well-defined occlusal crests; distal cusps on M^3 smaller; lower molars have low, rounded cusps and occlusal crests; crest connecting the entoconid and hypoconulid poorly developed or entirely lacking, and as a consequence the distal fovea is ill-defined and communicates directly with the talonid basin (a feature unique to *L. evansi*); M_3 smaller in size, and entoconid more distally positioned in relation to the hypoconid; mandibular corpus below the cheek teeth slightly higher (Harrison, 1981, 1982, 1988).

GENUS *Kalepithecus* Harrison, 1988

INCLUDED SPECIES *K. songhorensis*

SPECIES *Kalepithecus songhorensis*. (Andrews, 1978*a*) (Fig. 19.15)

TYPE SPECIMEN KNM-SO 378, right mandibular fragment

AGE AND GEOGRAPHIC RANGE Early–middle(?) Miocene, Kenya

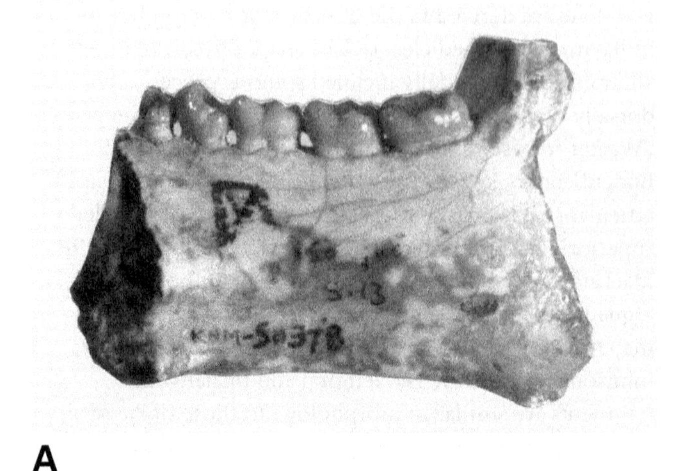

A

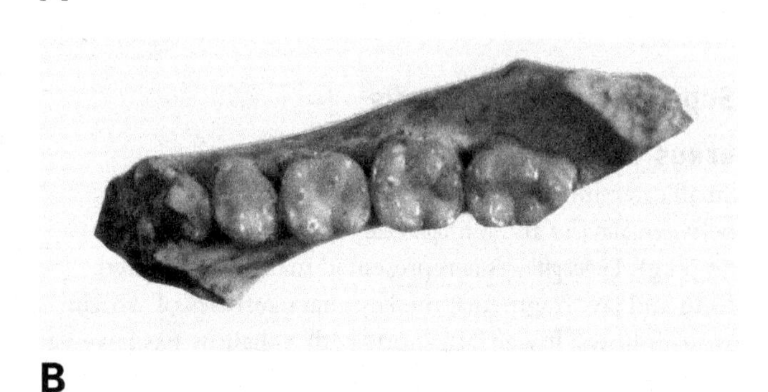

B

Fig. 19.15 *Kalepithecus songhorensis*. KNM-SO 378 (holotype), right mandibular fragment with P_3–M_3: (A) medial view; (B) occlusal view.

ANATOMICAL DEFINITION

A small catarrhine primate similar in dental size to *L. legetet* (probably $\geqslant$ 5 kg). Upper central incisor relatively broad and more spatulate compared with those in *Limnopithecus* or *Dendropithecus*; I^2 markedly bilaterally asymmetrical in shape, and relatively much smaller than I^1; lower incisors very high-crowned, slender and relatively bilaterally symmetrical in shape; canines moderately high-crowned, with only slightly bilaterally compressed; upper premolars relatively narrow, with well-developed transverse crest linking the buccal and lingual cusps; P_3 exhibits a moderate degree of specialization for sectoriality; P_4 relatively large and ovoid in shape, frequently being broader than long; upper molars relatively broad due to the strong development of a lingual cingulum; protocone voluminous and markedly buccally displaced away from the margin of the crown; breadth of the trigon only slightly greater than its length; lower molars short and broad, and rectangular to ovoid in shape; mesial fovea slightly oblique relative to the transverse axis of the crown; buccal cingulum broad, but rounded and poorly defined; $M_1 < M_2 \leqslant M_3$; molars have low, rounded and voluminous cusps that restrict the extent of the foveae and occlusal basins; occlusal crests low, rounded and poorly developed; anterior teeth large in relation to the size of the cheek teeth; unlike other early

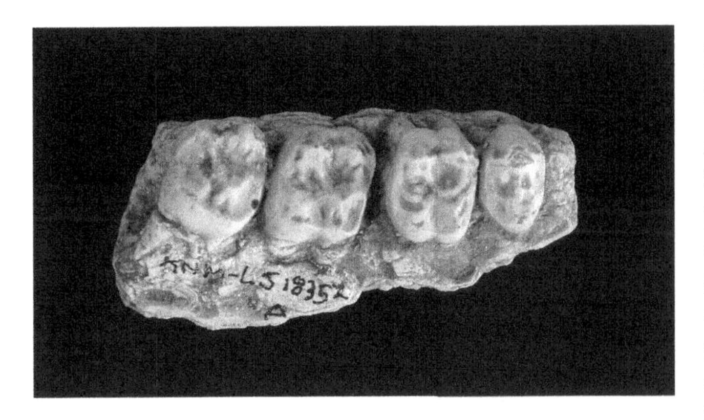

Fig. 19.16 *Kamoyapithecus hamiltoni*. KNM-LS 18352, right maxilla with P^4–M^3 (cast), occlusal view.

Miocene catarrhines, nasal aperture broad, particularly inferiorly, and nasoalveolar clivus relatively deep – a structural complex that superficially resembles that in great apes (Harrison, 1981, 1982, 1988). An isolated M^1 from the middle Miocene locality of Kipsaramon in Kenya has been provisionally identified as *Kalepithecus* cf. *songhorensis* (Hill *et al.*, 1991).

GENUS *Kamoyapithecus* Leakey *et al.*, 1995
INCLUDED SPECIES *K. hamiltoni*

SPECIES *Kamoyapithecus hamiltoni* (Madden, 1980) (Fig. 19.16)
TYPE SPECIMEN KNM-LS 7, right maxillary fragment previously listed as UCMP 52112 by Madden (1980)
AGE AND GEOGRAPHIC RANGE Late Oligocene (23.9–27.8 Ma), Kenya (Boschetto *et al.*, 1992; Leakey *et al.*, 1995*b*)
ANATOMICAL DEFINITION
A large catarrhine primate approximating *Proconsul nyanzae* in dental size. It differs from other East African catarrhines in the following combination of features: mandibular symphysis with large superior transverse torus and much smaller inferior transverse torus; mental foramen anteriorly facing and positioned just anterior to the mesial root of P_3; I_2 mesiodistally compressed without tapering at the cervix and with a robust root; upper and lower canines with stout roots; upper canine with relatively short and bilaterally compressed crown, deep mesial groove, and sharp distal crest; P_3 mesiodistally elongate, presumably with a long mesiobuccal honing face; P^4 ovoid with moderate degree of lingual flare and slight buccal flare; upper molars very broad and low-crowned, with distinct buccal and lingual basal flare, conical and bunodont cusps, a relatively narrow trigon, a broad lingual cingulum without crenulation (although subsidiary conules associated with the hypocone and protocone are situated on the lingual cingulum), and a narrow buccal cingulum; M^2 with large hypocone set close to the trigon, and crown distinctly shorter in buccal moiety than lingual moiety; M^3 with diminutive distal cusps; $M^1 < M^3 \leqslant M^2$ (Madden, 1980; Leakey *et al.*, 1995*b*).

Evolution of Afro-Arabian late Oligocene – middle Miocene catarrhines

Although our perspective on the relationships of the early Miocene catarrhines has been radically transformed in the 70 years since the first fossil primates were discovered in East Africa, their prime importance for understanding the early stages in the evolutionary history of the hominoids remains undisputed. The early Miocene taxa are members of a major radiation of catarrhines of modern aspect, some of which are undoubtedly stem catarrhines, while others are arguably identifiable as stem hominoids (Harrison, 1982, 1987, 1988, 1993; Andrews, 1985, 1992; Andrews & Martin, 1987*a*; Groves, 1989; Rose *et al.*, 1992; Begun *et al.*, 1997*a*; Fleagle, 1999). Current phylogenetic debate centers on which taxa have crossed the cladistic threshold to be considered stem hominoids rather than primitive stem catarrhines. The author's contention is that almost all of the fossil catarrhines discussed in this chapter are stem catarrhines. However, the majority of colleagues still favor recognizing the larger species as hominoids, while retaining most of the smaller species as stem catarrhines. This dichotomy based primarily on size is related to preconceived notions about the traditional relationships between extant hominoids and a lack of appreciation of the phenomenon of increasing organizational complexity as a consequence of scaling (Harrison, 1987, 1988, 1993). These ideas about catarrhine phylogenetic relationships pervaded even the earliest publications, and subsequently have become entrenched without adequate critical scrutiny.

The phylogenetic scheme presented in Fig. 19.17 represents the author's current views on the evolutionary relationships of the Miocene catarrhines discussed in this chapter. Four major clades are recognized: (1) Dendropithecoidea, (2) Proconsuloidea, (3) Cercopithecoidea and (4) Hominoidea. The first two clades are stem catarrhines of modern aspect composed exclusively of Oligo-Miocene Afro-Arabian taxa. Both clades are derived relative to propliopithecids and pliopithecids in the development of a fully formed tubular ectotympanic (at least in proconsuloids) and the loss of the entepicondylar foramen in the distal humerus (Harrison, 1987). The last two clades – the cercopithecoids and hominoids – are crown catarrhines that share a suite of synapomorphies (in the anatomy of the face, ear region, vertebral column and pelvis) not present in dendropithecoids and proconsuloids (Harrison & Sanders, 1999; T. Harrison, unpublished data). Of the early Miocene East African catarrhines, only *Morotopithecus* is included here as a stem hominoid.

The Dendropithecoidea includes *Dendropithecus*, *Micropithecus* and *Simiolus*. These taxa are linked by the possession of the following features: upper and lower canines strongly bilaterally compressed; P_3 shows a moderate to strong degree of specialization for sectoriality; the limb bones are slender; the humerus has a relatively straight shaft; the distal end of the

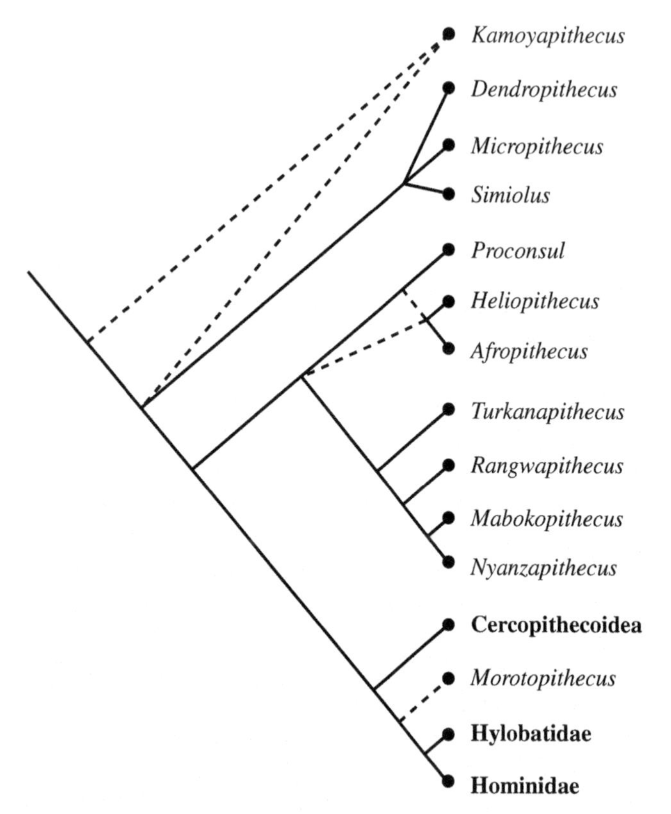

Fig. 19.17 Cladogram illustrating the phylogenetic relationships between the Oligo-Miocene Afro-Arabian catarrhines and extant cercopithecoids and hominoids. *Limnopithecus* and *Kalepithecus* are not shown because their relationships are uncertain. See text for further details.

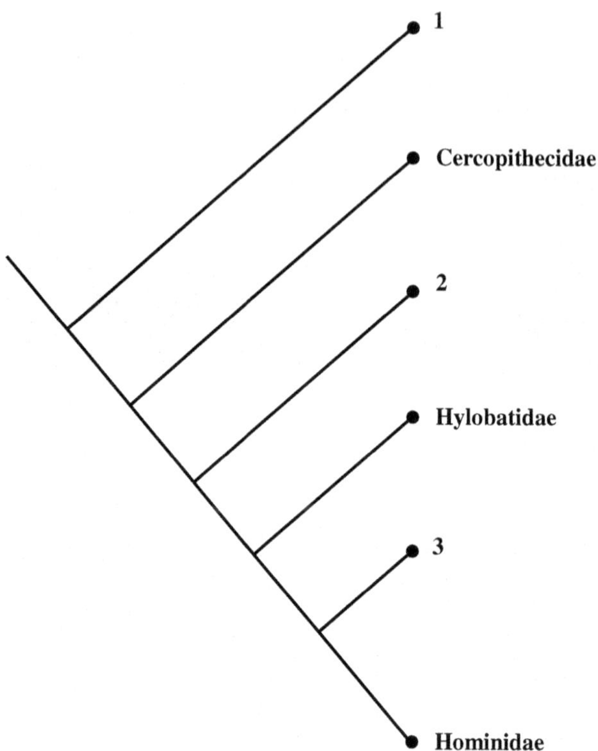

Fig. 19.18 Alternative hypotheses concerning the phylogenetic relationships of *Proconsul* to extant catarrhines: (1) *Proconsul* is a stem catarrhine that represents the sister group to cercopithecoids + hominoids (Harrison, 1987, 1988, 1993, and the hypothesis preferred here); (2) *Proconsul* is a stem hominoid that represents the sister group to hylobatids + hominids (e.g., Andrews, 1985, 1992; Rose, 1983, 1992, 1997; Begun, 1997; Kelley, 1997; Ward, 1997; Rae, 1999); (3) *Proconsul* is a stem hominid that represents the sister group to great apes and humans (e.g., Walker & Teaford, 1989; Walker, 1997; Rae, 1997).

humerus has a large medially directed medial epicondyle, a well-developed dorsal epitrochlear fossa, a broad and shallow zona conoidea, a weak lateral trochlear keel, a trochlear articular surface exhibiting minimal spooling, and a shallow olecranon fossa (Harrison, 1987, 1988; Rose *et al.*, 1992; Rose, 1997). Many of these characters can be interpreted as the primitive condition for catarrhines, so the dendropithecoids conceivably might be paraphyletic. However, the C–P_3 complex in dendropithecoids is not typical of propliopithecids or pliopithecids. It probably corresponds closely to the primitive condition for catarrhines of modern aspect (Harrison & Gu, 1999). The recognition of the Dendropithecoidea accords well with inferences derived from other recent studies (Harrison, 1987, 1993; Andrews, 1985; Rose *et al.*, 1992; Rose, 1997; Delson *et al.*, 2000). Rae (1997, 1999), in contrast, has presented a phylogenetic analysis of cranial data indicating that dendropithecoids are indeed paraphyletic, and that the individual taxa are more appropriately considered as stem hominoids (i.e., *Micropithecus*) and stem hominids (i.e., *Dendropithecus* and *Simiolus*). However, as discussed above, this hypothesis is not supported by other lines of evidence.

The Proconsuloidea is more derived than the Dendropithecoidea in many aspects of its postcranium, especially the forelimb. The current evidence supports recognizing the two groups as separate clades. In recent years, a number of different hypotheses have been proposed regarding the relationships of *Proconsul* to extant catarrhines (Fig. 19.18). Most authors contend that *Proconsul* shares derived characteristics linking it exclusively with extant hominoids (e.g., Rose, 1983, 1992, 1997; Andrews, 1985, 1992; Fleagle, 1986, 1999; Senut, 1989; Begun *et al.*, 1997*a*; Kelley, 1997; Ward, 1997; Rae, 1999). An alternative view recognizes *Proconsul* as a stem hominid – the sister taxon to extant great apes and humans (Walker & Teaford, 1989; Walker, 1997; Rae, 1997). As discussed above, Harrison (1987, 1988, 1993; Harrison & Rook, 1997; Harrison & Gu, 1999) has argued that *Proconsul* is a stem catarrhine, representing the sister group to cercopithecoids + hominoids. When the purported synapomorphies are critically scrutinized few provide convincing support for inferring a close phyletic relationship between *Proconsul* and extant hominoids. In fact, recent comparisons have shown that key synapomorphies of the face, ear region and postcranium shared by extant catarrhines are lacking in *Proconsul*, which thus retains a more primitive condition (Harrison & Sanders, 1999, unpublished data).

Three distinct clades are recognized within the Proconsuloidea: (1) the Proconsulinae, (2) the Afropithecinae and (3) the Nyanzapithecinae. The Proconsulinae includes four species belonging to a single genus, *Proconsul*. The Afropithecinae, comprising *Afropithecus* and *Heliopithecus*, may also include *Otavipithecus* (Andrews, 1992; Singleton, 2000), but the phylogenetic relationships of this taxon are difficult to establish because of the paucity of material. *Nacholapithecus* and *Equatorius* may also belong to this clade, but confirmation will have to await detailed comparisons of the recently published material, especially the partial skeletons from Kipsaramon and Nachola (Nakatsukasa *et al.*, 1998; Ishida *et al.*, 1999; S. Ward *et al.*, 1999). The postcranial morphology and basic anatomical plan of the face clearly link *Afropithecus* (and *Otavipithecus*) more closely to *Proconsul* than to extant hominoids (Rose, 1993*b*, 1997; Leakey & Walker, 1997; Ward, 1997, 1998; T. Harrison, unpublished data), and as a result the Afropithecinae is included here in the Proconsulidae.

The Nyanzapithecinae includes *Nyanzapithecus*, *Mabokopithecus*, *Rangwapithecus* and *Turkanapithecus*. The clade is recognized on the basis of synapomorphies of the premolars and molars. Precise assessment of the relationship between *Nyanzapithecus* and *Mabokopithecus* will have to await detailed comparisons of new finds from Maboko Island (Benefit & McCrossin, 1997*b*; Benefit *et al.*, 1998), but it is evident that the two taxa are closely related, and may even be congeneric. As noted previously by Harrison (1986*a*, 1986*b*), *Rangwapithecus* represents the conservative sister taxon of *Nyanzapithecus*. Since *Turkanapithecus* shares some derived features with them it can be inferred to represent the sister taxon of *Rangwapithecus* + (*Nyanzapithecus* + *Mabokopithecus*) (Fig. 19.17; see also Singleton, 2000).

The postcranial anatomy of *Turkanapithecus* is similar to that of *Proconsul* and more derived than that of *Dendropithecus* and *Simiolus* (see Rose, 1997). Features of the face and postcranium shared by *Turkanapithecus* and *Proconsul* indicate a close relationship. Isolated postcranials attributed to *Nyanzapithecus* and *Rangwapithecus* are consistent with this phylogenetic and taxonomic association (Harrison, 1982; Gebo *et al.*, 1988; McCrossin, 1992*a*).

Morotopithecus is the only early Miocene catarrhine included here in the Hominoidea, based on the presence of shared derived characteristics of the postcranium (particularly the lumbar vertebra) linking it with extant apes (Ward, 1993; Sanders & Bodenbender, 1994; Gebo *et al.*, 1997*b*; MacLatchy et al., 2000). The Moroto lumbar vertebra provides the first evidence of the evolutionary reorganization of the axial skeleton in the direction of hominoids (Sanders & Bodenbender, 1994) and contrasts with the condition seen in primitive catarrhines, such as *Proconsul*, which have long and flexible lower backs (Ward, 1993). Other postcranials firmly attributed to *Morotopithecus* indicate a general similarity to *Proconsul*, and imply that *Morotopithecus* may have retained the primitive catarrhine morphology, at least in its appendicular skeleton (Pickford *et al.*, 1999; MacLatchy *et al.*, 2000). The craniodental morphology of *Morotopithecus* does not appear to be more derived than that of proconsulids, and, as noted by Andrews & Martin (1987*b*) and Andrews (1992), its upper cheek teeth share derived features with afropithecines. At present, I favor including *Morotopithecus* as a member of the Hominoidea, based on the postcranial synapomorphies. However, I cannot discount the distinct possibility that the dental similarities with afropithecines are valid synapomorphies, and that *Morotopithecus* is merely a large orthograde proconsulid that has developed its own unique adaptations in the vertebral column in parallel with those of extant hominoids.

Acknowledgements

I would like to thank Walter Hartwig for inviting me to prepare a contribution for this volume, and for his good humor and patience in waiting for the final product. I thank the following colleagues and institutions for allowing me access to the fossil material in their care: M. G. Leakey, R. E. F. Leakey and E. Mbua, National Museums of Kenya, Nairobi; P. J. Andrews and J. Hooker, Natural History Museum, London; E. Delson, American Museum of Natural History, New York; J. G. Fleagle, State University of New York, Stony Brook, New York. The following individuals kindly provided me with photographs of specimens: P. Andrews, J. G. Fleagle, Y. Kunimatsu, M. G. Leakey, D. Pilbeam and L. Sarlo. Of the numerous colleagues and graduate students who have offered valuable help, advice and comments, the follow deserve special mention: P. Andrews, B. Benefit, E. Delson, W. Dirks, J. Fleagle, A. Hill, N. Jablonski, C. Jolly, J. Kelley, Y. Kunimatsu, J. Langdon, M. G. Leakey, L. MacLatchy, J. Manser, L. Martin, M. McCrossin, M. Pickford, D. Pilbeam, M. Rose, W. Sanders, L. Sarlo, B. Senut, A. Walker, C. Ward and S. Ward. I especially want to acknowledge with gratitude those individuals whose original contributions on the Miocene catarrhines of East Africa have been profoundly influential in shaping my own ideas, although they should not be held in any way accountable: Peter Andrews, John Fleagle, Martin Pickford, David Pilbeam, Michael Rose and Alan Walker.

Primary References

Afropithecus

Leakey, R. E. F. & Walker, A. C. (1985*b*). New higher primates from the early Miocene of Buluk, Kenya. *Nature*, **318**, 173–175.

Leakey, R. E. F. & Leakey, M .G. (1986*a*). A new Miocene hominoid from Kenya. *Nature*, **324**, 143–146.

Leakey, R. E. F., Leakey, M. G., & Walker, A. C. (1988*a*). Morphology of *Afropithecus turkanensis* from Kenya. *American Journal of Physical Anthropology*, **76**, 289–307.

Anyonge, W. (1991). Fauna from a new lower Miocene locality west of Lake Turkana, Kenya. *Journal of Vertebrate Paleontology*, **11**, 378–390.

Dendropithecus

MacInnes, D. G. (1943). Notes on the East African Miocene Primates. *Journal of the East Africa and Uganda Natural History Society*, **17**, 141–181.
Le Gros Clark, W. E. (1950). New palaeontological evidence bearing on the evolution of the Hominoidea. *Quarterly Journal of the Geological Society of London*, **105**, 225–259.
Le Gros Clark, W. E. & Leakey, L. S. B. (1950). Diagnoses of East African Miocene Hominoidea. *Quarterly Journal of the Geological Society of London*, **105**, 260–263.
Le Gros Clark, W. E. & Leakey, L. S. B. (1951). The Miocene Hominoidea of East Africa. *Fossil Mammals of Africa*, No. 1. London: British Museum (Natural History).
Le Gros Clark, W. E. & Thomas, D. P. (1951). Associated jaws and limb bones of *Limnopithecus macinnesi*. *Fossil Mammals of Africa*, No. 3. London: British Museum (Natural History).
Le Gros Clark, W. E. (1952). Report on fossil hominoid material collected by the British–Kenya Miocene Expedition, 1949–1951. *Proceedings of the Zoological Society of London*, **122**, 273–286.
Andrews, P. J. & Simons, E. L. (1977). A new Miocene gibbon-like genus, *Dendropithecus* (Hominoidea, Primates) with distinctive postcranial adaptations: its significance to origin of Hylobatidae. *Folia Primatologica*, **28**, 161–169.
Andrews, P. J. (1978). A revision of the Miocene Hominoidea of East Africa. *Bulletin of the British Museum (Natural History), Geology Series*, **30**, 85–225.
Harrison, T. (1981). New finds of small fossil apes from the Miocene locality of Koru in Kenya. *Journal of Human Evolution*, **10**, 129–137.
Langdon, J. (1986). Functional morphology of the Miocene hominoid foot. *Contributions to Primatology*, **22**, 1–225.
Harrison, T. (1988). A taxonomic revision of the small catarrhine primates from the early Miocene of East Africa. *Folia Primatologica*, **50**, 59–108.

Heliopithecus

Andrews, P. J., Hamilton, W. R., & Whybrow, P. J. (1978). Dryopithecines from the Miocene of Saudi Arabia. *Nature*, **274**, 249–250.
Andrews, P. J. & Martin, L. (1987). The phyletic position of the Ad Dabtiyah hominoid. *Bulletin of the British Museum (Natural History), Geology Series*, **41**, 383–393.

Kalepithecus

MacInnes, D. G. (1943). Notes on the East African Miocene Primates. *Journal of the East Africa and Uganda Natural History Society*, **17**, 141–181.
Le Gros Clark, W. E. & Leakey, L. S. B. (1951). The Miocene Hominoidea of East Africa. *Fossil Mammals of Africa*, No. 1. London: British Museum (Natural History).
Andrews, P. J. (1978). A revision of the Miocene Hominoidea of East Africa. *Bulletin of the British Museum (Natural History), Geology Series*, **30**, 85–225.
Harrison, T. (1988). A taxonomic revision of the small catarrhine primates from the early Miocene of East Africa. *Folia Primatologica*, **50**, 59–108.

Kamoyapithecus

Madden, C. T. (1980). New *Proconsul (Xenopithecus)* from the Miocene of Kenya. *Primates*, **21**, 241–252.
Leakey, M. G., Ungar, P. S., & Walker, A. C. (1995). A new genus of large primate from the late Oligocene of Lothidok, Turkana District, Kenya. *Journal of Human Evolution*, **28**, 519–531.

Limnopithecus

Hopwood, A. T. (1933*a*). Miocene primates from British East Africa. *Annals and Magazine of Natural History (Series 10)*, **11**, 96–98.
MacInnes, D. G. (1943). Notes on the East African Miocene Primates. *Journal of the East Africa and Uganda Natural History Society*, **17**, 141–181.
Le Gros Clark, W. E. & Leakey, L. S. B. (1950). Diagnoses of East African Miocene Hominoidea. *Quarterly Journal of the Geological Society of London*, **105**, 260–263.
Le Gros Clark, W. E. & Leakey, L. S. B. (1951). The Miocene Hominoidea of East Africa. *Fossil Mammals of Africa*, No. 1. London: British Museum (Natural History).
Le Gros Clark, W. E. (1952). Report on fossil hominoid material collected by the British-Kenya Miocene Expedition, 1949–1951. *Proceedings of the Zoological Society of London*, **122**, 273–286.
Walker, A. C. (1968). The Lower Miocene fossil site of Bukwa, Sebei. *Uganda Journal*, **32**, 149–156.
Andrews, P. J. (1978). A revision of the Miocene Hominoidea of East Africa. *Bulletin of the British Museum (Natural History), Geology Series*, **30**, 85–225.
Harrison, T. (1981). New finds of small fossil apes from the Miocene locality of Koru in Kenya. *Journal of Human Evolution*, **10**, 129–137.
Langdon, J. (1986). Functional morphology of the Miocene hominoid foot. *Contributions to Primatology*, **22**, 1–225.
Harrison, T. (1988). A taxonomic revision of the small catarrhine primates from the early Miocene of East Africa. *Folia Primatologica*, **50**, 59–108.

Mabokopithecus

Koenigswald, G. H. R. von (1969). Miocene Cercopithecoidea and Oreopithecoidea from the Miocene of East Africa. In *Fossil Vertebrates of Africa*, vol. 1, ed. L. S. B. Leakey, pp. 39–51. London: Academic Press.
Harrison, T. (1986). New fossil anthropoids from the middle Miocene of East Africa and their bearing on the origin of the Oreopithecidae. *American Journal of Physical Anthropology*, **71**, 265–284.

Micropithecus

Bishop, W. W. (1964). More fossil primates and other Miocene mammals from northeast Uganda. *Nature*, **203**, 1327–1331.
Pilbeam, D. R. & Walker, A. C. (1968). Fossil monkeys from the Miocene of Napak, northeast Uganda. *Nature*, **220**, 657–660.
Fleagle, J. G. (1975). A small gibbon-like hominoid from the Miocene of Uganda. *Folia Primatologica*, **24**, 1–15.
Andrews, P. J. (1978). A revision of the Miocene Hominoidea of

East Africa. *Bulletin of the British Museum (Natural History), Geology Series*, **30**, 85–225.

Fleagle, J. G. & Simons, E. L. (1978). *Micropithecus clarki*, a small ape from the Miocene of Uganda. *American Journal of Physical Anthropology*, **49**, 427–440.

Harrison, T. (1981). New finds of small fossil apes from the Miocene locality of Koru in Kenya. *Journal of Human Evolution*, **10**, 129–137.

Harrison, T. (1988). A taxonomic revision of the small catarrhine primates from the early Miocene of East Africa. *Folia Primatologica*, **50**, 59–108.

Harrison, T. (1989). A new species of *Micropithecus* from the middle Miocene of Kenya. *Journal of Human Evolution*, **18**, 537–557.

Morotopithecus

Allbrook, D. & Bishop, W. W. (1963). New fossil hominoid material from Uganda. *Nature*, **197**, 1187–1190.

Walker, A. C. & Rose, M. (1968). Fossil hominoid vertebra from the Miocene of Uganda. *Nature*, **217**, 980–981.

Pilbeam, D. R. (1969). Tertiary Pongidae of East Africa: evolutionary relationships and taxonomy. *Bulletin of the Peabody Museum of Natural History*, **31**, 1–185.

Gebo, D. L., MacLatchy, L., Kityo, R., Deino, A., Kingston, J., & Pilbeam, D. (1997). A hominoid genus from the Early Miocene of Uganda. *Science*, **276**, 401–404.

Pickford, M., Senut, B., & Gommery, D. (1999). Sexual dimorphism in *Morotopithecus bishopi*, an early Middle Miocene hominoid from Uganda, and a reassessment of its geological and biological contexts. In *Late Cenozoic Environments and Hominid Evolution: A Tribute to Bill Bishop*, eds. P. Andrews & P. Banham, pp. 27–38. London: Geological Society.

Nyanzapithecus

Andrews, P. J. (1974). New species of *Dryopithecus* from Kenya. *Nature*, **249**, 188–190.

Andrews, P. J. (1978*a*). A revision of the Miocene Hominoidea of East Africa. *Bulletin of the British Museum (Natural History), Geology Series*, **30**, 85–225.

Harrison, T. (1986). New fossil anthropoids from the middle Miocene of East Africa and their bearing on the origin of the Oreopithecidae. *American Journal of Physical Anthropology*, **71**, 265–284.

Gebo, D. L., Beard, K. C., Teaford, M. F., Walker, A. C., Larson, S. G., Jungers, W. L., & Fleagle, J. G. (1988). A hominoid proximal humerus from the Early Miocene of Rusinga Island, Kenya. *Journal of Human Evolution*, **17**, 393–401.

Kunimatsu, Y. (1992*a*). A revision of the hypodigm of *Nyanzapithecus vancouveringi*. *African Studies Monographs*, **14**, 231–235.

Kunimatsu, Y. (1992*b*). New finds of small anthropoid primate from Nachola, northern Kenya. *African Studies Monographs*, **14**, 237–249.

Kunimatsu, Y. (1997). New species of *Nyanzapithecus* from Nachola, northern Kenya. *Anthropological Science*, **105**, 117–141.

Proconsul

Hopwood, A. T. (1933*a*). Miocene primates from British East Africa. *Annals and Magazine of Natural History (Series 10)*, **11**, 96–98.

Hopwood, A. T. (1933*b*). Miocene primates from Kenya. *Journal of the Linnean Society of London – Zoology*, **38**, 437–464.

Leakey, L. S. B. (1943). A Miocene anthropoid mandible from Rusinga, Kenya. *Nature*, **152**, 319–320.

MacInnes, D. G. (1943). Notes on the East African Miocene Primates. *Journal of the East Africa and Uganda Natural History Society*, **17**, 141–181.

Le Gros Clark, W. E. (1949). Early Miocene apes from East Africa. *Advancement of Science*, **5**, 340–341.

Le Gros Clark, W. E. (1950). New palaeontological evidence bearing on the evolution of the Hominoidea. *Quarterly Journal of the Geological Society of London*, **105**, 225–259.

Le Gros Clark, W. E. & Leakey, L. S. B. (1950). Diagnoses of East African Miocene Hominoidea. *Quarterly Journal of the Geological Society of London*, **105**, 260–263.

Le Gros Clark, W. E. & Leakey, L. S. B. (1951). The Miocene Hominoidea of East Africa. *Fossil Mammals of Africa*, No. 1. London: British Museum (Natural History).

Le Gros Clark, W. E. (1952). Report on fossil hominoid material collected by the British–Kenya Miocene Expedition, 1949–1951. *Proceedings of the Zoological Society of London*, **122**, 273–286.

Napier, J. R. & Davis, P. R. (1959). The fore-limb skeleton and associated remains of *Proconsul africanus*. *Fossil Mammals of Africa*, No. 16. London: British Museum (Natural History).

Allbrook, D. & Bishop, W. W. (1963). New fossil hominoid material from Uganda. *Nature*, **197**, 1187–1190.

Bishop, W. W. (1964). More fossil primates and other Miocene mammals from northeast Uganda. *Nature*, **203**, 1327–1331.

Leakey, L. S. B. (1967). An early Miocene member of Hominidae. *Nature*, **213**, 155–163.

Pilbeam, D. R. (1969). Tertiary Pongidae of East Africa: evolutionary relationships and taxonomy. *Bulletin of the Peabody Museum of Natural History*, **31**, 1–185.

Andrews, P. J. (1978*a*). A revision of the Miocene Hominoidea of East Africa. *Bulletin of the British Museum (Natural History), Geology Series*, **30**, 85–225.

Martin, L. (1981). New specimens of *Proconsul* from Koru, Kenya. *Journal of Human Evolution*, **10**, 139–150.

Andrews, P. J., Harrison, T., Martin, L., & Pickford, M. (1981). Hominoid primates from a new Miocene locality named Meswa Bridge in Kenya. *Journal of Human Evolution*, **10**, 123–128.

Walker, A. C., Falk, D., Smith, R., & Pickford, M. (1983). The skull of *Proconsul africanus*: reconstruction and cranial capacity. *Nature*, **305**, 525–527.

Walker, A. C. & Pickford, M. (1983). New postcranial fossils of *Proconsul africanus* and *Proconsul nyanzae*. In *New Interpretations of Ape and Human Ancestry*, eds. R. L. Ciochon & R. S. Corruccini, pp. 325–351. New York: Plenum.

Langdon, J. (1986). Functional morphology of the Miocene hominoid foot. *Contributions to Primatology*, **22**, 1–225.

Beard, K. C., Teaford, M. F., & Walker, A. C. (1986). New wrist bones of *Proconsul africanus* and *nyanzae* from Rusinga Island, Kenya. *Folia Primatologica*, **47**, 97–118.

Teaford, M. F., Beard, K. C., Leakey, R. E., & Walker, A. C. (1988). New hominoid facial skeleton from the early Miocene of Rusinga Island, Kenya, and its bearing on the relationship between *Proconsul nyanzae* and *Proconsul africanus*. *Journal of Human Evolution*, **17**, 461–477.

Walker, A. C., Teaford, M. F., Martin, L., & Andrews, P. (1993). A new species of *Proconsul* from the early Miocene of Rusinga/Mfangano Islands, Kenya. *Journal of Human Evolution*, **25**, 43–56.

Ward, C. V., Walker, A. C., Teaford, M. F., & Odhiambo, I. (1993). Partial skeleton of *Proconsul nyanzae* from Mfangano Island, Kenya. *American Journal of Physical Anthropology*, **90**, 77–111.

Begun, D. R., Teaford, M. F., & Walker, A. C. (1994). Comparative and functional anatomy of *Proconsul* phalanges from the Kaswanga Primate Site, Rusinga Island, Kenya. *Journal of Human Evolution*, **26**, 89–165.

Rafferty, K. L., Walker, A. C., Ruff, C., Rose, M. D., & Andrews, P. J. (1995). Postcranial estimates of body weights in *Proconsul*, with a note on a distal tibia of *P. major* from Napak, Uganda. *American Journal of Physical Anthropology*, **97**, 391–402.

Ward, C. V., Ruff, C. B., Walker, A. C., Rose, M. D., Teaford, M. F., & Nengo, I. O. (1995). Functional morphology of *Proconsul* patellas from Rusinga Island, Kenya, with implications for other Miocene–Pliocene catarrhines. *Journal of Human Evolution*, **29**, 1–19.

Gommery, D., Senut, B., & Pickford, M. (1998). New hominoid postcranial remains from the early Miocene of Napak, Uganda. *Annales de paléontologie*, **84**, 287–306.

Rangwapithecus

Andrews, P. J. (1970). Two new fossil primates from the Lower Miocene of Kenya. *Nature*, **228**, 537–540.

Andrews, P. J. (1974). New species of *Dryopithecus* from Kenya. *Nature*, **249**, 188–190.

Andrews, P. J. (1978*a*). A revision of the Miocene Hominoidea of East Africa. *Bulletin of the British Museum (Natural History), Geology Series*, **30**, 85–225.

Nengo, I. O. & Rae, T. C. (1992). New hominoid fossils from the early Miocene site of Songhor, Kenya. *Journal of Human Evolution*, **23**, 423–429.

Simiolus

Leakey, L. S. B. (1968). Upper Miocene primates from Kenya. *Nature*, **218**, 527–528.

Andrews, P. J. & Walker, A. C. (1976). The primate and other fauna from Fort Ternan, Kenya. In *Human Origins: Louis Leakey and the East African Evidence*, eds. G. L. Isaac & E. R. McCown, pp. 279–304. Menlo Park, CA: W. A. Benjamin.

Leakey, R. E. F. & Leakey, M. G. (1987). A new Miocene small-bodied ape from Kenya. *Journal of Human Evolution*, **16**, 369–387.

Harrison, T. (1992). A reassessment of the taxonomic and phylogenetic affinities of the fossil catarrhines from Fort Ternan, Kenya. *Primates*, **33**, 501–522.

Rose, M. D., Leakey, M. G., Leakey, R. E. F., & Walker, A. C. (1992). Postcranial specimens of *Simiolus enjiessi* and other primitive catarrhines from the early Miocene of Lake Turkana, Kenya. *Journal of Human Evolution*, **22**, 171–237.

Turkanapithecus

Leakey, R. E. F. & Leakey, M. G. (1986*b*). A second new Miocene hominoid from Kenya. *Nature*, **324**, 146–148.

Leakey, R. E. F., Leakey, M. G., & Walker, A. C. (1988*b*). Morphology of *Turkanapithecus kalakolensis* from Kenya. *American Journal of Physical Anthropology*, **76**, 277–288.

20 | European hominoids

DAVID R. BEGUN

Introduction

Unlike their counterparts in Africa, European hominoids are restricted to the middle and late Miocene fossil record. Nevertheless, this group was highly diverse and successful. European hominoids persisted for about 8–9 million years, from the Pyrenees of France and Spain in the west to the Republic of Georgia in the east. Their paleobiological diversity arguably matches that of all other hominoids combined. Phylogenetically European hominoids span a broad spectrum of clades that include relatively primitive taxa, early members of the African ape and human clade, and those of uncertain affiliation. European hominoid diversity mirrors the Plio-Pleistocene hominoid record in Africa. The similarities in patterns of diversity and the evolutionary relations between European hominoids and their African descendants makes their fossil record especially relevant to analysis of great-ape and human evolutionary history.

History of discovery and debate

The history of the study of European hominoids parallels the intellectual history of evolutionary biology broadly and paleoanthropology more specifically. Both of these fields were born in Europe in the nineteenth century, and both were influenced fundamentally by the emergence of a fossil record of apes. Countless references to European anthropoids in the literature from the mid-nineteenth century to the present one include a huge number of named taxa and hypotheses about relations to living primates. This chaper reviews major historical trends following growth of the fossil record, beginning with the period preceding the recognition of any fossil anthropoids. By this I mean both the recognition that known specimens were indeed fossil anthropoids and the recognition that there could even be such a thing as a fossil anthropoid.

The pre-fossil period

Georges Cuvier, credited by many for developing the modern science of vertebrate paleontology, is also blamed for the slow recognition of the existence of fossil primates. This is a cruel historical irony, since he in fact described the first fossil primate ever discovered, *Adapis parisiensis*, from the Eocene age Montmartre gypsum quarry fossil assemblages (Cuvier, 1812). Cuvier thought *Adapis* was a primitive pachyderm, (probably a small perissodactyl in today's parlance). Although that seems to be a grotesque error, most Eocene mammalian orders resemble each other much more closely than do living members of different orders, and he lacked an extensive fossil record for comparison. However, Cuvier no doubt was cool to the idea of fossil primates, though he left the possibility open in his "Discours préliminaire", or Preface, to the first great work of vertebrate paleontology (Cuvier, 1812). In that defining work Cuvier says "Il n'y a point d'os humains fossiles" [there are no fossil humans]. After debunking several contemporary claims of associations between human fossils and extinct animals, Cuvier notes "mais je n'en veux pas conclure que l'homme n'existait point du tout avant cette époque" [but I do not want to conclude that man did not exist at all before this time].

Cuvier did not accept the potential for species to transform into new species, such as expressed at that time in the classic *Philosophie zoologique* (Lamarck, 1809). However, Cuvier's ideas on transformationism when viewed in the context of his time were quite reasonable. He meticulously documented assemblages of fossil vertebrates and recognized that they differed from each other. He surmised that repeated catastrophes, such as expansion and contraction of oceans, had extinguished previous assemblages of vertebrates, which were subsequently replaced by new ones. Essential to this view is the idea that successive faunas by definition can never be contemporaneous, hence his insistence on the absence of humans contemporaneous with extinct vertebrates, except in places from which they could have migrated following the last catastrophe. Cuvier's somewhat far-fetched explanation was criticized by Lamarck but only put to rest by Sir Charles Lyell (1830), in the first great text on modern geology, *Principles of Geology*.

Cuvier knew of many fossils claimed to represent human remains contemporaneous with fossil vertebrates. In most cases he recognized them as fraudulent, ridiculous reconstructions, or at the very least really bad identifications. Cuvier analyzed this "evidence" with the same basic principles of comparative and functional anatomy and geology central to paleontology today. For example, Cuvier rejected the antiquity of the Cannstadt human remains from Germany supposedly associated with Pleistocene fauna based on the quality of the "excavation" and loss of the geological contextual data he so highly valued (Cuvier, 1812). For these contributions we owe him much. However, Cuvier was also aware of a femur from Eppelsheim, in Germany, and this is the direct connection between Cuvier and the history of debate on European fossil anthropoids. A brief aside to explore the convoluted history of the Eppelsheim specimen follows, to illustrate the kind of bickering and cavalier systematics that has typified anthropoid evolutionary studies from the very beginning.

Eppelsheim is a late Miocene locality that is famous for complete specimens of the primitive and unusual proboscidean *Deinotherium*, an elephant with downwardly curved lower tusks. The locality is part of a complex of fossil sites called the "*Deinotherium* sands" that were already well known by the early 1800s. Cuvier, the leading authority on fossil vertebrates at the time, received a cast of an amazingly complete primate femur recovered there in 1820 by Schleichermacher (Pohlig, 1895; von Koenigswald, 1982). Schleichermacher suggested to Cuvier that the specimen was of a 12-year-old girl associated with *Deinotherium* and other fossil vertebrates (Pohlig, 1895). Le Grand Cuvier (as he is fondly remembered in Europe), did not comment on the specimen, despite repeated correspondence from Schleichermacher and his successor Johann Jakob Kaup (J. L. Franzen, pers. comm.; Hessisches Landesmuseum Archives, Darmstadt). The reasons for Cuvier's silence are lost to history, but he may have doubted the legitimacy of the find. The femur looks nothing like that of a human child at any age, and Cuvier surely would have known that it belonged to an adult anthropoid primate. Shortly after Cuvier's untimely death in 1832 Kaup tried to recover the cast, but was told that it no longer existed in the Paris collections.

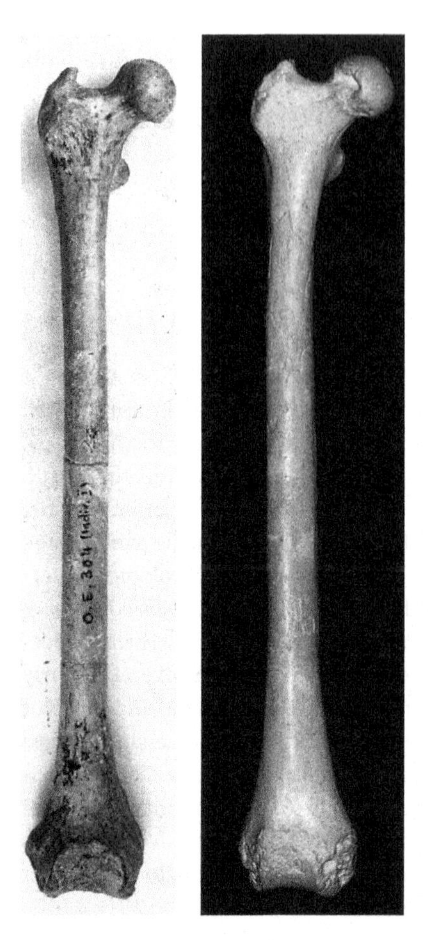

Fig. 20.1 The Eppelsheim femur (left) compared to a femur of *Epipliopithecus*, a pliopithecid from Slovakia, reduced to the same length.

In 1854 Kaup published the first volume of his famous series on the fossil mammals of the "*Deinotherium* sands" at Eppelsheim, and in 1861 (vol. 5) he finally described the Eppelsheim femur in detail (Fig. 20.1). In addition to two beautiful figures of the specimen, he includes a lengthy quote from the great English comparative anatomist and vertebrate paleontologist Richard Owen, a personal friend of the late Cuvier. Owen recognized the Eppelsheim femur as a fossil with similarities to gibbons, an accurate assessment given knowledge of comparative primate postcranial anatomy at the time. In Kaup (1861) Owen is quoted as referring the specimen to *Hylobates*. Another lengthy quote, from Lartet (see below), agrees in general with Owen, and even suggests that the Eppelsheim specimen may come from the same species as Lartet's newly described fossil ape *Dryopithecus fontani*. Hybridizing these ideas, Kaup referred the specimen to *Hylobates fontani* Owen. Hans Pohlig (1892) attributed it to *Dryopithecus*, but later changed his mind, citing characters that were supposedly advanced or more human-like than suggested by the more primitive looking jaws of *Dryopithecus*. So, comparing a femur to a mandible, Pohlig (1895) created the new nomen *Paidopithex rhenanus* for the Eppelsheim femur. Eugène Dubois, of *Pithecanthropus erectus* fame, in the very next paper of the same volume of the same journal in which Pohlig's publication appears, renames Pohlig's taxon *Pliohylobates eppelsheimensis*, citing strong affinities to gibbons, as noted by Owen, and errors of anatomy and interpretation on the part of Pohlig (Dubois, 1895). The argument does not stop there. In 1901 Schlosser published an important review of the evidence of fossil apes in Germany and gave *Pliohylobates* as a synonym of *Paidopithex* citing the principle of page priority. He then subsumed the genus name *Paidopithex* into *Dryopithecus*, following Pohlig's original argument, and retained the trivial name, thus establishing the taxon *Dryopithecus rhenanus* (Schlosser, 1901). Dubois wrote an emotional response, very critical of Schlosser, citing some relatively minor errors such as Schlosser's misspelling of *Paidopithex* (Dubois, 1901). Unfortunately, similar accounts can be given of the history of most other European hominoid taxa.

The first recognized "hominoids"

The first researcher to recognize in a scientific publication that a fossil, contemporaneous with obviously extinct species, represented an extinct hominoid was Édouard Lartet, just five short years after Cuvier's death. Naturally the history of this taxon is as confusing as any. In 1837 Lartet (1837*a*) announced the discovery of a "singe fossile", or fossil anthropoid, in the rich collection he was accumulating from

Sansan (Gers, France). Lartet (1837*b*) very quickly recognized affinities to gibbons in the mandible. Blainville (1840) also recognized that Lartet had a fossil ape, but was perhaps somewhat bolder than the more junior Lartet, and provided the first nomen for this specimen, *Pithecus antiquus*, or "ancient ape". The nomen *Pithecus* had been widely used by this time to designate so many different types of fossil and living primates that Gervais (1849*a*) rightly reassigned the Sansan fossils to the genus *Pliopithecus antiquus* Blainville 1840 (though *Pithecus* has since been suppressed, Blainville still gets the credit for naming the taxon). Lartet's discovery, the first of his two major non-human fossil primate discoveries, was arguably among the most important in the history of paleoanthropology. The discovery of *Pliopithecus* was very timely, showing convincingly that primates were also to be found in fossil faunas. Blainville in fact devoted a number of papers to impart the significance of this discovery to the scientific (and funding) communities (e.g., Blainville, 1837*a*, 1837*b*, 1838, 1840). Soon after the recognition of *Pithecus antiquus* (*Pliopithecus*) a flood of fossil primates would be recognized and named.

Once again it turns out that this fossil, of tremendous significance to the history of the study of fossil hominoids, is not a hominoid at all. For many years *Pliopithecus* and related taxa were thought to be fossil gibbons (see Chapter 15). Nearly all authorities in the nineteenth century concurred, and this view persisted well into the twentieth century, but new and more complete specimens show that *Pliopithecus* is not related to gibbons. It is quite clear from the morphology of highly informative anatomical areas, such as the ear region, the jaw joint, the elbow and the vertebral column, that pliopithecoids are not apes in the evolutionary sense of the term (Fleagle & Kay, 1983), even if they continue to be loosely referred to this group (e.g., Fleagle, 1999).

The early period (1856–1914)

As noted above, Lartet is remembered for two major fossil primate discoveries. In 1856 he published an analysis of fossils from the southern French site of St. Gaudens (Lartet, 1856). He recognized the fossils, brought to him by a local collector, Monsieur Fontan, as those of a great ape, which he called *Dryopithecus fontani*. The specimens consist of a humeral shaft with both ends missing and a juvenile mandible in three parts. Lartet was struck by their modern appearance, and he recognized arboreal features on the humerus.

Lartet noted strong affinities to living apes, particularly chimpanzees, and even some similarities to humans. The link between African apes and *Dryopithecus* eventually would fall out of favor, especially with the discovery of large samples of fossil apes from Africa, but the idea has been resurrected recently in light of new fossil evidence.

The next major discovery of a fossil ape in Europe took place in Tuscany. Fossils found around 1870 from a number of lignite (soft coal) localities in this region were examined by the French paleontologist Paul Gervais, who recognized them as hominoid. Gervais (1872) called this ape *Oreopithecus*, and while noting similarities to both gorillas and macaques that suggested to him a transitional form between the two, he nevertheless considered it to be hominoid. Schlosser (1887), in one of the first influential reviews of fossil primates, considered the fossils to belong to Old World monkeys, thus sparking a debate between the "hominoidists" and the "cercopithecoidists" that was not resolved until fairly recently (Harrison & Rook, 1997). Giuseppe Ristori (1890), who described a larger collection of upper and lower dentitions, assimilated a bit of both views, citing general agreement with Schlosser while suggesting that *Oreopithecus* is ancestral to hominoids (Ristori, 1890; Delson, 1986). Some current views of *Oreopithecus* as near the base of the modern hominoid radiation (Begun *et al.*, 1997*b*; Harrison & Rook, 1997) recall his conclusion. New discoveries sparked continued discussion of *Oreopithecus*, but it was not until the dedicated work of Johannes Hürzeler that intensive attention would be once again focused on this fossil primate (Hürzeler, 1949). At that time opinion ranged from the view that it was most closely related to humans (Hürzeler, 1956), that it was a primitive cercopithecid monkey (Robinson, 1956), or even a pig (Gregory, 1951).

The discovery of a new and more complete mandible from St. Gaudens, and a renewed awareness of isolated teeth from German localities in the Schwäbische Alb (Baden Württemberg), along with the "rediscovery" of the Eppelsheim femur, led to a flurry of new and important papers on fossil hominoids from Europe between 1887 and the beginning of the twentieth century. The first of these was Schlosser's (1887) monograph, in which the German specimens (initially published in monographs dating back to Jaeger, 1850) are revealed to a larger audience. Another influential work entitled simply "Le Dryopithèque" was published by Lartet's successor Albert Gaudry. It included much more detailed analysis of the original material described by Lartet and a new mandible from the site (Gaudry, 1890). Gaudry, who had earlier agreed with Lartet, modified his views to see *Dryopithecus* as a more primitive hominoid (possibly influenced by the "Hylobationists" cited above). Gaudry, best remembered for his work at the classic late Miocene locality of Pikermi in southern Greece (Gaudry, 1867), was a Darwin-era paleontologist, and is credited with making "transformationism" or evolution, fashionable in paleontology (Arambourg, 1937). His work departed from the more traditional, descriptive and taxonomic work of the day towards investigating patterns and processes of change. Fossil hominoid research in Europe at this time benefited from the attention of these eminent and respected researchers.

The last important paper from the nineteenth century is the review by Branco (1898), in which most of the specimens are figured together for the first time. Branco (1898) relied heavily on Schlosser and Gaudry, as well as other contemporary researchers, particularly Dubois. Following

Branco's review, Schlosser (1901, 1902) focused his study on European hominoids and recognized a new type of hominoid in Europe, *Dryopithecus brancoi* (originally his *Anthropodus brancoi*). Schlosser and others named additional species of *Dryopithecus*, but only Lartet's *D. fontani* and Schlosser's *D. brancoi* from this period have withstood the test of time.

As with the case of *Oreopithecus*, the intellectual history of *Dryopithecus* entered a period of relative quiet following the flurry of activity around the turn of the century. Attention to hominoid evolutionary history soon turned away from Europe east towards the rich deposits of the Siwalik Hills of India and Pakistan, and south to equally rich sites in the vicinity of Lake Victoria in Kenya.

One last discovery rounds out the first phase of discovery and debate in European fossil hominoid studies. In 1902 Othenio Abel published two fossil hominoid teeth from the Vienna Basin and also critically reviewed Schlosser (1901). Abel (1902) named the taxa *Griphopithecus suessi* and *Dryopithecus darwini*, each based on a single tooth. As noted by Steininger (1967), the left lower third molar that is the type of *Dryopithecus darwini* was first figured by Meyer in 1845, who took it to be from a plant-eating cetacean (whale). The type of *Griphopithecus suessi* is a heavily worn left upper last deciduous molar or dP^4. Two additional specimens later turned up: a worn upper left molar, probably an M^2, found in the Vienna Museum collections (Glässner, 1931), and a right M_3 from a private collection thought to come from the same site (Steininger, 1967). The only postcrania attributed to *Griphopithecus* and published in detail are from Klein Hadersdorf, Austria (Ehrensberg, 1938; Begun, 1992*a*). These are consistent with a large-bodied, above-branch arboreal quadruped, lacking the suspensory specializations of hominids and hylobatids. While this motley collection of teeth leaves much to be desired in terms of type specimens, they are probably from the same genus as the larger and more complete collection of fossils from Paşalar and Çandır, Turkey. But the names proposed for the isolated teeth are valid and must be used for the more informative samples so long as it cannot be demonstrated that they belong to different taxa. Thus, although based on relatively uninformative teeth from the Vienna Basin, *Griphopithecus* is mostly known from Turkey, and is discussed in another chapter of this volume (Kelley).

The middle years (1914–65)

The geographic range of European Miocene hominoids enlarged considerably during this time, with discoveries in Spain (Vidal, 1913) and Georgia (Burchak-Abramovich & Gabashvili, 1950). More postcranial fossils were described from Austria (Ehrensberg, 1938). William King Gregory, the enormously influential American vertebrate paleontologist of this time, wrote with Milo Hellman a still commonly cited review of fossil ape teeth. In it they discuss the importance of the "dryopithecus" lower molar cusp pattern that has come to be called "Y-5" for the pattern of fissures and number of cusps, for understanding hominoid evolutionary history (Gregory & Hellman, 1926). However, most of their conclusions actually are based on specimens now attributed to the South Asian hominoid *Sivapithecus*. Indeed, the Siwalik Hills fauna received considerably more attention from 1930 onwards than did European fossil faunas with regard to research on hominoid evolution. Of course, political instability and eventual global warfare seriously disrupted paleontology in Europe, which would take a generation to recover.

At about the same time fossils were being discovered in Africa, and a new genus of fossil ape, *Proconsul africanus*, was named in Hopwood (1933*b*). By the 1950s large samples had been collected from a number of localities in Kenya and were described in one of the most important monographs ever written on fossil hominoids, Le Gros Clark and Leakey's seminal 1951 work "The Miocene Hominoidea of East Africa." Le Gros Clark & Leakey (1951) suggested an African ancestry of great apes, with Europe as a side-stage, in keeping with Darwin's justified and probably correct pre-fossil-record speculation that the last common ancestor of African apes and humans would be found in Africa (Darwin, 1871).

The modern era (1958–2000)

There is some overlap between the middle and final phases of European hominoid evolutionary studies prior to two highly significant events, the discovery of a nearly complete skeleton of *Oreopithecus* (Hürzeler, 1960) and the seminal review by Simons & Pilbeam (1965). Back in Europe, Spanish paleontologist Miquel Crusafont-Pairo, one of the founders of the modern synthesis of European Neogene biostratigraphy, recovered and described numerous fossil hominoid jaw fragments and isolated teeth from a series of localities in northeastern Spain (Villalta & Crusafont, 1944; Crusafont, 1958; Crusafont & Hürzeler, 1961). Two of the species of *Dryopithecus* from these samples are among the few nomina that persist in the literature today, and one, *Dryopithecus laietanus*, has become central in discussions of hominoid evolution (see below). In 1954 the Swiss paleontologist Johannes Hürzeler hypothesized that *Oreopithecus* was a member of the human family (Hominidae in the sense of the term then) because of canine and premolar morphology, among other things (Hürzeler, 1951, 1954*b*, 1956, 1958). This conclusion sparked widespread reaction, much of it critical, but mostly from European colleagues of Hürzeler (e.g., Vallois, 1954; Remane, 1955; von Koenigswald, 1955; Viret, 1955; Meléndez, 1957; Butler & Mills, 1959; but see also Gregory, 1951 and Straus, 1957). However, the provincial appeal of *Oreopithecus* suddenly broadened on August 2, 1958, when Hürzeler and coal-miners from Baccinello, Italy, extracted a slab of lignite containing a skeleton of *Oreopithecus bambolii* (Delson, 1986). Before the year would end papers would be published in German, English, French and even Japanese on the discovery and its significance.

The skeleton offered unprecedented evidence of Miocene hominoid anatomy from the face, dentition, brain case, vertebral column and limbs, and yet, rather than clearing things up, the disagreements and controversies were stronger than ever. Disillusioned, Hürzeler was very slow in describing the skeleton and additional remains of *Oreopithecus* he had accumulated (e.g., Hürzeler, 1968). Decades later the sample is still being described (Delson, 1986; Harrison, 1986*a*; Szalay & Langdon, 1986; Jungers, 1987; Sarmiento, 1987; Harrison & Rook, 1997). At least in the modern era most researchers are satisfied that *Oreopithecus* is a hominoid and probably a great ape.

In the very important review of fossil apes by Simons & Pilbeam (1965) the European sample of "*Dryopithecus*", in their sense, was essentially reduced to a poorer version of that from South Asia and Africa. One or more of the smaller species of *Dryopithecus* was thought to be ancestral to chimpanzees, and the largest species was thought to be ancestral to gorillas (Simons & Pilbeam, 1965). Today it is widely accepted that specimens attributed to the seven species of the genus *Dryopithecus* in Simons & Pilbeam (1965) actually represent at least 19 species of 10 to 12 genera, including *Dryopithecus*, *Ouranopithecus* (and/or *Graecopithecus*; see below), *Ankarapithecus*, *Gigantopithecus*, *Sivapithecus*, *Griphopithecus*, *Proconsul*, *Lufengpithecus* and *Kenyapithecus*, and one or more pliopithecoids, *Paidopithex* and *Anapithecus*. The Simons & Pilbeam (1965) review illuminated the range of variation to be expected in fossil ape taxa, and eliminated an unbelievable number of superfluous and inappropriate fossil ape names. That they went a bit too far in cutting down the number of taxa reflects the poor quality of the fossil record at that time. A greater awareness of diversity and complexity in the fossil record comes largely from discoveries made since 1965.

In 1966 Gabor Hernyák, a chief mining geologist working near Rudabánya, in northern central Hungary, showed Miklos Kretzoi, Hungary's leading vertebrate paleontologist at the time, one of the fossils he had been collecting from unstable slopes of Neogene overburden. Kretzoi recognized the specimen as a hominoid mandible and published it as *Rudapithecus hungaricus* (Kretzoi, 1969). Kretzoi began work at Rudabánya and recovered numerous specimens of fossil apes that he eventually attributed to four genera, *Rudapithecus*, *Bodvapithecus*, *Pliopithecus* (*Anapithecus*) and *Ataxopithecus* (Kretzoi, 1969, 1975, 1984). In a much-cited paper in the journal *Nature*, Kretzoi interpreted *Rudapithecus* to be aligned with *Ramapithecus*, a genus of South Asian ape thought by most researchers for many years to be directly related to humans (Lewis, 1934; see Kelley, this volume, for a thorough review of this genus). Kretzoi published numerous papers on *Rudapithecus* and the earliest phases of "hominization", mostly in Hungarian and German, with the result that their impact, while significant, for the most part was confined to central Europe. After a hiatus of some years more controlled excavations were undertaken in several stages by László Kordos, director of the Geological Museum of Hungary. During this time more complete cranial and postcranial specimens were found, including two reasonably complete brain cases (Kordos, 1987, 1988; Kordos & Begun, 1997). Today Rudabánya is the richest hominoid locality in Europe. It contains nearly 300 fossils attributed to two primates, *Anapithecus* and *Dryopithecus*. Along with recent discoveries from Can Llobateres, Spain (Moyà-Solà & Köhler, 1995, 1996), the new finds at Rudabánya have sparked what one publication called "a renaissance of Europe's ape" (Martin & Andrews, 1993*a*). Current debate about *Dryopithecus* is concentrated on differing interpretations of the fossils from these two localities.

In 1972 a badly preserved mandible with a few severely damaged teeth was described as a new genus by von Koenigswald (1972). This is the only primate from the late Miocene site of Pygros in southern Greece, allegedly found during excavations for a swimming-pool during the Second World War (Freyberg, 1949, 1950; von Koenigswald, 1972). Von Koenigswald recognized the hominoid affinities of this specimen, at first attributed to *Mesopithecus* (a fossil monkey) by Dietrich (cited in Freyberg, 1949), and called it *Graecopithecus freybergi*. Shortly thereafter Louis de Bonis and colleagues described better-preserved specimens from Macedonia that they at first attributed to *Dryopithecus*, reflecting the decade-old influence of Simons and Pilbeam, but later changed to a new genus, *Ouranopithecus macedoniensis* (de Bonis, *et al.*, 1975; de Bonis & Melentis, 1977). Since then two issues continue to be debated. One is the dry but important question of the proper name for the taxon. Is it *Graecopithecus* or *Ouranopithecus*? The second, more interesting, debate concerns the evolutionary relations of *Ouranopithecus*. From the beginning, de Bonis and colleagues stressed the Pliocene hominid-like morphology of the jaws and teeth of *Ouranopithecus*, and consistently have argued that this primate is directly related to humans (see below). Others have seen these similarities as evolutionary parallelisms (Martin & Andrews, 1984; Begun & Kordos, 1997).

Finally, in 1957 a new type of fossil hominoid was described from Turkey (Ozansoy, 1957, 1965). The specimen, a partial mandible with most of the teeth but only parts of the symphysis preserved, showed affinities with *Sivapithecus* but with differences sufficient to warrant a new name, *Ankarapithecus* (Ozansoy, 1965). In the wake of the Simons & Pilbeam (1965) review this genus name received little support. New research, however, supports the original distinction. Re-examination and restoration of a second specimen described originally by Andrews & Tekkaya (1980) show the Turkish specimens to differ significantly from *Sivapithecus*, and raise new questions about the evolution of great apes in Asia (Begun & Güleç, 1998). More recent discoveries confirm this interpretation, and have led to other interpretations, to be discussed below (Alpagut *et al.*, 1996).

Taxonomy

Systematic framework

Hominoid higher-level systematics are as controversial as the more detailed taxonomy, due in part to concern over how to define Hominidae. Many anthropologists prefer the restricted definition that includes only humans and our immediate fossil relatives. Other hominoids are placed in two additional families, the Hylobatidae for gibbons and siamangs, and the Pongidae for great apes. This use stabilizes hominoid taxonomy but conflicts with universal practice in modern systematics of classifying according to evolutionary relations. Stability at the price of inaccuracy and misinformation is no bargain. Thus, in this chapter the systematic framework is based on the current consensus or near-consensus regarding hominoid higher-level relationships.

Order Primates Linnaeus, 1758
 Infraorder Catarrhini É. Geoffroy Saint-Hilaire, 1812
  Superfamily Hominoidea Gray, 1825
   Family Griphopithecidae (new rank)
    aff. *Griphopithecus*
    Subfamily Griphopithecinae (new rank)
     Genus *Griphopithecus* Abel, 1902
      Griphopithecus darwini Abel, 1902
      Griphopithecus alpani Tekkaya, 1974
      Griphopithecus africanus (new combination)
    Subfamily Kenyapithecinae Leakey, 1962
     Genus *Kenyapithecus* Leakey, 1962
   Family Hominidae Gray, 1825
    Subfamily Homininae Gray, 1825
     Tribe Dryopithecini Gregory & Hellman, 1939
      Genus *Dryopithecus* Lartet, 1856
       Dryopithecus fontani Lartet, 1856
       Dryopithecus brancoi Schlosser, 1901
       Dryopithecus laietanus Villalta & Crusafont, 1944
       Dryopithecus crusafonti Begun, 1992
      Genus *Ouranopithecus* Bonis & Melentis, 1977
       Ouranopithecus macedoniensis Bonis & Melentis, 1977
     Tribe Hominini Gray, 1825
      Genus *Gorilla* Geoffroy Saint-Hilaire, 1852
      Genus *Pan* Oken, 1816
      Genus *Homo* Linnaeus, 1758
    Subfamily Ponginae Elliot, 1913
      Genus *Pongo* Lacépède, 1799
      Genus *Sivapithecus* Pilgrim, 1910
      Genus *Ankarapithecus* Ozansoy, 1965
       Ankarapithecus meteai Ozansoy, 1965
    Subfamily Oreopithecinae Schwalbe, 1915
      Genus *Oreopithecus* Gervais, 1872
       Oreopithecus bambolii Gervais, 1872
    Subfamily *incertae sedis*
      Genus *Graecopithecus* von Koenigswald, 1972
       Graecopithecus freybergi von Koenigswald, 1972
   Family Hylobatidae Gray, 1877
    Genus *Hylobates* Illiger, 1811

INFRAORDER CATARRHINI

Superfamily Hominoidea

Family Griphopithecidae

GENUS *Griphopithecus* Abel, 1902
INCLUDED SPECIES *G. africanus, G. alpani, G. darwini*, aff. *Griphopithecus*

SPECIES *Griphopithecus darwini* Abel, 1902
TYPE SPECIMEN A single left M_3 (Abel, 1902: fig. 3) from Děvínská Nová Ves, in the collections of the Naturhistorische Museum, Wien, Austria
AGE AND GEOGRAPHIC RANGE The Sandberg locality at Děvínská Nová Ves was traditionally attributed to MN 6 but Mein (1986) suggests that it has a mixed fauna, with a few taxa such as the early pig *Bunolistriodon* reworked from older sediments. He thus dates the Sandberg locality to MN 7/8, a view also shared by van der Made (1999), whereas most other authors attribute it to late MN 6 (e.g., Andrews *et al.*, 1996; Rögl, 1999). Mein (1999) most recently has shifted his view to agree with the older age estimate based mainly on the geological arguments and micromammal evidence. Steininger (1999) correlates MN 6 to 13.5 to 15 Ma, and the locality would be toward the top of this range. Two postcranial specimens tentatively identified as *Griphopithecus darwini* come from Klein Hadersdorf, a locality in Austria roughly 60 km northwest of Děvínská Nová Ves in the same type of marine sand formation found at the type locality. Indications from the poor fauna of Klein Hadersdorf are consistent with a late MN 6 age (Steininger, 1986).
ANATOMICAL DEFINITION
Griphopithecus darwini is based on single lower molar, with three other isolated teeth and two fragmentary pieces of postcrania referred to it. As such, a definition of the species is necessarily short and relatively uninformative. Although the type species of the genus is *Griphopithecus darwini*, the samples attributed to *G. alpani* from Paşalar and Çandır in Turkey are much more complete and informative. A more comprehensive definition of the genus should be based on these samples (see Kelley, this volume). *Griphopithecus darwini* is a large-bodied hominoid dentally the size of a chimpanzee. The M_3 is tapered distally but the distal cusps are large and well formed, unlike many hominoids that have reduced last molars with small distal cusps. It has a well-developed buccal cingulum, broad, rounded cusps and a low crown. The other teeth from the type locality are

consistent with this morphology and almost certainly come from the same species. The dental morphology suggests thickly enameled molars, as is known to be the case for other *Griphopithecus* (Alpagut *et al.*, 1990). The humerus from Klein Hadersdorf is a portion of the shaft without the articular ends preserved. It is straight to slightly bent forwards (anteroflexed, or convex anteriorly and concave posteriorly) deep anteroposterioly compared to mediolateral breadth and has strong muscle attachment scars. The ulna from Klein Hadersdorf, which may be from the same individual, is more complete and has part of the proximal articular end and most of the shaft. It also has a deep shaft and strong muscle scars, and was probably short compared to the humerus length. The ulna has a narrow trochlear notch without the keel typical of modern hominoids and a more proximally projecting olecranon process than seen in living apes (see p. 362). These characters have been interpreted to indicate an arboreal adaptation involving powerful climbing capabilities and movement above branches, as opposed to suspensory positional behaviors (below-branch) typical of late Miocene and recent non-human hominoids (Begun, 1992*a*).

SPECIES aff. *Griphopithecus* Abel 1902
AGE AND GEOGRAPHIC RANGE MN 5 (European Land Mammal Zone) approximately 16.5–17 Ma (Heizmann, 1992; Heizmann, *et al.*, 1996), from Engelswies (Germany); this occurrence is included here because it may be the oldest evidence of hominoids anywhere outside of Africa
ANATOMICAL DEFINITION
The sole specimen is a worn right M^3 fragment that is diagnostically hominoid, with affinities to *Griphopithecus* (but not formally assigned). It has thick enamel, low dentine penetrance, shallow occlusal basins and broad, low cusps.

Family Hominidae

Subfamily Homininae

Tribe Dryopithecini

GENUS *Dryopithecus* Lartet, 1856
All species are known from partial upper and lower dentitions and share hominid characters such as robust incisors, compressed canines, mesiodistally elongated premolars and molars, and molars with smooth lingual (upper) or buccal (lower) surfaces (i.e., lacking cingula), P_4 with talonids close in height to the trigonids, and M^1 close in size but smaller than M^2. Ranges from 15 to 45 kg, with relatively strong sexual dimorphism varying among the species. Hominid characters known in all but *Dryopithecus fontani* include reduced P^3 cusp heteromorphy and a high root of the maxillary zygomatic process. All species share uniquely derived characters for the genus including narrow, tall-crowned upper central incisors and thinly

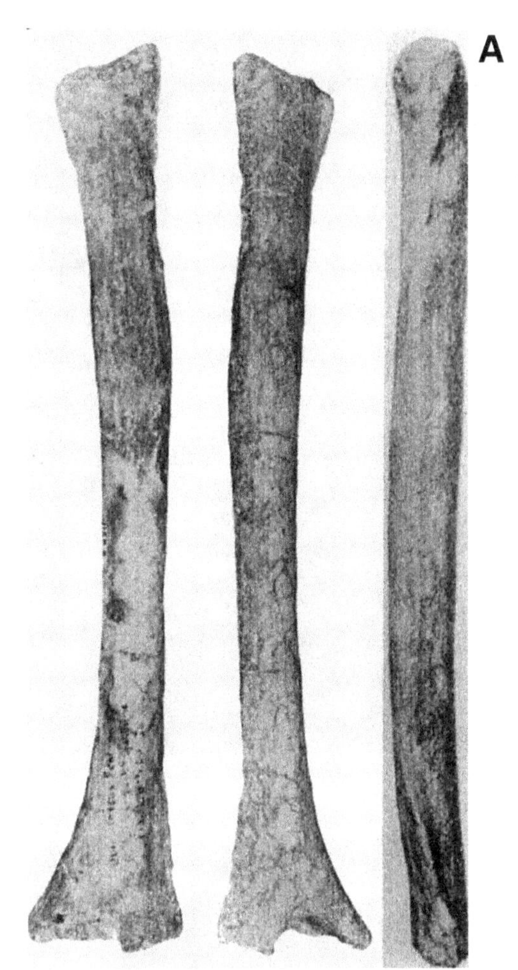

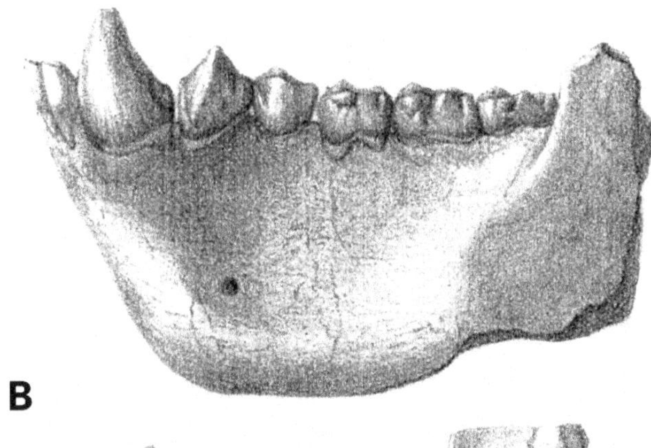

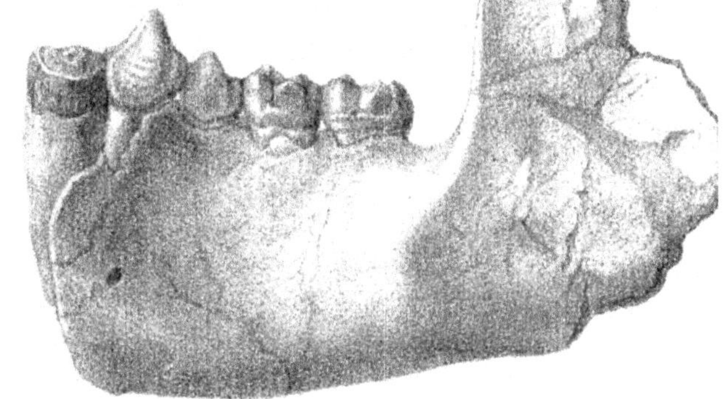

Fig. 20.2 *Dryopithecus fontani*. The humerus (A) and mandible (B) described by Lartet in 1856.

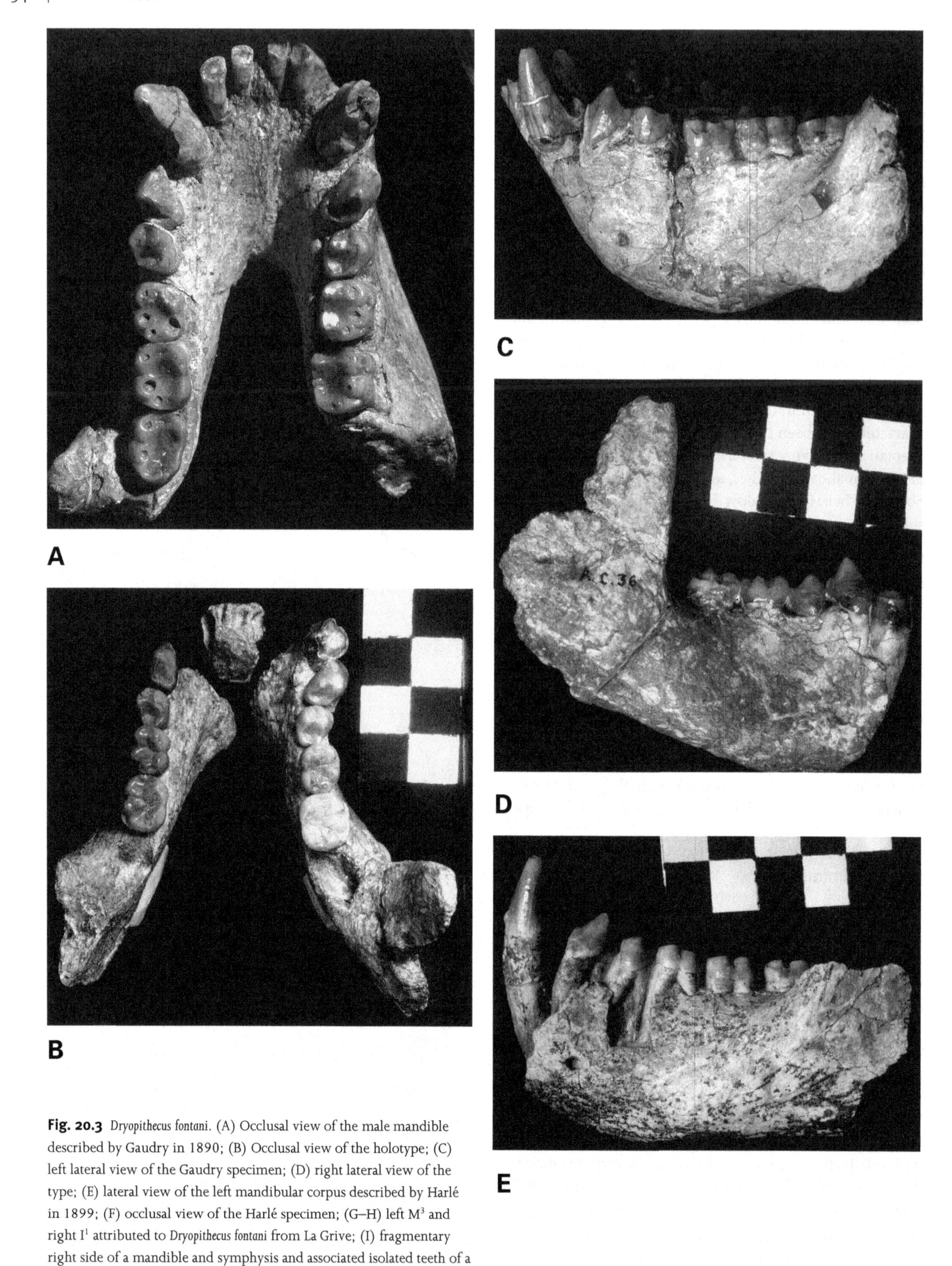

Fig. 20.3 *Dryopithecus fontani*. (A) Occlusal view of the male mandible described by Gaudry in 1890; (B) Occlusal view of the holotype; (C) left lateral view of the Gaudry specimen; (D) right lateral view of the type; (E) lateral view of the left mandibular corpus described by Harlé in 1899; (F) occlusal view of the Harlé specimen; (G–H) left M^3 and right I^1 attributed to *Dryopithecus fontani* from La Grive; (I) fragmentary right side of a mandible and symphysis and associated isolated teeth of a female *Dryopithecus fontani* from St. Stefan.

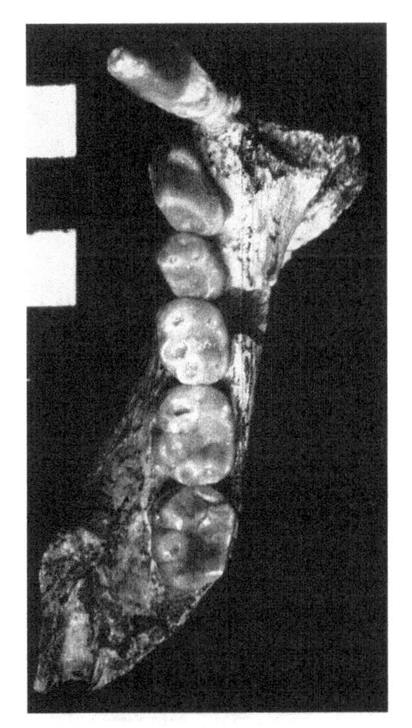

F

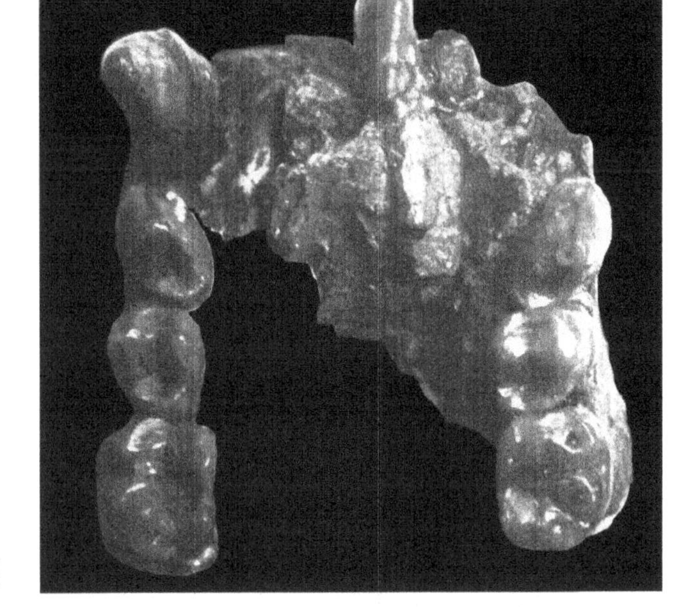

I

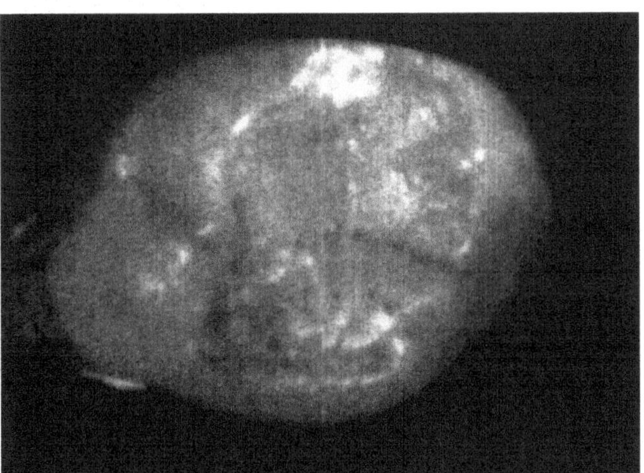

G

H

enameled molars with high dentine penetrance. Other uniquely derived dental characters (except in *D. crusafonti*) include narrow, tall-crowned lower central and lateral incisors, and strongly buccolingually compressed lower canines. Other traits known in all species of *Dryopithecus* include robust male mandibles, generally more robust than in early Miocene hominoids but less than in thickly enameled hominoids (e.g., *Griphopithecus*), molar cusps positioned at the peripheries of the crown, surrounding relatively broad, shallow basins. Most species are also known from cranial and postcranial characters shared exclusively with African apes alone or African apes and humans among living taxa (see species descriptions).

INCLUDED SPECIES *D. brancoi, D. crusafonti, D. fontani, D. laietanus*

SPECIES *Dryopithecus fontani* Lartet, 1856 (Figs. 20.2, 20.3)

TYPE SPECIMEN MNHN–AC 36 (from St. Gaudens, Haute Garonne, France), a subadult mandible in three fragments, preserving the canine to M_2 on both sides, and a fragment of the symphysis and now in the collections of the Musèum National d'Histoire Naturelle, Paris.

AGE AND GEOGRAPHIC RANGE From two localities in France (St. Gaudens and La Grive-St.-Alban M) and one in Austria (St. Stefan), and dated to the upper part of MN 7/8 (Andrews *et al.*, 1996; Mein, 1999), or about 11 to 12 Ma (Steininger, 1999)

ANATOMICAL DEFINITION

Dryopithecus fontani is only known from three fragmentary male mandibles, a fragmentary female mandible, a number of isolated teeth, and a humeral shaft. While all the lower teeth are known from at least one individual, only one $M^{2/3}$ and one I^1 are known from the upper dentition. The type mandible, a young male, and the female specimen from St.

Stefan (Mottl, 1957), are deep (tall corpus relative to corpus breadth), while other male specimens from St. Gaudens are more robust (thicker corpora). Male mandibles decrease in height strongly from the symphysis to the M_3, unlike other species in the genus. The male mandibles are among the largest of *Dryopithecus* and the female mandible and teeth are among the smallest, suggesting substantial sexual dimorphism in body mass. About 50% of the lower molars have partial cingula or buccal notches mesially or between the buccal cusps, and the canines are somewhat less compressed than in other species. The humeral shaft is long, straight and gracile, with a rounded shaft cross-section proximally and a flattened shaft distally. The olecranon fossa is broad and relatively deep. The position of the bicipital groove and the remnants of the greater and lesser tuberosities suggest that the humeral head was twisted medially as in living African apes and humans.

SPECIES *Dryopithecus brancoi* Schlosser, 1901 (Figs. 20.4, 20.5)
TYPE SPECIMEN A left M_3 figured in Branco (1898) now in the collections of the Institut für Geologie und Paläontologie, University of Tübingen
REFERRED SPECIMENS The hypodigm (the fossils on which the diagnosis of the taxon is based) was revised to include more informative specimens, primarily from Rudabánya, Hungary (Begun & Kordos, 1993)
AGE AND GEOGRAPHIC RANGE Known from at least two localities, the type locality of Salmendingen (Baden Württemberg, Germany) and Rudabánya (Hungary). This species may also be present at a number of localities in Germany (Melchingen, Ebingen, Trochtelfingen and Wissberg), Austria (Mariatal) and Georgia (Udabno), though these samples consist of isolated molars that cannot be compared directly to the type specimen. Rudabánya is dated to the upper half of MN 9 (Andrews *et al.*, 1996; Mein, 1999), or about 10 Ma (Steininger, 1999). Salmendingen is less securely dated, but is likely to be either late MN 9 or early MN 10 (approximately 9.5 Ma) (Franzen & Storch, 1999; Ginsburg, 1999; Mein, 1999) and not MN 11, as suggested in Andrews *et al.* (1996) and Andrews & Bernor (1999). Inclusion of the upper premolar and molar from Udabno, Georgia, currently attributed to *Udabnopithecus garedziensis* (Burchak-Abramovich & Gabashvili, 1950) would extend the range to as young as MN 11 (Ginsburg, 1999).

ANATOMICAL DEFINITION

Dryopithecus brancoi is known primarily from the large collection of fossils from Rudabánya, in northern central Hungary (Fig. 20.5). It is close in size to the type species, and similar in dental size to *Pan*. Lower incisors are tall-crowned, narrow and thick (labiolingually), and the male canines are tall and strongly compressed buccolingually. The P_3 usually has a well formed mesiolingual beak. The molars have reduced molar cingula compared to *D. fontani*, and an elongated, tapered M_3. The mandible is tall and gracile in females, shallow and robust in a single juvenile male, with narrow extramolar sulci, lateral eminences opposite M_{1-2}, and symphyses with poorly developed transverse tori. Upper incisors are also tall, narrow, buccolingually thick and heteromorphic. Male upper canines are small relative to M^2, female upper canines are relatively broader compared to *D. laietanus*, and the upper molars also lack cingula and have a simple occlusal morphology with large cusps and shallow basins. The cranium of *D. brancoi* is better known than that of most fossil apes. The maxilla have high alveolar processes, large maxillary sinuses (compared to early Miocene apes and *Oreopithecus*), and a nasal aperture with subvertical sides and broad flat base. The incisive foramen is reduced, the incisive canal is short and the subnasal floor is stepped. The premaxilla is biconvex, thick and elongated. The interorbital region is broad and the orbits squared or rounded. The frontal bone is relatively horizontal and has thick anterior temporal ridges, faint supraorbital ridges and a moderate-sized frontal air sinus extending below the level of nasion. The postorbital constriction is moderate, as in *Pan*. The mandibular fossa is deep transversely with prominent entoglenoid and vertical postglenoid processes. The articular and temporal portions of the temporal bone were probably fused, and the petrous portion of the temporal bone has a shallow subarcuate fossa.

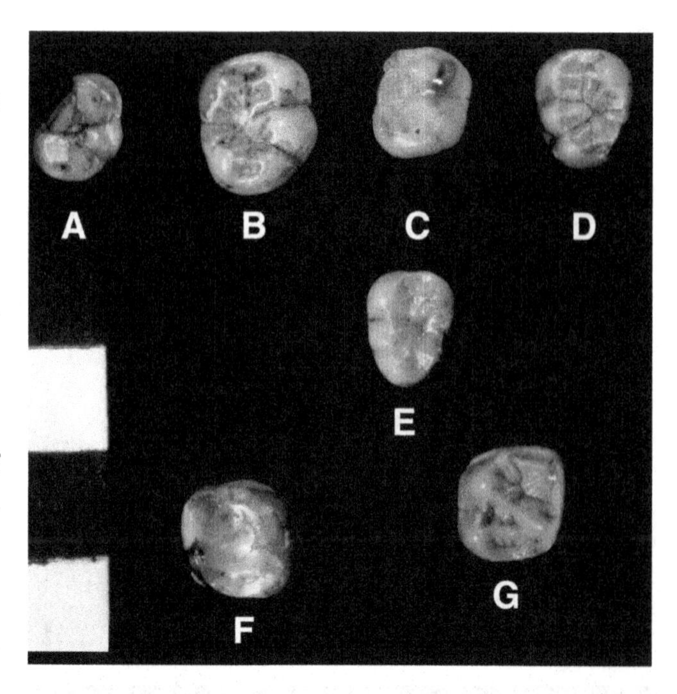

Fig. 20.4 European Miocene hominoid teeth known by the end of the nineteenth century. (A–C) Lower molars from Melchingen; (D) lower molar from Trochtelfingen; (E) lower molar from Salmendingen (the type specimen of *Dryopithecus brancoi*); (F–G) upper molars from Melchingen.

Overall the neurocranium is relatively elongated and the inion is placed at about the level of the superior orbital

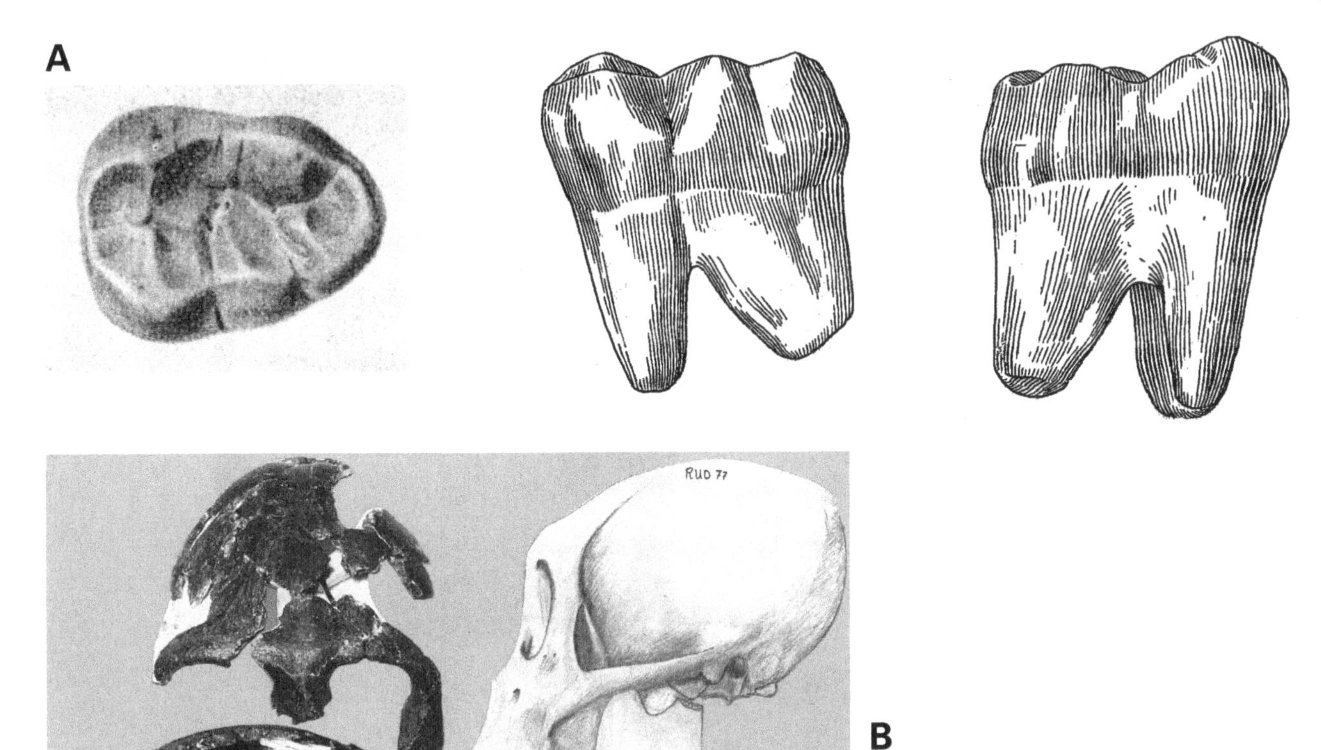

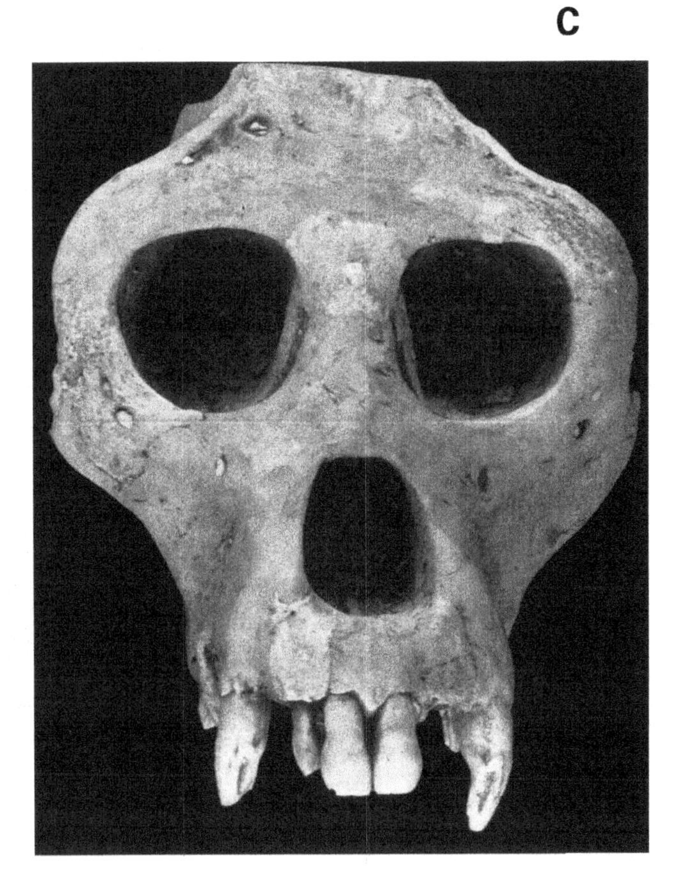

Fig. 20.5 *Dryopithecus brancoi*. (A) The type, an M_3 from Salmendingen, in occlusal, buccal and lingual views; (B) female cranium (RUD 77) on the left in frontal and lateral view, and a preliminary reconstruction of the skull on the right; (C) reconstructed face of a male (cast); (D) (*overleaf*) clockwise from upper right: restored palate of a male (cast), deciduous upper canine and premolars, male and female upper central incisors, occlusal view of RUD 15, lingual and occlusal views of RUD 14/70, juvenile mandible. (E) Postcrania from Rudabánya: to the left are anterior and lateral views of a fragment of proximal ulna. On the right are anterior, distal and posterior views of a distal humerus, from a different individual. The ulna shows characteristic hominoid features such as a median keel, robust shaft and large, flat, laterally facing radial facet. The humerus is also quite modern with a large capitulum and the deeply notched trochlea with prominent lateral trochlear keel and zona conoidea (arrows).

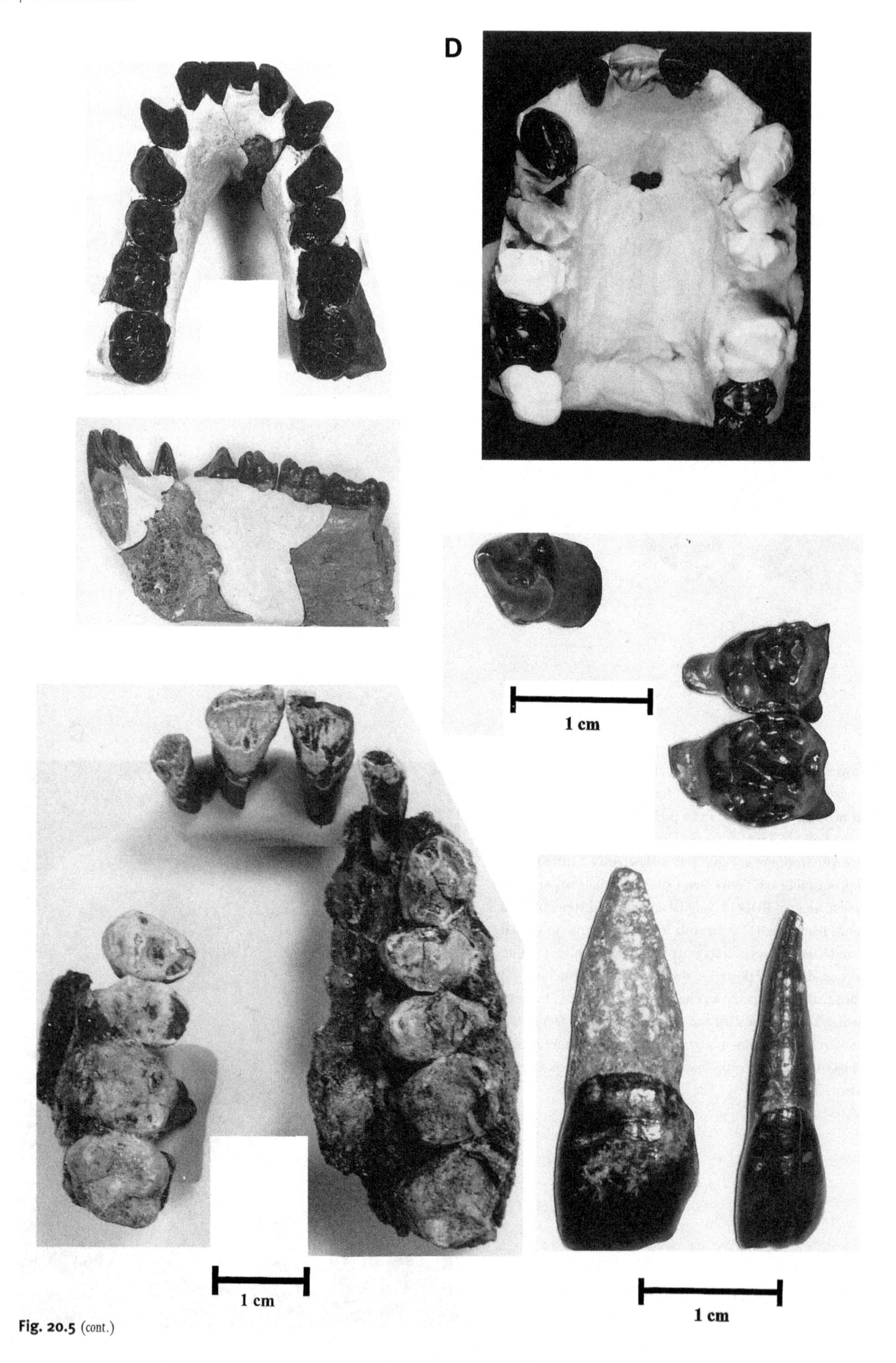

Fig. 20.5 (cont.)

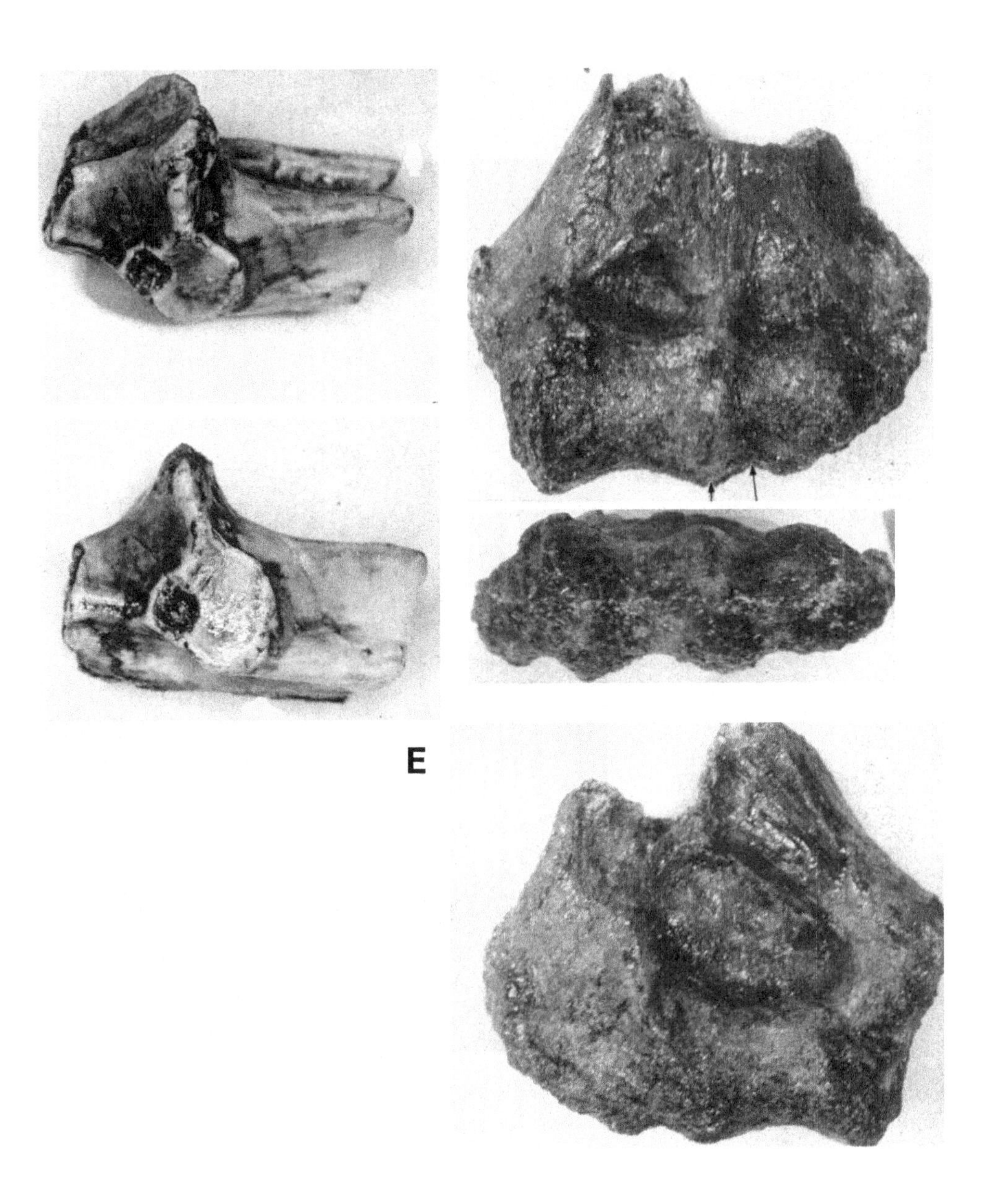

Fig. 20.5 (cont.)

margins in standard lateral view. The face projects moderately beyond the orbits and is downwardly deflected. A distal humerus has a broad, anteroposteriorly flat distal end, well-defined coronoid and radial fossa, prominent, medially-oriented medial epicondyle, a nearly symmetrical trochlea with a strongly developed lateral trochlear keel, a well-defined zona conoidea and large, spherical capitulum. The proximal ulna has a well-defined median keel, broad medial and lateral trochlear articular surfaces, a large, laterally oriented radial facet, a thick, prominent coronoid process and an anteroposteriorly flat proximal shaft. Several of the larger phalanges from Rudabánya, almost certainly attributable to *D. brancoi*, are long and have large articular ends with well-developed secondary shaft characters (curved shafts and strong ligament attachments). A thumb proximal phalanx is robust and a proximal articular surface of a proximal phalanx is oriented palmarly. A partial talus has a broad, low body, a broad but deep trochlea and long neck.

SPECIES *Dryopithecus laietanus* Villalta & Crusafont, 1944 (Figs. 20.6, 20.7)
TYPE SPECIMEN IPS 2, an associated set of lower teeth with P_3–M_3 now at the Institut de Paleontologia Miquel Crusafont, Sabadell, Spain (the definition of the taxon is based largely on the more complete sample from Can Llobateres)
AGE AND GEOGRAPHIC RANGE Roughly contemporaneous with *D. brancoi* but restricted to the Vallés Penedés of northeastern Spain (La Tarumba, Can Llobateres, Polinyá and Can Mata); La Tarumba is dated to MN 10 while the other localities occur in the upper half of MN 9 or span the MN 9–MN 10 boundary, from 9.5 to 10 Ma (Agustí et al., 1996)

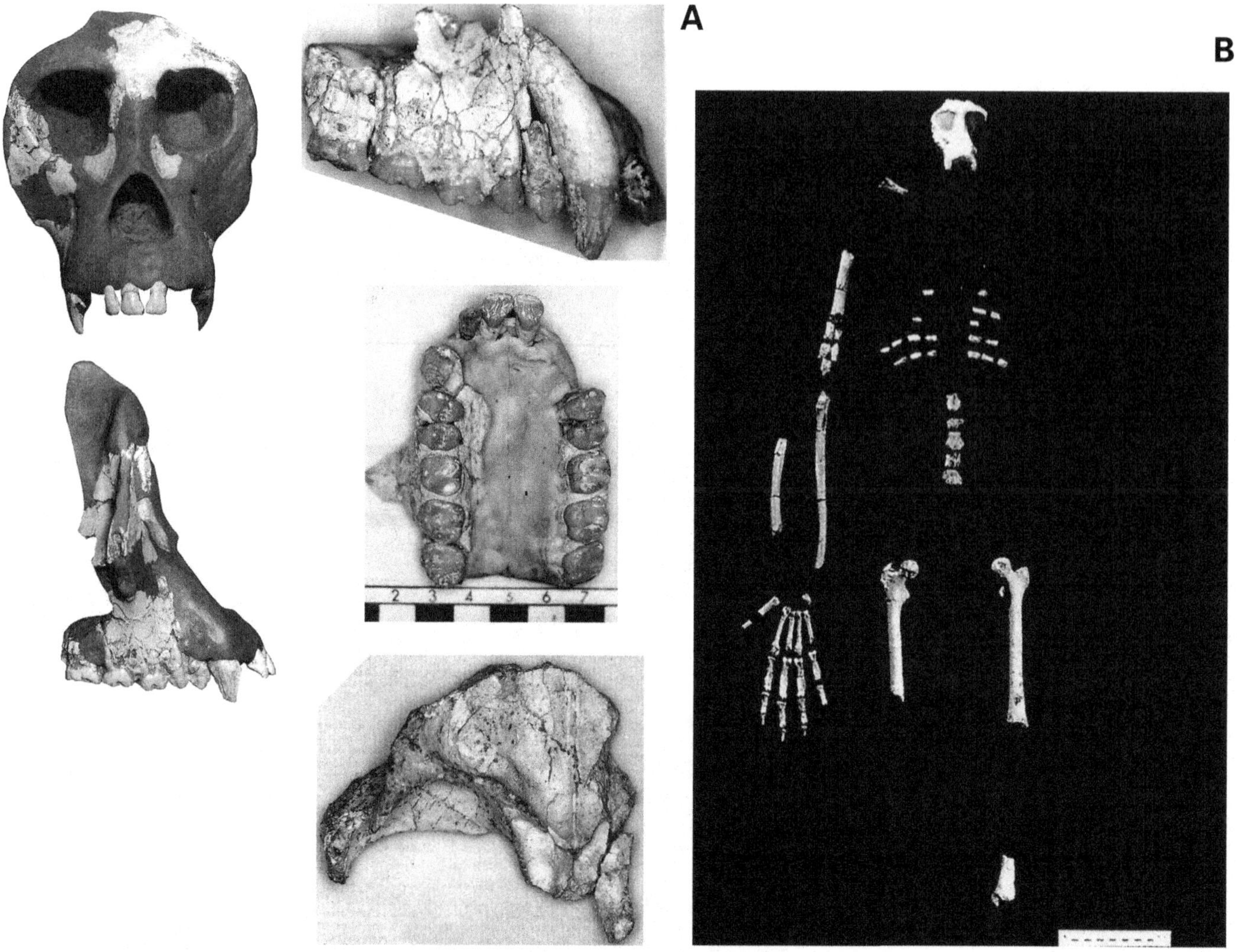

Fig. 20.6 *Dryopithecus laietanus*. (A) Left is a reconstructed male cranium in frontal and lateral views. Right are views of the palate in lateral and palatal view, and the frontal in posterior view, showing the large frontal sinuses. (B) CLl-18800, partial skeleton from Can Llobateres, Spain. The cranium was recovered approximately 50 m from the postcranial specimens and may not belong to the same individual. Photograph courtesy of Salvador Moyà-Solà.

ANATOMICAL DEFINITION

Several jaw fragments, isolated teeth, a partial cranium, and a partial postcranial skeleton indicate that this is the smallest species of the genus – perhaps 15–35 kg (Moyà-Solà & Köhler, 1995, 1996; Köhler *et al.*, 1999). Lower teeth are similar to *D. brancoi* but the premolars are relatively small and the molars have more rounded cusps that tend to fill the edges of the occlusal basins, and the M_3 is less tapered. Upper incisors are morphologically similar to *D. brancoi*. Mandibles tend to be relatively robust (low, thick corpora). In the single specimen with most of an associated upper dentition (CLl 18000) M^3 is the largest tooth. CLl 18000 shares with several specimens from Rudabánya detailed aspects of orbital, periorbital, maxillary and temporal morphology, frontal sinu2ses, supraorbital structures, and a shallow subarcuate fossa. Some differences in the Spanish specimen include a very thick and strongly expressed anterior temporal ridge, a deep and broad depression on the frontal bone superior to glabella, a very high root of the zygomatic process of the maxilla and a flatter zygomatic plane.

A few isolated postcranial specimens including a lunate and a few phalanges were supplemented recently by discovery of a partial skeleton with much of a forelimb including portions of the hand, femora, distal tibia, ribs and perhaps clavicle and vertebral column (Moyà-Solà & Köhler, 1996; Köhler *et al.*, 1999). While these postcrania are said to be associated with each other and with the cranium CLl 18000, individual specimens were found widely dispersed. The cranium and tibia were especially widely separated (author's personal observations at the time of discovery). Given these doubts, characters that depend on certainty of association must be considered cautiously. Proximally the ulna is similar to that from Rudabánya but its shaft is evidence of an absolutely long forearm. If the ulna and femora are associated then they

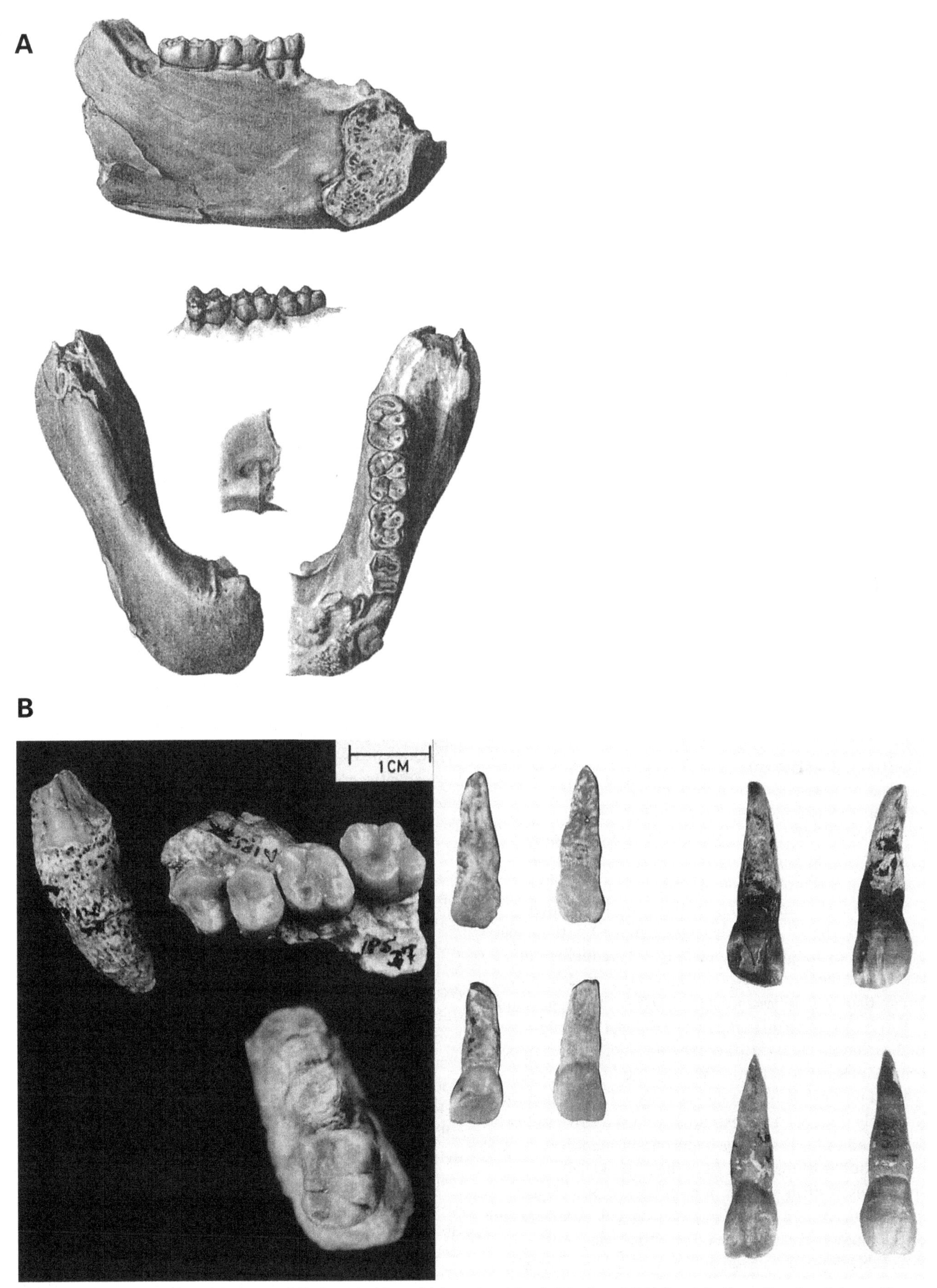

Fig. 20.7 *Dryopithecus crusafonti*. (A) The mandible described by Vidal (1913) and Smith-Woodward (1914), originally attributed to *D. fontani*. Note the massive mandible relative to dental size compared to *D. fontani*. (B) Top left, the type maxilla; bottom left, an M_2 in a mandibular fragment; right, tall, narrow I^1 in labial and lingual views compared to the shorter, broader I^1 of *Dryopithecus laietanus* (center).

provide the first direct evidence of ape-like relatively elongated forelimbs, especially forearms, in *Dryopithecus*, as known for some time in *Oreopithecus* (Hürzeler, 1960). The hands are also extremely elongated relative to the femora, and if they are associated then this is also a fundamental similarity to living great apes. Estimations of intermembral indices yielded results similar to those for chimpanzees and bonobos (Moyà-Solà & Köhler, 1996).

Fragmentary remains of the axial skeleton suggest other fundamental similarities to living great apes, including broad thoraxes and a dorsally positioned scapula (Moyà-Solà & Köhler, 1996). The femora are relatively short and have robust shafts, short thick necks and large, globular heads positioned well above the proximal end of the shaft. Manual phalanges resemble those described for *D. brancoi* (Begun, 1993*a*), but the metacarpals are short relative to them. The lunate is broad and robust with large joint surfaces and the wrist shows reduced stylotriquetral contact (Begun, 1994*a*; Moyà-Solà & Köhler, 1996).

SPECIES *Dryopithecus crusafonti* Begun, 1992 (Fig. 20.7)
TYPE SPECIMEN IPS 1798/1799, a badly preserved left maxilla with P^3–M^2 (1798) and a separate associated left canine fragment (1799) from Can Ponsic, in the collections of the Institut de Paleontologia Miquel Crusafont, Sabadell, Spain
AGE AND GEOGRAPHIC RANGE Confined to MN 9 and known only from the Vallés Penedés locality of Can Ponsic and probably El Firal (Seu d'Urgel) near Lerida, Spain; Can Ponsic and El Firal are both thought to be somewhat earlier in MN 9 than the localities with *D. laietanus*, and may date to about 10.5 Ma (Agustí *et al.*, 1996)
ANATOMICAL DEFINITION
Dryopithecus crusafonti is known from the type and an additional 15 isolated teeth from Can Ponsic, and a mandible from El Firal originally described as *D. fontani* (Vidal, 1913; Smith-Woodward, 1914). Its dental size indicates it was probably slightly larger than *D. laietanus*. Upper premolars and molars are quite similar to other *Dryopithecus*. The M^1 is slightly larger than the M^2, the upper premolars are longer compared to their breadth than in *D. brancoi*, and the upper molars are broader compared to their length than in *D. laietanus* (Begun, 1992*b*). The upper canine is also relatively broad compared to its length, though not to the extent seen in early Miocene forms. The three upper central incisors are extremely narrow and high-crowned, with an unusual pattern of lingual ridges. Tooth size in the mandible from El Firal is extremely robust relative to *D. fontani*, and its M_3 has a large hypoconulid and lacks a tuberculum sextum, unlike all of the St. Gaudens specimens. A small number of postcranial bones are also known, including a hamate that is elongated with large and distinct joint surfaces for the other wrist bones, and noticeable dorsal expansion of the proximal joint surfaces (Begun, 1994*a*). Its hamulus projects palmarly but not distally. A massive pit for a number of wrist ligaments marks the medial surface of the hamulus. The thumb phalanx is small and relatively long and gracile with a deep proximal articular surface. The metatarsals are robust with dorsal intermetatarsal articular surfaces, no ventral articular surfaces, and a medially oriented proximal articular surface.

GENUS *Ouranopithecus* Bonis & Melentis, 1977
INCLUDED SPECIES *O. macedoniensis*

SPECIES *Ouranopithecus macedoniensis* Bonis & Melentis, 1977 (Fig. 20.8)
TYPE SPECIMEN RPL 54, an adult female mandible from Ravin de la Pluie in Macedonia, now in the Department of Geology and Physical Geography of the University of Thessaloniki
AGE AND GEOGRAPHIC RANGE Known from three localities in northern Greece: Ravin de la Pluie, Xirochori and Nikiti, which can be dated with reasonable certainty to the end of the Vallesian, or MN 10, about 9 Ma (Bonis *et al.*, 1988*a*; Koufos, 1990; Mein, 1999; Steininger, 1999)
ANATOMICAL DEFINITION
Ouranopithecus macedoniensis is known from a large number of jaws and teeth, and two unpublished phalanges. It was the largest hominoid from Europe, with the male in the size range of female gorillas. Lower incisors are tall-crowned and slightly flared toward the incisive edges, with prominent lingual cingula and relatively symmetric I_2. Male canines are tall and compressed, but small in cross-section compared to molar size. Female canines are very low-crowned and broad, almost premolariform (Koufos, 1995). P_3 is triangular with a prominent mesial beak, and P_4 is elongated with a high talonid (Bonis & Koufos, 1993). Lower molars are elongated and lack cingula, though they often have buccal notches, as in *D. fontani*. $M_3 > M_1 \sim M^2$. The upper incisors are heteromorphic, with I^1 labiolingually thick and broad, and I^2 narrow and peg-shaped. Upper canines are relatively low-crowned, especially compared to molar size, but show significant sexual dimorphism compared to humans. P^3 is triangular, though less so than in the mandible, and P^4 is rectangular. M^2 is the largest upper molar. Molars are broad-cusped and relatively flat with simple occlusal morphology and very thick enamel. Dentine penetrance appears comparatively low based on the pattern of wear, in which crowns are worn nearly flat before dentine pits begin to form (Bonis & Koufos, 1993).

Female mandibles tend to be robust and male mandibles tall, but nevertheless transversely more massive than in living hominids. A broad and long planum reinforces the mandibular symphysis, which shows a shallow genioglossal fossa and a well-developed inferior transverse torus. The lateral eminence is situated opposite M_3 and the ramus ascends between M_2 and M_3. One specimen preserves a gonial region, which is extensive with strongly developed ridges for the medial pterygoid muscle. The same specimen

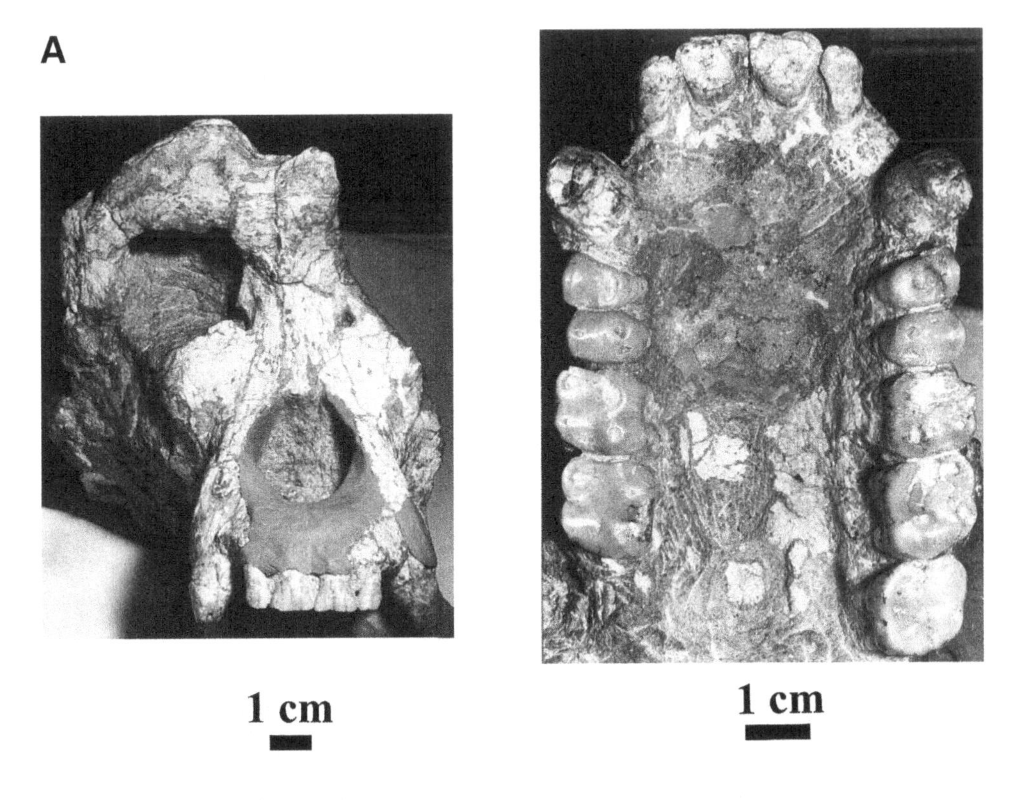

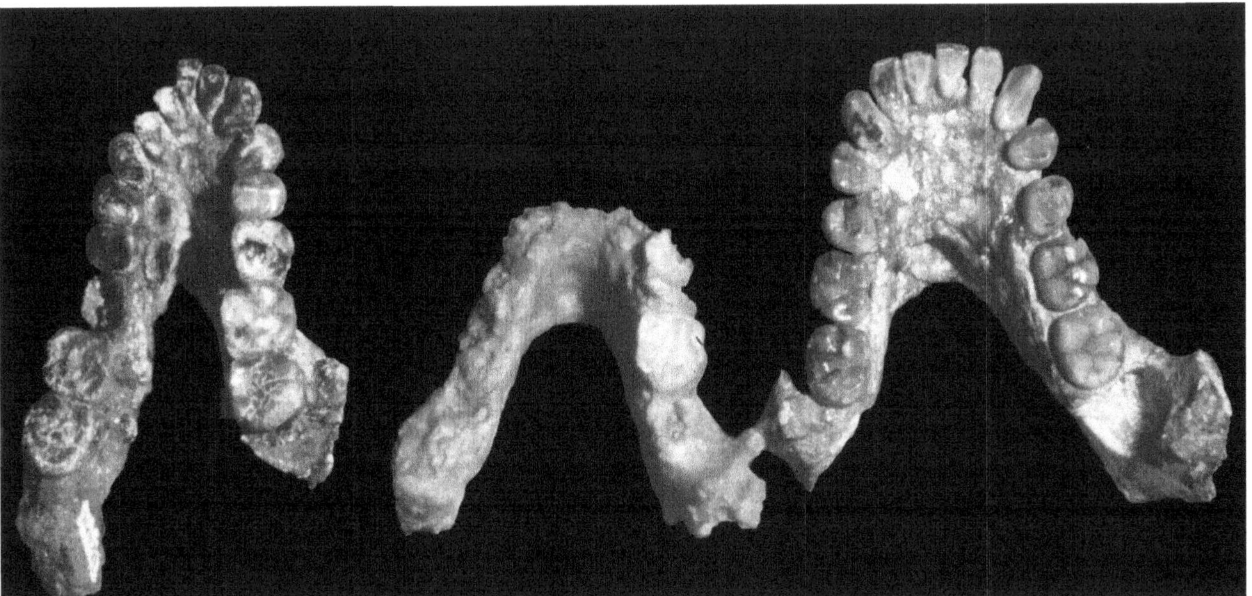

Fig. 20.8 *Ouranopithecus macedoniensis*. (A) Frontal (left) and palatal (right) views of a male cranium (not to the same scale). (B) Three mandibles from Greece; the specimen in the middle is a cast of the type of *Graecopithecus freybergi*, between female mandibles from Nikiti to the left and Ravin de la Pluie (the type) to the right.

preserves a condylar process and condyle, which, though damaged, can be described as relatively large but strongly convex anteroposteriorly (Bonis & Koufos, 1997). The premaxilla is robust, biconvex, and long compared to early Miocene apes and hylobatids but short compared to Asian great apes, chimpanzees and *Australopithecus*, being most like the premaxilla of gorillas and *Dryopithecus*. It is more vertically oriented than in Asian great apes, and has the same clivus and subnasal fossa as described for *Dryopithecus* (Bonis & Melentis, 1987; Begun, 1994a). Palates are relatively deep, broad anteriorly and parallel-sided. Unlike in most hominids, the zygomatic roots appear to arise from a low position on the maxilla. Canine fossae in males are broad and deep, but generally less so than in *Sivapithecus*. The nasal aperture is broad at its base. The midface is transversely flat and concave superoinferiorly, though less than in Asian great apes. The interorbital space is very thick, and the lateral orbital pillars are massive, with their surfaces oriented anterolaterally. The supraorbital tori are only modestly developed, again most like those of *Dryopithecus*. Glabella is also only mildly inflated and projecting. The anterior temporal lines are thick and strongly developed off the surface of the frontal bone behind the orbits, which are broad and rectangular in outline. A postglabellar depression is similar to that seen in *D. laietanus*. It does not continue across the surface of the frontal bone, an erroneous impression produced by postmortem damage. Instead the frontal squama was, like *Dryopithecus*, intermediate between the more vertical frontals of orangutans and the more horizontal frontals of African apes and early humans. A depression in the bone at the superomedial corner of the orbit, as well as a number of cracks between glabella and nasion, suggest the presence of a frontal sinus. *Ouranopithecus* was a comparatively large great ape, possibly 50 to 70 kg in size, though because it is mostly known from jaw fragments and appears to have been megadont, estimates are difficult.

Family Hominidae

Subfamily Ponginae

GENUS *Ankarapithecus* Ozansoy, 1965
INCLUDED SPECIES *A. meteai*

SPECIES *Ankarapithecus meteai* Ozansoy, 1965 (Fig. 20.9)
TYPE SPECIMEN MTA 2124, a male mandibular fragment in the collections of the Museum of the Maden Tetkik ve Arama Enstitüsü, Ankara
AGE AND GEOGRAPHIC RANGE Currently only known from Central Anatolia, and dated paleomagnetically to upper MN 9, about 10 Ma (Sen, 1991; Alpagut *et al.*, 1996)
ANATOMICAL DEFINITION
Ankarapithecus meteai is only known cranially from three specimens. Postcrania have been found but are unreported other than reference to their relative robusticity and adaptation to terrestrial locomotion (Köhler *et al.*, 1999). Lower incisors are labiolingually robust and relatively narrow and tall-crowned. The canine of the male is low-crowned but more massive in cross-section than in *Dryopithecus* or *Ouranopithecus*, while that of the female is more premolariform, as in *Ouranopithecus*. P_3 is large, oval and elongated, with a large mesial beak similar to *Dryopithecus* but small compared to *Ouranopithecus*, and the large P_4 is relatively broad. M_1 is small relative to M_2 compared to other late Miocene hominoids, with the possible exception of *Graecopithecus* (see below). Lower molars are also somewhat broader relative to their lengths than in other late Miocene hominoids. Their occlusal surfaces have broad, flat cusps, shallow basins, and no trace of cingula or buccal notches. Upper incisors are heteromorphic, with central ones labiolingually thick but also mesiodistally long and relatively low-crowned. Male upper canines are comparatively low-crowned, though not to the degree seen in *Ouranopithecus*, and flare cervically. Upper molars are somewhat less elongated than in other taxa described here, with simple occlusal surfaces lacking accessory ridging or cusps. The mandible is massive and strongly buttressed. The symphysis is very deep, narrow, vertical and "waisted" in anterior view, widening at its base. The incisors implant vertically and almost completely in line between the two canines. The sublingual plane extends to between the P_3–P_4 and the inferior transverse torus barely beyond that, around the level of M_1. The posterior corpus is massively thick transversely and so the ramus anteroposteriorly. The condyle appears to be large and convex, resembling *Ouranopithecus*.

The maxilla is massive, with a deep palate, roots of the zygomatic processes placed high on the robust alveolar process, a shallow canine fossa, compared to *Sivapithecus*, and a broad nasal aperture. Like *Ouranopithecus* the premaxilla is biconvex and less horizontally oriented than in *Sivapithecus*. The subnasal fossa is stepped and the incisive fossa is relatively large. Large maxillary sinuses invade the broad, laterally flared and frontated zygoma. The tall, prognathic midface is not flat as in *Sivapithecus* and *Pongo*. The orbits are squared, with a relatively narrow interorbital space. The nasal bones are extremely elongated. The lacrimal fossae are large and situated outside the orbits. The lateral orbital pillars are relatively broad and rounded compared to *Sivapithecus*, and the anterior temporal lines well marked. Supraciliary arches contour the superior edges of the orbits, but do not form a true torus (Alpagut *et al.*, 1996; Begun & Güleç, 1998).

Family Hominidae

Subfamily Oreopithecinae

GENUS *Oreopithecus* Gervais, 1872
INCLUDED SPECIES *O. bambolii*

SPECIES *Oreopithecus bambolii* Gervais, 1872 (Figs. 20.10–20.12)
TYPE SPECIMEN IGF 4335, a damaged juvenile mandible currently in the collections of the Instituto di Geologia University of Florence
AGE AND GEOGRAPHIC RANGE *Oreopithecus bambolii* is known only from Italy and possibly Moldova (Delson, 1986); fauna from the localities has been dated to MN12 and MN 13 (Hürzeler & Engesser, 1976; Andrews *et al.*, 1996; Harrison & Rook, 1997), or about 6 to 7 Ma (Steininger, 1999)
ANATOMICAL DEFINITION
Oreopithecus bambolii is the probably the best represented European fossil hominoid. Unusually small lower incisors

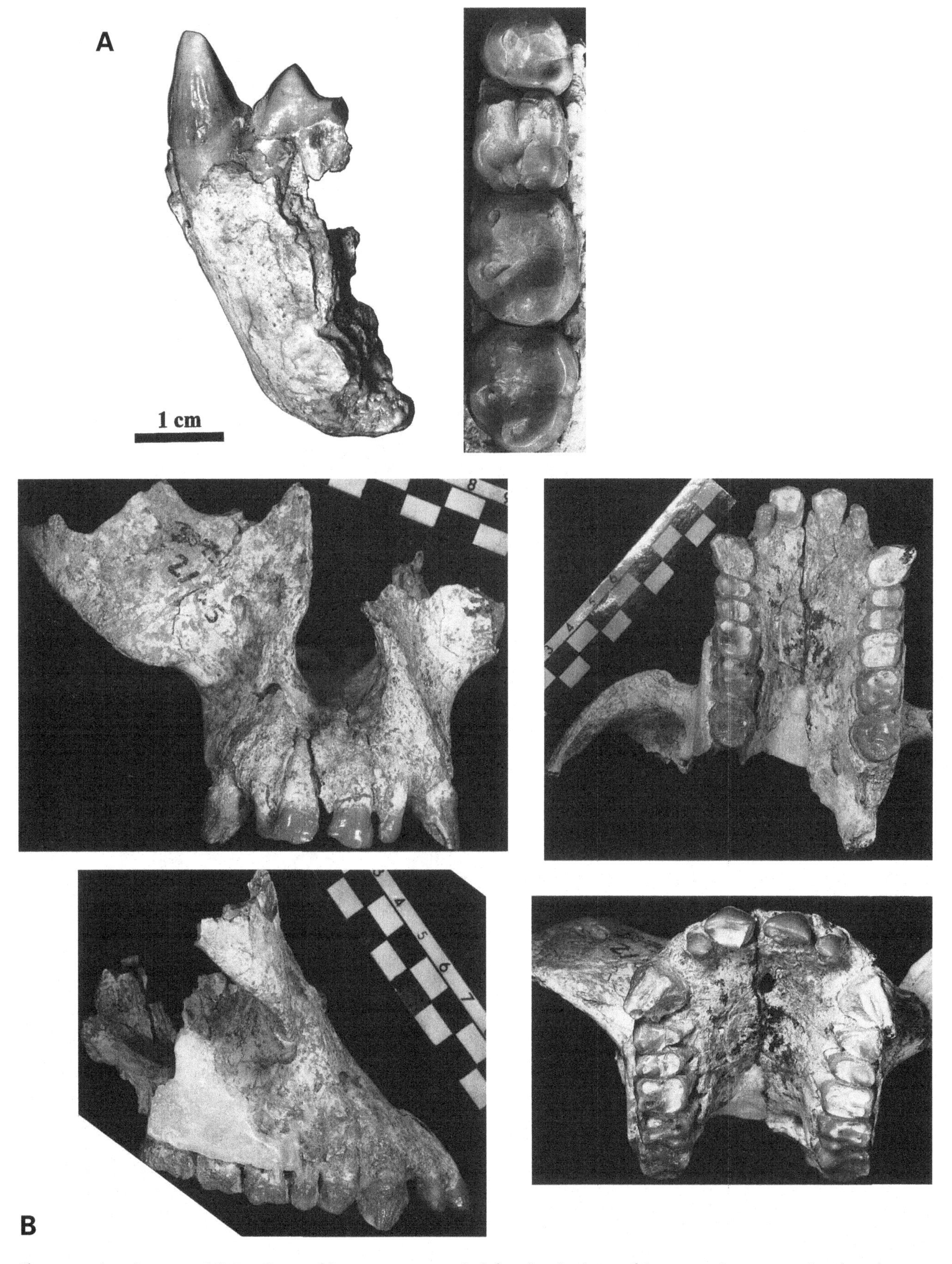

Fig. 20.9 *Ankarapithecus meteai*. (A) Lateral view of the type specimen to the left, and occlusal view of the associated posterior teeth to the right. (B) Clockwise from the top left, frontal, palatal, oblique palatal and right lateral views of a male *Ankarapithecus* partial face. The oblique palatal view shows a large incisive foramen behind the incisors, a character that distinguishes this taxon from *Pongo* and *Sivapithecus*.

have crowns compressed labiolingually. Male canines are small, rounded and relatively tall. P_3 is oval and elongated with a tall metaconid and sharp lingual and distal cingula. P_4 has two tall and distinct mesial cusps (protoconid–metaconid) and a low, short talonid. The molars are elongated and relatively low-crowned, but with tall compressed cusps. The buccal cusps are separated by a pronounced notch leading to a separate cusp in the center just distal to the mesial cusps, the centroconid, which is connected to them by sharp crests. The hypoconulid commonly is split into numerous separate cusps, and is thus difficult to identify. The molars also tend to be waisted. M_1 is close in size to M_2 and M_3 is the largest tooth. The four more mesial cusps (excluding the hypoconulid) all have crests that converge on the centroconids, making a very complicated and unique occlusal surface.

Upper incisors are heteromorphic, and very small but robust with marked lingual cingula. Female upper canines are low-crowned and morphologically similar to but larger than the lateral incisors. Male upper canines are large and strongly bilaterally compressed, with concave lingual surfaces and a long, sharp distal edge. Upper premolars are basically rectangular with nearly equal sized protocones and paracones. Upper molars are elongated and curiously similar to the lowers, which led many to attribute *Oreopithecus* to the Cercopithecoidea. Molar proportions are similar to the lower dentition.

The mandible is well known but distorted in nearly all specimens. Most appear to have been large, robust and tall relative to dental size. Mandibular height decreases slightly from P_3 to M_3. The corpora have well-developed lateral eminences and broad extramolar sulci. The rami are large, both in height and anteroposterior dimensions, and have strong markings for the masseter and the medial pterygoid muscles. The condylar processes also have deep scars for the lateral pterygoid muscle. The condyle itself is broad but convex mediolaterally.

The damaged palate has a low, anteriorly placed zygomatic root, deep but restricted maxillary sinuses, a very short, vertical premaxilla, shallow canine fossa, prominent, broad but not projecting canine jugae, a peculiar vertically oriented nasal aperture with the margin somewhat concave posteriorly in lateral view. The midface is projecting, and the nasal bones also probably relatively horizontally oriented. Around the orbits lacrimal fossae were obscured from frontal view by pronounced lacrimal crests.

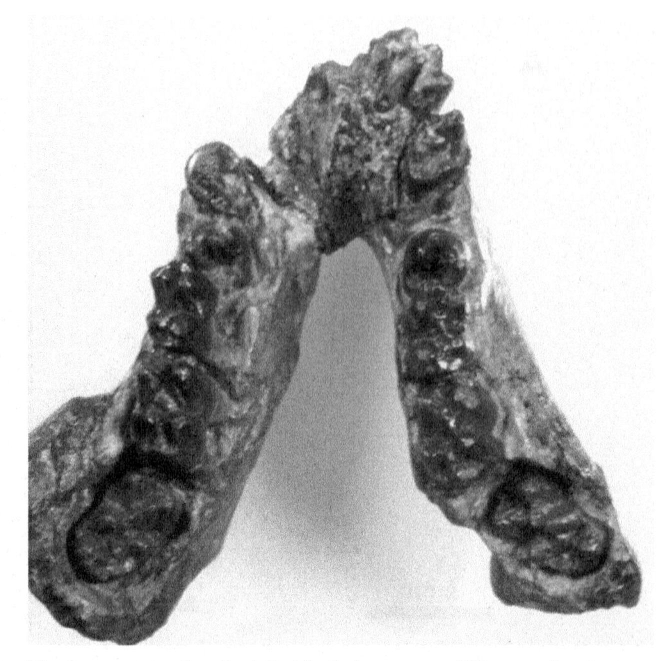

Fig. 20.10 *Oreopithecus bambolii*. The holotype mandible in occlusal view.

Fig. 20.11 *Oreopithecus bambolii*. The skeleton still in its slab.

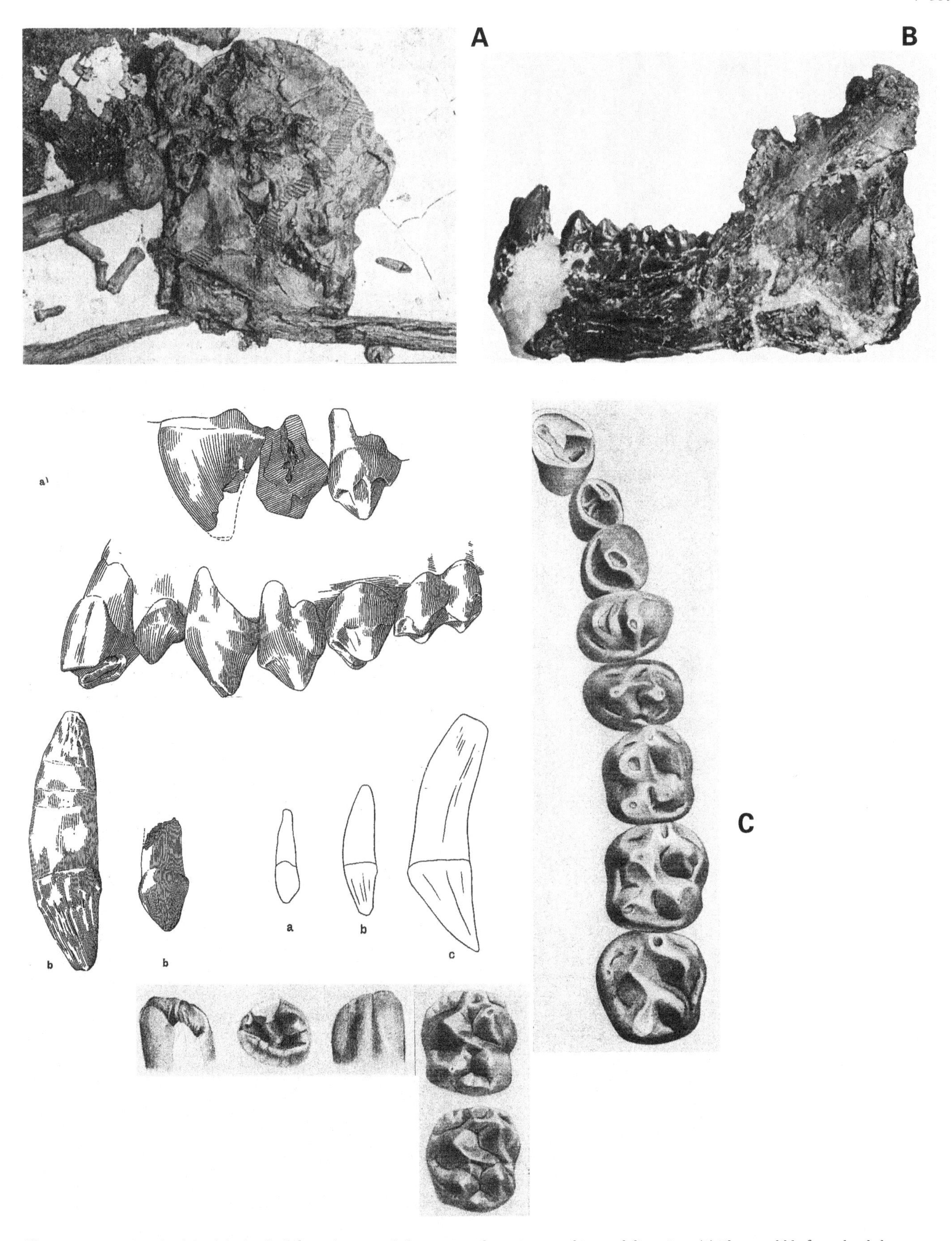

Fig. 20.12 *Oreopithecus bambolii*. (A) The skull from the 1958 skeleton. Note the serious crushing and distortion. (B) The mandible from the skeleton after preparation. (C) Upper dentition, first and second rows on the left, lateral views of a female palate and male and female canines; lower left, an upper central incisor in lingual, occlusal and labial views, and two unworn upper molars; right, occlusal view of an upper dentition; see also Fig. 20.10. (D) (*overleaf*) Postcrania. Top, lateral, inferior, anterior and medial views of a proximal ulna; bottom, posterior and superior views of a proximal femur. (E) Hand from the 1958 skeleton.

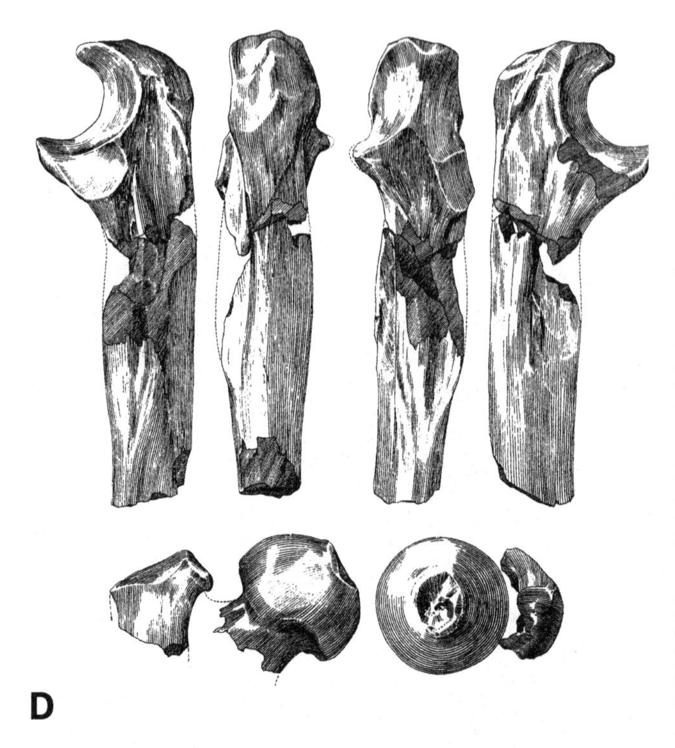

D

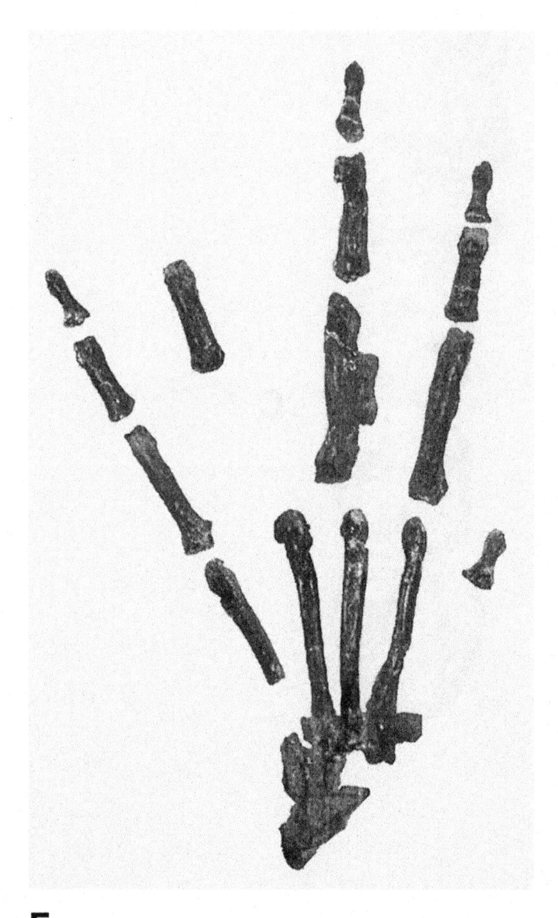

E

Fig. 20.12 (cont.)

The neurocranium and basicranium from the 1958 skeleton are preserved in large part, but are damaged and distorted. The articular eminence is mediolaterally broad, anteroposteriorly long, and saddle-shaped, and the entoglenoid process is large and inferiorly projecting. The articular and tympanic portions of the temporal are not fused, but the temporal petrous portion shares an important similarity with hominids in having a shallow and indistinct subarcuate fossa (Harrison & Rook, 1997). The mastoid process is broad and continuous with a large nuchal plane bearing scars for powerful nuchal muscles. Very strong nuchal and sagittal crests meet at inion. These crests form very deep surfaces for the temporalis muscles. The apparent flare of the root of the zygomatic process of the temporal, the flare of the temporal process of the zygomatic bone and the very large mandibular ramus all also indicate very large temporalis muscles. The temporal process of the zygomatic bone is oriented posterosuperiorly, suggesting that the temporomandibular joint was superiorly placed relative to the orbits. The orbital rims are smooth, relatively gracile and do not project far anteriorly beyond the anterior temporal lines. The anterior temporal lines are well marked and form the sagittal crest very far anteriorly.

Postcranially *Oreopithecus* is known from most bones, too many in fact to give a thorough review here. Broadly speaking, *Oreopithecus* postcrania resemble those of modern apes but with some primitive characters: broad thorax, deep glenoid fossae, long robust clavicle, a large humeral head, short lumbar region with robust vertebrae, short, broad ilia with prominent anterior inferior iliac spines, no tail, forelimbs much longer than hindlimbs, very deep humeral trochlea separated from the large, globular capitulum by a pronounced zona conoidea, very strong ulnar trochlear keel, large, deep, laterally oriented ulnar radial facet, very short olecranon process, robust ulnar shaft, circular radial head and long radial neck, large, globular femoral head with a very prominent, deep fovea, flattened distal femora with broad patellar surfaces, short, robust metacarpals and metatarsals but long, curved, gracile phalanges with strong ridges and proximally oriented articular surfaces.

The carpals indicate a wide range of movement potentials in having generally broad, flat articular surfaces, but they are also relatively gracile, unlike the blocky carpal bones of *Dryopithecus* and living great apes. The feet are short in the midfoot but long in the digital area, with a powerful hallux. The tarsal bones are similar in being comparatively flat with broad, continuous articular surfaces. The talus is low and broad with a short, divergent neck and a large head and the calcaneus is comparatively short with a strongly developed plantar process and sustentaculum. The cuboid has a well-developed beak for the calcaneus but little distinction between the 4th and 5th metatarsal facets. The navicular is short with distinct cuneiform facets and a rectangular facet for the cuboid, and the entocuneiform is also short but with a small facet for the navicular (Szalay & Langdon, 1986).

Family Hominidae

Subfamily *incertae sedis*

GENUS *Graecopithecus* von Koenigswald, 1972
INCLUDED SPECIES *G. freybergi*

SPECIES *Graecopithecus freybergi* von Koenigswald, 1972 (Fig. 20.8B)
TYPE SPECIMEN A badly preserved adult mandible now accessioned in the collections of the Geologische Institut Erlangen, Germany
AGE AND GEOGRAPHIC RANGE The exact age is difficult to determine; the presence of the mastodon *Mammut* at the site suggests contemporaneity with Greek sites dated to MN 12, between 6.6 and 8 Ma (Solounias, 1981; Mein, 1999; Steininger, 1999; *contra* Andrews *et al.*, 1996)
ANATOMICAL DEFINITION
Graecopithecus freybergi is the most poorly known European Miocene hominoid. It is difficult to distinguish from distinct Miocene hominoids such as *Sivapithecus*, *Ouranopithecus* and *Ankarapithecus*, as noted by Martin & Andrews (1984), who put all these genera in *Sivapithecus*. They considered *Ouranopithecus macedoniensis* and *Graecopithecus freybergi* to be the same species. *Graecopithecus* has large molars with thickly enameled crowns and an apparently robust mandibular corpus and is about the same size as females of *Ouranopithecus* (Martin & Andrews, 1984). Distinctive features of *Graecopithecus* include an M_2 breadth is actually greater than the breadth of the mandibular corpus at this level (Fig. 20.8B). Close and careful inspection of the original damaged corpus reveals this feature to be real, and not an artifact of poor preservation. The M_2 is also very broad relative to its length, though this has been reduced by interstitial wear. Overall the M_2 was a much larger tooth than in female *Ouranopithecus*. The M_1 is also small relative to the M_2, unlike other late Miocene to Recent great apes. Compared to *Ouranopithecus* the mandible is narrow, especially at the symphysis. Symphysis morphology differs from that of *Ouranopithecus* in having a relatively vertical lingual surface. Given these differences and uncertainties, as well as the substantial temporal difference between Pygros and the *Ouranopithecus* localities, *Graecopithecus* is recognized here as a distinct genus with unknown affinities to other hominids.

Evolution of European Hominoids

The earliest euhominoid?

Currently at least six great ape genera and 11 species are recognized in Europe and Western Asia spanning a temporal range of over 10 million years. There are about as many different ideas about phylogenetic relations among and between them and living great apes as there are researchers in this field. However, most researchers agree that all middle and late Miocene hominoids are more closely related to living hominoids than are early Miocene forms. Most also agree that most middle and late Miocene Eurasian hominoids are cladistically great apes, the exception being *Griphopithecus*, about which there is less consensus (Fig. 20.13). It is fairly clear that a thickly enameled hominoid made its way, probably from Africa, to Europe about 16 to 17 Ma, based on a tooth fragment from Engelswies, Germany (see above) that has received little attention thus far (Heizmann & Begun, 2001). The tooth fragment is clearly that of a hominoid with thick enamel, and is most similar to specimens attributed to *Griphopithecus* and *Kenyapithecus* from Europe, Western Asia and Africa. Most authorities now agree that these taxa, which some split into three genera, including *Equatorius africanus* (S. Ward *et al.*, 1999) (= *Griphopithecus africanus* here), are more primitive than any fossil or living great ape. There is also strong evidence to suggest that this group is more primitive than the hylobatid–hominid split (Begun *et al.*, 1997*b*; McCrossin & Benefit, 1997; Rose, 1997; Ward, 1997). Intriguingly, the German specimen is the oldest. An African taxon that predates *Griphopithecus* thus could be related to the taxon represented at Engelswies. *Afropithecus* and *Heliopithecus*, each over 17 million years old, are suitable candidates, but in the absence of more data from Engelswies the exact source of the invasion of Europe by the Hominoidea will have to remain a mystery. Nonetheless, even the more completely known Eurasian samples from this group of thickly enameled hominoids are at least as old if not older than *Griphopithecus* from East Africa, leading one to question the often assumed direction of dispersal of this group (Begun, 2000; Heizmann & Begun, 2001). If it is true that hominoids dispersed into Eurasia first about 17 Ma, from a thickly enameled East African ancestor like *Afropithecus*, and subsequently radiated throughout Eurasia and even back into Africa, then it is possible that thick enamel in hominoids represents a critical adaptation that made this success possible. Once established there, hominoids become highly diverse in Europe, but all appear to be traceable back to a large, thickly enameled, unspecialized arboreal species.

Concerning the evolutionary history of all other European hominoids there is widespread disagreement but no major camps. Nearly every worker has a different opinion. For the most part the conclusions presented here are updated from work presented in Begun *et al.* (1997*b*). This is the only comprehensive, total morphological pattern analysis evaluating the evolutionary history of Miocene hominoids using a computer-aided phylogenetic algorithm (Hennig86). Fig. 20.14 depicts cladistic relations among the taxa discussed here.

Oreopithecus

The bulk of the evidence suggests that *Oreopithecus* is the most primitive known great ape. This is the conclusion presented by Harrison (1986*a*) and Harrison & Rook (1997), in what

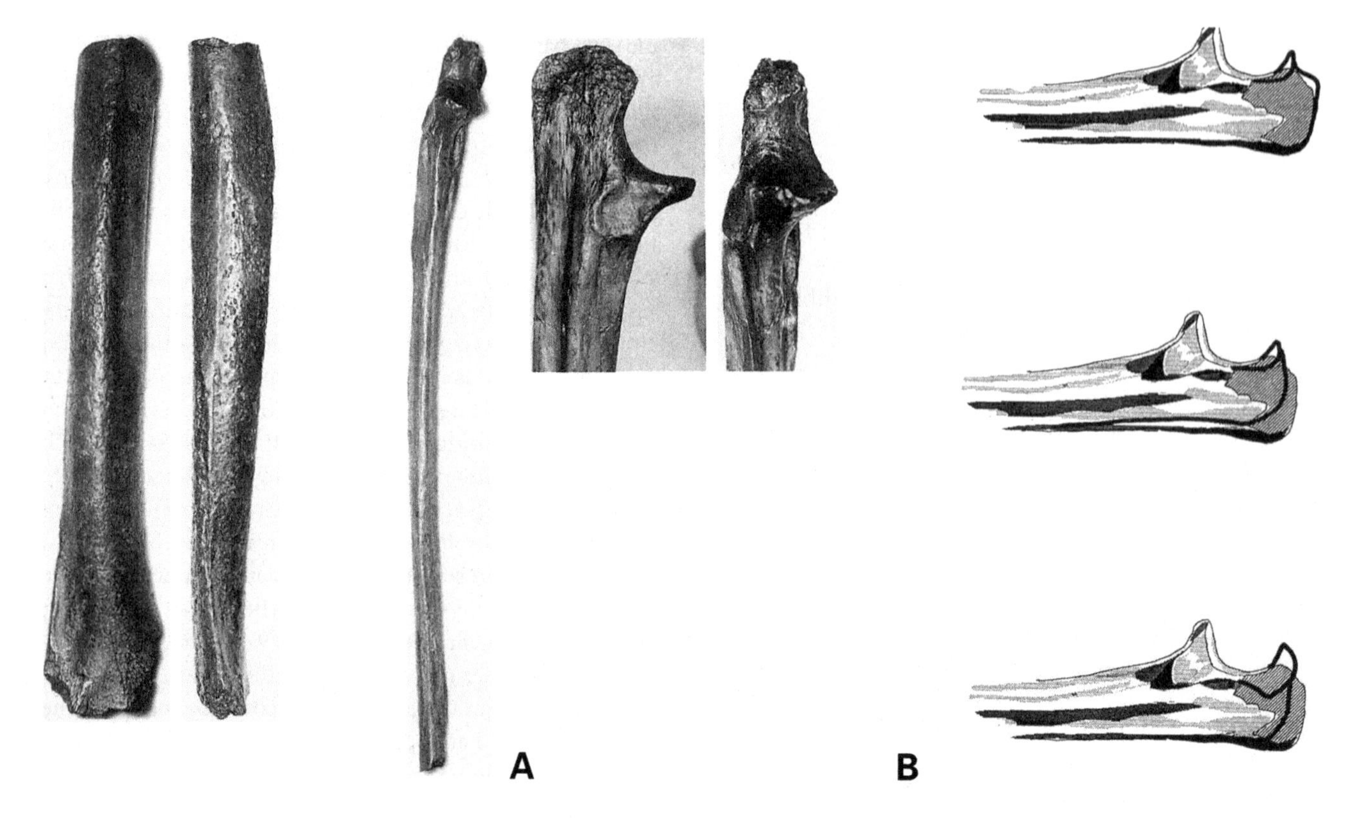

Fig. 20.13 *Griphopithecus* postcrania from Klein Hadersdorf. (A) Anterior and lateral views of a humeral shaft; middle, anterior view of an ulna, and close-ups of the proximal end in lateral and anterior views. (B) Drawings illustrating the probable presence of an olecranon process, a primitive feature not found in any living hominoid. The top drawing reconstructs the olecranon process (thick line), given the preserved anatomy of the fragment. The middle and bottom drawings illustrate a chimpanzee–like and orangutan–like olecranon superimposed on the specimen.

are the most comprehensive phylogenetic analyses of *Oreopithecus*. While sharing numerous characters with living apes, *Oreopithecus* retains several primitive characters of the face, neurocranium and postcranium (Harrison, 1986*a*). The main areas of disagreement concern other taxa that some researchers feel are closely related to *Oreopithecus*. Harrison & Rook (1997) consider *Dryopithecus* to be the sister clade to *Oreopithecus*, while Moyà-Solà & Köhler (1997) consider *Oreopithecus* to be part of a Eurasian great ape radiation, which excludes the African apes and humans. Moyà-Solà & Köhler further consider it to be a highly derived member of a clade that includes all Eurasian late Miocene hominoids (Moyà-Solà & Köhler, 1995, 1997). They argue that most of the cited primitive characters in *Oreopithecus* are actually autapomorphies that are merely homoplasious with similar characters in other taxa. They base this conclusion not on a phylogenetic analysis or a character analysis, which are the usual protocols for establishing homology and homoplasy (Ward *et al.*, 1997; Lockwood & Fleagle, 1999), but on various assumptions of developmental processes that could have contributed to craniofacial form in *Oreopithecus*. They focus on a small number of characters, such as facial foramina, that they consider to indicate a close relationship with *Dryopithecus*, which in turn they find to be closely related to *Sivapithecus*, following a similar methodology (Moyà-Solà & Köhler, 1995). However, when all the data are considered together, without a priori weighing of characters based on assumptions of ontogeny, *Oreopithecus* is placed unambiguously at the base of the large-bodied hominoid radiation (Begun *et al.*, 1997*b*; Harrison & Rook, 1997).

Harrison & Rook (1997) and Begun *et al.* (1997*b*) combined consider short, gracile premaxilla, long, narrow palate, vertical and narrow nasal aperture, projecting midface, low zygomatic root, low position of the orbits and a very small brain case to be primitive features retained in *Oreopithecus* and distinguishing this taxon from all other great apes. The brain of *Oreopithecus* is in fact so small relative to body mass that it may have undergone secondary reduction, a phenomenon often associated with extremely specialized forms of folivory (Clutton-Brock & Harvey, 1980; Harrison, 1989*b*). However, the common ancestor of hylobatids, *Oreopithecus* and all other hominids probably had a relative brain size similar to that of hylobatids and *Proconsul*, and below that of other fossil and living great apes (Kordos & Begun, 1998). *Oreopithecus* also shares numerous characters of the postcranium with hylobatids and hominids, related mostly to suspensory positional behavior, and with hominids to the exclusion of hylobatids, related mostly to large body mass in suspensory quadrupeds such as stable trunks, powerful grasping, slow moving, and high joint mobility (Harrison & Rook, 1997). The postcranial morphology of *Oreopithecus* accurately represents ancestral hominid

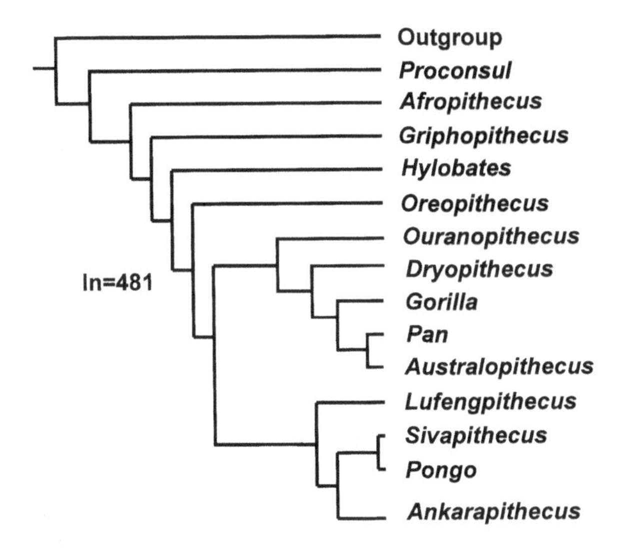

Fig. 20.14 The preferred cladogram of Miocene hominoids.

morphology, while the cranial and dental morphology presents features unique to that taxon.

Finally, an intriguing case has been made recently that *Oreopithecus* was a biped with a human-like hand capable of a precision grip (Köhler & Moyà-Solà, 1997; Moyà-Solà *et al.*, 1999; Rook *et al.*, 1999). However, most or all of the characters noted by these authors can be explained by a highly arboreal, slow-moving positional behavior most similar to that of living sloths, orangutans and prosimians, particularly the inferred positional behavior of some of the larger arboreal subfossil taxa from Madagascar (Wunderlich *et al.*, 1999). This latter explanation is also much more consistent with the overall anatomy of *Oreopithecus*, whose curved phalanges and greatly elongated arms are obvious suspensory arboreal characters.

Oreopithecus was a relatively large-bodied (30–35 kg) primitive great ape that was highly suspensory and a specialized folivore. Its huge ectocranial crests reflect both massive chewing muscles and a very small brain case, consistent with a diet relatively low in concentrated nutritional value, particularly calories and nutrients necessary to promote or allow for the development of a large brain. Its unusual morphology was probably the effect of a prolonged insularity (Köhler *et al.*, 1999) and a lengthy independent phylogenetic history.

The east–west divide

Figure 20.14 shows a principal separation between mostly Asian great apes on the one hand and Euro-African great apes on the other. On the Asian side, the results presented here are relatively uncontroversial and similar to those presented elsewhere (Begun & Güleç, 1998; Alpagut *et al.*, 1996; Köhler *et al.*, 1999). The only taxon in this group covered in this chapter is *Ankarapithecus*, which Begun & Güleç (1998) consider to be the most primitive member of the clade that includes *Sivapithecus* and *Pongo*, while Alpagut *et al.* (1996) consider *Ankarapithecus*, along with *Dryopithecus* and *Ouranopithecus*, to be the sister clade to all other great apes and humans (presumably excluding *Oreopithecus*). Both arguments agree that *Ankarapithecus* shares several characters with *Sivapithecus* and *Pongo*, but lacks some synapomorphies of this clade. Alpagut *et al.* (1996) also note features shared with *Dryopithecus* and *Ouranopithecus*, such as broad orbits, keeled nasals, exposed lacrimal fossae, extensive frontal sinus (in the interorbital region, not the frontal), supraorbital tori, flat facial profile, and more inclined frontal. Begun & Güleç (1998) also enumerate features shared with great apes other than *Sivapithecus* and *Pongo*, including a stepped subnasal fossa and comparatively large incisive canal and fossa. However, when all of the characters that have been used to reconstruct the evolutionary relations of *Ankarapithecus* are considered together (100 in all), it emerges as the sister to the *Sivapithecus*/*Pongo* clade. In summary, *Ankarapithecus* was a large (30–60 kg) great ape that is the primitive sister clade to *Sivapithecus* and *Pongo*. While little is known of its postcranial anatomy, craniodentally it was massive, suggesting the need for high occlusal loads and teeth comparatively resistant to abrasion. This is most consistent with a hard-object diet, and/or the presence of terrestrial foods that are relatively high in abrasives, or simply that it could process hard or terrestrial foods when necessary.

While the bulk of the evidence still favors an Asian great ape clade as depicted in Fig. 20.14, Alpagut *et al.* (1996) surmise that much of the anatomy of *Ankarapithecus* may be primitive for the great apes and humans. Taxa at the bases of the radiations of Asian and Afro-European great apes, such as *Ouranopithecus* and *Ankarapithecus*, while having synapomorphies of their respective clades, likely retain comparatively more features of their common ancestor than are found in their more derived relatives. If so, then the last common ancestor of the great apes and humans may have had many of the postcranial adaptations of *Oreopithecus* and *Dryopithecus* (see above), and craniodental characters typical of apes that generate high occlusal forces and high bite loads during mastication. These are usually associated with specialized forms of frugivory such as hard-object feeding, which has been suggested as a feeding strategy for many of these taxa (Kay, 1981; Kay & Ungar, 1997; Smith, 1999).

The Afro-European radiation

The hypothesis presented in Fig. 20.14 is actually three steps longer than a cladogram that places *Ouranopithecus* and *Lufengpithecus* as successive outgroups to the Asian great apes, two steps longer than the cladogram that places *Lufengpithecus* as a stem great ape after *Oreopithecus*, and one step longer than the cladogram that places *Ouranopithecus* among the Asian great apes excluding *Lufengpithecus* (Fig. 20.15). I prefer a less parsimonious hypothesis here because it makes more sense both functionally and biogeographically (see also Begun &

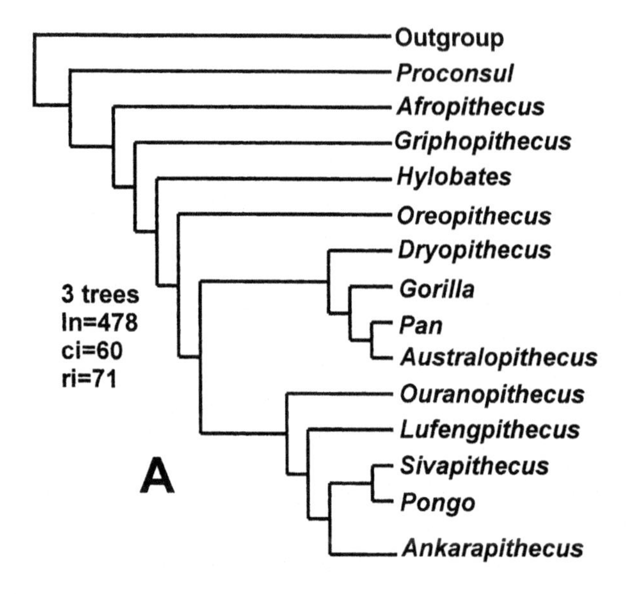

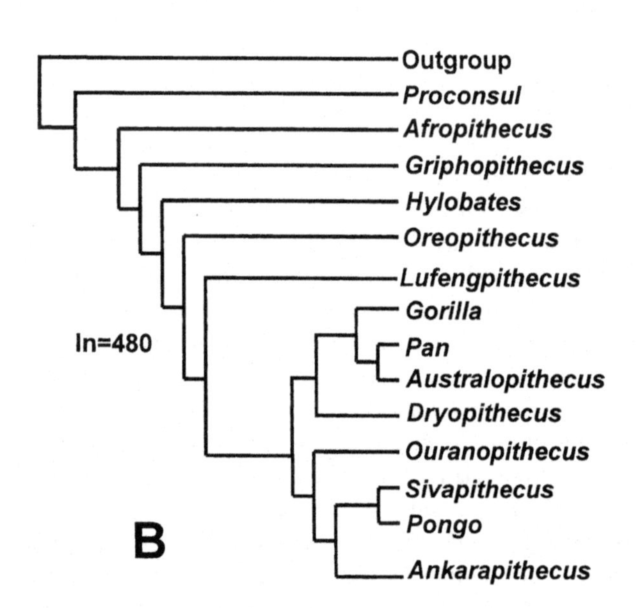

Fig. 20.15 (A) The most parsimonious cladogram generated from a data matrix of 14 taxa, one composite outgroup, and 247 characters. Data modified from Begun *et al.* (1997*b*) by inclusion of eight new forelimb characters, inclusion of *Ankarapithecus* and recoding of characters for *Griphopithecus* to exclude features known only for *Kenyapithecus*. The three most parsimonious trees differ only in the relative positions of *Afropithecus* and *Proconsul*. (B) Another more parsimonious cladogram with the positions of *Ouranopithecus* and *Lufengpithecus* altered.

Kordos, 1997). This illustrates what I believe to be the preferred use of computer parsimony algorithms, which are not designed to advise one to choose specific hypotheses, but simply to objectively describe the implications, in terms of inherent complexity, of a researcher's informed choice. In this case, while the total number of steps is smaller, the most parsimonious cladogram is not the preferred hypothesis here because the *types* of homoplasies are more diverse in nature across the entire dentition and skull, making them less easily explained by a single or comparatively small number of functional or developmental hypotheses. For example, *Ouranopithecus* is out of the Afro-European group in the most parsimonious cladogram because of the low position of its zygomatic root, which parallels the primitive condition, and because of its masticatory robusticity, which as noted earlier is probably primitive for the great apes and humans.

The slightly larger number of homoplasies involved in Fig. 20.14 is easier to explain by a single phenomenon, the independent evolution of masticatory robusticity, than are the slightly smaller number of homoplasies that would have to be posited between *Ouranopithecus* and the Afro-European taxa. These latter include characters of the incisors, canines, nasal aperture, midface, frontal and cranial architecture (facial hafting) that are not so obviously linked to a single, well-documented functional process. In addition, paleobiogeographically it makes more sense to link *Ouranopithecus* with *Dryopithecus* and *Lufengpithecus* with *Sivapithecus*/*Pongo*, on both temporal and geographic grounds.

One evolutionary implication of the Fig. 20.14 hypothesis is that *Ouranopithecus* retains a basic set of features related to heavy chewing that are also present in *Ankarapithecus*, *Sivapithecus*, *Lufengpithecus*, *Ardipithecus* and "australopithecines". To the limited extent that they are known, many of these features are also present in *Kenyapithecus* and *Griphopithecus*. Superimposed on the primitive set of gnathic attributes were specializations for enhanced hard object or terrestrial object feeding. Many of these traits are also found in "australopithecines" and, in varying degrees of development, in other primates such as *Ankarapithecus*, *Gigantopithecus*, *Sivapithecus*, *Lufengpithecus*, *Paranthropus*, *Afropithecus*, *Theropithecus*, *Cebus*, *Hadropithecus* and *Megaladapis*.

The hypothesis that *Ouranopithecus* is most closely related to *Australopithecus* (Bonis & Koufos, 1997; Bonis et al., 1998), which is based on these shared attributes, is nevertheless five steps less parsimonious. Once again the pattern of homoplasy required of this hypothesis does not lend itself easily to a functional or developmental explanation. Homo-plasies of *Australopithecus* and *Pan* would have to include elongated premaxilla, elongated, small-caliber incisive canal, spatulate upper lateral incisors, lower and more flared molar crowns, reduction in the thickness of the lateral orbital pillars, increased development of the supraorbital tori and supratoral sulci, more horizontal frontal bone and strong glabellar and supraorbital pneumatization. And once again, though not of overwhelming influence, associating the southeastern European Miocene hominid *Ouranopithecus* with a closely contemporaneous and anatomically similar taxon from central Europe (*Dryopithecus*) may be somewhat more straightforward than an association with terminal Miocene or even Pliocene hominids from East Africa.

The craniodental anatomy of *Ouranopithecus* recalls that of Plio-Pleistocene hominids and suggests parallel solutions to similar adaptive problems, namely, hard food objects and/or terrestrial food sources. From a structural morphological perspective there is a similarity in pattern between the northern great apes *Dryopithecus* and *Ouranopithecus*, and between *Pan* and *Australopithecus*, in the sense that both pairs diversified into soft-fruit and hard-object feeders, and lived in differing habitats that would predict such a pattern (closed forests vs. more open woodlands: Andrews *et al.*, 1997). The developmental biology of characters related to mastication probably is so constrained and similar in all hominids that a very small genetic change at any number of loci controlling the development of the jaws and teeth could result in the same morphological outcome, a change in masticatory robusticity.

Dryopithecus shares numerous characters with living African apes not found in any other fossil ape except *Ouranopithecus*. In the face these include detailed similarities in the premaxilla and subnasal fossa shared with gorillas, and nasal aperture morphology and periorbital morphology shared with Hominini more generally. *Dryopithecus* has an African-ape-like cranial form with an elongated cranium, relatively horizontal frontal, elongated temporal fossae, lower inion and many details of the temporal bone including partial fusion of the tympanic and articular portions and a large and projecting entoglenoid. These regions are not preserved in *Ouranopithecus*. *Dryopithecus* also shares a few postcranial characters with African apes including a robust, palmar hamate hamulus, an ulnar triquetral facet, and a shallow lunate scaphoid facet. However, *Dryopithecus* lacks a fused centrale and scaphoid (Begun, pers. observ.). The number and broad distribution of derived traits across several functional systems suggest that most are true synapomorphies. Although *Ouranopithecus* is the sister clade to the clade that includes *Dryopithecus* and the African apes and humans in Fig. 20.14, this may be because *Ouranopithecus* is known only from the face and teeth. Similarities to *Dryopithecus* in other anatomical regions, such as the neurocranium and postcranium would tend to falsify this hypothesis in favor of a *Dryopithecus–Ouranopithecus* clade (Begun & Kordos, 1997).

Dryopithecus is among the more derived Miocene hominoids, lacking many of the primitive masticatory adaptations of other taxa. *Dryopithecus* was also the most specious and widespread taxon and ranged in body mass across species from 18 to 45 kg. *Dryopithecus* was a highly arboreal hominid that may have ventured only rarely to the ground, and probably had a diet of soft fruits (Kay & Ungar, 1997; Smith, 1999). *Dryopithecus* had a large, *Pan*-sized brain, both in relative and absolute terms (Kordos & Begun, 1998), and was probably cognitively, and with respect to life-history variables, more great-ape-like than monkey-like.

It is widely assumed that the origin of Eurasian great apes was African, and that the fossil ancestors of African apes and humans remain to be found in the relatively less well-known late Miocene of Africa. In actual fact, none of the many late Miocene African fossil localities has any hominoids, while hominids that are closely related to African apes and humans are numerous in Eurasia. The straightforward explanation is that the ancestor of the African apes and humans migrated to Africa from Eurasia, probably Europe, about 9 million years ago. This is consistent with known patterns of biogeographic dispersal of many lineages of mammals in the late Miocene, with evidence of climatic change in the circum-Mediterranean region at the time, and with evidence of molecular evolution (Leakey *et al.*, 1996; Stewart & Disotell, 1998).

The *Gorilla*-like characters of both *Ouranopithecus* and *Dryopithecus* suggest that gorillas retain primitive features and that alternative character states shared by chimpanzees and humans are shared and derived between them (Begun, 1992*c*, 1994*a*). In this line of reasoning, it is not clear which taxon, *Dryopithecus* or *Ouranopithecus*, comes closer to the ancestral morphology of the African apes and humans. On the one hand *Ouranopithecus* is even more robust than early Pliocene hominids, especially *Ardipithecus*, which like *Dryopithecus* had thin enamel (White *et al.*, 1994). On the other hand, most living and extinct hominids have robust jaws and large, thickly enameled teeth. The evidence may favor *Dryopithecus* slightly over *Ouranopithecus* because all three lineages of the Hominini have relatively *Dryopithecus*-like gnathic structures (*Ardipithecus*, *Pan*, *Gorilla*). They also are more terrestrial than *Dryopithecus* (positional behavior is not known from *Ouranopithecus*), and this probably evolved after the dispersal into Africa. While no known Miocene hominid was either a biped or a knuckle-walker, the postcranial morphology of knuckle-walkers is closer to the general great ape pattern, and so increased terrestriality in the descendants of European hominids was most likely something similar to modern great ape knuckle-walking. Knuckle-walking probably evolved only once, before the split between the gorilla clade and the chimpanzee–human clade, and humans probably evolved from a knuckle-walker. This also explains numerous similarities in the forelimbs of African apes and humans quite plausibly related to the functional demands of knuckle-walking (Begun, 1993*b*, 1994*a*; Richmond & Strait, 2000).

In summary, Fig. 20.16 depicts a phylogenetic and paleobiogeographic hypothesis consistent with the cladogram in Fig. 20.14. All post-early Miocene hominoids may have evolved from a thickly enameled hominoid similar to *Afropithecus*, or possibly the less well-known *Heliopithecus*, following a migration to Eurasia before the Langhian transgression, about 16.5 Ma. The specimen from Engelswies may be that ancestor or its sister. The common ancestor of later hominoids, or Euhominoids (black box 1 in Fig. 20.16) would, like *Griphopithecus*, have had a robust masticatory apparatus but, unlike *Griphopithecus*, postcranial morphology permitting suspensory positional behavior. Combined these may have enabled this ancestor to exploit a wide range of terrestrial and arboreal dietary resources while maintaining an efficient positional behavior for a large-bodied arboreal quadruped. This may have been the key innovation that led to the explosive radiation of Eurasian euhominoids in the middle and late Miocene. Two of the most unusual hom-

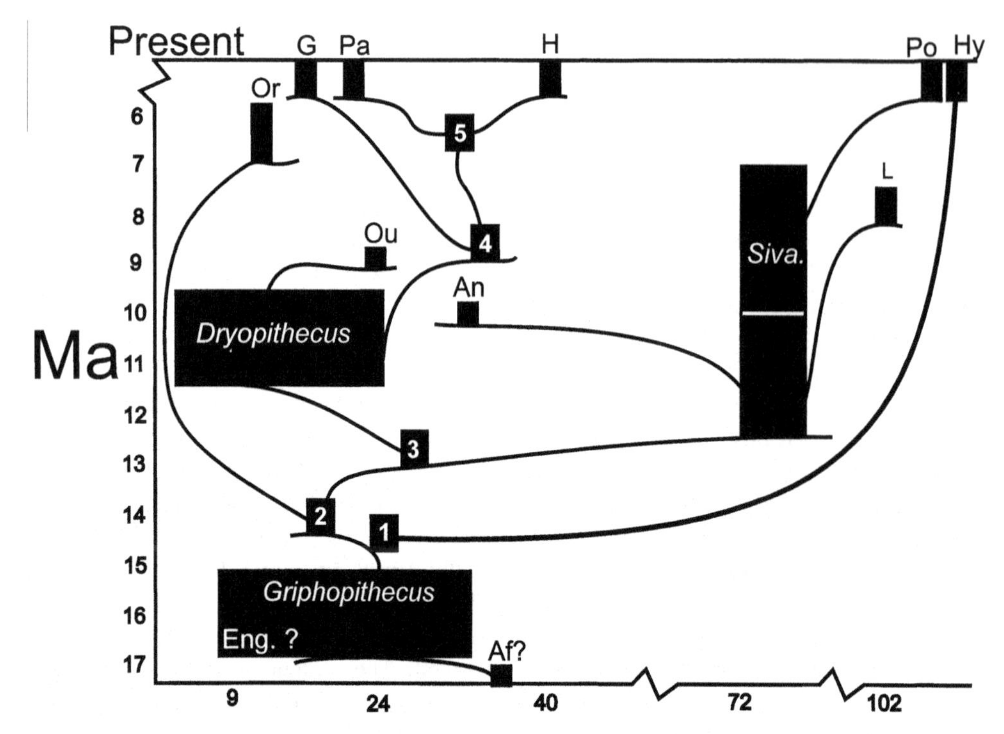

Fig. 20.16 One phylogeny consistent with Fig. 20.14. The *x* axis is longitude east of Greenwich and the *y* axis is time. Unnumbered polygons represent genus ranges in space and time, and numbered polygons represent hypothetical ancestors. G = *Gorilla*, Pa = *Pan*, H = humans, Po = *Pongo*, Hy = *Hylobates*, Or = *Oreopithecus*, Ou = *Ouranopithecus*, An = *Ankarapithecus*, L = *Lufengpithecus*, Af = *Afropithecus*, Eng. = Engelswies, Siva. = *Sivapithecus*. Boxes: 1. Common ancestor of all Eurasian and living hominoids; this could be the same as *Griphopithecus* or the taxon from Engelswies, if that proves to be different. 2. Common ancestor of all hominids. 3. Common ancestor of hominids excluding *Oreopithecus*. 4. Common ancestor of the African apes and humans. 5. Common ancestor of chimpanzees and humans. The line through the *Sivapithecus* bar marks a possible taxonomic distinction between Chinji and later *Sivapithecus* that would be more consistent with the hypothesis presented here (Begun & Güleç, 1998).

inoids evolve next, first *Hylobates* and then *Oreopithecus*, which share with hominids features including subtleties of the dentition, premaxilla and temporal bone and a more impressive list of postcranial features, mostly related to weight transmission through the axial skeleton and a powerful, grasping foot. These features probably were present in the taxon represented by black box 2 (but some may have been secondarily lost in hylobatids, which are probably phyletic dwarfs, given the evolutionary relations depicted here). Ancestor 2 may have been a large-bodied, slow-moving suspensory quadruped with feet capable of sustaining high loads and generating a powerful grip. The addition of specialized capabilities of the feet in this common ancestor may have been a prelude to further specialization including increased terrestriality in the African apes and humans and possibly at least one species of *Sivapithecus* (Begun *et al.*, 1997*b*; Harrison & Rook, 1997). The common ancestor of Afro-European and Asian hominids (black box 3) was probably characterized by a suite of characteristics of the face, neurocranium and postcranium related to a more intensive and complex use of diversified resources. The key innovation may have been the evolution of a large, great-ape-sized brain, or a great ape life-history strategy, or both. Large brains and the robust and expanded premaxilla, incisors and anterior portions of the chewing muscles in these taxa suggest an ability to exploit embedded or challenging food resources that require more complex cognitive and somatic strategies than other resources. Increased robusticity and stability in the wrist bones of these taxa could imply more loading in wider ranges of movement, suggesting either more diversified behavioral repertoires or more rapid suspensory positional behavior, like that of chimpanzees, as opposed to the possibly more sloth-like positional behavior of *Oreopithecus* (Wunderlich *et al.*, 1999). Ultimately this may have allowed for the substantial increase in body mass that characterizes most of the descendants of ancestor 3, and probably the increases in cognitive and behavioral complexity that are typical of all living great apes and humans.

With regard to the Asian radiation, only the western-most branch (*Ankarapithecus*) is reviewed. It appears to have arisen from a common ancestor shared by *Sivapithecus* and *Pongo*, characterized by more robust masticatory apparatuses, which may have been the key to the success of this clade as measured in species richness, and temporal and geographic range.

Dryopithecus and *Ouranopithecus* are represented here as an early radiation of the Afro-European clade of hominids (the Homininae) from which the African apes and humans arose. They share with African apes and humans an alternative strategy for enlarging the anterior jaw and dentition and providing muscle power toward anterior food processing, which may be related to differences in facial bone fusion and facial hafting, and cranial length (Begun, 1994*a*; Begun & Kordos, 1997). This *Bauplan* may have constrained descendants to certain stereotypical morphological responses to changes in diet and habitat, which may explain the parallelism in the development of soft-fruit frugivory and hard-object feeding in *Dryopithecus* and *Ouranopithecus* on the one hand and *Pan* and early humans on the other.

Many of the conclusions presented here are working hypotheses that will be tested and surely modified with the discovery of new fossils. Recent discoveries and new analyses have rescued the fossil record of European hominoids from relative obscurity. This new work has in fact revealed much about the complexity of this fossil record, but also about the evolutionary history of hominoids more generally.

Primary References

Ankarapithecus

Ozansoy, F. (1957). Faunes de mammifères du Tertiaire de Turquie et leurs révisions stratigraphiques. *Bulletin of the Mineral Research and Exploration Institute of Turkey*, **49**, 29–48.

Ozansoy, F. (1965). Étude des gisements continentaux et des mammifères du Cénozoique de Turquie. *Mémoires de la Sociéte géologique de France (nouvelle série)*, **44**, 1–92.

Andrews, P. J. & Tekkaya, I. (1980). A revision of the Turkish Miocene hominoid *Sivapthecus meteai*. *Paleontology*, **23**, 85–95.

Alpagut, B., Andrews, P., Fortelius, M., Kappelman, J., Temizsoy, I., Celebi, H., & Lindsay, W. (1996). A new specimen of *Ankarapithecus meteai* from the Sinap Formation of central Anatolia. *Nature*, **382**, 349–351.

Dryopithecus

Jaeger, G. F. (1850). Über die fossilen Säugetiere welche in Württemberg, als nachtag zu dem 1839 unter Gleichem htel erschienen Werke. *Nova Acta Academiae leopoldino carolinae germanicae naturae curiosum*, **22**, 765–924.

Lartet, E. (1856). Note sur un grand singe fossile qui se rattache au groupe des singes superieurs. *Comptes rendus de l'Académie des sciences de Paris*, **43**, 219–223.

Schlosser, M. (1902). Beiträge zur Kenntnis der Säugetierreste aus den süddeutschen Bohnerzen. *Geologische und paläontologische Abhandlungen*, **5**, 117–258.

Vidal, L. M. (1913). Nota sobre la presencia del "*Dryopithecus*" en el Mioceno superior del Pirineo catalán. *Boletín de la real Socieded española de historia natural, Sección geológica*, **13**, 499–507.

Villalta, J. F. & Crusafont, M. (1944). Dos nuevos antropomorfos del Mioceno español y su situación dentro del la moderna sistemática de los símidos. *Notas y Communicaciones, Instituto geológico y minero*, **13**, 91–139.

Crusafont, M. (1958). Neuvo hallazgo del pongido vallesiense *Hispanopithecus*. *Boletin informativo actividades europeas en paleontología de vertebrados, Sabadell, España*, **13–14**, 37–43.

Crusafont, M. & Hürzeler, J. (1961). Les Pongidés fossiles d'Espagne. *Comptes rendus de l'Académie des sciences de Paris*, **254**, 582–584.

Steininger, F. (1967). Ein weiterer Zahn von *Drypithecus* (*Dry.*) *fontani darwini* Abel, 1902 (Mammalia, Pongidae) aus dem Miozän des wiener Beckens. *Folia Primatologica*, **7**, 243–275.

Kretzoi, M. (1969). Geschichte der Primaten und der Hominisation. *Symposia Biologica Hungarica*, **9**, 3–11.

Kretzoi, M. (1975). New ramapithecines and *Pliopithecus* from the lower Pliocene of Rudabánya in north-eastern Hungary. *Nature*, **257**, 578–581.

Kordos, L. (1987). Description and reconstruction of the skull of *Rudapithecus hungaricus* Kretzoi (mammalia). *Annales historico naturales Musei nationalis Hungarici*, **79**, 77–88.

Begun, D. R. (1992*b*). *Dryopithecus crusafonti* sp. nov., a new Miocene hominid species fron Can Ponsic (Northeastern Spain). *American Journal of Physical Anthropology*, **87**, 291–310.

Begun, D. R. & Kordos, L. (1993). Revision of *Dryopithecus brancoi* Schlosser, 1901 based on the fossil hominoid material from Rudabánya. *Journal of Human Evolution*, **25**, 271–285.

Begun, D. R. (1994*a*). Relations among the great apes and humans: new interpretations based on the fossil great ape *Dryopithecus*. *Yearbook of Physical Anthropology*, **37**, 11–63.

Moyà-Solà, S. & Köhler, M. (1995). New partial cranium of *Dryopithecus* Lartet, 1863 (Hominoidea, Primates) from the upper Miocene of Can Llobateres, Barcelona, Spain. *Journal of Human Evolution*, **29**, 101–139.

Moyà-Solà, S. & Köhler, M. (1996). The first *Dryopithecus* skeleton: origins of great ape locomotion. *Nature*, **379**, 156–159.

Graecopithecus

Freyberg, B. V. (1949). Die Pikermifauna von Tour la Reine (Attika). *Annales géologiques des pays hellénoqies*, **3**, 7–10.

Koenigswald, G. H. R. von (1972). Ein Unterkiefer eines fossilen Hominoiden aus dem Unterpliozän Griechenlands. *Proceedings of the Koninklijke Nederlandse Akademie van Wetenschappen, Series B*, **75**, 385–394.

Griphopithecus darwini and aff. *Griphopithecus*

Abel, O. (1902). Zwei neue Menschenaffen aus den Leithakalkbildungen des wiener Beckens. *Sitzungsberichte-Akademie Wissenschaften in Wien, mathematisch-naturwissenschaftliche Klasse*, **1**, 1171–1207.

Glässner, M. F. (1931). Neue Zähne von Menschenaffen aus dem Miozän des wiener Beckens. *Annalen des naturhistorischens Museums in Wien*, **46**, 15–27.

Ehrensberg, K. (1938). *Austriacopithecus*, ein neuer menschen-affenartiger primate aus dem Miozan von klein-Hadersdorf bei Poysdorf in Niederosterreich (Nieder-Donau). *Sitzungsberichte-Akademie der Wissenschaften, Berlin, Klasse für Mathematik und allgemeine Naturwissenschaften*, **1**, 147.

Steininger, F. (1967). Ein weiterer Zahn von *Drypithecus* (*Dry.*) *fontani darwini* Abel, 1902 (Mammalia, Pongidae) aus dem Miozän des wiener Beckens. *Folia Primatologica*,**7**, 243–275.

Heizmann, E. (1992). Das Tertiär in südwestdeutschland. *Stuttgarter Beiträge zur Naturkunde, Serie C*, **33**, 1–90.

Heizmann, E., Duranthon, F., & Tassy, P. (1996). Miozäne Grossäugetiere. *Stuttgarter Beiträge zur Naturkunde, Serie C*, **39**, 1–60.

Heizmann, E. & Begun, D. R. (2001). The oldest Eurasian hominoid. *Journal of Human Evolution*, **41**, 463–481.

Oreopithecus

Gervais, P. (1872). Sur un singe fossile, d'espèce non encore décrite, qui a été découvert au Monte-Bamboli (Italie). *Comptes rendus de l'Académie des sciences de Paris*, **LXXIV**, 1217–1223.

Ristori, G. (1890). Le scimmie fossile italiane. *Bollettino del reale Comitato geologico d'Italia*, **XXI**, **178–196**, 225–237.

Hürzeler, J. (1949). Neubeschreibung von *Oreopithecus bambolii* Gervais. *Schweizerishe Paläontologische Abhandlungen*, **66**, 3–20.

Hürzeler, J. (1960). The significance of *Oreopithecus* in the genealogy of man. *Triangle*, **4**, 164–174.

Ouranopithecus

Bonis, L. de, Bouvrain, G., & Melentis, J. (1975). Nouveaux restes de primates hominoïdes dans le Vallésien de Macédoine (Grèce). *Comptes rendus de l'Académie des Sciences de Paris*, **182**, 379–382.

Bonis, L. de & Melentis, J. (1977). Les primates hominoïdes du Vallésien de Macédoine (Grèece): étude de la machoire inférieure. *Géobios*, **10**, 849–885.

Bonis, L. de & Koufos, G. (1993). The face and mandible of *Ouranopithecus macedoniensis*: description of new specimens and comparisons. *Journal of Human Evolution*, **24**, 469–491.

Koufos, G. D. (1995). The first female maxilla of the hominoid *Ouranopithecus macedoniensis* from the late Miocene of Macedonia, Greece. *Journal of Human Evolution*, **29**, 385–389.

Bonis, L. de, Koufos, G. D., Guy, F., Peigne, S., & Sylvestrou, I. (1998). Nouveaux restes du primate hominoïde *Ouranopithecus* dans les dépôts du Miocène supérieur de Macédoine (Grèc). *Comptes rendus de l'Académie des sciences de Paris*, **327**, 141–146.

21 | The hominoid radiation in Asia

JAY KELLEY

Introduction

The quality of the Asian hominoid record is generally relatively poor. This is somewhat ironic given the central role that Asian hominoids have played in some of the most contentious hominoid evolution debates during the twentieth century. The hominoid fossils from the Siwalik Sequence of Pakistan and India have by far the longest history among Asian hominoids, and for many decades after their initial discovery they constituted the most complete remains among all fossil hominoids. In contrast, fossil hominoids from western and eastern Asia are less well known overall. Discoveries at Shihuiba (Lufeng) and the Yuanmou Basin in China and Paşalar in Turkey are much more recent and the material has consequently been less intensively studied, but this is changing.

History of discovery and debate

Amongst the earliest reports on Siwalik vertebrates is a description by Dr. Hugh Falconer and Major Proby Thomas Cautley (1837) of an upper canine, purportedly of a large anthropoid primate. Comparisons by Mr. James Princep, editor of the *Journal of the Asiatic Society*, and later by Falconer himself, convinced both men that it came from an animal closely allied to the orangutan. Richard Lydekker (1885) assigned the canine to *Simia* sp. (*Pongo* sp.), by which time it already had been lost. If the tooth was of a fossil ape, it would rank as the first ever described. However, the quite worn canine was considerably larger than and morphologically different from any subsequently recovered fossil ape canine from the Siwaliks. Since no remains of *Pongo* have been reported from anywhere in the Indian subcontinent, the identification of this tooth as hominoid, or even anthropoid, remains in doubt.

The first unquestioned Asian fossil hominoid, a partial palate, was recovered in 1878 from the Siwalik sediments of the Potwar Plateau in what is now Pakistan by a Mr. Theobald. It was given the name *Palaeopithecus sivalensis* by Lydekker (1879), being only the second named fossil ape after *Dryopithecus fontani* (Lartet, 1856). It was said to have been collected from near the village of Jabi, but since Jabi (or Jabbi) is a common village name in the region, the provenance of the palate is unknown.

By the time of Lydekker's (1879) description of *P. sivalensis*, thousands of fossils of other mammalian taxa had already been recovered from the Siwaliks, revealing early on the extreme rarity of fossil apes there. More than 30 years passed before the next hominoids were recovered from the Siwaliks, from the vicinity of Chinji just north of the Salt Range on the Potwar Plateau. These were described by Guy Ellock Pilgrim (1910), Superintendent of the Geological Survey of India. One, a partial mandible with two teeth, was assigned to *Dryopithecus* (*D. punjabicus*), while the other, an isolated lower molar, was made the type of a new genus and species, *Sivapithecus indicus*.

During the next several years, additional specimens of fossil hominoids were collected from the Potwar Plateau, as well as from a new area near the village of Haritalyangar, in what is now Himachal Pradesh State in India. The new remains were described by Pilgrim in 1915, who, in addition to documenting more complete material of *D. punjabicus* and *S. indicus*, named one new genus and three new species, all based on single teeth. Thus was fully initiated what surely has been the most bewildering taxonomic proliferation ever for fossil primates recovered from any single region. Ultimately, Siwalik finds resulted in the naming of 11 new genera (and assignment to another five previously named genera) and 28 new species, with six genera and nine species having single teeth as their types. The unfortunate practice of naming species based on single teeth, common during the early period of discovery here as well as in Africa and Europe, has had repercussions for taxonomic nomenclature that continue to the present.

Pilgrim (1915) made detailed dental and gnathic comparisons to both extant apes and a small but growing number of recently recovered fossil anthropoids from several sites in Europe and the Fayum deposits in Egypt. Based on the occlusal crenulation patterns he regarded one of his new taxa, *Palaeosimia rugosidens*, as ancestral to extant *Pongo*. Species of *Dryopithecus* were suggested to be broadly related to African apes, particularly *Gorilla*. *Sivapithecus indicus* was placed within the Hominidae based primarily on Pilgrim's interpretation of a new partial mandible as having a shortened symphysis and an outward curvature of the corpus in the premolar region.

William King Gregory (1915, 1916) immediately challenged Pilgrim's reconstruction of the *S. indicus* mandible and its inclusion in the Hominidae. According to Gregory, the mandible fundamentally was that of an ape. In his view, *Sivapithecus* was closely related to *Palaeosimia*, both of which he placed in or close to orangutan ancestry. However, like

Pilgrim, he regarded the various Indian *Dryopithecus* species to be within the ancestry of African apes. He included *Palaeopithecus* in this group as well, which Pilgrim had regarded as a more primitive offshoot specifically related to *Gorilla*.

The growing proliferation of taxa and the phylogenetic speculations they generated established the backdrop for the next few decades of fossil discoveries in the Siwaliks. In the Pilgrim and Gregory debates are the seeds of subsequent controversies concerning Siwalik hominoids (particularly *Ramapithecus*), none of which is based on a record much better than that available to Pilgrim and Gregory.

In 1922, the American Museum of Natural History mounted an expedition to the Siwaliks led by Barnum Brown. Three partial mandibles were recovered from the new areas of Hasnot and Ramnagar and made the types of three species of *Dryopithecus* (Brown *et al.*, 1924). Gregory & Hellman (1926) described them more fully in the second major monograph on Siwalik fossil apes. Departing from Pilgrim's interpretation of substantial higher-level taxonomic diversity, Gregory & Hellman noted the general morphological similarity among all Siwalik hominoid taxa, and differences between these and European *Dryopithecus*. They speculated on consolidating genera to just those restricted to the Indian subcontinent. In this they clearly were influenced by Remane's (1921) demonstration of high levels of intraspecific variation in extant apes, and his and Schlosser's (1911) view that *Palaeopithecus* and *Sivapithecus* were congeneric. Also like Schlosser, they saw most or all of the Siwalik species as closer to *Pongo* than to African apes, based on overall morphology of the teeth.

The following year, Pilgrim (1927) described more Siwalik hominoid fossils, naming five new species and one new genus from a total of seven specimens, mostly from Haritalyangar. Having concluded based on one of his own new specimens that Gregory was correct about the ape-like morphology of the *Sivapithecus* mandible, Pilgrim acknowledged (1) that *Sivapithecus* and *Palaeopithecus* were closely related to each other (but not to *Pongo*), and (2) that *S. indicus* was not an early member of the Hominidae. However, influenced by the recent discovery of some very fragmentary remains from Moghara in Egypt (Fourteau, 1920), he still maintained that some earlier representative of *Sivapithecus* must have given rise to the Hominidae. Relying on the anatomical studies of Keith (1899), which he interpreted as showing a closer relationship of *Pan* to humans than to *Gorilla*, he reasoned that early *Sivapithecus* must be ancestral to *Pan* as well.

George Edward Lewis led the Yale North India Expedition to the Siwaliks in 1932. After an initial discourse on the need to appreciate variation and sex differences, Lewis (1934) proceeded to name four new genera and six new species from the hominoids collected. Among the genera were *Ramapithecus* and *Sugrivapithecus*, which Lewis tentatively placed within the Hominidae. *Ramapithecus* included two species and *Sugrivapithecus* one, all known only by their type specimens. The morphological features suggesting hominid status to Lewis included, among others, a presumed parabolic dental arcade, small, anteroposteriorly shortened canines, a lack of dental diastemata, slight facial prognathism and "dental crowding".

Aleš Hrdlička (1935) immediately challenged this interpretation and countered point-by-point each of the features proposed by Lewis to link the two Siwalik genera with the human lineage. In Hrdlička's view, Lewis's errors included simple misinterpretations of incomplete or damaged morphology, failure to take size into account (as, for example, when evaluating the degree of prognathism), and, importantly, the likelihood that the two most complete specimens were both female. In addition, he identified features in both the *Ramapithecus* maxilla and the *Sugrivapithecus* mandible that resembled apes.

Three years after his initial report – and one year after naming another new species of *Sugrivapithecus* (Lewis, 1936) – Lewis (1937) published the first comprehensive taxonomic revision of Siwalik hominoids, which by this time included more than 20 species in 10 genera. He reduced these numbers to four genera and nine species, with three of the genera (*Bramapithecus*, *Ramapithecus* and *Sugrivapithecus*) and five of the species having been named by him; the only holdover genus was *Sivapithecus*. *Dryopithecus* was restricted to European forms.

The following year, Gregory, Hellman & Lewis (1938) described additional specimens collected in 1935 by the Yale–Cambridge Expedition, headed by Helmut de Terra (most fossils collected by Nagamangalam Kesava Narasimha Aiyengar of the Geological Survey of India). For the first time in a report of new Siwalik hominoid material, no new taxa were named. Instead, the new specimens were used to further justify the many synonymies proposed by Lewis (1937). The growing number of fossils allowed Lewis and his colleagues to see how what formerly were thought to be diagnostic features could appear in various combinations in different specimens, or even in the different teeth of a single specimen. It also gave them a better sense of how dental wear influenced the appearance of morphology.

Perhaps stung by Hrdlička's critique of *Ramapithecus* and *Sugrivapithecus* as hominids, Gregory *et al.* (1938) appear to have backed away somewhat on this issue, stating that "while the Siwalik genus *Ramapithecus* and the South African *Australopithecus* were still simians by definition, they were almost at the human threshold, at least in respect to their known anatomical characteristics" (Gregory *et al.*, 1938: 25). They also continued to argue for a close relationship between *Sivapithecus* and *Pongo*, based on what they perceived to be numerous similarities between the two genera in the mandible and the mandibular dentition.

Also in 1938, Darashaw Nosherwan Wadia and Aiyengar published a comprehensive list of Siwalik primate fossils found up until that time. After more than 100 years of collecting in the Siwaliks, the total number of hominoid specimens recovered was a mere 60, of which just slightly more than half were single isolated teeth. There were no definitely associated upper and lower dentitions, no non-gnathic cranial remains and no postcrania. Nevertheless, this sample constituted the most abundant collection of fossil apes from anywhere in the world at the time and played a central role in most discussions of ape and human evolution.

This concludes what might be called the early period of collecting in the Siwaliks. Although Richard Dehm led another successful collecting expedition in 1939, the specimens were not described until nearly 45 years later.

Fossil hominoids from somewhere in Asia other than the Siwaliks were first described at about the time of the Yale and Yale–Cambridge Siwalik Expeditions. When Ralph von Koenigswald joined the Netherlands Geological Survey in the East Indies in 1931, he began to comb pharmacies throughout East Asia for "dragon bones" used as curatives in traditional Chinese medicine. The dragon bones were of course the remains – mostly teeth – of fossil mammals. Early in 1935 in Manila, he came across a number of Pleistocene teeth that he recognized must belong to fossil representatives of *Pongo*. Following up on this discovery led him to Hong Kong where he found hundreds of additional fossil orangutan teeth. Among these fossils was one giant lower molar that, in the paper describing the fossil orangutan teeth, von Koenigswald (1935) named a new genus and species, *Gigantopithecus blacki*. That same year, Pei (1935) also described some fossil orangutan teeth which he had obtained in the same manner. By 1939 von Koenigswald had purchased three additional teeth of *Gigantopithecus* but these, the type specimen and some additional anterior teeth were not described in detail until much later (von Koenigs-wald, 1952), because of his internment during the Second World War.

In the interim and to von Koenigswald's surprise (see von Koenigswald, 1958*b*), Franz Weidenreich (1945*a*) described the three additional teeth and reasoned that *Gigantopithecus* was ancestral to species now included in *Homo erectus*, and was thus at the root of human ancestry (see also Broom, 1939). At that time *Australopithecus* was not universally accepted as a hominid and Weidenreich viewed *Homo erectus* as near the beginning of the human lineage. Weidenreich's phylogenetic speculations initiated a debate on *Gigantopithecus* that would continue for another 30 years.

When von Koenigswald finally published his own descriptions in 1952, he challenged Weidenreich's contention. He did regard *Gigantopithecus* to be a hominid but, like *Australopithecus*, as a highly specialized side branch that paralleled the human lineage in many of its dental and gnathic adaptations. He also viewed the extreme size of *Gigantopithecus* as confirmation that it must be a terminal form. He saw *G. blacki* as the last survivor of an Asiatic stock that could be traced back to the Pliocene (now the latest Miocene) of the Siwaliks.

Two years earlier, von Koenigswald (1950) had assigned the type specimen of Pilgrim's (1915) *Dryopithecus giganteus*, a lower second or third molar, along with a large lower third premolar from Haritalyangar, to a new genus, *Indopithecus*. Singling out these two very large Siwalik teeth fit with his developing ideas about *Gigantopithecus* ancestry, and perhaps with the long-held notion that human ancestry could be traced to the Siwaliks as well. However, Dirk Hooijer (1951) challenged this proposal, prompted by his earlier questioning of Weidenreich's claims based on only a few molars (Hooijer, 1949).

The first more complete remains of *Gigantopithecus* were recovered in the late 1950s by Academia Sinica expeditions to southern Guangxi Province. The first of these was a mandible brought to the expedition by a local farmer digging in a cave for lime phosphate for fertilizer. This one cave eventually produced two more relatively complete mandibles as well as hundreds of isolated teeth (Pei & Woo, 1956; Pei & Li, 1959; Woo, 1962). Pei & Woo (1956) concluded that "the morphological pattern of *Gigantopithecus* indicates that it might belong to a side branch of the anthropoids, but to a point where it took off nearer to the hominid line than any other fossil anthropoids so far found." By the time of Woo's (1962) monograph, *Gigantopithecus* was more securely regarded as a primitive hominid. Von Koenigswald (1958*b*) particularly noted the "bicuspid" anterior lower premolar, which he reasoned must have been inherited from the last common ancestor of *Gigantopithecus*, *Australopithecus* and the Hominidae. In von Koenigswald's view, *Gigantopithecus* could not be an "anthropoid", that is an ape, because it did not have a sectorial lower third premolar.

When a joint Yale–Panjab University group recovered a relatively complete and very large mandible from the vicinity of Haritalyangar in 1968 (Simons & Chopra, 1969*a*), arguments about *Gigantopithecus* resurfaced and the focus shifted once more to the Siwaliks. The mandible was assigned to a new species of *Gigantopithecus*, *G. bilaspurensis*, by Simons & Chopra (1969*b*). It shared with *G. blacki* marked posterior divergence of the toothrows coupled with relatively small canines and incisors (Simons & Chopra, 1969*a*, 1969*b*). Simons & Chopra (1969*b*) concluded that it demonstrated ecological convergence between *Gigantopithecus* and early hominids. *Gigantopithecus* was in fact a pongid, with certain dental and gnathic specializations developed in parallel to humans and best interpreted as adaptations for foraging in open country. They arrived at this probably in large part because, among Siwalik taxa, *Ramapithecus* was viewed as the more likely human ancestor. Since the human lineage was still perceived to be strictly linear, the later and less human-like *G. bilaspurensis* could not possibly be a hominid. This conclusion was reinforced by what they correctly saw as links with earlier Siwalik apes in the "more primitive" features of *G. bilaspurensis* mandibular and molar morphology.

David Pilbeam (1970) expanded upon this in a more in-depth phylogenetic and ecological argument for excluding *Gigantopithecus* from the Hominidae. He traced the lineage from *Dryopithecus* (*Sivapithecus*) *indicus* through *D. giganteus* (the type molar) to *G. bilaspurensis*. Inclusion of the *D. giganteus* type in this phylogenetic sequence (suggested earlier by von

Koenigswald, 1950) was later formalized taxonomically by Szalay & Delson (1979), who made *G. bilaspurensis* a junior synonym of *D. giganteus*, which became *G. giganteus*.

Claims about the hominid status of *Gigantopithecus* continued in the 1970s (Eckhardt, 1973, 1975; Frayer, 1973). Protagonists on both sides of the argument viewed the *G. bilaspurensis* mandible as strong support, as either a phylogenetic link to earlier apes living alongside the early hominid *Ramapithecus*, or as a precursor to the earliest undoubted hominids of the African Plio-Pleistocene. In the end, the argument seems to have died of neglect in the avalanche of ever earlier and more complete early hominid remains from Africa. In the late 1980s to early 1990s, the geographic record of *Gigantopithecus* was extended into Vietnam (Cuong, 1984, 1985; Nisbett & Ciochon, 1993; Ciochon *et al.*, 1996).

Back in the Siwaliks, successful collecting efforts continued through the 1950s and 1960s: Kedar N. Prasad in 1951, 1954 and 1962 at Haritalyangar, Dehm in 1955–6 in the Potwar Plateau, Shri Uma Shanker in 1962–3, and later Vishwa Jit Gupta, in the Kangra foothills northwest of Haritalyangar in Himachal Pradesh, and von Koenigswald in1964–5, also in the Potwar Plateau (see Khatri, 1975). Three new species and one new genus were named: *Sivapithecus aiyengari* (Prasad, 1962) and *S. lewisi* (Pandey & Sastri, 1968) – both based on partial mandibles and diagnosed primarily on the basis of their very large size – and *Chinjipithecus atavus* (von Koenigswald, 1981), based on a single molar.

Also in the 1960s, Simons & Pilbeam (1965) published their seminal taxonomic revision of fossil apes reducing Lewis's four genera and ten species (plus the subsequently named *Indopithecus* and *S. aiyengari*) to two genera and four species. *Sivapithecus* was made one of three subgenera within *Dryopithecus*, with the latter accommodating most fossil hominoids known at that time. Size exclusively differentiated the two species of *D.* (*Sivapithecus*), *D.* (*S.*) *sivalensis* and *D.* (*S.*) *indicus*. The second genus, *Ramapithecus*, included one species, *R. punjabicus*. The hypodigms of all four species included material from outside the Siwaliks, including Europe, Africa, and West and East Asia, reflecting in part the strong emphasis on dental metric criteria to differentiate species viewed as having broadly similar morphology. With respect to phylogeny, Simons & Pilbeam (1965) noted that none of the *Sivapithecus* material significantly resembled *Pongo*.

The most significant matter of this period concerning Asian hominoids, and ultimately all Miocene hominoids, did not involve recovery of new fossils. In 1961 Elwyn Simons revived Lewis's argument that *Ramapithecus* was an early hominid. He – and later his student, David Pilbeam – championed this argument throughout most the next two decades. The evolution and unraveling of the case for the hominid status of *Ramapithecus* is a fascinating episode in the history of fossil primates (see Wolpoff, 1983; Frayer 1997*b*), only briefly summarized here with emphasis on new fossil discoveries and the debate.

According to Simons (1961*c*), a major part of the stimulus to re-examine *Ramapithecus* came from the astonishing new finds of early hominids by Mary and Louis Leakey at Olduvai Gorge. He mostly reiterated points made by Lewis (1934) regarding the perceived morphology of the type maxilla of *Ramapithecus brevirostrus*, GSP 13799, but in the context of an expanded African fossil record of what were now universally recognized as early hominids. He dismissed parallelism as an explanation largely because *Ramapithecus* was in the "proper place and time" to be a forerunner of the Pleistocene hominids. In 1963 Simons added two more partial maxillae to *Ramapithecus*, one from the Siwaliks and one from the middle Miocene site of Fort Ternan in Kenya, the latter described a year earlier as the type of a new genus, *Kenyapithecus* (Leakey, 1962*a*). It was also the only Miocene hominoid with a reasonably reliable minimum age, at about 14 Ma, thus pushing the origins of the human lineage back to at least this date.

Increasing numbers of Old World Miocene hominoid fossils were added to the *Ramapithecus* hypodigm during the next several years (see Simons, 1976). The list of presumed hominid characters also expanded, partly in response to changing perspectives regarding early hominid behavioral ecology, and a resulting shift in adaptive interpretations of the morphology (see Wolpoff, 1983). Following publication of Jolly's (1970*b*) *Theropithecus*-complex (or T-complex) model of hominid origins, hominid characters such as small, vertically implanted incisors, thickened molar enamel, heavy interproximal wear, a steep molar wear gradient, and heavily buttressed mandibles were viewed as additional evidence for inclusion in the human lineage. This character suite was linked to more powerful mastication, needed for ground feeding in open habitats (Simons, 1976; Simons & Pilbeam, 1972). Interestingly, the ecological backdrop for the T-complex was used to simultaneously argue for *parallel* evolution between hominids and *Gigantopithecus* (Pilbeam, 1970; Simons & Ettel, 1970).

During the 1970s, the hominid status of *Ramapithecus* came under attack, even as the hypodigm grew. The challenge came mostly from Milford Wolpoff and his students (Wolpoff, 1971*b*; Greenfield, 1974, 1975, 1978). David Frayer in particular offered a measured and pointed rebuttal to Simons and Pilbeam (Frayer, 1974). Like Hrdlička in 1935, he dissected the morphological arguments point-by-point. Focusing mostly on the original Siwalik material, he demonstrated that *Ramapithecus* was not fundamentally different from its contemporary, *Sivapithecus*, in those features that linked *Ramapithecus* to hominids (see also Frayer 1997*b*). By the late 1970s, Pilbeam, too, had come to recognize the essential similarity of all Siwalik hominoid material. This growing realization resulted partly from the many new fossils, including the first postcranial remains, produced by the Yale (later Harvard)–Geological Survey of Pakistan (GSP) project initiated in 1973 by him and still ongoing (Pilbeam *et al.*, 1977, 1980). Pilbeam maintained that *Ramapithecus* was a probable hominid ancestor, but admitted that there was insufficient evidence to demonstrate it. Significantly, he removed *Ramapithecus* from the Hominidae and instead associated it with *Sivapithecus* in the family Ramapithecidae. Shortly thereafter, Greenfield (1979) formally made *Ramapithecus* a junior synonym of *Sivapithecus*. For some, however, this simply meant that all species within this radiation were hominids (Kay, 1982; Kay & Simons, 1983).

Developments taking place in a discipline far removed from paleontology would bear directly on the phylogenetic interpretation of *Ramapithecus*. At the same time, ironically, that the 14-Ma Fort Ternan mandible was added to the *Ramapithecus* hypodigm, Morris Goodman (1962, 1963) published his ground-breaking work on the comparative immunochemistry of extant apes and humans. Vince Sarich and Allan Wilson extended this work and challenged the notion that the beginnings of the human lineage could be anywhere nearly so remote in time as the 14-Ma Fort Ternan material (Sarich & Wilson, 1966, 1967). According to their interpretation, the measured immunochemical distances among the extant great apes and humans could not accommodate a last common ancestor more distant in time than about 5 Ma. As biochemical studies multiplied and were refined (Goodman & Tashian, 1976; Cronin, 1977), it also became clear that, if their estimates of primate lineage divergence times were approximately accurate, even much younger *Ramapithecus* specimens from the Siwaliks (approximately 8 Ma: Tauxe, 1979) were likewise too old to belong to the human lineage. However, the paleontological community largely discounted the underlying premise of the regularity of genetic change and almost universally rejected this argument and the notion of 'molecular clocks" (Simons, 1976). For some, however, the comparative genetic data gradually began to erode their confidence in morphological arguments for *Ramapithecus* as a hominid (Pilbeam, 1979, 1983).

Peter Andrews (1982) was among the first paleoanthropologists to accept fully the validity of the molecular data. Andrews & Cronin (1982) combined this data with the morphology of a recently recovered partial skull from Turkey attributed to *Sivapithecus* (Andrews & Tekkaya, 1980) to argue that *Sivapithecus*, and by extension *Ramapithecus*, were not early hominids but were instead in the orangutan lineage. However, it was a spectacular new find from the Siwaliks that convinced the majority of workers – a relatively complete skull of *S. sivalensis* (GSP 15000) that bore a striking phenetic resemblance to living orangutans (Pilbeam, 1982). Detailed study of this skull and other more recent finds of the Yale–GSP Siwalik project by Steve Ward and colleagues produced an impressive list of presumed synapomorphies between *Sivapithecus* and *Pongo* (Ward & Kimbel, 1983; Ward & Pilbeam, 1983; Ward & Brown, 1986), elaborating and adding to those enumerated by Andrews & Cronin (1982).

It is one of those wonderful coincidences of fossil discovery that in 1879, exactly 100 years before the discovery of the *Sivapithecus* skull, Lydekker had written the following concerning the first described Siwalik hominoid specimen, the newly named *Sivapithecus sivalensis*:

> I can only hope that on some future occasion we may be fortunate enough to come across the cranium of this most interesting relic of the past, when we shall be able with some approach to certainty to assign to it its exact affinities, which with our present meagre specimens we can only vaguely guess at. (Lydekker, 1879: 39)

With the publication and subsequent analysis of GSP 15000, the position of *Sivapithecus* became so secure that, for many, the estimated age of divergence of the *Pongo* lineage based on the Siwalik record of *Sivapithecus* became the favored calibration point for molecular-based divergence times of other primate lineages.

Sporadic collecting in the Indian Siwaliks over the last few decades, mostly in the vicinity of Ramnagar, Haritalyangar and the Kangra Hills, has produced a small number of isolated teeth and jaws (Sahni *et al.*, 1974, 1980; Chopra & Kaul, 1975; Dutta *et al.*, 1976; Gupta *et al.*, 1982; Chopra, 1983; Verma & Gupta, 1997; Cameron *et al.*, 1999). The early 1980s also saw the belated publication of a number of *Sivapithecus* jaws and teeth collected from the Potwar Plateau in 1939 and 1955–6 (Dehm, 1983) and 1964–5 (von Koenigswald, 1983), some from Y311, which subsequently became the richest hominoid locality of the Yale–GSP project, known as the Nagri type locality to the earlier collectors. Numerous teeth and postcranial elements from Y311 (Pilbeam *et al.*, 1977, 1980, 1990; Rose 1983, 1986), are almost without exception the largest hominoid remains from the Siwaliks (Dehm, 1983); some were even assigned to *Gigantopithecus* initially (Pilbeam *et al.*, 1977). As a consequence, the hominoid from this locality was assigned to a new species, *Sivapithecus parvada*, which was nearly twice as large as any other *Sivapithecus* species (Kelley, 1988).

Unquestioned acceptance of *Sivapithecus* as an early orangutan was relatively short-lived. While none of the *Sivapithecus* postcrania collected up to this time closely resembled *Pongo*, neither were they so derived as to preclude this relationship. Most were sufficiently generalized to fit the hypothetical ancestral morphotype of the extant great apes and humans (Pilbeam *et al.*, 1980; Rose, 1983, 1984, 1986, 1989). This changed dramatically with the recovery of a humeral diaphysis of *S. parvada* from locality Y311 that more closely resembles more terrestrial, more strictly pronograde non-hominoid anthropoids (Pilbeam *et al.*, 1990). Since humeral features functionally linked to forelimb suspension universally were considered among the most definitive synapomorphies of the great ape and human clade (e.g., Straus, 1949; Tuttle, 1974), the *S. parvada* humerus suggested that *Sivapithecus* might be outside this clade altogether (Pilbeam et. al., 1990; Pilbeam, 1996). Others, however, have questioned earlier conceptions of the ancestral morphotype and instead have suggested that forelimb suspension evolved in parallel within the orangutan and great ape/human lineages from more pronograde ancestors (Andrews, 1992; Begun, 1993*a*, 1994*a*; Larson, 1998).

There were a number of significant hominoid discoveries elsewhere in Asia during the latter part of the twentieth century. In 1969, numerous isolated hominoid teeth were found near the village of Paşalar in western Turkey during a lignite exploration. Andrews & Tobien (1977) assigned them largely according to size to *Sivapithecus darwini* (now *Griphopithecus darwini*), known by a few isolated teeth from Děvínská Nová Ves, Slovakia (formerly Neudorf an der March), and *Ramapithecus wickeri*, known from Fort Ternan, Kenya. With the possible exception of a single molar from Engelswies in Germany (Heizmann, 1992), the Paşalar sample, and perhaps the teeth from Děvínská Nová Ves, represent the earliest known hominoids outside of Africa (approximately 15 Ma). Andrews and Berna Alpagut initiated a program of excavation at the site in 1983, which has recovered hundreds of teeth and several partial jaws (Alpagut *et al.*, 1990; Martin & Andrews, 1993*b*). Based on the larger collections, the latter authors now believe that the majority of the specimens belong to *Griphopithecus alpani* (based on a mandible from the somewhat younger Turkish site of Çandir: Tekkaya, 1974). The less common species remains unnamed and has not been fully characterized, but it shows many similarities to the Fort Ternan specimens (Ward *et al.*, 1999). Indeed, Eurasian–African hominoid taxonomy and phylogeny are in the midst of vigorous debate following recent discoveries in Kenya and the proposition of *Equatorius africanus* (see Ward *et al.* 1999; Begun, 2000; Benefit & McCrossin, 2000; Kelley *et al.*, 2000; Chapters 20, 22). Hominoid fossils have also been recovered from the later Miocene Sinap Formation in Turkey (*Ankarapithecus*; see Chapter 20).

Important discoveries of late Miocene, and perhaps early Pliocene, hominoids have also been made in China. The earliest of these, a partial mandible with all of the anterior teeth and the left molars (most tooth crowns damaged), was discovered in 1947–8 in southern Gansu province but first described in 1988, as a new species of *Dryopithecus*, *D. wuduensis* (Xue & Delson, 1988). Ten years later and approximately 1000 km further south, several teeth were recovered from the site of Xiaolungtan, near Kaiyuan in Yunnan (Woo, 1957). They initially also were described as a new species of *Dryopithecus*, *D. keiyuanensis*, but were later transferred to *Sivapithecus* by Simons & Pilbeam (1965). Additional fossils, including two partial maxillae, were discovered there in 1980 and 1982 (Zhang, 1987).

In the early 1970s, a second site in Yunnan was discovered near Shihuiba, Lufeng County. This late Miocene site produced, in addition to hundreds of teeth, numerous mandibles and maxillae and several relatively complete but badly crushed skulls. As was the fashion at the time, the larger specimens were assigned to *Sivapithecus* and the smaller to *Ramapithecus* (Xu *et al.*, 1978; Xu & Lu, 1979). However, the features distinguishing the two groups were those that typically distinguish males from females among living primates (Wu *et al.*, 1983, 1985, 1986; Kelley & Etler, 1989). Moreover, the Lufeng crania differ markedly from the *Sivapithecus* partial skull in a number of features. Wu (1987) concluded that the Lufeng sample represents a single, sexually dimorphic species belonging to a new genus, *Lufengpithecus lufengensis*.

It soon became apparent, however, that the Lufeng sample displayed an exceptional degree of overall metric variation in the dentition (Kelley & Etler, 1989; Wood & Xu, 1991). Kelley & Xu (1991) and Kelley (1993) demonstrated that this was due to a level of sexual size dimorphism in the dentition substantially greater than in any living hominoid. Several workers instead claimed that the variation must reflect the presence of two species in the sample (Martin, 1991; Cope & Lacy, 1992, 1994; Plavcan, 1993). However, in a study based on simulations specifically designed to reproduce various metric attributes of the Lufeng sample from two species, Kelley & Plavcan (1998) showed that any two-species solution was highly improbable. Thus, it appears that *Lufengpithecus lufengensis* was extremely dimorphic in tooth size, and perhaps in body size as well (Kelley & Xu, 1991), which has far-reaching implications for interpreting metric variation and determining species numbers in other Miocene hominoid samples (Kelley & Plavcan, 1998).

Lufengpithecus molar morphology is uncannily like extant *Pongo*, but its overall craniofacial morphology is very different (Wu *et al.*, 1983, 1985). This is the reverse of the situation in *Sivapithecus*. Schwartz (1990, 1997) later identified seemingly derived *Pongo*-like features in the Lufeng crania, but for many it has remained perplexing that *Sivapithecus* should have such a striking resemblance to *Pongo* in craniofacial but not dental morphology, while *Lufengpithecus* for the most part shows the reverse.

The most recent Chinese Miocene hominoid discoveries come from the Yuanmou Basin, less than 100 km north of Lufeng and include approximately 1500 teeth, nearly 20 upper and lower jaws and a well-preserved juvenile skull (Zhang *et al.*, 1987*a*, 1987*b*, 1988; Liu Wu, pers. comm.). The Yuanmou hominoids were assigned to two new species, *Ramapithecus hudienensis* and *Homo orientalis*, but Zheng & Zhang (1997) recognized that most specimens belonged to a single species, *Lufengpithecus hudienensis*, which they consider a sister species of *L. lufengensis*. They also named another new species from this collection, *L. yuanmouensis*, with the juvenile skull as the type specimen. Based on published accounts, however, it is probably best for the time being to regard *L. yuanmouensis* as a junior synonym of *L. hudienensis*. Most of the material from Yuanmou has yet to be described.

Finally, numerous mostly middle to late Pleistocene sites in China and elsewhere in southeast Asia have produced remains of fossil orangutans (von Koenigswald, 1935, 1940, 1957; Pei, 1935; Teilhard de Chardin *et al.*, 1935; Weidenreich, 1937; Hooijer, 1948, 1960; Gu *et al.*, 1987; Ciochon *et al.*, 1990, 1996; Nisbett & Ciochon, 1993; Schwartz *et al.*, 1994, 1995; Ho *et al.*, 1995) and gibbons (Matthew &

Granger, 1923; see also Delson, 1977 for a review of fossil gibbon sites). With few exceptions, these have been assigned to extant species of *Hylobates* and *Pongo*, with a number of new subspecies named for the latter. Schwartz *et al.* (1995) did name one new pongid genus, *Langsonia*, and one new species of *Pongo*, *P. hooijeri*, each based on a small number of isolated teeth. However, based on the published diagnoses and figures, it is not clear that the new taxa are warranted. Among the various fossil *Pongo* subspecies there is a fair amount of variation in tooth size, indicating an overall reduction in tooth size through the Pleistocene to the present (Hooijer, 1948; Ho *et al.*, 1995).

Taxonomy

Systematic framework

The taxonomy of Asian fossil hominoids that follows is very tentative. Not only is the species-level taxonomy uncertain in many cases, even some generic assignments are questionable or have not been fully explored. The Siwalik hominoids of Pakistan and India in particular badly need revision. Three species of *Sivapithecus* are described below but only *S. parvada* and *S. sivalensis* are unambiguously defined. Moreover, more species likely are present in the approximately 4-million-year Siwalik hominoid record (Badgley *et al.*, 1984; Kelley, 1986; Kelley & Pilbeam, 1986). Diagnosing them is complicated by the poor quality of the record at key time intervals and sites, and by the generally fragmentary nature of the fossils.

Three species of *Lufengpithecus* are recognized from the late Miocene of southern China. However, material from sites other than Lufeng has not been examined thoroughly. The small sample from Kaiyuan in particular (*L. keiyuanensis*) has been largely ignored in analyses of the much larger sample from Lufeng, and comparisons to *L. hudienensis* from Yuanmou are limited (Ho, 1988; Zheng & Zhang, 1997). It is not clear from published analyses that these samples represent three distinct species.

Order Primates Linnaeus, 1758
- Infraorder Catarrhini É. Geoffroy Saint-Hilaire, 1812
  - Superfamily Hominoidea Gray, 1825
    - Family Afropithecidae
      - Subfamily Griphopithecinae
        - Genus *Griphopithecus* Abel, 1902
          - *Griphopithecus alpani* Tekkaya, 1974
          - *Griphopithecus darwini* Abel, 1902
    - Family Hominidae Gray, 1825
      - Subfamily Homininae Gray, 1825
        - Tribe Dryopithecini Gregory & Hellman, 1939
          - Genus *Dryopithecus* Lartet, 1856
            - ?*Dryopithecus wuduensis* Xue & Delson, 1988
            - *Dryopithecus fontani* Lartet, 1856
            - *Dryopithecus brancoi* Schlosser, 1901
            - *Dryopithecus laietanus* Villalta & Crusafont, 1944
            - *Dryopithecus crusafonti* Begun, 1992
      - Subfamily Ponginae Elliot, 1913
        - Tribe Sivapithecini
          - Genus *Sivapithecus* Pilgrim, 1910
            - *Sivapithecus sivalensis* Lydekker, 1879
            - *Sivapithecus indicus* Pilgrim, 1910
            - *Sivapithecus parvada* Kelley, 1988
          - Genus *Gigantopithecus* von Koenigswald, 1935
            - *Gigantopithecus giganteus* Pilgrim, 1915
            - *Gigantopithecus blacki* von Koenigswald, 1935
        - Tribe Lufengpithecini
          - Genus *Lufengpithecus* Wu, 1987
            - *Lufengpithecus lufengensis* Xu *et al.*, 1978
            - *Lufengpithecus keiyuanensis* Woo, 1957
            - *Lufengpithecus hudienensis* Zhang *et al.*, 1987

Family Afropithecidae

Subfamily Griphopithecinae

GENUS *Griphopithecus* Abel, 1902
INCLUDED SPECIES *G. alpani*, *G. darwini*

SPECIES *Griphopithecus alpani* Tekkaya, 1974 (Fig. 21.1)
TYPE SPECIMEN MTA 2253, mandible with left P_3–M_3 and right P_4–M_3
AGE AND GEOGRAPHIC RANGE Middle Miocene (MN 6), Turkey
ANATOMICAL DEFINITION
Molars with bunodont, relatively thick-enameled cusps; lower molars with well-developed buccal cingula. Upper central incisors relatively narrow mesiodistally with a distinct, high-relief median lingual pillar; upper lateral incisors with highly asymmetric mesial and distal margins. Male canines, especially upper canines, robust and relatively low crowned. Mandible robust with well-developed superior, and especially inferior, transverse tori. Mandibular planum alveolare somewhat to extremely long and shallowly inclined.

Family Hominidae

Subfamily Homininae

Tribe Dryopithecini

GENUS *Dryopithecus* Lartet, 1856
INCLUDED SPECIES *D. brancoi*, *D. crusafonti*, *D. fontani*, *D. laietanus*, ?*D. wuduensis*

SPECIES ?*Dryopithecus wuduensis* Xue & Delson, 1988
TYPE SPECIMEN XD47Wd001, partial mandible with left P_{3-4} and several damaged teeth
AGE AND GEOGRAPHIC RANGE Late Miocene, southern Gansu Province, central China

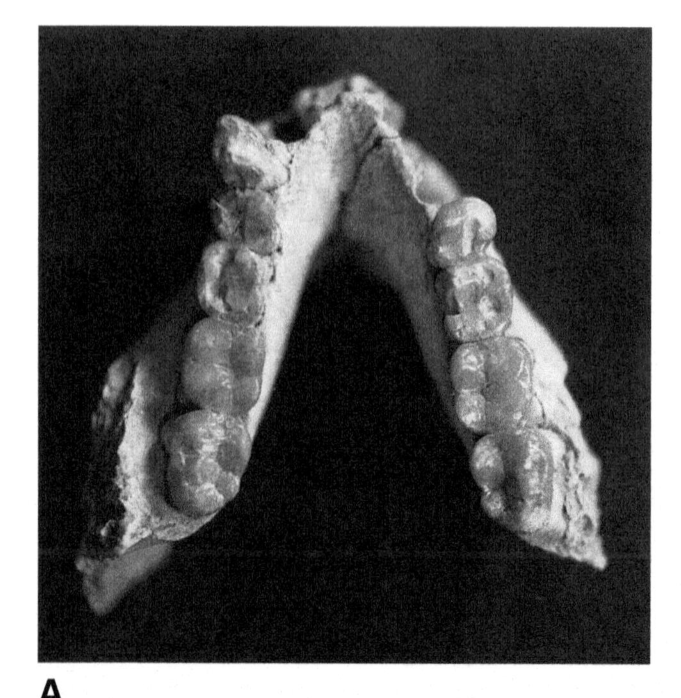

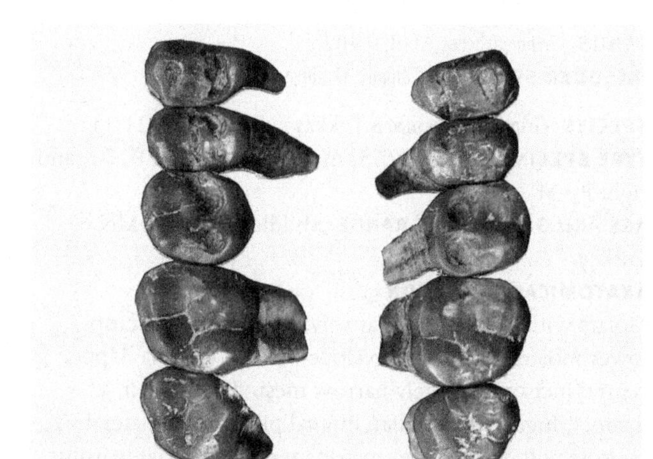

Fig. 21.1 *Griphopithecus alpani.* (A) Holotype (MTA 2253) from Çandir, Turkey; (B) associated maxillary tooth rows (R1636) from Paşalar, Turkey. Scale bars = 1 cm. (B) Courtesy of Peter Andrews.

ANATOMICAL DEFINITION

Dentally and gnathically small species of *Dryopithecus*. Mesiodistally elongate P_4 and relatively short P_3. Mandibular symphysis with steeply sloping planum alveolare and moderately developed superior and inferior transverse tori.

Subfamily Ponginae

Tribe Sivapithecini

GENUS *Sivapithecus* Pilgrim 1910

Molars with bunodont, relatively thick-enameled cusps and constricted basins; upper and lower molars both lack cingula. Upper central incisors relatively very wide mesiodistally, with a broad, low relief and heavily crenulated lingual tubercle that may extend from the mesial to the distal margins forming a continuous shelf. Upper central incisors very much wider mesiodistally than the lateral. Mandible robust with well-developed superior and inferior transverse tori. Orbits ovoid and taller than broad; distinct supraorbital costae; interorbital distance very narrow; inferior orbital margin well superior to the top of the nasal aperture. Lacks frontoethmoid sinus. Nasal floor smooth; incisive fossa (nasally) and incisive foramen (orally) both minute and indistinct; incisive canal very narrow. Nasoalveolar clivus long and strongly curved.

INCLUDED SPECIES: *S. indicus, S. parvada, S. sivalensis*

SPECIES *Sivapithecus sivalensis* Lydekker, 1879 (Figs. 21.2–21.4)

TYPE SPECIMEN GSI D-1, right maxilla with C and P^4–M^3

AGE AND GEOGRAPHIC RANGE Late Miocene (approximately 8.5–9.5 Ma), Siwaliks of India and Pakistan

ANATOMICAL DEFINITION

As for genus.

SPECIES *Sivapithecus indicus* Pilgrim, 1910 (Fig. 21.5)

TYPE SPECIMEN GSI D-176, right M_2 or M_3

AGE AND GEOGRAPHIC RANGE Late Miocene (approximately 10.5–12.5 Ma), Siwaliks of India and Pakistan

ANATOMICAL DEFINITION

Tooth size somewhat smaller on average than *S. sivalensis.* M_3 generally much larger than M_2. Nasoalveolar clivus perhaps somewhat shorter than in *S. sivalensis.* Humerus with prominent, anteriorly facing deltopectoral crest; shaft retroflexed anteroposteriorly and strongly internally curved mediolaterally (Fig. 21.5A).

SPECIES *Sivapithecus parvada* Kelley, 1988 (Fig. 21.5B, 21.6)

TYPE SPECIMEN BSPhG 1939 X 4, left and right mandibular corpora with left P_{3-4}, M_{2-3} and right C, P_3, M_2

AGE AND GEOGRAPHIC RANGE Late Miocene (approximately 10.0 Ma), Siwaliks of Pakistan

ANATOMICAL DEFINITION

Substantially larger dentally and postcranially than other *Sivapithecus* species. I^1 very wide mesiodistally in relation to breadth. M_3 much larger than M_2. Premolars, especially lower premolars, exceptionally large in relation to the molars. Symphyseal region and anterior corpora of mandible extremely deep.

GENUS *Gigantopithecus* von Koenigswald, 1935

Exceptionally large dentally and gnathically. Relatively small lower incisors. Canines low crowned in relation to mandibular size. Premolars somewhat to markedly molarized and relatively large in relation to the molars.

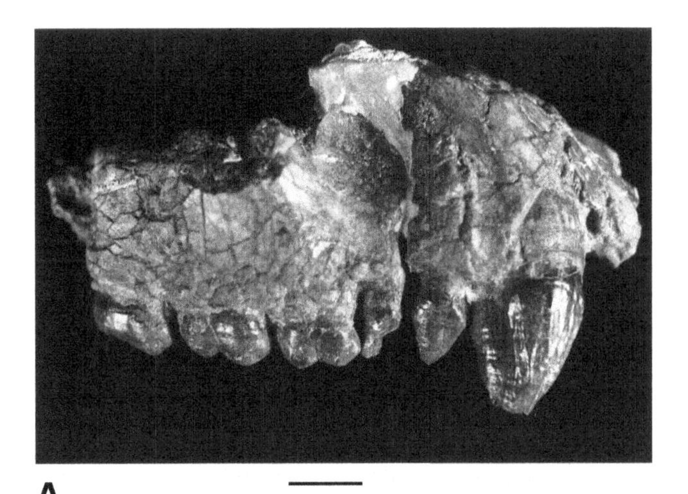

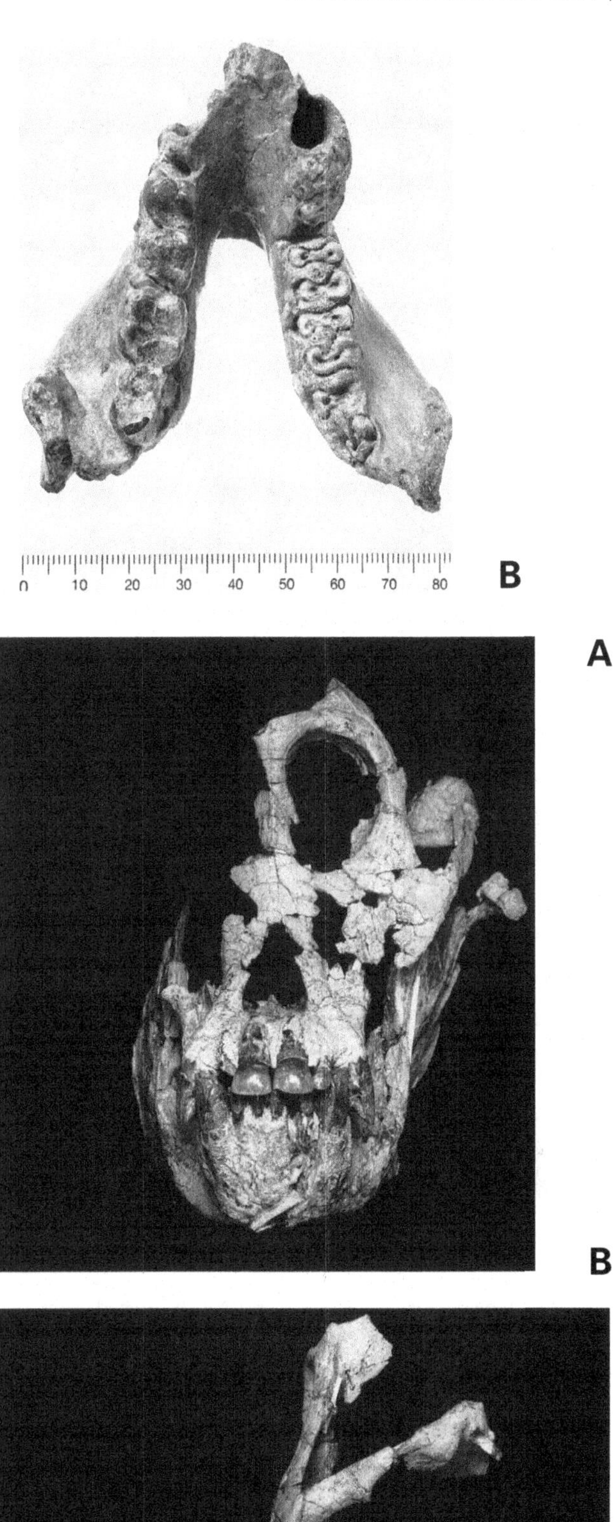

Fig. 21.2 *Sivapithecus* cf. *sivalensis* (A) Partial maxilla (GSI D-196) from the vicinity of Haritalyangar, India; one of the more complete specimens from the early collections. Scale bar = 1 cm; (B) male mandible of *Sivapithecus sivalensis* (GSP 9564) from the Potwar Plateau, Pakistan (photograph courtesy of David Pilbeam).

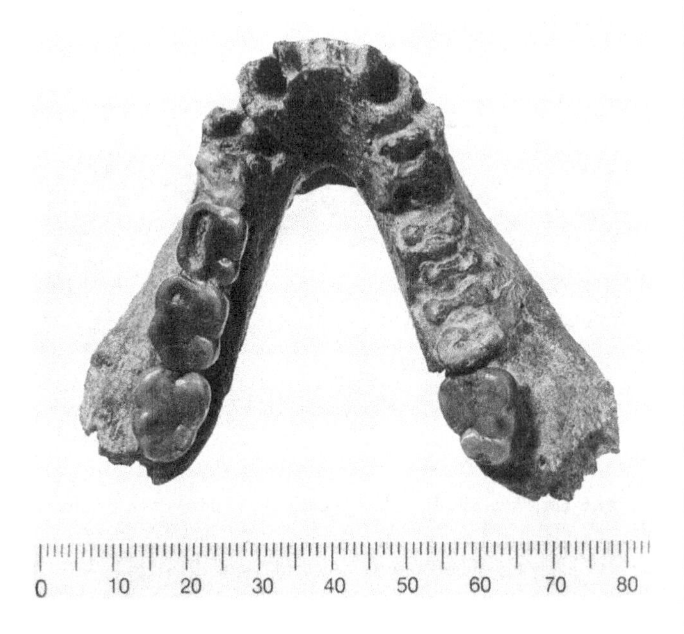

Fig. 21.3 *Sivapithecus sivalensis*. Female mandible (GSP 4622) from the Potwar Plateau, Pakistan; formerly assigned to *Ramapithecus*. Photograph courtesy of David Pilbeam.

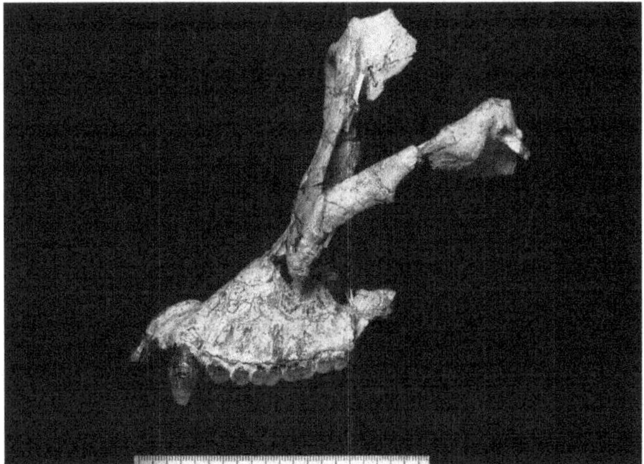

Fig. 21.4 *Sivapithecus sivalensis*. Partial skull (GSP 15000) from the Potwar Plateau, Pakistan: (A) anterior, (B) lateral views. Photographs courtesy of David Pilbeam.

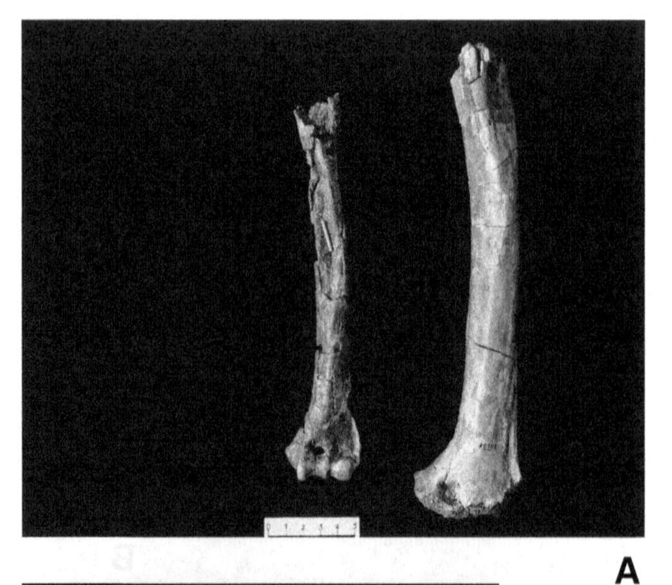

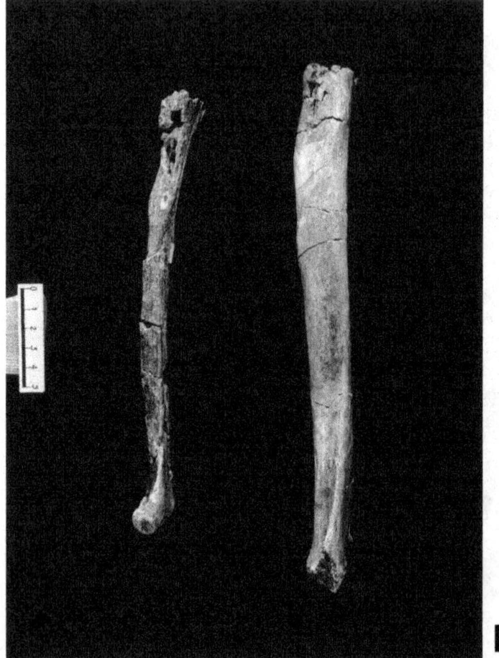

Fig. 21.5 Humeri of, left, *Sivapithecus indicus* (GSP 30730) and, right, *S. parvada* (GSP 30754) from the Potwar Plateau, Pakistan: (A) anterior, (B) lateral views. Photographs courtesy of David Pilbeam.

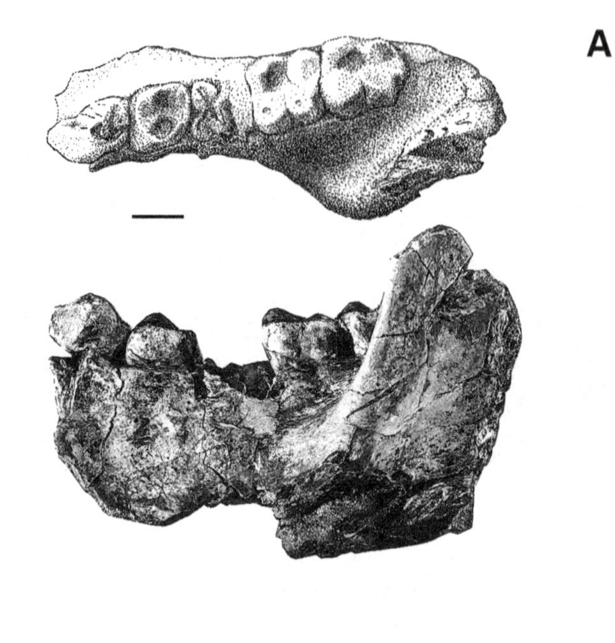

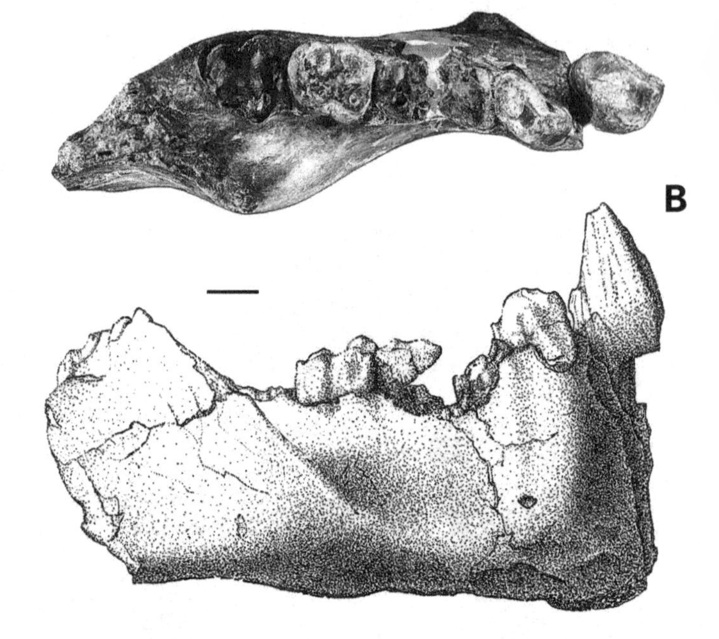

Fig. 21.6 *Sivapithecus parvada*. Holotype mandible (BSPhG 1939 X 4) from the Potwar Plateau, Pakistan: (A) left corpus, (B) right corpus. Scale bar = 1 cm. Photographs courtesy of Richard Dehm; drawings by Ken Upham.

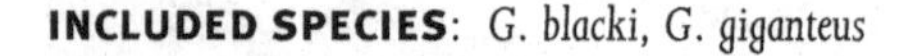

INCLUDED SPECIES: *G. blacki*, *G. giganteus*

SPECIES *Gigantopithecus giganteus* Pilgrim, 1915 (Fig. 21.7)
TYPE SPECIMEN GSI D-175, described as a right M_3, but probably M_2
AGE AND GEOGRAPHIC RANGE Late Miocene, Siwaliks of India and Pakistan
ANATOMICAL DEFINITION
As for genus.

SPECIES *Gigantopithecus blacki* von Koenigswald, 1935 (Fig. 21.8)
TYPE SPECIMEN Specimen 1, right M_3
AGE AND GEOGRAPHIC RANGE Pleistocene, southern China, Vietnam
ANATOMICAL DEFINITION
Larger dentally and gnathically than *G. giganteus*. Lower premolars more strongly molarized (P_3 more distinctly bicuspid, P_4 more elongated mesiodistally). Molars very high-crowned with low cusps and frequent development of accessory cuspules. Mandibular corpus extremely deep, especially posteriorly.

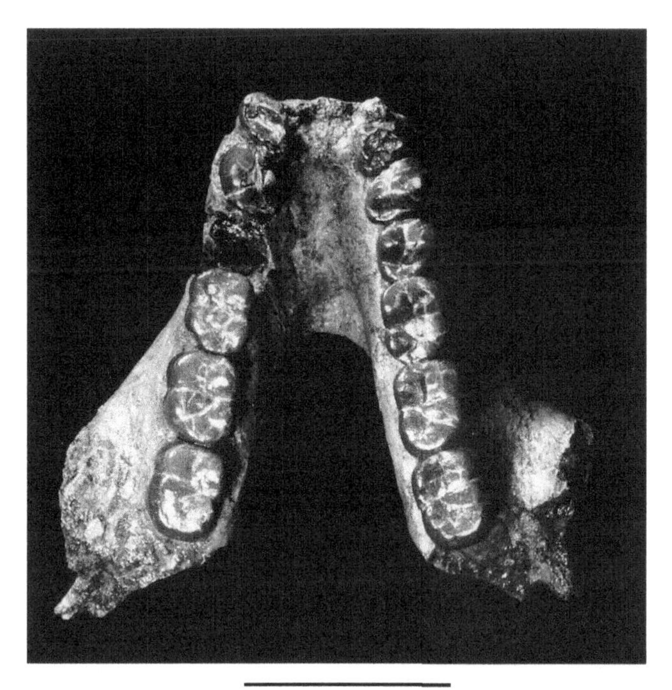

Fig. 21.7 *Gigantopithecus giganteus*. Mandible (CYP 359/68) from the vicinity of Haritalyangar, India. Scale bar = 4.5 cm. Photograph courtesy of Eric Delson.

Subfamily Ponginae

Tribe Lufengpithecini

GENUS *Lufengpithecus* Wu, 1987
Molars with relatively thick enamel, peripheralized cusp apices with expansive basins, and a dense, complex pattern of occlusal crenulations. Upper central incisors high-crowned and labiolingually thick in relation to mesiodistal length, with a distinct, high-relief median lingual pillar. Lower incisors high-crowned, relatively narrow mesiodistally and moderately procumbent. Male canines, especially lower canines, taper sharply toward the apex, appearing very gracile, and are relatively very high-crowned. Mandibular symphysis with a moderate superior transverse torus and prominent, robust inferior torus. Orbits approximately square in outline and interorbital region wide; glabella broad and depressed. Superior margin of the nasal aperture higher than inferior margin of the orbits. Nasoalveolar clivus relatively short.
INCLUDED SPECIES: *L. hudienensis*, *L. keiyuanensis*, *L. lufengensis*

SPECIES *Lufengpithecus lufengensis* Xu et al., 1978 (Figs. 21.9–21.10)
TYPE SPECIMEN PA 580, mandible with right and left I_2–M_3
AGE AND GEOGRAPHIC RANGE Late Miocene of Yunnan Province, southern China
ANATOMICAL DEFINITION
As for genus

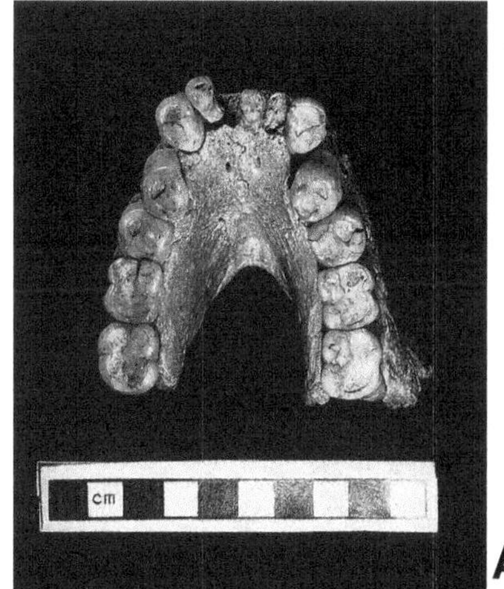

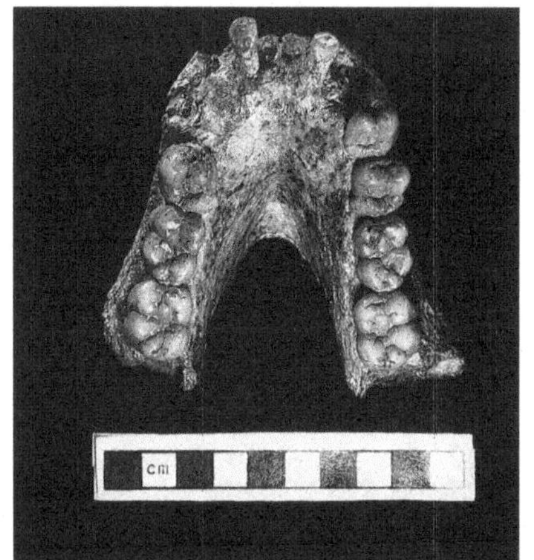

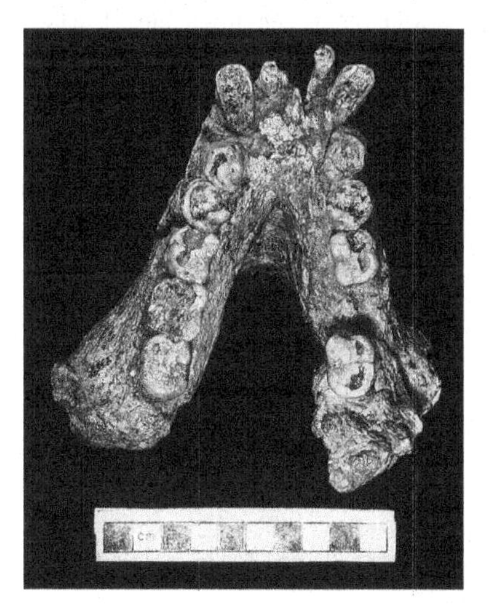

Fig. 21.8 *Gigantopithecus blacki*. Mandibles from Liucheng, Guangxi Province, China: (A) mandible 1, (B) mandible 2, (C) mandible 3. Scales in cm. Photographs courtesy of Eric Delson.

SPECIES *Lufengpithecus keiyuanensis* Woo, 1957
TYPE SPECIMEN IVPP An-613, left and right P_4 and M_2, right M_3
AGE AND GEOGRAPHIC RANGE Late Miocene of Yunnan Province, southern China

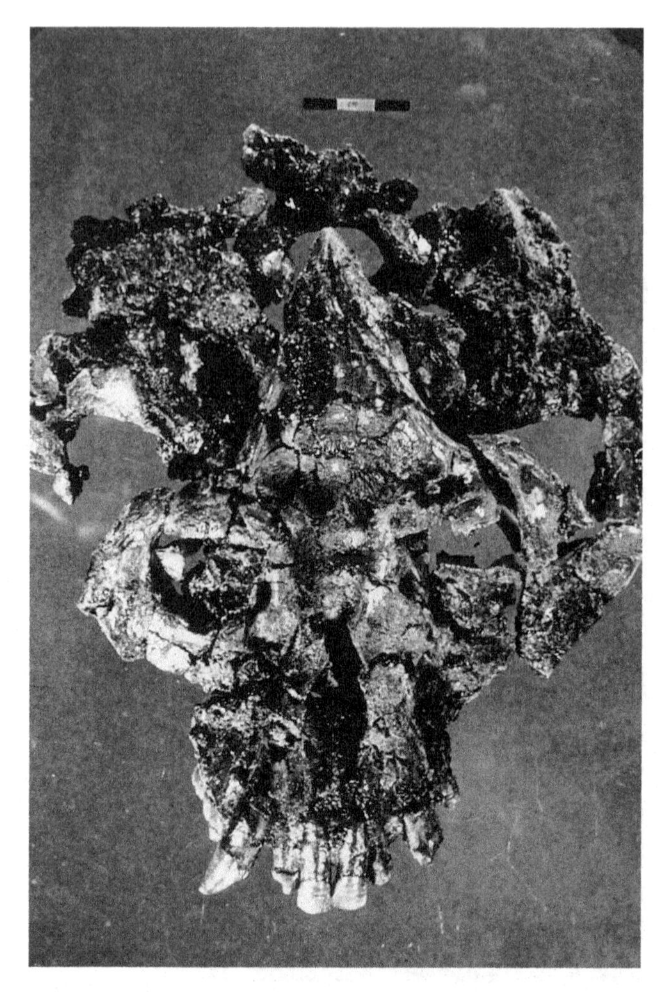

Fig. 21.9 *Lufengpithecus lufengensis*. Skull (PA 644) from Shihuiba (Lufeng), China. Photograph courtesy of Xu Qinghua.

Fig. 21.10 *Lufengpithecus lufengensis*. Unassociated lower molars from Shihuiba (Lufeng), China.

ANATOMICAL DEFINITION

Postcanine teeth smaller on average than those of *L. lufengensis*. Molar cingulum development perhaps greater than in *L. lufengensis*. Perhaps higher dentine penetrance and thinner enamel than in *L. lufengensis*.

SPECIES *Lufengpithecus hudienensis* Zhang *et al.*, 1987 (Fig. 21.11)

TYPE SPECIMEN YV 0916, partial maxilla with P^3–M^3

AGE AND GEOGRAPHIC RANGE Late Miocene (and early Pliocene?) of Yunnan Province, southern China

ANATOMICAL DEFINITION

Postcanine teeth smaller on average than those of *L. lufengensis*, with perhaps somewhat greater cingulum development.

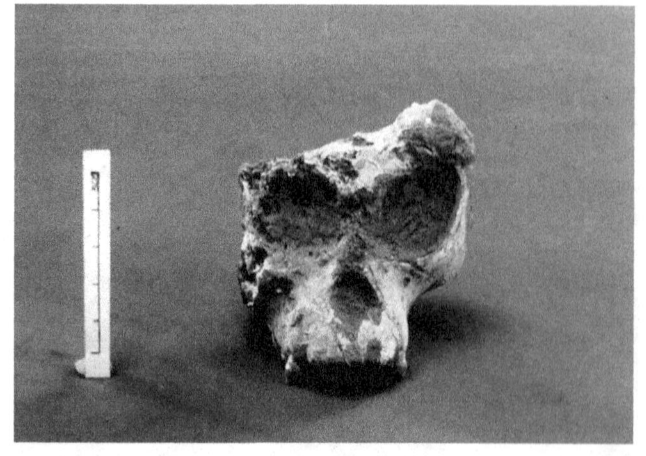

Fig. 21.11 *Lufengpithecus hudienensis*. Juvenile skull (YV0999) from Hudie Liangzi, Yuanmou Basin, China. Photograph courtesy of Dennis Etler.

Evolution of Asian hominoids in the Miocene–Pleistocene

As noted, the two species from the middle Miocene site of Paşalar in Turkey are among the oldest known hominoids outside of Africa. Recently, Ward *et al.* (1999) briefly noted a number of presumed synapomorphies between the unnamed species at Paşalar (see also Alpagut *et al.*, 1990; Martin & Andrews, 1993*b*) and *Kenyapithecus wickeri*, which appear to link them phylogenetically. Unfortunately, this possible relationship cannot be further explored until the Paşalar species has been fully described and named.

Begun (2000) has suggested that the more common species at Paşalar, assigned to *Griphopithecus alpani*, is closely related to another African middle Miocene species, *Equatorius africanus* (formerly *Kenyapithecus africanus*: Ward *et al.*, 1999). He in fact considers them to be congeneric, so the latter would take the name *G. africanus*. Kelley *et al.* (2000) acknowledged that the two species share many similarities, but also noted some differences that they feel warrant continued generic separation. What should not be lost in these disagreements, however, is that both Paşalar species can be linked plausibly to approximately contemporaneous African taxa.

A similarly plausible phylogenetic link between either Paşalar species and any later Eurasian hominoids cannot be established at this time (Andrews *et al.*, 1996). Non-dental remains from Paşalar are limited and display mostly primitive hominoid features. The highly distinctive upper central incisor morphology of *Griphopithecus alpani* is shared by *Lufengpithecus lufengensis* and some *Dryopithecus* species. Likewise, the second Paşalar species shares with the same later Miocene species a distinctive slender, high-crowned canine. However, it would be unwise to attach much phylogenetic significance to these or any other individual dental characters.

Thus, apart from an appropriate temporal and geographic position, no compelling evidence suggests a phylogenetic relationship between either Paşalar species and any later Miocene Eurasian genera (Begun *et al.*, 1997*b*). On the other hand, nothing in the known anatomy of either species precludes such relationships.

It has been suggested that *Ankarapithecus* and species of European *Dryopithecus* share derived features of the *Pongo* clade with *Sivapithecus* and *Lufengpithecus*, albeit as more distant relatives of *Pongo* (Schwartz, 1990, 1997; Moyà-Solà & Köhler, 1993*b*, 1995, 1996; Begun & Güleç, 1995; Alpagut *et al.*, 1996). All, or at least some, possibly are members of a single broad and diverse radiation, of which one (or both) Paşalar species is simply the earliest, most primitive member. The later Miocene Asian hominoids could also derive from a presently unknown or unrecognized African lineage that migrated separately into Eurasia, perhaps closer in time to the first appearance of *Sivapithecus* at around 12.5 Ma (Kappelman *et al.*, 1991; see also Begun *et al.*, 1997*b*).

Equally important is the question of how *Sivapithecus* and *Lufengpithecus* relate to the extant Asian great ape, *Pongo pygmaeus*. Virtually all cladistic and individual character analyses of cranial and dental features (e.g., Andrews & Cronin, 1982; Ward & Kimbel, 1983; Ward & Pilbeam, 1983; Ward & Brown, 1986; Andrews & Martin, 1987*a*; Schwartz, 1990, 1997; Begun, 1994*a*; Begun & Kordos, 1997; Begun *et al.*, 1997*b*; Ward, 1997) unequivocally place these genera within the *Pongo* clade, or, for *Lufengpithecus*, as its sister group. This is not to say that both genera share the same set of presumed synapomorphies with *Pongo* (Schwartz, 1997). Further, *Lufengpithecus* has a number of peculiar apomorphic cranial features (Wu *et al.*, 1983; Schwartz, 1990, 1997), but for both genera (*Sivapithecus* in particular) the list of credible *Pongo* synapomorphies is numerous.

As noted earlier, however, the morphology of the *Sivapithecus* humeral shaft (Pilbeam *et al.*, 1990) argues against placement within the *Pongo* clade if it is presumed that the last common ancestor of the extant great apes and humans possessed the basic hominoid body plan related to forelimb-dominated positional behaviors. In fact, a number of workers have come to question this ancestral morphotype and are comfortable with the idea of a more primitive morphotype and independent acquisition of suspensory-dominated behaviors and associated morphology (Andrews, 1992; Begun, 1993*b*, 1994*a*; Larson, 1998). Others believe the apparent craniofacial synapomorphies shared by *Sivapithecus* and *Pongo* were more likely to have evolved in parallel, perhaps in response to similar dietary demands (Pilbeam, 1996).

A third alternative can accommodate *Sivapithecus* within the *Pongo* clade without invoking pervasive homoplasy in either the cranium or postcranium. In this alternative, the apparently primitive morphology of the *Sivapithecus* humeral diaphysis is an apomorphic suite of character reversals associated with the reacquisition of predominantly pronograde behaviors from more suspensory ancestors (Andrews & Pilbeam, 1996; Ward, 1997). All such arguments lack universally accepted criteria for deciding among the competing hypotheses, particularly concerning the relative likelihood of homoplasy in the cranium versus postcranium.

This matter is still intensely debated and extends to other Miocene hominoid species as well. Recently described 20-Ma postcranial remains of *Morotopithecus bishopi* from Uganda suggest positional behaviors much closer to those of the extant great apes (Gebo *et al.*, 1997*b*). For some, this is corroboration that any Miocene hominoid with a more primitive morphology, or one derived in another direction, would lie outside the extant great ape and human radiation (Gebo *et al.*, 1997*b*). On the other hand, 20 Ma is well outside the range of all molecular-based estimates for the origin of the great-ape–human clade. This can be interpreted as evidence that suspensory-dominated positional behaviors evolved independently in *Morotopithecus*. This would lend further support to the argument that these behaviors have evolved repeatedly among anthropoids and that *Sivapithecus* humeral morphology does not exclude it from the *Pongo* clade.

Late Miocene Eurasian hominoids that show postcranial features related to suspensory positional behaviors further complicate this issue. These include *Oreopithecus bambolii* (Harrison, 1986*a*; Jungers, 1987; Sarmiento, 1987; Harrison & Rook, 1997), species of *Dryopithecus* (Begun, 1992*c*, 1994*a*; Moyà-Solà & Köhler, 1996), and very probably *Lufengpithecus* as well. Among the few postcranial elements of *L. lufengensis* is a proximal phalanx which, unlike that of *Sivapithecus*, resembles *Pongo* in articular morphology and shaft curvature and shape (Wu *et al.*, 1986; personal observation).

The majority but not consensus view is that both *Sivapithecus* and *Lufengpithecus* are part of the *Pongo* clade. If so, the members of this clade probably expressed a varied pattern of resemblances to extant *Pongo*, including an apparently diverse array of postural and locomotor behaviors and morphologies. A better late Miocene and Pliocene record from southern China surely will afford better understanding of evolu-tion within this clade.

Gigantopithecus also is most reasonably regarded as a member of the *Pongo* clade, especially if the attribution of Siwalik *G. giganteus* to this genus is correct. The type molar of *G. giganteus*, and molars of the mandible from Bilaspur, cannot be differentiated from *Sivapithecus* apart from their extremely large size. Also, the mandible itself is very similar to some *Sivapithecus* mandibles (Kelley, 1986*a*; Kelley & Pilbeam, 1986). Thus, assignment of this species to *Sivapithecus* might be more appropriate. On the other hand, as noted earlier, the mandible shares with Pleistocene *G. blacki* a relatively reduced anterior dentition and a tendency toward enlarged, molarized premolars. If the latter are shared-derived features, then this is perhaps the pattern one would expect if *Gigantopithecus* is derived from *Sivapithecus*, and *G. giganteus* is a very early representative of the genus.

Sivapithecus is the only Asian fossil hominoid whose paleobiology has been explored in any detail. It appears to be similar to extant great apes in two important respects. First, its extreme rarity in the densely sampled and well-collected Siwalik sequence, with more than 60 000 catalogued mammal specimens, argues for very low original population density (Kelley, 1994). Second, the calculated age at first molar emergence in a juvenile of *S. parvada* strongly suggests a life-history profile very similar to that of *Pan* (Kelley, 1997, in press). Life history is primarily about the scheduling of, and allocation of resources to, reproduction, which makes it one of the most fundamental aspects of a species' biology. Thus, the demonstration of modern ape life history in *Sivapithecus* implies that other features of biology and ecology were probably like those of extant apes as well.

Primary References

Dryopithecus wuduensis

Xue, X. & Delson, E. (1988). A new species of *Dryopithecus* from Gansu, China. *Kexue Tongbao*, **33**, 449–453.

Gigantopithecus

Pilgrim, G. E. (1915). New Siwalik primates and their bearing on the question of the evolution of man and the Anthropoidea. *Records of the Geological Survey of India*, **45**, 1–74.

Koenigswald, G. H. R. von (1935). Eine fossile Säugetierfauna mit *Simia* aus Südchine. *Proceedings of the Koninklijke Nederlandse Akademie Van Wetenschappen*, **38**, 872–79.

Weidenreich, F. (1945). Giant early man from Java and South China. *Anthropological Papers of the American Museum of Natural History*, **40**, 1–134.

Koenigswald, G. H. R. von. (1952). *Gigantopithecus blacki* von Koenigswald, a giant fossil hominoid from the Pleistocene of southern China. *Anthropological Papers of the American Museum of Natural History*, **43**, 292–325.

Pei, W. C. & Woo, J. K. (1956). New materials of *Gigantopithecus* teeth from South China. *Acta Palaeontologica*, **4**, 477–490.

Pei, W. C. (1957). Discovery of *Gigantopithecus* mandibles and other material in Liu-Cheng district of central Kwangsi in South China. *Vertebrata PalAsiatica*, **1**, 65–72.

Pei, W. C. & Li, Y. H. (1959). Discovery of a third mandible of *Gigantopithecus* in Liu-Cheng, Kwangsi, South China. *Vertebrata PalAsiatica*, **2**, 193–200.

Woo, J. K. (1962). The mandibles and dentition of *Gigantopithecus*. *Palaeontologica Sinica, new series* **11**, **146**, 1–94.

Simons, E. L. & Chopra, S. R. K. (1969*a*). A preliminary announcement of a new *Gigantopithecus* species from India. *Proceedings of the 2nd International Congress of Primatology*, **2**, 135–42.

Simons, E. L. & Chopra, S. R. K. (1969*b*). *Gigantopithecus* (Pongidae, Hominoidea) a new species from North India. *Postilla*, **138**, 1–18.

Chang, Y., Wu, M., & Liu, C. (1973). New discovery of *Gigantopithecus* teeth from Wuming, Kwangsi. *Kexue Tongbao*, **18**, 130–33.

Hsu, C., Han, K., & Wang, L. (1974). Discovery of *Gigantopithecus* teeth and associated fauna in western Hopei. *Vertebrata PalAsiatica*, **12**, 293–309.

Chang, Y., Wang, L., Dong, X., & Chen, W. (1975). Discovery of a *Gigantopithecus* tooth from Bama district in Kwangsi. *Vertebrata PalAsiatica*, **13**, 148–154.

Cuong, N. L. (1984). Paläeontologische Untersuchungen in Vietnam. *Zeitschrift für Archäeologie*, **18**, 247–251.

Cuong, N. L. (1985). Fossile Menschenfunde aus Nordvietnam. In *Menschwerdung-biotischer und gesellschaftlicher Entwicklungsprozess*, eds. J. Herrmann & H. Ullrich, pp. 96–102. Berlin: Akademieverlag.

Ciochon, R. L., Long, V. T., Larick, R., Gonzalez, L., Grun, R., Vos, J. de, Yonge, C., Taylor, L., Yoshida, H., & Reagan, M. (1996). Dated co-occurrence of *Homo erectus* and *Gigantopithecus* from Tham Khuyen Cave, Vietnam. *Proceedings of the National Academy of Sciences of the United States of America*, **93**, 3016–3020.

Griphopithecus alpani

Tekkaya, I. (1974). A new species of Tortonian anthropoid (Primates, Mammalia) from Anatolia. *Bulletin of the Mineral Research and Exploration Institute of Turkey*, **83**, 148–165.

Andrews, P. J. & Tobien, H. (1977). New Miocene locality in Turkey with evidence on the origin of *Ramapithecus* and *Sivapithecus*. *Nature*, **268**, 699–701.

Alpagut, B., Andrews, P., & Martin, L. (1990). New Miocene hominoid specimens from the Middle Miocene site at Paşalar, Turkey. *Journal of Human Evolution*, **19**, 397–422.

Martin, L. & Andrews, P. (1993*b*). Species recognition in Middle Miocene hominoids. In *Species, Species Concepts, and Primate Evolution*, eds. W. H. Kimbel & L. B. Martin, pp. 393–427. New York: Plenum Press.

Lufengpithecus

Woo, J. (1957). *Dryopithecus* teeth from Keiyuan, Yunnan Province. *Vertebrata PalAsiatica*, **1**, 25–32.

Woo, J. (1958). New materials of *Dryopithecus* from Keiyuan, Yunnan. *Vertebrata PalAsiatica*, **2**, 38–42.

Xu, Q., Lu, Q., Pan, Y., Zhang, X., & Zheng, L. (1978). Fossil mandible of the Lufeng *Ramapithecus*. *Kexue Tongbao*, **9**, 544–556.

Xu, Q. & Lu, Q. (1979). The mandibles of *Ramapithecus* and *Sivapithecus* from Lufeng, Yunnan. *Vertebrata PalAsiatica*, **17**, 1–13.

Xu, Q. & Lu, Q. (1980). The Lufeng ape skull and its significance. *China Reconstructs*, **29**, 56–57.

Wu, R., Han, D., Xu, Q., Lu, Q., Pan, Y., Zhang, X., Zheng, L., & Xiao, M. (1981). *Ramapithecus* skulls found first time in the world. *Kexue Tongbao*, **26**, 1018–1021.

Zhou, G. & Wang, Z. (1981). Preliminary observation on new found mandible and maxilla of *Sivapithecus yunnanensis*. *Memoirs of the Beijing Natural History Museum*, **10**, 10–17.

Wu, R., Han, D., Xu, Q., Qui, G., Lu, Q., Pan, Y., & Chen, W. (1982). More *Ramapithecus* skulls from Lufeng, Yunnan: report on the excavation of the site in 1981. *Acta Anthropologica Sinica*, **2**, 101–108.

Wu, R., Xu, Q., & Lu, Q. (1983). Morphological features of *Ramapithecus* and *Sivapithecus* and their phylogenetic relationships: morphology and comparison of the crania. *Acta Anthropologica Sinica*, **2**, 1–10.

Wu, R., Lu, Q., & Xu, Q. (1984). Morphological features of *Ramapithecus* and *Sivapithecus* and their phylogenetic relationships: morphology and comparison of the mandibles. *Acta Anthropologica*

Sinica, **3**, 1–10.

Wu, R., Xu, Q., & Lu, Q. (1985). Morphological features of *Ramapithecus* and *Sivapithecus* and their phylogenetic relationships: morphology and comparison of the teeth. *Acta Anthropologica Sinica*, **4**, 197–204.

Wu, R., Xu, Q., & Lu, Q. (1986). Relationship between Lufeng *Sivapithecus* and *Ramapithecus* and their phylogenetic position. *Acta Anthropologica Sinica*, **5**, 1–30.

Wu, R. (1987). A revision of the classification of the Lufeng great apes. *Acta Anthropologica Sinica*, **6**, 265–271.

Zhang, X. (1987). New materials of *Ramapithecus* from Kaiyuan, Yunnan. *Acta Anthropologica Sinica*, **6**, 81–86.

Zhang, X., Lin, Y., Jiang, C., & Xiao, L. (1987*a*). A new species of *Ramapithecus* from Yuanmou, Yunnan. *Journal of Yunnan University (Social Sciences)*, **3**, 54–56.

Zhang, X., Lin, Y., Jiang, C., & Xiao, L. (1987*b*). A new species of *Homo* from Yuanmou, Yunnan. *Journal of Yunnan University (Social Sciences)*, **3**, 57–60.

Zhang, X., Zheng, L., Gao, F., Jiang, C., & Zhang, J. (1988). A preliminary study of the skull of Lama ape unearthed at Hudie Hill of Yuanmou County. *Journal of Yunnan University (Social Sciences)*, **5**, 55–61.

Ho, C. K. (1990). A new Pliocene hominoid skull from Yuanmou southwest China. *Human Evolution*, **5**, 309–318.

Sivapithecus

Lydekker, R. (1879). Further notices of Siwalik Mammalia. *Records of the Geological Survey of India*, **12**, 33–52.

Pilgrim, G. E. (1910). Notices of new mammalian genera and species from the Tertiaries of India. *Records of the Geological Survey of India*, **40**, 63–71.

Pilgrim, G. E. (1915). New Siwalik primates and their bearing on the question of the evolution of man and the Anthropoidea. *Records of the Geological Survey of India*, **45**, 1–74.

Brown, B., Gregory, W. K., & Hellman, M. (1924). On three incomplete anthropoid jaws from the Siwaliks, India. *American Museum Novitates*, **130**, 1–9.

Gregory, W. K. & Hellman, M. (1926). The dentition of *Dryopithecus* and the origin of man. *American Museum of Natural History Anthropological Papers*, **28**, 1–123.

Pilgrim, G. E. (1927). A *Sivapithecus* palate and other primate fossils from India. *Memoirs Geological Survey of India (Palaeontologica Indica)*, **14**, 1–26.

Lewis, G. E. (1934). Preliminary notice of new man-like ape from India. *American Journal of Science*, **27**, 161–179.

Lewis, G. E. (1936). A new species of *Sugrivapithecus*. *American Journal of Science*, **31**, 450–452.

Gregory, W. K., Hellman, M., & Lewis, G. E. (1938). *Fossil Anthropoids of the Yale–Cambridge India Expedition of 1935*. Washington, DC: Carnegie Institute.

Prasad, K. N. (1962). Fossil primates from the Siwalik beds near Haritalyangar, Himachal Pradesh, India. *Journal of the Geological Society of India*, **3**, 86–96.

Prasad, K. N. (1964). Upper Miocene anthropoids from the Siwalik beds of Haritalyangar, Himachal Pradesh, India. *Palaeontology*, **7**, 124–134.

Pandey, J. & Sastri, V. V. (1968). On a new species of *Sivapithecus* from the Siwalik rocks of India. *Journal of the Geological Society of India*, **9**, 206–211.

Gupta, V. J. (1969). Fossil primates from the Lower Siwaliks of Kangra District, H.P. *Research Bulletin (new series), Panjab University, India*, **20**, 577–578.

Sahni, A., Kumar, V., & Srivastava, V. C. (1974). *Dryopithecus* (Subgenus: *Sivapithecus*) and associated vertebrates from the Lower Siwaliks of Uttar Pradesh. *Bulletin of the Indian Geological Association*, **7**, 54.

Chopra, S. R. K. & Kaul, S. (1975). New fossil *Dryopithecus* material from the Nagri beds at Haritalyangar (H.P.), India. In *Contemporary Primatology*, eds. S. Kondo, M. Kawai, & A. Ehara, pp. 2–11. Basel: Karger.

Dutta, A. K., Basu, P. K., & Sastry, M. V. A. (1976). On the new finds of hominids and additional finds of pongids from the Siwaliks of Ramnagar area, Udhampur District, J. & K. State. *Indian Journal of Earth Sciences*, **3**, 234–235.

Pilbeam, D. R., Meyer, G. E., Badgley, C., Rose, M. D., Pickford, M., Behrensmeyer, A. K., & Shah, S. M. I. (1977). New hominoid primates from the Siwaliks of Pakistan and their bearing on hominoid evolution. *Nature*, **270**, 689–695.

Pilbeam, D. R., Rose, M. D., Badgley, C., & Lipschutz, B. (1980). Miocene hominoids from Pakistan. *Postilla*, **181**, 1–94.

Koenigswald, G. H. R. von. (1981). A possible ancestral form of *Gigantopithecus* (Mammalia, Hominoidea) from the Chinji layers of Pakistan. *Journal of Human Evolution*, **10**, 511–515.

Gupta, S. S., Verma, B. C., & Tewari, A. P. (1982). New fossil hominoid material from the Siwaliks of Kangra District, Himachal Pradesh. *Journal of the Palaeontological Society of India*, **27**, 111–115.

Kay, R. F. (1982). *Sivapithecus simonsi*, a new species of Miocene hominoid with comments on the phylogenetic status of the Ramapithecinae. *International Journal of Primatology*, **3**, 113–173.

Pilbeam, D. R. (1982). New hominoid skull material from the Miocene of Pakistan. *Nature*, **295**, 232–234.

Chopra, S. R. K. (1983). Significance of recent hominoid discoveries from the Siwalik Hills of India. In *New Interpretations of Ape and Human Ancestry*, eds. R. L. Ciochon & R. S. Corruccini, pp. 539–557. New York: Plenum Press.

Dehm, R. (1983). Miocene hominoid primate dental remains from the Siwaliks of Pakistan. In *New Interpretations of Ape and Human Ancestry*, eds. R. L. Ciochon & R. S. Corruccini, pp. 527–537. New York: Plenum Press.

Koenigswald, G. H. R. von. (1983). The significance of hitherto undescribed Miocene hominoids from the Siwaliks of Pakistan in the Senckenberg Museum, Frankfurt. In *New Interpretations of Ape and Human Ancestry*, eds. R. L. Ciochon & R. S. Corruccini, pp. 517–526. New York: Plenum Press.

Raza, S. M., Barry, J. C., Pilbeam, D. R., Rose, M. D., Shah, S. M. I., & Ward, S. C. (1983). New hominoid primates from the middle Miocene Chinji Formation, Potwar Plateau, Pakistan. *Nature*, **306**, 52–54.

Rose, M. D. (1984). Hominoid specimens from the middle Miocene Chinji Formation, Pakistan. *Journal of Human Evolution*, **13**, 503–516.

Rose, M. D. (1986). Further hominoid postcranial specimens from the late Miocene Nagri Formation of Pakistan. *Journal of Human Evolution*, **15**, 333–367.

Kelley, J. (1988). A new large species of *Sivapithecus* from the Siwaliks of Pakistan. *Journal of Human Evolution*, **17**, 305–324.

Rose, M. D. (1989). New postcranial specimens of catarrhines from the middle Miocene Chinji Formation, Pakistan: descriptions and a discussion of proximal humeral functional morphology in anthropoids. *Journal of Human Evolution*, **18**, 131–162.

Pilbeam, D. R., Rose, M. D., Barry, J. C., & Shah, S. M. I. (1990). New *Sivapithecus* humeri from Pakistan and the relationship of

Sivapithecus and *Pongo*. *Nature*, **348**, 237–239.
Spoor, C. F., Sondaar, P. Y., & Hussain, S. T. (1991). A hominoid hamate and first metacarpal from the Late Miocene Nagri Formation of Pakistan. *Journal of Human Evolution*, **21**, 413–424.
Kelley, J., Anwar, M., McCollum, M., & Ward, S. C. (1995). The anterior dentition of *Sivapithecus parvada* with comments on the phylogenetic significance of incisor heteromorphy in hominoids. *Journal of Human Evolution*, **28**, 503–517.
Verma, B. C. & Gupta, S. S. (1997). New light on the antiquity of Siwalik great apes. *Current Science (India)*, **72**, 302–303.
Cameron, D. W., Patnaik, R., & Sahni, A. (1999). *Sivapithecus* dental specimens from Dhara locality, Kalgarh District, Uttar Pradesh, Siwaliks, Northern India. *Journal of Human Evolution*, **37**, 861–868.
Madar, S. I., Rose, M. D., Kelley, J., MacLatchy, L., & Pilbeam, D. (In press). New *Sivapithecus* postcranial specimens from the Siwaliks of Pakistan. *Journal of Human Evolution*.

22 | Middle and late Miocene African hominoids

STEVEN C. WARD AND DANA L. DUREN

Introduction

The fossil record of African middle to late Miocene hominoids documents a transition from taxa retaining primitive characters from their earlier Miocene antecedents to more derived forms, some of which may have affinities with contemporaneous and later hominoid genera in Eurasia. Recent discoveries in Kenya from sediments somewhat younger than 16 Ma and not older than approximately 14 Ma have significantly augmented what had been a relatively impoverished record of large hominoid taxa from the early middle Miocene (McCrossin & Benefit, 1997; Nakatsukasa *et al.*, 1998; S. Ward *et al.*, 1999). Sometime between 14 and 6 Ma hominoids dispersed beyond Afro-Arabia, and it is possible that hominoid genera re-entered Africa during this period although the fossil evidence for these events is all but non-existent.

Few new fossils from the later Miocene of Africa have come to light since the last major review of the hominoid record from this time period (Hill & Ward, 1988; Hill, 1994). Two quite notable exceptions are the discovery of *Otavipithecus namibiensis*, at Berg Aukas in Namibia (Conroy *et al.*, 1992), and the 6.0 Ma potentially bipedal hominoid *Orrorin tugenensis* from Kenya (Senut *et al.*, 2001). The existence of a large hominoid in southwestern Africa at an estimated age of around 13 Ma was unexpected, and suggested that by the end of the middle Miocene hominoids were distributed over much of Afro-Arabia. Nevertheless, African Miocene hominoids are still predominantly known from localities in Kenya and Uganda.

For much of the twentieth century, middle–later Miocene African hominoids have been regarded as the primary fossil evidence bearing on the ancestry of great apes and humans. Two developments in the latter part of the twentieth century cast the fossil apes of Africa in a rather different light: a dramatically expanded fossil record of Miocene hominoids from Europe and Asia, and the rise of molecular systematics as a coherent discipline. Radiations of Eurasian large hominoids between approximately 14 Ma and 8 Ma produced the greatest taxonomic diversity in the superfamily Hominoidea since the early Miocene, and the widest geographic distribution of the group in its history. Interestingly, this taxonomic fecundity was taking place at a time when the African record of large hominoids becomes very sparse. Not surprisingly, biogeographic arguments positing a Eurasian origin of the great-ape–human clade have become more frequent in discussions of hominid and African ape origins. Most of these hypotheses, bolstered by anatomical and as well as molecular evidence, suggest that the ancestors of great apes and humans emerged in Eurasia, probably from African ancestors, and "re-entered" Africa later in the Miocene (see Begun, Chapter 20, this volume).

Molecular biology also has had an increasingly prominent role in assessing the phylogenetic relationships of living and extinct hominoids, as well as attempting to calibrate lineage divergence events. Divergence estimates based on genetic data strongly suggest that a putative last common ancestor of the living great apes and humans appeared later than 14 Ma, and possibly much later, perhaps at the end of the Miocene. If true, this would exclude many middle–late Miocene taxa as candidate ancestors. At present, the middle–later Miocene hominoid fossil record is not sufficiently endowed to resolve these issues, but the picture has improved in recent years.

History of discovery and debate

Excellent summaries of early paleontological research in East Africa have been published by Le Gros Clark & Leakey (1951), Andrews (1981*a*) and Harrison (this volume). Three significant periods of hominoid fossil discovery in Africa can be identified – prior to the Second World War, the post-war interval up to and just following the colonial period, and an era of small to major expeditions that began in the late 1960s and extend to the present. Almost all that was known concerning middle Miocene hominoids until the 1960s were specimens recovered from Maboko Island in Lake Victoria. Archdeacon Walter Edwin Owen conducted excavations on Maboko in 1933, and recovered primate, including hominoid fossils. Donald Gordon MacInnes subsequently investigated the exposures at Maboko and among other specimens recovered a hominoid maxilla subsequently named as the type specimen of *Sivapithecus africanus* (Le Gros Clark & Leakey, 1950). It possessed a complex of dental and gnathic anatomy that satisfied its describers it was not referable to the genus *Proconsul*. This specimen has had a tortuous history in the annals of paleontology, and its provenance is not entirely certain. However, most agree that this specimen was indeed recovered at Maboko and mistakenly accessioned as having been found on Rusinga Island. Owen also discovered a series of middle Miocene localities outcropping on the mainland opposite Maboko Island. These localities, including Kaloma and Majiwa, produced hominoid specimens which were recovered by Martin Pickford in the late 1970s.

Significant new middle Miocene material was discovered first in the 1960s in Kenya. Several large hominoid specimens found in excavations at Fort Ternan beginning in 1961 were allocated by L. S. B. Leakey (1962*a*) to a new taxon, *Kenyapithecus wickeri*. This genus has played a prominent role in human origins debates, and as recently as 1999 has been the focus of intense interest in the evolutionary history of the great apes and humans (S. Ward *et al.*, 1999). Following the recovery of *K. wickeri*, a series of expeditions to western Kenya reinvestigated Maboko and Fort Ternan and searched for new localities. While many new early Miocene hominoids were recovered, middle Miocene hominoids remained elusive. These projects, led by David Pilbeam and Peter Andrews in 1973 to Maboko, Alan Walker to Fort Ternan in 1974 and Martin Pickford to Maboko in the early 1980s resulted in the recovery of important but fairly limited numbers of new hominoid specimens.

Middle–late Miocene hominoid research in the post-colonial era of East Africa resulted in a series of important discoveries in the middle 1980s that has continued into the new millennium. Three research projects were responsible for recovering hominoid material unprecedented in abundance or completeness. The Baringo Paleontological Research Project, directed by Andrew Hill, recovered a partial skeleton of a large hominoid at Kipsaramon in the Tugen Hills. Recovery of the skeleton clarified a long-standing taxonomic conundrum involving the genus *Kenyapithecus*. Analysis of the Kipsaramon skeleton demonstrated that a taxon previously known as *Kenyapithecus africanus* Leakey 1967 was morphologically incompatible with the genus *Kenyapithecus*. As a result Ward *et al.* (1999) erected the new nomen *Equatorius africanus* to accommodate the expanded fossil collections. Brenda Benefit and Monte McCrossin mounted a new expedition to Maboko in 1987 and, to date, have recovered over 100 specimens attributable to *Equatorius africanus* (their *Kenyapithecus africanus*). The array of anatomical regions represented in the new Maboko collections added considerable new information bearing on the locomotor and feeding behavior of *Equatorius*. Equally significant was the work of the Joint Japan–Kenya Samburu Hills Paleoanthropological Expedition lead by Hidemi Ishida. In the early 1980s his team found a hemimaxilla, now attributed to *Samburupithecus* (Ishida & Pickford, 1997), estimated to date from about 8.5 million years ago, a time in which few large hominoids are known from Africa. Throughout the late 1980s and 1990s, Ishida's group continued to recover large hominoids from the Nachola region in Samburu district, believed at that time to be *Kenyapithecus africanus*. However, discovery of an excellent partial skeleton as well as specimens from additional individuals provided the Japanese team with sufficient data to identify the Nachola hominoid as a new taxon, *Nacholapithecus kerioi* (Ishida *et al.*, 1984, 1999; Rose *et al.*, 1996; Nakatsukasa *et al.*, 1998, 2000).

In the meantime, a research team lead by Glenn Conroy and Martin Pickford discovered remains of a large hominoid in a breccia block at Berg Aukas, in Namibia. The new taxon, *Otavipithecus namibiensis*, represented an unexpected range extension of large Miocene hominoids to southwest Africa (Conroy *et al.*, 1992, 1993, 1995, 1996; Conroy, 1994, 1996). Its phylogenetic relationships are unclear (Begun, 1994*b*; Conroy, 1994; Singleton, 2000).

With the exception of *Otavipithecus* and *Samburupithecus*, few hominoid fossils are known from Africa after 14 Ma. Most of those are from the Lake Baringo basin. These include several teeth from the Ngorora sequence which date at approximately 12.5 Ma, and the Lukeino molar, which is between 5.6 and 6.0 Ma. Senut *et al.* (2001) have reported numerous fossils from the Baringo Basin that they align with the Lukeino molar as *Orrorin tugenensis*. Other localities in Africa that date between 6 and 14 Ma include Sahabi (Libya), the Chorora Formation (Ethiopia), Nakali and Aterir (Kenya), Semliki (Democratic Republic of the Congo) and the Manonga Valley in Tanzania. Unfortunately, hominoids are known from few of these localities, and these are limited to relatively uninformative specimens.

Three issues have dominated discussions of middle–later Miocene large hominoid evolutionary history: their role in great-ape and human origins, the extent of taxonomic diversity they represent, and their relationships to large hominoid taxa beyond continental Africa. All Miocene hominoids from Africa have historically been viewed as potential lineal ancestors of living great apes and humans. By the late 1960s two species of *Proconsul* were viewed as potential ancestors of chimpanzees and gorillas. The discovery and diagnosis of *Kenyapithecus wickeri* by Louis Leakey (1962*a*), followed in 1967 by his erection of the junior species *K. africanus*, heralded a period of almost 20 years during which hominid ancestry was traced to the middle Miocene. Moreover, some middle Miocene hominoids were more derived in certain elements of gnathic and dental anatomy when compared to earlier taxa. These factors, in conjunction with paleontologically rooted estimates of hominoid divergence dates, appeared to support the concept of a middle Miocene hominid origins.

However, *K. africanus* was to prove controversial. The type specimen was the "*Sivapithecus*" *africanus* maxilla originally named by Le Gros Clark & Leakey in 1950, and described in detail in 1951. It was more derived than *Proconsul* but was regarded as somewhat more primitive in certain features of its maxilla and teeth than the somewhat younger taxon *K. wickeri*. Overall though, the genus *Kenyapithecus*, with its two species, conformed anatomically to the prevailing operative paradigm defining hominid affinities. This paradigm was predicated on linking the robust dental and gnathic architecture of the early australopithecines with a middle Miocene molar megadont, robust jawed ape, that was not *too* ape-like.

These events reached their apotheosis in the decade of the *Ramapithecus* debate. Anatomical similarities between *Ramapithecus* and *Kenyapithecus wickeri* were noted, and for a time *K. wickeri* was allocated to *Ramapithecus* as the junior species *R. wickeri* (Simons & Pilbeam, 1972). The subsequent recognition that *Ramapithecus* was a junior synonym of the large South Asian hominoid *Sivapithecus*, and that *Sivapithecus* in turn was part of a hominoid radiation divorced from any role in hominid origins became a consensus view at the beginning of the 1980s following important discoveries by Pilbeam's group on the Potwar Plateau in Pakistan.

Despite recognition that *Kenyapithecus* was not closely related to the large hominoids of South Asia, there remained wide acceptance that it was ancestral to great apes and humans, based on details of maxillary, mandibular and dental anatomy (Ward & Pilbeam, 1983). Two developments transpired to change this view. First, the emergence of molecular "clocks" which indicated that a common ancestor for chimpanzees, gorillas and humans occurred much later in the Miocene, and second, a growing sense that the two included species of *Kenyapithecus* encompassed more anatomical variability than one genus could accommodate (Harrison, 1992). By the mid-1980s it was already evident that L. S. B. Leakey had erroneously included specimens attributable to *Proconsul* in the hypodigm of *K. africanus*, creating a taxonomic chimera. New material attributed to *K. africanus* from western Kenya and the Samburu District (Pickford, 1986*a*, 1986*b*) sustained the impression that this species was more primitive than *K. wickeri*.

Discovery of abundant new material from Maboko Island (e.g., McCrossin & Benefit 1993*a*, 1997), the Tugen Hills and the Samburu District permitted a detailed analysis of taxonomic diversity in the early middle Miocene hominoids of Kenya. Based on their analysis of the published hypodigm of *K. africanus* S. Ward *et al.* (1999) concluded that it was sufficiently distinctive in elements of gnathic and dental anatomy to warrant the erection of a new genus – *Equatorius*. At the same time, Ishida *et al.* (1999) erected the new genus and species *Nacholapithecus kerioi* to accommodate a large sample of fossils from the Samburu District that had previously been referred to *K. africanus*, largely on dental and gnathic characters. Partial skeletons recovered in the Tugen Hills and at Nachola proved to be distinct at the generic level from *Kenyapithecus wickeri*, and to be distinct in a number of postcranial features from each other. Thus, at a stroke, known hominoid taxonomic diversity in the early middle Miocene increased substantially. However, Benefit & McCrossin (2000) disputed the validity of *Equatorius*, asserting that the new genus was in fact a chimera composed of *K. africanus* and *Nacholapithecus*, and in addition that S. Ward *et al.* (1999) failed to take into account patterns of morphological and metric variation in the maxillae and dentition of *K. wickeri* and *K. africanus*. In response Kelley *et al.* (2000) noted that the postcranial anatomy of *Nacholapithecus* clearly warranted separation at the genus level, and that dental samples from multiple sedimentary levels at Maboko could include material attributable to *K. wickeri*.

The question of relationships of the early to later Miocene hominoids of Africa to their Eurasian counterparts remains unclear. S. Ward *et al.* (1999) noted presumptive shared and derived features of incisor and canine anatomy and proportions between *Kenyapithecus wickeri* and the rarer of the two species of large hominoid known from the middle Miocene locality of Paşalar in Turkey. This taxon is currently unnamed. If correct, this would represent the first known link between African and Eurasian Miocene large hominoids. However, in his response to the erection of *Equatorius*, Begun (2000) suggested that *Equatorius* was in fact congeneric with the more common of the two hominoid species at Paşalar in Turkey, and based on priority should be attributed to *Griphopithecus africanus*. Kelley *et al.* (2000) responded that similarities are indeed present but so are a number of differences that warrant the generic distinctiveness of *Equatorius*.

The remainder of the known middle Miocene fossil record of hominoid evolutionary history in Africa is an array of interesting but in many cases frustratingly enigmatic specimens. A fragmentary ulna from Fejej in Ethiopia has been reported to have a minimum age of approximately 16 Ma (Richmond *et al.*, 1998). This specimen is notable for several reasons, not the least of which is its occurrence outside of Kenya and Uganda. The Fejej ulna (F-18SB-68) consists of five fragments, preserving part of the proximal end as well as the proximal and central shaft. Based on anatomical and multivariate analysis Richmond *et al.* (1998) reported that the specimen showed similarities to the earlier Miocene hominid genera *Proconsul*, *Turkanapithecus* and *Pliopithecus* and asserted that it was functionally consistent with arboreal quadrupedality with a considerable range of forearm rotation.

Otavipithecus from Berg Aukas, Namibia, with an estimated age of approximately 13 Ma is known from a frontal bone, partial mandible, first cervical vertebra, proximal ulna and a manual phalanx (Conroy *et al.*, 1992, 1996; Senut & Gommery, 1997). At present the higher taxonomic affiliations of *Otavipithecus* are uncertain although Andrews (1992) suggested it might appropriately be allied with Afropithecinae. A hemimaxilla from the Samburu Hills in Kenya, and dated at around 8.5 Ma and recently referred to the new genus and species *Samburupithecus kiptalami* (Ishida & Pickford, 1997) is also impossible to link to any known large hominoid taxon or radiation.

Most recently, work in the Baringo Basin has recovered remains of a late Miocene hominoid that dates to potentially 6 Ma (Pickford & Senut, 2001). The fossils, named *Orrorin tugenensis* (Senut *et al.*, 2001), include portions of three femora that suggest bipedality. Senut *et al.* (2001) refer the enigmatic Lukeino molar (see Hill, 1994) to this taxon and argue that *Orrorin* displaces *Australopithecus* as a direct ancestor of *Homo*.

Taxonomy

Systematic framework

Since the publication of Andrews's important review of large hominoid systematics in 1992, the fossil record of Miocene catarrhines has expanded significantly. While the new specimens decisively have put to rest the formerly held perception that the middle Miocene documented a period of low hominoid diversity, there is still considerable uncertainty regarding the higher taxonomic affiliations of most, if not all hominoid taxa from that time. Andrews's taxonomic scheme recognized the early–middle Miocene non-cercopithecoid catarrhines as true hominoids. Alternatively, Harrison (1987, this volume) consistently has favored recognizing all early Miocene African and Arabian catarrhines as a paraphyletic grouping with respect to the Hominoidea. There are indeed compelling patterns of cranial, dental and postcranial anatomy supporting this view. However, assessing the phylogenetic relationships of middle Miocene large hominoids to earlier taxa as well as to later Miocene hominoids outside of Afro-Arabia is fraught with complexity. Most of these genera are derived in craniodental characters with respect to earlier genera, and new evidence suggests this may also be true of the postcranium as well. Thus, while Harrison's approach is elegant in its recognition of early Miocene stem hominoids, it does not directly address the evolutionary relationships of African middle Miocene large non-cercopithecoid catarrhines. Part of the difficulty lies in defining what a hominoid, especially a fossil hominoid, actually is (Kelley, 1997). Moreover, large collections of African middle Miocene hominoid fossils have been recovered but not described formally, making provisional any taxonomic hypotheses. At present, all we can say for certain is that there are more taxa present in Africa between 15.5 and 12.0 Ma than were thought probable as late as 1999.

The taxonomic scheme presented here follows Andrews's 1996 revision of his original 1992 contribution. Included in Afropithecinae are the early Miocene genus *Afropithecus*, as well as two recently named taxa, *Nacholapithecus* and *Equatorius*. *Kenyapithecus* is here referred to Kenyapithecinae. This intriguing subfamily is derived in several features of the dentition and maxilla with respect to *Equatorius* and *Nacholapithecus*. It also may have affinities with some middle Miocene hominoids from Turkey and the Vienna Basin (Andrews & Tekkaya, 1977; Andrews, 1992; Andrews *et al.*, 1996; S. Ward *et al.*, 1999; Kelley, this volume). Should further work indicate that the included genera of Afropithecinae were characterized by masticatory and locomotor anatomy incompatible with common patterns of feeding behavior and substrate use, it may become necessary to elevate some or all of them to family status (see Harrison, this volume). *Samburupithecus* from the later Miocene of Kenya is left *incertae sedis* although it seems possible that it is referable to Pongidae. The remaining individual and enigmatic fossils from Kenya and Ethiopia are left *incertae sedis*, including *Orrorin* because at this time it has not been evaluated independently.

Order Primates Linnaeus, 1758
- Infraorder Catarrhini É. Geoffroy Saint-Hilaire, 1812
  - Superfamily Hominoidea Gray, 1825
    - Family Hominidae Gray, 1825
      - Subfamily Afropithecinae Andrews, 1992
        - Genus *Nacholapithecus* Ishida *et al.*, 1999
          - *Nacholapithecus kerioi* Ishida *et al.*, 1999
        - Genus *Equatorius* Ward *et al.*, 1999
          - *Equatorius africanus* Ward *et al.*, 1999
      - Subfamily Kenyapithecinae Leakey, 1962
        - Genus *Kenyapithecus* Leakey, 1962
          - *Kenyapithecus wickeri* Leakey, 1962
      - Subfamily *incertae sedis*
        - Genus *Otavipithecus* Conroy *et al.*, 1992
          - *Otavipithecus namibiensis* Conroy *et al.*, 1992
        - Genus *Samburupithecus* Ishida & Pickford, 1997
          - *Samburupithecus kiptalami* Ishida & Pickford, 1997
        - Genus *Orrorin* Senut *et al.*, 2001
          - *Orrorin tugenensis* Senut *et al.*, 2001

Family Hominidae

Subfamily Afropithecinae

GENUS *Nacholapithecus* Ishida *et al.*, 1999

A large-bodied hominoid characterized by pronounced sexual dimorphism. Maxilla characterized by overlap of the posterior pole of the premaxilla and hard palate, small incisive foramen, pronounced canine pillars and deeply excavated canine fossae, and low origin of the zygomatic process with respect to the alveolar process. Mandibular corpus tall but thin, inferior transverse torus moderately expressed, vertical symphyseal region, and lateral corpus with excavated postcanine fossa below the premolars. Upper central incisors tall-crowned and robust. Upper canine relatively low-crowned and robust, upper premolars with pronounced buccal and lingual cusp heteromorphy, upper molars with reduced lingual cingula and relatively narrow crowns. Mandibular incisors tall and mesiodistally narrow. Lower canine low-crowned and buccolingually compressed. Lower premolars and molars with moderately expressed buccal cingula. Lower third molar with pronounced hypoconulid. Axial skeleton consisting of relatively large cervical vertebrae, thoracic vertebrae with moderately developed ventral keel. Lumbar vertebral bodies long with well-developed ventral keel and heart-shaped outline cranially, transverse process arising from the union of the pedicle and body, upper lumbar vertebrae with accessary processes. First sacral vertebra with mediolaterally narrow centrum, dorsoventrally shallow neural canal and robust articular process for the zygopophyseal joint. Clavicle relatively gracile with weak curvatures but large acromial facet. Scapula with buttressed axial margin and pear-shaped glenoid fossa. Humerus with flat deltoid plane, well-developed lateral supracondylar crest, large and projecting lateral epicondyle, large and posteromedially deflected medial epicondyle, deep olecranon fossa with well-developed lateral wall, tall and bulbous capitulum and well-defined groove of the zona conoidea. Ulna with proximally projecting and non-retroflected olecranon process, well-defined proximolateral extension of the trochlear notch, wide and projecting coronoid process, small and laterally facing radial notch. Distal radius robust with well-developed styloid process. Carpus with unfused os centrale. Ischium with large spine located close to the caudal rim of the acetabulum. Proximal femur with large head, positioned above the top of the greater trochanter, femoral neck short, robust and characterized by a high neck–shaft angle. Distal femur with patellar groove of relatively equal mediolateral and proximodistal dimensions, and limited asymmetry of the condyles. Patella anteroposteriorly shallow, with mediolateral and proximodistal dimensions approximately equal. Fibula with distal shaft triangular in cross-section, distal fibula expanded mediolaterally, fibular malleolus robust, and surfaces for attachment of talofibular and calcaneofibular ligaments expanded. Talus with trochlea broad with respect to length, lateral trochlear rim more projecting than its medial counterpart, trochlea with pronounced groove. Calcaneus generally shallow dorsoplantarly, moderately compressed calcaneal tuberosity, moderately large sustentaculum tali, long and deeply excavated groove for the flexor hallucis longus muscle. Pedal rays characterized by long and robust hallux. First metatarsal lacking prehallux facet, but evincing a pronounced insertion of the peroneus longus muscle. Proximal hallucial metatarsal markedly robust with well-developed tendon attachment surfaces. Terminal hallucial phalanx with broad proximal surface and deep ungual tuberosity. Second to fourth pedal rays relatively gracile, and lacking pronounced curvature. Fifth metatarsal with large and proximally projecting styloid process, gracile shaft and minimal curvature. Proximal phalanx with modest curvature, middle phalanx with straight shaft and terminal phalanx with deep but compressed ungual tuberosity. Limb proportions of *Nacholapithecus* said to be unusual for Miocene hominoids with the forelimb longer than the hindlimb. Proportions of pedal elements suggest the foot is relatively long with respect to the lower limb bones.

INCLUDED SPECIES *N. kerioi*

SPECIES *Nacholapithecus kerioi* Ishida *et al.*, 1999 (Fig. 22.1)
TYPE SPECIMEN KNM-BG 35250, partial skeleton
AGE AND GEOGRAPHIC RANGE Middle Miocene, approximately 15 Ma, from the Samburu District, Kenya (Sawada *et al.*, 1998)
ANATOMICAL DEFINITION
As for genus.

Subfamily Afropithecinae

GENUS *Equatorius* Ward *et al.*, 1999
A large-bodied hominoid characterized by pronounced sexual dimorphism. Maxillary incisors broad mesiodistally in relation to crown height, low lingual tubercle and continuous lingual cingulum and marginal ridges. Upper lateral incisors highly asymmetrical with a "spiraled" lingual cingulum. Maxillary premolars expanded mesiodistally and buccolingually with respect to M^1, reduction of buccal and lingual cusp heteromorphy, and reduction to absence of lingual cingula. Upper molars bunodont with reduced to absent lingual cingula. Lower incisors tall-crowned, mesiodistally narrow, labiolingually thick and characterized by procumbent implantation. Lower canine low-crowned relative to basal dimensions with lingual cingulum, lower canine roots converge strongly towards the symphyseal midline. Lower third premolar with sectorial morphology and variable expression of buccal cingulum. Lower fourth premolar with four cusps and minimal buccal cingulum. Lower molars showing size progression $M_1 < M_2 < M_3$. Lower molars generally bunodont with thickened enamel and minimal buccal cingulae. Deciduous third premolar with distinct metaconid. Deciduous fourth premolar molariform. Mandible with well-developed inferior transverse torus ("simian shelf"), long sublingual planum and robust corpus. Lower thoracic vertebra with ventral keel. Sternobrae relatively broad and flat. Scapula with axillary border longer than vertebral border, and acromion projecting beyond the glenoid region. Clavicle robust and not markedly curved or twisted along its long axis. Humerus with posteriorly oriented and flattened head, pronounced deltopectoral crest and retroflected shaft, posteriorly deflected medial epicondyle, deep olecranon fossa with keeled lateral border. Ulna with long olecranon process and proximolateral extension of the trochlear notch articulating with the lateral wall of the olecranon fossa, lateral facing radial notch, strongly buttressed coronoid process, strongly expressed supinator crest, and long styloid process articulating with the proximal carpus. Radius with circular head, relatively short lever arm for the biceps brachialis, and restricted distal articular surface. Carpal complex with proximally projecting pisiform, hamate with deep pit for the pisohamate ligament and pronounced triquetral groove. Morphology of the scaphoid indicates the presence of an os centrale. Metacarpal heads palmarly broad, variable expression of a dorsal transverse ridge on the distal metacarpals, pits for the metacarpophalangeal ligaments dorsally positioned, minimal curvature of all manual metacarpals and phalanges. Femur with relatively small head, high neck–shaft angle, relatively long neck, broad patellar groove. Patella relatively broad and anteroposteriorly thin. Distal tibia with anteroposteriorly and mediolaterally expanded articular

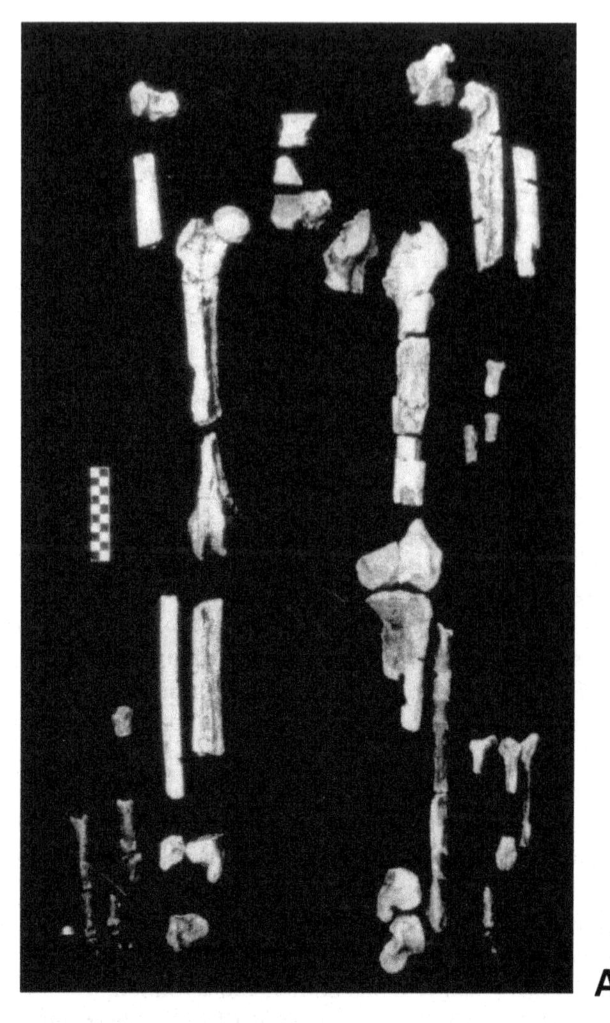

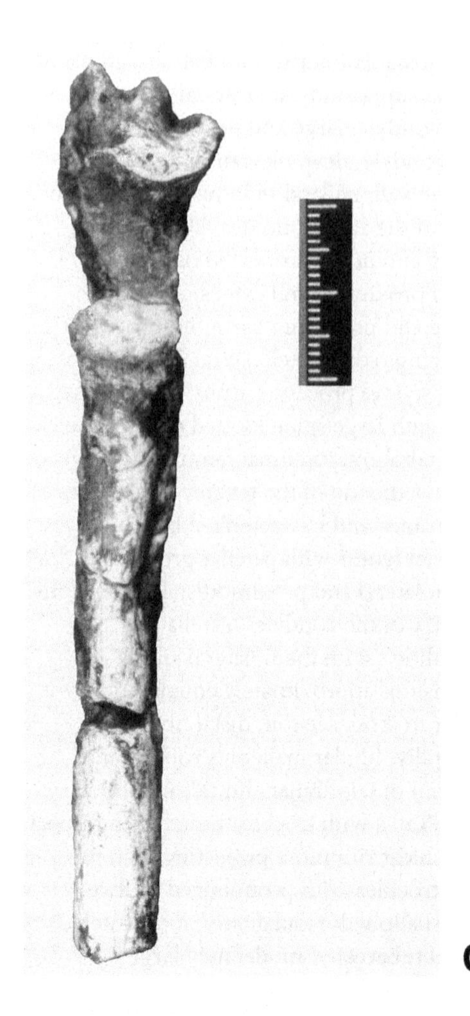

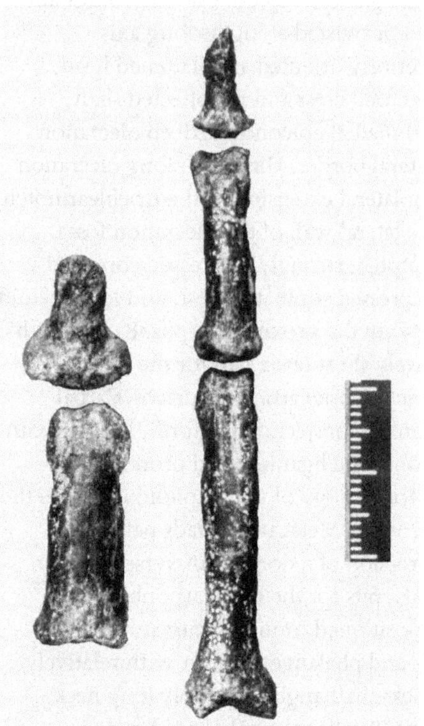

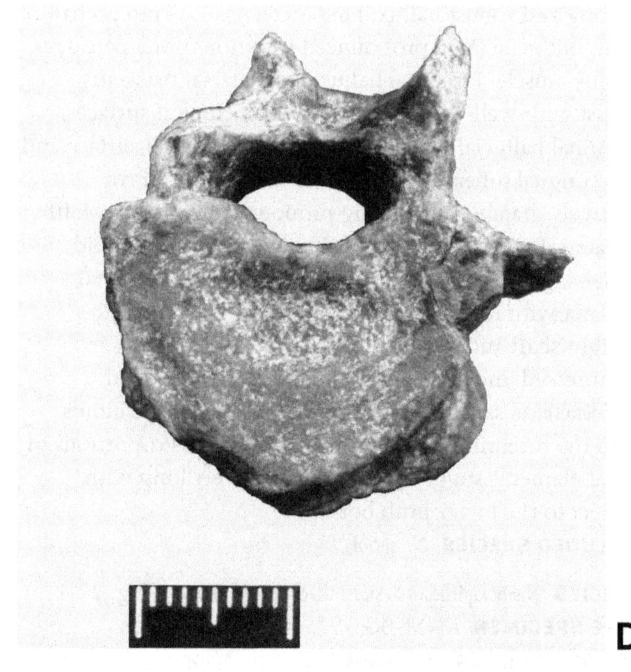

Fig. 22.1 *Nacholapithecus kerioi*. Type specimen KNM-BG 35250: (A) partial skeleton; (B) proximal and terminal hallucial phalanges and right pedal phalanges; (C) left proximal ulna in anterior view; (D) lumbar vertebra in superior view. Scales in cm. Photographs courtesy of Masato Nakatsukasa.

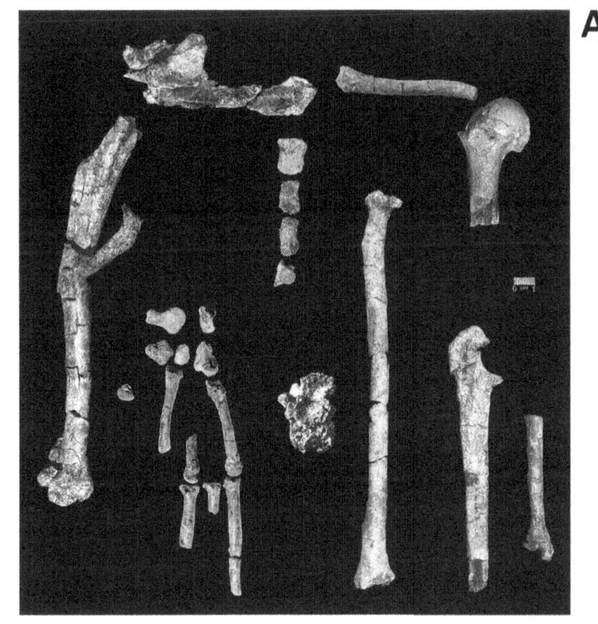

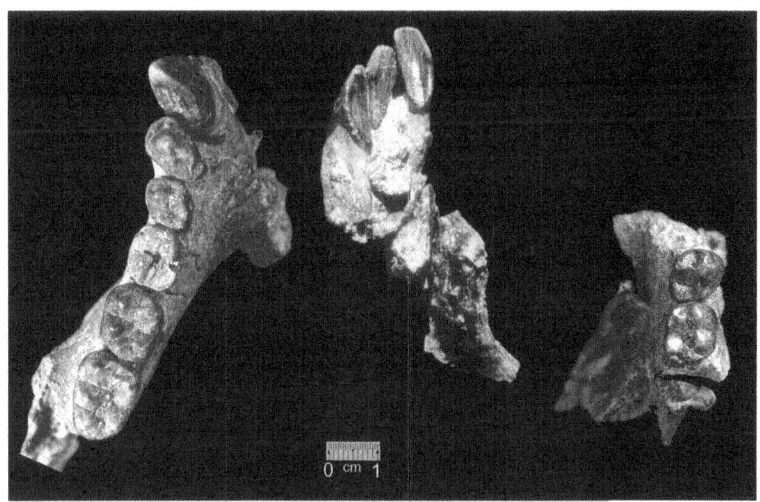

Fig. 22.2 *Equatorius africanus*. KNM-TH 28860: (A) partial skeleton; (B) mandibular dentition in occlusal view (left to right) left mandibular corpus with C–M_3, symphyseal fragment with incisors, right mandibular corpus fragment with P_4 and M_1. Scales in cm.

surface for the talus, with a median keel. Pedal elements characterized by a robust hallucial metatarsal. Entocuneiform with a flat, restricted hallucial facet, suggesting habitual adduction of the hallux. A prehallux facet is absent. Limb proportions calculated from a partial skeleton (KNM TH-28860) as well as possibly associated material from Maboko Island yield a humerus–radius index of 86, and a humerus–femur index of 95 (McCrossin & Benefit, 1997).

INCLUDED SPECIES *E. africanus*

SPECIES *Equatorius africanus* Ward *et al.*, 1999 (Fig. 22.2)

TYPE SPECIMEN BMNH 16649, partial left maxilla with P^3–M^1 and roots of M^2

AGE AND GEOGRAPHIC RANGE 14–15.5 Ma (Feibel & Brown, 1991; A. Deino, pers. comm.), from localities in western Kenya including Maboko Island, Nyacatch, Majiwa and Kaloma; and from the Tugen Hills, Baringo District, Kenya

ANATOMICAL DEFINITION

As for genus.

Subfamily Kenyapithecinae

GENUS *Kenyapithecus* Leakey, 1962

A large hominoid probably characterized by a significant degree of sexual dimorphism. Maxilla with pronounced canine fossae, high origin of the zygomatic process, minimal invasion of the maxillary sinus into the alveolar process. Upper central incisor with massively inflated lingual marginal ridges that envelop the basal region of the lingual crown surface, upper lateral incisor with symmetrical crown and lingual cingulum, and narrow median pillar on the lingual surface, upper canines highly size dimorphic: male canine robust overall, with tall and inflated crown, deep mesial developmental groove, root a flattened ovoid in section, and obliquely implanted (externally rotated). Female morph with conical crown, distinct lingual cingulum, lingual surface with moderately well-developed grooves and ridges, mesial groove well developed. Upper fourth premolar with minimal cusp heteromorphy, crown not compressed mesiodistally, central fovea between mesial and distal foveae, no definitive lingual cingulum. Upper first molar with four primary cusps, hypocone well separated from talon cusps, minimal crenulations and no definitive lingual cingulum. Upper second molar similar to M^1 except longer mesiodistally, and slightly broader. Mandible (female) with pronounced inferior transverse torus and long sublingual planum, mesiodistally compressed incisor portion of the alveolar process, robust corpus, reconstructed mandibular corpora diverging slightly posteriorly. Lower canine with tall crown relative mesiodistal length, reduced lingual cingulum and minimal crenulations. Lower third premolar sectorial, with partly to completely continuous buccal cingulum, lower fourth premolar with two mesial cusps, and poorly developed distal cusps, significant buccal flare of the crown. Lower molars poorly preserved but there is lack of significant buccal cingulum development. Distal humerus with wide trochlea relative to capitulum breadth, pronounced lateral trochlear keel, distinct zona conoidea and posteriorly deflected medial epicondyle.

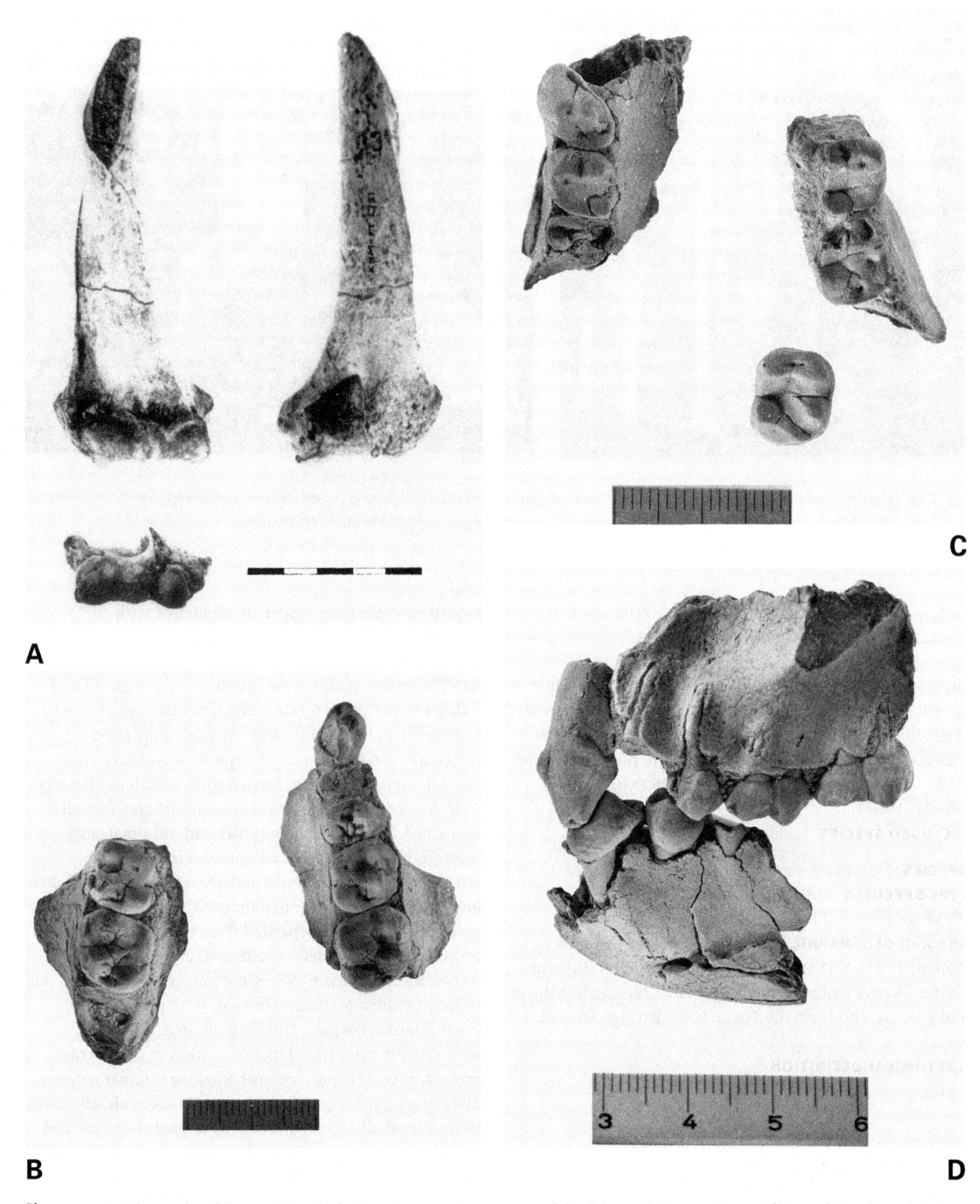

Fig. 22.3 *Kenyapithecus wickeri*. (A) KNM-FT 2751, humerus in anterior, posterior and distal views; (B) KNM-FT 47, right partial maxilla, and KNM-FT 46a, left partial maxilla with associated KNM-FT 46b canine (see Harrison (1992) for issues of association with KNM-FT 46a and b); (C) KNM-FT 45, left partial mandible with P3 and P4, KNM-FT 48, isolated right first molar, and KNM-FT 7, right partial mandible with P_4 and M_1; (D) composite of KNM-FT 46a/b and KNM-FT 45 in occlusion, lateral view. Photographs courtesy of Alan Walker.

INCLUDED SPECIES *K. wickeri*

SPECIES *Kenyapithecus wickeri* Leakey, 1962 (Fig. 22.3)
TYPE SPECIMEN KNM-FT 46, a left partial maxilla with P^4–M^2
AGE AND GEOGRAPHIC RANGE 14 Ma (Kelley & Pilbeam, 1986), Fort Ternan, Kenya
ANATOMICAL DEFINITION
As for genus.

Family Hominidae

Subfamily *incertae sedis*

GENUS *Otavipithecus* Conroy *et al.*, 1992
Frontal bone with superciliary ridges, rather than a transverse torus, postglabellar frontal sulcus temporoglabellar ridges, relatively wide interorbital dimensions, extensive invasion of the squamous portion of the frontal by the frontal sinus. First cervical vertebra with horizontal orientation of superior and inferior articular joints, transverse processes reduced in length relative to cercopithecoids. Mandible with robust corpus, and corpus depth does not increase appreciably posterior to anterior, well-developed postcanine fossa, inferior transverse torus extending only slightly more posteriorly that the superior torus, relatively vertical sublingual planum, and narrow incisive region, large retromolar trigone, posterior origin of ascending ramus. Premolar and molar cusps inflated and bunodont, restricted size of mesial and distal foveae, absence of buccal cingulae, but protostylar ridges present on second and third molars, molar size sequence: $M^2 > M^3 > M^1$, molar enamel relatively thin.
INCLUDED SPECIES *O. namibiensis*

SPECIES *Otavipithecus namibiensis* Conroy et al., 1992 (Fig. 22.4)
TYPE SPECIMEN BER I, 1'91, right partial mandible
AGE AND GEOGRAPHIC RANGE 13 ± 1 Ma, from Berg Aukas, Namibia
ANATOMICAL DEFINITION
As for genus.

GENUS *Samburupithecus* Ishida & Pickford, 1997
A large hominoid known only from one hemimaxilla. Palate with pronounced arch, shallow postcanine fossa, low origin of the zygomatic arch off the maxillary body, inflation of zygomatic process by the maxillary sinus, maxillary sinus extensive, margin of nasal floor with sharp edge, alveolar process straight from the canine alveolus to M^3. Premolars mesiodistally expanded, and posessing three roots, primary cusps similar in size, molar size sequence: $M^1 < M^2 < M^3$, molars with inflated bunodont cusps, well developed protocone cingula, thick enamel but high relief dentoenamel junctions.
INCLUDED SPECIES *S. kiptalami*

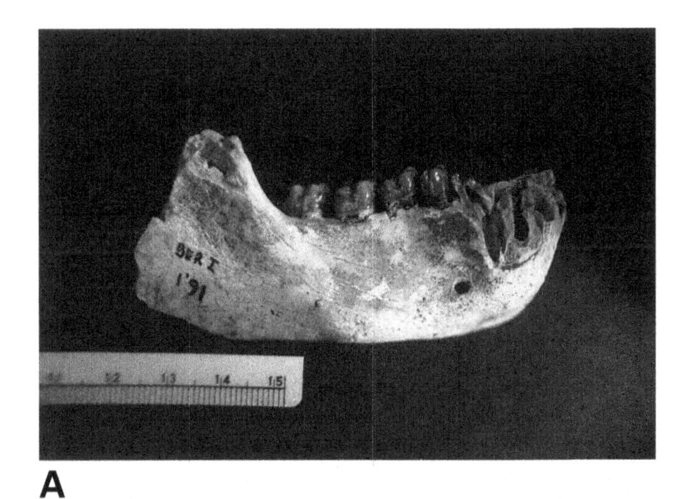

A

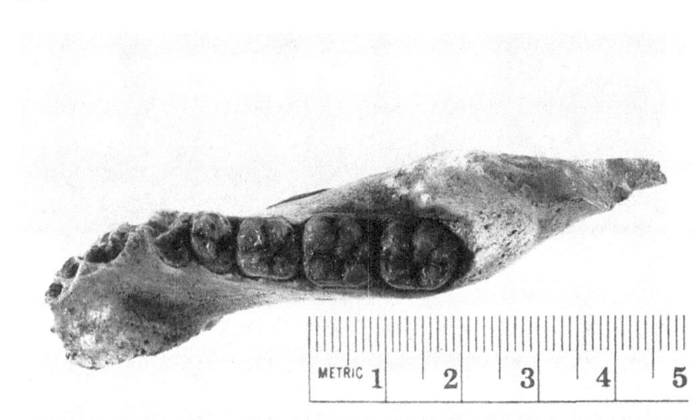

B

Fig. 22.4 *Otavipithecus namibiensis*. Type specimen BER I, right mandibular corpus in (A) lateral view, and (B) occlusal view. Scales in cm. Photographs courtesy of Glenn Conroy.

SPECIES *Samburupithecus kiptalami* Ishida & Pickford, 1997 (Fig. 22.5)
TYPE SPECIMEN KNM-SH 8531, a partial left maxilla and portion of palatine and premaxilla, with canine alveolus and complete P^3–M^3
AGE AND GEOGRAPHIC RANGE 9.5 Ma, Samburu Hills, north–central Kenya
ANATOMICAL DEFINITION
As for genus.

GENUS *Orrorin* Senut *et al.*, 2001
Jugal teeth smaller than australopithecines; large upper central incisor with thick enamel; mandibular corpus relatively deep below M_3; thick enamel on lower cheek teeth; femur with spherical head rotated anteriorly, neck elongated and oval in section; lesser trochanter medially salient; humerus with vertical brachioradialis crest; proximal manual phalanx curved; dentition small relative to body size.
INCLUDED SPECIES *O. tugenensis*

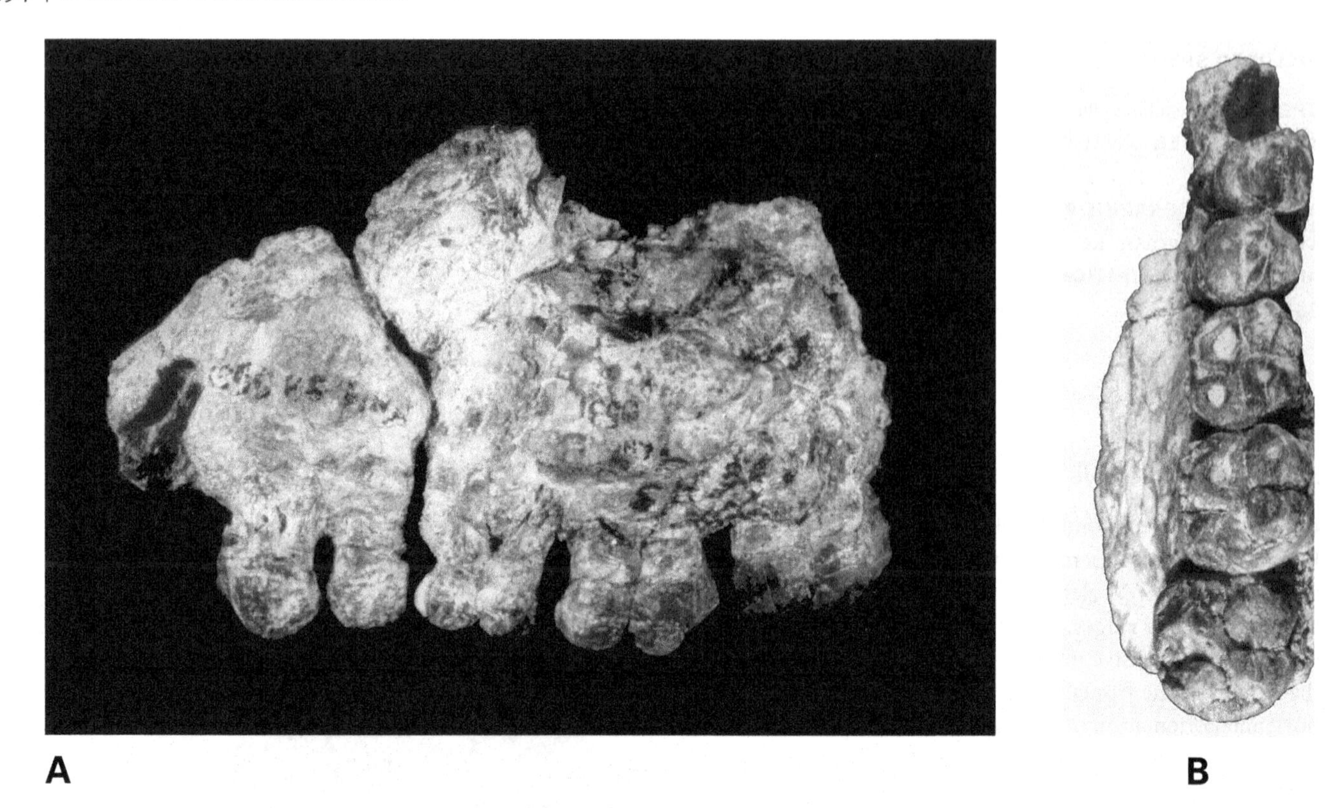

Fig. 22.5 *Samburupithecus kiptalami*. Type specimen KNM-SH 8531, left maxilla with P^3–M^3 and canine alveolus in (A) lateral view, and (B) occlusal view. Photographs courtesy of Hidemi Ishida.

SPECIES *Orrorin tugenensis* Senut *et al*., 2001 (Fig. 22.6)
TYPE SPECIMEN BAR 1000'00, 1000a'00, and 1000b'00, fragmentary mandibles
AGE AND GEOGRAPHIC RANGE 6 Ma, Baringo District, Kenya
ANATOMICAL DEFINITION
As for genus.

Evolution

As noted previously, the known middle–later Miocene hominoids span a time period in which fossil apes appear to become increasingly rare in Africa, while simultaneously diversifying in Europe and Asia. This conjunction of circumstances has increased interest in the hypothesis that hominoid radiations ancestral to great apes and humans occurred outside of Afro-Arabia, and some time after 10 million years ago the common ancestor of chimpanzees, gorillas and humans migrated into Africa (Stewart & Disotell, 1998). Evidence supporting "re-entry" scenarios is predicated on inferences about the fossil record that are now proving to be inaccurate. What was previously an impoverished fossil record of large hominoids in Africa between 15 and 6 million years ago is now becoming more robust with several new taxa being identified in recent years. Moreover, as has been shown by the discovery of *Otavipithecus* in Namibia, it is clear that hominoids were geographically widely dispersed on the African continent by the middle to later Miocene. Based on the results of recent field operations in Kenya, we can reasonably expect new hominoid taxa to be recovered as sediments ranging in age between 13 and 6 million years old are investigated systematically.

The relationships of *Equatorius* and *Nacholapithecus* may be closer to resolution. There is an emerging consensus that *Equatorius*, and probably *Nacholapithecus* as well, are derived representatives of a clade in which *Afropithecus* is the earliest known representative. If this is the case, the question arises as to whether or not the Afropithecinae represent a radiation of stem catarrhines, or are in fact stem hominoids. We agree with Harrison (this volume) that resolution revolves around determining when the "cladistic threshold" has been crossed from stem catarrhine to stem hominoid. Based on our assessment of cranial, dental and postcranial characters, we view *Equatorius* and *Nacholapithecus* as primitive hominoids, and by extension we regard the Proconsuloidea as defined by Harrison (this volume) as a stem hominoid clade.

The taxonomic affinities of *Kenyapithecus wickeri* remain an issue of intense interest. At present, three conflicting hypotheses concerning its relationships have been proposed. S. Ward *et al*. (1999) and Kelley *et al*. (2000) regard *K. wickeri* as more derived in elements of its incisor, canine and maxillary anatomy when compared to the slightly older taxon *Equatorius*. Further, S. Ward *et al*. (1999) and Kelley *et al*. (2000) tentatively suggested that *K. wickeri* manifested several characters in its anterior maxillary dental anatomy that are very similar to those seen in the rarer of the two large hominoid taxa known to be present at Paşalar, in Turkey. If the Turkish taxon proves to be closely related to *K. wickeri*, it would be the first evidence of a link between African and Eurasian hominoids. Another view (Benefit & McCrossin, 2000), holds that there is no generic separation between *Equatorius* and *Kenyapithecus*, and that the nomen *K. africanus* is still valid. This position is based on McCrossin & Benefit's analysis of variation in details of dental and gnathic morphology in specimens from Maboko and Fort Ternan. Finally, Begun (2000) also regarded the nomen *Equatorius* as invalid for other reasons – namely that it is a junior synonym of *Griphopithecus*, known from Paşalar and Slovakia. He further developed a biogeographic scenario positing an early (around 15.3 Ma) link between East African hominoids and those from Slovakia. In response to Begun (2000), Kelley *et al.* (2000) noted that the Slovakian sample of four teeth did not strongly support a taxonomic link between *Equatorius* and *Griphopithecus*. When the recent discoveries from Maboko, Nachola and Paşalar are published it may be possible to identify the taxonomic relationships of *Kenyapithecus*, *Griphopithecus* and other large hominoid taxa from Europe with a greater degree of certainty. However, it is our view that *K. wickeri* remains the best candidate as the earliest large hominoid taxon to have departed continental Africa and become established in the eastern Mediterranean region.

The affinities of two more enigmatic hominoid taxa from the African middle–later Miocene are also uncertain. *Otavipithecus* has been difficult to assign to any higher taxonomic affilation. Andrews (1992) and Singleton (2000) have suggested that *Otavipithecus* may be more closely related to *Afropithecus* than any other known Miocene hominoid, but Pickford *et al.* (1997) showed that the Berg Aukas frontal bone, which is assumed to be attibutable to *Otavipithecus*, possesses an array of characters that overall distinguishes it from *Afropithecus*. Pickford *et al.* (1997) concluded that *Otavipithecus* may be more closely related to the ancestry of African apes and humans than are the Miocene apes from Eurasia. It seems unlikely that the evolutionary relationships of *Otavipithecus* will be resolved any time soon. However, its geographic position in southern Africa underscores the need for caution when speculating on the relative lack of hominoid fossils in eastern Africa between 14 and 6 million years ago. It is entirely possible for example that the ancestry of chimpanzees and gorillas, which remains so elusive, will emerge from localities in regions heretofore thought to be devoid of hominoid fossils.

Similar to *Otavipithecus*, *Samburupithecus* remains an enigma. No additional specimens referable to this taxon have come to light since its discovery in the early 1980s and its formal description and naming by Ishida & Pickford in 1997. We can be sure that it represents a definitive hominoid of modern aspect, but its potential relationships to chimpanzees, gorillas, hominids, or any combination of these is uncertain.

The recently discovered *Orrorin tugenensis* includes intriguing suggestions of bipedal locomotion in a 6-Ma hominoid/hominid. The preserved femora, however, are proximal fragments and therefore lack the distal articular surfaces that are so indicative of bipedal locomotion. The specimen offers important data on the utility of enamel thickness and postcanine megadontia as evolutionary vectors of hominid origins and the australopithecine radiations. At the very least *Orrorin* signals a new initiative to recover primate fossils in this key temporal window of hominoid evolution. It can no longer be said that the record is devoid of fossils at the key place and time of supposed hominid divergence.

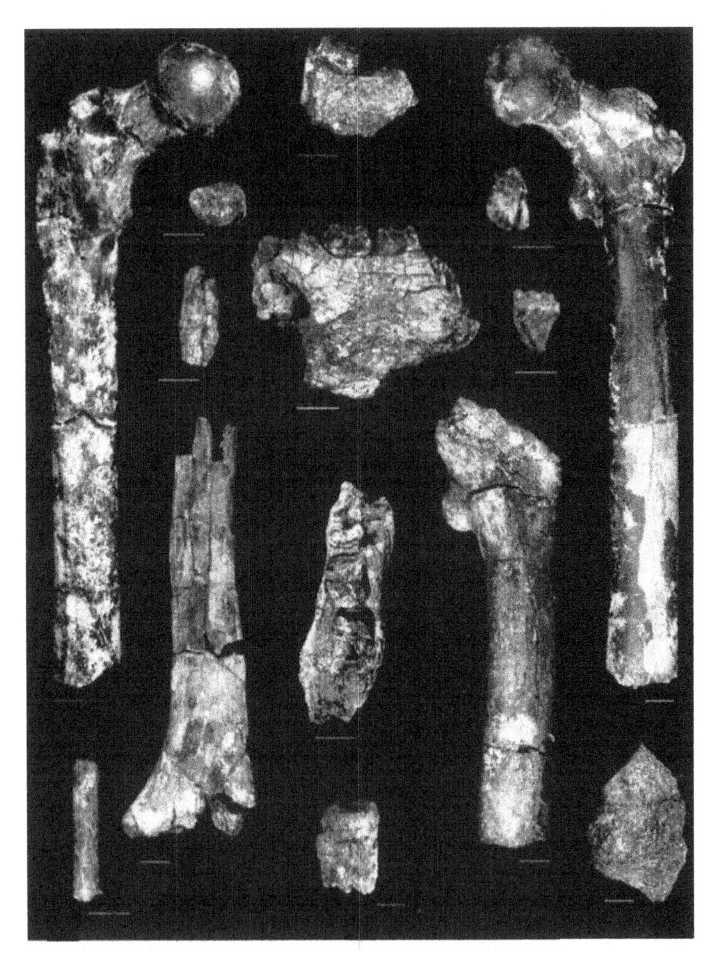

Fig. 22.6 *Orrorin tugenensis*. Composite of recent discoveries. Photographs courtesy of Brigitte Senut and Martin Pickford.

Significant events in the evolutionary history of the hominoids took place between 15.5 and 6 million years ago. The transition from more primitive "stem" catarrhines and hominoids is documented in the fossil record from many middle Miocene localities in Kenya. It is possible that the first large hominoid to migrate beyond Afro-Arabia did so around 14 million years ago. An array of diverse taxa manifesting differences in feeding behavior and locomotor characteristics emerged along the shores of Lake Victoria, the Baringo Basin and the Samburu region, as the Rift Valley was forming. After 13 million years ago, the history of hominoids in Africa is encumbered by a sparse but tantalizing body of fossil evidence. The question of whether the ancestors of African apes and humans evolved in Africa, or alternatively, migrated on to the continent from the north and east remain open questions, questions that the current body of evidence is poorly equipped to resolve. But whatever the ultimate answer to this pivotal event in human evolution turns out to be, the evidence is likely to come from eastern African Rift Valley sedimentary basins where long terrestrial sedimentary sequences are exposed.

Primary References

Equatorius

Le Gros Clark, W. E. & Leakey, L. S. B. (1950). Diagnoses of East African Miocene Hominoidea. *Quarterly Journal of the Geological Society of London*, **105**, 260–263.

Pickford, M. (1982). New higher primate fossils from the Middle Miocene deposits at Majiwa and Kaloma, western Kenya. *American Journal of Physical Anthropology*, **58**, 1–19.

Pickford, M. (1985). A new look at *Kenyapithecus* based on recent discoveries in western Kenya. *Journal of Human Evolution*, **14**, 113–143.

Ishida, H. (1986). Investigation in northern Kenya and new hominoid fossils. *Kagaku*, **56**, 220–226.

Benefit, B. R. & McCrossin, M. L. (1989). New primate fossils from the Middle Miocene of Maboko Island, Kenya. *Journal of Human Evolution*, **18**, 493–498.

Benefit, B. R. & McCrossin, M. L. (1993*c*). New *Kenyapithecus* postcrania and other primate fossils from Maboko Island, Kenya. *American Journal of Physical Anthropology*, Supplement **16**, 55–56.

McCrossin, M. L. & Benefit, B. R. (1993*a*). Recently recovered *Kenyapithecus* mandible and its implications for great ape and human origins. *Proceedings of the National Academy of Sciences of the United States of America*, **90**, 1962–1966.

McCrossin, M. L. & Benefit, B. R. (1993*b*). Clues to the relationships and adaptations of *Kenyapithecus africanus* from its mandibular and incisor morphology. *American Journal of Physical Anthropology*, Supplement **16**, 142.

Benefit, B. R. & McCrossin, M. L. (1994). Comparative study of the dentition of *Kenyapithecus africanus* and *K. wickeri*. *American Journal of Physical Anthropology*, Supplement **18**, 55.

McCrossin, M. L. (1994*b*). Semi-terrestrial adaptations of *Kenyapithecus*. *American Journal of Physical Anthropology*, Supplement **18**, 142–143.

McCrossin, M. L. & Benefit, B. R. (1994). Maboko Island and the evolutionary history of Old World monkeys and apes. In *Integrative Paths to the Past: Paleoanthropological Advances in Honor of F. C. Howell*, eds. R. S. Corruccini & R. L. Ciochon, pp. 95–122. Englewood Cliffs, NJ: Prentice-Hall.

McCrossin, M. L. (1997). New postcranial remains of *Kenyapithecus* and their implications for understanding the origins of hominoid terrestriality. *American Journal of Physical Anthropology*, Supplement **24**, 164.

Gitau, S. N., Benefit, B. R., McCrossin, M. L., & Roedl, T. (1998). Fossil primates and associated fauna from 1997 excavations at the middle Miocene site of Maboko Island, Kenya. *American Journal of Physical Anthropology*, Supplement **26**, 87.

McCrossin, M. L., Benefit, B. R., & Gitau, S. N. (1998). Functional and phylogenetic analysis of the distal radius of *Kenyapithecus*, with comments on the origin of the African great ape and human clade. *American Journal of Physical Anthropology*, Supplement **26**, 158.

Ward, S. C., Brown, B., Hill, A., Kelley, J., & Downs, W. (1999). *Equatorius*: A new hominoid genus from the middle Miocene of Kenya. *Science*, **285**, 1382–1386.

Gitau, S. N., Benefit, B. R., Johnson, K. B., & McCrossin, M. L. (2000). New dental remains of *Kenyapithecus africanus* from Maboko Island, Kenya support the congeneric status of *Kenyapithecus wickeri* and *K. africanus*. *American Journal of Physical Anthropology*, Supplement **30**, 59.

Kenyapithecus

Leakey, L. S. B. (1962*a*). A new lower Pliocene fossil primate from Kenya. *Annals of the Magazine of Natural History (Series 13)*, **4**, 689–696.

Andrews, P. J. & Walker, A. C. (1976). The primate and other fauna from Fort Ternan, Kenya. In *Human Origins: Louis Leakey and the East African Evidence*, eds. G. L. Isaac & E. R. McCown, pp. 279–304 Menlo Park, CA: W.A. Benjamin.

Harrison, T. (1992). A reassessment of the taxonomic and phylogenetic affinities of the fossil catarrhines from Fort Ternan, Kenya. *Primates*, **33**, 501–522.

Nacholapithecus

Ishida, H., Pickford, M., Nakaya, H., & Nakano, Y. (1984). Fossil anthropoids from Nachola and Samburu Hills, Samburu District, Northern Kenya. *African Study Monographs*, Supplementary Issue **2**, 73–86.

Rose, M. D., Nakano, Y., & Ishida, H. (1996). *Kenyapithecus* postcranial specimens from Nachola, Kenya. *African Studies Monographs*, Supplementary Issue **24**, 3–56.

Nakatsukasa, M., Yamanaka, A., Kunimatsu, Y., Shimizu, D., & Ishida, H. (1998). A newly discovered *Kenyapithecus* skeleton and its implications for the evolution of positional behavior in Miocene East African hominoids. *Journal of Human Evolution*, **34**, 657–664.

Ishida, H., Kunimatsu, Y., Nakatsukasa, M., & Nakano, Y. (1999). New hominoid genus from the middle Miocene of Nachola, Kenya. *Anthropological Science*, **107**(2), 189–191.

Nakatsukasa, M., Kunimatsu, Y., Nakano, Y., & Ishida, H. (2000). A new skeleton of the large hominoid from Nachola, northern Kenya. *American Journal of Physical Anthropology*, Supplement **30**, 235.

Orrorin

Andrews, P. J. (1978*b*). Stratigraphy and mammalian paleontology of the late-Miocene Lukeino Formation, Kenya. In *Geological Background to Fossil Man*, ed. M. Pickford, pp. 263–278. London: Academic Press.

Senut, B., Pickford, M., Gommery, D., Mein, P., Cheboi, K., & Coppens, Y. (2001). First hominid from the Miocene (Lukeino Formation, Kenya). *Comptes rendus de l'Académie des sciences de Paris*, **332**, 137–144.

Otavipithecus

Conroy, G. C., Pickford, M., Senut, B., Van Couvering, J., & Mein, P. (1992). *Otavipithecus namibiensis*, first Miocene hominoid from southern Africa. *Nature*, **356**, 144–148.

Pickford, M., Moyà-Solà, S., & Köhler, M. (1997). Phylogenetic implications of the first African Middle Miocene hominoid frontal bone from Otavi, Namibia. *Comptes rendus de l'Académie des sciences de Paris*, **325**, 459–466.

Senut, B. & Gommery, D. (1997). Postcranial skeleton of *Otavipithecus*, Hominoidea, from the Middle Miocene of Namibia. *Annales de Paléontologie*, **83**, 267–284.

Samburupithecus

Ishida, H. & Pickford, M. (1997). A new Late Miocene hominoid from Kenya: *Samburupithecus kiptalami* gen. et sp. nov. *Comptes rendus de l'Académie des sciences de Paris*, **325**, 823–829.

Unattributed

KNM-BN 1378 (Ngororo molar)

Bishop, W. W. & Chapman, G. R. (1970). Early Pliocene sediments and fossils from the northern Kenya Rift Valley. *Nature*, **226**, 914–918.

Bishop, W. W., Chapman, G. R., Hill, A., & Miller, J. A. (1971). Succession of Cainozoic vertebrate assemblages from the northern Kenya Rift Valley. *Nature*, **233**, 384–394.

KNM-BN 10489 (premolar)

Hill, A., Drake, R., Tauxe, L., Monaghan, M., Barry, J. C., Behrensmeyer, A. K., Curtis, G., Fine-Jacobs, B., Jacobs, L., Johnson, N., & Pilbeam, D. (1985). Neogene paleontology and geochronology of the Baringo Basin, Kenya. *Journal of Human Evolution*, **14**, 749–773.

KNM-BN 10556 (canine)

Hill, A. & Ward, S. C. (1988). Origin of the Hominidae: the record of African large hominoid evolution between 14 My and 4 My. *Yearbook of Physical Anthropology*, **31**, 49–83.

The fossil record of human ancestry

23 | Introduction to the fossil record of human ancestry

HENRY M. MCHENRY

In the last three decades human paleontology has been remarkably successful thanks to the skill, energy and persistence of the field researchers. They have expanded the sample enormously, discovered much about the habitats and biological communities in which early hominids lived, and established precise dates for most of the important steps in human evolution.

Public attention is so strong that new discoveries and even new analyses of well-known fossils regularly grace magazine covers and television screens. The public is captivated by the revelation of another distant ancestor, perhaps to the point of overindulging the analysis on how human evolution proceeded. In the context of this book the goal is to outline what the fossil record consistently and reliably has demonstrated about human evolution. A reasonably good picture of our evolution exists, but parts of it may always remain in darkness, due to several inherent limitations to the record.

Limitations of the record

The limitations of our knowledge include the fact that the sample is restricted geographically. For the first 3 million years of the record the sample derives almost entirely from sites in the African Rift Valley and limestone caves in northern South Africa. The exception is a maxillary fragment from Bahr el Ghazal in Chad that calls attention to the likelihood of hominids in many places as yet unsampled (Brunet *et al.*, 1995). Habitats favorable for hominid occupation undoubtedly appeared and disappeared throughout much of Africa with the drastic fluctuations in tropical climates that occurred during the Plio-Pleistocene (Vrba *et al.*, 1995).

Even with comparatively rich samples of paleospecies such as those of *Australopithecus afarensis* and *A. africanus*, there are limitations. Most specimens are very fragmentary and associated partial skeletons are rare. The shortage of associated skeletons makes reconstruction of body size and shape dependent on many assumptions. For example, the large morph that is presumably the male of *A. afarensis* is represented by unassociated elements of the fore- and hindlimb of different individuals (McHenry, 1992). The large and small morphs (male and female?) of *A. africanus* are known from two fragmentary associated partial skeletons and many isolated elements without complete fore- or hindlimbs (McHenry, 1992; McHenry & Berger, 1998*a*, 1998*b*). Body size and proportions are important parameters to know because they relate to how the organism moved about in its habitat and how relatively large were, for example, the brain and chewing mechanism.

Another dark side of the picture is that postcranial elements cannot be associated confidently with specific taxa in sites between 2.7 Ma (when two contemporary hominid species appear in the Shungura Formation of Omo) and about 1.2 Ma (when "robust" australopithecine species apparently go extinct). The wonderfully rich collections of postcranial elements in this time period with rare exceptions cannot be attributed to species. Craniodental characteristics cannot be evaluated confidently in reference to overall body size. Relative brain and tooth size, for example, can only be known approximately and evaluations potentially may be badly distorted.

Even with the relatively rich sample of craniodental remains, understanding is limited because homoplasy, the appearance of similarities in separate evolutionary lineages, was rampant in hominid evolution (Lieberman *et al.*, 1996; McHenry, 1996; Lockwood & Fleagle, 1999). Many evolutionary novelties must have evolved independently. For example, extreme cheek tooth expansion and all the structures involved in heavy chewing evolved in "robust" australopithecine species and, to a less extreme condition, in *A. afarensis*, *A. africanus* and *A. garhi*. Development of hypermastication links *A. aethiopicus*, *A. boisei* and *A. robustus* into an evolving lineage separate from the sequence of *A. afarensis* to *A. africanus* and *A. garhi* to *Homo*, but such an arrangement implies parallel evolution of facial flattening, cranial base flexing, and many other traits (Skelton & McHenry, 1992, 1998; Strait *et al.*, 1997).

Yet another limitation is the current state of research on character transformations. Recent cladistic analyses use 60 or more craniodental character transformations, but are these truly independent evolutionary novelties or are some traits part of complexes that change together? Alveolar thickness and tooth size are not independent, of course. Basicranial flexion and orthognathism are generally correlated. The strength of cladistic analysis is that it brings into focus these intercorrelations. Research on the developmental biology of characters will provide important clues of how to access the evolutionary independence of character states (Lovejoy *et al.*, 1999).

Accumulation of evolutionary novelties

Appreciation of recent successes balances the dark side of the

Table 23.1. Species, dates, body size, brain size and posterior tooth size in early hominids

Taxon[a]	Dates[b]	Mass[c]		Stature[d]		Cranial capacity[e]	Brain weight[f]	Postcanine tooth area[g]	Enaph-alization quotient[h]	Mega-dontial quotient[i]
		Male	Female	Male	Female					
Pan troglodytes	Extant	49	41	—	—	—	395	294	2.0	0.9
Ardipithecus ramidus	5.8–4.4									
Australopithecus anamensis	4.2–3.9	51	33	—	—	—	—	428	—	1.4
Australopithecus afarensis	3.6–2.9	45	29	151	105	438	434	460	2.4	1.7
Kenyanthropus platyops	3.5–3.3									
Australopithecus africanus	3.0–2.4	41	30	138	115	452	448	516	2.7	2.0
Australopithecus aethiopicus	2.7–2.2	—	—	—	—	—	—	688	—	—
Paranthropus boisei	2.3–1.4	49	34	137	124	521	514	756	2.7	2.7
Paranthropus robustus	1.9-1.4	40	32	132	110	530	523	588	3.0	2.2
Australopithecus garhi	2.5–?	—	—	—	—	450	446	—	—	—
Homo habilis	1.9–1.6	37	32	131	100	612	601	478	3.6	1.9
Homo rudolfensis	2.4–1.6	60	51	160	150	752	736	572	3.1	1.5
Homo ergaster	1.9–1.7	66	56	180	160	871	849	377	3.3	0.9
Homo sapiens	Extant	58	49	175	161	—	1350	334	5.8	0.9

[a] Taxonomy based on Klein (1999).
[b] Dates (in Ma) are from Klein (1999).
[c] Body mass estimates (in kg) are from McHenry (1992) except for the following: *A. anamensis* male is from Leakey *et al.* (1995*a*); *A. anamensis* female is calculated from the ratio of male and female in *A. afarensis*; *H. ergaster* is from Ruff *et al.* (1998).
[d] Stature estimates (in cm) are from McHenry (1991) except *H. ergaster* which is from Ruff & Walker (1993).
[e] Cranial capacity is in cm^3 from sources listed in McHenry (1994*a*, 1994*b*) with the addition of A.L. 444-2 (540 cm^3) to *A. afarensis* (W. H. Kimbel, pers. comm.); Stw 505 (515 cm^3) to *A. africanus* (Conroy *et al.*, 1998); KGA 10-525 (545 cm^3) to *P. boisei* (Suwa *et al.*, 1997); and BOU-VP-12/130 (450 cm^3) to *A. garhi* (Asfaw *et al.*, 1999).
[f] Brain weight is calculated from formula (6) in Ruff *et al.* (1998).
[g] Postcanine tooth area (in mm^2) is the sum of products of buccolingual and mesiodistal lengths of P_4, M_1 and M_2 and is taken from McHenry (1994*b*) with the addition of *A. anamensis* from Leakey *et al.* (1995*a*).
[h] Encephalization quotient is calculated as brain mass divided by ($11.22 \times$ body mass$^{0.76}$) from Martin (1981).
[i] Megadontial quotient is derived as postcanine tooth area divided by ($12.15 \times$ body mass$^{0.86}$) from McHenry (1988).

picture. One way of organizing them that expresses development of understanding most effectively is to follow through time the accumulation of evolutionary novelties. The subtlety of this approach is in appreciating that the known fossil sample represents the successful species at the time and not usually the direct ancestors of later species. Speciation probably occurred in small, isolated, peripheral populations that the fossil record did not sample. We collect what was successful at the time. We might expect, therefore, to find many unique characteristics of fossil species that exclude them from direct ancestry, but provide the keys to reconstructing the common ancestor between later species.

From this point of view, the record is superb. One can follow the hominid lineage step by step as the accumulation of human-like characteristics. Oddities are autapomorphies of the particular species that do not necessarily exclude the possibility that it and subsequent species shared a common ancestor. This accumulation has patterns but no predetermined direction. Orthogenesis has no place in this line of thinking. Table 23.1 summarizes one view of species characteristics, as a working hypothesis about body sizes and about how representative the brain and tooth sizes are.

Ardipithecus ramidus (5.8–4.4 Ma) shares key evolutionary novelties with later hominids, although it may not have some of the key traits traditionally regarded as diagnostic of the human family (White *et al.*, 1994; Haile-Selassie, 2001). Its apparently short cranial base, reduced canine projection, more incisiform canine, and other subtleties link it with later hominids relative to any known Miocene or modern ape. It is very primitive in the sense that it resembles what is best reconstructed as the last common ancestor of humans and apes, but it shares these novelties with the human family. It is the earliest and most primitive known member of Hominidae.

Australopithecus anamensis (4.2–3.9 Ma) is slightly later in time and has accumulated more novelties of the human lineage relative to *Ar. ramidus* (Leakey *et al.*, 1995*a*, 1998; C. Ward *et al.*, 1999). Most conspicuous is the thickening of enamel that characterizes all later hominids. The molars are expanded from side to side and the first and second molars are not markedly different in size. The tympanic tube is said to extend to the medial edge of the postglenoid process. Its deciduous lower first molar is reported to be intermediate between *Ar. ramidus* and *A. afarensis* in size and shape. Postcranially it is very much like later hominids. The original description of *A. anamensis* mentions that the lateral trochlear ridge on the distal humerus is weakly developed, unlike the condition in *Ar. ramidus*.

The sample of *A. afarensis* (3.6–2.9 Ma) postdates *A. anamensis* and shares derived characteristics with later hominids, including an external auditory meatus more rounded in outline, and less parallel and wider mandibular toothrows (Johanson *et al.*, 1982*b*; Leakey *et al.*, 1995*a*, 1998). The mandibular symphysis slopes less strongly posteroinferiorly. The canine root and crown are reduced. The upper canine root and associated facial skeleton are more posteriorly inclined. The trigons of the upper molars are relatively smaller. The lower molars have less sloping buccal sides and the upper molars more sloping lingual sides. The deciduous first lower molar is broadened and the talonid is well differentiated. The angle between the 2nd and 3rd metacarpal facets on the capitate is wider than in *A. anamensis* and more like later hominids.

Kenyanthropus platyops (3.5–3.3 Ma) overlaps in time with *A. afarensis*, but appears to be quite distinctive in its morphology (Leakey *et al.*, 2001). In some respects it is more primitive than its contemporary. For example, it lacks a petrous crest on the tympanic and has a narrow external acoustic meatus with a small aperture more like earlier hominids and not like *A. afarensis*. In other respects it resembles much later hominids, particularly *H. rudolfensis*, in having a relatively flat face, a tall malar region, small molars, and an anterior positioning of the zygomatic processes. These latter traits are related to mastication that, according to Leakey *et al.* (2001: 437) "suggests a diet-driven adaptive radiation among hominins in this time interval." It is *possible* that the resemblances between *H. rudofensis* and *K. platyops* are due to phylogenetically independent events driven by adaptations to similar situations and are, thereby, one more example of the pervasiveness of homoplasy. It is equally possible that the resemblances represent true homology and imply a clade linking *K. platyops* with *H. rudolfensis*. Leakey *et al.* (2001) favor the latter view that suggests transferring *H. rudolfensis* to the genus *Kenyanthropus* (i.e., *K. rudolfensis*).

The next oldest hominid species is *A. garhi* (2.5 Ma). It has some shared-derived traits with later and contemporary hominid species relative to *A. afarensis* (Asfaw *et al.*, 1999). Relative to *A. afarensis*, it shares with *A. africanus* and the "robust" australopithecines absolutely and relatively larger postcanine dentition. Its premolars show some molarization and its enamel is thicker. Its upper incisor roots are no longer lateral to the nasal aperture as they are in *A. afarensis*. The posterior temporalis and parietomastoid angles are reduced relatively, as in all post-*afarensis* hominids except *A. aethiopicus*.

Australopithecus africanus (3–2.4 Ma) is roughly contemporary with *A. garhi* and later than *A. afarensis*. Traits apparently derived and shared with *Homo* relative to these species include upper lateral incisor roots medial to the nasal aperture, a flatter clivus contour, reduced subnasal prognathism, reduced incisor procumbency and reduced bipartite lateral anterior facial contour (Asfaw *et al.*, 1999).

The earliest of the "robust" australopithecines, *A. aethiopicus* (2.7–2.3 Ma), shares very few derived characters with later "robusts" or *Homo* that are not related to hypermastication (Grine, 1988*b*). The later "robust" species, *A. robustus* (2-1 Ma) and *A. boisei* (2.3–1.4 Ma), do share apparent evolutionary novelties with *Homo* relative to other australopithecines such as a parabolic dental arcade shape, deep anterior depth of the palate, rare or absent diastema between the upper incisors and canine, a variable to weak canine jugum and fossa, reduced subnasal prognathism, and enlarged brain (Skelton & McHenry, 1992; Strait *et al.*, 1997).

One might despair at the apparently conflicting evidence for the origin of genus *Homo*, but from the perspective of the accumulation of shared-derived traits, the fossil record is less perplexing. *Brains expand and cheek teeth reduce*. In one species, *H. habilis* (2.3–1.6 Ma), the body appears to remain like that of *Australopithecus*, small with relatively large forelimbs and small hindlimbs (Johanson *et al.*, 1987). If the hindlimb and craniodental fossils of *H. rudolfensis* (approximately 1.9 Ma) truly are associated, then this is the first evidence of more human-like body proportions and hip architecture (Wood, 1992). Both of these species are transitional with some primitive and some derived characteristics of later *Homo*. Postcranial associations are critical here because body size appears to be very different. *Homo habilis* was very small-bodied (35 kg) and *H. rudolfensis* was large (55 kg) (McHenry, 1994*a*). Scaling cheek tooth size to body weight shows that they both had reversed the trend of ever-increasing cheek tooth size. Relative brain size expanded, especially in *H. habilis*.

Absolute brain size expands further with the appearance of *H. ergaster/erectus* by at least 1.8 Ma, but body size also increases (Walker & Leakey, 1993). The early African form of this species is often referred to the species *H. ergaster* to contrast it with the well-known Asian sample of *H. erectus* (Wood, 1991). Body size and especially hindlimb length reach modern proportions in this species (Ruff & Walker, 1993). Other synapomorphies with later *Homo* include reduction in prognathism, the postglenoid process, and the absolute and relative size of the cheek teeth. Brains continue to expand and cheek teeth get progressively smaller through the evolution of the genus *Homo*.

Brain evolution

The accumulation of evolutionary novelties reveals the continuity of human evolution. Demonstrating this continuity is a triumph of paleoanthropology. It further confirms Darwin's view of descent with modification. Hominid brain expansion is one of the most dramatic demonstrations of this, and follows time remarkably well (Table 23.1). The earliest species for which brain size is known, *A. afarensis*, has an average endocranial volume of 438 cm^3 (based on four specimens) compared to modern chimps at 400 cm^3 gorillas at 500 cm^3 and *H. sapiens* at 1350 cm^3. The endocranial volume for *A. garhi* (one specimen) is reported to be 450 cm^3, for *A. africanus* (seven specimens), 452 cm^3, for *A. boisei* (one specimen), 521 cm^3, for *A. robustus* (one specimen), 530 cm^3, for *H. habilis* (six specimens), 612 cm^3, for *H. rudolfensis* (one specimen), 752 cm^3, for *H. ergaster* (three specimens), 871 cm^3.

Brains expand through the Pleistocene from 914 cm^3 between 1.8 and 1.2 Ma to 1090 cm^3 between 550 000 and 400 000 years ago to 1186 cm^3 between 300 000 and 200 000 years ago, to above 1300 cm^3 after 150 000 years ago (Ruff *et al.*, 1998). Relative to estimated body weight, the fossil species go from slightly above the size seen in modern apes to 5.4 times the mammalian average (column 5 in Table 23.1). Unfortunately the fossil record is mute on the subject of brain reorganization which is a pity because there were presumably profound changes that accompanied the development of speech and other uniquely human faculties (Rilling & Insel, 1999).

Bipedalism

Bipedalism is the first major grade shift in hominid evolution, but the very first hominids were most likely generalized hominoids with rather ape-like bodies. They were part of the radiation of late Miocene hominoids that came, in Darwin's words, "to live somewhat less on trees and more on the ground", which was due to "a change in its manner of procuring subsistence, or to a change in the conditions of its native country" (Darwin, 1872). The first members of the human *clade* probably will be recognized in the fossil record by subtle evolutionary novelties such as the beginning of canine reduction. The major genetic/developmental transformations that produced the bipedally adapted trunk and hindlimb morphology defining the human grade probably appeared later. The precise sequence of trunk and hindlimb transformations may never be known from the fossil record. Parts of the bipedal body may have changed at different rates so that the length of the pelvic blades reduced earlier than the adduction of the hallux. The extinct and distinctly non-hominid Miocene ape *Oreopithecus* appears to have had a reduced pelvic length but retained a divergent big toe (Köhler & Moyà-Solà, 1997). The full analysis of *Ardipithecus ramidus* will clarify the transition to bipedality with our lineage.

Views about the meaning of the primitive postcranial traits retained in early species of *Australopithecus* have become polarized. One pole emphasizes the bipedal specializations (Latimer, 1991) and the other the many primitive characteristics (Stern, 2000). Both camps agree that all species of *Australopithecus* were bipedal. That implies these species did not climb like apes. However, they retained morphological features associated with arborealism for at least 1 million years and hip architecture different from later *Homo* species. That implies some difference from modern humans in gait and climbing ability, but they did not walk and climb like apes.

Discoveries of postcranial fossils associated with *H. ergaster/erectus* (Walker & Leakey, 1993) clarified the difference between *Australopithecus* positional and locomotor behavior and that seen in later species of *Homo*. The pelvis and femur of all species of *Australopithecus* and that of *H. ergaster/erectus* are strikingly different, with the latter appearing much more human-like. From this perspective, *Australopithecus* appears to be a biped with free hands for carrying, but best adapted to short distances and having a healthy appreciation of trees for safety, feeding and sleeping. The longer femora and more human-like pelves that appear by 1.9 Ma with *Homo* mark the beginning of an important change.

Megadontia

Although Darwin and his contemporaries predicted much of what the human fossil record would eventually reveal, no one anticipated the discovery of hominid megadonts. African apes and modern humans have small cheek teeth relative to body size. But early hominids had relatively huge molars and premolars with concomitantly gigantic jaws, alveolar bone, buttressed face and cheeks, and attachment areas for the chewing musculature. There is an approximate trend in successive species of increasing relative cheek-tooth size from small in *Ardipithecus ramidus* to moderate in *Australopithecus anamensis* and *A. afarensis*, to big in *A. africanus*, *A. robustus* and *A. garhi*, to huge in *A. boisei* and *A. aethiopicus*. The trend reverses in the *Homo* lineage so that from *H. habilis* and *H. rudolfensis* to modern *H. sapiens*, the relative size becomes progressively smaller (column 9 in Table 23.1).

Future surprises

The diversity of living members of the superfamily Hominoidea is impoverished relative to what it was in the past. Even with the limited fossil sample it is clear that there were many kinds of apes and humans long ago. More species await discovery. There were probably many evolutionary experiments in the varied habitats of the African Plio-Pleistocene. Although the current sample of fossil hominids leads some to the impression that there were only a few hominid lineages, it is far more likely that our family tree will turn out to be quite bushy. Species names may need to multiply to accommodate the diversity, although a balance needs to be maintained between excessive splitting and lumping.

It will be wonderful to resolve the question of knuckle-walking in human evolution. Living species of the African clade of Hominoidea include three knuckle-walkers and a biped. The odds are that the first hominid was also a knuckle-walker, yet the earliest hominids known so far are bipeds and climbers without specializations for knuckle-walking. Perhaps there are functional and developmental processes that predispose large-bodied apes to become knuckle-walkers when they adapt to habitats that require more terrestriality. If so, that apparently odd gait may have evolved in parallel in two or more lineages.

There may have been several population expansions out of Africa during favorable climatic conditions that then became extinct. It is seductive to imagine only one triumphant pulse of migration that spread humanity for the first time throughout Eurasia. The spread of anatomically modern *H. sapiens* appears to have happened in a short amount of geological time, but archaic species of *Homo* were probably much less adaptable and more vulnerable to the severe climatic shifts occurring everywhere over the last several million years. Local extinctions may have been common throughout the changing geographical range of early hominids. Before 40 000 years ago, humans were an insignificantly small part of the vertebrate fauna.

It is tempting to swing to extreme views of pessimism or optimism. The hominid fossil record is limited and incomplete, but it is also rich and consistent. Its overall quality is consistent with Darwin's view of descent with modification. It makes it very unlikely that, for example, significant encephalization began before the evolution of bipedalism in our lineage. The record is rich enough to make it unlikely that any hominid will be discovered that dates before 2 million years ago with a brain size of more than 600 cm^3. Conflicting interpretations will always be part of the science, but debate is a sign of intellectual health and helps to ensure ideas are grounded in accurate observations of material evidence and precise logic.

24 | Earliest hominids

TIM D. WHITE

Introduction

The fossil record of the earliest Hominidae is exclusively African, dating from the basal Pliocene. The first appearance of hominids outside of Africa, in Indonesia and Georgia, is suggested to be in the late Pliocene. The earliest genus that may be attributed with confidence to the Hominidae is the 5.8–4.4-Ma *Ardipithecus*, known from two subspecies. *Australopithecus* appeared later, was far more widespread in space and time, but was also exclusively African. Several *Australopithecus* species are recognized in at least two lineages. The genus spans approximately 4.1 to 1.4 Ma. *Australopithecus* is known from South Africa, Malawi, Tanzania, Kenya, Ethiopia, and Chad, but this geographic pattern owes itself to preservational conditions rather than true biological range. *Australopithecus* species exhibit postcanine megadontia and postcranial skeletons adapted to bipedality. Brain sizes were within ranges for the extant great apes. *Australopithecus* is widely recognized as the generic ancestor of relatively encephalized and mostly later species of *Homo*.

History of discovery and debate

There is no consensus regarding the definition of the family Hominidae. When only extant Hominoids were known during the 1800s, traditional, gradistic classifications were favored, and the Asian and African great apes were lumped in the family Pongidae. This conservative definition is still in wide use today. However, biomolecular studies show that the Asian orangutan *Pongo* is more distantly related to modern humans than to any of the three African apes (the chimpanzee, bonobo and gorilla). Some taxonomists therefore combine humans and the three modern African ape species in an expanded, cladistically defined Hominidae. Others subsume *all* extant great apes with humans in an even more inclusive Hominidae.

As increasingly earlier fossils were recovered from Africa, it became evident that bipedality arose among hominids before material culture or brain expansion. This has led most human paleontologists to adopt habitual bipedality as a marker for the Hominidae. This gradistically based familial classification has worked among the fossils so far described because all *Australopithecus* and early *Homo* species were habitually bipedal. However, as the late Miocene common ancestor to African apes and humans is approached in the fossil record, a familial distinction based on locomotor mode is likely to become more difficult to make. Fortunately, there is a phylogenetic means of classifying fossil forms at the family level that is consistent with traditional taxonomic practice. This defines Hominidae as the sister group of the African apes (Panidae (or Panidae and Gorillidae)). Hominids thus comprise all species derived in the human direction after the last common ancestor of African ape(s) (the chimpanzees based on molecular data) and humans. By this definition, Hominidae currently has one extant genus (*Homo*), and two extinct ones (*Ardipithecus* and *Australopithecus*). Several Old World Miocene ape fossils were mistakenly identified as cladistic hominids over the last century and are discussed in previous chapters. The recently proposed genus *Orrorin* may be a cladistic hominid (see Ward & Duren, this volume). However, further studies on additional materials are required to confirm this.

The first *Australopithecus* fossil was found in 1924. It was published the next year by Raymond Dart as *Australopithecus africanus* (Dart, 1925). Dart's claims that this immature child's cranium represented a cladistic hominid and a human ancestor were widely rejected for a variety of reasons. Ten years later Louis Leakey recovered the first adult *Australopithecus* specimen, and the first Pliocene hominid found in eastern Africa, at the site of Laetoli, Tanzania. Unfortunately, this lower canine was misidentified as a cercopithecid until the much later recognition of *A. afarensis* (White, 1981). In 1939 Ludwig Kohl-Larsen found a hominid maxilla with two premolars, an upper third molar, and an unrecognized lower incisor at Laetoli. It was not until the 1970s that these specimens were seen as specifically distinct from *A. africanus*.

In 1936 the first recognized adult *Australopithecus* specimen was recovered from Sterkfontein, South Africa. It was named *Australopithecus transvaalensis* by Robert Broom (Broom, 1936a). Two years later Broom placed this species in the genus *Plesianthropus* (Broom, 1938). In the same 1938 publication Broom published some adult remains from Kromdraai as *Paranthropus robustus*, calling attention to the many derived craniodental features of this species.

An important monographic treatment of the expanding South African samples was published by Broom & Schepers in 1946. Broom's work at Sterkfontein led to the 1947 discovery of a partial *Australopithecus* skeleton (Sts 14) that included a fairly complete pelvis. In 1948 Dart announced remains from another South African site, Makapansgat, as *Australopithecus prometheus* (Dart, 1948), In the very next year

Broom named another species of *Paranthropus* from Swartkrans (Broom, 1949), and recognized two contemporary hominids at that location (Broom & Robinson, 1950). Broom & Robinson's subsequent monograph (Broom & Robinson, 1952) represented an important milestone in the presentation and interpretation of these fossils.

By 1952 Le Gros Clark's work with the South African fossils had convinced him of their hominid status. The next year he joined Joseph Weiner and Kenneth Oakley in exposing Piltdown as a fraud, thereby assuring the widespread acceptance of *Australopithecus* as a hominid. By 1954 John Robinson had framed a dietary hypothesis in an attempt to explain the differences between what were widely, but in retrospect inappropriately, described as "gracile" and "robust" early hominid forms (Robinson, 1954, 1956). Robinson's work was followed by most authorities in interpreting *A. africanus* as an ancestor of *Homo*, and in considering *A.* ("*P.*") *robustus* as a distinct, but now extinct, lineage.

In 1959 the focus of early hominid recovery shifted from southern to eastern Africa with the recovery of a young adult male skull at Olduvai Gorge by Mary Leakey. Louis Leakey named the new form *Zinjanthropus boisei* and attributed to it the contemporary lithic technology that lay in close spatial proximity (Leakey, 1959). The application of the potassium–argon radioisotopic technique to nearby volcanic horizons dated this specimen at what was then a surprising 1.75 million years. The next year saw the discovery of more fragmentary hominid remains belonging to a second, contemporary hominid species from an equivalent stratigraphic horizon at Olduvai. This confirmed Broom and Robinson's earlier insight that at least two hominid lines had coexisted in the African Plio-Pleistocene.

Louis Leakey, Philip Tobias and John Napier described *Homo habilis* in 1964, based on an immature holotype specimen from Olduvai, (Leakey *et al.*, 1964). Robinson and many other workers initially considered this to represent a northern variety of *Australopithecus africanus*. It was not until the early 1970s that discoveries in Kenya confirmed that Leakey and colleagues were correct (see Tobias, 1991, and Dunsworth & Walker, this volume). The success of the Olduvai investigations focused attention on eastern Africa. By the end of the 1960s, this had resulted in discoveries of *Australopithecus* at Omo, Peninj, Kanapoi, the Baringo Basin, and Koobi Fora. The new species *Paraustralopithecus aethiopicus* was prematurely designated for an edentulous mandible from 2.5-Ma Omo Shungura deposits in southern Ethiopia (Arambourg & Coppens, 1968).

Work in the Turkana Basin during the 1970s dramatically increased sample sizes, and extended the record of early hominids more deeply in time. In 1975 a very early cranium of *H. erectus* (KNM-ER 3733) was recovered from Koobi Fora sediments that had also yielded *A. boisei*. This falsified the long-standing "single species" hypotheses about early human evolution (Leakey & Walker, 1976). Meanwhile, in efforts to avoid the taxonomic controversy that had surrounded the Olduvai discoveries a decade earlier, Richard Leakey's Koobi Fora Research Project dichotomously divided its recovered hominid fossils into *Homo* and *Australopithecus*. Species designations were withheld as Koobi Fora fossils were recovered in the 1970s. In eastern Africa, for these workers and this time, the name *Australopithecus* came to designate the species *A. boisei*.

When hominid fossils older than 3.0 Ma were found in the early and mid-1970s at Laetoli (Tanzania) and Hadar (Ethiopia), they were first considered to represent *Homo* according to this local Kenyan tradition (because they showed none of the specializations of *A. boisei*). However, subsequent analysis by Donald Johanson and Tim White led to the 1978 description of the combined Laetoli and Hadar samples as belonging to a separate species within the genus *Australopithecus*, *A. afarensis* (Johanson *et al.*, 1978). The discovery of a fairly complete partial skeleton from Hadar in 1974 (A.L. 288-1; "Lucy") and three sets of fossilized footprints at Laetoli in 1978 represented important new evidence from a time period that had been poorly represented before the mid-1970s (Johanson *et al.*, 1982*b*; Leakey & Harris, 1987).

Controversies about the 3.0 to 3.5 Ma fossils attributed to *Australopithecus afarensis* ensued during the 1980s. These debates involved the distinctiveness of *A. africanus* from *A. afarensis*; whether Hadar *A. afarensis* was a single species; whether Laetoli and Hadar sampled the same taxon; and whether the locomotor habits of these earliest hominids involved significant degrees of arboreality.

In 1986 Alan Walker and colleagues described a fairly complete cranium west of Lake Turkana as an early representative of *A. boisei* (the "black skull": Walker *et al.*, 1986). Most workers saw this fossil as validating the existence of a species *A. aethiopicus* distinct from *A. boisei*. An important volume on robust *Australopithecus* was published in 1988 (Grine, 1988*a*).

In the early 1990s work resumed at Hadar and in the Middle Awash of the Ethiopian Afar depression. This research generated greatly enlarged samples of *A. afarensis* (White *et al.*, 1993; Kimbel *et al.*, 1994). Work by Alun Hughes, Ron Clarke and Phillip Tobias at Sterkfontein in South Africa also led to dramatically enlarged samples of *A. africanus* (Clarke, 1998). The 1990s also saw the discovery of remains attributed to *A. boisei* in Malawi and at the site of Konso-Gardula (southern Ethiopia) where Gen Suwa and Yonas Beyene found the first associated mandible and cranium of this species (Suwa *et al.*, 1997). The important South African *A. robustus* site of Drimolen was opened by Andre Keyser in the 1990s (Keyser, 2000). Overshadowing recovery of these additional samples of previously recognized *Australopithecus* species in the 1990s was a burst of discovery of previously unknown hominid species in Kenya and Ethiopia.

In 1994 the Middle Awash paleoanthropological project described a new, 4.39-Ma hominid species, *A. ramidus*, based on its 1992 and 1993 discoveries at Aramis in the Afar depression of Ethiopia (White *et al.*, 1994). More fossils of this form were subsequently recovered, and the new genus name *Ardipithecus* was made available to receive this species pending comprehensive analysis and description of the remains (White *et al.*, 1995). Later in 1995 Meave Leakey and colleagues announced the new species *Australopithecus anamensis*, based largely on material from Kanapoi, Kenya (Leakey *et al.*, 1995*a*, 1998, 2001). In 1995 limited craniodental remains initially attributed to a new species of *Australopithecus* were recovered in Chad, central Africa by a team led by Michel Brunet (Brunet *et al.*, 1996, 1997). In 1997 Yohannes Haile-Selassie discovered the site of Galili, Mulu Basin, Afar, Ethiopia, and subsequently recovered teeth that may belong to *A. anamensis* (Haile-Selassie & Asfaw, 2000). In 1998 Ron Clarke announced the discovery of a partial skeleton of *Australopithecus* deep within the Sterkfontein cave complex (Clarke, 1998). In 1999 a 2.5-Ma partial cranium with dentition from Bouri in the Middle Awash study area of Ethiopia was named *Australopithecus garhi* by Asfaw and colleagues (Asfaw *et al.*, 1999), and evidence for contemporaneous butchery of large mammals with stone tools such as those found at Gona (Semaw *et al.*, 1997) was simultaneously announced (de Heinzelin, *et al.*, 1999). Early in 2001, Martin Pickford and B. Senut described the *Orrorin* fossils (Chapter 22, this volume). The discovery of *c.* 3.5-Ma hominid fossils west of Lake Turkana was announced in 2001 with the description of a putatively new hominid genus and species, "*Kenyanthropus platyops*" (Leakey *et al.*, 2001). In 2001 Haile-Selassie published the *c.* 5.7-Ma hominids from the western margin of the Middle Awash, naming them *Ardipithecus ramidus kadabba*.

Taxonomy

Systematic framework

In the early history of hominid paleontology, Linnean names were used to denote site-specific sets of fossils or even single specimens. As a result of misusing these taxonomic binomina instead of specimen numbers, early hominid taxonomy had become highly complicated, and was out of step with systematic biology. This began to change after the publication of Ernst Mayr's 1950 paper, "Taxonomic categories in fossil hominids." Mayr interpreted hominid evolution as unilineal and gradual, and sought to recognize this by lumping all of the then-known fossils into the genus *Homo*, with only three species. The important 1963 volume *Classification and Human Evolution*, edited by Sherwood Washburn, played a further important role in bringing hominid paleontology into conformity with the principles of zoological nomenclature, as did Bernard Campbell's 1965 publication of *The Nomenclature of the Hominidae* and Phillip Tobias' classic monograph, *Olduvai Gorge*, Vol. 2, on the *A. boisei* specimen from Olduvai Gorge (Tobias, 1967).

The rise of cladistic classification in paleoanthropology during the 1970s has once again turned the taxonomic tide. Parsimony cladistics has recently led to extensive splitting of hominid taxa at the genus level and below. This has resulted predictably in extreme nomenclatural instability at the level of the Hominidae, and in the taxonomic hierarchy below. Much of the cladistically based splitting among early hominid fossils has arisen from attempts to reconstruct phylogeny based on inadequate data sets and unbridled character atomization. Taxonomies based on these volatile phylogenetic hypotheses are highly unstable. Efforts to infer phylogeny through the cladistic analysis of biologically understood characters provide a solution but are just under way (Lovejoy *et al.*, 1999; McCollum, 1999). Meanwhile, in obsessional efforts to avoid paraphyletic taxa among hominids, cladists have produced an unwarranted proliferation of hominid genera and species. For example, the genus name "*Praeanthropus*" was employed by some workers to differentiate among a set of fossils thought to comprise a single species only a few years ago. Purely theoretical predictions of high early hominid species diversity and complex phylogeny that began in the 1970s have dramatically influenced expectations and interpretations of fossils found during the last decade. These expectations have been "met" by investigators who have mistaken intraspecific variation for species-level differences. The result has been phylogenetic and nomenclatorial confusion of an unprecedented magnitude.

Until the phylogenetic relationships among early hominids can be established with more reliability based on an expanded fossil record (see discussion below), it seems prudent to refrain from generic oversplitting of the known species. The genus "*Paranthropus*" might become a reasonable alternative to *Australopithecus* (in either gradistic or cladistic terms) but only after its monophyly has been demonstrated unequivocally. Thus, whether considered cladistically or gradistically (adaptively), "*Paranthropus*" and "*Praeanthropus*" are junior subjective synonyms of *Australopithecus*, as is the recently proposed "*Kenyanthropus*", which was based primarily on a fragmentary and highly distorted holotype.

The seven species in this genus of relatively small-brained, small-canined, postcanine megadont primate species show a modest amount of variation through time and across African space. Some of the species recognized here are similar and may ultimately be more appropriately considered as paleodemes; others appear to be ancestral-descendant representatives of single lineages. Compared to other mammalian genera, the amount of variation in the genus *Australopithecus* as defined here is not particularly marked. The following classification recognizes only two valid *Australopithecus* species in South Africa, the type species *A. africanus*, and *A. robustus*. The recently named *A. bahrelghazali* (Brunet *et al.*, 1996, 1997) is based on inadequate material and is here considered an invalid junior synonym of *A. afarensis*, with which it so far shares all of its allegedly diagnostic characters.

As noted above, there is no agreement on familial classification of extant and extinct hominoids, but the Hominidae is defined here as the sister group of the extant great apes.

Order Primates Linnaeus, 1758
  Infraorder Catarrhini É. Geoffroy Saint-Hilaire, 1812

Fig. 24.1 *Ardipithecus ramidus*. Juvenile mandible ARA-VP-1/129, from Aramis, Ethiopia. Scale in cm. Photograph by Tim White.

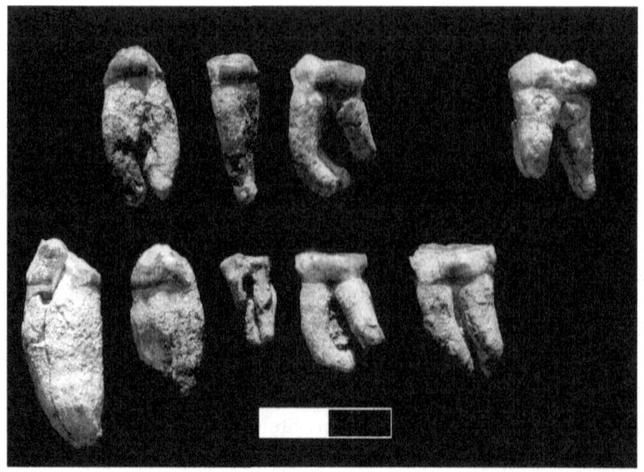

Fig. 24.2 *Ardipithecus ramidus*. Adult partial mandibular dentition ARA-VP-1/128, from Aramis, Ethiopia. Scale in cm. Photograph by Tim White.

Superfamily Hominoidea Gray, 1825
  Family Hominidae Gray, 1825
    Genus *Ardipithecus* White et al., 1995
      Ardipithecus ramidus ramidus White et al., 1994
      Ardipithecus ramidus kadabba Haile-Selassie, 2001
    Genus *Australopithecus* Dart, 1925
      Australopithecus afarensis Johanson et al., 1978
      Australopithecus aethiopicus Arambourg & Coppens, 1968
      Australopithecus africanus Dart, 1925
      Australopithecus anamensis Leakey et al., 1995
      Australopithecus boisei Leakey, 1959
      Australopithecus garhi Asfaw et al., 1999
      Australopithecus robustus Broom, 1938

Family Hominidae

GENUS *Ardipithecus* White et al., 1995
INCLUDED SPECIES *A. ramidus*

SPECIES *Ardipithecus ramidus* White et al., 1994 (Figs. 24.1–24.2)
TYPE SPECIMEN ARA-VP-6/1, partial dentition from Aramis
AGE AND GEOGRAPHIC RANGE 5.8–4.39 Ma, Saitune Dora, Alayba, Asa Koma, Digiba Dora, Kuseralee Dora, Amba, Sagantole and Aramis (all Middle Awash), Ethiopia

ANATOMICAL DEFINITION
Postcranial: not fully described; stature within *Australopithecus* range. Cranial: not fully described; temporomandibular joint more primitive than *Australopithecus*. Dental: not fully described; less postcanine megadontia than *Australopithecus*; incisors not hypertrophied; canine small and feminized, but large relative to posterior dentition; permanent third premolars and first deciduous premolar unmolarized relative to *Australopithecus*; non-honing C/P3 complex; enamel relatively and absolutely thin relative to *Australopithecus*.

GENUS *Australopithecus* Dart, 1925.
A genus of the family Hominidae. Postcranial: a postcranial skeleton adapted to habitual bipedal locomotion; in known species, forearm long relative to both upper arm and lower limb; lumbar column longer than in African or Asian apes; body size dimorphism greater than modern *Pan* and *Homo*, less than modern *Gorilla*, stature ranges from c. 1.2 to 1.5 m, body mass 30 to 55 kg. Cranial: brain size between 350 and 600 cm^3. Dental: incisors not hypertrophied, canine feminized; postcanine dentition megadont relative to body size; enamel thick relative to modern great apes.
INCLUDED SPECIES *A. aethiopicus, A. afarensis, A. africanus, A. anamensis, A. boisei, A. garhi, A. robustus*

SPECIES *Australopithecus afarensis* Johanson et al., 1978 (Fig. 24.3)
TYPE SPECIMEN Laetoli Hominid 4 (L.H. 4), adult mandible with teeth
AGE AND GEOGRAPHIC RANGE 3.6 to 2.9 Ma, Laetoli, Tanzania; Koobi Fora and West Turkana, Kenya; Omo, southern Ethiopia; Middle Awash (Maka) and Hadar, Afar depression, Ethiopia; ?Sterkfontein, South Africa
ANATOMICAL DEFINITION
Postcranial (and probably associated footprints): foot with

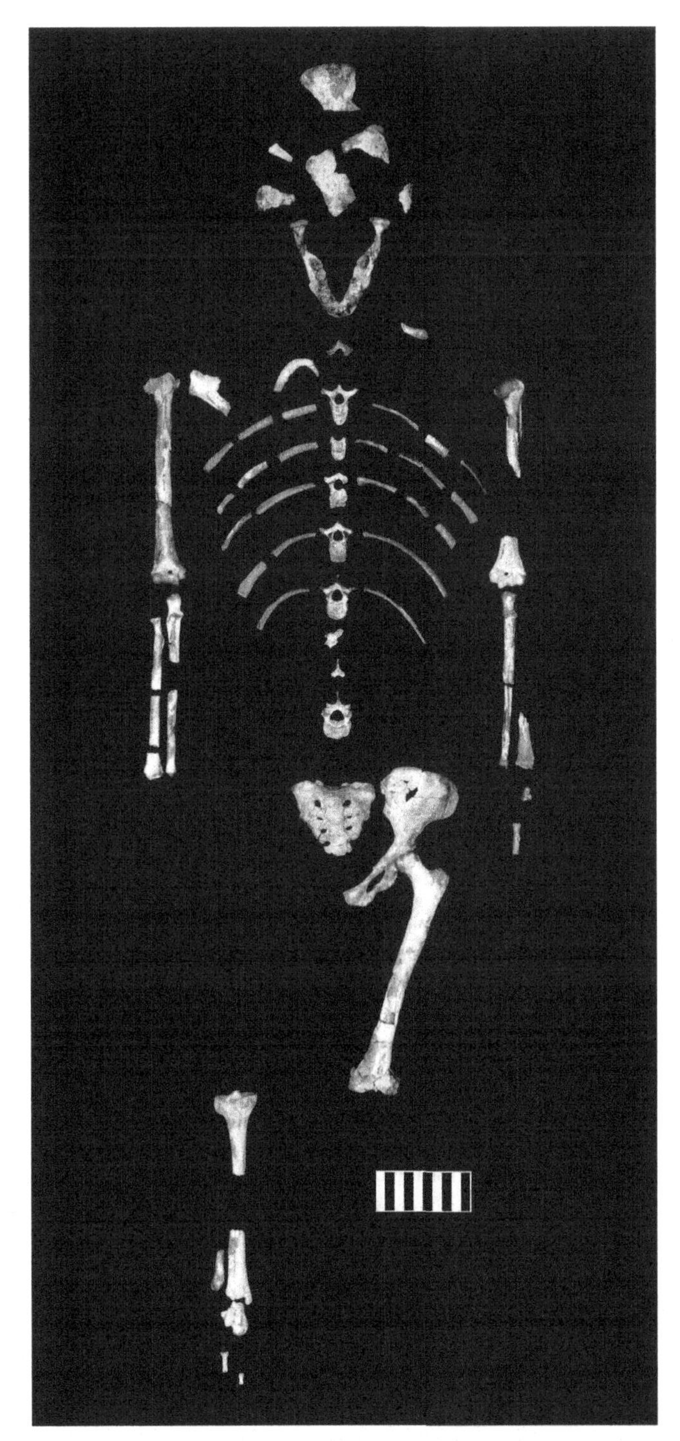

Fig. 24.3 *Australopithecus afarensis*. Partial skeleton A.L. 288-1, from Hadar, Ethiopia. Scale in cm. Photograph copyright Cleveland Museum of Natural History.

longitudinal and transverse arches, expanded calcaneal tuber, fully adducted hallux, moderately curved phalanges; human-like knee and pelvic girdle, with short, flared ilia, pubic symphysis and broad sacrum; and with gracile pollex, and long, moderately curved phalanges; limb proportions based on a single individual (A.L. 288-1) show long forearm, femur length intermediate between apes and humans; body size moderately sexually dimorphic, body size range large in both sexes. Cranial: prognathic face, unflexed external cranial base, marked posterior extension of cranial attachment of temporalis muscle, mandibular ramus tall relative to pongids but low relative to later *Australopithecus*; temporal with weak temporomandibular eminence; palate shallow with thin roof, maxillary toothrows range from parallel to convergent posteriorly (often with diastema); cranial capacity ranges from *c.* 380 to 430 cm^3 on a small sample, strong sexual dimorphism in cranial size, cresting and pneumatization. Dental: upper incisors heteromorphic; upper and lower canines large relative to postcanine teeth compared to later *Australopithecus* species, lower third premolar sometimes unicuspid.

SPECIES *Australopithecus aethiopicus* Arambourg & Coppens, 1968 (Fig. 24.4)
TYPE SPECIMEN Omo 18-(1967)-18, edentulous adult mandible
AGE AND GEOGRAPHIC RANGE *c.* 2.7 Ma – *c.* 2.3 Ma; West Turkana, Kenya, Omo Shungura, Ethiopia
ANATOMICAL DEFINITION
Postcranial: unknown. Cranial: cranium mostly known from single presumed male specimen (KNM-WT 17000) and edentulous holotype mandible; cranial base unflexed, cranial capacity *c.* 420 cm^3; mandible robust, face dished. Dental: anterior teeth larger relative to postcanine teeth than *A. boisei* and *A. robustus*; premolars molarized and molars with expanded talonid, wear flat, but lack specializations seen in *A. boisei*.

SPECIES *Australopithecus africanus* Dart, 1925 (Figs. 24.5, 24.6)
TYPE SPECIMEN Child's skull from Taung
AGE AND GEOGRAPHIC RANGE 3-2 Ma; South African sites of Taung, Makapansgat, Sterkfontein
ANATOMICAL DEFINITION
Postcranial: similar in known proportions and morphologies to *A. afarensis*. Cranial: average cranial capacity of *c.* 440 cm^3; cranium more globular, with less ectocranial cresting and pneumatization than other *Australopithecus* species; face prognathic with canine pillars framing nasal aperture. Dental: postcanine teeth relatively larger than in *A. afarensis*, but not disproportionally large as in *A. aethiopicus*, *A. robustus* or *A. boisei*; upper and lower canine and premolar morphology less primitive than earlier species. Pronounced variation in cranial and dental anatomy probably indicating that this taxon was sampled across a wide temporal span even at the Sterkfontein site in South Africa.

SPECIES *Australopithecus anamensis* Leakey *et al.*, 1995 (Fig. 24.7)
TYPE SPECIMEN KNM-KP 29281, adult mandible with dentition
AGE AND GEOGRAPHIC RANGE 4.17–3.9 Ma; Kanapoi and Allia Bay, northern Kenya, ?Fejej, southern Ethiopia, ?Galili, Ethiopia

ANATOMICAL DEFINITION
Postcranial: no known postcrania firmly associated with diagnostic cranial material; a proximal and distal tibia are assumed to represent this species and the latter shows adaptations to bipedal locomotion. Cranial: temporomandibular joint with very low eminence; shallow, procumbent palate with parallel toothrows; mandibular symphysis aligned at an oblique angle and strongly receding, with long postincisive planum. Dental: canines large relative to postcanine teeth; lower canines and asymmetrical lower third premolars more primitive than in *A. afarensis*; postcanine teeth larger than in *Ardipithecus*. Body size and craniodental sexual dimorphism marked.

SPECIES *Australopithecus boisei* Leakey, 1959 (Fig. 24.8)
TYPE SPECIMEN O.H. 5, cranium with dentition
AGE AND GEOGRAPHIC RANGE c. 2.3 Ma – c. 1.4 Ma; ?Chiwondo, Malawi; Olduvai Gorge and Peninj, Tanzania; Chesowanja, Koobi Fora, West Turkana, Kenya; Omo Shungura and Konso-Gardula, Ethiopia
ANATOMICAL DEFINITION
Postcranial: no known postcrania firmly associated with diagnostic cranial material. Cranial: face dished, with high, anteriorly placed zygoma, many specimens with evenly flared morphology of the anterior zygoma; ectocranial sagittal and compound temporo/nuchal crests common; frontal trigon and strong postorbital constriction; hard palate vertically thick, deeper than *A. robustus*; nasoalveolar clivus concave in coronal plane, with smooth entrance to nasal cavity, incisors set on bicanine line; mandibular ramus relatively and absolutely tall, corpus robust, with wide extramolar sulcus and strong posterior symphyseal buttressing; on average, larger than *A. robustus*. Dental: postcanine megadontia extreme, with highly molarized premolars and hyperthick enamel; lower first deciduous molar fully molarized.

SPECIES *Australopithecus garhi* Asfaw *et al.*, 1999 (Fig. 24.9)
TYPE SPECIMEN BOU-VP-12/130, cranium with vault and palate with dentition
AGE AND GEOGRAPHIC RANGE c. 2.5 Ma; Bouri, Afar Depression, Ethiopia; ?Baringo-Chemeron, Kenya; ?Omo Shungura, Ethiopia
ANATOMICAL DEFINITION
Postcranial: no known postcrania firmly associated with diagnostic cranial material. Cranial: only the holotype (presumably male) cranium, BOU-VP-12/130, is formally attributed to this taxon; vault crested, face as in *A. afarensis*; palate shallow, with thin hard palate; cranial capacity c. 450 cm^3 in single individual; frontal trigon present; dental arcade diverges posteriorly. Dental: canine very large; premolars and molars very large, but without molarization and flat wear seen in robust *Australopithecus*; upper premolars homomorphic relative to *A. afarensis*.

SPECIES *Australopithecus robustus* Broom, 1938 (Fig. 24.10)
TYPE SPECIMEN TM-1517, partial skull from Kromdraai, South Africa
AGE AND GEOGRAPHIC RANGE c. 1.7 Ma; Kromdraai, Swartkrans, Drimolen and Gondolin, South Africa
ANATOMICAL DEFINITION
Postcranial: no known postcrania uniquivocally associated with diagnostic cranial material, but most hand, foot, axial and appendicular fragments from Swartkrans are presumed to belong to the taxon and to indicate habitual bipedality. Cranial: face dished, with high, forwardly placed zygoma and weak pillars bounding the lateral nasal aperture; ectocranial sagittal and compound temporo/nuchal crests common; frontal trigon and strong postorbital constriction; hard palate vertically thick, shallower than *A. boisei*; nasoalveolar clivus concave in coronal plane, with smooth entrance to nasal cavity; incisors set on bicanine line; mandibular ramus relatively and absolutely tall, corpus robust, with wide extramolar sulcus and strong posterior symphyseal buttressing; on average, smaller than *A. boisei*. Dental: postcanine megadontia pronounced; enamel extremely thick; first deciduous molar fully molarized.

Evolution of the earliest hominids

Biochemically derived estimates for the split between the last common ancestor of humans and the African apes center around 5–6 million years, but with wide error ranges. There are no fossil representatives of the lineages leading to the chimpanzee, bonobo or gorilla. The earliest diagnostic fossil evidence of Hominidae is from the western margin of the Middle Awash, Ethiopia, and dates to approximately 5.7 Ma. These fossilized remains are assigned to *Ardipithecus ramidus kadabba*. This very generalized hominid inhabited a wooded setting. Another discovery site, Aramis, is a wider window into hominid evolution, at 4.4 Ma. Because of the singular nature of this data set, observations regarding the "bushy" nature of the earliest hominid family tree fall into the category of "X-files paleontology" (pure speculation – based on preconception – but unchecked by fossil data). Present data indicate that the hominid family tree was never particularly "bushy", or speciose in the normal mammalian sense, despite recent claims to the contrary (Lieberman, 2001).

The lack of a honing canine/premolar complex in *Ardipithecus* identifies it as a hominid. In-progress studies of its postcranial skeleton promise to elucidate its locomotor mode. Phylogenetically, *Ardipithecus* could represent the direct ancestor of its slightly younger, more derived close relative, *Australopithecus anamensis*. This interpretation would imply a period of rapid anatomical change during the 0.22-Ma period between 4.39 and 4.17 Ma. Alternatively, *Ardipithecus* may represent a relict species (persistent mother species) little changed from the ancestor of *Australopithecus*. Resolution of this phylogenetic issue will only be determined by

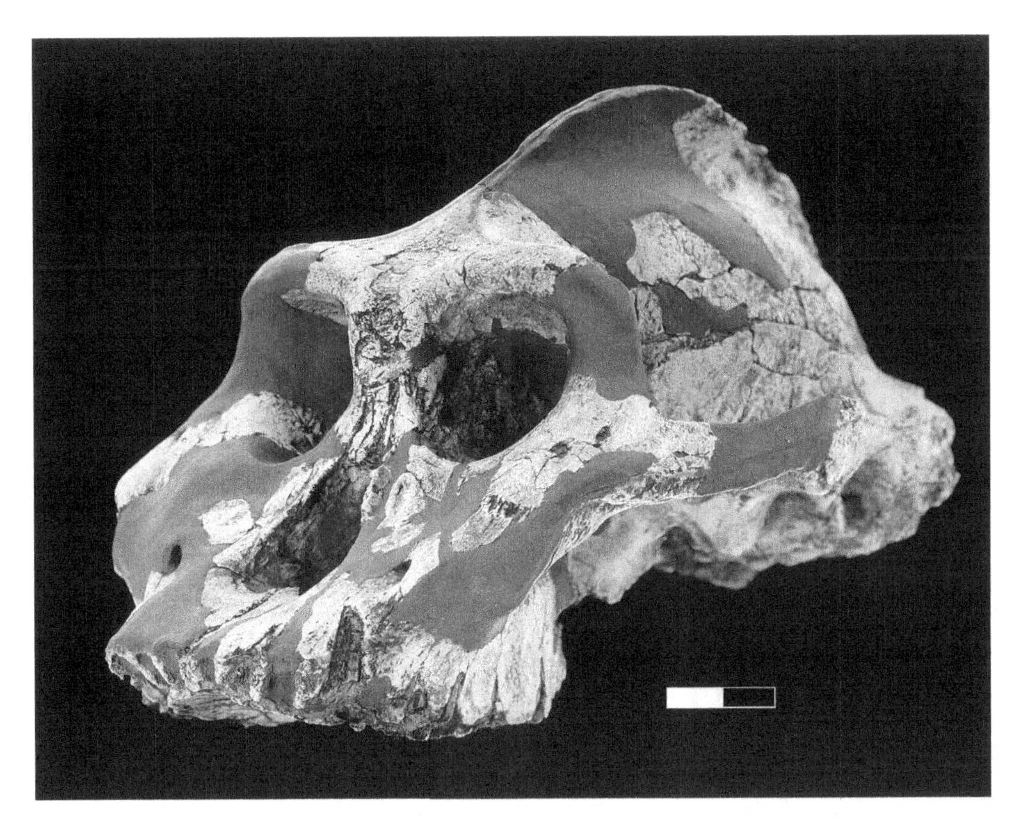

Fig. 24.4 *Australopithecus aethiopicus*. Cast of cranium KNM-WT 17000, from West Turkana, Kenya. Scale in cm. Photograph of cast copyright David L. Brill.

additional discoveries. There is currently nothing about the morphology of *Ar. ramidus* that would exclude this taxon as the ancestor of later hominids.

The earliest evidence of the genus *Australopithecus* comes in the form of limited craniodental and postcranial evidence from Kanapoi in Kenya, dated to 4.17-4.07 Ma. The fossils are very close to the younger taxon *Australopithecus afarensis*, but more pongid-like in details of the teeth and anterior mandibular corpus. The *A. anamensis* first deciduous molar is intermediate between *Ardipithecus* and *A. afarensis* (Leakey et al., 1998; C. Ward et al., 1999). This is also true of the lower third premolars and the canines. The known parts of the tibia match counterparts belonging to *A. afarensis* at Hadar. The postcanine dentition is expanded relative to *Ardipithecus* and the average enamel thickness may be greater. *Australopithecus anamensis* is considered here to represent the earliest known step in the adaptive trajectory of postcanine megadontia that characterized all later species of *Australopithecus*. It is likely that slightly younger fossils from Allia Bay, Fejej (Kappelman et al., 1996), and Belohdelie (Asfaw, 1987) belong to this taxon. However, the rate of what appears to be anagenetic evolution between *A. anamensis* and *A. afarensis* has not yet been established, and the samples are therefore currently inadequate to establish species boundaries.

In contrast to the limited sample of *A. anamensis*, the species *Australopithecus afarensis* is now the best known of its genus. It is well represented postcranially, with additional evidence of its habitually bipedal locomotor mode deriving from the

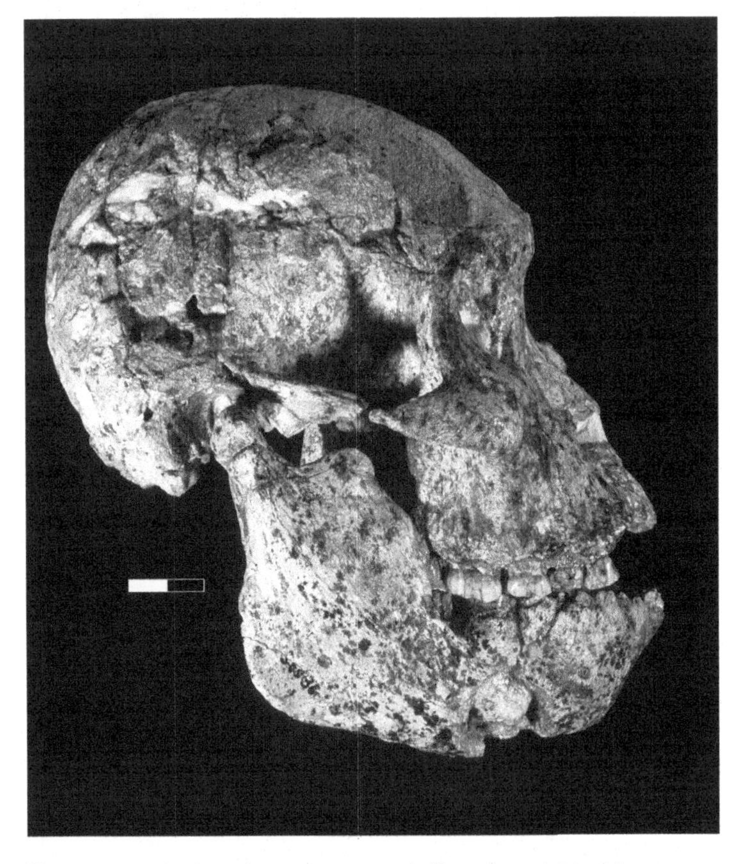

Fig. 24.5 *Australopithecus africanus*. Composite skull: cranium Sts 71 with mandible Sts 36, from Sterkfontein, South Africa. Scale in cm. Photograph by Tim White.

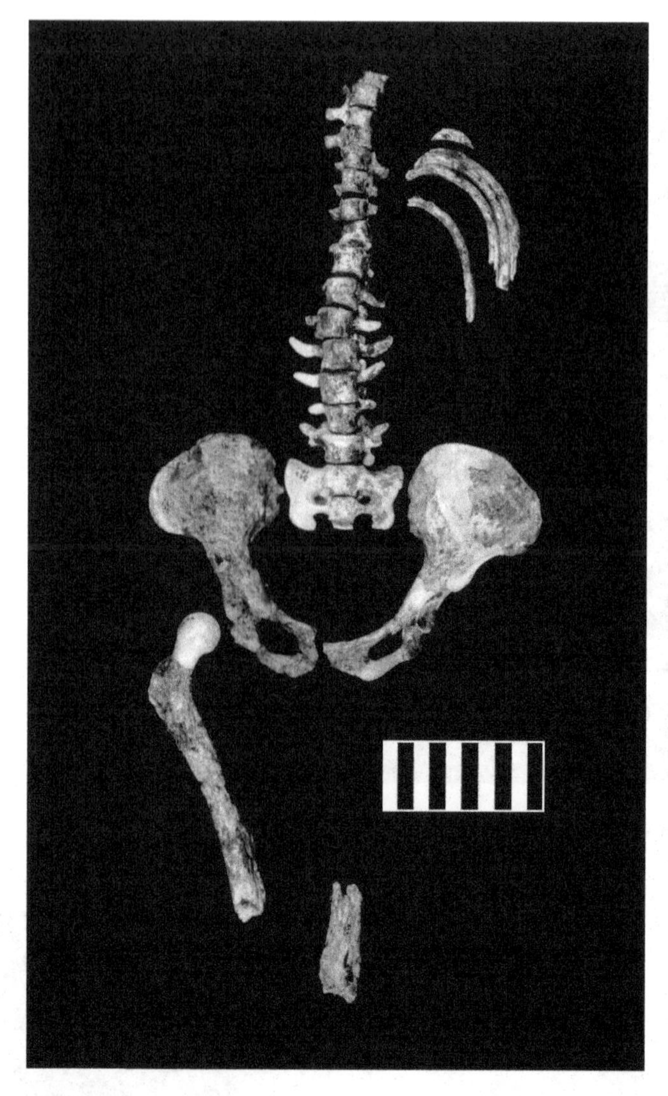

Fig. 24.6 *Australopithecus africanus*. Partial skeleton Sts 14, from Sterkfontein, South Africa. Scale in cm. Photograph by Tim White.

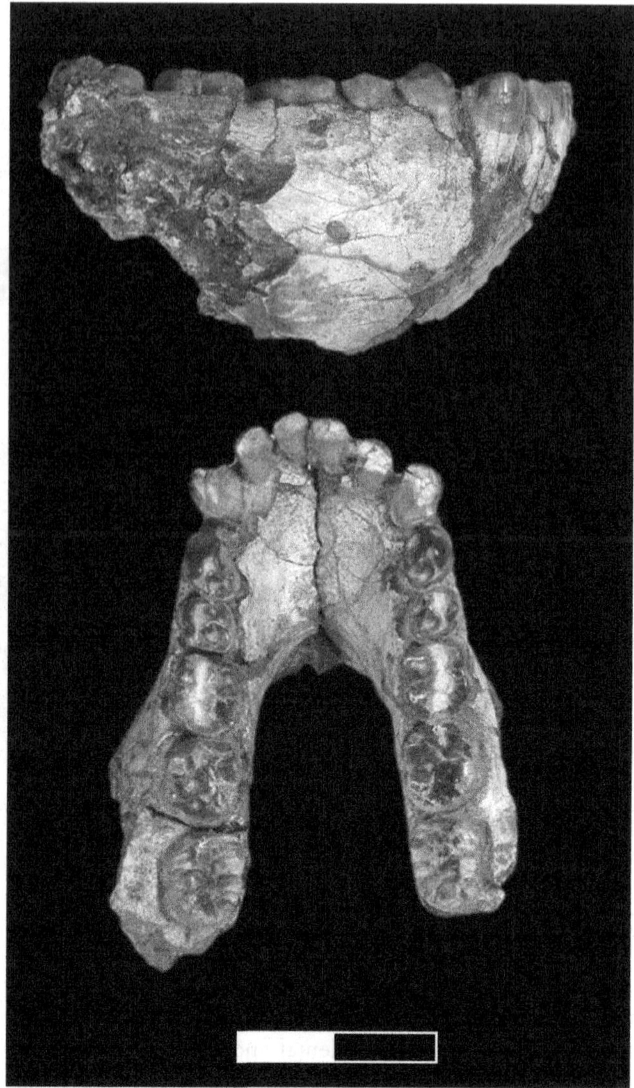

Fig. 24.7 *Australopithecus anamensis*. Holotype mandible KNM-KP 29281, from Kanapoi, Kenya. Scale in cm. Photograph by and courtesy of Alan Walker.

Laetoli footprint trails. Suggestions that the Laetoli footprints were made by an as-yet undiscovered hominid seem, in light of the current evidence, unwarranted. Debate persists about the taxonomic significance of the large morphological and metric variation observed among *A. afarensis* fossils. A variety of attempts have been made to divide the series into separate taxa, and to link one or more of these subdivisions with later hominid species. However, such interpretations have proven contradictory among themselves. Most often they have been found to be based on fallacies in statistical manipulation, or errors of basic observation or logic.

Australopithecus afarensis is known to have been distributed from Chad in the west, to Ethiopia on the east, and to Tanzania in the south. The recent discovery of Sterkfontein's oldest fossil hominid, the StW 573 partial skeleton, may dramatically extend the geographic range of *A. afarensis*. The published photographs of the partially exposed cranium (Clarke, 1998) suggest that this skeleton is not *A. africanus*, but rather belongs to *A. afarensis* or to a very close relative that may represent the immediate ancestor of *A. africanus*. Indeed, even before the discovery of this skeleton, it was widely held that *A. afarensis* was a suitable ancestor for all later hominid species.

The only stratigraphic succession that currently bears directly on the apparently limited early hominid diversification after 3.0 Ma is the Omo Shungura sequence (Suwa *et al.*, 1996). Unfortunately, this succession does not reveal many details about the nature of the evolutionary processes through which at least two species lineages arose from populations of *A. afarensis*. One of these lineages was the time-successive series *A. aethiopicus* to *A. boisei*. The other was a lineage ancestral to the genus *Homo*. However, even the question of how many contemporary early *Homo* species there were is controversial (see following chapter). Based on teeth, the cladogenesis between robust and non-robust lineages occurred before 2.7 Ma. Specimens belonging to the robust lineage (*A. aethiopicus*) appear for the first time in Omo Shungura Member C (Suwa *et al.*, 1996). The well-preserved cranium of this taxon, KNM-WT 17000, from West Turkana, is only slightly younger.

Many workers have adopted the position that the three robust *Australopithecus* species form a monophyletic group (sometimes called *Paranthropus*), with *A. aethiopicus* as the common ancestor. This position is based largely upon extensive trait lists used in numerical cladistic analyses. However, as noted by McCollum (1999), these traits, or characters, are not necessarily independent. Virtually all are related directly to basic masticatory adaptations involving megadontia, small anterior teeth, and large muscles of mastication. Similar dentognathic adaptations have arisen independently numerous times among primates, and even early *Homo* species evolved many of these characters in parallel. It remains a possibility that *A. robustus* evolved independently from *A. africanus* in southern Africa. It is also possible that *A. africanus* evolved from *A. afarensis* to become an endemic South African form that left no descendants.

The 1999 description of the megadont, non-robust species *Australopithecus garhi* from the Middle Awash emphasized the still-poor understanding of early hominid phylogenetic history after 3.0 Ma (Asfaw *et al.*, 1999). The species name "*garhi*" means "surprise" in the Afar language. Despite *post-hoc* comments to the contrary, *A. garhi* was completely unpredicted by any phylogenetic analysis that preceded it (including that of Wood & Collard (1999), published only weeks earlier in the same volume of *Science*). However, morphological evidence of the fragmentary 2.7 to 2.5 Ma Omo and Koobi Fora teeth, interpreted in light of stratophenetic constraints, did recognize the existence of a large-toothed, non-robust form allocated to *Homo/Australopithecus* indet. (Suwa *et al.*, 1996). Due to the discovery of *A. garhi* and the recognition of rampant homoplasy in early hominid masticatory systems, there is currently an unresolved polychotomy in early hominid phylogeny that involves *A. garhi*, *A. africanus*, *A. robustus*, *A. aethiopicus* and early *Homo* (whether one or more species). It is likely that this uncertainty will persist until the fossil record is substantially augmented.

Hominid fossils were recovered over the last century-and-a-half under expectations and assumptions that were usually unstated. The dominant model held that gradual evolution had produced a progression of time-successive hominid species that, when found, would link an ancestor resembling a modern chimpanzee with modern humans. As the relatively few fossils scattered through deep time were recovered, they were linked together under that model in satisfying imaginary chains.

We have since learned that hominid evolution was neither linear nor gradual. We know that the traditional model is inconsistent with perspectives from evolutionary theory and developmental biology. And it has been falsified by the fossil record. Abandonment of the model, however, brings a sobering realization regarding the inadequacies of the fossil record. It is now abundantly clear that we will need a very large and dense fossil record in order to obtain an accurate reading of the global experiment of hominid evolution. This is a frightening realization and a daunting challenge because the terrestrial data recorders which monitored this vast experiment were so poor. We have inherited only a few fossil hominid sites that inadequately sample deep time. We have

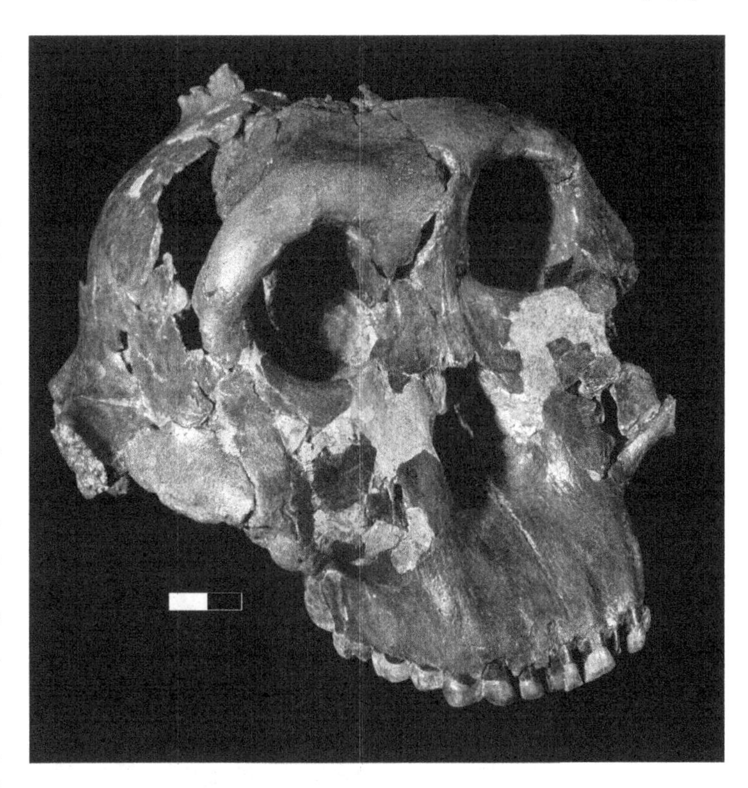

Fig. 24.8 *Australopithecus boisei*. Holotype cranium OH 5, from Olduvai Gorge, Tanzania. Scale in cm. Photograph by Tim White.

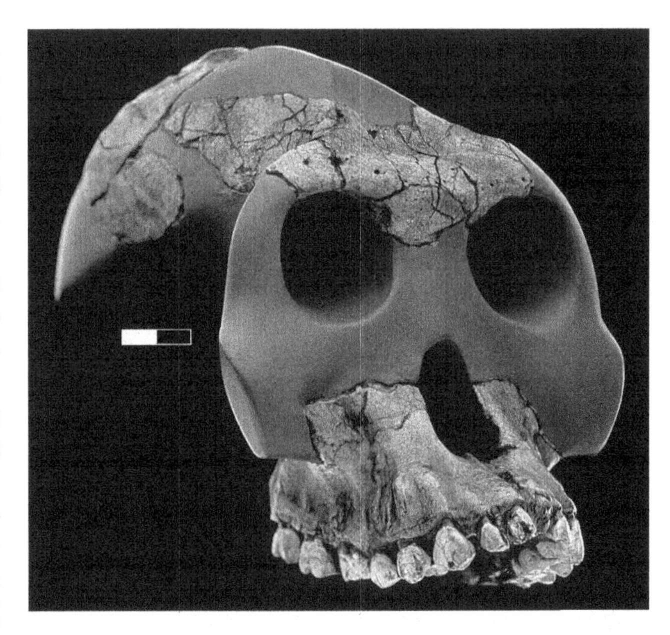

Fig. 24.9 *Australopithecus garhi*. Holotype partial cranium BOU-VP-12/131, from Bouri, Ethiopia. Scale in cm. Photograph of cast copyright David L. Brill.

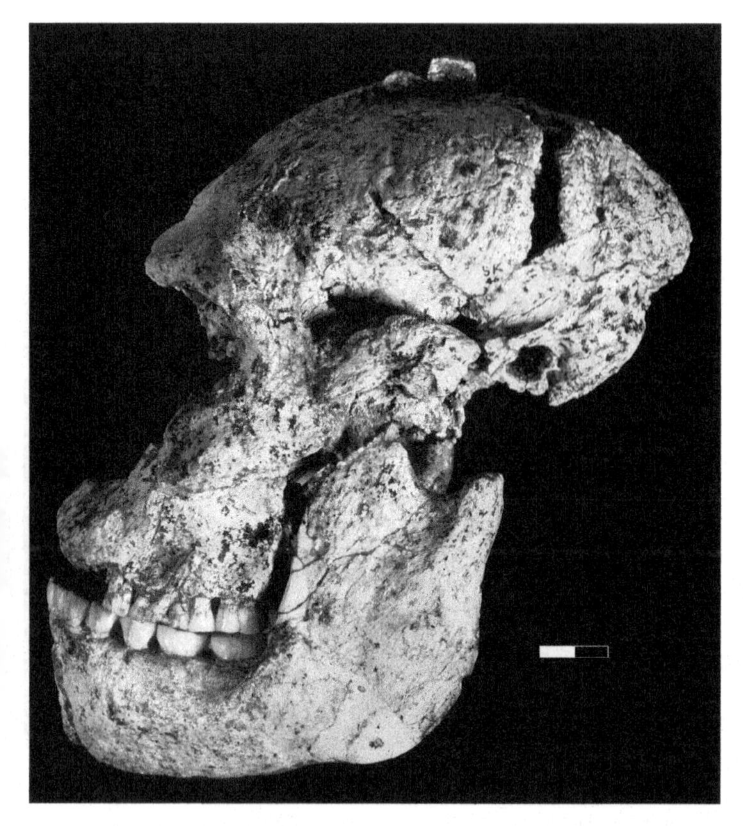

Fig. 24.10 *Australopithecus robustus*. Composite partial skull: cranium SK 52 with mandible SK 23, from Swartkrans, South Africa. Scale in cm. Photograph by Tim White.

probably already found the best of them. In working on them, we have learned that their largely surface accumulations are depleted by a few years of collection, but will require millennia to replenish (White, 2000). It is therefore unlikely that a comprehensive rendering of hominid evolution lies within our immediate grasp.

Primary References

Ardipithecus

White, T., Suwa, G., & Asfaw, B. (1994). *Australopithecus ramidus*: a new species of early hominid from Aramis, Ethiopia. *Nature*, **371**, 306–312.

White, T. D., Suwa, G., & Asfaw, B. (1995). Corrigendum: *Australopithecus ramidus*, a new species of early hominid from Aramis, Ethiopia. *Nature*, **375**, 88.

Haile-Selassie, Y. (2001). Late Miocene hominids from the Middle Awash, Ethiopia. *Nature*, **412**, 178–181.

Australopithecus

Dart, R.A. (1925). *Australopithecus africanus*: The man-ape of South Africa. *Nature*, **115**, 195–199.

Broom, R. (1936). A new fossil anthropoid skull from Sterkfontein, near Krugersdorp, South Africa. *Nature*, **138**, 486–488.

Broom, R. (1938). Pleistocene anthropoid apes of South Africa. *Nature*, **142**, 377–379.

Broom, R. & Schepers, G. W. H. (1946). The South African fossil ape-men: the Australopithecinae. *Transvaal Museum Memoir*, **2**, 1–272.

Dart, R. A. (1948). The Makapansgat protohuman *Australopithecus prometheus*. *American Journal of Physical Anthropology*, **6**, 259–283.

Broom, R. (1949). Another new type of fossil ape-man (*Paranthropus crassidens*). *Nature*, **163**, 57.

Broom, R. & Robinson, J. T. (1950). Man contemporaneous with the Swartkrans ape-man. *American Journal of Physical Anthropology*, **8**, 151–155.

Broom, R. & Robinson, J. T. (1952). Swartkrans ape-man: *Paranthropus crassidens*. *Transvaal Museum Memoir*, **6**, 1–123.

Robinson, J. T. (1954). The genera and species of the Australopithecines. *American Journal of Physical Anthropology*, **12**, 181–200.

Robinson, J. T. (1956). The dentition of the Australopithecinae. *Transvaal Museum Memoir*, **9**.

Leakey, L. S. B. (1959). A new fossil skull from Olduvai. *Nature*, **184**, 491–493.

Leakey, L. S. B., Tobias, P. V., & Napier, J. R. (1964). A new species of the genus *Homo* from Olduvai Gorge. *Nature*, **202**, 7–9.

Tobias, P. V. (1967). *Olduvai Gorge*, Vol. 2, *The Cranium and Maxillary Dentition of* Australopithecus (Zinjanthropus) boisei. Cambridge: Cambridge University Press.

Arambourg, C. & Coppens, Y. (1968). Découverte d'un australopithécien nouveau dans les gisements de l'Omo (Éthiopie). *South African Journal of Science*, **64**, 58–59.

Leakey, R. E. F. & Walker, A. C. (1976). *Australopithecus*, *Homo erectus* and the single species hypothesis. *Nature*, **261**, 572–574.

Johanson, D. C., White, T. D., & Coppens, Y. (1978). A new species of the genus *Australopithecus* (Primates: Hominidae) from the Pliocene of eastern Africa. *Kirtlandia*, **28**, 1–14.

White, T. D. (1981). Primitive hominid canine from Tanzania. *Science*, **213**, 348–349.

Johanson, D. C., Taieb, M., & Coppens, Y. (eds.) (1982). Pliocene hominids from Hadar, Ethiopia. *American Journal of Physical Anthropology*, **57**, 373–719.

Walker, A. C., Leakey, R. E. F., Harris, J. M., & Brown, F. H. (1986). 2.5-Myr *Australopithecus boisei* from west of Lake Turkana, Kenya. *Nature*, **322**, 517–522.

Leakey, M. D. & Harris, J. M. (eds.) (1987). *Laetoli: A Pliocene Site in Northern Tanzania*. Oxford: Clarendon Press.

Asfaw, B. (1987). The Belohdelie frontal: new evidence of early hominid cranial morphology from the Afar of Ethiopia. *Journal of Human Evolution*, **16**, 611–624.

Grine, F. E. (ed.) (1988). *Evolutionary History of the "Robust" Australopithecines*. New York: Aldine de Gruyter.

Tobias, P. V. (1991). *Olduvai Gorge*, Vol. 4, *The Skulls, Endocasts, and Teeth of* Homo habilis. Cambridge: Cambridge University Press.

Wood, B. A. (ed.) (1991). *Koobi Fora Research Project IV: Hominid Cranial Remains from Koobi Fora*. Oxford: Clarendon Press.

White, T. D., Suwa, G., Hart, W. K., Walter, R. C., WoldeGabriel, G., de Heinzelin, J., Clark, J. D., Asfaw, B., & Vrba, E. (1993). New discoveries of *Australopithecus* at Maka in Ethiopia. *Nature*, **366**, 261–265.

Kimbel, W. H., Johanson, D. C., & Rak, Y. (1994). The first skull and other new discoveries of *Australopithecus afarensis* at Hadar, Ethiopia. *Nature*, **368**, 449–451.

Leakey, M. G., Feibel, C. S., McDougall, I., & Walker, A. C. (1995). New four-million-year-old hominid species from Kanapoi and Allia Bay, Kenya. *Nature*, **376**, 565–571.

Brunet, M., Beauvilain, A., Coppens, Y., Heintz, E., Moutaye, A. H. E., & Pilbeam, D. (1996). *Australopithecus bahrelghazali*, a new species of early hominid from Koro Toro region, Chad. *Comptes rendus de l'Académie des sciences de Paris*, **322**, 907–913.

Kappelman, J., Swisher, C. C., Fleagle, J. G., Yirga, S., Bown, T. M., & Fesseha, M. (1996). Age of *Australopithecus afarensis* from Fejej, Ethiopia. *Journal of Human Evolution*, **30**, 139–146.

Suwa, G., White, T. D., & Howell, F. C. (1996). Mandibular postcanine dentition from the Shungura Formation, Ethiopia: crown morphology, taxonomic allocation, and Plio-Pleistocene hominid evolution. *American Journal of Physical Anthropology*, **101**, 247–282.

Brunet, M., Beauvilain, A., Geraads, D., Guy, F., Kasser, M., MacKaye, H. T., Maclatchy, L. M., Mouchelin, G., Sudre, J., & Vignaud, P. (1997). Tchad: un nouveau site à Hominidés Pliocène. *Comptes rendus de l'Académie des Sciences de Paris*, **324**, 341–345.

Semaw, S., Renne, P., Harris, J. W. K., Feibel, C. S., Bernor, R. L., Fesseha, N., & Mowbray, K. (1997). 2.5-million-year-old stone tools from Gona, Ethiopia. *Nature*, **385**, 333–336.

Suwa, G., Asfaw, B., Beyene, Y., White, T.D., Katoh, S., Nagaoka, S., Nakaya, H., Vzawa, K., Renne, P., & WoldeGabriel, G. (1997). The first skull of *Australopithecus boisei*. *Nature*, **389**, 489–492.

Leakey, M. G., Feibel, C. S., McDougall, I., Ward, C., & Walker, A. C. (1998). New specimens and confirmation of an early age for *Australopithecus anamensis*. *Nature*, **393**, 62–66.

Clarke, R. J. (1998). First ever discovery of a well-preserved skull and associated skeleton of *Australopithecus*. *South African Journal of Science*, **94**, 460–463.

Ward, C. V., Leakey, M. G., & Walker, A. C. (1999). The new hominid species *Australopithecus anamensis*. *Evolutionary Anthropology*, **7**, 197–205.

Heinzelin, J. de, Clark, J. D., White, T., Hart, W., Renne, P., WoldeGabriel, G., Beyene, Y., & Vrba, E. (1999). Environment and behavior of 2.5-million-year-old Bouri hominids. *Science*, **284**, 625–629.

Asfaw, B., White, T., Lovejoy, O., Latimer, B., Simpson, S., & Suwa, G. (1999). *Australopithecus garhi*: a new species of early hominid from Ethiopia. *Science*, **284**, 629–634.

Haile-Selassie, Y. & Asfaw, B. (2000). A newly discovered early Pliocene hominid bearing paleontological site in the Mulu Basin, Ethiopia. *American Journal of Physical Anthropology*, Supplement **30**, 170.

Keyser, A. W. (2000). The Drimolen skull: the most complete australopithecine cranium and mandible to date. *South African Journal of Science*, **96**, 189–193.

Leakey, M. G., Spoor, F., Brown, F. H., Gathogo, P. N., Kiarie, C., Leakey, L. N., & McDougall, I. (2001). New hominin genus from eastern Africa shows diverse middle Pliocene lineages. *Nature*, **410**, 433–440.

Ward, C. V., Leakey, M. G., & Walker, A. (2001). Morphology of *Australopithecus anamensis* from Kanapoi and Allia Bay, Kenya. *Journal of Human Evolution*, **41** 255–368.

25 | Early Genus *Homo*

HOLLY DUNSWORTH AND ALAN WALKER

Introduction

Fossils of early *Homo* were collected as early as 1891. From that beginning, the study of the origins of *Homo* has become increasingly volatile. Given the many different approaches ranging from morphometrics to cladistics, the taxonomy and phylogeny of *Homo* are difficult to resolve with any sort of consensus. While some experts disagree on the relevance of certain morphological features, others clash on the definition of a paleospecies, and so on.

In this account, early *Homo* includes three species, *H. habilis*, *H. rudolfensis* and *H. erectus*, that are often yet arguably distinguished from the preceding *Australopithecus* by morphological characteristics such as larger brains and reduced dentition, and behavioral ones such as the presumed ability to manufacture and use stone tools. *Homo habilis* and *H. rudolfensis* have limited ranges in eastern and/or southern Africa. However, *H. erectus* fossils have been found over much of the Old World marking the first hominid dispersal out of Africa into Eurasia. At least two of these species of *Homo* overlap temporally and span approximately 2 million years from the mid-Pliocene in Africa to the late Pleistocene in Indonesia. The earliest known member of the genus *Homo* dates to around 2.3 Ma (Kimbel *et al.*, 1996) and the latest known *H. erectus* suggests that the species possibly endured in Indonesia until as late as 30 000 BP (Swisher *et al.*, 1996).

The Plio-Pleistocene record of East Africa plays a major role in the story of human evolution, not only because it documents the emergence of the genus *Homo*, but also because it shows the probable contemporaneity of three hominid species (*A. boisei*, *H. habilis* and *H. rudolfensis*; then later *A. boisei*, *H. habilis* and *H. erectus*). The volume of literature covering the origins of our own genus is enormous. There are bound to be gaps and omissions in the following brief account.

History of discovery and debate

Eugène Dubois discovered the first known specimen of early *Homo* in Trinil, Java in October 1891 (Dubois, 1894; described in Dubois, 1924). The 1.0-Ma (Swisher *et al.*, 1998) adult skullcap, premolar, two molars, and femur were the focus of an intense scientific debate at the turn of the last century. The skullcap (Trinil 2) includes much of the frontal bone, both parietals and much of the superior portion of the occipital bone. It has a cranial capacity of 940 cm^3 (Holloway, 1981), a prominent supraorbital ridge and a receding narrow frontal with a keel at midline. Initially Dubois thought he had found a fossil chimpanzee since the only known hominoid fossils from Asia at the time were thought to be chimpanzees. He called the specimen *Anthropopithecus*, the name then used for chimpanzees. However, after more analysis and the discovery of a human-like femur, Dubois declared that the calotte was neither ape nor man yet shared characters with both. The combination of cranial and postcranial traits led Dubois to recognize it as an ape-man who walked erect, or *Pithecanthropus erectus* (Dubois, 1894). The genus was taken from Ernst Haeckel's hypothetical link between apes and humans, and so the new species was put about as the "missing link". According to Dubois, *Pithecanthropus erectus* was the immediate progenitor of humans.

Many researchers argued over the alleged human-like characteristics of the partial cranium from Indonesia. One skeptic, Richard Lydekker, accepted that the bones were from one individual yet contended that the Trinil remains were human, belonged to no "wild anthropoid", and could possibly be those of a "microcephalous idiot" (Lydekker, 1895). Some did, however, challenge the notion that the skullcap and femur belonged to the same individual since the femur was discovered several yards upstream from the skullcap. Although Dubois claimed that the stratum was the same for both the skullcap and the femur, the question of association placed even more doubt on his interpretation and the validity of the new species. As an interesting historical aside, Dubois was the first to ask whether a fossil was closer to humans or apes rather than assuming it to be the remains of an ancient human race. His work anticipated the modern paleoanthropological monograph (Shipman, 2000).

With similar education and aspirations to Dubois, Davidson Black, a Canadian anatomist, moved to Beijing, China, in search of the origins of humankind. Upon Black's arrival, a Swiss paleontological expedition led by Otto Zdansky uncovered some human-like teeth while excavating at a nearby cave site, Zhoukoudian (locally known as "Dragon Bone Hill"), which was discovered by geologist Johan Gunnar Andersson (Black, 1933). Convinced that these were the earliest remains of early humans known from China, Black named the new genus and species *Sinanthropus pekinensis* based on a single lower left molar (Black, 1927).

Between 1928 and 1937 Wen-chung Pei and the Cenozoic Research Laboratory unearthed several skulls and a few postcranial bones from Zhoukoudian Cave. Franz Weidenreich initially described the crania as *Sinanthropus pekinensis* Black, 1927 (Weidenreich, 1943). He argued that *Sinanthropus* shares almost all its essential characters with *Pithecanthropus* with only a few exceptions. For example, *Sinanthropus* has a larger cranial capacity than *Pithecanthropus*. Furthermore, the *Sinanthropus* frontal has a "bump-like protuberance" compared to the fairly flat frontal squama of *Pithecanthropus*. The obelion region of *Sinanthropus* is fairly depressed compared to the vaulted appearance in *Pithecanthropus*. The occiput of *Sinanthropus* is narrow and elongated while that of *Pithecanthropus* is broad and rounded. In 1939, von Koenigswald and Weidenreich synonomized *Sinanthropus pekinensis* Black, 1927 with *Pithecanthropus pekinensis* (Black, 1927) since the differences between the two genera are no more than the differences between "two different races of present mankind" (von Koenigswald & Weidenreich, 1939: 298). Today both genera are invariably integrated and almost always subsumed under the taxon *Homo erectus*.

Near Ngandong, just northeast of Trinil along the Solo River in Java, a team from the Dutch Geological Survey discovered eleven hominid skulls and two tibias between 1931 and 1933 that came to be known as "Solo Man". The taxonomic status of these skulls has been debated since their recovery, owing much to their location in geologically recent strata. Weidenreich himself initially thought that the skulls were Neanderthal-like but later argued that they were more like *Homo erectus* from Java and China (Weidenreich, 1951). C. Loring Brace (1967) placed them at the "Neanderthal stage" of evolution linking *H. erectus* and modern humans. Later Santa Luca (1980; see also Jacob, 1981) concluded that these specimens were all the same as the Peking, Sangiran and Trinil hominids. The cranial contours and pattern of contour development are similar in Ngandong adults to other *H. erectus* adults and do not show transitional vault shape between *H. erectus* and *H. sapiens* (Antón, 1999). The recent dating of the deposits that produced the Ngandong fossils indicate that *H. erectus* may have persisted in Indonesia as late as 30 000 BP (Swisher *et al.*, 1996), possibly overlapping with modern humans in the region.

In 1936, von Koenigswald, a member of the Dutch East Indies Geological Survey, named *Homo modjokertensis* based on a child's braincase (Perning 1) that was dubbed the "Modjokerto Child" after its location on Java. He also included several specimens from Sangiran, Java with similar presumed primitive dental characteristics in the new species that was later subsumed under the taxon *H. erectus*. The Modjokerto Child may be one of the oldest known *H. erectus* specimens ever found (Swisher *et al.*, 1994). Recent analysis of the cranium suggested a developmental age of four to six years for the individual who shares several cranial characteristics with representatives of Indonesian *H. erectus* (Antón, 1997). A year after the discovery of the Modjokerto Child, another skull, Sangiran 2, found its way into the hands of von Koenigswald. The long, low vault, sagittal keeling and pronounced occipital torus of the adult braincase mirrored Dubois's Trinil specimen.

Fossils of early *Homo* were not discovered on the African continent until 1949. At Swartkrans, South Africa, John T. Robinson uncovered a mandible of "a new type of man" (SK 15), that was considerably younger than *Paranthropus crassidens*, a species now known as *Australopithecus robustus*, from the same location (Broom & Robinson, 1949). Broom & Robinson labeled the nearly complete mandible and five well-preserved but slightly worn molars *Telanthropus capensis*, partly based on their structural analysis that it was intermediate between one of their ape-men, like *P. crassidens*, and *Homo*. However, they did not consider it to be in the genus *Homo* because it had well-developed third molars (Broom & Robinson, 1950). Later *Telanthropus* was subsumed under *H. erectus*.

Perceiving a trend towards taxonomic splitting in paleoanthropology, evolutionist Ernst Mayr attempted to simplify hominid taxonomy and align it with the standards of the rest of zoological nomenclature. Setting the stage for many workers to follow, he argued that the remains from Java and Peking (*Pithecanthropus* and *Sinanthropus*) belong to one species, *H. erectus* (Mayr, 1944, 1950). In this way, Mayr started a trend that has been followed by almost all paleoanthropologists since (e.g., Le Gros Clark, 1964; Campbell, 1963; Howell, 1978).

In 1954, Camille Arambourg (1955) excavated two hominid mandibles at Ternifine (now Tighennif), Algeria. He estimated the date of the site assemblage to be Middle Pleistocene based on artifacts and faunal remains. The two very robust mandibles share several features with South African and Asian representatives of *H. erectus* (Arambourg, 1955). For instance, the smaller specimen shares a low and broad ramus with *Sinanthropus*. They both have dental arcades nearly identical to those of *Sinanthropus*. Arambourg (1955) concluded that their overall dental and mandibular characteristics indicated a very close relationship to *Pithecanthropus* and *Sinanthropus*. However, since he could not align the two mandibles exactly with *Sinanthropus*, *Pithecanthropus* or *Telanthropus*, he assigned them the name *Atlanthropus mauritanicus*.

Later that year in North Africa, Pierre Biberson found a partial human mandible associated with Acheulean stone tools in Littorina Cave at the Schneider gravel pit in Sidi Abderrahman, Morocco (Arambourg & Biberson, 1956). Arambourg observed several mandibular and dental characteristics in common with the "pithecanthropines" and included the specimen in the North African genus he named a year earlier, *Atlanthropus* (*H. erectus*). The discovery of the *Atlanthropus* fossils brought evidence for the first time that *Homo erectus*, not *Australopithecus*, made an industrial transition from pebble tools and manufactured the first Acheulean or bifacial tools (Arambourg & Biberson, 1956). This was controversial evidence during a time when many workers argued against the presence of human-like capabilities or "human nature" in *H. erectus*.

In 1961 L. S. B. Leakey reported the 1960 discovery of a

cranium, now dated to about 1.4 Ma, from Upper Bed II at Olduvai Gorge, Tanzania. The skullcap known as OH 9 (Olduvai Hominid 9) has a flattened frontal region, a prominent nuchal crest, a flattened nuchal plane, and large, flaring brow ridges. The supramastoid crests are continuous with the nuchal crest and surmount small mastoid processes. For early *Homo*, OH 9 has one of the largest cranial capacities in Africa with an estimated brain size of 1067 cm^3 (Holloway, 1973). Originally Leakey referred to OH 9 as "Chellean Man" since it was associated with stone tools of the Chellean culture. At this time many paleoanthropologists agreed that hominid evolution in the Pleistocene comprised three successive stages: *Australopithecus*, *Pithecanthropus*, and then *Homo* or "true man" (Napier & Weiner, 1962).

Leakey (1961) likened the OH 9 cranium to fossils from Steinheim, Broken Hill, and Saldanha – crania considered early *Homo* then, but usually placed in *H. sapiens* today – because he noted that while the vault of OH 9 is low compared to modern humans, it is high in relation to the skulls from Java and China. However, John Napier and Joseph Weiner (1962) argued that the prominent supraorbital tori and flat frontal of OH 9 resemble *Pithecanthropus*. Arguing that OH 9 represented a new and distinct hominid, Gerhard Heberer (1963) proffered the name *Homo leakeyi*. After the initial interest in OH 9 dwindled, most paleoanthropologists, including L. S. B. Leakey (1966), adopted its inclusion in *H. erectus*. G. Philip Rightmire (1979) classified it authoritatively as *H. erectus* in a systematic description of it compared to other African and Asian fossils.

While continuing work at Olduvai Gorge in 1960, Jonathan Leakey, the oldest son of M.D. and L. S. B. Leakey, discovered a mandible with teeth, an upper molar, the parietals, and the hand bones of a juvenile individual from a putative living floor (FLK NN). The new hominid remains (OH 7) led Leakey, Philip Tobias and Napier to not only name a new species of early *Homo* that also includes OH 4, 6, 7, 8 and 13, but to also redefine the genus (Leakey *et al.*, 1964). Until the mid-1960s, cranial capacity was the major requirement used to distinguish *Homo* from *Australopithecus*. According to Leakey *et al.* (1964), OH 7 showed considerable brain expansion but did not quite reach the cerebral Rubicon that currently defined the genus *Homo*. Weidenreich, Keith and Vallois held slightly different standards for *Homo* of 700 cm^3, 750 cm^3, and 800 cm^3, respectively. Tobias estimated the cranial capacity of OH 7 to be around 657 cm^3 (Tobias, 1971) and in order to include the new species in *Homo*, Leakey *et al.* (1964) lowered the range of cranial capacity for *Homo* to 600 cm^3. Later Ralph Holloway (1980) re-estimated the endocranial volume of OH 7 to between 700 and 750 cm^3. These different estimates show how difficult it is to determine accurate cranial capacity from fragmentary specimens.

Leakey *et al.* (1964) based the new species, *Homo habilis*, on several morphological traits of OH 4, 6, 7, 8 and 13 that differ from *Australopithecus* and/or resembled *H. sapiens* (Tobias, 1964). The face and mandible of *H. habilis* are smaller than *Australopithecus* but similar in size to *H. erectus* and *H. sapiens*, and in certain respects the *H. habilis* clavicle, hand bones and foot bones resemble *H. sapiens* (Leakey *et al.*, 1964). Tobias (1966) proposed that *H. habilis* was an evolutionary link between *Australopithecus* and *H. erectus*.

The supposed human-like grip capabilities of *H. habilis* were a main focus of the debate over the new definition of *Homo*, since the new membership criteria for inclusion in the genus *Homo* involved tool-making abilities and behavior. *Homo habilis* means "able, handy, mentally skillful" hence the nickname "handy man'. It marked the first time a hominid species was defined according to cultural capacity instead of solely on anatomy. Since OH 7 was dubbed the first toolmaker, the contemporaneous *Zinjanthropus boisei* (now known as *Australopithecus boisei*) was no longer credited with making stone tools. Instead, it was reduced to an intruder or a victim on a *H. habilis* living site at Olduvai Gorge.

The naming of the new species of *Homo* did not sit well with many paleoanthropologists at the time. F.Clark Howell (1965), Robinson (1965) and others initiated an argument against *H. habilis* as basal *Homo* and against the validity of the new species designation altogether. Reflecting the popular paradigms of the time, they argued that Leakey et al. (1964) gave no evidence to support the few features used to separate the new species, that stone-tool manufacture was not relevant to the morphological characteristics required to separate a species, and that the new, expanded definition of the genus *Homo* was based on an invalid species. These arguments were one way or another based on biases regarding the genus *Homo* and hominid paleospecies in general. For instance, many at the time considered *A. africanus* as the ancestor of *Homo* (e.g., Campbell, 1963; Le Gros Clark, 1964) and few thought there was "enough morphological room" between *A. africanus* and *H. erectus* for another species. Later, Louis Leakey (1966) suggested that *H. habilis* might be the direct ancestor of *H. sapiens*, completely by-passing *H. erectus* phylogenetically.

Although the species *H. habilis* is widely accepted as valid today, the original hypodigm given by Leakey *et al.* (1964) is not supported as securely. Most workers agree on the classification of OH 13 in the *H. habilis* paratype. This small adolescent skull, affectionately known as "Cinderella", was recovered from MNK, Lower Bed II and dates to over 1.6 Ma (Leakey & Leakey, 1964; Delson *et al.*, 2000). The specimen has an estimated cranial capacity of 650 cm^3 (Tobias, 1971) and shares important features with OH 7. The fragmented skull OH 16, or "George", found in 1963 (Leakey & Leakey, 1964) was also placed within the *H. habilis* hypodigm (Tobias, 1991).

Assigning postcranial bones to *H. habilis* is difficult because those originally assigned to it were unassociated with teeth, jaws or crania. Although the juvenile hand bones and jaw and parietals were found on the same "living floor", a cursory examination of the site maps reveals alignment of bones by water. Furthermore, many of the bones, including hominids, show carnivore damage, so they could have been transported to the area. Adult hand phalanges were included with the juvenile ones in the early set of bones and there has been a dispute over the individual age of the foot bones. Indeed some workers have tried to bring together bones from two different sites and different geological horizons to make a single individual (Susman & Stern, 1982). A tibia and fibula, OH 35, from the same site as the type cranium of *A. boisei*, have been placed in *H. habilis*. They seem to be close matches with the same parts that are preserved in the Turkana Basin specimen, KNM-ER 1500, a partial skeleton that has a piece of mandible associated with it (Leakey *et al.*, 1989). This mandible fragment appears to be from a small *A. boisei*, but others have demurred (Wood, 1991). The foot skeleton, OH 8, was earlier argued by Wood (1974) on the basis of shape analysis to be from *A. boisei*, but he now includes it in *H. habilis*. The talus is, however, like one from the Turkana Basin that is associated with several *A. boisei* mandibular and dental specimens.

Despite these uncertainties, several workers have studied the postcranial anatomy of *H. habilis* from a functional perspective. Jack T. Stern and Randall L. Susman (1983) suggested that *H. habilis* postcrania indicated a habitual biped with retained primitive characters in the hand. They have argued that middle phalanges of the *H. habilis* hand are similar to apes and *Australopithecus* in robustness and curvature. However, others argue that the human-like length and overall morphology of the distal phalanges and the carpometacarpal joint outweigh those comparisons (Trinkaus, 1989). Wood (1974) and Lewis (1980) found primitive features in the OH 8 foot. Michael H. Day & Napier (1964) also analyzed the foot bones and documented a human-like metatarsal, arches and hallux. However, if any of the unassociated postcranial bones are not allocated to the correct species, then these functional interpretations are of course misleading.

Ronald J. Clarke has an extraordinary knack for finding hominids in museum drawers. Thirty years ago, on July 23, 1969, Clarke made a startling discovery while re-examining the fossil hominid collections at the Transvaal Museum in Pretoria, South Africa. He noticed that some hominid remains from Swartkrans looked different from the other "robust" *Australopithecus* remains already abundantly known at the site, and recognized that they may belong to *Telanthropus capensis* (*Homo erectus*) (Clarke *et al.*, 1970). Some of the fragments actually fit together to form the left side of a face and temporal region (SK 847). Several features that set SK 847 apart from *Australopithecus* include, for example, a pronounced supraorbital sulcus with thickening at the medial and lateral regions, a substantial frontal sinus, and moderate postorbital constriction. Clarke's museum recovery confirmed the previous minority view that two distinct hominids occurred at Swartkrans (Clarke & Howell, 1972). However, the taxonomic status of SK 847 is still somewhat disputed. Fred Grine and colleagues (1993) argued that significant morphological differences are apparent between SK 847 and two well-known early *H. erectus* crania. They suggested that the specimen does not belong to *H. erectus* and may indeed align closer with *H. habilis*.

In 1971 Sastrihamidjojo Sartono published observations on a new adult male cranium from Java labeled *Pithecanthropus VIII*, commonly known today as Sangiran 17. The somewhat distorted cranium came from the Middle Pleistocene Kabuh beds at Sangiran in 1969. Sartono (1971) placed Sangiran 17 within the same species as the other Asian hominids from Java that would later be subsumed under *H. erectus*. All the previously discovered hominid remains from Indonesia were faceless, so the rare preservation of the face and base of the vault in Sangiran 17 makes this specimen the most complete cranium of any published hominid from Indonesia. A new cranium with a facial skeleton, so far only published as an abstract, is said to resemble earlier African specimens (Tyler & Sartono, 1994).

Still at work in East Africa, M. D. Leakey, Clarke and L. S. B. Leakey reported the discovery and reconstruction of a crushed adult female *H. habilis* cranium. "Twiggy" or OH 24, came out of the eastern part of the gullies at Olduvai Gorge known as DK and differs from *Australopithecus* in that it does not exhibit marked postorbital constriction (Leakey *et al.*, 1971). Leakey *et al.* (1971) suggested that it resembles *H. habilis* from Bed II (OH 13) in both cranial and dental aspects, yet differs in some respects from the type specimen (OH 7) from Bed I. Until the discovery of OH 24, the differences in morphology and in size of cranial parts and dentition between *H. habilis* in Beds I and II at Olduvai were interpreted as the possible result of a long time interval. However, since OH 24 resembles the later morph (Bed II) yet aligns chronologically with the earlier type specimen, the time explanation was no longer suitable. Leakey et al. (1971) proposed that sexual dimorphism might be behind the variation with OH 13 and 24 representing females of *H. habilis* and OH 7 and 16 representing males.

In 1971, M. D. Leakey reported on the 1970 discovery of a left femur shaft and a hip bone (OH 28) of *H. erectus* from Olduvai Bed IV. The bones and an Acheulean industry were found together and represent the first ever *H. erectus* bones directly associated with a well-known tool assemblage (Leakey, 1971). Day (1971) reported on several features of the femur that are always present in *H. erectus* and pointed out the strong muscle markings, large acetabulum, massive iliac pillar, and medially rotated ischium of the hip bone.

The Koobi Fora Research Project initiated in 1968 at East Lake Turkana (Leakey & Leakey, 1978) produced large numbers of hominid fossils, some of them belonging to early *Homo*. In 1973, Richard E. F. Leakey published the reconstruction and analysis of an unusual cranium (Fig. 25.3) discovered by B. Ngeneo in Area 131, East Lake Turkana, Kenya. The maxilla and facial region of KNM-ER 1470 was said to be unlike those of any known form of hominid (Leakey, 1973*a*). None of the tooth crowns was preserved but the palate is large (Day *et al.*, 1975). However, since large cranial capacity had always been associated with the genus *Homo*, Leakey (1973*a*) proposed the taxon *Homo* sp. indet. for KNM-ER 1470. In fact, many of the features of the skull were argued to be more like those of *Australopithecus* than *Homo* (Walker, 1976), a suggestion that has recently been taken up again (Wood & Collard, 1999). At the time of discovery, KNM-ER 1470 was said to be about 2.9 Ma due to its location below the KBS tuff, which was then dated to about 2.6 Ma. However, redating of the KBS tuff puts KNM-ER 1470 at about 1.9 Ma (Feibel *et al.*, 1989), making this skull roughly contemporaneous with those of *H. habilis* from Olduvai Gorge and East Lake Turkana.

Continued explorations at Koobi Fora produced evidence of at least three contemporaneous hominid lineages in the Plio-Pleistocene of East Africa (Leakey, 1974). The cranium KNM-ER 1805, with a cranial capacity of 582 cm^3 (Holloway, 1983*a*), and an associated mandible were discovered *in situ* in the same horizon as a mandible of *A. boisei* (KNM-ER 1806). KNM-ER 1805 is a puzzling specimen. It has small sagittal and nuchal crests with a contrasting set of smallish teeth. Since sexual dimorphism is evident in *A. boisei* (e.g., KNM-ER 406 and 732), the possibility of the same phenomenon in early *Homo* is reasonable (Leakey, 1974). Leakey (1974) proposed the separation of these problematic specimens into three genera: *A. robustus* and *A. boisei* in one genus, the gracile specimens from Sterkfontein (*A. africanus*), KNM-ER 1813 (see below), and OH 24 in a second genus, and KNM-ER 1470, 1590 (another partial cranium), OH 7 and 16 and some from South Africa in the third genus, *Homo*. Most currently regard 1805 as an early representative of *Homo*.

Closer to the Koobi Fora base camp, Kamoya Kimeu located a diseased, partial *H. erectus* skeleton (KNM-ER 1808) with signs of extensive periostitis (Leakey, 1974). Walker *et al.* (1982) determined that the pathological condition was consistent with modern cases of hypervitaminosis A and that the individual likely acquired the disease from eating carnivore livers. Skinner (1991) proposed the vitamin A came from bee brood, but later found that bee brood has no vitamin A and retracted the idea (Skinner *et al.*, 1995). Rothschild and colleagues (1995) also examined the bones of KNM-ER 1808 and came to a different conclusion – that the distribution pattern of periosteal reaction is compatible with treponematosis in the form of yaws. The discovery of KNM-ER 1808 also had major implications on the possible social behaviors of early hominids since altruism must have been in place for this individual to survive as long as she did with the disease. Although the skeleton was severely pathological, Ruff & Walker (1993) were able to estimate stature from the length of the femur and show that this *H. erectus* individual was tall, unlike individuals of *H. habilis* or *Australopithecus*.

The cranium KNM-ER 1813 found in 1973 has a small braincase but *Homo*-like dentition (Leakey, 1974) and face (Rak, 1983). It has similar calvarial features to OH 24 and was later included in *H. habilis* (Howell, 1978; Clarke, 1985; Tobias, 1991; Wood, 1992). In 1995 Walter Ferguson provided an alternative classification for KNM-ER 1813. He suggested the name *Homo microcranous*, emphasizing that the small cranial capacity of KNM-ER 1813 was not due to sexual dimorphism since in his opinion the specimen is probably male (Ferguson, 1987). Today most other workers continue to include KNM-ER 1813 as a member of *H. habilis*. Also in 1973, a nearly 1.9-Ma mandible (KNM-ER 1802) was found (Leakey, 1973*b*; Feibel *et al.*, 1989) and regarded by Leakey to be a mandible of the same species as KNM-ER 1470 and 1590.

Colin P. Groves and Vratislav Mazak named a new hominid species in 1975 based on two associated hemimandibles with nearly complete dentition (KNM-ER 992) found by the Koobi Fora Research Project (Leakey & Wood, 1973). The new species was defined mainly by its dental differences from *H. habilis* and *A. africanus* but, oddly, not *H. erectus*. As originally defined, *Homo ergaster*, which can be translated as "working man", included many of the East Lake Turkana early Pleistocene *H. habilis* (KNM-ER 1805) and *Homo*-like specimens (Groves & Mazak, 1975). Later Groves also included KNM-ER 1813 in this species (Groves, 1989). Importantly, the new name later became available for those who wanted to split off the early African *H. erectus* from Asian ones.

In 1976 Richard Leakey reported more evidence from East Lake Turkana to further support the presence of early *Homo* during the Plio-Pleistocene of east Africa (Leakey, 1976). The well-preserved partial pelvic bone from a large and robust adult male (KNM-ER 3228) has similar acetabular buttressing and rotated ischium to that observed in the *H. erectus* OH 28 (Day, 1971). Together with the KNM-ER 1481 femur it provides critical evidence for early modification of the hip joint in *Homo* and this adds weight to the argument for the late Pliocene separation of *Homo* and *Australopithecus*.

The 1974–5 field season at East Lake Turkana also produced a relatively complete adult cranium. Bernard Ngeneo found KNM-ER 3733 in Area 104 in the upper member of the Koobi Fora Formation (Leakey, 1976) dated to approximately 1.78 Ma (Feibel *et al.*, 1989). The specimen was the best-preserved single *H. erectus* cranium known at the time (Leakey, 1976). *Homo erectus* features include prominent supraorbital tori, slight postorbital constriction and an enlarged cranial capacity of 848 cm^3 (Holloway, 1983*a*). Leakey & Walker (1976) argued that KNM-ER 3733 represents decisive evidence for two contemporaneous hominids in the East Lake Turkana region and was the evidence needed to falsify the single-species hypothesis (Wolpoff, 1971*a*). The accumulated cranial and postcranial east African evidence (e.g., KNM-ER 3733, 3228, 1470, 1590, 1481) now heavily favored a distinct *Homo* lineage in early Pleistocene levels (Leakey, 1976). By 1977 a second partial cranium had been found, KNM-ER 3883, a bit more robust and a little larger than KNM-ER 3733.

In 1977, Alun R. Hughes and Tobias excavated a cranium (Stw 53) and mandible fragment at Sterkfontein, South Africa. They aligned the remains with *H. habilis* because of the roundness of the skull, the pattern of the temporal lines, the thin brow ridges, and the age (Bed V) which is younger than *A. africanus* also found at the site (Hughes & Tobias, 1977). Based on material from East Lake Turkana (KNM-ER 1470, 1590, 1813, 3732) and Sterkfontein (Stw 53) they confirmed that *H. habilis* does indeed merit specific distinction from *A. africanus*.

In 1984, by spotting a mere fragment of a frontal bone, Kamoya Kimeu discovered his second hominid skeleton, this time the most complete early hominid skeleton ever collected. The bones of the juvenile *H. erectus* were situated on the south bank of the Nariokotome River on the western shore of Lake Turkana in 1.53 Ma sediments (Brown *et al.*, 1985). In-depth analysis of KNM-WT 15000 or the "Nariokotome Boy" brought new insights to the study of early *Homo* (Walker & Leakey, 1993). The limb proportions of this *H. erectus* were human-like and followed Allen's Rule regarding adaptation to heat stress. His stature was remarkable, especially compared to *Australopithecus* and early members of *Homo*. At an estimated age of 11 years old (Smith, 1993), he was already 160 cm with a projected adult height of 185 cm depending on the chosen growth model (Ruff & Walker, 1993). This height based on the length of the femur is consistent with KNM-ER 1808, the Trinil femur and the tallest populations of modern humans. An important comparison of brain size to body size is also possible here. It appears that, because this species had unexpectedly large body mass, relative brain size was fairly stable from about 1.8 Ma until late in the hominid fossil record, with encephalization quotients nearly half those of modern *H. sapiens*. The postcranial bones of KNM-WT 15000 were essentially like those of modern humans (Walker, 1993).

The new find also shed some light on the language debate. Tobias suggested that *H. habilis* possessed language capability (Tobias, 1980, 1983) since endocasts of *H. habilis* exhibit the earliest known evidence of an enlarged Broca's area (Holloway, 1983*b*). The narrow vertebral canal of the Nariokotome Boy suggests that the spinal cord was too small to contain the motor neurons needed to control breathing patterns in human speech, and thus modern language capabilities did not evolve until sometime after the middle Pleistocene (MacLarnon, 1993).

While surveying Bed I sediments at site FLK, Olduvai Gorge, Tim D. White made the most recent discovery of a partial skeleton of early *Homo*. The fossil included not only maxillary and mandibular pieces, but also radial, ulnar, humeral, femoral and tibial fragments representing the first limb bones securely assigned to *H. habilis*. Upon its discovery, OH 62 perplexed anthropologists because it exhibits *H. habilis* craniodental characteristics but was claimed to possess a postcranial anatomy similar to "Lucy", a much earlier *A. afarensis* (Johanson *et al.*, 1987). Its relatively wide palate, early *Homo* dental dimensions, and lack of facial and cranial specializations distinguish OH 62 from *Australopithecus* and align it with OH 24 and Stw 53. However, the upper limb proportions of OH 62 were said to be like *Australopithecus*. The OH 62 upper limb is longer and thinner than Lucy's, while the femur was claimed to be smaller and less robust than *Australopithecus* (Johanson *et al.*, 1987). OH 62 was claimed to be as small or smaller than any known fossil hominid.

Johanson *et al.* (1987) pointed out that since the taxon *H. habilis* was created on the basis of cranial remains, then the morphology of the face, palate and dentition of OH 62 merit its inclusion as a member of the taxon. Until this time, many researchers followed Tobias (1983) in placing all of the late Pliocene–early Pleistocene non-robust specimens that fall outside the range of *H. erectus* into a variable, polymorphic species, *H. habilis* (Johanson & White, 1979; Kimbel *et al.*, 1984; Skelton *et al.*, 1986). This view holds that morphological differences between individuals such as KNM-ER 1470 and 1813 reflect sexual, geographical and chronological effects on a single species. Johanson *et al.* (1987) agreed on this point and therefore viewed OH 62 as an elderly individual of a single species that was widespread in Africa. Until this discovery, most people thought *H. habilis* postcrania would resemble *H. erectus* or *H. sapiens* because of the hip bone (KNM-ER 3228) and leg bones such as KNM-ER 1471 and 1472 from the same time period as *H. habilis*. But Johanson *et al.* (1987) claimed that *H. habilis* postcranial morphology is more similar to *A. afarensis*, although OH 62 has hardly any joint surfaces with which to compare. The discovery of OH 62 and the claim made about its limb proportions seriously undermined the common practice of sorting late Pliocene–early Pleistocene postcrania into *Homo* or robust *Australopithecus* species on the basis of size and assumed limb proportions, making any previous taxonomic attributions of isolated postcranial bones highly tenuous.

Johanson *et al.* (1987) concluded that the long and powerful arms of *H. habilis* and *Australopithecus* emphasize the mosaic pattern of evolution in the early hominid postcranial skeleton. They also noted that the coincidence of otherwise relatively derived *H. erectus* postcrania at 1.5 Ma (KNM-WT 15000) and a postcranially primitive *H. habilis* at 1.8 Ma (OH 62) may imply an abrupt transition between these taxa in eastern Africa. Furthermore they postulated that perhaps the only distinguishing feature between small individuals of *Australopithecus* and *H. habilis* may have been cranial capacity (Johanson *et al.*, 1987). This reinforces the view that encephalization in the terminal Pliocene played a key role in hominid evolution (Johanson *et al.*, 1987). Note, however, the cranial capacity of OH 62 cannot be measured from the tiny fragments of skull.

KNM-ER 3735 is another early *Homo* partial skeleton (Leakey *et al.*, 1989). It has some skull bones including a temporal and zygomatic that show that it is not a robust *Australopithecus*. There are several pieces of the postcranial skeleton, including the distal humerus, forearm bones and sacrum. The limb proportions cannot be estimated from the short pieces of shaft that remain.

Considering the case of OH 62, and with new knowledge of hominid limb proportions from a recent discovery of a late Pliocene *Australopithecus*, Asfaw *et al.* (1999: 632) cautioned against using OH 62's femoral length:

> The past few years have witnessed a rash of attempts to estimate early hominid limb length proportions from fragmentary and unassociated specimens. These specimens have been used to generate a variety of functional and phylogenetic scenarios. Accurate estimates of the limb proportions of early hominids, however, must be confined to the very few specimens that actually preserve relevant elements, such as the A.L. 288-1 ("Lucy") specimens and KNM-WT 15000. The new Bouri VP-12/1 specimen [*Australopithecus garhi*] is only the third Plio-Pleistocene hominid to provide reasonably accurate limb length proportions. The Olduvai Hominid 62 specimen of *Homo habilis* has been erroneously argued to show humerus-to-femur proportions more primitive than those of "Lucy" (McHenry & Berger, 1998*a*; Hartwig-Scherer & Martin, 1991), but its femur length cannot be accurately estimated.

But several workers continue to use the original guesses of limb proportions of OH 62 to advance their taxonomic claims. Indeed, Wood & Collard (1999) have just done so to move *H. habilis* and *H. rudolfensis* into the genus *Australopithecus*.

Perhaps the oldest non-African remains of the genus *Homo* come from Dmanisi, Georgia, a medieval ruined city between the Black and Caspian Seas, built atop a Paleolithic site. A hominid mandible discovered there in 1991 (Gabunia *et al.*, 1991; Gabunia & Vekua, 1995) includes the greater portion of the body with complete dentition, but no rami. Recently two skulls have been discovered in the same bed (Bed V) that lies directly above a basalt flow dated to around 1.85 Ma (Gabunia *et al.*, 2000). Nearly 2 million years ago a basalt flow dammed a river forming a lake at Dmanisi. The basal basalt is unweathered with no signs of pedogenesis on its surface (Bräuer & Schultz, 1996). This indicates that the overlying layers were deposited rather quickly and makes a strong argument for the early date of the hominid remains from Bed V.

Since its discovery a small debate has surrounded the taxonomic status of the mandible. It has been compared to later *H. erectus* due to morphological similarities in the chin, anterior and posterior mandibular symphysis and mental foramina (Bräuer & Schultz, 1996). However, Rosas & de Castro (1998) pointed out that the mandible shares features with African *H. erectus*, or *H. ergaster*, and thus supported the taxonomic designation *Homo* sp. indet. (aff. *ergaster*).

Also in the early 1990s, workers in Malawi recovered a mandible that extended the range of early *Homo* to that part of eastern Africa (Schrenk *et al.*, 1993). This mandible has similarities to KNM-ER 1802 and has therefore been placed in *H. rudolfensis* (Bromage *et al.*, 1995).

In 1994 an Institute of Human Origins team working in Ethiopia discovered the earliest known member of the genus *Homo*, an adult maxilla (A.L. 666-1) found just below the BKT-3 marker tuff in the upper Kada Hadar Member of the Hadar Formation and dated to around 2.33 Ma (Kimbel *et al.*, 1996). This specimen was associated with Oldowan stone tools and a late Pliocene faunal assemblage. Based on its dissimilarity from *Australopithecus* and its *Homo*-like maxillary and dental morphology, Kimbel *et al.* (1996) aligned the new specimen with the genus *Homo*. Its *Homo*-like features include, for example, a relatively wide and deep palate, a parabolic dental arcade and slight subnasal prognathism. It has a flat nasoalveolar clivus sharply angled to the floor of the nasal cavity, which is a synapomorphy or shared-derived feature of *Homo*. Coupling this derived feature with the retained primitive *Australopithecus* condition of having an extensive intranasal platform that horizontally separates the anterior nasal spine from the vomeral insertion, ties A.L. 666-1 to other early *Homo* such as OH 62, SK 847 and Sangiran 4. The maxilla has a square anterior profile and frontal process morphology similar to KNM-ER 1805, 1470, 3733 and Sangiran 4. The teeth of the new hominid specimen also tie it to early *Homo*. It has a narrow upper first molar crown compared to *Australopithecus* and even to *H. habilis*. It also shows a rhomboid-shaped upper second molar typical of early *Homo* (e.g., most *H. habilis*, KNM-ER 1590, 3733, 807 and KMM-WT 15000) versus a typically square one in *A. afarensis*, *A. africanus* and "robust" *Australopithecus*. Initially according to Kimbel *et al.* (1996), none of these characters supports the assignment of A.L. 666-1 to any one species within the genus *Homo*, hence the designation of *Homo* sp. However, after further analysis Kimbel and colleagues (1997) attributed the specimen to *Homo* aff. *H. habilis*.

Since the accumulation of early *Homo* fossils from Koobi Fora in the 1970s, workers observed morphological differences between them and especially between those specimens attributed to *H. habilis*. Many researchers have noticed the size differences within *Homo habilis*, exemplified by KNM-ER 1470 and 1813. Some credited the cranial and facial size and shape differences to sexual dimorphism (Howell, 1978; Miller, 1991; Tobias, 1991). Others see a species-level difference between KNM-ER 1470 and 1813 and so recently the *H. habilis* hypodigm has been split into two different taxa. In separate accounts, Groves (1989) and Wood (1992) distinguished KNM-ER 1470 and 1590 as *H. rudolfensis*. They used the species name "*rudolfensis*" from *Pithecanthropus rudolfensis* Alexeev, 1986. *Homo rudolfensis* is distinguished by a large brain, a wide, flat face and large teeth, as opposed to the smaller brain, and more orthognathic face and teeth of *H. habilis*. However, several workers are not convinced that these features are adequate to distinguish a separate species (e.g., Suwa *et al.*, 1996).

Lieberman *et al.* (1996) used 48 frequently used cranial,

dental and mandibular characters in a cladistic study to see how homoplasy influences phylogenetic analyses of relationships among early *Homo* (Lieberman *et al.*, 1996). In other words, they investigated whether KNM-ER 1470 and KNM-ER 1813 are male and female representatives of *H. habilis sensu stricto* or are different species. They found the KNM-ER 1470 group to have an *Australopithecus*-like face with an expanded brain case and the KNM-ER 1813 group to have a smaller, more orthognathic, later *Homo*-like face and smaller brain. They postulated that perhaps increased encephalization occurred independently in more than one hominid clade and that the similarities of 1470 with *Australopithecus* could be either symplesiomorphies or homoplasies (Lieberman *et al.*, 1996). In 1999, Wood designated the cranium KNM-ER 1470 as the lectotype of *H. rudolfensis* after a minor debate over the validity of the taxon (Kennedy, 1999; Wood, 1999).

Currently paleoanthropologists also disagree on the validity of the taxon *H. ergaster*. Until the 1980s, anthropologists typically included early Pleistocene, large-brained, East African hominid remains in the same species as the well-known, contemporaneous Asian specimens: *H. erectus*. However, the vast geographic dispersal, long temporal duration, and polytypism of *H. erectus* make classifying the species difficult. *Homo erectus* is defined predominantly by several vault characteristics including a long and low vault, a pronounced bar-like supraorbital torus, an angular torus, cranial keeling and thick cranial vault bones (Rightmire, 1990). Peter Andrews (1984) suggested that the last three of these features are restricted to the Asian assemblage, and are autapomorphic to them. Others have also challenged the lumping of the African material into *H. erectus*, based on cranial observations, and support the taxon *H. ergaster* Groves and Mazak, 1975 for the African assemblage (e.g., Wood, in Bilsborough & Wood, 1988; Wood, 1992, 1994).

However, some workers have challenged the separation of the assemblages into two taxa. They have argued, based on cranial and facial analyses, that the differences between the two hominid assemblages are variable, slight, insignificant and unworthy of recognition of a separate species (Rightmire, 1986, 1998*a*; Bräuer & Mbua, 1992; Kramer, 1993; Walker, 1993; Bräuer, 1994). Rightmire (1998*a*) argued on the basis of facial morphology that the *H. erectus* material from Java and China is not diagnostically distinct from the Turkana Basin specimens. His argument supports Kramer (1993) on vault characteristics, suggesting that *H. erectus* shows comparable dispersal and polytypism to the one extant hominid species, *H. sapiens*.

Preliminary analysis of the crania discovered at Dmanisi, Georgia, in 1999 shed new light on the *H. erectus* problem. D2280 is a relatively complete calvaria that most likely represents an adult or a male individual, while D2282 is more complete preserving most of the calvaria and face with five teeth (Gabunia *et al.*, 2000). Size, structure, suture stage and dental wear differences between the cranium and the adult calvaria led to the suggestion that D2282 represents a subadult or young adult and possibly a female (Gabunia *et al.*, 2000). The overall cranial morphology of the specimens compares well with fossils from the early Pleistocene of East Africa. According to Gabunia et al. (2000: 1021) the similarity of the Dmanisi specimens to African ones is stronger than to Asian *H. erectus* specimens, leading them to assign the cranial remains to H. ex gr. *ergaster*.

The significance of the Dmanisi hominids reaches beyond their African-like morphology, however, to their estimated age. The Dmanisi hominids date to about 1.7 Ma, and are associated with artifacts that match those of east African pre-Acheulean assemblages from as early as 2.4 Ma (Semaw *et al.*, 1997). This is consistent with the earliest known appearance of the Acheulean industry in the fossil record at about 1.6 Ma (Asfaw *et al.*, 1992). More importantly, the age of these fossils (which coincides with, if not pre-dates, the appearance of *H. erectus* in East Africa) and the presence of some Asian *H. erectus* cranial, facial and dental features led to the suggestion that the Dmanisi remains represent the initial rapid dispersion of hominids out of Africa into the Caucasus (Gabunia *et al.*, 2000).

Taxonomy

Systematic framework

The definition of the genus *Homo* is a major point of contention among taxonomists. In 1964, Leakey, Tobias and Napier assigned a new definition for the genus, and since then others have also redefined *Homo*. As discussed above, Leakey *et al.* (1964) have been criticized for listing a set of characters or general trends of the genus, without emphasizing shared-derived features. More recently cladistic analysis redefined the genus by stressing character changes relative to *Australopithecus* (e.g., Wood, 1992). For our purposes we will follow Wood and include only those anatomical traits that distinguish *Homo* from *Australopithecus*. Some commonly used junior synonyms for the early genus *Homo* in the literature from the middle of the last century include *Pithecanthropus* Dubois, 1894, *Sinanthropus* Black, 1927, *Meganthropus* Weidenreich, 1944, *Telanthropus* Broom & Robinson, 1949, and *Atlanthropus* Arambourg, 1954.

The *Homo habilis* hypodigm has been split many different ways by taxonomists. For instance, in his latest Olduvai Gorge monograph, Tobias (1991) included the Olduvai and East Lake Turkana specimens in a single polytypic species, *H. habilis*. Wood (1992) agreed with placing the Olduvai remains in *H. habilis*, but added a twist to the taxonomy in his own Koobi Fora monograph. He concluded that two different species are represented on the eastern shore of Lake Turkana and designated *H. habilis* for the smaller specimens and *H. rudolfensis* for the larger ones. In this review we give what we think might be a consensus taxonomic scheme today by including *H. rudolfensis*. We are by no means convinced, however, that more than one species of early

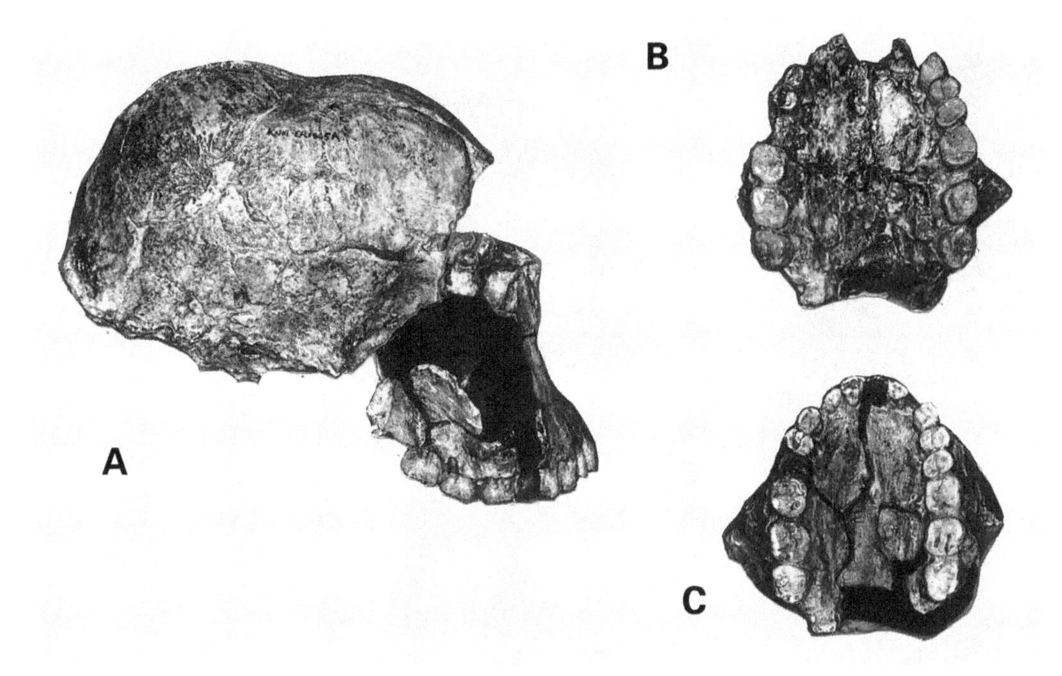

Fig. 25.1 *Homo habilis*. KNM-ER 1805, partial cranium, from Koobi Fora, Kenya: (A) right lateral view; (B) palatal view of unreconstructed palate and (C) of reconstructed palate (Walker, 1981).

Homo existed at about 2.0 Ma. Perhaps the best evidence in support of the two species is that a *H. habilis* specimen (OH 13) existed at about 1.6 Ma at Olduvai when early *H. erectus* (KNM-ER 3733) existed by 1.78 Ma at East Lake Turkana. This suggests either that *H. rudolfensis* alone might have been ancestral to *H. erectus* or that some populations of *H. habilis* persisted unchanged after *H. erectus* evolved from earlier ones.

Similar to the *H. habilis* conundrum, a present debate surrounds the taxonomic status of early Pleistocene African hominids (as discussed above). We chose to include them into one taxon *H. erectus*. However, the reader should be aware that *H. ergaster* is used increasingly more often when referring to African *erectus*-like fossils.

Order Primates Linnaeus, 1758
  Infraorder Catarrhini É. Geoffroy Saint-Hilaire, 1812
    Superfamily Hominoidea Gray, 1825
      Family Hominidae Gray, 1825
        Genus *Homo* Linnaeus, 1758
          Homo habilis Leakey et al., 1964
          Homo rudolfensis (Alexeev, 1986)
          Homo erectus (Dubois, 1892)

Family Hominidae

GENUS *Homo* Linnaeus, 1758
A genus of the family Hominidae. Cranial: highly variable cranial capacity, on average, larger than that of *Australopithecus*, cranial capacity large relative to body size; reduced postorbital constriction; reduced lower facial prognathism; more anteriorly situated foramen magnum; increased participation of the occipital bone in cranial sagittal arc length; increased cranial vault height. Dental: narrower tooth crowns, particularly mandibular premolars; reduction in length of the molar toothrow; variable molar size but smaller than *Australopithecus*; in general, teeth are not enlarged buccolingually as in *Australopithecus*.
INCLUDED SPECIES *H. erectus, H. habilis, H. rudolfensis*

SPECIES *Homo habilis* Leakey et al., 1964 (Figs. 25.1, 25.2)
TYPE SPECIMEN OH 7, mandible with teeth, upper molar, parietals, and (possibly) hand bones of a juvenile individual from Olduvai Gorge, Tanzania
AGE AND GEOGRAPHIC RANGE Approximately 2.33 to 1.6 Ma; Omo and Hadar, Ethiopia; Olduvai Gorge, Tanzania; East Lake Turkana, Kenya; Sterkfontein, South Africa
ANATOMICAL DEFINITION
Cranial: mean cranial capacity greater than *Australopithecus* but smaller than *H. erectus*; smaller mandibles and maxillae than those of *Australopithecus*, in the range of *H. erectus* and *H. sapiens*; incipient supraorbital torus; coronal chord greater than sagittal chord in the parietals; upper face breadth greater than midface breadth; nasal margins sharp and everted nasal sill. Dental: premolars narrower than *Australopithecus*, in the range of *H. erectus*. Molar size lies in the lower range of *Australopithecus* and in the upper range of *H. erectus*. Buccolingual narrowing and mesiodistal elongation of all teeth especially mandibular molars and premolars.

SPECIES *Homo rudolfensis* (Alexeev, 1986) (Figs. 25.3, 25.4)
TYPE SPECIMEN KNM-ER 1470, edentulous adult cranium from Koobi Fora, Kenya

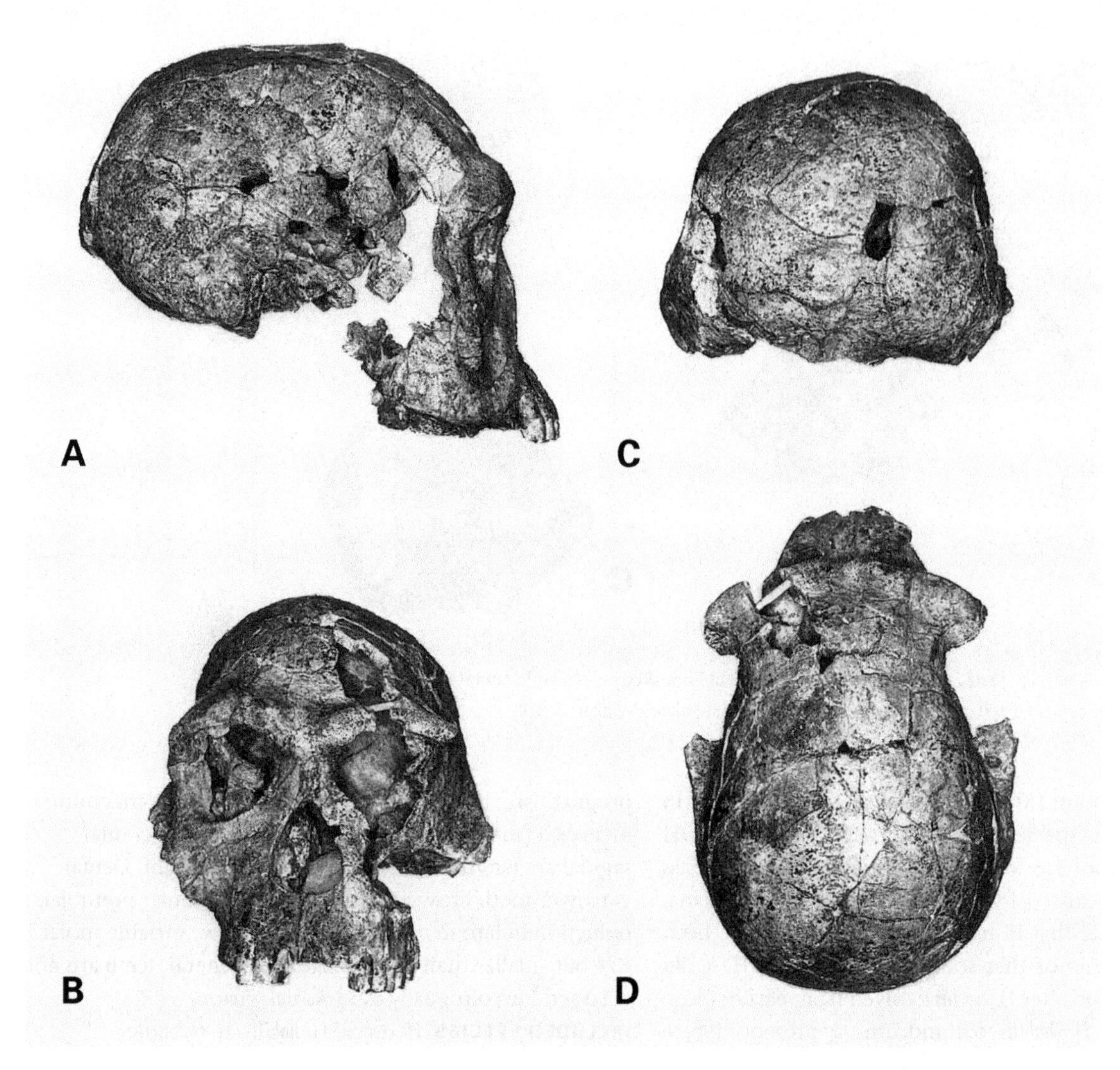

Fig. 25.2 *Homo habilis*. KNM-ER 1813, nearly complete adult cranium, from Koobi Fora, Kenya: (A) right lateral, (B) facial, (C) occipital and (D) superior views.

AGE AND GEOGRAPHIC RANGE Approximately 1.9 Ma; East Lake Turkana, Kenya; ?Uraha, Malawi
ANATOMICAL DEFINITION
Cranial: larger mean cranial capacity than *H. habilis*; prognathic overall with an orthognathic lower face; midface breadth greater than upper face breadth; less everted nasal margins than *H. habilis*; no nasal sill; flat superior surface of posterior zygomatic root, anteriorly inclined zygomatic surface; larger palate than *H. habilis*; rounded mandibular symphysis with no internal buttressing. Dental: absolutely and relatively large upper anterior alveoli compared to *H. habilis*.

SPECIES *Homo erectus* (Dubois, 1892) (Figs. 25.5–25.20)
TYPE SPECIMEN Trinil 2, adult partial cranium from Java, Indonesia
AGE AND GEOGRAPHIC RANGE 1.8–0.3 Ma; East and West LakeTurkana, Baringo Basin and ?Lainyamok, Kenya; Olduvai Gorge, Tanzania; Swartkrans, South Africa; Melka Kunture and Omo, Ethiopia; Tighennif, Algeria; Thomas Quarries, Salé and Sidi Abderrahman, Morocco; Ceprano, Italy; Dmanisi, Georgia; Narmada, India; Hexian, Gongwangling (Lantian), Zhoukoudian, Jianshi and Yunxian, China; Sambungmachan, Solo, Trinil, Ngandong, Modjokerto and Sangiran, Java, Indonesia
ANATOMICAL DEFINITION
Cranial: long and low skull with greatest breadth toward base of cranium; flat frontal bone characterized by pronounced supraorbital torus, supratoral sulcus and a frontal keel; significant degree of postorbital constriction compared to *H. sapiens*; parasagittal depressions on either side of the midsagittal keel of the parietals; strongly developed transverse occipital torus; superior border of the temporal bone is generally straight and low compared to *H. sapiens*; deep and narrow glenoid fossa; relatively small mastoid process; thick cranial vault bones averaging

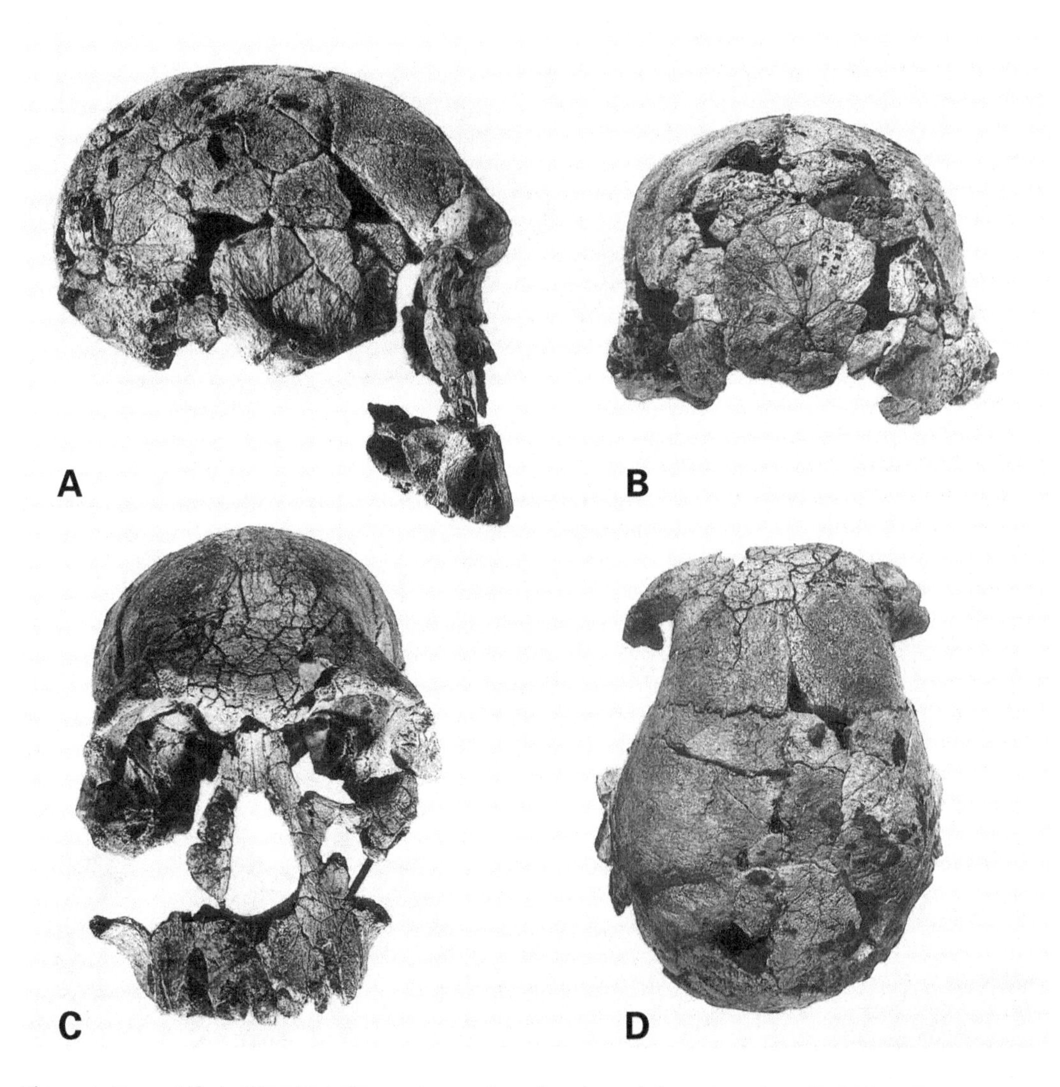

Fig. 25.3 *Homo rudolfensis*. KNM-ER 1470, nearly complete edentulous adult cranium, from Koobi Fora, Kenya: (A) right lateral, (B) facial, (C) occipital and superior views.

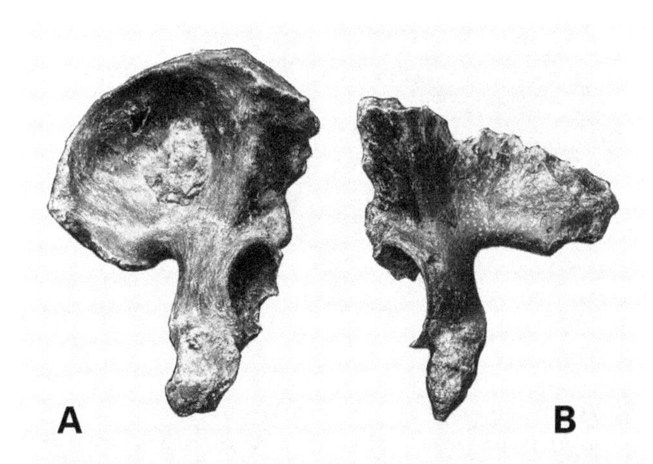

Fig. 25.4 (A) ?*Homo rudolfensis*, KNM-ER 3228, right hip bone; lateral view. (B) *Homo erectus*, OH 28, left hip bone, lateral view.

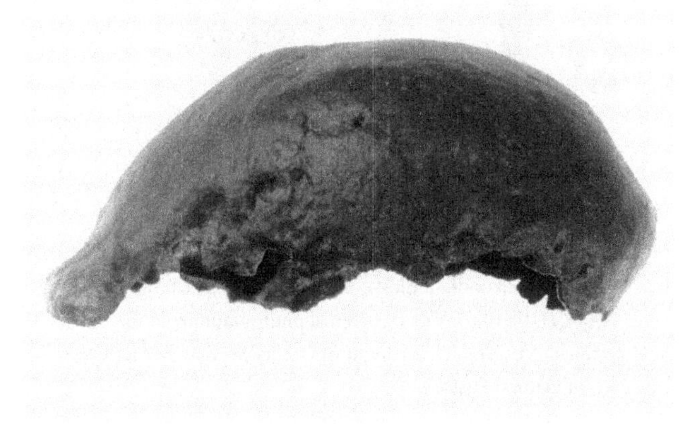

Fig. 25.5 *Homo erectus*. Trinil 2 (*Pithecanthropus erectus* of Dubois), from Trinil, Java: calotte, in left lateral view. Photograph copyright Nationaal Natuurhistorisch Museum, Leiden.

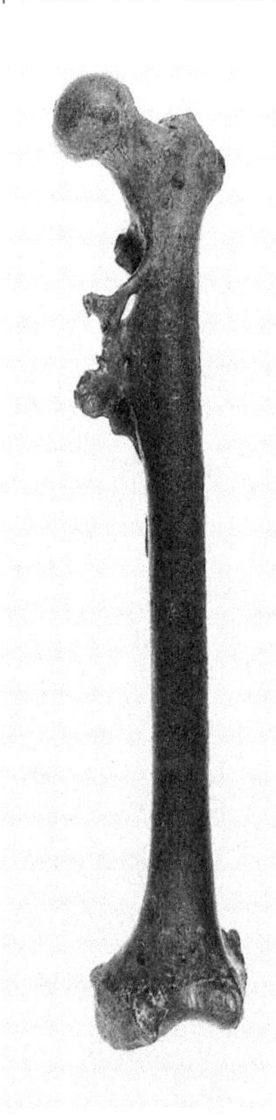

Fig. 25.6 *Homo erectus*. Trinil 2: complete left femur, anterior view. Photograph copyright Nationaal Natuurhistorisch Museum, Leiden.

Fig. 25.8 *Homo erectus*. OH 9, from Olduvai Gorge, Tanzania: calvaria in left lateral view.

Fig. 25.7 *Homo erectus*. Skull XII – Skull III Locus L (*Sinanthropus pekinensis* of Black), from Zhoukoudian, China: adult calvaria, left lateral view. Neg. no. 336412 (J. Coxe copy of an original photograph by Franz Weidenreich), courtesy Department of Library Services, American Museum of Natural History.

Fig. 25.9 *Homo erectus*. SK 847 (*Telanthropus capensis* of Broom & Robinson), from Swartkrans, South Africa: partial cranium including the left part of the face; facial view.

Fig. 25.10 *Homo erectus*. SK 847 (*Telanthropus capensis* of Broom & Robinson), from Swartkrans, South Africa: partial cranium including the left part of the face; left lateral view.

Fig. 25.12 *Homo erectus*. Lake Turkana, Kenya: (A) KNM-ER 1808, right femur; posterior view; (B) KNM-ER 737, left femur; posterior view.

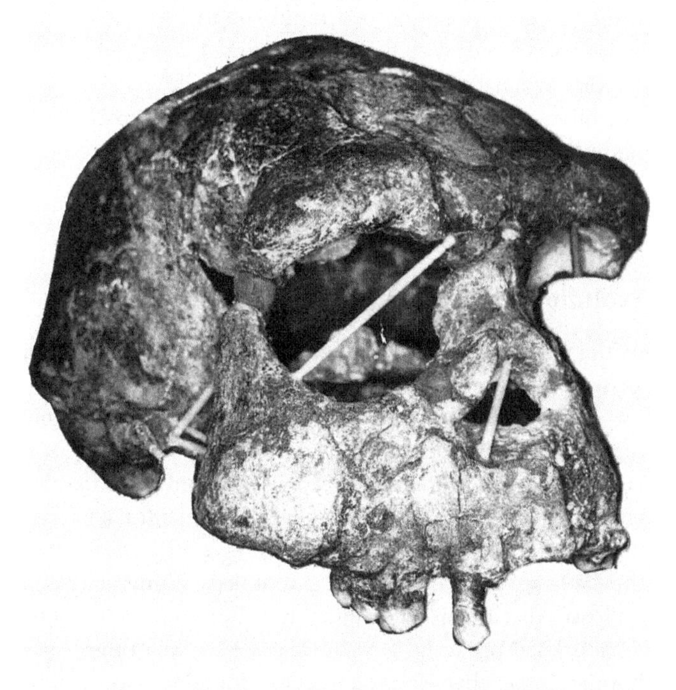

Fig. 25.11 *Homo erectus*. Sangiran 17, from Sangrian, Java: nearly complete cranium, in right three-quarters view. Photograph courtesy of Milford Wolpoff.

Fig. 25.13 *Homo erectus*. KNM-ER 992, from Koobi Fora, Kenya: left body and ramus of mandible, left lateral view.

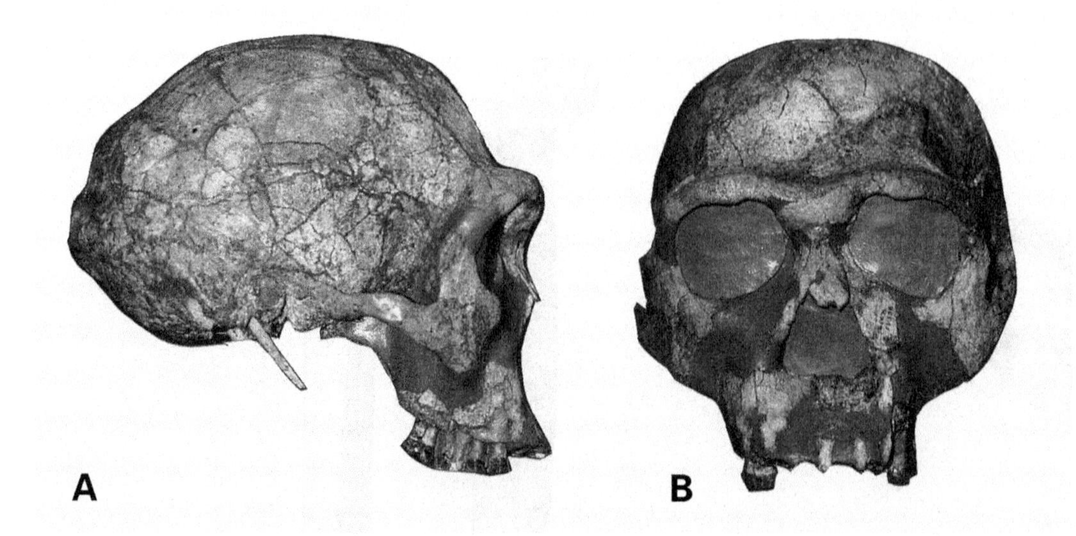

Fig. 25.14 *Homo erectus*. KNM-ER 3733, from Koobi Fora, Kenya: cranium, in (A) right lateral and (B) facial views.

9–10 mm – nearly twice as thick as modern humans; cranial capacity ranges from around 700 to 1300 cm^3; broad nasal bones with flat bridge; marked degree of alveolar prognathism; large bicondylar breadth of mandibles, no chin relative to modern humans, broad ramus. Dental: shovel-shaped upper central incisors. Postcranial: modern human upper and lower limb proportions.

Evolution of early genus *Homo*

At the moment we are plagued with more questions than answers concerning the evolution of early *Homo*. The majority of paleoanthropologists hold that the *Homo* clade probably originated in Africa sometime between 2.0 and 3.0 Ma. However, a gap in the African fossil hominid record at this crucial time period allows only general speculation concerning the origins of *Homo*. Often we read reports that *Homo* arose in Asia (e.g., Wanpo *et al.*, 1995), but these claims are based on fossils that are probably either non-hominid (Wu, 2000) or not dated properly. No *Australopithecus* or any other early hominid other than *H. erectus* has been found in Asia.

Homo may have been the first hominid to be adapted to open, arid environments (Reed, 1997). Archaeological evidence shows that early *H. erectus* in Africa were leaving the bones of large prey on their sites. Several predictions can be made about some physiological or life-history parameters when early hominids took up a more predatory role (Shipman & Walker, 1989). Increase in geographic range is one such prediction and the fossil record of the distribution of *H. erectus* confirms it. Before about 2.0 Ma hominids were confined to the African continent.

No one is certain from which species of *Australopithecus* the genus *Homo* came, although *A. africanus* has been a favorite (e.g., Tobias, 1991), and is sometimes itself placed in the genus *Homo* (e.g., Robinson, 1972). The new species, *A. garhi*

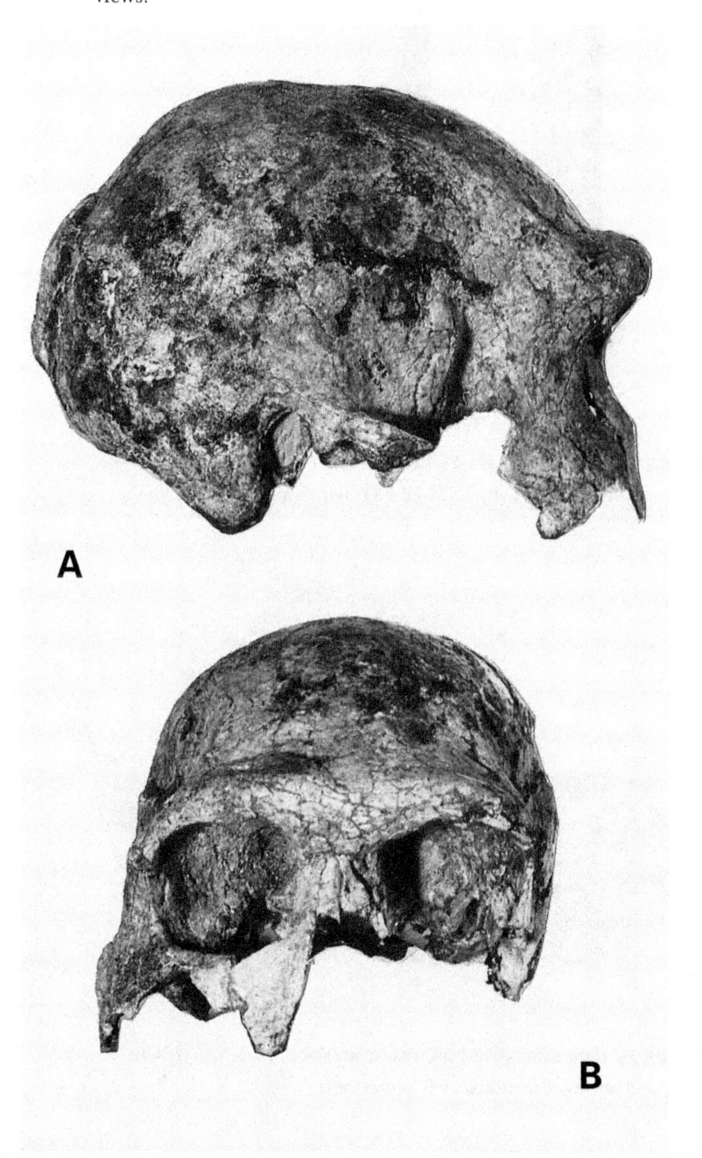

Fig. 25.15 *Homo erectus*. KNM-ER 3883, from Koobi Fora, Kenya: partial cranium, in (A) right lateral and (B) facial views.

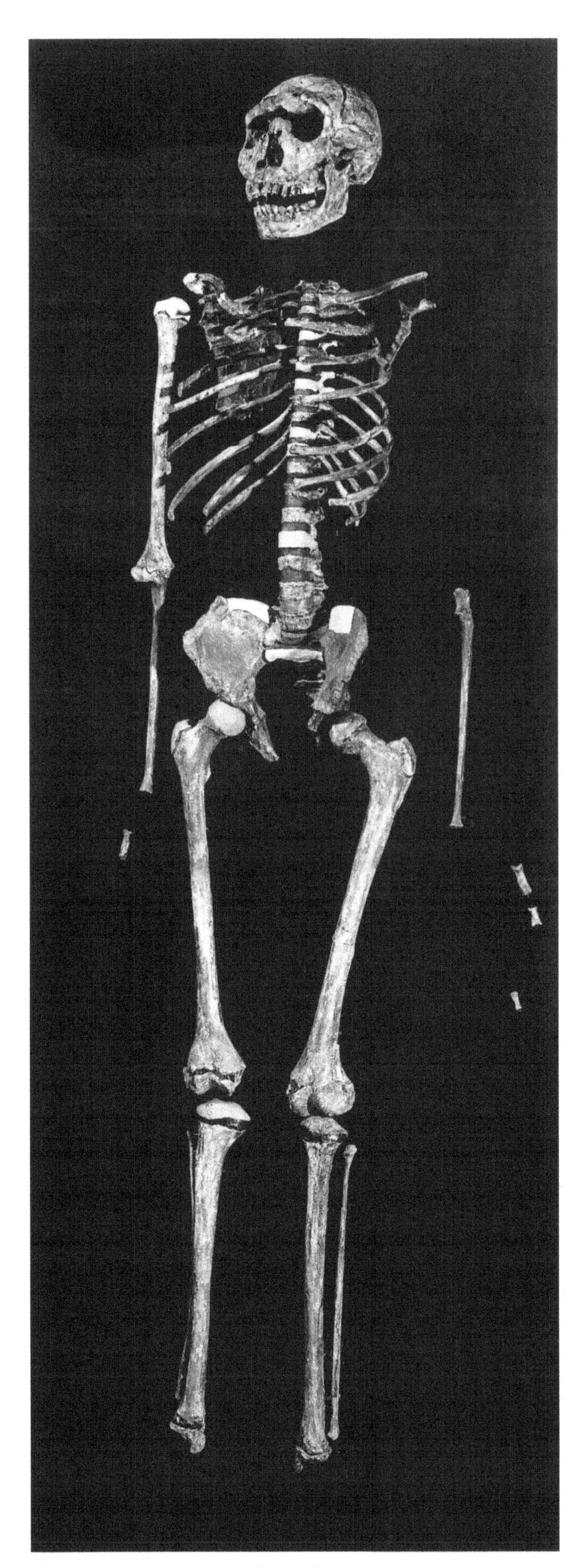

Fig. 25.16 *Homo erectus*. KNM-WT 15000, from Nariokotome, West Lake Turkana, Kenya: partial skeleton.

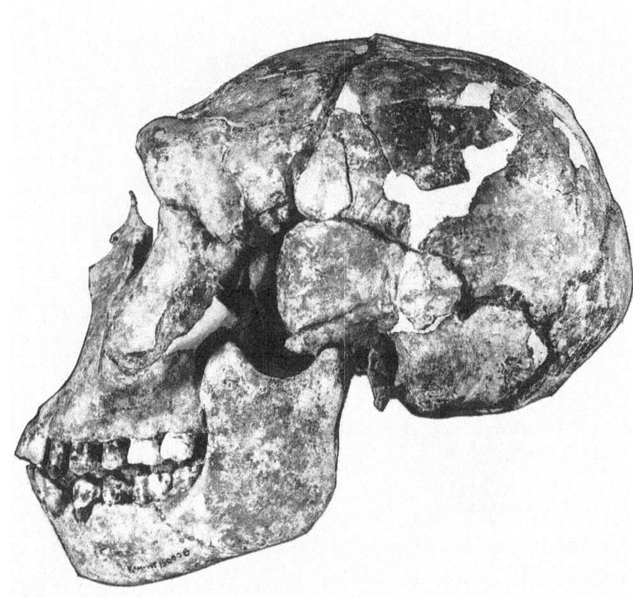

Fig. 25.17 *Homo erectus*. KNM-WT 15000, from Nariokotome, West Lake Turkana, Kenya: skull, left lateral view.

Asfaw *et al.*, 1999, is a possible candidate for the apical ancestor of *Homo* (Asfaw *et al.*, 1999). The *Homo*-like tooth and palate shape of *A. garhi* combined with its large teeth, in the size range of robust *Australopithecus*, suggests the following possible scenario. After the splitting event that led to robust *Australopithecus* in East Africa, another lineage, including *A. garhi*, presumably under similar ecological conditions, adapted with a similar increase in tooth size. However, if that second lineage included early *Homo*, then dental reduction would have been concomitant with the development of stone tools and the ingestion of more animal food.

Phylogenetic analysis of the morphology and chronology of the fossils we actually have of early *Homo* has produced four viable scenarios of the relationships of species in the genus. (1) The original unilinear phylogeny, which is still popular today, has *H. habilis* leading to *H. erectus*, which then leads to *H. sapiens* (e.g., Tobias, 1973). (2) Including *H. ergaster* slightly changes the phylogeny where *H. habilis* leads to *H. ergaster*, followed by a splitting event separating *H. erectus sensu stricto* from the direct ancestry of *H. sapiens* (e.g., Wood, 1994). The addition of *H. rudolfensis* to the hominid family tree creates further possibilities. (3) One scenario places *H. rudolfensis* and *H. habilis* together as a united sister group to *H. erectus/ergaster*. (4) Another places *H. rudolfensis* as the sister taxon to *H. erectus/ergaster* with *H. habilis* as the sister taxon to all three (see Wood & Collard (1999) for the various permutations).

Today paleoanthropologists commonly subscribe to one of three views concerning the evolution of *H. erectus*. The first states that the *H. erectus* hypodigm represents a single, widespread species that originated in Africa and dispersed throughout the Old World in the early Pleistocene and that

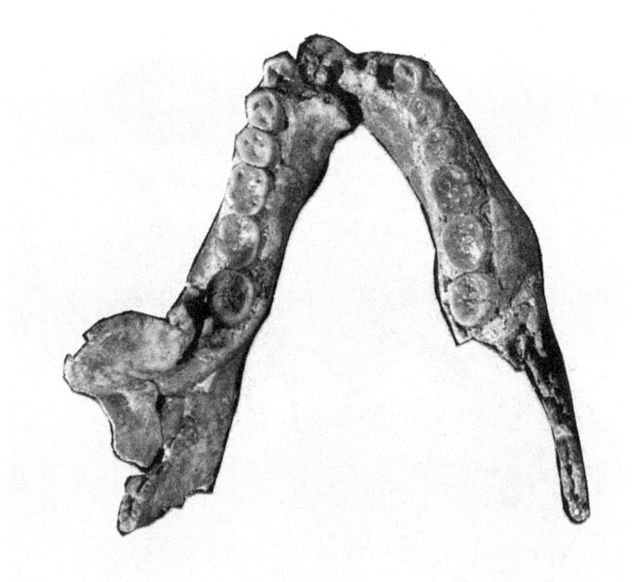

A

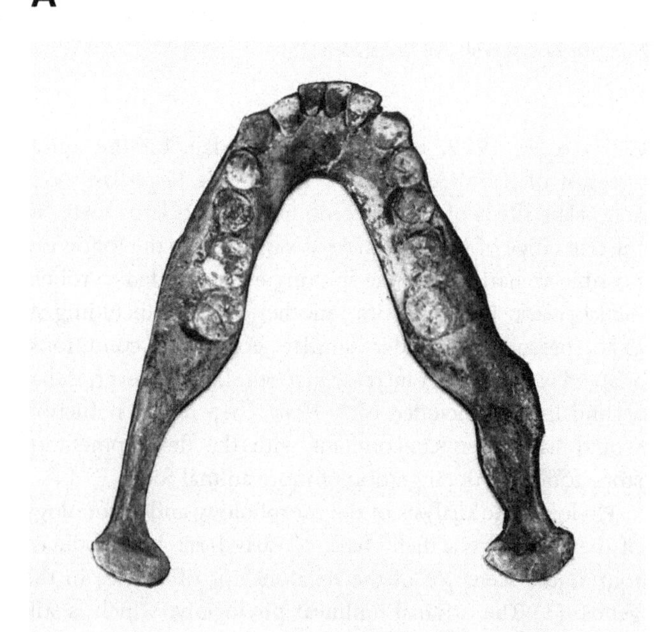

B

Fig. 25.18 *Homo erectus*. Lake Turkana, Kenya: (A) KNM-ER 992, mandible; occlusal view; (B) KNM-WT 15000, mandible, occlusal view.

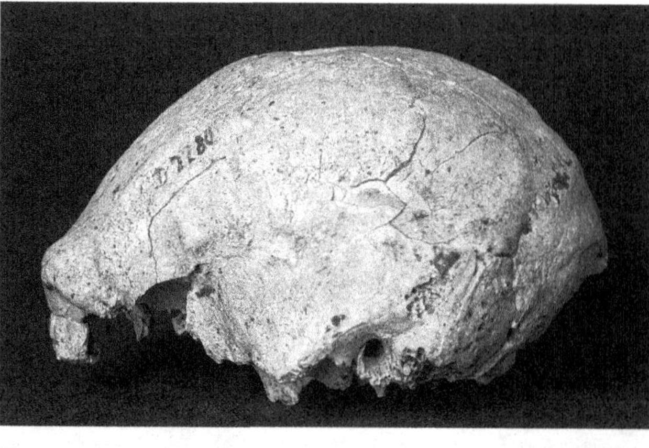

Fig. 25.19 *Homo erectus*. D2280, from Dmanisi, Georgia: calvaria, in left lateral view. Photograph courtesy of David Lordkipanidze.

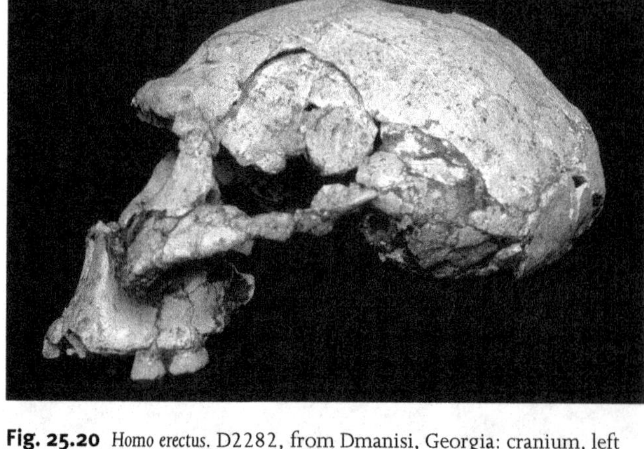

Fig. 25.20 *Homo erectus*. D2282, from Dmanisi, Georgia: cranium, left lateral view. Photograph courtesy of David Lordkipanidze.

one population of it was ancestral to modern humans. The second splits the fossil assemblage into several species with the Far East group (*H. erectus sensu stricto*) having no significant role in modern human ancestry. The third view subsumes the entire hypodigm under *H. sapiens* (e.g., Wolpoff *et al.*, 1994).

The controversy surrounding the origins of *Homo* may not be completely scientific in nature. Perhaps the problem lies in the elusive definition of what it means to be human. Clearly we need to collect more fossil representatives of early *Homo* in order to make better scientific sense of what little we have. We need complete or nearly complete adult skeletons of *Homo* and *Australopithecus* of around 2.5 Ma from east Africa. Augmentation of the fossil record is absolutely necessary if we are to proceed in any meaningful way in understanding the origin and evolution of our own genus.

Acknowledgements

We thank Walter Hartwig for inviting us to contribute to this volume. We thank all those paleoanthropologists both past and present for finding fossils of early *Homo* and trying to make sense of what they mean. We thank the following for their help: Susan Antón, Leslea Hlusko, David Lordkipanidze, Matthew Pavlick, Erica Phillips, Pat Shipman, John deVos, Milford Wolpoff, Fern and Ryan Wood.

Primary References

Early *Homo*

Dubois, E. (1894). Pithecanthropus erectus, *ein menschenähnliche Übergangsform aus Java*. Batavia: Landesdruckerei.

Dubois, E. (1924). Figures of the calvarium and endocranial cast, a fragment of the mandible and three teeth of *Pithecanthropus erectus*. *Proceedings of the Koninklijke Nederlandse Akademie van Wetenschappen*, **27**, 459–464.

Black, D. (1927). On a lower molar hominid tooth from the Chou Kou Tien deposit. *Palaeontologica Sinica, Series D*, **7**, 1.

Black, D. (1933). The Croonian Lecture: On the discovery, morphology and environment of *Sinanthropus pekinensis*. *Philosophical Transactions of the Royal Society, Series B*, **223**, 57–120.

Koenigswald, G. H. R. von (1936). Erste Mitteilungen über einen fossilen Hominiden aus dem Altpleistocän Ostjavas. *Proceedings of the Koninklijke Nederlandse Akademie van Wetenschappen*, **39**, 1000–1009.

Weidenreich, F. (1943). The skull of *Sinanthropus pekinensis*: a comparative study on a primitive hominid skull. *Palaeontologia Sinica, New Series D*, **10**, 1–484.

Weidenreich, F. (1944). Giant early man from Java and South China. *Science*, **99**, 479–482.

Broom, R. & Robinson, J. T. (1949). A new type of fossil man. *Nature*, **164**, 322–323.

Arambourg, C. (1954). L'hominien fossile de Ternifine (Algérie). *Comptes rendus de l'Académie des Sciences de Paris*, **239**, 893–895.

Arambourg, C. (1955). A recent discovery in human paleontology: *Atlanthropus* of Ternifine (Algeria). *American Journal of Physical Anthropology*, **13**, 191–201.

Arambourg, C. & Biberson, P. (1956). The fossil human remains from the Paleolithic site of Sidi Abderrahman (Morocco). *American Journal of Physical Anthropology*, **14**, 467–490.

Leakey, L. S. B. (1961). New finds at Olduvai Gorge. *Nature*, **189**, 649–650.

Heberer, G. (1963). Über einen neuen archanthropinen Typus aus der Oldoway-Schlucht. *Zeitschrift für Morphologie und Anthropologie*, **53**, 171–177.

Leakey, L. S. B. & Leakey, M. D. (1964). Recent discoveries of fossil hominids in Tanganyika: at Olduvai and near Lake Natron. *Nature*, **202**, 5–7.

Leakey, L. S. B., Tobias, P. V., & Napier, J. R. (1964). A new species of the genus *Homo* from Olduvai Gorge. *Nature*, **202**, 7–9.

Clarke, R. J., Howell, F. C., & Brain, C. K. (1970). More evidence of an advanced hominid at Swartkrans. *Nature*, **225**, 1219–1222.

Day, M. H. (1971). Postcranial remains of *Homo erectus* from Bed IV, Olduvai Gorge, Tanzania. *Nature*, **232**, 383–387.

Leakey, M. D. (1971). Discovery of postcranial remains of *Homo erectus* and associated artefacts in Bed IV at Olduvai Gorge, Tanzania. *Nature*, **232**, 380–383.

Leakey, M. D., Clarke, R.J., & Leakey, L. S. B. (1971). New hominid skull from Bed I, Olduvai Gorge, Tanzania. *Nature*, **232**, 308–312.

Sartono, S. (1971). Observations on a new skull of *Pithecanthropus erectus* (*Pithecanthropus VIII*) from Sangiran, Central Java. *Proceedings of the Academy of Sciences, Amsterdam* (B), **74**, 85–194.

Sartono, S. (1972). Discovery of another hominid skull at Sangiran, Central Java. *Current Anthropology*, **13**(1), 124–125.

Leakey, R. E. F. (1973*a*). Further evidence of Lower Pleistocene hominids from East Rudolf, North Kenya, 1972. *Nature*, **242**, 170–173.

Leakey, R. E. F. (1973*b*). Evidence for an advanced Plio-Pleistocene hominid from East Rudolf, Kenya. *Nature*, **242**, 447–450.

Leakey, R. E. F. & Wood, B. A. (1973). New evidence of the genus *Homo* from East Rudolf, Kenya. II. *American Journal of Physical Anthropology*, **39**, 355–368.

Day, M. H., & Leakey, R. E. F. (1974). New evidence of the genus *Homo* from East Rudolf, Kenya. III. *American Journal of Physical Anthropology*, **41**, 367–380.

Leakey, R. E. F. (1974). Further evidence of Lower Pleistocene hominids from East Rudolf, North Kenya, 1973. *Nature*, **248**, 653–656.

Day, M. H., Leakey, R. E. F., Walker, A. C., & Wood, B. A. (1975). New hominids from East Rudolf, Kenya. I. *American Journal of Physical Anthropology*, **42**, 461–476.

Groves, C. P. & Mazak, V. (1975). An approach to the taxonomy for the Hominidae: gracile Villafranchian hominids of Africa. *Casopis pro Mineralogii a Geologii*, **20**, 225–246.

Leakey, R. E. F. (1976). New hominid fossils from the Koobi Fora formation in Northern Kenya. *Nature*, **261**, 574–576.

Hughes, A. R. & Tobias, P. V. (1977). A fossil skull probably of the genus *Homo* from Sterkfontein, Transvaal. *Nature*, **265**, 310–312.

Brown F., Harris, J., Leakey, R., & Walker, A. C. (1985). Early *Homo erectus* skeleton from west Lake Turkana, Kenya. *Nature*, **316**, 788–792.

Alexeev, V. P. (1986). *The Origin of the Human Race*. Moscow: Progress Publishers.

Johanson, D. C., Masao, F. T., Eck, G. G., White, T. D., Walter, R. C., Kimbel, W. H., Asfaw, B., Manega, P., Ndessokia, P., & Suwa, G. (1987). New partial skeleton of *Homo habilis* from Olduvai Gorge, Tanzania. *Nature*, **327**, 205–209.

Leakey, R. E. F., Walker, A. C., Ward, C. V., & Grausz, H. M. (1989). A partial skeleton of a gracile hominid from the Upper Burgi Member of the Koobi Fora Formation, East Lake Turkana, Kenya. In *Hominidae: Proceedings of the 2nd International Congress of Human Paleontology, Turin, 1987*, ed. G. Giacobini, pp. 209–215. Milan: Jaca.

Gabunia, L. K., Justus, A., & Vekua, A. (1991). Der menschliche Unterkiefer. In *Dmanisi: Die Menschen der Altsteinzeit in Südgeorgient*, ed. Akademie der Wissenschaften Georgiens. Tblisi: Zentrum für Árchäologische Forschungen.

Wood, B. A. (1992). Origin and evolution of the genus *Homo*. *Nature*, **355**, 783–790.

Schrenk, R., Bromage, T. G., Betzler, C. G., Ring, U., & Juwayeyi, Y. M. (1993). Oldest *Homo* and Pliocene biogeography of the Malawi Rift. *Nature*, **365**, 833–835.

Tyler, D. E. & Sartono, S. (1994). A new *Homo erectus* skull from Sangiran, Java. *American Journal of Physical Anthropology*, Supplement **18**, 236.

Ferguson, W. W. (1995). A new species of the genus *Homo* (Primates: Hominidae) from the Plio-Pleistocene of Koobi Fora, in Kenya. *Primates*, **36**, 69–89.

Gabunia, L. K. & Vekua, A., (1995). A Plio-Pleistocene hominid from Dmanisi, East Georgia, Caucasus. *Nature*, **373**, 509–512.

Kimbel, W. H., Walter, R. C., Johanson, D. C., Aronson, J. L., Assefa, Z., Eck, G. G., Hovers, E., Marean, C. W., Rak, Y., Reed, K. E., Vondra, C., Yemane, T., York, D., Chen, Y., Evensen, N. M., & Smith, P. E. (1996). Late Pliocene *Homo* and Oldowan tools from the Hadar Formation (Kada Hadar Member), Ethiopia. *Journal of Human Evolution*, **31**, 549–561.

Gabunia, L. K., Vekua, A., Lordkipanidze, D., Swisher, C., Ferring, R., Justus, A., Nioradze, M., Tvalchrelidze, M., Antón, S., Bosinski, G., Jöris, O., De Lumley, M., Majsuradze, G., & Mouskhelishvili, A. (2000). Pleistocene hominid cranial remains from Dmanisi, Republic of Georgia: taxonomy, geological setting, and age. *Science*, **288**, 1019–1025.

26 | Migrations, radiations and continuity: Patterns in the evolution of Middle and Late Pleistocene humans

FRED H. SMITH

The biological history of the Primate Order is characterized by a series of ecologically diverse, multicontinental radiations, which have been detailed in the previous chapters of this volume. At their peaks such radiations extensively impacted the biogeography and biodiversity of the post-Paleocene Cenozoic. While products of all of these radiations continue to exist today, most are represented by only a few taxa, exhibiting relatively restricted ecological and/or geographic distributions. There is, however one exception to this pattern, at least in part – a single primate genus that covers virtually the entire Earth. This genus is, of course, our own – the genus *Homo*.

Extant human beings, all members of the taxon *Homo sapiens*, have managed to spread to every continent and climatic zone and to exploit to some extent every ecological region of the globe. While *Homo* epitomizes a colonizing taxon, one key aspect of its radiation – diversity – has declined. Living humans exhibit relatively limited amounts of both genetic and morphological diversity, certainly far less morphological diversity than existed throughout much of the Pleistocene (Wolpoff, 1999), and intense debates focus on this phenomenon (e.g., Wolpoff & Caspari, 1997 vs. Stringer & McKie, 1997). This chapter considers the historical basis for current debates on the evolution and taxonomy of *Homo* as evidenced in the fossil record. It also considers the radiations of the genus, beginning with its original expansion out of Africa and ending with the appearance of modern human morphology throughout the Old World. Finally this chapter assesses current debates on late Pleistocene human phylogeny and the models through which it is interpreted. This chapter explicitly does not attempt to diagnose the genus *Homo* or to place particular fossils in particular species – entire shelves of books have been and continue to be devoted to these goals.

Historical framework to *c.* 1900

Prior to 1856, the year of discovery of the Neander Valley skeleton, several specimens recovered in Europe were claimed to represent ancient humans. Few of these received serious scientific attention, and many turned out to be intrusive burials into old strata or not even human (Spencer, 1984). One exception was the Canstatt cranium (Germany) found in 1700 associated with the remains of extinct mammals (see Obermaier, 1905). This specimen subsequently became the "prototype" of the "Cannstatt race" (Quatrefages & Hamy, 1882), the presumed ancestor of modern Europeans. Another exception was the Paviland 1 skeleton, a burial covered in red ochre associated with a mammoth skull and bone and lithic tools, excavated by William Buckland from a cave in Wales in 1822–3. As both of these specimens were recovered at a time when the catastrophist paradigm denied the antiquity of humans, the geological age of both these specimens was vigorously attacked. For example, Buckland (1823), a staunch catastrophist, argued that the "Red Lady of Paviland" dated to the Roman occupation of Britain.

It was not until limestone quarrying in the Neander Valley exposed a partial skeleton that the possibility of ancient, pre-modern humans became a subject of serious scientific scrutiny. The remains (Fig. 26.1) were ultimately entrusted to Hermann Schaaffhausen, professor of anatomy at Bonn. Schaaffhausen (1858), a pre-Darwinian evolutionist, described the specimen and argued that it represented a normal individual of a "primitive race" of early Europeans. He pointed out the continuous supraorbital torus, receding frontal and projecting upper facial region of the specimen and noted that many features were relatively ape-like. Schaaffhausen argued further that the skull and other bones were ancient and probably contemporaries of diluvial (Pleistocene) animals. Because he believed the Neandertal specimen was part of a continuum linking recent human populations to the geological past, Schaaffhausen did not recognize the Neandertaler as a distinct archaic human species.

During the latter half of the nineteenth century the most common scholarly perspective corresponded closely to interpretations of Neandertals and other "barbarous races" as a racial succession sequence leading to living Europeans. In most forms, the racial succession model minimized or avoided evolutionary explanations for this sequence, focusing instead on concepts of cultural and linguistic diffusionism (Bowler, 1986; Smith, 1997*a*). An excellent example is the aforementioned "Cannstatt race". Because Neandertals and more modern specimens (Cannstatt, Brüx, Eguisheim and others) were included in it, the Canstatt Race concept supported the perception that Neandertals were only one pole on an unbroken morphological continuum leading to "civilized" races.

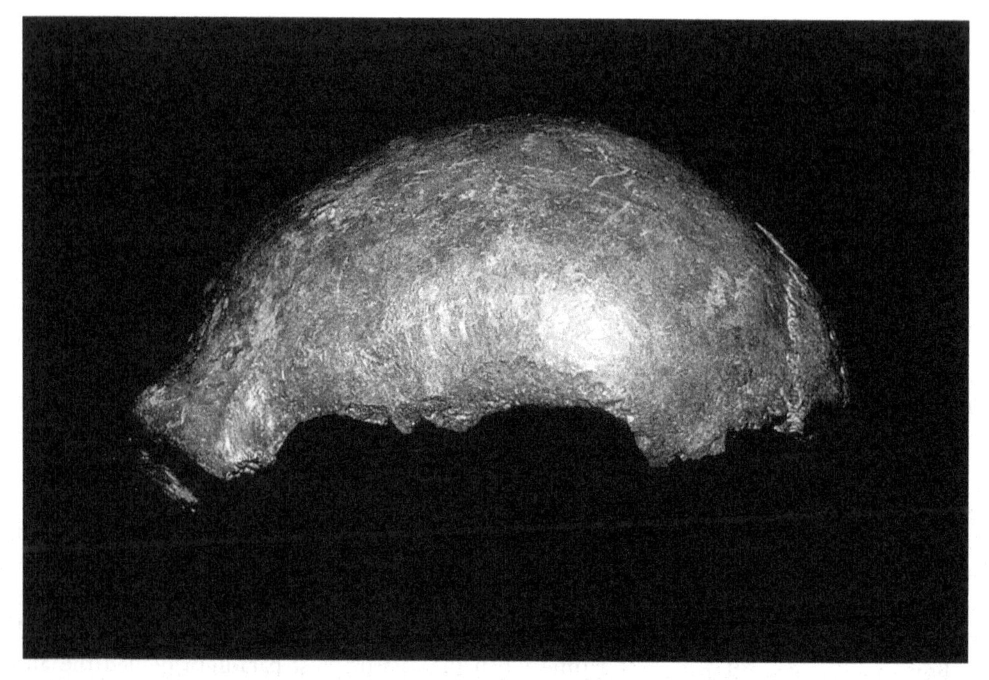

Fig. 26.1 The calotte from the original Neandertal specimen, discovered in 1856 in the Neander Valley near Düsseldorf, Germany. The projecting supraorbital torus and low, receding forehead played important roles in the early assessment of this specimen's taxonomic and phylogenetic position.

A second and strongly antievolutionary interpretation of Neandertals emerged in the 1870s. The noted anthropologist and pathologist Rudolf Virchow (1872) claimed that the original Neandertal had reached an advanced age (based on suture closure) and that this would only be possible in a sedentary, "civilized" race. He then argued that the unusual morphology of this specimen was due to a unique combination of disease and trauma (largely rickets and arthritis) in a recent human, not to membership in any ancient group or race.

A third view that emerged soon after the discovery and was first put forth by William King (1864) saw Neandertals as a distinct, extinct human "type". King emphasized that the specimen's primitive features lay outside the range of "primitive" recent human races in most respects and concluded that it resembled apes more closely than other studies recognized. To underscore its distinctions, King recommended classifying the Neandertal specimen at least in its own species – *Homo neanderthalensis* – and probably in its own genus (see also Brace, 1964; Spencer, 1984; Bowler, 1986).

While Virchow's pathological explanation was largely rejected in scientific circles by the early twentieth century, the other explanations have endured. Although modified in the light of accumulated knowledge, the specific question of whether Neandertals should be considered an extinct race or an extinct species of human still defines the modern debate on their evolutionary role (Smith, 1997*a*, 1998*b*).

A number of other important discoveries were made in Europe during the nineteenth century, including an 8–9-year-old child's calvarium from Engis, Belgium (Engis 2) and an adult cranium from Gibraltar (Gibraltar 1), that were actually discovered prior to the Neander Valley remains and subsequently recognized as Neandertals (Schwalbe, 1906; Sollas, 1908; Fraipont, 1936). Busk (1865) initially recognized the Forbes Quarry specimen as a very ancient skull, significantly different from *Homo sapiens* but not specifically as a Neandertal. Quatrefages & Hamy (1882) included it with the "Cannstatt Race", and its Neandertal affinities were detailed by Gustav Albert Schwalbe and William Johnson Sollas after 1900. In addition, two Neandertal mandibles were recovered from La Naulette (Belgium) and Šipka (now Czech Republic) in 1866 and 1880, respectively (Dupont, 1866; Maška, 1882). These specimens were associated securely with Mousterian tools and extinct fauna, but as fragmentary specimens they had little impact on the Neandertal debate.

In 1886, two relatively complete human skeletons were recovered from the cave of Spy (Belgium) in indisputable association with ancient fauna and Mousterian tools (Fig. 26.2). Carefully analyzed by Julien Fraipont and Maximin Lohest (Fraipont & Lohest, 1886), they were shown to be Neandertals in what constitutes the first systematic description of this group. The Spy site demonstrated that Neandertals made the Mousterian and derived from pre-modern times when different forms of animals inhabited Europe. Additionally, the Spy remains, along with the large Krapina sample (Croatia) discovered between 1899 and 1905 (Gorjanović-Kramberger, 1899, 1906), rendered Virchow's explanation untenable because they showed that Neandertal morphology was not a unique combination of pathologies (e.g., Schwalbe, 1901; Sollas, 1908; Keith, 1911).

The general consensus in the early 1900s viewed Neandertals as the direct, lineal ancestors of modern humans,

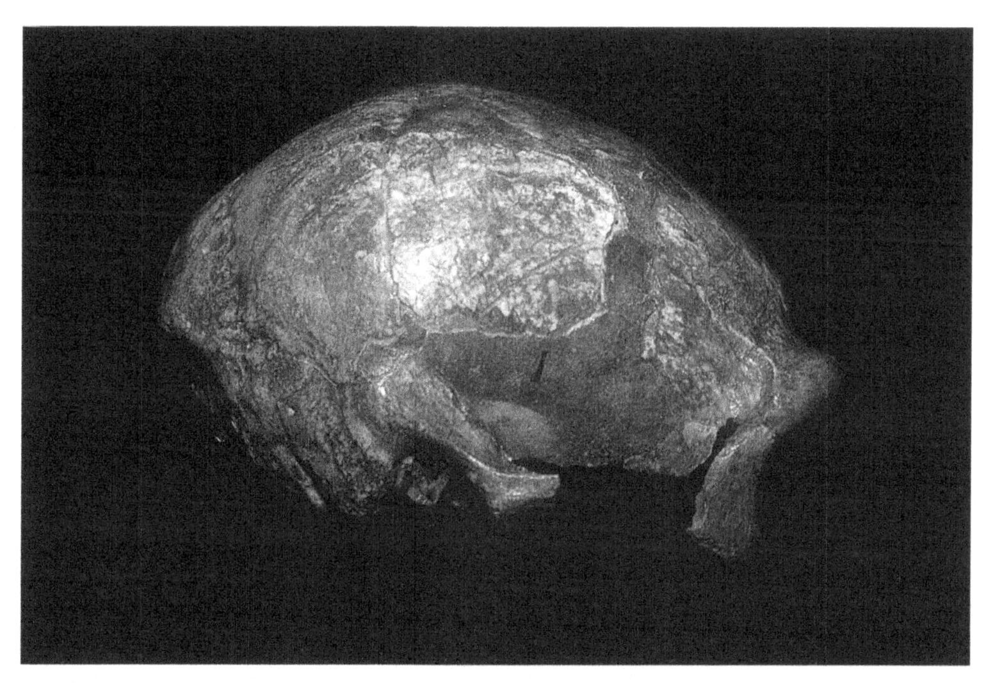

Fig. 26.2 The Spy 1 cranial vault from Belgium. In addition to the typically Neandertal form of the specimen, Spy 1 also exhibits a distinct occipital bun or "chignon".

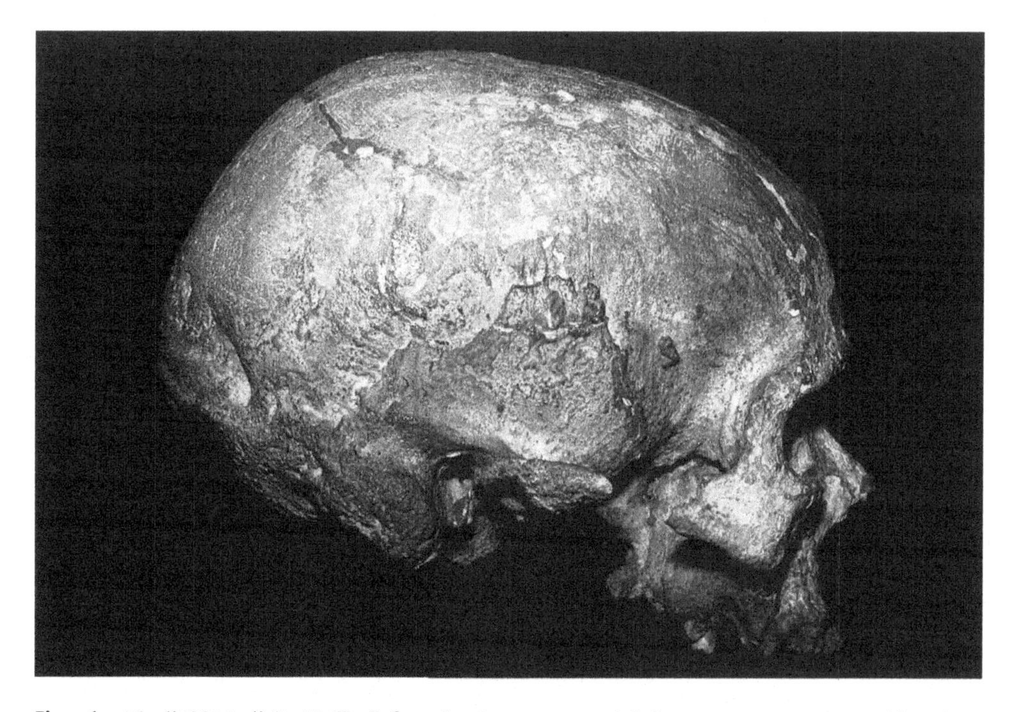

Fig. 26.3 The "Old Man" (Le Vieillard) from the Cro-Magnon rockshelter in Les Ezyies, France. Although often considered to represent the first early modern Europeans, Cro-Magnon 1 is actually not among the earliest representatives of this group. The high, rounded cranial vault, weak brow ridge development, and orthognathic (non-projecting) face are generally noted as key features distinguishing early modern Europeans from Neandertals.

fostered in part by the first discoveries of important human remains outside Europe. In 1890 and 1891, Eugène Dubois recovered fossil human remains from Kendung Brubus and Trinil in Java that were both earlier and more primitive (at least with respect to brain size) than Neandertals (Dubois, 1894). Considerable debate surrounded Dubois's *Pithecanthropus erectus* (see Chapter 25), but the remains provided a logical earlier stage in the evolution of humans which enhanced an intermediate position for Neandertals. Equally important is that these discoveries, along with the earlier ones at Wadjak (1889–90), demonstrated that human evolution was not restricted to Europe. Previously, consideration of the role of other continents in human evolution was largely theoretical (e.g., Haeckel, 1868).

The last half of the nineteenth century also witnessed the discovery of several specimens recognized as fundamentally modern but found in association with Paleolithic artifacts. Most influential were remains from the Abri de Cro-Magnon in France, excavated by Lartet in 1868 (Fig. 26.3). Lartet interpreted them as morphologically modern, with large brains, no protruding supraorbital tori, high vaults, and chins (Lartet, 1865–75). Initially, the Cro-Magnon specimens were interpreted in the light of the racial succession paradigm. Paul Broca (1869) considered them to show a mosaic of primitive and advanced features and concluded they represented the appearance of an early progressive race in Europe. Quatrefages & Hamy (1882) included them in the "Cannstatt Race". Several other specimens, for example Chancelade (1888) and Grotte des Enfants in France (Grimaldi – 1874, 1901) and Brünn in the Czech Republic (Brno 2 – 1891), were also initially interpreted within the race succession perspective. However by the beginning of the twentieth century, all of those fossils, as well as discoveries at Mladeč in the Czech Republic (1881) and Galley Hill in England (1884), were used to demonstrate the establishment of modern humans in Europe in a more evolutionary framework (Brace, 1964; Spencer, 1984; Bowler, 1986).

Changing evolutionary views: the early twentieth century

In the first decade of the twentieth century, Schwalbe (1904, 1906) argued that Neandertals exhibited a fundamentally different cranial form that necessitated their placement in a distinct species from *Homo sapiens*, such as *Homo primigenius* following Haeckel's (1874) terminology. He also made the same argument for *Pithecanthropus*, and its relationship to the Neandertals. Early unilinealists like Schwalbe and Dragutin Gorjanović–Kramberger (1906, 1918) viewed Neandertals as the logical ancestors of modern humans. However, two events early in the second decade of the century rapidly changed the predominant unilineal views. These were the publication of Marcelin Boule's monograph on the Neandertal skeleton from La Chapelle-aux-Saints (France) and the discovery of the Piltdown remains (England).

In 1908 and 1909, Neandertal skeletal remains were recovered from several significant sites in France: Le Moustier, La Chapelle-aux-Saints, La Quina and La Ferrassie (see Boule, 1921; Spencer, 1984). All but La Quina included partial skeletons and all received rapid preliminary assessments (e.g., Boule, 1908; Capitan & Peyrony, 1909; Klaatsch & Hauser, 1909; Martin, 1911), but La Chapelle-aux-Saints (Fig. 26.4) captured the detailed attention of Boule, who described and compared the specimen in a series of installments published between 1911 and 1913. Boule emphasized "simian-like" aspects of the postcranial skeleton, which he believed indicated a somewhat ape-like posture. He also noted the projecting supraorbital tours, receding forehead and flattened braincase. The latter feature, for Boule, reflected lack of development of the cerebral lobes indicating inferior mental abilities, despite the large cranial capacity (1620 cm^3). Boule concluded that Neandertals were unquestionably a separate species and placed them in King's taxon, *Homo neanderthalensis*. He also concluded they were far too primitive and ape-like to be ancestors of modern humans and relegated them to a side branch of human evolutionary history (Boule, 1911–13).

The 1912 announcement of the Piltdown discovery provided evidence that a lineage with a more modern human cranial form existed in Europe earlier than Neandertals. Piltdown was accepted widely (although not universally) as the root of a pre-*sapiens* lineage that bypassed the Neandertals and lead to modern humans (see Spencer, 1990). Keith (1915), Boule (1921) and others believed that such modern specimens as Galley Hill and Grotte des Enfants were contemporaries of Neandertals that represented the continuation of this lineage, and were the ancestors of the Cro-Magnon people (see Brace, 1964; Smith, 1997*b* for other historical reviews).

The major voice raised against this interpretation in the mid-twentieth century was that of Aleš Hrdlička. Hrdlička believed that early modern humans from Europe, many of which were former members of the "Cannstatt race", were intermediate morphologically between Neandertals and recent people and indicated an evolutionary succession (Hrdlička, 1927, 1930; Spencer & Smith, 1981). Despite Hrdlička's vigorous defense of a "Neanderthal phase of man", few scholars at this time viewed Neandertals as reasonable ancestors for modern people.

Several fossil specimens found in Europe during this span seemed to date earlier and exhibit somewhat different morphology than Neandertals. In 1907, workmen recovered a robust chinless mandible from Mauer, near Heidelberg (Germany) (Fig. 26.5). Schoetensack (1908) stressed the absence of a chin as indicating its primitive nature and designated it a new taxon: *Homo heidelbergensis*. The Steinheim cranium discovered in 1933 (Berckhemer, 1933) exhibited a relatively small cranial capacity (approximately 1100 cm^3), a low cranium and a well-developed supraorbital torus. Other features were suggested to be more modern than Neandertals. The same claims were made for the Swanscombe skull (Marston, 1936) found in England. Both specimens were assigned to their own individual species in the 1930s (see Table 26.1), and both were critical players in the arguments from the 1930s to the 1960s surrounding the existence of a pre-*sapiens* (non-Neandertal) lineage in Europe (Vallois, 1954; Howell, 1960; Wolpoff, 1999).

By the 1920s the perception that modern human evolution was a European phenomenon began to change. In 1921, workmen uncovered a massive skull (Fig. 26.6) and subsequently other bones at the Broken Hill mine near Kabwe, Northern Rhodesia (now Zambia) (Woodward, 1921). The cranium exhibited a suite of primitive features,

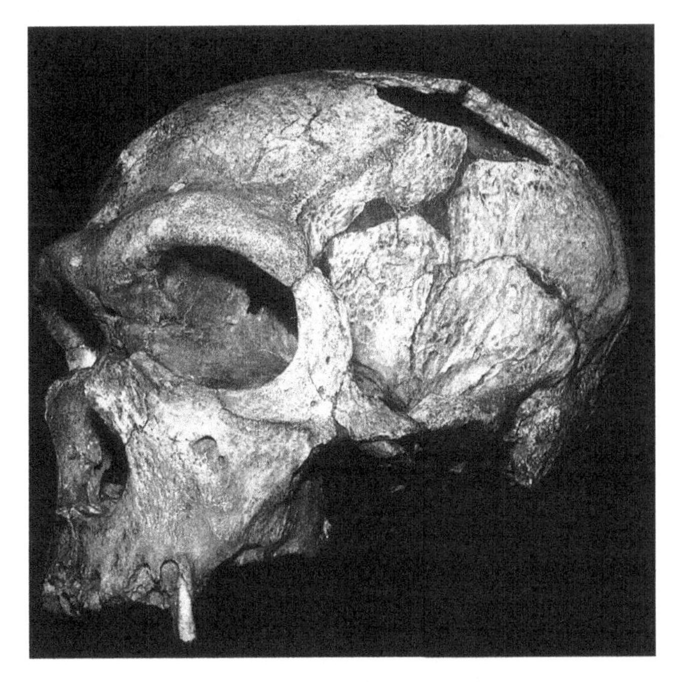

Fig. 26.4 The La Chapelle-aux-Saints cranium has become the typical example of a Neandertal skull. Note the supraorbital torus, large nasal aperture and oblique orientation of the inferior zygomaticoalveolar margin.

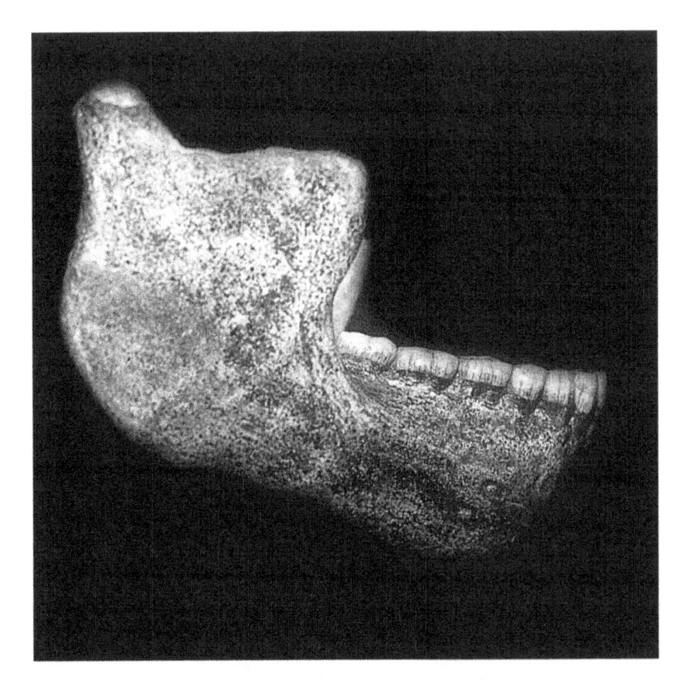

Fig. 26.5 The Mauer mandible, discovered in 1907, is the type specimen of *Homo heidelbergensis*. Note the receding mandibular symphysis and the broad mandibular ramus. There is no evidence of any chin structure or presence of a retromolar space.

including a large prognathic face with a tremendous supraorbital torus, a low cranial vault and a receding frontal squama. In many features "Rhodesian Man" compared favorably to the Neandertals, but in others the specimen seemed more primitive. Sir Arthur Smith Woodward placed the Broken Hill (Kabwe) remains in the taxon H. *rhodesiensis*, but it widely was considered an African equivalent of the Neandertal evolutionary stage (Hrdlička, 1930; Howells, 1959; Coon, 1962).

Additional discoveries in the Near East and Africa, such as the Galilee (Zuttiyeh) frontofacial specimen (described by Keith (1927)), supported the idea that Neandertals were geographically widespread. In 1932, Louis Leakey's East African Archaeological Research Expedition recovered cranial remains of five individuals at Kanjera and a mandibular symphysis fragment at Kanam, both in Kenya (Leakey, 1935, 1936). Leakey considered both to be Middle Pleistocene or earlier and indicative of a pre-*sapiens* lineage in Africa, separate from Neandertals, Broken Hill, and the remains known from Java and China. Leakey's perspective was the first cohesive argument for an initial evolution of modern people in Africa and it received support from initial interpretations of the Florisbad (1932) skull (cf. Howells, 1959).

From 1929 to 1934, excavations at the Mount Carmel caves of Skhūl and Tabūn provided a large fossil collection estimated to date to around 100 ka. Sir Arthur Keith and Theodore McCown (1937) initially identified two separate human "types", Neandertals from Tabūn (layer C and below) (Fig. 26.7) and more modern people from Skhūl (Fig. 26.8). However after completing their detailed analysis, they became convinced Skhūl/Tabūn was a single population, characterized by extensive variation (McCown & Keith, 1939). Although they were not certain of the exact meaning of the pattern, they interpreted the variation of the Mount Carmel people as reflecting a population in the "throes" of evolutionary change. Since the Mount Carmel people were thought to date earlier than European Neandertals, the line to modern humans could be viewed as emerging from a Neandertal-like ancestor outside Europe while "classic" Neandertals evolved in Europe (see Howell, 1951, 1957; and below). Excavations at Qafzeh in Israel (1933–5) supported the presence of Skhūl-like populations in the Near East at the same time, although the significance of Qafzeh only became apparent somewhat later (Howells, 1959).

Further discoveries of fossils now considered *Homo erectus* were made in Indonesia and in China during the 1920s and 1930s (see Chapter 25). In addition to the *Sinanthropus* remains, the site of Choukoutien (now Zhoukoudian) in north–central China yield three skulls and other fragments of modern humans from the Upper Cave deposits above locality 1 (Black, 1934). These specimens, along with Wadjak and Australian specimens like Talgai, Cohuna and others (Brown, 1997) demonstrated that modern humans were widespread in Asia by late Pleistocene times. On Java, a series of cranial vaults and tibial fragments were recovered at Ngandong, on a terrace of the Solo River between 1931 and 1933 (see von Koenigswald, 1958*a*). Geologically younger than the *Pithecanthropus* remains, the Ngandong sample exhibited many similar features but also a larger cranial capacity (average 1149 cm^3) and evidence of a braincase that is expanded

Table 26.1. Taxa (genera, species) created for selected Late and Middle Pleistocene human remains from 1864 to the present; for other Middle Pleistocene taxa see Dunsworth & Walker (this volume). Subspecies established within *Homo sapiens* (e.g., *Homo sapiens neanderthalensis* or *Homo sapiens soloensis*) are not included in this list

Taxon	Name/year[a]	Site[b]
Homo neanderthalensis	King, 1864	Neandertal, Germany
Homo primigenius	Haeckel, 1874	[Hypothetical]
Protoanthropus atavus	Haeckel, 1895	Neandertal, Germany
Homo europaeus primigenius	Wilser, 1898	Neandertal, Germany
Homo spelaeus	Lapouge, 1899	Cro-Magnon, Grimaldi Caves, France
Homo priscus	Lapouge, 1899	Laugerie-Basse, Chancelade, France
Homo niger	Wilser, 1903	Grotte des Enfants, France
Homo heidelbergensis	Schoetensack, 1908	Mauer, Germany
Homo grimaldi	Lapouge, 1905–6	Barma Grande, France
Homo transprimigenius mousteriensis	Forrer, 1908	Le Moustier, France
Homo antiquus	Adloff, 1908	Krapina, Croatia
Homo mousteriensis hauseri	Klaatsch & Hauser, 1909	Le Moustier, France
Homo sagensis	Krause, 1909	Spy, Belgium
Homo alpinus	Krause, 1909	Krapina, Croatia
Pseudohomo heidelbergensis	Ameghino, 1909	Mauer, Germany
Homo priscus	Krause, 1909	Spy, Belgium
Homo spyensis	Krause, 1909	Spy, Belgium
Homo aurignacensis hauseri	Klaatsch & Hauser, 1910	Combe Capelle, France
Palaeanthropus europaeus	Sergi, 1910	Neandertal, Germany
Homo breladensis	Marett, 1911	St. Brelade, England
Homo calpicus	Keith, 1911	Forbes Quarry, Gibraltar
Homo chapellensis	Buttel-Reepen, 1911	La Chapelle-aux-Saints, France
Palaeoanthropus krapinienis	Sergi, 1911	Krapina, Croatia
Notoanthropus eurafricanus recens	Sergi, 1911	Combe Capelle, Laugerie-Basse, Chancelade, France; Eguisheim, Germany
Notanthoropus eurafricanus archaius	Sergi, 1911	Předmostí, Czech Republic
Eoanthropus dawsoni	Dawson & Woodward, 1913	Piltdown, England
Homo fossilis proto-aethiopicus	Giuffrida-Ruggeri, 1915	Combe Capelle, France
Homo mediterraneus fossilis	Behm, 1915	Oberkassel, Germany
Protoanthropus heidelbergensis	Arldt, 1915	Mauer, Germany
Homo acheulensis moustieri	Wiegers, 1915	Le Moustier, France
Homo naulettensis	Baudouin, 1916	La Naulette, Belgium
Homo capensis	Broom, 1917	Boskop, South Africa
Archanthropus primigenius	Abel, 1920	La Chapelle-aux-Saints, France
Homo preědmostensis	Absolon, 1920	Prědmostí, Czech Republic
Homo rhodesiensis	Woodward, 1921	Broken Hill, Zambia
Homo wadjakensis	Dubois, 1921	Wadjak, Indonesia
Homo grimaldicus	Hilber, 1922	Grotte des Enfants, France
Homo gibraltarensis	Battaglia, 1924	Forbes Quarry, Gibraltar
Homo laterti	Pycraft, 1925	Cro-Magnon, France
Homo meridionalis protoaethiopicus	Giuffrida-Ruggeri, 1925	Combe Capelle, France
Anthropus neanderthalensis	Boyd-Dawkins, 1926	Neandertal, Germany
Cyphanthropus rhodesiensis	Pycraft, 1928	Broken Hill, Zambia
Homo primigenius africanus	Weidenreich, 1928	Broken Hill, Zambia
Homo ehringsdorfensis	Moller, 1928	Ehringsdorf, Germany
Homo australoideus africanus	Drennan, 1929	Cape Flats, South Africa
Homo galilensis	Joleaud, 1931	Zuttiyeh, Israel
Homo drennani	Kleinschmidt, 1931	Cape Flats, South Africa
Homo primigenius galilaeensis	Hennig, 1932	Zuttiyeh, Israel
Homo (Javanthropus) soloensis	Oppenoorth, 1932	Ngandong, Indonesia
Palaeanthropus palestinensis	Weidenreich, 1932	Skhūl, Zuttiyeh, Israel

Table 26.1. (*cont.*)

Taxon	Name/year[a]	Site[b]
Homo primigenius asiaticus	Weidenreich, 1932	Ngandong, Indonesia
Palaeanthropus palestinus	McCown & Keith, 1932	Mt. Carmel Caves, Israel
Praehomo heidelbergensis	Eikstedt, 1932	Mauer, Germany
Praehomo europaeus	Eikstedt, 1934	Mauer, Germany
Homo predmostensis	Matiegka, 1934	Předmostí, Czech Republic
Homo kanamensis	Leakey, 1935	Kanam, Kenya
Homo (Africanthropus) helmei	Dreyer, 1935	Florisbad, South Africa
Homo florisbadensis	Dreyer, 1935	Florisbad, South Africa
Palaeoanthropus njarasensis	Reck & Kohl-Larsen, 1936	Eyasi, Tanzania
Homo steinheimensis	Berckhemer, 1936	Steinheim, Germany
Homo murrensis	Weinert, 1936	Steinheim, Germany
Anthropus heidelbergensis	Weinert, 1937	Mauer, Germany
Africanthropus njarasensis	Weinert, 1938	Eyasi, Tanzania
Homo předmostí	Matiegka, 1938	Předmostí, Czech Republic
Homo marstoni	Patterson, 1940	Swanscombe, England
Homo ehringsdorfensis	Patterson, 1940	Ehringsdorf, Germany
Homo leakeyi	Paterson, 1940	Kanjera, Kenya
Homo kiik-kobiensis	Bontch-Osmolorski, 1941	Kiik Koba, Ukraine
Homo swanscombensis	Kennard, 1942	Swanscombe, England
Homo semiprimigenius palestinus	Montando, 1943	Qafzeh, Zuttiyeh, Skhūl, and Tabūn, Israel
Maueranthropus heidelbergensis	Montandon, 1943	Mauer, Germany
Homo sapiens proto-sapiens	Montandon, 1943	Swanscombe, England
Protanthropus tabunensis	Bonarelli, 1944	Tabūn, Israel
Nipponanthropus akasiensis	Hasebe, 1948	Nishiyagi, Japan
Homo saldanensis	Drennan, 1955	Hopefield, South Africa
Homo leakeyi	Heberer, 1963	Olduvai Gorge, Tanzania
Homo palaeohungaricus	Thoma, 1966	Vértesszöllös, Hungary
Homo antecessor	Bermúdez de Castro *et al.*, 1997	Atapuerca, Spain

[a]For complete references to authors of Latin names, see Campbell (1965).
[b]Countries given correspond to current political geography.

relative to the cranial base. For many scholars, the Ngandong sample represented a grade of human evolution in East Asia equivalent to the Neandertals of the west (von Koenigswald, 1958*a*), while other views placed them as late survivors of *Homo erectus* (Coon, 1962; Santa Luca, 1980).

One discovery made during the Second World War significantly impacted the debate on modern human origins. The Border Cave remains from South Africa (Border Cave 1, 2 and 3) are morphologically modern and were argued subsequently to be associated with an 80–100 ka Middle Stone Age context (Cooke *et al.*, 1945; Beaumont *et al.*, 1978; but see Sillen & Morris, 1996). They became important to the argument that modern humans were established in Africa at a very early date (Bräuer, 1984; Stringer & Andrews, 1988).

Emergence of modern views: discoveries and perspectives from the 1950s through the 1980s

The period just after the war's end ushered in the formulation of positions on Middle and Late Pleistocene human evolution that essentially characterize the current debates on this issue. Franz Weidenreich noticed strong morphological connections throughout the Middle and Late Pleistocene in both China and Australasia (Weidenreich, 1946, 1949). In China this continuity extended from the *Sinanthropus* remains through the "early" modern Upper Cave people, to modern east Asians – his "Mongoloid Group". He also noted morphological continuity from the Javan *Pithecanthropus* (including Ngandong), through the "early" modern Wadjak skulls, to the "Australian Group". Weidenreich's trellis model did not, however, maintain that modern humans arose independently in these areas. Rather he noted that variable levels of what is today termed gene flow (Fig. 26.9) constantly interconnected these regional lineages.

A variety of non-monocentric views followed from Weidenreich's perspectives. In the 1960s, both Carleton Stevens Coon and C. Loring Brace developed views with roots in the trellis model. Coon (1962) believed that regional lineages of *Homo erectus* evolved in "parallel" toward *Homo sapiens* in each region and that the root of major living races largely could be traced to *Homo erectus* in its region of origin. Coon based his ideas on a comprehensive assessment of the

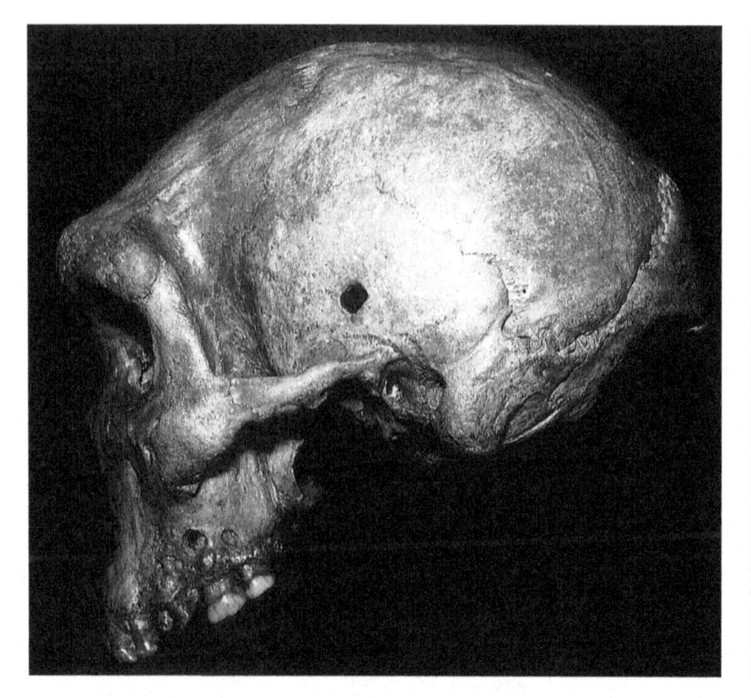

Fig. 26.6 The Kabwe 1 skull is often referred to as "Rhodesian Man". This specimen was the first post-*Homo erectus* archaic human recovered outside Europe. Found in 1921, the Kabwe material is often classified as *H. heidelbergensis* or *H. rhodesiensis* but may also be considered archaic *H. sapiens*.

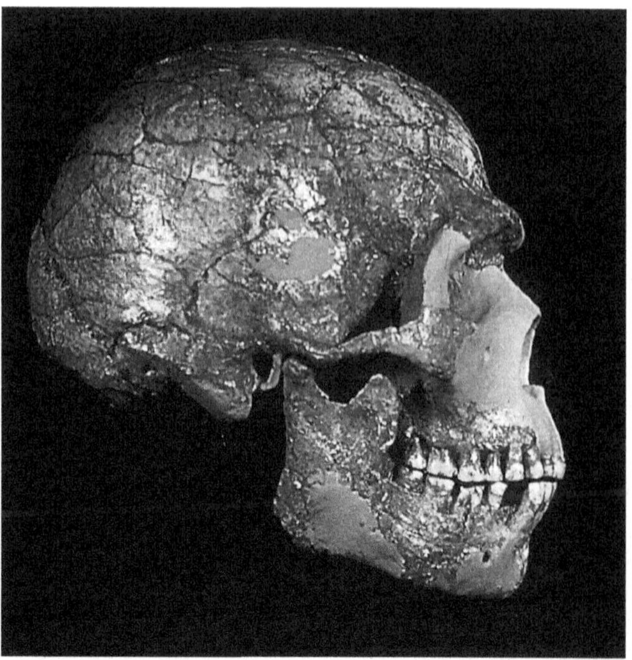

Fig. 26.8 The Skhūl 5 cranium exhibits a higher forehead and rounder cranial vault than is typical for Neandertals. Note the presence of a projecting chin and the maintenance of a supraorbital torus. Note also the absence of occipital bunning.

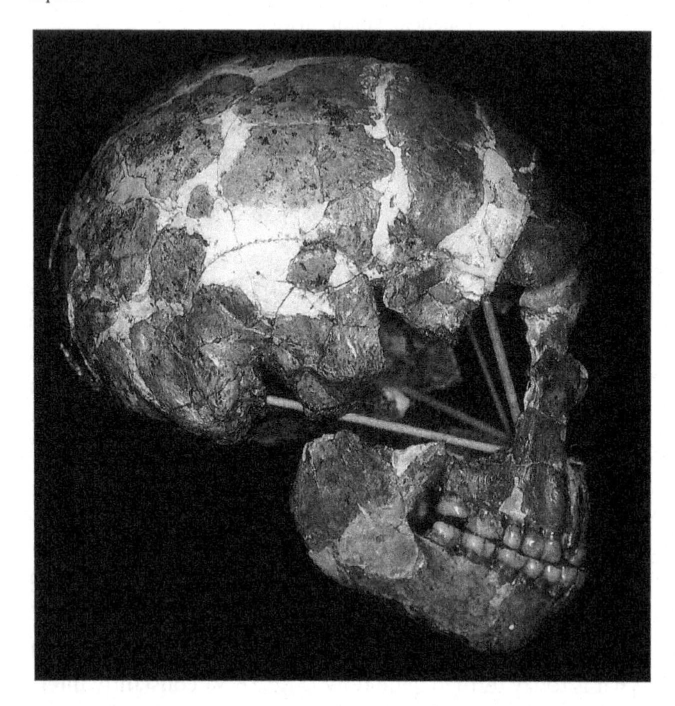

Fig. 26.7 The Tabūn C1 skull is considered to derive from a Near Eastern Neandertal. While most Neandertal features are present, note the absence of occipital bunning. The mandible exhibits a receding mandibular symphysis and a retromolar space.

known human fossil record and his own understanding of how a polytypic species should evolve. Brace, on the other hand, argued that race was an arbitrary construct in humans and thus could not be traced back in time. He emphasized the interconnection of human populations throughout the Pleistocene and argued that archaic forms of *Homo* evolved into modern ones due to relaxed selection on anatomy in the archaics (Brace, 1967). Thus similar modern human anatomical form throughout the Old World resulted from material culture sophistication that led to the relaxed selection (Brace, 1995).

The pre-*sapiens* model continued to be widely accepted from the 1940s until the 1960s. Despite the Piltdown hoax (Weiner, 1955) and other fossil context setbacks, the champion of pre-*sapiens* thinking during this period, Henri-Victor Vallois, simply substituted other specimens, particularly Swanscombe and Fontéchevade (Henri-Martin, 1957; Brace, 1964). Vallois (1949, 1958) argued that they were earlier than, or contemporary with, Neandertals but lacked most of their characteristic features, especially supraorbital tori and occipital buns. Although pre-*sapiens* status was implied for a few other specimens, even as late as 1966 by Andor Thoma for Vértesszöllös, the keys to the model's validity in the 1950s and 1960s were clearly Swanscombe and Fontéchevade.

During the 1950s through the 1970s, several fossil discoveries and analyses ultimately undermined the European pre-*sapiens* hypothesis. Studies by Sergio Sergi (1953), F. Clark Howell (1957), Thomas Dale Stewart (1964) and Joseph Weiner and Bernard Campbell (Weiner & Campbell, 1964) showed Swanscombe to have a primitive morphological pattern, with some features approaching the Neandertal condition. The same argument was made for

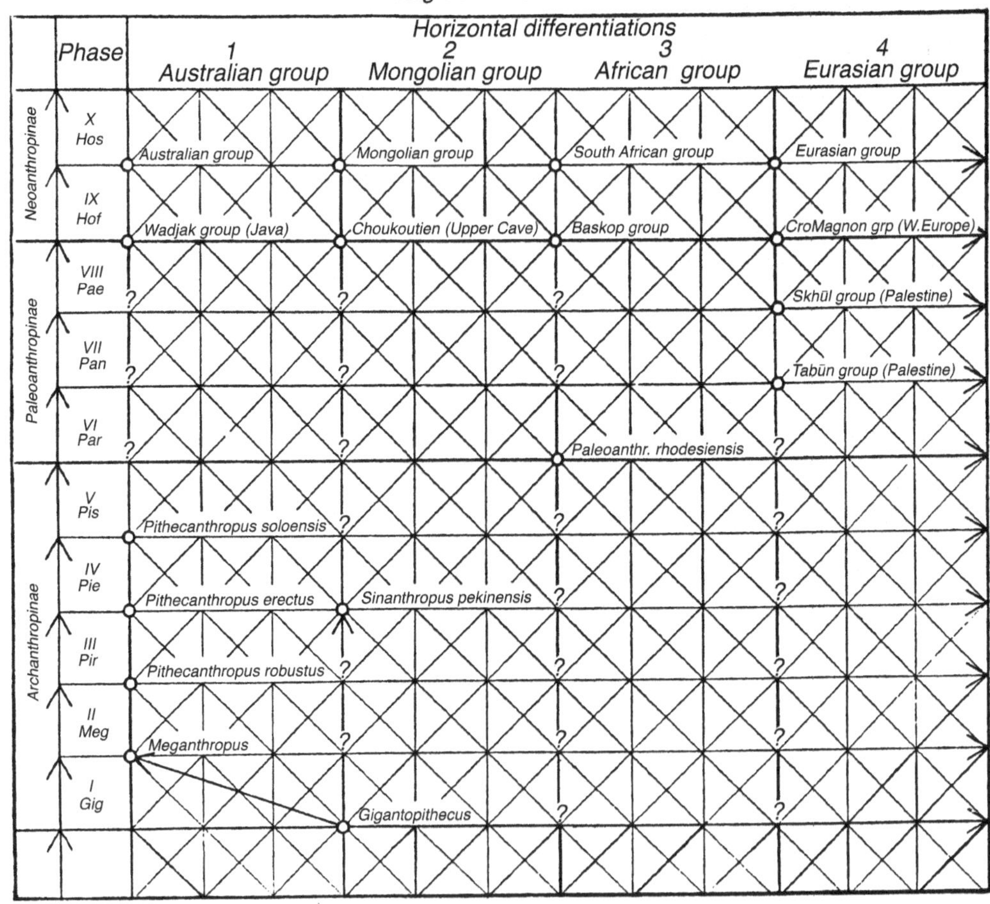

Fig. 26.9 Weidenreich's trellis model for human evolutionary history. Weidenreich's view emphasized continuity of humans over time in individual geographic regions (vertical lines) but also emphasized interconnections between these regional lineages (oblique and horizontal lines). This view explained the evolution of humans as a single polytypic species throughout the Middle and Late Pleistocene.

Fontéchevade (Brace, 1964; Trinkaus, 1973; Corruccini, 1975), and the pre-*sapiens* aspects of Vértesszöllös were strongly challenged (Brose & Wolpoff, 1971; Wolpoff, 1971c). Furthermore, discoveries of European "pre-Neandertal" remains from such sites as La Chaise (1949-60), Petralona (1959) and Arago (beginning 1964), demonstrated that Neandertal morphology was evolving slowly in Europe from at least 200 ka before the Riss–Würm Interglacial. Comprehensive analyses of the appropriate fossil record by Milford H. Wolpoff (1980) and Jean-Jacques Hublin (1982) demonstrated that all European remains prior to the Neandertals were a single lineage with a reasonable degree of variability. Moreover, recent analysis of the large Sima de los Huesos sample from Atapuerca (Spain), demonstrates considerable variation within a single sample of "pre-Neandertals" at around 300 ka (Arsuaga *et al.*, 1997; Rosas, 1997).

An alternative to pre-*sapiens* thinking emerged in the 1950s. Sergi (1953) argued that specimens like Fontéchevade, Swanscombe, Steinheim, Saccopastore and Ehringsdorf represented the common root for modern humans and Neandertals. At essentially the same time, Howell (1951) proposed that the same specimens, plus Krapina (Croatia) and Teshik Tash (Uzbekistan), represented a broadly distributed "progressive" Neandertal form which evolved into "classic" Neandertals in Europe but ultimately modern humans somewhere outside the continent. In 1957, Howell presented the now-popular view that early modern Europeans were derived from the Skhūl (and Qafzeh) sample.

Vallois (1958) designated the Sergi–Howell perspective as the pre-Neandertal hypothesis. This monocentric viewpoint, in the context of a pre-Neandertal model, is well represented in the work of William White Howells. In 1959, Howells strongly suggested that the center of modern human origins was somewhere in Asia. As modern humans spread from this hypothetical homeland, they likely hybridized with Neandertals and other archaic people, producing seemingly intermediate forms such as Skhūl. Howells continued to view Asia as the probable homeland for the origin of modern humans, although he was never specific about where in Asia this phenomenon took place.

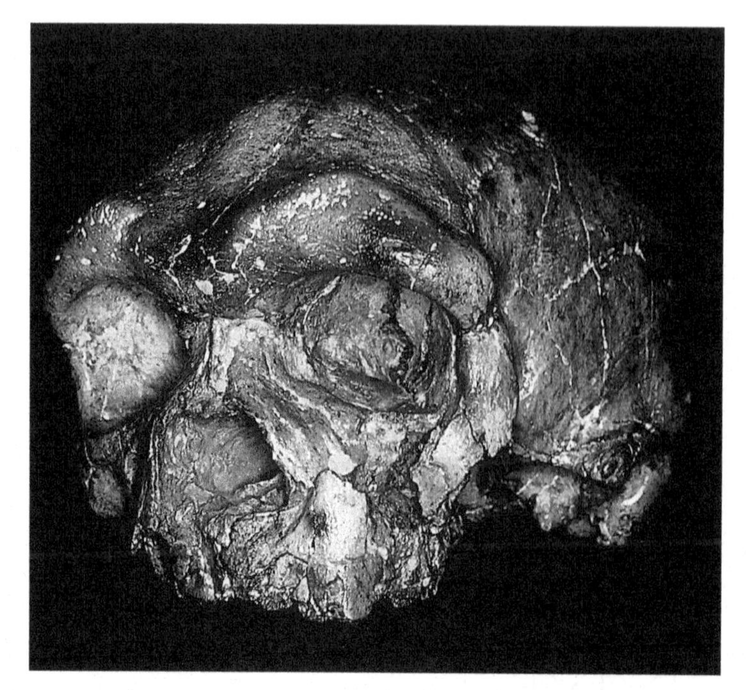

Fig. 26.10 Archaic (post-*Homo erectus* but pre-modern) human remains are relatively rare in East Asia. The Dali skull combines several primitive features (low forehead and cranial vault, well-developed supraorbital torus) with a shorter face, more angled zygomaticoalveolar margins and canine fossae.

Late Neandertal remains recovered in Europe in the early 1960s exhibited "transitional" features such as anterior dental reduction, an incipient mental eminence (Hortus, France) and a narrow nasal aperture and possible canine fossa (Kůlna, Czech Republic) (Jelínek, 1966; Smith, 1984; Wolpoff, 1999). Similar changes, in addition to reduced facial proportions and shape changes in the supraorbital tori, were also documented for the large, fragmentary series from Vindija (Croatia), excavated beginning in 1974 (Wolpoff *et al.*, 1981; Smith, 1984, 1994). A partial skeleton with clear Neandertal anatomy from St. Césaire (1979) and fragments of apparent Neandertals from Arcy-sur-Cure's Grotte du Renne (1950–8), both in France, were associated with the early Upper Paleolithic Chatelperronian culture (Leroi-Gourhan, 1958; Lévêque & Vandermeersch, 1981). These associations indicated a relatively late survival of Neandertals in Western Europe.

In western Asia during this time span, Ralph Solecki discovered a series of Neandertal skeletons from Shanidar Cave, Iraq (Solecki, 1963). The Shanidar skeletons demonstrated the expansion of Neandertals into western Asia and helped fill in the geographic gap between Levantine and Crimean Neandertals to the west and the Teshik Tash subadult Neandertal from Uzbekistan (Weidenreich, 1945*b*). Shanidar is one of the best sites to investigate patterns of Neandertal anatomy and variation, and the remains are the subject of one of the most thorough monographs available (Trinkaus, 1983). In the Crimea (Ukraine), the Staroselje 1 child was discovered in 1953 and considered to be a modern child, albeit with some archaic features, associated with the Mousterian (Ullrich, 1955). In addition, beginning in 1965, Bernard Vandermeersch continued excavations at Qafzeh and discovered additional modern humans, again with some archaic features, associated with the Mousterian (Vandermeersch, 1972, 1981*a*).

Farther to the east the most significant discovery at this time was the Mapa (Maba, China) calotte from early Late Pleistocene deposits. Woo & Peng (1959) interpreted it as intermediate between Chinese sinanthropines and Neandertals, although its only specific Neandertal-like feature is a high nasal angle (Wolpoff, 1999). However, Maba provided, for the first time, a fossil intermediate in age and possibly morphology between *Homo erectus* and Pleistocene modern humans in China. Further evidence would be found at Dali (1978) (Fig. 26.10) and Jinniushan (1984), which have been argued to demonstrate morphological continuity in China (Wu & Wu, 1985; see also Wolpoff *et al.*, 1984; Pope, 1992).

Notable discoveries also came from insular Southeast Asia and Australia. The claimed discovery of a > 40 ka adolescent modern skull at Niah Cave, Malaysia (Brothwell, 1960) provided one of the earliest dates for a modern human specimen then known; furthermore, discovery of the robust Kow Swamp remains in Australia (Thorne & Macumber, 1972) closed the morphological gap somewhat between modern Australia Aborigines and archaic populations like Ngandong. The Willandra Lakes hominids (see Webb, 1989) have also played an important role in perspectives on human evolution in Australia. WLH 50 has become a particularly important specimen for the controversy surrounding the nature of modern human origins in Australasia (see Stringer, 1998 vs. Hawks *et al.*, 2000; Adcock *et al.*, 2001; Relethford, 2001; Wolpoff *et al.*, 2001).

Historically, however, discoveries from Africa during this time have dominated thinking on modern human origins. Beginning with excavations in the Cave of Hearths in South Africa (1947), a series of presumably post-*Homo erectus* remains were recovered, including Hopefield in South Africa (1953), Ndutu in Tanzania (1973) and Bodo in Ethiopia (1976, 1983). Discoveries at Djebel Irhoud in Morocco (1961, 1963), Ngaloba in Tanzania (1976) and Eliya Springs in Kenya appeared to date later than the aforementioned early archaic Africans (see Rightmire, 1984). These, along with Florisbad in South Africa (Fig. 26.11), exhibited more modern-shaped faces and posterior cranial vaults than earlier Africans; and with dates generally falling between 100 ka and 200 ka, these fossils provided possible evidence of an early transition toward a modern human skull morphology (Bräuer, 1984). Similarly, dates potentially in excess of 80 ka were claimed for sites like Omo Kibish in Ethiopia (Day, 1969) and Border Cave and Klasies River Mouth in South Africa (Singer & Wymer, 1982), each of which provided fossils considered anatomically modern

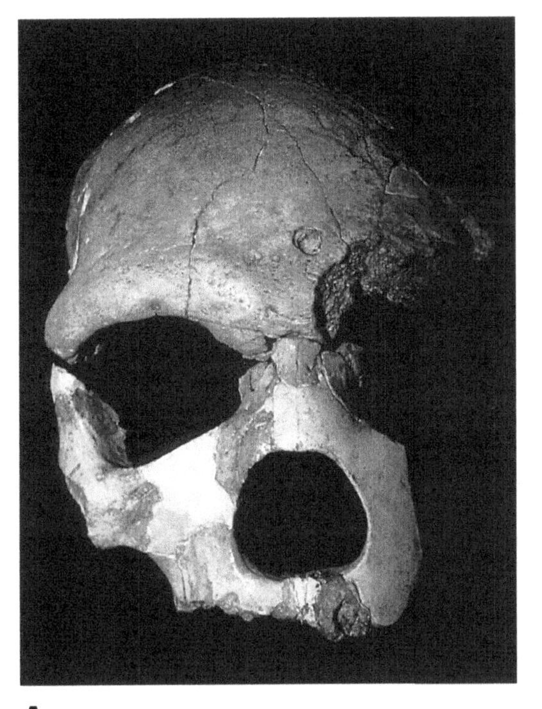
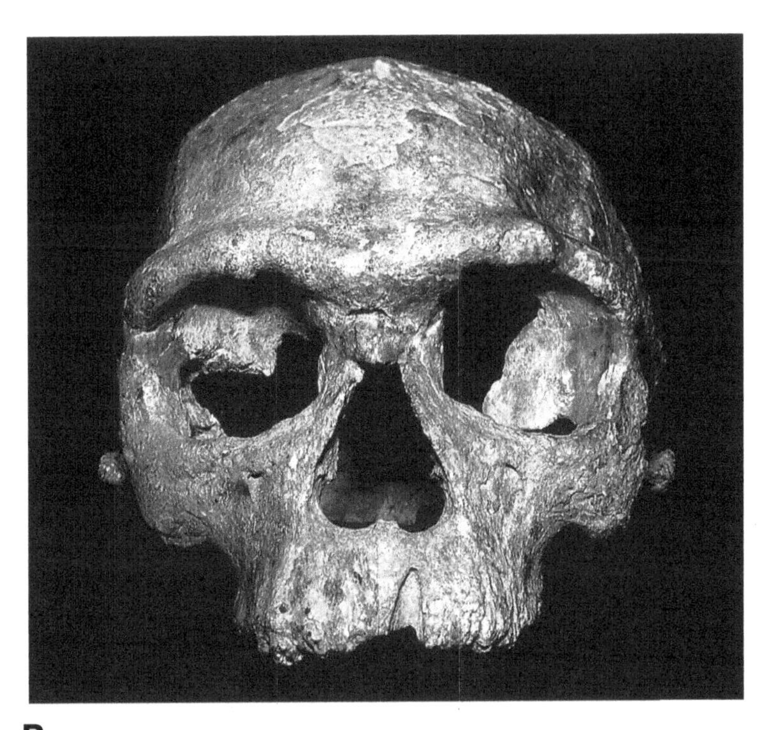

A **B**

Fig. 26.11 Facial views of (A) Florisbad (South Africa) and (B) Jebel Irhoud 1 (Morocco). Note the modern-like faces, exhibiting angled zygomaticoalveolar margins and canine fossae.

(Day & Stringer, 1982; Bräuer, 1984; Rightmire & Deacon, 1991).

In the last 10 to 15 years, relatively few fossil discoveries have impacted the debate on Middle and Late Pleistocene human evolution. In Asia, two crania from Yunxian are claimed to demonstrate continuity between *Homo erectus* and later Chinese hominids (Zhang, 1995), and the discovery of an archaic skull at Narmada, India (Sonakia, 1985) demonstrated late Middle Pleistocene humans in South Asia. In Europe, the Ceprano (Italy) calvaria is interpreted to exhibit distinctly *H. erectus* features, thus questioning the absence of this taxon in Europe (Ascenzi *et al.* 1996, 2000). The discoveries at Atapuerca and Dmanisi are discussed below. Finally, the 1998 excavation of the Lagar Velho child in Portugal has provided a specimen that is claimed to either demonstrate admixture between Neandertals and early modern humans (Duarte *et al.*, 1999) or simply a robust early modern with short limbs (Tattersall & Schwartz, 1999).

Finally, the impact of the synthetic theory of evolution to the study of early humans was somewhat formalized through the now-famous Cold Spring Harbor symposium in 1950, part of which considered the relationship of taxonomy and phylogeny in the human fossil record. As Table 26.1 illustrates, the taxonomy of fossil humans was largely an uninterpretable hodge-podge of names during most of the century. At the symposium McCown (1950) continued to suggest "superspecific" differentiation of Late Pleistocene humans, but the majority of papers favored a simpler taxonomy (Stewart, 1950; Mayr, 1950) in line with the implications of the synthetic theory. Wilton Marion Krogman (1950: 121) summarized the position that all Middle and Late Pleistocene humans should be classified either as *H. erectus* or *H. sapiens* (including Neandertals) but closed with the observation that still "there must be a certain amount of arbitrariness in any reconstructive taxonomic framework". The aftermath of the Cold Spring Harbor meeting was the tendency to simplify the taxonomic framework for human evolution during the following several decades. Thus, for most of the period from the 1950s through the 1980s, Middle and Late Pleistocene humans were generally classified into either *H. erectus* or *Homo sapiens*, but recent arguments (see below) have suggested a need to revise this practice.

Modern models in historical perspective

Numerous edited volumes over the last 20 years, from Smith & Spencer (1984) to Clark & Willermet (1997), review the issue of modern human origins and the implications of *Homo erectus* classification (e.g., Franzen, 1994*a*). Two major models have emerged since the late 1980s, known as the Multiregional Evolution (MRE) model (Wolpoff *et al.*, 1984) (Fig. 26.12) and the Recent African Evolution (RAE) model (Stringer & Andrews, 1988). Aiello (1993) has provided one of the few "independent" assessments of them, as well as two models that traditionally have received less attention – the African Hybridization and Replacement model (Bräuer, 1989) and the Assimilation model (Smith *et al.*, 1989).

The proximate history of RAE can be traced back to

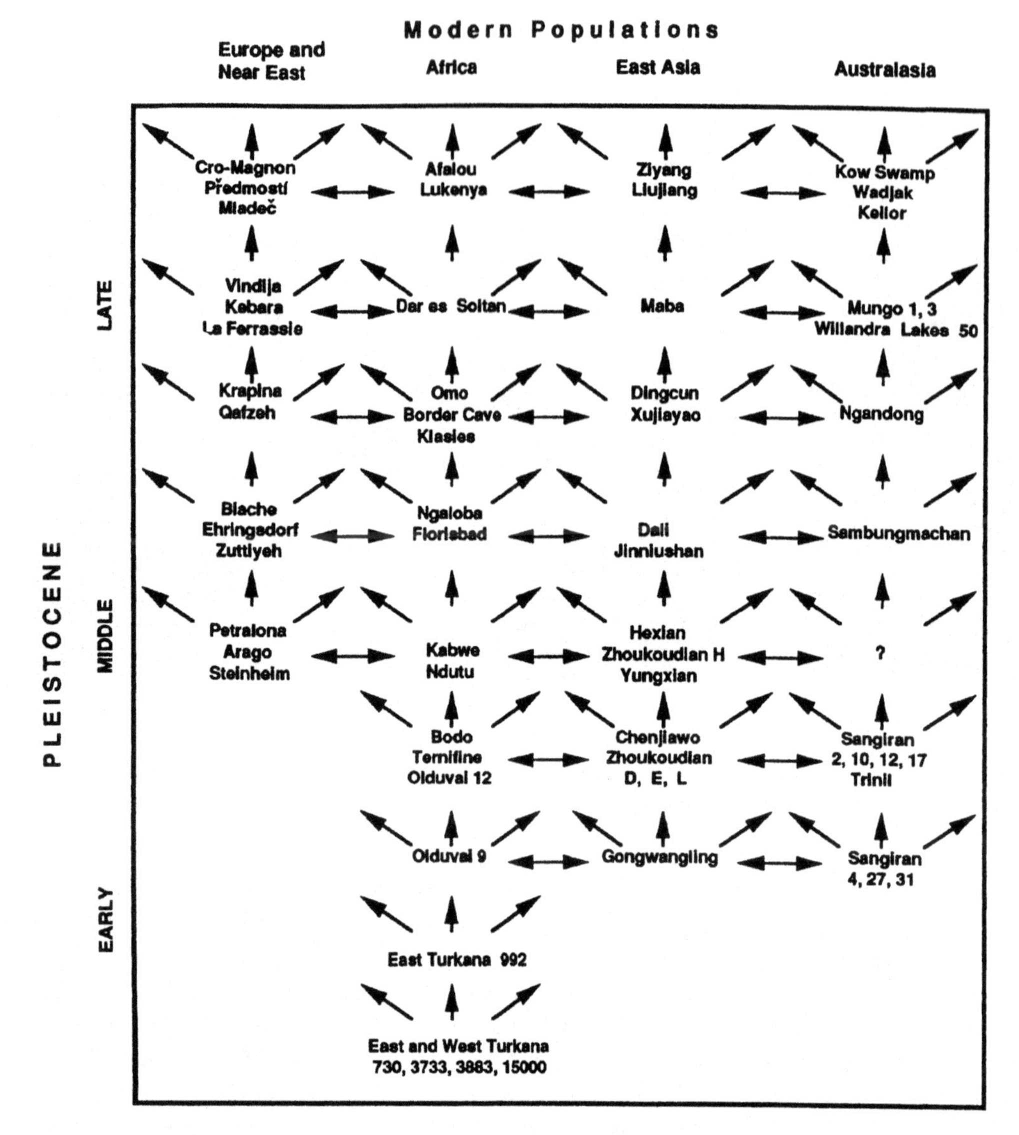

Fig. 26.12 A schematic representation of Multiregional Evolution (MRE). Note the broad similarity to Weidenreich's trellis model (Fig. 26.9), except for the addition of additional fossils discovered since Weidenreich's time. After Frayer *et al.* (1993).

Howells, who argued that modern humans had a single region of origin and that Neandertals could not represent their ancestry. The exclusion of Neandertals was based, in addition to earlier morphometric arguments, on initial multivariate cranial studies (Howells, 1974; Stringer, 1974) and the beginnings of a cladistic approach to defining Neandertal uniqueness (Santa Luca, 1978). These views ultimately formed the theoretical base of the RAE model.

Although Louis Leakey, Philip Rightmire (1976), and others had suggested Africa as the monocentric source for modern humans, Günter Bräuer's (1984, 1989) careful analyses of the pertinent African fossil record provided the most cogent argument. Bräuer based his Afro-European Sapiens hypothesis on evidence for an early transition to modern *H. sapiens* in Africa, that was, in turn, based on a series of new date estimations for critical African specimens (see Smith *et al.*, 1989). Like Howells, Howell and others before him, Bräuer (1989) accepted the possibility of limited hybridization with indigenous archaics, although he clearly viewed this to be of minor significance to the formation of modern Eurasian populations.

In addition to the fossil evidence from Africa and Eurasia, RAE drew heavily on mitochondrial DNA studies in living humans (e.g., Cann *et al.*, 1987). In addition to providing a time range of 90–180 ka for the last common ancestor of African and non-African mitochondrial (mt) DNA haplotypes, mt DNA data were interpreted to demonstrate a lack of interbreeding between expanding African modern and indigenous Eurasian archaic populations (Stoneking & Cann, 1989). Stringer & Andrews (1988) pointed out the apparent concurrence of the genetic data with the known fossil record and produced a RAE perspective (Fig. 26.13) that fundamentally supported a total replacement of all archaic Eurasian populations by a new species of human. In this context they

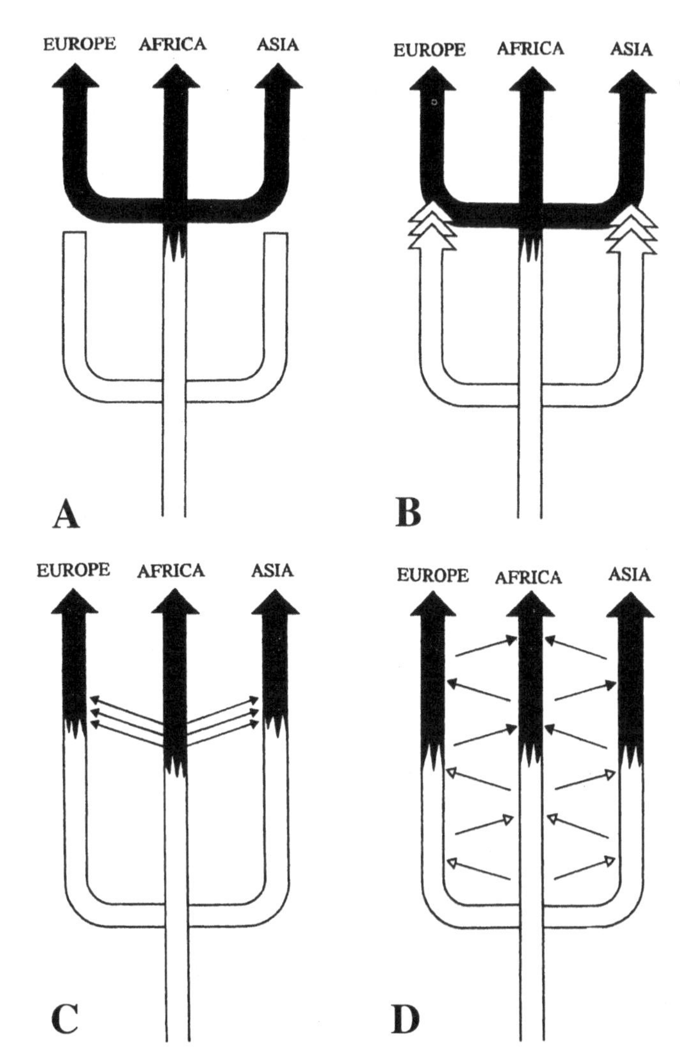

Fig. 26.13 Schematic representation of models for modern human origins. (A) The Recent African Origin model (RAE) emphasizing total replacement in Eurasia; (B) Bräuer's African Hybridization and Replacement model; (C) the Assimilation model; (D) the Multiregional Evolution (MRE) model. After Aiello (1993).

advocated restricting the taxon *H. sapiens* to specimens belonging to this new African species, which eventually spread throughout the entire world (see Stringer & McKie, 1997). The RAE model has received fundamental support in several subsequent morphological analyses (e.g., Rak ,1993; Waddle, 1994; Lahr, 1994, 1996; Bräuer & Stringer, 1997; Stringer *et al.*, 1997; Holliday, 2000) and genetic studies based on living humans (e.g., Cann *et al.*, 1994; Mountain, 1998; Sherry *et al.*, 1998; but see Spuhler, 1988; Excoffier, 1990; Templeton, 1993, 1999; Relethford, 1998, 1999).

Three other types of data significantly impacted the basis of this perspective. First, craniometric studies of modern human groups demonstrated the relative homogeneity of recent human cranial form and its distinctiveness *vis-à-vis* Neandertals (Howells, 1973, 1974; Lahr, 1996). Second, the application of a number of chronometric dating techniques, principally thermoluminescence and electron-spin resonance (ESR), to pertinent fossil human contexts (see Aitken *et al.*, 1993) produced a rather distinctive chronological pattern, at least in the western Old World. Prior to the later half of the 1980s, it was possible to present a strong case that modern human anatomical form appeared roughly at the same time (around 35 ka to 40 ka) throughout the Old World and that such a pattern matched well the expectations of MRE (Smith, 1985). But the emerging new chronology demonstrated a number of things (see reviews in Grün & Stringer, 1991 and Smith *et al.*, 1999): (1) an early appearance of near-modern morphology in the Near East and Africa; (2) a late survival of Neandertals and a late appearance of modern humans in Europe; and (3) an early date for transitional specimens between archaic and modern human morphology in Africa (Grün *et al.*, 1996). Third, ancient mt DNA from the original Feldhofer Grotte Neandertal provided a *c.* 380 base-pair sequence that fell on the extreme of the modern human range as judged by pair-wise comparisons (Krings *et al.*, 1997). This result, now supported by similar results on specimens from Mezmaiskaya in the Caucasus Mountains of Russia (Ovchinnikov *et al.*, 2000) and Vindija (Krings *et al.*, 2000) is interpreted to demonstrate that Neandertals were a distinctly separate lineage from the ancestors of modern humans for as much as 600 ka (but see Nordborg, 1998; Hawks *et al.*, 2000; Relethford, in press).

The punctuated equilibrium model of macroevolution (Eldredge & Gould, 1972) dovetailed smoothly with RAE. This model essentially required the recognition of lineage splits to explain macroevolutionary change, which in turn required the recognition of different taxa for each lineage. The practical meaning for paleoanthropology was that more distinct taxa had to be identified in the specimens traditionally called *H. erectus* and *H. sapiens* (Eldredge & Tattersall, 1982; Tattersall, 1986). This lead to the argument that *H. erectus* did not exist in Europe (Stringer, 1984) nor in Africa (see Chapter 25) but was restricted to Asia. Many specimens from Africa and Europe (previously considered *H. erectus* or early archaic *H. sapiens*) were reclassified in the resurrected taxon *H. heidelbergensis* (Stringer, 1984; Rightmire, 1990). In this view *H. heidelbergensis* was the common stock that split into *H. neanderthalensis* in Europe and *H. sapiens sensu stricto* in Africa (Rightmire, 1996, 1998*b*; Tattersall, 1992). Most recently, Juan-Luis Arsuaga and colleagues have named yet another taxon, based on the Gran Dolina site at Atapuerca, Spain (Bermúdez de Castro *et al.*, 1997). This taxon, *H. antecessor*, is argued to represent the last common ancestor of the line leading to Neandertals (through *H. heidelbergensis* in Europe) and the line leading to modern humans in Africa (Bermúdez de Castro *et al.*, 1999; Arsuaga *et al.*, 1999).

The most comprehensive non-monocentric model explaining Middle and Late Pleistocene human evolution is Multiregional Evolution (MRE), first articulated in 1984 by Milford H. Wolpoff, Alan G. Thorne and X. Z. Wu. MRE argues that modern human morphology is established, or consolidated, in any region by a complex interplay of gene flow (new characters being introduced) and continuity (regional characters being maintained) modulated by the nature of selective forces acting in a particular region (Wolpoff, 1989, 1999). The role of genetic drift can also be significant, particularly in the initial stages of regional diversification or other times when population sizes may be quite small in a particular region (Wolpoff, 1999). MRE recognizes variable degrees of morphological continuity in the fossil record of differing regions of Eurasia, as well as Africa (Frayer, 1992, 1997*a*; Pope, 1992; Frayer *et al.*, 1993) and purports that distinct morphological breaks and/or unequivocal apomorphies defining different species during the Middle and Late Pleistocene do not exist (Wolpoff, 1999). Recently, Wolpoff *et al.* (1994) argued that the evolutionary species concept permits classifying all members of the genus *Homo* (except fossils assigned to *H. habilis* or *H. rudolfensis*) into *H. sapiens*, which would extend this taxon back to around 1.9 Ma.

Supporters of MRE and others have also countered interpretations of genetic data that are cited in support of monocentric models, particularly RAE, of modern human origins. Spuhler (1988) detailed a number of specific objections to initial mt DNA studies supporting RAE. Excoffier (1990) noted that mt DNA haplotype distributions did not correspond to a neutral model, suggesting that selection is likely operating on this aspect of the genome. Templeton (1993, 1999) argued that patterning of mt DNA distribution actually fit what one would predict with MRE, not RAE. Relethford (1998) summarized both the pros and cons of RAE and noted that patterns interpreted as indicating an African origin for all mt DNA haplotypes might actually result from larger effective population sizes in Africa during the Pleistocene (Relenthford & Jorde, 1999).

Most recently, the interpretation that mt DNA sequences from three Neandertal specimens demonstrate that Neandertals are a distinct species from *Homo sapiens* (Krings *et al.*, 1997, 2000; Ovchinnikov *et al.*, 2000) has also been challenged (Nordborg, 1998; Hawks & Wolpoff, 2001; Relethford, 2001*b*). Indeed, Relethford has recently opined that "much of the genetic evidence is indeterminate and that both African replacement and multiregional models can explain observed patterns of genetic variation" (1999: 7). Finally, the recent analysis of ancient DNA from a series of Pleistocene and Holocene Australians (Adcock *et al.*, 2001) indicates that the deepest branch of mt DNA appears to be Australian, not African, thus casting doubt on one of the major genetic interpretations supporting the recent African origins (total replacement) model.

Among the several strong criticisms of MRE (e.g., Bräuer, 1989; Groves, 1989; Stringer, 1989, 1994; Lahr, 1994, 1996; Lieberman, 1995; Bräuer & Stringer, 1997), perhaps the most pointed are two assessments by F. Clark Howell in the mid-1990s. In comparing current models of modern human origins, Howell (1994: 306) wrote that "a far-reaching multiregional, continuity framework has, in fact, the least pragmatic and operational appeal as it stretches the bounds of credulity in respect to population distributions, affinities, and potentialities for gene flow". In 1996, he again asserts that MRE is a scenario vague in testable content and seriously flawed. In his view a "hard" MRE hypothesis "obscures rather than enlightens serious investigation into hominin phylogenetics" (Howell, 1996: 31).

The African Hybridization and Replacement model is essentially Bräuer's Afro-European Sapiens model with emphasis on the possible hybridization between indigenous archaic Eurasians and expanding moderns (Bräuer, 1992). However, the fact that Bräuer views hybridization to be basically insignificant is reflected in his recent dismissal of any evidence of continuity in the early modern Europeans from Mladeč (Bräuer & Broeg, 1998). The Assimilation model (Smith *et al.*, 1989) agrees with MRE that humans have evolved throughout the Middle and Late Pleistocene as a single polytypic species, but asserts that modern human morphology had a single, probably African, area of origin. Humans with this morphology spread and acted as a catalyst for the appearance of modern humans throughout Eurasia. This is the major difference between Assimilation and MRE *sensu stricto* (see Frayer *et al.*, 1993: 42), but both MRE and Assimilation reject total replacement (Fig. 26.13). The Assimilation model also argues that evidence of continuity is seen in details of morphology, not in the fundamental Gestalt of the modern human anatomical form, reflecting various degrees of assimilation of archaic peoples into the modern human gene pool in individual regions (Smith *et al.*, 1989; Smith, 1994; Churchill & Smith, 2001). However, unlike RAE or Replacement with Hybridization, the Assimilation model does not view the amount of gene flow from archaic Eurasians as "trivial".

Taxonomy

Taxonomy has always been a knotty issue in the history of human evolutionary studies. Historically, taxa were assigned to human fossils with reckless abandon, resulting in the impression that anthropological taxonomy was not in line with general zoological principles (Table 26.1). For example, Simpson (1963: 5) noted that "it is notorious that hominid nomenclature, particularly, has become chaotic." Under the influence of Simpson, Mayr (1950) and others (Campbell, 1963; Brace, 1967), the uninterpretable plethora of taxonomic names for fossil hominids reduced tremendously in the period from 1960 to 1980. For most, but certainly not all, researchers during this period, human fossil remains from the Middle and Late Pleistocene could be encompassed in two taxa, *Homo erectus* and *H. sapiens*. Subspecies categories designated regional and temporal variation, even patterns of variation (Campbell, 1963, 1965). Thus classifying Neandertals as *H. s. neanderthalensis* or the Ngandong remains as *H. s. soloensis* was common for several decades.

Growing dissatisfaction with this approach is reflected in

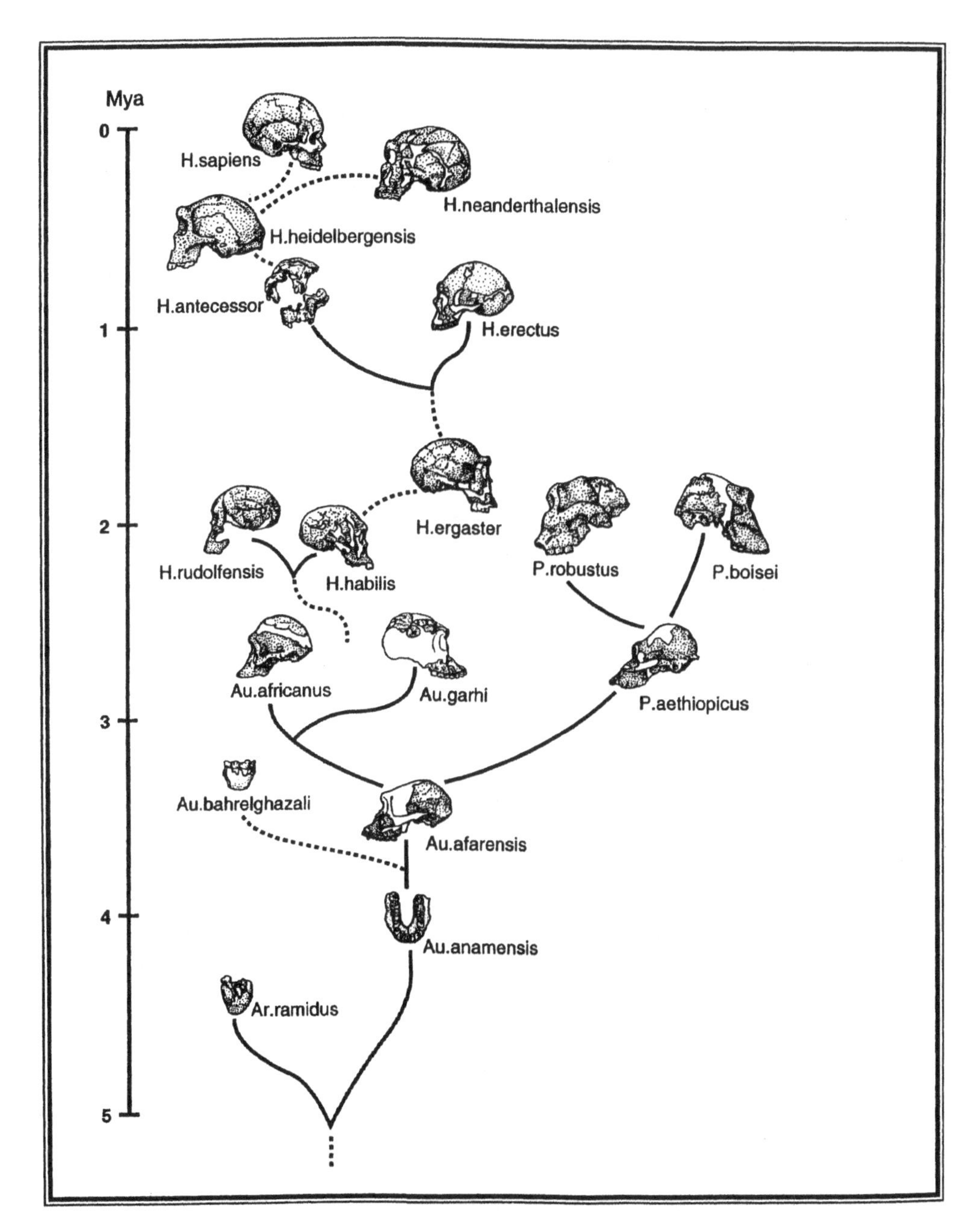

Fig. 26.14 Interpretation of human evolution recognizing the existence of numerous lineages/species. Courtesy of I. Tattersall.

several publications during this period (Stringer *et al.*, 1979; Eldredge & Tattersall, 1982). By the mid-1980s, the emergence of a cladistic perspective on fossil human history, intertwined with the impact of the punctuated equilibrium model of macroevolution, led to the insistence that both *H. erectus* and *H. sapiens* must each comprise multiple taxa (Tattersall, 1986). For specimens previously lumped under *H. sapiens*, Tattersall & Schwartz (2000) recognize three additional taxa: *H. neanderthalensis*, *H. heidelbergensis* (Rightmire, 1990, 1998b) and *H. antecessor* (Bermúdez de Castro *et al.*, 1997) (Fig. 26.14).

Defining "modern human", or what some term *H. sapiens sensu stricto* is a significant problem (e.g., LeGros Clark, 1964; Howell, 1978). Whereas most such definitions are rather general, Michael H. Day and Christopher B. Stringer attempted to quantify several specific features. For example, instead of simply stating "high, short cranial vault" or "relatively rounded occipital contour" they provided quantifiable values reflecting the parameters of such features that defined *H. sapiens* (Day & Stringer, 1982: 832).

However, Wolpoff (1986) noted that some recent humans, specifically subsamples of aboriginal Australians, would be excluded from *H. sapiens* following certain of Day and Stringer's criteria. For example, many recent human specimens in aboriginal Australasia and America have a continuous supraorbital torus not distinctly divided into medial and lateral parts and therefore would not conform to Day and Stringer's definition of modern. Furthermore, multivariate morphometric analysis does not unequivocally define modern *H. sapiens* (Kidder *et al.*, 1992). The problems involved in defining modern humans derive from several factors. Primary among these is that differential degrees of local continuity with archaic peoples in various regions influence the relative "homogeneity" of modern human anatomical form. For example, only a very few early modern European crania fall out of the recent human range morphometrically (Kidder *et al.*, 1992). On the other hand, in Australasia many specimens demonstrably do not conform to all of Day and Stringer's criteria. This could well indicate a greater contribution of archaic humans to early modern populations there, and what constitutes "modern" must involve a certain regional component. In other words, a recent Australian aborigine skull is easily identifiable as modern in the proper regional context even if not in a broader definition of modern *H. sapiens*.

Evolution

It is possible that the first humans migrated out of Africa about 1.7–1.8 Ma, very soon after *Homo erectus* appeared (Swisher *et al.*, 1994; Gabunia *et al.*, 1999; Dunsworth & Walker, this volume). But there are a number of reasons to doubt these early dates, or the association of dates with the human specimens (DeVos & Sondaar, 1994; Sémah *et al.*, 1997; Langbroek & Roebroeks, 2000). The vast majority of archaeological data support a more recent initial expansion of humans out of Africa. Oldowan-like artifacts, similar to those from Dmanisi, occur between 1.0 and 1.4 Ma at Ubeidiya in the Jordan Valley of Israel (Bar-Yosef & Goren-Imbar, 1993), and claims for earlier such tools, particularly at Riwat (Pakistan) are unconvincing (Klein, 1999). Further east, the earliest unequivocal artifacts in context come from the Nihewan Basin in China, which are paleomagnetically correlated to about 1 Ma (Schick *et al.*, 1991). In Europe, the earliest purported artifacts would only suggest a presence of humans at around 1 Ma. However, recent critical reviews of these sites (e.g., Soleihac, Vallonet Cave, Kärlich A, Stránska Skála and others) indicate that none is a valid demonstration of such antiquity and that the earliest occupation of Europe is likely on the order of 500 to 600 ka (Roebroeks & van Kolfschoten, 1994; Dennell & Roebroeks, 1996; Klein, 1999). The human exodus from Africa certainly should not be viewed as a single migration/radiation event, but rather a series of virtually continuous, overlapping "events".

Regional variation and archaic humans

Archaic Africans

Human remains from such sites as the Kapthurian Formation at Lake Baringo, the Masek Beds at Olduvai (OH 23 mandible) and possibly Bodo from Ethiopia are slightly younger than traditional African *Homo erectus* samples, but are continuous with them morphologically. The Bodo skull, estimated to be about 600 ka (Clark *et al.*, 1994), exhibits cranial vault features associated with *H. erectus* but other features more similar to later Africans (like Kabwe). Bodo's cranial capacity estimate of 1300 cm^3 (Rightmire, 1996) supports the suggestion made on the Tighenif parietal (Wolpoff, 1999) that human brain size had witnessed a significant increase by 600–700 ka.

Between roughly 400 ka and 300 ka, the African lineage continues with such specimens as the Kabwe sample, Ndutu (OH 11), Elandsfontein and North African specimens like Sidi Abderahman and Thomas Quarry. These specimens exhibit generally archaic cranial features but cranial capacities within the modern human range. The most recent archaic humans from Africa are dated from about 260 ka at Florisbad (Grün *et al.*, 1996) and perhaps Guomde (Bräuer *et al.*, 1997), and about 130 ka at Ngaloba (Leakey & Hay, 1982) to 90 ka at Jebel Irhoud (Amani & Geraads, 1993). These specimens (Fig. 26.15) exhibit relatively archaic cranial vaults with generally well-developed supraorbital tori combined with smaller faces and such modern facial features as canine fossae and angled zygomaticoalveolar margins (Bräuer, 1984; 1989; Rightmire, 1984; Stringer & Andrews, 1988; Smith *et al.*, 1989). Because of this mosaic morphological pattern, they often are referred to as the African Transitional Group and as a group strongly support an early transition toward modern human morphology in Africa.

Western Asia

The earliest definitive human remains from the Near East are two femoral fragments from Gesher Benot Ya'aqov in Israel. Associated with Acheulean tools and dated to *c.* 500–600 ka (Bar-Yosef, 1993, 1994), these femora illustrate the same robusticity and thick cortical bone characteristic of other roughly contemporary humans (Trinkaus, 1984). The Zuttiyeh specimen has been claimed to resemble modern humans (Vandermeersch, 1981) but its primitive supraorbital torus form and total facial prognathism demonstrate its archaic nature (Simmons *et al.*, 1991). The connections of these early western Asian specimens to subsequent Near Eastern people is far from clear.

Specimens from several sites including Tabûn, Amud, Kebara, Shanidar and Dederiyeh exhibit the craniofacial, dental and postcranial features that are generally typical of European Neandertals, although some features are described as less well developed in the Near Eastern variants (Akazawa *et al.*, 1998; Trinkaus, 1983, 1984, 1992; Kramer *et al.*, 2000). Bar-Yosef & Vandermeersch (1993) argue that Neandertals are late arrivals in the Levant, having migrated from Europe after 80 ka. However, others believe that some Neandertals date to well over 100 ka and reasonably have at least partial biological connections to earlier Levantine populations (Trinkaus, 1984; Quam & Smith, 1998; Wolpoff, 1999).

Eastern Asia

In China, several specimens fill the temporal gap between *H. erectus* and archaic *H. sapiens*. Cranial remains from Zhoukoudian level 3, Hexian, Nanjung and Yunxian are argued to morphologically connect *H. erectus* to later Chinese specimens such as Dali, Maba and Jinniushan (Li & Etler, 1992; Pope, 1997; Wolpoff, 1999). Exactly how clear the morphological connection to later humans really is continues to be a matter of some debate. For example, Johanson & Edgar (1996) classify Dali as *Homo sapiens sensu stricto*, meaning that this specimen represents a different lineage from earlier Chinese. Klein (1999) suggests that a second series of migrations into the Far East, antedating the original *H. erectus* migration, possibly explains the archaic sample from China.

The latest archaic human sample in China is best represented by the crania from Maba, Dali and Jinniushan (Yingkou), and several other more fragmentary remains (Wu & Wu, 1985; Pope, 1997; Wolpoff, 1999), spanning the same time as the African Transitional Group (100–230 ka). In general, they exhibit broad, low braincases with thick vaults and pronounced supraorbital tori. Faces, however, are broad but short and exhibit such modern-like features as canine fossae and angled zygomaticoalveolar margins (Pope, 1992; Klein, 1999; Wolpoff, 1999). As is noted elsewhere (Smith, 1993) these are the same general features that characterize the African Transitional Group and are cited in support an early transition to modern anatomy in Africa. Cranial capacity ranges from 1120 cm^3 (Dali) to 1260 cm^3 (Jinniushan). The Jinniushan innominate is very robust, with a prominent anterior pillar and an elongated pubis reminiscent of Neandertals (Rosenberg, 1998).

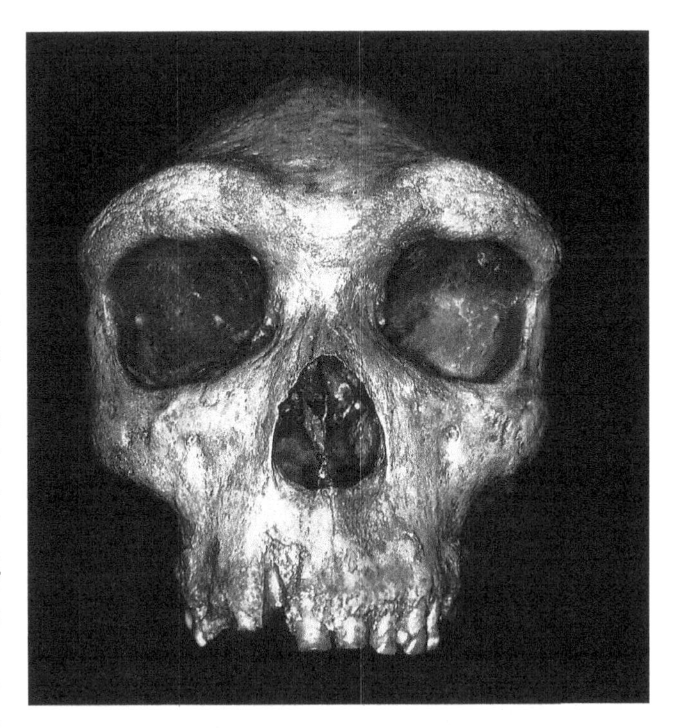

Fig. 26.15 Facial view of the Kabwe 1 cranium.

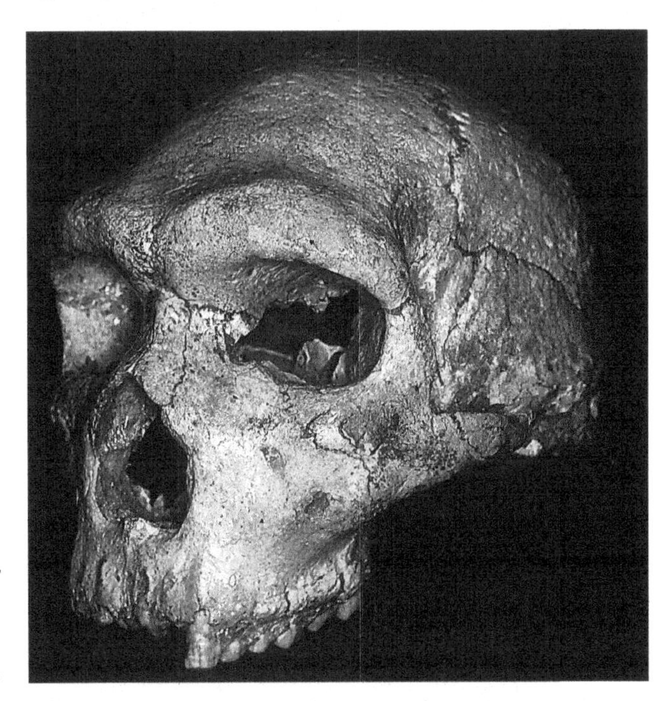

Fig. 26.16 The Petralona cranium from Greece exhibits facial and other features that approach the Neandertal condition. The exact data of the specimen is unknown but indirect evidence suggests an age between 300 ka and 700 ka.

Australasia

As the only Late Pleistocene archaic fossils from this part of the world, the ones comprising the Ngandong, or Solo, sample are particularly critical to interpreting regional Late Pleistocene human evolution. Their morphology is clearly archaic with large supraorbital tori, low cranial vaults with broad cranial bases, primitive occipital bone morphology, and a cranial capacity averaging 1149 cm^3 ($n = 6$). Both dating and taxonomy of the sample are controversial.

The Ngandong specimens were recovered in assumed association with Late Pleistocene fauna on the High Terrace deposit of the Solo River. Bartstra *et al.* (1988) dated animal remains from the lowest levels of this deposit to 101 ± 10 ka using uranium thorium, but Swisher *et al.* (1996) dated other animal teeth found near the Ngandong site to between 27 ka and 53 ka on the basis of ESR and uranium-series dates. However, as Grün & Thorne (1997) note, the human fossils appear to have fossilized in very different environments from the fauna, which suggests that the samples are not contemporaneous. Much older dates are reported based on a modified ESR date on one of the human skulls and other dates on High Terrace fauna (see Wolpoff, 1999). The best assumption is that the Ngandong sample is broadly Late Pleistocene

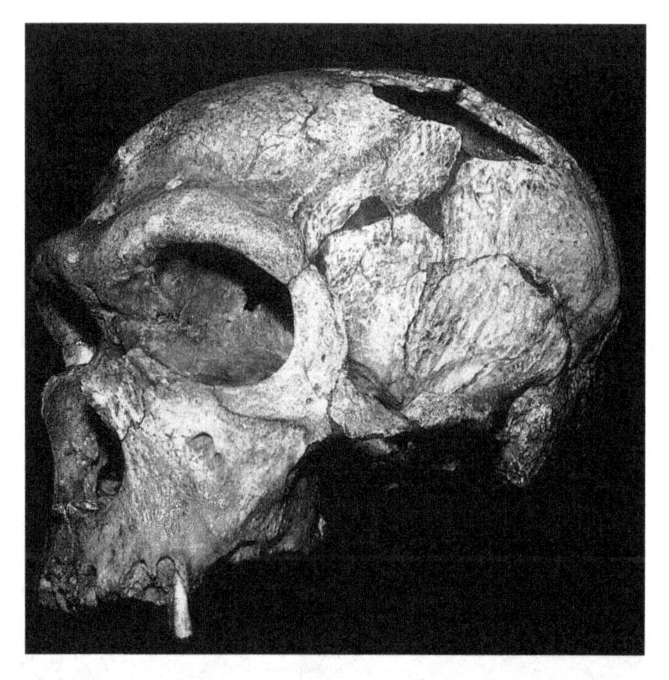

Fig. 26.17 The La Chapelle-aux-Saints Neandertal cranium in oblique view.

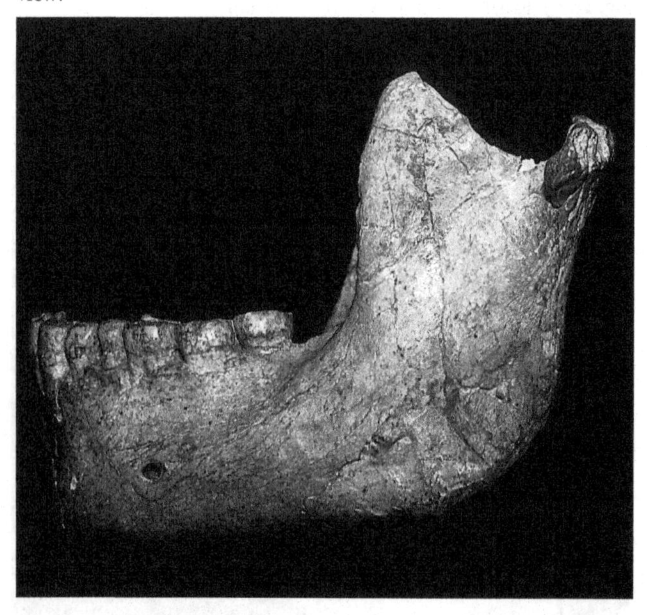

Fig. 26.18 The Amud Neandertal mandible. Although from Israel, not Europe, this specimen exhibits typically Neandertal morphology (receding symphysis, retromolar space, lack of chin projection, etc.).

because all other remains from the High Terrace appear to derive from this period.

Traditionally, the Ngandong sample generally has been considered archaic *H. sapiens* (Weidenreich, 1946), but both Coon (1962) and Howells (1959) argued that classification as *H. erectus* was more appropriate. Santa Luca (1980) noted several features connecting the Ngandong skulls to Asian *H. erectus* (see also Rightmire, 1990; Antón & Weinstein, 1999; Klein, 1999). On the other hand, Smith (1985) and Wolpoff (1999) have noted numerous features intermediate between Kabuh *H. erectus* from Indonesia and modern Aboriginal samples from Australia. These include expansion of cranial capacity and the cranial vault on the cranial base. Tattersall & Schwartz (2000) view Ngandong as *H. erectus*, unlike those who accept that late Pleistocene regional populations tended to be interconnected by gene flow (Frayer *et al.*, 1993). Wolpoff (1999) has noted that the cladistic approach, which places Ngandong in *H. erectus*, confuses Australasian regional features with *H. erectus* features. Others (Habgood, 1989; Brown, 1993, 1997) have questioned the validity of many of these characters as distinct Australasian regional features.

Europe

Of all regions of the Old World, Europe still has the most complete record of Late Pleistocene human evolution. The earliest possible human fossils from Europe derive from Ceprano in Italy and Gran Dolina (Atapuerca) in Spain, both of which may date to around 800 ka (Bermúdez de Castro *et al.*, 1997; Ascenzi *et al.*, 2000; but see Roebroeks & van Kolfschoten, 1994; Dennell & Roebroeks, 1996). Ceprano exhibits features suggesting classification in *Homo erectus* (Ascenzi *et al.*, 1996, 2000), although some scholars argue this taxon is not present in Europe. The Gran Dolina remains have been classified as *Homo antecessor* (see above), but such purportedly unique features as the canine fossae and angled zygomaticoalveolar margins can also be found in archaic *H. sapiens* (or *H. heidelbeigensis*) specimens in Europe.

Most agree that a single lineage is recognizable in Europe and that this line evolved gradually into Neandertals (Wolpoff, 1980, 1999; Hublin, 1982, 1998; Stringer, 1983; Smith, 1985; Arsuaga *et al.*, 1993; Bermúdez de Castro *et al.*, 1997). Specimens from Arago, Swanscombe, Steinheum and Sima de los Huesos and others exhibit emergent Neandertal features in the vault and face. This pattern may extend even earlier depending upon the age of the Petralona (Fig. 26.16) specimen from Greece. Every Neandertal did not exhibit all characteristic Neandertal features (Frayer, 1992), but rather a complex including almost all of them (Figs. 26.17, 26.18). Late Neandertals, particularly at sites like Vindija (Croatia), Kùlna (Czech Republic) and Hortus (France), more closely resemble modern humans in selected features that some regard as parallelisms (e.g., Stringer & Bräuer, 1994) but others as connections to modern populations (Frayer *et al.*, 1993; Smith, 1994; Wolpoff, 1999).

In Europe, Neandertals are associated with Mousterian or "transitional" industries like the Chatelperronian (Churchill & Smith, 2001). Mousterian tools have been dated to as recently as 28 ka on the Iberian Peninsula, and until recently, the latest dates for Neandertal fossils derived from Zafarraya, Spain at 32 ka (Hublin *et al.*, 1995; Klein, 1999). Such discoveries led to the interpretation that expanding modern humans pushed Neandertals into refugial areas such as the Iberian peninsula. However, dating of Neandertals at the site

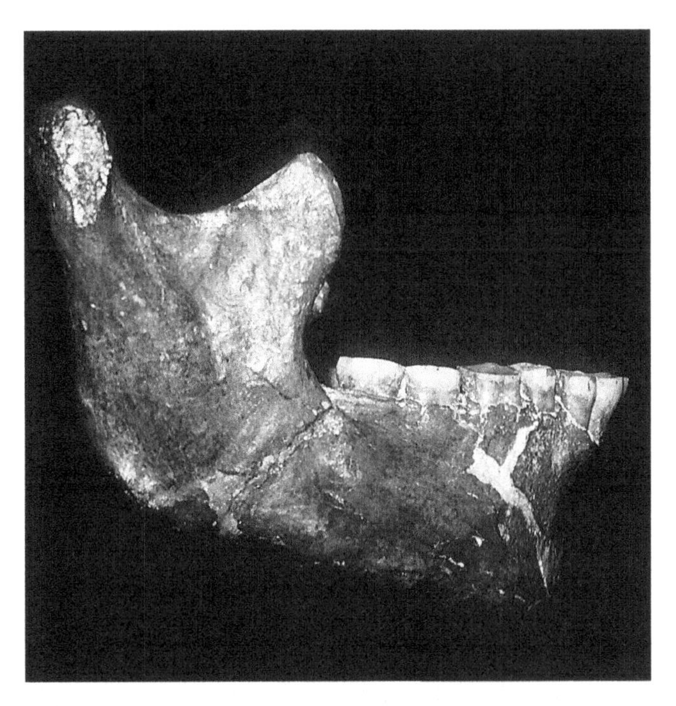

Fig. 26.19 The Tabūn C2 mandible exhibits a mosaic of archaic and possibly modern-like features.

of Vindija (Croatia) to 28–29 ka (Smith *et al.*, 1999) places Neandertals in prime European territory at a very late date and indicates that Neandertals were not so readily displaced. Furthermore, the mixture of Mousterian and more modern tools, including bone tools and ornaments at such sites as Arcy-sur-Cure (Hublin *et al.*, 1996) and Vindija (Karavanić & Smith, 1998) suggests a complex interaction between late Neandertals and early modern humans in Europe.

The pattern of modern human origins: models and meanings

Despite lack of agreement on a precise definition of "modern human", human remains can be characterized as fundamentally modern or archaic, at least within a specific region. For example, Thorne & Macumber (1972) noted that the Kow Swamp skeletons from Australia exhibited archaic, even *Homo erectus*-like, features. Yet any researcher familiar with Australasian skeletal biology would have no problem identifying these as modern Australians morphologically. Similarly, male crania from the European site of Mladeč exhibit some similarities to Neandertals (Smith, 1982; Kidder *et al.*, 1992). Still the overall morphological pattern of the Mladeč sample is certainly modern (Smith & Trinkaus, 1991). These examples demonstrate that while there is a definite homogeneity in the basic morphological form of modern human skeletal remains (Howells, 1973; Lahr, 1996), details of that pattern clearly exhibit regional differences probably due to different degrees of assimilation of archaic humans into modern populations in different regions.

In west Asia, the Skhūl (Fig. 26.8) and Qafzeh humans exhibit archaic features (Kidder *et al.*, 1992; Wolpoff, 1999; Kramer *et al.*, 2000), but their overall morphology (Trinkaus, 1984, 1992), including their body form and limb proportions (Holliday, 2000), is far more modern-like than Neandertal-like. The exact origin of the Skhūl/Qafzeh people is not certain. For example, the 130-ka (or 170-ka) Tabūn C2 mandible (Fig. 26.19) indicates an early "transitional" morphology in the Near East (Quam & Smith, 1998). But whether the Tabūn C2 anatomy resulted from assimilation of African migrants into Near East Neandertals (Quam & Smith, 1998), represents a parallelism (Stefan & Trinkaus, 1998), or indicates an even earlier appearance of modern Near Easterners (Rak, 1998) is unknown. It is generally assumed that the Skhūl/Qafzeh people represent migrants from Africa into the Near East, but no strict cladistic cranial traits connect them to Africa. Holliday's (1997, 2000) argument for tropical origin of the Skhūl/Qafzeh body form provides data for such a connection, but cranial similarities are basically phenetic. Thus, it is not clear to what extent the Skhūl/Qafzeh morphological pattern reflects continuity with local archaic populations, immigration from Africa or another region, or some combination of both.

Australasia lacks large samples of early modern specimens but is a region in which models of modern human emergence recently have been tested (Adcock *et al.*, 2001; Wolpoff *et al.*, 2001). Archaeological evidence and recent dating of the Willandra Lakes (Mungo) 3 skeleton suggest that people reached Australia by 60 ka (Thorne *et al.*, 1999). Many specimens of early Australasians, including Willandra Lakes 3 and the Niah Cave adolescent from Borneo (estimated to date at *c.* 40 ka), are rather gracile (cf. Webb, 1989) and exhibit no strong similarity to samples like Ngandong. However, other specimens like Willandra Lakes (WLH) 50 and the much more recent Kow Swamp remains exhibit features of the frontal bone and supraorbital torus, among others, that indicate continuity with archaic Indonesians (Hawks *et al.*, 2000; Wolpoff, 1999). Again, however, evidence of strong continuity is far from ubiquitous in early modern Australasians, even for specimens like WLH 50. For example, multivariate analysis by Stringer (1998) suggests that the specimen is modern and not closely related to samples like Ngandong (but see Hawks *et al.*, 2000). While it seems reasonable to exclude total replacement of archaic Australasians, the precise balance of immigration and local continuity that resulted in early modern Australasians is not evident from the existing biological record.

The evidence summarized above suggests that models implying a *total replacement* of all archaic humans in Eurasia by expanding populations out of Africa are not likely to be correct. Indeed, virtually all researchers now allow for at least some degree of archaic human contributions to Eurasian early modern populations. Proponents of MRE argue for high levels of continuity in East Asia and Australasia. But MRE does not demand strong continuity in all regions, and proponents recently have suggested that external gene flow might have played a larger role than continuity in the origin of modern Europeans.

The Assimilation model shares much with MRE, but differs on two points. First, modern human morphology likely does have a single region of origin, probably in Africa, but does not result from a speciation event (Smith *et al.*, 1989; Smith, 1994). Both Assimilation and MRE models hold that the development of a distinctive morphological trajectory in a region does not necessarily lead to the creation of a new species if reproductive isolation is never accomplished (Simpson, 1949). Second, evidence of continuity in regions of Eurasia is seen largely in details of morphology, not in fundamental form. Thus European early modern crania exhibit a general shape that makes them easily identifiable as modern humans but also have features that are difficult to explain if Neandertals are totally excluded from their ancestry. Perhaps the difference between these closely related models can best be illustrated by the following quotation.

> A compromise hypothesis, or really a range of hypotheses, would argue that while there has been widespread, long-term genic exchange between human populations, modern humans originated from the expansion of a single, once-localized population within a species. Since this hypothesis allows for admixture, it is not a speciation event. Nevertheless the ultimate origin of unique modern features is traced to a single (yet to be discovered) population. There is no theoretical reason why the notion of partial regional continuity could not have been the case, but we (except for Smith) believe that paleontological, archaeological and genetic evidence contradicts it. (Frayer *et al.*, 1993: 41–42)

There are, of course, legitimate potential problems with Assimilation that may ultimately prove its undoing. Specifically, if future fossil discoveries demonstrate that the appearance of modern humans in East Asia and Australasia exhibits a stronger pattern of continuity with their regional archaic predecessors than is the case in Europe and West Asia, classic MRE would be a better universal explanatory model than Assimilation. This is because robust patterns of continuity (as opposed to continuity mainly in morphological details) would be established as the predominant pattern of modern human origins.

It is also possible that the traits noted to indicate continuity in both MRE and Assimilation may turn out to be unreliable indicators of biological relationships. For example, Lieberman (1995) has pointed out that the genetic basis of most paleoanthropologically relevant features is poorly known and that many traits used to support continuity may not be homologous between archaic and modern samples. However, at the current time Assimilation does remain a viable alternative perspective that recognizes the strength of some aspects of the African origin models, while at the same time noting evidence for not-insignificant patterns of regional continuity.

There continues to be considerable debate surrounding models of modern human origins and how well each model explains the phenomenon as it is reflected in fossils, recent skeletal biology and genetics. These debates are focused on theoretical issues pertaining to how these data should be properly interpreted, and each model represents perspectives well grounded in differing views on the evolutionary process. While these debates are not likely to end soon, there is clear evidence on several fronts that many participants practice good science. Earlier tenets have been revised as new data have become available, both on the paleontological and genetic sides of the debate, and these shifts have tended to migrate toward a middle ground. Along with the ever-increasing amount of data pertinent to the pattern of modern human origins, these evolving models are certain to enhance greatly our understanding of the radiations and evolutionary history of the genus *Homo*.

Acknowledgements

I am indebted to Walter Hartwig for inviting me to write this chapter and for his help in many aspects of its preparation. This paper has benefited both from my interaction with many colleagues over the years and the opportunities afforded me to study much of the original fossil material pertinent to later human evolution. It would be impossible to individually acknowledge each person and institution here, but I am very grateful for all of the cooperation and collaboration I have received in the discipline. Because I was asked to write this paper with a strong historical focus, I want to acknowledge my debt to my late colleague and close friend, Frank Spencer, by dedicating this paper to his memory.

References

Abel, O. (1902). Zwei neue Menschenaffen aus den Leithakalkbildungen des wiener Beckens. *Sitzungsberichte-Akademie der Wissenschaften in Wien, mathematisch-naturwissenschaftliche Klasse*, **1**, 1171–1207.

Abel, O. (1931). Die Stellung des Menschen im Rahmen der Wirbeltiere. Jena: Fischer Verlag.

Adcock, G. J., Dennis, E. S., Easteal, S., Huttley, G. A., Jermiin, L. S., Peacock, W. J. & Thorne, A. (2001). Mitochondrial DNA sequences in ancient Australians: implications for modern human origins. *Proceedings of the National Academy of Sciences of the United States of America*, **98**, 537–542.

Adkins, R. M. & Honeycutt, R. L. (1991). Molecular phylogeny of the superorder Archonta. *Proceedings of the National Academy of Sciences of the United States of America*, **88**, 10317–10321.

Adkins, R. M. & Honeycutt, R. L. (1994). Evolution of the primate cytochrome-c-oxidase subunit-II gene. *Journal of Molecular Evolution*, **38**, 215–231.

Aguirre, A. C. (1971). O mono *Brachyteles arachnoides* (E. Geoffroy). Rio de Janeiro: Academia brasileira de ciências.

Agustí, J., Köhler, M., & Moyà-Solà, S. (1996). Can Llobateres: the pattern and timing of the Vallesian hominoid radiation reconsidered. *Journal of Human Evolution*, **31**, 143–155.

Aiello, L. C. (1993). The fossil evidence for modern human origins in Africa: a revised view. *American Anthropologist*, **95**, 73–96.

Aimi, M. (1981). Fossil *Macaca nemestrina* (Linnaeus, 1766) from Java, Indonesia. *Primates*, **22**, 409–413.

Aitken, M. J., Stringer, C. B., & Mellars, P. A. (eds) (1993). *The Origin of Modern Humans and The Impact of Chronometric Dating*. Princeton, NJ: Princeton University Press.

Akazawa, T. Aoki, K., & Bar-Yosef, O. (eds.) (1998). *Neandertals and Modern Humans in Western Asia*. New York: Plenum Press.

Alexander, J. P. (1994). Sexual dimorphism in notharctid primates. *Folia Primatologica*, **63**, 59–62.

Alexander, J. P. & Macphee, R. D. E. (1999*a*). Skull of *Omomys carteri*, an Eocene omomyid primate. *American Journal of Physical Anthropology*, Supplement **28**, 83.

Alexander, J. P. & Macphee, R. D. E. (1999b). Skull morphology of *Omomys carteri*. *Journal of Vertebrate Paleontology*, Supplement **3**, 29A.

Alexeev, V. P. (1986). *The Origin of the Human Race*. Moscow: Progress Publishers.

Allard, M. W., McNiff, B. E., & Miyamoto, M. M. (1996). Support for interordinal eutherian relationships with an emphasis on primates and their archontan relatives. *Molecular Phylogenetics and Evolution*, **5**, 78–88.

Allbrook, D. & Bishop, W. W. (1963). New fossil hominoid material from Uganda. *Nature*, **197**, 1187–1190.

Alpagut, B., Andrews, P., & Martin, L. (1990). New Miocene hominoid specimens from the Middle Miocene site at Paşalar, Turkey. *Journal of Human Evolution*, **19**, 397–422.

Alpagut, B., Andrews, P., Fortelius, M., Kappelman, J., Temizsoy, I., Celebi, H. & Lindsay, W. (1996). A new specimen of *Ankarapithecus meteai* from the Sinap Formation of central Anatolia. *Nature*, **382**, 349–351.

Alterman, L., Doyle, D. A., & Izard, M. K. (eds.) (1995). *Creatures of the Dark: The Nocturnal Prosimians*. New York: Plenum Press.

Amadon, D. (1947). An estimated weight of the largest known bird. *Condor*, **4**, 159–164.

Amani, F. & Geraads, D. (1993). Le gisement moustérien du Djebel Irhoud, Maroc: précisions sur la fauna et la biochronologie et description d'un nouveau reste humain. *Comptes rendus de l'Académie des sciences de Paris*, **316**, 847–852.

Ameghino, F. (1891*a*). Los monos fosiles del Eoceno de la Republica Argentina. *Revista argentina de historia natural*, **1**, 383–397.

Ameghino, F. (1891*b*). Nuevos restos de mamiferos fosiles descubiertos por C. Ameghino en el Eoceno inferior de la Patagonia austral: especies nuevas, adiciones y correcciones. *Revista argentina de historia natural*, **1**, 289–328.

Ameghino, F. (1910). *Montaneia anthropomorpha*: un género de monos hoy extinguido de la isla de Cuba: nota preliminar. *Anales del Museo nacional de historia natural de Buenos Aires* (ser. 3), **13**, 317–318.

Anapol, F. & Fleagle, J. G. (1988). Fossil platyrrhine forelimb bones from the early Miocene of Argentina. *American Journal of Physical Anthropology*, **76**, 417–428.

Andrews, C. W. (1906). *A Descriptive Catalog of the Tertiary Vertebrata of the Fayum, Egypt*. London: British Museum (Natural History).

Andrews, C. W. (1916). Note on a new baboon (*Simopithecus oswaldi*, gen. et sp. n.) from the Pliocene of British East Africa. *The Annals and Magazine of Natural History, including Zoology, Botany, and Geology*, **18**, 410–419.

Andrews, P. J. (1970). Two new fossil primates from the Lower Miocene of Kenya. *Nature*, **228**, 537–540.

Andrews, P. J. (1973). Miocene primates (Pongidae, Hylobatidae) of East Africa. PhD dissertation, University of Cambridge.

Andrews, P. J. (1974). New species of *Dryopithecus* from Kenya. *Nature*, **249**, 188–190.

Andrews, P. J. (1978*a*). A revision of the Miocene Hominoidea of East Africa. *Bulletin of the British Museum (Natural History), Geology Series*, **30**, 85–225.

Andrews, P. J. (1978*b*). Stratigraphy and mammalian paleontology of the late-Miocene Lukeino Formation, Kenya. In *Geological Background to Fossil Man*, ed. M. Pickford, pp 263–278. London: Academic Press.

Andrews, P. J. (1980). Ecological adaptations of the smaller fossil apes. *Zeitschrift für Morphologie und Anthropologie*, **71**, 164–173.

Andrews, P. J. (1981*a*). Species diversity and diet in monkeys and apes during the Miocene. In *Aspects of Human Evolution*, ed. C. B. Stringer, pp. 25–61. London: Taylor & Francis.

Andrews, P. J. (1981*b*). A short history of Miocene field palaeontology in western Kenya. *Journal of Human Evolution*, **10**, 3–9.

Andrews, P. J. (1982). Hominoid evolution. *Nature*, **295**, 185–86.

Andrews, P. J. (1984). An alternative interpretation of the characters used to define *Homo erectus*. *Courier Forschungsinstitut Senckenberg*, **69**, 167–175.

Andrews, P. J. (1985). Family group systematics and evolution among catarrhine primates. In *Ancestors: The Hard Evidence*, ed. E. Delson, pp. 14–22. New York: Alan R. Liss.

Andrews, P. J. (1990). Paleoecology of the Miocene fauna from Paşalar, Turkey. *Journal of Human Evolution*, **19**, 569–582.

Andrews, P. J. (1992). Evolution and environment in the Hominoidea. *Nature*, **360**, 641–647.

Andrews, P. J., Begun, D. R., & Zylstra, M. (1997). Interrelationships between functional morphology and paleoenvironments in Miocene hominoids. In *Function, Phylogeny and Fossils: Miocene Hominoid Evolution and Adaptations*, eds. D. R. Begun, C. V. Ward, & M. D. Rose, pp. 29–58. New York: Plenum Press.

Andrews, P. J. & Bernor, R. L. (1999). Vicariance biogeography and paleoecology of Eurasian Miocene hominoid primates. In *The Evolution of Neogene Terrestrial Ecosystems in Europe*, eds. J. Agusti, L. Rook, & P. Andrews, pp. 454–487. Cambridge: Cambridge University Press.

Andrews, P. J. & Cronin, J. (1982). The relationships of *Sivapithecus* and *Ramapithecus* and the evolution of the orang-utan. *Nature*, **297**, 541–46.

Andrews, P. J., Hamilton, W. R., & Whybrow, P. J. (1978). Dryopithecines from the Miocene of Saudi Arabia. *Nature*, **274**, 249–250.

Andrews, P. J., Harrison, T., Delson, E., Bernor, R. L., & Martin, L. (1996). Distribution and biochronology of European and Southwest Asian Miocene catarrhines. In *The Evolution of Western Eurasian Neogene Mammal Faunas*, eds. R. L. Bernor, V. Fahlbusch, & H.-W. Mittmann, pp. 168–295. New York: Columbia University Press.

Andrews, P. J., Harrison, T., Martin, L., & Pickford, M. (1981*a*). Hominoid primates from a new Miocene locality named Meswa Bridge in Kenya. *Journal of Human Evolution*, **10**, 123–128.

Andrews, P. J. & Martin, L. (1987*a*). Cladistic relationships of extant and fossil hominoids. *Journal of Human Evolution*, **16**, 101–118.

Andrews, P. J. & Martin, L. (1987*b*). The phyletic position of the Ad Dabtiyah hominoid. *Bulletin of the British Museum (Natural History), Geology Series*, **41**, 383–393.

Andrews, P. J. & Martin, L. (1991). Hominoid dietary evolution. *Philosophical Transactions of the Royal Society of London, Series B*, **334**, 199–209.

Andrews, P. J., Martin, L., & Whybrow, P. J. (1987). Earliest known member of the great ape and human clade. *American Journal of Physical Anthropology*, **72**, 174–175.

Andrews, P. J., Meyer, G. E., Pilbeam, D. R., Van Couvering, J. A., & Van Couvering, J. A. H. (1981*b*). The Miocene fossil beds of Maboko Island, Kenya: geology, age, taphonomy and palaeontology. *Journal of Human Evolution*, **10**, 35–48.

Andrews, P. J. & Pilbeam, D. (1996). The nature of the evidence. *Nature*, **379**, 123–24.

Andrews, P. J. & Simons, E. (1977). A new Miocene gibbon-like genus, *Dendropithecus* (Hominoidea, Primates), with distinctive postcranial adaptations: its significance to origin of Hylobatidae. *Folia Primatologica*, **28**, 161–169.

Andrews, P. J. & Tekkaya, I. (1980). A revision of the Turkish Miocene hominoid *Sivapthecus meteai*. *Paleontology*, **23**, 85–95.

Andrews, P. J. & Tobien, H. (1977). New Miocene locality in Turkey with evidence on the origin of *Ramapithecus* and *Sivapithecus*. *Nature*, **268**, 699–701.

Andrews, P. J. & Van Couvering, J. H. (1975). Palaeoenvironments in the East African Miocene. In *Approaches to Primate Paleobiology*, ed. F. S. Szalay, pp. 62–103. Basel: Karger.

Andrews, P. J. & Walker, A. C. (1976). The primate and other fauna from Fort Ternan, Kenya. In *Human Origins: Louis Leakey and the East African Evidence*, eds. G. L. Isaac & E. R. McCown, pp. 279–304. Menlo Park, CA: W. A. Benjamin.

Andrews, T. D., Jermiin, L. S., & Easteal, S. (1998). Accelerated evolution of cytochrome *b* in simian primates: adaptive evolution in concert with other mitochondrial proteins? *Journal of Molecular Evolution*, **47**, 249–257.

Anemone, R. L. (1990). The VCL hypothesis revisited: patterns of femoral morphology among quadrupedal and saltatorial prosimian primates. *American Journal of Physical Anthropology*, **83**, 373–393.

Anemone, R. L. (1993). The functional anatomy of the hip and thigh in primates. In *Postcranial Adaptation in Nonhuman Primates*, ed. D. L. Gebo, pp. 150–174. DeKalb, IL: Northern Illinois University Press.

Anemone, R. L. & Covert, H. H. (2000). New skeletal remains of *Omomys* (Primates, Omomyidae): functional morphology of the hindlimbs and locomotor behavior of a middle Eocene primate. *Journal of Human Evolution*, **38**, 607–633.

Anemone, R. L., Covert, H. H., & Nachman, B. A. (1997). Functional anatomy and positional behavior of *Omomys carteri*. *Journal of Vertebrate Paleontology*, **17**, 29A.

Ankel, F. (1965). Der Canalis sacralis als Indikator für die Länge der Caudalregion der Primaten. *Folia Primatologica*, **3**, 263–276.

Ankel-Simons, F. (1996). Deciduous dentition of the aye aye, *Daubentonia madagascariensis*. *American Journal of Primatology*, **39**, 87–97.

Ankel-Simons, F., Fleagle, J. G., & Chatrath, P. S. (1998). Femoral anatomy of *Aegyptopithecus zeuxis*, an early Oligocene anthropoid. *American Journal of Physical Anthropology*, **106**, 413–424.

Anthony, H E. (n.d.). *Daily Journal of Expedition to Jamaica, Nov. 18, 1919 to Mar. 19, 1920* [title on first interior page]. New York: Department of Mammalogy Library and Archives, American Museum of Natural History.

Anthony, M. R. L. & Kay, R. F. (1993). Tooth form and diet in ateline and alouattine primates: reflections on the comparative method. *American Journal of Science*, **293-A**, 356–382.

Antón, S. C. (1994). Mechanical and other perspectives on Neandertal craniofacial morphology. In *Integrative Paths to the Past: Paleoanthropological Advances in Honor of F. Clark Howell*, eds R. Corruccini & R. Ciochon, pp. 677–695. Englewood Cliffs, NJ: Prentice-Hall.

Antón, S. C. (1997). Developmental age and taxonomic affinity of the Modjokerto child, Java, Indonesia. *American Journal of Physical Anthropology*, **102**, 497–514.

Antón S. C. (1999). Cranial growth in *Homo erectus*: how credible are the Ngandong juveniles? *American Journal of Physical Anthropology*, **108**, 223–236.

Antón, S. C. & Weinstein, K. J. (1999). Artificial cranial deformation and fossil Australians revisited. *Journal of Human Evolution*, **36**, 195–209.

Anyonge, W. (1991). Fauna from a new lower Miocene locality west of Lake Turkana, Kenya. *Journal of Vertebrate Paleontology*, **11**, 378–390.

Arambourg, C. (1937). Paléontologie générale et paléontologie humaine. *Revue générale des sciences pures et appliquées*, **114**, 2–11.

Arambourg, C. (1943*a*). Sur les affinités de quelques anthropoïdes fossiles d'Afrique et leur relations avec la lignée humaine. *Comptes* rendus de l'Académie des sciences de Paris, **216**, 593–595.

Arambourg, C. (1943*b*). L'État actuel de nos connaissances sur les origines l'homme. *L'Anthropologie*, **52**, 193.

Arambourg. C. (1947). *Mission scientifique à l'Omo*, vol. 1, Géologie, anthropologie. Paris: Muséum national d'histoire naturelle.

Arambourg, C. (1948). La classification des primates et particulièrement des hominiens. *Mammalia*, **12**, 123–135.

Arambourg, C. (1952). Observations sur la phylogénie des primates et l'origine des hominiens. *Proceedings of the 1st Pan-African Congress on Prehistory*, **1947**, 116–119.

Arambourg, C. (1954). L'Hominien fossile de Ternifine (Algérie). *Comptes rendus de l'Académie des sciences de Paris*, **239**, 893–895.

Arambourg, C. (1955). A recent discovery in human paleontology: *Atlanthropus* of Ternifine (Algeria). *American Journal of Physical Anthropology*, **13**, 191–201.

Arambourg, C. & Biberson, P. (1956). The fossil human remains from the Paleolithic site of Sidi Abderrahman (Morocco). *American Journal of Physical Anthropology*, **14**, 467–490.

Arambourg, C. & Coppens, Y. (1968). Découverte d'un australopithécien nouveau dans les gisements de l'Omo (Éthiopie). *South African Journal of Science*, **64**, 58–59.

Archibald, J. D. (1996). Fossil evidence for a Late Cretaceous origin of "hoofed" mammals. *Science*, **272**, 1150–1153.

Arctander, P., Johansen, C., & Coutellec-Vreto, M. A. (1999). Phylogeography of three closely related African bovids (Tribe Alcelaphini). *Molecular Biology and Evolution*, **16**, 1724–1739.

Ardito, G., & Mottura, A. (1987). An overview of the geographic and chronologic distribution of west European cercopithecoids. *Human Evolution*, **2**, 29–45.

Arredondo, O. & Varona, L. (1983). Sobre la validez de *Montaneia anthropomorpha* Ameghino, (1910) (Primates: Cebidae). *Poeyana*, **255**, 1–21.

Arsuaga, J. L., Martínez, I.,Gracia, A., Carretero, J., & Carbonell, E. (1993). Three new human skulls from the Sima de los Huesos Middle Pleistocene site in Sierra de Atapuerca, Spain. *Nature*, **362**, 534–537.

Arsuaga, J. L., Martínez, I., Gracia, A., & Lorenzo, C. (1997). The Sima de los Huesos crania (Sierra de Atapuerca, Spain): a comparative study. *Journal of Human Evolution*, **33**, 219–281.

Arsuaga, J. L., Martínez, I., Lorenzo, C., Gracia, A., Muñoz, A., Alonso, O., & Gallego, J. (1999). The human cranial remains from Gran Dolina Lower Pleistocene site (Sierra de Atapuerca, Spain). *Journal of Human Evolution*, **37**, 431–457.

Ascenzi, A., Biddittu, I., Cassoli, P., Serge, A., & Serge-Naldini, E. (1996). A calvarium of late *Homo erectus* from Ceprano, Italy. *Journal of Human Evolution*, **31**, 409–423.

Ascenzi, A., Mallegni, F., Manzi, G., Serge, A., & Serge-Naldini, E. (2000). A reappraisal of Ceprano calvarial affinities with *Homo erectus* after the new reconstruction. *Journal of Human Evolution*, **39**, 443–450.

Asfaw, B. (1987). The Belohdelie frontal: new evidence of early hominid cranial morphology from the Afar of Ethiopia. *Journal of Human Evolution*, **16**, 611–624.

Asfaw, B., Beyene, Y., Semaw, S., Suwa, G., White, T., & WoldeGabriel, G. (1991). Fejej: a new paleoanthropological research area in Ethiopia. *Journal of Human Evolution*, **21**, 137–144.

Asfaw, B., White, T., Lovejoy, O., Latimer, B., Simpson, S., & Suwa, G. (1999). *Australopithecus garhi*: a new species of early hominid from Ethiopia. *Science*, **284**, 629–634.

Ashton, E. H., Flinn, R. M., Oxnard, C. E., & Spence, T. F. (1976). The adaptive and classificatory significance of certain quantitative features of the forelimb in primates. *Journal of Zoology*, **179**, 515–556.

Ashton, E. H., Healy, M. J. R., Oxnard, C. E., & Spence, T. F. (1965). The combination of locomotor features of the primate shoulder girdle by canonical analysis. *Journal of Zoology*, **147**, 406–429.

Atchley, W. R. & Hall, B. K (1991). A model for development and evolution of complex morphological structures. *Biological Reviews*, **66**, 101–157.

Azzaroli, A. (1946). La scimmia fossile della Sardegna. *Rivista di scienze preistoriche*, **1**, 68–76.

Bacon, A. M. & Godinot, M. (1998). Analyse morphofonctionelle des fémurs et des tibias des "*Adapis*" du Quercy: mise en évidence de cinq type morphologiques. *Folia Primatologica*, **69**, 1–21.

Badgley, C., Kelley, J., Pilbeam, D., & Ward, S. (1983). The paleobiology of South Asian Miocene Hominoidea. In *The Peoples of South Asia: The Biological Anthropology of India, Pakistan and Nepal*, ed. J. Lucacs, pp. 3–27. New York: Plenum Press.

Badoux, D. M. (1959). *Fossil Mammals from Two Fissure Deposits at Punung (Java)*. Utrecht: Keminck & Zoon.

Bailey, W. J., Fitch, D. H. A., Tagle, D. A., Czelusniak, J., Slightom, J. L., & Goodman, M. (1991). Molecular evolution of the γ-globin gene locus: gibbon phylogeny and the hominoid slowdown. *Molecular Biology and Evolution*, **8**, 155–184.

Baker, B. H., Williams, L. A. J., Miller, J. A., & Fitch, F. J. (1971). Sequence and geochronology of the Kenya Rift volcanics. *Tectonophysics*, **11**, 191–215.

Baker, R. J., Honeycutt, R. L., & Van den Bussche, R. A. (1991). Examination of monophyly in bats: restriction map of the ribosomal DNA cistron. *Bulletin of the American Museum of Natural History*, **206**, 42–53.

Baker, W. E., & Durand, H. M. (1836). Sub-Himálayan fossil remains of the Dádupur collection. (Quadrumana). *Journal of the Asiatic Society of Bengal*, **5**, 739–741.

Ba Maw, Ciochon, R. L., & Savage, D. E. (1979). Late Eocene of Burma yields earliest anthropoid primate, *Pondaungia cotteri*. *Nature*, **282**, 65–67.

Barry, J. C. (1987). The history and chronology of Siwalik cercopithecids. *Human Evolution*, **2**, 47–58.

Barry, J. C., Morgan, M. E., Flynn, L. J., Pilbeam, D., Jacobs, L. L., Lindsay, E. H., Raza, S. M., & Solounias, N. (1995). Patterns of faunal turnover and diversity in the Neogene Siwaliks of Northern Pakistan. *Palaeogeography, Palaeoclimatology, Palaeoecology*, **115**, 209–226.

Bartlett, T. (1999). The gibbons. In *The Nonhuman Primates*, eds. P. Dolhinow & A. Fuentes, pp. 44–49. Mountain View, CA: Mayfield Publishing Company.

Bartstra, G. J., Soegondho, S., & Van Der Wijk, A. (1988). Ngandong man: age and artifacts. *Journal of Human Evolution*, **17**, 325–337.

Bar-Yosef, O. (1993). The role of western Asia in modern human origins. In *The Origin of Modern Humans and the Impact of Chronometric Dating*, eds. M. Aitken, C. Stringer, & P. Mellars, pp. 132–147. Princeton, NJ: Princeton University Press.

Bar-Yosef, O. (1994). Contributions of Southwest Asia to the study of the origin of modern humans. In *Origins of Anatomically Modern Humans*, eds. M. H. Nitecki & D. V. Nitecki, pp. 23–66. New York: Plenum Press.

Bar-Yosef, O. & Goren-Imbar, N. (1993). The lithic assemblages of 'Ubeidiya: a Lower Paleolithic site in the Jordan Valley. *Oedem*, **45**, 1–266.

Bar-Yosef, O. & Vandermeersch, B. (1993). Modern humans in the Levant. *Scientific American*, **267**, 94–100.

Basilici, G., Faraone, A. G., & Gentili, S. (1991). Un nuovo reperto di *Macaca* nelle brecce ossifere pleistoceniche di Monte Peglia (Terni, Italia centrale). *Bollettino della Società paleontologica italiana*, **30**, 251–254.

Battistini, R. & Vérin, P. (1966). Les transformations écologiques à Madagascar à l'époque protohistorique. *Bulletin de Madagascar*, **16**, 845–856.

Beard, K. C. (1987). *Jemezius*, a new omomyid primate from the early Eocene of northwestern New Mexico. *Journal of Human Evolution*, **16**, 457–468.

Beard, K. C. (1988*a*). New notharctine primate fossils from the early Eocene of New Mexico and southern Wyoming [USA] and the phylogeny of Notharctinae. *American Journal of Physical Anthropology*, **75**, 439–470.

Beard, K. C. (1988*b*). The phylogenetic significance of strepsirhinism in Paleogene primates. *International Journal of Primatology*, **9**, 83–96.

Beard, K. C. (1990). Gliding behavior and paleoecology of the alleged Primate Family Paromomyidae (Mammalia: Dermoptera). *Nature*, **345**, 340–341.

Beard, K. C. (1993). Phylogenetic systematics of the Primatomorpha, with special reference to Dermoptera. In *Mammal Phylogeny: Placentals*, eds. F. S. Szalay, M. J. Novacek, & M. C. McKenna, pp. 129–150. New York: Springer-Verlag.

Beard, K. C. (1998*a*). A new genus of Tarsiidae (Mammalia: Primates) from the middle Eocene of Shanxi Province, China, with notes on the historical biogeography of tarsiers. *Bulletin of the Carnegie Museum of Natural History*, **34**, 260–277.

Beard, K. C. (1998*b*). East of Eden: Asia as an important center of taxonomic origination in mammalian evolution. *Bulletin of the Carnegie Museum of Natural History*, **34**, 5–39.

Beard, K. C. (1998*c*). Unmasking an Eocene primate enigma: the true identity of *Hoanghonius stehlinii*. *American Journal of Physical Anthropology*, Supplement, **26**, 69.

Beard, K. C., Dagosto, M., Gebo, D. L., & Godinot, M. (1988). Interrelationships among primate higher taxa. *Nature*, **331**, 712–714.

Beard, K. C. & Godinot, M. (1988). Carpal anatomy of *Smilodectes gracilis* (Adapiformes, Notharctinae) and its significance for lemuriform phylogeny. *Journal of Human Evolution*, **17**, 71–92.

Beard, K. C., Krishtalka, L., & Stucky, R. K. (1991). First skulls of the early Eocene primate *Shoshonius cooperi* and the anthropoid–tarsier dichotomy. *Nature*, **349**, 64–67.

Beard, K. C., Krishtalka, L., & Stucky, R. K. (1992). Revision of the Wind River faunas, Early Eocene of central Wyoming. XII. New species of omomyid primates (Mammalia: Primates: Omomyidae) and omomyid taxonomic composition across the Early–Middle Eocene boundary. *Annals of Carnegie Museum*, **61**, 39–62.

Beard, K. C. & MacPhee, R. D. E. (1994). Cranial anatomy of *Shoshonius* and the antiquity of Anthropoidea. In *Anthropoid Origins*, eds. J. G. Fleagle & R. F. Kay, pp. 55–97. New York: Plenum Press.

Beard, K. C., Qi, T., Dawson, M. R., Wang, B., & Li, C. (1994). A diverse new primate fauna from middle Eocene fissure-fillings in southeastern China. *Nature*, **368**, 604–609.

Beard, K. C. & Tabrum, A. R. (1990). The first early Eocene mammal from eastern North America: an omomyid primate from the Bashi Formation, Lauderdale County, Mississippi. *Mississippi Geology*, **11**, 1–6.

Beard, K. C., Teaford, M. F., & Walker, A. C. (1986). New wrist bones of *Proconsul africanus* and *nyanzae* from Rusinga Island, Kenya. *Folia Primatologica*, **47**, 97–118.

Beard, K. C., Tong, Y., Dawson, M. R., Wang, J., & Huang, X. (1996). Earliest complete dentition of an anthropoid primate from the late middle Eocene of Shanxi Province, China. *Science*, **272**, 82–85.

Beard, K. C. & Wang, B. (1991). Phylogenetic and biogeographic significance of the tarsiiform primate *Asiomomys changbaicus* from the Eocene of Jilin Province, People's Republic of China. *American Journal of Physical Anthropology*, **85**, 159–166.

Bearder, S. K. (1987). Lorises, bushbabies, and tarsiers: diverse societies in solitary foragers. In *Primate Societies*, eds. B. Smuts, D. Cheney, R. Seyfarth, R. Wrangham, & T. Struhsaker, pp. 11–24. Chicago, IL: University of Chicago Press.

Bearder, S. K. (1999). Physical and social diversity among nocturnal primates: a new view based on long-term research. *Primates*, **40**, 267–282.

Bearder, S. K., Honess, P. E., & Ambrose, L. (1995). Species diversity among galagos with special reference to mate recognition. In *Creatures of the Dark: The Nocturnal Prosimians*, eds. L. Alterman, G. A. Doyle, & M. I. Izard, pp. 331–352. New York: Plenum Press.

Beaumont, P., De Villiers, H., & Vogel, J. (1978). Modern man in sub-Saharan Africa prior to 49,000 year BP: a review and evaluation with particular reference to Border Cave, South Africa. *South African Journal of Science*, **74**, 409–419.

Begun, D. R. (1988*a*). Catarrhine phalanges from the Late Miocene (Vallesian) of Rudabánya, Hungary. *Journal of Human Evolution*, **17**, 413–438.

Begun, D. R. (1988*b*). Diversity and affinities of *Dryopithecus*. *American Journal of Physical Anthropology*, **75**, 185.

Begun, D. R. (1989*a*). New species of *Dryopithecus* from the Vallesian of Can Ponsic. *American Journal of Physical Anthropology*, **78**, 191.

Begun, D. R. (1989*b*). A large pliopithecine molar from Germany and some notes on the Pliopithecinae. *Folia Primatologica*, **52**, 156–166.

Begun, D. R. (1990). Postcranial anatomy and positional behavior in European Miocene catarrhines. *American Journal of Physical Anthropology*, **81**, 192–193.

Begun, D. R. (1991). European Miocene catarrhine diversity. *Journal of Human Evolution*, **20**, 521–526.

Begun, D. R. (1992*a*). Phyletic diversity and locomotion in primitive European hominids. *American Journal of Physical Anthropology*, **87**, 311–340.

Begun, D. R. (1992*b*). *Dryopithecus crusafonti* sp. nov., a new Miocene hominid species from Can Ponsic (Northeastern Spain). *American Journal of Physical Anthropology*, **87**, 291–310.

Begun, D. R. (1992*c*). Miocene fossil hominids and the chimp–human clade. *Science*, **257**, 1929–1933.

Begun, D. R. (1993*a*). New catarrhine phalanges from Rudabánya (Northeastern Hungary) and the problem of parallelism and convergence in hominoid postcranial morphology. *Journal of Human Evolution*, **24**, 373–402.

Begun, D. R. (1993*b*). Knuckle walking ancestors. *Science*, **259**, 294.

Begun, D. R. (1994*a*). Relations among the great apes and humans: new interpretations based on the fossil great ape *Dryopithecus*. *Yearbook of Physical Anthropology*, **37**, 11–63.

Begun, D. R. (1994*b*). The significance of *Otavipithecus namibiensis* to interpretations of hominoid evolution. *Journal of Human Evolution*, **27**, 385–394.

Begun, D. R. (1995). Late Miocene European orang-utans, gorillas, humans, or none of the above? *Journal of Human Evolution*, **29**, 169–180.

Begun, D. R. (1997). A Eurasian origin of the Hominidae. *American Journal of Physical Anthropology*, Supplement **24**, 73–74.

Begun, D. R. (2000). Middle Miocene hominoid origins. *Science*, **287**, 2375a.

Begun, D. R. & Güleç, E. (1995). Restoration and interpretation of the facial specimen attributed to *Sivapithecus meteai* from Kayincak. *American Journal of Physical Anthropology*, Supplement **20**, 63–64.

Begun, D. R. & Güleç, E. (1998). Restoration of the type and palate of *Ankarapithecus meteai*: taxonomic and phylogenetic implications. *American Journal of Physical Anthropology*, **105**, 279–314.

Begun, D. R. & Kordos, L. (1993). Revision of *Dryopithecus brancoi* Schlosser, 1901 based on the fossil hominoid material from Rudabanya. *Journal of Human Evolution*, **25**, 271–285.

Begun, D. R. & Kordos, L. (1997). Phyletic affinities and functional convergence in *Dryopithecus* and other Miocene and living hominids. In *Function, Phylogeny, and Fossils: Miocene Hominoid Evolution and Adaptation*, eds. D. R. Begun, C. V. Ward, & M. D. Rose, pp. 291–316. New York: Plenum Press.

Begun, D. R., Moyà-Solà, S., & Köhler, M. (1990). New Miocene hominoid specimens from Can Llobateres (Valles Penedes, Spain) and their geological and paleoecological context. *Journal of Human Evolution*, **19**, 255–268.

Begun, D. R., Teaford, M. F., & Walker, A. C. (1994). Comparative and functional anatomy of *Proconsul* phalanges from the Kaswanga Primate Site, Rusinga Island, Kenya. *Journal of Human Evolution*, **26**, 89–165.

Begun, D. R., Ward, C. V., & Rose, M. D. (eds.) (1997*a*). *Function, Phylogeny, and Fossils: Miocene Hominoid Evolution and Adaptation*. New York: Plenum Press.

Begun, D. R., Ward, C. V., & Rose, M. D. (1997*b*). Events in hominoid evolution. In *Function, Phylogeny and Fossils: Miocene Hominoid Evolution and Adaptation*, eds. D. R. Begun, C. V., Ward & M. D. Rose, pp. 389–415. New York: Plenum Press.

Behrensmeyer, A. K., Damuth, J. D., DiMichele, W. A., Potts, R., Sues, H. D., & Wing, S. L. (1992). *Terrestrial Ecosystems Through Time: Evolutionary Paleoecology of Terrestrial Plants and Animals*. Chicago, IL: University of Chicago Press.

Behrensmeyer, A. K., Todd, N. E., Potts, R., & McBrinn, G. E. (1997). Late Pliocene faunal turnover in the Turkana Basin, Kenya and Ethiopia. *Science*, **278**, 1589–1594.

Benefit, B. R. (1987). The molar morphology, natural history, and phylogenetic position of the middle Miocene monkey *Victoriapithecus*. PhD dissertation, New York University.

Benefit, B. R. (1991). The taxonomic status of Maboko small apes. *American Journal of Physical Anthropology*, Supplement **12**, 50–51.

Benefit, B. R. (1993). The permanent dentition and phylogenetic position of *Victoriapithecus* from Maboko Island, Kenya. *Journal of Human Evolution*, **25**, 83–172.

Benefit, B. R. (1994). Phylogenetic, paleodemographic and taphonomic implications of *Victoriapithecus* deciduous teeth from Maboko, Kenya. *American Journal of Physical Anthropology*, **95**, 277–331.

Benefit, B. R. (1999*a*). *Victoriapithecus*, the key to Old World monkey and catarrhine origins. *Evolutionary Anthropology*, **7**, 155–174.

Benefit, B. R. (1999*b*). Mandibular evidence bearing on the phylogenetic position of *Victoriapithecus* and its relationship to *Prohylobates*. *American Journal of Physical Anthropology*, Supplement **28**, 90.

Benefit, B. R. (1999*c*). Biogeography, dietary specialization and the diversification of African Plio-Pleistocene monkeys. In *African Biogeography, Climate Change, and Human Evolution*, eds. T. G. Bromage & F. Schrenk, pp. 172–188. Oxford: Oxford University Press.

Benefit, B. R (2000). Old World monkey origins and diversification: an evolutionary study of diet and dentition. In *Old World Monkeys*, eds. P. F. Whitehead & C. J. Jolly, pp. 133–179. Cambridge: Cambridge University Press.

Benefit, B. R., Gitau, S. N., McCrossin, M. L., & Palmer, A. K. (1998). A mandible of *Mabokopithecus clarki* sheds new light on oreopithecid evolution. *American Journal of Physical Anthropology*, Supplement **26**, 109.

Benefit, B. R., & McCrossin, M. L. (1989). New primate fossils from the Middle Miocene of Maboko Island Kenya. *Journal of Human Evolution*, **18**, 493–498.

Benefit, B. R., & McCrossin, M. L. (1989). The facial morphology of *Victoriapithecus*. *American Journal of Physical Anthropology*, **78**, 191.

Benefit, B. R. & McCrossin, M. L. (1990). Diet, species diversity and distribution of African fossil baboons. *Kroeber Anthropological Society Papers*, **71/72**, 77–93.

Benefit, B. R., & McCrossin, M. L. (1991). Ancestral facial morphology of Old World higher primates. *Proceedings of the National Academy of Sciences of the United States of America*, **88**, 5267–5271.

Benefit, B. R. & McCrossin, M. L. (1993*a*). The facial anatomy of *Victoriapithecus* and its relevance to the ancestral cranial morphology of Old World monkeys and apes. *American Journal of Physical Anthropology*, **92**, 329–370.

Benefit, B. R. & McCrossin, M. L. (1993*b*). The lacrimal fossa of Cercopithecoidea, with special reference to cladistic analysis of Old World monkey relationships. *Folia Primatologica*, **60**, 133–145.

Benefit, B. R. & McCrossin, M. L. (1993*c*). New *Kenyapithecus* postcrania and other primate fossils from Maboko Island, Kenya. *American Journal of Physical Anthropology*, Supplement **16**, 55–56.

Benefit, B. R. & McCrossin, M. L. (1994). Comparative study of the dentition of *Kenyapithecus africanus* and *K. wickeri*. *American Journal of Physical Anthropology*, Supplement **18**, 55.

Benefit, B. R. & McCrossin, M. L. (1995). Miocene hominoids and hominid origins. *Annual Review of Anthropology*, **24**, 237–256.

Benefit, B. R. & McCrossin, M. L. (1997*a*). Earliest known Old World monkey skull. *Nature*, **388**, 368–371.

Benefit, B. R. & McCrossin, M. L. (1997*b*). New fossil evidence bearing on the relationship of *Nyanzapithecus* and *Oreopithecus*. *American Journal of Physical Anthropology*, Supplement **24**, 74.

Benefit, B. R. & McCrossin, M. L. (2000). Middle Miocene hominoid origins. *Science*, **287**, 2375a.

Benefit, B. R. & Pickford, M. (1986). Miocene fossil cercopithecoids from Kenya. *American Journal of Physical Anthropology*, **69**, 441–464.

Benton, M. J. (1998). Molecular and morphological phylogenies of mammals: congruence with stratigraphic data. *Molecular Phylogenetics and Evolution*, **9**, 398–407.

Benveniste, R. E. & Todaro, G. J. (1976). Evolution of type C viral genes: evidence for an Asian origin of man. *Nature*, **261**, 101–108.

Berckhemer, F. (1933). Ein Menschen-Schädel aus den diluvalen Schottern von Steinheim a. d. Murr. *Anthropologischer Anzeiger*, **10**, 318–321.

Bergounioux, F. M. & Crouzel, F. (1965). Les Pliopithèques de France. *Annale de paléontologie*, **51**, 45–65.

Bermúdez de Castro, J., Arsuaga, J. L., Carbonell, E., Rosas, A., Martínez, I., & Mosquera, M. (1997). A hominid from the Lower Pleistocene of Atapuerca, Spain: a possible ancestor to Neandertals and modern humans. *Science*, **276**, 1392–1395.

Bermúdez de Castro, J., Rosas, A., & Nicolás, M. (1999). Dental remains from Atapuerca TD 6 (Gran Dolina site, Burgos, Spain). *Journal of Human Evolution*, **37**, 523–566.

Bernor, R. L. (1983). Geochronology and zoogeographic relationships of Miocene Hominoidea. In *New Interpretations of Ape and Human Ancestry*, eds. R. L. Ciochon & R. S. Corruccini, pp. 21–64. New York: Plenum Press.

Bernor, R. L. (1984). A zoogeographic theater and biochronologic play: the time/biofacies phenomena of Eurasian and African Miocene mammal provinces. *Contributions in Paleobiology*, **18**, 121–142.

Bernor, R. L., Flynn, L. J., Harrison, T., Hussain, S. T., & Kelley, J. (1988). *Dionysopithecus* from Southern Pakistan and the biochronology and biogeography of early Eurasian catarrhines. *Journal of Human Evolution*, **17**, 339–358.

Beynon, A. D., Dean, M. C., Leakey, M. G., Reid, D. J., & Walker, A. C. (1998). Comparative dental development and microstructure of *Proconsul* teeth from Rusinga Island, Kenya. *Journal of Human Evolution*, **35**, 163–209.

Biedermann, W. G. A. (1863). *Petrefacten aus der Umgegend von Winterthur*, vol. 2, *Die Braunkohlen von Elgg*. Winterthur: Bleuler-Hausheer.

Biegert, J. (1957). Der Formwandel des Primatenschadels. *Morphologisches Jahrbuch*, **98**, 77–199.

Biggers, J. D. (1967). Notes on the reproduction of the woolly opossum (*Caluromys derbianus*) in Nicaragua. *Journal of Mammalogy*, **48**, 678–680.

Bilsborough, A. & Wood, B. A. (1988). Cranial morphometry of early hominids: facial region. *American Journal of Physical Anthropology*, **76**, 61–86.

Bishop, W. W. (1964). More fossil primates and other Miocene mammals from north-east Uganda. *Nature*, **203**, 1327–1331.

Bishop, W. W. (1968). The evolution of fossil environments in East Africa. *Transactions of the Leicester Literary and Philosophical Society*, **62**, 22–44.

Bishop, W. W. & Chapman, G. R. (1970). Early Pliocene sediments and fossils from the northern Kenya Rift Valley. *Nature*, **226**, 914–918.

Bishop, W. W., Chapman, G. R., Hill, A., & Miller, J. A. (1971) Succession of Cainozoic vertebrate assemblages from the northern Kenya Rift Valley. *Nature*, **233**, 384–394.

Bishop, W. W., Miller, J. A., & Fitch, F. J. (1969). New potassium-argon age determinations relevant to the Miocene fossil mammal sequence in East Africa. *American Journal of Science*, **267**, 669–699.

Black, D. (1927). On a lower molar hominid tooth from the Chou Kou Tien deposit. *Palaeontologica Sinica, Series D*, **7**, 1.

Black, D. (1933). The Croonian Lecture: On the discovery, morphology and environment of *Sinanthropus pekinensis*. *Philosophical Transactions of the Royal Society of London, Series B*, **223**, 57–120.

Black, D. (1934). Recent discoveries at Choukoutien. *Nature*, **133**, 89–90.

Blainville, H. de (1836). Prétendues empreintes de pieds d'un quadrupède dans le grès bigarré de Hildburghausen en Saxe. *Comptes rendus de l'Académie des sciences de Paris*, **II**, 454–455.

Blainville, H. de (1837*a*). Sur un nouvel envoi de fossiles provenant du dépôt de Sansan. *Comptes rendus de l'Académie des sciences de Paris*, **V**, 417–427.

Blainville, H. de (1837*b*). Rapport sur la découverte de plusieurs ossements fossiles de quadrumanes, dans le dépôt tertiaire de Sansan, près d'Auch. *Annales des sciences naturelles*, **13**, 232–247.

Blainville, H. de (1838). Sur l'importance des résultats obtenus par M. Lartet dans les fouilles qu'il a entreprises pour recherches des ossements fossiles. *Comptes rendus de l'Académie des sciences de Paris*, **VII**, 100–106.

Blainville, H. de (1839). *Ostéographie des mammifères*, vol. 1, *Primates et Secundates*. Paris: Baillière.

Blainville, H. de (1840). De l'ancienneté des primates à la surface de la terre. In *Téographie ou description iconographique comparée du squelette et du système dentaire des mammifères récents et fossiles pour servir de base à la zoologie et à la géologie*, ed. H. de Blainville. Paris.

Bloch, J. I., & Silcox, M. T. (2001). New basicrania of Paleocene–Eocene *Ignacius*: re-evaluation of the pleisiadapiform–dermopteran link. *American Journal of Physical Anthropology*, **116**, 184–198.

Block, J. I., Fisher, D. C., Gingerich, P. D., Gunnell, G. F., Simons, E. L., Uhen, M. D., Kay, R. F., Ross, C., Williams, B. A., & Johnson, D. (1997). Cladistic analysis and anthropoid origins (and reply). *Science*, **278**, 2134–2136.

Bluntschli, H. (1913). Die fossilen Affen patagoniens und der Ursprung der playtrrhinen Affen. *Anatomischer Anzeiger*, **44**, 33–43.

Bluntschli, H. (1931). *Homunculus patagonicus* und die ihm zugereihten Fossilfunde aus den Santa-Cruz-Schichten patagoniens: eine morphologische Rivision an Hand der Originalstucke in der Sammlung Ameghino zu La Plata. *Genenbaurs morphologisches Jahrbuch*, **67**, 811–892.

Boaz, N. T., Bernor, R. L., Brooks, A. S., Cooke, H. B. S., De Heinzelin, J., Dechamps, R., Delson, E., Gentry, A. W., Harris, J. W. K., Meylan, P., Pavlakis, P. P., Sanders, W. J., Stewart, K. M., Verniers, J., Williamson, P. G., & Winkler, A. J. (1992). A new evaluation of the significance of the Late Neogene Lusso Beds, Upper Semliki Valley, Zaïre. *Journal of Human Evolution*, **22**, 505–517.

Bonis, L. de, Bouvrain, G., Geraads, D., & Koufos, G. D. (1997). New material of *Mesopithecus* (Mammalia, Cercopithecidae) from the late Miocene of Macedonia, Greece. *Neues Jahrbuch für Mineralogie, Geologie und Paläontologie 1997*, **5**, 255–265.

Bonis, L. de, Bouvrain, G., & Koufos, G. (1988*a*). Late Miocene mammal localities of the lower Axios valley (Macedonia, Greece) and their stratigraphic significance. *Modern Geology*, **13**, 141–147.

Bonis, L. de, Bouvrain, G., & Melentis, J. (1975). Nouveaux restes de primates hominoïdes dans le Vallésien de Macédoine (Grèce). *Comptes rendus de l'Académie des sciences de Paris*, **182**, 379–382.

Bonis, L. de, Jaeger, J. J., Coiffat, B., & Coiffait, P. E. (1988*b*). Découverte du plus ancien primate catarrhinen connu dans l'Éocène supérieur d'Afrique du Nord. *Comptes rendus de l'Académie des sciences de Paris*, **306**, 929–934.

Bonis, L. de & Koufos, G. (1993). The face and mandible of

Ouranopithecus macedoniensis: description of new specimens and comparisons. *Journal of Human Evolution*, **24**, 469–491.

Bonis, L. de & Koufos, G. (1997). The phylogenetic and functional implications of *Ouranopithecus macedoniensis*. In *Function, Phylogeny, and Fossils: Miocene Hominoid Evolution and Adaptations*, eds. D. R. Begun, C. V. Ward, & M. D. Rose, pp. 317–326. New York: Plenum Press.

Bonis, L. de, Koufos, G. D., Guy, F., Peigne, S., & Sylvestrou, I. (1998). Nouveaux restes du primate hominoïde *Ouranopithecus* dans les dépôts du Miocène supérieur de Macédoine (Grèce). *Comptes rendus de l'Académie des science de Paris*, **327**, 141–146.

Bonis, L. de & Melentis, J. (1977). Les Primates hominoïdes du Vallésien de Macédoine (Grèce): étude de la machoire inférieure. *Géobios*, **10**, 849–885.

Bonis, L. de & Melentis, J. (1987). Intérêt de l'anatomie naso-maxillaire pour la phylogénie de Hominoidea. *Comptes rendus de l'Académie des sciences de paris*, **304**, 767–769.

Bonner, T. I., Heinemann, R., & Todaro, G. J. (1980). Evolution of DNA sequences has been retarded in Malagasy primates. *Nature*, **286**, 470–423.

Bordas, A. F. (1942). Anotaciones sobre un "Cebidae" fosil de Patagonia. *Physis*, **19**, 265–269.

Borissoglebskaya, M. B. (1981). New species of monkey (Mammalia, Primates) from the Pliocene of N. Mongolia. *Iskopaemye pozvonochnye Mongolii*, **15**, 95–108.

Boschetto, H. B., Brown, F. H., & McDougall, I. (1992). Stratigraphy of the Lothidok Range, northern Kenya, and K-Ar ages of its Miocene primates. *Journal of Human Evolution*, **22**, 47–71.

Bosler, W. (1981). Species groupings of Early Miocene dryopithecine teeth from East Africa. *Journal of Human Evolution*, **10**, 151–158.

Boule, M. (1908). L'homme fossile de La Chapelle-aux-Saints. *L'Anthropologie*, **20**, 260–264.

Boule, M. (1911–13). L'homme fossile de la Chapelle-aux-Saints. *Annales de paléontologie*, **6**, 111–172; **7**, 21–192; **8**, 1–70, 209–278.

Boule, M. (1921). *Les Hommes fossiles: éléments de paléontologie humaine*. Paris: Masson.

Bowler, P. (1986). *Theories of Human Evolution: A Century of Debate, 1844–1944*. Baltimore, MD: Johns Hopkins University Press.

Bown, T. M. (1974). Notes on some early Eocene anaptomorphine primates. *Contributions in Geology, University of Wyoming*, **13**, 19–26.

Bown, T. M. (1976). Affinities of *Teilhardina* (Primates, Omomyidae) with description of a new species from North America. *Folia Primatologica*, **25**, 62–72.

Bown, T. M. (1979). New omomyid primates (Haplorhini, Tarsiiformes) from middle Eocene rocks of west-central Hot Springs County, Wyoming. *Folia Primatologica*, **31**, 48–73.

Bown, T. M. & Fleagle, J. G. (1993). Systematics, biostratigraphy, and dental evolution of the Palaeothentidae, later Oligocene to early–middle Miocene (Deseadan–Santacrucian) caenolestoid marsupials of South America. *Journal of Paleontology*, **67** (**2** Supplement), 1–176.

Bown, T. M. & Kraus, M. D. (1988). Geology and paleoenvironment of the Oligocene Jebel Qatrani Formation and adjacent rocks, Fayum Depression, Egypt. *United States Geological Survey Special Papers*, **1452**, 1–60.

Bown, T. M., Kraus, M. J., Wing, S. L., Fleagle, J. G., Tiffney, B. H., Simons, E. L., & Vondra, C. F. (1982). The Fayum primate forest revisited. *Journal of Human Evolution*, **11**, 603–632.

Bown, T. M. & Rose, K. D. (1976). New Early Tertiary primates and a reappraisal of some Plesiadapiformes. *Folia Primatologica*, **26**, 109–138.

Bown, T. M. & Rose, K. D. (1984). Reassessment of some early Eocene Omomyidae, with description of a new genus and three new species. *Folia Primatologica*, **43**, 97–112.

Bown, T. M. & Rose, K. D. (1987). Patterns of dental evolution in early Eocene anaptomorphine primates (Omomyidae) from the Bighorn Basin, Wyoming. *Journal of Paleontology*, **61**(**5** Supplement), 1–162.

Bown, T. M. & Rose, K. D. (1991). Evolutionary relationships of a new genus and three new species of omomyid primates (Willwood Formation, Lower Eocene, Bighorn Basin, Wyoming). *Journal of Human Evolution*, **20**, 465–480.

Brace, C. L. (1964). The fate of the "classic" Neanderthals: a consideration of hominid catastrophism. *Current Anthropology*, **5**, 3–43.

Brace, C. L. (1967). *The Stages of Human Evolution*. Englewood Cliffs, NJ: Prentice-Hall.

Brace, C. L. (1995). *The Stages of Human Evolution*, 5th edn. Englewood Cliffs, NJ: Prentice-Hall.

Braga, L. (1995). Définition de certains caractères discrets crâniens chez *Pongo*, *Gorilla* et *Pan*: perspectives taxonomiques et phylogénétiques. PhD dissertation, University of Bordeaux I.

Branco, W. (1898). Die menschenähnlichen Zähne aus dem Bohnertz der schwäbischen Alp. *Jahreshefte des Vereins für vaterländische Naturkunde in Württemberg*, **54**, 1–144.

Brandon-Jones, D. (1995). A revision of the Asian pied leaf monkeys (Mammalia: Cercopithecidae: Superspecies: *Semnopithecus auratus*), with a description of a new subspecies. *Raffles Bulletin of Zoology*, **43**, 3–43.

Bräuer, G. (1984). A craniological approach to the origin of anatomically modern *Homo sapiens* in Africa and implications for the origin of modern Europeans. In *The Origins of Modern Humans: A World Survey of the Fossil Evidence*, eds. F. H. Smith & F. Spencer, pp. 327–410. New York: Alan R. Liss.

Bräuer, G. (1989). The evolution of modern humans: a comparison of the African and non- African evidence. In *The Human Revolution*, eds. P. Mellars & C. Stringer, pp. 123–154. Edinburgh: Edinburgh University Press.

Bräuer, G. (1992). Africa's place in the evolution of *Homo sapiens*. In *Continuity or Replacement: Controversies in* Homo sapiens *Evolution*, eds. G. Bräuer & F. H. Smith, pp. 83–98. Rotterdam: Balkema.

Bräuer, G. (1994). How different are Asian and African *Homo erectus*? *Courier Forschungsinstitut Senckenberg*, **171**, 301–318.

Bräuer, G. & Broeg, H. (1998). On the degree of Neandertal–modern continuity in the earliest Upper Palaeolithic crania from the Czech Republic: evidence from non-metrical features. In *The Origins of Past Modern Humans: Towards Reconciliation*, eds. K. Omoto & P. V. Tobias, pp. 106–125. Singapore: World Scientific.

Bräuer, G. & Mbua, E. (1992). *Homo erectus* features used in cladistics and their variability in Asian and African hominids. *Journal of Human Evolution*, **22**, 79–108.

Bräuer, G. & Schultz, M. (1996). The morphological affinities of the Plio-Pleistocene mandible from Dmanisi, Georgia. *Journal of Human Evolution*, **30**, 445–481.

Bräuer, G. & Stringer, C. B. (1997). Models, polarization and perspectives on modern human origins. In: *Conceptual Issues in Modern Human Origins Research*, eds. G. Clark & C. Willermet, pp.

191–201. New York: Aldine de Gruyter.

Bräuer, G., Yokoyama, Y., Falguéres, C., & Mbua, E. (1997). Modern human origins backdated. *Nature*, **386**, 337–338.

Brenoe, F. T. (2000). A total evidence phylogeny of fossil and extant callitrichids using morphology, mitochondrial, and nuclear genes. *Journal of Vertebrate Paleontology*, **20**, 31A.

Broca, P. (1869). On the crania and bones of Les Eyzies. In: *Transactions of the International Congress of Anthropology and Prehistoric Archaeology* (Norwich, England, 1868), pp. 168–175. London: Longmans Green.

Bromage, T. G. & Schrenk, F. (1986). A cercopithecoid tooth from the Pliocene of Malawi. *Journal of Human Evolution*, **15**, 497–500.

Bromage, T. G., Schrenk, F., & Zonneveld, F. W. (1995). Paleoanthropology of the Malawi Rift: an early hominid mandible from the Chiwondo Beds, northern Malawi. *Journal of Human Evolution*, **28**, 71–108.

Broom, R. (1936*a*). A new fossil anthropoid skull from Sterkfontein, near Krugersdorp, South Africa. *Nature*, **138**, 486–488.

Broom, R. (1936*b*). A new fossil baboon from the Transvaal. *Annals of the Transvaal Museum*, **18**, 393–396.

Broom, R. (1937). On some new Pleistocene mammals from limestone caves of the Transvaal. *South African Journal of Science*, **33**, 750–768.

Broom, R. (1938). Pleistocene anthropoid apes of South Africa. *Nature*, **142**, 377–379.

Broom, R. (1939). On the affinities of the South African Pleistocene anthropoids. *South African Journal of Science*, **36**, 408–411.

Broom, R. (1940). The South African Pleistocene cercopithecid apes. *Annals of the Transvaal Museum*, **20**, 89–100.

Broom, R. (1949). Another new type of fossil ape-man (*Paranthropus crassidens*). *Nature*, **163**, 57.

Broom, R. & Jensen, J. S. (1946). A new fossil baboon from the caves at Potgietersrust. *Annals of the Transvaal Museum*, **20**, 337–340.

Broom, R. & Robinson, J. T. (1949). A new type of fossil man. *Nature*, **164**, 322–323.

Broom, R. & Robinson, J. T. (1950). Man contemporaneous with the Swartkrans ape-man. *American Journal of Physical Anthropology*, **8**, 151–155.

Broom, R. & Robinson, J. T. (1952). Swartkrans ape-man: *Paranthropus crassidens*. *Transvaal Museum Memoir*, **6**, 1–123.

Broom, R. & Schepers, G. W. H. (1946). The South African fossil ape-men: the Australopithecinae. *Transvaal Museum Memoir*, **2**, 1–272.

Brose, D. & Wolpoff, M. (1971). Early Upper Paleolithic man and late Middle Paleolithic tools. *American Anthropologist*, **73**, 1156–1194.

Brothwell, D. (1960). Upper Pleistocene human skull from Niah Caves, Sarawak. *Sarawak Museum Journal*, **9**, 323–349.

Brown, B., Gregory, W. K., & Hellman, M. (1924). On three incomplete anthropoid jaws from the Siwaliks, India. *American Museum Novitates*, **130**, 1–9.

Brown, F., Harris, J., Leakey, R., & Walker, A. C. (1985). Early *Homo erectus* skeleton from west Lake Turkana, Kenya. *Nature*, **316**, 788–792.

Brown, P. (1993). Recent human evolution in East Asia and Australasia. In *The Origin of Modern Humans and the Impact of Chronometric Dating*, eds. M. Aitken, C. Stringer, & P. Mellars, pp. 217–233. Princeton, NJ: Princeton University Press.

Brown, P. (1997). Australian paleoanthropology. In *History of Physical Anthropology: An Encyclopedia*, ed. F. Spencer, pp. 138–145. New York: Garland.

Brunet, M., Beauvilain, A., Coppens, Y., Heintz, E., Moutaye, A. H. E., & Pilbeam, D. (1995). The first australopithecine 2500 kilometres west of the Rift Valley (Chad). *Nature*, **378**, 273–275.

Brunet, M., Beauvilain, A., Coppens, Y., Heintz, E., Moutaye, A. H. E., & Pilbeam, D. (1996). *Australopithecus bahrelghazali*, a new species of early hominid from Koro Toro region, Chad. *Comptes rendus de l'Académie des sciences de Paris*, **322**, 907–913.

Brunet, M., Beauvilain, A., Geraads, D., Guy, F., Kasser, M., Mackaye, H. T., MacLatchy, L. M., Mouchelin, G., Sudre, J., & Vignaud, P. (1997). Tchad: un nouveau site à hominidés pliocène. *Comptes rendus de l'Académie des sciences de Paris*, **324**, 341–345.

Buckland, W. (1823). *Reliquiae Diluvianae; or: Observations on the Organic Remains Contained in Caves, Fissures, and Diluvial Gravel and on Other Geological Phenomena Attesting the Action of an Universal Deluge*. London: John Murray.

Buckley, G. A. (1997). A new species of *Purgatorius* (Mammalia; Primatomorpha) from the Lower Paleocene Bear Formation, Crazy Mountains Basin, South-Central Montana. *Journal of Palaeontology*, **71**, 149–155.

Bugge, J. (1974). The cephalic arterial system in insectivores, primates, rodents, and lagomorphs, with special reference to the systematic classification. *Acta Anatomica*, **101**, 45–61.

Burchak-Abramovitsch, N. O. & Gabashvili, E. G. (1950). Discovery of a fossil anthropoid in Georgia (in Russian). *Priroda, Moscow*, **9**, 70–72.

Burney, D. A. (1993). Late Holocene environmental changes in arid southwestern Madagascar. *Quaternary Research*, **40**, 98–106.

Burney, D. A. (1997). Theories and facts regarding Holocene environmental change before and after human colonization. In *Natural Change and Human Impact in Madagascar*, eds S. M. Goodman & B. D. Patterson, pp. 75–89. Washington, DC: Smithsonian Institution Press.

Burney, D. A. (1999). Rates, patterns, and the processes of landscape transformation and extinction in Madagascar. In *Extinctions in Near Time: Causes, Contexts, and Consequences*, ed. R. D. E. MacPhee, pp. 145–164. New York: Plenum Press.

Burney, D. A., James, H. F., Grady, F. V., Rafamantanantsoa, J. G., Ramilisonina, Wright, H. T., & Cowart, J. B. (1997). Environmental change, extinction, and human activity: evidence from caves in NW Madagascar. *Journal of Biogeography*, **24**, 755–767.

Burney, D. A. & Ramilisonina (1998). The kilopilopitsofy, kidoky, and bokyboky: accounts of strange animals from Belo-sur-mer, Madagascar, and the megafaunal "extinction window". *American Anthropologist*, **100**, 957–966.

Burr, D. B., Ruff, C. B., & Johnson, C. (1989). Structural adaptations of the femur and humerus to arboreal and terrestrial environments in three species of macaque. *American Journal of Physical Anthropology*, **79**, 357–367.

Busk, G. (1865). On a very ancient human cranium from Gibraltar. *Report of the 34th Meeting of the British Association for the Advancement of Science* (Notices and Abstracts), 91–92.

Butler, P. M. (1972). The problem of insectivore classification. In *Studies in Vertebrate Evolution*, eds. J. A. Joysey & T. S. Kemp, pp. 253–265. Edinburgh: Oliver & Boyd.

Butler, P. M. & Mills, J. R. E. (1959). A contribution to the odontology of *Oreopithecus*. *Bulletin of the British Museum (Natural History), Geology Series*, **4**, 1–26.

Caccone, A. & Powell, J. R. (1989). DNA divergence among hominoids. *Evolution*, **43**, 925–942.

Cachel, S. M. (1979*a*). A functional analysis of the primate masticatory system and the origin of the anthropoid post-orbital septum. *American Journal of Physical Anthropology*, **50**, 1–18.

Cachel, S. (1979*b*). A paleoecological model for the origin of higher primates. *Journal of Human Evolution*, **8**, 351–359.

Cachel, S. (1981). Plate tectonics and the problem of anthropoid origins. *Yearbook of Physical Anthropology*, **24**, 139–172.

Cameron, D. W. (1991). Sexual dimorphism in the early Miocene species of *Proconsul* from the Kisingiri Formation of East Africa: a morphometric examination using multivariate statistics. *Primates*, **32**, 329–343.

Cameron, D. W. (1992). A morphometric analysis of extant and Early Miocene fossil hominoid maxillo-dental specimens. *Primates*, **33**, 377–390.

Cameron, D. W. (1997). A revised systematic scheme for the Eurasian Miocene fossil Hominidae. *Journal of Human Evolution*, **33**, 449–477.

Cameron, D. W., Patnaik, R., & Sahni, A. (1999). *Sivapithecus* dental specimens from Dhara locality, Kalgarh District, Uttar Pradesh, Siwaliks, Northern India. *Journal of Human Evolution*, **37**, 861–868.

Campbell, B. G. (1963). Quantitative taxonomy and human evolution. In *Classification and Human Evolution*, ed. S. L. Washburn, pp. 50–74. Chicago, IL: Aldine Press.

Campbell, B. G. (1965). *The Nomenclature of the Hominidae*. Royal Anthropological Institute, Occasional Paper, **22**, 1–33.

Campbell, B. G., Amini, M. H., Bernor, R. L., Dickinson, W., Drake, R., Morris, R., Van Couvering, J. A., & Van Couvering, J. A. H. (1980). Maragheh: a classical late Miocene vertebrate locality in northwestern Iran. *Nature*, **287**, 837–841.

Canavez, F. C., Moreira, M. M., Ladasky, J. J., Pissinatti, A., Parham, P., & Seuánez, H. (1999). Molecular phylogeny of New World primates (Platyrrhini) based on β2-microglobulin DNA sequences. *Molecular Phylogeny and Evolution*, **12**, 74–82.

Cann, R., Richards, O., & Lum, K. (1994). Mitochondrial DNA and human evolution: our one lucky mother. In *Origins of Anatomically Modern Humans*, eds. M. H. Nitecki & D. V. Nitecki, pp. 135–148. New York: Plenum Press.

Cann, R., Stoneking, M., & Wilson, A. (1987). Mitochondrial DNA and human evolution. *Nature*, **325**, 31–36.

Capitan, L. & Peyrony, D. (1909). Deux squelettes humains au milieu de foyer de l'époque moustérienne. *Revue d'anthropologie*, **22**, 76–99.

Carleton, A. (1936). The limb-bones and vertebrae of the extinct lemurs of Madagascar. *Proceedings of the Zoological Society of London*, **106**, 281–307.

Cartelle, C. (1993). Achado de *Brachyteles* do Pleistoceno Final. *Neotropical Primates*, **1**, 8.

Cartelle, C. & Hartwig, W. C. (1996*a*). A new extinct primate among the Pleistocene megafauna of Bahia, Brazil. *Proceedings of the National Academy of Sciences of the United States of America*, **93**, 6405–6409.

Cartelle, C. & Hartwig, W. C. (1996*b*). Updating the two Pleistocene primates from Bahia, Brasil. *Neotropical Primates*, **4**, 46–48.

Cartmill, M. (1972). Arboreal adaptations and the origin of the order Primates. In *The Functional and Evolutionary Biology of Primates*, ed. R. Tuttle, pp. 97–122. Chicago, IL: Aldine-Atherton.

Cartmill, M. (1974*a*). Pads and claws in arboreal locomotion. In *Primate Locomotion*, ed. F. A. Jenkins, Jr., pp. 45–83. New York: Academic Press.

Cartmill, M. (1974*b*). *Daubentonia*, *Dactylopsila*, woodpeckers, and klinorhynchy. In *Prosimian Biology*, eds. R. D. Martin, G. A. Doyle, & A. C. Walker, pp. 655–670. London: Duckworth.

Cartmill, M. (1974*c*). Rethinking primate origins. *Science*, **184**, 436–443.

Cartmill, M. (1975). Strepsirhine basicranial structures and the affinities of the Cheirogaleidae. In *Phylogeny of the Primates*, eds. W. Luckett & F. Szalay, pp. 313–356. New York: Plenum Press.

Cartmill, M. (1980). Morphology, function, and evolution of the anthropoid postorbital septum. In *Evolutionary Biology of the New World Monkeys and Continental Drift*, eds. R. L. Ciochon & A. B. Chiarelli, pp. 243–274. New York: Plenum Press.

Cartmill, M. (1992). New views on primate origins. *Evolutionary Anthropology*, **1**, 105–111.

Cartmill, M., & Kay, R. F. (1978). Cranio-dental morphology, tarsier affinities, and primate suborders. In *Recent Advances in Primatology*, vol. 3, eds. D. J. Chivers & K. A. Joysey, pp. 205–214. London: Academic Press.

Cartmill, M., MacPhee, R. D. E., & Simons, E. L. (1981). Anatomy of the temporal bone in early anthropoids, with remarks on the problem of anthropoid origins. *American Journal of Physical Anthropology*, **56**, 3–21.

Castenholtz, A. (1984). The eye of *Tarsius*. In *The Biology of Tarsiers*, ed. C. Niemitz, pp. 127–134. New York: Gustav Fischer.

Caton, J. M. (1999). Digestive strategy of the Asian colobine genus *Trachypithecus*. *Primates*, **40**, 311–325.

Cautley, P. T. (1837). On the finding of the remains of a quadrumanous animal in the Sewaliks, or Sub-Himalayan range of mountains. *Proceedings of the Geological Society of London*, **II**, 544–545.

Cautley, P. T. & Falconer, H. (1840). Notice on the remains of a fossil monkey from the Tertiary strata of the Sewalik Hills in the North of Hindoostan. *Transactions of the Geological Society of London (Series 2)*, **5**, 499–504.

Cave, A. J. E. (1967). Observations on the platyrrhine nasal fossa. *American Journal of Physical Anthropology*, **26**, 277–288.

Cerling, T. E., Quade, J., Ambrose, S. H., & Sikes, N. E. (1991). Fossil soils from Fort Ternan, Kenya: grassland or woodland? *Journal of Human Evolution*, **21**, 295–306.

Chaimanee, Y., Suteethorn, V., Jaeger, J. J., & Ducrocq, S. (1997). A new late Eocene anthropoid from Thailand. *Nature*, **385**, 429–431.

Chaimanee, Y., Thein, T., Ducrocq, S., Soe, A. N., Benammi, M., Tun, T., Lwin, T., Wai, S., & Jaeger, J. J. (2000*a*). A lower jaw of *Pondaungia cotteri* from the late middle Eocene Pondaung Formation (Myanmar) confirms its anthropoid status. *Proceedings of the National Academy of Sciences of the United States of America*, **97**, 4102–4105.

Chaimanee, Y., Khansubha, S., & Jaeger, J. J. (2000*b*). A new lower jaw of *Siamopithecus eocaenus* from the late Eocene of Thailand. *Comptes rendus de l'Académie des sciences de Paris*, **323**, 235–241.

Chang, Y., Wang, L., Dong, X., & Chen, W. (1975). Discovery of a *Gigantopithecus* tooth from Bama district in Kwangsi. *Vertebrata PalAsiatica*, **13**, 148–154.

Chang, Y., Wu, M., & Liu, C. (1973). New discovery of *Gigantopithecus* teeth from Wuming, Kwangsi. *Kexue Tongbao*, **18**, 130–133.

Chantre, E. & Gaillard, C. (1897). Sur la faune du gisement sidérolithique éocène de Lissieu (Rhône). *Comptes rendus de l'Académie des sciences de Paris*, **125**, 986–987.

Charles-Dominique, P., Cooper, H. M., Hladik, A., Hladik, C. M., Pages, E., Pariente, G. F., Petter-Rousseaux, A., Petter, J. J., & Schilling, A. (eds.) (1980). *Nocturnal Malagasy Primates*. New York: Academic Press.

Charles-Dominique, P. & Martin, R. D. (1970). Evolution of lorises and lemurs. *Nature*, **227**, 257–260.

Chopra, S. R. K. (1983). Significance of recent hominoid discoveries from the Siwalik Hills of India. In *New Interpretations of Ape and Human Ancestry*, eds. R. L. Ciochon & R. S. Corruccini, pp. 539–557. New York: Plenum Press.

Chopra, S. R. K. & Kaul, S. (1975). New fossil *Dryopithecus* material from the Nagri beds at Haritalyangar (H. P.), India. In *Contemporary Primatology*, eds S. Kondo, M. Kawai, & A. Ehara, pp. 2–11. Basel: Karger.

Chopra, S. R. K. & Vasishat, R. N. (1980). A new Mio-Pliocene *Indraloris* (Primate) with comments on the taxonomic status of *Sivanasua* (Carnivora) from the Siwaliks of the Indian subcontinent. *Journal of Human Evolution*, **9**, 129–132.

Chow, M. (1961). A new tarsioid primate from the Lushi Eocene, Hunan. *Vertebrata PalAsiatica*, **5**, 1–5.

Churcher, C. S. (1970). Two new Upper Miocene giraffids from Fort Ternan, Kenya, East Africa: *Palaeotragus primaevus* n. sp. and *Samotherium africanum* n. sp. In *Fossil Vertebrates of Africa*, vol. 1, ed. L. S. B. Leakey, pp. 1–105. London: Academic Press.

Churchill, S. E. & Smith, F. H. (2001). The makers of the early Aurignacian of Europe. *Yearbook of Physical Anthropology*, **44**, 61–116.

Ciochon, R. L. & Chiarelli, A. B. (eds.) (1980*a*). *Evolutionary Biology of New World Monkeys and Continental Drift*. New York: Plenum Press.

Ciochon, R. L. & Chiarelli, A. B. (1980*b*). Concluding remarks. In *Evolutionary Biology of the New World Monkeys and Continental Drift*, eds R. L. Ciochon & A. B. Chiarelli, pp. 495–501. New York: Plenum Press.

Ciochon, R. L. & Corruccini, R. S. (1975). Morphometric analysis of platyrrhine femora with taxonomic implications and notes on two fossil forms. *Journal of Human Evolution*, **4**, 193–217.

Ciochon, R. L. & Corruccini, R. S. (1977). The phenetic position of *Pliopithecus* and its phylogenetic relationship to the Hominoidea. *Systematic Zoology*, **26**, 290–299.

Ciochon, R. L. & Corruccini, R. S. (eds.) (1983). *New Interpretations of Ape and Human Ancestry*. New York: Plenum Press.

Ciochon, R. L. & Holroyd, P. A. (1994). The Asian origin of Anthropoidea revisited. In *Anthropoid Origins*, eds. J. G. Fleagle & R. F. Kay, pp. 143–162. New York: Plenum Press.

Ciochon, R. L., Holroyd, P. A., & Thein, T. (1999). New *Amphipithecus* from the middle Eocene of Myanmar: implications for stem anthropoid origins. *American Journal of Physical Anthropology*, Supplement **28**, 108.

Ciochon, R. L., Long, V. T., Larick, R., Gonzalez, L., Grun, R., Vos, J. de, Yonge, C., Taylor, L., Yoshida, H., & Reagan, M. (1996). Dated co-occurrence of *Homo erectus* and *Gigantopithecus* from Tham Khuyen Cave, Vietnam. *Proceedings of the National Academy of Sciences of the United States of America*, **93**, 3016–3020.

Ciochon, R. L., Olsen, J., & James, J. (1990). *Other Origins*. New York: Bantam.

Ciochon, R. L., Savage, D. E., Tint, T., & Ba Maw (1985). Anthropoid origins in Asia? New discovery of *Amphipithecus* from the Eocene of Burma. *Science*, **229**, 756–759.

Clark, G. A. & Willermet, C. M. (eds.) (1997). *Conceptual Issues in Modern Human Origins Research*. New York: Aldine de Gruyter.

Clark, J. (1941). An anaptomorphid primate from the Oligocene of Montana. *Journal of Paleontology*, **15**, 562–563.

Clark, J. D., De Heinzelin, J., Schick, K., Hart, W., White, T., WoldeGabriel, G., Walter, R., Suwa, G., Asfaw, B., Vrba, E., & Haile-Selassie, Y. (1994). African *Homo erectus*: old radiometric ages and young Oldowan assemblages in the Middle Awash Valley, Ethiopia. *Nature*, **264**, 1907–1910.

Clarke, R. J. (1985). *Australopithecus* and early *Homo* in southern Africa. In *The Hard Evidence*, ed. E. Delson, pp. 171–177. New York: Alan R. Liss.

Clarke, R. J. (1998). First ever discovery of a well-preserved skull and associated skeleton of *Australopithecus*. *South African Journal of Science*, **94**, 460–463.

Clarke, R. J. (1999). Discovery of complete arm and hand of the 3.3 million-year-old *Australopithecus* skeleton from Sterkfontein. *South African Journal of Science*, **96**, 477–480.

Clarke, R. J. & Howell, F. C. (1972). Affinities of the Swarkrans 847 hominid cranium. *American Journal of Physical Anthropology*, **37**, 319–336.

Clarke, R. J., Howell, F. C., & Brain, C. K. (1970). More evidence of an advanced hominid at Swartkrans. *Nature*, **225**, 1219–1222.

Clemens, W. A. (1974). *Purgatorius*, an early paromomyid primate (Mammalia). *Science*, **184**, 903–905.

Clutton-Brock, T. H. & Harvey, P. H. (1980). Primates, brains and ecology. *Journal of Zoology, London*, **190**, 309–323.

Cocchi, I. (1872). Su di due scimmie fossili italiane. *Bollettino del reale Comitata geologico d'Italia*, **3**, 59–71.

Colbert, E. H. (1935). Siwalik mammals in the American Museum of Natural History. *Transactions of the American Philosophical Society*, **26**, 1–401.

Colbert, E. H. (1937). A new primate from the upper Eocene Pondaung Formation of Burma. *American Museum Novitates*, **951**, 1–18.

Colbert, E. H. & Hooijer, D. A. (1953). Pleistocene mammals from the limestone fissures of Szeschwan, China. *Bulletin of the American Museum of Natural History*, **102**, 1–134.

Collard, M. & Wood, B. (2000). How reliable are human phylogenetic hypotheses? *Proceedings of the National Academy of Sciences of the United States of America*, **97**, 5003–5006.

Conklin-Brittain, N., Wrangham, R. W., & Hunt, K. (1998). Dietary response of chimpanzees and cercopithecines to seasonal variation in fruit abundance. II. Macronutrients. *International Journal of Primatology*, **19**, 971–997.

Conroy, G. C. (1976*a*). Hallucial tarsometatarsal joint in an Oligocene anthropoid, *Aegyptopithecus zeuxis*. *Nature*, **262**, 684–686.

Conroy, G. C. (1976*b*). Primate postcranial remains from the Oligocene of Egypt. *Contributions in Primatology*, **8**, 1–134.

Conroy, G. C. (1987). Problems of body-weight estimation in fossil primates. *International Journal of Primatology*, **8**, 115–137.

Conroy, G. C. (1990). *Primate Evolution*. New York: W. W. Norton.

Conroy, G. C. (1994). *Otavipithecus*: or how to build a better hominid-Not. *Journal of Human Evolution*, **27**, 373–383.

Conroy, G. C. (1996). The cave breccias of Berg Aukas, Namibia: a clustering approach to mine dump paleontology. *Journal of Human Evolution*, **30**, 349–355.

Conroy, G. C., Lichtman, J. W., & Martin, L. B. (1995). Brief communication: some observations on enamel thickness and enamel prism packing in the Miocene hominoid *Otavipithecus namibiensis*. *American Journal of Physical Anthropology*, **98**, 595–600.

Conroy, G. C., Pickford, M., Senut, B., van Couvering, J. & Mein, P. (1992). *Otavipithecus namibiensis*, first Miocene hominoid from southern Africa. *Nature*, **356**, 144–148.

Conroy, G. C., Pickford, M., Senut, B., & Mein, P. (1993). Diamonds in the desert: the discovery of *Otavipithecus namibiensis*. *Evolutionary Anthropology*, **2**, 46–52.

Conroy, G. C. & Rose, M. D. (1983). The evolution of the primate foot from the earliest primates to the Miocene hominoids. *Foot and Ankle*, **3**, 342–364.

Conroy, G. C., Senut, B., Gommery, D., Pickford, M., & Mein, P. (1996). Brief communication: new primate remains from the Miocene of Namibia, Southern Africa. *American Journal of Physical Anthropology*, **99**, 487–492.

Conroy, G. C., Weber, G. W., Seidler, H., Tobias, P. V., Kane, A., & Brunsden, B. (1998). Endocranial capacity in an early hominid cranium from Sterkfontein, South Africa. *Science*, **280**, 1730–1731.

Cooke, H. B. S., Malan, B. D., & Wells, L. H. (1945). Fossil man in the Lebombo Mountains, South Africa: the Border cave, Inwavuma District, Zululand. *Man*, **45**, 6–13.

Coon, C. S. (1962). *The Origin of Races*. New York: Knopf.

Cooper, C. F. (1932). On some mammalian remains from the lower Eocene of the London clay. *Annals of the Magazine of Natural History, Series 10*, **9**, 458–467.

Cope, D. A. & Lacy, M. G. (1992). Falsification of a single species hypothesis using the coefficient of variation: a simulation approach. *American Journal of Physical Anthropology*, **89**, 359–378.

Cope, D. A. & Lacy, M. G. (1994). Testing single species hypotheses using the combined referent CV: applications to fossil hominoid dental samples. *American Journal of Physical Anthropology*, Supplement **16**, 71.

Cope, E. D. (1872). On a new vertebrate genus from the northern part of the Tertiary basin of Green River. *Proceedings of the American Philosophical Society*, **12**, 554.

Cope, E. D. (1873). On the extinct Vertebrata of Wyoming observed by the expedition of 1872, with notes on the geology. 6th *Annual Report of the United States Geological Survey of Territories*, 545–649.

Cope, E. D. (1875). Systematic catalogue of Vertebrata of the Eocene of New Mexico, collected in 1874. In *Geographical Expansion and Survey West of 100th Meridian*, ed. G. M. Wheeler, pp. 5–37. Washington, DC: Government Printing Office.

Cope, E. D. (1877). Report upon the extinct Vertebrata obtained in New Mexico by parties of the expedition of 1874. Chapter XII. Fossils of the Eocene period. In *Geographical Surveys West of 100th Meridian*, ed. G. M. Wheeler, pp. 37–282. Washington, DC: Government Printing Office.

Cope, E. D. (1882*a*). An anthropomorphous lemur. *American Naturalist*, **16**, 73–74.

Cope, E. D. (1882*b*). Contributions to the history of the Vertebrata of the lower Eocene of Wyoming and New Mexico, made during 1881. I. The fauna of the Wasatch beds of the basin of the Bighorn River. II. The fauna of the *Catathlaeus* beds, or lowest Eocene, New Mexico. *Proceedings of the American Philosophical Society*, **20**, 139–197.

Cope, E. D. (1883). On the mutual relations of the bunotherian Mammalia. *Proceedings of the National Academy of Sciences of the United States of America*, **35**, 77–83.

Cope, E. D. (1885). The Vertebrata of the Tertiary Formations of the West. *Reports of the United States Geological Survey of Territories*, **3**, 1–1009.

Corruccini, R. S (1975). Metrical analysis of Fontéchevade II. *American Journal of Physical Anthropology*, **42**, 95–98.

Corruccini, R. S. & Ciochon, R. L. (1976). Morphometric affinities of the human shoulder. *American Journal of Physical Anthropology*, **45**, 19–38.

Corruccini, R. S. & Ciochon, R. L. (1978). Morphoclinal variation in the anthropoid shoulder. *American Journal of Physical Anthropology*, **48**, 539–542.

Corruccini, R. S. & Henderson, A. M. (1978). Palatofacial comparison of *Dryopithecus (Proconsul)* with extant catarrhines. *Primates*, **19**, 35–44.

Covert, H. H . (1985). Adaptations and evolutionary relationships of the Eocene primate family Notharctidae. PhD dissertation, Duke University.

Covert, H. H. (1986). Biology of early Cenozoic primates. In *Comparative Primate Biology*, vol. 1, *Systematics, Evolution, and Anatomy*, eds. D. W. Swindler & J. Erwin, pp. 335–359. New York: Alan R. Liss.

Covert, H. H. (1988). Ankle and foot morphology of *Cantius mckennai*: adaptations and phylogenetic implications. *Journal of Human Evolution*, **17**, 57–70.

Covert, H. H. (1990). Phylogenetic relationships among the Notharctinae of North America. *American Journal of Physical Anthropology*, **81**, 381–398.

Covert, H. H. (1995). Locomotor adaptations of Eocene primates: adaptive diversity among the earliest prosimians. In *Creatures of the Dark: The Nocturnal Prosimians*, eds. L. Alterman, G. A. Doyle, & M. Z. Izard, pp. 495–509. New York: Plenum Press.

Covert, H. H. (1997). The early primate adaptive radiations and new evidence about anthropoid origins. In *Biological Anthropology: The State of the Science*, eds. N. T. Boaz & L. D. Wolfe, pp. 1–24. Corvallis, OR: Oregon State University Press.

Covert, H. H. & Hamrick, M. W. (1993). Description of new skeletal remains of the early Eocene anaptomorphine primate *Absarokius* (Omomyidae) and a discussion about its adaptive profile. *Journal of Human Evolution*, **25**, 351–362.

Covert, H. H. & Williams, B. A. (1991). The anterior lower dentition of *Washakius insignis* and adapid–anthropoidean affinities. *Journal of Human Evolution*, **21**, 463–467.

Covert, H. H. & Williams, B. A. (1994). Recently recovered specimens of North American Eocene omomyids and adapids and their bearing on debates about anthropoid origins. In *Anthropoid Origins*, eds. J. G. Fleagle & R. F. Kay, pp. 29–54. New York: Plenum Press.

Crompton, R. H. (1995). "Visual predation", habitat structure, and the ancestral primate niche. In *Creatures of the Dark: The Nocturnal Prosimians*, eds. L. Alterman, G. A. Doyle, & M. K. Izard, pp. 11–30. New York: Plenum Press.

Cronin, J. E. (1977). Anthropoid evolution: the molecular evidence. *Kroeber Anthropological Society Papers*, **50**, 75–84.

Cronin, J. E. & Meikle, W. E. (1982). Hominid and gelada baboon evolution: agreement between molecular and fossil time scales. *International Journal of Primatology*, **3**, 469–482.

Crovella, S., Masters, J. C., & Rumpler, Y. (1994*a*). Highly repeated DNA sequences as phylogenetic markers among the Galaginae.

American Journal of Primatology, **3**, 177–185.

Crovella, S., Montagnon, D., Rakotosamimanana, B., & Rumpler, Y. (1994*b*). Molecular biology and systematics of an extinct lemur: *Pachylemur insignis*. *Primates*, **35**, 519–522.

Crusafont, M. (1958). Neuvo hallazgo del pongido vallesiense *Hispanopithecus*. *Boletín informative actividades europeas paleontología vertebrados, Sabadell, España*, **13–14**, 37–43.

Crusafont, M. & Hürzeler, J. (1961). Les Pongidés fossiles d'Espagne. *Comptes rendus de l'Académie des sciences de Paris*, **254**, 582–584.

Crusafont-Pairo, M. (1967). Sur quelques prosimiens de l'Éocène de la zone préaxiale pyrénaique et un essai provisoire de reclassification. *Colloques internationaux du Centre National de la recherche scientifique Problèmes actuels de paléontologie*, **163**, 611–632.

Crusafont-Pairo, M. & Golpe-Posse, J. M. (1973). Yacimientos del Eoceno prepirenaico (nuevas localidades del Cuisiense). *Acta Geologica Hispanica*, **8**(5), 145–147.

Cuong, N. L. (1984). Paläeontologische Untersuchungen in Vietnam. *Zeitschrift für Archäologie*, **18**, 247–251.

Cuong, N. L. (1985). Fossile Menschenfunde aus Nordvietnam. In *Menschwerdung-biotischer und gesellschaftlicher Entwicklungsprozess*, eds. J. Herrmann & H. Ullrich, pp. 96–102. Berlin: Akademieverlag.

Cuvier, G. (1812). *Recherches sur les ossements fossiles, où l'on rétablit les caractères de plusieurs animaux dont les révolutions du globe ont détruit les espèces*. Paris: Déterville.

Cuvier, G. (1821). *Discours sur la théorie de la terre, servant d'introduction aux recherches sur les ossements fossiles*. Paris.

Dagosto, M. (1983). Postcranium of *Adapis parisiensis* and *Leptadapis magnus* (Adapiformes, Primates): adaptations and phylogenetic significance. *Folia Primatologica*, **41**, 49–101.

Dagosto, M. (1985). The distal tibia of Primates with special reference to the Omomyidae. *International Journal of Primatology*, **6**, 45–76.

Dagosto, M. (1986). The joints of the tarsus in the strepsirhine primates. PhD dissertation, City University of New York.

Dagosto, M. (1988). Implications of postcranial evidence for the origin of euprimates. *Journal of Human Evolution*, **17**, 35–56.

Dagosto, M. (1990). Models for the origin of the anthropoid postcranium. *Journal of Human Evolution*, **19**, 121–140.

Dagosto, M. (1993). Postcranial anatomy and locomotor behavior in Eocene primates. *Postcranial Adaptation in Nonhuman Primates*, ed. D. L. Gebo, pp. 150–174. DeKalb, IL: Northern Illinois University Press.

Dagosto, M. & Gebo, D. L. (1994). Postcranial anatomy and the origin of the Anthropoidea. In *Anthropoid Origins*, eds. J. G. Fleagle & R. F. Kay, pp. 567–593. New York: Plenum Press.

Dagosto, M., Gebo, D. L., & Beard, K. C. (1999). Revision of the Wind River faunas, early Eocene of central Wyoming. Part XIV. Postcranium of *Shoshonius cooperi* (Mammalia: Primates). *Annals of Carnegie Museum*, **68**, 175–211.

Dagosto, M., Gebo, D. L., Beard, C., & Qi, T. (1996). New primate postcranial remains from the middle Eocene Shanghuang fissures, southeastern China. *American Journal of Physical Anthropology*, Supplement **22**, 92–93.

Dagosto, M. & Schmid, P. (1996). Proximal femoral anatomy of omomyiform primates. *Journal of Human Evolution*, **30**, 29–56.

Dagosto, M. & Terranova, C. J. (1992). Estimating the body size of Eocene primates: a comparison of results from dental and postcranial variables. *International Journal of Primatology*, **13**, 307–344.

Dandouau, A. (1922). *Contes populaires des Sakalava et des Tsimihety de la région d'Analalava*. Publications de la Faculté des Lettres d'Alger. Algiers: Jules Carbonel.

Dart, R. A. (1925). *Australopithecus africanus*: the man–ape of South Africa. *Nature*, **115**, 195–199.

Dart, R. A. (1948). The Makapansgat protohuman *Australopithecus prometheus*. *American Journal of Physical Anthropology*, **6**, 259–283.

Darwin, C. (1859). *On the Origin of Species by Means of Natural Selection*. London: John Murray.

Darwin, C. (1871). *The Descent of Man and Selection in Relation to Sex*. London: John Murray.

Darwin, C. (1872). *The Descent of Man and Selection in Relation to Sex*. New York: D. Appleton & Co.

Dashzeveg, D. & McKenna, M. C. (1977). Tarsioid primate from the early Tertiary of the Mongolian People's Republic. *Acta Palaeontologica Polonica*, **22**, 119–137.

Davis, L. C. (1987). Morphological evidence of positional behavior in the hindlimb of *Cebupithecia sarmientoi* (Primates: Platyrrhini). MA thesis, Arizona State University.

Davis, L. C. (1988). Morphological evidence of locomotor behavior in a fossil platyrrhine. *American Journal of Physical Anthropology*, **75**, 202.

Davis, P. R. & Napier, J. (1963). A reconstruction of the skull of *Proconsul africanus* (R. S. 51). *Folia Primatologica*, **1**, 20–28.

Day, M. H. (1969). Omo human skeletal remains. *Nature*, **222**, 1140–1143.

Day, M. H. (1971). Postcranial remains of *Homo erectus* from Bed IV, Olduvai Gorge, Tanzania. *Nature*, **232**, 383–387.

Day, M. H. & Leakey, R. E. F. (1974). New evidence of the genus *Homo* from East Rudolf, Kenya (III). *American Journal of Physical Anthropology*, **41**, 367–380.

Day, M. H., Leakey, R. E. F., Walker, A. C., & Wood, B. A. (1975). New hominids from East Rudolf, Kenya, I. *American Journal of Physical Anthropology*, **42**, 461–476.

Day, M. H. & Napier, J. R. (1964). Hominid fossils from Bed I, Olduvai Gorge, Tanganyika. Fossil foot bones. *Nature*, **201**, 967–970.

Day, M. H. & Stringer, C. B. (1982). A reconsideration of the Omo Kibish remains and the *erectus-sapiens* transition. In: L'Homo erectus *et la place de l'homme de Tautavel parmi les hominidés fossiles*, ed. M. A. de Lumley, pp. 814–846. Nice: Centre national de la recherche scientifique.

Dechow, P. C. & Singer, R. (1984). Additional fossil *Theropithecus* from Hopefield, South Africa: a comparison with other African sites and a reevaluation of its taxonomic status. *American Journal of Physical Anthropology*, **63**, 405–435.

Decker, R. L. & Szalay, F. S. (1974). Origins and function of the pes in the Eocene Adapidae (Lemuriformes, Primates). In *Primate Locomotion*, ed. F. A. Jenkins, Jr., pp. 261–291. New York: Academic Press.

Dehm, R. (1983). Miocene hominoid primate dental remains from the Siwaliks of Pakistan. In *New Interpretations of Ape and Human Ancestry*, eds R. L. Ciochon & R. S. Corruccini, pp. 527–537. New York: Plenum Press.

Deinard, A. S., Sirugo, G., & Kidd, K. K. (1998). Hominoid phylogeny: inferences from a sub-terminal minisatellite analyzed by repeat expansion detection (RED). *Journal of Human Evolution*, **35**, 313–318.

De Jong, W. W. & Goodman, M. (1988). Anthropoid affinities of *Tarsius* supported by lens αA-crystallin sequences. *Journal of Human*

Evolution, **17**, 575–82.

Delfortrie, E. (1873). Un singe de la familie des Lémuriens. *Actes de la Société linnéenne de Bordeaux*, **24**, 87–95.

DelPero, M., Crovella, S., Cervella, P., Ardito, G., & Rumpler, Y. (1995). Phylogenetic relationships among Malagasy lemurs as revealed by mitochondrial-DNA sequence-analysis. *Primates*, **36**, 431–440.

DelPero, M., Masters, J. C., Cervella, P., Crovella, S., Ardito, G., & Rumpler, Y. (2001). Phylogenetic relationships among the Malagasy lemuriforms (Primates: Strepsirhini) as indicated by mitochondrial sequence data from the 12S rRNA gene. *Zoological Journal of the Linnean Society*, **133**, 83–103.

Delson, E. (1973). Fossil colobine monkeys of the circum-Mediterranean region and the evolutionary history of the Cercopithecidae (Primates, Mammalia). PhD dissertation, Columbia University.

Delson, E. (1975*a*). Evolutionary history of the Cercopithecidae. In *Approaches to Primate Paleobiology: Contributions to Primatology*, vol. 5, ed. F. S. Szalay, pp. 167–217. Basel: Karger.

Delson, E. (1975*b*). Paleoecology and zoogeography of the Old World Monkeys. In *Primate Functional Morphology and Evolution*, ed. R. H. Tuttle, pp. 37–64. Paris: Mouton.

Delson, E. (1975*c*). Toward the origin of the Old World Monkeys. *Colloques internationaux du Centre national de la recherche scientifique*, **218**, 839–850.

Delson, E. (1977). Vertebrate paleontology, especially of non-human primates in China. In *Paleoanthropology in the People's Republic of China*, eds. W. W. Howells & P. J. Tsuchitani, pp. 40–65. Washington, DC: National Academy of Sciences.

Delson, E. (1979). *Prohylobates* (Primates) from the Early Miocene of Libya: a new species and its implications for cercopithecid origins. *Geobios*, **12**, 725–733.

Delson, E. (1980). Fossil macaques, phyletic relationships and a scenario of deployment. In *The Macaques: Studies in Ecology, Behavior and Evolution*, ed. D. G. Lindburg, pp. 10–30. New York: Van Nostrand Reinhold.

Delson, E. (1986). An anthropoid enigma: historical introduction to the study of *Oreopithecus bambolii*. *Journal of Human Evolution*, **15**, 523–531.

Delson, E. (1993). *Theropithecus* fossils from Africa and India and the taxonomy of the genus. In Theropithecus: *The Rise and Fall of a Primate Genus*, ed. N. G. Jablonski, pp. 157–189. Cambridge: Cambridge University Press.

Delson, E. (1994). Evolutionary history of the colobine monkeys in paleoenvironmental perspective. In *Colobine Monkeys: Their Ecology, Behaviour and Evolution*, eds. A. G. Davies & J. F. Oates, pp. 11–43. Cambridge: Cambridge University Press.

Delson, E. & Dean, D. (1993). Are *Papio baringensis* R. Leakey, 1969, and *P. quadratirostris* Iwamoto, 1982, species of *Papio* or *Theropithecus*? In Theropithecus: *The Rise and Fall of a Primate Genus*, ed. N. G. Jablonski, pp. 125–156. Cambridge: Cambridge University Press.

Delson, E., Eck, G. G., Leakey, M. G., & Jablonski, N. G. (1993). Appendix I: A partial catalogue of fossil remains of *Theropithecus*. In Theropithecus: *The Rise and Fall of a Primate Genus*, ed. N. G. Jablonski, pp. 499–525. Cambridge: Cambridge University Press.

Delson, E. & Hoffstetter, R. (1993). *Theropithecus* from Ternifine, Algeria. In Theropithecus: *The Rise and Fall of a Primate Genus*, ed. N. G. Jablonski, pp. 191–208. Cambridge: Cambridge University Press.

Delson, E. & Rosenberger, A. L. (1980). Phyletic perspectives on platyrrhine origins and anthropoid relationships. In *Evolutionary Biology of the New World Monkeys and Continental Drift*, eds. R. L. Ciochon & A. B. Chiarelli, pp. 445–458. New York: Plenum Press.

Delson, E. & Rosenberger, A. L. (1984). Are there any anthropoid primate "living fossils'? In *Living Fossils*, eds. N. Eldredge & S. Stanley, pp. 50–61. New York: Springer-Verlag.

Delson, E., Tattersall, I., Van Couvering, J., & Brooks, A. S. (2000). *Encyclopedia of Human Evolution and Prehistory*, 2nd edn. New York: Garland.

Delson, E., Terranova, C., Jungers, W. L., Sargis, E. J., Jablonski, N. G. & Dechow, P. C. (2000). Body mass in Cercopithecidae (Primates, Mammalia): estimation and scaling in extinct and extant taxa. *Anthropological Papers, American Museum of Natural History*, **83**, 1–159.

Demes, B. & Jungers, W. L. (1993). Long bone cross-sectional geometry of extant and subfossil indrioid primates. *American Journal of Physical Anthropology*, Supplement **16**, 80–81.

Dene, H. T., Goodman, M., & Prychodko, W. (1976*a*). Immunodiffusion evidence on the phylogeny of the primates. In *Molecular Anthropology*, eds. M. Goodman & R. E. Tashian, pp. 141–170. New York: Plenum.

Dene, H. T., Goodman, M. & Prychodko, W. (1980). Immunodiffusion systematics of the primates. IV. Lemuriformes. *Mammalia*, **44**, 211–223.

Dene, H. T., Goodman, M., Prychodko, W., & Moore, G. W. (1976*b*). Immunodiffusion systematics of the primates: the Strepsirhini. *Folia Primatologica*, **25**, 35–61.

Denham, W. W. (1987). West Indian green monkeys: problems in historical biogeography. *Contributions to Primatology*, **24**, 1–79.

Dennell, R. & Roebroeks, W. (1996). The earliest colonization of Europe, the short chronology revisited. *Antiquity*, **70**, 535–542.

Depéret, C. (1887). Sur la faune vertébrés miocène de La Grive-Saint Alban (Isère). *Archives du Muséum d'histoire naturelle de Lyon*, **5**.

Depéret, C. (1889). Sur le *Dolichopithecus ruscinensis* nouveau singe fossile du Pliocène du Roussillion. *Comptes rendus de l'Académie des sciences de Paris*, **109**, 982–983.

Depéret, C. (1929). Nouveau singe du Pliocène supérieur de Sénèze (Hte-Loire). *Travaux du Laboratoire de géologie de la Faculté des sciences de Lyon*, **12**, 5–12.

DeVos, J. & Sondaar, P. (1994). Dating hominid sites in Indonesia. *Science*, **266**, 1726.

Dewar, R. E. (1984). Recent extinctions in Madagascar: the loss of the subfossil fauna. In *Quaternary Extinctions: A Prehistoric Revolution*, eds. P. S. Martin & R. G. Klein, pp. 574–593. Tucson, AZ: University of Arizona Press.

Dewar, R. E. (1997). Were people responsible for the extinction of Madagascar's subfossils, and how will we ever know? In *Natural Change and Human Impact in Madagascar*, eds. S. M. Goodman & B. D. Patterson, pp. 364–377. Washington, DC: Smithsonian Institution Press.

Dietrich, W. O. (1942). Altestquartiare Säugetiere aus der sudlichen Serengeti, Deutsch-Ostafrika. *Palaeontographica (A)*, **94**, 43–133.

Disotell, T. (1994). Generic level relationships of the Papionini (Cercopithecoidea). *American Journal of Physical Anthropology*, **94**,

47–57.

Disotell, T. (1996). Phylogeny of Old World monkeys. *Evolutionary Anthropology*, **5**, 18–24.

Disotell, T. R., Honeycutt, R. L., & Ruvolo, M. (1992). Mitochondrial DNA phylogeny of the Old World monkey Tribe Papionini. *Molecular Biology and Evolution*, **9**, 1–13.

Djian, P. & Green, H. (1991). Involucrin gene of tarsioids and other primates: alternatives in evolution of the segment of repeats. *Proceedings of the National Academy of Sciences of the United States of America*, **88**, 5321–5325.

Dolhinow, P. & Fuentes, A. (eds.) (1999). *The Nonhuman Primates*. Mountainview, CA: Mayfield Publishing Company.

Domning, D. P., Emry, R. J., Portell, R. W., Donovan, S. K., & Schindler, K. S. (1997). Oldest West Indian land mammal: rhinocerotoid ungulate from the Eocene of Jamaica. *Journal of Vertebrate Paleontology*, **17**, 638–641.

Donoghue, M. J., Doyle, J. A., Gauthier, J., & Kluge, A. G. (1989). The importance of fossils in phylogeny reconstruction. *Annual Reviews in Ecology and Systematics*, **20**, 43–60.

Dornum, M. von & Ruvolo, M. (1999). Phylogenetic relationships of the New World monkeys (Primates, Platyrrhini) based on nuclear G6PD DNA sequences. *Molecular Phylogenetics and Evolution*, **11**, 459–476.

Doyle, G. A. & Martin, R. D. (eds.) (1979). *The Study of Prosimian Behavior*. New York: Academic Press.

Drake, R., Van Couvering, J. A., Pickford, M., Curtis, G., & Harris, J. A. (1988). New chronology for the early Miocene mammalian faunas of Kisingiri, western Kenya. *Journal of the Geological Society of London*, **145**, 479–491.

Duarte, C., Maurício, J., Pettitt, P., Souto, P., Trinkaus, E., van der Plicht, H. & Zilhão, J. (1999). The early Upper Paleolithic human skeleton from the Abrigo do Lagar Velho (Portugal) and modern human emergence in Iberia. *Proceedings of the National Academy of Sciences of the United States of America*, **96**, 7604–7609.

Dubois, E. (1894). Pithecanthropus erectus, *eine menschenähnliche Übergangsform aus Java*. Batavia: Landesdruckerei.

Dubois, E. (1895). Sur le *Pithecanthropus erectus* de Pliocène de Java. *Bulletin de la Societé belge de geologie de paléontologie et d'hydrologie*, **9**, 151–160.

Dubois, E. (1901). Zur systematischen Stellung der ausgestorbenen Menschenaffen. *Zoologisches Anzeiger*, **24**, 556–560.

Dubois, E. (1924). Figures of the calvarium and endocranial cast, a fragment of the mandible and three teeth of *Pithecanthropus erectus*. *Proceedings of the Koninklijke Nederlandse Akademie van Wetenschappen*, **27**, 459–464.

Ducrocq, S. (1998). Eocene primates from Thailand: are Asian anthropoideans related to African ones? *Evolutionary Anthropology*, **7**, 97–104.

Ducrocq, S. (1999). *Siamopithecus eocaenus*, a late Eocene anthropoid primate from Thailand: its contribution to the evolution of anthropoids in Southeast Asia. *Journal of Human Evolution*, **36**, 613–636.

Ducrocq, S., Chaimanee, Y., Suteethorn, V., & Jaeger, J. J. (1994). Ages and paleoenvironment of Miocene mammalian faunas from Thailand. *Palaeogeography, Palaeoclimatology, Palaeoecology*, **108**, 149–163.

Ducrocq, S., Jaeger, J. J., Chaimanee, Y. & Suteethorn, V. (1995). New primate from the Palaeogene of Thailand, and the biogeographical origin of anthropoids. *Journal of Human Evolution*, **28**, 477–485.

Dunbar, R. I. M. (1993). Socioecology of the extinct theropiths: a modelling approach. In Theropithecus: *The Rise and Fall of a Primate Genus*, ed. N. G. Jablonski, pp. 465–486. Cambridge: Cambridge University Press.

Dupont, E. (1866). Étude sur les fouilles scientifiques exécutées pendant l'hiver de 1865–1866 dans les cavernes des bordes de la Lesse. *Bulletin d'Académie royale des sciences, des lettres, et des beaux-arts de Belgiques*, (Série 2) **22**, 31–54.

Dutrillaux, B. (1988). Chromosome evolution in primates. *Folia Primatologica*, **50**, 134–135.

Dutta, A. K., Basu, P. K., & Sastry, M. V. A. (1976). On the new finds of hominids and additional finds of pongids from the Siwaliks of Ramnagar area, Udhampur District, J & K State. *Indian Journal of Earth Sciences*, **3**, 234–235.

Earle, C. (1897). On the affinities of *Tarsius*: a contribution to the phylogeny of the Primates. *American Naturalist*, **31**, 569–575.

Eck, G. G. (1976*a*). Diversity and frequency distribution of Omo Group Cercopithecoidea. *Journal of Human Evolution*, **6**, 55–63.

Eck, G. G. (1976*b*). Cercopithecoidea from Omo Group Deposits. In *Earliest Man and Environments in the Lake Rudolf Basin*, eds. Y. Coppens, F. C. Howell, G. L. Isaac & R. E. F. Leakey, pp. 332–344. Chicago, IL: University of Chicago Press.

Eck, G. G. (1987). *Theropithecus oswaldi* from the Shungura Formation, Lower Omo Basin, Southwestern Ethiopia. In *Les Faunes Plio-Pleistocènes de la vallée de l'Omo (Éthiopie)*, eds. G. G. Eck, N. G. Jablonski, & M. Leakey, pp. 123–140. Paris: Centre national de la recherche scientifique.

Eck, G. G. (1993). *Theropithecus darti* from the Hadar Formation, Ethiopia. In Theropithecus: *The Rise and Fall of a Primate Genus*, ed. N. G. Jablonski, pp. 15–83. Cambridge: Cambridge University Press.

Eck, G. G. & Howell, F. C. (1972). New fossil *Cercopithecus* material from the Lower Omo Basin, Ethiopia. *Folia Primatologica*, **18**, 325–355.

Eck, G. G. & Jablonski, N. G. (1984). A reassessment of the taxonomic status and phyletic relationships of *Papio baringensis* and *Papio quadratirostris* (Primates: Cercopithecidae). *American Journal of Physical Anthropology*, **65**, 109–134.

Eck, G. G. & Jablonski, N. G. (1987). The skull of *Theropithecus brumpti* compared with those of other species of the genus *Theropithecus*. In *Les Faunes Plio-Pleistocènes de la vallée de l'Omo (Éthiopie)*, eds. G. G. Eck, N. G. Jablonski, & M. Leakey, pp. 10–122. Paris: Centre national de la recherche scientifique.

Eckhardt, R. B. (1973). *Gigantopithecus* as a hominid ancestor. *Anthropologisches Anzeiger*, **34**, 1–8.

Eckhardt, R. B. (1975). *Gigantopithecus* as a hominid. In *Paleoanthropology, Morphology and Paleoecology*, ed. R. L. Tuttle, pp. 105–127. The Hague: Mouton.

Ehrensberg, K. (1938). *Austriacopithecus*, ein neuer menschen-affenartiger Primate aus dem Miozän von klein-Hadersdorf bei Poysdorf in Niederosterreich (Nieder-Donau). *Sitzungsberichte-Akademie der Wissenschaften, Berlin, Klasse für Mathematik und allgemeine Naturwissenschaften*, **1**, 147.

Eisenberg, J. F., & Wilson, D. E., (1981). Relative brain size and demographic strategies in didelphid marsupials. *American Naturalist*, **118**, 1–15.

Eisenhart, B. (1974). The fossil cercopithecoids of Makapansgat and Sterkfontein. BA thesis, Harvard College, Cambridge, MA.

Ekblom, T. (1951). Studien über subfossile Lemuren von Madagaskar. *Bulletin of the Geological Institute of Uppsala*, **34**, 123–190.

Eldredge, N. & Gould, S. J. (1972). Punctuated equilibrium: an alternative to phyletic gradualism. In *Models in Paleobiology*, ed. T. Schopf, pp. 82–115. San Francisco, CA: W. H. Freeman.

Eldredge, N. & Tattersall, I. (1982). *The Myths of Human Evolution*. New York: Columbia University Press.

Emry, R. J. (1990). Mammals of the Bridgerian (middle Eocene) Elderberry Canyon local fauna of eastern Nevada. *Geological Society of America Special Papers*, **243**, 187–210.

Erikson, G. E. (1963). Brachiation in New World monkeys and in anthropoid apes. *Symposium of the Zoological Society, London*, **10**, 135–164.

Eudey, A. A. (1980). Pleistocene glacial phenomena and the evolution of Asian macaques. In *The Macaques: Studies in Ecology, Behavior and Evolution*, ed. D. G. Lindburgh, pp. 52–83. New York: Van Nostrand Reinhold.

Evans, E. M. N., Van Couvering, J. A. H., & Andrews, P. (1981). Palaeoecology of Miocene sites in western Kenya. *Journal of Human Evolution*, **10**, 35–48.

Evernden, J., Curtis, G., & James, G. (1965). The potassium-argon dating of late Cenozoic rocks in East Africa and Italy. *Current Anthropology*, **6**, 343–384.

Excoffier, L. (1990). Evolution of human mitochondrial DNA: evidence for departure from a pure neutral model of populations at equilibrium. *Journal of Molecular Evolution*, **30**, 125–139.

Falconer, H. & Cautley, P. T. (1837). On additional fossil species of the order Quadrumana from the Sewalik Hills. *Journal of the Asiatic Society of Bengal*, **VI**, 354–360.

Falk, D. (1983). A reconsideration of the endocast of *Proconsul africanus*: implications for primate brain evolution. In *New Interpretations of Ape and Human Ancestry*, eds. R. L. Ciochon & R. S. Corruccini, pp. 239–248. New York: Plenum Press.

Feibel, C. S. & Brown, F. H. (1991). Age of the primate-bearing deposits on Maboko Island, Kenya. *Journal of Human Evolution*, **21**, 221–225.

Feibel, C. S., Brown, F. H., & McDougall, I. (1989). Stratigraphic context of fossil hominids from the Omo Group deposits: Northern Turkana Basin, Kenya and Ethiopia. *American Journal of Physical Anthropology*, **78**, 595–622.

Feldesman, M. R. (1986). The forelimb of the newly "rediscovered" *Proconsul africanus* from Rusinga Island, Kenya: morphometrics and implications for catarrhine evolution. In *Variation, Culture and Evolution in African Populations*, eds. R. Singer & J. K. Lundy, pp. 179–193. Johannesburg: Witwatersrand University Press.

Felsenstein, J. (1978). Cases in which parsimony or compatibility methods will be positively misleading. *Systematic Zoology*, **27**, 401–410.

Ferembach, D. (1953). Affinités et mode de vie de *Limnopithecus macinnesi*, Le Gros Clark et Leakey. *Comptes rendus de l'Académie des sciences de Paris*, **236**, 2101–2103.

Ferembach, D. (1954). La Mandible et les dents inférieures des Limnopithèques. *Comptes rendus de l'Académie des sciences de Paris*, **239**, 1656–1658.

Ferembach, D. (1958). Les Limnopithèques du Kenya. *Annales de paléontologie*, **44**, 149–249.

Ferguson, W. W. (1987). Taxonomic status of the hominine cranium KNM-ER 1813 (Primates: Homininae) from the Plio/Pleistocene of Koobi Fora. *Primates*, **28**, 423–438.

Ferguson, W. W. (1995). A new species of the genus *Homo* (Primates: Hominidae) from the Plio-Pleistocene of Koobi Fora, in Kenya. *Primates*, **36**, 69–89.

Filhol, H. (1873). Sur un nouveau genre de Lémurien fossile récemment découvert dans les gisements de phosphate de chaux du Quercy. *Comptes rendus de l'Académie des sciences de Paris*, **77**, 1111–1112.

Filhol, H. (1874). Nouvelles observations sur les mammifères des gisements de phosphates de chaux, Lémuriens et Pachylémuriens. *Bibliothèque de l'École des hautes études, sciences naturelles*, **9**.

Filhol, H. (1880). Note sur des mammifères fossiles nouveaux provenant des phosphorites du Quercy. *Bulletin sociéte philomathique de Paris*, **7**(4).

Filhol, H. (1882). Mémoires sur quelques mammifères fossiles des phosphorites du Quercy. *Annales de la Société des sciences physiques et naturelles de Toulouse*, **5**,19–156.

Filhol, H. (1889–90). Description d'une nouvelle espèce de Lémurien fossile (*Necrolemur parvulus*). *Bulletin de la Société philomathique de Paris*, **8**, 39–40.

Filhol, H. (1895). Observations concernant les mammifères contemporains de *Aepyornis* à Madagascar. *Bulletin du Muséum national d'histoire naturelle, Paris*, **1**, 12–14.

Flacourt, É. de (1658). *Histoire de la grande Isle Madagascar composée par le Sieur de Flacourt*, 2 vols. Paris: G. de Lvyne.

Fladerer, F. A. (1987). *Macaca* (Cercopithecidae, Primates) in the lower Pleistocene from Deutsch-Altenberg, lower Austria. *Beitrage zur Paläontologie von Österreich*, **13**, 1–24.

Fladerer, F. A. (1989). Hohlenschutz und Eiszeitforschung Erstnachweis von Affen (Gattung *Macaca*) im Jungpleistozän Mitteleuropas. *Mitteilungen des naturwissenshaftlichen Vereines für Steiemark*, **119**, 23–26.

Fladerer, F. A. (1991). Der erste Fund von *Macaca* (Cercopithecidae, Primates) im Jungpleistozän von Mitteleuropa. *Zeitschrift für Säugetierkunde*, **56**, 272–283.

Fleagle, J. G. (1975). A small gibbon-like hominoid from the Miocene of Uganda. *Folia Primatologica*, **24**, 1–15.

Fleagle, J. G. (1983). Locomotor adaptations of Oligocene and Miocene hominoids and their phyletic implications. In *New Interpretations of Ape and Human Ancestry*, eds. R. L. Ciochon & R. S. Corruccini, pp. 301–324. New York: Plenum Press.

Fleagle, J. G. (1986). The fossil record of early catarrhine evolution. In *Major Topics in Primate and Human Evolution*, eds. B. Wood, L. Martin & P. Andrews, pp. 130–149. Cambridge: Cambridge University Press.

Fleagle, J. G. (1988). *Primate Adaptation and Evolution*. San Diego, CA: Academic Press.

Fleagle, J. G. (1990). New fossil platyrrhines from the Pinturas Formation, southern Argentina. *Journal of Human Evolution*, **19**, 61–85.

Fleagle, J. G. (1999). *Primate Adaptation and Evolution*, 2nd edn. San Diego, CA: Academic Press.

Fleagle, J. G. (2000). The century of the past: one hundred years in the study of primate evolution. *Evolutionary Anthropology*, **9**, 87–100.

Fleagle, J. G. & Anapol, F. (1992). The indriid ischium and the hominid hip. *Journal of Human Evolution*, **22**, 285–306.

Fleagle, J. G. & Bown, T. M. (1983). New primate fossils from late Oligocene (Colhuehuapian) localities of Chubut, Province, Argentina. *Folia Primatologica*, **41**, 240–266.

Fleagle, J. G., Bown, T. M., Harris, J. M., Watkins, R. W., & Leakey, M. G. (1997*a*). Fossil monkeys from Northern Kenya. *American Journal of Physical Anthropology*, Supplement **24**, 111.

Fleagle, J. G., Bown, T. M., Swisher, C., & Buckley, G. (1995). Age of the Pinturas and Santa Cruz formation. *Actas VI Congreso argentino de paleontologia y biostratigraffa*, 129–135.

Fleagle, J. G. & Hartwig , W. C. (1997). Paleoprimatology. In *History of Physical Anthropology: An Encyclopedia*, ed. F. Spencer, pp. 796–810. New York: Garland.

Fleagle, J. G., Janson, C. H., & Reed, K. (eds.) (1999). *Primate Communities*. Cambridge: Cambridge University Press.

Fleagle, J. G. & Kay, R. F. (1983). New interpretations of the phyletic positions of Oligocene hominoids. In *New Interpretations of Ape and Human Ancestry*, eds. R. L. Ciochon & R. S. Corruccini, pp. 181–210. New York: Plenum Press.

Fleagle, J. G. & Kay, R. F. (1985). The paleobiology of catarrhines. In *Ancestors: The Hard Evidence*, ed. E. Delson, pp. 23–36. New York: Alan R. Liss.

Fleagle, J. G. & Kay, R. F. (1987). The phyletic position of the Parapithecidae. *Journal of Human Evolution*, **16**, 483–532.

Fleagle, J. G. & Kay, R. F. (1989). The dental morphology of *Dolichocebus gaimanensis*, a fossil monkey from Argentina. *American Journal of Physical Anthropology*, **78**, 221.

Fleagle, J. G. & Kay, R. F. (eds.) (1994). *Anthropoid Origins*. New York: Plenum Press.

Fleagle, J. G. & Kay, R. F. (1997). Platyrrhines, catarrhines, and the fossil record. In *New World Primates: Ecology, Evolution, and Behavior*, ed. W. G. Kinzey, pp. 3–23. New York: Aldine de Gruyter.

Fleagle, J. G., Kay, R. F., & Anthony, M. R. L. (1997*b*). Fossil New World monkeys. In *Vertebrate Paleontology in the Neotropics: The Miocene Fauna of La Venta, Colombia*, eds. R. F. Kay, R. H. Madden, R. L. Cifelli, & J. J. Flynn, pp. 473–496. Washington, DC: Smithsonian Institution Press.

Fleagle, J. G., Kay, R. F. & Simons, E. L. (1980). Sexual dimorphism in early anthropoids. *Nature*, **287**, 328–330.

Fleagle, J. G. & McGraw, W. S. (1999). Skeletal and dental morphology supports diphyletic origin of baboons and mandrills. *Proceedings of the National Academy of Sciences of the United States of America*, **96**, 1157–1161.

Fleagle, J. G. & Meldrum, D. J. (1988). Locomotor behavior and skeletal morphology of two sympatric pithecine monkeys, *Pithecia pithecia* and *Chiropotes satanas*. *American Journal of Primatology*, **16**, 227–250.

Fleagle, J. G., Powers, D. W., Conroy, G. C., & Watters, J. P. (1987). New fossil platyrrhines from Santa Cruz Province, Argentina. *Folia Primatologica*, **48**, 65–77.

Fleagle, J. G. & Rae, T. C. (1992). Primate cranial remains from the Pinturas Formation, Argentina. *American Journal of Physical Anthropology*, Supplement **14**, 75–76.

Fleagle, J. G. & Reed, K. E. (1996). Comparing primate communities: a multivariate approach. *Journal of Human Evolution*, **30**, 489–510.

Fleagle, J. G. & Reed, K. E. (1999). Phylogenetic and temporal perspectives on primate ecology. In *Primate Communities*, eds. J. G. Fleagle, C. H. Janson, & K. Reed, pp. 92–115. Cambridge: Cambridge University Press.

Fleagle, J. G. & Rosenberger, A. L. (1983). Cranial morphology of the earliest anthropoids. In *Morphologie, évolutive, morphogenèse du crâne et anthropogenèse*, ed. M. Sakka, pp. 141–153. Paris: Centre national de la recherche scientifique.

Fleagle, J. G. & Rosenberger, A. L. (eds.) (1990). The platyrrhine fossil record. *Journal of Human Evolution*, **19**, 1–254.

Fleagle, J. G. & Simons, E. L. (1978*a*). Humeral morphology of the earliest apes. *Nature*, **276**, 705–707.

Fleagle, J. G. & Simons, E. L. (1978*b*). *Micropithecus clarki*, a small ape from the Miocene of Uganda. *American Journal of Physical Anthropology*, **49**, 427–440.

Fleagle, J. G. & Simons, E. L. (1979). Anatomy of the bony pelvis in parapithecid primates. *Folia Primatologica*, **31**, 176–186.

Fleagle, J. G. & Simons, E. L. (1982*a*). Skeletal remains of *Propliopithecus chirobates* from the Egyptian Oligocene. *Folia Primatologica*, **39**, 161–177.

Fleagle, J. G. & Simons, E. L. (1982*b*). The humerus of *Aegyptopithecus zeuxis*, a primitive anthropoid. *American Journal of Physical Anthropology*, **59**, 175–194.

Fleagle, J. G. & Simons, E. L. (1983). The tibio-fibular articulation in *Apidium phiomense*, an Oligocene anthropoid. *Nature*, **301**, 238–239.

Fleagle, J. G. & Simons, E. L. (1995). Limb skeleton and locomotor adaptations of *Apidium phiomense*, an Oligocene anthropoid from Egypt. *American Journal of Physical Anthropology*, **97**, 235–289.

Fleagle, J. G., Simons, E. L., & Conroy, G. C. (1975). Ape limb bone from the Oligocene of Egypt. *Science*, **189**, 135–137.

Flynn, J. J., Guerrero, J., & Swisher, C. C. (1997). Geochronology of the Honda Group. In *Vertebrate Paleontology in the Neotropics: The Miocene Fauna of La Venta, Colombia*, eds. R. F. Kay, R. H. Madden, R. L. Cifelli, & J. J. Flynn, pp. 44–60. Washington, DC: Smithsonian Institution Press.

Flynn, J. J., Parrish, J. M., Rakotosamimanana, B., Simpson, W., & Wyss, A. R. (1999). A Middle Jurassic mammal from Madagascar. *Nature*, **401**, 57–60.

Flynn, J. J., Wyss, A. R., Charrier, R., & Swisher, C. C. (1995). An early Miocene anthropoid skull from the Chilean Andes. *Nature*, **373**, 603–607.

Flynn, L. J., Barry, J. C., Morgan, M. E., Pilbeam, D., Jacobs, L. L., & Lindsay, E. H. (1995). Neogene Siwalik mammalian lineages: species longevities, rates of change, and modes of speciation. *Palaeogeography, Palaeoclimatology, Palaeoecology*, **115**, 249–264.

Flynn, L. J., Jacobs, L. L., & Cheema, I. U. (1986). Baluchimyinae: a new ctenodactyloid rodent subfamily from the Miocene of Baluchistan. *American Museum Novitates*, **2841**, 1–58.

Fooden, J. (1990). The bear macaque, *Macaca arctoides*: a systematic review. *Journal of Human Evolution*, **19**, 607–686.

Fooden, J. (1995). Systematic review of southeast Asian longtail macaques, *Macaca fascicularis* (Raffles, [1821]). *Fieldiana*, **81**, 1–205.

Fooden, J. (2000). Systematic review of the rhesus macaque, *Macaca mulatta* (Zimmermann, 1780). *Fieldiana*, **96**, 1–180.

Foote, M., Hunter, J. P., Janis, C. M., & Sepkoski, J. J. (1999). Evolutionary and preservational constraints on origins of biological groups: divergence times of Eutherian mammals. *Science*, **283**, 1310–1314.

Ford, S. M. (1980*a*). Affinities and interpretations of the primate fossil postcranials from South America. *American Journal of Physical Anthropology*, **52**, 227.

Ford, S. M. (1980*b*). *A Systematic Revision of the Platyrrhini Based on Features of the Postcranium*. Ann Arbor, MI: University Microfilms.

Ford, S. M. (1986*a*). Subfossil platyrrhine tibia (Primates: Callitrichidae) from Hispaniola: a possible further example of island gigantism. *American Journal of Physical Anthropology*, **70**, 47–62.

Ford, S. M. (1986*b*). Systematics of the New World monkeys. In *Comparative Primate Biology*, vol. 1, *Systematics, Evolution and Anatomy*, eds. D. R. Swindler & J. Erwin, pp. 73–135. New York: Alan R. Liss.

Ford, S. M. (1988). Postcranial adaptations of the earliest platyrrhine. *Journal of Human Evolution*, **17**, 155–192.

Ford, S. M. (1990*a*). Locomotor adaptations of fossil platyrrhines. *Journal of Human Evolution*, **19**, 141–174.

Ford, S. M. (1990*b*). Platyrrhine evolution in the West Indies. *Journal of Human Evolution*, **19**, 237–254.

Ford, S. M. (1994). Primitive platyrrhines? Perspectives on anthropoid origins from platyrrhine, parapithecid, and preanthropoid postcrania. In *Anthropoid Origins*, eds. J. G. Fleagle & R. F. Kay, pp. 595–673. New York: Plenum Press.

Ford, S. M., Davis, L. C., & Kay, R. F. (1991). New platyrrhine astragalus from the Miocene of Colombia. *American Journal of Physical Anthropology*, Supplement **12**, 73–74.

Ford, S. M. & Morgan, G. S. (1986). A new ceboid femur from the Late Pleistocene of Jamaica. *Journal of Vertebrate Paleontology*, **6**, 281–289.

Ford, S. M. & Morgan, G. S. (1988). Earliest primate fossil from the West Indies. *American Journal of Physical Anthropology*, **75**, 209.

Forsyth-Major, C. I. (1893). Verbal report on an exhibition of a specimen of a subfossil lemuroid skull from Madagascar. *Proceedings of the Zoological Society of London*, **36**, 532–535.

Forsyth-Major, C. I. (1894). On *Megaladapis madagascariensis*, an extinct gigantic lemuroid from Madagascar, with remarks on the associated fauna, and on its geologic age. *Philosophical Transactions of the Royal Society of London, Series B*, **185**, 15–38.

Forsyth-Major, C. I. (1896). Preliminary notice on fossil monkeys from Madagascar. *Geological Magazine, New Series, Decade 4*, **3**, 433–436.

Forsyth-Major, C. I. (1897). On the brains of two sub-fossil Malagasy lemuroids. *Proceedings of the Royal Society of London*, **62**, 46–50.

Forsyth-Major, C. I. (1900*a*). Exhibition of, and remarks upon, specimens of two subfossil mammals from Madagascar. *Proceedings of the Zoological Society of London* (for 1899), 988–989.

Forsyth-Major, C. I. (1900*b*). Extinct mammalia from Madagascar. 1. *Megaladapis insignis*. sp. n. *Philosophical Transactions of the Royal Society, Series B*, **193**, 47–50.

Forsyth-Major, C. I. (1900*c*). A summary of our present knowledge of extinct primates from Madagascar. *Geological Magazine*, **7**, 492–499.

Forsyth-Major, C. I. (1901). On some characters of the skull in the lemurs and monkeys. *Proceedings of the Zoological Society of London* (for 1900), 129–153.

Fourtau, R. (1918). *Contribution à l'étude des vertèbrés miocènes de l'Égypte*. Paris: Survey Department, Ministry of Finance.

Fourtau, R. (1920). *Contribution à l'étude vertébrés miocènes de l'Égypte*. Cairo: Government Press.

Frailey, C. D. (1986). Late Miocene and Holocene mammals, exclusive of the Notoungulata, of Rio Acre region, western Amazonia. *Contributions in Science of the Los Angeles County Museum*, **373**, 1–46.

Fraipont, C. (1936). Les hommes fossiles d'Engis. *Archives de l'Institut de Paléontologie humaine*, **16**, 1–52.

Fraipont, J. & Lohest, M. (1886). La race de Néanderthal ou de Canstatt, en Belgique: recherches ethnographiques sur des ossements humains découverts dans dépôts quaternaires d'une grotte à Spy et détermination de leur âge géologique: note préliminaire. *Bulletin de l'Académie royale des sciences, des lettres, et des beaux-arts de Belgique, Série 3*, **12**, 741–784.

Franzen, J. L. (1987). Ein neuer Primate aus dem Mitteleozän der Grube Messel (Deutschlend, S-Hessen). *Courier Forschungsinstitut Senckenberg*, **91**, 151–187.

Franzen, J. L. (1988). Ein weiterer Primatenfund aus der Grube Messel bei Darmstadt. *Courier Forschungsinstitut Senckenberg*, **107**, 275–289.

Franzen, J. L. (ed.) (1994*a*). *100 Years of* Pithecanthropus: *The* Homo erectus *Problem*. Frankfurt: Forschungsinstitut Senckenberg.

Franzen, J. L. (1994*b*). The Messel primates and anthropoid origins. In *Anthropoid Origins*, eds. J. G. Fleagle & R. F. Kay, pp. 99–122. New York: Plenum Press.

Franzen, J. L. & Storch, G. (1999). Late Miocene mammals from Central Europe. In *Hominoid Evolution and Climate Change in Europe*, vol. 1, *The Evolution of Neogene Terrestrial Ecosystems in Europe*, eds. J. Agustí, L. Rook, & P. Andrews, pp. 165–190. Cambridge: Cambridge University Press.

Frayer, D. W. (1973). *Gigantopithecus* and its relationship to *Australopithecus*. *American Journal of Physical Anthropology*, **39**, 413–426.

Frayer, D. W. (1974). A reappraisal of *Ramapithecus*. *Yearbook of Physical Anthropology*, **18**, 19–30.

Frayer, D. W. (1992). Evolution at the European edge: Neanderthal and Upper Paleolithic relationships. *Préhistoire européen*, **2**, 9–69.

Frayer, D. W. (1997*a*). Perspectives on Neanderthals as ancestors. In *Conceptual Issues in Modern Human Origins Research*, eds. G. Clark & K. Willermet, pp. 220–234. New York: Aldine de Gruyter.

Frayer, D. W. (1997*b*). *Ramapithecus*. In *History of Physical Anthropology: An Encyclopedia*, ed. F. Spencer, pp. 868–870. New York: Garland.

Frayer, D. W., Wolpoff, M. H., Thorne, A. G., Smith, F. H., & Pope, G. G. (1993). Theories of modern human origins: the paleontological test. *American Anthropologist*, **95**, 14–50.

Freedman, L. (1957). The fossil Cercopithecoidea of South Africa. *Annals of the Transvaal Museum*, **23**, 8–262.

Freedman, L. (1960). Some new cercopithecoid specimens from Makapansgat, South Africa. *Palaeontologica Africana*, **7**, 7–45.

Freedman, L. (1965). Fossil and subfossil primates from the limestone deposits at Taung, Bolt's Farm and Witkrans, South Africa. *Palaeontologica Africana*, **9**, 19–48.

Freedman, L. (1970). A new check-list of fossil Cercopithecoidea of South Africa. *Palaeontologica Africana*, **13**, 109–110.

Freedman, L. (1976). South African fossil Cercopithecoidea: a re-assessment including a description of new material from Makapansgat, Sterkfontein and Taung. *Journal of Human Evolution*, **5**, 297–315.

Freedman, L. & Brain, C. K. (1972). Fossil cercopithecoid remains from the Kromdraai australopithecine site (Mammalia: Primates). *Annals of the Transvaal Museum*, **28**, 1–16.

Freyberg, B. V. (1949). Die Pikermifauna von Tour la Reine (Attika). *Annales géologiques des pays helléniques*, **3**, 7–10.

Freyberg, B. V. (1950). Das neogen-gebiet Nordwestlich Athen. *Annales géologiques des pays Helléniques*, **3**, 65–86.

Frisch, J. E. (1965). Trends in the evolution of the hominoid dentition. *Bibliotheca Primatologica*, **3**, 1–130.

Gabunia, L., Jöris, A., Justus, A., Lordkipanidze, D., Muschelišvili,

A., Nioradze, M., Swisher, C., & Vekua, A. (1999). Neue Hominidenfunde des altpaläolithischen Fundplatzes Dmanisi (Georgien, Kaukasus im Kontext actueller Grabungsergebnisse). *Archäologisches Korrespondenzblatt*, **29**, 451–488.

Gabunia, L. K., Justus, A., & Vekua, A. (1991). Der menscheliche Unterkiefer. In *Dmanisi: Die Menschen der Altsteinzeit in Südgeorgient*, ed. Akademie der wissenschaften Georgiens. Tblisi: zentrum für archäologische Forschungen.

Gabunia, L. K. & Vekua, A., (1995). A Plio-Pleistocene hominid from Dmanisi, East Georgia, Caucasus. *Nature*, **373**, 509–512.

Gabunia, L. K, Vekua, A., Lordkipanidze, D., Swisher, C., Ferring, R., Justus, A., Nioradze, M., Tvalchrelidze, M., Antón, S., Bosinski, G., Jöris, O., de Lumley, M., Majsuradze, G., & Mouskhelishvili, A. (2000). Pleistocene hominid cranial remains from Dmanisi, Republic of Georgia: taxonomy, geological setting, and age. *Science*, **288**, 1019–1025.

Gade, D. W. (1996). Deforestation and its effects in highland Madagascar. *Mountain Research and Development*, **16**, 101–116.

Gagneux, P., Wills, C., Gerloff, U., Tautz, D., Morin, P. A., Boesch, C., Fruth, B., Hohmann, G., Ryder, O. A., & Woodruff, D. S. (1999). Mitochondrial sequences show diverse evolutionary histories of African hominoids. *Proceedings of the National Academy of Sciences of the United States of America*, **96**, 5077–5082.

Gagnon, M. (1997). Ecological diversity and community ecology in the Fayum sequence (Egypt). *Journal of Human Evolution*, **32**, 133–160.

Ganzhorn, J. U. (1997). Test of Fox's assembly rule for functional groups in lemur communities in Madagascar. *Journal of the Zoological Society of London*, **241**, 533–542.

Ganzhorn, J. U. (1998). Nested patterns of species composition and their implications for lemur biogeography in Madagascar. *Folia Primatologica*, Supplement **1**, 332–341.

Ganzhorn, J. U., Wright, P. C., & Ratsimbazafy, J. (1999). Primate communities: Madagascar. In *Primate Communities*, eds. J. G. Fleagle, C. H. Janson, & K. Reed, pp. 75–89. Cambridge: Cambridge University Press.

Garbutt, N. (1999). *Mammals of Madagascar*. New Haven, CT: Yale University Press.

Gaudry, A. (1867). *Animaux fossiles et géologie de l'Attique*. Paris.

Gaudry, A. (1890). Le Dryopithèque. *Mémoires de la Société geologique de France*, **1**, 1–11.

Gazin, C. L. (1952). The lower Eocene Knight Formation of western Wyoming and its mammalian faunas. *Smithsonian Miscellaneous Collections*, **117**, 1–82.

Gazin, C. L. (1958). A review of the middle and upper Eocene Primates of North America. *Smithsonian Miscellaneous Collections*, **136**, 1–112.

Gazin, C. L. (1962). A further study of the lower Eocene mammalian faunas of southwestern Wyoming. *Smithsonian Miscellaneous Collections*, **144**, 1–98.

Gear, J. H. S. (1926). A preliminary account of the baboon remains from Taungs. *South African Journal of Science*, **23**, 731–747.

Gebo, D. L. (1985). The nature of the primate grasping foot. *American Journal of Physical Anthropology*, **67**, 269–278.

Gebo, D. L. (1986). Anthropoid origins: the foot evidence. *Journal of Human Evolution*, **15**, 421–430.

Gebo, D. L. (1987*a*). Humeral morphology of *Cantius*, an early Eocene adapid. *Folia Primatologica*, **49**, 52–56.

Gebo, D. L. (1987*b*). Locomotor diversity in prosimian primates. *American Journal of Primatology*, **13**, 271–281.

Gebo, D. L. (1988). Foot morphology and locomotor adaptations in Eocene primates. *Folia Primatologica*, **50**, 3–41.

Gebo, D. L. (1989). Miocene lorisids: the foot evidence. *Folia Primatologica*, **47**, 217–225.

Gebo, D. L. (1993). Postcranial anatomy and locomotor adaptation in early African anthropoids. In *Postcranial Adaptation in Nonhuman Primates*, ed. D. L. Gebo, pp. 220–234. De Kalb, IL: Northern Illinois University Press.

Gebo, D. L. (1996). Climbing, brachiation, and terrestrial quadrupedalism: historical precursors of hominid bipedalism. *American Journal of Physical Anthropology*, **101**, 55–92.

Gebo, D. L., Beard, K. C., Teaford, M. F., Walker, A. C., Larson, S. G., Jungers, W. L., & Fleagle, J. G. (1988). A hominoid proximal humerus from the Early Miocene of Rusinga Island, Kenya. *Journal of Human Evolution*, **17**, 393–401.

Gebo, D. L., Dagosto, M., Beard, K. C., & Qi, T. (2000*a*). The smallest primates. *Journal of Human Evolution*, **38**, 585–594.

Gebo, D. L., Dagosto, M., Beard, K. C., Qi, T., & Wang, J. (2000*b*). The oldest known anthropoid postcranial fossils and the early evolution of higher primates. *Nature*, **404**, 276–278.

Gebo, D. L., Dagosto, M., Beard, K. C., & Wang, J. (1999). A first metatarsal of *Hoanghonius stehlini* from the Late Middle Eocene of Shanxi Province, China. *Journal of Human Evolution*, **37**, 801–806.

Gebo, D. L., Dagosto, M., & Rose, K. D. (1991). Foot morphology and evolution in early Eocene *Cantius*. *American Journal of Physical Anthropology*, **86**, 51–73.

Gebo, D. L., Dagosto, M., Rosenberger, A. L., & Setoguchi, T. (1990). New platyrrhine tali from La Venta, Colombia. *Journal of Human Evolution*, **19**, 737–746.

Gebo, D. L., Maclatchy, L., & Kityo, R. (1997*a*). A new lorisid humerus from the Early Miocene of Uganda. *Primates*, **38**, 423–427.

Gebo, D. L., MacLatchy, L., Kityo, R., Deino, A., Kingston, J., & Pilbeam, D. (1997*b*). A hominoid genus from the Early Miocene of Uganda. *Science*, **276**, 401–404.

Gebo, D. L. & Simons, E. L. (1987). Morphology and locomotor adaptations of the foot in early Oligocene anthropoids. *American Journal of Physical Anthropology*, **74**, 83–101.

Gebo, D. L., Simons, E. L., Rasmussen, D. T., & Dagosto, M. (1994). Eocene anthropoid postcrania from the Fayum, Egypt. In *Anthropoid Origins*, eds. J. G. Fleagle & R. F. Kay, pp. 203–233. New York: Plenum Press.

Genet-Varcin, E. (1963). *Les Singes actuels et fossiles*. Paris: N. Boubée.

Gentili, S., Mottura, A., & Rook, L. (1998). The Italian fossil primate record: recent finds and their geological context. *Geobios*, **31**, 675–686.

Gentry, A. W. (1970). The Bovidae (Mammalia) of the Fort Ternan fossil fauna. In *Fossil Vertebrates of Africa*, vol. 1, ed. L. S. B. Leakey, pp. 243–342. London: Academic Press.

Gentry, A. W. & Heizmann, E. P. J. (1996). Miocene ruminants of the Central and Eastern Paratethys. In *The Evolution of Western Eurasian Neogene Mammal Faunas*, eds. R. L. Bernor, V. Farlbusch, & H.-W. Mittmann, pp. 378–391. New York: Columbia University Press.

Gentry, A. W., Rössner, G. E., & Heizmann, E. P. J. (1999). Suborder Ruminantia. In *The Miocene Land Mammals of Europe: The Continental European Miocene*, eds. G. E. Rössner & K. Heissig, pp. 225–258. Munich: Dr. Friedrich Pfeil.

Geoffroy, É. (1812). Tableau des quadrumanes. I. Ordre Quadrumanes. *Annales du Muséum national d'Histoire naturelle, Paris*, **19**, 85–122.

Gervais, P. (1849*a*). Note sur une nouvelle espèce de singe fossile. *Comptes rendus de l'Académie des sciences de Paris*, **XXVIII**, 699–700, 789.

Gervais, P. (1849*b*). *Zoologie et paléontologie françaises*. Paris: A. Bertrand.

Gervais, P. (1859). *Zoologie et paléontologie françaises*, 2nd edn, 2 vols, texte et atlas. Paris: A. Bertrand.

Gervais, P. (1867). *Zoologie et paléontologie générales*. Paris: Arthus.

Gervais, P. (1872). Sur un singe fossile, d'espèce non encore décrite, qui a été découvert au Monte-Bamboli (Italie). *Comptes rendus de l'Académie des sciences de Paris*, **LXXIV**, 1217–1223.

Gervais, P. (1876). *Zoologie et paléontologie générale*, 2nd edn. Paris: A. Bertrand.

Gervais, P. & Ameghino, F. (1880). Les mammifères fossiles de l'Amérique du Sud. Paris: F. Savy.

Gheerbrant, E., Thomas, H., Roger, J., Sen, S., & Al-Sulaimani, Z. (1993). Deux nouveaux primates dans l'Oligocène inférieur de Taqah (Sultanat d'Oman): premiers adapiformes (?Anchomomyini) de la Péninsula Arabique. *Palaeovertebrata*, **22**, 141–196.

Gheerbrant, E., Thomas, H., Sen, S. & Al-Sulaimani, Z. (1995). Nouveau primate Oligopithecinae (Simiiformes) de l'Oligocène inférieur de Taqah, Sultanat d'Oman. *Comptes rendus de l'Académie des sciences de Paris*, 425–432.

Gibbard, P. L. (1994). *Pleistocene History of the Lower Thames Valley*. Cambridge: Cambridge University Press.

Gibbs, S. (1999). Comparative soft tissue morphology of the extant Hominoidea, including Man. PhD dissertation, University of Liverpool.

Gibbs, S., Collard, M., & Wood, B. (2000). Soft-tissue characters in higher primate phylogenetics. *Proceedings of the National Academy of Sciences of the United States of America*, **97**, 11130–11132.

Gibert, J., Ribot, F., Gibert, L., Leakey, M., Arribas, A., & Martinez, B. (1995). Presence of the cercopithecid genus *Theropithecus* in Cueva Victoria (Murcia, Spain). *Journal of Human Evolution*, **28**, 487–493.

Gidley, J. W. (1923). Paleocene primates of the Fort Union, with discussion of relationships of Eocene primates. *Proceedings of the United States National Museum*, **63**, 1–38.

Gingerich, P. D. (1973). Anatomy of the temporal bone in the Oligocene anthropoid *Apidium* and the origin of Anthropoidea. *Folia Primatologica*, **19**, 329–337.

Gingerich, P. D. (1975*a*). Dentition of *Adapis parisiensis* and the evolution of lemuriform primates. In *Lemur Biology*, eds. I. Tattersall & R. W. Sussman, pp. 65–79. New York: Plenum Press.

Gingerich, P. D. (1975*b*). A new genus of Adapidae (Mammalia, Primates) from the late Eocene of southern France, and its significance for the origin of higher primates. *Contributions from the Museum of Paleontology, University of Michigan*, **24**, 163–170.

Gingerich, P. D. (1976*a*). Cranial anatomy and evolution of early Tertiary Plesiadapidae (Mammalia, Primates). *Museum of Paleontology, University of Michigan, Papers on Paleontology*, **15**, 1–140.

Gingerich, P. D. (1976*b*). Paleontology and phylogeny: patterns of evolution at the species level in early Tertiary mammals. *American Journal of Science*, **276**, 1–28.

Gingerich, P. D. (1977*a*). Radiation of Eocene Adapidae in Europe. *Géobios, Mémoire spécial*, **1**, 165–182.

Gingerich, P. D. (1977*b*). New species of Eocene primates and the phylogeny of European Adapidae. *Folia Primatologica*, **28**, 60–80.

Gingerich, P. D. (1979). Phylogeny of middle Eocene Adapidae (Mammalia, Primates) in North America: *Smilodectes* and *Notharctus*. *Journal of Paleontology*, **53**, 153–163.

Gingerich, P. D. (1980*a*). Dental and cranial adaptations in Eocene Adapidae. *Zeitschrift für Morphologie und Anthropologie*, **71**, 135–142.

Gingerich, P. D. (1980*b*). Eocene Adapidae, paleobiogeography, and the origin of South American Platyrrhini. In *Evolutionary Biology of the New World Monkeys and Continental Drift*, eds. R. L. Ciochon & A. B. Chiarelli, pp. 123–138. New York: Plenum Press.

Gingerich, P. D. (1981*a*). Cranial morphology and adaptations in Eocene Adapidae. I. Sexual dimorphism in *Adapis magnus* and *Adapis parisiensis*. *American Journal of Physical Anthropology*, **56**, 217–234.

Gingerich, P. D. (1981*b*). Early Cenozoic Omomyidae and the evolutionary history of tarsiiform primates. *Journal of Human Evolution*, **10**, 345–374.

Gingerich, P. D. (1984*a*). Paleobiology of tarsiiform primates. In *Biology of Tarsiers*, ed. C. Niemitz, pp. 33–44. Stuttgart: Gustav Fischer.

Gingerich, P. D. (1984*b*). Primate evolution: evidence from the fossil record, comparative morphology, and molecular biology. *Yearbook of Physical Anthropology*, **27**, 57–72.

Gingerich, P. D. (1986). Early Eocene *Cantius torresi*: oldest primate of modern aspect from North America. *Nature*, **320**, 319–322.

Gingerich, P. D. (1990). African dawn for primates. *Nature*, **346**, 411.

Gingerich, P. D. (1993*a*). Early Eocene *Teilhardina brandti*: oldest omomyid primate from North America. *Contributions from the Museum of Paleontology, University of Michigan*, **28**, 321–326.

Gingerich, P.D. (1993*b*). Oligocene age of the Gebel Qatrani Formation, Fayum, Egypt. *Journal of Human Evolution*, **24**, 207–218.

Gingerich, P. D. (1995). Sexual dimorphism in earliest Eocene *Cantius torresi* (Mammalia, Primates, Adapoidea). *Contributions from the Museum of Paleontology, University of Michigan*, **29**, 185–199.

Gingerich, P. D., Dashzeveg, D., & Russell, D. E. (1991). Dentition and systematic relationships of *Altanius orlovi* (Mammalia, Primates) from the early Eocene of Mongolia. *Geobios*, **24**, 637–646.

Gingerich, P. D. & Haskin, R. A. (1981). Dentition of early Eocene *Pelycodus jarrovii* (Mammalia, Primates) and the generic attribution of species formerly referred to *Pelycodus*. *Contributions from the Museum of Paleontology, University of Michigan*, **25**, 327–337.

Gingerich, P. D., Holroyd, P. A., & Ciochon, R. L. (1994). *Rencunius zhoui*, a new primate from the late middle Eocene of Henan, China, and a comparison with some early Anthropoidea. In *Anthropoid Origins*, eds. J. G. Fleagle & R. F. Kay, pp. 163–177. New York: Plenum Press.

Gingerich, P. D. & Martin, R. D. (1981). Cranial morphology and adaptations in Eocene Adapidae. II. The Cambridge skull of *Adapis parisiensis*. *American Journal of Physical Anthropology*, **56**, 235–257.

Gingerich, P. D. & Sahni, A. (1979). *Indraloris* and *Sivaladapis*: Miocene adapid primates from the Siwaliks of India and Pakistan. *Nature*, **279**, 415–416.

Gingerich, P. D. & Sahni, A. (1984). Dentition of *Sivaladapis nagrii* (Adapidae) from the late Miocene of India. *International Journal of Primatology*, **5**, 63–79.

Gingerich, P. D. & Schoeninger, M. (1977). The fossil record and primate phylogeny. *Journal of Human Evolution*, **6**, 483–505.

Gingerich, P. D. & Simons, E. L. (1977). Systematics, phylogeny and evolution of early Eocene Adapidae (Mammalia, Primates) in North America. *Contributions from the Museum of Paleontology, University* of *Michigan*, **24**, 245–279.

Gingerich, P. D., Smith, B. H., & Rosenberg, K. (1982). Allometric scaling in the dentition of primates and prediction of body weight from tooth size in fossils. *American Journal of Physical Anthropology*, **58**, 81–100.

Ginsburg, L. (1975). Les Pliopithèques des faluns helvétiens de la Touraine et de l'Anjou. *Colloques internationaux du Centre national de la recherche scientifique*, **218**, 877–886.

Ginsburg, L. (1986). Chronology of the European pliopithecids. In *Primate Evolution*, eds. J. G. Else & P. C. Lee, pp. 47–57. Cambridge: Cambridge University Press.

Ginsburg, L. (1999). Order Carnivora. In *The Miocene Land Mammals of Europe: The Continental European Miocene*, eds. G. E. Rössner & K. Heissig, pp. 109–148. Munich: Dr. Friedrich Pfeil.

Ginsburg, L. & Mein, P. (1980). *Crouzelia rhondanica*,nouvelle espèce de primate catarrhinien, et essai sur la position systématique de Pliopithécidae. *Bulletin du Muséum national d'histoire naturelle, Paris*, **4**, 57–85.

Ginsburg, L. & Mein, P. (1987). *Tarsius thailandica* nov. sp., premier Tarsiidae (Primates, Mammalia) fossile d'Asie. *Comptes rendus de l'Académie des sciences de Paris*, **304**, 1213–1215.

Gitau, S. N. & Benefit, B. R. (1995). New evidence concerning the facial morphology of *Simiolus leakeyorum* from Maboko Island. *American Journal of Physical Anthropology*, Supplement **20**, 99.

Gitau, S. N., Benefit, B. R., Johnson, K. B., & McCrossin, M. L. (2000). New dental remains of *Kenyapithecus africanus* from Maboko Island, Kenya support the congeneric status of *Kenyapithecus wickeri* and *K. africanus*. *American Journal of Physical Anthropology*, Supplement **30**, 59.

Gitau, S. N., Benefit, B. R., McCrossin, M. L., & Roedl, T. (1998). Fossil primates and associated fauna from 1997 excavations at the middle Miocene site of Maboko Island, Kenya. *American Journal of Physical Anthropology*, Supplement **26**, 87.

Glässner, M. F. (1931). Neue Zähne von Menschenaffen aus dem Miozän des wiener Beckens. *Annalen des naturhistorischens Museums in Wien*, **46**, 15–27.

Godfrey, L. R. (1977). Structure and function in *Archaeolemur* and *Hadropithecus* (subfossil Malagasy lemurs): the postcranial evidence. PhD dissertation, Harvard University.

Godfrey, L. R. (1986*a*). The tale of the tsy-aomby-aomby. *The Sciences*, **1986**, 49–51.

Godfrey, L. R. (1986*b*). What were the subfossil indriids of Madagascar up to? *American Journal of Physical Anthropology*, **69**, 205–206.

Godfrey, L. R. (1988). Adaptive diversification of Malagasy strepsirhines. *Journal of Human Evolution*, **17**, 93–134.

Godfrey, L. R., Jungers, W. L., Reed, K. E., Simons, E. L., & Chatrath, P. S. (1997*a*). Subfossil lemurs: inferences about past and present primate communities In *Natural Change and Human Impact in Madagascar*, eds. S. M. Goodman & B. D. Patterson, pp. 218–256. Washington, DC: Smithsonian Institution Press.

Godfrey, L. R., Jungers, W. L., Simons, E. L., Chatrath, P. S., & Rakotosamimanana, B. (1999). Past and present distributions of lemurs in Madagascar. In *New Directions in Lemur Studies*, eds. B. Rakotosamimanana, H. Rasamimanana, J. U. Ganzhorn, & S. M. Goodman, pp. 19–53. New York: Kluwer.

Godfrey, L. R., Jungers, W. L., Wunderlich, R. E., & Richmond, B. G. (1997*b*). Reappraisal of the postcranium of *Hadropithecus* (Primates, Indroidea). *American Journal of Physical Anthropology*, **103**, 529–556.

Godfrey, L. R., Petto, A. J., & Sutherland, M. R. (2001). Dental ontogeny and life history strategies: the case of the giant extinct indroids of Madagascar. In *Reconstructing Behavior in the Primate Fossil Record*, eds. J. M. Plavcan, R. Kay, C. P. van Schaik, & W. L. Jungers, pp. 113–157. New York: Kluwer.

Godfrey, L. R., Simons, E. L., Chatrath, P. S., & Rakotosamimanana, B. (1990). A new fossil lemur (*Babakotia*, Primates) from northern Madagascar. *Comptes rendus de l'Académie des sciences de Paris (Série 2)*, **310**, 81–87.

Godfrey, L. R., Sutherland, M. R., Paine, R. R., Williams, F. L., Boy, D. S., & Vuillaume-Randriamanantena, M. (1995). Limb joint surface areas and their ratios in Malagasy lemurs and other mammals. *American Journal of Physical Anthropology*, **97**, 11–36.

Godinot, M. (1978). Un nouvel Adapidé (Primate) de l'Éocène inférieur de Provence. *Comptes rendus de l'Académie des sciences de Paris*, **286**, 1869–1872.

Godinot, M. (1984). Un nouveau genre témoignant de la diversité des Adapinés (Primates, Adapidae) à l'Éocène terminal. *Comptes rendus de l'Académie des sciences de Paris*, **299**, 1291–1296.

Godinot, M. (1986). Evolution of the Adapina during the late Eocene. *Primate Report*, **14**, 86.

Godinot, M. (1988). Les Primates adapidés de Bouxwiller (Éocène Moyen, Alsace) et leur apport à comprehension de la faune de Messel et à l'évolution des Anchomomyini. *Courier Forschungsinstitut Senckenberg*, **107**, 383–407.

Godinot, M. (1991). Toward the locomotion of two contemporaneous *Adapis* species. *Zeitschrift für Morphologie und Anthropologie*, **78**, 387–405.

Godinot, M. (1992*a*). Apport à la systématique de quatre genres d'Adapiformes (Primates, Éocène). *Comptes rendus de l'Académie des sciences de Paris*, **314**, 237–242.

Godinot, M. (1992*b*). Early euprimate hands in evolutionary perspective. *Journal of Human Evolution*, **22**, 267–283.

Godinot, M. (1994). Early North African primates and their significance for the origin of Simiiformes (= Anthropoidea). In *Anthropoid Origins*, eds. J. G. Fleagle & R. F. Kay, pp. 235–295. New York: Plenum Press.

Godinot, M. (1998). A summary of adapiform systematics and phylogeny. *Folia Primatologica*, **69**, 218–249.

Godinot, M., Crochet, J. Y., Hartenberger, J. L., Lange-Badre, B., Russell, D. E., & Sigé, B. (1987). Nouvelles données sur les mammifères de Palette (Éocène inférieur, Provence). *Munchner geowissen Abhandlungen A*, **10**, 273–288.

Godinot, M. & Dagosto, M. (1983). The astragalus of *Necrolemur* (Primates, Microchoerinae). *Journal of Paleontology*, **57**, 1321–24.

Godinot, M. & Jouffroy, F. K. (1982). La main d'*Adapis* (Primate, Adapidae). In *Actes du Symposium paléontologique G. Cuvier*, Montbeliard, pp. 221–242.

Godinot, M. & Mahboubi, M. (1992). Earliest known simian primate found in Algeria. *Nature*, **357**, 324–326.

Godinot, M. & Mahboubi, M. (1994). Les petits primates simiiformes de Glib Zegdou (Éocène inférieur à moyen d'Algérie). *Comptes rendus de l'Académie des sciences de Paris*, **319**, 357–364.

Godinot, M., Russell, D. E., & Louis, P. (1992). Oldest known *Nannopithex* (Primates, Omomyiformes) from the early Eocene of France. *Folia Primatologica*, **58**, 32–40.

Golpe Posse, J. M. (1982). Un pliopitécido persistente en el Vallesience Medio-Superior de los alrededores de Terrassa (cuenca del

Vallés, España) y problemas de su adaptación. *Boletín di geología y minería*, **93**, 287–296.

Gommery, D., Senut, B., & Pickford, M. (1998). New hominoid postcranial remains from the early Miocene of Napak, Uganda. *Annales de paléontologie*, **84**, 287–306.

Gommery, D., Ziegle, P., Ramanivosoa, B., & Cauvin, J. (1998). Découverte d'un nouveau site à lémuriens subfossiles dans les karsts malgaches. *Comptes rendus de l'Académie des sciences, de Paris*, **326**, 823–826.

Goodfriend, G. A. & Mitterer, R. M. (1987). Age of the ceboid femur from Coco Ree, Jamaica. *Journal of Vertebrate Paleontology*, **7**, 344–5.

Goodman, M. (1962). Immunochemistry of the primates and primate evolution. *Annals of the New York Academy of Sciences*, **102**, 219–234.

Goodman, M. (1963). Man's place in the phylogeny of the primates as reflected in serum proteins. In *Classification and Human Evolution*, ed. S. L. Washburn, pp. 204–234. Chicago, IL: Aldine.

Goodman, M., Bailey, W. J., Hayasaka, K., Stanhope, M. J., Slightom, J. & Czelusniak, J. (1994). Molecular evidence on primate phylogeny from DNA sequences. *American Journal of Physical Anthropology*, **94**, 3–24.

Goodman, M., Porter, C. A., Czelusniak, J., Page, S. L., Schneider, H., Shoshani, J., Gunnell, G. F., & Groves, C. P. (1998). Toward a phylogenetic classification of primates based on DNA evidence complemented by fossil evidence. *Molecular Phylogenetics and Evolution*, **9**, 585–598.

Goodman, M. & Tashian, R. E. (eds.) (1976). *Molecular Anthropology*. New York: Plenum Press.

Goodman, S. M. (1994*a*). Description of a new species of subfossil eagle from Madagascar: *Stephanoaetus* (Aves, Falconiformes) from the deposits of Ampasambazimba. *Proceedings of the Biological Society of Washington*, **107**, 421–428.

Goodman, S. M. (1994*b*). The enigma of antipredator behavior in lemurs: evidence of a large extinct eagle on Madagascar. *International Journal of Primatology*, **15**, 129–134.

Goodman, S. M. & Patterson, B. D. (eds.) (1997). *Natural Change and Human Impact in Madagascar*. Washington, DC: Smithsonian Institution Press.

Goodman, S. M., Pidgeon, M., Hawkins, A. F. A., & Schulenberg, T. S. (1997). *The Birds of Southeastern Madagascar*. Chicago, IL: Field Museum of Natural History.

Goodman, S. M. & Rakotozafy, L. M. A. (1995). Evidence for the existence of two species of *Aquila* on Madagascar during the Quaternary. *Géobios*, **28**, 241–246.

Goodman, S. M., & Rakotozafy, L. M. A. (1997). Subfossil birds from coastal sites in Western and Southwestern Madagascar: a paleoenvironmental reconstruction. In *Natural Change and Human Impact in Madagascar*, eds. S. M. Goodman & B. D. Patterson, pp. 257–279. Washington, DC: Smithsonian Institution Press.

Gorjanović-Kramberger, K. (1899). Der paläolithische Mensch und seine Zeitgenossen aus dem Diluvium von Krapina in Kroatien. *Mitteilungen der anthropologischen Gesellschaft in Wien*, **29**, 65–68.

Gorjanović-Kramberger, K. (1906). *Der diluviale Mensch von Krapina in Kroatien: Ein Beitrag zur Paläoanthropologie*. Wiesbaden: Kriedel.

Gorjanović-Kramberger, K. (1918). *Pračovjek iz Krapine*. Zagreb: Hrvatsko Prirodoslovno Društvo.

Grandidier, A. (1868*a*). Lettre rectifiant la position géographique des principales rivières de la côte sud-est et annonçant la découverte d'ossements fossiles d'un nouvel *Aepyornis* et d'un hippopotame. *Bulletin de la Société de géographie, Paris*, nov.-déc. **1868**, 508–510.

Grandidier, A. (1868*b*). Sur des découvertes zoologiques faites récemment à Madagascar (description d'un hippopotame nouveau, de tortues colossales et d'une espèce de *Chirogale*). *Comptes rendus de l'Académie des sciences de Paris*, **63**, 1165–1167 and *Annals of Science and Nature (Zoology)*, **10**, 375–378.

Grandidier, G. (1899*a*). Description d'ossements de lémuriens disparus. *Bulletin du Muséum national d'histoire naturelle, Paris*, **5**, 272–276.

Grandidier, G. (1899*b*). Description d'ossements de lémuriens disparus. *Bulletin du Muséum national d'histoire naturelle, Paris*, **5**, 344–348.

Grandidier, G. (1900). Sur les lémuriens subfossiles de Madagascar. *Comptes rendus de l'Académie des sciences de Paris*, **130**, 1482–1485.

Grandidier, G. (1901). Un nouvel édenté subfossile de Madagascar. *Bulletin du Muséum national d'histoire naturelle, Paris*, **7**, 54–56.

Grandidier, G. (1902*a*). Observations sur les lémuriens disparus de Madagascar: Collections Alluaud, Gaubert, Grandidier. *Bulletin du Muséum national d'histoire naturelle, Paris*, **8**, 497–505.

Grandidier, G. (1902*b*). Observations sur les lémuriens disparus de Madagascar: Collections Alluaud, Gaubert, Grandidier (suite). *Bulletin du Muséum national d'histoire naturelle, Paris*, **8**, 587–592.

Grandidier, G. (1904). Un nouveau lémurien fossile de France, le *Pronycticebus gaudryi*. *Bulletin du Muséum national d'histoire naturelle, Paris*, **10**, 9–13.

Grandidier, G. (1905). Recherches sur les lémuriens disparus et en particulier sur ceux qui vivaient à Madagascar. *Nouvelles archives du Muséum national d'histoire naturelle, Paris*, **7**, 1–142.

Grandidier, G. (1929, for the year 1928). Une variété du *Cheiromys madagascariensis* actuel et un nouveau *Cheiromys* subfossile. *Bulletin de l'Académie* (nouvelle série), **11**, 101–107.

Granger, W. (1910). Tertiary faunal horizons in the Wind River Basin, Wyoming, with descriptions of new Eocene mammals. *Bulletin of the American Museum of Natural History*, **28**, 235–251.

Graybeal, A. (1998). Is it better to add taxa or characters to a difficult taxonomic problem? *Systematic Biology*, **47**, 9–17.

Greenfield, L. O. (1972). Sexual dimorphism in *Dryopithecus africanus*. *Primates*, **13**, 395–410.

Greenfield, L. O. (1973). Note on the placement of the most complete "*Kenyapithecus africanus*" mandible. *Folia Primatologica*, **20**, 274–279.

Greenfield, L. O. (1974). Taxonomic reassessment of two *Ramapithecus* specimens. *Folia Primatologica*, **22**, 97–115.

Greenfield, L. O. (1975). A comment on relative molar breadth in *Ramapithecus*. *Journal of Human Evolution*, **4**, 267–273.

Greenfield, L. O. (1978). On the dental arcade reconstructions of *Ramapithecus*. *Journal of Human Evolution*, **7**, 345–359.

Greenfield, L. O. (1979). On the adaptive pattern in *Ramapithecus*. *American Journal of Physical Anthropology*, **50**, 527–548.

Gregory, W. K. (1910). The orders of mammals. *Bulletin of the American Museum of Natural History*, **27**, 1–524.

Gregory, W. K. (1913). Relationship of the Tupaiidae and of Eocene lemurs, especially *Notharctus*. *Bulletin of the Geological Society of America*, **24**, 247–252.

Gregory, W. K. (1915). On the classification and phylogeny of the Lemuroidea. *Bulletin of the Geological Society of America*, **26**, 419–446.

Gregory, W. K. (1916). Studies on the evolution of the primates. *Bulletin of the American Museum of Natural History*, **35**, 239–355.

Gregory, W. K. (1920). On the structure and relations of *Notharctus*, an American Eocene primate. *Memoirs of the American Museum of Natural History*, new series, **3**(2), 49–243.

Gregory, W. K. (1951). *Evolution Emerging*. New York: Macmillan.

Gregory, W. K. & Hellman, M. (1926). The dentition of *Dryopithecus* and the origin of man. *American Museum of Natural History Anthropological Papers*, **28**, 1–123.

Gregory, W. K., Hellman, M., & Lewis, G. E. (1938). *Fossil Anthropoids of the Yale–Cambridge India Expedition of 1935*. Washington, DC: Carnegie Institute.

Grine, F. E. (ed.) (1988*a*). *Evolutionary History of the "Robust" Australopithecines*. New York: Aldine de Gruyter.

Grine, F. E. (1988*b*). Evolutionary history of the robust australopithecines: a summary and the historical perspective. In *Evolutionary History of the "Robust" Australopithecines*, ed. F. E. Grine, pp. 509–510. New York: Aldine de Gruyer.

Grine, F. E., Demes, B., Jungers, W. L., & Cole, T. M. III (1993). Taxonomic affinity of the early *Homo* cranium from Swartkrans, South Africa. *American Journal of Physical Anthropology*, **92**, 411–426.

Grine, F. E. & Hendey, Q. B. (1981). Earliest primate remains from South Africa. *South African Journal of Science*, **77**, 374–376.

Groves, C. P. (1972). Systematics and phylogeny of the gibbons. In *Gibbon and Siamang*, vol. 1, ed. D. M. Rumbaugh, pp. 1–89. Basel: Karger.

Groves, C. P. (1974*a*). New evidence on the evolution of apes and man. *Vestnik Ústredniho Ústavu Geologickeho*, **49**, 53–56.

Groves, C. P. (1974*b*). Taxonomy and phylogeny of prosimians. In *Prosimian Biology*, eds. R. D. Martin, G. A. Doyle, & A. C. Walker, pp. 449–473. London: Duckworth.

Groves, C. P. (1986). Systematics of the Great Apes. In *Comparative Primate Biology*, vol. 1, *Systematics, Evolution and Anatomy*, eds D. Swindler & S. Erwin, pp. 187–217. New York: Alan R. Liss.

Groves, C. P. (1988). The evolutionary ecology of the Hominoidea. *Anales de psicología*, **39**, 87–98.

Groves, C. P. (1989). *A Theory of Human and Primate Evolution*. Oxford: Clarendon Press.

Groves, C. P. (1993). Order Primates. In *Mammal Species of the World*, eds. D. E. Wilson & D. M. Reeder, pp. 243–277. Washington, DC: Smithsonian Institution Press.

Groves, C. P. (1997). Taxonomy and phylogeny of primates. In *Molecular Biology and Evolution of Blood Group and MHC Antigens in Primates*, ed. A. Blancher, pp. 3–23. Berlin: Springer-Verlag.

Groves, C. P. (2000). The phylogeny of the Cercopithecoidea. In *Old World Monkeys*, eds. P. F. Whitehead & C. J. Jolly, pp. 77–98. Cambridge: Cambridge University Press.

Groves, C. P. & Mazak, V. (1975). An approach to the taxonomy for the Hominidae: gracile Villafranchian hominids of Africa. *Casopis pro Mineralogii a Geologii*, **20**, 225–46.

Groves, C. P. & Paterson, J. D. (1991). Testing hominoid phylogeny with the PHYLIP programs. *Journal of Human Evolution*, **20**, 167–183.

Grün, R., Brink, J., Spooner, N., Taylor, A., Stringer, C., Franciscus, R., & Murray, A. (1996). Direct dating of Florisbad hominid. *Nature*, **382**, 500–501.

Grün, R. & Stringer, C. (1991). Electron spin resonance dating and the evolution of modern humans. *Archaeometry*, **33**, 153–199.

Grün, R. & Thorne, A. (1997). Dating the Ngandong humans. *Science*, **276**, 1575–1576.

Gu, Y. & Hu, C. (1991). A fossil cranium of *Rhinopithecus* found in Xinan, Henan Province. *Vertebrata PalAsiatica*, **29**, 55–58.

Gu, Y., Huang, W., Chen, D., Guo, X., & Jablonski, N. G. (1996). Pleistocene fossil primates from Luoding, Guangdong. *Vertebrata PalAsiatica*, **34**, 235–250.

Gu, Y., Huang, W., Song, F., Guo, X., & Chen, D. (1987). The study of some fossil orang-utans from Guangdong and Guangxi. *Acta Anthropologica Sinica*, **6**, 272–83.

Gu, Y. & Lin, Y. (1983). First discovery of *Dryopithecus* in East China. *Acta Anthropologica Sinica*, **2**, 305–314.

Gundling, T. & Hill, A. (2000). Geological context of fossil Cercopithecoidea from eastern Africa. In *Old World Monkeys*, eds. P. F. Whitehead & C. J. Jolly, pp. 180–213. Cambridge: Cambridge University Press.

Gunnell, G. F. (1989). Evolutionary history of Microsyopoidea (Mammalia, ?Primates) and the relationship between Plesiadapiformes and Primates. *Museum of Paleontology, University of Michigan, Papers on Paleontology*, **27**, 1–154.

Gunnell, G. F. (1995*a*). Omomyid primates (Tarsiiformes) from the Bridger Formation, middle Eocene, southern Green River Basin, Wyoming. *Journal of Human Evolution*, **28**, 147–187.

Gunnell, G. F. (1995*b*). New notharctine (Primates, Adapiformes) skull from the Uintan (Middle Eocene) of San Diego County, California. *American Journal of Physical Anthropology*, **98**, 447–470.

Gunnell, G. F. (1997). Wasatchian–Bridgerian (Eocene) paleoecology of the western interior of North America: changing paleoenvironments and taxonomic composition of omomyid (Tarsiiformes) primates. *Journal of Human Evolution*, **32**, 105–132.

Gunnell, G. F., Bartels, W. S., Gingerich P. D., & Torres, V. (1992). Wapiti Valley faunas: early and middle Eocene fossil vertebrates from the North Fork of the Shoshone River, Park County, Wyoming. *Contributions from the Museum of Paleontology, University of Michigan*, **29**, 247–287.

Gunnell, G. F. & Miller, E. R. (2001). Origin of Anthropoidea: dental evidence and recognition of early anthropoids in the fossil record, with comments on the Asian anthropoid radiation. *American Journal of Physical Anthropology*, **114**, 177–191.

Gupta, S. S., Verma, B. C., & Tewari, A. P. (1982). New fossil hominoid material from the Siwaliks of Kangra District, Himachal Pradesh. *Journal of the Palaeontological Society of India*, **27**, 111–115.

Gupta, V. J. (1969). Fossil primates from the Lower Siwaliks of Kangra District, H. P. *Research Bulletin (new series), Panjab University, India*, **20**, 577–578.

Gupta, V. J. & Sahni, A. (1981). *Theropithecus delsoni*, a new cercopithecine species from the Upper Siwaliks of India. *Bulletin of the Indian Geological Association*, **14**, 69–71.

Habgood, P. (1989). The origin of anatomically modern humans in Australasia. In *The Human Revolution*, eds. P. Mellars & C. Stringer, pp. 245–273. Edinburgh: University of Edinburgh Press.

Haeckel, E. (1868). *Natürliche Schöpfunsgeschichte*. Berlin: Reimer.

Haeckel, E. (1874). *Anthropogenie, oder Entwicklungsgeschichte des Menschen*. Leipzig: Engelmann.

Haile-Selassie, Y. (2001). Late Miocene hominids from the Middle Awash. *Nature*, **412**, 178–181.

Haines, R. (1950). The interorbital septum in mammals. *Journal of the Linnean Society of London*, **41**, 585–607.

Hamilton, A. C. (1988). Guenon evolution and forest history. In *A Primate Radiation: Evolutionary Biology of the African Guenons*, eds. A. Gautier, F. Bourlière, J.-P. Gautier, & J. Kingdon, pp. 13–34.

Cambridge: Cambridge University Press.

Hamrick, M. W. (1996). Locomotor adaptations reflected in the wrist joints of early Tertiary primates (Adapiformes). *American Journal of Physical Anthropology*, **100**, 585–604.

Hamrick, M. W. (1998). Functional and adaptive significance of primate pads and claws: evidence from New World anthropoids. *American Journal of Physical Anthropology*, **106**, 113–127.

Hamrick, M. W. (1999). First carpals of the Eocene primate family Omomyidae. *Contributions from the Museum of Paleontology, University of Michigan*, **30**, 191–198.

Hamrick, M. W. (2001). Primate origins: evolutionary change in digital ray patterning and segmentation. *Journal of Human Evolution*, **40**, 339–351.

Hamrick, M. W. & Alexander, J. P. (1996). The hand skeleton of *Notharctus tenebrosus* (Primates, Notharctidae) and its significance for the origin of the primate hand. *American Museum Novitates*, **3182**, 1–20.

Hamrick, M. W., Meldrum, D. J., & Simons, E. L. (1995). Anthropoid phalanges from the Oligocene of Egypt. *Journal of Human Evolution*, **28**, 121–145.

Hamrick, M. W., Rosenman, B. A., & Brush, J. A. (1999). Phalangeal morphology of the Paromomyidae (?Primates, Plesiadapiformes): the evidence for gliding behavior reconsidered. *American Journal of Physical Anthropology*, **109**, 397–413.

Hamrick, M. W., Simons, E. L., & Jungers, W. L. (2000). New wrist bones of the Malagasy giant subfossil lemurs. *Journal of Human Evolution*, **38**, 635–650.

Harada, M. L., Schneider, H., Schneider, M. P. C., Sampaio, I. M., Czelusniak, J., & Goodman, M. (1995). DNA evidence on the phylogenetic systematics of the New World monkeys: support for the sister-grouping of *Cebus* and *Saimiri* from two unlinked nuclear genes. *Molecular Phylogenetics and Evolution*, **4**, 331–349.

Harris, E. E. & Disotell, T. R. (1998). Nuclear gene trees and the phylogenetic relationships of the mangabeys (Primates: Papionini). *Molecular Biology and Evolution*, **15**, 892–900.

Harris, J. M., Brown, F. H., & Leakey, M. G. (1988*a*). Stratigraphy and paleontology of Pliocene and Pleistocene localities west of Lake Turkana, Kenya. *Contributions in Science*, **399**, 1–128.

Harris, J. M., Brown, F. H., Leakey, M. G., Walker, A. C., & Leakey, R. E. (1988*b*). Pliocene and Pleistocene hominid-bearing sites from west of Lake Turkana, Kenya. *Science*, **239**, 27–33.

Harrison, T. (1981). New finds of small fossil apes from the Miocene locality of Koru in Kenya. *Journal of Human Evolution*, **10**, 129–137.

Harrison, T. (1982). Small-bodied apes from the Miocene of East Africa. PhD dissertation, University of London.

Harrison, T. (1986*a*). A reassessment of the phylogenetic relationship of *Oreopithecus bambolii* Gervais. *Journal of Human Evolution*, **15**, 541–583.

Harrison, T. (1986*b*). New fossil anthropoids from the middle Miocene of East Africa and their bearing on the origin of the Oreopithecidae. *American Journal of Physical Anthropology*, **71**, 265–284.

Harrison, T. (1987). The phylogenetic relationships of the early catarrhine primates: a review of the current evidence. *Journal of Human Evolution*, **16**, 41–80.

Harrison, T. (1988). A taxonomic revision of the small catarrhine primates from the early Miocene of East Africa. *Folia Primatologica*, **50**, 59–108.

Harrison, T. (1989*a*). A new species of *Micropithecus* from the middle Miocene of Kenya. *Journal of Human Evolution*, **18**, 537–557.

Harrison, T. (1989*b*). New estimates of cranial capacity, body size, and encephalization in *Oreopithecus bambolii*. *American Journal of Physical Anthropology*, **78**, 237.

Harrison, T. (1989*c*). New postcranial remains of *Victoriapithecus* from the middle Miocene of Kenya. *Journal of Human Evolution*, **18**, 3–54.

Harrison, T. (1992). A reassessment of the taxonomic and phylogenetic affinities of the fossil catarrhines from Fort Ternan, Kenya. *Primates*, **33**, 501–522.

Harrison, T. (1993). Cladistic concepts and the species problem in hominoid evolution. In *Species, Species Concepts, and Primate Evolution*, eds. W. H. Kimbel & L. B. Martin, pp. 345–371. New York: Plenum Press.

Harrison, T. (1998). Evidence for a tail in *Proconsul heseloni*. *American Journal of Physical Anthropology*, Supplement **26**, 93–94.

Harrison, T. (1999). A reconsideration of the taxonomic and phylogenetic relationships of *Nyanzapithecus*. *International Symposium on Evolution of Middle and Late Miocene Hominoids in Africa, Kyoto* (abstracts).

Harrison, T., Delson, E., & Jian, G. (1991). A new species of *Pliopithecus* from the middle Miocene of China and its implications for early catarrhine zoogeography. *Journal of Human Evolution*, **21**, 329–362.

Harrison, T. & Gu, Y. (1999). Taxonomy and phylogenetic relationships of early Miocene catarrhines from Sihong, China. *Journal of Human Evolution*, **37**, 225–277.

Harrison, T. & Harris, E. E. (1996). Plio-Pleistocene cercopithecids from Kanam, western Kenya. *Journal of Human Evolution*, **30**, 539–561.

Harrison, T. & Rook, L. (1997). Enigmatic anthropoid or misunderstood ape? The phylogenetic status of *Oreopithecus bambolii* reconsidered. In *Function, Phylogeny and Fossils: Miocene Hominoid Evolution and Adaptation*, eds. D. R. Begun, C. V. Ward, & M. D. Rose, pp. 327–362. New York: Plenum Press.

Harrison, T. & Sanders, W. J. (1999). Scaling of lumbar vertebrae in anthropoid primates: its implications for the positional behavior and phylogenetic affinities of *Proconsul*. *American Journal of Physical Anthropology*, Supplement **28**, 146.

Hartenberger, J. L., Crochet, J. Y., Martinez, C., Feist, M., Godinot, M., Mannai Tayech, B., Marandat, B., & Sigé, B. (1997). Le gisement de mammifères de Chambi (Éocène, Tunisie Centrale) dans son contexte géologique apport à la connaissance de l'évolution des mammifères en Afrique. *Mémoires et travaux de l'Institut de Montpellier*, **21**, 263–274.

Hartenberger, J. L. & Marandat, B. (1992). A new genus and species of an early Eocene primate from North Africa. *Human Evolution*, **7**, 9–16.

Hartwig, W. C. (1994). Patterns, puzzles and perspectives on platyrrhine origins. In *Integrative Paths to the Past: Paleoanthropological Essays in Honor of F. Clark Howell*, eds. R. S. Corruccini & R. L. Ciochon, pp. 69–94. Englewood Cliffs, NJ: Prentice-Hall.

Hartwig, W. C. (1995*a*). A giant New World monkey from the Pleistocene of Brazil. *Journal of Human Evolution*, **28**, 189–195.

Hartwig, W. C. (1995*b*). *Protopithecus*: rediscovering the first fossil primate. *History and Philosophy of the Life Sciences*, **17**, 447–460.

Hartwig, W. C. (1996). Perinatal life history traits in New World monkeys. *American Journal of Primatology*, **40**, 99–130.

Hartwig, W. C. & Cartelle, C. (1996). A complete skeleton of the giant South American primate *Protopithecus*. *Nature*, **381**, 307–311.

Hartwig-Scherer, S. (1993). Allometry in hominoids: a comparative study of skeletal growth trends. PhD dissertation, University of Zürich.

Hartwig-Scherer, S. & Martin, R. D. (1991). Was "Lucy" more human than her "Child'? Observations on early hominid postcranial skeletons. *Journal of Human Evolution*, **21**, 439–449.

Haughton, R. H. (1925). A note on the occurrence of a species of baboon in limestone deposits near Taungs. *Transactions of the Royal Society of South Africa*, **12**, 68.

Hawks, J., Oh, S., Hunley, K., Dobson, S., Cabana, G., Dayalu, P., & Wolpoff, M. (2000). An Australian test of the recent African origin theory using the WLH-50 calvarium. *Journal of Human Evolution*, **39**, 1–22.

Hawks, J. & Wolpoff, M. (2001). Brief communication: Paleoanthropology and the population genetics of ancient genes. *American Journal of Physical Anthropology*, **114**, 269–272.

Hayasaka, K., Fujii, K., & Horai, S. (1996). Molecular phylogeny of macaques: implications of nucleotide sequences from an 896-base pair region of mitochondrial DNA. *Molecular Biology and Evolution*, **13**, 1044–1053.

Hayashida, A. (1984). Paleomagnetic study of the Miocene continental deposits in La Venta badlands, Colombia. *Kyoto University Overseas Research Reports of New World Monkeys*, **4**, 77–88.

Heberer, G. (1963). Über einen neuen archanthropinen Typus aus der Oldoway-Schlucht. *Zeitschrift für Morphologie und Anthropologie*, **53**, 171–177.

Heinrich, R. E., Rose, M., Leakey, R. E., & Walker, A. C. (1993). A new hominid radius from the Middle Pliocene of West Turkana, Kenya. *American Journal of Physical Anthropology*, **93**, 139–148.

Heintz, E., Brunet, M., & Battail, B. (1981). A cercopithecid primate from the late Miocene of Molayan, Afghanistan, with remarks on *Mesopithecus*. *International Journal of Primatology*, **2**, 273–284.

Heinzelin, J. de, Clark, J. D., White, T., Hart, W., Renne, P., WoldeGabriel, G., Beyene, Y., & Vrba, E. (1999). Environment and behavior of 2.5-million-year-old Bouri hominids. *Science*, **284**, 625–629.

Heizmann, E. (1992). Das Tertiär in Südwestdeutschland. *Stuttgarter Beiträge zur Naturkunde, Serie C*, **33**, 1–90.

Heizman, E. & Begun, D. R. (2001). The oldest Eurasian hominoid. *Journal of Human Evolution*, **41**, 463–481.

Heizmann, E., Duranthon, F., & Tassy, P. (1996). Miozäne Grossäugetiere. *Stuttgarter Beiträge zur Naturkunde, Serie C*, **39**, 1–60.

Heller, F. (1930). Die Säugetierfauna der mitteleozänen Braunkohle des Geiseltales bei Halle a. *Stuttgarter Jahrbuch Halleschen Verbandes*, **9**, 13–14.

Henri-Martin, G. (1957). La grotte de Fontéchevade. I. Historique, fouilles, stratigraphie, archéologie. *Archives de l'Institut de paléontologie humaine*, **28**, 1–247.

Hershkovitz, P. (1970). Notes on Tertiary platyrrhine monkeys and description of a new genus from the late Miocene of Colombia. *Folia Primatologica*, **12**, 1–37.

Hershkovitz, P. (1974). A new genus of late Oligocene monkey (Cebidae, Platyrrhini) with notes on postorbital closure and platyrrhine evolution. *Folia Primatologica*, **21**, 1–35.

Hershkovitz, P. (1977). *Living New World Monkeys (Platyrrhini) with an Introduction to Primates*. Chicago, IL: University of Chicago Press.

Hershkovitz, P. (1981). Comparative anatomy of platyrrhine mandibular cheek teeth dpm4, pm4, m1, with particular reference to those of *Homunculus* (Cebidae), and comments on platyrrhine origins. *Folia Primatologica*, **35**, 179–217.

Hershkovitz, P. (1982). Supposed squirrel monkey affinities of the late Oligocene *Dolichocebus gaimanensis*. *Nature*, **298**, 201–202.

Hershkovitz, P. (1984). More on *Homunculus* dpm4 and M1 and comparisons with *Alouatta* and *Stirtonia* (Primates, Platyrrhini, Cebidae). *American Journal of Primatology*, **7**, 261–283.

Hershkovitz, P. (1985). A preliminary taxonomic review of the South American bearded saki monkeys genus *Chiropotes* (Cebidae, Platyrrhini), with the description of a new species. *Fieldiana: Zoology*, n.s., **27**, 1–46.

Hershkovitz, P. (1987). The taxonomy of South American sakis, genus *Pithecia* (Cebidae, Platyrrhini): a preliminary report and critical review with the description of a new species and a new subspecies. *American Journal of Primatology*, **12**, 387–468.

Hershkovitz, P. (1988). The subfossil monkey femur and subfossil monkey tibia of the Antilles: a review. *International Journal of Primatology*, **9**, 365–384.

Hershkovitz, P. (1990). Titis, New World monkeys of the genus *Callicebus* (Cebidae, Platyrrhini): a preliminary taxonomic review. *Fieldiana: Zoology*, n.s., **55**, 1–109.

Hewitt, G. (2000). The genetic legacy of the Quaternary ice ages. *Nature*, **405**, 907–913.

Hill, A. (1994). Late Miocene and early Pliocene hominoids from Africa. In *Integrative Paths to the Past: Paleoanthropological Advances in Honor of F. Clark Howell*, eds. R. S. Corruccini & R. L. Ciochon, pp. 123–146. Englewood Cliffs, NJ: Prentice-Hall.

Hill, A. (1995). Faunal and environmental change in the Neogene of east Africa: evidence from the Tugen Hills sequence, Baringo District, Kenya. In *Paleoclimate and Evolution, with Emphasis on Human Origins*, eds. E. S. Vrba, G. H. Denton, T. C. Partridge, & L. H. Burkle, pp. 178–193. New Haven, CT: Yale University Press.

Hill, A., Behrensmeyer, K., Brown, B., Deino, A., Rose, M., Saunders, J., Ward, S.,, & Winkler, A. (1991). Kipsaramon: a lower Miocene hominoid site in the Tugen Hills, Baringo District, Kenya. *Journal of Human Evolution*, **20**, 67–75.

Hill, A., Drake, R., Tauxe, L., Monaghan, M., Barry, J. C., Behrensmeyer, A. K., Curtis, G., Fine-Jacobs, B., Jacobs, L., Johnson, N., & Pilbeam, D. (1985). Neogene paleontology and geochronology of the Baringo Basin, Kenya. *Journal of Human Evolution*, **14**, 749–773.

Hill, A. & Odhiambo, I. (1987). New mandible of *Rangwapithecus* from Songhor, Kenya. *American Journal of Physical Anthropology*, **72**, 210.

Hill, A. & Ward, S. C. (1988). Origin of the Hominidae: the record of African large hominoid evolution between 14 My and 4 My. *Yearbook of Physical Anthropology*, **31**, 49–83.

Hill, W. C. O. (1953). *Primates: Comparative Anatomy and Taxonomy*. vol. 1. *Strepsirhini*. Edinburgh: Edinburgh University Press.

Hill, W. C. O. (1962). *Primates: Comparative Anatomy and Taxonomy*, vol. 5. *Cebidae*. Edinburgh: Edinburgh University Press.

Hill, W. C. O. (1966). *Primates: Comparative Anatomy and Taxonomy*, vol. 6. *Catarrhini, Cercopithecoidea, Cercopithecinae*. Edinburgh: Edinburgh University Press.

Ho, C. K. (1988). Human origins in Asia? *Human Evolution*, **3**, 357–365.

Ho, C. K. (1990). A new Pliocene hominoid skull from Yuanmou southwest China. *Human Evolution*, **5**, 309–318.

Ho, C. K., Zhou, G. X., & Swindler, D. R. (1995). Dental evolution of the orangutan in China. *Human Evolution*, **10**, 249–64.

Hoelzer, G. A., Hoelzer, M. A., & Melnick, D. J. (1992). The evolutionary history of the *sinica*-group of macaque monkeys as revealed by mtDNA restriction site analysis. *Molecular Phylogenetics and Evolution*, **1**, 215–222.

Hoelzer, G. A. & Melnick, D. J. (1996). Evolutionary relationships of the macaques. In *Evolution and Ecology of Macaque Societies*, eds J. E. Fa & D. G. Lindburg, pp. 3–19. New York: Cambridge University Press.

Hofer, H. O. (1953). Über Gehirn und Schadel von *Megaladapis edwardsi* G. Grandidier (Lemuroidea). *Zeitschrift für wissenschaftliche Zoologie*, **157**, 220–284.

Hofer, H. O. (1979). The external nose of *Tarsius bancanus borneanus* Horsfield, 1821 (Primates, Tarsiiformes). *Folia Primatologica*, **32**, 180–192.

Hoffstetter, R. (1969). Un primate de l'Oligocène inférieur sud-americain: *Branisella boliviana* gen. et. sp. nov. *Comptes rendus de l'Académie des sciences de Paris*, **269**, 434–437.

Hoffstetter, R. (1972). Relationships, origins, and history of the ceboid monkeys and caviomorph rodents: A modern reinterpretation. In *Evolutionary Biology*, vol. 6, eds. T. Dobzhansky, M. K. Hecht, & W. C. Steere, pp. 323–347. New York: Appleton-Century-Crofts.

Hoffstetter, R. (1974). Phylogeny and geographical deployment of the primates. *Journal of Human Evolution*, **3**, 327–350.

Hoffstetter, R. (1977). Phylogénie des Primates: confrontation des résultats obtenus par les diverses voies d'approche du problème. *Bulletin et mémoires de la Société d'anthropologie de Paris* (Série 13), **4**, 327–346.

Hoffstetter, R. (1980). Origin and deployment of New World monkeys emphasizing the southern continents route. In *Evolutionary Biology of the New World Monkeys and Continental Drift*, eds. R. L. Ciochon & A. B. Chiarelli, pp. 103–122. New York: Plenum Press.

Hoffstetter, R. (1982). Les Primates Simiiformes (= Anthropoidea): compréhension, phylogénie, histoire biogéographique. *Annales de paléontologie*, **68**, 241–290.

Hofmann, A. (1863). Die Fauna von Göriach. *Abhandlungen der kaiserlich-königlichen geologischen Reichsansalt Wien*, **15**, 1–87.

Holliday, T. W. (1997). Postcranial evidence of cold adaptation in European Neandertals. *American Journal of Physical Anthropology*, **104**, 245–258.

Holliday, T. W. (2000). Evolution at the crossroads: modern human emergence in western Asia. *American Anthropologist*, **102**, 54–68.

Holloway, R. L. (1973). New endocranial values for the East African early hominids. *Nature*, **243**, 97–99.

Holloway, R. L. (1980). The OH 7 (Olduvai Gorge, Tanzania) hominid partial endocast revisited. *American Journal of Physical Anthropology*, **53**, 267–274.

Holloway, R. L. (1981). The Indonesian *Homo erectus* brain endocasts revisited. *American Journal of Physical Anthropology*, **55**, 503–521.

Holloway, R. L. (1983*a*). Human brain evolution: a search for units, models and synthesis. *Canadian Journal of Anthropology*, **3**, 215–230.

Holloway, R. L. (1983*b*). Human paleontological evidence relevant to language behavior. *Human Neurobiology*, **2**,105–114.

Holroyd, P. A. & Maas, M. C. (1994). Paleogeography, paleobiogeography, and anthropoid origins. In *Anthropoid Origins*, eds. J. G. Fleagle & R. F. Kay, pp. 297–333. New York: Plenum Press.

Honey, J. G. (1990). New washakiin primates (Omomyidae) from the Eocene of Wyoming and Colorado, and comments on the evolution of the Washakiini. *Journal of Vertebrate Paleontology*, **10**, 206–221.

Hooijer, D. A. (1948). Prehistoric teeth of man and the orangutan from central Sumatra with notes on the fossil orangutan from Java and southern China. *Zoologische mededelingen, Rijksmuseum van Natuurlijke Historie*, **29**, 175–301.

Hooijer, D. A. (1949). Some notes on the *Gigantopithecus* question. *American Journal of Physical Anthropology*, **7**, 513–518.

Hooijer, D. A. (1951). Questions relating to a new large anthropoid ape from the Mio-Pliocene of the Siwaliks. *American Journal of Physical Anthropology*, **9**, 79–95.

Hooijer, D. A. (1960). Quaternary gibbons from the Malay archipelago. *Zoologische Verhandelingen*, **46**, 1–42.

Hooijer, D. A. (1962*a*). The Middle Pleistocene fauna of the Near East. In *Evolution und Hominization*, ed. G. Kurth, pp. 81–83. Stuttgart: Gustav Fischer.

Hooijer, D. A. (1962*b*). Prehistoric bone: the gibbons and monkeys of Niah Great Cave. *Sarawak Museum Journal*, **11**, 428–449.

Hooijer, D. A. (1962*c*). Quaternary langurs and macaques from the Malay Archipelago. *Zoololigsche Verhandelingen*, **55**, 3–64.

Hooijer, D. A. (1962*d*). Report upon a collection of Pleistocene mammals from tin-bearing deposits in a limestone cave near Ipoh, Kinta Valley, Perak. *Federation Museums Journal*, **VII**, 1–5.

Hooijier, D. A. (1963). Miocene Mammalia of the Congo. *Annales du Muséum royal d'Afrique Centrale* (*Série 8, Sciences geologiques*), **46**, 1–71.

Hooker, J. J. (1986). Mammals from the Bartonian (middle/late Eocene) of the Hampshire Basin, southern England. *Bulletin of the British Museum (Natural History)*, **39**, 191–478.

Hooker, J. J., Russell, D. E., & Phélizon, A. (1999). A new family of Plesiadapiformes (Mammalia) from the Old World lower Paleogene. *Palaeontologia*, **42**, 377–407.

Hopwood, A. T. (1933*a*). Miocene primates from British East Africa. *Annals and Magazine of Natural History, Series 10*, **11**, 96–98.

Hopwood, A. T. (1933*b*). Miocene primates from Kenya. *Journal of the Linnean Society of London–Zoology*, **38**, 437–464.

Hopwood, A. T. (1934). New fossil mammals from Olduvai, Tanganyika Territory. *Annals and Magazine of Natural History Series 10*, **14**, 546–550.

Hopwood, A. T. (1936). New and little-known fossil mammals from the Pleistocene of Kenya Colony and Tanganyika Territory. *Annals and Magazine of Natural History, Series 10*, **17**, 636–641.

Horai, S., Hayasaka, K., Kondo, R., Tsugane, K., & Takahata, N. (1995). Recent African origin of modern humans revealed by complete sequences of hominoid mitochondrial DNAs. *Proceedings of the National Academy of Sciences of the United States of America*, **92**, 532–536.

Horovitz, I., (1997). Platyrrhine systematics and the origin of Greater Antillean monkeys. PhD dissertation, State University of New York, Stony Brook.

Horovitz, I. (1999). A phylogenetic study of living and fossil platyrrhines. *American Museum Novitates*, **3269**, 1–40.

Horovitz, I. & MacPhee, R. D. E. (1996). New materials of *Paralouatta* and a new hypothesis for Antillean monkey systematics. *Journal of Vertebrate Paleontology*, **16**, 42A.

Horovitz, I. & MacPhee, R. D. E. (1999). The Quaternary Cuban platyrrhine *Paralouatta varonai* and the origin of Antillean monkeys. *Journal of Human Evolution*, **36**, 33–68.

Horovitz, I. & MacPhee, R. D. E. (in press). The primate fossil record of the Greater Antilles. In *Contributions to the Land Mammal Fauna of the West Indies*, eds. C. A. Woods, R. Borroto-Pàez, & J. A. Ottenwalder. Gainesville, FL: University of Florida.

Horovitz, I., MacPhee, R. D. E., Flemming, C., & McFarlane, D. A. (1997). Cranial remains of *Xenothrix* and their bearing on the question of Antillean monkey origins. *Journal of Vertebrate Paleontology*, **17**, 54A.

Horovitz, I. & Meyer, A. (1995). Systematics of the New World monkeys (Platyrrhini, Primates) based on 16S mitochondrial DNA sequences: a comparative analysis of different weighting methods in cladistic analysis. *Molecular Phylogenetics and Evolution*, **4**, 448–456.

Horovitz, I., Zardoya, R., & Meyer A. (1998). Platyrrhine systematics: a simultaneous analysis of molecular and morphological data. *American Journal of Physical Anthropology*, **106**, 261–287.

Houle, A. (1999). The origin of platyrrhines: an evaluation of the Antarctic scenario and the floating island model. *American Journal of Physical Anthropology*, **109**, 541–559.

Howell, F. C. (1951). The place of Neanderthal man in human evolution. *American Journal of Physical Anthropology*, **9**, 376–416.

Howell, F. C. (1957). The evolutionary significance of variation and varieties of "Neanderthal" man. *Quarterly Review of Biology*, **32**, 330–347.

Howell, F. C. (1958). Upper Pleistocene men of the Southwest Asian Mousterian. In *Hundert Jahre Neanderthaler*, ed. G. H. R. von Koenigswald, pp. 185–198. Utrecht: Keminck & Zoon.

Howell, F. C. (1960). European and northwest African Middle Pleistocene hominids. *Current Anthropology*, **1**, 195–232.

Howell, F. C. (1965). New discoveries in Tanganyika: their bearing on hominid evolution (reply to Tobias). *Current Anthropology*, **6**, 399–401.

Howell, F. C. (1978). Hominidae. In *Evolution of African Mammals*, eds. V. J. Maglio & H. B. S. Cooke, pp. 154–248. Cambridge, MA: Harvard University Press.

Howell, F. C. (1994). A chronostratigraphic and taxonomic framework of the origins of modern humans. In *Origins of Anatomically Modern Humans*, eds. M. H. Nitecki & D. V. Nitecki, pp. 253–319. New York: Plenum Press.

Howell, F. C. (1996). Thoughts on the study and interpretation of the human fossil record. In *Contemporary Issues in Human Evolution*, eds. W. E. Meikle, F. C. Howell, & N. G. Jablonski, pp. 1–45. San Francisco, CA: California Academy of Sciences.

Howells, W. W. (1959). *Mankind in the Making*. Garden City, MI: Doubleday.

Howells, W. W. (1973). *Cranial Variation in Man*. Papers of the Peabody Museum of Archaeology and Ethnology, **67**. Cambridge, MA: Harvard University.

Howells, W. W. (1974). Neanderthals: names, hypotheses and scientific method. *American Anthropologist*, **76**, 24–38.

Howells, W. W. (1976). Explaining modern man: evolutionists versus migrationalists. *Journal of Human Evolution*, **5**, 577–596.

Hrdlička, A. (1927). The Neanderthal phase of man. *Journal of the Royal Anthropological Institute*, **57**, 249–274.

Hrdlička, A. (1930). The skeletal remains of early man. *Smithsonian Miscellaneous Collections*, **83**, 1–379.

Hrdlička, A. (1935). The Yale fossils of anthropoid apes. *American Journal of Science*, **29**, 34–40.

Hsu, C., Han, K., & Wang, L. (1974). Discovery of *Gigantopithecus* teeth and associated fauna in western Hopei. *Vertebrata PalAsiatica*, **12**, 293–309.

Hu, C., & Qi, T. (1978). Gongwangling Pleistocene mammalian fauna of Lantian, Shaanxi. *Chinese Paleontology*, **155**, 1–64.

Hublin, J. J. (1982). Les anténéandertaliens: présapiens ou prénéandertaliens. *Géobios, Mémoire special*, **6**, 345–357.

Hublin, J. J. (1998). Climatic changes, paleogeography, and the evolution of the Neandertals. In *Neandertals and Modern Humans in Western Asia*, eds. T. Akazawa, K. Aoki, & O. Bar-Yosef, pp. 295–310. New York: Plenum Press.

Hublin, J. J., Barroso Ruiz, C., Medina Lara, P., Fontugne, M., & Reyss, J. L. (1995). The Mousterian site of Zafarraya (Andalucia, Spain): dating and implications on the Paleolithic peopling processes of western Europe. *Comptes rendus de l'Académie des sciences de Paris*, **321**, 931–937.

Hublin, J. J., Spoor, F., Braun, M., & Zonnenveld, F. (1996). A late Neanderthal associated with Upper Paleolithic artifacts. *Nature*, **381**, 224–226.

Hubrecht, A. A. W. (1908). Early ontogenetic phenomena in mammals and their bearing on our interpretation of the phylogeny of the vertebrates. *Quarterly Journal of Microscope Science*, **53**, 1–181.

Hughes, A. R. & Tobias, P. V. (1977). A fossil skull probably of the genus *Homo* from Sterkfontein, Transvaal. *Nature*, **265**, 310–312.

Hunsaker, D. & Shupe, D. (1977). Behavior of New World marsupials. In *The Biology of Marsupials*, ed. D. Hunsaker, pp. 279–347. New York: Academic Press.

Hürzeler, J. (1946). Zur Charakteristik, systematischen Stellung, Phylogenese und Verbreitung der Necrolemuriden aus demeuropäischen Eocaen. *Schweizerische paläontologische Abhandlungen*, **10**, 352–354.

Hürzeler, J. (1948). Zur Stammesgeschichte der Necrolemuriden. *Schweizerische paläontologische Abhandlungen*, **55**, 1–46.

Hürzeler, J. (1949). Neubeschreibung von *Oreopithecus bambolii* Gervais. *Schweizerische paläontologische Abhandlungen*, **66**, 3–20.

Hürzeler, J. (1951). Contribution à l'étude de la dentition de lait d'*Oreopithecus bambolii* Gervais. *Eclogae geologicae Helvetiae*, **44**, 404–411.

Hürzeler, J. (1954*a*). Contribution à l'odontologie et à la phylogénèse du genre *Pliopithecus* Gervais. *Annales de paléontologie*, **40**, 1–63.

Hürzeler, J. (1954*b*). Zur systematischen Stellung von *Oreopithecus*. *Verhandlungen der naturforschenden Gesellschaft, Basel*, **65**, 85–95.

Hürzeler, J. (1956). *Oreopithecus*, un point de repère pour l'histoire de l'humanité à l'ére tertiaire. *Colloques internationaux, Centre national de la recherche scientifique*, **60**, 115–121.

Hürzeler, J. (1958). *Oreopithecus bambolii* Gervais: a preliminary report. *Verhandlungen der naturforschenden Gesellschaft, Basel*, **69**, 1–47.

Hürzeler, J. (1960). The significance of *Oreopithecus* in the genealogy of man. *Triangle*, **4**, 164–174.

Hürzeler, J. (1968). Questions et réflexions sur l'histoire des anthropomorphes. *Annales de paléontologie*, **44**, 13–233.

Hürzeler, J. & Engesser, B. (1976). Les faunes de mammifères néogènes du Bassin de Baccinello (Grosseto, Italie). *Comptes rendus de l'Académie des sciences de Paris*, **283**, 333–336.

Isaac, G. L. (1977). *Olorgesailie: Archeological Studies of a Middle Pleistocene Lake Basin in Kenya*. Chicago, IL: University of Chicago Press.

Ishida, H. (1986). Investigation in northern Kenya and new hominoid fossils. *Kagaku*, **56**, 220–226.

Ishida, H., Kunimatsu, Y., Nakatsukasa, M., & Nakano, Y. (1999). New hominoid genus from the middle Miocene of Nachola, Kenya. *Anthropological Science*, **107**, 189–191.

Ishida, H. & Pickford, M. (1997). A new Late Miocene hominoid from Kenya: *Samburupithecus kiptalami* gen. et sp. nov. *Comptes rendus de l'Académie des sciences de Paris*, **325**, 823–829.

Ishida, H., Pickford, M., Nakaya, H., & Nakano, Y. (1984). Fossil anthropoids from Nachola and Samburu Hills, Samburu District, Northern Kenya. *African Study Monographs*, Supplementary Issue **2**, 73–86.

Iturralde-Vinent, M. A. & MacPhee, R. D. E. (1999). Paleogeography of the Caribbean region: implications for Cenozoic biogeography. *Bulletin of the American Museum of Natural History*, **238**, 1–95.

Iwamoto, M. (1975). On a skull of a fossil macaque from the Shikimizu Limestone Quarry in the Shikoku District, Japan. *Primates*, **16**, 83–94.

Iwamoto, M. (1982). A fossil baboon skull from the Lower Omo Basin, Southwest Ethiopia. *Primates*, **23**, 533–541.

Iwamoto, M. & Hasegawa, Y. (1972). Two macaque fossil teeth from the Japanese Pleistocene. *Primates*, **13**, 77–81.

Jablonski, N. G. (1986). The hand of *Theropithecus brumpti*. In *Proceedings of the 10th Congress of the International Primatological Society*, eds. P. C. Lee & J. Else, pp. 173–182. Cambridge: Cambridge University Press.

Jablonski, N. G. (ed.) (1993*a*). Theropithecus: *The Rise and Fall of a Primate Genus*. Cambridge: Cambridge University Press.

Jablonski, N. G. (1993*b*). Quaternary environments and the evolution of primates in East Asia, with notes on two new specimens of fossil Cercopithecidae from China. *Folia Primatologica*, **60**, 118–132.

Jablonski, N. G. (1993*c*). Evolution of the masticatory apparatus in *Theropithecus*. In Theropithecus: *The Rise and Fall of a Primate Genus*, ed. N. G. Jablonski, pp. 299–329. Cambridge: Cambridge University Press.

Jablonski, N. G. (1994). New fossil cercopithecid remains from the Humpata Plateau, southern Angola. *American Journal of Physical Anthropology*, **94**, 435–464.

Jablonski, N. G. (1998). The evolution of the doucs and snub-nosed monkeys and the question of the phyletic unity of the odd-nosed colobines. In *The Natural History of the Doucs and Snub-Nosed Monkeys*, ed. N. G. Jablonski, pp. 13–52. Singapore: World Scientific Press.

Jablonski, N. G. & Kelley, J. (1997). Did a major immunological event shape the evolutionary histories of apes and Old World monkeys? *Journal of Human Evolution*, **33**, 513–520.

Jablonski, N. G., Pan, Y. R., & Zhang, X. Y. (1994). New cercopithecid fossils from Yunnan Province, People's Republic of China. In *Current Primatology*, eds. B. Thierry, J. R. Anderson, J. J. Roeder, & N. Herrenschmidt, pp. 303–311. Strasbourg: Université Louis Pasteur.

Jablonski, N. G. & Peng, Y. Z. (1993). The phylogenetic relationships and classification of the doucs and snub-nosed langurs of China and Vietnam. *Folia Primatologica*, **60**, 36–55.

Jablonski, N. G. & Tyler, D. E. (1999). *Trachypithecus auratus sangiranensis*, a new fossil monkey from Sangiran, Central Java, Indonesia. *International Journal of Primatology*, **20**, 319–326.

Jablonski, N. G., Whitfort, M. G., Roberts-Smith, N., & Xu, Q. Q. (2000). The influence of life history and diet on the distribution of catarrhine primates during the Pleistocene in eastern Asia. *Journal of Human Evolution*, **39**, 131–157.

Jablonski, N. G. & Yumin, G. (1991). A reassessment of *Megamacaca lantianensis*, a large monkey from the Pleistocene of north-central China. *Journal of Human Evolution*, **20**, 51–66.

Jacob, T. (1981). Solo man and Peking man. In Homo erectus: *Papers in Honor of Davidson Black*, eds. B. A. Sigmon & J. S. Cybulski, pp. 87–104. Toronto: University of Toronto Press.

Jacobs, L. L. (1981). Miocene lorisid primates from the Pakistan Siwaliks. *Nature*, **289**, 585–587.

Jaeger, G. F. (1850). Über die fossilen Säugetiere welche in Württemberg, als nachtag zu dem 1839 unter Gleichem htel erschienen Werke. *Nova Acta academiae leopoldino-carolinae germanicae naturae curiosorum*, **22**, 765–924.

Jaeger, J. J., Chaimanee, Y. & Ducrocq, S. (1998*a*). Origin and evolution of Asian hominoid primates: paleontological data versus molecular data. *Comptes rendus de l'Académie des sciences de Paris*, **321**, 73–78.

Jaeger, J. J., Soe, A. N., Aung, A. K., Benammi, M., Chaimanee, Y., Ducrocq, R. M., Tun, T., Thein, T., & Ducrocq, S. (1998*b*). New Myanmar middle Eocene anthropoids: an Asian origin for catarrhines? *Comptes rendus de l'Académie des sciences de Paris*, **321**, 953–959.

Jaeger, J. J., Thein, T., Benammi, M., Chaimanee, Y., Soe, A. N., Lwin, T., Tun, T., Wai, S., & Ducrocq, S. (1999). A new primate from the middle Eocene of Myanmar and the Asian early origin of anthropoids. *Science*, **286**, 528–530.

Jaimez Salgado, E., Gutiérrez Calvache, D., MacPhee, R. D. E., & Gould, G. C. (1992). The monkey caves of Cuba. *Cave Science*, **19**, 25–28.

Janis, C. M. (1993). Tertiary mammal evolution in the context of changing climates, vegetation, and tectonic events. *Annual Review of Ecology and Systematics*, **24**, 467–500.

Jelínek, J. (1966). Jaw of an intermediate type of Neanderthal man from Czechoslovakia. *Nature*, **212**, 701–702.

Jenkins, F. A., Jr. (1974). *Primate Locomotion*. New York: Academic Press.

Jenkins, P. D. (1987). *Catalogue of Primates in the British Museum (Natural History) and Elsewhere in the British Isles, part 4, Suborder Strepsirhini, including the Subfossil Madagascan Lemurs and Family Tarsiidae*. London: British Museum (Natural History).

Jepsen, G. L. (1930). New vertebrate fossils from the lower Eocene of the Bighorn Basin, Wyoming. *Proceedings of the American Philosophical Society*, **69**, 117–131.

Jia, L. & Huang, W. (1990). *The Story of Peking Man: From Archaeology to Mystery*. Beijing: Foreign Languages Press.

Johanson, D. & Edgar, B. (1996). *From Lucy to Language*. New York: Nevraumont.

Johanson, D. C., Masao, F. T., Eck, G. G., White, T. D., Walter, R. C., Kimbel, W. H., Asfaw, B., Manega, P., Ndessokia, P., & Suwa, G. (1987). New partial skeleton of *Homo habilis* from Olduvai Gorge, Tanzania. *Nature*, **327**, 205–209.

Johanson, D. C., Taieb, M., & Coppens, Y. (eds) (1982*a*). Pliocene hominids from Hadar, Ethiopia. *American Journal of Physical Anthropology*, **57**, 373–719.

Johanson, D. C., Taieb, M., & Coppens, Y. (1982*b*). Pliocene hominids from the Hadar Formation, Ethiopia (1973–1977): stratigraphic, chronologic, and paleoenvironmental contexts, with notes on hominid morphology and systematics. *American Journal of Physical Anthropology*, **57**, 373–402.

Johanson, D. C. & White, T. D. (1979). A systematic assessment of early African hominids. *Science*, **203**, 321–330.

Johanson, D. C., White, T. D., & Coppens, Y. (1978). A new species of the genus *Australopithecus* (Primates: Hominidae) from the Pliocene of eastern Africa. *Kirtlandia*, **28**, 1–14.

Johnson, K. B., McCrossin, M. L., & Benefit, B. R. (2000). Circular shapes do not an ape make: comments on interpretation of the inferred *Morotopithecus* scapula. *American Journal of Physical Anthropology*, Supplement **30**, 189–190.

Jolly, C. J. (1965). The origins and specialization of the long-faced Cercopithecoidea, PhD dissertation, University of London.

Jolly, C. J. (1967). The evolution of baboons. In *The Baboon in Medical Research*, vol. 2, ed. H. Vagtborg, pp. 23–50. Austin, TX: University of Texas Press.

Jolly, C. J. (1970*a*). *Hadropithecus*: a lemuroid small-object feeder. *Man*, **5**, 619–626.

Jolly, C. J. (1970*b*). The seed-eaters: a new model of hominid differentiation based on a baboon analogy. *Man*, **5**, 5–26.

Jolly, C. J. (1972). The classification and natural history of *Theropithecus* (*Simopithecus*) (Andrews, 1916), baboons of the African Plio-Pleistocene. *Bulletin of the British Museum (Natural History), Geology Series*, **22**, 1–123.

Jones, F. W. (1916). *Arboreal Man*. London: Edward Arnold.

Jones, T. R. (1937). A new fossil primate from Sterkfontein, Krugersdorp, Transvaal. *South African Journal of Science*, **33**, 709–728.

Jouffroy, F. K. (1959). Un crâne subfossile de macaque du Pleistocène du Viet Nam. *Bulletin du Muséum national d'histoire naturelle, Paris*, **31**, 309–316.

Jouffroy, F. K. (1960). Caractères adaptatifs dans les proportions des membres chez les Lémurs fossiles. *Comptes rendus de l'Académie des sciences de Paris*, **251**, 2756–2757.

Jouffroy, F. K. (1963). Contribution à la connaissance du genre *Archaeolemur* Filhol, 1895. *Annales de paléontologie*, **49**, 129–155.

Jouffroy, F. K. (1974). Biomechanics of vertical leaping from the ground in *Galago alleni*: a cineradiographic analysis. In *Prosimian Biology*, eds. R. D. Martin, G. A. Doyle, & A. C. Walker, pp. 817–827. London: Duckworth.

Jouffroy, F. K., Godinot, M., & Nakano, Y. (1991). Biometrical characteristics of primate hands. *Human Evolution*, **6**, 269–306.

Jully, A. & Standing, H. F. (1904). Les gisements fossilifères d'Ampasambazimba. *Bulletin de l'Académie malgache*, **3**, 87–94.

Jungers, W. L. (1976). Osteological form and function: the appendicular skeleton of *Megaladapis*, a subfossil prosimian from Madagascar (Primates, Lemuroidea). PhD dissertation. University of Michigan, Ann Arbor.

Jungers, W. L. (1977). Hindlimb and pelvic adaptations to vertical climbing and clinging in *Megaladapis*, a giant subfossil prosimian from Madagascar. *Yearbook of Physical Anthropology*, **20**, 508–524.

Jungers, W. L. (1978). The functional significance of skeletal allometry in *Megaladapis* in comparison to living prosimians. *American Journal of Physical Anthropology*, **49**, 303–314.

Jungers, W. L. (1980). Adaptive diversity in subfossil Malagasy prosimians. *Zeitschrift für Morphologie und Anthropologie*, **71**, 177–186.

Jungers, W. L. (1985). Body size and scaling of limb proportions in primates. In *Size and Scaling in Primate Biology*, ed. W. L. Jungers, pp. 345–381. New York: Plenum Press.

Jungers, W. L. (1987). Body size and morphometric affinities of the appendicular skeleton in *Oreopithecus bambolii* (IGF 11778). *Journal of Human Evolution*, **16**, 445–456.

Jungers, W. L., Godfrey, L. R, Simons, E. L., & Chatrath, P. S. (1995). Subfossil *Indri indri* from the Ankarana Massif of Northern Madagascar. *American Journal of Physical Anthropology*, **97**, 357–366.

Jungers, W. L., Godfrey, L. R, Simons, E. L., & Chatrath, P. S. (1997). Phalangeal curvature and positional behavior in extinct sloth lemurs (Primates, Palaeopropithecidae). *Proceedings of the National Academy of Sciences of the United States of America*, **94**, 11998–12001.

Jungers, W. L., Godfrey, L. R., Simons, E. L., Chatrath, P. S., & Rakotosamimanana, B. (1991). Phylogenetic and functional affinities of *Babakotia radofilai*, a new fossil lemur from Madagascar. *Proceedings of the National Academy of Sciences of the United States of America*, **88**, 9082–9086.

Jungers, W. L., Godfrey, L. R., Simons, E. L., Wunderlich, R. E., Richmond, B. G., & Chatrath, P. S. (2001). Ecomorphology and behavior of giant extinct lemurs from Madagascar. In *Reconstructing Behavior in the Primate Fossil Record*, eds. J. M. Plavcan, R. F. Kay, C. P. van Schaik, & W. L. Jungers, pp. 371–411. New York: Kluwer.

Jungers, W. L., Jouffroy, F. K., & Stern, J. T. (1980). Gross structure and function of the quadriceps femoris in *Lemur fulvus*: an analysis based on telemetered electromyography. *Journal of Morphology*, **164**, 287–299.

Jungers, W. L., Simons, E. L., Godfrey, L. R., Chatrath, P. S., & Rakotosamimanana, B. (in prep.). A new sloth lemur (Palaeopropithecidae) from Northwest Madagascar.

Jungers, W. L., Wunderlich, R. E., Lemelin, P., Godfrey, L. R., Burney, D. A., Simons, E. L., Chatrath, P. S., & James, H. F. (1998). New hands and feet for an old lemur (*Archaeolemur*). *American Journal of Physical Anthropology*, Supplement **26**, 131–132.

Kaessmann, H., Wiebe, V., & Paabo, S. (1999). Extensive nuclear DNA sequence diversity among chimpanzees. *Science*, **286**, 1159–1162.

Kalb, J. E., Jolly, C. J., Mebrate, A., Tebedge, S., Smart, C., Oswald, E. B., Cramer, D., Whitehead, P., Wood, C. B., Conroy, G. C., Adefris, T., Sperling, L., & Kana, B. (1982*a*). Fossil mammals and artifacts from the Middle Awash Valley, Ethiopia. *Nature*, **298**, 25–29.

Kalb, J. E., Jolly, C. J., Tebedge, S., Mebrate, A., Smart, C., Oswald, E. B., Whitehead, P. F., Wood, C. B., Adefris, T., & Rawn-Schatzinger, V. (1982*b*). Vertebrate faunas from the Awash Group, Middle Awash Valley, Afar, Ethiopia. *Journal of Vertebrate Paleontology*, **2**, 237–258.

Kälin, J. (1961). Sur les primates de l'Oligocène inférieur d'Egypte. *Annales de paléontologie*, **47**, 1–48.

Kälin, J. (1962). Über *Moeripithecus markgrafi* Schlosser und die phyletischen Vorstufien der Bilophodontie der Cercopithecoidea. *Biblioteca Primatologica*, **1**, 32–42.

Kalmykov, N. P. & Maschenko, E. N. (1992). The most northern representative of early Pliocene Cercopithecidae from Asia. *Paleontologicheskifi Zhurnal*, **4**, 136–138.

Kalmykov, N. P. & Maschenko, E. N. (1995). *Parapresbytis eohanuman*, a cercopithecid monkey (Primates, Cercopithecidae) from Pliocene of Baikal. *Voprosy Antropologii*, **88**, 91–116.

Kappeler, P. M. & Ganzhorn, J. U. (1993). The evolution of primate communities and societies in Madagascar. *Evolutionary Anthropology*, **2**, 159–171.

Kappelman, J. (1992). The age of the Fayum primates as determined by paleomagnetic reversal stratigraphy. *Journal of Human Evolution*, **22**, 495–503.

Kappelman, J., Kelley, J., Pilbeam, D., Sheikh, K., Ward, S., Anwar, M., Barry, J., Brown, B., Hake, P., Johnson, N., Raza, S. M., & Shah, S. I. (1991). The earliest occurrence of *Sivapithecus* from the middle Miocene Chinji Formation of Pakistan. *Journal of Human Evolution*, **21**, 61–73.

Kappelman, J., Simons, E. L., & Swisher, C. C. III (1992). New age determinations for the Eocene–Oligocene boundary sediments in the Fayum depression, northern Egypt. *Journal of Geology*, **100**, 647–667.

Kappelman, J., Swisher, C. C., Fleagle, J. G., Yirga, S., Bown, T. M., & Fesseha, M. (1996). Age of *Australopithecus afarensis* from Fejej, Ethiopia. *Journal of Human Evolution*, **30**, 139–146.

Karavanić, I. & Smith, F. H. (1998). The Middle/Upper Paleolithic interface and the relationship of Neanderthals and early modern humans in the Hrvatsko Zagorje, Croatia. *Journal of Human Evolution*, **34**, 223–248.

Karpanty, S. & Goodman, S. M. (1999). Prey profile of the Madagascar harrier-hawk, *Polyboroides radiatus*, in southeastern Madagascar. *Raptor Research*, **33**, 313–316.

Kaup, J. J. (1835). *Description d'ossements fossiles de mammifères inconnus jusqu" à présent qui se trouvent au Muséum grand-ducal de Darmstadt*. Darmstadt.

Kaup, J. J. (1861). *Beitrage zur naheren Kenntniss der urweltlichen Säugethiere*, vol. 5. Darmstadt and Leipzig.

Kay, R. F. (1977). The evolution of molar occlusion in the Cercopithecidae and early catarrhines. *American Journal of Physical Anthropology*, **46**, 327–352.

Kay, R. F. (1980). Platyrrhine origins: a reappraisal of the dental evidence. In *Evolutionary Biology of the New World Monkeys and Continental Drift*, eds. R. L. Ciochon & A. B. Chiarelli, pp. 159–188. New York: Plenum Press.

Kay, R. F. (1981). The nut-crackers: a new theory of the adaptations of the Ramapithecinae. *American Journal of Physical Anthropology*, **55**, 141–151.

Kay, R. F. (1982). *Sivapithecus simonsi*, a new species of Miocene hominoid with comments on the phylogenetic status of the Ramapithecinae. *International Journal of Primatology*, **3**, 113–173.

Kay, R. F. (1984). On the use of anatomical features to infer foraging behavior in extinct primates. In *Adaptations for Foraging in Nonhuman Primates: Contributions to an Organismal Biology of Prosimians, Monkeys and Apes*, eds. P. S. Rodman & J. G. H. Cant, pp. 21–53. New York: Columbia University Press.

Kay, R. F. (1990). The phyletic relationships of extant and fossil Pitheciinae (Platyrrhini, Anthropoidea). *Journal of Human Evolution*, **19**, 175–208.

Kay, R. F. (1994). "Giant" tamarin from the Miocene of Colombia. *American Journal of Physical Anthropology*, **95**, 333–353.

Kay, R. F. & Cartmill, M. (1977). Cranial morphology and adaptation of *Palaechthon nacimienti* and other Paromomyidae, with a description of a new genus and species. *Journal of Human Evolution*, **6**, 19–53.

Kay, R. F. & Covert, H. H. (1984). Anatomy and the behavior of extinct primates. In *Food Acquisition and Processing in Primates*, eds. D. J. Chivers, B. A. Wood, & A. Bilsborough, pp. 467–508. New York: Plenum Press.

Kay, R. F., Fleagle, J. G., & Simons, E. L. (1981). A revision of the Oligocene apes from the Fayum Province, Egypt. *American Journal of Physical Anthropology*, **55**, 293–322.

Kay, R. F. & Frailey, C. D. (1993). Large fossil platyrrhines from the Rio Acre local fauna, late Miocene, western Amazonia. *Journal of Human Evolution*, **25**, 319–327.

Kay, R. F., Johnson, D., & Meldrum, D. J. (1998*a*). A new pitheciin primate from the middle Miocene of Argentina. *American Journal of Primatology*, **45**, 317–336.

Kay, R. F., Johnson, D., & Meldrum, D. J. (1999*a*). Corrigendum. *American Journal of Primatology*, **47**, 347.

Kay, R. F., MacFadden, B. J., Madden, R. H., Sandeman, H., & Anaya, F. (1998*b*). Revised age of the Salla Beds, Bolivia, and its bearing on the age of the Deseadan South American Land Mammal "Age". *Journal of Vertebrate Paleontology*, **18**, 189–199.

Kay, R. F. & Madden, R. H. (1997). Paleogeography and paleoecology. In *Vertebrate Paleontology in the Neotropics: The Miocene Fauna of La Venta, Colombia*, eds. R. F. Kay, R. H. Madden, R. L. Cifelli, & J. J. Flynn, pp. 520–550. Washington, DC: Smithsonian Institution Press.

Kay, R. F., Madden, R. H., Cifelli, R. L., & Flynn, J. J. (eds.) (1997*a*). *Vertebrate Paleontology in the Neotropics: The Miocene Fauna of La Venta, Colombia*. Washington, DC: Smithsonian Institution Press.

Kay, R. F., Madden, R. H., & Guerrero-Diaz, J. (1988). Newly recovered remains of monkeys from the Miocene of Colombia. *Ameghiniana*, **25**, 203–212.

Kay, R. F., Madden, R. H., Mazzoni, M., Vucetich, M. G., Heizler, M., & Sandeman, H. (1999*b*). The oldest Argentine primates: first age determinations for the Colhuehuapian South American Land Mammal "Age". *American Journal of Physical Anthropology*, Supplement **28**, 166.

Kay, R. F., Madden, R. H., Plavcan, J. M., Cifelli, R. L., & Diaz, J. G. (1987). *Stirtonia victoriae*, a new species of Miocene Colombian primate. *Journal of Human Evolution*, **16**, 173–196.

Kay, R. F. & Meldrum, D. J. (1997). A new small platyrrhine and the phyletic position of Callitrichinae. In *Vertebrate Paleontology in the Neotropics: The Miocene Fauna of La Venta, Colombia*, eds. R. F. Kay, R. H. Madden, R. L. Cifelli, & J. J. Flynn, pp. 435–458. Washington, DC: Smithsonian Institution Press.

Kay, R. F., Ross, C., & Williams, B. A. (1997*b*). Anthropoid origins. *Science*, **275**, 797–803.

Kay, R. F. & Simons, E. L. (1980). The ecology of Oligocene African Anthropoidea. *International Journal of Primatology*, **1**, 21-37.

Kay, R. F. & Simons, E. L. (1983). A reassessment of the relationships between later Miocene and subsequent Hominoidea. In *New Interpretations of Ape and Human Ancestry*, eds. R. L. Ciochon & R. S. Corruccini, pp. 577–624. New York: Plenum Press.

Kay, R. F., Thewissen, J. G. M., & Yoder, A. D. (1992). Cranial anatomy of *Ignacius graybullianus* and the affinities of the Plesiadapiformes. *American Journal of Physical Anthropology*, **89**, 477–498.

Kay, R. F., Thorington, R. W., Jr., & Houde, P. (1990). Eocene plesiadapiform shows affinities with flying lemurs not primates. *Nature*, **345**, 342–344.

Kay, R. F. & Ungar, P. S. (1997). Dental evidence for diet in some Miocene catarrhines with comments on the effects of phylogeny on the interpretation of adaptation. In *Function, Phylogeny, and Fossils: Miocene Hominoid Evolution and Adaptations*, eds. D. R. Begun, C. V. Ward, & M. D. Rose, pp. 131–151. New York: Plenum Press.

Kay, R. F. & Williams, B. A. (1994). Dental evidence for anthropoid origins. In *Anthropoid Origins*, eds. J. G. Fleagle & R. F. Kay, pp. 361–445. New York: Plenum Press.

Kay, R. F. & Williams, B. A. (1995). Recent finds of monkeys from the Oligocene/Miocene of Salla, Bolivia. *American Journal of Physical Anthropology*, Supplement **20**, 124.

Keith, A. (1899). On the chimpanzees and their relationship to the gorilla. *Proceedings of the Zoological Society of London*, **1**, 296–312.

Keith, A. (1911). *Ancient Types of Man*. New York: Harper.

Keith, A. (1915). *The Antiquity of Man*. London: Williams & Norgate.

Keith, A. (1927). A report on the Galilee skull. In *Reserches in Prehistoric Galilee*, ed. F. Turville-Petre, pp. 53–106. Jerusalem: British School of Archeology in Jerusalem.

Keith, A. & McCown, T. D. (1937). Mount Carmel man: his bearing on the ancestry of modern races. In *Early Man*, ed. G. G. MacCurdy, pp. 44–52. Philadelphia, PA: Lippincott.

Kelley, J. (1986*a*). Paleobiology of Miocene hominoids. PhD dissertation, Yale University.

Kelley, J. (1986*b*). Species recognition and sexual dimorphism in *Proconsul* and *Rangwapithecus*. *Journal of Human Evolution*, **15**, 461–495.

Kelley, J. (1988). A new large species of *Sivapithecus* from the Siwaliks of Pakistan. *Journal of Human Evolution*, **17**, 305–324.

Kelley, J. (1993). Taxonomic implication of sexual dimorphism in *Lufengpithecus*. In *Species, Species Concepts and Primate Evolution*, eds. W. H. Kimbel & L. B. Martin, pp. 429–458. New York: Plenum Press.

Kelley, J. (1994). A biological hypothesis of ape species density. In *Current Primatology*, eds. B. Thierry, J. R. Anderson, J. J. Roeder, & N. Herrenschmidt, pp. 11–18. Strasbourg: Université Louis Pasteur.

Kelley, J. (1995). Sex determination in Miocene catarrhine primates. *American Journal of Physical Anthropology*, **96**, 391–417.

Kelley, J. (1997). Paleobiological and phylogenetic significance of life history in Miocene hominoids. In *Function, Phylogeny and Fossils: Miocene Hominoid Evolution and Adaptation*, eds. D. R. Begun, C. V. Ward, & M. D. Rose, pp. 173–208. New York: Plenum Press.

Kelley, J. (in press). Life-history evolution in Miocene and extant apes. In *Human Evolution through Developmental Change*, eds. N. Minugh-Purvis & K. J. McNamara. Baltimore, MD: Johns Hopkins University Press.

Kelley, J., Anwar, M., McCollum, M., & Ward, S. C. (1995). The anterior dentition of *Sivapithecus parvada* with comments on the phylogenetic significance of incisor heteromorphy in hominoids. *Journal of Human Evolution*, **28**, 503–517.

Kelley, J. & Etler, D. (1989). Hominoid dental variability and species number at the late Miocene site of Lufeng, China. *American Journal of Primatology*, **18**, 15–34.

Kelley, J. & Pilbeam, D. (1985). Miocene hominoid cladistics and the demise of the "thick enameled" apes. *American Journal of Physical Anthropology*, **66**, 188.

Kelley, J. & Pilbeam, D. R. (1986). The dryopithecines: taxonomy, comparative anatomy and phylogeny of Miocene large hominoids. In *Comparative Primate Biology*, eds. D. R. Swindler & J. Erwin, pp. 361–411. New York: Alan R. Liss.

Kelley, J. & Plavcan, M. (1998). A simulation test of species number at Lufeng, China: implications for the use of the coefficient of variation in paleotaxonomy. *Journal of Human Evolution*, **35**, 577–596.

Kelley, J., Ward, S., Brown, B., Hill, A., & Downs, W. (2000). Middle Miocene hominoid origins: Response. *Science*, **287**, 2375a.

Kelley, J. & Xu, Q. (1991). Extreme sexual dimorphism in a Miocene hominoid. *Nature*, **352**, 151–153.

Kelly, T. S. (1990). Biostratigraphy of Uintan and Duchesnean land mammal assemblages from the middle member of the Sespe Formation, Simi Valley, California. *Contributions in Science of the Natural History Museum of Los Angeles*, **419**, 1–42.

Kennedy, G. E. (1999). Is "*Homo rudolfensis*" a valid species? *Journal of Human Evolution*, **36**, 119–121.

Keyser, A. W. (1991). The palaeontology of Haasgat: a preliminary account. *Palaeontologica Africana*, **28**, 29–33.

Keyser, A. W. (2000). The Drimolen skull: the most complete australopithecine cranium and mandible to date. *South African Journal of Science*, **96**, 189–193.

Khatri, A. P. (1975). The early fossil hominids and related apes of the Siwalik foothills of the Himalayas: recent discoveries and new interpretations. In *Paleoanthropology: Morphology and Paleoecology*, ed. R. H. Tuttle, pp. 31–58. The Hague: Mouton.

Kidder, J., Jantz, R., & Smith, F. (1992). Defining modern humans: a multivariate approach. In *Continuity or Replacement: Controversies in* Homo sapiens *Evolution*, eds. G. Bräuer & F. H. Smith, pp. 157–177. Rotterdam: Balkema.

Kimbel, W. H., Johanson, D. C., & Rak, Y. (1994). The first skull and other new discoveries of *Australopithecus afarensis* at Hadar, Ethiopia. *Nature*, **368**, 449–451.

Kimbel, W. H., Johanson, D. C. & Rak, Y. (1997). Systematic assessment of a maxilla of *Homo* from Hadar, Ethiopia. *American Journal of Physical Anthropology*, **103**, 235–262.

Kimbel, W. H., Walter, R. C., Johanson, D. C., Aronson, J. L., Assefa, Z., Eck, G. G., Hovers, E., Marean, C. W., Rak, Y., Reed, K. E., Vondra, C., Yemane, T., York, D., Chen, Y., Evensen, N. M., & Smith, P. E. (1996). Late Pliocene Homo and Oldowan tools from the Hadar Formation (Kada Hadar Member), Ethiopia. *Journal of Human Evolution*, **31**, 549–561.

Kimbel, W. H., White, T. D., & Johanson, D. C. (1984). Cranial morphology of *Australopithecus afarensis*: a comparative study based on a composite reconstruction of the adult skull. *American Journal of Physical Anthropology*, **64**, 337–388.

King, T., Aiello, L. C., & Andrews, P. (1999). Dental microwear of *Griphopithecus alpani*. *Journal of Human Evolution*, **36**, 3–31.

King, W. (1864). The reputed fossil man of the Neanderthal. *Quaternary Journal of Science*, **1**, 88–97.

Kingston, J. D. & Hill, A. (1999). Late Miocene palaeoenvironments in Arabia: a synthesis. In *Fossil Vertebrates of Arabia: With Emphasis on the Late Miocene Faunas, Geology, and Palaeoenvironments of the Emirate of Abu Dhabi, United Arab Emirates*, eds. P. J. Whybrow & A. Hill, pp. 389–407. New Haven, CT: Yale University Press.

Kinzey, W. G. (1992). Dietary and dental adaptations in the Pitheciinae. *American Journal of Physical Anthropology*, **88**, 499–514.

Kitching, J. W. (1952). A new type of fossil baboon: *Brachygnathopithecus peppercorni*, gen. et sp. . *South African Journal of Science*, **Nov. 1952**, 15–17.

Klaatsch, H. & Hauser, O. (1909). *Homo mousteriensis* Hauseri: ein altdiluvialer Skelettfund im Department Dordogne und seine Zuhörogkeit zum Neandertaltypus. *Archiv für Anthropologie*, **7**, 287–297.

Klein, R. G. (1999). *The Human Career: Human Biological and Cultural Origins*. Chicago, IL: University of Chicago Press.

Kluge, A. G. (1983). Cladistics and the classification of the great apes. In *New Interpretations of Ape and Human Ancestry*, eds. R. Ciochon & R. Corruccini, pp. 151–177. New York: Plenum Press.

Knott, C. D. (1998). Changes in orangutan caloric intake, energy balance, and ketones in response to fluctuating fruit availability. *International Journal of Primatology*, **19**, 1061–1079.

Koenigswald, G. H. R. von (1935). Eine fossile Säugetierfauna mit *Simia* aus Südchine. *Proceedings of the Koninklijke Nederlandse Akademie van Wetenschappen*, **38**, 872–79.

Koenigswald, G. H. R. von (1936). Erste mitteilungen über einen fossilen Hominiden aus dem Altpleistocän Ostjavas. *Proceedings of the Koninklijke Nederlandse Akademie van Wetenschappen*, **39**, 1000–1009.

Koenigswald, G. H. R. von (1940). Neue *Pithecanthropus*-Funde 1936–1938: ein Beitrag zur Kenntnis der Praehominiden. *Wetenschappelijke Mededelingen – Dienst van den Mijnbouw in Nederlandsch-Oost Indië*, **28**, 1–232.

Koenigswald, G. H. R. von (1950). Bemerkungen zu *Dryopithecus giganteus* Pilgrim. *Eclogae geologicae Helvetiae*, **42**, 515–519.

Koenigswald, G. H. R. von. (1952). *Gigantopithecus blacki* von Koenigswald, a giant fossil hominoid from the Pleistocene of southern China. *Anthropological Papers of the American Museum of Natural History*, **43**, 292–325.

Koenigswald, G. H. R. von (1955). Remarks on *Oreopithecus*. *Rivista di Scienze Preistoriche*, **10**, 1–11.

Koenigswald, G. H. R. von (1957). Remarks on *Gigantopithecus* and other hominoid remains from southern China. *Proceedings of the Koninklijke Nederlandse Akademie van Wetenschappen, Series B*, **60**, 153–159.

Koenigswald, G. H. R. von (1958*a*). Der Solo-Mensch von Java: Ein tropischer Neanderthaler. In *Hundert Jahre Neanderthaler*, ed. G. H. R. von Koenigswald, pp. 21–26. Utrecht: Keminck & Zoon.

Koenigswald, G. H. R. von (1958*b*). *Gigantopithecus* and *Australopithecus*. *The Leech*, **28**, 101–105.

Koenigswald, G. H. R. von (1969). Miocene Cercopithecoidea and Oreopithecoidea from the Miocene of East Africa. In *Fossil Vertebrates of Africa*, vol. 1, ed. L. S. B. Leakey, pp. 39–51. London: Academic Press.

Koenigswald, G. H. R. von (1972). Ein Unterkiefer eines fossilen-Hominoiden aus dem Unterpliozän Griechenlands. *Proceedings of the Koninklijke Nederlandse Akademie van Wetenschappen, Series B*, **75**, 385–394.

Koenigswald, G. H. R. von (1981). A possible ancestral form of *Gigantopithecus* (Mammalia, Hominoidea) from the Chinji layers of Pakistan. *Journal of Human Evolution*, **10**, 511–515.

Koenigswald, G. H. R. von (1983). The significance of hitherto undescribed Miocene hominoids from the Siwaliks of Pakistan in the Senckenberg Museum, Frankfurt. In *New Interpretations of Ape and Human Ancestry*, eds. R. L. Ciochon & R. S. Corruccini, pp. 517–526. New York: Plenum Press.

Koenigswald, G. H. R. von & Weidenreich, F. (1939). The relationship between *Pithecanthropus* and *Sinanthropus*. *Nature*, **144**, 926–929.

Koenigswald, W. von (1979). Ein Lemurenrest aus dem eozänen Ölschiefer der Grube Darmstadt. *Paläontologische Zeitschrift*, **53**, 63–76.

Koenigswald, W. von (1982). Das Dinotherium von Eppelsheim. *Alzeyer Geschichtsblätter*, **8**, 17–29.

Köhler, M. & Moyà-Solà, S. (1997). Ape-like or hominid-like? The positional behavior of *Oreopithecus bambolii* reconsidered. *Proceedings of the National Academy of Sciences of the United States of America*, **94**, 11747–11750.

Köhler, M. & Moyà-Solà, S. (1999). A finding of Oligocene primates on the European continent. *Proceedings of the National Academy of Sciences of the United States of America*, **96**, 14664–14667.

Köhler, M., Moyà-Solà, S., & Alba, D. M. (2000). *Macaca* (Primates, Cercopithecidae) from the Late Miocene of Spain. *Journal of Human Evolution*, **38**, 447–452.

Köhler, M., Moyà-Solà, S., & Andrews, P. (1999). Order Primates. In *The Miocene Land Mammals of Europe: The Continental European Miocene*, eds. G. Rössner & K. Heissig, pp. 91–104. Munich: Dr. Friedrich Pfeil.

Kolnicki, R. L. (1999). Karyotypic fission theory applied: kinetochore reproduction and lemur evolution. *Symbiosis*, **26**, 123–141.

Koop, B. F., Siemieniak, D., Slightom, J. L., Goodman, M., Dunbar, J., Wright, P. C., & Simons, E. L. (1989*a*). *Tarsius* d- and b-globin genes: conversions, evolution, and systematic implications. *Journal of Biological Chemistry*, **264**, 68–79.

Koop, B. F., Tagle, D. A., Goodman, M. & Slightom, J. L. (1989*b*). A molecular view of primate phylogeny and important systematic and evolutionary questions. *Molecular Biology and Evolution*, **6**, 580–612.

Kordos, L. (1987). Description and reconstruction of the skull of *Rudapithecus hungaricus* Kretzoi (Mammalia). *Annales historico-naturales Musei nationalis hungarici*, **79**, 77–88.

Kordos, L. (1988). Comparison of early primate skulls from Rudabánya and China. *Anthropologia hungarica*, **20**, 9–22.

Kordos, L. & Begun, D. R. (1997). A new reconstruction of RUD 77, a partial cranium of *Dryopithecus brancoi* from Rudábanya, Hungary. *American Journal of Physical Anthropology*, **103**, 277–294.

Kordos, L. & Begun, D. R. (1998). Encephalization and endocranial morphology in *Dryopithecus brancoi*: implications for brain evolution in early hominids. *American Journal of Physical Anthropology*, Supplement **26**, 141–142.

Kordos, L. & Begun, D. R. (1999). Femora of *Anapithecus* from Rudabánya. *American Journal of Physical Anthropology*, Supplement **28**, 173.

Kordos, L. & Begun, D. R. (2000). Four catarrhine crania from Rudabánya. *American Journal of Physical Anthropology*, Supplement **30**, 193.

Kordos, L. & Begun, D. R. (2001). Primates from Rudabánya: allocation of specimens to individuals, sex and age categories. *Journal of Human Evolution*, **40**, 17–39.

Koufos, G. D. (1990). The hipparions of the lower Axios valley (Macedonia, Greece): implications for the Neogene stratigraphy and the evolution of *Hipparion*. In *European Neogene Mammal Chronology*, eds E. Lindsay, V. Fahlbusch, & P. Mein, pp. 321–338. New York: Plenum Press.

Koufos, G. D. (1995). The first female maxilla of the hominoid *Ouranopithecus macedoniensis* from the late Miocene of Macedonia, Greece. *Journal of Human Evolution*, **29**, 385–389.

Koufos, G. D., Syrides, G. E., & Koliadiou, K. K. (1991). A Pliocene primate from Macedonia (Greece). *Journal of Human Evolution*, **21**, 283–294.

Kowalski, K. & Zapfe, H. (1974). *Pliopithecus antiquus* (Blainville, 1839) Primates, (Mammalia) from the Miocene of Przeworno in Silesia (Poland). *Acta Zoologica Cracoviensia*, **19**, 19–30.

Kraglievich, J. L. (1951). Contribuciones al conocimiento de los primates fosiles da la Patagonia. I. Diagnosis previa de un nuevo primate fosil del Oligocene superior (Colhuehuapiano) de Gaiman, Chubut. *Comunicaciones, Instituto nacional investigación de las ciencias de naturales*, **2**, 57–82.

Kramer, A. (1993). Human taxonomic diversity in the Pleistocene:

does *Homo erectus* represent multiple hominid species? *American Journal of Physical Anthropology*, **91**, 161–171.

Kramer, A., Crummett, T., & Wolpoff, M. (2000). Out of Africa and into the Levant: replacement or admixture? *Quaternary International*, **75**, 51–63.

Krause, D. W. (1991). Were paromomyids gliders? Maybe, maybe not. *Journal of Human Evolution*, **21**, 177–188.

Krause, D. W., Hartman, J. H., & Wells, N. A. (1997). Late Cretaceous vertebrates from Madagascar: implications for biotic change in deep time. In *Natural Change and Human Impact in Madagascar*, eds. S. M. Goodman & B. D. Patterson, pp. 3–43. Washington, DC: Smithsonian Institution Press.

Krause, D. W. & Maas, M. C. (1990). The biogeographic origins of late Paleocene–early Eocene mammalian immigrants to the western interior of North America. *Geological Society of America, Special Papers*, **243**, 71–105.

Krentz, H. B. (1993). Postcranial anatomy of extant and extinct species of *Theropithecus*. In *Theropithecus: The Rise and Fall of a Primate Genus*, ed. N. G. Jablonski, pp. 383–424. Cambridge: Cambridge University Press.

Kretzoi, M. (1969). Geschichte der Primaten und der Hominisation. *Symposia biologica hungarica*, **9**, 3–11.

Kretzoi, M. (1975). New ramapithecines and *Pliopithecus* from the lower Pliocene of Rudabánya in north-eastern Hungary. *Nature*, **257**, 578–581.

Kretzoi, M. (1984). Uj Hominid Lelet Rudabáyáról. *Anthropológia közlemények*, **28**, 91–96.

Krings, M., Stone, A., Schmitz, R., Krainitzki, H., Stoneking, M., & Pääbo, S. (1997). Neanderthal DNA sequences and the origin of modern humans. *Cell*, **90**, 19–30.

Krings, M., Capelli, C., Tschentscher, F., Geisert, H., Meyer, S., von Haeseler, A., Grossschmidt, K., Possnet, G., Paunović, M., & Pääbo, S. (2000). A view of Neandertal genetic diversity. *Nature Genetics*, **26**, 145–146.

Krishtalka, L. (1978). Paleontology and geology of the Badwater Creek area, central Wyoming. XV. Review of the late Eocene primates from Wyoming and Utah, and the Plesitarsiiformes. *Annals of the Carnegie Museum of Natural History*, **47**, 335–360.

Krishtalka, L. & Schwartz, J. H. (1978). Phylogenetic relationships of plesiadapiform–tarsiiform primates. *Annals of the Carnegie Museum of Natural History*, **47**, 515–540.

Krishtalka, L., Stucky, R. K., & Beard, K. C. (1990). The earliest fossil evidence for sexual dimorphism in primates. *Proceedings of the National Academy of Sciences of the United States of America*, **87**, 5223–5226.

Krogman, W. M. (1950). Classification of fossil men: concluding remarks of the chairman. *Cold Spring Harbor Symposia on Quantitative Biology*, **15**, 119–121.

Kumar, S. & Hedges, B. (1998). A molecular timescale for vertebrate evolution. *Nature*, **392**, 917–920.

Kunimatsu, Y. (1992*a*). A revision of the hypodigm of *Nyanzapithecus vancouveringi*. *African Studies Monographs*, **14**, 231–235.

Kunimatsu, Y. (1992*b*). New finds of small anthropoid primate from Nachola, northern Kenya. *African Studies Monographs*, **14**, 237–249.

Kunimatsu, Y. (1997). New species of *Nyanzapithecus* from Nachola, northern Kenya. *Anthropological Science*, **105**, 117–141.

Lahr, M. M. (1994). The multiregional model of modern human origins: a reassessment of its morphological basis. *Journal of Human Evolution*, **26**, 23–56.

Lahr, M. M. (1996). *The Evolution of Modern Human Diversity*. Cambridge: Cambridge University Press.

Lamarck, J. B. (1809). *Philosophie zoologique, ou exposition des considérations relative à l'histoire naturelle des animaux*. Paris: Dentu et L'Auteur.

Lamberton, C. (1934*a*). Contribution à la connaissance de la faune subfossile de Madagascar: lémuriens et ratites: *Archaeoindris fontoynonti* Stand. *Mémoires de l'Académie malgache*, **17**, 9–39 plus plates.

Lamberton, C. (1934*b*). Contribution à la connaissance de la faune subfossile de Madagascar: lémuriens et ratites. *Chiromys robustus* sp. nov. Lamb. *Mémoires de l'Académie malgache*, **17**, 40–46 plus plates.

Lamberton, C. (1934*c*). Contribution à la connaissance de la faune subfossile de Madagascar: lémuriens et ratites. Les *Megaladapis*. *Mémoires de l'Académie malgache*, **17**, 47–105 plus plates.

Lamberton, C. (1934*d*). Contribution à la connaissance de la faune subfossile de Madagascar: lémuriens et ratites: ratites subfossiles de Madagascar. *Mémoires de l'Académie malgache*, **17**, 123–168 plus plates.

Lamberton, C. (1936*a*). Nouveaux lémuriens fossiles du groupe des Propithèques et l'intérêt de leur découverte. *Bulletin du Muséum national d'histoire naturelle, Paris, Série 2*, **8**, 370–373.

Lamberton, C. (1936*b*). Fouilles paléontologiques faites en 1936. *Bulletin de l'Académie malgache* (nouvelle série), **19**, 1–19.

Lamberton, C. (1938, for the year 1937). Contribution à la connaissance de la faune subfossile de Madagascar. Note III. Les Hadropithèques. *Bulletin de l'Académie malgache* (nouvelle série), **20**, 127–170 plus plates.

Lamberton, C. (1939*a*). Contribution à la connaissance de la faune subfossile de Madagascar: lémuriens et cryptoproctes. Note V. Petits lémuriens subfossiles. *Mémoires de l'Académie malgache*, **27**, 51–73 plus plates.

Lamberton, C. (1939*b*). Contribution à la connaissance de la faune subfossile de Madagascar: lémuriens et cryptoproctes. Note VI. Des os du pied de quelques lémuriens subfossiles malgaches. *Mémoires de l'Académie malgache*, **27**, 75–139 plus plates.

Lamberton, C. (1939*c*). Contribution à la connaissance de la faune subfossile de Madagascar: lémuriens et cryptoproctes. Note VIII. Les Cryptoprocta fossiles. *Mémoires de l'Académie malgache*, **27**, 155–193 plus plates.

Lamberton, C. (1946, for the years 1942–43). Contribution à la connaissance de la faune subfossile de Madagascar. Note XV. *Plesiorycteropus madagascariensis* Filhol. *Bulletin de l'Académie malgache* (nouvelle série), **25**, 25–53 plus plates.

Lamberton, C. (1947, for the years 1944–45). Contribution à la connaissance de la faune subfossile de Madagascar. Note XVI. *Bradytherium* ou Palaeopropithèque? *Bulletin de l'Académie malgache* (nouvelle série), **26**, 89–140 plus plates.

Lamberton, C. (1948*a*, for the year 1946). Contribution à la connaissance de la faune subfossile de Madagascar. Note XVII. Les Pachylemurs. *Bulletin de l'Académie malgache* (nouvelle série), **27**, 7–22 plus plates.

Lamberton, C. (1948*b*, for the year 1946). Contribution à la connaissance de la faune subfossile de Madagascar. Note XX. Membre posterieur des Neopropithèques et des Mesopropithèques. *Bulletin de l'Académie malgache* (nouvelle série), **27**, 30–32 plus plates.

Lamberton, C. (1957, for the year 1956). Examen de quelques hypothèses de Sera concernant les lémuriens fossiles et actuels. *Bulletin de l'Académie malgache* (nouvelle série), **34**, 51–65.

Lanèque, L. (1992*a*). Analyse de matrice de distance euclidienne de la région du museau chez *Adapis* (Adapiforme, Eocène). *Comptes rendus de l'Académie des sciences de Paris*, **314**, 1387–1393.

Lanèque, L. (1992*b*). Variation in the shape of the palate in *Adapis* (Eocene, Adapiformes) compared with living primates. *Human Evolution*, **7**, 1–16.

Lanèque, L. (1993). Variation of orbital features in adapine skulls. *Journal of Human Evolution*, **25**, 287–318.

Langbroek, M. & Roebroeks, W. (2000). Extraterrestrial evidence on the age of the hominids from Java. *Journal of Human Evolution*, **38**, 595–600.

Langdon, J. (1986). Functional morphology of the Miocene hominoid foot. *Contributions to Primatology*, **22**, 1–225.

Larson, S. G. (1998). Parallel evolution in the hominoid trunk and forelimb. *Evolutionary Anthropology*, **6**, 87–99.

Larson, S. G., Schmitt, D., Lemelin, P., & Hamrick, M. (2000). Uniqueness of primate forelimb posture during quadrupedal locomotion. *American Journal of Physical Anthropology*, **112**, 87–101.

Lartet, É. (1837*a*). Note sur les ossements fossiles des terrains tertiaires de Simorre, de Sansan, etc., dans le département du Gers, et sur la découverte récente d'une mâchoire de singe fossile. *Annales des sciences naturelles*, **13**, 116–123. [Also published in *Comptes rendus de l'Académie des sciences de Paris*, **IV**, 85–93.

Lartet, É. (1837*b*). Nouvelles observations sur une mâchoire inférieure fossile, crue d'un singe voisin du gibbon, et sur quelques dents et ossements attribués à d'autres quadrumanes. *Comptes rendus de l'Academie de sciences de Paris*, **IV**, 583–584.

Lartet, É. (1856). Note sur un grand singe fossile qui se rattache au groupe des singes supérieurs. *Comptes rendus de l'Académie des sciences de Paris*, **43**, 219–223.

Lartet, L. (1865–75). A burial place of the cave-dwellers of the Périgord. In *Reliquiae Acquitanicae; Being Contributions to the Archaeology and Paleontology of Périgord and the Adjoining Provinces*, eds. E. Lartet & H. Christy, pp. 62–72. London: Williams & Norgate.

Latimer, B. M. (1991). Locomotor adaptations in *Australopithecus afarensis*: the issue of arboreality. In *Origine(s) de la bipédie chez les hominides*, eds. Y. Coppens & B. Senut, pp. 169–176. Paris: Cahiers de Paleoanthropologie, Centre national de la recherche scientifique.

Leakey, L. S. B. (1935). *The Stone Age Races of Kenya*. Oxford: Oxford University Press.

Leakey, L. S. B. (1936). Fossil human remains from Kanam and Kanjera, Kenya Colony. *Nature*, **138**, 643.

Leakey, L. S. B. (1943). A Miocene anthropoid mandible from Rusinga, Kenya. *Nature*, **152**, 319–320.

Leakey, L. S. B. (1946). Fossil finds in Kenya: ape or primitive man? *Antiquity*, **20**, 201–204.

Leakey, L. S. B. (1959). A new fossil skull from Olduvai. *Nature*, **184**, 491–493.

Leakey, L. S. B. (1961). New Finds at Olduvai Gorge. *Nature*, **189**, 649–50.

Leakey, L. S. B. (1962*a*). A new lower Pliocene fossil primate from Kenya. *Annals of the Magazine of Natural History, Series 13*, **4**, 689–696.

Leakey, L. S. B. (1962*b*). Primates. In *The Mammalian Fauna and Geomorphological Relations of the Napak Volcanics, Karamoja*, ed. W. W. Bishop, pp. 1–18. Entebbe: Recent Geological Survey of Uganda (1957–58).

Leakey, L. S. B. (1963). East African fossil Hominoidea and the classification within this super-family. In *Classification and Human Evolution*, ed. S. L. Washburn, pp. 32–49. Chicago, IL: Aldine.

Leakey, L. S. B. (1966). *Homo habilis*, *Homo erectus* and the australopithecines. *Nature*, **209**, 1279–1281.

Leakey, L. S. B. (1967*a*). An early Miocene member of Hominidae. *Nature*, **213**, 155–163.

Leakey, L. S. B. (1967*b*). Notes on the mammalian faunas from the Miocene and Pleistocene of East Africa. In *Background to Evolution in Africa*, eds. W. W. Bishop & J. D. Clark, pp. 7–29. Chicago, IL: University of Chicago Press.

Leakey, L. S. B. (1968). Upper Miocene primates from Kenya. *Nature*, **218**, 527–528.

Leakey, L. S. B. (1970). *The Stone Age Races of Kenya*, 2nd edn. Oxford: Oxford University Press.

Leakey, L. S. B. & Leakey, M. D. (1964). Recent discoveries of fossil hominids in Tanganyika: at Olduvai and near Lake Natron. *Nature*, **202**, 5–7.

Leakey, L. S. B., Tobias, P. V., & Napier, J. R. (1964). A new species of the genus *Homo* from Olduvai Gorge. *Nature*, **202**, 7–9.

Leakey, L. S. B. & Whitworth, T. (1958). Notes on the genus *Simopithecus* with a description of a new species from Olduvai. *Occasional Papers of the Coryndon Museum*, **6**, 1–14.

Leakey, M. D. (1971). Discovery of postcranial remains of *Homo erectus* and associated artefacts in Bed IV at Olduvai Gorge, Tanzania. *Nature*, **232**, 380–383.

Leakey, M. D., Clarke, R. J., & Leakey, L. S. B. (1971). New hominid skull from Bed I, Olduvai Gorge, Tanzania. *Nature*, **232**, 308–312.

Leakey, M. D. & Harris, J. M. (eds) (1987). *Laetoli: A Pliocene Site in Northern Tanzania*. Oxford: Clarendon Press.

Leakey, M. D. & Hay, R. (1982). The chronological position of the fossil hominids of Tanzania. In: *L'Homo erectus et la place de l'homme de Tautavel parmi les hominidés fossiles*, ed. M. A. de Lumley, pp. 753–765. Nice: Centre national de la recherche scientifique

Leakey, M. G. (1982). Extinct large colobines from the Plio-Pleistocene of Africa. *American Journal of Physical Anthropology*, **58**, 153–172.

Leakey, M. G. (1985). Early Miocene cercopithecids from Buluk, Northern Kenya. *Folia Primatologica*, **44**, 1–14.

Leakey, M. G. (1987). Colobinae (Mammalia, Primates) from the Omo Valley, Ethiopia: Cercopithecidae de la formation de Shungura. *Faunes de Plio-Pleistocene, Vallée Omo*, **3** 147–169.

Leakey, M. G. (1988). Fossil evidence for the volution of the guenons. In *A Primate Radiation: Evolutionary Biology of the African Guenons*, eds. A. Gautier-Hion, F. Boulière, & J.-P. Gautier, pp. 7–12. Cambridge: Cambridge University Press.

Leakey, M. G. (1993). Evolution of *Theropithecus* in the Turkana Basin. In Theropithecus: *The Rise and Fall of a Primate Genus*, ed. N. G. Jablonski, pp. 85–123. Cambridge: Cambridge University Press.

Leakey, M. G. & Delson, E. (1987). Fossil Cercopithecidae from the Laetolil Beds. In *Laetoli: A Pliocene Site in Northern Tanzania*, eds. M. D. Leakey & J. M. Harris, pp. 91–107. Oxford: Clarendon Press.

Leakey, M. G., Feibel, C. S., Bernor, R. L., Harris, J. M., Cerling, T. E., Stewart, K. M. Storrs, G. W., Walker, A., Werdelin, L., & Winkler, A. J. (1996). Lothagam: a record of faunal change in the late Miocene of East Africa. *Journal of Vertebrate Paleontology*, **16**, 556–570.

Leakey, M. G., Feibel, C. S., McDougall, I., & Walker, A. C. (1995*a*). New four-million-year-old hominid species from Kanapoi and Allia Bay, Kenya. *Nature*, **376**, 565–571.

Leakey, M. G., Feibel, C. S., McDougall, I., Ward, C., & Walker, A.

C. (1998). New specimens and confirmation of an early age for *Australopithecus anamensis*. *Nature*, **393**, 62–66.

Leakey, M. G. & Leakey, R. E. F. (1973). Further evidence of *Simopithecus* (Mammalia, Primates) from Olduvai and Olorgesailie. *Fossil Vertebrates of Africa*, **3**, 101–120.

Leakey, M. G. & Leakey, R. E. F. (1976). Further Cercopithecinae (Mammalia, Primates) from the Plio/Pleistocene of East Africa. *Fossil Vertebrates of Africa*, **4**, 121–146.

Leakey, M. G., Leakey, R. E. F., Richtsmeier, J. T., Simons, E. L., & Walker, A. C. (1991). Similarities in *Aegyptopithecus* and *Afropithecus* facial morphology. *Folia Primatologica*, **56**, 65–85.

Leakey, M. G., Spoor, F., Brown, F. H., Gathogo, P. N., Kiarie, C., Leakey, L. N., & McDougall, I. (2001). New hominin genus from eastern Africa shows diverse middle Pliocene lineages. *Nature*, **410**, 433–440.

Leakey, M. G., Ungar, P. S., & Walker, A. C. (1995*b*). A new genus of large primate from the late Oligocene of Lothidok, Turkana District, Kenya. *Journal of Human Evolution*, **28**, 519–531.

Leakey, M. G. & Walker, A. C. (1997). *Afropithecus*: function and phylogeny. In *Function, Phylogeny, and Fossils: Miocene Hominoid Evolution and Adaptations*, eds. D. R. Begun, C. V. Ward, & M. D. Rose, pp. 225–239. New York: Plenum Press.

Leakey, R. E. F. (1969). New Cercopithecidae from the Chemeron Beds of Lake Baringo, Kenya. *Fossil Vertebrates of Africa*, **1**, 53–69.

Leakey, R. E. F. (1973*a*). Further evidence of Lower Pleistocene hominids from East Rudolf, North Kenya, 1972. *Nature*, **242**, 170–173.

Leakey, R. E. F. (1973*b*). Evidence for an advanced Plio-Pleistocene hominid from East Rudolf, Kenya. *Nature*, **242**, 447–450.

Leakey, R. E. F. (1974). Further evidence of Lower Pleistocene hominids from East Rudolf, North Kenya, 1973. *Nature*, **248**, 653–656.

Leakey, R. E. F. (1976). New hominid fossils from the Koobi Fora formation in Northern Kenya. *Nature*, **261**, 574–576.

Leakey, R. E. F. & Leakey, M. G. (1978). *Koobi Fora Research Project*, vol. 1, *The Fossil Hominids and an Introduction to Their Context, 1968–1974*. Oxford: Clarendon Press.

Leakey, R. E. F. & Leakey, M. G. (1986*a*). A new Miocene hominoid from Kenya. *Nature*, **324**, 143–146.

Leakey, R. E. F. & Leakey, M. G. (1986*b*). A second new Miocene hominoid from Kenya. *Nature*, **324**, 146–148.

Leakey, R. E. F. & Leakey, M. G. (1987). A new Miocene small-bodied ape from Kenya. *Journal of Human Evolution*, **16**, 369–387.

Leakey, R. E. F., Leakey, M. G., & Walker, A. C. (1988*a*). Morphology of *Afropithecus turkanensis* from Kenya. *American Journal of Physical Anthropology*, **76**, 289–307.

Leakey, R. E. F., Leakey, M. G., & Walker, A. C. (1988*b*). Morphology of *Turkanapithecus kalakolensis* from Kenya. *American Journal of Physical Anthropology*, **76**, 277–288.

Leakey, R. E. F. & Walker, A. C. (1976). *Australopithecus, Homo erectus* and the single species hypothesis. *Nature*, **261**, 572–574.

Leakey, R. E. F. & Walker, A. C. (1985*a*). Further hominids from the Plio-Pleistocene of Koobi Fora, Kenya. *American Journal of Physical Anthropology*, **67**, 135–164.

Leakey, R. E. F. & Walker, A. C. (1985*b*). New higher primates from the early Miocene of Buluk, Kenya. *Nature*, **318**, 173–175.

Leakey, R. E. F., Walker, A. C., Ward, C. V., & Grausz, H. M. (1989). A partial skeleton of a gracile hominid from the Upper Burgi Member of the Koobi Fora Formation, East Lake Turkana, Kenya. In *Hominidae: Proceedings of the 2nd International Congress of Human Paleontology, Turin, 1987*, ed. G. Giacobini, pp. 209–215. Milan: Jaca.

Leakey, R. E. F. & Wood, B. A. (1973). New evidence of the genus *Homo* from East Rudolf, Kenya. II. *American Journal of Physical Anthropology*, **39**, 355–368.

Lecointre, G. (1912). *Les Faluns de Touraine*. Tours: Société d'agriculture, science, arts et belles-lettres de département Indre-et-Loire.

Lee, P. C. & Foley, R. A. (1993). Ecological energetics and extinction of giant gelada baboons. In Theropithecus: *The Rise and Fall of a Primate Genus*, ed. N. G. Jablonski, pp. 487–498. Cambridge: Cambridge University Press.

Le Gros Clark, W. E. (1934). *Early Fore-Runners of Man*. London: Baillière.

Le Gros Clark, W. E. (1945). A note on the palaeontology of the lemuroid brain. *Journal of Anatomy*, **79**, 123–126.

Le Gros Clark, W. E. (1949). Early Miocene apes from East Africa. *Advancement of Science*, **5**, 340–341.

Le Gros Clark, W. E. (1950). New palaeontological evidence bearing on the evolution of the Hominoidea. *Quarterly Journal of the Geological Society of London*, **105**, 225–259.

Le Gros Clark, W. E. (1952). Report on fossil hominoid material collected by the British-Kenya Miocene Expedition, 1949–1951. *Proceedings of the Zoological Society of London*, **122**, 273–286.

Le Gros Clark, W. E. (1956). A Miocene lemuroid skull from East Africa. *Fossil Mammals of Africa*, No. 9. London: British Museum (Natural History).

Le Gros Clark, W. E. (1959). *The Antecedents of Man*. Edinburgh: Edinburgh University Press.

Le Gros Clark, W. E. (1964). *The Fossil Evidence for Human Evolution: An Introduction to the Study of Paleoanthropology*, 2nd edn., Chicago, IL: University of Chicago Press.

Le Gros Clark, W. E. & Leakey, L. S. B. (1950). Diagnoses of East African Miocene Hominoidea. *Quarterly Journal of the Geological Society of London*, **105**, 260–263.

Le Gros Clark, W. E. & Leakey, L. S. B. (1951). The Miocene Hominoidea of East Africa. *Fossil Mammals of Africa*, **1**, 1–117.

Le Gros Clark, W. E. & Thomas, D. P. (1951). Associated jaws and limb bones of *Limnopithecus macinnesi*. *Fossil Mammals of Africa*, **3**, 1–27.

Le Gros Clark, W. E. & Thomas, D. P. (1952). The Miocene lemuroids of East Africa. *Fossil Mammals of Africa*, **5**, 1–20.

Leidy, J. (1869). Notice of some extinct vertebrates from Wyoming and Dakota. *Proceedings of the Academy of Natural Sciences of the United States of America, Philadelphia*, **21**, 63–67.

Leidy, J. (1870). Descriptions of *Palaeosyops paludosus*, *Microsus cuspidatus* and *Notharctus tenebrosus*. *Proceedings of the Academy of Natural Sciences of the United States of America, Philadelphia*, **22**,111–114.

Leidy, J. (1872). Remarks on some extinct mammals. *Proceedings of the Academy of Natural Sciences of the United States of America, Philadelphia*, **24**, 37–38.

Leidy, J. (1873). Contribution to the extinct vertebrate fauna of the western territories. *Reports of the United States Geological Survey of Territory*, Part 1.

Leighton, M. (1993). Modeling dietary selectivity by Bornean orangutans: evidence for integration of multiple criteria in fruit selection. *International Journal of Primatology*, **14**, 257–313.

Lemelin, P. (1996). The evolution of prehensility in primates: a comparative study of prosimians and didelphid marsupials. PhD dissertation, State University of New York, Stony Brook.

Lemelin, P. (1999). Morphological correlates of substrate use in didelphid marsupials: implications for primate origins. *Journal of Zoology*, **247**, 165–175.

Lemelin, P. (2000). Micro-anatomy of the volar skin and interordinal relationships of primates. *Journal of Human Evolution*, **38**, 257–267.

Lemoine, V. (1878). Communication sur les ossements fossiles des terrains tertiaires inférieures des environs de Reims à la Société d'étude d'histoire naturelle de Reims, 1–24.

Leroi-Gourhan, A. (1958). Étude des restes humains fossiles provenant des Grottes d'Arcy-sur-Cure. *Annales de paléontologie*, **44**, 87–148.

Lévêque, F. & Vandermeersch, B. (1981). Le néandertalien de Saint-Césaire. *La Recherche*, **12**, 242–244.

Lewis, G. E. (1933). Preliminary notice of a new genus of lemuroid from the Siwaliks. *American Journal of Science*, **26**, 134–138.

Lewis, G. E. (1934). Preliminary notice of new man-like ape from India. *American Journal of Science*, **27**, 161–179.

Lewis, G. E. (1936). A new species of *Sugrivapithecus*. *American Journal of Science*, **31**, 450–452.

Lewis, G. E. (1937). Taxonomic syllabus of Siwalik fossil anthropoids. *American Journal of Science*, **34**, 139–142.

Lewis, O. J. (1972). The evolution of the hallucial tarsometatarsal joint in the Anthropoidea. *American Journal of Physical Anthropology*, **37**, 13–34.

Lewis, O. J. (1980). The joints of the evolving foot. III. The fossil evidence. *Journal of Anatomy*, **131**, 275–298.

Li, C. (1978). A Miocene gibbon-like primate from Shihhung, Kiangsu Province. *Vertebrata PalAsiatica*, **16**, 187–192.

Li, T. & Etler, D. (1992). New Middle Pleistocene hominid crania from Yunxian in China. *Nature*, **357**, 404–407.

Li, W., Zhang, Z., Gu, Y., Luin, Y., & Yan, F. (1982). A fauna from Lianhua Cave, Dantu, Jiangsu. *Acta Anthropologica Sinica*, **1**, 169–179.

Lieberman, D. E. (1995). Testing hypotheses about recent human evolution from skulls: integrating morphology, function, development, and phylogeny. *Current Anthropology*, **36**, 159–197.

Lieberman, D. E. (1999). Homology and hominid phylogeny: problems and potential solutions. *Evolutionary Anthropology*, **7**, 142–151.

Lieberman, D. E. (2001). Another face in our family tree. *Nature*, **410**, 419–420.

Lieberman, D. E., Wood, B. A., & Pilbeam, D. R. (1996). Homoplasy and early *Homo*: an analysis of the evolutionary relationships of *H. habilis* sensu stricto and *H. rudolfensis*. *Journal of Human Evolution*, **30**, 97–120.

Lillegraven, J. A. (1980). Primates from the later Eocene rocks of southern California. *Journal of Mammalogy*, **61**, 181–204.

Linnaeus, C. (1758). *Systema naturae per regna tria naturae, secundum classes, ordines, genera, species cum characteribus, differentiis, synonymis, locis*, 10th edn, revised. Stockholm: Laurentius Salvius.

Lockwood, C. A. & Fleagle, J. G. (1999). The recognition and evaluation of homoplasy in primate and human evolution. *Yearbook of Physical Anthropology*, **42**, 189–232.

Lönnberg, E. (1936). On some fossil mammalian remains from East Afrika. *Archives für Zoologische*, **27**, 1–23.

Loomis, F. B. (1906). Wasatch and Wind River primates. *American Journal of Science*, **21**, 277–284.

Lorenz von Liburnau, L. (1899). Über einen fossilen Anthropoidenvon Madagaskar. *Anzeiger der kaiserlichen Akademie der Wissenschaften in Wien*, **37**, 8–9.

Lorenz von Liburnau, L. (1900*a*). Über einige Reste ausgestorbener Primaten von Madagaskar. *Denkschriften der kaiserlichen Akademie der Wissenschaften in Wien*, **70**, 1–15.

Lorenz von Liburnau, L. (1900*b*). *Paleolemur destructus*. *Anzeiger der kaiserlichen Akademie der Wissenschaften in Wien*, **1**, 8.

Lorenz von Liburnau, L. (1902). Über *Hadropithecus stenognathus* Lz. Nebst bemerkungen zu einigen anderen austestorbenen Primaten von Madagascar. *Denkschriften der Kaiserlichen Akademie der Wissenschaften in Wien*, **72**, 243–254.

Lorenz von Liburnau, L. (1905). *Megaladapis edwardsi* G. Grandidier. *Denkshriften der Kaiserlichen Akademie der Wissenschaften in Wien*, **77**, 451–490.

Louis, P. & Sudre, J. (1975). Nouvelles données sur les primates de l'Éocène supérieur européen. *Colloques internationaux dei Centre national de la recherche scientifique*, **218**, 805–828.

Lovejoy, C. O., Cohn, M. J., & White, T. D. (1999). Morphological analysis of the mammalian postcranium: a developmental perspective. *Proceedings of the National Academy of Sciences of the United States of America*, **96**, 13247–13252.

Luchterhand, K., Kay, R. F., & Madden, R. H. (1986). *Mohanamico hershkovitzi*, gen. et sp. nov., un primate du Miocene moyen d'Amérique du Sud. *Comptes rendus de l'Académie des sciences de Paris*, **303**, 1753–1758.

Luckett, W. P. (1975). Ontogeny of the fetal membranes and placenta: their bearing on primate phylogeny. In *Phylogeny of the Primates*, eds. W. Luckett & F. S. Szalay, pp. 157–82. New York: Plenum Press.

Luckett, W. P. (1980). *Comparative Biology and Evolutionary Relationships of Tree Shrews*. New York: Plenum Press.

Luckett, W. P. & Maier, W. (1986). Developmental evidence for anterior tooth homologies in the aye aye, *Daubentonia*. *American Journal of Physical Anthropology*, **69**, 233.

Lund, P. W. (1838). Blik paa Brasiliens dyreverden for sidste jordomvaeltning. *Det Kongelige Danske Videnskabernes Selskabs Naturvidenskabelige og Mathematiske Afhandlinger*, **8**, 61–144.

Lund, P. W. (1839). Coup d'oeil sur les espèces éteintes de mammifères du Brésil. *Annuaire des sciences naturelles (Paris)*, **11**, 214–234.

Lund, P. W. (1840*a*). Nouvelles recherches sur la faune fossile du Brésil. *Annuaire des sciences naturelles (Paris)*, **13**, 310–319.

Lund, P. W. (1840*b*). View of the fauna of Brazil, previous to the last geological revolution. *Charlesworth's Magazine of Natural History*, **4**, 1–8; 49–57; 105–112; 153–161; 207–213; 251–259; 307–317; 373–389.

Lydekker, R. (1877). Notices of new and rare mammals from the Siwaliks. *Records of the Geological Survey of India*, **10**, 76–83.

Lydekker, R. (1878). Notices of Siwalik mammals. *Records of the Geological Survey of India*, **11**, 64–104.

Lydekker, R. (1879). Further notices of Siwalik mammalia. *Records of the Geological Survey of India*, **12**, 33–52.

Lydekker, R. (1884). Indian Tertiary and post-Tertiary Vertebrata: rodents, ruminants and synopsis of Mammalia. *Memoirs of the Geological Survey of India (Palaeontologica Indica)*, **3**, 105–134.

Lydekker, R. (1885). *Catalogue of the Fossil Mammalia in the British Museum, Part I, Containing the Orders Primates, Chrioptera, Insectivora, Carnivora, and Rodentia*, London.

Lydekker, R. (1887). *Catalogue of the Fossil Mammalia in the British Museum*, Part V. London: British Museum (Natural History).

Lydekker, R. (1895). Review of Dubois: Pithecanthropus erectus, eine Menschenähnliche Übergangsform aus Java. *Nature*, **51**, 291.

Lyell, C. (1830). *Principles of Geology: Being an Attempt to Explain the Former Changes of the Earth's Surface, by Reference to Causes Now in Operation.* London: John Murray.

Maas, M. C. & O'Leary, M. (1996). Evolution of molar enamel microstructure in North American Notharctidae (Primates). *Journal of Human Evolution*, **31**, 293–310.

Macdonald, J. R. (1963). The Miocene faunas from the Wounded Knee area of western South Dakota. *Bulletin of the American Museum of Natural History*, **125**, 141–238.

MacFadden, B. J. (1990). Chronology of Cenozoic primate localities in South America. *Journal of Human Evolution*, **19**, 7–21.

MacInnes, D. G. (1943). Notes on the East African Miocene Primates. *Journal of the East Africa and Uganda Natural History Society*, **17**, 141–181.

MacLarnon, A. (1993). The vertebral canal. In *The Nariokotome* Homo erectus *Skeleton*, eds. A. C. Walker & R. E. F. Leakey, pp. 359–390. Cambridge, MA: Harvard University Press.

MacLarnon, A. (1995). The distribution of spinal cord tissues and locomotor adaptation in primates. *Journal of Human Evolution*, **29**, 463–482.

MacLarnon, A. (1996). The evolution of the spinal cord in primates: evidence from the foramen magnum and the vertebral canal. *Journal of Human Evolution*, **30**, 121–138.

MacLatchy, L., Gebo, D., Kityo, R., & Pilbeam, D. (2000). Postcranial functional morphology of *Morotopithecus bishopi*, with implications for the evolution of modern ape locomotion. *Journal of Human Evolution*, **39**, 159–183.

MacLatchy, L. & Pilbeam, D. (1999). Renewed research in the Ugandan early Miocene. In *Late Cenozoic Environments and Hominid Evolution: A Tribute to Bill Bishop*, eds. P. Andrews & P. Banham, pp. 15–25. London: Geological Society.

MacPhee, R. D. E. (1977). Ontogeny of the ectotympanic–petrosal plate relationships in strepsirhine prosimians. *Folia Primatologica*, **27**, 245–283.

MacPhee, R. D. E. (1981). Auditory regions of primates and eutherian insectivores: morphology, ontogeny, and character analysis. *Contributions in Primatology*, **18**, 1–282.

MacPhee, R. D. E. (ed.) (1993*a*). *Primates and their Relatives in Phylogenetic Perspective.* New York: Plenum Press.

MacPhee, R. D. E. (1993*b*). Summary. In *Primates and their Relatives in Phylogenetic Perspective*, ed. R. D. E. MacPhee, pp. 363–373. New York: Plenum Press.

MacPhee, R. D. E. (1994). Morphology, adaptations and relationships of *Plesiorycteropus*, and a diagnosis of a new order of eutherian mammals. *Bulletin of the American Museum of Natural History*, **220**, 1, 3–214.

MacPhee, R. D. E. (1996). The Greater Antillean monkeys. *Revista de Ciencia*, **18**, 13–32.

MacPhee, R. D. E., Beard, K. C., & Qi, T. (1995*a*). Significance of primate petrosal from Middle Eocene fissure-fillings at Shanghuang, Jiangsu Province, People's Republic of China. *Journal of Human Evolution*, **29**, 501–513.

MacPhee, R. D. E. & Burney, D. A. (1991). Dating of modified femora of extinct dwarf *Hippopotamus* from southern Madagascar: implications for constraining human colonization and vertebrate extinction events. *Journal of Archaeological Science*, **18**, 695–706.

MacPhee, R. D. E. & Cartmill, M. (1986). Basicranial structures and primate systematics. In *Comparative Primate Biology*, Vol. 1, eds. D. Swindler & J. Erwin, pp. 219–275. New York: Alan R. Liss.

MacPhee, R. D. E., Cartmill, M., & Rose, K. D. (1989). Craniodental morphology and relationships of the supposed Eocene dermopteran *Plagiomene* (Mammalia). *Journal of Vertebrate Paleontology*, **9**, 329–349.

MacPhee, R. D. E. & Fleagle, J. G. (1991). Postcranial remains of *Xenothrix mcgregori* (Primates, Xenotrichidae) and other late Quaternary mammals from Long Mile Cave, Jamaica. *Bulletin of the American Museum of Natural History*, **206**, 287–321.

MacPhee, R. D. E. & Flemming, C. (1999). *Requiem aeternum*: the last five hundred years of mammalian species extinctions. In *Extinctions in Near Time: Causes, Contexts, and Consequences*, ed. R. D. E. MacPhee, pp. 333–372. New York: Kluwer.

MacPhee, R. D. E., Horovitz, I., Arredondo, O., & Jiménez Vasquez, O. (1995*b*). A new genus for the extinct Hispaniolan monkey *Saimiri bernensis* (Rímoli, 1977), with notes on its systematic position. *American Museum Novitates*, **3134**, 1–21.

MacPhee, R. D. E. & Iturralde-Vinent, M. A. (1994). First Tertiary land mammal from Greater Antilles: an Early Miocene sloth (Megalonychidae, Xenarthra) from Cuba. *American Museum Novitates*, **3094**, 1–13.

MacPhee, R. D. E. & Iturralde-Vinent, M. A. (1995*a*). Earliest monkey from Greater Antilles. *Journal of Human Evolution*, **28**, 197–200.

MacPhee, R. D. E. & Iturralde-Vinent, M. A. (1995*b*). Origin of the Greater Antillean land mammal fauna. I. New Tertiary fossils from Cuba and Puerto Rico. *American Museum Novitates*, **3141**, 1–31.

MacPhee, R. D. E. & Jacobs, L. L. (1986). *Nycticeboides simpsoni* and the morphology, adaptations, and relationship of Miocene Siwalik Lorisidae. *Contributions to Geology, University of Wyoming, Special Paper* **3**, 131–161.

MacPhee, R. D. E. & Marx, P. A. (1997). The 40 000 year plague: humans, hyperdisease, and first-contact extinctions. In *Natural Change and Human Impact in Madagascar*, eds. S. M. Goodman & B. D. Patterson, pp. 169–217. Washington, DC: Smithsonian Institution Press.

MacPhee, R. D. E., McFarlane, D. A., & Ford, D. F. (1989). Pre-Wisconsinan land mammals from Jamaica and models of late Quaternary extinction in the Greater Antilles. *Quaternary Research*, **31**, 94–106.

MacPhee, R. D. E. & Raholimavo, E. M. (1988). Modified subfossil aye-aye incisors from southwestern Madagascar: species allocation and paleoecological significance. *Folia Primatologica*, **51**, 126–142.

MacPhee, R. D. E. & Rivero de la Calle, M. (1996). Accelerator mass spectrometry 14C determination for the alleged "Cuban spider monkey", *Ateles* (= *Montaneia*) *anthropomorphus*. *Journal of Human Evolution*, **30**, 89–94.

MacPhee, R. D. E, Simons, E. L., Wells, N. A., & Vuillaume-Randriamanantena, M. (1984). Team finds giant lemur skeleton. *Geotimes*, **29**, 10–11.

MacPhee, R. D. E., White, J., & Woods, C. A. (2000). New species of Megalonychidae (Xenarthra, Phyllophaga) from the Quaternary of Hispaniola. *American Museum Novitates*, **3303**, 1–32.

MacPhee, R. D. E. & Woods, C. A. (1982). A new fossil cebine from

Hispaniola. *American Journal of Physical Anthropology*, **58**, 419–436.

Madden, C. T. (1980). New *Proconsul (Xenopithecus)* from the Miocene of Kenya. *Primates*, **21**, 241–252.

Madden, R. H., Savage, D., & Fields, R. W. (1997). A history of vertebrate paleontology in the Magdalena Valley. In *Vertebrate Paleontology in the Neotropics: The Miocene Fauna of La Venta, Colombia*, eds. R. F. Kay, R. H. Madden, R. L. Cifelli, & J. J. Flynn, pp. 1–11. Washington, DC: Smithsonian Institution Press.

Made, J. van der (1999). Intercontinental relationship Europe-Africa and the Indian subcontinent. In *The Miocene Land Mammals of Europe: The Continental European Miocene*, eds G. E. Rössner & K. Heissig, pp. 457–472. Munich: Dr. Friedrich Pfeil.

Mahé, J. (1965). Un gisement nouveau de subfossiles à Madagascar. *Compte rendu sommaire des séances de la Société géologique de France*, **2**, 66.

Mahé, J. (1976). Craniométrie des lémuriens: analyses multivariables–phylogénie. *Mémoires du Muséum national d'histoire naturelle, Paris, Série C*, **32**, 1–342.

Mahé, J. & Sourdat, M. (1972). Sur l'extinction des vertébrés subfossiles et l'aridification du climat dans le sud-ouest de Madagascar. *Bulletin de la Société géologique de France, Série 7*, **14**, 295–309.

Maier, W. (1970). New fossil Cercopithecoidea from the lower Pleistocene cave deposits of the Makapansgat Limeworks, South Africa. *Palaeontologica Africana*, **13**, 69–108.

Maier, W. (1971). Two new skulls of *Parapapio antiquus* from Taung and a suggested phylogenetic arrangement of the genus *Parapapio*. *Annals of the South African Museum*, **59**, 1–16.

Maier, W. (1972). The first complete skull of *Simopithecus darti* from Makapansgat, South Africa, and its systematic position. *Journal of Human Evolution*, **1**, 395–405.

Maier, W. (1980). Konstruktions-morphologische Untersuchungen am Gebiss der rezente Prosimiae (Primates). *Abhandlungen der senckenbergischen naturforschenden Gesellschaft*, **538**, 1–158.

Manser, J. & Harrison, T. (1999). Estimates of cranial capacity and encephalization in *Proconsul* and *Turkanapithecus*. *American Journal of Physical Anthropology*, **28**, 189.

Marsh, O. C. (1871). Notice of some fossil mammals from the Tertiary formation. *American Journal of Science*, **2**, 35–44, 120–127.

Marsh, O. C. (1872). Preliminary description of new Tertiary mammals. Parts I through IV. *American Journal of Science*, **4**, 22–28, 202–224.

Marston, A. (1936). Preliminary note on a new fossil human skull from Swanscombe, Kent. *Nature*, **138**, 200–201.

Martin, H. (1911). Sur un squelette humain de l'époque moustérienne trouvé en Charente. *Comptes rendus de l'Académie des sciences de Paris*, **153**, 728.

Martin, L. (1981). New specimens of *Proconsul* from Koru, Kenya. *Journal of Human Evolution*, **10**, 139–150.

Martin, L. (1991). Teeth, sex, and species. *Nature*, **352**, 111–112.

Martin, L. & Andrews, P. (1984). The phyletic positon of *Graecopithecus freyberi* Koenigswald. *Courier Forschungsinstitut Senckenberg*, **69**, 25–40.

Martin, L. & Andrews, P. (1993*a*). Renaissance of Europe's ape. *Nature*, **365**, 494.

Martin, L. & Andrews, P. (1993*b*). Species recognition in Middle Miocene hominoids. In *Species, Species Concepts, and Primate Evolution*, eds. W. H. Kimbel & L. B. Martin, pp. 393–427. New York: Plenum Press.

Martin, P. S. (1984). Prehistoric overkill: the global model. In *Quaternary Extinctions: A Prehistoric Revolution*, eds. P. S. Martin & R. G. Klein, pp. 354–403. Tucson, AZ: University of Arizona Press.

Martin, R. D. (1968*a*). Towards a new definition of Primates. *Man*, **3**, 377–401.

Martin, R. D. (1968*b*). Reproduction and ontogeny in tree-shrews (*Tupaia belangeri*) with reference to their general behaviour and taxonomic relationships. *Zeitschrift für Tierpsychologie*, **25**, 409–532.

Martin, R. D. (1979). Phylogenetic aspects of prosimian behavior. In *The Study of Prosimian Behavior*, eds. G. A. Doyle & R. D. Martin, pp. 45–77. New York:Academic Press.

Martin, R. D. (1981). Relative brain size and basal metabolic rate in terrestrial vertebrates. *Nature*, **293**, 57–60.

Martin, R. D. (1986). Are fruit-bats primates? *Nature*, **320**, 482–483.

Martin, R. D. (1990). *Primate Origins and Evolution: A Phylogenetic Reconstruction*. London: Chapman & Hall.

Martin, R. D. (1993). Primate origins: plugging the gaps. *Nature*, **363**, 223–234.

Martin, R. D. (1998). Comparative aspects of human brain evolution: scaling, energy costs and confounding variables. In *The Origin and Diversification of Language*, eds. N. G. Jablonski & L. C. Aiello, pp. 35–68. San Francisco, CA: California Academy of Sciences.

Martin, R. D., Doyle, G. A., & Walker, A. C. (eds.) (1974). *Prosimian Biology*. London: Duckworth.

Maška, K. (1882). Über den diluvialen Menschen in Stramberg. *Mitteilungen der anthropologischen Gesellschaft in Wien*, **12**, 32–38.

Mason, M. A. (1990). New fossil primate from the Uintan (Eocene) of southern California. *PaleoBios*, **13**, 1–7.

Masters, J. C., Rayner, R. J., Ludewick, H., Zimmermann, E., Molez-Verriere, N., Vincent, F., & Nash, L. T. (1994). Phylogenetic relationships among the Galaginae as indicated by erythrocytic allozymes. *Primates*, **35**, 177–190.

Masters, J. C., Rayner, R. J., & Tattersall, I. (1995). Pattern and process in strepsirhine phylogeny. In *Creatures in the Dark: The Nocturnal Prosimians*, eds. L. Alterman, G. A. Doyle, & M. K. Izard, pp. 31–44. New York: Plenum Press.

Matthew, W. D. (1909). The Carnivora and Insectivora of the Bridger Basin, middle Eocene. *Memoirs of the American Museum of Natural History*, **9**, 291–567.

Matthew, W. D. (1915*a*). A revision of the lower Eocene Wasatch and Wind River faunas. IV. Entelonychia, Primates, Insectivora (part). *Bulletin of the American Museum of Natural History*, **34**, 429–483.

Matthew, W. D. (1915*b*). Climate and evolution. *Annals of the New York Academy of Sciences*, **24**, 171–318.

Matthew, W. D. & Granger, W. (1923). New fossil mammals from the Pliocene of Sze-Chuan, China. *Bulletin of the American Museum of Natural History*, **48**, 563–598.

Mayr, E. (1944). On the concepts and terminology of vertical subspecies and species. *National Research Council Bulletin*, **2**, 11–16.

Mayr, E. (1950). Taxonomic categories in fossil hominids. *Cold Spring Harbor Symposia on Quantitative Biology*, **15**, 109–118.

McArdle, J. E. (1981). Functional morphology of the hip and thigh of the Lorisiformes. *Contributions in Primatology*, **17**, 1–132.

McCollum, M. A. (1999). The robust australopithecine face: a morphogenetic perspective. *Science*, **284**, 301–305.

McCown, T. D. (1950). The genus *Palaeoanthropus* and the problem of superspecific differentiation among the Hominidae. *Cold Spring*

Harbor Symposia on Quantitative Biology, **15**, 87–96.

McCown, T. D. & Keith, A. (1939). *The Stone Age of Mount Carmel*, vol. 2. *The Fossil Human Remains from the Levalloiso-Mousterian*. Oxford: Clarendon Press.

McCrossin, M. L. (1992*a*). An oreopithecid proximal humerus from the middle Miocene of Maboko Island, Kenya. *International Journal of Primatology*, **13**, 659–677.

McCrossin, M. L. (1992*b*). New species of bushbaby from the middle Miocene of Maboko Island, Kenya. *American Journal of Physical Anthropology*, **89**, 215–233.

McCrossin, M. L. (1994*a*). The phylogenetic relationships, adaptations, and ecology of *Kenyapithecus*. PhD dissertation, University of California, Berkeley.

McCrossin, M. L. (1994*b*). Semi-terrestrial adaptations of *Kenyapithecus*. *American Journal of Physical Anthropology*, Supplement **18**, 142–143.

McCrossin, M. L. (1995). New perspectives on the origins of terrestriality among Old World higher primates. *American Journal of Physical Anthropology*, Supplement **20**, 147.

McCrossin, M. L. (1996). A reassessment of forelimb evidence for the phylogenetic relationships of *Kenyapithecus* and other large-bodied hominoids of the middle–late Miocene. *American Journal of Physical Anthropology*, Supplement **22**, 161–162.

McCrossin, M. L. (1997). New postcranial remains of *Kenyapithecus* and their implications for understanding the origins of hominoid terrestriality. *American Journal of Physical Anthropology*, Supplement **24**, 164.

McCrossin, M. L. (1999*a*). New postcranial remains of *Kenyapithecus* and their implications for understanding the origins of hominoid terrestriality. Paper presented at *Evolution of Middle and Late Miocene Hominoids in Africa*, Kyoto University, July 1999.

McCrossin, M. L. (1999*b*). Phylogenetic relationships and paleoecological adaptations of a new bushbaby from the middle Miocene of Kenya. *American Journal of Physical Anthropology*, Supplement **28**, 195–196.

McCrossin, M. L. & Benefit, B. R. (1992). Comparative assessment of the ischial morphology of *Victoriapithecus macinnesi*. *American Journal of Physical Anthropology*, **87**, 277–290.

McCrossin, M. L. & Benefit, B. R. (1993*a*). Recently recovered *Kenyapithecus* mandible and its implications for great ape and human origins. *Proceedings of the National Academy of Sciences of the United States of America*, **90**, 1962–1966.

McCrossin, M. L. & Benefit, B. R. (1993*b*). Clues to the relationships and adaptations of *Kenyapithecus africanus* from its mandibular and incisor morphology. *American Journal of Physical Anthropology*, Supplement **16**, 142.

McCrossin, M. L. & Benefit, B. R. (1994). Maboko Island and the evolutionary history of Old World monkeys and apes. In *Integrative Paths to the Past: Paleoanthropological Advances in Honor of F. C. Howell*, eds. R. S. Corruccini & R. L. Ciochon, pp. 95–122. Englewood Cliffs, NJ: Prentice-Hall.

McCrossin, M. L. & Benefit, B. R. (1997). On the relationships and adaptations of *Kenyapithecus*, a large-bodied hominoid from the middle Miocene of eastern Africa. In *Function, Phylogeny and Fossils: Miocene Hominoid Origins and Adaptations*, eds. D. R. Begun, C. V. Ward, & M. D. Rose, pp. 241–267. New York: Plenum Press.

McCrossin, M. L., Benefit, B. R., & Gitau, S. N. (1998). Functional and phylogenetic analysis of the distal radius of *Kenyapithecus*, with comments on the origin of the African great ape and human clade. *American Journal of Physical Anthropology*, Supplement **26**, 158.

McCrossin, M. L., Benefit, B. R., Gitau, S., & Blue, K. T. (1998). Fossil evidence for the origins of terrestriality among Old World higher primates. In *Primate Locomotion: Recent Advances*, eds. E. L. Strasser, J. G. Fleagle, A. L. Rosenberger, & H. M. McHenry, pp. 353–396. New York: Plenum Press.

McDougall, I. & Watkins, R. (1985). Age of hominoid-bearing sequence at Buluk, northern Kenya. *Nature*, **318**, 175–178.

McHenry, H. M. (1986). The first bipeds: a comparison of the *Australopithecus afarensis* and *Australopithecus africanus* postcranium and implications for the evolution of bipedalism. *Journal of Human Evolution*, **15**, 177–192.

McHenry, H. M. (1988). New estimates of body weights in early hominids and their significance to encephalization and megadontia in "robust" australopithecines. In *Evolutionary History of the "Robust" Australopithecines*, ed. F. E. Grine, pp. 133–148. New York: Aldine de Gruyter.

McHenry, H. M. (1991). Femoral lengths and stature in Plio-Pleistocene hominids. *American Journal of Physical Anthropology*, **85**, 149–158.

McHenry, H. M. (1992). Body size and proportions in early hominids. *American Journal of Physical Anthropology*, **87**, 407–431.

McHenry, H. M. (1994*a*). Behavioral ecological implications of early hominid body size. *Journal of Human Evolution*, **27**, 77–87.

McHenry, H. M. (1994*b*). Tempo and mode in human evolution. *Proceedings of the National Academy of Sciences of the United States of America*, **91**, 6780–6786.

McHenry, H. M. (1996). Homoplasy, clades and hominid phylogeny. In *Contemporary Issues in Human Evolution*, eds. W. E. Meikle, F. C. Howell, & N. G. Jablonski, pp. 77–92. San Francisco, CA: California Academy of Sciences.

McHenry, H. M., Andrews, P., & Corruccini, R. S. (1980). Miocene hominoid palatofacial morphology. *Folia Primatologica*, **33**, 241–252.

McHenry, H. M. & Berger, L. R. (1998*a*). Body proportions in *Australopithecus afarensis* and *A. africanus* and the origin of the genus *Homo*. *Journal of Human Evolution*, **35**, 1–22.

McHenry, H. M. & Berger, L. R. (1998*b*). Limb lengths in *Australopithecus* and the origin of the genus *Homo*. *South African Journal of Science*, **94**, 447–450.

McHenry, H. M. & Corruccini, R. S. (1983). The wrist of *Proconsul africanus* and the origin of hominoid postcranial adaptations. In *New Interpretations of Ape and Human Ancestry*, eds. R. L. Ciochon & R. S. Corruccini, pp. 353–367. New York: Plenum Press.

McKee, J. K. (1993). Taxonomic and evolutionary affinities of *Papio izodi* fossils from Taung and Sterkfontein. *Palaeontologica Africana*, **30**, 43–49.

McKee, J. K. & Keyser, A. W. (1994). Craniodental remains of *Papio angusticeps* from the Haasgat Cave Site, South Africa. *International Journal of Primatology*, **15**, 823–841.

McKenna, M. C. (1990). Plagiomenids (Mammalia: ?Dermoptera) from the Oligocene of Oregon, Montana, and South Dakota, and middle Eocene of northwestern Wyoming. *Geological Society of America, Special Papers*, **243**, 211–234.

McKenna, M. C. & Bell, S. K. (1997). *Classification of Mammals above the Species Level*. New York: Columbia University Press.

Meikle, W. E. (1987). Fossil Cercopithecidae from the Sahabi Formation. In *Neogene Paleontology and Geology of Sahabi*, eds. N. T. Boaz, A. El-Arnauti, A. W. Gaziry, J. de Heinzelin, & D. D. Boaz, pp. 119–127. New York: Alan R. Liss.

Mein, P. (1986). Chronological succession of hominoids in the European Neogene. In *Primate Evolution*, eds. J. G. Else & P. C. Lee, pp. 59–70. Cambridge: Cambridge University Press.

Mein, P. (1989). Updating of MN zones. In *European Neogene Mammal Chronology*, eds. E. H. Lindsay, V. Fahlbusch, & P. Mein, pp. 73–90. New York: Plenum Press.

Mein, P. (1999). European Miocene mammal biochronology. In *The Miocene Land Mammals of Europe: The Continental European Miocene*, eds. G. E. Rössner & K. Heissig, pp. 25–38. Munich: Dr. Friedrich Pfeil.

Meireles, C. M., Czelusniak, J., Schneider, M. P. C., Muniz, J. A. P. C., Brigido, M. C., Ferreira, H. S., & Goodman, M. (1999). Molecular phylogeny of ateline New World monkeys (Platyrrhini, Atelinae) based on g-globin gene sequences: evidence that *Brachyteles* is the sister group of *Lagothrix*. *Molecular Phylogenetics and Evolution*, **12**, 10–30.

Meldrum, D. J. (1990). New fossil platyrrhine tali from the early Miocene of Argentina. *American Journal of Physical Anthropology*, **83**, 403–418.

Meldrum, D. J. (1993). Postcranial adaptations and positional behavior in fossil platyrrhines. In *Postcranial Adaptation in Nonhuman Primates*, ed. D. L. Gebo, pp. 235–251. DeKalb, IL: Northern Illinois University Press.

Meldrum, D. J., Dagosto, M., & White, J. (1997). Hindlimb suspension and hind foot reversal in *Varecia variegata* and other arboreal mammals. *American Journal of Physical Anthropology*, **103**, 85–102.

Meldrum, D. J., Fleagle, J. G., & Kay, R. F. (1990). Partial humeri of two Miocene Colombian primates. *American Journal of Physical Anthropology*, **81**, 413–422.

Meldrum, D. J. & Kay, R. F. (1997*a*). *Nuciruptor rubricae*, a new pitheciin seed predator from the Miocene of Colombia. *American Journal of Physical Anthropology*, **102**, 407–427.

Meldrum, D. J. & Kay, R. F. (1997*b*). Postcranial skeleton of Laventan platyrrhines. In *Vertebrate Paleontology in the Neotropics: The Miocene Fauna of La Venta, Colombia*, eds. R. F. Kay, R. H. Madden, R. L. Cifelli, & J. J. Flynn, pp. 459–472. Washington, DC: Smithsonian Institution Press.

Meldrum, D. J. & Lemelin, P. (1991). Axial skeleton of *Cebupithecia sarmientoi* (Pitheciinae, Platyrrhini) from the middle Miocene of La Venta, Colombia. *American Journal of Primatology*, **25**, 69–90.

Meldrum, D. J. & Pan, Y. (1988). Manual proximal phalanx of *Laccopithecus robustus* from the latest Miocene site of Lufeng. *Journal of Human Evolution*, **17**, 719–731.

Meléndez, B. (1957). Sobre el significado del "*Oreopithecus*.'. *Boletín de la real Sociedad española de historia natural Seción geologica*, **54**, 141–145.

Melnick, D. J., Hoelzer, G. A., Absher, R., & Ashley, M. V. (1993). mtDNA diversity in rhesus monkeys reveals overestimates of divergence time and paraphyly with neighboring species. *Molecular Biology and Evolution*, **10**, 282–295.

Melnick, D. J. & Kidd, K. K. (1985). Genetic and evolutionary relationships among Asian macaques. *International Journal of Primatology*, **6**, 123–160.

Mercerat, A. (1891). Sobre la presencia de restos de monos en el Eoceno de Patagonia. *Revista Museo La Plata*, **2**, 73–74.

Meyer, H. von (1848). In *Index Palaeontologicus*, ed. H. G. Bronn. Stuttgart: E. Schweizerbart.

Miller, E. R. & E. L. Simons (1997). Dentition of *Proteopithecus sylviae*, an archaic anthropoid from the Fayum, Egypt. *Proceedings of the National Academy of Sciences of the United States of America*, **94**, 13760–13764.

Miller, G. S. (1916). The teeth of a monkey found in Cuba. *Smithsonian Miscellaneous Collections*, **66**(13), 1–3.

Miller, G. S. (1929). Mammals eaten by Indians, owls, and Spaniards in the coast region of the Dominican Republic. *Smithsonian Miscellaneous Collections*, **82**(5), 1–16.

Miller, J. A. (1991). Does brain size variability provide evidence of multiple species in *Homo habilis*? *American Journal of Physical Anthropology*, **84**, 385–398.

Mittermeier, R. A., Tattersall, I., Konstant, B., Meyers, D., & Mast, R. B. (1994). *Lemurs of Madagascar*. Washington, DC: Conservation International.

Mivart, S. G. (1864). Notes on the crania and dentition of the Lemuridae. *Proceedings of the Zoological Society of London*, **1864**, 611–648.

Mollett, O. (1947). Fossil mammals from the Makapan Valley, Potgietersrust. I. Primates. *South African Journal of Science*, **43**, 295–303.

Montagnon, D., Ravaoarimanana, B., Rakotosamimanana, B., & Rumpler, Y. (2001). Ancient DNA from *Megaladapis edwardsi* (Malagasy subfossil): Preliminary results using cytochrome *b* sequence. *Folia Primatologica*, **72**, 30–32.

Morales, J. C. & Melnick, D. J. (1998). Phylogenetic relationships of the macaques (Cercopithecidae: *Macaca*), as revealed by high resolution restriction site mapping of mitochondrial ribosomal genes. *Journal of Human Evolution*, **34**, 1–23.

Morris, W. J. (1954). An Eocene fauna from the Cathedral Bluffs Tongue of the Washakie Basin, Wyoming. *Journal of Paleontology*, **28**, 195–203.

Mottl, M. (1957). Bericht über die neuen Menschenaffenfunde aus Österreich, von Sankt Steven in Lavanttal, Karnten. *Carinthia II*, **67**, 39–84.

Mottura, A. & Ardito, G. (1992). Observations on the Turin specimen of *Mesopithecus pentelici* (Wagner, 1839). *Human Evolution*, **7**, 67–73.

Mountain, J. (1998). Molecular evolution and modern human origins. *Evolutionary Anthropology*, **7**, 21–37.

Moyà-Soyà, S. & Köhler, M. (1993*a*). Middle Bartonian locality with *Anchomomys* (Adapidae, Primates) in the Spanish Pyrenees: preliminary report. *Folia Primatologica*, **60**, 158–163.

Moyà-Solà, S. & Köhler, M. (1993*b*). Recent discoveries of *Dryopithecus* shed new light on evolution of great apes. *Nature*, **365**, 543–545.

Moyà-Solà, S. & Köhler, M. (1995). New partial cranium of *Dryopithecus* Lartet, 1863 (Hominoidea, Primates) from the upper Miocene of Can Llobateres, Barcelona, Spain. *Journal of Human Evolution*, **29**, 101–139.

Moyà-Solà, S. & Köhler, M. (1996). The first *Dryopithecus* skeleton: origins of great ape locomotion. *Nature*, **379**, 156–159.

Moyà-Solà, S. & Köhler, M. (1997). The phylogenetic relationships of *Oreopithecus bambolii* Gervais, 1872. *Comptes rendus de l'Académie des sciences de Paris*, **324**, 141–148.

Moyà-Solà, S., Köhler, M., & Alba, D. M. (2001). *Egarapithecus narcisoi*, a new genus of Pliopithecidae (Primates, Catarrhini) from the late Miocene of Spain. *American Journal of Physical Anthropology*, **114**, 312–324.

Moyà-Solà, S., Köhler, M., & Rook, L. (1999). Evidence of hominid-like precision grip capability in the hand of the Miocene ape

Oreopithecus. *Proceedings of the National Academy of Sciences of the United States of America*, **96**, 313–317.

Moyà-Solà, S., Moyà, J. P., & Köhler, M. (1992). Primates catarrinos (Mammalia) del Neogeno de la peninsula Iberica. *Paleontologia i evolució*, **23**, 41–45.

Murphey, P. C., Torick, L., Bray, E., Chandler, R., & Evanoff, E. (1998). Taphonomy and paleoecology of the *Omomys* Quarry, an unusual fossil accumulation from the Bridger Formation, Wyoming. *Journal of Vertebrate Paleontology*, **18**, 65A.

Musser, G. G. & Dagosto, M. (1987). The identity of *Tarsius pumilus*, a pygmy species endemic to the montane mossy forests of central Sulawesi. *American Museum Novitates*, **2867**, 1–53.

Nakatsukasa, M., Kunimatsu, Y., Nakano, Y., & Ishida, H. (2000). A new skeleton of the large hominoid from Nachola, northern Kenya. *American Journal of Physical Anthropology*, Supplement **30**, 235.

Nakatsukasa, M., Takai, M., & Setoguchi, T. (1997). Functional morphology of the postcranium and locomotor behavior of *Neosaimiri fieldsi*, a *Saimiri*-like middle Miocene platyrrhine. *American Journal of Physical Anthropology*, **102**, 515–544.

Nakatsukasa, M., Yamanaka, A., Kunimatsu, Y., Shimizu, D., & Ishida, H. (1998). A newly discovered *Kenyapithecus* skeleton and its implications for the evolution of positional behavior in Miocene East African hominoids. *Journal of Human Evolution*, **34**, 657–664.

Napier, J. R. & Davis, P. R. (1959). The fore-limb skeleton and associated remains of *Proconsul africanus*. *Fossil Mammals of Africa*, No. 16. London: British Museum (Natural History).

Napier, J. R. & Napier, P. H. (1967). *A Handbook of Living Primates*. London: Academic Press.

Napier, J. R. & Walker, A. C. (1967). Vertical clinging and leaping: a newly recognized category of primate locomotion. *Folia Primatologica*, **6**, 204–219.

Napier, J. R. & Weiner, J. S. (1962). Olduvai Gorge and human origins. *Antiquity*, **36**, 41–47.

Napier, P. H. (1981). *Catalogue of Primates in the British Museum (Natural History) and Elsewhere in the British Isles, part 2, Family Cercopithecidae, Subfamily Cercopithecinae*. London: British Museum (Natural History).

Napier, P. H. (1985). *Catalogue of Primates in the British Museum (Natural History) and Elsewhere in the British Isles, part 3, Family Cercopithecidae, Subfamily Colobinae*. London: British Museum (Natural History).

Nash, L. T., Bearder, S. K., & Olson, T. R. (1989). Synopsis of *Galago* species characteristics. *International Journal of Primatology*, **10**, 57–80.

Necrasov, O., Samson, P., & Radulesco, C. (1961). Sur un nouveau singe catarrhinien fossile, découvert dans un nid fossilifère d'Olténie (R. P. R.). *Analele stiintifice ale Universitatii "Al. I. Cuza" din Iasi*, **7**, 401–416.

Nei, M. & Roychoudhury, A. (1982). Genetic relationships and evolution of human races. *Evolutionary Biology*, **15**, 1–59.

Nengo, I. O. & Rae, T. C. (1992). New hominoid fossils from the early Miocene site of Songhor, Kenya. *Journal of Human Evolution*, **23**, 423–429.

Niemitz, C., Nietsch, A., Warter, S., & Rumpler, Y. (1991). *Tarsius dianae*: a new primate species from central Sulawesi (Indonesia). *Folia Primatologica*, **56**, 105–116.

Nisbett, R. A. & Ciochon, R. L. (1993). Primates in northern Viet Nam: a review of the ecology and conservation status of extant species, with notes on Pleistocene localities. *International Journal of Primatology*, **14**, 765–795.

Nordborg, M. (1998). On the probability of Neanderthal ancestry. *American Journal of Human Genetics*, **63**, 1237–1240.

Novacek, M. J. (1992*a*). Fossils, topologies, missing data, and the higher level phylogeny of eutherian mammals. *Systematic Biology*, **41**, 58–73.

Novacek, M. J. (1992*b*). Mammalian phylogeny: shaking the tree. *Nature*, **356**, 121–125.

Nowak, R. M. (1999). *Walker's Mammals of the World*. Baltimore, MD: Johns Hopkins University Press.

Oates, J. F. (1994). The natural history of African colobines. In *Colobine Monkeys: Their Ecology, Behavior and Evolution*, eds. A. G. Davies & J. F. Oates, pp. 75–128. Cambridge: Cambridge University Press.

Obermaier, H. (1905). Les restes humains quaternaires dans l'Europe centrale. *L'Anthropologie*, **41**, 385–410.

O'Brien, S. J., Nash, W. G., Wildt, D. E., Bush, M. E., & Benveniste, R. E. (1985). A molecular solution to the riddle of the giant panda's phylogeny. *Nature*, **317**, 140–144.

O'Leary, M. (1996). Dental evolution in the Early Eocene Notharctinae (Primates:Adapiformes) from the Bighorn Basin, Wyoming: documentation of gradual evolution in the oldest true primates. PhD dissertation, Johns Hopkins University.

Olsen, S. J. (1982). An osteology of some Maya mammals. *Papers of the Peabody Museum of Archeology and Ethnology, Harvard University*, **73**, 1-91.

Olson, S. L. & Rasmussen, D. T. (1986). Paleoenvironment of the earliest hominoids: new evidence from the Oligocene avifauna of Egypt. *Science*, **233**, 1202-1204.

Olson, T. R. (1979). Studies on aspects of the morphology and systematics of the genus *Otolemur*, 1859 (Primates: Galagidae). PhD dissertation, University of London.

Olson, T. R. (1986). Species diversity and zoogeography in the Galagidae. *Primate Reports*, **14**, 213.

Orlosky, F. J. (1973). *Comparative Dental Morphology of Extant and Extinct Cebidae*. Ann Arbor, MI: University Microfilms.

Orlosky, F. J. & Swindler, D. R. (1975). Origins of New World monkeys. *Journal of Human Evolution*, **4**, 77–83.

Osborn, H. F. (1895). Fossil mammals of the Uinta Basin: expedition of 1894. *Bulletin of the American Museum of Natural History*, **7**, 71–105.

Osborn, H. F. (1902). American Eocene primates, and the supposed rodent family Mixodectidae. *Bulletin of the American Museum of Natural History*, **16**, 169–214.

Osborn, H. F. (1908). New fossil mammals from the Fayum Oligocene, Egypt. *Bulletin of the American Museum of Natural History*, **24**, 265–272.

Ovchinnikov, I. V., Götherström, A., Romanova, G., Kharitonov, V., Lindén, K., & Goodwin, W. (2000). Molecular analysis of Neanderthal DNA from the Northern Caucasus. *Nature*, **404**, 490–493.

Owen, R. (1845). Notice sur la découverte, faite en Angleterre, de restes fossiles d'un quadrumane du genre macaque, dans une formation d'eau douce appartenant au nouveau Pliocene. *Comptes rendus de l'Académie des sciences de Paris*, **21**, 573–575.

Oxnard, C. E. (1977). Morphometric affinities of the human shoulder. *American Journal of Physical Anthropology*, **46**, 367–374.

Oxnard, C. E. (1981). The uniqueness of *Daubentonia*. *American Journal of Physical Anthropology*, **54**, 1–21.

Oxnard, C. E. (1984). *The Order of Man: A Biomathematical Anatomy of the*

Primates. New Haven, CT: Yale University Press.

Ozansoy, F. (1957). Faunes de mammifères du Tertiaire de Turquie et leurs révisions stratigraphiques. *Bulletin of the Mineral Research and Exploration Institute of Turkey*, **49**, 29–48.

Ozansoy, F. (1965). Étude des gisements continentaux et des mammifères du Cénozoique de Turquie. *Mémoires de al Société geologique de France (nouvelle série)*, **44**, 1–92.

Page, S. L., Chiu, C. H., & Goodman, M. (1999). Molecular phylogeny of Old World Monkeys (Cercopithecidae) as inferred from γ-globin DNA sequences. *Molecular Phylogenetics and Evolution*, **13**, 348–359.

Palmer, A. K., Benefit, B. R., McCrossin, M. L., & Gitau, S. N. (1998). Paleoecological implications of dental microwear analysis for the middle Miocene primate fauna from Maboko Island, Kenya. *American Journal of Physical Anthropology*, Supplement **26**, 175.

Palmer, A. K., Benefit, B. R., & McCrossin, M. L. (1999). Was *Kenyapithecus africanus* a sclerocarp feeder? An exploration of the dietary adaptations of a middle Miocene hominoid through anterior dental microwear analysis. *American Journal of Physical Anthropology*, Supplement **28**, 217.

Palmer, A. K., Benefit, B. R., & McCrossin, M. L. (2000). Does dental microwear analysis confirm or reject dietary predictions based on functional dental morphology? A comparative test case for fossil primate sutilizing the middle Miocene primates from Maboko Island, Kenya. *American Journal of Physical Anthropology*, Supplement **30**, 244.

Pan, Y. (1988). Small fossil primates from Lufeng, a latest Miocene site in Yunnan Province, China. *Journal of Human Evolution*, **17**, 359–366.

Pan, Y. (1996). A small-sized ape from the Xiaohe area hominoid sites, Yuanmou, Yunnan. *Acta Anthropologica Sinica*, **15**, 102–106.

Pan, Y. (1998). Middle-small bodied apes from Neogene in China and their significance. *Acta Anthropologica Sinica*, **17**, 283–292.

Pan, Y. R. & Jablonski, N. G. (1987). The age and geographical distribution of fossil cercopithecids in China. *Human Evolution*, **2**, 59–69.

Pan, Y. R., Peng, Y. Z., Zhang, X. Y., & Pan, R. l. (1992). Cercopithecid fossils discovered in Yunnan and their stratigraphical significance. *Acta Anthropologica Sinica*, **11**, 303–311.

Pan, Y. R., Waddle, D. M., & Fleagle, J. G. (1989). Sexual dimorphism in *Laccopithecus robustus*, a late Miocene hominoid from China. *American Journal of Physical Anthropology*, **79**, 137–158.

Pandey, J. & Sastri, V. V. (1968). On a new species of *Sivapithecus* from the Siwalik rocks of India. *Journal of the Geological Society of India*, **9**, 206–211.

Patterson, B. (1954). The geologic history of non-hominid primates in the Old World. *Human Biology*, **26**, 191–209.

Patterson, B. & Pascual, R. (1972). The fossil mammal fauna of South America. In *Evolution, Mammals, and Southern Continents*, eds. A. Keast, F. C. Erk, & B. Glass, pp. 247–309. Albany, NY: State Univesity of New York Press.

Paula-Couto, C. de (1979). *Tratado de Paleomastozoologia*. Rio de Janeiro: Academia Brasileira de Ciências.

Payseur, B. A., Covert, H. H., Vinyard, C. J., & Dagosto, M. (1999). New body mass estimates for *Omomys carteri*, a middle Eocene primate from North America. *American Journal of Physical Anthropology*, **109**, 41–52.

Pei, W. C. (1935). Fossil mammals from the Kwangsi caves. *Bulletin of the Geological Society of China*, **14**, 413–435.

Pei, W. C. (1957). Discovery of *Gigantopithecus* mandibles and other material in Liu-Cheng district of central Kwangsi in South China. *Vertebrata PalAsiatica*, **1**, 65–72.

Pei, W. C. & Li, Y. H. (1959). Discovery of a third mandible of *Gigantopithecus* in Liu-Cheng, Kwangsi, South China. *Vertebrata PalAsiatica*, **2**, 193–200.

Pei, W. C. & Woo, J. K. (1956). New materials of *Gigantopithecus* teeth from South China. *Acta Palaeontologica*, **4**, 477–490.

Penela, A. M. (1983). Presence of the genus *Macaca* in the Pleistocene site of the Solana del Zamborino (Fonelas, Granada, Spain): preliminary study. *Boletín de la real Sociedad española de historia Natural, Sección geológia*, **81**, 187–195.

Peng, Y. Z., Pan, R. L., & Jablonski, N. G. (1993). Classification and evolution of Asian colobines. *Folia Primatologica*, **60**, 107–117.

Perret, M. (1995). Chemocommunication in the reproductive function of mouse lemurs. In *Creatures of the Dark: The Nocturnal Prosimians*, eds. L. Alterman, G. A. Doyle, & M. K. Izard, pp. 377–392. New York: Plenum Press.

Petter, J. J. & Petter-Rousseaux, A. (1979). Classification of the Prosimians. In *The Study of Prosimian Behavior*, eds. G. A. Doyle & R. D. Martin, pp. 1–44. New York: Academic Press.

Pettigrew, J. D. (1989). Phylogenetic relationships between [sic] microbats, megabats and primates (Mammalia: Chiroptera and Primates). *Philosophical Transactions of the Royal Society of London, Series B*, **325**, 489–559.

Pettigrew, J. D. (1991). Wings or brain? Convergent evolution in the origins of bats. *Systematic Biology*, **40**, 231–239.

Phillips, C. J., & Jones, J. K., Jr. (1968). Additional comments on reproduction in the woolly opossum (*Caluromys derbianus*) in Nicaragua. *Journal of Mammalogy*, **49**, 320–321.

Phillips, E. M. & Walker, A. C. (2000). A new species of fossil lorisid from the Miocene of East Africa. *Primates*, **41**, 367–372.

Pickford, M. (1981). Preliminary Miocene mammalian biostratigraphy for western Kenya. *Journal of Human Evolution*, **10**, 73–97.

Pickford, M. (1982). New higher primate fossils from the Middle Miocene deposits at Majiwa and Kaloma, western Kenya. *American Journal of Physical Anthropology*, **58**, 1–19.

Pickford, M. (1983). Sequence and environment of the lower and middle Miocene hominoids of western Kenya. In *New Interpretations of Ape and Human Ancestry*, eds. R. L. Ciochon & R. S. Corruccini, pp. 421–440. New York: Plenum Press.

Pickford, M. (1985). A new look at *Kenyapithecus* based on recent discoveries in western Kenya. *Journal of Human Evolution*, **14**, 113–143.

Pickford, M. (1986*a*). Cainozoic paleontological sites of western Kenya. *Müncher geowissenschaftliche Abhandlungen*, (*a*), **8**, 1–151.

Pickford, M. (1986*b*). Geochronology of the Hominoidea: a summary. In *Primate Evolution*, eds. J. G. Else & P. C. Lee, pp. 123–128. Cambridge: Cambridge University Press.

Pickford, M. (1986*c*). Première decouverte d'une faune mammalienne terrestre paléogene d'Afrique sub-saharienne. *Comptes rendus de l'Académie des sciences de Paris*, **302**, 1205–1210.

Pickford, M. (1986*d*). Sexual dimorphism in *Proconsul*. *Human Evolution*, **1**, 111–148.

Pickford, M. (1987). The chronology of the Cercopithecoidea of East Africa. *Human Evolution*, **2**, 1–17.

Pickford, M. (1993*a*). Climatic change, biogeography, and *Theropithecus*. In Theropithecus: *The Rise and Fall of a Primate Genus*, ed. N. G. Jablonski, pp. 227–243. Cambridge: Cambridge University

Press.

Pickford, M. (1993*b*). Old World suoid systematics, phylogeny, biogeography, and biostratigraphy. *Paleontologia i evolució*, **26–27**, 237–269.

Pickford, M. (1998). A new genus of Tayassuidae (Mammalia) from the middle Miocene of Uganda and Kenya. *Annales de paléontologie*, **84**, 275–285.

Pickford, M. & Andrews, P. J. (1981). The Tinderet Miocene sequence in Kenya. *Journal of Human Evolution*, **10**, 11–33.

Pickford, M., Moyà-Solà, S., & Köhler, M. (1997). Phylogenetic implications of the first African Middle Miocene hominoid frontal bone from Otavi, Namibia. *Comptes rendus de l'Académie des sciences de Paris*, **325**, 459–466

Pickford, M. & Senut, B. (2001). The geological and faunal context of Late Miocene hominid remains from Lukeino, Kenya. *Comptes rendus de l'Académie des sciences de Paris*, **332**, 145–152.

Pickford, M., Senut, B., & Gommery, D. (1999). Sexual dimorphism in *Morotopithecus bishopi*, an early Middle Miocene hominoid from Uganda, and a reassessment of its geological and biological contexts. In *Late Cenozoic Environments and Hominid Evolution: A Tribute to Bill Bishop*, eds. P. Andrews & P. Banham, pp. 27–38. London: Geological Society.

Pickford, M., Senut, B., Hadoto, D., Musisi, J. & Kariira, C. (1986*a*). Nouvelles découvertes dans le Miocène inférieur de Napak, Ouganda oriental. *Comptes rendus de l'Académie des sciences de Paris*, **302**, 47–52.

Pickford, M., Senut, B., Hadoto, D., Musisi, J., & Kariira, C. (1986*b*). Découvertes récentes dans les sites Miocènes de Moroto (Ouganda oriental): aspects biostratigraphiques et paléoécologiques. *Comptes rendus de l'Académie des sciences de Paris*, **302**, 681–686.

Pilbeam, D. R. (1969). Tertiary Pongidae of East Africa: evolutionary relationships and taxonomy. *Bulletin of the Peabody Museum of Natural History*, **31**, 1–185.

Pilbeam, D. R. (1970). *Gigantopithecus* and the origins of Hominidae. *Nature*, **225**, 1093–1094.

Pilbeam, D. R. (1972). *The Ascent of Man*. New York: Macmillan.

Pilbeam, D. R. (1979). Recent finds and interpretations of Miocene hominoids. *Annual Review of Anthropology*, **8**, 333–353.

Pilbeam, D. R. (1982). New hominoid skull material from the Miocene of Pakistan. *Nature*, **295**, 232–234.

Pilbeam, D. R. (1983). Hominoid evolution and hominid origins. In *Recent Advances in the Evolution of Primates*, ed. C. Chagas, pp. 43–61. Vatican City: Pontificiae Scripta Varia.

Pilbeam, D. R. (1986). Distinguished lecture: Hominoid evolution and hominid origins. *American Anthropologist*, **88**, 295–312.

Pilbeam, D. R. (1996). Genetic and morphological records of the Hominoidea and hominid origins: a synthesis. *Molecular Phylogenetics and Evolution*, **5**, 155–168.

Pilbeam, D. R. (1997). Research on Miocene hominoids and hominid origins: the last three decades. In *Function, Phylogeny and Fossils: Miocene Hominoid Evolution and Adaptation*, eds. D. R. Begun, C. V. Ward, & M. D. Rose, pp. 13–28. New York: Plenum Press.

Pilbeam, D. R., Meyer, G. E., Badgley, C., Rose, M. D., Pickford, M., Behrensmeyer, A. K., & Shah, S. M. I. (1977). New hominoid primates from the Siwaliks of Pakistan and their bearing on hominoid evolution. *Nature*, **270**, 689–695.

Pilbeam, D. R., Rose, M. D., Badgley, C., & Lipschutz, B. (1980). Miocene hominoids from Pakistan. *Postilla*, **181**, 1–94.

Pilbeam, D. R., Rose, M. D., Barry, J. C., & Shah, S. M. I. (1990). New *Sivapithecus* humeri from Pakistan and the relationship of *Sivapithecus* and *Pongo*. *Nature*, **348**, 237–239.

Pilbeam, D. R. & Simons, E. L. (1971). Humerus of *Dryopithecus* from Saint-Gaudens, France. *Nature*, **229**, 408–409.

Pilbeam, D. R. & Walker, A. C. (1968). Fossil monkeys from the Miocene of Napak, north-east Uganda. *Nature*, **220**, 657–660.

Pilgrim, G. E. (1910). Notices of new mammalian genera and species from the Tertiaries of India. *Records of the Geological Survey of India*, **40**, 63–71.

Pilgrim, G. E. (1915). New Siwalik primates and their bearing on the question of the evolution of man and the Anthropoidea. *Records of the Geological Survey of India*, **45**, 1–74.

Pilgrim, G. E. (1927). A *Sivapithecus* palate and other primate fossils from India. *Memoirs of the Geological Survey of India (Palaeontologica Indica)*, **14**, 1–26.

Pilgrim, G. E. (1932). The fossil Carnivora of India. *Memoirs of the Geological Survey of India (Palaeontologica Indica)*, **18**, 1–232.

Pinto da Silveira, E. K. (1985). O subgenero *A.* (*Brachyteles*) e a evolucao dos Atelini Gray, 1825. *Brasil Departamento nacional da producão mineral, Serie geologia, Secão paleontologia e estratigrafia*, **27**, 179–182.

Piveteau, J. (1950). Recherches anatomique sur l'encéphale de lémuriens disparus. *Annales de paléontologie*, **36**, 87–103.

Piveteau, J. (1956). L'encéphale d'*Hadropithecus*, lémurien subfossile de Madagascar. *Annales de paléontologie*, **42**, 141–150.

Piveteau, J. (1957). *Traité de paléontologie*, vol. 7, *Primates*. Paris: Masson.

Piveteau, J. (1961). Behavior and ways of life of the fossil primates. In *Social Life of Early Man*, ed. S. L. Washburn, pp. 11–16. Chicago, IL: Aldine.

Plavcan, J. M. (1993). Catarrhine dental variability and species recognition in the fossil record. In *Species, Species Concepts and Primate Evolution*, eds. W. H. Kimbel & L. B. Martin, pp. 239–263. New York: Plenum Press.

Pocock, R. I. (1918). On the external characters of the lemurs and of *Tarsius*. *Proceedings of the Zoological Society of London*, **1918**, 19–53.

Pohlig, H. (1892). *Dryopithecus*. *Sitzungsberichte niederrheinische Gesellschaft für Natur- und Heilkunde*, **1891**, 42–43.

Pohlig, H. (1895). *Paidopithex rhenanus*, n.g.n.s., le singe anthropomorphe du Pliocene rhenan. *Bulletin de la Société belge de géologie, de paléontologie et d'hydrologie*, **9**, 149–151.

Pollock, J. I. (1987). Vitamin C biosynthesis in prosimians: evidence for the anthropoid affinities of *Tarsius*. *American Journal of Physical Anthropology*, **73**, 65–70.

Pope, G. G. (1992). Craniofacial evidence for the origin of modern humans in China. *Yearbook of Physical Anthropology*, **35**, 243–298.

Pope, G. G. (1997). Paleoanthropological research traditions in the Far East. In *Conceptual Issues in Modern Human Origins Research*, eds. G. Clark & K. Willermet, pp. 269–282. New York: Aldine de Gruyter.

Porter, C. A., Czelusniak, J., Schneider, H., Schneider, M. P. C., Sampaio, I., & Goodman, M. (1999). Sequences from the 5' flanking region of the ε-globin gene support the relationship of *Callicebus* with the pitheciins. *American Journal of Primatology*, **48**, 69–75.

Porter, C. A., Page, S. L., Czelusniak, J., Schneider, H., Schneider, M. P. C., Sampaio, I., & Goodman, M. (1997). Phylogeny and evolution of selected primates as determined by sequences of the ε-globin locus and 5' flanking regions. *International Journal of Primatology*, **18**, 261–295.

Prasad, K. N. (1962). Fossil primates from the Siwalik beds near Haritalyangar, Himachal Pradesh, India. *Journal of the Geological Society of India*, **3**, 86–96.

Prasad, K. N. (1964). Upper Miocene anthropoids from the Siwalik beds of Haritalyangar, Himachal Pradesh, India. *Palaeontology*, **7**, 124–134.

Preuschoft, H. (1971). Mode of locomotion in subfossil giant lemuroids from Madagascar. *Proceedings of the 3rd International Congress of Primatology*, Zurich 1970, **1**, 79–90.

Preuschoft, H. (1973). Body posture and locomotion in some East African Miocene Dryopithecinae. In *Human Evolution*, ed. M. Day, pp. 13–46. London: Taylor & Francis.

Qi, T. & Beard, K. C. (1998). Late Eocene sivaladapid primate from Guangxi Zhuang Autonomous Region, People's Republic of China. *Journal of Human Evolution*, **35**, 211–220.

Qiu, Z. (1990). The Chinese Neogene mammalian biochronology: its correlation with the European Neogene. In *European Neogene Mammal Chronology*, eds. E. H. Lindsay, V. Fahlbusch, & P. Mein, pp. 527–556. New York: Plenum Press.

Qiu, Z. & Qiu, Z. (1995). Chronological sequence and subdivision of Chinese Neogene mammalian faunas. *Palaeogeography, Palaeoclimatology, Palaeoecology*, **116**, 41–70.

Qiu, Z. & Storch, G. (1990). New murids (Mammalia: Rodentia) from the Lufeng hominoid locality, late Miocene of China. *Journal of Vertebrate Paleontology*, **10**, 467–472.

Qiu, Z., Wu, W., & Qiu, Z. (1999). Miocene mammal faunal sequence of China: palaeozoogeography and Eurasian relationships. In *The Miocene Land Mammals of Europe: The Continental European Miocene*, eds. G. E. Rössner & K. Heissig, pp. 443–455. Munich: Dr. Friedrich Pfeil.

Quade, J., Cerling, T. E., Andrews, P., & Alpagut, B. (1995). Paleodietary reconstruction of Miocene faunas from Paşalar, Turkey using stable carbon and oxygen isotopes of fossil tooth enamel. *Journal of Human Evolution*, **28**, 373–384.

Quam, R. & Smith, F. H. (1998). A reassessment of the Tabun C2 mandible. In *Neandertals and Modern Humans in Western Asia*, eds. T. Akazawa, K. Aoki, & O. Bar-Yosef, pp. 405–421. New York: Plenum Press.

Quatrefages de Breau, J. L. A. & Hamy, J. T. E. (1882). *Craina ethnica: Les Crânes des races humains*. Paris: Ballière.

Radinsky, L. B. (1970). The fossil evidence of prosimian brain evolution. In *The Primate Brain*, eds. C. R. Noback & W. Montagna, pp. 209–224. New York: Appleton-Century-Crofts.

Radinsky, L. (1973). *Aegyptopithecus* endocast: oldest record of a pongid brain. *American Journal of Physical Anthropology*, **39**, 239–247.

Radinsky, L. B. (1975). Primate brain evolution. *American Scientist*, **63**, 656–663.

Rae, T. C. (1997). The early evolution of the hominoid face. In *Function, Phylogeny and Fossils: Miocene Hominoid Evolution and Adaptation*, eds. D. R. Begun, C. V. Ward, & M. D. Rose, pp. 59–77. New York: Plenum Press.

Rae, T. C. (1999). Mosaic evolution in the origin of the Hominoidea. *Folia Primatologica*, **70**, 125–135.

Rafferty, K. L. (1990). The functional and phylogenetic significance of the carpometacarpal joint of the thumb in anthropoid primates. MA thesis, New York University.

Rafferty, K. L., Walker, A. C., Ruff, C., Rose, M. D. & Andrews, P. J. (1995). Postcranial estimates of body weights in *Proconsul*, with a note on a distal tibia of *P. major* from Napak, Uganda. *American Journal of Physical Anthropology*, **97**, 391–402.

Rak, Y. (1983). *The Australopithecine Face*. New York: Academic Press.

Rak, Y. (1993). Morphological variation in *Homo neanderthalensis* and *Homo sapiens* in the Levant: a biogeographic analysis. In *Species, Species Concepts, and Primate Evolution*, eds. W. Kimbel & L. Martin, pp. 523–536. New York: Plenum Press.

Rak, Y. (1998). Does any Mousterian cave present evidence of two hominid species? In *Neandertals and Modern Humans in Western Asia*, eds. T. Akazawa, K. Aoki, & O. Bar-Yosef, pp. 353–366. New York: Plenum Press.

Rakotoarisoa, J. A. (1997). A cultural history of Madagascar: evolution and interpretation of the archaeological evidence. In *Natural Change and Human Impact in Madagascar*, eds. S. M. Goodman & B. D. Patterson, pp. 331–341. Washington, DC: Smithsonian Institution Press.

Rasmussen, D. T. (1986). Anthropoid origins: a possible solution to the Adapidae–Omomyidae paradox. *Journal of Human Evolution*, **15**, 1–12.

Rasmussen, D. T. (1990*a*). Primate origins: lessons from a Neotropical marsupial. *American Journal of Primatology*, **22**, 263–278.

Rasmussen, D. T. (1990*b*). The phylogenetic position of *Mahgarita stevensi*: protoanthropoid or lemuroid? *International Journal of Primatology*, **11**, 439–469.

Rasmussen, D. T. (1994). The different meanings of a tarsioid–anthropoid clade and a new model of anthropoid origin. In *Anthropoid Origins*, eds. J. G. Fleagle & R. F. Kay, pp. 335–360. New York: Plenum Press.

Rasmussen, D. T. (1996). A new Middle Eocene omomyine primate from the Uinta Basin, Utah. *Journal of Human Evolution*, **31**, 75–87.

Rasmussen, D. T., Bown, T. M. & Simons, E. L. (1992). The Eocene–Oligocene transition in continental Africa. In *Eocene–Oligocene Climatic and Biotic Evolution*, eds. D. R. Prothero & W. A. Berggren, pp. 548–566. Princeton, NJ: Princeton University Press.

Rasmussen, D. T., Conroy, G. C., & Simons, E. L. (1998). Tarsier-like locomotor specializations in the Oligocene primate *Afrotarsius*. *Proceedings of the National Academy of Sciences of the United States of America*, **95**, 14848–14850.

Rasmussen, D. T. & Nekaris, K. A. (1998). Evolutionary history of lorisiform primates. *Folia Primatologica*, **69**, 250–285.

Rasmussen, D. T., Olson, S. L., & Simons, E. L. (1987). Fossil birds from the Oligocene Jebel Qatrani Formation, Fayum Province. Egypt. *Smithsonian Contributions to Paleobiology*, **62**, 1–18.

Rasmussen, D. T. & Simons, E. L. (1988). New specimens of *Oligopithecus savagei*, early Oligocene primate from the Fayum, Egypt. *Folia Primatologica*, **51**, 182–208.

Rasmussen, D. T. & Simons, E. L. (1992). Paleobiology of the oligopithecines, the earliest known anthropoid primates. *International Journal of Primatology*, **13**, 477–508.

Rasoloharijaona, S. (1999). Contribution à l'étude du genre *Archaeolemur* sp. (Archaeolemuridae): un lémurien subfossil provenant de la région de l'Ankarana: essai de reconstitution du paléoenvironnement de la région de l'Ankarana. *Lemur News*, **4**, 7–10.

Ravololonarivo, G. (1990). Contribution à l'étude de la colonne vertébrale du genre *Pachylemur* (Lamberton, 1946): anatomie et analyse cladistique. 3rd Cycle thesis, University of Antananarivo, Madagascar.

Ravosa, M. J. (1991). Structural allometry of the prosimian mandibular corpus and symphysis. *Journal of Human Evolution*, **20**, 3–20.

Ravosa, M. J. & Hylander, W. L. (1994). Function and fusion of the mandibular symphysis in primates: stiffness or strength? In *Anthropoid Origins*, eds. J. G. Fleagle & R. F. Kay, pp. 447–468. New York: Plenum Press.

Ravosa, M. J. & Simons, E. L. (1994). Mandibular growth and function in *Archaeolemur*. *American Journal of Physical Anthropology*, **95**, 63–76.

Raza, S. M., Barry, J. C., Pilbeam, D., Rose, M. D., Shah, S. M. I., & Ward, S. C. (1983). New hominoid primates from the middle Miocene Chinji Formation, Potwar Plateau, Pakistan. *Nature*, **306**, 52–54.

Reed, K. E. (1997). Early hominid evolution and ecological change through the African Plio-Pleistocene. *Journal of Human Evolution*, **32**, 289–322.

Reeser, L. A. (1984). Morphological affinities of new fossil talus of *Dolichocebus gaimanensis*. *American Journal of Physical Anthropology*, **63**, 206–207.

Relethford, J. (1998). Genetics of modern human origins and diversity. *Annual Reviews of Anthropology*, **27**, 1–23.

Relethford, J. (1999). Models, predictions and the fossil record of modern human origins. *Evolutionary Anthropology*, **8**, 7–10.

Relethford, J. (2001*a*). Ancient DNA and the origin of modern humans. *Proceedings of the National Academy of Sciences of the United States of America*, **98**, 390–391.

Relethford, J. (2001*b*). Regional affinities of Neandertal DNA with living humans do not reject multiregional evolution. *American Journal of Physical Anthropology*, **115**, 95–98.

Relethford, J. & Jorde, L. (1999). Genetic evidence for larger African population size during recent human evolution. *American Journal of Physical Anthropology*, **108**, 251–260.

Remane, A. (1921). Zur Beurteilung der fossilen Anthropoiden. *Zentralblatt für Mineralogie Geologie und Paläontologie*, **14**, **11**, 335–339.

Remane, A. (1925). Der fossile Pavian (*Papio* Sp.) von Oldoway nebst Bemerkungen über die Gattung *Simopithcus* C. W. Andrews. *Wissenschaftliche Ergebnisse der Oldoway Expedition*, **2**, 83–90.

Remane, A. (1955). Ist *Oreopithecus* ein Hominide? *Abhnadlungen Akademie der Wissenschaften und der Literatur, mathematisch-naturwissenschaftliche Klasse, Mainz*, **1955**, 467–497.

Remane, A. (1965). Die Geschichte der Menschenaffen. In *Menschliche Abstammungslehre*, ed. G. Heberer, pp. 249–309. Göttingen: Gustav Fischer.

Remis, M. J. (1997). Western lowland gorillas (*Gorilla gorilla gorilla*) as seasonal frugivores: use of variable resources. *American Journal of Physical Anthropology*, **43**, 87–109.

Retallack, G. J. (1992). Middle Miocene fossil plants from Fort Ternan (Kenya) and evolution of African grasslands. *Paleobiology*, **18**, 383–400.

Richmond, B. G., Fleagle, J. G., Kappelman, J., & Swisher, C. C. (1998). First hominoid from the Miocene of Ethiopia and the evolution of the catarrhine elbow. *American Journal of Physical Anthropology*, **105**, 257–277.

Richmond, B. G. & Strait, D. S. (2000). Evidence that humans evolved from a knuckle-walking ancestor. *Nature*, **404**, 382–385.

Rightmire, G. P. (1976). Relationships of Middle and Upper Pleistocene hominids from sub-Saharan Africa. *Nature*, **260**, 238–240.

Rightmire, G. P. (1979). Cranial remains of *Homo erectus* from Beds II and IV, Olduvai Gorge, Tanzania. *American Journal of Physical Anthropology*, **51**, 99–115.

Rightmire, G. P. (1984). *Homo sapiens* in sub-Saharan Africa. In *The Origins of Modern Humans: A World Survey of the Fossil Evidence*, eds. F. H. Smith & F. Spencer, pp. 295–325. New York: Alan R. Liss.

Rightmire, G. P. (1986). Species recognition and *Homo erectus*. *Journal of Human Evolution*, **15**, 823–826.

Rightmire, G. P. (1990). *The Evolution of* Homo erectus: *Comparative Anatomical Studies of an Extinct Species*. Cambridge: Cambridge University Press.

Rightmire, G. P. (1996). The human cranium from Bodo, Ethiopia: evidence for speciation in the Middle Pleistocene? *Journal of Human Evolution*, **31**, 21–39.

Rightmire, G. P. (1998*a*). Evidence from facial morphology for similarity of Asian and African representatives of *Homo erectus*. *American Journal of Physical Anthropology*, **106**, 61–85.

Rightmire, G. P. (1998*b*). Human evolution in the Middle Pleistocene: the role of *Homo heidelbergensis*. *Evolutionary Anthropology*, **6**, 218–227.

Rightmire, G. P. & Deacon, H. J. (1991). Comparative studies of late Pleistocene human remains from Klasies River, South Africa. *Journal of Human Evolution*, **20**, 131–150.

Rilling, J. K. & Insel, T. R. (1999). The primate neocortex in comparative perspective using magnetic resonance imaging. *Journal of Human Evolution*, **37**, 191–223.

Rímoli, R. (1977). Una nueva especie de mono (Cebidae: Saimirinae: *Saimiri*) de La Hispaniola. *Cuadernos de CENDIA, Universidad Autónoma de Santo Domingo*, **242**, 5–14.

Ristori, G. (1890). Le scimmie fossile italiane. *Bollettino del reale Comitato geolico d'Italia* **XXI, 178–196**, 225–237.

Rivero de la Calle, M. & Arredondo, O. (1991). *Paralouatta varonai*, a new Quaternary platyrrhine from Cuba. *Journal of Human Evolution*, **21**, 1–11.

Robinson, J. T. (1954). The genera and species of the Australopithecines. *American Journal of Physical Anthropology*, **12**, 181–200.

Robinson, J. T. (1956). The dentition of the Australopithecinae. *Transvaal Museum Memoir*, **9**.

Robinson, J. T. (1965). *Homo habilis* and the australopithecines. *Nature*, **205**, 121–124.

Robinson, J. T. (1972). *Early Hominid Posture and Locomotion*. Chicago, IL: University of Chicago Press.

Robinson, P. (1966). Fossil Mammalia of the Huerfano Formation, Eocene, of Colorado. *Bulletin of the Peabody Museum of Natural History*, **21**, 1–95.

Robinson, P. (1968). The paleontology and geology of the Badwater Creek area, central Wyoming. IV. Late Eocene primates from Badwater, Wyoming, with a discussion of material from Utah. *Annals of the Carnegie Museum*, **39**, 307–326.

Roebroeks, W. & van Kolfschoten, T. (1994). The earliest occupation of Europe: a short chronology. *Antiquity*, **68**, 489–503.

Roger, J., Sen, S., Thomas, H., Cavelier, C., & Al-Sulaimani, Z. (1993). Stratigraphic, palaeomagnetic and palaeoenvironmental study of the Early Oligocene vertebrate locality of Taqah (Dhofar, Sultanate of Oman). *Newsletters on Stratigraphy*, **28**, 93–119.

Roger, O. (1898). Wirbeltierreste aus dem Dinotheriensande der bayerisch-schwäbischen Hochebene. *Bericht des naturwissenschaftlichen Vereins für Schwaben und Neuburg in Augsburg*, **33**, 1–46.

Rögl, V. F., Zapfe, H., Bernor, R. L., Brzobohaty, R. L., Daxner-

Höck, G., Draxler, I., Fehar, O., Gaudant, J., Hermann, P., Rabeder, G., Schultz, O., & Zetter, R. (1993). Die Primatenfundstelle Götzendorf an der Leitha (Obermiozän des Winer Beckens, Niederösterreich). *Jahrbuch für Geologie*, **136**, 503–526.

Rögl, F. (1999). Circum-Mediterranean Miocene paleogeography. In *The Miocene Land Mammals of Europe: The Continental European Miocene*, eds. G. E. Rössner & K. Heissig, pp. 39–48. Munich: Dr. Friedrich Pfeil.

Rook, L. (1999). Late Turolian *Mesopithecus* (Mammalia, Primates, Colobinae) from Italy. *Journal of Human Evolution*, **36**, 535–547.

Rook, L., Bondioli, L., Köhler, M., Moyà-Solà, S., & Macchiarelli, R. (1999). *Oreopithecus* was a bipedal ape after all: evidence from the iliac cancellous architecture. *Proceedings of the National Academy of Sciences of the United States of America*, **96**, 8795–8799.

Rook, L., Harrison, T., & Engesser, B. (1996). The taxonomic status and biochronological implications of new finds of *Oreopithecus* from Baccinello (Tuscany, Italy). *Journal of Human Evolution*, **30**, 3–27.

Rook, L., Mottura, A., & Gentili, S. (2001). Fossil *Macaca* remains from RDB quarry (Villafranca d'Asti, Italy): new data and overview. *Journal of Human Evolution*, **40**, 187–202.

Rosas, A. (1997). A gradient of size and shape for the Atapuerca sample and Middle Pleistocene hominid variability. *Journal of Human Evolution*, **33**, 319–331.

Rosas, A. & de Castro, J. M. B. (1998). On the taxonomic affinities of the Dmanisi mandible (Georgia). *American Journal of Physical Anthropology*, **107**, 145–162.

Rose, K. D. (1995*a*). The earliest primates. *Evolutionary Anthropology*, **3**, 159–173.

Rose, K. D. (1995*b*). Anterior dentition and relationships of the early Eocene omomyids *Arapahovius advena* and *Teilhardina demissa*, sp. nov. *Journal of Human Evolution*, **28**, 231–244.

Rose, K. D. & Bown, T. M. (1984). Early Eocene *Pelycodus jarrovii* (Primates: Adapidae) from Wyoming: phylogenetic and biostratigraphic implications. *Journal of Paleontology*, **58**, 1532–1535.

Rose, K. D. & Bown, T. M. (1991). Additional fossil evidence on the differentiation of the earliest euprimates. *Proceedings of the National Academy of Sciences of the United States of America*, **88**, 98–101.

Rose, K. D. & Fleagle, J. G. (1981). The fossil history of nonhuman primates in the Americas. In *Ecology and Behavior of Neotropical Primates*, vol. 1, eds. A. F. Coimbra-Filho & R. A. Mittermeier, pp. 111–167. Rio de Janeiro: Academia Brasileira de Ciências.

Rose, K. D., Godinot, M., & Bown, T. M. (1994). The early radiation of euprimates and the initial diversification of Omomyidae. In *Anthropoid Origins*, eds. J. G. Fleagle & R. F. Kay, pp. 1–28. New York: Plenum Press.

Rose, K. D. & Krause, D. W. (1984). Affinities of the primate *Altanius* from the Early Tertiary of Mongolia. *Journal of Mammalogy*, **65**, 721–726.

Rose, K. D., MacPhee, R. D. E., & Alexander, J. P. (1999). Skull of early Eocene *Cantius abditus* (Primates: Adapiformes) and its phylogenetic implications, with a reevaluation of "*Hesperolemur*" *actius*. *American Journal of Physical Anthropology*, **109**, 523–539.

Rose, K. D. & Rensberger, J. M. (1983). Upper dentition of *Ekgmowechashala* (Omomyidae, Primates) from the John Day Formation, Oligo-Miocene of Oregon. *Folia Primatologica*, **41**, 102–111.

Rose, K. D. & Walker, A. C. (1985). The skeleton of early *Cantius*, oldest lemuriform primate. *American Journal of Physical Anthropology*, **66**, 73–89.

Rose, M. D. (1983). Miocene hominoid postcranial morphology: monkey-like, ape-like, neither, or both? In *New Interpretations of Ape and Human Ancestry*, eds. R. L. Ciochon & R. S. Corruccini, pp. 405–420. New York: Plenum Press.

Rose, M. D. (1984). Hominoid specimens from the middle Miocene Chinji Formation, Pakistan. *Journal of Human Evolution*, **13**, 503–516.

Rose, M. D. (1986). Further hominoid postcranial specimens from the late Miocene Nagri Formation of Pakistan. *Journal of Human Evolution*, **15**, 333–367.

Rose, M. D. (1988). Another look at the anthropoid elbow. *Journal of Human Evolution*, **17**, 193–224.

Rose, M. D. (1989). New postcranial specimens of catarrhines from the middle Miocene Chinji Formation, Pakistan: descriptions and a discussion of proximal humeral functional morphology in anthropoids. *Journal of Human Evolution*, **18**, 131–162.

Rose, M. D. (1992). Kinematics of the trapezium–1st metacarpal joint in extant anthropoids and Miocene hominoids. *Journal of Human Evolution*, **22**, 255–266.

Rose, M. D. (1993*a*). Functional anatomy of the elbow and forearm in primates. In *Postcranial Adaptation in Nonhuman Primates*, ed. D. L. Gebo, pp. 175–198. DeKalb, IL: Northern Illinois University Press.

Rose, M. D. (1993*b*). Locomotor anatomy of Miocene hominoids. In *Postcranial Adaptation in Nonhuman Primates*, ed. D. L. Gebo, pp. 252–272. DeKalb, IL: Northern Illinois University Press.

Rose, M. D. (1994). Quadrupedalism in some Miocene catarrhines. *Journal of Human Evolution*, **26**, 387–411.

Rose, M. D. (1997). Functional and phylogenetic features of the forelimb in Miocene hominoids. In *Function, Phylogeny and Fossils: Miocene Hominoid Evolution and Adaptation*, eds. D. R. Begun, C. V. Ward, & M. D. Rose, pp. 79–100. New York: Plenum Press.

Rose, M. D., Leakey, M. G., Leakey, R. E. F., & Walker, A. C. (1992). Postcranial specimens of *Simiolus enjiessi* and other primitive catarrhines from the early Miocene of Lake Turkana, Kenya. *Journal of Human Evolution*, **22**, 171–237.

Rose, M. D., Nakano, Y., & Ishida, H. (1996). *Kenyapithecus* postcranial specimens from Nachola, Kenya. *African Studies Monographs*, Supplementary Issue **24**, 3–56.

Rosenberg, K. (1998). Morphological variation in West Asian postcrania: implications for obstetric and locomotor behavior. In *Neandertals and Modern Humans in West Asia*, eds. T. Akazawa, K. Aoki, & O. Bar-Yosef, pp. 367–379. New York: Plenum Press.

Rosenberger, A. L. (1977). *Xenothrix* and ceboid phylogeny. *Journal of Human Evolution*, **6**, 461–481.

Rosenberger, A. L. (1978). New species of Hispaniolan monkey. *Anales cientificas, Universidade Central Este de Dominican Republic*, **3**, 249–251.

Rosenberger, A. L. (1979*a*). Cranial anatomy and implications of *Dolichocebus*, a late Oligocene ceboid primate. *Nature*, **279**, 416–418.

Rosenberger, A. L. (1979*b*). *Phylogeny, Evolution and Classification of New World Monkeys (Platyrrhini, Primates)*. Ann Arbor, MI: University Microfilms.

Rosenberger, A. L. (1980). Gradistic views and adaptive radiation of platyrrhine primates. *Zeitschrift für Morphologie und Anthropologie*, **71**, 157–163.

Rosenberger, A. L. (1981*a*). A mandible of *Branisella boliviana* (Platyrrhini, Primates) from the Oligocene of South America. *Interna-*

tional Journal of Primatology, **2**, 1–7.

Rosenberger, A. L. (1981*b*). Systematics: the higher taxa. In *Ecology and Behavior of Neotropical Primates*, vol. 1, eds. A. F. Coimbra-Filho & R. A. Mittermeier, pp. 9–27. Rio de Janeiro: Academia Brasileira de Ciências.

Rosenberger, A. L. (1982). Supposed squirrel monkey affinities of the late Oligocene *Dolichocebus gaimanensis*. *Nature*, **298**, 202.

Rosenberger, A. L. (1983). Aspects of the systematics and evolution of marmosets. In *A Primatologia no Brasil*, ed. M. T. de Mello, Brasilia: Universidad federal Districto federal.

Rosenberger, A. L. (1984). Fossil New World monkeys dispute the molecular clock. *Journal of Human Evolution*, **13**, 737–742.

Rosenberger, A. L. (1985). In favor of the *Necrolemur*–tarsier hypothesis. *Folia Primatologica*, **45**, 179–194.

Rosenberger, A. L. (1992). Evolution of feeding niches in New World monkeys. *American Journal of Physical Anthropology*, **88**, 525–562.

Rosenberger, A. L. & Dagosto, M. (1992). New craniodental and postcranial evidence of fossil tarsiiforms. In *Topics in Primatology*, vol. 3, eds. R. H. Tuttle, H. Ishida, & M. Goodman, pp. 37–51. Kyoto: University of Kyoto Press.

Rosenberger, A. L., Hartwig, W. C., Takai, M., Setoguchi, T., & Shigehara, N. (1991*a*). Dental variability in *Saimiri* and the taxonomic status of *Neosaimiri fieldsi*, an early squirrel monkey from Colombia, South America. *International Journal of Primatology*, **12**, 291–301.

Rosenberger, A. L., Hartwig, W. C., & Wolff, R. G. (1991*b*). *Szalatavus attricuspis*, an early platyrrhine primate. *Folia Primatologica*, **56**, 225–233.

Rosenberger, A. L., Setoguchi, T., & Hartwig, W. C. (1991*c*). *Laventiana annectens*, new genus and species: fossil evidence for the origins of callitrichine monkeys. *Proceedings of the National Academy of Sciences of the United States of America*, **88**, 2137–2140.

Rosenberger, A. L., Setoguchi, T., & Shigehara, N. (1990). The fossil record of callitrichine primates. *Journal of Human Evolution*, **19**, 209–236.

Rosenberger, A. L. & Strasser, E. (1988). Toothcomb origins: support for the grooming hypothesis. *Primates*, **26**, 73–84.

Rosenberger, A. L., Strasser, E., & Delson, E. (1985). Anterior dentition of *Notharctus* and the adapid–anthropoid hypothesis. *Folia Primatologica*, **44**, 15–39.

Rosenberger, A. L. & Strier, K. B. (1989). Adaptive radiation of the ateline primates. *Journal of Human Evolution*, **18**, 717–750.

Rosenberger, A. L. & Szalay, F. S. (1980). On the tarsiiform origins of the Anthropoidea. In *Evolutionary Biology of the New World Monkeys and Continental Drift*, eds. R. L. Ciochon & A. B. Chiarelli, pp. 139–157. New York: Plenum Press.

Ross, C. F. (1994). The craniofacial evidence for anthropoid and tarsier relationships. In *Anthropoid Origins*, eds. J. G. Fleagle & R. F. Kay, pp. 469–547. New York: Plenum Press.

Ross, C. F., Williams, B., & Kay, R. F. (1998). Phylogenetic analysis of anthropoid relationships. *Journal of Human Evolution*, **35**, 221–306.

Rothschild, B. M., Hershkovitz, I., & Rothschild, C. (1995). Origin of yaws in the Pleistocene. *Nature*, **378**, 343–344.

Rowe, T. (1999). At the roots of the mammalian family tree. *Nature*, **398**, 283–284.

Ruff, C. B. (1994). Morphological adaptation to climate in modern and fossil hominids. *Yearbook of Physical Anthropology*, **37**, 65–107.

Ruff, C. B. (1998). Evolution of the hominid hip. In *Primate Locomotion: Recent Advances*, eds. E. Strasser, J. G. Fleagle, A. L. Rosenberger, & H. M. McHenry, pp. 449–470. New York: Plenum Press.

Ruff, C. B. & Runestad, J. A. (1992). Primate limb bone structural adaptations. *Annual Reviews in Anthropology*, **21**, 407–433.

Ruff, C. B., Trinkaus, E., & Holliday, T. W. (1998). Body mass and encephalization in Pleistocene *Homo*. *Nature*, **387**, 173–176.

Ruff, C. B. & Walker, A. C. (1993). Body size and body shape. In *The Nariokotome* Homo erectus *Skeleton*, eds. A. C. Walker & R. E. F. Leakey, pp. 234–265. Cambridge, MA: Harvard University Press.

Ruff, C. B., Walker, A. C., & Teaford, M. F. (1989). Body mass, sexual dimorphism and femoral proportions of *Proconsul* from Rusinga and Mfangano Islands, Kenya. *Journal of Human Evolution*, **18**, 515–536.

Runestad, J. A. (1994). Humeral and femoral diaphyseal cross-sectional geometry and articular dimensions in Prosimii and Platyrrhini (Primates) with application for reconstruction of body mass and locomotor behavior in Adapidae (Primates, Eocene). PhD dissertation, Johns Hopkins University School of Medicine.

Runestad, J. A. & Ruff, C. B. (1995). Structural adaptations for gliding in mammals with implications for locomotor behavior in paromomyids. *American Journal of Physical Anthropology*, **98**, 101–119.

Rusconi, C. (1933). Nuevos restos de monos fosiles del terciario antiguo de la Patagonia. *Anales de ciencias Argentina*, **116**, 286–289.

Rusconi, C. (1935*a*) Sobre morfogenesis basicraneana de algunos primates actuales y fosiles. *Revista argentina de paleontologia y antropologia ameghinia*, **1**, 3–23.

Rusconi, C. (1935*b*). Los especies de primates del oligoceno de Patagonia (gen. *Homunculus*). *Revista argentina de paleontologia y antropologia ameghinia*, **1**, 39–126.

Russell, D. E. & Gingerich P. D. (1980). Un nouveau primate omomyide de l'Éocène du Pakistan. *Comptes rendus de l'Académie des sciences de Paris*, **291**, 621–624.

Russell, D. E. & Gingerich, P. D. (1987). Nouveaux primates de l'Éocène du Pakistan. *Comptes rendus de l'Académie des sciences de Paris*, **304**, 209–214.

Russell, D. E., Louis, P., & Savage, D. E. (1967). Primates of the French early Eocene. *University of California Publications in the Geological Sciences*, **73**, 1–46.

Rütimeyer, L. (1862). Eocäne Säugetiere aus dem Gebiet des schweizerischen Jura. *Allgemeine schweizerische Gesellschaft für die gesamten naturwissenschaften Denkschriften*, **19**, 1–98.

Rütimeyer, L. (1890). Übersicht der eocänen Fauna von Egerkingen nebst einer Erwiderung an Prof. E. D. Cope. *Abhandlungen der schweizerischen paläontologischen Gesellschaft*, **17**, 1–24.

Ruvolo, M. (1988). Genetic evolution in the African guenons. In *A Primate Radiation: Evolutionary Biology of the African Guenons*, eds. A. Gautier-Hion, F. Bourlière, & J.-P. Gautier, pp. 127–139. Cambridge: Cambridge University Press.

Ruvolo, M. (1994). Molecular evolutionary processes and conflicting gene trees: the hominoid case. *American Journal of Physical Anthropology*, **94**, 89–114.

Ruvolo, M. (1997). Molecular phylogeny of the hominoids: inference from multiple independent DNA sequence data sets. *Molecular Biology and Evolution*, **14**, 248–265.

Saban, R. (1956). L'os temporal et ses rapports chez les lémuriens subfossiles de Madagascar. I. Types à molaires quadrituberculées formes archaïques. *Mémoire de l'Institut de recherche scientifique de*

Madagascar, Tananarive (Série A), **10**, 251–297.

Saban, R. (1963). Contribution à l'étude de l'os temporal des Primates. *Mémoires du Muséum national d'histoire naturelle, Paris*, **29**, 1–378 plus plates.

Sahni, A., Kumar, V., & Srivastava, V. C. (1974). *Dryopithecus* (subgenus: *Sivapithecus*) and associated vertebrates from the Lower Siwaliks of Uttar Pradesh. *Bulletin of the Indian Geological Association*, **7**, 54.

Sahni, A., Tiwari, B. N., & Kumar, K. (1980). An additional Lower Siwalik vertebrate fauna from the Kalgarh area, District Puari Garhwal, Uttar Pradesh. *Proceedings of the Indian Geological Congress*, **3**, 81–90.

Sanders, W. J. & Bodenbender, B. E. (1994). Morphometric analysis of lumbar vertebra UMP 67–28: implications for spinal function and phylogeny of the Miocene Moroto hominoid. *Journal of Human Evolution*, **26**, 203–237.

Santa Luca, A. (1978). A re-examination of presumed Neanderthal fossils. *Journal of Human Evolution*, **7**, 619–636.

Santa Luca, A. (1980). The Ngandong fossil hominids. *Yale University Publications in Anthropology*, **78**, 1–175.

Sarich, V. M. (1971). A molecular approach to the question of human origins. In *Background for Man*, ed. P. Dolhinow, pp. 60–81. Boston, MA: Little, Brown.

Sarich, V. M. (1993). Mammalian systematics: twenty-five years among their albumins and transferrins. In *Mammal Phylogeny: Placentals*, eds. F. S. Szalay, M. Novacek, & M. McKenna, pp. 103–114. New York: Springer-Verlag.

Sarich, V. M. & Cronin, J. E. (1976). Molecular systematics of the primates. In *Molecular Anthropology*, eds. M. Goodman & R. E. Tashian, pp. 141–170. New York: Plenum Press.

Sarich, V. M., Schmid, C. W., & Marks, J. (1989). DNA hybridization as a guide to phylogeny: a critical analysis. *Cladistics*, **5**, 3–32.

Sarich, V. M. & Wilson, A. C. (1966). Quantitative immunochemistry and the evolution of primate albumins: microcomplement fixation. *Science*, **154**, 1563–1566.

Sarich, V. M. & Wilson, A. C. (1967). Immunological time scale for hominoid evolution. *Science*, **158**,1200–1203.

Sarmiento, E. (1987). The phylogenetic position of *Oreopithecus* and its significance in the origin of the Hominoidea. *American Museum Novitates*, **2881**, 1–44.

Sartono, S. (1971). Observations on a new skull of *Pithecanthropus erectus* (*Pithecanthropus VIII*) from Sangiran, Central Java. *Proceedings of the Academy of Sciences, Amsterdam (B)*, **74**, 85–194.

Sartono, S. (1972). Discovery of another hominid skull at Sangiran, Central Java. *Current Anthropology*, **13**, 124–125.

Savage, D. E., Russell, D. E., & Waters, B. T. (1977). Critique of certain Eocene primate taxa. *Géobios, Mémoire spéciale*, **1**, 159–164.

Savage, D. E. & Waters, B. T. (1978). A new omomyid primate from the Wasatch Formation of southern Wyoming. *Folia Primatologica*, **30**, 1–29.

Sawada, Y., Pickford, M., Itaya, T., Makinouchi, T., Tateishi, M., Kabeto, K., Ishida, S., & Ishida, H. (1998). K–Ar ages of Miocene Hominoidea (*Kenyapithecus* and *Samburupithecus*) from Samburu Hills, Northern Kenya. *Comptes rendus de l'Académie des sciences de Paris*, **326**, 445–451.

Schaaffhausen, H. (1858). Zur Kenntnis der ältesten Rassenschädel. *Müller's Archiv*, **1858**, 453–478.

Schick, K., Toth, N., Qi, W., Clark, J., & Etler, D. (1991). Archaeological perspectives in the Nihewan Basin, China. *Journal of Human Evolution*, **21**, 13–26.

Schlosser, M. (1887). Die Affen, Lemuren, Chiropteren, Insectivoren, Marsupialier, Creodonten, und Carnivoren des europaischen Tertiars und deren Beziehungen zu ihren lebenden und fossilen aussereuropaischen Verwandten. *Beitrage zur Paläontologie und Geologie Öesterreich–Ungarns und des Orients*, **6**, 1–227; **7**, 1–117.

Schlosser, M. (1900). Die neueste Litteratur über die ausgestorbenen Anthropomorphen. *Zoologischer Anzeiger*, **23**, 289–301.

Schlosser, M. (1901). Die menshenähnlichen Zähne aus dem Bohnerz der schwäbischen Alb. *Zoologischer Anzeiger*, **24**, 261–271.

Schlosser, M. (1902). Beiträge zur Kenntnis der Säugetierreste aus den süddeutschen Bohnerzen. *Geologische und paläontologische Abhandlungen*, **5**, 117–258.

Schlosser, M. (1907). Bietrag zur Osteologie un systematischen Stellung der Gattung *Necrolemur*, sowie zur Stammesgeschichte der Primaten überhaupt. *Neue Jahrbuch für Mineralogie, Geologie, und Paläontologie*, **1907**, 197–226.

Schlosser, M. (1910). Über einige fossile Säugetiere aus dem Oligocän von Ägypten. *Zoologischer Anzeiger*, **35**, 500–508.

Schlosser, M. (1911). Beiträge zur Kenntnis der Oligozänen Landsäugetiere aus dem Fayum (Ägypten). *Beiträge zur Paläontologie und Geologie Österreich–Ungarns und des Orients*, **24**, 51–167.

Schlosser, M. (1924). Fossil primates from China. *Palaeontologica Sinica*, **1**, 1–16.

Schmid, P. (1979). Evidence of microchoerine evolution from Dielsdorf (Zurich region, Switzerland): a preliminary report. *Folia Primatologica*, **31**, 301–311.

Schmid, P. (1983). Front dentition of the Omomyiformes (Primates). *Folia Primatologica*, **40**, 1–10.

Schmidt, E. & Meldrum, D. J. (2000). Significance of the postentoconid notch in platyrrhine systematics and dental evolution. In *Program of the 42nd Annual Meeting of the Idaho Academy of Science*, p. 26.

Schmitt, D. (1996). Humeral head shape as an indicator of locomotor behavior in extant strepsirhines and Eocene adapids. *Folia Primatologica*, **67**, 137–151.

Schneider, H. (2000). The current status of the New World monkey phylogeny. *Anais da Academia brasileira de ciências*, **72**, 165–172.

Schneider, H. & Rosenberger, A. L. (1996). Molecules, morphology and platyrrhine systematics. In *Adaptive Radiations of Neotropical Primates*, eds. M. A. Norconk, A. L. Rosenberger, & P. A. Garber, pp. 3–20. New York: Plenum Press.

Schneider, H., Sampaio, I., Harada, M. L., Barroso, C. M. L., Schneider, M. P. C., Czelusniak, J., & Goodman, M. (1996). Molecular phylogeny of the New World monkeys (Platyrrhini, Primates) based on two unlinked nuclear genes: IRBP intron I and ε-globin sequences. *American Journal of Physical Anthropology*, **100**, 153–180.

Schoetensack, O. (1908). *Der Unterkiefer des* Homo heidelbergensis *aus den Sanden von Mauer bei Heidelberg*. Leipzig: Engelmann.

Schrenk, R., Bromage, T. G., Betzler, C. G., Ring, U., & Juwayeyi, Y. M. (1993). Oldest *Homo* and Pliocene biogeography of the Malawi Rift. *Nature*, **365**, 833–835.

Schultz, A. (1968). The recent hominoid primates. In *Perspectives on Human Evolution*, eds. S. L. Washburn & P. C. Jay, pp. 122–195. New York: Holt, Rinehart, & Winston.

Schultz, A. (1969). Observations of the acetabulum of primates. *Folia Primatologica*, **11**, 181–199.

Schwalbe, G. (1901). Der Neanderthalschädel. *Bonner Jahrbücher*,

106, 1–72.

Schwalbe, G. (1904). *Die Vorgeschichte des Menschen*. Braunschweig: Friedrich Vieweg und Sohn.

Schwalbe, G. (1906). *Studien zur Vorgeschichte des Menschen*. Stuttgart: E. Schweitzer-bartsche Verlagsbuchhandlung.

Schwartz, G. T. & Conroy, G. C. (1996). Cross-sectional geometric properties of the *Otavipithecus* mandible. *American Journal of Physical Anthropology*, **99**, 613–623.

Schwartz, G. T., Samonds, K. E., Jungers, W. L., & Godfrey, L. R. (2000). Dental microstructure and life history in subfossil lemurs. *Journal of Vertebrate Paleontology*, **20**, 68A.

Schwartz, J. H. (1974). Dental development and eruption in the prosimians and its bearing on their evolution. PhD dissertation, Columbia University.

Schwartz, J. H. (1975). Development and eruption of the premolar region of prosimians and its bearing on their evolution. In *Lemur Biology*, eds. I. Tattersall & R. W. Sussman, pp. 41–63. New York: Plenum Press.

Schwartz, J. H. (1984). The evolutionary relationships of man and orang-utans. *Nature*, **308**, 501–505.

Schwartz, J. H. (1990). *Lufengpithecus* and its potential relationship to an orang-utan clade. *Journal of Human Evolution*, **19**, 591–605.

Schwartz, J. H. (1992). Phylogenetic relationships of African and Asian lorisids. In *Topics in Primatology: Evolutionary Biology, Reproductive Endrocrinology and Virology*, vol. 3, ed. S. Mantana, pp. 65–81. Tokyo: University of Tokyo Press.

Schwartz, J. H. (1996). *Pseudopotto martini*: a new genus and species of extant lorisiform primate. *Anthropological Papers of the American Museum of Natural History*, **78**, 1–14.

Schwartz, J. H. (1997). *Lufengpithecus* and hominoid phylogeny: problems in delineating and evaluating phylogenetically relevant characters. In *Funtion, Phylogeny, and Fossils: Miocene Hominoid Evolution and Adaptations*, eds. D. R. Begun, C. V. Ward, & M. D. Rose, pp. 363–388. New York: Plenum Press.

Schwartz, J. H., Long, V. T., Cuong, N. L., Kha, L. T., & Tattersall, I. (1994). A diverse hominoid fauna from the late middle Pleistocene breccia cave of Tham Khuyen, Socialist Republic of Vietnam. *Anthropological Papers of the American Museum of Natural History*, **73**, 1–11.

Schwartz, J. H., Long, V. T., Cuong, N. L., Kha, L. T., & Tattersall, I. (1995). A review of the Pleistocene hominoid fauna of the Socialist Republic of Vietnam (excluding Hylobatidae). *Anthropological Papers of the American Museum of Natural History*, **76**, 1–24.

Schwartz, J. H., Shoshani, J., Tattersall, I., Simons, E. L., & Gunnell, G. F. (1998). Lorisidae Gray 1821 and Galagidae Gray, 1825 (Mammalia, Primates): proposed conservation as the correct original spellings. *Bulletin of Zoological Nomenclature*, **55(3)**, 165–168.

Schwartz, J. H. & Tattersall, I. (1982). A note on the status of "*Adapis priscus*" Stehlin, 1916. *American Journal of Primatology*, **3**, 295–298.

Schwartz, J. H. & Tattersall, I. (1985). Evolutionary relationships of living lemurs and lorises (Mammalia, Primates) and their potential affinities with European Eocene Adapidae. *Anthropological Papers of the American Museum of Natural History*, **60**, 1–100.

Schwartz, J. H. & Tattersall, I. (1987). Tarsiers, adapids and the integrity of the Strepsirhini. *Journal of Human Evolution*, **16**, 23–40.

Schwartz, J. H., Tattersall, I., & Eldredge, N. (1978). Phylogeny and classification of the primates revisited. *Yearbook of Physical Anthropology*, **21**, 95–133.

Scott, R. S., Kappelman, J., & Kelley, J. (1999). The paleoenvironment of *Sivapithecus parvada*. *Journal of Human Evolution*, **36**, 245–274.

Scott, W. B. (1928). Astropotheria of the Santa Cruz beds: primates of the Santa Cruz beds. Reports of the Princeton University Expeditions to Patagonia, **4 & 5**, 301–351.

Seiffert, E. R., Simons, E. L., & Fleagle, J. G. (2000). Anthropoid humeri from the late Eocene of Egypt. *Proceedings of the National Academy of Sciences of the United States of America*, **97**, 10062–10067.

Seligsohn, D. & Szalay, F. S. (1974). Dental occlusion and the masticatory apparatus in *Lemur* and *Varecia*: their bearing on the systematics of living and fossil primates. In *Prosimian Biology*, eds R. D. Martin, G. A. Doyle, & A. C. Walker, pp. 543–561. London: Duckworth.

Sémah, E., Falgueres, C., Yokayama, Y., Féraud, G., Saleki, H., & Djubiantono, T. (1997). Arrivée et disparation des *Homo erectus* à Java, les données actuelles. In *Abstracts of the 3rd Meeting of the European Association of Archaeologists*, pp. 11–12.

Semaw, S., Renne, P., Harris, J. W. K., Feibel, C. S., Bernor, R. L., Fesseha, N., & Mowbray, K. (1997). 2. 5-million-year-old stone tools from Gona, Ethiopia. *Nature*, **385**, 333–336.

Sen, S. (1991). Stratigraphie, faunes de mammifères et magnétostratigraphie du Néogène du Sinap Tepe, Province d'Ankara, Turquie. *Bulletin du Muséum national d'histoire naturelle, Paris, Série 4*, **12**, 243–277.

Senut, B. (1986). Nouvelle découvertes de restes postcraniens de primates (Hominoidea et Cercopithecoidea) sur le site Maboko au Kenya occidental. *Comptes rendus de l'Académie des sciences de Paris*, **303**, 1359–1362.

Senut, B. (1989). *Le Coude chez les Primates Hominoïdes: Anatomie, fonction, taxonomie et évolution*. Paris: Cahiers de Paléoanthropologie. Centre national de al recherche scientifique.

Senut, B. (1994). Cercopithecoidea néogènes et quaternaires du Rift Occidental (Ouganda). In *Geology and Palaeobiology of the Albertine Rift Valley, Uganda–Zäire*, vol. 2, *Palaeobiology*, pp. 195–205. Orléans, France: CIFEG.

Senut, B. (1996). Plio-Pleistocene Cercopithecoidea from the Koanaka Hills (Ngamiland, Botswana). *Comptes rendus de l'Académie des sciences de Paris*, **322**, 423–428.

Senut, B. & Gommery, D. (1997). Postcranial skeleton of *Otavipithecus*, Hominoidea, from the Middle Miocene of Namibia. *Annales de paléontologie*, **83**, 267–284.

Senut, B., Pickford, M., Gommery, D., & Kinematsu, Y. (2000). A new genus of Early Miocene hominoid from East Africa: *Ugandapithecus major* (Le Gros Clark & Leakey, 1950). *Comptes rendus de l'Académie des sciences de Paris*, **331**, 227–233.

Senut, B., Pickford, M., Gommery, D., Mein, P., Cheboi, K., & Coppens, Y. (2001). First hominid from the Miocene (Lukeino Formation, Kenya). *Comptes rendus de l'Académie des sciences de Paris*, **332**, 137–144.

Senut, B. & Thomas, H. (1994). First discoveries of anthropoid postcranial remains from Taqah (early Oligocene, Sultanate of Oman). In *Current Primatology*, eds. B. Thierry, J. R. Anderson, J. J. Roeder, & N. Herrenschmidt, pp. 255–260. Strasbourg: Université Louis Pasteur.

Sera, G. L. (1917). La testimianza dei fossili di anthropomorfi per la questione dell'origine dell'uomo. *Attidella Società italiana di scienze naturali*, **56**, 1–156.

Sera, G. L. (1935). I caratteri morfologici di "*Paleopropithecus*" e l'adattamento acquatico primitivo dei Mammiferi e dei Primati in particolare: contributo alla morfologia, all filogenesi e alla paleobiologia dei Mammifera. *Archivio italiano di anatomia e di embriologia*, **35**, 229–370.

Sera, G. L. (1938). Alcuni cratteri scheletrici di importanza ecologica e filletica nei Lemuri fossili ed attuali: studi sulla paleobiologia e sulla filogenesi dei Primati. *Rivista italiana di paleontologia e stratigrafia, Memorie*, **38** (n.s. 8), 1–113.

Sera, G. L. (1950). Ulteriori osservazioni sui lemuri fossili ed attuali significato di alcuni caratteri in rapporto con l'evoluzione dei Primati. *Rivista italiana di paleontologia e stratigrafia, Memorie*, **47** (n.s. 17), 1–97.

Sergi, S. (1953). Morphological position of the "Prophaneranthropi" (Swanscombe and Fontéchevade). In *Actes du Congrès international du Quaternaire* (1953), pp. 651–665.

Setoguchi, T. (1980). Discovery of a fossil primate from the Miocene of Colombia. *Monkey*, **24**, 64–69.

Setoguchi, T. (1985). *Kondous laventicus*, a new ceboid primate from the Miocene of La Venta, Colombia, South America. *Folia Primatologica*, **44**, 96–101.

Setoguchi, T. & Rosenberger, A. L. (1985*a*). Miocene marmosets: first evidence. *International Journal of Primatology*, **6**, 615–625.

Setoguchi, T. & Rosenberger, A. L. (1985*b*). Some new ceboid primates from the La Venta, Colombia, South America. *Memorias VI Congreso latinoamericano de geologica*, **1**, 187–198.

Setoguchi, T. & Rosenberger, A. L. (1987). A fossil owl monkey from La Venta, Colombia. *Nature*, **326**, 692–694.

Setoguchi, T., Takai, M., & Shigehara, N. (1990). A new ceboid primate, closely related to *Neosaimiri*, found in the Upper Red Bed in the La Venta badlands, middle Miocene of Colombia, South America. *Kyoto University Overseas Research Reports on New World Monkeys*, **7**, 9–14.

Setoguchi, T., Takai, M., Villarroel, A. C., Shigehara, N., & Rosenberger, A. L. (1988). New specimen of *Cebupithecia* from La Venta, Miocene of Colombia, South America. *Kyoto University Special Publications*, **1988**, 7–9.

Setoguchi, T., Watanabe, T., & Mouri, T. (1981). The upper dentition of *Stirtonia* (Ceboidea, Primates) from the Miocene of Colombia, South America, and the origins of the posterointernal cusp of upper molars of howler monkeys. *Kyoto University Overseas Research Reports on New World Monkeys*, **3**, 51–60.

Shapiro, L., Jungers, W. L., Godfrey, L. R., & Simons, E. L. (1994). Vertebral morphology of extinct lemurs. *American Journal of Physical Anthropology*, Supplement **18**, 179–180.

Shea, B. T. (1984). An allometric perspective on the morphological and evolutionary relationships between pygmy (*Pan paniscus*) and common (*Pan troglodytes*) chimpanzees. In *The Pygmy Chimpanzee*, ed. R Susman, pp. 89–130. New York: Plenum Press.

Shea, B. T. (1985). On aspects of skull form in African apes and orangutans, with implications for hominoid evolution. *American Journal of Physical Anthropology*, **68**, 329–342.

Shen, P., Wang, F., Underhill, P. A., Franco, C., Yang, W. H., Roxas, A., Sung, R., Lin, A. A., Hyman, R. W., Vollrath, D., Davis, R. W., Cavalli-Sforza, L. L., & Oefner, P. J. (2000). Population genetic implications from sequence variation in four Y chromosome genes. *Proceedings of the National Academy of Sciences of the United States of America*, **97**, 7354–7359.

Sherry, S. T., Batzer, M. A., & Harpending, H. C. (1998). Modeling the genetic architecture of modern populations. *Reviews in Anthropology*, **27**, 153–169.

Shipman, P. (2000). *The Man Who Found the Missing Link*. London: Weidenfeld & Nicolson.

Shipman, P. & Walker, A. C. (1989). The costs of becoming a predator. *Journal of Human Evolution*, **18**, 373–392.

Shipman, P., Walker, A. C., Van Couvering, J. A., Hooker. P. J., & Miller, J. A. (1981). The Fort Ternan hominoid site, Kenya: geology, age, taphonomy and paleoecology. *Journal of Human Evolution*, **10**, 49–72.

Shoshani, J., Groves, C. P., Simons, E. L., & Gunnell, G. (1996). Primate phylogeny: morphological vs molecular results. *Molecular Phylogenetics and Evolution*, **5**, 102–154.

Shoshani, J. & McKenna, M. (1998). Higher taxonomic relationships among extant mammals based on morphology, with selected comparisons of results from molecular data. *Molecular Phylogenetics and Evolution*, **9**, 572–584.

Sibley, C. G. & Ahlquist, J. E. (1984). The phylogeny of the hominoid primates, as indicated by DNA–DNA hybridization. *Journal of Molecular Evolution*, **20**, 2–15.

Sibley, C. G., Comstock, J. A., & Ahlquist, J. E. (1990). DNA hybridization evidence of hominoid phylogeny: a reanalysis of the data. *Journal of Molecular Evolution*, **30**, 202–236.

Sigé, B., Jaeger, J. J., Sudre, J., & Vianey-Liaud, M. (1990). *Altiatlasius koulchii* n. gen. et sp., primate omomyidé du Paléocène supérieur du Maroc, et les origines des euprimates. *Palaeontographica Abteilung A*, **214**, 31–56.

Sillen, A. & Morris, A. (1996). Diagenesis of bone from Border Cave: implications for the age of the Border Cave hominids. *Journal of Human Evolution*, **31**, 499–506.

Simmons, T., Falsetti, A., & Smith, F. (1991). Frontal bone morphometrics of Southwest Asian Pleistocene hominids. *Journal of Human Evolution*, **20**, 249–269.

Simons, C. V. M. (1997). Diet, dental eruption and dental variation in *Archaeolemur* specimens from northwestern Madagascar. *American Journal of Physical Anthropology*, Supplement **24**, 211–212.

Simons, E. L. (1959). An anthropoid frontal bone from the Fayum Oligocene of Egypt: the oldest skull fragment of a higher primate. *American Museum Novitates*, **1976**, 1–16.

Simons, E. L. (1960). *Apidium* and *Oreopithecus*. *Nature*, **186**, 824–826.

Simons, E. L. (1961*a*). Notes on Eocene tarsioids and a revision of some Necrolemurinae. *Bulletin of the British Museum (Natural History), Geology Series*, **5**, 43–70.

Simons, E. L. (1961*b*). The dentition of *Ourayia*: its bearing on relationships of omomyid primates. *Postilla*, **54**, 1–20.

Simons, E. L. (1961*c*). The phyletic position of *Ramapithecus*. *Postilla*, **57**, 1–9.

Simons, E. L. (1961*d*). An anthropoid mandible from the Oligocene Fayum beds of Egypt. *American Museum Novitates*, **2051**, 1–20.

Simons, E. L. (1962*a*). A new Eocene primate genus, *Cantius*, and a revision of some allied European lemuroids. *Bulletin of the British Museum (Natural History), Geology Series*, **7**, 1–30.

Simons, E. L. (1962*b*). Two new primate species from the African Oligocene. *Postilla*, **64**, 1–12.

Simons, E. L. (1965). New fossil apes from Egypt and the initial differentiation of Hominoidea. *Nature*, **205**, 135–139.

Simons, E. L. (1967*a*). A fossil *Colobus* skull from the Sudan (Primates, Cercopithecidae). *Postilla*, **111**, 1–12.

Simons, E. L (1967*b*). The earliest apes. *Scientific American*, **217**, 28–35.

Simons, E. L. (1967*c*). Review of the phyletic interrelationships of

Oligocene and Miocene Old World Anthropoidea. *Colloqes internationaux du Centre national de la recherche scientifique*, **163**, 597–602.

Simons, E. L. (1968). Early Cenozoic mammalian faunas, Fayum Province, Egypt: introduction. *Bulletin of the Peabody Museum of Natural History*, **28**, 1–21.

Simons, E. L (1969). Miocene monkey (*Prohylobates*) from northern Egypt. *Nature*, **223**, 687–689.

Simons, E. L. (1970). The deployment and history of Old World Monkeys (Cercopithecidae, Primates). In *Old World Monkeys: Evolution, Systematics, and Behavior*, eds. J. R. Napier & P. H. Napier, pp. 99–137. New York: Academic Press.

Simons, E. L. (1972). *Primate Evolution: An Introduction to Man's Place in Nature*. New York: Macmillan.

Simons, E. L. (1974). *Parapithecus grangeri* (Parapithecidae, Old World higher primates): new species from the Oligocene of Egypt and the initial differentiation of Cercopithecoidea. *Postilla*, **166**, 1–12.

Simons, E. L. (1976). The fossil record of primate phylogeny. In *Molecular Anthropology*, eds. M. Goodman, R. E. Tashian, & J. H. Tashian, pp. 35–61. New York: Plenum Press.

Simons, E. L. (1986). *Parapithecus grangeri* of the African Oligocene: An archaic catarrhine without lower incisors. *Journal of Human Evolution*, **15**, 205–214.

Simons, E. L. (1987). New faces of *Aegyptopithecus* from the Oligocene of Egypt. *Journal of Human Evolution*, **16**, 273–290.

Simons, E. L. (1989). Description of two genera and species of late Eocene Anthropoidea from Egypt. *Proceedings of the National Academy of Sciences of the United States of America*, **86**, 9956–9960.

Simons, E. L. (1990). Discovery of the oldest known anthropoidean skull from the Paleogene of Egypt. *Science*, **247**, 1567–1569.

Simons, E. L. (1992). Diversity in the early Tertiary anthropoidean radiation in Africa. *Proceedings of the National Academy of Sciences of the United States of America*, **89**, 10743–10747.

Simons, E. L. (1993). New endocasts of *Aegyptopithecus*: oldest well-preserved record of the brain in Anthropoidea. *American Journal of Science*, **293**, 383–390.

Simons, E. L. (1994*a*). New monkeys (*Prohylobates*) and an ape humerus from the Miocene Moghara Formation of Northern Egypt. In *Current Primatology*, eds B. Thierry, J. R. Anderson, J. Roeder, & N. Herrenschmidt, pp. 247–253. Strasbourg: Université Louis Pasteur.

Simons, E. L. (1994*b*). The giant aye-aye *Daubentonia robusta*. *Folia Primatologica*, **62**, 14–21.

Simons, E. L. (1995*a*). Crania of *Apidium*: primitive anthropoidean (Primates, Parapithecidae) from the Egyptian Oligocene. *American Museum Novitates*, **3124**, 1–10.

Simons, E. L. (1995*b*). Egyptian Oligocene primates: a review. *Yearbook of Physical Anthropology*, **38**, 199–238.

Simons, E. L. (1995*c*). Skulls and anterior teeth of *Catopithecus* (Primates: Anthropoidea) from the Eocene and anthropoid origins. *Science*, **268**, 1885–1888.

Simons, E. L. (1997*a*). Discovery of the smallest Fayum Egyptian primates (Anchomomyini, Adapidae). *Proceedings of the National Academy of Sciences of the United States of America*, **94**, 180–184.

Simons, E. L. (1997*b*). Lemurs: old and new. In *Natural Change and Human Impact in Madagascar*, eds. S. M. Goodman & B. D. Patterson, pp. 142–166. Washington, DC: Smithsonian Institution Press.

Simons, E. L. (1997*c*). Preliminary description of the cranium of *Proteopithecus sylviae*, an Egyptian late Eocene anthropoidean primate. *Proceedings of the National Academy of Sciences of the United States of America*, **94**, 14970–14975.

Simons, E. L. (1998). The prosimian fauna of the Fayum Eocene/Oligocene deposits of Egypt. *Folia Primatologica*, **69**, 286–294.

Simons, E. L. & Bown, T. M. (1985). *Afrotarsius chatrathi*, new genus, new species: first tarsiiform primate (Tarsiidae?) from Africa. *Nature*, **313**, 475–477.

Simons, E. L., Bown, T. M., & Rasmussen, D. T. (1986). Discovery of two additional prosimian primate families (Omomyidae, Lorisidae) in the African Oligocene. *Journal of Human Evolution*, **15**, 431–438.

Simons, E. L., Burney, D. A., Chatrath, P. S., Godfrey, L. R., Jungers, W. L., & Rakotosamimanana, B. (1995*a*). AMS ^{14}C dates for extinct lemurs from caves in the Ankarana Massif, Northern Madagascar. *Quaternary Research*, **43**, 249–254.

Simons, E. L. & Chopra, S. R. K. (1969*a*). A preliminary announcement of a new *Gigantopithecus* species from India. In *Proceedings of the 2nd International Congress of Primatology*, vol. 2, pp. 135–42.

Simons, E. L. & Chopra, S. R. K. (1969*b*). *Gigantopithecus* (Pongidae, Hominoidea), a new species from North India. *Postilla*, **138**, 1–18.

Simons, E. L. & Delson, E. (1978). Cercopithecidae and Parapithecidae. In *Evolution of African Mammals*, eds. V. J. Maglio & H. B. S. Cooke, pp. 100–119. Cambridge, MA: Harvard University Press.

Simons, E. L. & Ettel, P. C. (1970). *Gigantopithecus*. *Scientific American*, **222**, 77–85.

Simons, E. L. & Fleagle, J. G. (1973). The history of extinct gibbon-like primates. In *Gibbon and Siamang*, vol. 2, ed. D. M. Rumbaugh, pp. 121–148. Basel: Karger.

Simons, E. L., Godfrey, L. R., Jungers, W. L., Chatrath, P. S., & Rakotosamimanana, B. (1992). A new giant subfossil lemur *Babakotia* and the evolution of the sloth lemurs. *Folia Primatologica*, **58**, 190–196.

Simons, E. L., Godfrey, L. R., Jungers, W. L., Chatrath, P. S., & Ravaoarisoa, J. (1995*b*). A new species of *Mesopropithecus* (Primates, Palaeopropithecidae) from Northern Madagascar. *International Journal of Primatology*, **16**, 653–682.

Simons, E. L., Godfrey, L. R., Vuillaume-Randriamanantena, M., Chatrath, P. S., & Gagnon, M. (1990). Discovery of new giant subfossil lemurs in the Ankarana Mountains of northern Madagascar. *Journal of Human Evolution*, **19**, 311–320.

Simons, E. L. & Kay, R. F. (1983). *Qatrania*, new basal anthropoid primate from the Fayum, Oligocene of Egypt. *Nature*, **304**, 624–626.

Simons, E. L. & Kay, R. F. (1988). New material of *Qatrania* from Egypt with comments on the phylogenetic position of the Parapithecidae (Primates, Anthropoidea). *American Journal of Primatology*, **15**, 337–348.

Simons, E. L. & Miller, E. R. (1997). An upper dentition of *Aframonius dieides* (Primates) from the Fayum, Egyptian Eocene. *Proceedings of the National Academy of Sciences of the United States of America*, **94**, 7993–7996.

Simons, E. L. & Pilbeam, D. R. (1965). Preliminary revision of the Dryopithecinae (Pongidae, Anthropoidea). *Folia Primatologica*, **3**, 81–152.

Simons, E. L. & Pilbeam, D. R. (1971). A gorilla-sized ape from the Miocene of India. *Science*, **173**, 23–27.

Simons, E. L. & Pilbeam, D. R. (1972). Hominoid paleoprimatology. In *Functional and Evolutionary Biology of Primates*, ed. R. L. Tuttle, pp. 36–62. Chicago, IL: Aldine.

Simons, E. L., Plavcan, J. M., & Fleagle, J. G. (1999). Canine sexual dimorphism in Egyptian Eocene anthropoid primates: *Catopithecus* and *Proteopithecus*. *Proceedings of the National Academy of Sciences of the United States of America*, **96**, 2559–2562.

Simons, E. L. & Rasmussen, D. T. (1989). Cranial morphology of *Aegyptopithecus* and *Tarsius* and the question of the tarsier–anthropoidean clade. *American Journal of Physical Anthropology*, **79**, 1–23.

Simons, E. L. & Rasmussen, D. T. (1991). The generic classification of Fayum Anthropoidea. *International Journal of Primatology*, **12**, 163–178.

Simons, E. L. & Rasmussen, D. T. (1994*a*). A remarkable cranium of *Plesiopithecus teras* (Primates, Prosimii) from the Eocene of Egypt. *Proceedings of the National Academy of Sciences of the United States of America*, **91**, 9946–9950.

Simons, E. L. & Rasmussen, D. T. (1994*b*). A whole new world of ancestors: Eocene anthropoideans from Africa. *Evolutionary Anthropology*, **3**, 128–139.

Simons, E. L. & Rasmussen, D. T. (1996). Skull of *Catopithecus browni*, an Early Tertiary catarrhine. *American Journal of Physical Anthropology*, **100**, 261–292.

Simons, E. L., Rasmussen, D. T., Bown, T. M., & Chatrath, P. S. (1994). The Eocene origin of anthropoid primates: adaptation, evolution, and diversity. In *Anthropoid Origins*, eds. J. G. Fleagle & R. F. Kay, pp. 179–201. New York: Plenum Press.

Simons, E. L., Rasmussen, D. T., & Gebo, D. L. (1987). A new species of *Propliopithecus* from the Fayum, Egypt. *American Journal of Physical Anthropology*, **73**, 139–148.

Simons, E. L., Rasmussen, D. T., & Gingerich, P. D. (1995*c*). New cercamoniine adapid from Fayum, Egypt. *Journal of Human Evolution*, **29**, 577–589.

Simons, E. L. & Russell, D. (1960). Notes on the cranial anatomy of *Necrolemur*. *Breviora*, **127**, 1–14.

Simons, E. L. & Seiffert, E. R. (1999). A partial skeleton of *Proteopithecus sylviae* (Primates, Anthropoidea): first associated dental and postcranial remains of an Eocene anthropoidean. *Comptes rendus de l'Académie des sciences de Paris*, **329**, 921–927.

Simpson, G. G. (1940). Studies of the earliest primates. *Bulletin of the American Museum of Natural History*, **77**, 185–212.

Simpson, G. G. (1945). The principles of classification and a classification of mammals. *Bulletin of the American Museum of Natural History*, **85**, 1–350.

Simpson, G. G. (1949). *The Meaning of Evolution*. New Haven, CT: Yale University Press.

Simpson, G. G. (1959). Primates. *Bulletin of the American Museum of Natural History*, **117**, 152–157.

Simpson, G. G. (1961). *Principles of Animal Taxonomy*. New York: Columbia University Press.

Simpson, G. G. (1963). The meaning of taxonomic statements. In *Classification and Human Evolution*, ed. S. L. Washburn, pp. 1–31. New York: Aldine de Gruyter.

Simpson, G. G. (1965). Mammalian fauna other than Bovidae. In *Olduvai Gorge 1951–1961*, ed. L. S. B. Leakey, pp. 15–16. Cambridge: Cambridge University Press.

Simpson, G. G. (1967). The Tertiary lorisiform primates of Africa. *Bulletin of the Museum of Comparative Zoology*, **136**, 39–61.

Simpson, G. G. (1980). *Splendid Isolation: The Curious History of South American Mammals*. New Haven, CT: Yale University Press.

Singer, R., Wolff, R. G., Gladfelter, B. G., & Wymer, J. J. (1982). Pleistocene *Macaca* from Hoxne, Suffolk, England. *Folia Primatologica*, **37**, 141–152.

Singer, R. & Wymer, J. (1982). *The Middle Stone Age at Klasies River Mouth in South Africa*. Chicago, IL: University of Chicago Press.

Singleton, M. (2000). The phylogenetic affinities of *Otavipithecus namibiensis*. *Journal of Human Evolution*, **38**, 537–573.

Skelton, R. R. & McHenry, H. M. (1992). Evolutionary relationships among early hominids. *Journal of Human Evolution*, **23**, 309–349.

Skelton, R. R. & McHenry, H. M. (1998). Trait list bias and a reappraisal of early hominid phylogeny. *Journal of Human Evolution*, **34**, 109–113.

Skelton, R. R., McHenry, H. M., & Drawhorn, G. M. (1986). Phylogenetic analysis of early hominids. *Current Anthropology*, **27**, 21–43.

Skinner, M. (1991). Bee brood consumption: an alternative explanation for hypervitaminosis A in KNM-ER 1808 (*Homo erectus*) from Koobi Fora, Kenya. *Journal of Human Evolution*, **20**, 493–503.

Skinner, M., Jones, K. E., & Dunn, B. P. (1995). Undetectability of vitamin A in bee brood. *Apidologie*, **26**, 407–414.

Smith, B. H. (1993). The physiological age of KNM-WT 15000. In *The Nariokotome* Homo erectus *Skeleton*, eds. A. C. Walker & R. E. F. Leakey, pp. 195–220. Cambridge, MA: Harvard University Press.

Smith, E. J. (1999). A functional analysis of molar morphometrics in living and fossil hominoids using 2-D digitized images. Unpublished manuscript, Department of Anthropology, University of Toronto.

Smith, F. H. (1982). Upper Pleistocene hominid evolution in south–central Europe: a review of the evidence and analysis of trends. *Current Anthropology*, **23**, 667–703.

Smith, F. H. (1984). Fossil hominids from the Upper Pleistocene of Central Europe and the origin of modern Europeans. In *The Origins of Modern Humans: A World Survey of the Fossil Evidence*, eds. F. H. Smith & F. Spencer, pp. 137–209. New York: Alan R. Liss.

Smith, F. H. (1985). Continuity and change in the origin of modern *Homo sapiens*. *Zeitschrift für Morphologie und Anthropologie*, **75**, 197–222.

Smith, F. H. (1993). Models and realities in modern human origins: the African fossil evidence. In *The Origin of Modern Humans and the Impact of Chronometric Dating*, eds. M. Aitken, C. Stringer, & P. Mellars, pp. 234–248. Princeton, NJ: Princeton University Press.

Smith, F. H. (1994). Samples, species, and speculations in the study of modern human origins. In *Origins of Anatomically Modern Humans*, eds. M. H. Nitecki & D. V. Nitecki, pp. 227–249. New York: Plenum Press.

Smith, F. H. (1997*a*). Neandertals. In *History of Physical Anthropology: An Encyclopedia*, ed. F. Spencer, pp. 711–722. New York: Garland.

Smith, F. H. (1997*b*). Modern human origins. In *History of Physical Anthropology: An Encyclopedia*, ed. F. Spencer, pp. 661–672. New York: Garland.

Smith, F. H., Falsetti, A. B., & Donnelly, S. M. (1989). Modern human origins. *Yearbook of Physical Anthropology*, **32**, 35–68.

Smith, F. H. & F. Spencer (eds.) (1984). *The Origins of Modern Humans: A World Survey of the Fossil Evidence*. New York: Alan R. Liss.

Smith, F. H. & Trinkaus, E. (1991). Les origines de l'homme moderne en Europe centrale: un cas de continuité. In *Aux Origines d'Homo sapiens, Nouvelle Encyclopédie Diderot*, eds. J. J. Hublin & A. M. Tillier, pp. 251–290. Paris: Presses Universitaires de France.

Smith, F. H., Trinkaus, E., Pettitt, P., Karavanić, I. & Paunović, M.

(1999). Direct radiocarbon dates for Vindija G_1 and Velika Pećina Late Pleistocene hominid remains. *Proceeding of the National Academy of Sciences of the United States of America*, **96**, 12281–12286.

Smith, G. E. (1908). Appendix: On the brain in the extinct lemurs of Madagascar, with some remarks on the affinities of the Indrisinae. *Transactions of the Zoological Society of London*, **18**, 163–177.

Smith, G. E. (1913). The evolution of man. *Annual Report of the Board of Regents of the Smithsonian Institution*, **1912**, 553–572.

Smith, J. & Savage, R. (1956). Some locomotory adaptations in mammals. *Zoological Journal of the Linnean Society*, **42**, 603–622.

Smith, R. J. & Jungers, W. L. (1997). Body mass in comparative primatology. *Journal of Human Evolution*, **32**, 523–559.

Smith-Woodward, A. (1914). On the lower jaw of an anthropoid ape (*Dryopithecus*) from the upper Miocene of Lerida (Spain). *Quarterly Journal of the Geological Society*, **70**, 316–320.

Solecki, R. (1963). Prehistory in Shanidar valley, northern Iraq. *Science*, **139**, 179–193.

Soligo, C. & Müller, A. E. (1999). Nails and claws in primate evolution. *Journal of Human Evolution*, **36**, 97–114.

Sollas, W. J. (1908). On the cranial and facial characters of the Neandertal race. *Philosophical Transactions of the Royal Society*, **199**, 281–339.

Solounias, N. (1981). Mammalian fossils of Samos and Pikermi. II. Resurrection of a classic Turolian fauna. *Annals of Carnegie Museum of Natural History*, **50**, 231–269.

Sonakia, A. (1985). Early *Homo* from Narmada Valley, India. In *Ancestors: The Hard Evidence*, ed. E. Delson, pp. 334–338. New York: Alan R. Liss.

Spencer, F. (1984). The Neandertals and their evolutionary significance: a brief historical survey. In *The Origins of Modern Humans: A World Survey of the Fossil Evidence*, eds. F. H. Smith & F. Spencer, pp. 1–49. New York: Alan R. Liss.

Spencer, F. (1990). *Piltdown: A Scientific Forgery*. Oxford: Oxford University Press.

Spencer, F. & Smith, F. H. (1981). The significance of Aleš Hrdlička's "Neanderthal phase of man": A historical and current assessment. *American Journal of Physical Anthropology*, **56**, 435–459.

Spoor, C. F., Sondaar, P. Y., & Hussain, S. T. (1991). A hominoid hamate and first metacarpal from the Late Miocene Nagri Formation of Pakistan. *Journal of Human Evolution*, **21**, 413–424.

Spoor, F., Walker, A. C., Lynch, J., Liepins, P., & Zonneveld, F. (1998). Primate locomotion and vestibular morphology, with special reference to *Adapis*, *Necrolemur*, and *Megaladapis*. *American Journal of Physical Anthropology*, Supplement **26**, 207.

Spuhler, J. (1988). Evolution of mitochondrial DNA in monkeys, apes, and humans. *Yearbook of Physical Anthropology*, **31**, 15–48.

Standing, H. F. (1903). Rapport sur des ossements subfossiles provenant d'Ampasambazimba. *Bulletin de l'Académie-malgache*, **2**, 227–235.

Standing, H. F. (1904). Rapport sur des ossements subfossiles provenant d'Ampasambazimba. *Bulletin de l'Académie-malgache*, **3**, 305–310.

Standing, H. F. (1905). Rapport sur des ossements subfossiles provenant d'Ampasambazimba. *Bulletin de l'Académie-malgache*, **4**, 95–100.

Standing, H. F. (1908). On recently discovered subfossil primates from Madagascar. *Transactions of the Zoological Society of London*, **18**, 69–162.

Standing, H. F. (1909, for the year 1908). Subfossiles provenant des fouilles d'Ampasambazimba. *Bulletin de l'Académie-malgache*, **6**, 9–11.

Standing, H. F. (1910, for the year 1909). Note sur les ossements subfossiles provenant des fouilles d'Ampasambazimba. *Bulletin de l'Académie-malgache*, **7**, 61–64.

Standing, H. F. (1913, for the year 1912). Procès-verbal de la séance du 28 novembre 1912. (Reported by M. Fontoynont.) *Bulletin de l'Académie-malgache*, **10**, 41–44.

Stanhope, M. J., Waddell, V. G., Madsen, O., de Jong, W., Hedges, S. B., Cleven, G. C., Kao, D., & Springer, M. S. (1998). Molecular evidence for multiple origins of Insectivora and for a new order of endemic African insectivore mammals. *Proceedings of the National Academy of Sciences of the United States of America*, **95**, 9967–9972.

Stefan, V. & Trinkaus, E. (1998). Discrete trait and dental morphometric affinities of the Tabun 2 mandible. *Journal of Human Evolution*, **34**, 443–468.

Stehlin, H. G. (1912). Die Säugetiere des schweizerschen Eocäns: critischer Catalog der Materialen. Part 7, first half. *Abhandlungen der schweizerischen paläontologischen Gesellschaft*, **38**,1165–1298.

Stehlin, H. G. (1914). Übersicht über die Säugetiere der schweizerischen Molasseformation, ihre Fundorte und ihre stratigraphische Verbreitung. *Verhandlungen der naturforwchenden Gesellschaft*, **25**,

Stehlin, H. G. (1916). Die Säugetiere des schweizerischen Eocäns: critischer Catalog der Materialen. VIIb, *Abhandlungen der schweizerischen paläontologischen Gesellschaft*, **41**, 1299–1552.

Steiner, K. E. (1981). Nectarivory and potential pollination by a neotropical opossum. *Annals of the Missouri Botanical Garden*, **68**, 505–513.

Steininger, F. (1967). Ein weiterer Zahn von *Drypithecus* (*Dry.*) *fontani darwini* Abel, 1902 (Mammalia, Pongidae) aus dem Miozän des wiener Beckens. *Folia Primatologica*,**7**, 243–275.

Steininger, F. (1986). Dating the Paratethys Miocene hominoid record. In *Primate Evolution*, eds. J. G. Else & P. C. Lee, pp. 71–84. Cambridge: Cambridge University Press.

Steininger, F. (1999). Chronostratigraphy, geochronology and biochronology of the Miocene "European Land Mammal Mega-Zones" (ELMMZ)and the Miocene "Mammal-Zones (MN-Zones)". In *The Miocene Land Mammals of Europe: The Continental European Miocene*, eds. G. E. Rössner & K. Heissig, pp. 9–24. Munich: Dr. Friedrich Pfeil.

Sterling, E. (1994*a*). Taxonomy and distribution of *Daubentonia*: a historical perspective. *Folia Primatologica*, **62**, 8–13.

Sterling, E. (1994*b*). Aye-ayes: specialists on structurally defended resources. *Folia Primatologica*, **62**, 142–154.

Stern, J. T. (2000). Climbing to the top: a personal memoir of *Australopithecus afarensis*. *Evolutionary Anthropology*, **9**, 113–133.

Stern, J. T. & Susman, R. L. (1983). The locomotor anatomy of *Australopithecus afarensis*. *American Journal of Physical Anthropology*, **60**, 279–317.

Steunes, S. (1989). Taxonomy, habits, and relationships of the subfossil Madagascan hippopotami *Hippopotamus lemerlei* and *H. madagascariensis*. *Journal of Vertebrate Paleontology*, **9**, 241–268.

Stevens, N. J. & Fleagle, J. G. (1998). A new fossil platyrrhine femur from the Pinturas Formation of Argentina. *American Journal of Physical Anthropology*, Supplement **26**, 209.

Stewart, C. B. & Disotell, T. R. (1998). Primate evolution: in and out of Africa. *Current Biology*, **8**, 582–588.

Stewart, T. D. (1950). The problem of the earliest claimed representatives of *Homo sapiens*. *Cold Spring Harbor Symposia on Quantitative Biology*, **15**, 97–107.

Stewart, T. D. (1964). A neglected primitive feature of the Swanscombe skull. *Occasional Papers of the Royal Anthropological Institute*, **20**, 151–160.

Stirton, R. A. (1951). Ceboid monkeys from the Miocene of Colombia. *Bulletin of the University of California Publications in the Geological Sciences*, **28**, 315–356.

Stirton, R. A. & Savage, D. E. (1951). A new monkey from the La Venta late Miocene of Colombia. *Compilación de los estudios geológicos oficiales en Colombia*, **7**, 345–346.

Stock, C. (1933). An Eocene primate from California. *Proceedings of the National Academy of Sciences of the United States of America*, **19**, 954–959.

Stock, C. (1934). A second Eocene primate from California. *Proceedings of the National Academy of Sciences of the United States of America*, **20**, 150–154.

Stock, C. (1938). A tarsiid primate and a mixodectid from the Poway Eocene, California. *Proceedings of the National Academy of Sciences of the United States of America*, **24**, 288–293.

Stoneking, M. & Cann, R. (1989). African origins of human mitochondrial DNA. In *The Human Revolution*, eds. P. Mellars & C. Stringer, pp. 17–30. Edinburgh: Edinburgh University Press.

Storer, J. E. (1990). Primates of the Lac Pelletier Lower Fauna (Eocene: Duchesnean), Saskatchewan. *Canadian Journal of Earth Sciences*, **27**, 520–524.

Strait, D. S., Grine, F. E., & Moniz, M. A. (1997). A reappraisal of early hominid phylogeny. *Journal of Human Evolution*, **32**, 17–82.

Strait, S. G. (1991). Dietary reconstruction in small-bodied fossil primates. PhD dissertation, State University of New York, Stony Brook.

Strait, S. G. (1993). Differences in occlusal morphology and molar size in frugivores and faunivores. *Journal of Human Evolution*, **25**, 471–484.

Strait, S. G. (1997). Tooth use and the physical properties of food. *Evolutionary Anthropology*, **6**, 199–211.

Strasser, E. (1993). Kaswanga *Proconsul* foot proportions. *American Journal of Physical Anthropology*, Supplement **16**, 191.

Strasser, E. (1997). Cladistic analysis of the cercopithecoid foot. *American Journal of Physical Anthropology*, Supplement **24**, 222.

Strasser, E. & Delson, E. (1987). Cladistic analysis of cercopithecid relationships. *Journal of Human Evolution*, **16**, 81–99.

Straus, W. L. Jr. (1949). The riddle of man's ancestry. *Quarterly Review of Biology*, **24**, 200–23.

Straus, W. L. Jr. (1957). *Oreopithecus bambolii*. *Science*, **126**, 345–346.

Straus, W. L. Jr. & Wislocki, G. B. (1932). On certain similarities between sloths and slow lemurs. *Bulletin of the Museum of Comparative Zoology*, **74** (3), 45–56.

Stringer, C. B. (1974). Population relationships of later Pleistocene hominids: a multivariate study of available crania. *Journal of Archaeological Science*, **1**, 317–342.

Stringer, C. B. (1983). Some further notes on the morphology and dating of the Petralona hominid. *Journal of Human Evolution*, **12**, 731–742.

Stringer, C. B. (1984). The definition of *Homo erectus* and the existence of the species in Africa and Europe. *Courier Forschungsinstitut Senckenberg*, **69**, 131–144.

Stringer, C. B. (1989). The origin of modern Europeans: a comparison of the African and the non-African evidence. In *The Human Revolution*, eds. P. Mellars & C. Stringer, pp. 123–154. Edinburgh: Edinburgh University Press.

Stringer, C. B. (1994). Out of Africa: a personal history. In *Origins of Anatomically Modern Humans*, eds. M. H. Nitecki & D. V. Nitecki, pp. 149–174. New York: Plenum Press.

Stringer, C. B. (1998). A metrical study of the WLH–50 calvaria. *Journal of Human Evolution*, **34**, 327–332.

Stringer, C. B. & Andrews, P. (1988). Genetic and fossil evidence for the origin of modern humans. *Science*, **239**, 1263–1268.

Stringer, C. B. & Bräuer, G. (1994). Models, misreading, and bias. *American Anthropologist*, **96**, 416–424.

Stringer, C. B., Howell, F. C., & Melentis, J. (1979). The significance of the fossil hominid skull from Petralona, Greece. *Journal of Archaeological Science*, **6**, 235–253.

Stringer, C. B., Hublin, J. J., & Vandermeersch, B. (1984). The origin of anatomically modern humans in western Europe. In *The Origins of Modern Humans: A World Survey of the Fossil Evidence*, eds. F. H. Smith & F. Spencer, pp. 51–135. New York: Alan R. Liss.

Stringer, C. B., Humphrey, L. T., & Compton, T. (1997). Cladistic analysis of dental traits in recent humans using a fossil outgroup. *Journal of Human Evolution*, **32**, 389–402.

Stringer, C. B. & McKie, R. (1997). *African Exodus*. New York: Henry Holt.

Stringer, C. B., Trinkaus, E., Roberts, M., Parfitt, S., & Macphail, R. (1998). The Middle Pleistocene human tibia from Boxgrove. *Journal of Human Evolution*, **34**, 509–547.

Stromer, E. (1913). Mitteilungen über Wirbeltierreste aus dem Mittelpliocän des Natrontales (Ägypten). *Zeitschrift deutsche geologische Gesellschaft*, **65**, 350–372.

Stucky, R. K. (1984). The Wasatchian–Bridgerian Land Mammal Age boundary (early to middle Eocene) in western North America. *Annals of the Carnegie Museum of Natural History*, **53**, 347–382.

Suckling, J. A., Suckling, E. E., & Walker, A. C. (1969). Suggested function of the vascular bundles in the limbs of *Perodicticus potto*. *Nature*, **221**, 379–380.

Susman, R. L. (1979). Comparative and functional morphology of hominoid fingers. *American Journal of Physical Anthropology*, **50**, 215–236.

Susman, R. L. & Stern, J. T. (1982). Functional morphology of *Homo habilis*. Science, **217**, 931–934.

Susman, R. L., Stern, J. T., & Jungers, W. L. (1984). Arboreality and bipedality in the Hadar [Ethiopia] hominids. *Folia Primatologica*, **43**, 113–156.

Sussman, R. W. (1977). Feeding behavior of *Lemur catta* and *Lemur fulvus*. In *Primate Ecology: Studies of Feeding and Ranging Behavior in Lemurs, Monkeys, and Apes*, ed. T. H., Clutton-Brock, pp. 1–37. London: Academic Press.

Sussman, R. W. (1991). Primate origins and the evolution of angiosperms. *American Journal of Primatology*, **23**, 209–223.

Sussman, R. W. (1995). How primates invented the rainforest and vice versa. In *Creatures of the Dark: The Nocturnal Prosimians*, eds. L. Alterman, G. A. Doyle, & M. K. Izard, pp. 1–10. New York: Plenum Press.

Sussman, R. W. (1999). *Primate Ecology and Social Structure*, vol. 1, *Lorises, Lemurs, and Tarsiers*. Needham Heights, MA: Pearson Custom Publishing.

Sussman, R. W. & Raven, P. H. (1978). Pollination by lemurs and marsupials: an archaic coevolutionary system. *Science*, **200**, 731–736.

Suteethorn, V., Buffetaut, E., Buffetaut-Tong, H., Ducrocq, S.,

Helmcke-Ingavat, R., Jaeger, J. J., & Jongkanjanasoontorn, Y. (1990). A hominoid locality in the Middle Miocene of Thailand. *Comptes rendus de l'Académie des sciences de Paris*, **311**, 1449–1454.

Suwa, G., Asfaw, B., Beyene, Y., White, T. D., Katoh, S., Nagaoka, S., Nakaya, H., Uzawa, K., Renne, P., & WoldeGabriel, G. (1997). The first skull of *Australopithecus boisei*. *Nature*, **389**, 489–492.

Suwa, G., White, T. D., & Howell, F. C. (1996). Mandibular post-canine dentition from the Shungura Formation, Ethiopia: crown morphology, taxonomic allocation, and Plio-Pleistocene hominid evolution. *American Journal of Physical Anthropology*, **101**, 247–282.

Swisher, C. C. III, Curtis, G. H., Jacob, T., Getty, A. G., Suprijo, A., Widiasmoro (1994). Age of the earliest known hominids in Java, Indonesia. *Science*, **263**, 1118–1121.

Swisher, C. C. III, Rink, W. J., Antón, S. C., Schwarcz, H. P., Curtis, G. H., Suprijo, A., & Widiasmoro (1996). Latest *H. erectus* of Java: potential contemporaneity with *H. sapiens* in Southeast Asia. *Science*, **274**, 1870–1874.

Swisher, C. C. III, Scott, G. R., Curtis, G. H., Butterworth, J., Antón, S. C., Jacob, T., Suprijo, A., Widiasmoro, Sukandar, Koeshardjono. (1998). Antiquity of *Homo erectus* in Java: ^{40}Ar/^{39}Ar dating and paleomagnetic study of the Sangiran area. In *Abstracts of Contributions to the Dual Congress*, Sun City, South Africa.

Swofford, D. L. (1993). *PAUP: Phylogenetic Analysis Under Parsimony*, 3.1.1. Champaign–Urbana, IL: Illinois Natural History Survey.

Szalay, F. S. (1968). The beginnings of primates. *Evolution*, **22**, 19–36.

Szalay, F. S. (1970). Late Eocene *Amphipithecus* and the origins of catarrhine primates. *Nature*, **227**, 355–357.

Szalay, F. S. (1972). *Amphipithecus* revisited. *Nature*, **236**,179.

Szalay, F. S. (1974). New genera of European Eocene adapid primates. *Folia Primatologica*, **22**,116–133.

Szalay, F. S. (1975*a*). Phylogeny of primate higher taxa: the basicranial evidence. In *Phylogeny of the Primates*, eds. W. Luckett & F. Szalay, pp. 91–125. New York: Plenum Press.

Szalay, F. S. (1975*b*). Phylogeny, adaptations and the dispersal of the tarsiiform primates. In *Phylogeny of the Primates*, eds. W. Luckett & F. Szalay, pp. 357–404. New York: Plenum Press.

Szalay, F. S. (1976). Systematics of the Omomyidae (Tarsiiformes, Primates): taxonomy, phylogeny, and adaptations. *Bulletin of the American Museum of Natural History*, **156**, 157–450.

Szalay, F. S. (1981). Phylogeny and the problems of adaptive significance: the case of the earliest primates. *Folia Primatologica*, **36**, 157–182.

Szalay, F. S. & Dagosto, M. (1980). Locomotor adaptations as reflected in the humerus of Paleogene primates. *Folia Primatologica*, **34**, 1–45.

Szalay, F. S. & Dagosto, M. (1988). Evolution of hallucial grasping in the primates. *Journal of Human Evolution*, **17**, 1–34.

Szalay, F. S. & Delson, E. (1979). *Evolutionary History of the Primates*. New York: Academic Press.

Szalay, F. S. & Katz, C. C. (1973). Phylogeny of lemurs, galagos, and lorises. *Folia Primatologica*, **19**, 88–103.

Szalay, F. S. & Langdon, J. H. (1986). The foot of *Oreopithecus*: an evolutionary assessment. *Journal of Human Evolution*, **15**, 585–621.

Szalay, F. S., Rosenberger, A. L., & Dagosto, M. (1987). Diagnosis and differentiation of the Order Primates. *Yearbook of Physical Anthropology*, **30**, 75–106.

Takahata, N. (1995). A genetic perspective on the origin and history of humans. *Annual Reviews in Ecology and Systematics*, **26**, 343–472.

Takai, M. (1994). New specimens of *Neosaimiri fieldsi* from La Venta, Colombia: a middle Miocene ancestor of the living squirrel monkeys. *Journal of Human Evolution*, **27**, 329–360.

Takai, M. & Anaya, F. (1996). New specimens of the oldest fossil platyrrhine, *Branisella boliviana*, from Salla, Bolivia. *American Journal of Physical Anthropology*, **99**, 301–317.

Takai, M., Anaya, F., Shigehara, N., & Setoguchi, T. (2000*a*). New fossil materials of the earliest New World Monkey, *Branisella boliviana*, and the problem of platyrrhine origins. *American Journal of Physical Anthropology*, **111**, 263–282.

Takai, M., Shigehara, N., Tsubamoto, T., Egi, N., Aung, A. K., Thein, T., Soe, A. N., & Tun, S. T. (2000*b*). The latest middle Eocene primate fauna in Pondaung area, Myanmar. *Asian Paleoprimatology*, **1**, 7–28.

Takemura, K. & Danhara, T. (1986). Fission-track dating of the upper part of Miocene Honda Group in La Venta badlands, Colombia. *Kyoto University Overseas Research Reports on New World Monkeys*, **5**, 31–37.

Tardieu, C. & Jouffroy, F. K. (1979). Les surfaces articulaires fémorales du genou chez les Primates: étude préliminaire. *Annales des sciences naturelles, Zoologie, Paris, Série 13*, **1**, 23–38.

Tattersall, I. (1971). Revision of the subfossil Indriinae. *Folia Primatologica*, **15**, 257–269.

Tattersall, I. (1973). Cranial anatomy of the Archaeolemurinae (Lemuroidea, Primates). *Anthropological Papers of the American Museum of Natural History*, **52**, 1–110.

Tattersall, I. (1975). Notes on the cranial anatomy of the subfossil Malagasy lemurs. In *Lemur Biology*, eds. I. Tattersall & R. W. Sussman, pp. 111–124. New York: Plenum Press.

Tattersall, I. (1982). *The Primates of Madagascar*. New York: Columbia University Press.

Tattersall, I. (1986). Species recognition in human paleontology. *Journal of Human Evolution*, **15**, 165–175.

Tattersall, I. (1992). Species concepts and species identification in human evolution. *Journal of Human Evolution*, **22**, 341–349.

Tattersall, I. (1999). Patterns of origin and extinction in the mammal fauna of Madagascar. In *Elephants have a Snorkel! Papers in Honour of Paul Y. Sondaar*, eds. J. W. F. Reumer & J. DeVos, pp. 303–311. Rotterdam: Het Natuurmuseum.

Tattersall, I. & Schwartz, J. H. (1974). Craniodental morphology and the systematics of the Malagasy lemurs (Primates, Prosimii). *Anthropological Papers of the American Museum of Natural History*, **52**, 137–192.

Tattersall, I. & Schwartz, J. H. (1999). Hominids and hybrids: the place of Neanderthals in human evolution. *Proceedings of the National Academy of Sciences of the United States of America*, **96**, 7117–7119.

Tattersall, I. & Schwartz, J. H. (2000). *Extinct Humans*. Boulder, CO: Westview.

Tattersall, I., Simons, E. L., & Vuillaume-Randriamanantena, M. (1992). Case 2785, *Palaeopropithecus ingens* G. Grandidier, 1899 (Mammalia, Primates): proposed conservation of both generic and specific names. *Bulletin of Zoological Nomenclature*, **49**, 55–57.

Tauber, A. (1991). *Homunculus patagonicus* Ameghino, 1891 (Primates, Cebidae), Mioceno medio, de la costa Atlantica austral, Provincia de Santa Cruz, Republica Argentina. *Academia nacional de ciencias (Cordoba, Argentina), Miscelanea*, **82**, 1–32.

Tauxe, L. (1979). A new date for *Ramapithecus*. *Nature*, **282**, 399–401.

Teaford, M. F., Beard, K. C., Leakey, R. E., & Walker, A. C. (1988). New hominoid facial skeleton from the early Miocene of Rusinga Island, Kenya, and its bearing on the relationship between *Proconsul nyanzae* and *Proconsul africanus*. *Journal of Human Evolution*, **17**, 461–477.

Teaford, M. F., Maas, M. C., & Simons, E. L. (1996). Dental microwear and microstructure in early Oligocene primates from the Fayum, Egypt: Implications for diet. *American Journal of Physical Anthropology*, **101**, 527–543.

Teaford, M. F. & Walker, A. C. (1984). Quantitative differences in dental microwear between primate species with different diets and a comment on the presumed diet of *Sivapithecus*. *American Journal of Physical Anthropology*, **64**, 191–200.

Teaford, M. F., Walker, A. C., & Mugaisi, G. S. (1993). Species discrimination in *Proconsul* from Rusinga and Mfwangano Islands, Kenya. In *Species, Species Concepts, and Primate Evolution*, eds. W. H. Kimbel & L. B. Martin, pp. 373–392. New York: Plenum Press.

Teilhard de Chardin, P. (1927). Les mammifères de l'Éocène inférieur de la Belgique. *Memoires du Muséum royal d'histoire naturelle de Belgique*, **36**, 1–33.

Teilhard de Chardin, P. (1938) Fossils from Locality 12, Choukoutien. *Palaeontologica Sinica*, **5**, 1–50.

Teilhard de Chardin, P., Young, C. C., Pei, W. C., & Chang, H. C. (1935). On the Cenozoic formations of Kwangsi and Kwangtung. *Bulletin of the Geological Society of China*, **14**, 179–205.

Tejedor, M. F. (1995*a*). Descripción de nuevos restos dentarios asignados a *Homunculus patagonicus* (Primates, Platyrrhini) procedentes de la localidad de Monte Observación (Santacrucense), Provincia de Santa Cruz. *Resumenes XI Jornadas argentinas de paleontologia de vertebrados Tucuman*, mayo de 1995.

Tejedor, M. F. (1995*b*). La diversidad de platirrinos fosiles en la Patagonia. *Neotropical Primates*, **3**, 1–4.

Tejedor, M. F. (1997). La dentición de *Callicebus* y el morfotipo ancestral de los platirrinos. *Neotropical Primates*, **5**, 43–46.

Tejedor, M. F. (1998*a*). La posición de *Aotus* y *Callicebus* en la filogenia de los primates platirrinos. *Boletin Primatologico Latinamericano*, **7**, 13–29.

Tejedor, M. F. (1998*b*). The evolutionary history of platyrrhines:old controversies and new interpretations. *Neotropical Primates*, **6**, 77–82.

Tejedor, M. F. (in press). New material of *Homunculus patagonicus* (Primates, Platyrrhini) from the Santacrucian of Argentina. *American Journal of Physical Anthropology*.

Tekkaya, I. (1974). A new species of Tortonian anthropoid (Primates, Mammalia) from Anatolia. *Bulletin of the Mineral Research and Exploration Institute of Turkey*, **83**, 148–65.

Templeton, A. (1993). The "Eve" hypothesis: a genetic critique and reanalysis. *American Anthropologist*, **95**, 51–72.

Templeton, A. (1999). Human races: a genetic and evolutionary perspective. *American Anthropologist*, **100**, 632–650.

Tesakov, A. S. & Maschenko, E. N. (1992). The first reliable finding of a macaque (Cercopithecidae, Primates) in the Pliocene of Ukraine. *Paleontologicheskiaei Zhurnal*, **4**, 47–52.

Thalmann, U. (1994). Die Primaten aus dem eozänen Geiseltal bei Halle/Saale (Deutschland). *Courier Forschungsinstitut Senckenberg*, **175**, 1–161.

Thalmann, U., Haubold, H., & Martin, R. D. (1989). *Pronycticebus neglectus*: an almost complete adapid primate specimen from the Geiseltal (GDR). *Palaeovertebrata*, **19**, 115–130.

Thenius, E. (1953). Zur Gebiss-Analyse von *Megaladapis edwardsi* (Lemur, Mammal.). *Zoologischer Anzeiger*, **150**, 251–260.

Thenius, E. (1965). Ein Primaten-Rest aus dem Altpleistozan von Voigtstedt in Thuringen. *Paläontologische Abhandlungen*, **2**, 681–686.

Thenius, E. (1970). Zum Problem der airorhynchie des Säugetierschadels. *Ein Zoologische Anzeiger*, **185**, 159–172.

Thewissen, J. G. M., Hussain, S. T., & Arif, M. (1997). New *Kohatius* (Omomyidae) from the Eocene of Pakistan. *Journal of Human Evolution*, **32**, 473–477.

Thoma, A. (1966). L'occipital de l'homme Mindelian de Vértesszőllős. *L'Anthropologie*, **70**, 499–533.

Thomas, H. (1985). The early and middle Miocene land connection of the Afro-Arabian plate and Asia: a major event for hominoid dispersal? In *Ancestors: The Hard Evidence*, ed. E. Delson, pp. 42–50. New York: Alan R. Liss.

Thomas, H., Roger, J., Sen, S., & Al-Sulaimani, Z. (1988). Découverte des plus anciens "anthropoïdes" du continent araboafricain et d'un primate tarsiiforme dans l'Oligocène du Sultanat d'Oman. *Comptes rendus de l'Académie des sciences de Paris*, **306**, 823–829.

Thomas, H., Roger, J., Sen, S., & Al-Sulaimani, Z. (1991). The discovery of *Moeripithecus markgrafi* Schlosser (Propliopithecidae, Anthropoidea, Primates) in the Ashawq Formation (early Oligocene of Dhofar Province, Sultanate of Oman). *Journal of Human Evolution*, **20**, 33–49.

Thomas, H., Roger, J., Sen, S., Bourdillon-de-Grissac, C., & Al-Sulaimani, Z. (1989). Découverte de vertébrés fossiles dans l'Oligocène inférieur de Dhofar (Sultanat d'Oman). *Géobios*, **22**, 101–120.

Thomas, H., Roger, R., Sen, S., Pickford, M., Gheerbrant, E., Al-Sulaimani, Z., & Al-Busaidi, S. (1999). Oligocene and Miocene terrestrial vertebrates in the southern Arabian Peninsula (Sultanate of Oman) and their geodynamic and palaeogeographic settings. In *Fossil Vertebrates of Arabia*, eds. P. J. Whybrow & A. Hill, pp. 430–442. New Haven, CT: Yale University Press.

Thomas, H. & Verna, S. N. (1979). Découverte d'un primate adapiforme (Sivaladapinae subfam. nov.) dans le Miocène moyen des Siwaliks de la région de Ramnager (Jammu et Cachemire, Inde). *Comptes rendus de l'Académie des sciences de Paris*, **289**, 833–836.

Thorington, R. W. & Groves, C. P. (1970). An annotated classification of the Cercopithecoidea. In *Old World Monkeys: Evolution, Systematics, and Behavior*, eds. J. R. Napier & P. H. Napier, pp. 629–647. New York: Academic Press.

Thorne, A., Grün, R., Mortimer, G., Spooner, N., Simpson, J., McCulloch, M., Taylor, L., & Curnoe, D. (1999). Australia's oldest human remains: age of the Lake Mungo 3 skeleton. *Journal of Human Evolution*, **36**, 591–612.

Thorne, A. & Macumber, P. (1972). Discoveries of late Pleistocene man at Kow Swamp, Australia. *Nature*, **238**, 316–319.

Tobias, P. V. (1964). The Olduvai Bed 1 Hominine with special reference to its cranial capacity. *Nature*, **202**, 3–5.

Tobias, P. V. (1966). The distinctiveness of *Homo habilis*. *Nature*, **209**, 953–960.

Tobias, P. V. (1967). *Olduvai Gorge*, vol. 2, *The Cranium and Maxillary Dentition of Australopithecus (Zinjanthropus) boisei*. Cambridge: Cambridge University Press.

Tobias, P. V. (1971). *The Brain in Hominid Evolution*. New York:

Columbia University Press.

Tobias, P. V. (1973). Implications of the new age estimates of the early South African hominids. *Nature*, **246**, 79–83.

Tobias, P. V. (1980). L'évolution du cerveau humain. *La Recherche*, **11(109)**, 282–292.

Tobias, P. V. (1983). Hominid evolution in Africa. *Canadian Journal of Anthropology*, **3**, 163–185.

Tobias, P. V. (1991). *Olduvai Gorge*, vol. 4, *The Skulls, Endocasts, and Teeth of* Homo habilis. Cambridge: Cambridge University Press.

Tobien, H. (1986). An Early Upper Miocene (Vallesian) *Mesopithecus* molar from Rheinhessen. *Primate Report*, **14**, 49.

Todaro, G. J. (1980). Evidence using viral gene sequences suggesting an Asian origin of man. In *Current Argument on Early Man*, ed. L.-K. Konigsson, pp. 252–260. Oxford: Pergamon Press.

Tong, H. & Jaeger, J. J. (1993). Muroid rodents from the middle Miocene Fort Ternan locality (Kenya) and their contribution to the phylogeny of muroids. *Palaeontographica*, **229**, 51–73.

Tong, Y. (1997). Middle Eocene small mammals from Liguanqiao Basin of Henan Province and Yuanqu Basin of Shanxi Province, central China. *Palaeontologica Sinica* (Series C), **26**, 1–256.

Trinkaus, E. (1973). A reconsideration of the Fontéchevade fossils. *American Journal of Physical Anthropology*, **39**, 25–35.

Trinkaus, E. (1983). *The Shanidar Neandertals*. New York: Academic Press.

Trinkaus, E. (1984). Western Asia. In *The Origins of Modern Humans: A World Survey of the Fossil Evidence*, eds. F. H. Smith & F. Spencer, pp. 251–293. New York: Alan R. Liss.

Trinkaus, E. (1987). The Neandertal face: evolutionary and functional perspectives on a recent human face. *Journal of Human Evolution*, **16**, 429–443.

Trinkaus, E. (1989). Olduvai hominid 7 trapezial metacarpal 1 articular morphology: contrasts with recent humans. *American Journal of Physical Anthropology*, **80**, 411–416.

Trinkaus, E. (1992). Morphological contrasts between the Near Eastern Qafzeh–Skhūl and late archaic human samples: grounds for a behavioral difference? In *The Evolution and Dispersal of Modern Humans in Asia*, eds. T. Akazawa, K. Aoki, & T. Kimura, pp. 277–294. Tokyo: Hokusen-Sha.

Trofimov, B. A. (1977). Primate *Paradolichopithecus pushkini* sp. nov. from Upper Pliocene of the Pamirs Piedmont. *Journal of the Paleontological Society of India*, **20**, 26–32.

Trouessart, E. L. (1879). Catalogue des mammifères vivants et fossiles. *Revue et magasin de zoologie* 7, 223–230.

Trouessart, E. L. (1897). *Catalogus mammalium tam viventium quam fossilium*, vol. 1, 2nd edn. Berlin: Friedlander & Sohn.

Turnbull, W. D. (1972). The Washakie Formation of Bridgerian–Uintan ages, and the related faunas. In *Field Conference on Tertiary Biostratigraphy of Southern and Western Wyoming*, pp. 20–31. Privately printed by R. M. West.

Tuttle, R. H. (1974). Darwin's apes, dental apes and the descent of man: normal science in evolutionary anthropology. *Current Anthropology*, **15**, 389–426.

Tyler, D. E. & Sartono, S. (1994). A new *Homo erectus* skull from Sangiran, Java. *American Journal of Physical Anthropology*, Supplement **18**, 236.

Ullrich, H. (1955). Paläolithische Menschenreste aus der Sowjetunion. I. Das Mousterian-Kind von Staroselje (Krim). *Zeitschrift für Morphologie und Anthropologie*, **47**, 99–112.

Ungar, P. S. (1996). Dental microwear of European Miocene catarrhines: evidence for diets and tooth use. *Journal of Human Evolution*, **31**, 335–366.

Ungar, P. S. & Kay, R. F. (1995). The dietary adaptations of European Miocene catarrhines. *Proceedings of the National Academy of Sciences of the United States of America*, **92**, 5479–5481.

Ungar, P. S. & Teaford, M. F. (1996). Preliminary examination of non-occlusal dental microwear in anthropoids:implications for the study of fossil primates. *American Journal of Physical Anthropology*, **100**, 101–114.

Vallois, H. V. (1949). The Fontéchevade fossil man. *American Journal of Physical Anthropology*, **7**, 339–362.

Vallois, H. V. (1954). L'Oréopithèque, Hominidé tertiare primitif? *L'Anthropologie*, **58**, 349–351.

Vallois, H. V. (1958). La grotte de Fontéchevade. I. Anthropologie. *Archives de L'Institut de paléontologie humaine*, **29**, 7–164.

Van Couvering, J. A. (1972). Geology of Rusinga Island and correlation of the Kenya mid-Tertiary fauna. PhD dissertation, University of Cambridge.

Van Couvering, J. A. & Harris, J. A. (1991). Late Eocene age of Fayum mammal faunas. *Journal of Human Evolution*, **21**, 241–260.

Van Couvering, J. A. & Miller, J. A. (1969). Miocene stratigraphy and age determinations, Rusinga Island, Kenya. *Nature*, **221**, 628–632.

Van Couvering, J. H. & Van Couvering, J. A. (1976). Early Miocene mammal fossils from East Africa: aspects of geology, faunistics and paleoecology. In *Human Origins*, eds. G. L. Isaac & E. R. McCown, pp. 155–207. Menlo Park, CA: W. A. Benjamin.

Vandermeersch, B. (1972). Récentes découvertes de squelettes humains á Qafzeh (Israël): essai d'interprétation. In *The Origin of* Homo sapiens, ed. F. Bordes, pp. 49–53. Paris: UNESCO.

Vandermeersch, B. (1981*a*). *Les Hommes fossiles de Qafzeh (Israël)*. Paris: Centre national de la recherche scientifique.

Vandermeersch, B. (1981*b*). Les premiers *Homo sapiens* au Proche-Orient. In *Les Processus de l'Hominization*, ed. D. Ferembach, pp. 97–100. Paris: Centre national de la recherche scientifique.

Van Valen, L. M. (1994). The origin of the plesiadapid primates and the nature of *Purgatorius*. *Evolutionary Monographs*, **15**, 1–79.

Van Valen, L. & Sloan, R. E. (1965). The earliest primates. *Science*, **150**, 743–745.

Verma, B. C. & Gupta, S. S. (1997). New light on the antiquity of Siwalik great apes. *Current Science (India)*, **72**, 302–3.

Vidal, L. M. (1913). Nota sobre la presencia del "*Dryopithecus*' en el Mioceno superior del Pirineo catalán. *Boletín de la real Sociedad española de historia natural, Sección geológic*, **13**, 499–507.

Villalta, J. F. & Crusafont, M. (1944). Dos nuevos antropomorfos del Mioceno español y su situación dentro del la moderna sistemática de los símidos. *Notas y comunicaciones, Instituto geológica y minero*, **13**, 91–139.

Viranta, S. & Andrews, P. (1995). Carnivore guild structure in the Paşalar Miocene fauna. *Journal of Human Evolution*, **28**, 359–372.

Virchow, R. (1872). Untersuchung des Neanderthal-Schädels. *Zeitschrift für Ethnologie*, **4**, 157–165.

Virchow, R. (1882). Der Kiefer aus der Schipka-Höhle und der Kiefer von la Naulette. *Zeitschrift für Ethnologie*, **14**, 277–310.

Viret, J. (1955). À propos de l'Oréopithèque. *Mammalia*, **19**, 320–324.

Vogel, C. (1968). The phylogenetical evaluation of some characters and some morphological trends in the evolution of the skull in catarrhine primates. In *Taxonomy and Phylogeny of Old World Primates with References to the Origin of Man*, ed. B. Chiarelli, pp. 21–55. Turin: Rosenberg & Sellier.

Voigt, F. S. (1835). Thier-Fährten in Hildburghauser Sandstein (Palaeopithecus). *Neues Jahrbuch für Mineralogie, Geologie und Paläontologie*, **1835**, 322–326.

von Dornum, M. & Ruvolo, M. (1999). Phylogenetic relationships of the New World monkeys (Primates, Platyrrhini) based on nuclear G6PD DNA sequences. *Molecular Phylogenetics and Evolution*, **11**, 459–476.

Vrba, E. S., Denton, G. H., Partridge, T. C., & Burckle, L. H. (eds.) (1995). *Paleoclimate and Evolution, With Emphasis on Human Origins*. New Haven, CT: Yale University Press.

Vuillaume-Randriamanantena, M. (1982). Contribution à l'étude des os longs des Lémuriens subfossiles malgaches. 3rd Cycle thesis. Uiversity of Madagascar.

Vuillaume-Randriamanantena, M. (1988). The taxonomic attributions of giant subfossil lemur bones from Ampasambazimba: *Archaeoindris* and *Lemuridotherium*. *Journal of Human Evolution*, **17**, 379–391.

Vuillaume-Randriamanantena, M. (1990). *Palaeopropithecus ingens* Grandidier, 1899 synonyme de *Thaumastolemur grandidieri* Filhol, 1895. *Comptes rendus de l'Académie des sciences de Paris*, **310**, 1307–1313.

Vuillaume-Randriamanantena, M., Godfrey, L. R., Jungers, W. L., & Simons, E. L. (1992). Morphology, taxonomy and distribution of *Megaladapis*: giant subfossil lemur from Madagascar. *Comptes rendus de l'Académie des sciences de Paris*, **315**, 1835–1842.

Vuillaume-Randriamanantena, M., Godfrey, L. R., & Sutherland, M. R. (1985). Revision of *Hapalemur (Prohapalemur) gallieni* (Standing, 1905). *Folia Primatologica*, **45**, 89–116.

Wadia, D. N. & Aiyengar, N. K. N. (1938). Fossil anthropoids of India: a list of the fossil material hitherto discovered from the Tertiary deposits of India. *Records of the Geological Survey of India*, **72**, 467–494.

Waddle, D. (1994). Matrix correlation tests support a single origin for modern humans. *Nature*, **368**, 452–455.

Wagner, A. (1839). Fossile Überreste von einem Affenschadel und anderen Säugtieren aus Griechenland. *Gelehrte Anziegen bayerisches Akademie Wissenschaften*, **38**, 306–311.

Wagner, P. J. (2000). Exhaustion of morphologic character states among fossil taxa. *Evolution*, **54**, 365–386.

Walker, A. C. (1967*a*). Locomotor adaptation in recent and fossil Madagascar lemurs. PhD dissertation, University of London.

Walker, A. C. (1967*b*). Patterns of extinction among the subfossil Madagascar lemuroids. In *Pleistocene Extinctions: The Search for a Cause*, eds. P. S. Martin & H. E. Wright, pp. 407–424. New Haven, CT: Yale University Press.

Walker, A. C. (1968). The Lower Miocene fossil site of Bukwa, Sebei. *Uganda Journal*, **32**, 149–156.

Walker, A. C. (1969*a*). True affinities of *Propotto leakeyi* Simpson, 1967. *Nature*, **223**, 647–648.

Walker, A. C. (1969*b*). New evidence from Uganda regarding the dentition of Miocene Lorisidae. *Uganda Journal*, **33**, 90–91.

Walker, A. C. (1970). Post-cranial remains of the Miocene Lorisidae of East Africa. *American Journal of Physical Anthropology*, **33**, 249–261.

Walker, A. C. (1974*a*). Locomotor adaptations in past and present prosimian primates. In *Primate Locomotion*, ed. F. A. Jenkins, Jr., pp. 349–381. New York: Academic Press.

Walker, A. C. (1974*b*). A review of the Miocene Lorisidae of East Africa. In *Prosimian Biology*, eds. R. D. Martin, G. A. Doyle, & A. C. Walker, pp. 435–447. London: Duckworth.

Walker, A. C. (1976). Remains attributable to *Australopithecus* in the East Rudolf succession. In *Earliest Man and Environments in the Lake Rudolf Basin*, eds. Y. Coppens, F. C. Howell, G. L. Isaac, & R. E. F. Leakey, pp. 484–489. Chicago, IL: University of Chicago Press.

Walker, A. C. (1978). Prosimian primates. In *Evolution of African Mammals*, eds. V. J. Maglio & H. B. S. Cooke, pp. 90–99. Cambridge, MA: Harvard University Press.

Walker, A. C. (1981). The Koobi Fora hominids and their bearing on the origins of the genus *Homo*. In *Homo erectus: Papers in Honor of Davidson Black*, eds. B. A. Sigmon & J. S. Cybulski, pp. 193–215. Toronto: University of Toronto Press.

Walker, A. C. (1987). Fossil Galaginae from Lateoli. In *Laetoli: A Pliocene Site in Northern Tanzania*, eds. M. D. Leakey & J. M. Harris, pp. 88–91. Oxford: Clarendon Press.

Walker, A. C. (1993). Perspectives on the Nariokotome discovery. In *The Nariokotome Homo erectus Skeleton*, eds. A. C. Walker & R. E. F. Leakey, pp. 410–430. Cambridge, MA: Harvard University Press.

Walker, A. C. (1997). *Proconsul* function and phylogeny. In *Function, Phylogeny and Fossils: Miocene Hominoid Evolution and Adaptation*, eds. D. R. Begun, C. V. Ward, & M. D. Rose, pp. 209–224. New York: Plenum Press.

Walker, A. C., Falk, D., Smith, R., & Pickford, M. (1983). The skull of *Proconsul africanus*: reconstruction and cranial capacity. *Nature*, **305**, 525–527.

Walker, A. C. & Leakey, R. E. F. (eds.) (1993). *The Nariokotome Homo erectus Skeleton*. Cambridge, MA: Harvard University Press.

Walker, A. C., Leakey, R. E., Harris, J. M., & Brown, F. H. (1986). 2.5- Myr *Australopithecus boisei* from west of Lake Turkana, Kenya. *Nature*, **322**, 517–522.

Walker, A. C. & Pickford, M. (1983). New postcranial fossils of *Proconsul africanus* and *Proconsul nyanzae*. In *New Interpretations of Ape and Human Ancestry*, eds. R. L. Ciochon & R. S. Corruccini, pp. 325–351. New York: Plenum Press.

Walker, A. C. & Rose, M. (1968). Fossil hominoid vertebra from the Miocene of Uganda. *Nature*, **217**, 980–981.

Walker, A. C. & Teaford, M. (1988). The Kaswanga Primate Site: an early Miocene hominoid site on Rusinga Island, Kenya. *Journal of Human Evolution*, **17**, 539–544.

Walker, A. C. & Teaford, M. (1989). The hunt for *Proconsul*. *Scientific American*, **260**, 76–82.

Walker, A. C., Teaford, M. F., & Leakey, R. E. F. (1985). New information concerning the R114 *Proconsul* site, Rusinga Island, Kenya. In *Primate Evolution*, eds. J. G. Else & P. C. Lee, pp. 143–149. Cambridge: Cambridge University Press.

Walker, A. C., Teaford, M. F., Martin, L., & Andrews, P. (1993). A new species of *Proconsul* from the early Miocene of Rusinga/Mfangano Islands, Kenya. *Journal of Human Evolution*, **25**, 43–56.

Walker, A. C., Zimmerman, M., & Leakey, R. E. F. (1982). A possible case of hypervitaminosis A in *Homo erectus*. *Nature*, **296**, 248–250.

Wall, C. E. (1997). The expanded mandibular condyle of the Megaladapidae. *American Journal of Physical Anthropology*, **103**, 263–276.

Wang, B. & Li, C. (1990). First Paleogene mammalian fauna from northeast China. *Vertebrata PalAsiatica*, **28**, 165–205.

Wanpo, H., Ciochon, R., Yumin, G., Larick, R., Qiren, F., Schwarcz, H., Yonge, C., deVos, J., & Rink, W. (1995). Early *Homo* and associated artifacts from Asia. *Nature*, **378**, 275–278.

Ward, C. V. (1993). Torso morphology and locomotion in *Proconsul nyanzae*. *American Journal of Physical Anthropology*, **92**, 291–328.

Ward, C. V. (1997). Functional anatomy and phylogenetic implications of the hominoid trunk and hindlimb. In *Function, Phylogeny, and Fossils: Miocene Hominoid Evolution and Adaptations*, eds. D. R. Begun, C. V. Ward, & M. D. Rose, pp. 101–130. New York: Plenum Press.

Ward, C. V. (1998). *Afropithecus, Proconsul*, and the primitive hominoid skeleton. In *Primate Locomotion: Recent Advances*, eds. E. Strasser, J. Fleagle, A. Rosenberger, & H. McHenry, pp. 337–352. New York: Plenum Press.

Ward, C. V., Begun, D. R., & Rose, M. D. (1997). Function and phylogeny in Miocene hominoids. In *Function, Phylogeny and Fossils: Miocene Hominoid Evolution and Adaptation*, eds. D. R. Begun, C. V. Ward, & M. D. Rose, pp. 1–12. New York: Plenum Press.

Ward, C. V, Leakey, M. G., & Walker, A. (2001). Morphology of *Australopithecus anamensis* from Kanapoi and Allia Bay, Kenya. *Journal of Human Evolution*, **41**, 255–368.

Ward, C. V., Ruff, C. B., Walker, A. C., Rose, M. D., Teaford, M. F., & Nengo, I. O. (1995). Functional morphology of *Proconsul* patellas from Rusinga Island, Kenya, with implications for other Miocene–Pliocene catarrhines. *Journal of Human Evolution*, **29**, 1–19.

Ward, C. V., Walker, A. C., & Teaford, M. F. (1991). *Proconsul* did not have a tail. *Journal of Human Evolution*, **21**, 215–220.

Ward, C. V., Walker, A. C., & Teaford, M. F. (1999). Still no evidence for a tail in *Proconsul heseloni*. *American Journal of Physical Anthropology*, Supplement **28**, 273.

Ward, C. V., Walker, A. C., Teaford, M. F., & Odhiambo, I. (1993). Partial skeleton of *Proconsul nyanzae* from Mfangano Island, Kenya. *American Journal of Physical Anthropology*, **90**, 77–111.

Ward, C. V., Leakey, M. G., & Walker, A. C. (1999). The new hominid species *Australopithecus anamensis*. *Evolutionary Anthropology*, **7**, 197–205.

Ward, S. C. (1997). The taxonomy and phylogenetic relationships of *Sivapithecus* revisited. In *Funtion, Phylogeny, and Fossils: Miocene Hominoid Evolution and Adaptations*, eds. D. R. Begun, C. V. Ward, & M. D. Rose, pp. 269–290. New York: Plenum Press.

Ward, S. C. & Brown, B. (1986). The facial skeleton of *Sivapithecus indicus*. In *Comparative Primate Biology*, vol. 1, *Systematics, Evolution, and Anatomy*, eds. D. Swindler & S. Erwin, pp. 413–452. New York: Alan R. Liss.

Ward, S. C., Brown, B., Hill, A., Kelley, J., & Downs, W. (1999). *Equatorius*: a new hominoid genus from the middle Miocene of Kenya. *Science*, **285**, 1382–1386.

Ward, S. C. & Kimbel, W. H. (1983). Subnasal morphology and the systematic position of *Sivapithecus*. *American Journal of Physical Anthropology*, **61**, 157–171.

Ward, S. C. & Pilbeam, D. (1983). Maxillofacial morphology of Miocene hominoids from Africa and Indo-Pakistan. In *New Interpretations of Ape and Human Ancestry*, eds. R. L. Ciochon & R. S. Corruccini, pp. 211–238. New York: Plenum Press.

Washburn, S. L. (ed.) (1963). *Classification and Human Evolution*. Chicago, IL: Aldine.

Washburn, S. L. (1968). Condon Lectures: The study of human evolution. Eugene, OR: Oregon State System of Higher Education.

Watkins, R. T. (1989). The Buluk Member, a fossil hominoid-bearing sedimentary sequence of Miocene age from Northern Kenya. *Journal of African Earth Sciences*, **8**, 107–112.

Wayne, R. K., Benveniste, R. E., Janczewski, D. N., & O'Brien, S. J. (1989). Molecular and biochemical evolution of the Carnivora. In *Carnivore Behavior, Ecology, and Evolution*, ed. J. L. Gittleman, pp. 465–494. Ithaca, NY: Comstock.

Webb, S. G. (1989). *The Willandra Lakes Hominids*. Canberra: Department of Prehistory, School of Pacific Studies, Australian National University.

Wegner, R. N. (1913). Teriär und umgelagerte Kreide bei Oppeln (Oberschlesien). *Palaeontographica*, **60**, 175–274.

Weidenreich, F. (1937). The dentition of *Sinanthropus pekinensis*. *Palaeontologica Sinica*, whole series **101**, new series D, **1**, 1–180.

Weidenreich, F. (1943). The skull of *Sinanthropus pekinensis*: a comparative study on a primitive hominid skull. *Palaeontologia Sinica*, new series D, **10**, 1–484.

Weidenreich, F. (1944). Giant early man from Java and South China. *Science*, **99**, 479–482.

Weidenreich, F. (1945*a*). Giant early man from Java and South China. *Anthropological Papers of the American Museum of Natural History*, **40**, 1–134.

Weidenreich, F. (1945*b*). The Paleolithic child from the Teshik-Tash cave in Southern Uzbekistan (Central Asia). *American Journal of Physical Anthropology*, **3**, 21–32.

Weidenreich, F. (1946). *Apes, Giants and Man*. Chicago, IL: University of Chicago Press.

Weidenreich, F. (1949). Interpretations of the fossil material. In *Early Man in the Far East: Studies in Physical Anthropology*, ed. W. W. Howells, pp. 149–157. Detroit, MI: American Association of Physical Anthropologists.

Weidenreich, F. (1951). Morphology of Solo man. *Anthropological Papers of the American Museum of Natural History*, **43**, 205–290.

Weigelt, J. (1933). Neue Primaten aus der mitteleozänen (oberlutetischen) Braunkohle des Geiseltals. *Nova Acta Leopoldense*, **1**, 97–156.

Weiner, J. S. (1955). *The Piltdown Forgery*. Oxford: Oxford University Press.

Weiner, J. S. & Campbell, B. G. (1964). The taxonomic status of the Swanscombe skull. *Occasional Papers of the Royal Anthropology Institute*, **20**, 175–209.

Weitzel, V., Yang, C. M., & Groves, C. (1988). *The Raffles Bulletin of Zoology*. Singapore: National University of Singapore.

Welcomme, J. L., Aguelar, J. P., & Ginsburg, L. (1991). Découverte d'un nouveau Pliopithèque (Primates, Mammalia) associé à des rongeurs dans les sables du Miocène supérieur de Priay (Ain, France) et remarques sur la paléogéographie de la Bresse au Vallésien. *Comptes rendus de l'Académie des sciences de Paris*, **313**, 723–729.

Werdelin, L. & Solounias, N. (1996). The evolutionary history of hyaenas in Europe and Western Asia during the Miocene. In *The Evolution of Western Eurasian Neogene Mammal Faunas*, eds. R. L. Bernor, V. Fahlbusch, & H.-W. Mittmann, pp. 290–306. New York: Columbia University Press.

Wesselman, H. B. (1984). The Omo micromammals: systematics and paleoecology of early man sites from Ethiopia. In *Contributions to Vertebrate Evolution*, vol. 7, eds. M. K. Hecht & F. S. Szalay, pp. 64–82. Basel: Karger.

White, T. D. (1981). Primitive hominid canine from Tanzania. *Science*, **213**, 348–349.

White, T. D. (2000). A view on the science: physical anthropology at the millennium. *American Journal of Physical Anthropology*, **113**, 287–292.

White, T. D. & Suwa, G. (1987). Hominid footprints at Laetoli [Tanzania]: facts and interpretations. *American Journal of Physical Anthropology*, **72**, 485–514.

White, T. D., Suwa, G., Hart, W. K., Walter, R. C., WoldeGabriel, G., De Heinzelin, J., Clark, J. D., Asfaw, B., & Vrba, E. (1993). New discoveries of *Australopithecus* at Maka in Ethiopia. *Nature*, **366**, 261–265.

White, T. D., Suwa, G., & Asfaw, B. (1994). *Australopithecus ramidus*: a new species of early hominid from Aramis, Ethiopia. *Nature*, **371**, 306–312.

White, T. D., Suwa, G., & Asfaw, B. (1995). Corrigendum: *Australopithecus ramidus*, a new species of early hominid from Aramis, Ethiopia. *Nature*, **375**, 88.

Whittaker, R. J. (1998). *Island Biogeography: Ecology, Evolution, and Conservation*. New York: Oxford University Press.

Whybrow, P. J. & Andrews, P. J. (1978). Restoration of the holotype of *Proconsul nyanzae*. *Folia Primatologica*, **30**, 115–125.

Wible, J. R. & Covert, H. H. (1987). Primates: cladistic diagnosis and relationships. *Journal of Human Evolution*, **16**, 1–20.

Wible, J. R. & Martin, J. R. (1993). Ontogeny of the tympanic floor and roof in archontans. In *Primates and their Relatives in Phylogenetic Perspective*, ed. R. D. E. MacPhee, pp. 111–148. New York: Plenum Press.

Wible, J. R. & Novacek, M. J. (1988). Cranial evidence for the monophyletic origin of bats. *American Museum Novitates*, **2911**, 1–19.

Wible, J. R. & Zeller, U. (1994). Cranial circulation of the pen-tailed tree shrew *Ptilocercus lowii* and relationships of Scandentia. *Journal of Mammalian Evolution*, **2**, 209–230.

Wiley, E. O. (1981). *Phylogenetics*. New York: Wiley.

Williams, B. A. & Covert, H. H. (1994). New early Eocene anaptomorphine primate (Omomyidae) from the Washakie Basin, Wyoming, with comments on the phylogeny and paleobiology of anaptomorphines. *American Journal of Physical Anthropology*, **93**, 323–340.

Williams, E. E. & Koopman, K. F. (1952). West Indian fossil monkeys. *American Museum Novitates*, **1546**, 1–16.

Wilson, J. A. (1966). A new primate from the earliest Oligocene, west Texas, preliminary report. *Folia Primatologica*, **4**, 227–248.

Wilson, J. A. & Szalay, F. S. (1976). New adapid primate of European affinities from Texas. *Folia Primatologica*, **25**, 294–312.

Winge, H. (1895). Jordfundne og nulevende aber (primates) fra Lagoa Santa, Minas Gerais, Brasilien. *E Museo Lundii*, **2**, 1–57.

WoldeGabriel, G., White, T. D., Suwa, G., Renne, P., De Heinzelin, J., Hart, W. K., & Heiken, G. (1994). Ecological and temporal placement of early Pliocene hominids at Aramis, Ethiopia. *Nature*, **371**, 330–333.

Wolff, R. (1984). New fossil specimens of *Branisella boliviana* from the Early Oligocene of Salla, Bolivia. *Journal of Vertebrate Paleontology*, **4**, 570–574.

Wolin, L. R. & Massopust, L. C. (1970). Morphology of the primate retina. In *The Primate Brain*, eds. C. R. Noback & W. Montagna, pp. 1–27. New York: Appleton-Century-Crofts.

Wolpoff, M. H. (1971*a*). Competitive exclusion among Lower Pleistocene hominids: the single species hypothesis. *Man*, **6**, 601–614.

Wolpoff, M. H. (1971*b*). Metric trends in hominid dental evolution. *Case Western Reserve University Studies in Anthropology*, **2**, 1–244.

Wolpoff, M. H. (1971*c*). Vérteszőllős and the presapiens theory. *American Journal of Physical Anthropology*, **35**, 209–216.

Wolpoff, M. H. (1980). Cranial remains of Middle Pleistocene European hominids. *Journal of Human Evolution*, **9**, 339–358.

Wolpoff, M. H. (1983). *Ramapithecus* and human origins: an anthropologist's perspective of changing perceptions. In *New Interpretations of Ape and Human Ancestry*, eds. R. L. Ciochon & R. S. Corruccini, pp. 651–676. New York: Plenum Press.

Wolpoff, M. H. (1986). Describing anatomically modern *Homo sapiens*: a distinction without a definable difference. *Anthropos (Brno)*, **23**, 41–53.

Wolpoff, M. H. (1989). Multiregional evolution: the fossil alternative to Eden. In *The Human Revolution*, eds. P. Mellars & C. Stringer, pp. 62–108. Edinburgh: University of Edinburgh Press.

Wolpoff, M. H. (1999). *Paleoanthropology*. 2nd edn. New York: McGraw-Hill.

Wolpoff, M. H. & Caspari, R. (1997). *Race and Human Evolution*. New York: Simon & Schuster.

Wolpoff, M. H., Hawks, J., Frayer, D. W., & Hunley, K. (2001). Modern human ancestry at the peripheries: a test of the replacement theory. *Science*, **291**, 293–297.

Wolpoff, M. H., Smith, F. H., Malez, M., Radovčić, J., & Rukavina, D. (1981). Upper Pleistocene hominid remains from Vindija cave, Croatia, Yugoslavia. *American Journal of Physical Anthropology*, **54**, 499–545.

Wolpoff, M. H., Thorne, A. G., Jelínek, J., & Yinyum, Z. (1994). The case for sinking *Homo erectus*: 100 years of *Pithecanthropus* is enough! *Courier Forschungsinstitut Senckenberg*, **171**, 341–361.

Wolpoff, M. H., Wu, X. Z. & Thorne, A. G. (1984). Modern *Homo sapiens* origins: a general theory of hominid evolution involving the fossil evidence from East Asia. In *The Origins of Modern Humans: A World Survey of the Fossil Evidence*, eds. F. H. Smith & F. Spencer, pp. 411–483. New York: Alan R. Liss.

Woo, J. (1957). *Dryopithecus* teeth from Keiyuan, Yunnan Province. *Vertebrata PalAsiatica*, **1**, 25–32.

Woo, J. (1958). New materials of *Dryopithecus* from Keiyuan, Yunnan. *Vertebrata PalAsiatica*, **2**, 38–42.

Woo, J. K. (1962). The mandibles and dentition of *Gigantopithecus*. *Palaeontologica Sinica*, **146**, new series **11**, 1–94.

Woo, J. & Peng, R. (1959). Fossil skull of early paleoanthropic stage found at Mapa, Shaoquan, Kwangtung Province. *Vertebrata PalAsiatica*, **3**, 176–182.

Wood, B. A. (1974). Olduvai Bed I post-cranial fossils: a reassessment. *Journal of Human Evolution*, **3**, 373–378.

Wood, B. A. (ed.) (1991). *Koobi Fora Research Project IV: Hominid Cranial Remains from Koobi Fora*. Oxford: Clarendon Press.

Wood, B. A. (1992). Origin and evolution of the genus *Homo*. *Nature*, **355**, 783–790.

Wood, B. A. (1994). Taxonomy and evolutionary relationships of *Homo erectus*. *Courier Forschungsinstitut Senckenberg*, **171**, 159–165.

Wood, B. A. (1999). "*Homo rudolfensis*" Alexeev, 1986: fact or phantom? *Journal of Human Evolution*, **36**, 115–118.

Wood, B. A. & Collard, M. (1999). The human genus. *Science*, **284**, 65–71.

Wood, B. A. & Xu, Q. (1991). Variation in the Lufeng dental

remains. *Journal of Human Evolution*, **20**, 291–311.

Wood, S. (1844). Record of the discovery of an alligator with several new Mammalia in the freshwater strata at Hordwell. *Annals of the Magazine of Natural History, London*, **14**, 349–351.

Wood, S. (1846). On the discovery of an alligator and of several new Mammalia in the Hordwell Cliff, with observations on the geological phenomena of that locality. *Geological Journal of London*, **1**, 1–7.

Woodward, A. S. (1921). A new cave man from Rhodesia, South Africa. *Nature*, **108**, 371–372.

Wortman, J. L. (1903–1904). Studies of Eocene Mammalia in the Marsh Collection, Peabody Museum. *American Journal of Science*, **15**, 163–176, 399–414, 419–436; **16**, 345–368; **17**, 23–33, 133–140, 203–214.

Wrangham, R. W., Conklin-Brittain, N., & Hunt, K. (1998). Dietary response of chimpanzees and cercopithecines to seasonal variation in fruit abundance. I. Antifeedants. *International Journal of Primatology*, **19**, 949–970.

Wright, P. C. & Martin, L. B. (1995). Predation, pollination and torpor in two nocturnal prosimians: *Cheirogaleus major* and *Microcebus rufus* in the rainforest of Madagascar. In *Creatures of the Dark: The Nocturnal Prosimians*, eds. L. Alterman, G. A. Doyle, & M. K. Izard, pp. 45–60. New York: Plenum Press.

Wu, R. (1987). A revision of the classification of the Lufeng great apes. *Acta Anthropologica Sinica*, **6**, 265–71.

Wu, R., Han, D., Xu, Q., Lu, Q., Pan, Y., Zhang, X., Zheng, L., & Xiao, M. (1981). *Ramapithecus* skulls found first time in the world. *Kexue Tongbao*, **26**, 1018–1021.

Wu, R., Han, D., Xu, Q., Qui, G., Lu, Q., Pan, Y., & Chen, W. (1982). More *Ramapithecus* skulls from Lufeng, Yunnan: report on the excavation of the site in 1981. *Acta Anthropologica Sinica*, **2**, 101–108.

Wu, R., Lu, Q., & Xu, Q. (1984). Morphological features of *Ramapithecus* and *Sivapithecus* and their phylogenetic relationships: morphology and comparison of the mandibles. *Acta Anthropologica Sinica*, **3**, 1–10.

Wu, R. & Pan, Y. (1984). A late Miocene gibbon-like primate from Lufeng, Yunnan Province. *Acta Anthropologica Sinica*, **3**, 185–194.

Wu, R. & Pan, Y. (1985). A new adapid primate from the Lufeng Miocene, Yunnan Province. *Acta Anthropologica Sinica*, **4**, 1–6.

Wu, R. & Wang, L. (1988). Single species and sexual dimorphism in *Sinoadapis*. *Acta Anthropologica Sinica*, **7**, 1–8.

Wu, R., Xu, Q., & Lu, Q. (1983). Morphological features of *Ramapithecus* and *Sivapithecus* and their phylogenetic relationships: morphology and comparison of the crania. *Acta Anthropologica Sinica*, **2**, 1–10.

Wu, R., Xu, Q., & Lu, Q. (1985). Morphological features of *Ramapithecus* and *Sivapithecus* and their phylogenetic relationships: morphology and comparison of the teeth. *Acta Anthropologica Sinica*, **4**, 197–204.

Wu, R., Xu, Q., & Lu, Q. (1986). Relationship between Lufeng *Sivapithecus* and *Ramapithecus* and their phylogenetic position. *Acta Anthropologica Sinica*, **5**, 1–30.

Wu, X. (2000). Longgupo hominoid mandible belongs to ape. *Acta Anthropologica Sinica*, **19**, 1–10.

Wu, X. & Wu, M. (1985). Early *Homo sapiens* in China. In *Paleoanthropology and Paleolithic Archaeology in the People's Republic of China*, eds. R. Wu & J. Olsen, pp. 91–106. Orlando, FL: Academic Press.

Wunderlich, R. E., Simons, E. L., & Jungers, W. L. (1996). New pedal remains of *Megaladapis* and their functional significance. *American Journal of Physical Anthropology*, **100**, 115–138.

Wunderlich, R. E., Walker, A. C., & Jungers, W. L. (1999). Rethinking the positional repertoire of *Oreopithecus*. *American Journal of Physical Anthropology*, Supplement **28**, 282.

Xiao, M. (1981). Discovery of fossil hominoid scapula at Lufeng, Yunnan. *Journal of the Yunnan Provincial Museum*, **30**, 41–44.

Xu, Q. & Lu, Q. (1979). The mandibles of *Ramapithecus* and *Sivapithecus* from Lufeng, Yunnan. *Vertebrata PalAsiatica*, **17**, 1–13.

Xu, Q. & Lu, Q. (1980). The Lufeng ape skull and its significance. *China Reconstructs*, **29**, 56–57.

Xu, Q., Lu, Q., Pan, Y., Zhang, X., & Zheng, L. (1978). Fossil mandible of the Lufeng *Ramapithecus*. *Kexue Tongbao*, **9**, 544–556.

Xue, X. & Delson, E. (1988). A new species of *Dryopithecus* from Gansu, China. *Kexue Tongbao*, **33**, 449–453.

Yang, X. (1999). *The Golden Age of Chinese Archaeology: Celebrated Discoveries from the People's Republic of China*. Washington, DC: National Gallery of Art.

Yoder, A. D. (1994). Relative position of the Cheirogaleidae in strepsirhine phylogeny: a comparison of morphological and molecular methods and results. *American Journal of Physical Anthropology*, **94**, 25–46.

Yoder, A. D. (1997). Back to the future: a synthesis of strepsirhine systematics. *Evolutionary Anthropology*, **6**, 11–22.

Yoder, A. D. (2000). Implications of a multigene phylogeny for strepsirhine biogeography and evolution. *American Journal of Physical Anthropology*, Supplement **30**, 329.

Yoder, A. D. Cartmill, M., Ruvolo, M., Smith, K., & Vilgalys, R. (1996). Ancient single origin for Malagasy primates. *Proceedings of the National Academy of Sciences of the United States of America*, **93**, 5122–5126.

Yoder, A. D., Rakotosamimanana, B., & Parsons, T. (1999). Ancient DNA in subfossil lemurs: methodological challenges and their solutions. In *New Directions in Lemur Studies*, eds. B. Rakotosamimanana, H. Rasaminanana, J. U. Ganzhorn, & S. M. Goodman, pp. 1–17. New York: Plenum Press.

Yoder, A. D., Rasoloarison, R. M., Goodman, S. M., Irwin, J. A., Atsalis, S., Ravosa, M. J., & Ganzhorn, J. U. (2000). Remarkable species diversity in Malagasy mouse lemurs (Primates, *Microcebus*). *Proceedings of the National Academy of Sciences of the United States of America*, **97**, 11325–11330.

Young, N. & MacLatchy, L. (2000). Is *Morotopithecus* a great ape? *American Journal of Physical Anthropology*, Supplement **30**, 329–330.

Young, C. C., & P'ei, W. C. (1933). On the fissure deposits of Chinghsinghsien with remarks on the Cenozoic geology of the same area. *Bulletin of the Geological Society of China*, **13**, 63–71.

Yuerong, P., Waddle, D. M., & Fleagle, J. G. (1989). Sexual dimorphism in *Laccopithecus rubustus*, a late Miocene hominoid from China. *American Journal of Physical Anthropology*, **79**, 137–158.

Zambon, S. N., McCrossin, M. L. & Benefit, B. B. (1999). Estimated body weight and degree of sexual dimorphism for *Victoriapithecus-macinnesi*, a Miocene cercopithecoid. *American Journal of Physical Anthropology*, Supplement **28**, 284.

Zapfe, H. (1952). Die *Pliopithecus*-Funde aus der Spaltenfüllung von Neudorf an der March (CRS). Vienna: *Verhandlungen der geologischen bundesanstalt*, Sonderheft **C**, 126–130.

Zapfe, H. (1958). The skeleton of *Pliopithecus* (*Epipliopithecus*) *vindobonensis* Zapfe & Hürzeler. *American Journal of Physical Anthropology*, **16**, 441–455.

Zapfe, H. (1960). Die Primatenfunde aus der miozänen Spaltenfüllung von Neudorf an der March (Děvínská Nová Ves), Tschechoslovakei. Mit anhang: Der Primatenfund aus dem Miozän von Klein Hadersdorf in Niederosterreich. *Abhandlungen der schweizerischen paläontologischen Gesellschaft*, **78**, 1–293.

Zapfe, H. (1961). Ein Primatenfund aus der miozänen Molasse von Oberösterreich. *Zeitschrift für Morphologie und Anthropologie*, **51**, 247–267.

Zapfe, H. (1963). Lebensbild von *Megaladapis edwardsi* (Grandidier). *Folia Primatologica*, **1**, 178–187.

Zapfe, H. (1991). Mesopithecus pentelicus *Wagner aus dem Turolien von Pikermi bei Athen, Odontologie und Osteologie (Eine Dokumentation)*. Vienna. Verlag Ferdinand Berger & Sohne.

Zapfe, H. & Hürzeler, J. (1957). Die Fauna der miozänen Spaltenfüllung von Neudorf a. d. March (CSR). *Sitzungsberichte Österreich der Akademie der Wissenschaften, Mathematik, Naturwissenschaften Klasse*, **166**, 113–123.

Zdansky, O. (1928). Die Säugetiere der Quarterfauna von Chou-K'ou-Tien. *Geological Survey of China, series C*, **5**, 123–141.

Zdansky, O. (1930). Die alttertiären Säugetiere Chinas nebst stratigraphischen Bemerkungen. *Palaeontologica Sinica, series C*, **6**,1–87.

Zehr, S. (1999). A nuclear and mitochondrial phylogeny of the lesser apes (Primates, genus *Hylobates*). PhD dissertation, Harvard University.

Zhang, X. (1987). New materials of *Ramapithecus* from Kaiyuan, Yunnan. *Acta Anthropologica Sinica*, **6**, 81–86.

Zhang, X., Lin, Y., Jiang, C., & Xiao, L. (1987*a*). A new species of *Ramapithecus* from Yuanmou, Yunnan. *Journal of Yunnan University (Social Sciences)*, **3**, 54–56.

Zhang, X., Lin, Y., Jiang, C., & Xiao, L. (1987*b*). A new species of *Homo* from Yuanmou, Yunnan. *Journal of Yunnan University (Social Sciences)*, **3**, 57–60.

Zhang, X., Zheng, L., Gao, F., Jiang, C., & Zhang, J. (1988). A preliminary study of the skull of Lama ape unearthed at Hudie Hill of Yuanmou County. *Journal of Yunnan University (Social Sciences)*, **5**, 55–61.

Zhang, Y. (1995). Human fossil crania from Yunxian: morphological comparison with *Homo erectus* crania from Zhoukoudian. Acta Anthropologica Sinica, **14**, 1–7.

Zhang, Y.P., & Ryder, O.A. (1998). Mitochondrial cytochrome *b* gene seqences of Old World monkeys, with special reference on evolution of Asian colobines. *Primates*, **39**, 39–49.

Zheng, L. & Zhang, X. (1997). Hominoid fossils. In *Yuanmou Hominoid Fauna*, ed. Z. He, pp. 1–21. Kunming: Yunnan Science Press.

Zhou, G. & Wang, Z. (1981). Preliminary observation on new found mandible and maxilla of *Sivapithecus yunnanensis*. *Memoirs of the Beijing Natural History Museum*, **10**, 10–17.

Zietkiewicz, E., Richer, C., & Labuda, D. (1999). Phylogenetic affinities of *Tarsius* in the context of primate Alu repeats. *Molecular Phylogeny and Evolution*, **11**, 77–83.

Zihlman, A. L. (1979). Pygmy chimpanzee morphology and the interpretation of early hominids. *South African Journal of Science*, **75**, 165–167.

Zong, G., Pan, Y., Chu, J., & Lin, X. (1991). Stratigraphic subdivision of hominoid fossil localities of Yuanmou, Yunnan. *Acta Anthropologica Sinica*, **10**, 165–166.

Historical figures index

This index lists the individuals whose contributions to primate paleontology date to at least 20 or more years ago and who are recognized by the chapter authors for the historical context of their contribution. The index does not account for all instances of citation of the indicated individuals, nor does it comprehensively account the biography of primate paleontology, which is particularly rich during the most recent 20 years.

Taxonomic index

This index lists the fossil taxa recognized or addressed by chapter authors, as well as select higher-order categories that are formally defined or applied in the text. Extant genera are indicated for instances in which the formal genus name is used in comparison to extinct genera. Informal usages of the terms below are not accounted in this index; i.e., this index finds *Hominidae*, but not *hominid*. Pages on which the taxon is illustrated are indicated in bold, italic typeface.